# THE ELEMENTS

| Element | Symbol | Atomic Number | Atomic Mass* |
|---|---|---|---|
| Actinium | Ac | 89 | (227) |
| Aluminum | Al | 13 | 26.98 |
| Americium | Am | 95 | (243) |
| Antimony | Sb | 51 | 121.8 |
| Argon | Ar | 18 | 39.95 |
| Arsenic | As | 33 | 74.92 |
| Astatine | At | 85 | (210) |
| Barium | Ba | 56 | 137.3 |
| Berkelium | Bk | 97 | (247) |
| Beryllium | Be | 4 | 9.012 |
| Bismuth | Bi | 83 | 209.0 |
| Boron | B | 5 | 10.81 |
| Bromine | Br | 35 | 79.90 |
| Cadmium | Cd | 48 | 112.4 |
| Calcium | Ca | 20 | 40.08 |
| Californium | Cf | 98 | (249) |
| Carbon | C | 6 | 12.01 |
| Cerium | Ce | 58 | 140.1 |
| Cesium | Cs | 55 | 132.9 |
| Chlorine | Cl | 17 | 35.45 |
| Chromium | Cr | 24 | 52.00 |
| Cobalt | Co | 27 | 58.93 |
| Copper | Cu | 29 | 63.55 |
| Curium | Cm | 96 | (247) |
| Dysprosium | Dy | 66 | 162.5 |
| Einsteinium | Es | 99 | (254) |
| Erbium | Er | 68 | 167.3 |
| Europium | Eu | 63 | 152.0 |
| Fermium | Fm | 100 | (253) |
| Fluorine | F | 9 | 19.00 |
| Francium | Fr | 87 | (223) |
| Gadolinium | Gd | 64 | 157.3 |
| Gallium | Ga | 31 | 69.72 |
| Germanium | Ge | 32 | 72.59 |
| Gold | Au | 79 | 197.0 |
| Hafnium | Hf | 72 | 178.5 |
| Hahnium** | Ha | 105 | (262) |
| Hassium | Hs | 108 | (265) |
| Helium | He | 2 | 4.003 |
| Holmium | Ho | 67 | 164.9 |
| Hydrogen | H | 1 | 1.008 |
| Indium | In | 49 | 114.8 |
| Iodine | I | 53 | 126.9 |
| Iridium | Ir | 77 | 192.2 |
| Iron | Fe | 26 | 55.85 |
| Krypton | Kr | 36 | 83.80 |
| Lanthanum | La | 57 | 138.9 |
| Lawrencium | Lr | 103 | (257) |
| Lead | Pb | 82 | 207.2 |
| Lithium | Li | 3 | 6.941 |
| Lutetium | Lu | 71 | 175.0 |
| Magnesium | Mg | 12 | 24.31 |
| Manganese | Mn | 25 | 54.94 |
| Meitnerium | Mt | 109 | (267) |
| Mendelevium | Md | 101 | (256) |
| Mercury | Hg | 80 | 200.6 |
| Molybdenum | Mo | 42 | 95.94 |
| Neodymium | Nd | 60 | 144.2 |
| Neon | Ne | 10 | 20.18 |
| Neptunium | Np | 93 | (244) |
| Nickel | Ni | 28 | 58.70 |
| Nielsbohrium | Ns | 107 | (262) |
| Niobium | Nb | 41 | 92.91 |
| Nitrogen | N | 7 | 14.01 |
| Nobelium | No | 102 | (253) |
| Osmium | Os | 76 | 190.2 |
| Oxygen | O | 8 | 16.00 |
| Palladium | Pd | 46 | 106.4 |
| Phosphorus | P | 15 | 30.97 |
| Platinum | Pt | 78 | 195.1 |
| Plutonium | Pu | 94 | (242) |
| Polonium | Po | 84 | (209) |
| Potassium | K | 19 | 39.10 |
| Praseodymium | Pr | 59 | 140.9 |
| Promethium | Pm | 61 | (145) |
| Protactinium | Pa | 91 | (231) |
| Radium | Ra | 88 | (226) |
| Radon | Rn | 86 | (222) |
| Rhenium | Re | 75 | 186.2 |
| Rhodium | Rh | 45 | 102.9 |
| Rubidium | Rb | 37 | 85.47 |
| Ruthenium | Ru | 44 | 101.1 |
| Rutherfordium | Rf | 104 | (261) |
| Samarium | Sm | 62 | 150.4 |
| Scandium | Sc | 21 | 44.96 |
| Seaborgium | Sg | 106 | (263) |
| Selenium | Se | 34 | 78.96 |
| Silicon | Si | 14 | 28.09 |
| Silver | Ag | 47 | 107.9 |
| Sodium | Na | 11 | 22.99 |
| Strontium | Sr | 38 | 87.62 |
| Sulfur | S | 16 | 32.07 |
| Tantalum | Ta | 73 | 180.9 |
| Technetium | Tc | 43 | (98) |
| Tellurium | Te | 52 | 127.6 |
| Terbium | Tb | 65 | 158.9 |
| Thallium | Tl | 81 | 204.4 |
| Thorium | Th | 90 | 232.0 |
| Thulium | Tm | 69 | 168.9 |
| Tin | Sn | 50 | 118.7 |
| Titanium | Ti | 22 | 47.90 |
| Tungsten | W | 74 | 183.9 |
| Uranium | U | 92 | 238.0 |
| Vanadium | V | 23 | 50.94 |
| Xenon | Xe | 54 | 131.3 |
| Ytterbium | Yb | 70 | 173.0 |
| Yttrium | Y | 39 | 88.91 |
| Zinc | Zn | 30 | 65.39 |
| Zirconium | Zr | 40 | 91.22 |

* All atomic masses have four significant figures. Values in parentheses represent the mass number of the most stable isotope.

** Although final approval has not yet been obtained, the American Chemical Society has recommended the following names and symbols for elements 104 through 109: 104, Rutherfordium (Rf); 105, Hahnium (Ha); 106, Seaborgium (Sg); 107, Nielsbohrium (Ns); 108, Hassium (Hs); 109, Meitnerium (Mt).

# Introductory Chemistry

## Investigating the Molecular Nature of Matter

# Introductory Chemistry

## Investigating the Molecular Nature of Matter

**John P. Sevenair**
*Xavier University of Louisiana*

**Allan R. Burkett**
*Dillard University*

WCB Wm. C. Brown Publishers
Dubuque, IA Bogota Boston Buenos Aires Caracas Chicago Guilford, CT
London Madrid Mexico City Seoul Singapore Sydney Taipei Tokyo Toronto

WCB Wm. C. Brown Publishers

Vice President and Publisher: James M. Smith
Executive Editor: Lloyd W. Black
Senior Managing Editor: Judith Hauck
Photo Researcher: Donata Dettbarn
Project Manager: Chris Baumle
Production Editor: Stacy M. Loonstyn
Design Manager: Nancy McDonald
Design: Ox & Company, Inc.
Illustrations: ArtScribe, Inc., Greg and Carolyn Duffy
Manufacturing Manager: William A. Winneberger, Jr.

First Edition



A Times Mirror Company



Printed in the U.S.

Composition by Progressive Information Technologies
Printing/binding by Von Hoffman Press

Wm. C. Brown Communications, Inc.
2460 Kerper Boulevard
Dubuque, IA 52001

**Library of Congress Cataloging-in-Publication Data**
Sevenair, John.
Introductory chemistry: investigating the molecular nature of matter/John P. Sevenair, Allan R. Burkett.—1st ed.
p. cm.
Includes bibliographical references (p.—) and index.
ISBN 0-8016-6558-2
1. Chemistry. I. Burkett, Allan R. II. Title.
QD31.2.S468 1996 96-38217
540—dc20 CIP

97 98 99 00 01/9 8 7 6 5 4 3 2 1

# Preface

This textbook is written for students who either have no plans to take or are unprepared for General Chemistry for science majors, but who need an understanding of the sciences in general and/or chemistry in particular. It is intended to help instructors advance their students from their existing knowledge of the world to a deeper, more rigorous and lasting understanding of the science of chemistry.

In all of the chapters, facts and experimental evidence are presented first, followed by a theoretical interpretation of the facts. This approach follows Piaget's ideas. Piaget was a Swiss psychologist who investigated intellectual development, and his concepts have been widely applied by science educators. Piaget wrote about the learning transition from a concrete thinking style (thinking in terms of familiar, tangible things) to a formal operational style (thinking in more abstract or theoretical terms). In his original formulation, Piaget proposed that these styles were typical of adolescence and adulthood respectively.

## KEY INSTRUCTIONAL FEATURES

We have made the writing style welcoming, informal, and colloquial rather than stiff and formal. We have developed explanations that anticipate student difficulties and forestall them. At the beginning of discussions, there are analogies to familiarize students with chemical processes and events.

**Experimental Exercises** and **Experimental Problems** are a unique feature of this textbook. They appear in the body of the text and at the end of the chapter. Each is a short description of an experiment, followed by a series of qualitative and quantitative questions that test the student's comprehension of the chemistry described, and their grasp of the principles at work in the reactions or processes. The goal of these exercises is to develop the students' critical thinking skills, to promote the students' ability to analyze chemical reactions, and to improve their skills in finding the relationship between theory and real phenomena. In some cases, the Experimental Exercises reproduce historical events. For example, one Experimental Exercise is based on Ben Franklin's kite experiment. The answers to the experimental exercise questions are at the end of each chapter. All of the Experimental Exercises are available on CD-ROM, and several are reproduced on transparencies.

The text includes **Tool Boxes** in the main body rather than relegating reviews of math to appendices. Tool Boxes expand on the origins and uses of mathematical tools that can be used for problem solving or representing information. There is a Tool Box on percentages in Chapter 2, working with exponents in Chapter 3, reading graphs in Chapter 4, and logarithms in Chapter 15. Every Tool Box includes examples followed by exercises. In addition, Tool Box–related questions are given in the end of chapter problems where appropriate. The answers to all Tool Box exercises are given at the end of each chapter.

The text also includes **Chapter Boxes**. The Chapter Boxes and the body of the text often describe the historical origin of ideas and techniques. In some ways, early chemists faced the same problems of learning that introductory chemistry students face today. Students can use these specific sections to follow the processes that led to "how we know what we know"—and to some instructive errors. Chapter Boxes also supply further interesting details about subjects discussed only briefly in the text.

This textbook emphasizes helping students visualize the chemistry they are learning at a molecular level. Students are encouraged to translate a problem statement into a picture of the physical reality that occurs at the molecular level. In every possible instance, the art program and text provide molecular views of events. In some instances, students are asked to draw their own versions of molecular events.

**Problem solving** is also emphasized in this textbook. It is well-known among chemistry educators that everyone solves problems one step at a time in five steps or less. Solving a problem in more than five steps is inordinately difficult. It is useful to remember that when you say, "calcium chloride" to a novice, it can take several problem-solving steps to get to the formula $CaCl_2$, but an experienced problem solver is there in far fewer. The textbook's problem-solving approach first shows students how to handle individual steps, and then leads them toward the grouping of steps, which is the technique experienced problem solvers use to reduce problems to five or fewer cognitive steps.

The authors and editors have provided a large quantity and quality of line drawings exceeding competing books in both respects. Art is used in the text, examples, problem sets, and Experimental Exercises. Graphic depiction of concepts and mathematical relationships is strongly emphasized. The line art that illustrates mathematical relationships is color screened to match words and phrases in the legends and text.

## TOOLS FOR INSTRUCTORS AND STUDENTS

**Key terms** are given in each chapter opener, and each section begins with a simply stated **Goal**. Margin notes called **Study Alerts** draw the students' attention to skills to use in problem solving, refresh skills they have already learned, offer analogies to aid comprehension of skills, and direct the student to specific sections that apply to the current material. Other **margin notes** call attention to applications of concepts learned, suggest demonstrations of concepts that students can perform, supply historical anecdotes, mention familiar applications of text concepts, and give further interesting details. **Chapter Summaries** are included at the end of each chapter.

Each section includes at least one **worked sample problem** (called an **Example**). The Examples are immediately followed by **Exercises,** and the answers to the Exercises are given at the end of each chapter. End-of-Chapter Problems are given by section, followed by "mixed" problems. Each chapter has approximately 100 end-of-chapter problems at several levels of difficulty. The answers to selected end-of-chapter problems are given in Appendix B.

The Instructor's Resource Manual that accompanies this textbook contains several unique features. A list of behavioral objectives is given for each chapter. These objectives list operations that students should be able to perform after the completion of each chapter. These objectives can also be used in preparing course syllabi and planning a lecture strategy. A short summary statement is given for each chapter and each chapter section. A list of relevant experiments taken from volumes 1 and 2 of the second edition of *Chemical Demonstrations: A Sourcebook for Teachers* (available from the American Chemical Society in Washington, D.C.) is provided where appropriate. There is a list of films and video material that can be rented or purchased from the Penn State Audio-Visual Service as well as a list of interesting and pertinent web sites. Finally, a list of end-of-chapter problems that have the solutions included in Appendix B of the text is given. The solutions to all of the end-of-chapter problems are also available in the Instructor's Solutions Manual that accompanies the textbook.

# Acknowledgments

These faculty served as reviewers. Each of them gave valuable expert advice and assistance for this book.

Dr. Walter Volland
*Bellevue Community College, WA*

Dr. Kathleen M. Trahanovsky
*Iowa State University*

Dr. J.E. Hardcastle
*Texas Woman's University*

Dr. Wayne Hiller
*California State University, Chico*

Dr. Caroline Ayers
*East Carolina University*

Dr. Mahmoud Jawad
*El Paso Community College*

Dr. Gerald Swanson
*Daytona Beach Community College*

Dr. Gordon J. Ewing
*New Mexico State University*

Dr. James A. Petrich
*San Antonio College*

Dr. Joseph Asire
*Cuesta College*

Dr. David L. Dozark
*Kirkwood Community College*

Dr. Samuel Rieger
*Mattatuck Community College*

Dr. Carleton P. Stinchfield
*Greenfield Community College*

Dr. Joel Shelton
*Bakersfield College*

Dr. Vernon Theilman
*Southwest Missouri State University*

Dr. Mary Ellen Schaff
*University of Wisconsin-Milwaukee*

Dr. Sandor Reichman
*California State University, Northridge*

Dr. Linda Woodward
*University of Southwestern Louisiana*

Dr. T.C. Ichniowski
*Illinois State University*

Dr. Henry Gehrke
*South Dakota State University*

Dr. John J. Aklonis
*University of Southern California*

Dr. Joseph Kanamueller
*Western Michigan University*

Dr. Vernon Thielmann
*Southwest Missouri State University*

Floyd W. Kelly
*Casper College*

Erwin W. Richter
*The University of Northern Iowa*

Pamela Coffin
*The University of Michigan, Flint*

Alton Hassell
*Baylor University*

Gerald Berkowitz
*Erie Community College*

Sharmaine S. Cady
*East Stroudsburg University*

Galen G. George
*Santa Rosa Jr. College*

Katrina Hartman
*Aquinas College*

Stan Cherim
*Delaware County Community College*

Jim Smith, Vice-President and Publisher, signed us on for this project and provided excellent support even after he was promoted to a position supervising an entire line of textbooks. We have worked closely with Jim on two other projects and he is unquestionably one of the finest, most competent working professionals we have encountered.

Judy Hauck, Senior Managing Editor, supervised the review process and the art program, and provided useful, workable suggestions and considerable encouragement during those trying times. Many of the strengths of this textbook are due to Judy. (The weaknesses are ours.)

Jay Freedman, Developmental Editor, worked with us to integrate this flood of expert advice and assistance into the text. Carolyn and Greg Duffy, of ArtScribe, created the wonderful molecular views and other computer artwork for the book. Donata Dettbarn searched the world to find photographs and illustrations from historical sources. When nothing could be found that suited our needs, new photos were taken by Chris Sorensen and Pat Watson.

Once we had our "final" version of the text, Jane Hoover, Copyeditor, edited the manuscript. Dorothy Kurland of the West Virginia Institute of Technology, Accuracy Checker, meticulously read the page proofs, and found and corrected any remaining errors, both minor and not-so-minor. Lloyd Black served as Executive Editor through the entire production phase. Finally, Stacy Loonstyn, Production Editor, supervised the integration of text, images, and the final layout of the manuscript to make this book a coherent whole.

Christopher Rawlings, Assistant Editor, supervised the production of all the ancillary materials. We are grateful for the painstaking work of the authors of these ancillaries. Our thanks go to Daniele Peters, who solved all of the problems for the Solutions Manual; Jerry C. Swanson, who wrote the Laboratory Manual; Joe Ledbetter, for the Study Guide; Donna Friedman, who wrote the Instructor's Resource Manual; and Mitchel Fedak, who wrote the Test Bank.

On another level, one of us, Allan Burkett, thanks Roland Rose and William Stack for their enthusiastic support, and is particularly grateful to Dr. James J. Prestage for his encouragement and understanding during the entire project. The other, John Sevenair, thanks JW Carmichael and the members of the Science Education Research Group at Xavier University of Louisiana for their ideas and inspiration over the years, Sr. Monica Loughlin, Dr. Deidre Labat, Dr. Antoine Garibaldi, and Dr. Norman C. Francis at Xavier for their support. We both wish to thank our students for their indispensable help in teaching us how to teach chemistry.

As you can see, a textbook isn't just something that two authors sit down and write. Any textbook is a major team effort, and we wholeheartedly and gratefully thank every single member of our team.

# Contents in Brief

# Contents

## CHAPTER 2
## Mixtures, Compounds, and Elements 29



## CHAPTER 3
## Measurements and Units 69

## CHAPTER 6
## Atomic Structure In More Detail 189



## CHAPTER 7
## Simple Inorganic Compounds: Names and Properties 221

CHAPTER 9

## Formulas and Reactions: Stoichiometry 299



CHAPTER 10

## Chemical Bonding and Molecular Structure 339

## CHAPTER 12
## Liquids and Solids 425



## CHAPTER 13
## Solutions 455

## CHAPTER 15
## Acids and Bases 543



## CHAPTER 16
## Oxidation-Reduction Reactions 593

1

# AN INTRODUCTION TO THE STUDY OF CHEMISTRY



**KEY TERMS**

Curiosity is one of the defining characteristics of humanity. People have always tried to understand the universe, often starting with things they could see, hear, touch, smell, and taste. As human beings from earlier eras accumulated knowledge and developed ideas, they evolved specialized areas of interest. One of these is chemistry.

In the beginning of this chapter we define chemistry and introduce the philosophy of science from a historical perspective. The main study points focus on the methods and some of the specialized language used by scientists in their work. This chapter also presents some useful guidelines on how to study for this and similar courses.

## THE HISTORY OF CHEMISTRY AND THE PHILOSOPHY OF SCIENCE

### 1.1 What Is Chemistry?

GOAL: To define chemistry and to discuss its boundaries with other areas of study.

**Matter is typically defined as being anything that has mass and occupies space.**

**Chemistry** is the study of matter and the changes it undergoes. People who study and use chemistry have played an essential part in building our society, with all its strengths and weaknesses. A knowledge of chemistry is useful, and frequently necessary, for individuals who want to improve health care, increase food production and energy supplies, and provide clothing, shelter, and consumer goods. Any material that has been altered from the way it occurs in nature has been worked on by chemists or chemical engineers.

FIGURE 1.1
Chemistry-based industries use chemical processes to convert copper-containing minerals (an example is shown above) to copper metal for electronics and plumbing and copper alloys for sculpture (right). The method of conversion is called *smelting* (left), and it poses environmental problems.

Chemistry is in action all around you and within you. The paper this book is made of, the ink on its pages, and the workings of your eyes that make them sensitive to brightness and color are all firmly rooted in chemistry. Though you are much more than the sum of your parts, your physical being is made up entirely and exclusively of chemicals. Instructors sometimes capitalize on this. One trick is to give students a small sign that says "This isn't made of chemicals" and tell them to hang it on something that fits the description. It can't be done. If you can put a sign on something, it consists of chemicals.

### EXAMPLE 1.1 Is It Chemical?

Name something that doesn't consist of chemicals.

**SOLUTION**

Neither emotions nor ideas are made up of chemicals. Hope, love, fear, intelligence, capitalism, and evolution are nonchemical. Physicists study things that don't consist of chemicals, though they may be properties of chemicals. Heat, radiation, and energy fall in this category.

**EXERCISE 1.1**

Which of these are made up of chemicals?
**a.** Birds **b.** Hatred **c.** Air **d.** Food **e.** Speed

Your automobile operates because of chemical reactions. The car starts because the battery (a mixture of chemicals) provides electrical energy to turn the motor. The car moves when you step on the accelerator because gasoline (a chemical) mixes with oxygen (a chemical) from the air and then burns (a chemical reaction). This provides energy to move the pistons, which, after a series of mechanically linked steps, turn the wheels of your car. The car speeds up when you "step on it" because the gas feeds, and thus burns, faster. This moves both the pistons and the wheels faster. Essentially the same thing is happening inside you at this very moment. Oxygen in the air reacts with the food you have eaten to keep your "motor running."

FIGURE 1.2
Chemical reactions provide the energy necessary for both automobiles and athletes to run.

Chemists generally think in terms of fundamental building blocks of matter—atoms and molecules. Some scientists work with larger systems at more complex levels. Biologists study the complexities of living systems, for instance. Because living systems consist of atoms and molecules, however, the boundary between chemistry and biology is not clear-cut and absolute. Chemists who deal with the molecules found in living systems are called *biochemists.* Biologists interested in the same molecules are called *molecular biologists.* These two groups of scientists study essentially the same thing.

Physics is the science of forces, motion, and the internal structure of atoms. Knowledge of these things is often important to the study of atoms and molecules. Chemists who study this aspect of their subject are called *physical chemists.* Some physicists, called *chemical physicists,* study the application of atomic and molecular ideas to their own subject.

Chemistry and physics together make up the *physical sciences.* Along with such disciplines as biology, geology, and astronomy, they form a larger collection called the *natural sciences.* Each individual science has its own obvious subject of study, but the sciences aren't isolated from each other. They have loose and overlapping boundaries, and call on each others' expertise sometimes.

### EXAMPLE 1.2 Identifying Overlapping Areas of Science

On what two fields is each of the following research efforts based?

**a.** Discovering what processes form different types of petroleum
**b.** Studying the flow of blood in arteries
**c.** Finding out what meteorites are made of

**SOLUTIONS**

**a.** Chemists study the kinds of molecules in petroleum. Petroleum (crude oil) comes from the ground and is studied by geologists. (This field is called *geochemistry.*)
**b.** Blood and arteries are studied by biologists. Flow is a type of motion, studied by physicists. (This field is called *biophysics.*)
**c.** Meteorites are pieces of metal or rock that have fallen from space and are interesting to astronomers. Finding out what they're made of is a job for chemists.

**EXERCISE 1.2**

In which of the following areas is a knowledge of chemistry useful?

**a.** Cooking
**b.** Office work
**c.** Nursing
**d.** Panning for gold

## 1.2 The Historical Origins of Modern Chemistry

**GOAL:** To explain how the modern science of chemistry arose from earlier philosophies and technologies.

People have used chemical reactions to make their lives easier for hundreds of thousands of years. Fire was almost certainly the first reaction to be used this way. The people who discovered it were seeking light and warmth, of course, not knowledge. Other reactions that have been known for thousands of years include copper smelting (heating copper compounds to make the metal), fermentation (combining fruits or grains with yeast to produce alcoholic beverages), saponification (heating fats and oils with wood ashes to make soap), and papermaking (converting plant fibers to paper).

FIGURE 1.3
In this Renaissance distillery the operator uses the brick furnace to heat containers of mixed liquids. Some of the liquids evaporate, condense above the furnace, and flow down into flasks.

All of these processes are still in use today. Think about paper. The first papermaking process was developed by the Chinese court official Ts'ai Lun in about 50 AD. Since then paper has been the primary medium for sharing, distributing, and storing information; this book is an example. Communication became easier—the world shrank—as paper became more available.

One goal of early scientists was to figure out what matter is. Such questions as "What is a rock?" and "How can I convert lead into gold?" were taken seriously. The essential tools of science, the methods of asking and answering questions (see Section 1.3), took more than 2500 years to develop. Early attempts were deeply rooted in philosophy. In fact, *natural philosopher* was the term used for those interested in science right up to a century ago. Even today the highest degree you can earn in science is the Ph.D.—Doctor of Philosophy.

**The classification of some scientists as chemists did not occur until the 1600's.**

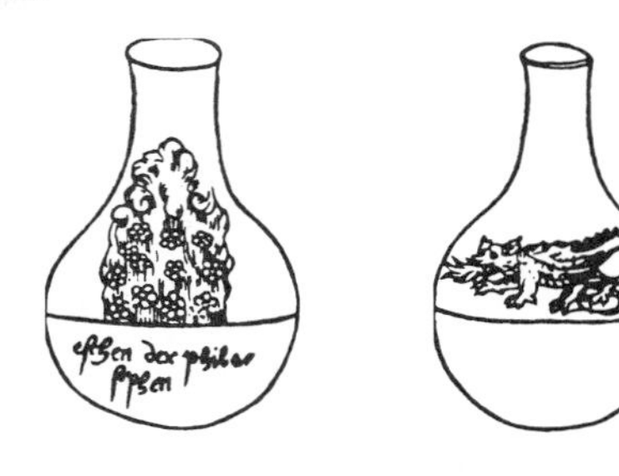

FIGURE 1.4
The elements air, fire, and earth, from illustrations by the Swiss physician and alchemist Paracelsus (Theophrastus Bombastus von Hohenheim, 1493–1541).

What follows is a necessarily brief and oversimplified summary of the earliest chemical theory and practice. As you read through the rest of this section, note the shift in emphasis from the abstract to the concrete as people learned more and more about the world around them.

The Babylonians carefully observed the sky and developed a complex mythology loosely based on what they saw. Some of those myths came down to us through the Greeks in the form we call *astrology*. Those myths had a greater scope in ancient times than they do now. The Greeks thought that not only human lives but also substances were controlled by the stars and planets. The sun controlled gold, and the moon, silver. Mars was associated with iron, and Venus with copper.

The Greek philosopher Thales (640–546 BC) traveled to Egypt. There he learned from both the intensely mystical Babylonian culture and the more practical culture of his North African hosts. Later he proposed that water was the basis of the universe. This seemed to be a perfectly reasonable conclusion, because water is found everywhere—in plants, in animals, and in the ground. Furthermore, water can be converted from its liquid form into either a solid or a gas. Since all matter on earth is in one of these three states, water makes a very sensible choice as the basic element. According to Babylonian myth, the universe was created from water.

Other philosophers followed with competing ideas. Empedocles (about 492–432 BC) combined several of these proposals into one. He theorized that there were four basic elements: earth, air, water, and fire. He viewed transformations of matter as being governed by such things as love and strife.

Aristotle (384–322 BC) adopted much of this system. He thought that the four elements postulated by Empedocles were not truly fundamental but arose from the combination of characteristics that were even more basic. Fire was hot and dry, air was hot and moist, water was cold and moist, and earth was cold and dry. Aristotle also maintained that gold was the most perfect metal.

**Alchemy might be considered the earliest form of chemistry. Historically it was deeply rooted in the philosophies discussed in this section. It also had magical associations. The chief aims of the alchemists were the transmutation of matter (turning common metals, such as lead, into gold) and the discovery of the proverbial elixir of life and perpetual youth.**

Alchemists used this kind of thinking for 2000 years after Aristotle. While trying to remove the imperfections from baser substances to make gold, they developed some useful recipes for making other things. As European civilization collapsed into the Dark Ages, the young and vigorous Arabic civilization took up alchemy. The Arabs borrowed ideas and methods from the Chinese (including papermaking) and made substantial contributions of their own. Europeans were recovering their own cast-aside heritage from the Arabs (with Arabic modifications and improvements) by about the thirteenth century.

In later ages in Europe natural philosophers and mathematicians modified, extended, and in many cases replaced the ideas of the earlier philosophers and alchemists to produce modern science. The details of this fascinating transformation are beyond the scope of this book, but one historian of chemistry, Eduard Farber, has identified three important philosophical threads.

**The ancient Chinese developed a system similar to that of Empedocles but with five elements: water, metal, wood, fire, and earth. The five elements combine with the twelve signs of the Chinese zodiac (which correspond to years, not months) to produce a calendar cycle of 60 years.**

First, substances had to be viewed as independent of the gods and planetary influences before modern chemistry could arise. This was difficult, because the same alchemical symbols were used for some of the planets and elements. Alchemy finally withered away not long before 1800.

Second, words like *love* and *strife* had to be limited to human affairs and not applied to substances and their actions. Modern chemists view reactions as arising from impersonal forces. Physicists, in particular Sir Isaac Newton, established this viewpoint during the 1600's by using such emotionless terms as *gravity* to describe natural forces.

Finally, the chemical elements had to be viewed as specific, distinct substances in their own right, rather than as manifestations of universal principles. Natural philosophers had fully accepted this by the end of the 1600's, and dozens of elements were quickly discovered after that.

FIGURE 1.5
This is both the astrologers' symbol for Venus and the alchemists' symbol for copper. You may recognize it as the symbol used for *female* in modern biology.

Another essential change involved the distribution of knowledge. Alchemists often kept their methods and discoveries secret. They usually passed on their methods only within their families, and if they wrote them down, they deliberately made their texts hard to understand. Leonardo da Vinci did his work in a scientific spirit, but wrote his notes backwards—an effective code in an era when few people could read well. Furthermore, most of da Vinci's work was not published until long after his death. Most modern chemists and other scientists see their obligations very differently. They see themselves as having a duty to publish their results, to make them as widely known as possible. This does not mean that chemists do not use a specialized vocabulary of their own (see Section 1.8). All chemists are taught this vocabulary, and though it may be difficult to master, it isn't secret.

As time went on, chemists refined their methods and made them more precise. Often this meant making more numerical measurements and making them more carefully. One of the earliest practitioners of careful measurement was the French chemist Antoine Lavoisier (1743–1794). Lavoisier was affiliated with the tax collection department and had access to the best available balances. More importantly, he understood the need for accurate weighing. He performed experiments and measurements to test the theory that when a substance is burned, a fluid called *phlogiston* is transferred from one place to another.

Although the phlogiston theory was widely accepted in the eighteenth century and had much to recommend it, Lavoisier was disturbed about some of the properties of this proposed fluid. It apparently had a positive mass some of the time and a negative mass at other times. His experiments demonstrated either that phlogiston did not exist or that, if it did, it had no mass at all. He found out that the substance essential to burning is simply

## EXPERIMENTAL EXERCISE 1.1

The alchemist Jan van Helmont (1577–1644) grew a tree in a weighed amount of soil. He added only water to the soil for 5 years. He found that the tree gained 164 pounds while the soil lost only 2 ounces. Since van Helmont had added only water to the soil, he concluded that the tree's substance was basically water and argued that water is the basic element of the universe.

**a.** What aspect of van Helmont's work goes back to earlier times?

**b.** What did van Helmont do that was pioneering—that is, before its time?

**c.** What assumption did van Helmont make in interpreting his results?

oxygen. These experiments helped transform the young science of chemistry into something recognizable by today's standards. Because of this and other contributions, Lavoisier is sometimes called the Father of Modern Chemistry.

FIGURE 1.6
Antoine Lavoisier is known as the Father of Modern Chemistry for good reason. He was the first to declare that air is a mixture of elements, to establish a workable theory for combustion, and to prepare a reasonably accurate list of elements. He also developed the first system for naming compounds, which made it easier for scientists to talk to one another.

## THE EXPERIMENTAL BASIS OF CHEMISTRY

### 1.3 Experiments and the Scientific Method

GOAL: To explain the scientific method and the role of experiments in it, and to distinguish between what science is and what it is not.

An **experiment** is a test or a trial. It is almost always directed toward answering a specific question or series of questions. For example, at one point in his career the inventor Thomas Edison (1847–1931) needed to dissolve rubber. This was a serious problem because, as far as anyone knew at the time, rubber was insoluble in everything. So Edison had a question, "What dissolves rubber?" To obtain an answer he did a set of experiments. He cut a piece of rubber into small chunks and added one chunk to every bottle in his laboratory that contained a liquid. He succeeded in finding his solvent in this way.

Experiments can be as simple as Edison's. They can also be extremely complicated, taking many days and dollars to perform. They share one goal—to answer questions.

How do you know what experiments to do? How do you figure out what questions need to be answered? All of this is related to a systematic way to ask and answer questions called the *scientific method*. The scientific method is widely known and respected, almost to the point of being a cliché. However, too few people understand what it really is and how it works. One way to explain it is to follow a chemist through a research project.

First, the chemist must decide what to do, what experiments to perform. What line of thought might lead to something new, interesting, or useful? Inspiration can come from knowledge and theory learned in class, from experiences both inside and outside the laboratory, from comments of co-workers, family, and friends, or from one's own cultural background. One chemist reported that one of his best ideas about the structure of compounds came to him in a dream about snakes! Scientists are no different from historians or artists or anyone else in this respect. Good ideas can come from anywhere.

Many ideas in science come from *observations* of the natural world. "Why is the sky blue?" and "Why are diamonds hard?" are questions based on observations. One of the more interesting myths about the scientific method is that observations are necessary to generate ideas. Observations are certainly useful for that, and a thorough knowledge of "what's out there" is tremendously valuable to scientists. However, observations are necessary only in veri-

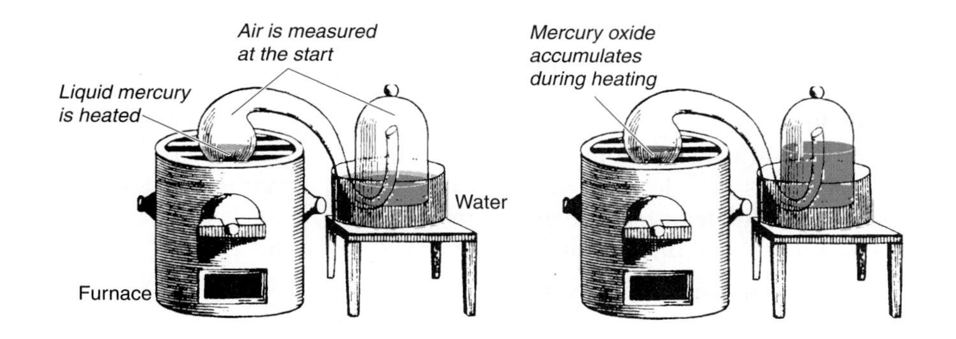


FIGURE 1.7
This apparatus was used for an experiment performed by Antoine Lavoisier to test his theories on combustion. For more details about the experiment, see Experimental Problem 1.1 at the end of the chapter.

fying explanations. The atomic theory (Chapter 4) was almost universally accepted by chemists long before atoms were even indirectly observable.

After generating an idea, the chemist then runs experiments. Some experiments are simple explorations, such as Edison's search for a substance that dissolved rubber. Other experiments address more complex questions: "What will happen if I mix these two compounds together? What new compounds will be formed, how much heat will be given off or taken up, and how fast will the reaction go?" Some experiments are analytical: "What compounds are present in this sample from the environment? Are any of them dangerous?" "How much vitamin C does this fruit contain?" "How much sugar is present in the blood of a healthy individual?" Experiments can also be used to test hypotheses. This is often done by making a prediction, then performing an experiment to determine the accuracy of the prediction: "If the Whoozis Hypothesis is true, I will get water from this reaction. Let's see if Whoozis was right."

Finally, the chemist finishes the experiment, thinks over the results, draws (or jumps to) a conclusion, and writes all this down for other chemists to read. If the results are interesting and important, other chemists will repeat the experiment or do similar ones. Only if others get the same results—that is, if the results are *reproducible*—will the conclusion finally be accepted.

The other chemists will also closely analyze the original experimenter's conclusion. Do the experimental results support the original conclusion? (The answer is not always yes—look at Experimental Exercise 1.1). What other conclusions are possible? If the original conclusion is valid, then what new experiments can be done to test its validity? If the experimental results don't support the hypothesis or could be interpreted to support more than one explanation, then it's back to the drawing board for both experimenters and theoreticians. If the original chemist and the others get no water from the Whoozis experiment mentioned above, then the Whoozis Hypothesis must be thrown out or modified.

How do you tell whether an idea is scientific or not? Physicist and philosopher Sir Karl Popper proposed that for an idea to be scientific, it must be **falsifiable.** In practical terms this means that the idea must be testable by experiment and/or observation. For example, the idea that the dinosaurs were killed off by undetectable rays is not scientific. Undetectable rays, by definition, are not subject to testing. The idea that the dinosaurs were killed by gold cannonballs can be tested, so it can be thought of as scientific. Just be-

FIGURE 1.8
Experiments are the soul of science. Only through experiments and observations can scientific ideas be tested. (Note: The lab coat and safety goggles are worn for safety reasons; they're not just a uniform.)

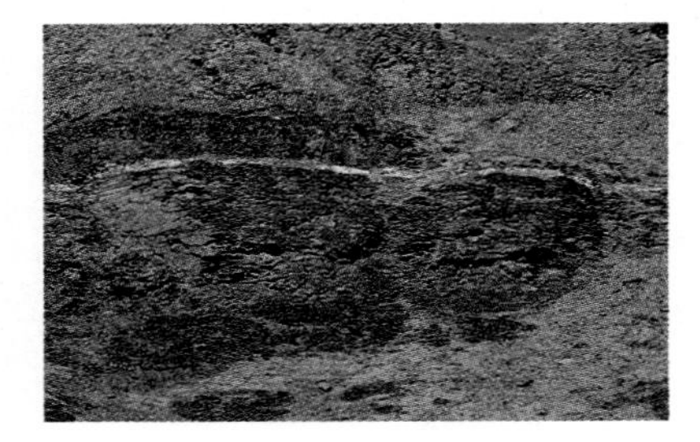

FIGURE 1.9
The narrow pale layer of rock in the middle of the picture was deposited at the very end of the Cretaceous Era, 65 million years ago. It contains relatively high concentrations of the element iridium, which is rare in the earth's crust but more abundant in meteors. Layers like this, deposited at the end of the Age of Dinosaurs, have been found all around the world.

cause an idea is scientific doesn't necessarily mean that it's good, valid, or even plausible! The possibility that the dinosaurs were killed by the effects of a large meteor that hit the earth is also scientific because it can be tested and potentially *falsified* (proven wrong). Chemical and physical traces of a meteoric impact at the correct time have been found in the earth's rocks. This evidence supports the idea that meteors, an asteroid, or a comet might be responsible for the demise of the dinosaurs.

The scientific method, then, has three basic elements:

1. First, you come up with an idea.
2. Next, you perform experiments and make observations to test the idea.
3. Finally, you interpret your results, and discard or modify your original idea if it is in conflict with the results.

Scientists' interpretations very often lead to new ideas, new experiments, and refinements and changes in the original ideas. The basic theme that scientists should go where their results and observations lead them—that they should not blindly accept the ideas of authority figures such as Aristotle—is the keystone of modern science. When the Royal Society of London, one of the earliest scientific organizations, was chartered in 1663, its founders adopted a Latin motto that echoed this idea. The motto in turn was inspired by these lines of the ancient Roman poet Horace:

> I am not bound to swear allegiance
> to the word of any master.
> Where the storm carries me,
> I put into port and make myself at home.

Most scientists try to work in this spirit. They don't always succeed. On the other hand, nobody ever succeeded at anything worthwhile without trying.

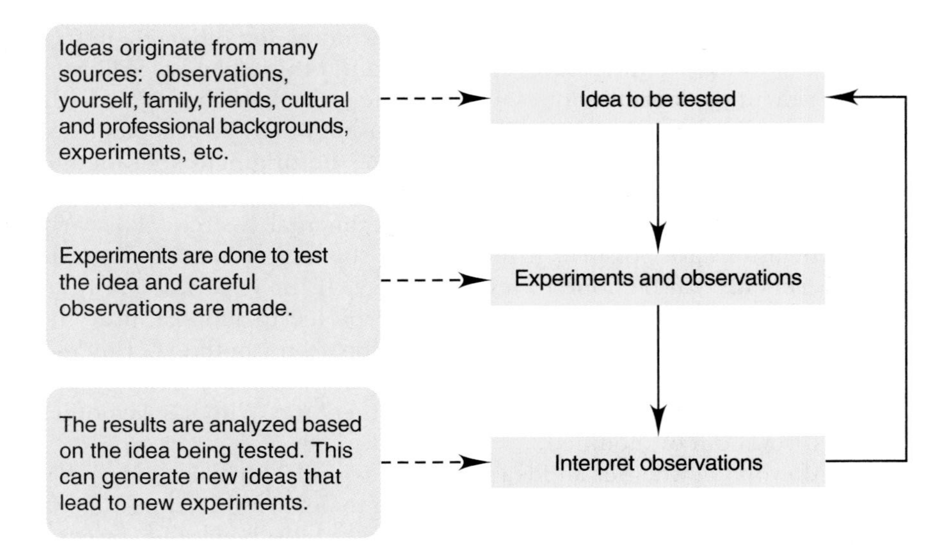


FIGURE 1.10
The scientific method is an endless cycle of ideas, experiments, and interpretations. Most of the technological and scientific advances of the twentieth century came from this systematic approach to problem solving.

## 1.4 Data, Results, and Conclusions

**GOAL:** To define data, results, and conclusions and explain the relationship among them.

To describe the information they obtain from experiments, scientists use three terms—data, results, and conclusions. These words have specific meanings to scientists, and the distinctions are important.

**Data** are the raw *information* you get from experiments. Examples of data include a description of the appearance of a reaction mixture, the melting point of a reaction product, and pairs of numbers used to make points on a graph.

**Results** are data presented in organized form, typically to describe the more important outcomes of experiments. Results might be presented as a table summarizing the appearance of several reaction mixtures, a list of the melting points of several groups of compounds, or a graph.

**Conclusions** are the deductions and inferences that you draw from the results. You might *conclude* that acids form a gaseous substance when added to limestone, or that the melting points of compounds containing metals are higher than the melting points of compounds that do not contain metals.

A key criterion of the scientific method is that data (experimental information) and results (the presentation of the data) should be reproducible if experiments are done in the same way. Anyone who repeats an experiment exactly should be able to get the same information and present it the same

**STUDY ALERT**

**Beginners in science frequently have difficulties distinguishing among data, results, and conclusions. Since the distinctions are both useful and important, you should make certain that you have them straight.**

**The Latin word *data* is plural; *datum* is the singular form. Many people now use the word *data* in English as the singular, the plural, and the name of a character in *Star Trek: The Next Generation.* It is unlikely that your instructor will dip you in boiling oil if you use *data* instead of *datum*.**

Suppose that some engineers heat an object in a furnace and then allow it to cool down again in contact with air. Here is a sample of the data that they might acquire.

| Time (minutes) | Temperature (Celsius) |
|---|---|
| 0 | 20 |
| 6 | 130 |
| 10 | 185 |
| 14 | 230 |
| 16 | 241 |
| 20 | 175 |
| 24 | 131 |
| 36 | 60 |
| 45 | 40 |

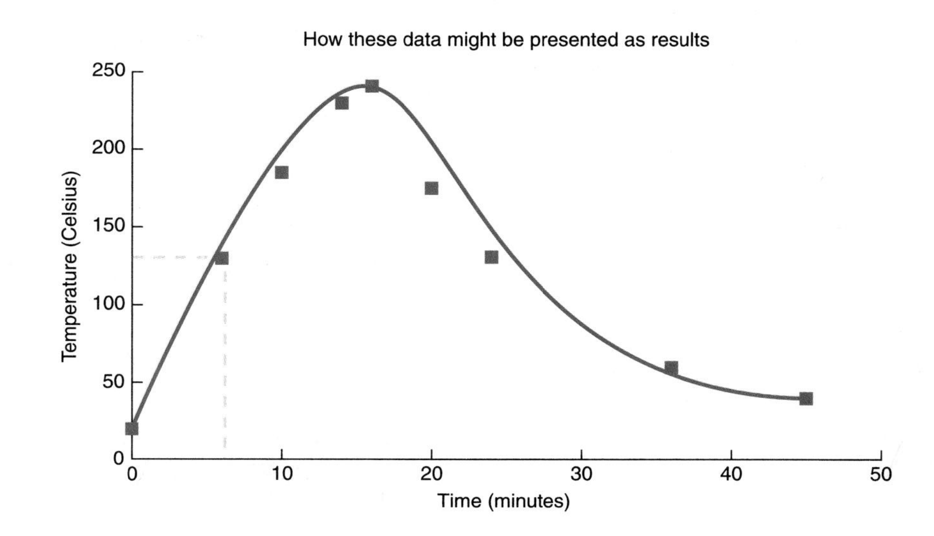


FIGURE 1.11
Each point (shown as a square) on the graph corresponds to two measurements: a temperature and a time. Note that you can read the individual data points for each measurement from the presentation. The dashed lines show one pair of observations, 130°C at the 6-minute mark. The results (the graph itself) are organized to show the trends in the data.

way. Conclusions, however, are attempts to classify or interpret results. The conclusions of the original experimenter and those of later experimenters, even if they get the same data and results, may not coincide.

Here's an example, an imaginary test of the proposal that the dinosaurs were killed off by gold cannonballs. First, scientists find a deposit of dinosaur skeletons and excavate it. The data are their detailed descriptions of the deposit and what was found there, including the precise location of every bone and rock. The results are a summary of the data, telling (among other things) how many skeletons (and cannonballs, if any) were found, what kind of dinosaurs the skeletons came from, and how complete and well-preserved the skeletons were. The conclusions will no doubt include the statement that these particular dinosaurs weren't killed by cannonballs of any kind. Someone might even point out that this conclusion is reasonable because other experiments have indicated that the dinosaurs were dead about 75 million years before there were any cannonball makers.

## 1.5 Hypothesis, Law, and Theory

GOAL: To define hypothesis, law, and theory and explain the relationship among them.

The words *hypothesis, law,* and *theory* describe ideas and are not associated with a particular experiment or series of experiments. All three words are frequently misunderstood and misapplied.

A **hypothesis** is an idea put forth for testing. To a scientist the word suggests an idea that hasn't been carefully examined yet. (The idea that the dinosaurs were killed off by gold cannonballs is a particularly inept hypothesis.)

To scientists **laws** are observations or experimental results that have been confirmed time after time. Laws are usually simple verbal statements or equations. Examples include Boyle's Law (the pressure of a gas multiplied by its volume is a constant in a closed system at constant temperature), Newton's Third Law (to every action there is an equal and opposite reaction), and Murphy's Law (if anything can go wrong, it will).

**Theories** are more general constructions, put forth to explain either laws or behavior that seems to follow some law. Generally laws tell what happens, and theories explain why. Laws are simple statements, and theories are more complex.

You may have been told that laws are more accurate and more widely accepted than theories, but the opposite is true fairly often. Newton's *laws* have been replaced and extended by the *theory of relativity,* because relativity gives predictions that are more accurate in many cases. The most accurate numerical predictions in the history of science have been made using a theory called *quantum electrodynamics.*

**EXPERIMENTAL EXERCISE 1.2**

In the latter half of the 1700's many people still believed in the four ancient Greek elements and in transmutation. Supporters of transmutation pointed out that if water is heated in a glass vessel for many days, a sediment forms. Lavoisier decided to look more closely at this observation. He had a completely closed glass vessel made, so that he could boil water without letting the steam escape. He carefully weighed both the vessel and the water and then boiled the water in the vessel for 100 days. By the end of the experiment a sediment had formed at the bottom of the vessel. Lavoisier weighed the water, the sediment, and the glass.

The mass of the water remained unchanged. The mass of the glass decreased by an amount exactly equal to the mass of the sediment. Lavoisier decided that the sediment was not water that had been converted into another substance, but rather material that had been removed from the vessel itself during the experiment.

**a.** What hypothesis was Lavoisier testing?
**b.** What data were collected?
**c.** What conclusion did Lavoisier draw from the results?
**d.** Do the results support Lavoisier's conclusion?
**e.** If the water had been transmuted into sediment, how would Lavoisier's results have been different?

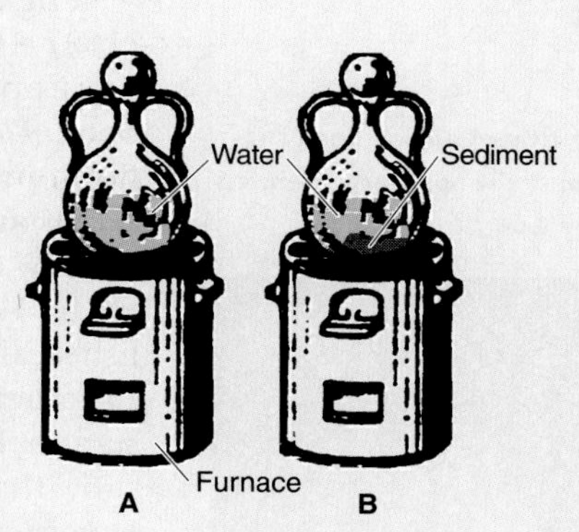


FIGURE 1.12
The vessel Lavoisier used in this experiment was called a *pelican* because its shape was thought to resemble that of the bird. Part (A) shows the vessel before 100 hours of heating, and part (B) shows the same vessel after the experiment.

## THE DEVELOPMENT OF THE PERIODIC TABLE

### 1.6 What is an Element? An Experimental Definition

**GOAL:** To define the term *element.*

An exciting time for science had emerged in Europe by the early seventeenth century. The Dark Ages were following the dinosaurs into extinction (once more without the use of gold cannonballs), and creative thinking was encouraged on all fronts (though not by everybody). These were the times of Copernicus and Galileo, and major advances were being made in astronomy, physics, and chemistry. The philosophical and intellectual climate was receptive to scientific advance.

England was a hotbed of work on new ideas, and Robert Boyle (1627–1691) was a leader in what we now call physics and chemistry. In 1661 he proposed an experimentally based definition of the word *element* in his book *The Skeptical Chemist:* an element is a substance that cannot be broken down into simpler substances. Boyle continued to believe in the ancient elements and in the possibility that other metals could be turned into gold, but his new definition paved the way for progress.

**What does the title of Boyle's book suggest to you about the approach scientists take when they work?**

FIGURE 1.13
By defining the word *element* (and in many other ways), Robert Boyle moved science toward the modern era.

In 1789 Lavoisier revived Boyle's definition of an element, and he published a list of substances he believed were elements based on that definition. Lavoisier's list wasn't perfect, and he himself wasn't certain that all the substances listed were in fact elements. He recognized that it is much easier to show that a substance is not an element than it is to verify that it really is one. Lavoisier's doubts were justified, because his list of elements included some compounds called oxides that are very hard to break down. But the list served as a basis for more research.

FIGURE 1.14
These ten elements are, from left to right, carbon, copper, gold, iron, lead, mercury, silver, sulfur, tin, and antimony. Alchemists were familiar with all of these substances (though they did not think of them as elements) long before the modern concept of the elements was developed by Boyle.

Since Lavoisier's time, of course, Boyle's definition of the word *element* has been revised. Physicists have succeeded in breaking atoms apart and putting new ones together, and almost everyone knows that nuclear weapons work this way. You have almost certainly heard of protons, neutrons, electrons, and perhaps other subatomic particles. Physicists, however, use much more energy to build and break up atoms than chemists use for anything they do.

Here is a modern definition: an **element** is a substance that cannot be broken down into simpler substances by chemical methods.

Based on this modern definition, there are currently 112 known elements. Physicists have made 21 of these in their laboratories, and the other 91 occur in nature.

The early alchemists knew 10 of the elements, although they didn't recognize their elemental characteristics at the time. Nine of these—carbon, copper, gold, iron, lead, mercury, silver, sulfur, and tin—are more or less familiar to you, and the tenth is antimony. Two more elements—arsenic and phosphorus—were added to the list by the last half of the seventeenth century. About 50 elements were known by 1850, and the naturally occurring elements were all known by the first quarter of this century. Then technological developments made it possible for physicists to prepare new elements (frequently called artificial elements, or transuranium elements). All of these new elements are notoriously unstable. They decompose into more familiar elements, usually very rapidly.

**The differences between the *nuclear* energy studied and used by physicists and the *chemical* energy studied and used by chemists are discussed in Chapter 17.**

FIGURE 1.15
Elements 110 and 111 were first made in late 1994 by an international group of scientists at Darmstadt in Germany using the instrument shown here, a heavy-ion accelerator.

EXAMPLE 1.3 **Identifying Elements and Compounds From Their Properties**

Which fact is evidence that water is *not* an element?

**a.** When water is chilled, it freezes.
**b.** Water forms two gases when an electric current is run through it.
**c.** When water is added to a certain pink solid, the solid turns blue.
**d.** Water is clear and colorless in small quantities and blue in large amounts.

SOLUTION

In (b) water is chemically broken down into two substances. This could not happen if water were an element.

EXERCISE 1.3

Which of these facts about aluminum supports the conclusion that aluminum is an element?

**a.** Heating samples of aluminum in a vacuum, even to 500°C, doesn't change them.
**b.** Aluminum samples gain weight and form a white solid when heated with oxygen.
**c.** Foil made from aluminum has a shiny, metallic luster.
**d.** Adding aluminum to water gives no observable change.

## 1.7 Dmitri Mendeleev and the Periodic Table

**GOAL:** To explain how the modern organization of elements came about.

The German chemist Julius Lothar Meyer independently developed a periodic table very similar to Mendeleev's and deserves partial credit for the idea. He worked with atomic volumes rather than atomic masses, published his table a year after Mendeleev did, and did not predict new elements.

By the late 1860s more than 60 elements were known, even after the removal of the oxides from Lavoisier's original list. Some of the elements resembled others, forming families of sorts. A few pioneers were trying to organize the elements into a larger scheme. The Russian chemist Dmitri Mendeleev succeeded.

Mendeleev's method was simple. He wrote the names, atomic masses, and properties of the elements on cards. He formed a table by arranging the elements in rows from lightest to heaviest and creating columns of elements that had visible and uniform trends in their chemical behavior. Once he was satisfied, he printed his table on a single piece of paper and had it published. Unlike most Russian work of that era, it was quickly translated and made known in Western Europe.

Mendeleev made some bold predictions based on his table. It seemed that placing some elements a little out of the strict order of atomic masses would give a better organization of rows and columns, and he did that. He predicted that the accepted atomic masses would be proven wrong in those cases. Where there seemed to be a gap in the table, he left it and predicted that a previously unknown element would be found to fill it. Chemists quickly found new elements to fill the gaps, and Mendeleev's table was accepted with enthusiasm.

THE PERIODIC ARRANGEMENT OF THE ELEMENTS

| Periods | Group 0 | | Group I | | Group II | | Group III | | Group IV | | Group V | | Group VI | | Group VII | | Group VIII |
|---|---|---|---|---|---|---|---|---|---|---|---|---|---|---|---|---|---|
| | A | B | A | B | A | B | A | B | A | B | A | B | A | B | A | B | |
| 1 | He = 3.99 | | Li = 6.94 | | Gl = 9.1 | | B = 11 | | C = 12 | | N = 14.01 | | O = 16 | | F= 19 | | |
| 2 | | Ne = 20.2 | | Na = 23 | | Mg = 24.32 | | Al = 27.1 | | Si = 28.3 | | P = 31.04 | | S = 32.07 | | Cl = 35.46 | |
| 3 | A = 39.88 | | K = 39.1 | | Ca = 40.07 | | Sc = 44.1 | | Ti = 48.1 | | V = 51 | | Cr = 52 | | Mn = 54.93 | | Fe = 55.84<br>Ni = 58.68<br>Co = 58.97 |
| 4 | | | | Cu = 63.57 | | Zn = 65.37 | | Ga = 69.9 | | Ge = 72.5 | | As = 74.96 | | Se = 79.2 | | Br = 79.92 | |
| 5 | Kr = 82.92 | | Rb = 85.45 | | Sr = 87.63 | | Y = 89 | | Zr = 90.6 | | Cb = 93.5 | | Mo = 96 | | | | Ru = 101.7<br>Rh = 102.9<br>Pd = 106.7 |
| 6 | | | | Ag = 107.88 | | Cd = 112.4 | | In = 114.8 | | Sn = 119 | | Sb = 120.2 | | Te = 127.5 | | I = 126.92 | |
| 7 | X = 130.2 | | Cs = 132.81 | | Ba = 137.37 | | La = 139 | | Ce–Yb*<br>140.25 – 172 | | Ta = 181.5 | | W = 184 | | | | Os = 190.9<br>Ir = 193.1<br>Pt = 195.2 |
| 8 | | | | Au = 197.2 | | Hg = 200.6 | | Tl = 204 | | Pb = 207.1 | | Bi = 208 | | | | | |
| 9 | Nt = 222.4 | | | | Ra = 226.4 | | | | Th = 232.4 | | U = 238.5 | | | | | | |
| Formula of oxide<br>Formula of hydride | | | $R_2O$<br>RH | | RO<br>$RH_3$ | | $R_2O_3$<br>$RH_3$ | | $RO_3$<br>$RH_4$ | | $R_2O_5$<br>$RH_3$ | | $RO_3$<br>$RH_2$ | | $R_2O_7$<br>RH | | $RO_4$ |

* This includes a number of elements whose atomic weights lie between 140 and 173, but which have not been accurately studied, and so their proper arrangement is uncertain.

FIGURE 1.16
Early periodic tables, such as the 1915 version shown here, did not look quite like modern periodic tables (such as the one shown in Figure 1.17). The idea behind this and all other periodic tables is the same as Mendeleev's original idea—to group the elements by similarities in their properties.

| | 1A | 2A | 3B | 4B | 5B | 6B | 7B | 8B | 8B | 8B | 1B | 2B | 3A | 4A | 5A | 6A | 7A | 8A |
|---|---|---|---|---|---|---|---|---|---|---|---|---|---|---|---|---|---|---|
| 1 | 1 **H** 1.008 | | | | | | | | | | | | | | | | | 2 **He** 4.003 |
| 2 | 3 **Li** 6.941 | 4 **Be** 9.012 | | | | | | | | | | | 5 **B** 10.81 | 6 **C** 12.01 | 7 **N** 14.01 | 8 **O** 16.00 | 9 **F** 19.00 | 10 **Ne** 20.18 |
| 3 | 11 **Na** 22.99 | 12 **Mg** 24.31 | | | | | | | | | | | 13 **Al** 26.98 | 14 **Si** 28.09 | 15 **P** 30.97 | 16 **S** 32.07 | 17 **Cl** 35.45 | 18 **Ar** 39.95 |
| 4 | 19 **K** 39.10 | 20 **Ca** 40.08 | 21 **Sc** 44.96 | 22 **Ti** 47.90 | 23 **V** 50.94 | 24 **Cr** 52.00 | 25 **Mn** 54.94 | 26 **Fe** 55.85 | 27 **Co** 58.93 | 28 **Ni** 58.70 | 29 **Cu** 63.55 | 30 **Zn** 65.39 | 31 **Ga** 69.72 | 32 **Ge** 72.59 | 33 **As** 74.92 | 34 **Se** 78.96 | 35 **Br** 79.90 | 36 **Kr** 83.80 |
| 5 | 37 **Rb** 85.47 | 38 **Sr** 87.62 | 39 **Y** 88.91 | 40 **Zr** 91.22 | 41 **Nb** 92.91 | 42 **Mo** 95.94 | 43 **Tc** (98) | 44 **Ru** 101.1 | 45 **Rh** 102.9 | 46 **Pd** 106.4 | 47 **Ag** 107.9 | 48 **Cd** 112.4 | 49 **In** 114.8 | 50 **Sn** 118.7 | 51 **Sb** 121.8 | 52 **Te** 127.6 | 53 **I** 126.9 | 54 **Xe** 131.3 |
| 6 | 55 **Cs** 132.9 | 56 **Ba** 137.3 | 71 **Lu** 175.0 | 72 **Hf** 178.5 | 73 **Ta** 180.9 | 74 **W** 183.9 | 75 **Re** 186.2 | 76 **Os** 190.2 | 77 **Ir** 192.2 | 78 **Pt** 195.1 | 79 **Au** 197.0 | 80 **Hg** 200.6 | 81 **Tl** 204.4 | 82 **Pb** 207.2 | 83 **Bi** 209.0 | 84 **Po** (209) | 85 **At** (210) | 86 **Rn** (222) |
| 7 | 87 **Fr** (223) | 88 **Ra** (226) | 103 **Lr** (260) | 104 **Rf** (Unq) | 105 **Ha** (Unp) | 106 **Sg** (Unh) | 107 **Ns** (Uns) | 108 **Hs** (Uno) | 109 **Mt** (Une) | 110 | 111 | 112 | | | | | | |

| | | | | | | | | | | | | | | | |
|---|---|---|---|---|---|---|---|---|---|---|---|---|---|---|---|
| 6 | Lanthanides | 57 **La** 138.9 | 58 **Ce** 140.1 | 59 **Pr** 140.9 | 60 **Nd** 144.2 | 61 **Pm** (145) | 62 **Sm** 150.4 | 63 **Eu** 152.0 | 64 **Gd** 157.3 | 65 **Tb** 158.9 | 66 **Dy** 162.5 | 67 **Ho** 164.9 | 68 **Er** 167.3 | 69 **Tm** 168.9 | 70 **Yb** 173.0 |
| 7 | Actinides | 89 **Ac** (227) | 90 **Th** 232.0 | 91 **Pa** (231) | 92 **U** 238.0 | 93 **Np** (244) | 94 **Pu** (242) | 95 **Am** (243) | 96 **Cm** (247) | 97 **Bk** (247) | 98 **Cf** (251) | 99 **Es** (252) | 100 **Fm** (257) | 101 **Md** (258) | 102 **No** (259) |

FIGURE 1.17

This is a modern periodic table. Elements in vertical columns on this table are similar to one another. The elements in a column belong to the same *group*. The elements in two of the groups are highlighted. Members of the group on the left, highlighted in brown, are called the *alkaline earth metals,* and members of the group on the right, highlighted in green, are called the *halogens*. The elements in a group have similar properties, and the properties vary systematically within groups.

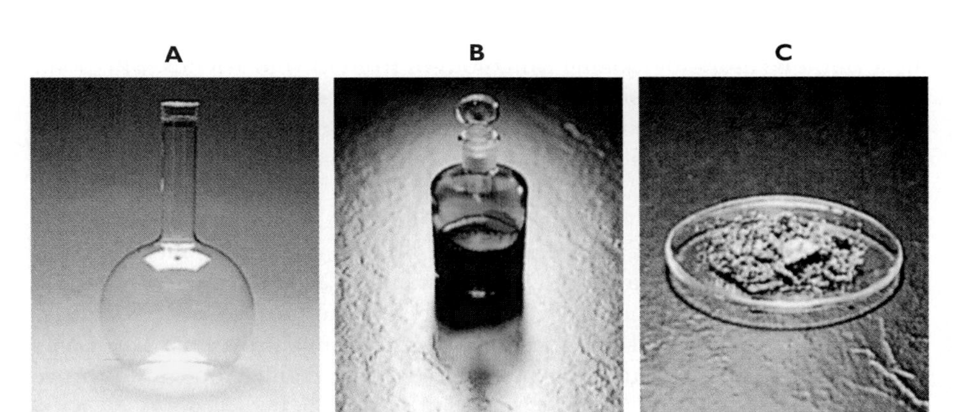


FIGURE 1.18

The three elements shown here are (A) chlorine, (B) bromine, and (C) iodine. These are members of a group of elements collectively known as the halogens (see Figure 1.17).

FIGURE 1.19
Dmitri Ivanovich Mendeleev (1834–1907). Mendeleev's development of the periodic table provided scientists with a systematic grouping of the elements on the basis of their properties. This meant that instead of having to memorize the chemistry of the elements one at a time, scientists could learn trends and predict the outcome of thousands of experiments. It is still necessary to memorize details in chemistry, but the periodic table makes a chemist's life much easier.

Mendeleev's other prediction—that certain atomic masses were wrong—proved to be incorrect. Fortunately for Mendeleev, atomic masses are not fundamentally important in the organization of the table. We'll go into this issue in more depth in Chapter 5.

## HOW TO STUDY CHEMISTRY

### 1.8 The Language of Science

**GOAL:** To emphasize the importance of using specialized language carefully and precisely when dealing with chemistry and other sciences.

The word *jargon* can be defined as the specialized vocabulary of a profession. Another definition is baffling speech. These two definitions are certainly connected—if you don't know the specialized vocabulary, then speech must be baffling! During this course, you will be introduced to a large number of words that at least some people would classify as jargon. Much of this jargon can simply be memorized. For example, in Chapter 2 you must learn the names and symbols of some of the elements.

Difficulty with the language of science, however, cannot be overcome merely by memorizing the meanings of words you don't know. (Both your instructor and the authors of this book assume you will do that.) You may encounter communication problems involving words that you already know, but that are used differently by scientists. Words such as *mass, weight, energy,* and *temperature* have very precise and specific definitions in the scientific community. A phrase such as "I'm full of energy today" is fine and dandy, as long as you are not trying to be scientific. If you said that to scientists in a scientific context, they might want to know what pharmaceutical you took with breakfast!

**STUDY ALERT**

**One of the greatest barriers to learning is the fear of looking silly in front of your fellow students. When was the last time you gave a term paper to your roommate to proofread? Were you too embarrassed to share your paper with a friend? Did you turn it in anyway, even if you knew it wasn't your best work? Students have an exaggerated idea of the consequences of looking silly. Your parents won't disinherit you and your dog won't abandon you if you practice using scientific vocabulary regularly.**

A very good way to learn jargon is to use it. When you discuss various aspects of this course with your friends, use the language of the course. (You have formed a study group, haven't you? See Section 1.9, on problem solving.) Don't be afraid of looking ridiculous in front of your friends. The point is to get the best grade you can and learn some chemistry along the way. Ten years from now your grade will still be on your transcript, and not a single soul in your class will remember that you "talked funny" during the study sessions for this course.

Precision in scientific language is not limited to definitions of words. Entire paragraphs, articles, chapters, and books are written to mean exactly what they say, no more and no less. For example, in Experimental Exercise 1.2 you were asked the question "Do the results support Lavoisier's conclusion?" You were *not* asked the question "Do the results prove that Lavoisier's conclusion is true?" because this question is unanswerable. You cannot prove a hypothesis on the basis of one experiment, although you can disprove one. When you read in science, you must be very careful not to make inferences that are not supported by what is written.

The word *can,* for example, causes students no end of trouble. All too frequently, *can* means "maybe." "This object can turn blue" does not mean that the object is going to turn blue. It means that the object may or may not turn blue sometime between now and the end of time.

A good way to practice identifying the precise meanings of words is to listen carefully to TV commercials. Although federal law prohibits outright lies in ads, you might hear, for example, Company X promise, "These are the best tires we ever made!" What does that tell you about the quality of the tires? Nothing. The best tires Company X ever made could be worse than cheaper tires from another manufacturer. The slogan is an attempt to seduce you into drawing conclusions that are not supported—a serious "no-know" in science.

Here is another example. Company Y advertises that its toothpaste is "unsurpassed in fighting cavities." Does this mean that Company Y manufactures the best toothpaste available for fighting cavities? Nope! "Unsurpassed" means that the toothpaste is at least as good as any other toothpaste. It does not mean that it is uniquely the best toothpaste.

Try listening to commercials in this light. You will find that your appreciation for the precise use of language will grow rapidly.

## 1.9 An Introduction to Problem Solving

**GOAL:** To emphasize the importance of problem solving and to describe a method of learning how to do it.

Everybody has problems to solve. They can be simple ("What do I have for lunch?"), profound ("What can I do to improve the environment?"), or *really* profound ("How can I get a date for Friday night?"). You have been solving problems all your life, and whether you know it or not, you have developed a variety of techniques for doing so. Solving problems in science, however, often requires a slightly different approach from the one you use to solve everyday problems. To solve science problems you must often apply logic and grammar in a rigorous and unfamiliar way. Don't panic! You can learn to solve problems the same way you learn any other skill—through practice.

FIGURE 1.20
Golf and problem solving can both be learned.

Think about how you might acquire a new skill. Suppose you were trying to learn how to hit a golf ball. First, a golf pro would show you the proper grip and swing. Then you'd practice your swing while the pro was watching and making suggestions. Finally, you'd practice on your own, often for hours at a time. When you had difficulties, you'd go back to your pro for more lessons or ask a fellow golfer for suggestions.

You might ask what golf has to do with learning to solve problems in chemistry. Golf swings are visible, but problem solving happens inside your

FIGURE 1.21
If you solve problems with your friends, you (and they) will probably get better at it faster.

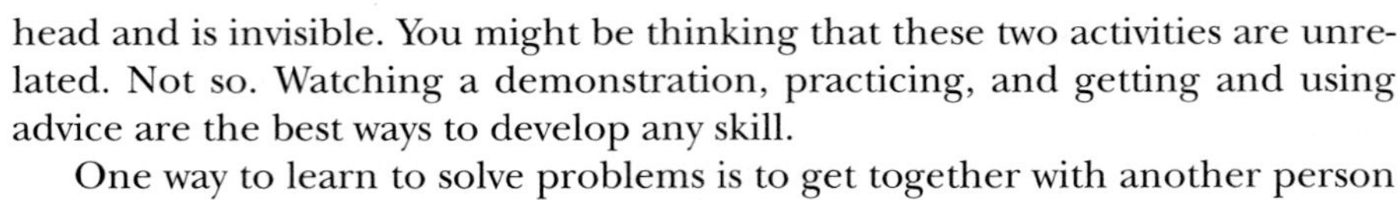

head and is invisible. You might be thinking that these two activities are unrelated. Not so. Watching a demonstration, practicing, and getting and using advice are the best ways to develop any skill.

One way to learn to solve problems is to get together with another person or a small group. One person tries to solve the problem, describing the process *out loud* while the others listen. If the problem solver doesn't express an idea clearly, a listener should ask for clarification. If the problem solver makes a mistake, a listener should make corrections gently, knowing that the roles of problem solver and listener will be exchanged later. If it's followed carefully, this process can make you a good problem solver.

Courses that involve problem solving, including this one, generally require more time than other courses. Problem solving is time-consuming because you are involved in a creative activity. You are also learning and developing reasoning skills when you solve problems in science, and this too takes time. The more problems you solve the easier it gets—but those first hundred problems seem like a mountain!

You'll make things easier if you start working problems early. It is very useful to work chemistry problem sets in small chunks—say, two or three a day, every day. This approach will help you get the correct answer more frequently, because you are allowing yourself enough time to think seriously about the problems. Furthermore, you will certainly get better scores on your examinations if you work a lot of problems (and remember, your transcript will follow you around). When you read a section of this book, work the problems that go with it and/or the assigned end-of-chapter problems. *Don't* wait until you finish a chapter before you start working the problems. The more problems you work, the better your chemistry grade will be.

## 1.10 The Best Way to Use Your Time

**GOAL:** To emphasize the importance of using time wisely and to suggest ways to do this.

FIGURE 1.22

One of the most important of all study skills is using your time efficiently and wisely. The best way to illustrate this is by practical example: How long do you take for lunch? If you're like most people, lunchtime is 1 hour, so that's how long you take. You finish eating in 30 minutes and then discuss whatever comes to mind with your friends for the other 30 minutes. What will happen if you give up the 30 minutes of gossip and read your textbook instead? In half an hour the average student can read four pages. If your semester lasts 15 weeks, and you read at lunchtime for 5 days a week, you will finish 300 pages of the book by the end of the semester. You will in fact have done most of the required reading over lunch!

Once you have read the book and produced a good set of notes (see Section 1.11), what do you do next? You may answer, "Go over the book (or notes) again." It would be more effective if you start to solve problems as soon as possible. The fact that you have gotten this far in your education strongly suggests that you can memorize facts and definitions. Problem solving is harder for almost all students, and exams in chemistry are made up mostly of problems.

The important point is to use your time wisely, and only you can decide how to do that. One trap to avoid, however, is allocating the same amount of study time to each of your courses. Some courses are more difficult than others or place greater demands on your concentration. All courses are not created equal in this respect. There is no general guideline—what's difficult for you may be easy for Lolina, and what's difficult for Eduard may be easy for you. You have to allocate time according to your specific needs and talents.

Another point to consider is *when* you study *what.* It is a good idea to study your most difficult courses at the time of day you are most ready to learn. This varies widely among individuals. One of the authors of this book thinks that 6:30 AM is the best time of day to work on creative projects and that midafternoon should be dedicated to taking a nap, if possible. The other author has just the opposite opinion. He thinks that 6:30 AM is a good time to sleep and that afternoons and evenings are best for creative endeavors.

## 1.11 Taking Complete and Accurate Notes

**GOAL:** To describe methods for making sure that your notes are as complete and accurate as possible.

You cannot take a complete and accurate set of notes in class. Social scientists have verified this in experiments specifically designed to measure the completeness and accuracy of student notes. The results suggest that the *best* note takers get about 80% of the material. In other words, if you are really good at taking notes, you will miss a fifth of the material presented in class! The average person misses about 40% of the material. Don't be discouraged by this; you're in good company. It has been seriously suggested that if Albert Einstein had not had a friend who took good notes, he wouldn't have graduated from college. After graduation Einstein tried to find a university position but finally had to take a job in the Swiss patent office. Of course, he didn't waste his time there—he worked hard to refine his ability to visualize as he reviewed inventions. He spent most of his spare time thinking about theoretical physics. After four years in this job Einstein finished his Ph.D. requirements (in 1905) and wrote five papers, all of which were of Nobel Prize quality.

FIGURE 1.23
Writing down what is said in class, and later rewriting your notes to make sure they are accurate and that you can understand them, is an important method of studying.

The key to getting a good set of notes to study from is to recognize that paying attention in class isn't enough. There are basically two options, and both of them are useful. You can get together with your fellow class members, discuss the day's lecture, and compile a set of notes. Or you can read this book and embellish your notes with what you find here. Both of these methods work. Not only does either method ensure that you will end up with a good set of notes, but applying either one requires you to think seriously about the content of the course. Combined with practice at problem solving, thoughtful note taking improves your learning, enhances your GPA, and puts you in line to acquire more marketable skills.

## CHAPTER SUMMARY

1. Chemistry is the study of matter and the changes that matter undergoes.
2. The subject areas studied by scientists in various fields often overlap.
3. An idea is considered scientific only if there is some possible way of proving it invalid (that is, of falsifying it) by observation and/or experiment.
4. The scientific method is a systematic approach to asking and answering questions. It involves three basic steps: (1) framing an idea or hypothesis, which can come from almost anywhere; (2) testing the hypothesis by experiment and observation; and (3) modifying the idea if the experimental results conflict with it.
5. The raw information garnered from experiments is classified as data. Results are an organized presentation of data, and conclusions are deductions and inferences based on results.
6. A hypothesis is a method used for testing ideas. Laws are confirmed observations or experimental results. Theories are general explanations of laws or behaviors.
7. Elements are substances that cannot be broken down by chemical methods.
8. The periodic table is used to characterize the elements on the basis of their chemical similarities. Elements with similar properties are grouped vertically in the periodic table.
9. Mastering the language of science is essential to communicating effectively with your instructor and other scientists.
10. Problem solving can be a time-consuming and often frustrating process, but you can learn it if you work at it. A very good method is to work problems aloud while other students or a tutor listen and comment, and then to exchange roles.
11. To use time wisely you must plan carefully and think ahead.
12. No individual, no matter how skilled, can compile a complete set of notes during a classroom lecture. To obtain a complete and accurate set of notes, you must make efforts outside the classroom.

## PROBLEMS

### SECTION 1.1 What Is Chemistry?

1. Give three specific professions (other than chemist) in which a knowledge of chemistry is useful and important.

*2. Give three specific activities in your life in which chemicals play an important role.

3. List three things not already named in this chapter that do not consist of chemicals.

4. List three things not already named in this chapter that consist exclusively of chemicals.

5. On what two fields are each of the following research efforts based?
   a. Studying the composition of the sun
   b. Finding out what pollutants are in the air
   c. Determining the composition of chromosomes

* You will find solutions to the problems numbered in red in Appendix B.

6. On what two fields are each of the following research efforts based?
   a. Determining the composition of Jupiter's atmosphere
   b. Measuring the force required to bend a steel beam
   c. Studying the composition of minerals
7. Name a subject of study that would involve both of the fields named. Do not use any of the examples given in the previous problems.
   a. Chemistry and astronomy
   b. Chemistry and physics
   c. Chemistry and medicine
   d. Biology and physics
8. Name a subject of study that would involve both of the fields named. Do not use any of the examples given in the previous problems.
   a. Chemistry and biology
   b. Chemistry and geology
   c. Chemistry and engineering
   d. Geology and physics

### SECTION 1.2 The Historical Origins of Modern Chemistry

9. Name or describe three chemical reactions that were extensively used by many societies in the year Jesus was born.
10. Name or describe three chemical reactions, not including your answers to the previous problem, that were extensively used by many societies around 1800.
11. Present an argument that Empedocles might use to justify classifying earth, air, water, and fire as elements.
12. Present an argument that Aristotle might use to justify his conclusion that earth, air, water, and fire arose from more fundamental characteristics.
13. Without using examples, describe the three philosophical threads that Eduard Farber used to describe the emergence of modern science from older versions.
14. Each of the following sentences illustrates one of Farber's three philosophical threads in the evolution of modern chemistry. Which thread is represented by each sentence?
   a. White phosphorus loves to react with oxygen; if you put the two together you get an immediate fire.
   b. The properties of iron arise from its essential masculinity. (Yes, people did talk like that once.)
   c. Iron and Mars have a mystical association; they have the same symbol. [If you think about part (b) of this question you might be able to figure out what the symbol is.]

### SECTION 1.3 Experiments and the Scientific Method

Describe an experiment that you can perform to test each of the following statements.

15. Hot water freezes faster than cold water.
16. A open bottle of ketchup absorbs water from the air.
17. Table salt is more soluble in water than sugar is.
18. Vinegar dissolves in water and not in cooking oil.
19. Diamonds are harder than sugar cubes.
20. Air can be compressed.

Which of these ideas can be regarded as scientific? Explain briefly.

21. All men are created equal.
22. A little learning is a dangerous thing.
23. Two objects cannot occupy the same space at the same time.
24. There are 112 chemical elements.
25. Overeating is bad for your health.
26. Texas is larger than Japan.
27. You have not worked enough end-of-chapter problems.
28. A good note taker misses 20% of the material presented in class.
29. To love learning is to be near knowledge.
30. Sexism in a society hinders that society's productivity.
31. The love of money is the root of all evil.
32. Air is an element.
33. Smoking is bad for your health.
34. There are 91 naturally occurring elements.
35. Automobiles made in Japan are better than U.S.-made automobiles.
36. The more problems you work, the better your exam scores will be.
37. Intelligence can be taught.
38. No two sets of fingerprints are identical.
39. Strong emotions are caused by chemical reactions.
40. Competent modern biologists know a lot of chemistry.
41. What you learn is more important than the grade you get.

### SECTION 1.4 Data, Results, and Conclusions

42. Classify each of these statements about a specific experiment as data, results, or conclusion, and briefly explain your answer.
    a. A sample of salt water had a mass of 1000 grams.
    b. After the sample was evaporated, 192 grams of solid remained.
    c. The sample contained 19.2% dissolved solids.
    d. The sample is much more salty than ordinary seawater.

43. Classify each of these statements about a specific experiment as data, results, or conclusion, and briefly explain your answer.
    a. A sample of pure calcium metal had a mass of 1.0 gram.
    b. After the sample was heated in the presence of air, it had a mass of 1.4 grams.
    c. The mass of the sample increased by 40% during the experiment.
    d. The calcium combines with something in the air when heated.

44. Classify each of these statements as data, results, or conclusion, and briefly explain your answer.
    a. Acids taste sour.
    b. Chlorine is more abundant than bromine in seawater.
    c. Pure table salt melts at 801°C.
    d. Vinegar reacts with baking soda to produce a gas.

45. Classify each of these statements as data, results, or conclusion, and briefly explain your answer.
    a. If you run an electric current through a sample of water and 18 grams of the water disappear, you will make 16 grams of oxygen.
    b. Samples of pure calcium have a metallic luster.
    c. Samples of fool's gold (iron pyrite) have a metallic luster.
    d. Calcium is a metal, and fool's gold is not.

What conclusions can you draw from each of these experimental results?

46. Acids react with iron to produce a certain gas. The same gas evolves when vinegar is mixed with iron.

47. Powerful acids react rapidly with iron to produce a certain gas. Vinegar slowly produces the same gas when added to iron.

48. Baking soda turns a certain test solution blue. Bases turn the same test solution blue.

49. Table salt yields no visible change when added to the test solution of the previous problem. Bases turn this test solution blue, and acids give no visible color change. (Be careful what you conclude!)

50. When certain metals are heated strongly in open containers, they form substances that were called *calxes* in the eighteenth century. Scientists concluded that a calx had a larger mass than the metal sample from which it was formed. Design an experiment that tests this conclusion.

51. A Parisian jeweler who dabbled in science claimed that diamonds would not burn in the absence of air, and the jeweler was willing to donate a diamond to confirm the fact. The jeweler asked Lavoisier to perform the experiment. How would you design the experiment if you were Lavoisier? (Remember, you are only going to get one crack at this—diamonds were even harder to find in the 1770's than they are today.)

### SECTION 1.5 Hypothesis, Law, and Theory

52. You have heard of the Theory of Relativity and the Theory of Evolution. Are relativity and evolution theories? Explain your answer.

53. The statement "Matter is neither created nor destroyed in a chemical reaction" is sometimes called the Law of Conservation of Mass. According to Section 1.5, is this really a law? Explain your answer.

Formulate a hypothesis based on each of these sets of facts.

54. Burning a candle or a chunk of wood produces water vapor and carbon dioxide. Water vapor and carbon dioxide are also the material products of respiration.

55. When a large amount of phosphorus was burned in a closed vessel filled with air, the phosphorus stopped burning when one fifth of the air was consumed. When iron, tin, or candles are burned, the combustion also stops after one fifth of the air is consumed.

### SECTION 1.6 What is an Element? An Experimental Definition

56. Suggest a method other than strong heating that might be used to decompose a substance into its elements.

57. Does the fact that a substance undergoes no change on heating demonstrate conclusively that it is an element? Explain your answer.

58. Aluminum is more abundant in the earth's crust than iron, yet iron has been used for many centuries longer than aluminum has. How do you account for this?

59. Table salt has the chemical name *sodium chloride.* People have been familiar with salt for thousands of years, but the elements sodium and chlorine have been known for only about 200 years. How do you account for this?

60. Can an element be defined as a substance that contains only one kind of matter? Explain your answer.

61. What is the difference between Boyle's definition of an element and the modern definition given in the text?

### SECTION 1.7 Dmitri Mendeleev and the Periodic Table

62. Referring to the periodic table and the list of elements inside the front cover, write out the names and symbols of the alkaline earth metals (Group 2A).

63. Using the periodic table and the list of elements inside the front cover, write out the names and the symbols of the halogens (Group 7A).

64. Using the periodic table and the list of elements inside the front cover, write out the names and the symbols of the alkali metals (Group 1A).

65. Using the periodic table and the list of elements inside the front cover, write out the names and the symbols of the noble gases (Group 8A).

66. Why was the development of the periodic table an important event in the history of chemistry?

67. Mendeleev made two predictions based on his periodic table. What were they, and which one was wrong?

## ADDITIONAL PROBLEMS

## EXPERIMENTAL PROBLEM 1.1

Lavoisier was intensely interested in understanding combustion (how and why objects burn). His ultimate goal was a practical one: he wanted to light the streets of Paris. In the eighteenth century there were several theories about combustion (the phlogiston theory described in Section 1.2 was one). Lavoisier was convinced that the presence of air was essential to the burning of anything. He performed the following experiment.

He placed a sample of mercury in a closed vessel with a limited supply of air and weighed this apparatus carefully. Then he heated it for several days. A red substance called a calx formed on the surface of the mercury. (For details of a related experiment, see Figure 1.7). The chemists of the day already knew that mercury calx prepared in this way has a greater mass than the mercury from which it is formed. Lavoisier weighed his closed vessel and its contents again after heating. The total mass remained unchanged. When the vessel was opened to the atmosphere, air rushed in, and the mass of vessel and its contents increased because of the added air. Lavoisier reasoned that the mercury reacted with the air, removing some of it as the calx formed.

### QUESTIONS

68. What was Lavoisier's hypothesis?

**69.** Further experiments showed that not all of the air in the vessel will react, regardless of how much metal is used. Does this fact disprove Lavoisier's conclusion?

70. What would the results have been if the metal lost phlogiston when it was heated, assuming that phlogiston is a gas?

**71.** List the data collected by Lavoisier.

**72.** How should these data be organized for presentation to fellow scientists?

**73.** Suppose Lavoisier had used a larger vessel in the experiment. How would the data compare with the original data? How would the results compare with the original results? How would his conclusion compare with the original conclusion?

## EXPERIMENTAL PROBLEM 1.2

A scientist named A. L. Chemist (you will often read about A. L. Chemist in this book) decided to repeat the van Helmont experiment (Experimental Exercise 1.1). A tree weighing 1 pound was planted in 5 pounds of soil and watered regularly. At the end of the growing period, the tree weighed 170 pounds and the weight of the soil was reduced by 2 ounces. The tree, including the roots, was desiccated (that is, all the water was removed), and the resulting material had a mass of 40 pounds. A. L. Chemist and his team decided that these results showed that some substance in addition to water was responsible for the increase in the tree's weight. They further concluded that the other substance was sunlight absorbed by the tree and offered these observations in support of this conclusion: (1) trees do not grow where there is no sunlight (there are no trees deep in caves, for example); (2) when chemists place a sample of a metal in the flames of a fire, they observe an increase in weight.

### QUESTIONS

**74.** What data did A. L. Chemist and his team collect?

**75.** How should they organize their results for presentation to fellow scientists?

**76.** Suppose A. L. Chemist had started with a smaller tree. How would the data compare with the original data? How would the results compare with the original results?

**77.** Do the results support A. L. Chemist's conclusion?

**78.** When chemists heat a limestone rock in a flame, the amount of rock present decreases. Does this fact disprove A. L. Chemist's sunlight hypothesis?

**79.** Do the results support the conclusion that trees are composed mostly of water?

## SOLUTIONS TO EXERCISES

### Exercise 1.1

1 Items (a), (c), and (d) are material objects and are therefore made up of chemicals. Item (b) is an emotion and is not chemical. As for (e), it depends—as a rate of movement, speed isn't chemical, but as a drug of abuse, it is.

### Exercise 1.2

2 A knowledge of chemistry is useful for everything listed. When you cook, you carry out chemical reactions. Office workers use copying toner, correction fluid, etc., and some knowledge of their flammability and ability to act as poisons is useful. Nursing is, in part, the application of chemistry to the human body. When panning for gold, you must be able to tell fool's gold from the genuine article.

### Exercise 1.3

3 If aluminum is the only substance present in (a), it cannot be broken down into other substances by ordinary methods. Note that this is not a conclusive experiment. Aluminum oxide is not an element, but samples of it do not change even at a temperature of 2000°C.

### Experimental Exercise 1.1

1 **a.** In arguing that water is the basic element of the universe, van Helmont agreed with the ancient Greek philosopher Thales, who lived 2000 years earlier.
**b.** The fact that van Helmont performed an experiment at all is unusual for the seventeenth century. Even more unusual is the fact that he weighed the soil and the tree—that is, he performed numerical measurements.
**c.** The tree was in contact with the air as well as with soil and water, but van Helmont assumed that the air was not a factor in the experiment results (a semireasonable assumption for the time). We now know that plants make most of their substance by using energy from the sun to combine water with carbon dioxide from the air.

### Experimental Exercise 1.2

2 **a.** Hypothesis: The sediment is formed from the transmutation of water.
**b.** Data: The masses of the vessel and the water both before and after the experiment, the mass of the sediment formed in the experiment, and the number of days the water was boiled.
**c.** Conclusion: Water was not converted into other substances.
**d.** Yes, because the total mass of the sediment exactly equaled the mass lost by the vessel.
**e.** If the transmutation of water to sediment had occurred, Lavoisier would have found that the mass of the water decreased by an amount exactly equal to the mass of the sediment. The mass of the glass vessel would have remained unchanged.

2

# MIXTURES, COMPOUNDS, AND ELEMENTS

Materials in the World Around Us



## KEY TERMS

In Chapter 1 you learned that chemists are scientists who study matter and the changes it undergoes. The way to begin any such study is to get some samples of material things and start examining them. Chemists as well as many other scientists start with samples. An archaeologist studying the ruins of an ancient city might begin by digging in a small area and collecting pieces of building materials, pottery, and food remains. A botanist (plant biologist) studying a forest might collect some leaves and flowers. A chemist might find some use for samples like these but would probably try to break them down further. You probably remember the term *element* from Chapter 1. If the material in a sample cannot be broken down into simpler substances by chemical methods, it is an element. Individual elements differ from one another in many other respects, but they all have this one property in common. In this chapter we look more closely at material substances (including the elements) to see how they differ from one another.

Where can we begin? When scientists begin to explore something unknown, classification—putting things into categories—is almost always the first step.

A close look at the world reveals that many different things can be grouped on the basis of their properties. Your own observations and experiences tell you that solids, liquids, and gases do not behave the same. Furthermore, any gas has properties in common with all other gases. All gases are easy to compress, for example; this is one way gases differ from liquids and solids. Squeezing a liquid or a solid into a smaller volume is much more difficult than compressing a gas. You can see that classifying a material as a solid, a liquid, or a gas tells you something about how that substance behaves.

Chemists use classification schemes as tools for understanding why objects in the material world behave as they do. The names of many classes of substances are part of the special language of chemistry (see Section 1.8). Chemists use them carefully, to make sure that other scientists and students know exactly what they are saying.

Classification systems can confuse you, but that is not their purpose. Chemists classify substances to make them easier to understand and remember. Take care to learn the classification schemes presented in this book as soon as they are introduced. You will find it easier to understand the concepts in this course if you treat its language seriously. Think of the vocabulary as a foreign language (which it certainly is for beginners) and study it accordingly.

## MATERIALS IN THE WORLD AROUND US

People generally don't think about the materials they look at. They usually see rocks, oceans, and wedding rings without thinking about minerals, salts dissolved in water, and metals. Chemists, on the other hand, are scientists who think about the composition of samples they encounter.

Most things you encounter in daily life are composed of several different materials. The ink in your pen contains a dye (or several dyes) dissolved in some liquid. Vinegar consists of water and a substance called acetic acid. In the first two sections of this chapter, we look at the similarities and differences between various types of substances. (Ink and vinegar are similar in some respects and different in others.) Next, we consider some of the methods used to separate mixtures of substances into their component parts.

Though most of the items you use in everyday life are composed of mixtures, not all of them are. Baking soda, table sugar, and ordinary salt have only one component each. The properties of one-component substances differ from the properties of mixtures in several important ways. We will explore these differences and then present the symbols chemists use to represent elements and compounds. Finally, we take a closer look at the differences between the states of matter and see how these differences apply to some common substances.

## 2.1 Mixtures

**GOAL:** To define and explain the two types of mixtures: homogeneous and heterogeneous.

If you inspect some objects around you, you can come to some conclusions about how to classify what you see. Where can you start looking? You might begin with a close inspection of the piece of granite shown in Figure 2.1. The granite is made up of grains of different substances cemented together. Some of the grains are transparent and some are opaque. The light reflects off the grains in different ways, so they have various colors. Granite is a mixture containing several substances called *minerals*. If you smashed a chunk of granite into small enough pieces, you'd be able to pick out grains of the individual minerals. (You might need a magnifying glass to see them, but it can be done.)

A **mixture** is a sample of variable composition. Mixtures are very common in nature, and you encounter them every day. The air you breathe is a mixture of gases, and the food you eat is an amazingly complex mixture of liquids and solids. Products you buy are almost always mixtures. For example, any perfume or cologne is a mixture of fragrant substances, water, and alcohol. Rubbing alcohol is a relatively simple mixture that contains 30% water and 70% of a chemical commonly called isopropyl alcohol.

A

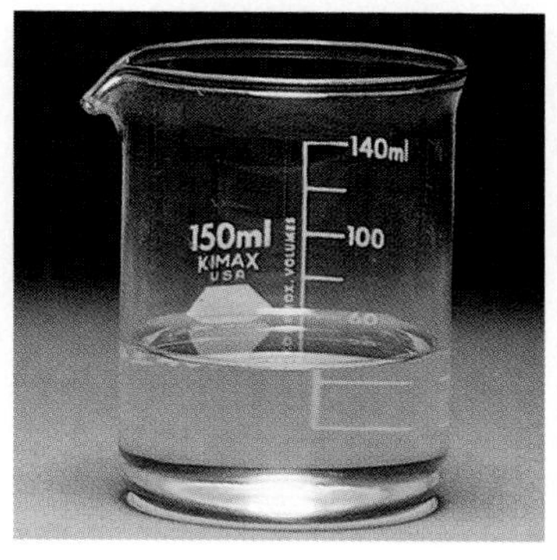


B

FIGURE 2.1
Homogeneous and heterogeneous mixtures. (A) Note the grains of clearly different substances in granite. These substances are different minerals that have crystallized together. This piece of rock is a heterogeneous mixture. (B) Salt water is clear and transparent. Since it is impossible to see any boundaries between the salt and the water, salt water is classified as a homogeneous mixture.

A

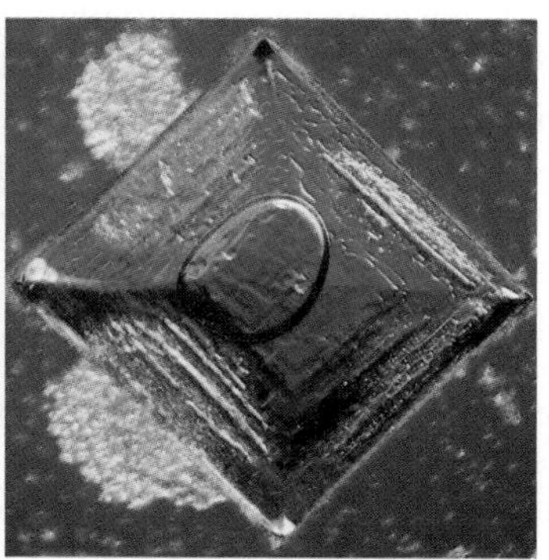

B

FIGURE 2.2
(A) When salt water is evaporated, the salt is left behind. Sea salt consists of several substances, though common table salt (called *sodium chloride* by chemists) is the most abundant. (B) A crystal of sodium chloride.

## CHAPTER BOX 2.1 ALLOYS

**Alloys are** often very useful. Most pure metals have some properties that make them less than ideal for use in industry or everyday life. Consider iron. You have probably never seen a sample of pure iron metal. Typically what you see is steel, a solid mixture of iron, carbon, and usually other elements such as nickel, magnesium, and chromium. Steel has many uses, and there are many different, specialized steel compositions for these uses.

Pure gold is much too soft to use in jewelry. A 24-karat gold wedding ring could be squeezed shut with two fingers, so jewelers add silver and copper to pure gold to increase the metal's strength. Metalworkers have known how to do this since ancient times—most of the ancient "gold" artifacts you see in museums are made of alloys of gold and copper.

Alloys are by no means a modern invention. The Bronze Age (3500–1000 BC) was given its name because of the impact bronze metal had on the development of its cultures. Bronze is an alloy of copper, zinc, and tin. Tools and weapons made of this alloy are much stronger and harder than those made of pure copper, which was used before the discovery of bronze.

FIGURE 2.3
This golden bird from the north coast of Peru is made of an alloy of gold and copper.

FIGURE 2.4
Bronze Princess of Benin

Look closely at a sample of salt water. Its name tells you it's a mixture. If you get a clean sample, it looks completely clear and uniform. If you boil the salt water or let it stand in the sun, the water evaporates and white salt crystals are left behind (see Figure 2.2).

What about a wedding ring? That's just gold, isn't it? But you've heard of 14-karat gold and 18-karat gold. Do these numbers indicate different grades of gold, as "prime" and "choice" do for the beef you buy? The answer is no. The "gold" jewelers use is actually a mixture of gold, silver, and copper. The designations 14-karat and 18-karat tell how much of the mixture is gold, with pure gold defined as 24-karat. Almost all of the metals you encounter in everyday life are actually mixtures of several metallic elements. These mixtures are called **alloys.** In most alloys you can't see the boundaries between one metal and another, even under a microscope.

**The prefix *hetero* means "different," and the prefix *homo* means "same." When chemists use the word *homogeneous* or *heterogeneous* to refer to a mixture, *homo* and *hetero* indicate whether the component parts of the mixture can be separated visually. If you can see the components and the boundaries between them, the substances are different and the mixture is heterogeneous. If not, it's homogeneous.**

As you can see, mixtures can be classified into two broad groups. One type of mixture has visible boundaries between the substances that make it up, as granite does. This kind is called a **heterogeneous mixture.** If you can't see the boundaries between the substances, as in salt water and 18-karat gold, the mixture is called a **homogeneous mixture.**

Let's look at two other examples: red wine and blood. Red wine is transparent, though it does have an intense color. You can see all the way through a glass of red wine. Blood is also red, but it is opaque; you cannot see through a sample of blood. Are these liquids homogeneous or heterogeneous? The clue to answering this question is provided by what happens to light when it passes through the two different substances. The fact that red wine is transparent is evidence that it is a homogeneous mixture. Nothing (that is, no boundary between substances) prevents the light from traveling directly from one side to the other. Light doesn't pass directly through blood, however. This suggests that there are boundaries between different parts of the blood, and they block or scatter the light. Blood is a heterogeneous mixture. If you look at blood under a microscope, you'll see that blood cells are dispersed throughout the sample.

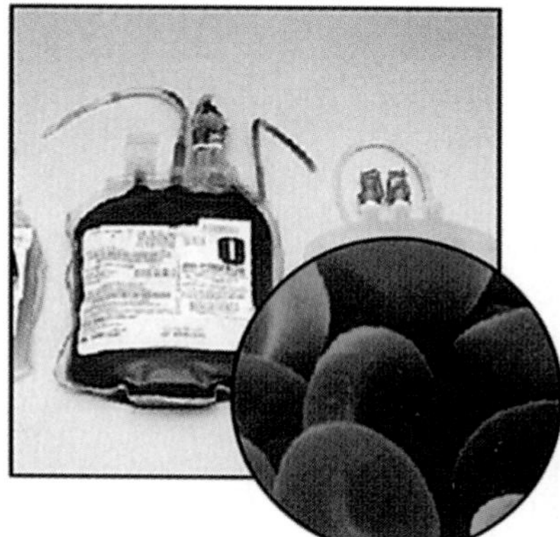

FIGURE 2.5
Light passes through the glass of red wine (top), which indicates that the mixture is homogeneous. Blood is opaque (center) because the cells in it prevent the light from passing directly through. The blood cells are visible under a microscope (right), providing positive evidence that blood is a heterogeneous mixture.

### EXAMPLE 2.1 Classifying Mixtures as Heterogeneous or Homogeneous

Classify each of the following mixtures as either homogeneous or heterogeneous.

**a.** T-bone steak
**b.** Oil and vinegar salad dressing
**c.** Homogenized milk
**d.** Green glass (as in disposable bottles)

SOLUTIONS

**a.** The bone is a giveaway: a steak is heterogeneous. Even without the bone, you can easily distinguish the fat from the flesh.
**b.** The boundaries between the two parts of oil and vinegar salad dressing are easy to see even when the mixture has been shaken. The dressing is a heterogeneous mixture.
**c.** Beware of common names. You can't see boundaries in homogenized milk, but the fact that the milk isn't transparent is strong evidence that it is a heterogeneous mixture. The word *homogenized* means that the milk has been processed so that droplets of cream won't float to the surface to form a separate layer. The cream stays dispersed.
**d.** Green glass is transparent, so it's a homogeneous mixture.

EXERCISE 2.1

Classify each of the following mixtures as either homogeneous or heterogeneous.

**a.** Cola freshly poured into a glass
**b.** Tears
**c.** Beef stew
**d.** A carrot

TOOLBOX

## TOOL BOX: PERCENTAGES

**GOAL:** To work numerical problems involving percentages.

Jewelers use the word *karat* to indicate the fraction of gold present in jewelry. Pure gold is defined as being 24-karat. If a bracelet is made of 18-karat gold, $^{18}/_{24}$ of the total mass of the bracelet is gold, and the rest ($^{6}/_{24}$ of the total mass) is a mixture of copper and silver. This system is fine for jewelers, but most of the rest of the world uses **percentages** to indicate relative amounts.

You find percentages in all sorts of places. The evening news almost always gives information about the economy—inflation rates, profits and losses, change in almost anything—as percentages. Coaches of sports teams talk about "giving 100%." You'll find percentages throughout chemistry, too.

What is a percentage? The word literally means "per hundred." Here's an example. If 20% of the voters cast ballots in favor of candidate Smith, that means that Smith received 20 out of every 100 votes. Is this a good total? Well, 80 out of every 100 voters cast their ballots for someone else. Smith probably lost, unless there were five or more other candidates and each of them got less than 20% of the votes.

This situation is typical of what you encounter in chemistry. All the percentages (parts per 100) add up to a whole of 100%. If you compare numbers of the same thing at two different times, you can get percentages of more than 100 ("Last month I had two rabbits, but now I have twelve—an increase of 600%!"). Physical scientists rarely use percentages this way.

Most scientists use and interpret numbers continually in their work. One very common way to express relationships among numbers is an equation. You will find this equation useful in calculating the makeup of substances or groups:

$$\text{Sum of parts} = 100\%\ (\text{Whole})$$

The fact that 80% of the voters didn't cast their ballots for candidate Smith can be determined using this equation. Here's how to use it.

$$\text{Sum of parts} = 100\%$$

$$20\%\ (\text{Smith's part}) + X\ (\text{all other parts}) = 100\%\ (\text{all of the votes: the whole})$$

$$20\% + X = 100\%$$

To solve this equation for $X$, subtract 20% from each side of the equation:

$$20\% + X - 20\% = 100\% - 20\%$$

This gives

$$X = 100\% - 20\%$$

$$X = 80\%$$

The percentage of votes cast for candidates other than Smith is 80%.

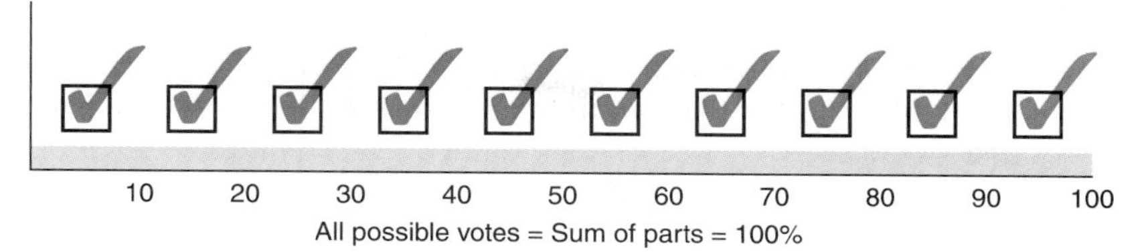


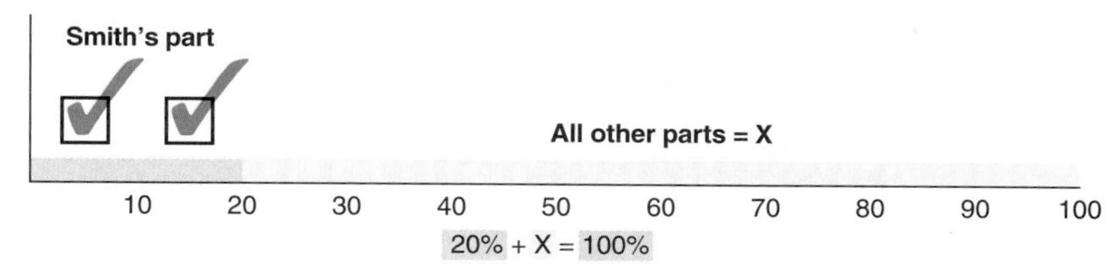


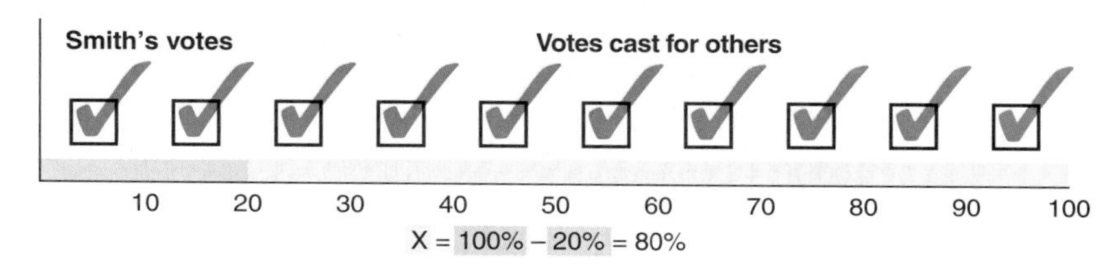


FIGURE 2.6

**TOOL BOX EXAMPLE 1 Finding the Unknown Percentage From the Known.**

Brass is an alloy of copper and zinc. A sample of brass contains 22% zinc. What is the percentage of copper in this sample?

**SOLUTION**

$$\text{Sum of parts} = 100\% \text{ (Whole)}$$

$$22\% \text{ (zinc)} + X \text{ (copper)} = 100\% \text{ (all of the brass: the whole)}$$

$$22\% + X = 100\%$$

Subtract 22% from both sides of this equation to get

$$X = 100\% - 22\% = 78\% \text{ copper in this sample of brass}$$

**TOOL BOX EXERCISE 1**

**a.** A sample of steel contains 2% carbon; the rest is iron. What is the percentage of iron in this sample?

**b.** A sample of bronze contains 5% zinc and 72% copper. The rest is tin. What is the percentage of tin in this sample?

**c.** When an election was over, it was reported that candidate Smith got 20% of the vote, candidate Jones got 25%, and candidate Crook got 65%. A. L. Chemist, who was a member of the Board of Elections, argued that the election results should be thrown out and the election investigated. This recommendation was followed simply on the basis of the vote totals. The police later discovered that candidate Crook had bribed one of the election officials. Candidate Smith won the runoff. What made A. L. Chemist think there was something wrong with the election?

**TOOL BOX EXAMPLE 2 Calculating Percentages.**

The "sum of parts" equation is useful if you already have percentages, but how do you calculate percentages in the first place? You use this equation:

$$\frac{\text{Part}}{\text{Whole}} \times 100\% = \text{Percentage}$$

**a.** What is the percentage of gold in 18-karat gold?
**b.** A metallurgist made bronze using 5 pounds of tin, 15 pounds of copper, and 5 pounds of zinc. What is the percentage of copper in the bronze?

**SOLUTIONS**

**a.** This gold sample contains 18 parts gold and 6 parts everything else. (Remember, jewelers divide gold alloys into 24 equal parts. If 18 parts are gold, then the rest must be another metal. The total of all parts equals 24—that's where the 6 comes from.)

$$\frac{\text{Part}}{\text{Whole}} \times 100\% = \text{Percentage} = \frac{18}{24} \times 100\% = 75\% \text{ gold}$$

**b.** The question asks for "the percentage of copper in the bronze." This means that the weight of the copper is "Part" in the equation, and the final weight of the bronze is "Whole." So what is the final weight of the bronze? It's the weight of everything—5 pounds of tin plus 15 pounds of copper plus 5 pounds of zinc, or 25 pounds. This is what you get when you put these numbers in the equation:

$$\frac{\text{Part}}{\text{Whole}} \times 100\% = \frac{15 \text{ lb}}{25 \text{ lb}} \times 100\% = 60\% \text{ copper}$$

**TOOL BOX EXERCISE 2**

**a.** A sample of an experimental alloy was made by adding 100 pounds of titanium to 300 pounds of iron. What is the percentage of titanium in this sample?
**b.** What is the percentage of gold in 14-karat gold?

**STUDY ALERT**
**To solve problems involving percentages, you will need to solve simple equations (in mathematical terms, linear first-order equations). For a brief review of the simple algebra involved, see Appendix A.**

**TOOL BOX EXAMPLE 3 Using Percentages.**

A very common calculation, particularly in science, requires you to determine "how much" when you are given the percentage. This procedure is demonstrated in this example.

A sample of bronze contains 5% zinc. If the sample weighs 40 pounds, what is the weight of the zinc in it?

**SOLUTION**

The percentage is given here (5%). The weight of the bronze is the "Whole," and it's given too. The unknown *X* is the "Part," the amount of zinc in the bronze. The equation is:

$$\frac{\text{Part}}{\text{Whole}} \times 100\% = \text{Percentage}$$

Substitute $X$ for the "Part," 40 lb for the "Whole," and 5% for the "Percentage":

$$\frac{X}{40 \text{ lb}} \times 100\% = 5\%$$

Multiply both sides of the equation by 40 lb:

$$40 \text{ lb} \times \frac{X}{40 \text{ lb}} \times 100\% = 5\% \times 40 \text{ lb}$$

After you cancel the 40 lb on the left you get:

$$X \times 100\% = 40 \text{ lb} \times 5\%$$

Dividing both sides by 100% and solving the equation gives you:

$$X = \frac{40 \text{ lb} \times 5\%}{100\%} = 2 \text{ lb, the amount of zinc in the sample of bronze}$$

**TOOL BOX EXERCISE 3**

**a.** A metallurgist wants to make 50 kilograms of bronze that contains 70% copper. How much copper should the metallurgist use?

**b.** Another metallurgist has 21 kilograms of copper and more than enough of the other ingredients for bronze. How much bronze (which is 70% copper) can the metallurgist make?

## 2.2 Solids, Liquids, and Gases

**GOAL:** To describe the three common states of matter.

You are already familiar with the three forms that matter can take under ordinary circumstances. Water, as you know, exists in all three forms. All of them are colorless. The liquid is the most familiar form to almost everyone. As ice or snow, water is a solid. Gaseous water, called water vapor, is invisible. Air with water vapor in it looks just the same as air without it. Both fog and steam consist of small droplets of *liquid* water floating in air. All three forms of water—liquid, ice, and vapor—are the same substance. Only the physical form differs.

FIGURE 2.7
Water exists on earth in all three states. You can see both liquid water and ice in this photo, but water vapor is invisible.

The three **states** of matter are no mystery to you, but it might be worthwhile to review what they're like. Solids are rigid. If you drop a pencil into a cup, the pencil doesn't collapse and flow to the bottom. Solids aren't very compressible. If you squeeze a solid, it resists the pressure and, as far as you can tell, keeps the same volume. This incompressibility is an important *property* (characteristic) of solids. What would happen to a skyscraper if the steel beams used in its construction were compressible?

**There is a fourth state of matter called *plasma*. It occurs only in gaseous substances at very high temperatures. It exists in lightning bolts, the sun, and the laboratories of physicists studying *nuclear fusion* (see Chapter 17).**

Even a casual inspection of the solids around you demonstrates that there are many different kinds. The paper in this textbook is certainly solid, and it's quite flexible. So is a rubber ball. Almost all metals can be pounded into thin sheets like aluminum foil, but if you try to pound a diamond flat, you'll

FIGURE 2.8
Some substances can exist in more than one solid form. Diamond and graphite (left) are both pure carbon. Hard candy and table sugar (right) are both essentially pure sugar.

end up with a very expensive powder. We will explore the reasons behind these variations in properties in later chapters. At this point these examples simply demonstrate that the properties of solids depend (at least in part) on their chemical composition.

Many substances can exist in more than one solid phase (or form). One example is the element carbon, which has several solid forms. The two most familiar of these are graphite and diamond. Graphite is the primary component of the "lead" in a pencil (which contains no lead at all), and you know what diamonds look like. It's remarkable that both diamond and graphite are 100% carbon since their physical properties aren't similar at all. Diamond is the hardest substance that occurs naturally, and graphite is so soft that tiny pieces of it come off if you rub it on paper (that's why you can see the mark a pencil makes). Pure diamonds are colorless, and graphite is black.

Another example of a substance that has more than one solid phase (or form) is sugar. The crystalline form is the stuff you put in or on food to flavor it. When making lollipops and hard candies, candy makers use a transparent form of sugar that lacks a regular crystal structure. Both of these are forms of the same chemical substance. You might conclude that the properties of solids also depend on how the units that make up the solids are connected, and you would be correct.

Liquids flow. Put a liquid into a container and it will assume the shape of the container. Liquids, like solids, are not easy to compress. The brakes on your car use this property to transmit the pressure of your foot on the pedal through the brake fluid to the braking mechanism. In general, most substances have only one liquid form. Only one substance, helium, is known to have two different liquid forms.

**In general, substances that flow are called *fluids*. Scientists classify both gases and liquids as fluids.**

Gases also flow, but unlike liquids, they expand and contract as well. Put a gas sample in a rigid container and it will expand to fill the whole volume. Gases are much easier to compress than liquids and solids are. Scuba divers take advantage of this property of gases when they carry tanks of compressed air on their backs. These tanks can hold an amount of air that would fill a closet-sized space at normal pressure. All mixtures of gases are homogeneous (if they are not disturbed by outside forces). Both liquid and solid mixtures can be either homogeneous or heterogeneous.

FIGURE 2.9
The compressibility of gases allows a large quantity of air to be stored in a small volume of space. The compressed air in these divers' tanks will keep them going for some time, but the same volume of uncompressed air would provide only a few moments' worth of oxygen.

EXAMPLE 2.2 **Properties of Solids**

A common kitchen sponge is solid (it certainly doesn't flow), but it's compressible. Why doesn't this property contradict the information in this section?

SOLUTION

When you compress a sponge, you're not compressing a solid; you're forcing air or water out of the pores (holes) in the sponge. When the solid fibers that make up a sponge have been pushed into contact, it suddenly becomes much harder to compress the sponge any further. Chemistry is an experimental science—try compressing a wet sponge and observe closely what occurs.

EXERCISE 2.2

Many people can crush an aluminum can with one hand. Why doesn't the compressibility of the can contradict the information in this section?

## 2.3 Separating Mixtures

**GOAL:** To identify, illustrate, and explain methods for separating various mixtures.

**STUDY ALERT**

**Chemists separate a mixture by using the differences between the properties of the substances that make up the mixture. As you study this section, be sure to identify the property being used in each of the separation techniques. Later you will be asked to explain how to separate various mixtures, and the answers will depend on what property is easiest to use.**

Every pure substance has its own unique properties, or characteristics. This means that the properties of the individual substances that make up a given mixture are different. Chemists (and many others) use these differences to separate the components of mixtures.

Farmers were faced with one of the first separation problems. When they harvested rice or wheat, they got a mixture of the grain and chaff (stems, leaves, and husks), not the grain by itself. Their method of separation, called *winnowing,* is older than written history. The farmers dried the mixture of grain and chaff in the sun and then threw it up in the air on a breezy day. The feathery stems, leaves, and husks blew away with the wind, while the compact grain fell back to earth. Modern harvesting machines (called *combines,* because they combine all the individual harvesting steps into one operation) use the same principle, blowing the chaff away with powerful fans.

Prospectors who pan for gold apply the same basic idea. They put a mixture of water, sand, gravel, and (they hope) gold into a pan. Then they slosh the mixture back and forth, which washes the sand and some of the gravel over the side of the pan. Gold nuggets are very heavy for their size and aren't easily washed away by the movement of the water. Most of the sand and gravel wash away, leaving the gold behind.

FIGURE 2.10

Early farmers (and some farmers today in less wealthy nations) used the difference in the properties of grain and chaff. The chaff is light and blows away on the wind. The heavier grain falls back to the ground.

When different phases (solid and liquid or solid and gas) are present in a heterogeneous mixture, **filtration** is a common method of separation. A mixture is filtered by passing it through something that has openings designed to trap the solid and let the liquid (and anything dissolved in the liquid) pass through. Cooks use filtration when they pour spaghetti into a colander. The spaghetti is trapped, and the water used to cook the spaghetti pours through the holes. Air conditioners use filters to take dust out of the air. The dust particles are trapped against the fibers of the filter, while the air flows through.

Chemists use filters that are much finer than colanders or air conditioner filters in their separations. They use paper (or a membrane) precisely made so that it has holes of uniform size. This *filter paper* traps small crystals and lets a liquid flow through.

Another separation technique used since ancient times is **extraction.** This method takes advantage of differences in the solubility of a mixture's components in water (or some other liquid). You use extraction yourself every time you use a tea bag to make tea. The flavoring agents and a few other substances in the tea are soluble in water, and the rest is not. The tea bag is a filter used to separate the undissolved portion from the drinkable stuff. When you make coffee in a coffee maker, you also use extraction. The hot water poured in the top flows through the coffee and a filter, which extracts the drinkable part and leaves the grounds behind. Cooks use vanilla and almond extracts. These flavorings are called *extracts* because they are prepared from vanilla beans or almonds by similar methods.

FIGURE 2.11
It takes energy to pull air through a dirty, plugged-up air conditioner filter. That's why these filters should be cleaned fairly frequently.

FIGURE 2.12
Here a chemist is using filter paper in a funnel to remove solid impurities from a solution. Coffee filters work the same way, trapping the grounds while letting the drinkable liquid through.

## CHAPTER BOX 2.2 SUBLIMATION

**The process** of transforming a solid directly into a gas without melting is called **sublimation.** You may have seen this with dry ice. Dry ice is solid carbon dioxide ($CO_2$) and is very cold. When removed from a storage container, the dry ice sublimes and white "smoke" appears. This smoke is not carbon dioxide gas (which is invisible); rather, it consists of droplets of water vapor condensed from the air by the cold carbon dioxide.

Most solids don't sublime easily. Although sublimation is a fairly common laboratory technique, it doesn't occur often in the natural world.

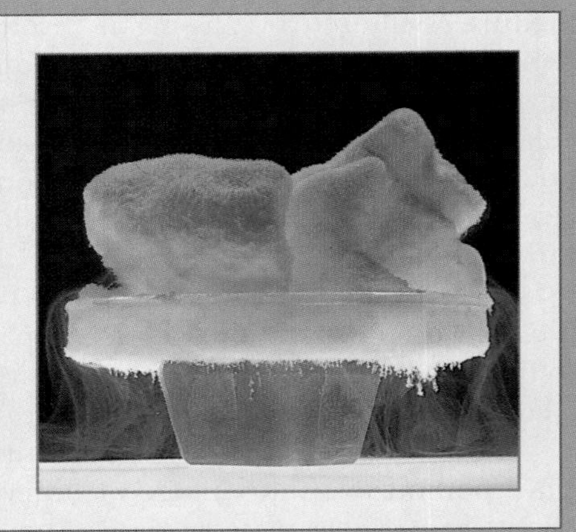

FIGURE 2.13
Dry ice (solid carbon dioxide) sublimes—that is, goes directly from the solid to the gaseous state. The visible smoke is not carbon dioxide gas but condensed water vapor.

FIGURE 2.14
In the coldest areas of the world, such as the top of Mt. Everest (shown here), snow that lands on the ground evaporates directly into the air without melting. This is an example of sublimation.

**Evaporation** is the conversion of a liquid into the gaseous state. People have used this method since ancient times to separate salt from seawater. To do this, they put the seawater in a large, shallow vat (or even a shallow pond). The heat of the sun causes the water to evaporate, leaving the salt behind. Note how the different properties of the substances in the mixture come into play in the separation process. The water evaporates from the vat, and the salt doesn't.

When a glass holding an iced drink sits out on a hot day, water vapor *condenses* on the glass. **Condensation** is the process of converting a gas into a liquid or a solid. Dehumidifiers use condensation to remove water vapor from the air. This is relatively easy to do because water condenses at a much higher temperature than anything else in the air. If you have gleaned from this paragraph the notion that condensation is the opposite of evaporation, you are right on track. Figure 2.16 summarizes the terms used for phase changes.

FIGURE 2.15
Condensation can mean either the change from gas to liquid (droplets on the glass, left) or the change from gas to solid (frost on the window, right).

What do you do if you have salt water and you want the water but not the salt? You evaporate the water away from the salt by boiling, move the water vapor away, and condense it. This combined process is called **distillation.** This technique is based on the difference in **volatility** between the components of a mixture—that is, the difference in how easily the substances evaporate. In this example the more volatile water evaporates much more easily (at a much lower temperature) than the less volatile salt.

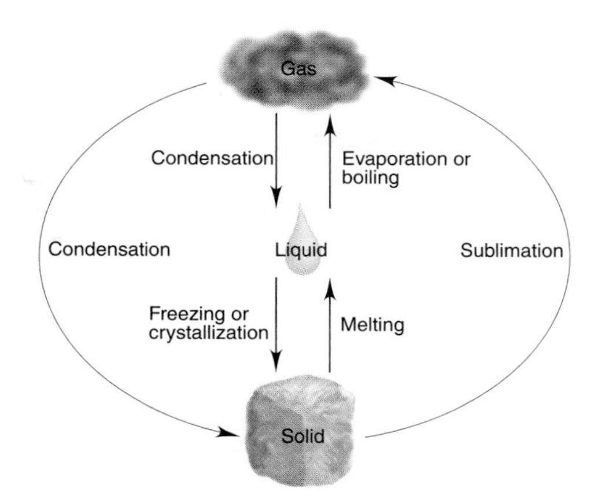


FIGURE 2.16
The phase changes

Chemists use distillation to separate many other things besides salt and water. It can be used if any component of a mixture is a liquid that can be evaporated. Distillation is commonly used to separate a homogeneous mixture that contains two or more liquids. For example, the first step in refining crude oil is to distill it.

Figure 2.18 shows an apparatus chemists often use in laboratory distillations. The chemist heats a mixture until a liquid component of the mixture starts to boil and evaporate. The gas that forms expands out of the sample flask and into the *condenser.* As the name implies, the condenser cools the gas and converts it back into liquid. Then this liquid flows down to the collection flask. The less volatile components of the mixture remain in the original flask.

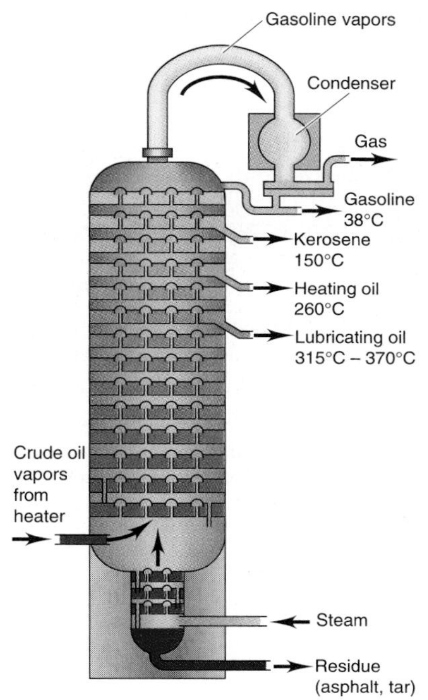


FIGURE 2.17
Oil refineries use an apparatus like this to distill crude oil. The oil is an extremely complex mixture, and the fractions (gasoline, kerosene, etc.) are mixtures themselves. The kerosene fraction contains jet engine fuel, fuel oil, and fuel for diesel engines. The lubricating oil fraction includes thick oils, greases, and petroleum jelly.

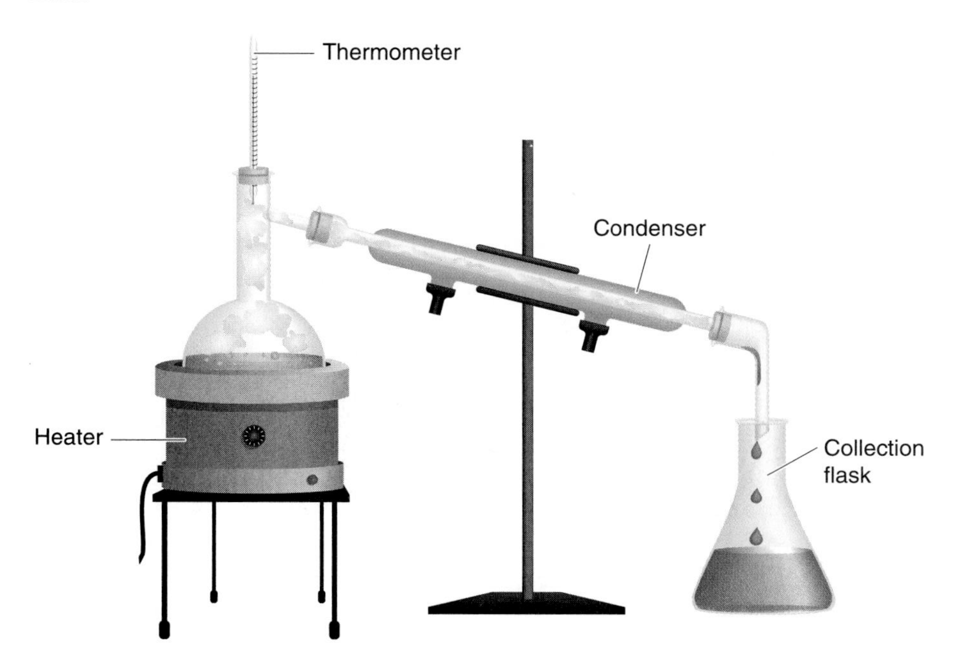


FIGURE 2.18
A typical laboratory setup for distillation. If the solution being distilled is salt water, the salt remains in the distilling flask. The water boils there, returns to the liquid phase in the condenser, and flows down to the collection flask.

CHAPTER BOX

## 2.3 WHY WAS BRONZE DEVELOPED BEFORE BRASS?

**Brass**—a mixture of copper and zinc—was very important to the development of technology. Brass is much harder than bronze and can be used in heavy-duty applications—cannons and steamship propellers, for instance.

Why did the discovery of bronze—a mixture of copper, tin, and zinc—precede the discovery of brass, which contains only two of these metals? Early metallurgists knew how to obtain both copper and tin from the minerals (ores) that contained them: heating the ore in a wood fire worked just fine. Alloys made of only copper and tin are too soft to be very useful. However, if you throw some tin ore into a batch of molten copper, heat well, stir vigorously, maybe throw in a chunk of wood (which burns), and let the molten mixture cool, you will get the softer alloy bronze. So where did the zinc come from? The ancients never knew that they were using zinc. They simply had a recipe for making something useful.

The confusion arose from the properties of tin and zinc. Tin deposits usually contain zinc. Tossing "tin ore" into molten copper actually adds both tin and zinc, forming bronze. Why didn't anyone notice the presence of zinc in the ore when the tin was isolated from it? The answer comes from the difference in volatility between the two metals. Zinc boils at 970°C, tin at 2270°C. When the mixture of ores was thrown into a container over a hot wood fire, the tin melted and collected at the bottom of the container. The zinc boiled away and was never seen at all.

When the ancient metalworkers made bronze, the molten copper was hot enough to melt both the tin and the zinc, but not hot enough to evaporate the zinc. Industries using brass could not develop until zinc was isolated as an element in the late 1700s.

**When something is *absorbed*, it permeates the entire absorbing material, as a sponge absorbs water. When something is *adsorbed*, it sticks only to the surface of the adsorbing material, and does not penetrate to the interior.**

To separate more difficult homogeneous mixtures, chemists often use a method called **chromatography.** The mixture being purified moves, together with a mobile phase, along a stationary material that adsorbs some parts of the mixture more strongly than others (see Figure 2.19). Their different adsorbencies make the substances in the mixture travel along the stationary material at different rates. Some substances are adsorbed so strongly that they don't travel at all.

You can try the process yourself with a felt-tip pen, a drinking glass, some rubbing alcohol, a coffee filter, and water. Put 3 tablespoons of the rubbing

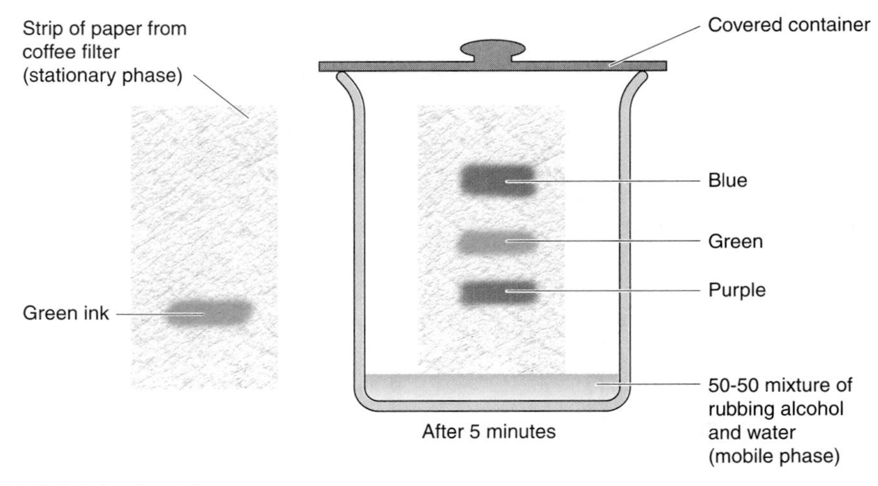


FIGURE 2.19

The filter paper adsorbs the solvent, which travels up the paper by capillary action. The various components of the ink have different adsorbencies and move up the paper at different rates. This green ink is a mixture of three dyes.

alcohol and 3 tablespoons of water in the glass. Cut a rectangle out of the coffee filter about 1 inch wide and 4 inches long. With the felt-tip pen, draw a single line about 1 inch from the bottom of the filter paper and place the paper into the glass. The end of the paper with the ink line must be the bottom end, and the ink line must be above the level of the liquid. Cover the glass with some aluminum foil or plastic wrap. In a few minutes you will observe several "bands," each representing a different component of the ink. The results of one such experiment are illustrated in Figure 2.19.

You can buy devices that purify your drinking water using chromatography. Two of these devices are activated charcoal "filters" and ion-exchange purifiers.

FIGURE 2.20
Activated charcoal (the black material in this fish tank water purifier) adsorbs unwanted substances dissolved in the water. Glass wool (the white material) filters particles from the water. Water itself is not adsorbed and passes through the filter. This apparatus uses both chromatography and filtration.

**The French chemist Joseph Proust (1754–1826) was the first to distinguish clearly between chemical compounds and mixtures. He decided that the substances in mixtures can be separated by making use of their physical properties, as the methods given in Section 2.3 do. Compounds can be broken down into their component parts only by chemical reactions.**

### EXAMPLE 2.3 Using Separation Methods

Which of the following four methods can be used to separate each mixture?

**1.** Winnowing **2.** Filtration
**3.** Distillation **4.** Chromatography

**a.** Live fish in water
**b.** Sugar dissolved in water
**c.** Sand mixed with gravel

**SOLUTIONS**

**a.** You can use a net to remove the fish. This is a type of filtration (2).
**b.** Sugar water is a homogeneous mixture. Since it can be distilled, distillation (3) will work.
**c.** A blower will move the sand and leave the gravel behind. This is a winnowing method (1). A filter with a large mesh (method 2) will also work.

**EXERCISE 2.3**

Which of the following four methods can be used to carry out each separation?

**1.** Extraction **2.** Filtration
**3.** Evaporation **4.** Chromatography

**a.** Separating egg yolks from whites
**b.** Recovering sugar from sugar water
**c.** Separating salt from sand

## 2.4 Properties

GOAL: To classify properties in two ways: (1) physical or chemical and (2) intensive or extensive.

**Properties** is the word scientists use for the typical characteristics of substances. A sample of any substance has many properties.

FIGURE 2.21
A chemical reaction occurs when sodium metal is placed in water. Two observations support this idea. Bubbles of gas form where the sodium touches the water. Also, if you evaporate the water afterwards, you will obtain a white powder—something that clearly isn't sodium metal.

**Aqua regia is prepared by mixing two common laboratory acids: hydrochloric acid (three parts) and nitric acid (one part). Its most useful property is its ability to dissolve relatively unreactive metals such as platinum and gold.**

Let's consider water. An ordinary glass holds about 250 grams (½ pound) of the stuff. Water is transparent and apparently colorless. If you put it in a freezer, it solidifies at a temperature of 0°C (32°F). Salt dissolves in water, and if you evaporate the water, the salt is left behind.

If you add a pea-sized piece of sodium metal to water (taking stringent safety precautions as you do so!), you can make all sorts of interesting observations. The sodium bounces along the surface of the water like a bug and melts to form a metallic bead that quickly starts to shrink. If you look closely at the point of contact between the water and the sodium (before the sodium disappears), you will see gas bubbles being formed. Sometimes this experiment produces small explosions and sparks (so if you do look closely, do it through a safety shield). If you evaporate the water after all the sodium has disappeared, you will be left with a white solid that is obviously neither water nor sodium metal. A different substance has been formed; a chemical reaction has taken place.

Next let's think about gasoline. An ordinary glass holds about 200 grams of gasoline. If you add it to water, the gasoline floats on top. If you touch a lighted match to gasoline in air, it burns, and it burns even faster if you use pure oxygen instead of air. If you add a pea-sized piece of sodium metal to gasoline, the metal just sinks to the bottom and sits there.

Now consider gold. An ordinary glass holds about 5000 grams (about 11 pounds) of it. Gold is soft, and its surface is shiny and yellow. It doesn't corrode or tarnish. You can dissolve it in a mixture of very strong acids called *aqua regia.* If you do that and then remove the aqua regia, what's left behind isn't gold metal.

What can you make of this collection of properties? Some of them seem to be of different kinds. Chemists distinguish among properties in two important ways.

If a property changes when the amount of the substance changes, it's called an **extensive property.** The volume occupied by a substance is an extensive property. If you try to put 5000 grams of water in a regular glass—well, you'd better have some paper towels handy! The mass of a substance is also an extensive property. Which would you rather have, a gold nugget the size of a golf ball or one the size of a tennis ball? If you plan to convert your gold to cash, the answer is obvious.

If a property doesn't change when the amount of the substance changes, it's an **intensive property.** An ounce of gold is yellow, and so is a ton. An ice cube freezes at 0°C, and so does a lake.

The other classification of properties depends on whether the substance changes into something else when you determine the property. When you freeze water, it's still water. If you float gasoline on top of water, it is still gasoline. If the substance remains the same before and after you demonstrate a property, as water and gasoline do in these cases, the property is called a **physical property.** On the other hand, if you dissolve sodium in water or burn gasoline, the original substance becomes something different. If the identity of the substance changes when a property is demonstrated, the property is a **chemical property.**

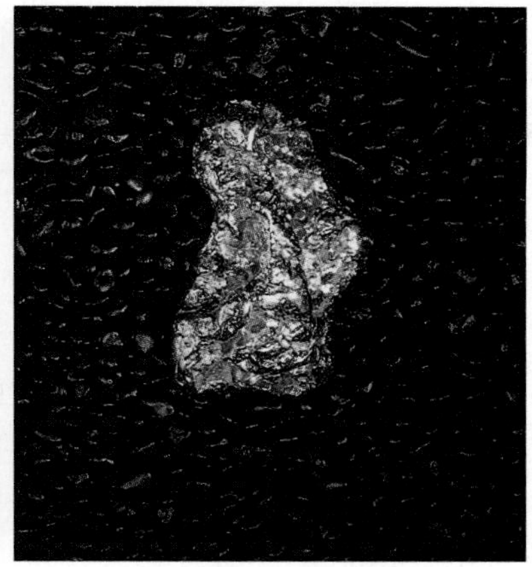

FIGURE 2.22
The color of gold is an intensive property. The coin and the nugget are both yellow.

EXAMPLE 2.4 **Classifying Properties as Physical or Chemical**

Classify each of the properties described as either physical or chemical.

**a.** Steel forms a red solid (rust) when in contact with moist air.
**b.** Ordinary liquid helium turns into superfluid helium at very low temperatures (near −273°C).
**c.** Sugar dissolves in water.

SOLUTIONS

**a.** The steel turns into a different substance with different properties (both physical and chemical). Rusting is a chemical property of steel.
**b.** Changing from a liquid to a superfluid is a physical property. Liquid helium in any form is still helium.
**c.** The sugar retains its identity; if you want the sugar back, you can evaporate the water. Dissolving in water is a physical property of sugar.

EXERCISE 2.4

Classify each property described as either physical or chemical.

**a.** An open can of cola goes flat.
**b.** A stick of dynamite explodes.
**c.** Gasoline is distilled in an oil refinery.

A **pure substance** is one that has its own unique properties. These properties distinguish it from all other substances. The properties of a mixture change when the amounts of its component parts change, but the intensive properties of pure substances are unvarying. For example, pure water boils at 100.0°C under normal conditions. This is a property of water that can be used to identify it. A mixture of salt and water boils at higher temperatures;

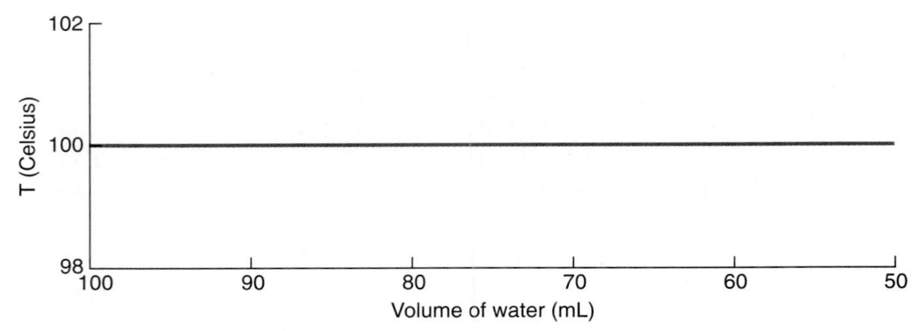


**A** The temperature of boiling water as the volume is reduced from 100 to 50 mL

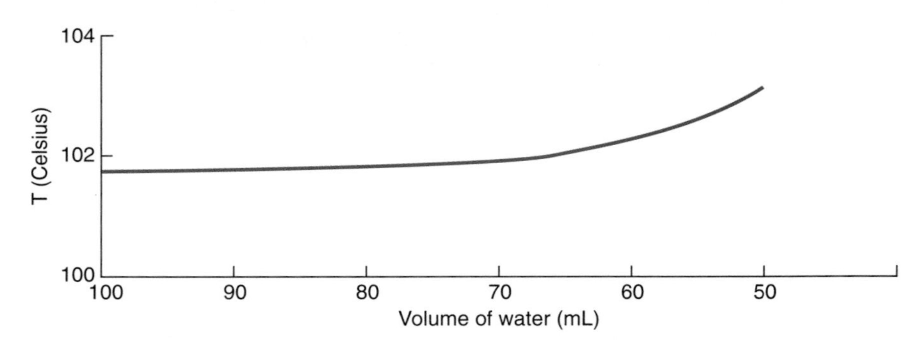


**B** The temperature of boiling salt water as the volume is reduced from 100 to 50 mL

FIGURE 2.23

(A) This graph shows that the boiling point of pure water remains constant as the water evaporates. (B) This graph shows that the boiling point of a salt-water mixture increases as the water evaporates.

the exact temperature depends on the amounts of salt and water present. The higher the percentage of salt in the water, the higher the boiling point will be.

## 2.5 The Atomic Theory

GOAL: To explain how chemists know that matter is composed of atoms.

Suppose you take a sample of a pure substance and divide it in half, and then in half again, and so on indefinitely. What will finally happen? There are only two possibilities. One of these is that the substance might be made up of particles that are too small to see. After a very large number of divisions, you might get down to one particle of a substance that can't be divided any further. The other possibility is that the substance might be infinitely divisible.

Ancient Greek philosophers put forth both of these ideas. The philosopher Democritus (about 470–380 BC) was the first to propose the existence of indivisible tiny particles, and he called them **atoms.** Aristotle, on the other hand, adopted the continuous fluid idea. Since real atoms are too small to be seen, Aristotle at least had common sense on his side, and among the few people who thought about the subject at all, his viewpoint was more widely believed. Neither philosophical camp had any evidence.

**The word *atom* comes from the Greek and means "not cuttable."**

Atomist ideas enjoyed a revival after the Middle Ages, but the necessary experimental basis was still lacking. The French chemist Joseph Proust found a key link. By his time it was known that most substances are composed of more than one element, and these substances were called *compounds* (see Section 2.7). Proust performed experiments to show that the weights of the elements making up any given compound always have the same ratio, no matter how the compound was made.

The English schoolteacher John Dalton (1766-1844) extended these ideas and published them in detail in 1808. He presented his original theory in essay form, but chemists often summarize it in five statements:

1. Elements are composed of extremely small individual particles called atoms.
2. All atoms of a given element have identical properties, which differ from the properties of atoms of other elements.
3. Atoms cannot be created from nothing, transformed into atoms of other elements, or destroyed.
4. Compounds are formed when atoms of different elements combine in whole-number ratios.
5. The relative numbers and kinds of atoms are constant in a given compound.

Chemists today know that there are problems with two of these five points. First, there *are* differences between atoms in most elements. Though these differences have only a small effect on the chemistry of the elements, they do exist. Second, some atoms (of radioactive elements) *do* transform themselves into others, and a wide variety of such changes are possible given enough energy. To many chemists both of these differences are subtle and usually unimportant.

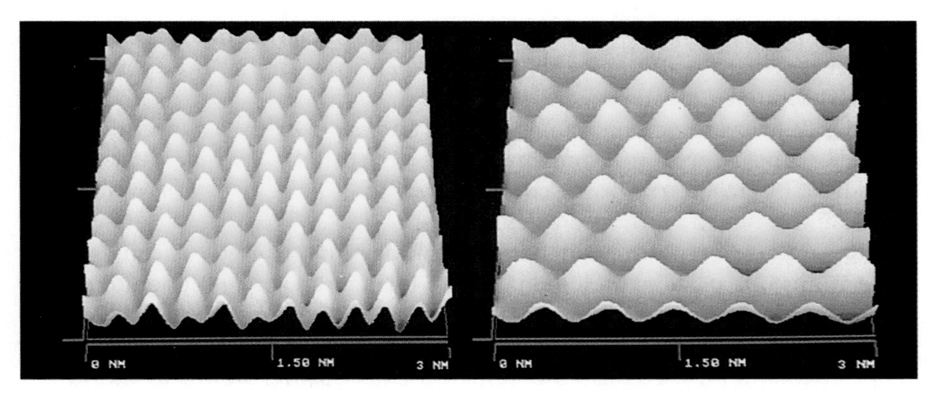


FIGURE 2.24

Scientists have obtained visible images of atoms only recently. This image from a scanning tunneling microscope shows the surface of a piece of gold foil.

If you look through a very strong microscope at bits of dust suspended in water, you can see the dust move. This movement is called Brownian motion. Einstein proposed that the dust is being bounced around by the motions of water molecules (composed of atoms) that are too small to see.

Most chemists of the time quickly accepted Dalton's proposals. A few holdouts said that they couldn't be sure atoms existed because nobody had ever seen any. The only evidence for the existence of atoms was that it explained so many observations so well. A century later Albert Einstein (1879–1955) was the first to explain a visible phenomenon—Brownian motion—in terms of atomic theory. The first visible images of actual atoms date from the 1950s. On the whole, Dalton's theory has survived two centuries of scientific advances remarkably well.

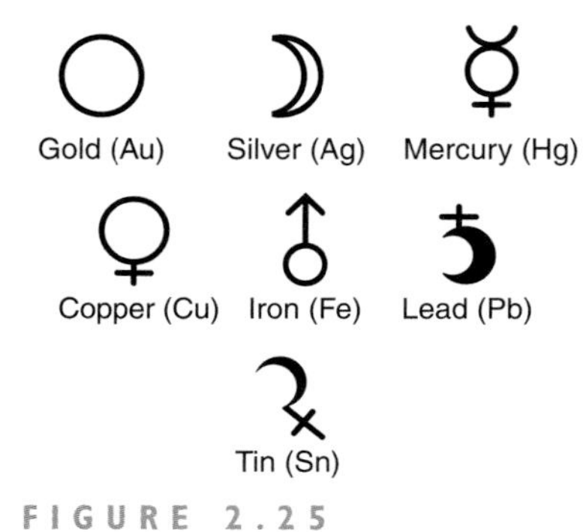


FIGURE 2.25
Alchemical symbols for elements and their modern equivalents

## 2.6 Symbols of the Elements

GOAL: To identify an element by name when given its symbol, and vice versa.

As you may remember from Chapter 1, the alchemists used exotic symbols for the elements. Modern scientists use simpler symbols made from one, two, or (rarely) three letters to represent the elements. These symbols are an essential part of the language of chemistry. You need to memorize at least some of them.

A dozen elements—including some that are most important to life—have one-letter symbols. For instance, carbon is C, hydrogen is H, nitrogen is N, and oxygen is O. Some elements' symbols consist of the two letters that begin their names, which makes these symbols easy to remember. For instance, helium is He, aluminum is Al, and bromine is Br. For pairs of elements with names that begin with the same letters, the symbols may be a bit harder to remember. Magnesium is Mg and manganese is Mn; cobalt is Co and copper is Cu (from the Latin *cuprum*). The symbols of many elements that were known in ancient times were taken from the Latin names originally used for the elements: for example, iron is Fe *(ferrum)*, tin is Sn *(stannum)*, and gold is Au *(aurum)*. Sodium is Na from the Latin word for common table salt *(natrium)*, which is a compound of sodium and chlorine. Three-letter symbols are used only for the newest of the artificial elements. These elements are primarily of interest to specialists, and you can ignore them in this course without a guilty conscience.

**STUDY ALERT**
**When writing the symbols for the elements, the first letter is always uppercase and the other letter is lowercase. You must be able to distinguish between symbols such as Co (the element cobalt) and CO (a compound called carbon monoxide, which consists of carbon and oxygen).**

Table 2.1 lists some common elements and their symbols. To understand the rest of this course, you must memorize this list. A list of elements in the periodic table inside the front cover of this book gives the names and symbols of all of the currently known elements.

Very few elements exist in nature as individual atoms. Instead, two or more atoms of a particular element join to form a discrete, independent unit called a **molecule.** A molecule is a group of atoms held together by electrical forces of attraction called *bonds.* The nature of bonding is discussed in more detail in Chapter 10. For the moment you can consider bonds as the force, or glue, that links atoms tightly to form molecules.

**TABLE 2.1 Common Elements and Their Symbols***

| Element | Symbol | Element | Symbol | Element | Symbol |
|---|---|---|---|---|---|
| aluminum | Al | hydrogen | H | platinum | Pt |
| arsenic | As | iodine | I | potassium (*kalium*) | K |
| barium | Ba | iron (*ferrum*) | Fe | silicon | Si |
| boron | B | lead (*plumbum*) | Pb | silver (*argentum*) | Ag |
| bromine | Br | lithium | Li | sodium (*natrium*) | Na |
| calcium | Ca | magnesium | Mg | sulfur | S |
| carbon | C | manganese | Mn | tin (*stannum*) | Sn |
| chlorine | Cl | mercury (*hydrargyrum*) | Hg | titanium | Ti |
| chromium | Cr | neon | Ne | uranium | U |
| copper (*cuprum*) | Cu | nickel | Ni | zinc | Zn |
| fluorine | F | nitrogen | N | | |
| gold (*aurum*) | Au | oxygen | O | | |
| helium | He | phosphorus | P | | |

* The symbols of some elements are based on Latin or Greek names, which are given in parentheses.

The elements display a wide variety of molecular compositions. Several of them exist in nature as molecules with two atoms. These are known as **diatomic molecules** (*di* is a prefix meaning "two"). More specifically, the molecules of these elements are homonuclear (same-nucleus) diatomic (two-atom) molecules.

A molecule of nitrogen contains two nitrogen atoms (whose symbol is N). How do chemists represent nitrogen molecules in formulas? Chemists use a numerical subscript to indicate the number of a particular type of atom present in a molecule. The formula for nitrogen molecules is $N_2$. The subscript 2 tells you that there are two nitrogen atoms in one molecule of nitrogen. Be sure to remember that the subscript is written *after* the symbol for the element. The symbol $N_2$ stands for one nitrogen molecule containing two atoms; in contrast, 2 N means two separate nitrogen atoms.

The elements that form homonuclear diatomic molecules are hydrogen, nitrogen, oxygen, fluorine, chlorine, bromine, and iodine. These molecules have the formulas $H_2$, $N_2$, $O_2$, $F_2$, $Cl_2$, $Br_2$, and $I_2$, respectively.

Most elements form molecules that contain more than two atoms. Phosphorus and arsenic, for instance, are usually found as $P_4$ and $As_4$ molecules, respectively. Sulfur molecules have the formula $S_8$. All metallic elements form molecules of indeterminate size containing a very large number of atoms. A piece of aluminum foil or a seamless aluminum beverage can can be thought of as a single molecule of aluminum. A single aluminum atom in a piece of foil is bonded to all of its immediate neighbor atoms in exactly the same way (see Figure 2.26). The only beginning or end to such an array is at the surface of the metal. Carbon atoms do the same thing. A diamond is a single gigantic carbon molecule, and a piece of graphite is made up of a relatively small number of very large molecules. By convention, scientists represent the formula of a substance of this kind simply by writing the symbol of the element.

**The only elements that exist in nature as separate atoms are helium, neon, argon, krypton, xenon, and radon. Collectively these were formerly known as the *inert gases*, because chemists thought they were completely unreactive. Chemists discovered, however, that krypton, xenon, and radon form a few compounds under exceptional circumstances, so they now call these elements the *noble gases*.**

**STUDY ALERT**

**Memorize the names and formulas of the homonuclear diatomic molecules.**

**STUDY ALERT**

**Note that the symbol for an element can be used to represent a single-atom species (such as He) or an element whose "molecule" consists of an incredibly large number of atoms (Al).**

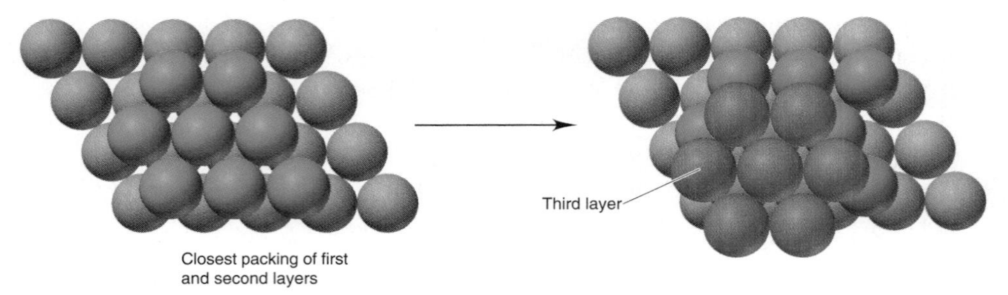


FIGURE 2.26
Each of the spheres represents an aluminum atom. Only three layers of a very small cross section of a sheet of aluminum foil are shown. In a piece of aluminum foil large enough to see, the number of layers is in the millions, and the number of atoms present is larger than the number of seconds in the lifetime of the earth. Every atom is identical to every other atom, and there are no distinct molecules present. In a sense the entire array is a single molecule. Scientists simply use the symbol of the element to represent arrays of this kind. As a matter of convenience and convention, the formula for aluminum is written Al.

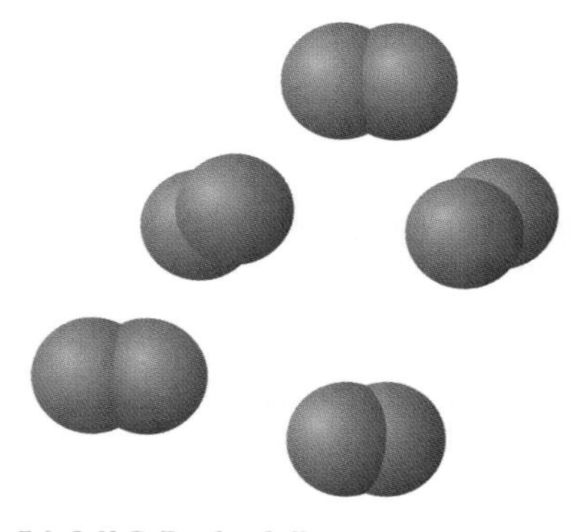

FIGURE 2.27
Nitrogen in the air exists as diatomic molecules. Here the individual nitrogen atoms are represented by spheres, and the fact that they have formed molecules is indicated by the contact between the spheres. There are ten nitrogen atoms and five nitrogen molecules in this figure.

### EXAMPLE 2.5 Element Names and Symbols

**a.** Name the elements that have the symbols Si, Ag, and F.
**b.** Give the symbols for the elements potassium and phosphorus.

**SOLUTIONS**

**a.** Si is silicon, which is used in computers; do not confuse it with silver. Ag is silver (from the Latin word *argentum*); do not confuse it with silicon. F is fluorine (beware of the spelling of this name—it does not contain "flour"!).
**b.** Potassium is K (from the Latin word *kalium*), *not* P. Phosphorus is P. (The answer is not $P_4$. You were asked for the symbol for the *element*, not the molecule.)

**EXERCISE 2.5**

**a.** First, cover the symbols of the elements in Table 2.1 and write the names. Then cover the names and write the symbols.
**b.** Write the symbols for molecules of oxygen, phosphorus, and fluorine.

## 2.7 Compounds

GOAL: To determine how many atoms of each element are present in a formula unit of a compound.

A sample of an element contains only one kind of atom. A **compound** is a pure substance that contains atoms of more than one element. If you look at samples of elements and compounds, you won't be able to see differences between the two kinds of substances. The ancients never discovered the distinction, and even the careful Lavoisier included a few compounds that are hard to break down in his list of elements. Today's list of elements is the result of 200 years of experiments and research since Lavoisier's time.

If chemists know how many atoms of each kind of element are present in a compound, they can create a kind of accounting of the atoms, called a **chemical formula.** This section describes how to interpret these formulas. Constructing the formula from the name of a compound or the name from the formula is the main topic of Chapter 7.

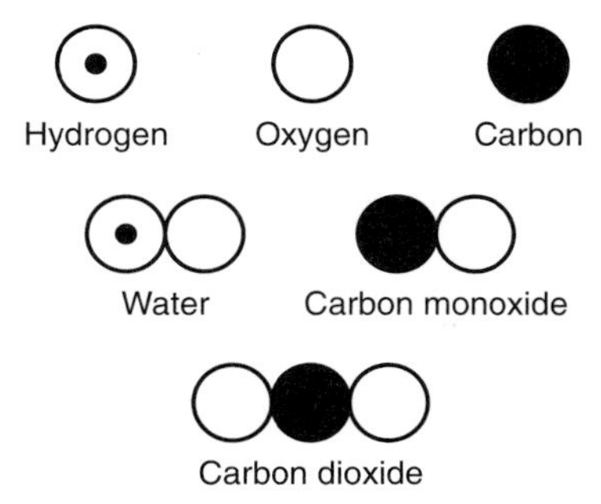


FIGURE 2.28
Dalton proposed circular symbols for the elements, some of which are shown here. The system using letters was proposed by the Swedish chemist Jons Jakob Berzelius in 1813. It's easier to use, and chemists soon adopted it.

Chemical formulas use subscripts to indicate the number of each kind of element in a compound. This use of subscripts is the same as their use in the formulas for the molecular elements. The subscript number is written *after* the symbol for each element. For example, a water molecule contains two hydrogen atoms and one oxygen atom. The formula for water, then, is $H_2O$. When only one atom of a particular type is present (such as oxygen in $H_2O$), there is no subscript. That is, the number of atoms is 1 when no subscript appears.

The formula for table salt is NaCl. There is one atom of each element in one unit of the compound. Sodium hydrogen carbonate, commonly called sodium bicarbonate, is an ingredient in both stomach antacids and baking soda, and it has the formula $NaHCO_3$. The formula for sodium bicarbonate indicates one atom each of sodium, hydrogen, and carbon, as well as three atoms of oxygen in a formula unit.

Some formulas are written with parentheses. Aluminum hydroxide is another ingredient sometimes used in stomach antacids, and its formula is $Al(OH)_3$. This means that one unit of this compound contains one aluminum atom and three OH units. The three OH units, in turn, contain a total of three oxygen atoms and three hydrogen atoms. A formula unit of $Al(OH)_3$ therefore contains a total of one aluminum atom, three oxygen atoms, and three hydrogen atoms.

$Al(OH)_3 = Al + OH + OH + OH$

**This formula shows that one unit of the compound contains one aluminum atom. Three oxygen and three hydrogen atoms form three OH units.**

FIGURE 2.29
Potassium metal (left) (stored here under an inert liquid to prevent it from reacting with oxygen in the air) does not resemble greenish yellow chlorine gas (in the round-bottomed flask). The two combine in a vigorous reaction (center) to form the compound potassium chloride (right). The fact that two of these substances are elements and one is a compound is not obvious from the way they look.

**$Fe_3(PO_4)_2 = Fe + Fe + Fe + PO_4 + PO_4$**
**One unit of the compound contains three iron atoms. Two phosphorus atoms and eight oxygen atoms form two $PO_4$ units.**

A similar example is a unit of $Fe_3(PO_4)_2$, which has three iron atoms and two $PO_4$ units. Each $PO_4$ unit contains one phosphorus and four oxygen atoms. Therefore one unit of the compound contains a total of three iron atoms, two phosphorus atoms, and eight oxygen atoms.

You are probably wondering why the parentheses are necessary. Why are these formulas written as $Al(OH)_3$ and $Fe_3(PO_4)_2$ instead of $AlO_3H_3$ and $Fe_3P_2O_8$? Figure 2.30 provides the key to answering this question. Many chemical formulas do not give information about how the atoms of each element in the compound are connected to one another. On the other hand, one of the most important principles in chemistry is that **the physical and chemical properties of substances depend on their structures.** Certain collections of atoms, including the OH and $PO_4$ units, have distinct and well-known structures and properties. To indicate the presence of such structural units, chemists group them in formulas using parentheses.

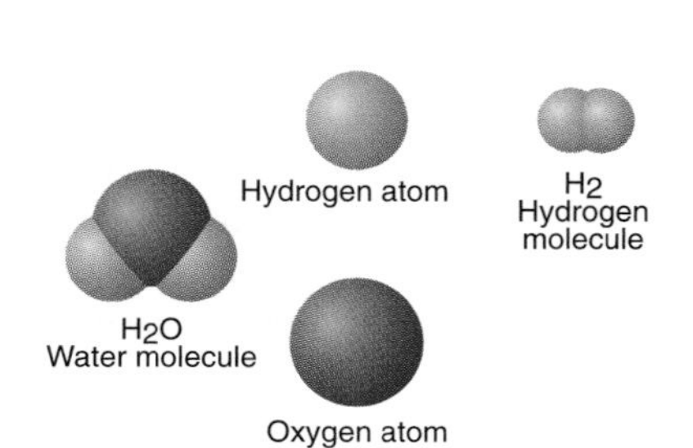


FIGURE 2.30
A water molecule, $H_2O$, contains two atoms of hydrogen and one atom of oxygen. A hydrogen molecule, $H_2$, contains two atoms of hydrogen. In a hydrogen molecule the two atoms of hydrogen are bonded (joined) to each other. In a water molecule they are not—each is bonded to the oxygen. Both formulas contain the designation $H_2$, but this symbol implies nothing at all about how the atoms are arranged with respect to each other. You must recognize that chemical formulas are atom counts and nothing more. The arrangements of atoms cannot be determined from the formulas without considerable experience, and often not even then!

### EXAMPLE 2.6 Counting Atoms in Formulas

How many atoms of each element are present in a formula unit of each of these compounds?

**a.** $CoCl_2$ **b.** $COCl_2$ **c.** $C_{12}H_{22}O_{11}$ **d.** $Ca_3(PO_4)_2$

SOLUTIONS

**a.** $CoCl_2$ has one cobalt atom (Co) and two chlorine atoms (Cl).
**b.** Look closely at the two formulas in (a) and (b), which demonstrate a small but important difference. Molecules of $COCl_2$, the poisonous gas phosgene, contain one carbon, one oxygen, and two chlorine atoms.
**c.** One molecule of this compound (table sugar, or sucrose) contains twelve carbon atoms, twenty-two hydrogen atoms, and eleven oxygen atoms.
**d.** A formula unit of this compound contains three calcium atoms and two $PO_4$ units. Each $PO_4$ unit in turn contains one phosphorus atom and four oxygen atoms. Add these atoms together to get the final answer: three calcium, two phosphorus, and eight oxygen atoms.

EXERCISE 2.6

How many atoms of each element are present in a formula unit of each of the following compounds?

**a.** $Ca(OH)_2$ **b.** $C_6H_{12}$ **c.** $Al_2(SO_4)_3$ **d.** $CO_2$

## 2.8 Microscopic Views of Matter

GOAL: To describe what you would see if you could view the atoms and molecules of solids, liquids, and gases.

In Section 2.2 we described the familiar properties of solids, liquids, and gases. Solids and liquids aren't compressible, and gases are. Liquids and gases flow, and solids don't. Some solids are flexible and without much form, and some have rigid structures. Why do substances act this way? What is it about the atoms and molecules in them that causes solids, liquids, and gases to be the way they are?

Some solids are crystalline. Look closely at some grains of table salt. You can sometimes make out a regular cubic shape without the aid of a magnifying glass. In crystalline solids the atoms and molecules have an orderly, rigid arrangement. The atoms that make up a sodium chloride crystal, for example, are organized in a structure resembling that of the aluminum atoms in

FIGURE 2.31
The atoms and molecules in a crystal have a regular, repeating arrangement. Different compounds can have different arrangements of atoms. Note that crystals of salt (left), quartz (center), and pyrite (right) have different shapes. The differences are caused partly by variations in the arrangement of the atoms in the individual crystals.

## EXPERIMENTAL EXERCISE 2.1

A. L. Chemist and his team were given a sample of gunpowder to analyze. A. L. knew that gunpowder is a mixture of sulfur, potassium nitrate ($KNO_3$), and carbon. $KNO_3$ dissolves in water, and carbon and sulfur do not. Sulfur dissolves in carbon disulfide, and $KNO_3$ and carbon do not.

After measuring out 40 grams of gunpowder, A. L. placed it in a large amount of water, stirred vigorously for 10 minutes, and then removed the undissolved residue (solid 1) from the water. Solid 1 had a dry weight of 8.8 grams.

A. L. then put Solid 1 in excess carbon disulfide and stirred for 10 minutes. A second solid residue (solid 2) was collected and dried. It had a mass of 4.0 grams.

A. L. Chemist and his team concluded that the gunpowder was 22% $KNO_3$ ($^{8.8}/_{40} \times 100\%$), 45.45% sulfur ($^{4.0}/_{8.8} \times 100\%$), and 32.55% carbon (100% − 22% − 45.45%).

1. Name the elements present in $KNO_3$.
2. Name the separation technique that the team used to isolate solid 1 from the water.
3. What are the elements present in solid 1?
4. What elements are present in solid 2?
5. What was left dissolved in the water after solid 1 was isolated?
6. What was left dissolved in the carbon disulfide after solid 2 was isolated?
7. Assuming that the procedure for this experiment is stated correctly (and it is), are the team's conclusions supported by the results? (Justify your answer—do not respond simply "yes" or "no.")

Figure 2.26. In crystalline solids the orderly arrangement of atoms is repeated all the way to the surface of the sample in three dimensions. This does not mean that the molecules and atoms in crystals (or any other solids) are not moving. They are, in fact, vibrating constantly about their fixed positions. Chapter 12 explores this point in more detail.

FIGURE 2.32
The long-chain molecules in this amorphous solid are stretched easily in any direction. This property gives rise to the shapeless nature of these solids.

FIGURE 2.33
Crystal glass is amorphous at the molecular level. The "crystal" in the name refers to the shape the glass was formed into when it was made. Once again, scientists' use of a term is somewhat different from common usage.

Many solids lack the rigid structures that crystals have. Rubber and most plastics are examples of these **amorphous solids.** The word *amorphous* (from a Greek word that means "without definite shape") is quite descriptive, because the molecules are arranged more randomly in these solids than they are in crystalline solids. Many amorphous solids are composed of long molecules that coil up and bunch together something like spaghetti. Figure 2.32 shows an arrangement of this kind. Because the molecules are coiled, they are easily stretched in any direction—a particularly useful property of plastic wrap and rubber bands.

Appearances can be deceiving. Glass is an amorphous solid. But this fact is far from obvious—it's not uncommon for students to guess (incorrectly) that glass is crystalline. However, close examination of old windows (such as those found in cathedrals in Europe) shows that the panes of glass are thicker near the bottom than they are near the top. The relentless force of gravity over time has caused the glass to flow.

In both liquids and solids the atoms or molecules are in close contact with one another. It is possible, but difficult, to compress a liquid or a solid by applying pressure and forcing the atoms or molecules to fill the gaps more efficiently. It almost always requires more pressure than you could apply by hand to do this. The solids that you can compress by hand usually have pores or holes in them, as sponges do. The solid substance is not actually compressed; instead, air is forced out of the holes.

Adjacent atoms or molecules in liquids are held together by fairly strong forces, though the forces are usually weaker than those in solids. The molecular arrangement in liquids is more random than it is in solids. Molecules

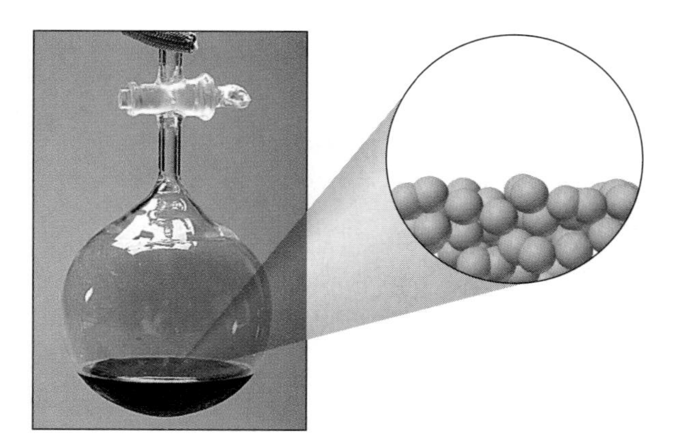

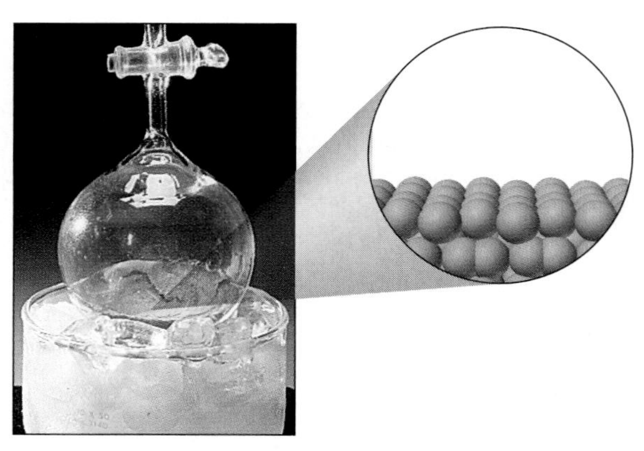


FIGURE 2.34
The molecules in liquids are oriented more randomly than the ones in solids. Also, the molecules can slide past one another rather easily. This property allows liquids to flow. Note that in liquids, as in solids, the molecules are close together.

can exchange places with adjacent ones rather easily. This permits liquids to flow, which most solids (except amorphous ones) don't do.

In gases the individual atoms or molecules are widely separated. The forces that attract these atoms or molecules to one another are so weak that the atoms or molecules bounce around freely. Gases apply pressure to their containers because of the force of the collisions of their atoms or molecules with the bottom, sides, and top of a container. When you increase the pressure on a sample of gas and compress it, what actually becomes smaller is the volume of empty space surrounding the molecules.

A gallon of water at room temperature weighs 800 times as much as a gallon of air (yes, air does weigh something). This is because there are very many more molecules present in a given volume of a liquid than in an identical volume of a gas. Even at the high pressure inside a scuba diver's tank, air is still mostly empty space.

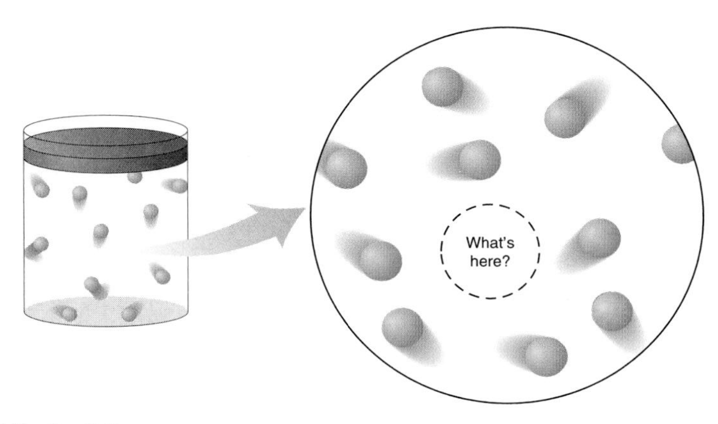


FIGURE 2.35
Gas molecules are widely separated and have little attraction for each other. These molecules are constantly in motion and move past each other very easily. This is why gases flow. Gas molecules exert pressure by colliding with the walls of their container.

### EXAMPLE 2.7 Microscopic Descriptions of Matter

In Figure 2.35, what is at the location indicated by the dashed circle?

**SOLUTION**

There's a vacuum there—nothing at all. (You didn't say "air," did you?)

**EXERCISE 2.7**

In each of these pairs, select the item with the larger number of molecules.

**a.** A cup of solid gold metal or a cup of molten gold
**b.** A cup of liquid water or a cup of steam
**c.** A gallon of ice or a gallon of water vapor

## 2.9 Some Typical Substances

**GOAL:** To describe some familiar substances in terms of the atoms and molecules that make them up.

Let's start with that most familiar of substances, water. On cold days $H_2O$ molecules lock together to form a rigid crystalline solid. The beauty of frost crystals on windows and of snowflakes is proverbial. Ice crystals have a rather open molecular structure; this is why ice floats on water rather than sinking. This property is very unusual for a solid—samples of almost all solid substances will sink in the liquid form of the same substance. When ice melts, the spaces between the molecules close up a little. If you heat the liquid water enough, it boils. The heat provides the energy to push the molecules of water apart. The molecules go off into space as widely separated gas molecules. Figure 2.36 shows what water molecules do as a sample of water goes from solid to liquid to gas.

Table salt is sodium chloride, NaCl. As we pointed out earlier, if you examine grains of salt closely, you'll see that many of them are small cubes. NaCl is a crystalline solid with a high melting point. It won't melt even at the highest temperatures possible on laboratory Bunsen burners. This shows that the attraction between the atoms in salt crystals is very strong. Many similar compounds behave the same way.

The most common compound in natural gas (used for cooking and heating) is methane, $CH_4$. Like all gases, methane has widely separated molecules. Methane is odorless; power companies add a strong-smelling com-

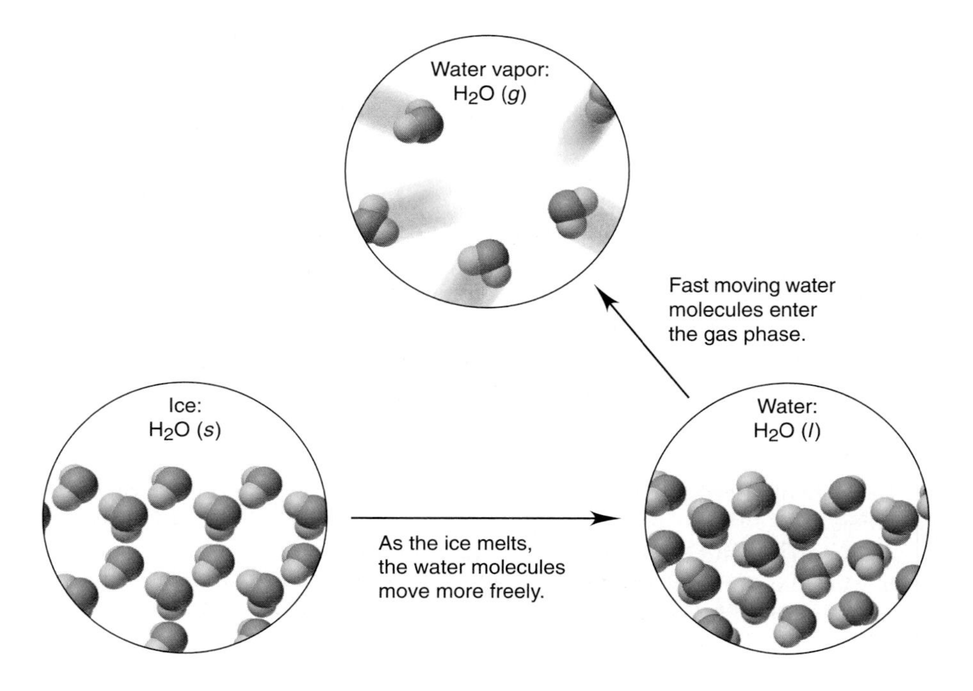
Water vapor:
H2O (g)
Fast moving water
molecules enter
the gas phase.
Ice:
H2O (s)
As the ice melts,
the water molecules
move more freely.
Water:
H2O (l)

FIGURE 2.36
Molecules of water in the solid, liquid, and gas states.

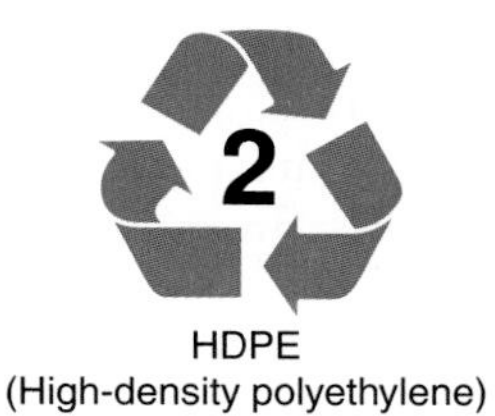


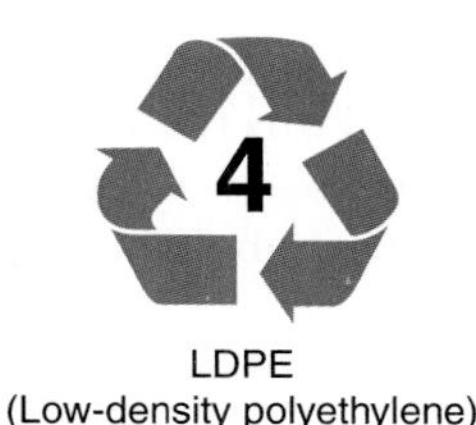


FIGURE 2.37
Polyethylene can be recycled effectively and efficiently. High-density polyethylene (HDPE), the form with longer chains of $CH_2$ groups, should be recycled separately from the low-density form (LDPE).

pound to natural gas to alert people to gas leaks. The most useful property of methane is a chemical property: it burns—reacts with oxygen, a chemist would say. This reaction gives off heat, which is typically used to cook or to generate electricity.

Polyethylene is one of the most common amorphous solids. The plastic jugs that milk comes in and the lightweight plastic bags that many stores use are made of polyethylene. A typical polyethylene molecule is a very long string of $CH_2$ units. In the best and purest polyethylene each molecule contains hundreds of thousands of these units. The long, flexible chains don't fall into the rigid patterns typical of crystalline solids, which is why objects made of polyethylene are flexible.

Like methane, polyethylene burns with a relatively clean flame and reacts with few compounds other than oxygen. Also, it takes considerable heat to start the reaction with oxygen. This is both a blessing and a curse. Objects made of polyethylene last a long time—while you're using them and in the dump after you've thrown them away.

Ammonia is another common substance. It's used in fertilizers and in household cleaners. Once again, names can be deceiving. Pure ammonia, $NH_3$, is a foul-smelling gas at room temperature. Household ammonia is a homogeneous mixture of $NH_3$ and water.

**STUDY ALERT**

**Of course, you should learn the formulas for water, methane, and ammonia.**

## CHAPTER SUMMARY

1. Mixtures can be classified as homogeneous or heterogeneous.
2. There are three states of matter: solids, liquids, and gases. These differ primarily in the amount of empty space between the particles that make up the substance. The amount of space depends on the strength of the attraction between the particles.
3. The properties of solids depend on both their chemical composition and their structure.
4. Gases and liquids are fluids.
5. All gaseous mixtures are homogeneous.
6. Methods used to separate mixtures of substances are based on the differences in the physical and/or chemical properties of the substances.
7. The common separation techniques used in chemistry are filtration, distillation, extraction, and chromatography.
8. Extensive properties depend on the amount of substance present, and intensive properties do not.
9. Physical properties can be investigated without changing the identity of a substance, but demonstrating chemical properties does change the identity of the substance.

10. Elements are composed of one type of atom. An atom is the smallest unit of matter that retains the identity of the element.
11. Compounds are made up of atoms of two or more elements.
12. The names and symbols of certain elements, given in Table 2.1, must be memorized.
13. The homonuclear diatomic molecules are $H_2$, $N_2$, $O_2$, $F_2$, $Cl_2$, $Br_2$, and $I_2$.
14. The formulas for water, methane, and ammonia are $H_2O$, $CH_4$, and $NH_3$, respectively.

## PROBLEMS

### SECTION 2.1 Mixtures

1. Give an example of each of these:
   a. A homogeneous mixture containing three or more substances
   b. A heterogeneous mixture containing two substances
   c. An alloy *not* mentioned in this chapter
   d. A homogeneous mixture that is also a fluid
2. Give an example of each of these:
   a. A homogeneous mixture containing two substances
   b. A heterogeneous mixture containing three or more substances
   c. An alloy you use or see daily
   d. A homogeneous mixture that is liquid
3. Classify each item as homogeneous or heterogeneous. Briefly explain your answers.
   a. Vinegar
   b. A peach
   c. Mushroom soup
   d. A cloud
4. Classify each item as homogeneous or heterogeneous. Briefly explain your answers.
   a. White wine
   b. Your instructor
   c. Ice cream
   d. Cooking oil
5. Are these items homogeneous or heterogeneous? Classify each one and briefly explain your answers.
   a. A goldfish
   b. Air
   c. Water vapor
   d. A quartz crystal
6. Are these items homogeneous or heterogeneous? Classify each one and briefly explain your answers.
   a. A razor blade
   b. A piece of wood
   c. Gasoline
   d. An icicle

### TOOL BOX: Percentages

7. A ring made from gold and copper is 91.6% gold. What is the percentage of copper in the ring?
8. A sample of bronze contains 10% zinc and 73% copper. The rest of the sample is tin. What is the percentage of tin in this sample of bronze?
9. Gunpowder consists of carbon, sulfur, and potassium nitrate. A sample of gunpowder contains 15% carbon and 75% potassium nitrate. What is the percentage of sulfur in this sample?
10. Another sample of gunpowder contains 17% carbon and 72% potassium nitrate. What is the percentage of sulfur in this sample?
11. Before 1920 British silver coins contained 92.5% silver. What was the percentage of other metals in this alloy (which was called "sterling silver")?
12. The last British gold coins to be minted contained 8.3% other metals. What is the percentage of gold in this alloy?
13. What is the percentage of gold in each of these samples?
    a. 14-karat gold
    b. 17-karat gold
14. What is the percentage of gold in each of these samples?
    a. 12-karat gold
    b. 15-karat gold
15. Players on the Green Leaf Bookstore basketball team had the following shooting records over a five-game span. What was the shooting percentage of each player (that is, the percentage of shots attempted that were successful)?
    a. 47 attempts, 22 successful
    b. 12 attempts, 2 successful
    c. 75 attempts, 24 successful
    d. 41 attempts, 23 successful
    e. 36 attempts, 13 successful

16. The shooting records of the players on the Heavy Metal Audio basketball team are as follows. Once again, what was the shooting percentage of each player?
    a. 9 attempts, 3 successful
    b. 113 attempts, 31 successful
    c. 35 attempts, 26 successful
    d. 29 attempts, 9 successful
    e. 44 attempts, 20 successful
17. Use the following shooting records of the players on the Poor Boy Sandwich Shop basketball team to determine the shooting percentage of each player.
    a. 37 attempts, 5 successful
    b. 72 attempts, 45 successful
    c. 54 attempts, 25 successful
    d. 15 attempts, 10 successful
    e. 39 attempts, 19 successful
18. Use the following shooting records of the players on the Shiny Auto Repair basketball team to determine the shooting percentage of each player.
    a. 8 attempts, 2 successful
    b. 25 attempts, 12 successful
    c. 155 attempts, 86 successful
    d. 47 attempts, 17 successful
    e. 77 attempts, 35 successful
19. A biologist on a small island in the South Atlantic woke up every morning, went to a nearby beach, and counted all the seals there. Here are the records for five days. Calculate the percentage of each kind of seal on the beach for each day.
    a. Monday: 199 elephant seals, 245 fur seals, 2 crabeater seals
    b. Tuesday: 212 elephant seals, 301 fur seals, 1 crabeater seal
    c. Wednesday: 129 elephant seals, 387 fur seals, 2 crabeater seals
    d. Thursday: 154 elephant seals, 355 fur seals, 3 crabeater seals
    e. Friday: 234 elephant seals, 414 fur seals, no crabeater seals
20. Here are some more records from the seal-counting biologist. Once again, calculate the percentage of each kind of seal on the beach for each day.
    a. Saturday: 185 elephant seals, 420 fur seals, 5 crabeater seals
    b. Sunday: 121 elephant seals, 424 fur seals, 1 crabeater seal
    c. Monday: 83 elephant seals, 397 fur seals, no crabeater seals
    d. Tuesday: 65 elephant seals, 440 fur seals, 1 crabeater seal
    e. Wednesday: 73 elephant seals, 433 fur seals, 4 crabeater seals
21. A metallurgist prepared a series of alloy samples for some tests. The amounts of metals used to make five samples of alloy are as follows. Calculate the percentage of each metal in each of the samples the metallurgist made.
    a. Trial 1: 50 grams copper, 20 grams tin, 5 grams zinc
    b. Trial 2: 50 grams copper, 20 grams tin, 10 grams zinc
    c. Trial 3: 50 grams copper, 20 grams tin, 15 grams zinc
    d. Trial 4: 50 grams copper, 20 grams tin, 20 grams zinc
    e. Trial 5: 50 grams copper, 20 grams tin, 25 grams zinc
22. The same metallurgist prepared more alloy samples later in the week. The amounts of metals used to make five samples of alloy are as follows. Calculate the percentage of each metal in each of the samples the metallurgist made.
    a. Trial 1: 40 grams copper, 10 grams tin, 10 grams zinc
    b. Trial 2: 40 grams copper, 20 grams tin, 10 grams zinc
    c. Trial 3: 40 grams copper, 30 grams tin, 10 grams zinc
    d. Trial 4: 40 grams copper, 40 grams tin, 10 grams zinc
    e. Trial 5: 40 grams copper, 50 grams tin, 10 grams zinc
23. A sample of steel was made from 20 pounds of iron, 2 pounds of nickel, 1.5 pounds of chromium, and 0.5 pounds of carbon. What is the percentage of each substance in the sample of steel?
24. A sample of steel was made from 45 pounds of iron, 4 pounds of nickel, 2 pounds of chromium, and 1.5 pounds of carbon. What is the percentage of each substance in this sample of steel?
25. How many pounds of tin are in a 250-pound sample of tin alloy that is 27% tin?
26. How many pounds of tin are in a 30-pound sample of tin alloy that is 42% tin?
27. A chemist was asked to make some 100-gram samples of mixtures of aluminum hydroxide, magnesium carbonate, and calcium carbonate, to be tested as antacids. Here are the desired percentages. How many grams of each substance should the chemist use?
    a. 10% aluminum hydroxide, 30% magnesium carbonate, 60% calcium carbonate
    b. 10% aluminum hydroxide, 35% magnesium carbonate, 55% calcium carbonate
    c. 10% aluminum hydroxide, 40% magnesium carbonate, 50% calcium carbonate
    d. 10% aluminum hydroxide, 45% magnesium carbonate, 45% calcium carbonate

28. The same chemist was asked to make other 100-gram samples of mixtures of the same three compounds. Here are the desired percentages. How many grams of each substance should the chemist use?
    a. 35% aluminum hydroxide, 15% magnesium carbonate, 50% calcium carbonate
    b. 25% aluminum hydroxide, 25% magnesium carbonate, 50% calcium carbonate
    c. 15% aluminum hydroxide, 35% magnesium carbonate, 50% calcium carbonate
    d. 5% aluminum hydroxide, 45% magnesium carbonate, 50% calcium carbonate

29. Next our chemist was asked to make 250-gram samples of other mixtures of the three compounds. Here are the desired percentages. How many grams of each substance should the chemist use?
    a. 15% aluminum hydroxide, 30% magnesium carbonate, 55% calcium carbonate
    b. 20% aluminum hydroxide, 30% magnesium carbonate, 50% calcium carbonate
    c. 25% aluminum hydroxide, 30% magnesium carbonate, 45% calcium carbonate
    d. 30% aluminum hydroxide, 30% magnesium carbonate, 40% calcium carbonate

30. Finally, the chemist was asked to make 750-gram samples of mixtures of the three compounds. Here are the desired percentages. How many grams of each substance should the chemist use?
    a. 25% aluminum hydroxide, 20% magnesium carbonate, 55% calcium carbonate
    b. 30% aluminum hydroxide, 20% magnesium carbonate, 50% calcium carbonate
    c. 35% aluminum hydroxide, 20% magnesium carbonate, 45% calcium carbonate
    d. 40% aluminum hydroxide, 20% magnesium carbonate, 40% calcium carbonate

31. A metallurgist wants to make 1000 kilograms of an alloy that is 30% magnesium and 70% aluminum. How much magnesium and aluminum should the metallurgist use?

32. A metallurgist wants to make 500 kilograms of an alloy that is 35% magnesium and 65% aluminum. How much magnesium and aluminum should the metallurgist use?

## SECTION 2.2 Solids, Liquids, and Gases

33. Give examples of three substances other than air that are gases.

34. Give examples of three substances not mentioned in this chapter that are liquids.

35. Give two examples of situations not mentioned in the chapter in which the incompressibility of solids is an important property.

36. Give two examples of situations not mentioned in the chapter in which the incompressibility of liquids is an important property.

37. Name the state or states of matter that fit each of these descriptions.
    a. Expands to fill its container
    b. Does not flow
    c. Assumes the shape of its container
    d. The most compressible state

38. Name the state or states of matter that fit each of these descriptions.
    a. The substance is not easily compressible.
    b. The substance always has the same volume as its container.
    c. Mixtures are always homogeneous.
    d. Many substances have more than one form in this state.

## SECTION 2.3 Separating Mixtures

39. Give two examples of mixtures that can be separated by each of these methods.
    a. Winnowing b. Filtration c. Distillation

40. Give three examples of mixtures that can be separated by each of these methods.
    a. Extraction b. Chromatography

41. State the difference in physical properties that is utilized by each of the following separation techniques.
    a. Filtration b. Chromatography

42. State the difference in physical properties that is utilized by each of the following separation techniques.
    a. Extraction b. Distillation

43. Name a separation technique that can be used to separate each of the following mixtures. Justify your answer by explaining what happens to each substance during the separation.
    a. Sugar and gravel
    b. Kerosene and crude petroleum
    c. Two coloring agents found in leaves

44. Name a separation technique that can be used to separate each of the following mixtures. Justify your answer by explaining what happens to each substance during the separation.
    a. Salt and pepper
    b. Sand and seawater
    c. Popped popcorn and unpopped kernels

45. What separation technique uses differences in each of these physical properties to separate the components of mixtures?
    a. Adsorptivity    b. Boiling point    c. Particle size

46. What separation technique uses differences in each of these physical properties to separate the components of mixtures?
    a. Adsorption on the fibers of paper
    b. Physical trapping by the fibers of filter paper
    c. Ability to be blown in the wind

47. What term describes each of these transformations?
    a. Ice to liquid water
    b. Ice to water vapor
    c. Liquid water to water vapor

48. What term describes each of these transformations?
    a. Water vapor to liquid water    c. Liquid water to ice
    b. Water vapor to ice

Alchemists and early chemists used the device shown in Figure 2.38, called a *retort,* to carry out distillations.

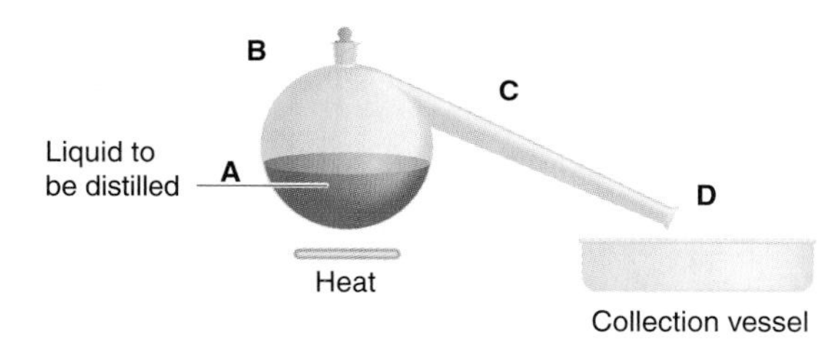


FIGURE 2.38

49. At which of the labeled points does evaporation occur?

50. Which component of the liquid mixture evaporates first, the more volatile or the less volatile one?

51. At which of the labeled points does condensation occur?

52. What happens to the boiling point of the liquid mixture as the distillation proceeds?

### SECTION 2.4 Properties

53. Classify each of these properties as either extensive or intensive.
    a. Color    b. Volume

54. Classify each of these properties as either extensive or intensive.
    a. Mass    b. Odor

55. Tell whether each of the following is describing a chemical or a physical property of the substance.
    a. Baking soda dissolves in water.
    b. Adding vinegar to baking soda produces gas bubbles.
    c. Ammonia is colorless.
    d. Polyethylene burns with a clean, clear flame.

56. Tell whether each of the following is describing a chemical or a physical property of the substance.
    a. Ammonia gas is extremely soluble in water.
    b. Mercury metal is a liquid at room temperature.
    c. Helium forms no compounds with any of the other elements.
    d. Sodium azide explodes in air bags to produce nitrogen gas.

57. A sample of barium metal weighs 5 grams. The metal is bright and shiny. When added to a sample of gasoline, the piece of barium sinks. When heated in air, the metal burns with a green flame and gives a white powder as a product. From this description, give a property of barium in each of the following classifications.
    a. Extensive    c. Chemical
    b. Intensive    d. Physical

58. A sample of bromine has a volume of 0.12 liters. It is a brown liquid that gives off red fumes. When a sample of bromine is added to a piece of potassium metal, an explosion occurs and a white solid is found among the debris. From this description, give a property of bromine in each of the following classifications.
    a. Extensive    c. Chemical
    b. Intensive    d. Physical

59. A sample of hydrogen has a volume of 1.2 liters. The sample is a colorless, odorless gas. When a sample of hydrogen is added to a sample of oxygen and a spark is struck, an explosion occurs and (if the container doesn't break) water can be found as a product of the explosion. From this description, give a property of hydrogen in each of the following classifications.
    a. Extensive    c. Chemical
    b. Intensive    d. Physical

60. A sample of sodium hydroxide weighs 85 grams. It is a white solid that dissolves easily in water. When hydrogen chloride is added to sodium hydroxide, water and a different white solid form. From this description, give a property of sodium hydroxide in each of the following classifications.
    a. Extensive    c. Chemical
    b. Intensive    d. Physical

### SECTION 2.5 The Atomic Theory

61. Give examples of four pure substances not mentioned in this chapter that are not elements.

62. Give examples of four mixtures (impure substances) that are not mentioned in this chapter.

63. Is it possible to have a pure compound? Justify your answer.

64. Is it possible to have a pure mixture? Justify your answer.

65. Comment on the often-heard phrase "pure air."

66. Comment on the phrase "pure mountain spring water."

67. List the five statements usually given to summarize Dalton's atomic theory.

68. Which two of the statements in the previous problem are not true?

## SECTION 2.6 Symbols of the Elements

69. List the names and symbols of all the elements you know. Compare this list with Table 2.1. Do you need to study more?

70. Give the symbols and names for all the diatomic elements.

71. Draw a picture similar to Figure 2.27 that shows six oxygen *molecules*.

72. Draw a picture similar to Figure 2.27 that shows fourteen helium *atoms*.

73. Give the symbol and the name of three elements that form molecules containing more than two atoms.

74. Give the symbol and the name of four elements that exist in nature as single atoms.

75. Give the symbol of each of these elements.
   a. Aluminum
   b. Arsenic
   c. Bromine
   d. Calcium
   e. Carbon
   f. Chlorine
   g. Chromium
   h. Copper

76. Give the symbol of each of these elements.
   a. Magnesium
   b. Manganese
   c. Mercury
   d. Lead
   e. Lithium
   f. Neon
   g. Nickel
   h. Nitrogen

77. Give the name of each of these elements.
   a. Ba
   b. B
   c. F
   d. Au
   e. He
   f. H
   g. I
   h. Fe
   i. O
   j. P
   k. Pt

78. Give the name of each of these elements.
   a. Pu
   b. K
   c. Si
   d. Ag
   e. Na
   f. Sr
   g. S
   h. Sn
   i. Ti
   j. U
   k. Zn

## SECTION 2.7 Compounds

79. Write the chemical formula for each compound described.
   a. One iron atom and three chlorine atoms
   b. One copper atom and two OH units
   c. Six carbon atoms, ten hydrogen atoms, and two bromine atoms
   d. Two sodium atoms and one $SO_3$ unit
   e. Two $NH_4$ units and one $CO_3$ unit
   f. Two iron atoms and nine CO units
   g. One copper atom and one sulfur atom

80. Write the chemical formula for each compound described.
   a. Two copper atoms and one sulfur atom
   b. Two hydrogen atoms, one sulfur atom, and four oxygen atoms
   c. One hydrogen atom, one nitrogen atom, and three oxygen atoms
   d. One sodium atom, one nitrogen atom, and three oxygen atoms
   e. Two nitrogen atoms and four hydrogen atoms
   f. Two nitrogen atoms and four oxygen atoms
   g. Two silver atoms and one oxygen atom

81. Write the molecular formula for each of the following substances.
   a. Oxygen
   b. Methane
   c. Phosphorus
   d. Nitrogen
   e. Chlorine

82. Write the molecular formula for each of the following substances.
   a. Sulfur
   b. Bromine
   c. Helium
   d. Arsenic
   e. Ammonia

83. How many atoms of each element are in each of the following formula units?
   a. $Fe(OH)_3$
   b. $CaCl_2$
   c. $K_2SO_4$
   d. $Ag_2Cr_2O_7$
   e. $CH_3Cl$
   f. $CF_2Cl_2$
   g. $Ca_3(PO_4)_2$
   h. $(NH_4)_2SO_4$

84. How many atoms of each element are in each of the following formula units?
   a. $Ca(NO_3)_2$
   b. $(NH_4)_3PO_4$
   c. $NaHSO_4$
   d. $NH_4NO_3$
   e. $Fe(CN)_2$
   f. $KMnO_4$
   g. $Al_2(SO_4)_3$
   h. $CoCO_3$

### SECTION 2.8 Microscopic Views of Matter

85. In what state of matter are the particles closest together?

86. In what state of matter are the particles farthest apart?

87. Draw figures that represent $N_2$ as a solid, a liquid, and a gas. (You might use Figure 2.27 as a model.)

88. Draw figures that represent Al as a solid, a liquid, and a gas.

89. Explain the difference between an amorphous solid and a crystalline solid.

90. What is the primary difference between solids and liquids with respect to the arrangement of molecules?

91. Name a solid not mentioned in this chapter that is flexible.

92. Name a solid not mentioned in this chapter that is rigid.

### SECTION 2.9 Some Typical Substances

93. Give the molecular formula of ammonia. How many hydrogen atoms are present in a molecule of ammonia? How many hydrogen molecules are present in a molecule of ammonia?

94. Give the molecular formula of methane. How many hydrogen atoms are present in a molecule of methane? How many hydrogen molecules are present in a molecule of methane?

95. Which boils at the highest temperature—sodium chloride, methane, or water?

96. Which freezes at the highest temperature—sodium chloride, methane, or water?

### ADDITIONAL PROBLEMS

97. List facts and/or observations you have made that support each of these conclusions.
   a. The properties of solids depend on their chemical composition.
   b. Molecules in liquids are closer together than molecules in gases.
   c. Gases are more easily compressed than liquids.
   d. Ketchup is a heterogeneous mixture.

98. List facts and/or observations you have made that support each of these conclusions.
   a. The properties of solids depend on how their structural units are connected.
   b. Gases are fluids.
   c. Water vapor is easily condensed.
   d. Ammonia gas is very soluble in water.

99. Give an example of each of the following:
   a. A liquid more volatile than water
   b. A chemical property of gasoline
   c. A physical property of methane
   d. An element with an indeterminate molecular formula

100. Give an example of each of the following:
   a. A liquid less volatile than water
   b. A chemical property of methane
   c. A physical property of gasoline
   d. An element that forms homonuclear diatomic molecules

## EXPERIMENTAL PROBLEM 2.1

An ink manufacturer wanted to know how many components there were in an ink sample obtained from a rival company. The company called on A. L. Chemist for an analysis. A. L. took a piece of chromatographic paper and drew a line across the bottom, as shown in Figure 2.39(a), with a pen that used the ink to be analyzed. The paper was then placed in a covered jar with a small amount of water at the bottom and allowed to sit for 20 minutes. The results are shown in Figure 2.39(b). A. L. repeated the experiment using isopropyl alcohol instead of water and got the results shown in Figure 2.39(c). A. L. reported that the ink sample contained five substances.

In answering each of the following questions, explain or justify your answer in 20 words or less.

**101.** At what level (A, B, or C) in Figure 2.39 should the water be at the beginning of the experiment?

**102.** Is water or isopropyl alcohol better for this analysis?

**103.** Assume that the lowest lines in Figure 2.39(b) and (c) are the same substance. In which of these two experiments is this substance adsorbed more readily?

**104.** Assume that the topmost lines in Figure 2.39(b) and (c) are the same substance. In which of these two experiments is this substance adsorbed more readily?

**105.** What differences are there between the results for water [Figure 2.39(b)] and isopropyl alcohol [Figure 2.39(c)]?

**106.** Based on these results, was A. L.'s conclusion that the ink sample contained five substances correct?

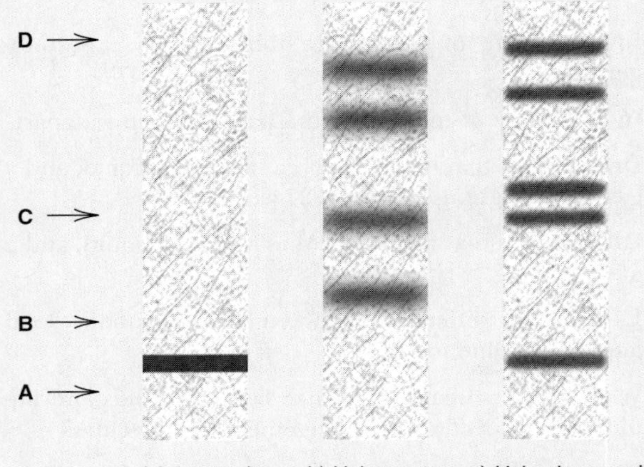


FIGURE 2.39
(A) A line is drawn on a piece of chromatography paper, using the ink to be analyzed. (B) The paper is placed in a covered jar containing some water. After 20 minutes the paper shows four blurred lines. (C) This paper shows what happens when isopropyl alcohol is used instead of water with a sample of the same ink.

## SOLUTIONS TO EXERCISES

### Exercise 2.1

**a.** The bubbles (carbon dioxide gas) in the cola indicate that the mixture is heterogeneous. Cola that has gone flat (without bubbles) or is still in its original bottle (unshaken) is a homogeneous mixture.

**b.** Tears are clear, but they taste salty. They are salt water—a homogeneous mixture.

**c.** You can see pieces of carrots, potatoes, and beef. Stew is a heterogeneous mixture.

**d.** A cross-section of a carrot clearly shows two distinct rings, making it heterogeneous.

### TOOL BOX Exercise 1

**a.** Sum of parts = 100% (Whole)

$2\%\ (\text{carbon}) + X\ (\text{iron}) = 100\%$ (all of the steel: the whole)

$2\% + X = 100\%$

$X = 100\% - 2\% = 98\%$ iron in this sample of steel

**b.** This mixture contains three components, but the basic technique for obtaining the answer is the same.

Sum of parts = 100% (Whole)

$5\%\ (\text{zinc}) + 72\%\ (\text{copper}) + X\ (\text{tin}) = 100\%$ (all of the bronze: the whole)

$5\% + 72\% + X = 77\% + X = 100\%$

$X = 100\% - 77\% = 23\%$ tin in the bronze

**c.** Add up the vote totals:
20% (Smith) + 25% (Jones) + 65% (Crook) = 110%
This is impossible because by definition, the whole (the total percentage of votes) is 100%. Therefore at least one of the three parts (20, 25, or 65) had to be too large. (Never swallow statistics whole, no matter what their source.)

**TOOL BOX Exercise 2**

**a.** The whole is the sum of the parts: 100 lb titanium + 300 lb iron = 400 lb of alloy.

$$\frac{\text{Part}}{\text{Whole}} \times 100\% = \frac{100 \text{ lb}}{400 \text{ lb}} \times 100 = 25\% \text{ titanium in the alloy}$$

**b.** $$\frac{\text{Part}}{\text{Whole}} \times 100\% = \text{Percentage}$$

$$\frac{14}{24} \times 100\% = 58\% \text{ gold}$$

**TOOL BOX Exercise 3**

**a.** $$\frac{\text{Part}}{\text{Whole}} \times 100\% = 70\%$$ The part is unknown; call it *X*.

$$\frac{X}{50 \text{ kg}} \times 100\% = 70\%$$ Substitute 50 kg for the whole.

$$X = \frac{50 \text{ kg} \times 70\%}{100\%} = 35 \text{ kg} \quad \text{copper}$$

**b.** $$\frac{\text{Part}}{\text{Whole}} \times 100\% = 70\%$$ The whole is the unknown, *X*.

$$\frac{21 \text{ kg}}{X} \times 100\% = 70\%$$ Substitute 21 kg for the part. Multiply both sides by *X*, then divide both sides by 70%.

$$X = \frac{21 \text{ kg} \times 100\%}{70\%} = 30 \text{ kg of bronze}$$

**Exercise 2.2**

The answer is identical to that for Example 2.2. The air inside an aluminum can is forced out as the can collapses. The aluminum itself is not compressible.

**Exercise 2.3**

**a.** Using an egg separator is a type of filtration (2). You can use one to isolate both the yolk and the egg white.
**b.** Evaporation (3) is the method to use in this case.
**c.** If you wash the mixture with water, you can recover both the salt (which dissolves in water) and the sand (which does not). This is an extraction procedure (1).

**Exercise 2.4**

**a.** Effervescence (bubbliness) is a physical property. The cola is still cola even after the carbon dioxide gas is removed.
**b.** Explosiveness is a chemical property. Once the dynamite reacts (as a chemist would say), it no longer exists.
**c.** The gasoline is still gasoline after it is purified by distillation; its volatility allows it to be separated by this method and is a physical property.

**Exercise 2.5**

**a.** The answers are in Table 2.1.
**b.** An oxygen molecule is $O_2$ (O is one of the diatomic elements). Phosphorus is $P_4$, and fluorine is $F_2$ (F is another of the diatomic elements).

**Exercise 2.6**

**a.** One atom of calcium, two atoms of oxygen, and two atoms of hydrogen
**b.** Six atoms of carbon, twelve atoms of hydrogen
**c.** Two atoms of aluminum, three atoms of sulfur, twelve atoms of oxygen
**d.** One atom of carbon, two atoms of oxygen

**EXPERIMENTAL EXERCISE 2.1**

**1.** Potassium (K), nitrogen (N), and oxygen (O)
**2.** Extraction followed by filtration
**3.** Carbon and sulfur
**4.** Carbon only
**5.** $KNO_3$
**6.** Sulfur
**7.** No, A. L. Chemist and his team messed up. If the $KNO_3$ dissolved in water and the other two substances were left, the percentage of $KNO_3$ was

$$\frac{40 - 8.8}{40} \times 100\% = 78\%$$

The sulfur dissolved in the carbon disulfide, so the amount of sulfur present was 8.8 g (carbon plus sulfur) minus 4.0 g (carbon only) = 4.8 g, and the percentage of sulfur was

$$\frac{8.8 - 4.0}{40} \times 100\% = 12\%$$

Solid 2, which remained after the $KNO_3$ and the sulfur were removed, was carbon. Its percentage was

$$\frac{4.0}{40} \times 100\% = 10\%$$

**Exercise 2.7**

The answers are (a) solid gold metal, (b) liquid water, and (c) ice. In each instance the more condensed phase contains the larger number of molecules. The only exception to this guideline is ice. Ice floats on water, and therefore is less dense than water (see Section 2.9).

5
2

3

# MEASUREMENTS AND UNITS



KEY TERMS

Many important discoveries in the history of chemistry were made by chemists who performed their experiments more carefully and precisely than those who went before. Some scientists say that modern chemistry originated with the careful experiments and proposals of Lavoisier, although (as you saw in Chapter 1) the evolution of the scientific method as the approach to answering questions about the natural world was a long process.

Simple descriptions are important, and the great ideas in chemistry can almost always be expressed in simple sentences. Scientists use careful numerical measurements as their main tool for acquiring the insights that lead to these simple laws and for getting closer to the truth.

## MEASUREMENTS AND CALCULATIONS

Most people in modern society use numbers and calculations in their everyday lives. A great many people are accustomed to making calculations when balancing a checkbook or figuring out how much change they should get. Sports is full of numbers, of course. This fullback weighs 248 (big), and that basketball player is five-ten (short for a professional basketball player). The usual way of using numbers—omitting the units and assuming that the listener will understand—can cause difficulties in a chemistry course. It sounds stuffy to say that the fullback weighs 248 *pounds* and that the basketball player is five *feet,* ten *inches* tall, but in science this sort of exactness is necessary for effective communication. The way scientists view numbers and work with them is the primary theme of the next four sections.

### 3.1 An Introduction to Measurement

GOAL: To define and illustrate terms involved in measurement.

**STUDY ALERT**

**With few exceptions, numbers (magnitudes) used in this course have units. You must include the unit every time you write a number that has one. You have already met one kind of number without units: percentages have no units (because they are ratios of quantities that have the same units). Later in this chapter you will encounter logarithms, and in Chapter 6 you will use quantum numbers; neither of these has units because of the way they are defined.**

A large majority of the numbers that scientists use come from measurements. **Measurement** is defined as the act of *determining the size or amount* of something. Measurements always have two parts. One part is the **magnitude,** or size, of a measurement; this is the number. The other part is the **unit,** which compares a measurement to a standard. Suppose that a person weighs 95 kilograms; the 95 is the magnitude and the kilogram is the unit. The same person weighs 209 pounds; here, 209 is the magnitude and pounds is the unit. Both measurements describe the same person.

Note that the magnitude (such as 95 or 209) depends on the unit that's used in the measurement. To a chemist, the number (the magnitude) is meaningless without the unit. This is the key point here. Scientists don't think in terms of numbers by themselves, they think in terms of numbers *with units.* Very few quantities consist of numbers without units.

When scientists make measurements, they are seeking the true value of the quantity they are measuring. A typical procedure for doing this is to repeat the measurement several times and then average the results. This process tends to even out random errors and fluctuations in the measurements and can also point out where a mistake has been made. As you may already know, the **average** of a series of measurements is their sum divided by the number of measurements that were made.

$$\text{Average} = \frac{\text{Measurement 1} + \text{Measurement 2} + \cdots}{\text{Number of measurements}}$$

If you know the true value of a measured quantity, you can compare the average to that value. If there is reasonable agreement between the average and the true value, then the measurements have **accuracy.** If the average and the true value don't agree well, then an error (or possibly more than one) was made in the technique or methods used in making the measurements. (Often, of course, scientists have no idea what the true value is other than what their measurements tell them. This is the main reason why they repeat their own experiments and those of their colleagues.)

Similarly, you can compare the individually measured values to the average result. If the differences are relatively small (that is, if the values cluster tightly around the average), then measurements have **precision.** If the measurements are scattered all over the map, they are not precise, and there may be something wrong with the procedures, measuring instruments, or calculations you used.

As an example, suppose that the 95-kilogram person mentioned earlier stepped on three different bathroom scales and got measurements of 91, 94, and 97 kilograms. The average of these numbers is

$$\frac{91 \text{ kg} + 94 \text{ kg} + 97 \text{ kg}}{3} = 94 \text{ kg}$$

The person's true mass is 95 kilograms, and the average is only 1 kilogram away from that, so the measurements are reasonably accurate. However, the individual values are scattered over a range of 6 kilograms from lowest to highest. This means that the precision of these measurements is poor.

Now suppose the same person was weighed three times on a balance in a doctor's office, and the measurements were 96.8, 96.9, and 96.9 kilograms. The average mass is 96.9 kilograms, which—if the true value is 95—is further away (less accurate) than the measurements made on the bathroom scale. However, the measurements are more tightly clustered. All of them are within 0.1 kilogram of the average, so they are more precise than the bathroom-scale measurements. It is likely that some constant, repeatable error is making these measurements less accurate than the ones made on the bathroom scale. Perhaps the person being weighed forgot to remove an item of clothing, or perhaps the manufacturer of the balance simply put the numbers in the wrong place. This kind of mistake is called a *systematic error.*

A set of measurements can have one of four possible combinations of accuracy and precision:

- A set of measurements can be both accurate and precise (the most desirable combination).

**There is another difference between 95 kilograms and 209 pounds. Kilogram is a unit of *mass,* and a pound is a unit of *weight.* An object's mass is independent of the gravitational field it experiences, but its weight is not. This means that weight changes when the gravitational field changes, but mass remains the same. A 95-kilogram person still has a mass of 95 kilograms on the moon but has a weight of only about 35 pounds (rather than 209 pounds) because the force of gravity on the moon is about $\frac{1}{6}$ of that on earth. The difference between mass and weight can be very important in physics and astronomy but is somewhat less important in chemistry.**

FIGURE 3.1
This astronaut has the same mass on the earth's surface as in orbit. However, the astronaut's weight in orbit is 0 pounds, while the astronaut's weight on the earth's surface (including equipment) is more than 200 pounds.

- A set of measurements can be accurate but not precise (sometimes acceptable).
- A set of measurements can be precise but not accurate (usually the result of a consistent mistake).
- A set of measurements can be neither accurate nor precise (not useful for much of anything).

Example 3.1 provides a visual representation of these possibilities.

### EXAMPLE 3.1 Accuracy and Precision

In a horseshoe pitching contest four people tossed three horseshoes each. Their results are shown in Figure 3.2. (a) Which person's tosses illustrate most clearly neither precision nor accuracy? (b) Which person's tosses illustrate most clearly accuracy but not precision?

**SOLUTIONS**

**a.** Measurements that are not precise are far apart; measurements that are not accurate are far from the correct value. Person 4's tosses best fit this description.

**b.** When measurements are accurate, their average is close to the correct value. They can be accurate without being close together (that is, they do not have to be precise in order to be accurate). Person 1's results best fit this description.

**EXERCISE 3.1**

Which of the results in Example 3.1 best illustrates (a) precision but not accuracy and (b) both accuracy and precision?

FIGURE 3.2

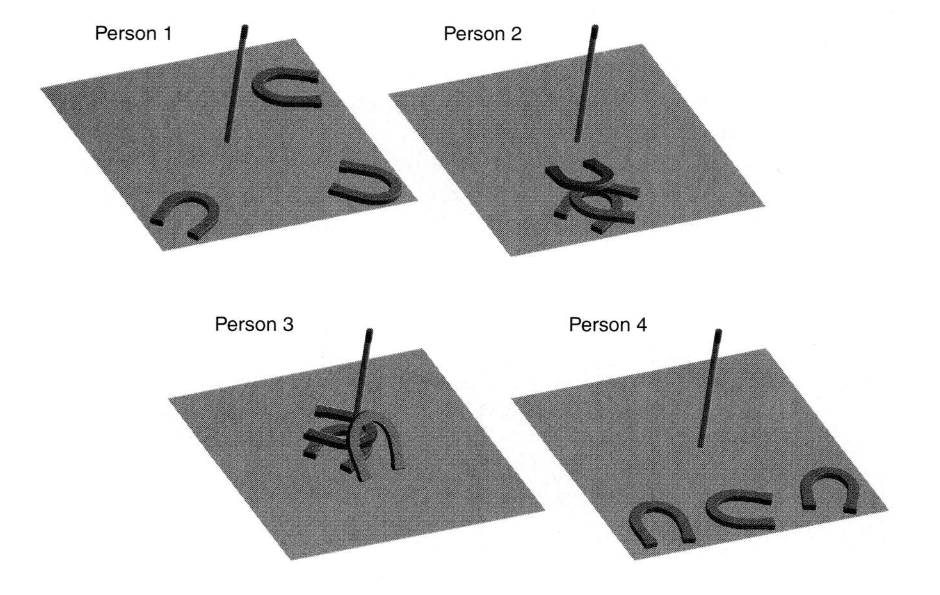

## 3.2 Uncertainty in Measurement: Significant Figures

**GOAL:** To understand what a significant figure is and to be able to state the number of significant figures in a measurement or other number.

It is impossible to make perfectly exact measurements. The use of any measuring device (for example, a thermometer or balance) is subject to some error, and thus all measurements have some uncertainty associated with them. This is illustrated by the bathroom scale and the doctor's office balance as shown in Figure 3.3.

Roughly speaking, scientists record all the digits of a measurement that they know exactly, plus the first one that is at all uncertain—that is, the first one they have to guess about. A scientist might record the mass shown on the bathroom scale in Figure 3.3 as 95.1 kilograms. The 95 is known exactly, and the next digit (the 1) is estimated. For the doctor's balance in Figure 3.3, a scientist might record the result as 95.16 kilograms. The 95.1 is known exactly. (Note that the 1 is known even though you can't see it on the measuring device. The pointer is clearly more than halfway between 95.0 and 95.2 kg.) The scientist who records the value of 95.16 kilograms has estimated the last digit in the sequence.

The digits that you can read exactly, plus the digit that you must guess at, are called **significant figures.** The number of significant figures in 95.1 is three, and 95.16 has four significant figures. Some additional rules apply when counting the number of significant figures in a number that contains zeros. The box on the following page lists all of the rules for counting significant figures.

Bathroom scale:
Note that the measured mass is slightly greater than 95 kg. Exactly how much greater is impossible to determine from this. Is it 95.1, 95.05, or 95.02 kg? Since you cannot tell, the measurement is inexact; that is, it has an error associated with it.

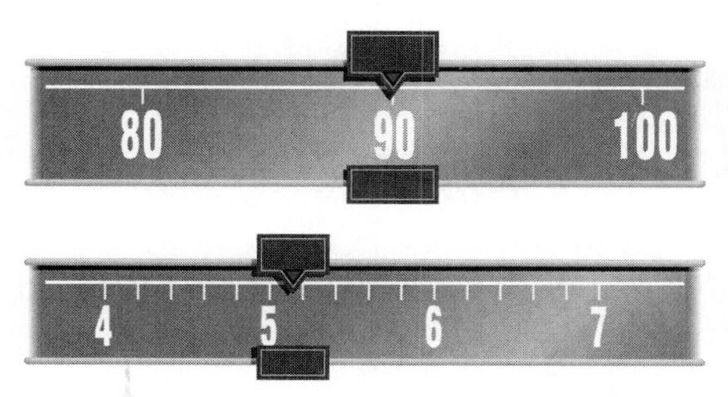


Doctor's balance:
This suggests that the mass is greater than 95.1 kg and less than 95.2 kg. You must still guess about the last digit. This measurement, too, has an error associated with it.

**FIGURE 3.3**
Every measurement in science is subject to uncertainty. There is always a limit to the number of significant figures that can be reported for a measurement.

**Rules for Counting Significant Figures**

- Nonzero digits are *always* significant.
- Zeros between nonzero digits are *always* significant. For example, the number 2.002 has four significant figures.
- Zeros to the left of nonzero digits (called *leading zeros*) are *never* significant. The number 0.00065 has only two significant figures.
- Zeros to the right of nonzero digits (called *trailing zeros*) are *sometimes* significant. If a number contains a decimal point, the zeros to the right of it are always significant. The number 21.00 has four significant figures. In a number without a decimal point, you can't tell for sure whether any of the trailing zeros are significant. For example, the number 265,000 has at least three and at most six significant figures.

**Most calculators don't display trailing zeros at all. You must remember the rules for significant figures and add trailing zeros to the calculator results when they are needed. On the other hand, calculators can also give you too many nonzero digits. Section 3.3 discusses what to do in this situation.**

Is a measured value of 14.0 different from a measured value of 14.00? Yes, it is. The number 14.00 is known more precisely, to a greater number of significant figures. (This does not necessarily mean that it is known more accurately.) Do not neglect significant figures that are zeros.

The number of significant figures in a measured number depends on the precision of the measuring instrument. If you measure the same sample or quantity repeatedly and the results fall very close together, then you are justified in using the largest number of significant figures you can read from the instrument. To determine if these measurements are also accurate, you have to compare them to some "true" value, or **standard.** This comparison procedure is called *calibration.* Sometimes the "true" value is unknown—after all, the purpose of measurement is usually to find an unknown value. All you can do in that case is assume that the measurements are accurate until you can find other ways of confirming them.

**This assumes that the number is small enough to count without human error. If you asked 10 different people to count the number of words in this chapter, you would probably get 10 different answers.**

There is one more point to be made here. Consider the following question. How many words are in this sentence? (The answer, of course, is seven words. You didn't just say "seven," did you? *All* measured quantities have units.) You made a measurement by counting the words in the sentence. The answer is an example of an *exact number.* Exact numbers typically are determined by counting (or represent a defined quantity such as one dozen, which is 12 things) and are always known to an infinite number of significant figures. The formal answer to the question about the sentence is 7.000 . . . words, but the magnitude is always written as 7.

### EXAMPLE 3.2 The Number of Significant Figures

How many significant figures are there in each of these values?

**a.** 21.05 centimeters
**b.** 0.0120 liter
**c.** 1.00115965221
**d.** 18 eggs

**SOLUTIONS**

**a.** The zero between the nonzero digits is significant. This measurement has four significant figures.

b. The zeros to the left are leading zeros and are not significant. The trailing zero on the extreme right, following the decimal point, is significant. This number has three significant figures.
c. This number has twelve significant figures. It is the value of a unitless quantity called *Dirac's number,* determined after very careful measurements made by physicists in the early 1980s.
d. 18.00 . . . eggs is an exact number and is known to an infinite number of significant figures.

**EXERCISE 3.2**

How many significant figures are there in each of these values?

a. 0.00006 gram
b. 10.005 yards
c. 789.030 kilometers
d. 0.001050 kg

## 3.3 Significant Figures in Calculations

**GOAL:** To perform simple calculations and give answers with the correct number of significant figures.

Many quantities are not measured directly but are calculated by combining two or more measured values. For example, in Chapter 2 you learned to calculate percentages using this equation:

$$\frac{\text{Part}}{\text{Whole}} \times 100\% = \text{Percentage}$$

How do you decide how many significant figures there are in a calculated percentage when you use this equation? The rules presented in this section show you how to make this decision. They give you a means to establish an uncertainty for calculated answers that is consistent with the uncertainty in the starting numbers.

When you multiply or divide two numbers, the decision is relatively simple. You count the number of significant figures in each number used in the calculation. *The number of significant figures in the answer is the same as the number of significant figures in the least precise number used in the calculation* (that is, the one that has the fewest number of significant figures). Remember that exact numbers have an infinite number of significant figures, so you can ignore them in this process.

Here is an example. A 50.20-gram sample of an alloy contains 16 grams of copper. What is the percentage of copper in the alloy? Plugging the numbers into a hand calculator gives you a result with plenty of figures, something like this:

$$\frac{\text{Part}}{\text{Whole}} \times 100\% = \frac{16 \text{ g}}{50.20 \text{ g}} \times 100\% = 31.87251\%$$

To decide how many of the digits in 31.87251% are significant, look at the numbers involved in the calculation: 16 has two significant figures, 50.20 has four significant figures, and 100 is an exact number that has an infinite number of significant figures. The answer, then, should contain two significant figures. So how do you reduce the seven digits obtained from the calculator to the proper two-digit answer? You *round off* to get the final answer. The rules for rounding off, shown in the box, are straightforward.

**Rules for Rounding Off**

- If the digit you remove from the right-hand end of the number is 0, 1, 2, 3, or 4, then the last digit you keep is unchanged. For example, 1.24 rounds off to 1.2.
- If the digit you remove from the right side of the number is 5, 6, 7, 8, or 9, then the last number kept is increased by one unit. For example, 1.28 rounds off to 1.3, and 1.25 also rounds off to 1.3.

For the alloy example above, the calculated number was 31.87251%. To round this number off to *two* significant figures, you must consider the first *three* digits and you can ignore the rest (7251). Following the rounding rules, 31.8% rounds off to 32%, which has the correct number of significant figures.

The rules for deciding on significant figures in addition and subtraction are a little trickier to use. Only digits to the right of the decimal point are important in this rule. *The answer has the same number of digits to the right of the decimal point as the number in the calculation with the fewest digits to the right of the decimal point.*

Here's an example. Suppose you want to add some masses (given in grams) on your calculator; you enter these numbers:

| | |
|---|---|
| 100.1 | (one digit to the right of the decimal point) |
| 25.656 | (three digits to the right of the decimal point) |
| 0.0006 | (four digits to the right of the decimal point, three zeros and one digit) |
| 125.7566 | (calculator result) |

What is the correct answer? It is 125.8 grams. (Note that the answer has units of grams—though a calculator doesn't keep track of units, they are part of the answer.) The answer has one digit to the right of the decimal point, just as 100.1 does.

**Significant Figure Calculations: A Summary**

| | |
|---|---|
| Multiplication and Division | The number of significant figures in the calculated answer is identical to the number of significant figures in the *least precise* number used in the calculation. |
| Addition and Subtraction | The calculated answer has the same number of digits to the right of the decimal point as the number with the *fewest digits to the right of the decimal point* used in the calculation. |

| Rounding Off | When using a calculator to do a series of calculations (sometimes called a chain operation), *round off only at the end* of the series of calculations. |
|---|---|

**EXAMPLE 3.3** **Using Significant Figures in Calculations**

Perform each of these calculations and give the answer with the correct number of significant figures.

**a.** 5.4771 lb × 3.1 lb
**b.** 45.95 ton ÷ 43.2 ton
**c.** 3.8 lb + 32 lb
**d.** 153.1 lb − 0.002 lb
**e.** An alloy contains 3.2 grams of copper and 1.27 grams of tin. What is the percentage of copper in this alloy?

SOLUTIONS

**a.** You round off the calculator answer of 16.97901 $lb^2$ to 17 $lb^2$, which has two significant figures, as 3.1 lb has. Remember that your calculator doesn't keep track of the units, so you have to. Here lb × lb = $lb^2$, the proper units for the answer.
**b.** 1.0636574 rounds off to 1.06, with three significant figures (as 43.2 ton has). Here the units cancel (ton/ton = 1).
**c.** 35.8 lb rounds up to 36 lb. The number 32 lb has no numbers to the right of the decimal point, so the answer can't have any either.
**d.** The calculator answer is 153.098 which rounds off to 153.1 lb. Suppose you weigh yourself on a bathroom scale, cut off a few hairs, and weigh yourself again. Your scale won't be able to tell the difference in your weight, will it?
**e.** Using the equation for percentage, you get

$$\frac{\text{Part}}{\text{Whole}} \times 100\% = \text{Percentage}$$

$$\frac{3.2\ \text{g}}{3.2\ \text{g} + 1.27\ \text{g}} \times 100\% = \text{Percentage of copper}$$

Some calculators can perform such a calculation all at once, giving you a displayed result something like 71.58836689. Rounding off this number to two significant figures gives 72% as the answer. However, some calculators work with only two numbers at a time. If you are using one of these calculators, you first have to compute the denominator of the fraction and then do the division. Computing the denominator gives (3.2 g + 1.27 g) = 4.47 g. Following the rule about rounding off a chain operation only at the end, you should use 4.47 g as the value of the denominator; this will give you the same answer as above. If you round off the denominator to 4.5 g and then finish the calculation, the final display will be 71.11111111, which rounds off to 71%.

**STUDY ALERT**
**Remember that numbers with positive exponents are larger than 1 and those with negative exponents are smaller than 1.**

To convert from scientific notation back to conventional, nonexponential notation, you simply reverse the process. For $3.5 \times 10^5$, you move the decimal point five spaces to the right, which gives you 350,000. For $4.2 \times 10^{-3}$, you move the decimal point three spaces to the left and get 0.0042. Note that for numbers less than 1 it is conventional to precede the decimal point with a zero, as in 0.0042. The zero is added to prevent a reader from overlooking the decimal point when reading the number. You will find your problem-solving life easier if you develop this habit.

### EXAMPLE 3.5 Converting from Scientific to Nonexponential Notation

Convert each of the following numbers to nonexponential notation.

**a.** $2.7 \times 10^5$ grams
**b.** $2.3 \times 10^{-2}$ centimeter

**SOLUTIONS**

**a.** Move the decimal point five places to the right and add four zeros. This gives 270,000 grams.
**b.** Move the decimal point two places to the left. This yields 0.023 centimeter.

**EXERCISE 3.5**

Convert each of the following numbers to nonexponential notation.

**a.** $4 \times 10^7$ meters
**b.** $4 \times 10^{-7}$ gram

**The × button (the "times" button) is *never* used to enter numbers into a calculator in scientific notation. Also, don't confuse the exponent in scientific notation with the exponential function keys (these keys may be labeled $10^x$ or $e^x$).**

A scientific calculator—one that can accept and display numbers in scientific notation—can make your work in this course much easier. Such calculators have several advantages. They can do chain calculations rather easily, and they contain functions such as logarithms that will be useful later in this course. Each brand of calculator has its own means for entering exponents, often a button labeled Exp or EE. Check the manual that comes with your calculator to be certain you know how to enter and read numbers displayed in scientific notation.

TOOLBOX

## TOOL BOX: WORKING WITH EXPONENTS

In the late sixteenth century, a Scottish mathematician named John Napier (1550–1617) came to the conclusion that *all* numbers can be written in exponential form. You already know that $1 = 3^0$, $3 = 3^1$, $9 = 3^2$, and $27 = 3^3$. Napier went further. Since the number 15 lies between $3^2$ and $3^3$, the number 15 can be written exponentially as 3 to some fractional power between 2 and 3. (To be more exact, $15 = 3^{2.465}$.)

Actually, 3 is not a particularly useful base. Napier wrote his exponential numbers using the base *e*, the so-called natural logarithm, or exponential function, which appears quite frequently in mathematics. The value of *e* is approximately 2.718281828. Using $e$ as a base, Napier calculated such numbers as $1 = e^0$, $1.5 = e^{0.406}$, $2 = e^{0.693}$, $3 = e^{1.099}$, $6 = e^{1.792}$, $14{,}000 = e^{9.547}$, and $0.006 = e^{-5.116}$. Napier spent about 20 years working out formulas and obtaining exponential expressions for various numbers. His work was incredibly useful because it allowed people to multiply and divide more easily. Napier published his work a long time before the long division and multiplication methods you learned in school were developed, and hundreds of years before the invention of the electronic calculator.

To multiply numbers using exponential notation, you look up the exponents in a standard table, add them together, and look up the number whose exponent you now have. To divide, you subtract the exponents. You probably know what $3 \times 2$ and $3 \div 2$ are, but here's how you calculate these values with exponents:

$$3 = e^{1.099} \text{ and } 2 = e^{0.693}$$

$$3 \times 2 = (e^{1.099})(e^{0.693}) = e^{(1.099\,+\,0.693)} = e^{1.792} = 6$$

$$3 \div 2 = \frac{3}{2} = \frac{e^{1.099}}{e^{0.693}} = e^{(1.099\,-\,0.693)} = e^{0.406} = 1.5$$

As soon as Napier published the results of his calculations, the English mathematician Henry Briggs adopted the system. Briggs suggested that the more familiar 10 be used as the base instead of $e$. This approach gives such results as $1 = 10^0$, $2 = 10^{0.301}$, $3 = 10^{0.477}$, and $1000 = 10^3$. Both Napier's and Briggs's exponents are used today; you know them as natural logarithms (ln x on your calculator) and common logarithms (log x on your calculator). The rest of this story is given in the Chapter 15 Tool Box on logarithms.

The rules for multiplying and dividing numbers written in scientific notation follow the pattern established here. A number written in scientific notation contains two parts, the number between 1 and 10 and the exponential part. When you multiply numbers written in scientific notation, you multiply the first parts and add the exponents in the second parts. After you have done this, you rewrite the result in scientific notation if necessary.

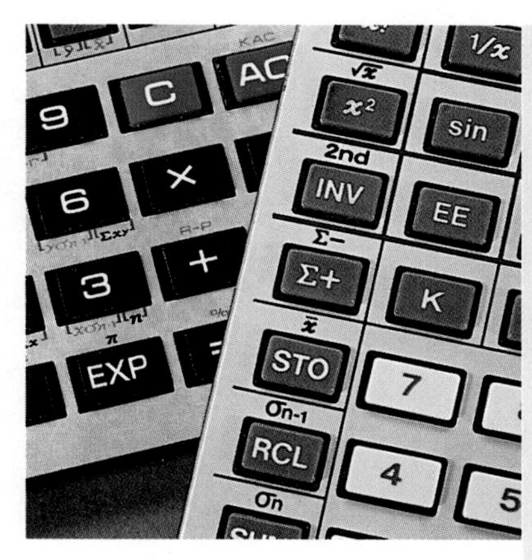


FIGURE 3.7
Scientific calculators allow you to enter numbers in scientific notation, in most cases using a key labeled Exp or EE. If you want to enter the number $3 \times 10^6$ into such a calculator, the typical method is to punch in 3, then hit the Exp or EE button, then punch in 6. Be sure to read the manual for your calculator and practice.

TOOL BOX EXAMPLE 1 **Multiplying Numbers in Scientific Notation**

Write the product of each of the following pairs in scientific notation with the correct number of significant figures.

**a.** $(3 \times 10^2\text{ ft})(3 \times 10^5\text{ ft})$
**b.** $(1.4 \times 10^{-8}\text{ in})(2.55 \times 10^2\text{ in})$
**c.** $(6.00 \times 10^{-5}\text{ ft})(5.443 \times 10^{-2}\text{ ft})$

SOLUTIONS

**a.** First, multiply the first parts:

$$3 \times 3 = 9$$

Second, add the exponents (which multiplies the exponential parts):

$$(10^2)(10^5) = 10^{(2\,+\,5)} = 10^7$$

This gives $9 \times 10^7$ (in scientific notation with one significant figure)
This gives the number; next we take care of the units.

$$\text{ft} \times \text{ft} = \text{ft}^2$$

The final answer is $9 \times 10^7\text{ ft}^2$.

**b.** First, multiply the first parts:

$$1.4 \times 2.55 = 3.57$$

Add the exponents:

$$(10^{-8})(10^{2}) = 10^{(-8+2)} = 10^{-6}$$

Finally work out the units:

$$\text{in} \times \text{in} = \text{in}^2$$

The answer is $3.6 \times 10^{-6}$ in$^2$ (two significant figures and correct scientific notation).

**c.** Multiply the first parts:

$$(6.00)(5.443) = 32.658$$

Add the exponents:

$$(10^{-5})(10^{-2}) = 10^{[-5+(-2)]} = 10^{-7}$$

This gives $32.7 \times 10^{-7}$, which is the correct number, but it is not written in proper scientific notation. The solution is to note that $32.7 = 32.7 \times 10^{1}$. Therefore, $32.7 \times 10^{-7} = 3.27 \times 10^{1} \times 10^{-7} = 3.27 \times 10^{[1+(-7)]} = 3.27 \times 10^{-6}$, which is the number in scientific notation. Finally, the units are:

$$\text{ft} \times \text{ft} = \text{ft}^2$$

This gives $3.27 \times 10^{-6}$ ft$^2$, which is the correct answer.

**STUDY ALERT**

**Decide the number of significant figures based on only the first parts of the original number. The exponential parts represent trailing zeros or leading zeros and are not significant.**

**TOOL BOX EXERCISE 1**

Write the product of each of these in scientific notation with the correct number of significant figures.

**a.** $(3.3 \times 10^{2}$ ft$)(3 \times 10^{2}$ ft$)$
**b.** $(2.6 \times 10^{-2}$ lb$)(9.111 \times 10^{5}$ lb$)$
**c.** $(3.99 \times 10^{5}$ lb$)(8 \times 10^{-5}$ lb$)$

When dividing numbers written in scientific notation, you divide the two first parts as usual. You then subtract the exponent of the denominator (the number below the fraction bar) from the exponent of the numerator (the number above the fraction bar). Finally, you change the result to correct scientific notation if necessary. This process is illustrated in Tool Box Example 2.

**TOOL BOX EXAMPLE 2** **Dividing Numbers in Scientific Notation**

Write the result of each of the following operations in scientific notation with the correct number of significant figures.

**a.** $\dfrac{5.1 \times 10^{2}\ \text{grams}}{4.6 \times 10^{3}\ \text{grams}}$

**b.** $\dfrac{3.6 \times 10^{2}\ \text{lb}}{8.33 \times 10^{-2}\ \text{in}^2}$

**c.** $\dfrac{1.00 \times 10^{-6}\ \text{kilograms}}{5.00 \times 10^{3}\ \text{kilograms}}$

SOLUTIONS

**a.** Divide the first parts of the numbers:

$$\frac{5.1}{4.6} = 1.1086957$$

Determine the exponent:

$$\frac{(10^2)}{(10^3)} = 10^{(2-3)} = 10^{-1}$$

Round off and put the numbers together:

$1.1 \times 10^{-1}$ (to two significant figures: the units cancel)

**b.** Divide the first parts of the numbers:

$$\frac{3.6}{8.33} = 0.4321729$$

Determine the exponent:

$$\frac{(10^2)}{(10^{-2})} = 10^{[2-(-2)]} = 10^{(2+2)} = 10^4$$

Put these together and get $0.43 \times 10^4$. This is the correct number, but it is not written in correct scientific notation. Again, you apply the rules you learned earlier: $0.43 = 4.3 \times 10^{-1}$, so $0.43 \times 10^4 = 4.3 \times 10^{-1} \times 10^4 = 4.3 \times 10^3$.

Work out the units:

$$\frac{\text{lb}}{\text{in}^2}$$

And the final answer is $4.3 \times 10^3$ lb/in$^2$

**c.** Divide the first parts of the numbers:

$$\frac{1.00}{5.00} = 0.200 \text{ (Both of the zeros on the end are significant!)}$$

Determine the exponent:

$$\frac{(10^{-6})}{(10^3)} = 10^{(-6-3)} = 10^{-9}$$

This gives $0.200 \times 10^{-9} = 2.00 \times 10^{-1} \times 10^{-9} = 2.00 \times 10^{-10}$ as the final answer because the units cancel.

TOOL BOX EXERCISE 2

Perform each of the following calculations, giving the answer with the correct number of significant figures.

**a.** $\dfrac{2.3 \text{ lb}}{4.1 \times 10^6 \text{ in}^2}$ (Note: In scientific notation 2.3 is $2.3 \times 10^0$, although it is rarely written that way.)

**b.** $\dfrac{1.5 \times 10^3 \text{grams}}{3.0 \times 10^{-6} \text{grams}}$

**c.** $\dfrac{2.00 \times 10^{-2} \text{ lb}}{1.00 \times 10^{10} \text{ gallon}}$

## THE METRIC SYSTEM

Jefferson had one success in his attempts to persuade Congress to use the metric system. The monetary system in the United States—with 10 cents to the dime and 10 dimes to the dollar and so forth—is directly attributable to Jefferson. It was relatively easy to persuade Congress on this one point because the United States was still using European currency and was not yet printing its own money.

There is one major system of measurement in the world today, along with one minor system and some historical curiosities. The major one is the **metric system,** which was created in the wake of the French Revolution. In 1799 Thomas Jefferson tried to persuade the U.S. Congress to adopt the metric system. He said that it would eventually become the standard of the world. He was right, but Congress rejected his proposal, and the metric system has still not been adopted for general use in the United States. Scientists use it universally, primarily because the relationships among the units are convenient. Most of the world's people use the metric system in their everyday lives.

You must become familiar with the metric system of measurement to succeed in science. This means you must be able to use the base units as well as some of the prefixes that are associated with these units. Metric measurements are used almost exclusively in the rest of this book and in all other science books.

### 3.5 Metric Prefixes and Base Units

GOAL: To learn the base units of the metric system and the prefixes used with them.

FIGURE 3.8
Thomas Jefferson was a surveyor, an architect, a scientist, and the major author of the Declaration of Independence, in addition to being President.

Before delving into the fundamental units of the metric system, let us first see why it is superior to the English system that you are probably more familiar with. The key feature of the metric system is that it uses factors of 10 to link units of the same kind together and uses the same system of prefixes with all units. If you know that 1 milliliter equals 0.001 liter, you also know that 1 millimeter is 0.001 meter and 1 milligram is 0.001 gram. Clearly, the prefix *milli* must be related to the base unit by a factor of 0.001 (three factors of 10: $\frac{1}{10} \times \frac{1}{10} \times \frac{1}{10}$, or $10^{-3}$).

In the English system the relationships between units are irregular. You know that 12 inches is a foot, 3 feet is a yard, 5280 feet is a mile, and so on. These measurements have nothing to do with the fact that there are 2 pints in a quart and 4 quarts in a gallon. The merits of the metric system over this jumble of relationships are clear. The United States will continue to have problems competing in (or even communicating with) global markets as long as it sticks with the cumbersome English system of measurement.

The key to using metric system prefixes is that **each prefix indicates the factor of 10 by which to multiply the base unit.** You have to know what the prefixes mean. The most common and useful of the metric prefixes are listed in Table 3.1. A complete list is given inside the back cover of this book.

All of the units in the metric system are constructed from the prefixes listed in Table 3.1 (and some less common ones not listed in the table) and the base units. The next example illustrates how the prefixes are used.

**TABLE 3.1 Common Prefixes**

| Base Unit (meter, gram, second, kelvin, etc.) | Prefix (Symbol) | Factor by Which to Multiply Base Unit |
|---|---|---|
| Quantities Smaller than Base | **pico-** (p) | $10^{-12}$ (one trillionth: 0.000 000 000 001) |
| | **micro-** (μ) | $10^{-6}$ (one millionth: 0.000 001) |
| | **milli-** (m) | $10^{-3}$ (one thousandth: 1/1000 or .001) |
| | **centi-** (c) | $10^{-2}$ (one hundredth: 1/100 or .01) |
| Quantities Larger than Base | **kilo-** (k) | $10^{3}$ (one thousand: 1000) |
| | **mega-** (M) | $10^{6}$ (one million: 1,000,000) |

FIGURE 3.9
The labels of many products in the United States use both metric and English system units.

### EXAMPLE 3.6 Using Metric Prefixes

Fill in the blanks without looking at Table 3.1.
**a.** 1 centimeter = _____ meter
**b.** 1 megameter = _____ meters
**c.** 1 millimeter = _____ meter

**SOLUTIONS**

**a.** 1 centimeter = $10^{-2}$ meter = 0.01 meter = 1/100 meter
**b.** 1 megameter = $10^{6}$ meters = 1,000,000 meters
**c.** 1 millimeter = $10^{-3}$ meter = 0.001 meter = 1/1000 meter

**EXERCISE 3.6**

Fill in the blanks without looking at Table 3.1.
**a.** 1 microliter = _____ liter
**b.** 1 picometer = _____ meter
**c.** 1 kilogram = _____ grams
**d.** 1 megaton = _____ tons

**STUDY ALERT**
**It is essential that you memorize the prefixes listed in Table 3.1. All of them are used extensively in the rest of this book.**

Now that you have seen how prefixes are used in the metric system, let's turn our attention to the base units used by scientists when they make measurements. Seven base units are given in Table 3.2. For the moment the first four of these—those for length, mass, time, and temperature—are the important ones.

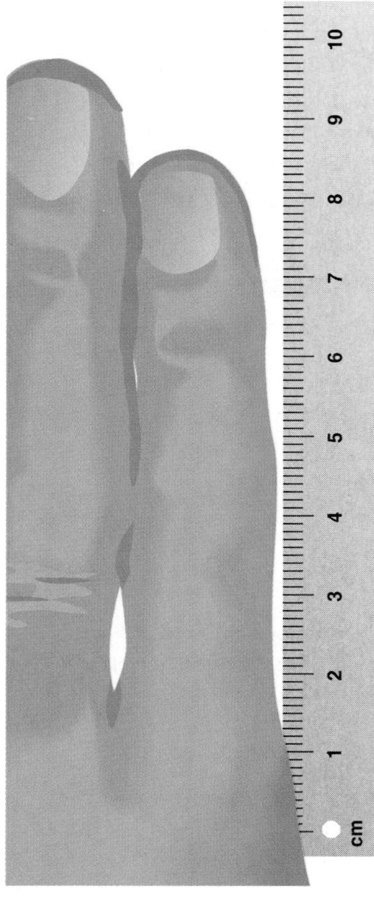


FIGURE 3.10
This ruler is graduated in centimeters and millimeters. Fingers are about 8 to 10 cm (80 to 100 mm) long. It is useful to develop a sense of sizes in the metric system.

FIGURE 3.11
Merchants of Pompeii used balances like this one to "weigh" goods offered for sale.

**TABLE 3.2 The Seven Base Units**

| Physical quantity | Unit | Abbreviation | Discussed in |
|---|---|---|---|
| Length | meter | m | Chapter 3 |
| Mass | kilogram | kg | Chapter 3 |
| Time | second | s | Chapter 3 |
| Temperature | kelvin | K | Chapter 3 |
| Amount of substance | mole | mol | Chapter 5 |
| Electric current | ampere | A | Chapter 16 |
| Luminous intensity | candela | cd | — |

**LENGTH.** The base unit used in measuring lengths and distances in the metric system is the **meter.** A meter is a little more than 3 feet in length. If the length or distance being measured is much larger or smaller than this, a prefix is used to make a unit that has a more convenient size. Scientists regularly use prefixes to indicate units of different sizes, and you will find it helpful to try to develop a feel for how large or small the various units are.

Here are some examples of how scientists use prefixes to size units appropriately. Chemists and physicists use *picometer* (1 pm $= 10^{-12}$ m) to measure the sizes of atoms and molecules. An oxygen atom is about 140 pm in diameter (it is much easier to write and say 140 pm than $1.40 \times 10^{-10}$ m). Biologists find the *micrometer* (1 $\mu$m $= 10^{-6}$ m) useful for measuring bacteria and other microorganisms. You can find *millimeters* (1 mm = 0.001 m) and *centimeters* (1 cm = 0.01 m) on most rulers. The *kilometer* (km) is 1000 meters, or about 0.6 of a mile.

## EXAMPLE 3.7 Using Metric Units of Length

Which of the following measures closest to a centimeter?

**a.** Thickness of a dime
**b.** Diameter of a penny
**c.** Height of a 4-year-old child
**d.** Distance between cities

SOLUTION

The answer is (b). You will calculate many answers in this course, and if you have some feel for the size of the units, you can help yourself avoid mistakes.

EXERCISE 3.7

Match each of the given sizes (left column) with the appropriate object (right column)

| Approximate Size | Object |
|---|---|
| 2 m | Length of a pencil |
| 15 cm | Length of a fly |
| 10 mm | Height of a skyscraper |
| 0.3 km | Width of an automobile |

**MASS.** Mass is often confused with weight (see the margin note on p. 71). Even scientists occasionally do it; in fact, they still use the verb *to weigh* when they mean *to find the mass of* in almost all laboratory settings. Almost all older chemistry books refer to *atomic weights* instead of the correct *atomic masses.* In recent years scientists (as well as textbook authors) have become much more careful about terminology.

The names of the mass units in the metric system are based on the gram, which is the mass of about 20 drops of water. For many everyday applications this unit of measurement is too small, so the kilogram (about 2.2 lb) is more widely used. Scientists use the *milligram* (1 mg $= 10^{-3}$ g) and the *microgram* (1 $\mu$g $= 10^{-6}$ g) when measuring the masses of small samples. The metric ton is 1000 kg (which is $10^6$ g, so the metric ton might be called a megagram, but it isn't). It is used in commerce, and is about 10% larger than the English system ton.

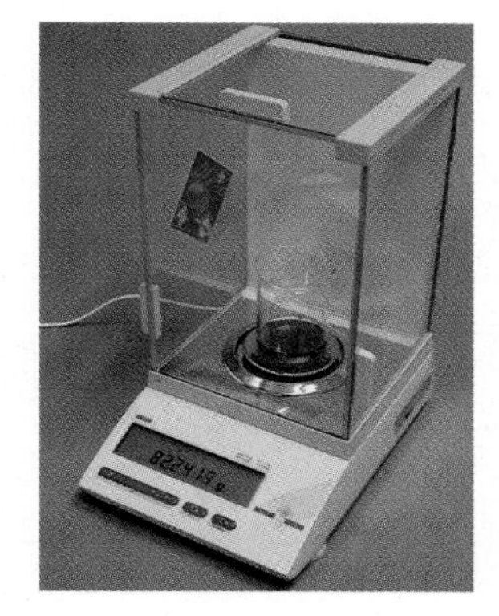

FIGURE 3.12
Modern balances are quite easy to use and feature a digital readout of the measured mass.

### EXAMPLE 3.8 Using Metric Units of Mass

Guess the mass of each of the following in conveniently sized metric units.

**a.** A paper clip
**b.** An automobile
**c.** A 4-year-old child
**d.** A strand of hair

**SOLUTIONS**

**a.** About 1 g
**b.** A metric ton ($10^6$ g, or about 2200 lb)
**c.** About 20 kg (about 44 lb)
**d.** About 1 mg ($10^{-3}$ g)

**EXERCISE 3.8**

Guess the mass of each of the following in conveniently sized metric units.

**a.** A nickel
**b.** A fly
**c.** A pail of water
**d.** A speck of dust

**TIME.** For scientists the base unit of time is the familiar second. They use the metric system prefixes to make units to measure time spans shorter than a second, such as *millisecond* and *microsecond.* For longer time spans they usually revert to units that are not related to seconds by factors of 10, including minutes, hours, days, and years.

FIGURE 3.13
The light-year is sometimes mistaken for a unit of time, which it most certainly is not. It is a unit of length equal to the distance that light travels in a year, about $9.5 \times 10^{15}$ m. This is a huge distance; the radius of the solar system is only a small fraction of this value. The light-year is used extensively by astronomers to measure the distances separating stars, nebulae, and galaxies.

## 3.6 Temperature, Volume, and Density

**GOAL:** To learn the commonly used units of temperature, volume, and density.

The **temperature** of an object is a measure of its hotness or coldness. A warmer object in contact with a colder one cools off, and the colder one warms up. At its most basic level, you can view temperature as a measure of the amount of motion of the atoms and molecules that make up an object. Objects at high temperatures have more molecular motion than they do at lower temperatures.

The temperature scale in common use in the United States is the **Fahrenheit scale.** No doubt you know that the freezing point of water on this scale is 32°F and that water boils at 212°F. The temperature scale in common use around the world is called the **centigrade scale** in most places, but scientists call it the **Celsius scale.** You probably learned in high school or even earlier that water freezes at 0°C and boils at 100°C.

There are 180 "steps" between the freezing and boiling points of water in the Fahrenheit scale, and the same temperature difference is covered by only 100 "steps" in the Celsius scale. In other words, the Celsius degree is larger than the Fahrenheit degree by a factor of exactly $^{180}/_{100}$, or 1.8. Also, of course, 0°C = 32°F, because water freezes at this temperature no matter how you measure it. You can combine these two pieces of information to make Equation 3.1, which converts temperatures from one scale to the other.

$$°\mathrm{F} = (1.8)(°\mathrm{C}) + 32° \qquad \text{(Equation 3.1)}$$

The factor of 1.8 compensates for the larger size of the Celsius degree ($^{180}/_{100} = 1.8$), and the 32° compensates for the different zero points.

## CHAPTER BOX 3.1 SI UNITS

In 1960 the scientific community adopted a worldwide standard for units (that is, everybody agreed to use them). The international committee responsible for preparing the agreement used the seven units given in Table 3.2 as a standard. Two more units were adopted for angular measurements. (These units are for special applications, and you will not encounter them in this course.)

The prefixes given in Table 3.1, plus several others, are part of the International System of Units, known as **SI units** (from the initials of its French name, *Système International d'Unités*). As you already know, the prefixes define fractions or multiples of the base units.

Scientists (and others) often use approximately fifteen *derived units* in addition to the base units. Like the original seven, the derived units are measures of physical quantities. For example, *volume* is not included in the base list, though it certainly is a physical quantity that scientists measure (see Experimental Exercise 3.1 in Section 3.8). If you want to calculate the volume of a box, you multiply its length times its width times its height. Length, width, and height are distances, and the base unit for distance is the meter. Therefore you can *derive* (calculate from the bases) an SI unit for volume, the cubic meter ($\mathrm{m} \times \mathrm{m} \times \mathrm{m} = \mathrm{m}^3$). Some of the derived units look quite bizarre to beginners. For instance, the unit for energy, called a *joule,* is equivalent to $\mathrm{kg\ m^2/s^2}$.

In theory, the SI units seemed fine, but scientists had some practical objections. One difficulty was that two centuries of tradition and convention are hard to overturn by the mere decree of a committee. The volume unit used by most chemists today is not the cubic meter but the liter (a little more than a quart, equal to $10^{-3}\ \mathrm{m}^3$). The liter is a more practical size for laboratory work, and most of the laboratory glassware that chemists use to measure volumes is labeled in units of liters or milliliters. Scientists had similar objections to some of the other SI derived units.

The point of all this is that a standard set of units, called SI units, is used extensively by practicing scientists. In this book we used the units most commonly employed by chemists.

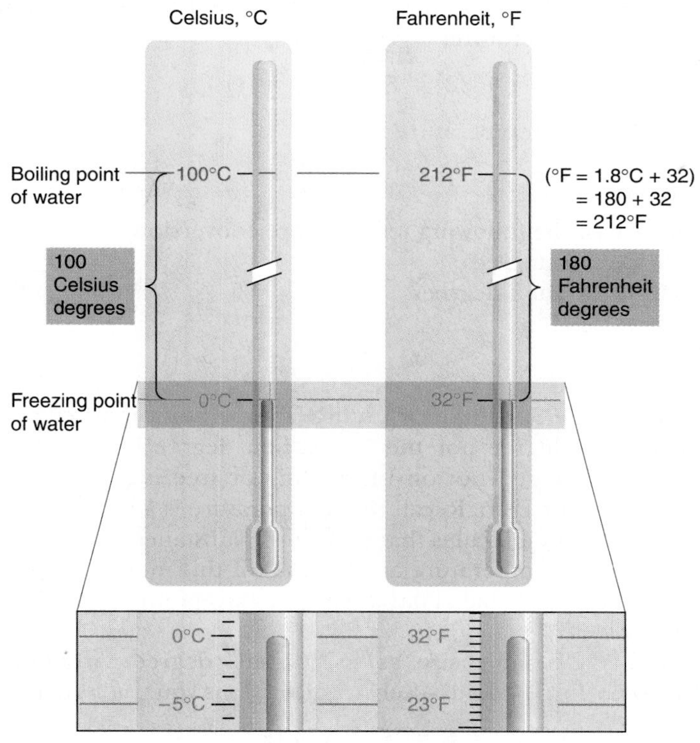


FIGURE 3.14

**EXAMPLE 3.9** **Celsius–Fahrenheit Conversions**

Perform each of the following temperature conversions:

**a.** 68°F to Celsius degrees
**b.** 45°C to Fahrenheit degrees

**SOLUTIONS**

Equation 3.1 can be used to do both conversions. You simply plug the numbers in and calculate. (If you have any difficulty following the solutions given below, look at Appendix A, Algebra and Solving Numerical Problems.)

**a.** You know the Fahrenheit temperature: 68°F. The Celsius temperature is what you are looking for:

$$68^\circ = (1.8)(^\circ\mathrm{C}) + 32^\circ$$

$$(1.8)(^\circ\mathrm{C}) = 68^\circ - 32^\circ = 36^\circ$$

$$^\circ\mathrm{C} = \frac{36^\circ}{1.8} = 20^\circ$$

The Celsius temperature is 20°C.

**b.** This problem is even more straightforward than the one in part (a). You know the Celsius temperature (45°C), and you are looking for the Fahrenheit temperature:

$$°F = (1.8)(45°) + 32° = 81° + 32° = 113°$$

The Fahrenheit temperature is 113°F.

**EXERCISE 3.9**

Perform each of the following temperature conversions:

**a.** 25°F to Celsius degrees

**b.** 25°C to Fahrenheit degrees

**The K is written without a degree mark, for example, 200 K. You should read 200 K as "two hundred kelvins," *not* "two hundred *degrees* kelvin."**

**STUDY ALERT**

**When you write a temperature in this course, always indicate what scale you're using. That is, write 100°C or 100°F or 100 K, not just 100°.**

As shown in Table 3.2, the base unit for temperature that scientists use is neither the Celsius degree nor the Fahrenheit degree but the **kelvin** (K). The **Kelvin scale** is based not on the boiling or freezing point of any substance, but on absolute zero. Recall that temperature is a measure of the motion of the atoms and molecules that make up a substance. Absolute zero can be thought of as the temperature at which all of this motion stops; and you can't get any colder than that. That is why the zero point on the Kelvin scale is called *absolute zero.*

The kelvin is the same size as the Celsius degree. Absolute zero is −273.15°C. To perform Celsius–Kelvin conversions you use Equation 3.2:

$$\mathbf{K = °C + 273.15} \qquad \text{(Equation 3.2)}$$

Since the kelvin and the Celsius degree are the same size, the 273.15 accounts for the difference in the zero points between the two temperature scales.

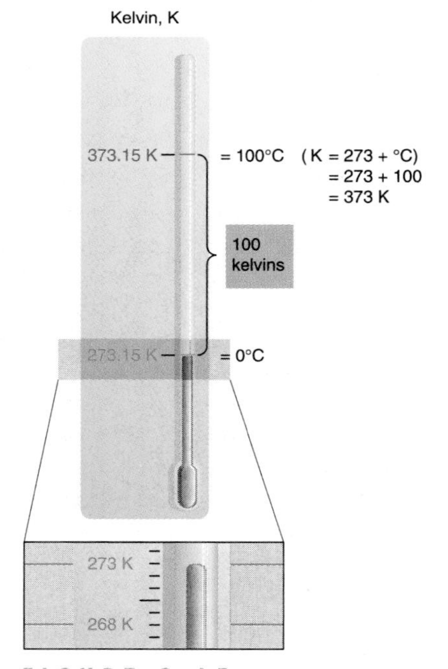


FIGURE 3.15

## EXAMPLE 3.10 Celsius–Kelvin Conversions

Perform each of the following temperature conversions:

**a.** 20°C to kelvins

**b.** 20 K to Celsius degrees

**SOLUTIONS**

**a.** Use Equation 3.2, which gives

$$K = °C + 273.15 = 20 + 273.15 = 293.15$$

Don't forget significant figures. Since the 20°C has no digits to the right of the decimal point, your answer can't have any either. Therefore, 20°C = 293 K.

**b.** Once again, plug the numbers into Equation 3.2:

$$20 = °C + 273.15$$

$$20 - 273.15 = °C = -253.15$$

The answer to the correct number of significant figures is −253°C.

**EXERCISE 3.10**

Perform each of the following temperature conversions:

**a.** 100°C to kelvins

**b.** 200 K to Celsius degrees

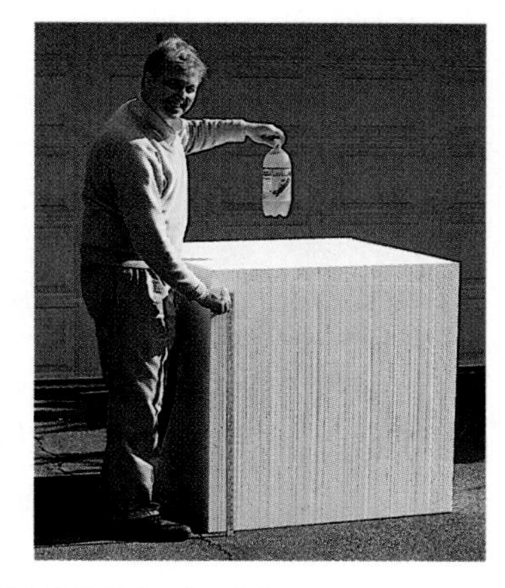

FIGURE 3.16
The large box has a volume of 1 cubic meter ($m^3$). The bottle has a volume of $2 \times 10^{-3}$ $m^3$, or 2 L. The liter is the more widely used unit of volume.

Perhaps because it's used on soda bottles, the most familiar metric system unit of volume is the **liter** (L). This unit is a little larger than the English system quart. Units of volume are derived units that come from the meter, the base unit of length (see Chapter Box 3.1). The derived unit of area is the square meter ($m^2$), and that of volume is a cubic meter ($m^3$).

Unfortunately, the cubic meter is such a large volume (1000 L, about 250 gal) that it is inconvenient to use around the house or in a laboratory. For practical reasons, then, the liter and the milliliter (mL) are the volume units used most frequently by chemists. The volume of a small grape is approximately one milliliter. Another unit you will often see is the cubic centimeter (cc, or $cm^3$), which is exactly equal to the milliliter (1 mL ≡ 1 cc = 1 $cm^3$). This unit is used extensively in medicine and biology. Medical hypodermic syringes are usually graduated in cubic centimeters, for instance.

The more of a substance you have, the greater its mass and the larger the volume it occupies. For example, a 2-L sample of water has twice the mass of a 1-L sample. In other words, volume is an extensive property. This means that the volume occupied by a substance cannot be used to identify that substance. However, the relationship between volume and mass can be combined to define the property known as the **density.** Density is an intensive

## CHAPTER BOX 3.2 REASONABLE ANSWERS

At some point you might get tangled up while using either Equation 3.1 or Equation 3.2. Perhaps you'll get a sign wrong and subtract 32 when you should have added it, or maybe you'll divide by 1.8 instead of multiplying, or vice versa. Some folks write down the wrong answer and continue on their merry way without even looking to see if their calculated result makes sense (a terrible "know-no" when working any kind of problem).

How can you spot a wrong answer? The best way is to have a sense of what the answer should be. Suppose you subtracted 32 instead of adding it in Example 3.9(b) (as one of the authors did while writing this book). The answer you would get is 45°C = 49°F. You should know that temperatures above zero in °F are larger numbers than those in °C. You also should realize that 45°C is above the human body temperature of 37°C, while 49°F is well below that temperature. Thus the answer 45°C = 49°F has to be wrong, so you would recalculate.

Everybody makes errors in calculations from time to time. That's not a problem; the crime is failing to check to see whether your answer makes sense. This is an extremely important problem-solving tool. The trick is to recognize that you are not finished with a problem once you have formulated an answer. *There is one more step to perform: you must ask yourself whether your answer is reasonable.*

If you expect to have trouble checking your answers because you are "not a scientist," don't worry. Your ability to check answers for reasonableness is much greater than you probably think it is—you just haven't done it very much, so making a negative judgment like that is premature. You know a lot more than you may think you do. Trust yourself!

property that is often used to help identify substances. The definition of density is

$$d = \frac{m}{V} \qquad \text{(Equation 3.3)}$$

Note that $mL^{-1} = \frac{1}{mL}$, just as $10^{-1} = \frac{1}{10}$.

where $m$ is the mass (typically in grams), $V$ is the volume (in mL for solids and liquids, in L for gases), and $d$ is the density. This equation gives grams divided by milliliters as the commonly used units of density for solids and liquids, which can be written as grams per milliliter, g/mL, or as g $mL^{-1}$. For gases the units of density are usually g $L^{-1}$.

Suppose someone hands you a small painted object and asks whether it's made of balsa wood or lead. That's an easy question to answer. You lift the object. If it feels heavy, it's made of lead, and if it feels light, it's made of balsa wood. Many people will tell you, "Lead is heavy and balsa wood is light." However, a large board of balsa wood is heavier than a small lead fishing sinker. The correct statement is that lead is *more dense* than balsa wood. (The hidden lesson here is to be careful in what you say—your instructor will insist that you use precise wording in this course.)

FIGURE 3.17
A small grape has a volume of about 1 mL (1 $cm^3$).

This kind of observation can be useful. For example, once while on vacation one of your authors was approached by a street vendor. The vendor offered a handsome silver religious medallion for sale. "For you, a special price," he said, and dropped the medallion to the pavement, where it gave a nice metallic ring. The author picked up the medallion and guessed from its light weight that it was aluminum rather than silver. He decided to save his money.

In science it is generally useful to measure densities more precisely than this. The relatively vague concept of "heaviness" is too imprecise to be useful in most situations. The mass of an object is easy to determine, and has been since ancient times. (The merchants of any civilization could not have worked without being able to weigh things accurately.) The more difficult experimental problem is to measure the volume, because most interesting objects have irregular shapes.

Suppose you are given a shiny yellow metal crown and are told that it's pure gold. How do you tell whether it's pure gold or an alloy of gold and silver? (Both the pure gold and the alloy are "heavy," so an informal observation of the weight of the crown won't work.) The ancient Greek philosopher Archimedes was given just this problem by the king, who suspected that he had been swindled by the person who made his crown. Archimedes was not allowed to damage the crown while testing it. He wanted to determine the density, because he knew that pure gold is more dense than an alloy of gold and silver.

Archimedes didn't know what to do, until one day he got into his bath and it overflowed. He realized that he could measure the volume of the crown by submerging it in water and measuring how much the level of the water rose. The story goes that Archimedes leaped out of his bath and ran through the streets naked, shouting "Eureka!" ("I've got it!"). (The ancient Greeks were much more relaxed about public nudity than we are, so Archimedes wasn't arrested and taken away.)

**Densities of Some Familiar Substances (g $mL^{-1}$)**

| Substance | Density |
|---|---|
| Balsa wood | 0.12 |
| Ethanol (grain alcohol) | 0.79 |
| Water | 1.00 |
| Aluminum | 2.70 |
| Diamond | 3.51 |
| Silver | 10.5 |
| Lead | 11.3 |
| Mercury | 13.6 |
| Gold | 19.3 |

### EXAMPLE 3.11 Calculating Density from Experimental Data

**a.** What is the density in g $mL^{-1}$ of a piece of metal that has a volume of 25 mL and a mass of 125 g?

**b.** A chemist put 30.0 mL of water in a graduated cylinder and then submerged a 127-g piece of metal in the water. The water level rose to 52.2 mL. What is the density of the metal?

**SOLUTIONS**

**a.** To solve this one, use Equation 3.3:

$$d = \frac{m}{V} = \frac{125 \text{ g}}{25 \text{ mL}} = 5.0 \text{ g mL}^{-1} \text{ (Don't forget significant figures!)}$$

**b.** Since you are asked to find the density, you need both the mass and the volume of the metal. You were given the mass; you must calculate the volume. The water level rose from 30.0 mL to 52.2 mL, so the volume of the metal is

$$52.2 \text{ mL (total volume)} - 30.0 \text{ mL (initial volume of water)} = 22.2 \text{ mL (volume of metal)}$$

The density is then

$$d = \frac{m}{V} = \frac{127 \text{ g}}{22.2 \text{ mL}} = 5.72 \text{ g mL}^{-1}$$

**EXERCISE 3.11**

**a.** What is the density of a rock that has a volume of 12.2 mL and a mass of 89.3 g?

**b.** A chemist places a sample of a pure liquid in a flask with a capacity of 12.1 mL. The dry flask has a mass of 25.1 g, and the filled flask has a mass of 35.1 g. What is the density of the liquid?

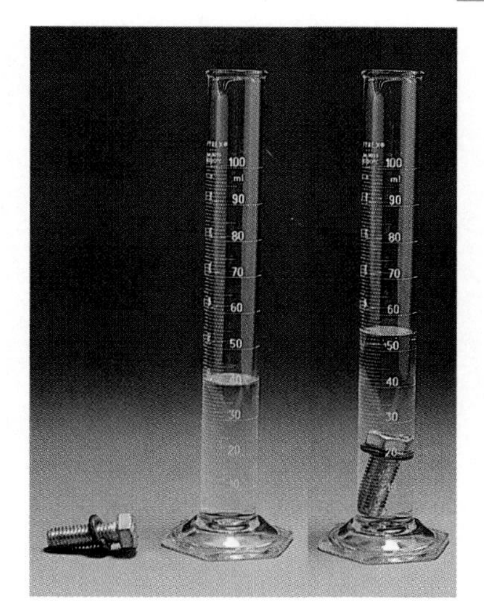

FIGURE 3.18
You can measure the volume occupied by an irregularly shaped piece of metal by measuring the amount of water the submerged piece displaces. See Example 3.11(b).

## 3.7 Working with Units: Dimensional Analysis

GOAL: To convert one unit to another using dimensional analysis.

In Section 3.1 you learned that you *must* keep the unit with the magnitude when writing a measurement. A magnitude without a unit usually doesn't make any sense. Suppose someone told you, "It's 50 between these two places." If the unit is the picometer, you're talking about two parts of the same atom. If it's the light-year, then you probably have two stars.

Remember that metric system units relate to one another by factors of 10. You can tell what the connections are from the prefixes. For instance,

0.001 L is equivalent to 1 mL. We can write this relationship as an algebraic equation:

$$1 \text{ mL} = 0.001 \text{ L}$$

This equation can be written another way. If 0.001 L and 1 mL are equal, then either quantity divided by the other equals 1:

$$\underset{\text{One}}{\frac{0.001 \text{ L}}{1 \text{ mL}}} = \underset{\text{One}}{1} = \underset{\text{One}}{\frac{1 \text{ mL}}{0.001 \text{ L}}}$$

Each of these fractions is called a **unit factor.** Any time you have an equality involving units, you can construct a pair of these unit factors in exactly this way. Such unit factors are the tools you use to convert from units with prefixes to base units or to units with another set of prefixes. For this reason, these ratios are more commonly called **conversion factors.**

Suppose you want to convert $2.3 \times 10^4$ mL into liters. You can do this conversion, and all like it, in three steps. You need the conversion factor that contains the *dimensions* of both the *quantity you know* and the *quantity you want to get.* That is why this general technique is called dimensional analysis: your task is to pick the conversion factor with the correct dimensions.

**Step 1** First, you write your starting point above a line like this. Don't forget the units.

$$\frac{2.3 \times 10^4 \text{ mL}}{1}$$

**Step 2** Write the appropriate conversion factor next to it. What is the appropriate conversion factor? You choose the one that places the unit you want to remove (mL in this example) in the denominator, and the unit you want to convert to in the numerator (L). You do this because the unwanted unit cancels when you calculate (in Step 3), leaving the unit you want in the numerator.

$$\frac{2.3 \times 10^4 \cancel{\text{mL}}}{1} \times \frac{0.001 \text{ L}}{1 \cancel{\text{mL}}}$$

**Step 3** You then multiply all this together. The overall quantity cannot change because the conversion factor itself is equal to 1, and multiplying by 1 changes nothing. The only unit left is the one you wanted (L here).

$$\frac{2.3 \times 10^4 \cancel{\text{mL}}}{1} \times \frac{0.001 \text{ L}}{1 \cancel{\text{mL}}} = \frac{(2.3 \times 10^4)(0.001 \text{ L})}{1}$$
$$= 2.3 \times 10^1 \text{ L} = 23 \text{ L}$$

So $2.3 \times 10^4$ mL = 23 L is the answer to the problem.

The fact that the answer does have the correct units is strong evidence that the conversion was done correctly assuming that the conversion factor itself was correct. If you make an error, you'll be able to tell instantly. Suppose, for example, you wrote the conversion factor incorrectly in Step 2 (a very common error—and not only for beginners). The result would have looked like this:

$$\frac{2.3 \times 10^4 \text{ mL}}{1} \times \frac{1 \text{ mL}}{0.001 \text{ L}}$$

Note that the unwanted units (mL) do not cancel. What you get from this calculation is $2.3 \times 10^7 \text{ (mL)}^2 \text{ L}^{-1}$. These are not the units you are looking for, and they don't make sense anyway. You know that there is a mistake because the answer isn't reasonable.

Dimensional analysis is a powerful tool for working a vast array of numerical problems. Its great strength is that if you get the wrong dimensions in the answer, you know you worked the problem wrong and can remedy the situation immediately. Furthermore, as the next example shows, the technique is not limited to making conversions between metric units.

### EXAMPLE 3.12 Using Dimensional Analysis

**a.** How many picometers are there in 5.00 cm?
**b.** How many centimeters are there in 5.000 pm?
**c.** What is the volume occupied by a 240-g piece of wood that has a density of $0.90 \text{ g mL}^{-1}$?
**d.** A chemist measured the volume of a sample of hydrogen gas and found it to be $1.7 \times 10^9 \text{ cm}^3$. What is the volume in cubic meters?

**SOLUTIONS**

**a.** Here you want to convert centimeters to picometers. The problem is that no appropriate single conversion factor is listed in Table 3.1. You must first convert centimeters into something that exists in the table (meters) and then convert the meters to picometers. In other words, do these operations: cm → m → pm. From Table 3.1, $1 \text{ pm} = 1 \times 10^{-12} \text{ m}$ and $1 \text{ cm} = 1 \times 10^{-2} \text{ m}$. The first conversion is

$$\frac{5.00 \cancel{\text{cm}}}{1} \times \frac{1 \times 10^{-2} \text{ m}}{1 \cancel{\text{cm}}} = 5.00 \times 10^{-2} \text{ m}$$

The second conversion is

$$\frac{5.00 \times 10^{-2} \cancel{\text{m}}}{1} \times \frac{1 \text{ pm}}{1 \times 10^{-12} \cancel{\text{m}}} = \frac{5.00 \times 10^{-2} \text{ pm}}{1 \times 10^{-12}}$$
$$= 5.00 \times 10^{10} \text{ pm}$$

**b.** This conversion is done much the same way as that in part (a). Here the conversion sequence is pm → m → cm. Rather than writing the conversions as two separate equations, you can combine them:

$$\frac{5.000 \cancel{\text{pm}}}{1} \times \frac{1 \times 10^{-12} \cancel{\text{m}}}{1 \cancel{\text{pm}}} \times \frac{1 \text{ cm}}{1 \times 10^{-2} \cancel{\text{m}}} =$$
$$\frac{(5.000)(1 \times 10^{-12}) \text{ cm}}{1 \times 10^{-2}} = 5.000 \times 10^{-10} \text{ cm}$$

**c.** Despite its appearance, this is a dimensional analysis problem. The task is to convert the mass of wood into the volume of wood. Density is defined as $m/V$, or g $mL^{-1}$ (remember that this means g/mL), and this provides the key to the solution. Look carefully at what 0.90 g $mL^{-1}$ really means—that 0.90 g of wood has a volume of 1 mL. In equation form, 0.9 g wood $\approx$ 1 mL (the equals sign cannot be used, since a mass can never equal a volume, it can only be equivalent to one). Once you recognize this, you can proceed as in a normal dimensional analysis. Since you want to cancel units of grams, the conversion factor is written with grams in the denominator:

$$\frac{240\ \cancel{g}}{1} \times \frac{1\ \text{mL}}{0.90\ \cancel{g}} = 267\ \text{mL}$$

$$= 2.7 \times 10^2\ \text{mL (to two significant figures)}$$

**d.** There is a special problem involved in converting volumes. You start off as usual by writing the given measurement, $1.7 \times 10^9$ $cm^3$. But how do you convert $cm^3$ to $m^3$? Well, you know that 1 cm = 0.01 m. You also know that 1 $cm^3$ = 1 cm × 1 cm × 1 cm. You have to use the unit factor 1 cm = 0.01 m three times to get the units to cancel:

$$\frac{1.7 \times 10^9\ \cancel{cm}^3}{1} \times \frac{0.01\ \text{m}}{1\ \cancel{cm}} \times \frac{0.01\ \text{m}}{1\ \cancel{cm}} \times \frac{0.01\ \text{m}}{1\ \cancel{cm}} =$$

Then you do the arithmetic to get

$1.7 \times 10^9$ $cm^3$ = $1.7 \times 10^3$ $m^3$

EXERCISE 3.12

**a.** An aquarium manager was told that one of the tanks contained $1.8 \times 10^5$ mL of water. How many liters is that?
**b.** A chemist jokingly told a friend that they were $1.7 \times 10^5$ cm from home. How many kilometers is this?
**c.** What is the mass of 30 cc of gold? Gold has a density of 19.31 g $mL^{-1}$.
**d.** A piece of paper has an area of 550 $cm^2$. How many square meters is this?

## 3.8 Converting English System and Metric System Units

**GOAL:** To convert measurements and units from the English system to the metric system, and vice versa.

Most scientists born in the United States get a feel for metric system units in their work. They measure amounts of compounds in milligrams or grams, volume in milliliters or liters, and temperatures in Celsius degrees, and they know what the numbers they get mean. Then they go home, buying a quart or a gallon of milk on the way, and cook with recipes that call for cups and

## EXPERIMENTAL EXERCISE 3.1

A. L. Chemist was given what looked like a small golden statue to identify. The donor said, "A. L., we know that this thing has a microscopically thin gold layer over another metal. We also know that the interior metal is either copper or lead, but we don't know which. Can you find out for us?"

A. L. looked up the densities of the two metals. The density of copper is 8.96 g $mL^{-1}$ and the density of lead is 11.3 g $mL^{-1}$. A. L. placed exactly 10.00 mL of water in a graduated cylinder and then submerged the statue, which had a mass of 15.5 g. The total volume rose to 11.73 mL, so A. L. calculated that the volume of the metal was

$$11.73 \text{ mL} - 10.00 \text{ mL} = 1.73 \text{ mL}$$

A. L. concluded that the density is

$$\frac{15.5 \text{ g}}{1.73 \text{ mL}} = 9 \text{ g mL}^{-1}$$

Thus the unknown metal was copper.

1. What would the total volume have been if the unknown metal had been lead?
2. Is the volume of the unknown metal reported to the correct number of significant figures? If not, what is the correct volume?
3. Is the density of the unknown metal reported to the correct number of significant figures? If not, what should the answer be?
4. Do the experimental results support A. L. Chemist's conclusion?

teaspoons. They check their weight in pounds and hear weather reports and read recipes with temperatures in Fahrenheit degrees. They have a feel for those numbers, too. It would be simpler if they could use just one set of units, but it's too late for that.

Sometimes, though, scientists get information in one system and have to convert it to the other. Some chemicals are still sold by the gallon or the pound, and engineers often use these English system units in their work. Scientists or nonscientists who vacation abroad, of course, have these conversion problems all the time. "The clerk says it's 4.7 kilometers to the embassy, and we have to be there in 20 minutes—should we walk?" "The guide said that the mountain over there on the horizon is 8848 meters high. Is that big enough to warrant a visit?"

You can convert between the two systems using the dimensional analysis method discussed in Section 3.7. You will also need conversion factors between the systems. Some of these are given in Table 3.3.

FIGURE 3.19
The largest fish in this figure has a mass of about 1 kg (about 2.2 lb).

**TABLE 3.3 Conversion Factors for Some English and Metric System Units**

| | |
|---|---|
| 1 qt = 0.9464 L | 1 mi = 5280 ft (exact) |
| 1 lb = 453.6 g | 1 yd = 3 ft = 36 in (exact) |
| 1 in = 2.54 cm (exact) | 4 qt = 1 gal (exact) |
| | 1 lb = 16 oz (exact) |

FIGURE 3.20
Mt. Everest is 29,028 ft high.

EXAMPLE 3.13 **Converting Metric and English System Measurements**

If it's 4.7 km to the embassy, how far is that in miles? (Is 20 minutes enough time to walk there?)

**SOLUTION**

The only conversion factor in Table 3.3 for converting metric distances to English is 1 in. = 2.54 cm. First you must convert the kilometers to meters, then meters to centimeters, then centimeters to inches, inches to feet, and finally feet to miles (km→m→cm→in→ft→mi). Here is the equation:

$$\frac{4.7\ \cancel{\text{km}}}{1} \times \frac{1000\ \cancel{\text{m}}}{1\ \cancel{\text{km}}} \times \frac{1\ \cancel{\text{cm}}}{0.01\ \cancel{\text{m}}} \times \frac{1\ \cancel{\text{in}}}{2.54\ \cancel{\text{cm}}} \times \frac{1\ \cancel{\text{ft}}}{12\ \cancel{\text{in}}} \times \frac{1\ \text{mi}}{5280\ \cancel{\text{ft}}} = 2.9\ \text{mi}$$

So 4.7 km is 2.9 mi, and you'd better take a taxi.

**EXERCISE 3.13**

Mt. Everest is 29,028 ft high. How high is this in meters?

## CHAPTER SUMMARY

1. All measured numbers consist of two parts: the magnitude and the unit. The unit must be included in all reported measurements.
2. Precision is a characteristic of a set of measured numbers. The closer the numbers are to each other, the greater the precision.
3. Accuracy is a comparison of a measured quantity to a known value.
4. All measurements have some uncertainty associated with them.
5. All digits known exactly plus the first digit that is estimated are called significant figures.
6. For multiplying or dividing, the number of significant figures in the result must be identical to the number of significant figures in the least precise number.
7. For adding or subtracting, the result must have the same number of digits to the right of the decimal point as the number that has the fewest digits to the right of the decimal point.
8. In scientific notation, numbers are written in the form $A \times 10^x$, where $A$ is a number between 1 and 10 and $x$ is any positive or negative integer.
9. The metric system uses base units and a set of prefixes reflecting factors of 10 to record measured or calculated results.
10. The prefixes in the metric system indicate the factor of 10 by which you must multiply the measurement to obtain the number in base units. (You must memorize the prefixes given in Table 3.1. A complete list is given in the back cover of this book.)

11. The temperature scales used most frequently by scientists are the Kelvin scale and the Celsius scale; K = °C + 273.15 is the relationship that interconverts the two systems of measurement.
12. Density is defined as mass per unit volume ($d = m/V$). Densities of solids and liquids are reported in g $mL^{-1}$; densities of gases are reported in g $L^{-1}$.
13. Dimensional analysis is a problem-solving method that helps eliminate calculation errors by ensuring that the answer has the correct units. It has broad applications in science.

## PROBLEMS

### SECTION 3.1 An Introduction to Measurement

1. Name five measuring devices.

2. Name a device typically found around a house, garage, or automobile that is used to measure each of the following:
   a. Pressure
   b. Temperature
   c. Volume
   d. Weight

3. Explain why a unitless number is almost always worthless in conveying scientific information to another person.

4. Explain the importance of both the magnitude and the unit in a number.

5. What is the major difference between mass and weight?

6. Why is the distinction between mass and weight more important for experiments done in a space shuttle than those done on the earth's surface?

7. To a scientist, something is wrong with each of the following statements. Identify the problems (there may be more than one) and suggest improved statements.
   a. Under normal conditions water boils at 100 degrees.
   b. The quarterback weighs 210.
   c. A sprinter ran the hundred in 10 flat.

8. To a scientist, something is wrong with each of the following statements. Identify the problems (there may be more than one) and suggest improved statements.
   a. Normal body temperature is 98.6.
   b. The guard is five ten.
   c. A. L. Chemist got a speeding ticket for doing 50 in a 35 zone.

9. What effect does each of the following actions have on accuracy and precision when you weigh yourself? These are the possible answers: lowers both accuracy and precision, lowers accuracy only, lowers precision only, or lowers neither accuracy nor precision. Briefly justify each answer.
   a. You wear a cast.
   b. You eat a chocolate bar afterward.
   c. You bounce up and down on the scale and read your weight at random times.
   d. You bounce up and down on the scale and read the lowest weight during several bounces.

10. What effect does each of the following actions have on accuracy and precision when you take your temperature with a normal thermometer? These are the possible answers: lowers both accuracy and precision, lowers accuracy only, lowers precision only, or lowers neither accuracy nor precision. Briefly justify each answer.
    a. You put the thermometer under your arm instead of in your mouth.
    b. You put an ice cube in your mouth along with the thermometer.
    c. The directions say to keep the thermometer in your mouth for 2 minutes, but you wait 5 minutes.
    d. You put the thermometer in your mouth for 2 seconds and then read the temperature.

11. What is the average of each of these sets of numbers?
    a. 25.6 mL, 24.9 mL, 30.0 mL
    b. 4.6 kg, 4.6 kg, 4.7 kg, 4.5 kg, 4.5 kg
    c. 12.165 m, 14.555 m, 16.970 m, 18.299 m, 13.343 m
    d. 4099 s, 5001 s, 5677 s, 5069 s
    e. 100.2 mL, 37.9 mL, 45.6 mL
    f. 102.1 kg, 100.9 kg, 101.3 kg, 102.0 kg, 102.2 kg
    g. 12 m, 16 m, 19 m, 12 m
    h. 300.0 s, 299.1 s, 299.6 s, 299.9 s
    i. 28 h, 35 h, 122 h, 5 h, 9 h, 12 h
    j. 0.222 g, 0.983 g, 0.199 g, 0.456 g, 0.335 g

12. What is the average of each of these sets of numbers?
    a. 123.6 mL, 133.9 mL, 128.8 mL, 122.0 mL
    b. 5.6 kg, 10.2 kg, 13.7 kg
    c. 12.26 m, 14.99 m, 19.02 m, 14.99 m, 13.05 m
    d. 40 s, 53 s, 58 s, 43 s, 45 s, 47 s, 44 s
    e. 10.5 mL, 47.8 mL, 45.9 mL, 35.7 mL
    f. 1002.1 kg, 1019.9 kg, 1221.3 kg
    g. 12.335 m, 29.386 m, 21.009 m, 12.224 m
    h. 470.0 s, 286.7 s, 200.5 s, 354.7 s
    i. 48 h, 35 h, 22 h, 57 h, 34 h
    j. 0.022 g, 0.083 g, 0.099 g, 0.056 g

13. A scientist weighed an object four times and obtained masses of 10.23 g, 10.21 g, 10.23 g, and 10.21 g.
    a. What is the average mass of the object?
    b. If the true value of the mass is 9.00 g, what can you say about the accuracy and the precision of these results?

14. A scientist measured the volume of a piece of metal four times and obtained volumes of 1.26 mL, 1.06 mL, 1.37 mL, and 1.11 mL.
    a. What is the average volume of this piece of metal?
    b. If the true volume is 1.19 mL, what can you say about the accuracy and the precision of these results?

15. A scientist measured the maximum temperature achieved by a chemical reaction four times and obtained temperatures of 53.21°C, 53.22°C, 53.22°C, and 53.22°C.
    a. What is the average maximum temperature of this reaction?
    b. If the true maximum temperature is 53.22°C, what can you say about the accuracy and the precision of these results?

16. Four times a scientist measured the time it takes for a particular reaction to end and obtained times of 53 s, 62 s, 46 s, and 102 s.
    a. What is the average time it takes for this reaction to end?
    b. If the true maximum time is 52 s, what can you say about the accuracy and the precision of these results?

## SECTION 3.2 Uncertainty in Measurement: Significant Figures

17. How many significant figures does each of these numbers have?
    a. 12.10 mi
    b. 0.00023 in
    c. 5001 ft
    d. 9.003°C
    e. 0.00780 ton
    f. 2 words
    g. 19.01 mL
    h. 0.0090°C
    i. 30.0 g
    j. 6 spoons

18. How many significant figures does each of these numbers have?
    a. 14.606 kg
    b. 0.02133 ft
    c. 100.01000 in
    d. 0.00004 lb
    e. 46 g
    f. 46.000 g
    g. 1000 mi
    h. 400 paragraphs
    i. 400.000 m
    j. 9835.1 cm

19. A ruler graduated in centimeters is shown in Figure 3.21.
    a. To how many significant figures can the number be read at the point marked by the arrow?
    b. To what number is the arrow pointing?

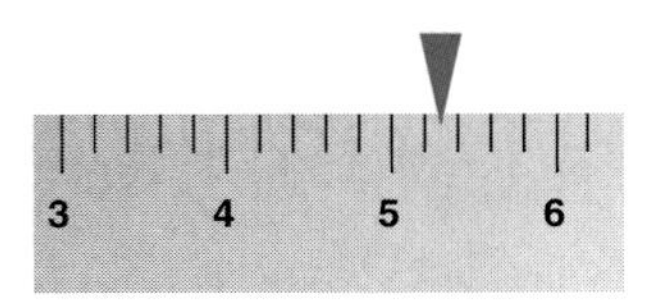


FIGURE 3.21

20. A ruler graduated in centimeters is shown in Figure 3.22.
    a. To how many significant figures can the number be read at the point marked by the arrow?
    b. To what number is the arrow pointing?

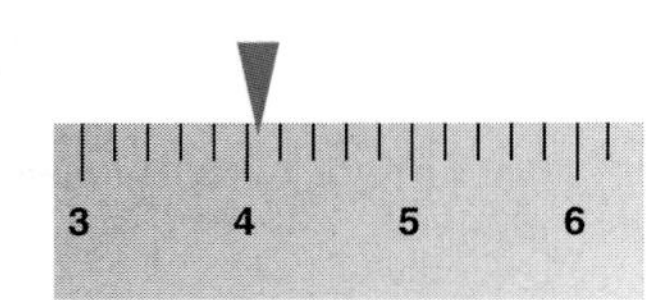


FIGURE 3.22

21. A bathroom scale graduated in pounds is shown in Figure 3.23.
    a. To how many significant figures can the number be read at the point marked?
    b. What is the weight indicated on the scale?

FIGURE 3.23

22. A bathroom scale graduated in pounds is shown in Figure 3.24.
    a. To how many significant figures can the number be read at the point marked?
    b. What is the weight indicated on the scale?

FIGURE 3.24

23. A doctor's office balance, graduated in pounds, is shown in Figure 3.25.
    a. To how many significant figures can the number be read at the point marked?
    b. What is the weight of this patient?

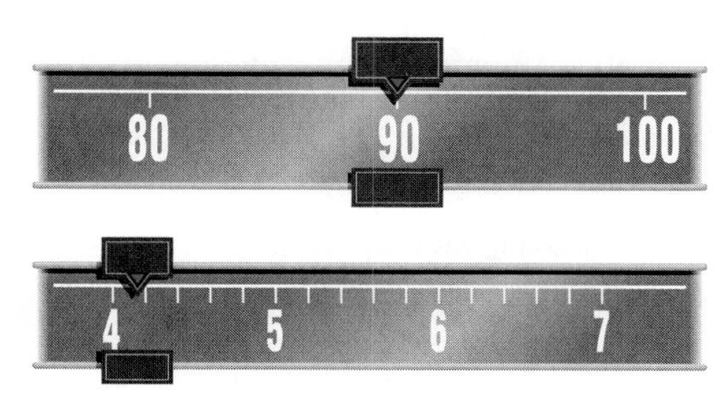


FIGURE 3.25

24. A doctor's office balance, graduated in pounds, is shown in Figure 3.26.
    a. To how many significant figures can the number be read at the point marked?
    b. What is the weight of this patient?

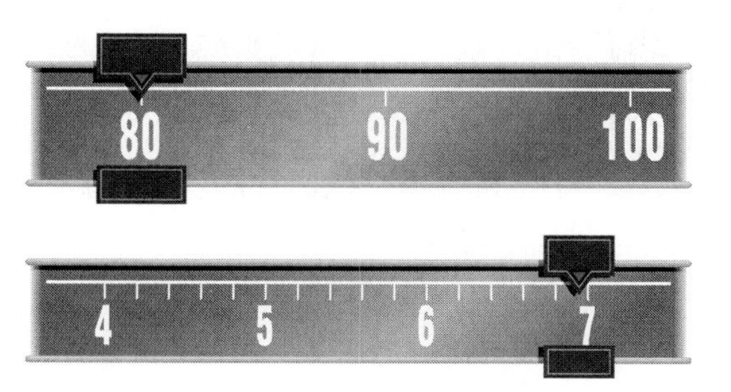


FIGURE 3.26

25. A buret is a piece of glassware used to measure volumes in milliliters. A portion of one is shown in Figure 3.27.
    a. To how many significant figures can the number be read at the point marked by the arrow?
    b. To what number is the arrow pointing?

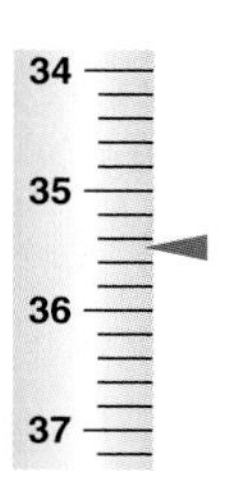


FIGURE 3.27

26. A portion of a buret is shown in Figure 3.28.
    a. To how many significant figures can the number be read at the point marked by the arrow?
    b. To what number is the arrow pointing?

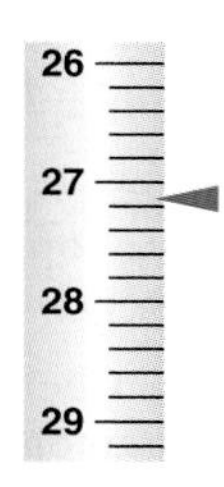


FIGURE 3.28

## SECTION 3.3 Significant Figures in Calculations

27. Perform each of the following calculations, giving each answer with the correct number of significant figures.
    a. 2.100 g + 0.01 g + 21.1 g
    b. 123.2 ft − 0.07 ft
    c. 55.6 m × 1.2 m
    d. $\frac{34.0 \text{ L}}{1.235 \text{ L}}$
    e. 4.0 lb + 324 lb + 23.78 lb
    f. 120.07 mi − 12.5 mi
    g. 98 s × 5.156 s
    h. $\frac{55.34 \text{ lb}}{61.2 \text{ in}^2}$
    i. 154 m + 0.001 m + 0.100 m
    j. 345.0 kg − 45 kg

28. Perform each of the following calculations, giving each answer with the correct number of significant figures.
   a. $2.399\ m + 0.01\ m + 45\ m$
   b. $89.444\ g - 0.010\ g - 76\ g$
   c. $67.2\ ft \times 13.3\ ft$
   d. $\frac{898\ m}{2.001\ h}$
   e. $4.0\ s + 444\ s + 44.44\ s$
   f. $56.55\ mi + 100.2\ mi - 0.02\ mi$
   g. $14.556\ in^3 \times \pi$
   h. $398.00\ lb - 98.0\ lb$
   i. $\frac{50.00\ ft}{2.000\ s}$
   j. $500.0\ hr + 55.12\ hr + 5.333\ hr$
29. Calculate the number of grams of Cu in 9.3345 g of an alloy that is 5.24% Cu. Report the answer to the correct number of significant figures.
30. Calculate the number of grams of nickel in 19.4 g of an alloy that is 0.750% nickel. Report the answer to the correct number of significant figures.
31. What is the percentage of silver in an alloy made from 10.22 g Ag and 1.3334 g Cu? Report the answer to the correct number of significant figures.
32. What is the percentage of copper in an alloy made from 14.99 g tin, 0.674 g copper, and 2.0 g zinc? Report the answer to the correct number of significant figures.
33. The laboratory of A. L. Chemist contained 1000 mL of a stock solution of ammonia. On successive experiments A. L. used 137.24 mL, 167.00 mL, 156 mL, and .0566 mL of this solution. How much stock solution was left after these experiments? Report the answer to the correct number of significant figures.
34. The laboratory of another scientist contained 250.00 mL of a stock solution of ammonia. On successive experiments this scientist used 0.0977 mL, 60.00 mL, 10.222 mL, and 150 mL of this solution. How much stock solution was left after these experiments? Report the answer to the correct number of significant figures.
35. Give two examples of exact numbers that do not appear in this chapter.
36. Give two examples of numbers that are *not* exact numbers.

## SECTION 3.4 Scientific Notation

37. Convert each of these numbers to scientific notation. Assume that any zeros between the decimal point and the nonzero portion of the entire number are not significant.
   a. 234,000,000,000 y
   b. $230.1\ s^2$
   c. 0.0034 g
   d. 0.000000000787 s
   e. 52,100,000 g
   f. $5240\ m^3$
   g. $0.02456\ g\ L^{-1}$
   h. $0.00000004\ s^2$
   i. 345,000,000 g
38. Convert each of these numbers to scientific notation. Assume that any zeros between the decimal point and the nonzero portion of the entire number are not significant.
   a. $100\ pm^3$
   b. 0.1000 mg
   c. 234.001 km
   d. 4,000,000,000,000 y
   e. $5,980,000,000\ ft^3$
   f. 0.000388 km
   g. $1000\ L^3$
   h. 51.11 mL
   i. 0.4 mg
   j. 2,387,000 pm
   k. 0.004001 m
39. Convert each of these to nonexponential notation.
   a. $2.3 \times 10^5\ g$
   b. $5.456 \times 10^2\ kg$
   c. $1.23 \times 10^{-4}\ L$
   d. $7 \times 10^{-12}\ m$
   e. $9.19 \times 10^7\ ft^3$
   f. $4.876 \times 10^1\ mg$
   g. $7.7 \times 10^{-6}\ g$
   h. $2 \times 10^{-17}\ kg$
   i. $4.0 \times 10^{-1}\ g\ mL^{-1}$
   j. $9.110 \times 10^2\ cm^3$
40. Convert each of these to nonexponential notation.
   a. $9.66 \times 10^{-3}\ g$
   b. $8.001 \times 10^3\ m^3$
   c. $6.991 \times 10^4\ m^2$
   d. $7.7371 \times 10^{-2}\ kg$
   e. $8.48 \times 10^{-4}\ m$
   f. $1.414 \times 10^{-2}\ s$
   g. $5.01 \times 10^{-7}\ g$
   h. $9.78 \times 10^5\ ft^3$
   i. $2.37 \times 10^7\ m^2$
   j. $1.11 \times 10^{12}\ s$

### TOOL BOX Working with Exponents

41. Perform each of these operations giving the answers to the correct number of significant figures.
   a. $(1.237 \times 10^{-17}\ m)(8.7 \times 10^5\ m^2)$
   b. $\frac{5.5 \times 10^{12}\ s}{8.992 \times 10^3\ s}$
   c. $(4.93 \times 10^{22}\ ft^2)(6.119 \times 10^3\ ft)$
   d. $\frac{6 \times 10^{-8}\ g^2}{1.559 \times 10^{-4}\ g}$
   e. $(1.88 \times 10^{-3}\ pm)(2.4 \times 10^{-9}\ pm^2)$
   f. $\frac{2.983 \times 10^6\ m^3}{9.983 \times 10^{-12}\ m}$
42. Perform each operation and give the answer to the correct number of significant figures.
   a. $\frac{9.44 \times 10^5\ cm^3}{6.7 \times 10^{-6}\ cm}$
   b. $(4.9 \times 10^6\ m^3)(2.12 \times 10^{-5}\ m)$
   c. $\frac{7.2359 \times 10^{-7}\ m}{4.73 \times 10^7\ s}$
   d. $(4.75 \times 10^3\ m^2)(3.56 \times 10^{-2}\ m)$
   e. $\frac{6.7066 \times 10^4\ lb}{3.1 \times 10^{15}\ in^2}$
   f. $(8.2 \times 10^{22}\ m^3)(9.166 \times 10^8\ m^3)$

43. Perform each operation and give the answer to the correct number of significant figures.
    a. $\dfrac{(3.556 \times 10^{-5}\ \text{s})(6.3000 \times 10^{-2}\ \text{s})}{2.6634 \times 10^{7}\ \text{s}^2}$
    b. $\dfrac{(3.792\ \text{g})(4.566\ \text{g})}{4.4\ \text{g} + 3.6557\ \text{g}}$
    c. $\dfrac{5.9800\ \text{m}^2 + 9.001\ \text{m}^2}{(2.63 \times 10^{2}\ \text{m})(3.1 \times 10^{-5}\ \text{m})}$
    d. $\dfrac{(5 \times 10^{5}\ \text{s}) + 3.55\ \text{s}}{5 \times 10^{5}\ \text{s}}$
    e. $\dfrac{(5.00 \times 10^{5}\ \text{m}^2) + (3.1 \times 10^{4}\ \text{m}^2)}{5 \times 10^{5}\ \text{m}}$
    f. $\dfrac{1.5111 \times 10^{8}\ \text{ft}^3}{(5.23 \times 10^{4}\ \text{ft}) - (2.3 \times 10^{3}\ \text{ft})}$

44. Perform each operation and give the answer to the correct number of significant figures.
    a. $\dfrac{(2.00 \times 10^{8}\ \text{m})(1.7378 \times 10^{-4}\ \text{m})}{8.337 \times 10^{-3}\ \text{m}}$
    b. $\dfrac{(1.0 \times 10^{33})(4.566)\text{m}^3}{2.7\ \text{m}^2 - 1.2310\ \text{m}^2}$
    c. $\dfrac{7.100\ \text{g} + 8.72\ \text{g}}{(1.912 \times 10^{4})(2.9899 \times 10^{-4}\ \text{mL})}$
    d. $\dfrac{6.1 \times 10^{5}\ \text{s}^2}{(6.1 \times 10^{5}\ \text{s}) + (6.233\ \text{s})}$
    e. $\dfrac{(1.2377 \times 10^{4}\ \text{g}) + (9.0 \times 10^{2}\ \text{g})}{8 \times 10^{-12}\ \text{g}}$
    f. $\dfrac{3.9 \times 10^{5}\ \text{m}^3}{(2.99 \times 10^{4}\ \text{m}^3) - (9.3 \times 10^{3}\ \text{m}^3)}$

## SECTION 3.5 Metric Prefixes and Base Units

45. Without sneaking a peek at Table 3.1, write down the factor by which the base unit is multiplied to perform each conversion.
    a. Meters to micrometers
    b. Meters to centimeters
    c. Meters to picometers

46. Write down the factor by which each base unit is multiplied to do the conversion.
    a. Grams to kilograms
    b. Grams to milligrams
    c. Bucks to megabucks

47. Fill in each blank.
    a. 1 micrometer = ______ meter
    b. 1 milligram = ______ gram
    c. 1 metric ton = ______ kilograms

48. Fill in each blank.
    a. 1 millimeter = ______ meter
    b. 1 microgram = ______ gram
    c. 1 cubic centimeter = ______ milliliter

49. Estimate the mass of each object in conveniently sized metric units.
    a. A penny
    b. A grapefruit
    c. This book
    d. Your instructor
    e. A mouse

50. Estimate the mass of each object in conveniently sized metric units.
    a. An automobile
    b. A grape
    c. A ball-point pen
    d. A glass of water
    e. An iron skillet

51. Estimate the length of each object in conveniently sized metric units.
    a. This book
    b. Yourself
    c. A pickup truck
    d. Your classroom
    e. A laboratory bench

52. Estimate the length or height of each object in conveniently sized metric units.
    a. A city block
    b. A house
    c. A fly
    d. A penny
    e. A coffee cup

53. Name an object that has a mass approximately equal to:
    a. Metric ton
    b. Kilogram
    c. Gram
    d. Milligram

54. Name an object typically found in a kitchen that has a mass approximately equal to:
    a. 500 g
    b. 10 kg
    c. 30 g
    d. 1 kg

55. Name an object that has a volume in the given range.
    a. 500–1000 L
    b. 40–50 L
    c. 0.5–1.0 L
    d. 1–5 mL

56. Name an object typically found in a kitchen that has a volume approximately equal to:
    a. 500 mL
    b. 2 L
    c. 15 mL
    d. 0.5 $\text{m}^3$

## SECTION 3.6 Temperature, Volume, and Density

57. Perform each of the following temperature conversions.
    a. 98.6°F to °C
    b. − 80°C to °F
    c. − 78°C to K
    d. 100 K to °C
    e. − 12°F to °C
    f. 56°C to °F
    g. − 221°C to K
    h. 77 K to °C
    i. 0°F to °C
    j. 273 K to °C

58. Perform each of the following temperature conversions.
    a. 100°C to K
    b. − 200°C to °F
    c. 80°F to °C
    d. 373 K to °C
    e. 1000°C to K
    f. 1000°C to °F
    g. 100°F to °C
    h. 100°C to K
    i. − 40°F to °C
    j. − 40°C to °F

59. Estimate the volume of each object in conveniently sized metric units.
    a. A pumpkin
    b. This book
    c. A beer mug
    d. An atom
    e. A sauce pan

60. Estimate the volume of each object in conveniently sized metric units.
    a. A large coffee pot
    b. A refrigerator
    c. A jar of mayonnaise
    d. A can of tuna
    e. A gallon of water

61. What is the density of a sample of liquid that has a mass of 12.3 g and a volume of 15.0 mL?

62. What is the density of a sample that has a mass of 51.66 g and a volume of 58.99 mL?

63. A sample of a liquid had a volume of 21.5 mL, measured in a graduated cylinder. The cylinder together with the liquid had a mass of 123.2 g. After the liquid was poured out, the mass of the cylinder was found to be 103.6 g. What is the density of the liquid?

64. A graduated cylinder has a mass of 122.5 g. After 26.1 mL of a liquid was added to it, the graduated cylinder together with the liquid had a mass of 145.8 g. What is the density of the liquid?

65. A sample of an alloy had a mass of 55.4 g. When this sample was submerged in water in a graduated cylinder, the water level rose from 12.0 mL to 16.9 mL. What is the density of this alloy?

66. A sample of a mineral has a volume of 86 mL and a mass of 540.5 g. What is the density of this mineral?

67. What observation suggests that ice is less dense than water?

68. What observation suggests that lead has a higher density than water?

69. All the substances listed in the following table are common inside the earth. Select three substances you would expect to find relatively close to the center of the earth, and justify your selections.

| Substance | Density (g $mL^{-1}$) |
|---|---|
| Basalt | 2.4–3.1 |
| Granite | 2.64–2.76 |
| Iron | 7.87 |
| Marble | 2.6–2.84 |
| Nickel | 8.90 |
| Olivine | 3.26–3.40 |
| Quartz | 2.65 |
| Sandstone | 2.12–2.36 |
| Talc | 2.8–2.8 |
| Water | 1.00 |

70. From the table in problem 69, select three substances you would expect to find relatively close to the earth's surface, and justify your selections.

## SECTION 3.7 Working with Units: Dimensional Analysis

71. A carver ordered a block of the mineral dolomite (density = 2.84 g $mL^{-1}$) that was 10 cm on a side. What is the mass of this block?

72. A collector had a cube of quartz (d = 2.65 g $mL^{-1}$) that was 15 cm on a side. What is the mass of this cube?

73. What is the mass of 100 mL of each of these substances? The densities in g $mL^{-1}$ are given in parentheses.
    a. Balsa wood (0.12)
    b. Ethanol (0.785)
    c. Water (1.00)
    d. Mercury (13.55)
    e. Aluminum (2.70)

74. What is the mass of 100 mL of each of these substances? The densities in g $mL^{-1}$ are given in parentheses.
    a. Gold (19.3)
    b. Lithium (0.534)
    c. Iron (7.9)
    d. Magnesium (1.74)
    e. Graphite (2.3)

75. An ice cube melted, giving 12 $cm^3$ of water. What is this volume in liters?

76. A goldsmith needed 0.022 L of liquid gold. What is this volume in milliliters?

77. The velocity of light is $3.00 \times 10^8$ m $s^{-1}$. What is its velocity in km $s^{-1}$?

78. The velocity of light is $3.00 \times 10^8$ m $s^{-1}$. What is its velocity in cm $s^{-1}$?

79. The diameter of a zinc atom is 270 pm. What is its diameter in meters?

80. The diameter of a copper atom is 240 pm. What is its diameter in meters?

81. How many seconds are there in a day?

82. How many seconds are there in a week?

83. A jeweler has $2.5 \times 10^5$ mg of gold. How many kilograms is this?

84. A chemist told a friend that they were $4.7 \times 10^9$ mm from home. How many kilometers is this?

85. Write the conversion factor (including units, as always) that can be used to convert:
    a. pm to m
    b. g to pg
    c. m to mm
    d. mg to g
    e. kg to g

86. Write the conversion factor (including units, as always) that can be used to convert:
    a. g to kg
    b. $\mu$m to m
    c. g to $\mu$g
    d. m to cm
    e. cm to m

87. Convert each of these quantities to the unit indicated. Use proper scientific notation and give the answer to the correct number of significant figures.
    a. 23,022 m to km
    b. 1299 g to kg
    c. 55 cm to m
    d. 278 mg to g
    e. 655.9 $\mu$g to g
    f. 455,701 km to m
    g. 250.0 mL to L
    h. 509 L to mL
    i. 345 pm to m
    j. 0.079 mg to g

88. Convert each of these quantities to the unit indicated. Use proper scientific notation and give the answer to the correct number of significant figures.
    a. 988 mm to m
    b. 7.21 Mm to m
    c. 98.66 kg to g
    d. 454 $\mu$g to g
    e. 98 m to pm
    f. 0.783 km to m
    g. 622.5 cm to m
    h. 75.0 m to cm
    i. 100.0 mg to g
    j. 0.0065 g to mg

89. Convert each quantity as indicated. Use proper scientific notation and give the answer to the correct number of significant figures.
    a. $1.23 \times 10^3$ pg to mg
    b. $9.30 \times 10^{-5}$ m to $\mu$m
    c. $3.00 \times 10^{10}$ cm to m
    d. $4.55 \times 10^{-6}$ kg to mg
    e. $4.67 \times 10^{12}$ pm to cm
    f. $4.6 \times 10^3$ mg to kg
    g. 155 mm to cm
    h. 155 cm to mm
    i. $8.00 \times 10^3$ $\mu$m to mm
    j. 927 mg to $\mu$g

90. Convert each quantity as indicated. Use proper scientific notation and give the answer to the correct number of significant figures.
    a. 5.00 mL to L
    b. $6.78 \times 10^{-2}$ L to mL
    c. $8.99 \times 10^{-5}$ kg to mg
    d. 981 mg to kg
    e. 6.52 kg to mg
    f. 74 $\mu$s to ms
    g. 1.45 mg to $\mu$g
    h. 56 Mg to kg
    i. 155 pm to mm
    j. $9.88 \times 10^{-6}$ mm to pm

91. Convert each of these quantities as indicated. Use proper scientific notation and give the answer to the correct number of significant figures.
    a. 14.6 mg $mL^{-1}$ to g $L^{-1}$
    b. 1.79 g $L^{-1}$ to g $mL^{-1}$
    c. $3 \times 10^{10}$ cm $s^{-1}$ to m $s^{-1}$
    d. 4.59 g $L^{-1}$ to kg $m^{-3}$
    e. 18.6 $\mu$g $mL^{-1}$ to g $L^{-1}$

92. Convert each of these quantities as indicated. Use proper scientific notation and give the answer to the correct number of significant figures.
    a. $1.72 \times 10^2$ mg $cm^{-3}$ to g $mL^{-1}$
    b. 46.00 mg $mL^{-1}$ to g $L^{-1}$
    c. 45 mm $s^{-1}$ to m $h^{-1}$
    d. $4.60 \times 10^2$ $\mu$g $\mu L^{-1}$ to mg $mL^{-1}$
    e. $3 \times 10^8$ m $s^{-1}$ to km $h^{-1}$

## SECTION 3.8 Converting English System and Metric System Units

93. Perform each of these conversions. Don't forget significant figures, even when your instructor or authors don't mention them.
    a. 77 kg to lb
    b. 77 in to cm
    c. 77 $in^2$ to $cm^2$
    d. 77 $in^3$ to $cm^3$
    e. 77 lb to kg
    f. 77 cm to in
    g. 77 $cm^2$ to $in^2$
    h. 77 $cm^3$ to $in^3$
    i. 77 $in^2$ to $m^2$

94. Perform each of these conversions. Don't forget significant figures, even when your instructor or authors don't mention them.
    a. 6.55 lb to kg
    b. 14 lb $in^{-2}$ to kg $m^{-2}$
    c. $6.57 \times 10^5$ $in^3$ to $m^3$
    d. 500 mi $h^{-1}$ to km $h^{-1}$
    e. 0.005 lb $L^{-1}$ to g $mL^{-1}$
    f. $2.0 \times 10^3$ km to mi
    g. 761 mm to in
    h. 33.00 ft to m
    i. 25 $ft^3$ to L
    j. 6.10 lb to kg

95. The speed of sound in air is 1116 ft $s^{-1}$. Express this quantity in each of the following units (to four significant figures):
    a. mi $h^{-1}$
    b. ft $min^{-1}$
    c. m $s^{-1}$
    d. cm $s^{-1}$

96. Under certain circumstances the density of nitrogen gas is 1.251 g $L^{-1}$. Express this quantity in each of the following units.
    a. g $mL^{-1}$
    b. lb $ft^{-3}$
    c. kg $L^{-1}$
    d. mg $L^{-1}$

97. A furlong is $\frac{1}{8}$ of a mile. How many meters is this?

98. The 100-yd dash is how long in meters?

99. The speed limit on a certain road is 35 mi $h^{-1}$. What is this speed in m $s^{-1}$?

100. The speed limit on another road is 55 km $h^{-1}$. What is this speed in m $h^{-1}$?

101. Sirloin steak is selling for $10.25 for 1 kg. How much is this per pound?

102. Bananas are selling for 79¢ for 500 g. How much is this per pound?

103. The velocity of light is $3.00 \times 10^8$ m $s^{-1}$. What is its velocity in m $h^{-1}$?

104. The velocity of light is $3.00 \times 10^8$ m $s^{-1}$. What is its velocity in mi $y^{-1}$? (In other words, how many miles are there in a light-year?)

## ADDITIONAL PROBLEMS

105. Scientists have determined that the radius of the earth is $3.96 \times 10^3$ mi and its mass is $5.97 \times 10^{24}$ kg. Perform each of the following calculations to the correct number of significant figures.
    a. What is the radius of the earth in meters?
    b. Assume that the earth is a perfect sphere. (This isn't true, but it is a reasonable approximation.) The equation for computing the volume of a sphere is $V = \frac{4}{3}\pi r^3$, where $V$ is the volume and $r$ is the radius. Compute the volume of the earth in cubic meters.
    c. Compute the density of the earth in g $cm^{-3}$.

106. Scientists have determined that the radius of the planet Mars is $2.11 \times 10^3$ mi and its mass is $6.42 \times 10^{23}$ kg. Perform each of the following calculations to the correct number of significant figures.
    a. What is the radius of Mars in meters?
    b. Assume that Mars is a perfect sphere. (Once again this is a reasonable approximation, though it is not exactly true.) The equation for computing the volume of a sphere is $V = \frac{4}{3}\pi r^3$, where $V$ is the volume and $r$ is the radius. Calculate the volume of Mars in cubic meters.
    c. Compute the density of Mars in g $cm^{-3}$.

107. How many significant figures does each of these numbers have?
    a. $2.03 \times 10^{23}$ molecules
    b. $1.100 \times 10^{-12}$ kg
    c. $6.770 \times 10^3$ lb
    d. $2.091 \times 10^{-8}$ kg
    e. $9.00 \times 10^{-3}$ g

108. How many significant figures does each of these numbers have?
    a. $1.400 \times 10^5$ m
    b. $3.00 \times 10^{-2}$ L
    c. $3.07 \times 10^3$ L
    d. $2.00001 \times 10^6$ m
    e. $4.8 \times 10^4$ g

109. Compute the mass of each metal present in 5.0 kg of each of these alloys.
    a. Gun metal: 10% tin, 90% copper (once used to make cannon)
    b. Aluminum bronze: 3.5% aluminum, the rest is copper
    c. German silver: 20% zinc, 25% nickel, and 55% copper (formerly used to make resistance coils)
    d. Bell metal: 20% zinc, the rest is copper (once used to make bells)
    e. Brass: 27% zinc, the rest is copper

110. Compute the mass of each metal present in 5.0 kg of each of these alloys.
    a. Bronze: 71% copper, 12% zinc, the rest is tin (one of many recipes for bronze)
    b. Bronze: 82% copper, 12% zinc, the rest is tin (another bronze recipe)
    c. Gold coin: 90% gold, the rest is copper (very common in gold coins used in the nineteenth century)
    d. Common brass: 33% zinc, 67% copper
    e. British copper coins: 95% copper, 4% tin, 1% zinc (used in the nineteenth century)

111. Name five objects that are about 15 cm in length.

112. Name five objects that are about 1 m in length.

113. Name five objects that have a mass of about 200 g.

114. Name five objects that have a mass of about 1 kg.

## EXPERIMENTAL PROBLEM

### 3.1

A. L. Chemist headed a scientific team that measured the percentage of tin in an unknown sample. The results obtained from four separate experiments were 24.67%, 24.67%, 24.64%, and 26.44%. There are two ways to report this type of experimental results. (Theoretically, they are equally valid.) The most familiar method is to take the average value of all results, which in this case is 25.11%. Another possible method is to report the *median* result—the number in the middle when the results are listed from highest to lowest. The median here is 24.67%. What percentage do you think A. L. Chemist should report for the tin in this sample? Explain your answer. (Chemists seldom use either of these methods. Instead they use a rigorous mathematical approach—an error analysis—to compute the reliability of their data.)

## SOLUTIONS TO EXERCISES

**Exercise 3.1**

**a.** Precise measurements are close together. Measurements that are not accurate have an average value far from the center. Person 2's results best fit this description.
**b.** The average of accurate measurements is close to the true value. Precise measurements are close together. Person 3's results best fit this description.

**Exercise 3.2**

**a.** One significant figure—none of the zeros are significant, because they are all to the left of the nonzero digit (6).
**b.** Five significant figures—all of the zeros lie between two nonzero digits.
**c.** Six significant figures—trailing zeros in a number with a decimal point are significant.
**d.** Four significant figures—the first three zeros are to the left of all the nonzero digits and are not significant, the next zero is between nonzero digits and is significant, and the last one is a trailing zero in a number with a decimal point and is significant.

**Exercise 3.3**

**a.** 0.02115 $in^2$ rounds off to 0.02 $in^2$ (one significant figure)
**b.** 22.046908 rounds off to 22.0 (three significant figures)
**c.** 43.49 lb rounds off to 43.5 lb (one decimal place in the answer)
**d.** 1.61 lb rounds off to 1.6 lb (one decimal place in the answer)

**Exercise 3.4**

**a.** $2.3034 \times 10^2$ grams (Remember to include the units.)
**b.** $2.001 \times 10^3$ kilograms
**c.** $6.6 \times 10^{-3}$ meters
**d.** $5.550 \times 10^{-6}$ meters (Be careful: the last zero is significant.)

**Exercise 3.5**

**a.** 40,000,000 meters
**b.** 0.0000004 grams

**TOOL BOX Exercise 1**

**a.** $9.9 \times 10^4\ ft^2 = 10 \times 10^4\ ft^2$ (after rounding off) $= 1 \times 10^1 \times 10^4 = 1 \times 10^5\ ft^2$ (with one significant figure)
**b.** $23.6886 \times 10^3\ lb^2 = 2.4 \times 10^1\ lb^2 \times 10^3\ lb^2 = 2.4 \times 10^4\ lb^2$ (two significant figures)
**c.** $31.92 \times 10^0\ lb^2 = 3.192 \times 10^1\ lb^2 = 3 \times 10^1\ lb^2$ (one significant figure)

**TOOL BOX Exercise 2**

**a.** $0.5609756 \times 10^{(0-6)}\ lb/in^2 = 0.56 \times 10^{-6}\ lb/in^2 = 5.6 \times 10^{-1} \times 10^{-6}\ lb/in^2 = 5.6 \times 10^{-7}\ lb/in^2$
**b.** $0.50 \times 10^{[3-(-6)]} = 0.50 \times 10^9 = 5.0 \times 10^{-1} \times 10^9 = 5.0 \times 10^8$
**c.** $2.00 \times 10^{(-2-10)}$ lb/gallon $= 2.00 \times 10^{-12}$ lb/gallon

**Exercise 3.6**

**a.** $10^{-6}$ liter
**b.** $10^{-12}$ meter
**c.** $10^3$ grams
**d.** $10^6$ tons

**Exercise 3.7**

**a.** 2 m: width of an automobile
**b.** 15 cm: length of a pencil
**c.** 10 mm: length of a fly
**d.** 0.3 km: height of a skyscraper

**Exercise 3.8**

**a.** A nickel: about 5 g
**b.** A fly: 10 to 20 mg
**c.** A pail of water: about 5 kg
**d.** A speck of dust: 20 $\mu$g

**Exercise 3.9**

**a.** $°F = (1.8)(°C) + 32° = 25°$
$(1.8)(°C) = 25° - 32° = -7$
$°C = \frac{-7°}{1.8} = -3.9° \simeq -4°C$
If the Fahrenheit temperature is below 32°F, the Celsius temperature is negative. (Can you state why this is true?)
**b.** $°F = (1.8)(°C) + 32° = (1.8)(25°) + 32° = 45° + 32° = 77°$

**Exercise 3.10**

**a.** $K = °C + 273.15 = 100 + 273.15 = 373.15$; so the answer is 373 K.
**b.** $200 = °C + 273.15$
$200 - 273.15 = °C = -73.15$, or $-73°C$ to the correct number of significant figures

**Exercise 3.11**

**a.** $d = \frac{m}{V} = \frac{89.3\ \text{g}}{12.2\ \text{mL}} = 7.32\ \text{g mL}^{-1}$
**b.** Here you know the volume, 12.1 mL, and must get the mass. The mass of the liquid is the difference between the mass of the cylinder when it's filled and the mass when it's empty. The mass of the liquid is
$35.1\ \text{g} - 25.1\ \text{g} = 10.0\ \text{g}$
Then $d = \frac{m}{V} = \frac{10.0\ \text{g}}{12.1\ \text{mL}} = 0.826\ \text{g mL}^{-1}$

**Exercise 3.12**

**a.** This is a one-stage problem. You convert milliliters to liters using 1 mL = 0.001 L.

$$\frac{1.8 \times 10^5\ \cancel{\text{mL}}}{1} \times \frac{0.001\ \text{L}}{1\ \cancel{\text{mL}}} = 1.8 \times 10^2\ \text{L}$$

**b.** Here you have to use two conversion factors. First, you convert centimeters to meters and then meters to kilometers. The equation is

$$\frac{1.7 \times 10^5\ \cancel{\text{cm}}}{1} \times \frac{0.01\ \cancel{\text{m}}}{1\ \cancel{\text{cm}}} \times \frac{1\ \text{km}}{1000\ \cancel{\text{m}}} = 1.7\ \text{km}$$

**c.** This mass is calculated the same way as the volume of the wood was calculated in part (c) of Example 3.12. Here you must also recognize that 1 cc = 1 mL.

$$\frac{30\ \cancel{\text{cc}}}{1} \times \frac{1\ \cancel{\text{mL}}}{1\ \cancel{\text{cc}}} \times \frac{19.31\ \text{g}}{1\ \cancel{\text{mL}}} = 5.8 \times 10^2\ \text{g}$$

**d.** Here the conversion factor from centimeters to meters must be used twice:

$$\frac{550\ \cancel{\text{cm}^2}}{1} \times \frac{0.01\ \text{m}}{1\ \cancel{\text{cm}}} \times \frac{0.01\ \text{m}}{1\ \cancel{\text{cm}}} = 0.055\ \text{m}^2$$

$$= 5.5 \times 10^{-2}\ \text{m}^2$$

### Exercise 3.13

The only conversion factor in Table 3.3 for converting metric distances to English is 1 in = 2.54 cm. Your strategy should be as follows: convert 29,028 ft to inches, inches to centimeters, and centimeters to meters (ft → in → cm → m). The equation is

$$\frac{29{,}028\ \cancel{\text{ft}}}{1} \times \frac{12\ \cancel{\text{in}}}{1\ \cancel{\text{ft}}} \times \frac{2.54\ \cancel{\text{cm}}}{1\ \cancel{\text{in}}} \times \frac{0.01\ \text{m}}{1\ \text{cm}}$$

$$= 8847.7344\ \text{m, which is rounded to } 8847.7\ \text{m}$$

The numbers in all of the conversion factors you used are exact, so they don't have any effect on the number of significant figures. Since 29,028 has five significant figures, your answer should, too: Mt. Everest is 8847.7 m high. (Yes, that 8848-m mountain mentioned earlier in the section was a big one!)

### EXPERIMENTAL EXERCISE 3.1

1. The total volume is the amount of water used (10.00 mL) plus the volume occupied by 15.5 g of lead, which would be

$$\frac{15.5\ \cancel{\text{g Pb}}}{1} \times \frac{1\ \text{mL}}{11.3\ \cancel{\text{g Pb}}} = 1.37\ \text{mL}$$

   Therefore the total volume would have been 10.00 mL + 1.37 mL = 11.37 mL.
2. The key operation is subtraction, and both numbers are known to two digits after the decimal point. Thus the answer must also be reported to two digits past the decimal point and is given correctly.
3. Both the mass and the volume are known to three significant figures. They key operation is division, so the density must also be reported to three significant figures. The answer was reported incorrectly and should be $8.96\ \text{g mL}^{-1}$.
4. Yes, even a calculated density reported to one significant figure is enough to allow you to distinguish between these two metals.

4

# MASS, ENERGY, FORCES, AND ATOMS



## KEY TERMS

Chemistry is the study of matter and the changes it undergoes. When a sample of matter changes, there is almost always a transfer of energy to or from the sample. Forces give the push that makes the changes happen.

Energy in the form of light from the sun pours down on the surface of the earth in vast quantities. Green plants and algae on land and in the sea absorb some of this energy and use it to convert carbon dioxide and water into more complicated molecules such as sugars and starches. These compounds are the fuel of living organisms, the stores of energy we all use for movement and other changes. Later, the plants (and the animals that eat the plants) combine the starches and sugars with oxygen from the air, releasing the stored energy to support other life processes. This sequence of storing and releasing energy from the sun supports almost all life on earth. (A few minor energy sources, primarily the earth's internal radioactivity, also play a part.)

Of course, the changes that matter undergoes do not have to be on a scale this grand. Whether you like your eggs scrambled or fried, you supply energy (in the form of heat) to raw eggs in a skillet to cause the eggs to change. In this chapter we will consider the interactions of matter, energy, and forces in more detail. In particular, we will study these interactions in atoms and discuss how scientists learned about atomic structure.

## AN INTRODUCTION TO ENERGY

Energy is in the news all the time, especially when gasoline and fuel oil prices go up or when there is a crisis in a part of the world that produces petroleum. Informed citizens in every country worry about the cost of energy, conserving energy resources, and using energy wisely.

You have some idea of what energy is from your everyday experiences. It takes gasoline to move your automobile, electrical current to run your flashlight, natural gas or electricity to operate your oven, and food to keep your body going. Gasoline, batteries, natural gas, and food are sources of energy. Chemical reactions release this energy in a form you can use.

Like many other words, the word *energy* has different popular and scientific meanings. The differences this time, however, are not profound.

FIGURE 4.1

## 4.1 Energy and Its Forms

GOAL: To define and give examples of kinetic and potential energy.

**Energy** is defined as the ability to do work. To a scientist, this definition means that the energy of anything is related to its ability to move an object some distance. The energy sources that are topics on the evening news—coal, oil, natural gas, nuclear power, wind—are important for their ability to move things around. Exactly what moves depends on the particular application. Some movement, such as car and truck traffic, is obvious. Some is less clear. Refrigerators, for example, operate by using a *compressor* to move gas molecules through their cooling coils.

**Energy moves not only industrialized society but also other societies. Computers, planes and cars and trains, and a horse-drawn plow are all moved by energy (see Figure 4.1). Understanding how scientists think about energy is an important part of understanding the world around you.**

Let's take a closer look at the implications of the concept of doing work. A gallon of gasoline contains energy that you can use somehow. This must be true, because you use gasoline to make your car move. In other words, you can do work with gasoline. A moving car can also do work. If it hits a wooden fence, it will send pieces of wood flying all over the place. Since the automobile did work (not constructive work, but work nonetheless), it must have energy. Is the energy of the gasoline the same as the energy of the moving auto? The answer to this question seems to be no, suggesting that energy can take more than one form.

Let's start thinking about the different forms of energy by taking an imaginary ride on a roller coaster. The coaster's machinery pulls the cars up the tracks to the top of the first hill. Then the cars start to roll down an incline. You start slowly but go faster and faster as you move farther down the slope. When you reach the bottom, you're going fast. As you climb up the next hill, the car slows until it crosses the next summit, which is lower than the first. You start down again, and you feel the rush of wind in your face getting stronger and hear the rumble of the wheels getting louder. The car goes around curves and through loops, but if it's a true roller coaster (with no engine to drive the cars), you will never again go as high as you did at first. Finally you come to a stop.

FIGURE 4.2
Note that the highest hill this roller coaster climbs is near the beginning of the ride.

Here is how a scientist looks at what happens to the energy of the roller coaster as you go through the ride. First, the machinery that pulls you up to the high point supplies the energy that moves you from start to finish. At the top of the first hill, almost all the energy for the ride is in the form of what is called **potential energy.**

**It may not be obvious, but these two ways to view potential energy are really equivalent (that is, they are two ways of saying the same thing). Sometimes it is easier to view potential energy as stored energy (in a battery or a gallon of gasoline, for example), and sometimes it is easier to view it as energy of position (as in the roller coaster car).**

There are two ways to think about potential energy. First, it is the energy an object possesses *by virtue of its position in relation to another object that exerts a force on it.* For the roller coaster the potential energy is provided by the car's position above the earth. The car's elevation gives it the ability to move downward because of the earth's force of gravity. A second, perhaps more general, way to view potential energy is that it is *energy stored by a system.* In the roller coaster the mechanism added potential energy to the cars by moving them from the starting point to the summit of the first hill. The cars are barely moving at the summit, and most of the added energy is stored in the cars. This added energy is potential energy.

**Scientists use the word *velocity* instead of *speed* because it is more precise. Speed is simply rate of progress, while velocity is rate of progress in a given direction.**

As the roller coaster car starts down the slope, you move faster and faster. The potential energy you had at the top of the hill changes into *energy of motion,* or **kinetic energy.** As you are gaining speed (and your kinetic energy is increasing), you are moving down to a lower position (so your potential energy is decreasing). The more massive an object is, the more kinetic energy it has at any given velocity. An 80-kg adult has twice as much kinetic energy as a 40-kg child traveling at the same speed. Kinetic energy also increases as objects go faster, and it increases at a much faster rate than the velocity increases. Both adult and child have four times as much energy at 20 m $s^{-1}$ than they have at 10 m $s^{-1}$.

The equation that relates kinetic energy, mass, and velocity for a given object is

**Kinetic energy equation**

$$\mathbf{KE} = \tfrac{1}{2}\boldsymbol{mv}^2 \qquad \text{Equation 4.1}$$

where KE is the kinetic energy of the object, $m$ is its mass, and $v$ is its velocity. Scientists almost always use SI units in this equation. Thus the mass is in kilograms (kg) and the velocity (a derived unit) is in meters per second (m $s^{-1}$), giving KE units of kg $m^2$ $s^{-2}$—called the **joule** (J):

$$1\ \text{J} = 1\ \text{kg m}^2\ \text{s}^{-2}$$

Scientists often use the joule, and we will use it extensively in this book (in conjunction with the usual metric system prefixes). The joule is popular among scientists primarily because it is defined in SI units, which makes calculations easier to do. Such a calculation is illustrated in the next example.

## EXAMPLE 4.1 Calculating the Kinetic Energy

What is the kinetic energy of an 80-kg adult traveling at 10 m $s^{-1}$? (Don't forget significant figures)

**SOLUTION**

Use Equation 4.1:

$$\text{KE} = \tfrac{1}{2}mv^2 = \tfrac{1}{2}(80\ \text{kg})(10\ \text{m s}^{-1})^2$$

$$= 4.0 \times 10^3\ \text{kg m}^2\ \text{s}^{-2} = 4.0 \times 10^3\ \text{J} = 4.0\ \text{kJ}$$

The answer is expressed in *kilojoules:* 1 kJ = $1 \times 10^3$ J.

**EXERCISE 4.1**

What is the kinetic energy of an 80-kg adult traveling at 20 m $s^{-1}$?

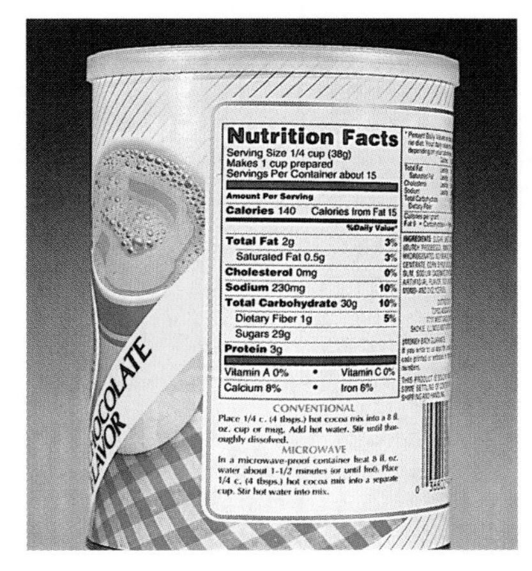


FIGURE 4.3
The "energy value" of food is typically reported in units of Calories: 1 Cal = 1 kcal.

Other units of energy are used in many different fields. A **calorie** (abbreviated cal) is the amount of heat it takes to raise the temperature of 1 gram of water by 1 degree Celsius. To convert calories to joules you use the relationship 1 cal ≡ 4.184 J. When a package of food gives the energy content in terms of *Calories* (with a capital C), it's referring to kilocalories—the amount of heat it takes to raise the temperature of 1 kilogram of water 1 degree Celsius. Many people confuse calories with Calories—don't be one of them.

Engineers sometimes use BTUs (British Thermal Units) to define energy. A **BTU** is $\frac{1}{180}$ the heat needed to raise the temperature of 1 pound of water from 0°C (32°F) to 100°C (212°F). One BTU is equal to 1055.8 J. The BTU is often used to describe the energy requirements of heating and air-conditioning equipment.

The **kilowatt-hour** is another energy unit. The utility company uses this unit to charge you for the electrical energy you use. The rate at which energy is produced or used is called **power,** and the SI unit of power is the **watt:** 1 watt (W) is defined as 1 joule per second (J $s^{-1}$). A kilowatt-hour, then, is 1000 watt-hours of energy use. This is equal to $3.6 \times 10^6$ J.

FIGURE 4.4

**Summary of Common Energy Units**

| Unit | Abbreviation | Equivalent in Joules |
|---|---|---|
| calorie | cal | 1 cal = 4.184 J |
| Calorie | Cal | 1 Cal = 4184 J = 4.184 kJ |
| British Thermal Unit | BTU | 1 BTU = 1055.8 J = 1.0558 kJ |

FIGURE 4.5

Now that we've gone over all that, think some more about the roller coaster ride. Suppose you are moving past the bottom of the first slope and traveling very rapidly. This means that you have a lot of kinetic energy. As you go up the second slope, your car slows down. The kinetic energy you had at the bottom of the slope changes back to potential energy as you go upward. The second high point in the roller coaster can't be as high as the first, because some of the energy you started with has left you.

A moment's reflection should tell you that you *cannot* have all the energy you started with. It takes energy to push the air aside, to produce the rumbling of the wheels, and to vibrate the structure. There is friction between the wheels and the tracks, and this slows you down and heats up the tracks and the surrounding air.

The "missing" energy doesn't disappear from the universe. Like potential and kinetic energy, it changes form. What happens as the roller coaster car plows through the air is pretty much the same thing that happens when a

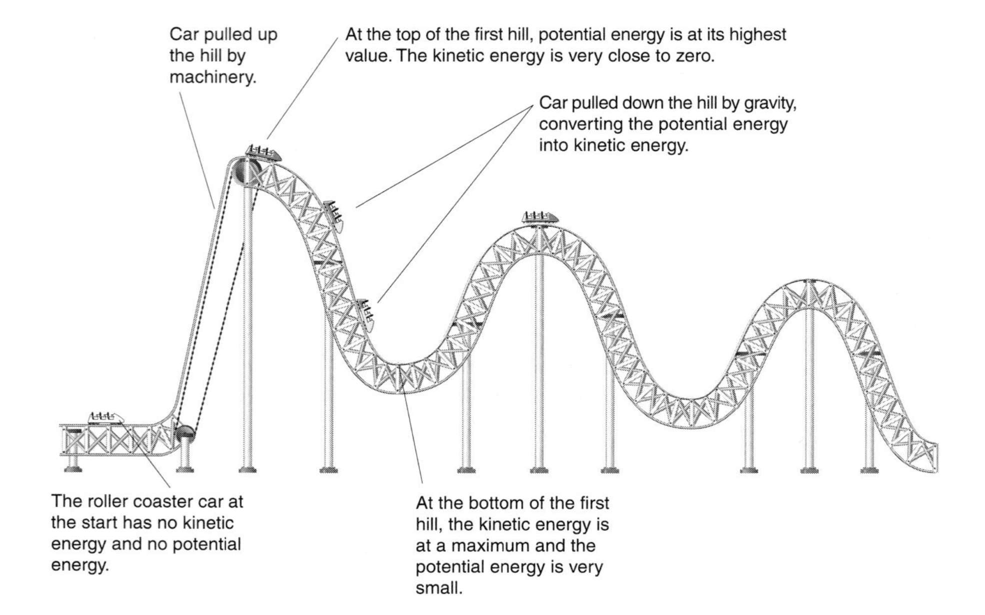
Car pulled up the hill by machinery.
At the top of the first hill, potential energy is at its highest value. The kinetic energy is very close to zero.
Car pulled down the hill by gravity, converting the potential energy into kinetic energy.
The roller coaster car at the start has no kinetic energy and no potential energy.
At the bottom of the first hill, the kinetic energy is at a maximum and the potential energy is very small.

FIGURE 4.6
Energy and a roller coaster ride.

**To demonstrate for yourself how much heat can be generated by friction, place your hands together and rub them very quickly. You will feel your hands warming up. If this doesn't convince you, feel the tires on your automobile after a trip at high speed.**

moving automobile smacks into a wooden fence. The roller coaster car collides with air molecules and sends them flying in all directions.

The air molecules gain kinetic energy because the collision increases their velocity. (Remember Equation 4.1, in which kinetic energy increases as the square of the velocity increases.) As the kinetic energy of the air molecules increases, the kinetic energy of the roller coaster car decreases. As the car loses kinetic energy, it slows down. In formal terms, energy transfers from the car to the air; speaking less formally, the air heats up. **Heat** is the kinetic energy of moving molecules. Near the end of the ride most of the potential energy the car had at the top of the first summit has become heat.

The brakes at the end of the ride finish the process—when they are applied, they absorb any kinetic energy remaining in the roller coaster car, and heat up as they do so. All of the potential energy the roller coaster car had at the top of the first slope has become heat at the end of the ride.

## TOOL BOX: READING GRAPHS

Scientists frequently present experimental data and results in the form of graphs. Graphs serve several useful purposes:

1. They show trends in the data visually ($y$ increases as $x$ increases, or $y$ increases as $x$ decreases, etc.). The eye can discern trends from graphs much more easily than it can from tables of numbers.
2. They make it possible for you to predict results of experiments not yet done (and frequently of experiments you have no intention of doing).
3. They can be used to find the quantitative relationships between variables. That is, they can be used to write an equation relating the variables to each other.

4. They allow scientists to share large amounts of data with interested readers in a concise way.

Reading and interpreting graphs is an important skill. A very common mistake that beginning students make when they study is to ignore anything that is not text. Most books on science, from beginning to very advanced texts, contain many graphs. Graphs give information in a concise, clear way. To ignore graphs when you study science is to miss the point.

The thing to keep in mind about reading graphs is that they were created to convey information to you. You must pay close attention to the labels of the axes and the magnitudes and units on them. In Graph 1, note that the units of velocity are cm $s^{-1}$ and the unit for time is s. The graduations on the velocity axis (the vertical axis) are 20 cm $s^{-1}$, and on the time axis (the horizontal axis) they are 1 s.

**All measured numbers have units. You should find units on every graph you see.**

Graphs 1 and 2 show experimental data as dots. (The data points could be crosses, triangles, circles, etc.) Note that not all the dots are directly on the line drawn in Graph 2. The line in a graph is typically the best representation of the *trend* in the data. Scientists seldom, if ever, play the game "connect the dots."

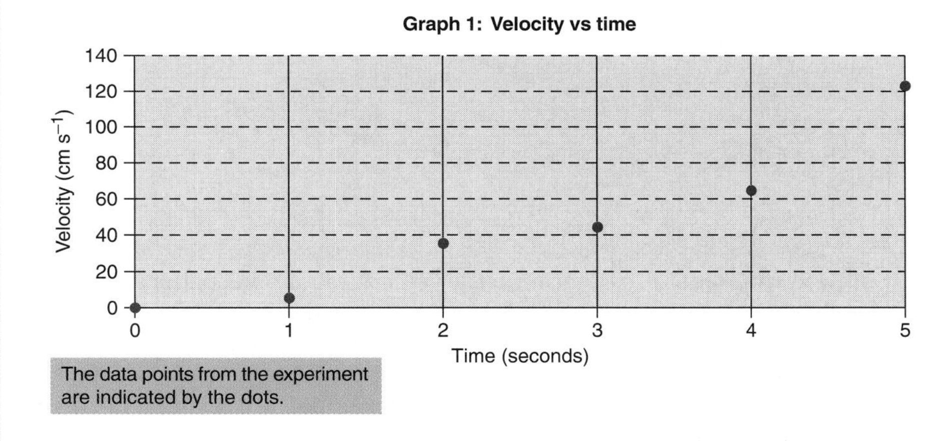


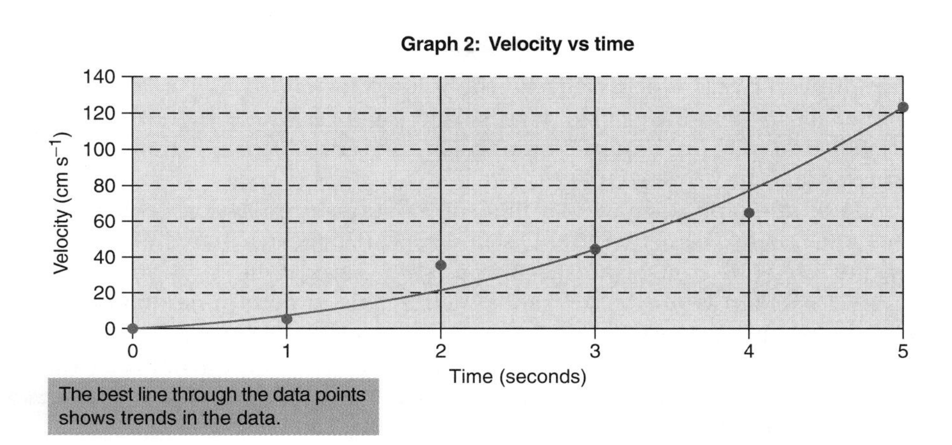

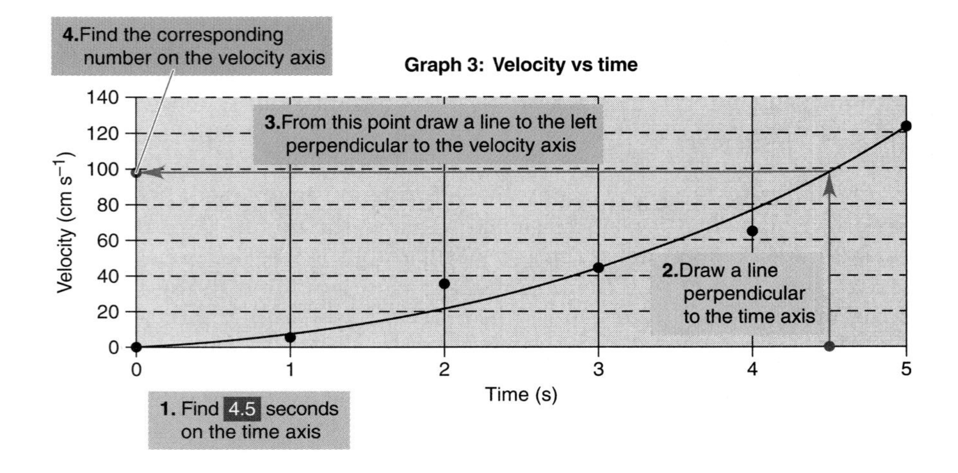


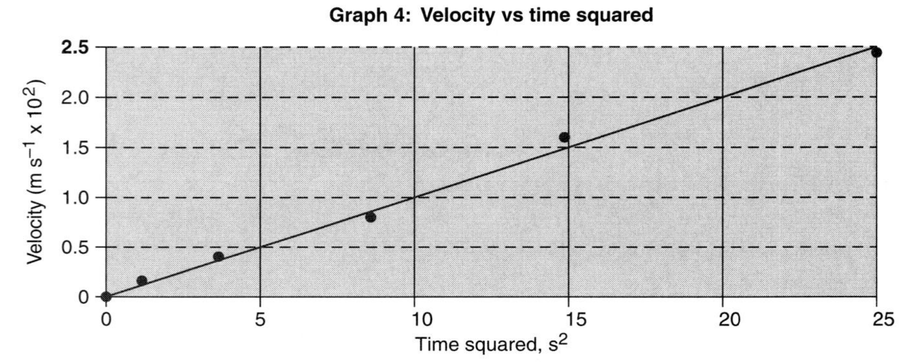


Suppose you wanted to determine the velocity of an object after 4.5 s. You first find 4.5 on the time axis. (It's halfway between the points marked 4 and 5.) Next, draw a vertical line from the time axis up to the curving line. The point you want is the point where your vertical line intersects this curve. Finally, you find the velocity corresponding to this point by drawing a horizontal line to the left until it reaches the velocity axis. Graph 3 illustrates these steps. You can see that the velocity after 4.5 s is about 99 cm $s^{-1}$.

Some care must be taken in reporting the numbers on graphs when exponents are involved. Graph 4 plots the velocity of an object against the square of the time. Note that the vertical axis is labeled "Velocity (m $s^{-1} \times 10^2$)." This means that each set of digits on the velocity axis must be multiplied by $10^2$ when you read the velocity. Scientists use units with exponents in order to have conveniently sized numbers with which to label the axis. For example, the boldface number on the vertical axis in Graph 4 is 2.5; it corresponds to $2.5 \times 10^2$ m $s^{-1}$.

A look at Graph 4 shows you that the relationship between velocity and time squared in this experiment is linear, which means that the plot is a straight line. (Of course, not all graphs show straight lines, as you saw in Graph 2.) Scientists often use graphs such as this to determine quantitative relationships between variables.

#### TOOL BOX EXERCISE 1

Using Graph 4, determine the velocity of the object at 4.5 s.

## 4.2 The First and Second Laws of Thermodynamics

**GOAL:** To understand that (1) energy can change from one form to another but can never be created or destroyed and (2) energy can never change from less useful to more useful forms.

The story of a roller coaster ride in Section 4.1 embodies two rules. To scientists these are ideas so fundamental that they have achieved the status of "laws of nature." Both of them have been tested by experiment thousands of times and have passed every one of those tests.

**Recall what a law is from Chapter 1.**

As the roller coaster car rolled along its rails, the energy in the system changed from potential to kinetic and back again, and from kinetic energy to heat energy. Energy changed from one form to another but was neither created nor destroyed. This is the **First Law of Thermodynamics,** which states that energy can change from one form into another but can *never* be created or destroyed. A shorter way of saying this is to say that energy is *conserved.*

The second of the two laws in the roller coaster story is less obvious. Potential and kinetic energy changed from one to the other and back again, but once energy was transformed to sound or heat, it never changed back into potential or kinetic energy to help move the cars along. One way of stating the law underlying this observation is to say that energy may be changed from one useful form to another or from more useful forms to less useful forms, but never *(never!)* from less useful to more useful forms. This is a statement of the **Second Law of Thermodynamics,** which is one of the more frustrating laws of nature.

**The *perpetual motion machine* is a classic tool of swindlers and con artists, though sometimes honest people fool themselves into believing they have invented one that works. Such machines supposedly either create energy out of nothing or use the same energy to do work over and over again. This is thermodynamically quite impossible.**

Unfortunately, this law is easy to misinterpret, because what is defined as "useful" can change—the heat in the roller coaster tracks is not useful for anything, but heat can be very useful on a cold day. There are several other ways to state the Second Law, but as the statements grow more precise, they also become harder to interpret in ordinary terms. Here are two more statements of the Second Law: (1) Heat flows from warmer objects to colder objects, but never the reverse. (2) Isolated systems become more disordered as time passes.

FIGURE 4.7
It takes energy and effort to create order (as shown in the photo on the left), but disorder can be made from order relatively easily (as shown in the photo on the right). This is another way of stating the Second Law of Thermodynamics.

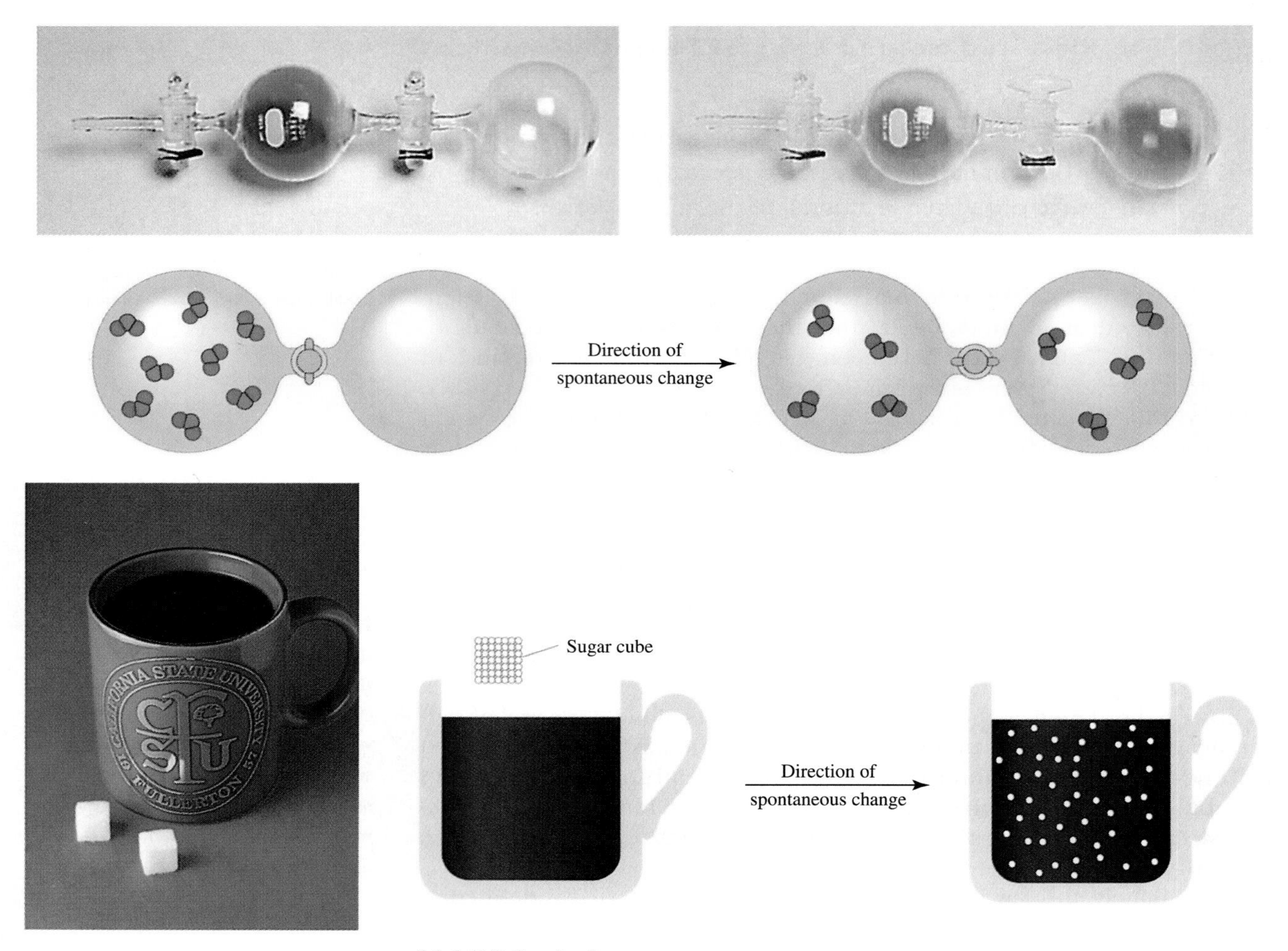


FIGURE 4.8
These two experiments illustrate the Second Law of Thermodynamics. Above when the valve between the two bulbs opens, the nitrogen dioxide gas in the full bulb expands into the empty bulb. The gas will never flow back to a single bulb, any more than heat from a pair of objects in contact will flow to one of them. Below, the sugar cube dissolves, but will never reconstruct itself from the coffee.

### EXAMPLE 4.2 The Laws of Thermodynamics

Show where the two laws of thermodynamics were assumed in the account of the roller coaster ride given in Section 4.1.

**SOLUTION**

The account describes energy as being converted from one form into another (potential to kinetic and back again, and kinetic to heat) but not as being either created or destroyed. This is the first law. Kinetic and potential energy become heat energy, but the reverse never occurs. The heat dissipates into the environment and cannot be used to do anything useful. This is the second law.

**EXERCISE 4.2**

A driver steps on the gas, and a car accelerates from a complete stop. Then the driver steps on the brakes, and the car comes to a complete stop again. Describe what happens to the potential energy of the fuel during these events.

## EXAMPLE 4.3 Reading a Graph of Energy Versus Distance

The graph in Figure 4.9 shows the amounts of three different types of energy contained in roller coaster cars during a ride. (a) How much potential energy, kinetic energy, and heat do the cars have when they have traveled 150 m? What is their total energy? (b) What is happening to the cars at this point in the ride?

**SOLUTIONS**

**a.** The potential energy is represented by the distance from the mark for 150 m at the bottom of the graph to the curvy line; this is approximately 120 kJ. The kinetic energy is the distance from the curvy line (at 120 kJ) to the sloping straight line above it (at 410 kJ), or about 290 kJ. The heat is the distance from the sloping line to the horizontal line at the top of the graph, or approximately 40 kJ. The total energy is represented by this top line, at 450 kJ.

**b.** The kinetic energy is large but has just begun to decrease at 150 m, and potential energy is small but has just begun to increase. This is just past the first low point of the roller coaster ride.

**EXERCISE 4.3**

Use the graph in Figure 4.9 to answer the following questions. (a) How much potential energy, kinetic energy, and heat do the cars have when they have traveled 450 m? What is their total energy? (b) What is happening at this point in the ride? (c) How far into the ride are the brakes applied?

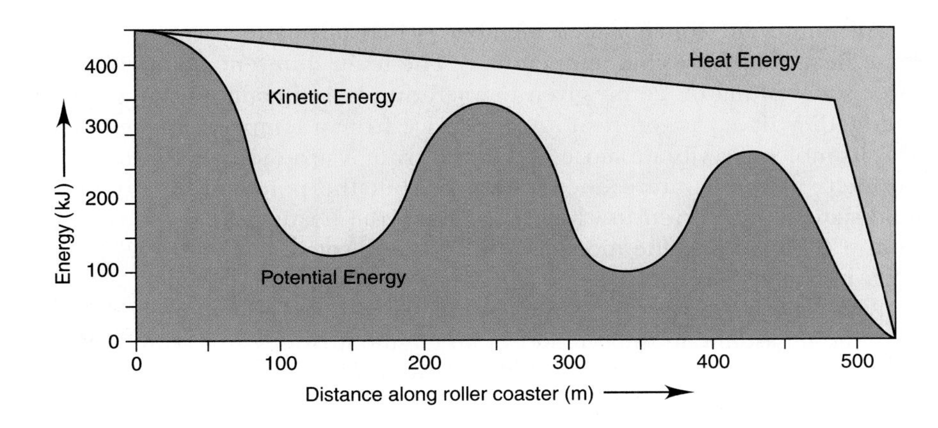


FIGURE 4.9

FIGURE 4.10
Dams like this are used to produce hydroelectric power. (*Power* is defined as energy per unit time.)

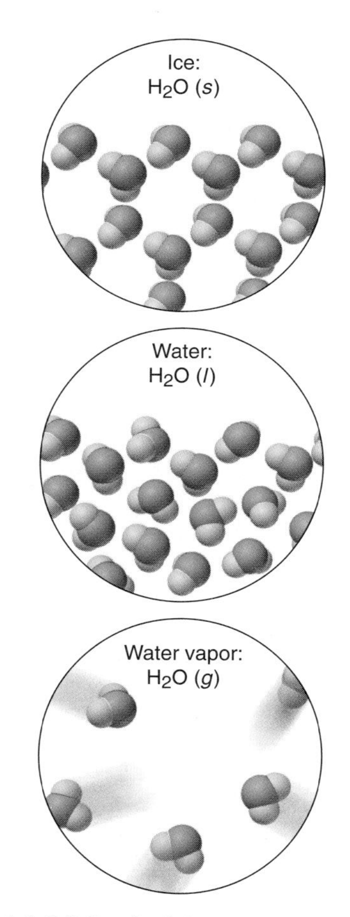


FIGURE 4.11
The molecules in solid water (ice) have relatively low kinetic energies. The molecules in warmer liquid water have more kinetic energy than those in ice. The molecules in hot water vapor have relatively large kinetic energies.

## 4.3 Potential and Kinetic Energy and Their Forms

GOAL: To give examples of potential and kinetic energy.

In Section 4.1 you encountered the two major types of energy: potential and kinetic. Potential energy is the energy an object has because of its position. A roller coaster car at the top of a slope has potential energy. Water behind a high dam has potential energy, too. The water can be released to pass through a powerhouse. Energy of position (the water is above the river level downstream) is converted into kinetic energy (the water moves rapidly, spinning a turbine) and then into electrical energy. Hydroelectric dams produce a large amount of electrical energy for the western United States.

As you have seen, a gallon of gasoline also has potential energy. The gasoline burns in an engine that in turn moves a car or cuts the grass. In other words, the engine converts the potential energy of the gasoline (the energy stored in the gasoline) into physical work. Remember that energy is defined as the ability to do work. The potential energy that fossil fuels of all kinds have is frequently referred to as **chemical energy.** (The fuels used in nuclear power plants have another form of potential energy called nuclear energy; see Chapter 17.)

Kinetic energy is the energy of motion, as you already know. The moving roller coaster car at the bottom of the slope, the moving water in the powerhouse of a dam, and the spinning wheels inside the mechanism of the powerhouse all have kinetic energy and are therefore capable of doing work. Sound is a form of kinetic energy in which molecules vibrate back and forth in an organized way. As you listen to your instructor speak, your ears are detecting the movement of air molecules that are oscillating back and forth. These vibrations of the atmosphere do work by striking your eardrum, causing this membrane to vibrate in turn. These vibrations are ultimately decoded by the brain as sound.

Heat is also a form of kinetic energy. The sensation we call heat is caused by molecular motion. When you feel the heat on a hot day at the beach, your nerve endings are detecting the rapid motion of air molecules. Your feet on the hot sand detect the motion of rapidly vibrating molecules in the hot solid.

An important concept to remember is that all matter contains at least some heat at any possible temperature. The main difference between these three states of matter at any given temperature is in the amount of potential energy they have. As you heat a piece of ice from a temperature well below 0°C, its molecules vibrate more and more strongly around fixed positions. As ice melts its temperature doesn't change, but the potential energy of the solid state is transferred to whatever is doing the heating. As you heat water from 0°C to boiling, the molecules will vibrate, rotate, and move past each other more vigorously.

A hot substance next to a colder substance also has potential energy. The burning of gasoline in the cylinder of an engine provides an example. When the gasoline burns, the hot molecules of the burned gases push against the

## CHAPTER BOX 4.1 HEAT, TEMPERATURE, AND KINETIC ENERGY

**You learned** in Chapter 3 that temperature and the motion of molecules (more specifically, their kinetic energy) are related to each other. In fact, absolute zero (0 K) can be defined as the temperature at which there is no molecular motion (which is theoretically impossible to attain, for reasons that are beyond the scope of this text). You have also learned that the amount of heat a substance contains is related to the amount of motion of its molecules.

One important conclusion that can be drawn from these facts is that the amount of heat a substance contains depends on the number of molecules present. That is, a ton of water contains more heat than a drop of water at the same temperature because there are a lot more molecules in a ton. The heat content of a substance is an extensive property.

The **specific heat capacity** of a substance is the amount of heat (in joules) it takes to raise the temperature of 1 gram of the substance by 1 degree Celsius. Specific heat capacity, then, has units of $J\ g^{-1}\ °C^{-1}$, and it is an intensive property. The values of the specific heat capacities of substances vary quite widely. The specific heat capacity of water is $4.2\ J\ g^{-1}\ °C^{-1}$, that of alcohol (the kind in beverages) is $2.5\ J\ g^{-1}\ °C^{-1}$, and that of metallic zinc is $0.39\ J\ g^{-1}\ °C^{-1}$. You can see that it takes ten times as much heat to raise the temperature of 1 g of water by 1°C than it does to do the same for zinc (4 J for water and 0.4 J for zinc).

Now, let's do a thought experiment. First, take 1 g of water at 0°C and raise its temperature 20°C by adding 80 J of heat. Then, take 1 g of zinc at 0°C and raise its temperature the same 20°C by adding 8 J of heat. Clearly the water contains more heat than the zinc, even though both substances have the same temperature. Equal masses of two different substances at the same temperature typically contain different amounts of heat.

piston and (through a mechanical linkage) move the car. The colder gas molecules outside the piston provide little resistance to the piston's movement because they have considerably less kinetic energy than the fast-moving molecules inside the cylinder.

The heat in a substance can be used to do useful work only when the hot substance or object is working against something that has a lower temperature. Furthermore, heat always flows from warmer substances to colder ones. The molecules in hotter substances have more kinetic energy than the molecules in cooler substances. When the hot molecules collide with the cold ones, they transfer some of their kinetic energy. This process continues until the molecules in the two samples, originally hotter and colder, have the same average kinetic energy. At this point the two samples have the same temperature. A moment's reflection may reveal to you that this is an example of the Second Law of Thermodynamics in action.

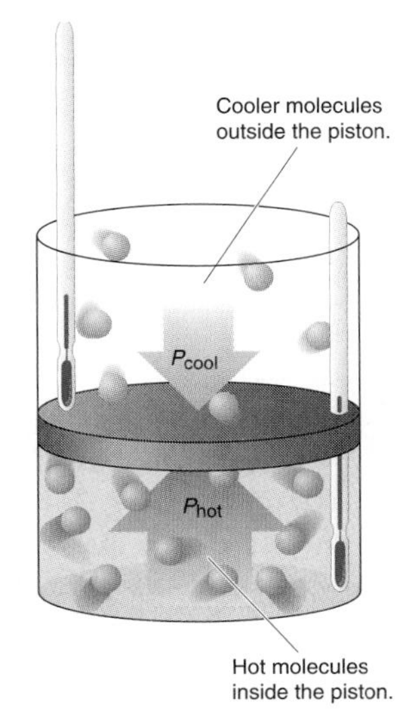


FIGURE 4.12
The slower-moving, colder molecules outside the piston have less kinetic energy and can do less work than the faster, hotter molecules inside the cylinder. This is why the piston moves when the fuel inside the cylinder is mixed with air and burned.

### EXAMPLE 4.4 The Laws of Thermodynamics in Action

You put a refrigerator in the middle of a well-insulated room, close the door of the refrigerator, plug it in, and leave. It gets cold inside the refrigerator. What happens to the temperature in the room outside the refrigerator—does it increase, decrease, or remain the same? Explain your answer.

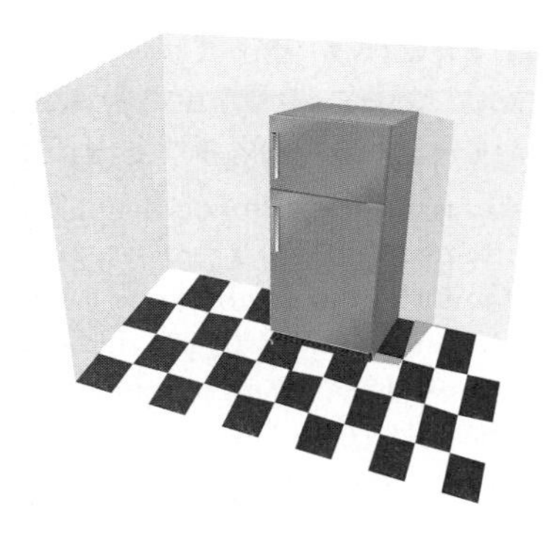

SOLUTION

Some of the heat that was inside the refrigerator is no longer there. The First Law of Thermodynamics says that energy can be moved around or changed in form but is neither created nor destroyed. Therefore, the heat originally inside the refrigerator must be somewhere outside it—in the room. The temperature in the room *increases*.

EXERCISE 4.4

You put a refrigerator in the middle of a well-insulated room, open the door of the refrigerator, plug it in, and leave. What happens to the temperature in the room?

## THE STRUCTURE AND COMPOSITION OF THE ATOM

In the next few sections you will learn how scientists arrived at the currently accepted ideas of the composition and structure of the atom. Atoms are the building blocks of chemistry—all molecules are composed of atoms of various sorts. Each type of atom has its own unique properties, many of which will be discussed in the next two chapters.

All atomic properties are related to the structure of the atom, and many of them are quite predictable if you understand how atoms are put together. Chemists think about matter in terms of atoms and molecules. It is therefore important for anyone studying chemistry to know something about what atoms and the forces that govern their behavior are like.

### CHAPTER BOX 4.2 MASS AS A FORM OF ENERGY

**Scientists in** the nineteenth century believed that neither mass nor energy changed in the course of any process. That is, both mass and energy were individually conserved, as far as anyone could tell from experiments done up to that time. Albert Einstein proposed that these "laws"—known as the *Law of Conservation of Mass and the Law of Conservation of Energy*—were not strictly true. In a paper published in 1905 he said that mass can be converted into energy (and vice versa). To calculate how much potential energy there is in a given mass (or vice versa), you can use Einstein's most famous equation, $E = mc^2$, where $E$ is the energy (in J) in an object, $m$ is the object's mass (in kg), and $c$ is the velocity of light ($3 \times 10^8$ m s$^{-1}$).

Einstein's theory was untested by experiment when it was first published, but high-energy physicists have confirmed it countless times since then. A very small mass can be converted into a huge amount of energy—only a few grams of mass is converted in the average nuclear explosion.

Despite Einstein's theory, chemists can still use the laws of conservation of mass and energy separately and be confident that their results will be accurate within any limit they can measure. High-energy physicists, on the other hand, consider mass to be just another form of energy. Thus they can say "energy is conserved" with no inaccuracy.

## 4.4 Electricity and Chemistry

**GOAL:** To learn about the history of electricity and how its connection with chemistry was first established.

FIGURE 4.13
Lightning is a form of electric current, and it has been around as long as the earth has. However, lightning is too violent and dangerous to serve as an easy object of study. It played almost no part in the scientific investigation of electricity either before or since Benjamin Franklin's famous, and extremely dangerous, experiment.

You use electricity without thinking about it very much. You push a switch and the light comes on. You plug in appliances, push some buttons, and they work. Your local utility company measures how much electricity you use and sends a bill. Electricity is used in transportation, communication, and almost any other part of modern life you can think of.

Electricity is a vital part of chemistry. We have mentioned it several times earlier in this book. Several of the elements were isolated for the first time when a chemist ran an electric current through a compound of the element. Aluminum is one of many substances manufactured in this way today. The connection between electricity and chemistry is an intimate one. In this section we explore how physicists and chemists discovered this connection.

According to the ancient Greeks, Thales of Miletus was the first person to investigate static electricity, which is safer and more accessible than lightning. Thales found that if you rub a piece of fossilized pine resin, which the Greeks called *elektron* and is now known as *amber,* with cloth (a silk cloth works very well), the piece can attract small, lightweight objects. Human understanding of electricity advanced very little over the next 2000 years.

The next notable discovery about electricity was made by the English physicist and physician William Gilbert (1544–1603). Gilbert investigated the power of objects to attract other small objects after being rubbed, and found that many substances besides amber show the same behavior. He pointed out that this effect was different from the attractive powers of magnets, which were made only from iron in those days. This may sound like a simple and obvious discovery, but both scientific knowledge and experimentalists were rare in the sixteenth century. In fact, Galileo considered Gilbert to be the pioneer of experimental science.

FIGURE 4.14
The red-cockaded woodpecker (left) of the southeastern United States drills its nest hole in old pine trees. It pecks small holes above the nest opening to let sticky pine sap—resin—flow around the hole, presumably to discourage predators. Fossilized resin (right) is called *amber*. Rare pieces of amber contain insects that were trapped by the sticky sap millions of years ago.

FIGURE 4.15
This apparatus (called an *electroscope*) shows that electric charges of the same kind repel one another. The rod above transfers the charge to the metal (frequently gold), and the light piece of foil on the right pushes away from the other piece of metal.

Beginning in the seventeenth century a number of people looked at static electricity. The next few paragraphs may seem like a blizzard of names and dates. You needn't memorize them. Our purpose is to give you a sense of history and an idea of how science advances and to convey the impression of an age of activity. Natural philosophers (as scientists were called in those days) were experimenting, writing each other letters, and publishing articles and books. It was a lively age for science.

In 1729 the Englishman Stephen Gray electrified a glass tube by rubbing it and found that the corks that closed the ends of the tube became electrified, too. Shortly afterwards and also in England, John Théophile Desaguliers coined the term **conductors** for substances that could transfer this strange attraction from one object to another and the term **insulators** for those that could not.

Charles François de Cisternay Du Fay took things further in France. He found that two corks electrified by a rubbed amber or resin rod repelled each other. Two corks electrified by a rubbed glass rod also repelled each other, but corks electrified by glass attracted corks electrified by resin. He proposed in 1733 that two different fluids were involved, which he called *vitreous* (glassy) and *resinous.* He thought that corks electrified by glass contain excess vitreous fluid, and corks electrified by resin or amber contain an excess of the resinous fluid.

**Gilbert pointed out in the late sixteenth century that electricity and magnetism were two different phenomena. In the late nineteenth century the English physicist James Clerk Maxwell concluded that the two forces, though different in their effects, were aspects of the same underlying phenomenon. See your friendly local physicist for more details.**

The next theoretical step was taken by the only American natural philosopher of the colonial era to have a European reputation. Benjamin Franklin proposed that there was only one "fluid", and that an excess or deficit of it in a substance accounted for the attraction and repulsion. This was reasonable—why should such a symmetrical phenomenon be caused by two different fluids? Franklin's idea turned out to be essentially correct.

Others were making practical advances, too. The Leyden jar, invented at the University of Leyden in the Netherlands in 1745, made it possible to store electricity. Franklin and others made good use of it. The invention of what we now call the battery by Alessandro Volta of Italy made it possible to generate electrical current. In 1800 Volta wrote of his discovery to the president of the Royal Society of England.

Within a month of reading this letter, William Nicholson and Anthony Carlisle in England had made their own *voltaic pile,* dropped a wire from

***Don't* try the Franklin Kite Experiment on your own. Franklin was lucky to survive it, and several people have been killed by lightning while trying to repeat it.**

EXPERIMENTAL EXERCISE

## 4.1
## The Franklin Kite Experiment

Very early in the history of electricity, natural philosophers noted the similarity between electric sparks and lightning. In 1752 Benjamin Franklin showed that these were identical except for size.

Franklin's original plan was to place a metal rod in the spire of a newly constructed church; his intention was to withdraw electricity from the clouds through the rod. The sight of a boy's kite caused Franklin to modify his plans, primarily because the kite could be flown higher in the sky than the spire of the church could reach.

Franklin stretched a silk handkerchief over a light wooden frame to form a kite. He attached a hemp string to the kite and tied a key to the end of the string with a silk cord. He flew the kite on a stormy night with lightning bolts flashing all over the sky. For a while nothing happened, but as the string became wet, Franklin observed string filaments repelling each other. When Franklin placed his knuckle near the key, he saw a spark and felt an electric shock.

Based on this and other experiments Franklin proposed that an elastic fluid called the *electric fluid* exists throughout all space. This electric fluid has the property of repelling its own particles and attracting particles of other matter. All objects in their natural state possess a specific quantity of this fluid. If electric fluid is added to a body, giving that body an excess of the fluid, the body becomes positively charged. If some of the electric fluid is removed from a body, the body becomes negatively charged.

FIGURE 4.16

An alternative theory to explain electricity was proposed by Du Fay. Du Fay's theory of electricity also proposed that an electric fluid was found throughout all space. The fluid existed as two types: a vitreous fluid and resinous fluid (also referred to as positive and negative). Two samples of vitreous fluid repelled each other, as did two samples of resinous fluid, but vitreous fluid attracted resinous fluid, and vice versa. All neutral bodies possessed equal amounts of the two fluids.

### QUESTIONS

1. Suggest a reason why Franklin observed nothing happening in his experiment until after the hemp string became wet.
2. Use Franklin's theory to explain why he observed string filaments repelling each other.
3. Use Du Fay's theory of electricity to explain why Franklin observed the filaments repelling each other.
4. Which of these theories supports the conclusion that one kind of electricity (say, positive) cannot be produced without the other appearing?
5. What experimental evidence described in Section 4.4 supports Du Fay's theory?
6. A voltaic pile generates electricity by chemical reactions. Use Franklin's theory to explain how a chemical reaction can generate electricity.
7. Apply Du Fay's theory to interpret Franklin's results, and suggest an experiment that could verify whether the key contained an excess of vitreous or resinous fluid at the end of his experiment.

each end of it into a container of water, and produced hydrogen and oxygen. This experiment showed that there is a strong connection between electricity and chemistry. Ever since then chemists have used electricity as a tool in investigating and manufacturing substances. (You can learn more about the details of these activities in the sections about *electrochemistry* in Chapter 16.)

FIGURE 4.17
The Leyden jar, invented in 1745 as a way to store electricity, consisted of a glass jar, coated both inside and outside with tin foil to within an inch of its upper edge. The mouth of the jar was fitted with a cork. A brass rod, with a brass ball on the outside end, was passed through the cork and arranged so that it touched the foil inside the jar. The jar was charged with electricity by attaching the brass ball to a source of electricity. Early experimenters discharged the jar simply by touching the brass ball and receiving an electric shock. They measured the amount of electricity stored in the jar by how much they got hurt by the electric shock. (This is not a recommended procedure.)

## 4.5 From the Indivisible Atom to Details of Atomic Structure

GOAL: To explain how scientists arrived at the modern concept of the structure and composition of the atom.

By the late nineteenth century the atomic theory was accepted by almost everyone, though there were some holdouts among the physicists and philosophers. Part of John Dalton's original statement of the atomic theory said that atoms were structureless and indivisible. A young Swedish chemist named Svante Arrhenius (1859–1927) brought this notion into question when he wrote up his analysis of the properties of solutions.

**Dalton's atomic theory was first presented in Section 2.5. You might want to review it before reading this section.**

One of the phenomena Arrhenius examined was the conduction of electricity by a solution of table salt (sodium chloride) in water. He proposed that sodium chloride separates into electrically charged particles called **ions** when it dissolves. He included this idea in his 1884 Ph.D. thesis, and his skeptical examiners gave him a barely passing grade. If atoms could have charges, then there must be something wrong with Dalton's theory as it was then understood.

**Some of the same people who sat on the review board for Arrhenius's Ph.D. thesis—and barely gave him a passing grade—were also on the Nobel committee that awarded him the prize in 1903. By this time there was no argument about the merit of Arrhenius's ideas.**

Several of chemistry's rising stars thought that Arrhenius's ideas were sensible, but without more evidence the issue was in doubt. An avalanche of discoveries began in France in 1895. Antoine-Henri Becquerel discovered radioactivity, and Marie Curie found that radioactivity was a property of atoms. (For more about this topic, see Chapter 17.) In England J. J. Thomson studied the properties of the tiny, negatively charged particle that was called a cathode ray (see Chapter Box 4.3). Thomson found that the mass of a cathode ray was a tiny fraction of the mass of the hydrogen atom, and he called this particle the **electron.**

Since his electron was so much smaller than an atom, Thomson proposed that atoms had structure. He saw them as containing electrons embedded in a positively charged matrix, just as raisins are embedded in raisin bread. Following Thomson's discovery, Arrhenius's ion proposal was quickly accepted, and he won a Nobel Prize in 1903 for the thesis that had barely passed in 1884. (Thomson's Nobel Prize followed in 1906.)

The New Zealand native Ernest Rutherford (1871–1937) filled an experimental void. He was studying alpha particles, which he knew were helium atoms with their electrons removed. He looked at what happened when these small, positively charged particles were fired at a thin piece of gold foil (see Figure 4.20). Rutherford chose gold for this experiment primarily because gold can easily be pounded into extremely thin sheets.

## CHAPTER BOX 4.3 CATHODE RAYS

**Michael Faraday** (1791–1867) was one of the first to investigate the effects of an electric current on gases. To study these effects, he placed a gas inside a glass tube and then created an electric field inside the tube by running current from the positive electrode at one end to the negative electrode at the other. When most of the gas in a tube like this is pumped out, the remaining gas glows. (This is the same effect seen in neon signs). When almost all of the gas is pumped out, the glass around the positive electrode begins to glow.

Figure 4.18 shows a more modern version of Faraday's apparatus. If you place an object between the negative electrode (the *cathode*) and the positive electrode (the *anode*), its shadow appears on the anode end. Apparently something is given off by the cathode, and it travels in a straight line toward the anode as if it were a beam of light. The German physicist Eugen Goldstein was one of several physicists who investigated these rays, and he coined the term *cathode ray* in 1876.

If you place a cathode ray tube in a second electric field that crosses the tube (as in Figure 4.19), the stream of rays is deflected toward the positive side of the field. In 1895 the French physicist Jean-Baptiste Perrin aimed cathode rays at a cylinder placed inside the tube. After the cylinder absorbed many of the rays, he found that it had a large negative charge. Therefore, Perrin decided, cathode rays are particles with a negative charge.

If you think this process might be refined to give pictures inside the tube, you're right. Television was developed from this basic idea much later.

Rutherford knew from Thomson's work that alpha particles are more than 7000 times as massive as electrons, so the electrons in the gold should provide almost no resistance to the path of the alpha particles. Furthermore, Thomson's positively charged matrix could not have much effect on the path of the alpha particles, because the positive charge of the atom was assumed to be spread evenly throughout the atom (Figure 4.20(A)).

Most of the alpha particles shot straight through the gold foil, as Rutherford expected. A few of them bounced to the side (some at very wide angles, which was a surprise), and a very small number of them came straight back toward their source. This last result was completely unexpected. Rutherford compared the situation to firing an artillery shell at a newspaper and having it bounce back!

He decided that Thomson's model of the atom must be incorrect, because the experimental results that that model predicted were not observed. Rutherford concluded that all the positive charge in an atom must be concentrated in a massive center he called the *nucleus,* which is surrounded by the lighter negatively charged electrons (see Figure 4.20(C)), and he published this conclusion in 1911. This is the model of the atom scientists use today.

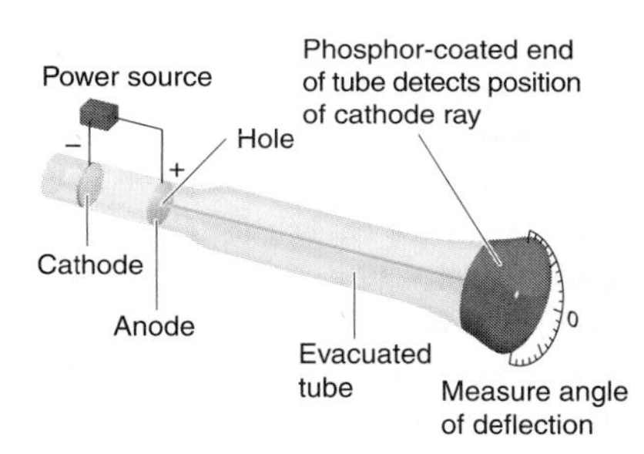


FIGURE 4.18
Cathode ray tube

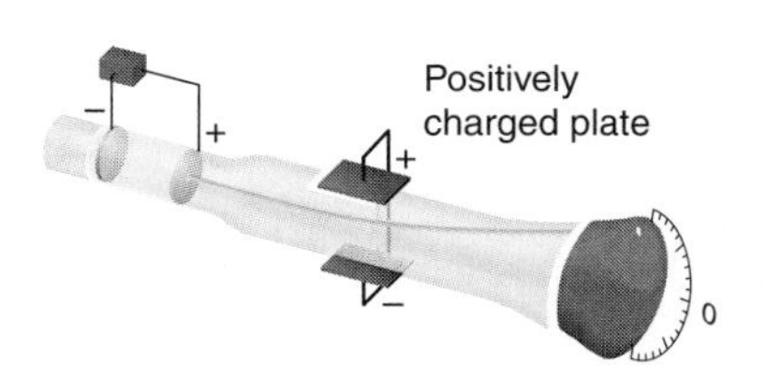


FIGURE 4.19
Cathode ray tube in an external electric field

## 4.6 The Periodic Table, Part I

**GOAL:** To understand the relationship between the structure of the atomic nucleus and the periodic table.

By 1911 Mendeleev's periodic table was well established. (For the beginning of this story, see Sections 1.6 and 1.7.) Mendeleev had correctly predicted

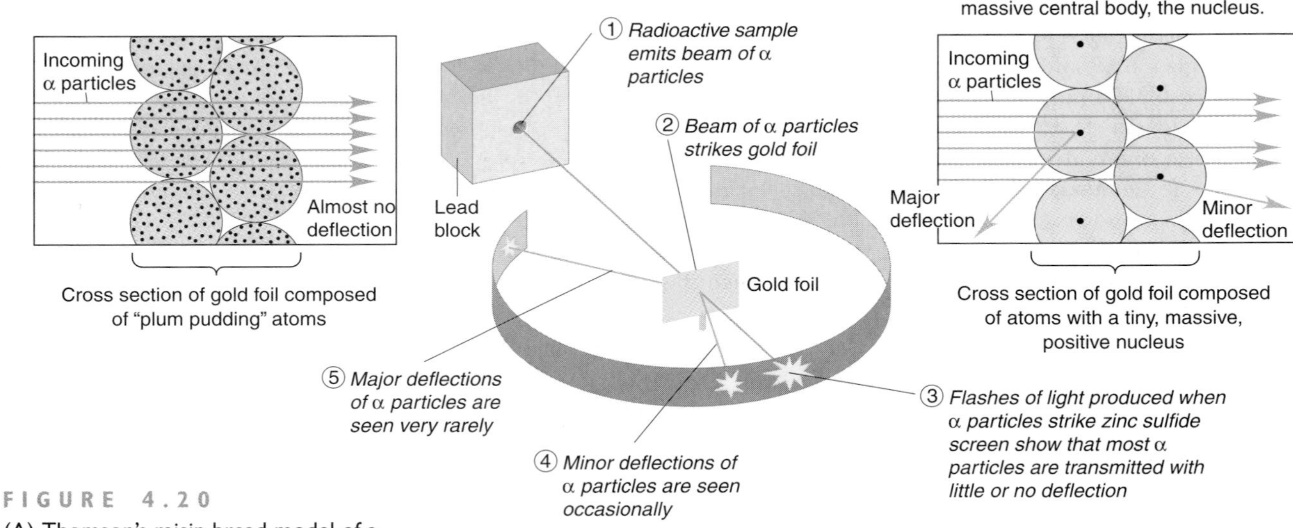
A. Thomson's "plum pudding" or "raisin bread" model of a gold atom. If this were a good representation of atomic structure, the alpha particles should pass right through the gold atoms with very little observed deflection.
Incoming α particles
Almost no deflection
Cross section of gold foil composed of "plum pudding" atoms
B. In Rutherford's actual experiment, alpha particles aimed at gold foil emit a flash of light when they hit a detector screen.
① Radioactive sample emits beam of α particles
② Beam of α particles strikes gold foil
Lead block
Gold foil
③ Flashes of light produced when α particles strike zinc sulfide screen show that most α particles are transmitted with little or no deflection
④ Minor deflections of α particles are seen occasionally
⑤ Major deflections of α particles are seen very rarely
C. The actual results show that most particles are not noticeably deflected, but a few are deflected at wide angles. This led Rutherford to conclude that the positive charge in atoms is concentrated in a small, relatively massive central body, the nucleus.
Incoming α particles
Major deflection
Minor deflection
Cross section of gold foil composed of atoms with a tiny, massive, positive nucleus

FIGURE 4.20
(A) Thomson's raisin bread model of a gold atom. If this were a good representation of atomic structure, the alpha particles should pass right through the gold with very little observed deflection. (B) In Rutherford's experiment alpha particles aimed at gold foil emit a flash of light when they hit a detector screen. (C) The results show that most particles are not noticeably deflected, but a few are deflected at wide angles. This led Rutherford to conclude that the positive charge in atoms is concentrated in a small, relatively massive central body—the nucleus.

that new elements would be discovered. As you may remember, Mendeleev also predicted that some atomic masses would be proven wrong. It turned out that this prediction was wrong, not the masses. Atomic mass doesn't increase smoothly along rows of the periodic table. What is the principle underlying the order of elements in the table if it isn't atomic mass?

Soon after Rutherford discovered how mass and charge are distributed in the atom, one of his students did some research on X rays. Atoms give off X rays when electrons change position near the nucleus. The student, Henry Moseley, found that the energy of the X rays was greater in atoms whose nuclei had more positive charges. He listed the atoms in order of this increasing charge, and he found that his order matched the periodic table order exactly.

In the meantime Rutherford kept on investigating. He decided that the simplest particle with a positive charge, which he called the **proton,** was the nucleus of the hydrogen atom. The most common type of hydrogen atom consists of two subatomic particles: a proton and an electron.

Masses at the atomic level are quite small. A typical atom has a mass of about $10^{-26}$ kg. Chemists almost always use a more convenient unit for atomic masses, the **atomic mass unit** (amu). The mass of a proton is approximately 1 amu.

**When an atom gains or loses one or more electrons, the number of protons no longer equals the number of electrons and the atom has an overall charge. These charged atoms are called *ions* (see Section 4.8).**

Chemists first calculated the relative masses of atoms from the ratios of masses of substances that react. The actual masses of atoms and the particles that make them up were measured much later. The proton is now known to have a mass of $1.673 \times 10^{-27}$ kg, for instance, and the electron has a mass of $9.109 \times 10^{-31}$ kg. As you can calculate, the mass of a proton is 1837 times greater than the mass of an electron. Therefore, the mass of a hydrogen atom is essentially the same as the mass of a proton. This is not a surprise in view of Rutherford's experimental results.

**TABLE 4.1 Subatomic Particles**

| Name | Mass (amu) | Mass (kg) | Charge | Charge (C) |
|---|---|---|---|---|
| Proton | 1 | $1.673 \times 10^{-27}$ | +1 | $+1.602 \times 10^{-19}$ |
| Electron | 0.00054 | $9.109 \times 10^{-31}$ | −1 | $-1.602 \times 10^{-19}$ |
| Neutron | 1 | $1.675 \times 10^{-27}$ | 0 | 0 |

The charges of the proton and the electron are identical in magnitude but opposite in sign. Primarily as a matter of convenience the charge of a proton is almost always given as + 1 and that of an electron as − 1. In SI units the charge of a proton is actually $+1.602 \times 10^{-19}$ **coulomb** (C). (For more information see Chapter Box 4.4.) The total charge of a hydrogen atom or any other atom is zero, because the number of protons equals the number of electrons in a neutral atom.

**Remember that every measurement has a magnitude and a unit (Chapter 3).**

All other atoms, including a small fraction of hydrogen atoms, contain one or more additional subatomic particles called neutrons. A **neutron** has no charge, and its mass of $1.675 \times 10^{-27}$ kg is almost identical to the mass of a proton. The discovery of the neutron in the early 1930s completed the modern chemists' model of the atom. (For more about neutrons, see Section 4.9.) Atoms and ions contain these particles—protons, neutrons, and electrons—in various numbers. The masses and charges of these three subatomic particles are summarized in Table 4.1.

All the protons and neutrons in an atom are in the nucleus, and the nucleus is surrounded by a cloud of electrons. The **atomic number** of an atom is the number of protons in the nucleus. Every atom of a given element has the same number of protons in its nucleus, although the number of neutrons may vary. Hydrogen, for example, comes in three varieties: a hydrogen atom can contain zero, one, or two neutrons.

**Hydrogen atoms that have no neutrons make up 99.985% of all naturally occurring hydrogen on earth. Hydrogen atoms with one neutron (which are given a special name, *deuterium*) constitute only 0.015% of the earth's hydrogen. Hydrogen atoms with two neutrons (called *tritium*) are unstable and decompose within a few years (see Chapter 17). The tritium used by scientists in experimental work is produced artificially.**

The atomic number of an element usually appears at the top of the box designating that element in the periodic table (see Figure 4.21). Because the atomic number gives the number of protons in the nucleus of an element, it can be used to identify the element. Since the number of protons is equal to the number of electrons in a neutral atom, the atomic number also gives the number of electrons.

## EXAMPLE 4.5 The Number of Protons in a Nucleus

You may refer to a periodic table in answering these questions.

**a.** How many protons are there in the nucleus of an atom of each of the following elements? (1) C (2) Ca (3) Cl (4) U

**b.** Name the elements that have nuclei containing (1) 10 protons (2) 35 protons.

**SOLUTION**

**a.** (1) On the periodic table the atomic number of carbon is 6, so all carbon nuclei contain 6 protons. (2) Calcium: 20 protons. (3) Chlorine: 17 protons. (4) Uranium: 92 protons.

**b.** (1) Neon (atomic number 10). (2) Bromine (atomic number 35).

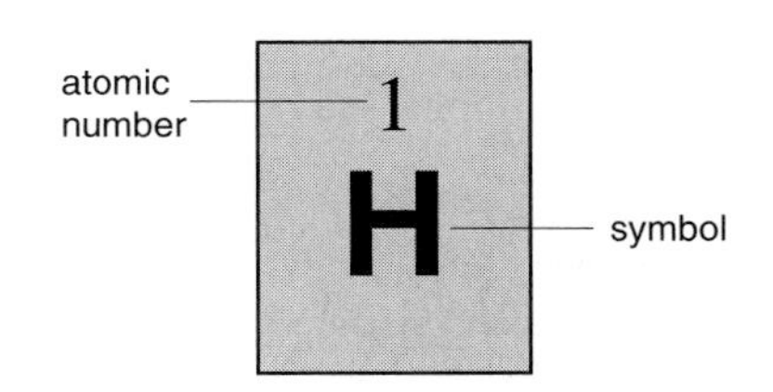


FIGURE 4.21
The box for hydrogen on a periodic table

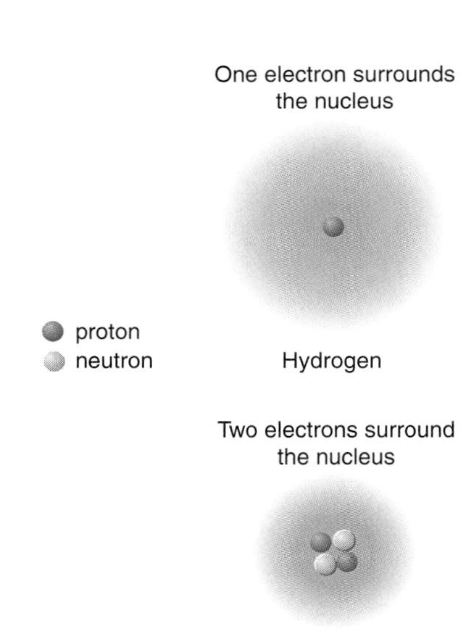


FIGURE 4.22
These representations of a hydrogen atom and a helium atom are not drawn to scale. The nuclei are actually much smaller than the volume of the atom—the hydrogen and helium nuclei here would be invisible if they were drawn to scale.

**EXERCISE 4.5**

**a.** How many protons are there in the nucleus of an atom of each of the following elements? (1) N (2) Fe (3) He (4) S

**b.** Name the elements that have nuclei containing (1) 50 protons (2) 79 protons.

## 4.7 Picturing the Atom

**GOAL:** To get a sense of the sizes of the components of atoms.

Figure 4.22 illustrates a hydrogen atom and a helium atom. The volume of an atom is essentially the volume occupied by its electrons. The nucleus is extremely tiny, having a radius of about 1/100,000 of that of the atom as a whole. This size difference makes it difficult to represent the nucleus and the atomic volume on the same scale. If you drew a nucleus as a dot like this (•), the whole atom would be nearly the size of a football field.

Another difficulty is how to show the electrons. The electrons don't really travel in circular orbits about the nucleus, although this is the classical picture (see Figure 4.23). Instead, you should think of the electrons as occupying the entire volume of the atom. There are several good reasons why this picture is better. We will explore this in more detail in Chapter 6.

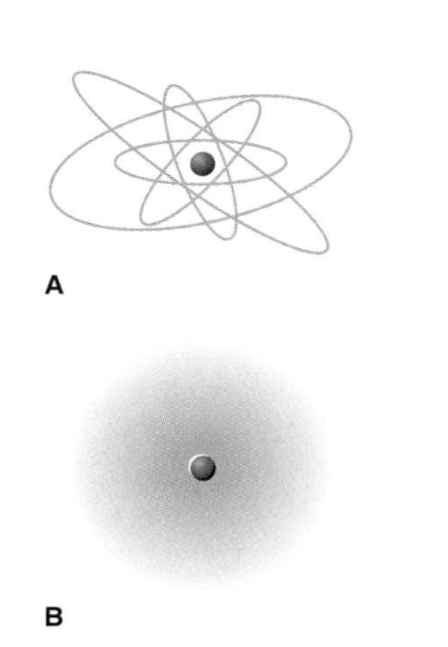


FIGURE 4.23
(A) Traditional symbol of the atom. (B) Picture of an atom as a fuzzy ball. The entire spherical volume is occupied by the electrons. This is probably the best way for chemists to model an atom in their minds.

## 4.8 Electrical Charges and Ions

**GOAL:** To calculate the number of electrons, the number of protons, and the charge of atoms and ions, given the necessary information.

Sections 4.5 through 4.7 tied the modern picture of the atom up in a package. Each atom has a small nucleus at its center that contains almost all of the atomic mass. The nucleus, in turn, is made up of the same number of protons as the element's atomic number, plus some neutrons. (The nucleus of the most common form of hydrogen atom is the only nucleus with no neutrons.) The nucleus is surrounded by a cloud of electrons.

In this section we look at the electrons in more detail. You already know that the total mass of the electrons in an atom is very small compared with the mass of the nucleus. In a **neutral atom** (that is, an atom with no charge) the number of electrons is equal to the number of protons. Remember that the charges of a proton and an electron are equal in magnitude and opposite in sign (see Table 4.1).

The fact that electrons are outside the nucleus and protons are inside the

## CHAPTER BOX 4.4 THE CHARGE OF AN ELECTRON

**In this** and any other chemistry course, you will see charges written as +1 or −3; that is, as magnitudes without units. Charges do have units, however. The SI unit of charge is the *coulomb* (C), the amount of electrical charge carried by a current of 1 ampere (A) in 1 second.

What is the charge on an electron? This question was answered early in the twentieth century by the American physicist Robert A. Millikan. His famous experiment is called the Millikan Oil Drop Experiment.

First, Millikan produced small droplets of oil by spraying them from a nozzle. (He tried using water first, but it tended to evaporate.) The oil droplets drifted slowly downward, pulled by gravity. Millikan shot X rays toward the drops. The high-energy rays knocked electrons loose from air molecules, and sometimes the electrons stuck to the drops of oil.

While all this was happening the oil drops were falling between two metal plates (see Figure 4.24). The top plate had a positive charge and the bottom one had a negative charge. The droplets with extra electrons had negative charges, of course, and were attracted to the positively charged plate. They stopped falling and began to rise. By measuring the motion of the drops against gravity, Millikan was able to calculate the charge on the drops. Millikan found that the charge on each drop was $-1.602 \times 10^{-19}$ C or a multiple of that.

You will almost always see the charge of an electron given as −1. This is a shorthand notation that chemists and other scientists (and this book) use in writing formulas. A charge given as −1 is actually a charge of $-1.602 \times 10^{-19}$ C.

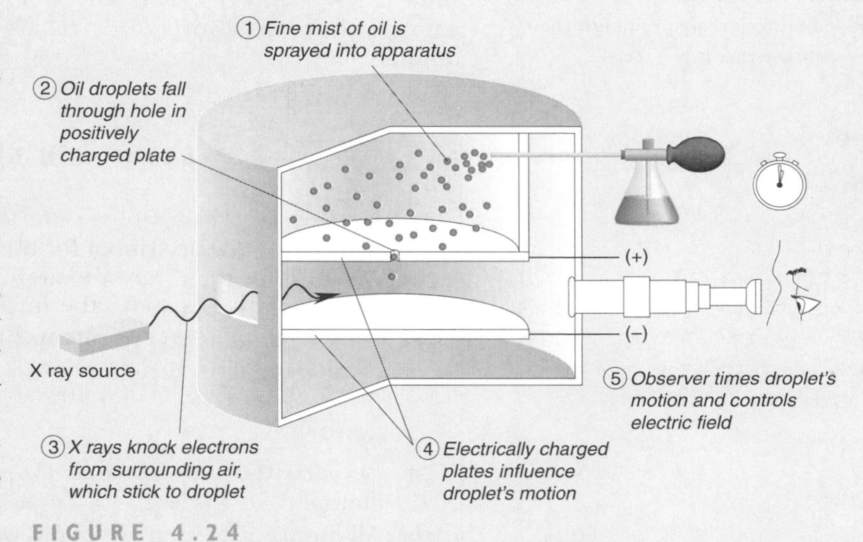
1 Fine mist of oil is sprayed into apparatus
2 Oil droplets fall through hole in positively charged plate
(+)
(–)
X ray source
3 X rays knock electrons from surrounding air, which stick to droplet
4 Electrically charged plates influence droplet's motion
5 Observer times droplet's motion and controls electric field

**FIGURE 4.24**
Millikan's oil-drop experiment for measuring an electron's charge.

nucleus turns out to be important. The charge of an electron has the same magnitude as that of a proton, so the forces that push charges around push as hard on the electron as they do on the proton (see Section 4.10 for more on this subject). Because electrons have a very small mass and are located outside the nucleus, any forces acting on an atom can remove electrons from or add them to the atom relatively easily, while the protons and neutrons stay at home. Exceptions occur only when the forces involved are truly enormous, as in nuclear reactions.

How do the charged species called *ions* come to exist? Consider a neutral sodium atom, for example. The atom contains 11 protons (sodium's atomic number) and 11 electrons. If you remove an electron from this atom, you get a species with 11 protons and only 10 electrons. (It's still sodium; remember

In older books the symbol for the calcium ion is written as $Ca^{+2}$. This convention is seldom used today.

that you can tell what element you have by the number of protons its atoms contain.) This species has a charge of $(+11) + (-10) = +1$.

The symbol for sodium is Na, and the symbol for the sodium ion is $Na^+$. The superscript plus sign following the atomic symbol tells you that the species is ionic with a $+1$ charge. $Ca^{2+}$ tells you that the calcium ion has a $+2$ charge; that is, that two electrons were removed from a neutral calcium atom to form it. Note that in this kind of superscript the number comes first, followed by the sign.

Positively charged ions are called **cations.** Both $Na^+$ and $Ca^{2+}$ are cations. When the charge is $+1$ (or $-1$, as in the next paragraph), the number 1 is not written in the superscript. It is simply understood to be there.

**STUDY ALERT**

**You are accustomed to seeing positive numbers without a plus sign. When you are indicating a charge you *must* show the sign, whether it is + or −.**

A neutral fluorine atom contains 9 electrons and 9 protons. Adding one more electron to one of these atoms gives it 10 electrons, and the charge is then $(+9) + (-10) = -1$. The symbol for this ion of fluorine is $F^-$, indicating that it has an overall $-1$ charge. Negatively charged ions are called **anions.**

### EXAMPLE 4.6 Protons and Electrons in Atoms and Ions

**a.** What is the charge of an ion that contains 17 protons and 18 electrons? What is the symbol for this ion? Is it a cation or an anion?
**b.** An ion of oxygen has a charge of $-2$. How many protons and electrons does it contain?
**c.** An ion contains 20 protons and 18 electrons. What element is it? What is the symbol for this ion?

**SOLUTION**

**a.** Because the ion contains 17 protons it's an ion of chlorine. The charge is $(+17) + (-18) = -1$. Its symbol is $Cl^-$ and it is an anion.
**b.** All atoms of oxygen contain 8 protons. Since this ion has a charge of $-2$, it contains two extra electrons, for a total of 10 electrons.
**c.** Because the ion contains 20 protons, it is calcium (atomic number 20). Its charge is $(+20) + (-18) = +2$. The symbol is $Ca^{2+}$.

**EXERCISE 4.6**

**a.** What is the charge on an ion that contains 7 protons and 10 electrons? What is the symbol for this ion? Is it a cation or an anion?
**b.** An ion of aluminum has a charge of $+3$. How many protons and electrons does it contain and what is its symbol?
**c.** An ion contains 53 protons and 54 electrons. What element is it? What is the charge on this ion and what is its symbol?

## 4.9 Isotopes

GOAL: To learn the terminology for atoms that have different numbers of neutrons and to calculate the numbers of protons, neutrons, and electrons in an atom or ion.

In 1912 J. J. Thomson (the discoverer of the electron) was studying a phenomenon called *canal rays,* which had been discovered some time earlier. Finding that these rays were made of positively charged particles, he called them *positive rays.* He thought that they had something to do with the nuclei of atoms, since Rutherford's discovery of the atomic nucleus had just been announced.

Thomson fired these positive rays (which turned out to be nothing more than cations) through a magnetic field and measured how much their paths were deflected as they passed through the field and toward a piece of photographic film. Thomson's knowledge of physics told him that charged particles travel a curved path in a magnetic field and the mass and the charge of a particle determine the amount of curvature in its path. To his surprise, Thomson found that canal rays made from neon produced two spots on the film. This meant that there were two kinds of neon cations. He knew that the cations had to have either different masses or different charges, or both.

Henry Moseley's work soon established that all nuclei of the same element have the same positive charge. This probably meant that Thomson's neon cations had different masses. The British chemist Francis William Aston improved on Thomson's apparatus, and by 1919 he had experimental apparatus that could separate the positive ions and find their masses. His instrument was called a **mass spectrometer,** and modern versions are widely used today. Almost all the masses of the elements that you see on periodic tables and in reference books were measured using a mass spectrometer, which can measure masses with pinpoint accuracy.

Aston determined that almost all of the positive rays had the same charge, + 1. He found neon cations that had two different masses: 20 and 22 amu. These neon ions, then, had the same positive charge in the nucleus,

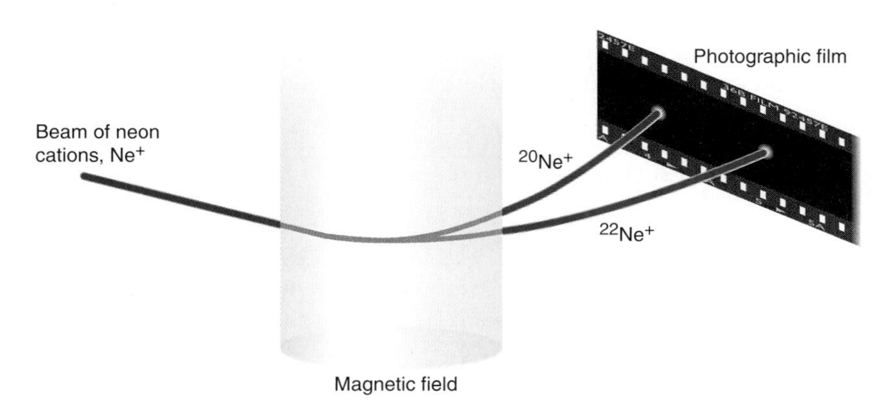


FIGURE 4.25
Charged particles travel a curved path in a magnetic field. The radius of curvature depends on the particle's mass and its charge. J. J. Thomson ionized some neon and fired the ions through a magnetic field toward a photographic plate. He was quite surprised to find the plate exposed in two places, suggesting that there were neon ions with two different masses. This type of experimental evidence demonstrated the existence of isotopes. (More sensitive modern detectors have been able to find low concentrations of a third neon ion with a mass of 21 amu).

the same number of electrons around the nucleus, and different masses. When the neutron was discovered a few years later, the evidence was complete. The two forms of neon had the same number of protons in the nucleus but different numbers of neutrons.

The balance of protons and electrons in its atoms is the primary factor that determines the chemical behavior of an element. Aston showed that some atoms of the same elements have different masses than others. Two atoms that have the same number of protons but different numbers of neutrons are called **isotopes.**

**In recent years physicists have discovered that protons and neutrons have structures of their own. You may know, for instance, that they are composed of *quarks*. For more information about this, once again, please see your friendly local physicist.**

The mass of a proton is very close to that of a neutron, that is, very close to 1 amu. The total mass of an atom is almost entirely due to the masses of its protons and neutrons, with the very light electrons making up about $\frac{1}{2000}$ of the total mass at most.

### EXAMPLE 4.7 Protons, Neutrons, and Electrons in Atoms

Consider the following descriptions of atoms:

1. 9 protons, 10 neutrons, 10 electrons
2. 10 protons, 8 neutrons, 10 electrons
3. 11 protons, 12 neutrons, 10 electrons
4. 12 protons, 9 neutrons, 10 electrons

a. Which of these atoms has the largest charge in the nucleus?
b. Which has the largest overall positive charge?
c. Which has the largest overall negative charge?
d. Which has the largest mass?
e. Which is sodium?

**SOLUTIONS**

To work problems of this type, it is useful to summarize all the information you know or can deduce. Here is a table that does this:

| Atom | Nuclear Charge | Overall Charge | Total Mass, amu |
|---|---|---|---|
| (1) | +9 | $9(+1) + 10(-1) = -1$ | $9 + 10 = 19$ |
| (2) | +10 | $10(+1) + 10(-1) = 0$ | $10 + 8 = 18$ |
| (3) | +11 | $11(+1) + 10(-1) = +1$ | $11 + 12 = 23$ |
| (4) | +12 | $12(+1) + 10(-1) = +2$ | $12 + 9 = 21$ |

a. The proton is the only charged particle in the nucleus. The atom with the most protons has the largest nuclear charge: atom (4).
b. The overall charge is equal to the number of protons (which have a positive charge) times +1, plus the number of electrons (negative charge) times −1, as shown in the table. Atom (4) has the largest overall positive charge.

**c.** Atom (1) has the largest negative charge.
**d.** To find the mass, add the number of protons to the number of neutrons. The mass of the electrons is so small it can be ignored. As shown in the table, atom (3) has the largest mass.
**e.** The atomic number is equal to the number of protons. The periodic table shows that the atomic number of sodium is 11. Atom (3) has 11 protons, so it is sodium.

**EXERCISE 4.7**

An ion contains 15 protons, 16 neutrons, and 18 electrons. Answer each of the following questions about this ion.
**a.** What is the charge in the nucleus of this ion?
**b.** What is the total charge of this ion?
**c.** What is the mass of this ion in atomic mass units?
**d.** What element is this ion?

## FORCES AND MOTION

### 4.10 Force

**GOAL:** To identify the forces known to science and explain their relative importance to chemists.

What kinds of motion are there? Scientists know of only two kinds, uniform and accelerated. If you are sitting in your chair at home or drifting along in a spaceship at 40,000 miles an hour, you're in uniform motion. If you are speeding up, slowing down, or changing direction, you're in accelerated motion. You probably recognize that, once again, scientists are using a definition that is a little different from the everyday one.

A **force** is a physical agency that causes acceleration. When a rock falls off a cliff, the force of gravity accelerates it, making it fall faster and faster. When you step on the gas pedal of your car, the engine applies a force to the wheels that makes the car speed up. When you step on the brake pedal of your car, a force is applied to the wheels that makes the car slow down. To scientists all of these are accelerations (changes in the rate of motion).

Scientists have identified four basic forces that operate in nature. The only one that chemists use to explain the actions of atoms and molecules is the *electromagnetic force.* This is the force that moves electrons and protons around. When particles the size of atoms and molecules are involved, it is by far the strongest force. Gravity is much too weak a force to compete with the electromagnetic force at the atomic level. The other forces operate only at nuclear distances (that is, within the nuclei of atoms). Since atoms have 10,000 times the diameter of their nuclei, nuclear forces are not useful or interesting to most chemists.

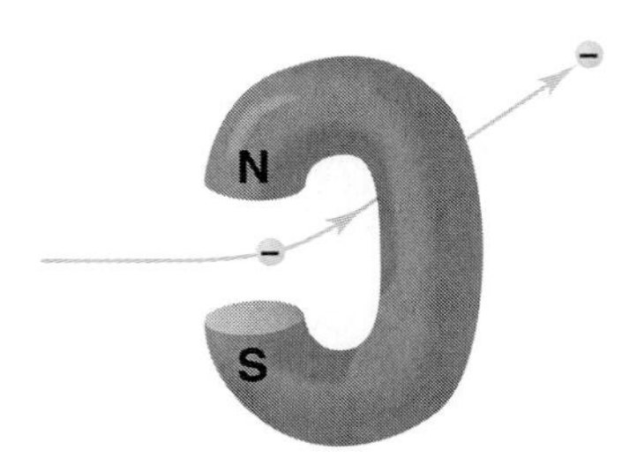


Electromagnetic force

Gravitational force

FIGURE 4.26

**The four forces are *gravity*, the *electromagnetic force* (important in chemistry), the *strong nuclear force*, and the *weak nuclear force*. The last two forces are negligible at distances larger than the size of an atomic nucleus.**

## 4.11 The Electromagnetic Force and Its Importance in Chemistry

GOAL: To understand how the electromagnetic force works in simple situations.

Earlier in this book chemistry was defined as the study of matter and the changes it undergoes. Chemistry can also be defined as the study of electrons and nuclei and their interactions by way of the electromagnetic force. These are two ways of saying the same thing.

Chemists often use the term *electrostatic forces,* which are electromagnetic forces that operate between two charged species (specifically, between two point charges). This is a subclassification of the more general electromagnetic forces.

As far as chemists are concerned, only two basic properties of the electromagnetic force are important. First, opposite charges attract and like charges repel. For example, a positive ion will attract a negative ion but will repel another positive ion (see Figure 4.27). The force of attraction or repulsion is proportional to the product of the charges on the two ions. For example, the attraction of a $+2$ ion for a $-1$ ion is twice as strong as the attraction of a $+1$ ion for a $-1$ ion.

FIGURE 4.27

Second, the strength of the electromagnetic force decreases with the square of the distance. The attraction of ion A for ion B in Figure 4.28 is four times as strong as the attraction of C for B, because ion A is half as far from B.

FIGURE 4.28

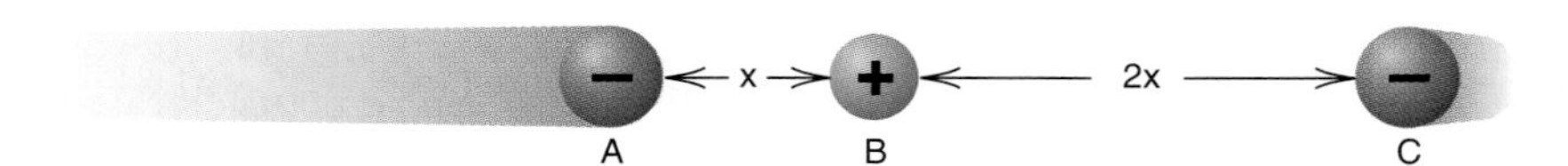


The purpose of this section is to give you a feel for the operation of the electromagnetic force. Most chemists and many other people prefer to think in terms of equations. The previous two paragraphs are summarized by the equation called **Coulomb's Law:**

$$F = k\left(\frac{Q_1Q_2}{r^2}\right) \qquad \text{Equation 4.2}$$

where $F$ is the force of attraction or repulsion, $k$ is a constant known as the *Coulomb constant,* $Q_1$ and $Q_2$ are the magnitudes of the charges, and $r$ is the distance between the charges.

STUDY ALERT
Coulomb's Law is of fundamental importance in understanding the interactions of molecules, ions, and atoms. We refer to it frequently in subsequent chapters, and understanding what it means in qualitative terms will be quite helpful to you.

The ideas behind Equation 4.2 are among the most important you will encounter in this book. You may never need to solve the equation describing Coulomb's Law, but understanding what it says about electromagnetic force is crucial in understanding how ions, molecules, and atoms interact with one another.

## EXAMPLE 4.8 Attraction and Repulsion of Charges

For each of these questions, explain your answer.

**a.** A polar molecule (that is, a molecule with a positive end and a negative end) approaches an ion with a positive charge. Which end of the molecule is attracted to the ion?

**b.** Two ions having charges of + 1 and − 1 are in contact with each other. What is the relationship between the size of the ions and the force of attraction? (Do larger ions attract each other more strongly, for instance?)

### SOLUTIONS

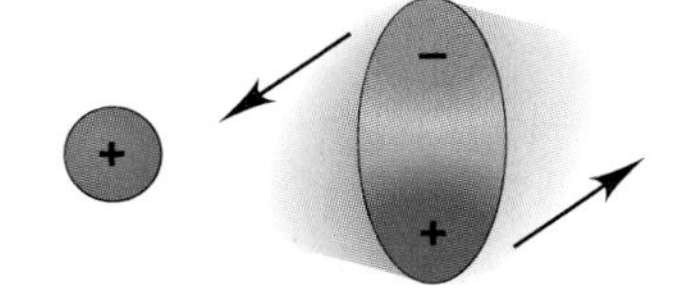

**a.** The negative end of the polar molecule (right) is attracted to the positively charged ion because opposite charges attract each other.

**b.** Smaller ions attract each other more strongly. For smaller ions the distance between their centers of charge is smaller ($d_2 < d_1$), and the force of attraction is greater at shorter distances.

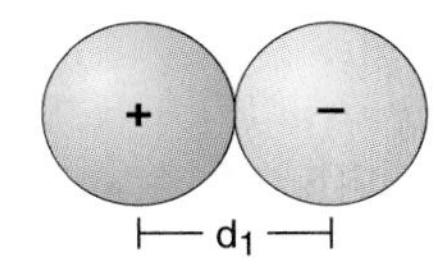


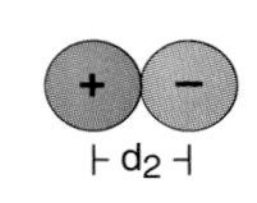


### EXERCISE 4.8

For each of these questions, explain your answer.

**a.** Which of the four other ions shown is attracted most strongly to ion Q?

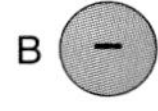


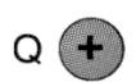


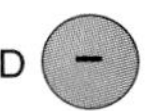


**b.** The positive end of a polar molecule (a molecule that has a positive end and a negative end) is in contact with an ion having a negative charge. Does the ion attract or repel the molecule, or is there no effect at all?

## CHAPTER SUMMARY

1. Energy is defined as the ability to do work. This means that the energy of anything is related to its ability to move an object some distance.
2. Potential energy is the energy an object possesses by virtue of its position in relation to another object that exerts a force on it. A second, perhaps more general, way to view potential energy is that it is energy stored by a system. Chemical energy is a form of potential energy.
3. Kinetic energy is the energy of motion. For objects, the kinetic energy is equal to one-half the mass times the square of the velocity

($KE = \frac{1}{2}mv^2$). Heat is a form of kinetic energy arising from atoms and molecules in motion.

4. The joule ($1\ J = 1\ kg\ m^2\ s^{-2}$) is a unit of energy commonly used by scientists. Other widely used units of energy include the calorie, the kilocalorie (or Calorie used by dieters), the BTU (British Thermal Unit), and the kilowatt-hour.
5. The First Law of Thermodynamics states that energy can change from one form into another but can never be created or destroyed. "Energy is conserved" is a shorter way of saying this.
6. One way of stating the Second Law of Thermodynamics is to say that energy may be changed from one useful form into another, or from more useful forms into less useful forms, but never from less useful into more useful forms. Here are two other, more precise statements of the Second Law: (1) Heat flows from warmer objects to colder objects, but never the reverse. (2) Isolated systems become more disordered as time passes.
7. Electrons have a negative charge. They surround the nuclei of atoms.
8. Protons have a positive charge and are found inside the nuclei of atoms. Neutrons have no charge and are also found in nuclei. Protons and neutrons are almost 2000 times as massive as electrons.
9. An atom contains a small heavy central body called a nucleus, which has an overall positive charge. The magnitude of the charge of the nucleus is equal to the atomic number of the element, and all nuclei of the same element have the same positive charge. Nuclei in atoms are surrounded by clouds of electrons. In neutral (uncharged) atoms, the number of electrons in the clouds equals the number of protons in the nuclei.
10. Ions form when electrons are added to or removed from atoms. Cations have positive charges (that is, they have fewer electrons than protons). Anions have negative charges (more electrons than protons). Chemists write the formulas of ions by writing the charge as a superscript following the symbol of the atoms: for example, $Na^+$ and $O^{2-}$.
11. Isotopes of an element all have the same number of protons but different numbers of neutrons in their nuclei.
12. A force is a physical agency that causes acceleration (change in the rate of motion). There are four basic forces known to scientists. The electromagnetic force, the attraction of opposite charges and the repulsion of like charges, is by far the most important force in chemistry.
13. The force of attraction or repulsion of two charged objects is equal to the product of a constant (the Coulomb constant) and the two charges divided by the square of the distance separating them. The equation expressing this force ($F = kQ_1Q_2/r^2$) is called Coulomb's Law.

## IMPORTANT EQUATIONS

Kinetic Energy: $\mathbf{KE} = \frac{1}{2}mv^2$

Coulomb's Law: $F = k\left(\frac{Q_1Q_2}{r^2}\right)$

## PROBLEMS

### SECTION 4.1 Energy and Its Forms

1. A golfer picks up a golf ball from the ground, places it on a tee, and hits it to an elevated green (15 ft higher than the golfer); the ball rolls into the cup for a hole-in-one. Describe the changes in the kinetic and potential energy of the golf ball during each of these steps.

2. A baseball player picks up a baseball from the ground, tosses it into the air, and hits it with a bat; the ball lands in the upper deck of the baseball stadium. Describe the changes in the kinetic and potential energy of the baseball during each of these steps.

3. Compute the kinetic energy of each of these:
   a. 1000-kg automobile traveling at 27 m $s^{-1}$ (60 mi $h^{-1}$, or mph)
   b. 50-kg person traveling in an automobile at 25 m $s^{-1}$
   c. 200-g object traveling at 50 m $s^{-1}$
   d. $2.50 \times 10^2$ g object traveling at 50 m $s^{-1}$
   e. 450-kg bull traveling at 0.200 m $s^{-1}$

4. Compute the kinetic energy of each of these:
   a. 115-g baseball traveling at 45 m $s^{-1}$ (100 mi $h^{-1}$)
   b. $5.00 \times 10^3$ g meteorite traveling at 225 m $s^{-1}$
   c. $3.00 \times 10^{-26}$ kg molecule traveling at $1.00 \times 10^3$ m $s^{-1}$
   d. $2.57 \times 10^{-22}$ g molecule traveling at $1.10 \times 10^3$ m $s^{-1}$
   e. 80-kg person walking at 0.811 m $s^{-1}$

5. What is the mass (in kilograms) of an object that has each velocity and kinetic energy?
   a. Velocity of $2.00 \times 10^2$ m $s^{-1}$ and kinetic energy of 2.30 kJ
   b. Velocity of $5.00 \times 10^2$ m $s^{-1}$ and kinetic energy of $2.71 \times 10^2$ kJ
   c. Velocity of $1.00 \times 10^3$ m $s^{-1}$ and kinetic energy of $2.56 \times 10^{-20}$ kJ
   d. Velocity of 5.00 cm $s^{-1}$ and kinetic energy of 2.33 J

6. What is the mass (in kilograms) of an object that has each velocity and kinetic energy?
   a. Velocity of $5.02 \times 10^2$ cm $s^{-1}$ and kinetic energy of $5.44 \times 10^2$ J
   b. Velocity of 0.523 m $s^{-1}$ and kinetic energy of 4.22 kJ
   c. Velocity of 8.99 m $h^{-1}$ and kinetic energy of $8.11 \times 10^{-3}$ kJ
   d. Velocity of $2.13 \times 10^3$ m $s^{-1}$ and kinetic energy of 4.00 kJ

7. The energy value of food items is usually reported in Calories. Compute the energy value in joules and kilojoules of these typical portions of common food items.
   a. 12-oz soda, 160 Cal
   b. 1 oz of cheddar cheese, 115 Cal
   c. 1 cup of whole milk, 150 Cal
   d. 10-oz chocolate shake, 335 Cal
   e. 1 egg, 80 Cal
   f. 1 tbsp of butter, 100 Cal

8. Compute the energy value in joules and kilojoules of these typical portions of common food items.
   a. 1 tbsp of olive oil, 135 Cal
   b. 3 oz of pickled herring, 190 Cal
   c. 3 oz of fried shrimp, 200 Cal
   d. 1 apple, 80 Cal
   e. 1 banana, 105 Cal
   f. 1 cup of cherries, 90 Cal

9. Show that 1 kilowatt-hour is equal to $3.6 \times 10^6$ J (1 watt = 1 J $s^{-1}$).

10. Show that 1 BTU is equal to 1.06 kJ.

11. How many joules of energy are consumed when a 100-watt incandescent light bulb burns for 30 min?

12. How many joules of energy are consumed when a 15-watt compact fluorescent light bulb burns for 30 min?

13. A person holds a ball at arm's length and drops it. The ball bounces a few times and rolls on the floor a couple of feet and then comes to rest (see the figure below).
    a. At which of the indicated positions is the potential energy of the ball at its minimum value?
    b. At which of the indicated positions is the potential energy of the ball decreasing?
    c. At which of the indicated positions is the kinetic energy of the ball at its minimum value?
    d. At which of the indicated positions is the kinetic energy of the ball decreasing?

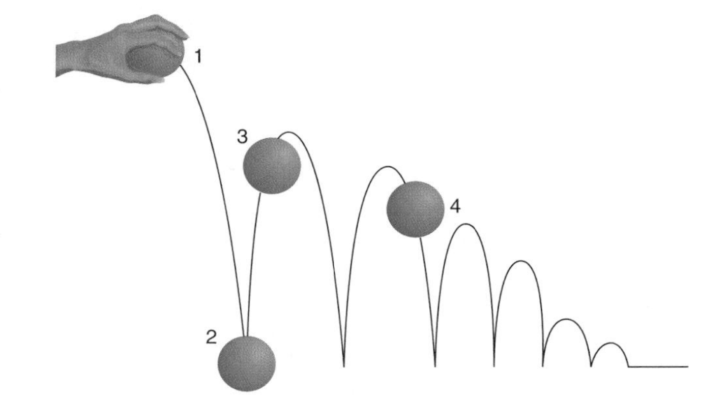

14. Consider the figure in Problem 13.
    a. At which of the indicated positions is the potential energy of the ball at its maximum value?
    b. At which of the indicated positions is the potential energy of the ball increasing?
    c. At which of the indicated positions is the kinetic energy of the ball at its maximum value?
    d. At which of the indicated positions is the kinetic energy of the ball increasing?

**TOOL BOX Reading Graphs**

15. The graph shows how the volume of a gas changes as the temperature changes at constant pressure.

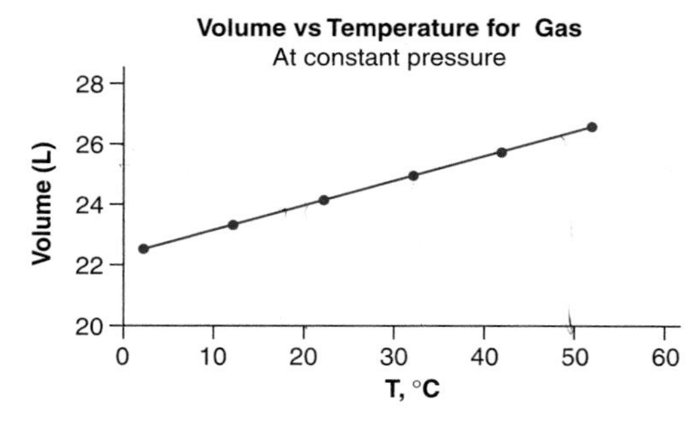


    a. At what temperature does the gas have a volume of 25 L?
    b. At what temperature does the gas have a volume of $2.4 \times 10^4$ mL?
    c. What is the volume of the gas at 50°C?
    d. What is the volume of the gas at 300 K?
    e. Is there a linear relationship between temperature and volume?

16. The graph gives some physical properties (melting points, boiling points, and formula masses) of certain compounds of the element potassium.

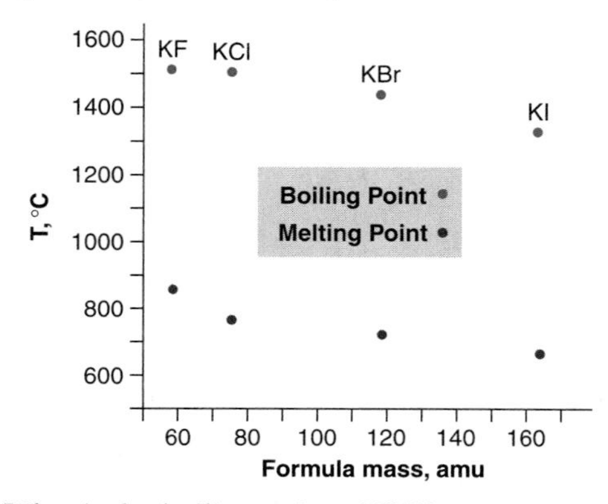


    a. What is the boiling point of KCl?
    b. What is the melting point of KBr?
    c. What is the formula mass of KI?
    d. Is there any relationship between the melting point and the formula mass for these compounds?

17. The graph gives the melting and boiling points of some hydrocarbons (compounds containing only hydrogen and carbon).

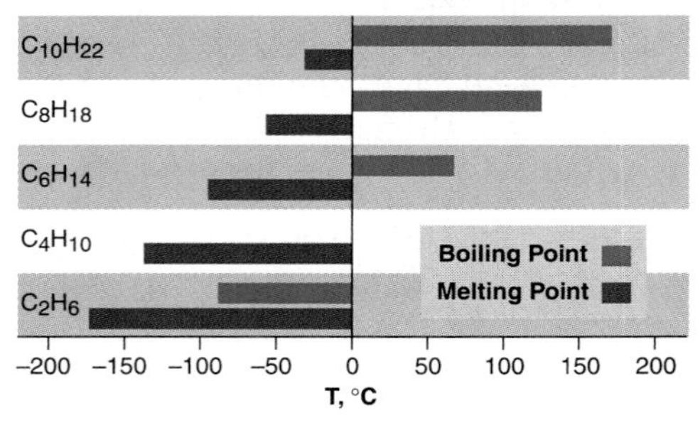


    a. What is the boiling point of $C_{10}H_{22}$?
    b. What is the melting point of $C_8H_{18}$?
    c. Make a reasonable estimate of the melting point of $C_9H_{20}$.
    d. Make a reasonable estimate of the boiling point of $C_9H_{20}$.
    e. What trend do the data suggest in the relationship between the number of atoms the hydrocarbon contains and its boiling point?
    f. What trend do the data suggest in the relationship between the number of atoms the hydrocarbon contains and its melting point?
    g. Is there a linear relationship between the number of atoms in a hydrocarbon and its boiling point?

18. For various metals the graph gives some physical properties: specific heat capacities (see Chapter Box 4.1), atomic radii, and densities.

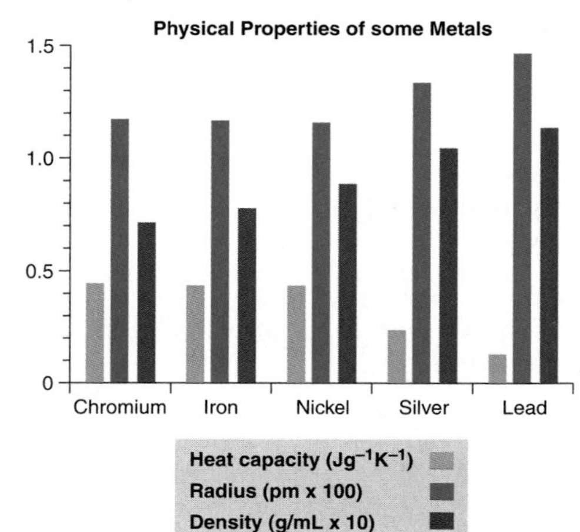


    a. What is the specific heat capacity of chromium? (Don't forget units.)
    b. What is the density of lead?
    c. What is the radius of a nickel atom?
    d. Look up the atomic numbers of these elements. Is there any relationship between the atomic number and any of the listed variables?

## SECTION 4.2 The First and Second Laws of Thermodynamics

19. When you drive an automobile, only a small part of the energy obtained from the combustion of the gasoline is used to make the car move. What happens to the rest of the energy?
20. A book is dropped to the floor from a height of 1 m. The book loses potential energy. Where does this "lost" energy go?
21. When scientists are in a playful mood, they sometimes restate the First Law of Thermodynamics as "You cannot win." What do they mean by that?
22. Scientists in a playful mood may restate the Second Law of Thermodynamics as "You cannot break even." What do they mean by that?

## SECTION 4.3 Potential and Kinetic Energy and Their Forms

23. Potential energy can be defined as the energy of an object's position or as the energy stored by an object. Which of the definitions most accurately describes the following?
    a. A coiled spring
    b. An airplane 2 km in the air
    c. A flashlight battery
    d. A parked automobile
    e. A tank of fuel oil

24. Potential energy can be defined as the energy of an object's position or as the energy stored by an object. Which of the definitions most accurately describes each of the following?
    a. An unburned coal
    b. A quiet volcano
    c. A hot spring
    d. A parachutist before jumping from a plane
    e. A mountain climber at the summit of a mountain

25. Give five examples (not mentioned in this chapter) in which it is convenient to view potential energy as energy of position.

26. Give five examples (not mentioned in this chapter) in which it is convenient to view potential energy as energy stored by a system.

27. What is the relationship between kinetic energy and heat?

28. Compare the average kinetic energy of liquid water molecules at 10°C to the average kinetic energy of liquid water molecules at 90°C.

29. The figures represent bromine ($Br_2$) as a solid, liquid, and a gas. Which one represents solid bromine?

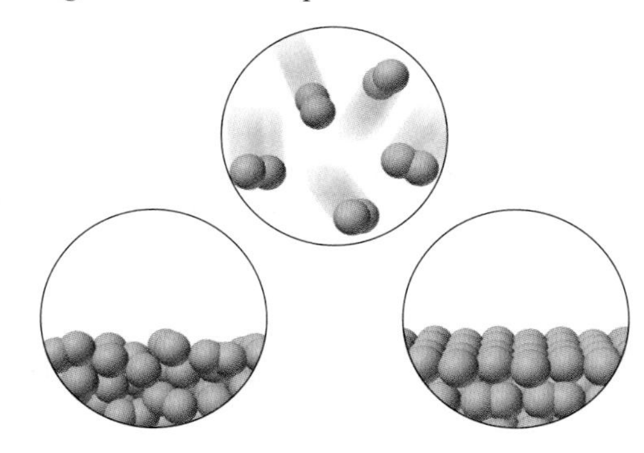

30. Which of the diagrams in Problem 29 represents gaseous bromine?

31. Describe the difference between liquid bromine and gaseous bromine in terms of the kinetic energy of the bromine molecules.

32. Explain why heat always "flows" from a hot substance to a cold substance when they are in contact.

33. For each substance listed below, how much heat (in joules) must be supplied to raise the temperature of 50 g of the substance from 20°C to 40°C?

| Substance | Specific Heat Capacity ($J\ g^{-1}\ °C^{-1}$) |
|---|---|
| a) Graphite | 0.711 |
| b) Water | 4.184 |
| c) Cement | 0.88 |
| d) Steel | 0.45 |
| e) Granite | 0.79 |

34. For each substance listed below, how much heat (in joules) must be supplied to raise the temperature of 100 g of the substance from 10°C to 25°C?

| Substance | Specific Heat Capacity ($J\ g^{-1}\ °C^{-1}$) |
|---|---|
| a) Gold | 0.128 |
| b) Copper | 0.387 |
| c) Ammonia | 4.70 |
| d) Lithium | 3.6 |
| e) Glass | 0.84 |

35. Compute the amount of energy released in kJ when 1.00 g of a substance is converted completely into energy.

36. Compute the amount of energy released in kJ when 0.0100 g of a substance is converted completely into energy.

37. Chemists frequently say, "Mass is conserved." Why is this statement not completely accurate?

38. High-energy physicists frequently say, "Energy is conserved." Why is this statement completely accurate?

### SECTION 4.4 Electricity and Chemistry

39. Describe an experiment that links chemistry and electricity.

40. Name a device that converts chemical energy into electricity.

41. Use Franklin's theory of electricity to interpret the gold leaf electroscope experiment shown in Figure 4.15.

42. Use Du Fay's theory of electricity to interpret the gold leaf electroscope experiment shown in Figure 4.15.

43. In the early part of the eighteenth century, scientists used tin foil to make Leyden jars because tin was readily available. What kind of foil would you use if you were making a Leyden jar? Why?

44. In the early part of the eighteenth century, scientists used a brass rod in Leyden jars because brass was readily available. What kind of metal rod would you use if you were making a Leyden jar? Why?

45. How would you go about placing an electrical charge in a Leyden jar?

46. Look at the Leyden jar in Figure 4.17. Which parts are insulating, and which parts are conducting?

### SECTION 4.5 From the Indivisible Atom to Details of Atomic Structure

47. Why was a large part of the scientific community skeptical about Arrhenius's proposal that table salt forms ions when dissolved in solution?

48. What aspect of Arrhenius's proposal supported the idea that chemistry and electricity are closely linked?

49. What experimental result led Thomson to conclude that atoms had structure, contrary to Dalton's atomic theory?

50. Explain why Thomson's model of the atom is much more sensible than Rutherford's in light of the electrical forces of attraction and repulsion.

### SECTION 4.6 The Periodic Table, Part I

51. Using the information in Table 4.1, show by calculation that the mass of a proton is 1837 times greater than that of an electron.

52. Using the information in Table 4.1, show by calculation that the mass of a neutron is 1839 times greater than that of an electron.

53. How many protons are there in the nucleus of one atom of each of the following elements?
a. He b. O c. Au d. U e. Mg

54. How many protons are there in the nucleus of one atom of each of the following elements?
a. Cr b. Ag c. Pu d. Li e. Ni

55. How many electrons are in a neutral atom of each of the elements listed in problem 53?

56. How many electrons are in a neutral atom of each of the elements listed in problem 54?

57. What element has atoms whose nuclei contain the number of protons given?
a. 5 b. 15 c. 24 d. 30 e. 80

58. What element has atoms whose nuclei contain the number of protons given?
a. 9 b. 19 c. 29 d. 35 e. 82

59. For what element do neutral atoms contain the number of electrons given?
a. 1 b. 3 c. 12 d. 13 e. 16

60. For what element do neutral atoms contain the number of electrons given?
a. 18 b. 38 c. 47 d. 50 e. 53

### SECTION 4.7 Picturing the Atom

61. Draw a sketch of a carbon atom similar to those in Figure 4.22.

62. Draw a sketch of an oxygen atom similar to those in Figure 4.22.

### SECTION 4.8 Electrical Charges and Ions

63. How many electrons are present in each of the following ions?
a. $Na^+$ b. $Mg^{2+}$ c. $O^{2-}$ d. $F^-$ e. $Fe^{3+}$ f. $N^{3-}$ g. $Br^-$ h. $Pb^{2+}$ i. $Al^{3+}$ j. $Ag^+$

64. How many electrons are present in each of the following ions?
a. $Ca^{2+}$ b. $K^+$ c. $Cl^-$ d. $S^{2-}$ e. $Fe^{2+}$ f. $I^-$ g. $Li^+$ h. $Sc^{3+}$ i. $P^{3-}$ j. $Zn^{2+}$

65. Fill in any missing information in the table below: the symbol of an element, its number of protons, its number of electrons, its total charge, and whether it is a cation, an anion, or a neutral species.

| | Symbol | Number of Protons | Number of Electrons | Charge | Cation, Anion, or Neutral |
|---|---|---|---|---|---|
| (a) | He | | | 0 | |
| (b) | K | | 18 | | |
| (c) | P | | 18 | | |
| (d) | O | | 10 | | |
| (e) | | 9 | 10 | | |
| (f) | | 35 | | −1 | |
| (g) | Na | | | | Neutral |
| (h) | | | 18 | −2 | |
| (i) | | 53 | | −1 | |
| (j) | | 11 | 10 | | |
| (k) | Ti | | 22 | | |
| | | | | | |

66. Fill in any missing information in the table below: the symbol of an element, its number of protons, its number of electrons, its total charge, and whether it is a cation, an anion, or a neutral species.

| | Symbol | Number of Protons | Number of Electrons | Charge | Cation, Anion, or Neutral |
|---|---|---|---|---|---|
| (a) | | 3 | 2 | | |
| (b) | | 20 | | +2 | |
| (c) | | 26 | 23 | | |
| (d) | | 10 | 10 | | |
| (e) | Ca | | 20 | | |
| (f) | | 1 | | −1 | |
| (g) | | 24 | 21 | | |
| (h) | | | 10 | +3 | |
| (i) | Zn | 53 | | +2 | |
| (j) | | 7 | 10 | | |
| (k) | | 81 | | +1 | |

## SECTION 4.9 Isotopes

67. Consider the following descriptions of atoms.
    1. 6 protons, 6 neutrons, 6 electrons
    2. 7 protons, 7 neutrons, 10 electrons
    3. 6 protons, 7 neutrons, 8 electrons
    4. 5 protons, 6 neutrons, 2 electrons

    a. Which of these atoms are isotopes of each other?
    b. Which atom has a $-2$ charge?
    c. Which atom has a $+3$ charge?
    d. Which atom is boron?
    e. Which atom is neutral?

68. Consider the following descriptions of atoms:
    1. 7 protons, 7 neutrons, 10 electrons
    2. 7 protons, 8 neutrons, 7 electrons
    3. 8 protons, 8 neutrons, 10 electrons
    4. 6 protons, 7 neutrons, 6 electrons

    a. Which of these atoms are isotopes of each other?
    b. Which atom has a $-2$ charge?
    c. Which atom has a $-3$ charge?
    d. Which atom is carbon?
    e. Which atoms are neutral?

69. Consider the following descriptions of atoms:
    1. 19 protons, 20 neutrons, 18 electrons
    2. 35 protons, 44 neutrons, 36 electrons
    3. 30 protons, 35 neutrons, 28 electrons
    4. 35 protons, 46 neutrons, 35 electrons

    a. Which of these atoms are isotopes of each other?
    b. Which atom has a $-1$ charge?
    c. Which atom has a $+2$ charge?
    d. Which atom is zinc?
    e. Which atom is neutral?

70. Consider the following descriptions of atoms:
    1. 11 protons, 12 neutrons, 10 electrons
    2. 11 protons, 14 neutrons, 11 electrons
    3. 17 protons, 20 neutrons, 18 electrons
    4. 16 protons, 16 neutrons, 18 electrons

    a. Which of these atoms are isotopes of each other?
    b. Which atom has a $+1$ charge?
    c. Which atom has a $-2$ charge?
    d. Which atom is sulfur?
    e. Which atom is neutral?

## SECTION 4.10 Force

71. Give an example of uniform motion that is not mentioned in this chapter.

72. Give an example of accelerated motion that is not mentioned in this chapter.

73. For each of these activities, describe the motion as either uniform or accelerated.
 a. Taking a train ride
 b. Traveling in a space shuttle orbiting the earth
 c. Traveling in a space shuttle during a landing
 d. Riding in an elevator

74. For each situation described, classify the motion as either uniform or accelerated.
 a. Standing on the moon
 b. Playing basketball
 c. Riding a bus going 50 mi $h^{-1}$
 d. Riding a bus in the city

**SECTION 4.11 The Electromagnetic Force and Its Importance in Chemistry**

75. a. Which of the four ions shown in the figure is attracted most strongly to ion Q?
 b. Which of the four ions shown is attracted least strongly to ion Q?

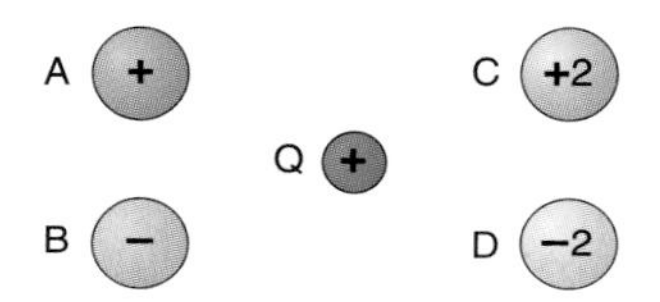


76. a. Which of the four ions in the figure in Problem 75 most strongly repels ion Q?
 b. Which of the four ions shown repels ion Q, but less strongly than the ion in part (a)?

77. Why does it take approximately twice as much energy to separate $Mg^{2+}$ ions from $Cl^-$ ions as it does to separate $Na^+$ ions from $Cl^-$ ions?

78. Why does it take approximately four times as much energy to separate $Mg^{2+}$ ions from $O^{2-}$ ions as it does to separate $Na^+$ ions from $Cl^-$ ions?

79. Show how three polar molecules (like the one shown below) would align themselves vertically in a solid.

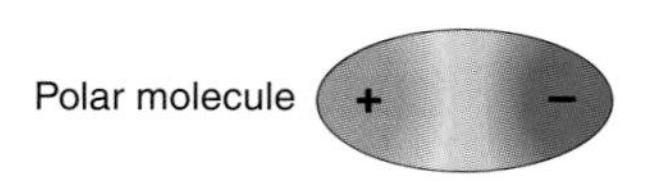


80. Show how three polar molecules (shown in Problem 79) would align themselves horizontally in a solid.

**ADDITIONAL PROBLEMS**

81. Experimental Exercise 4.1 pointed out that Franklin's original idea for studying electricity was to place a metal rod in the spire of a church. Franklin decided to perform the kite experiment instead, but it still bore fruit as one of his most important inventions. What was the invention?

82. Suppose you had an experimental apparatus similar to the cathode ray tube shown in Figure 4.19 except that you were studying protons instead of electrons. Show the direction the protons would travel down the tube. How would the protons be deflected under the influence of the external electric field?

83. Tritium does not occur naturally anywhere on earth. Why not? (See the marginal note on p. 131 for a description of tritium).

84. Restate Coulomb's Law in your own words.

## SOLUTIONS TO EXERCISES

**Exercise 4.1**

You use Equation 4.1 to solve this problem.

$$KE = \tfrac{1}{2}mv^2 = \tfrac{1}{2}(80 \text{ kg})(20 \text{ m s}^{-1})^2$$
$$= 1.6 \times 10^4 \text{ kg m}^2 \text{ s}^{-2} = 1.6 \times 10^4 \text{ J} = 16 \text{ kJ}$$

If you compare Example 4.1 with Exercise 4.1, you will notice that doubling the velocity increases the kinetic energy by a factor of 4.

**TOOL BOX Exercise 1**

The key to solving this problem is to remember that the units on the time axis are $s^2$. Since you were given a time of 4.5 s, you must calculate the value of $(4.5 \text{ s})^2$, which is 20 $s^2$. The answer is about $2 \times 10^2$ m $s^{-1}$.

**Exercise 4.2**

As the car accelerates, the potential energy of the fuel is converted partly into heat and partly into kinetic energy. (More specifically, the potential energy of the fuel plus oxygen from the atmosphere is converted to concentrated heat in the engine, and the engine converts some of this concentrated heat into the car's kinetic energy and dissipates the rest). When the car brakes, its kinetic energy is converted to heat, which dissipates into the atmosphere.

**Exercise 4.3**

**a.** The potential energy is approximately 275 kJ. The kinetic energy is about 100 kJ. The heat energy is approximately 75 kJ. The total energy is still 450 kJ.

**b.** The potential energy is large but has just begun to decrease at 450 m, and the kinetic energy is small and has just begun to increase. This is just past the last high point of the roller coaster ride.
**c.** The brakes are applied at about 500 m into the ride; the amount of heat energy suddenly begins to increase rapidly.

### Exercise 4.4

A refrigerator's motor normally pumps heat from the inside to the outside of the closed refrigerator. Since both the inside and the outside of the open refrigerator are in the room, they will tend to reach room temperature. The motor will run constantly trying to cool the inside, giving off heat. The room warms up (in the long run) if you open the refrigerator door.

### EXPERIMENTAL EXERCISE 4. I

1. The dry hemp string was not a good electrical conductor. When the string became wet, the rain water increased the electrical conductivity of the system and allowed an electric current to pass down the string.
2. The electric fluid was extracted from the air by the kite, moved down the string, and concentrated in its filaments. Since the filaments contained an excess of the electric fluid (and were therefore positively charged, according to Franklin's theory), they repelled each other.
3. The explanation is similar in this case. The kite extracted either vitreous or resinous fluid from the atmosphere, which concentrated in the filaments. The excess fluid concentrated in the filaments caused them to repel each other.
4. Both theories do. In Franklin's theory, the electric fluid flows from one object to another. This suggests that one object acquires an excess of the fluid and is positively charged, while the other is left with a deficit of the fluid and is negatively charged. In Du Fay's theory, if one fluid is removed from an object, an excess of the other will remain.
5. Corks electrified by glass attract corks electrified by resin and repel other corks electrified by glass.
6. The simple explanation is that the electric fluid is a product of the chemical reaction.
7. Check to see if the key attracts or repels a cork electrified by glass (or resin). A glass-electrified cork contains the vitreous fluid, according to Du Fay; so if the key attracts the cork, then the key contains resinous fluid. If it repels the cork, the key contains vitreous fluid.

### Exercise 4.5

**a.** (1) 7; (2) 26; (3) 2; (4) 16
**b.** (1) Tin (Sn); (2) gold (Au)

### Exercise 4.6

**a.** The charge is $(+7) + (-10) = -3$. The symbol is $N^{3-}$; it is an anion.
**b.** All atoms of aluminum contain 13 protons. If this one has a charge of $+3$, it contains 3 fewer electrons than protons, or $13 - 3 = 10$ electrons; its symbol is $Al^{3+}$.
**c.** Because the atom contains 53 protons, it is an iodine ion. Its charge is $(+53) + (-54) = -1$; its symbol is $I^-$.

### Exercise 4.7

**a.** The nuclear charge is $+15$; the nucleus contains 15 protons.
**b.** $15 - 18 = -3$.
**c.** 15 amu + 16 amu = 31 amu
**d.** Phosphorus

### Exercise 4.8

**a.** Ion B is most strongly attracted to ion Q. Q has a positive charge, so it attracts negative ions and repels positive ions. The force of attraction is stronger at shorter distances, so ion B is attracted more strongly because it is closer to Q than ion D.
**b.** The positive end of the molecule attracts the negative ion. The negative end of the molecule repels the negative ion. (Opposite charges attract; like charges repel.) The positive end of the molecule is closer to the ion, so its force of attraction is stronger than the force of repulsion exerted by the two negative charges. (Forces of attraction and repulsion are stronger at shorter distances.) Overall, the ion attracts the molecule.

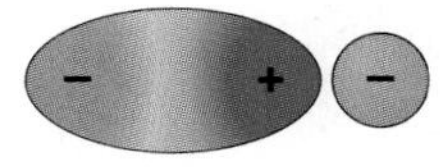

40 42
44 46
48 50

# THE PERIODIC TABLE



KEY TERMS

In Chapters 1 and 4 you were introduced to the history of the periodic table and to some of the atomic structure underlying it. In this chapter you will learn more about how to use it. The periodic table is one of the most important tools chemistry students have at their command. If you know the position of an element on the table, you can predict quite a few things about its physical and chemical properties, even if you have never seen a sample of it. This is possible because the periodic table classifies the elements by similarities in their chemical properties and by trends in their physical properties.

## 5.1 Metals, Nonmetals, and Semimetals

**GOAL:** To classify elements as metals, nonmetals, and semimetals from their positions on the periodic table, and to predict some of their properties based on these classifications.

**STUDY ALERT**

**In chemistry courses students often encounter lists of characteristics of certain types of substances—metals, nonmetals, acids, alcohols, etc. It is important to keep in mind that these are just tendencies. Exceptions can and do occur. This book will tell you the important exceptions (the fact that mercury is a liquid is an example).**

By the middle of the nineteenth century, chemists had isolated and studied several dozen elements. These elements vary considerably in their properties. For example, most of the known elements are solids, but some are gases, and two of them (bromine and mercury) are liquids at room temperature. Some of these elements (for example, sodium) are extremely reactive. Others (platinum among them) don't react with much of anything.

Once an element was isolated, its discoverer and others studied its physical state, color, and odor, its melting and boiling points, and its chemical reactions with a variety of other substances. The result was an enormous amount of information. Mendeleev's publication of the periodic table (see Chapter 1) was greeted with some skepticism, but when the usefulness of the table as a predictor was confirmed by experiment, there must have been a great feeling of relief among chemists. The periodic table makes the study of the elements a lot simpler.

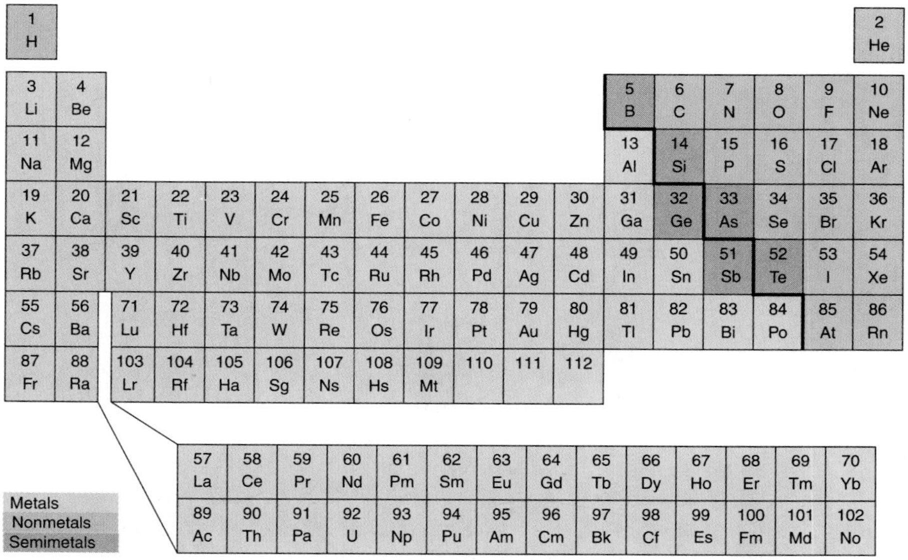

| | | | | | | | | | | | | | | | | | |
|---|---|---|---|---|---|---|---|---|---|---|---|---|---|---|---|---|---|
| 1 H | | | | | | | | | | | | | | | | | 2 He |
| 3 Li | 4 Be | | | | | | | | | | | 5 B | 6 C | 7 N | 8 O | 9 F | 10 Ne |
| 11 Na | 12 Mg | | | | | | | | | | | 13 Al | 14 Si | 15 P | 16 S | 17 Cl | 18 Ar |
| 19 K | 20 Ca | 21 Sc | 22 Ti | 23 V | 24 Cr | 25 Mn | 26 Fe | 27 Co | 28 Ni | 29 Cu | 30 Zn | 31 Ga | 32 Ge | 33 As | 34 Se | 35 Br | 36 Kr |
| 37 Rb | 38 Sr | 39 Y | 40 Zr | 41 Nb | 42 Mo | 43 Tc | 44 Ru | 45 Rh | 46 Pd | 47 Ag | 48 Cd | 49 In | 50 Sn | 51 Sb | 52 Te | 53 I | 54 Xe |
| 55 Cs | 56 Ba | 71 Lu | 72 Hf | 73 Ta | 74 W | 75 Re | 76 Os | 77 Ir | 78 Pt | 79 Au | 80 Hg | 81 Tl | 82 Pb | 83 Bi | 84 Po | 85 At | 86 Rn |
| 87 Fr | 88 Ra | 103 Lr | 104 Rf | 105 Ha | 106 Sg | 107 Ns | 108 Hs | 109 Mt | 110 | 111 | 112 | | | | | | |

| | | | | | | | | | | | | | |
|---|---|---|---|---|---|---|---|---|---|---|---|---|---|
| 57 La | 58 Ce | 59 Pr | 60 Nd | 61 Pm | 62 Sm | 63 Eu | 64 Gd | 65 Tb | 66 Dy | 67 Ho | 68 Er | 69 Tm | 70 Yb |
| 89 Ac | 90 Th | 91 Pa | 92 U | 93 Np | 94 Pu | 95 Am | 96 Cm | 97 Bk | 98 Cf | 99 Es | 100 Fm | 101 Md | 102 No |

Metals
Nonmetals
Semimetals

FIGURE 5.1
The metals, nonmetals, and semimetals are indicated by different colors on this periodic table.

One way the periodic table simplifies the study of the elements is that it clearly separates the metals from the nonmetals. Figure 5.1 shows that the metallic elements (blue) are all on the left side of the table. The nonmetals (purple) are grouped to the right, except for hydrogen, which is perched at the top of the table on the far left. The reason that hydrogen appears out of place is partly historical (it is the smallest element in both mass and size and was therefore placed in the table first) and, as you'll see later, partly theoretical. The semimetals (greenish blue) are located along the boundary between the metals and the nonmetals. These elements have some properties of metals and some of nonmetals. In Figure 5.1 a dividing line like a staircase appears between the metals and nonmetals.

FIGURE 5.2
These typical metals are aluminum, copper, and iron.

Now that you know how to identify an element as a metal, a nonmetal, or a semimetal from the periodic table, what do these terms mean? Each of them tells you about many of the properties of any element in its class. Here are some of the typical properties of **metals:**

- Metals reflect light well, a property that is called *metallic luster.*
- Metals conduct heat well. Put a metal spoon and a plastic one in a refrigerator overnight. When you remove them from the refrigerator, their temperatures will be the same. The metal spoon will feel colder, however, because it conducts the heat away from your hand faster than the plastic spoon can.
- Metals are **malleable** and **ductile.** That is, you can pound or squeeze them into sheets (aluminum foil is an example) and draw them out into wires (copper wire, for instance).
- Metals conduct electricity well. You've probably noticed metal wires (most often copper or aluminum) at the core of electrical wiring.
- Metals are typically solids at room temperature.

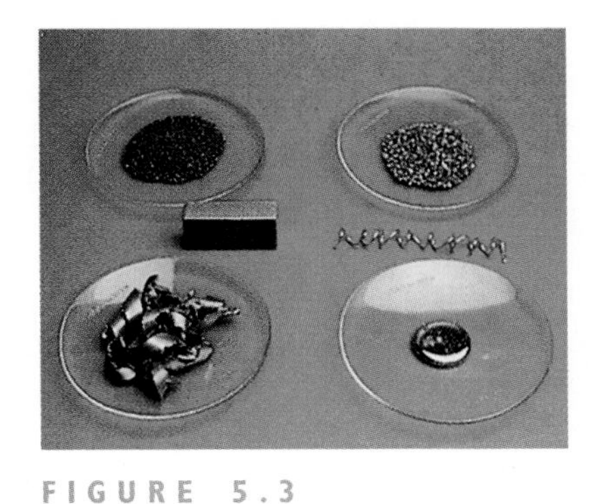

FIGURE 5.3
Mercury (bottom right) is the only metallic element that is a liquid at room temperature. The others, clockwise from mercury, are copper, iron, lead, nickel, and silver.

It is important to note that not all metals have every one of these properties. For example, chromium is a rather brittle metal that cannot be pounded into flat sheets. Mercury metal is a liquid at room temperature; it has a melting point of −39°C. Metals *tend* to have these properties, but the list is not a collection of ironclad rules.

The **nonmetals** also have a set of typical properties that you can use to identify them. As usual, there are exceptions.

- Nonmetallic elements do not have a shiny metallic appearance. They don't reflect light the way metals do.
- Nonmetals are poor conductors of heat. You can see this when you cook hamburgers too close to the coals on a charcoal grill. The outside layer of the hamburger burns, making what is essentially pure carbon. The carbon acts as a thermal insulator that prevents heat from reaching the middle of the hamburger, which takes a long time to cook.
- In solid form nonmetallic elements are **brittle** (that is, they crack and break easily under stress) rather than being malleable and ductile.
- Nonmetals do not conduct electricity well. Graphite (a form of pure carbon) does conduct electricity and is an exception among the nonmetallic elements.
- Nonmetals may be gases, liquids, or solids at room temperature. All of the gaseous elements are nonmetals. There is also one liquid (bromine) and several solids (sulfur and phosphorus among others) on the list of nonmetallic elements.

FIGURE 5.4
The carbon (a nonmetal) on the outside of a burger is a thermal insulator and hinders the flow of heat to the middle of the burger. This increases the cooking time.

FIGURE 5.5
These nonmetals, from left to right, are carbon (graphite), sulfur, chlorine, bromine, and iodine.

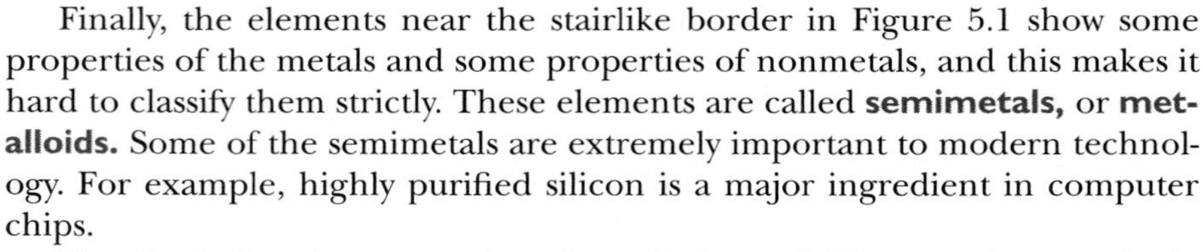

Finally, the elements near the stairlike border in Figure 5.1 show some properties of the metals and some properties of nonmetals, and this makes it hard to classify them strictly. These elements are called **semimetals,** or **metalloids.** Some of the semimetals are extremely important to modern technology. For example, highly purified silicon is a major ingredient in computer chips.

Usually tin has the properties of a typical metal. However, it can exist in another form. Organists during the extreme cold of Russian winters found that this was a serious problem. Their tin organ pipes sometimes crumbled suddenly and collapsed because of a change they called *tin disease.* Chemists eventually found that the atomic structure of tin rearranges itself from a typical metal form to a diamond-like nonmetal form in deep cold. When this happened, the tin organ pipes became brittle, as nonmetals typically are, and shattered under the vibrations caused by the sound.

FIGURE 5.6
These semimetals (metalloids), from left to right, are boron, silicon, arsenic, antimony, and tellurium.

### EXAMPLE 5.1 Classifying Elements from the Periodic Table

Which of the following elements is a metal?
**1.** Co **2.** B **3.** I **4.** As

**SOLUTION**

Only (1), Co (cobalt), lies to the left of the stairstep line in Figure 5.1, which indicates the boundary between metals and nonmetals.

**EXERCISE 5.1**

Which of the following elements is a nonmetal?
**1.** Cu **2.** Ba **3.** Na **4.** Ar

## 5.2 Columns and Rows: Groups and Periods

GOAL: To identify groups and periods in the periodic table

Mendeleev's organizing principle for the periodic table was fairly simple. Elements above and below one another in the table display similar chemical properties. Each set of elements in a vertical column is called a **group.** As you'll see later in this section, the chemical and physical properties of the elements in a particular group vary in a systematic way. The properties of the elements within horizontal rows, called **periods,** also vary systematically.

In Figure 5.8 note that the periods are numbered 1 to 7 on the left side of the periodic table and that the groups are numbered 1A, 2A, 3B, etc., across the top. You can use these designations to find elements in the table. For example, the second-period element in Group 4A is carbon.

FIGURE 5.7
Computer chips made from the semimetal silicon are key components of all computers: (left) a memory chip on a computer keyboard; (right) section of a computer chip magnified 300 times.

Group Numbers / Period Numbers

| Period | 1A | 2A | 3B | 4B | 5B | 6B | 7B | 8B | 8B | 8B | 1B | 2B | 3A | 4A | 5A | 6A | 7A | 8A |
|---|---|---|---|---|---|---|---|---|---|---|---|---|---|---|---|---|---|---|
| 1 | H | | | | | | | | | | | | | | | | | He |
| 2 | Li | Be | | | | | | | | | | | B | C | N | O | F | Ne |
| 3 | Na | Mg | | | | | | | | | | | Al | Si | P | S | Cl | Ar |
| 4 | K | Ca | Sc | Ti | V | Cr | Mn | Fe | Co | Ni | Cu | Zn | Ga | Ge | As | Se | Br | Kr |
| 5 | Rb | Sr | Y | Zr | Nb | Mo | Tc | Ru | Rh | Pd | Ag | Cd | In | Sn | Sb | Te | I | Xe |
| 6 | Cs | Ba | Lu | Hf | Ta | W | Re | Os | Ir | Pt | Au | Hg | Tl | Pb | Bi | Po | At | Rn |
| 7 | Fr | Ra | Lr | Rf | Ha | Sg | Ns | Hs | Mt | | | | | | | | | |

| La | Ce | Pr | Nd | Pm | Sm | Eu | Gd | Tb | Dy | Ho | Er | Tm | Yb |
|---|---|---|---|---|---|---|---|---|---|---|---|---|---|
| Ac | Th | Pa | U | Np | Pu | Am | Cm | Bk | Cf | Es | Fm | Md | No |

FIGURE 5.8
Groups and periods in the periodic table. The four highlighted groups are called the alkali metals (blue), the alkaline earth metals (brown), the halogens (green), and the noble gases (purple).

The groups nearest the left and right sides of the periodic table show the strongest periodic relationships. The elements in Groups 1A, 2A, 7A, and 8A are so distinctive that these groups have special names. Let's look more closely at these groups.

### The Alkali Metals

The metallic members of Group 1A (Li, Na, K, Rb, Cs, and Fr) are called the **alkali metals.** These elements are extremely reactive. No one has ever found any of them in an uncombined state in nature. Very early in the formation of the earth or even before, all the alkali metals reacted with other elements to form compounds. Alkali metals are so reactive that they must be protected from air, water, and most other substances to prevent them from forming compounds.

**Hydrogen is usually shown in Group 1A but sometimes in Group 7A for theoretical reasons (see Section 5.4). Actually it shares very few chemical or physical properties with either group. Like the alkali metals, it reacts with many elements; like the halogens, it forms a diatomic molecule.**

Like most metals, alkali metals are shiny solids when pure and conduct heat and electricity well. They are among the softest metals. You could cut a sample of any of them easily with an ordinary kitchen knife (but don't try it; these elements are not safe to handle without professional training). They also have relatively low melting points. Cesium, with a melting point of 28°C, is a liquid on a warm day.

**All alkali metals react similarly with chlorine and water. Here are sodium's reactions as an example:**

$$2\ Na(s) + Cl_2(g) \longrightarrow 2\ NaCl(s)$$

$$2\ Na(s) + 2\ H_2O(l) \longrightarrow 2\ NaOH(aq) + H_2(g)$$

**A detailed description of reaction equations is given in Chapters 7 and 8.**

The alkali metals react with chlorine to form salts. If you use the letter M as a general symbol for a metal atom, then the formula of all of the alkali metal chlorides is MCl. Specific examples are NaCl, KCl, and CsCl. Samples of these elements react strongly with water to form hydrogen gas and solutions of compounds called *hydroxides.* These hydroxides all have the formula MOH (see Chapters 7 and 8 for more details). We will explore how to predict the formulas of these and many other compounds in Section 5.6.

### The Alkaline Earth Metals

The **alkaline earth metals** (Be, Mg, Ca, Sr, Ba, and Ra) are the members of Group 2A. As typical metals, they are shiny solids that conduct heat and electricity well. They are reactive, although not as reactive as the alkali metals. For example, sodium reacts very strongly with water at room temperature, but magnesium reacts only if the water is boiling (see Figure 5.11). These metals are harder than the alkali metals and have higher melting points. The alkaline earth hydroxides all have the general formula $M(OH)_2$, and the chlorides have the formula $MCl_2$.

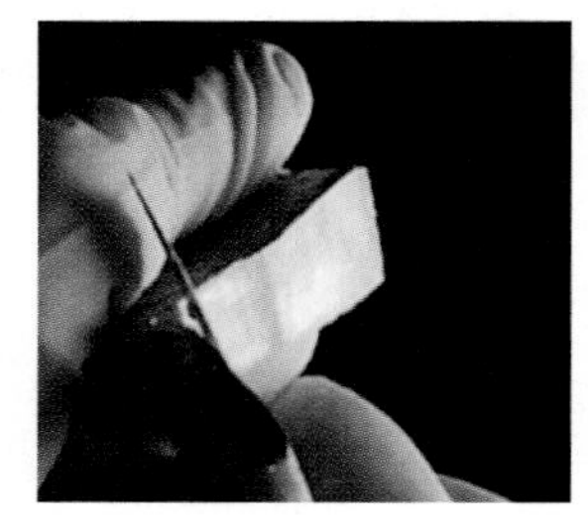

FIGURE 5.9
Chemists store sodium under an inert liquid to isolate the metal from atmospheric oxygen and water vapor. Sodium metal is very soft and easily cut with a knife.

### The Halogens

The members of Group 7A (F, Cl, Br, and I), near the right side of the periodic table, are called the **halogens.** These nonmetals are quite reactive, and none of them exist in nature in an uncombined form. They all form diatomic molecules: $F_2$, $Cl_2$, $Br_2$, $I_2$. Astatine, At, is a highly radioactive, artificial element, and its chemical properties are not well known.

The halogens have different physical states at room temperature. Fluorine is a pale yellow gas, chlorine is a greenish-yellow gas, bromine is a reddish-brown liquid that gives off red vapor, and iodine is a bluish-black solid that gives off purple fumes when you heat it (see Figure 5.12). All of the halogens boil at temperatures lower than the boiling point of mercury, which has the lowest boiling temperature of any metal. The melting and boiling points of the halogens are given in Table 5.1. These illustrate the trends in properties evident in the groups of the periodic table.

All of the halogens form compounds with the Group 1A and Group 2A metals. If the letter X is used as a general symbol for a halogen atom, the formula for all of the compounds of alkali metals and halogens is MX (NaF, NaCl, KBr, CsI, etc.). The formula for all of the compounds of alkaline earth metals and halogens is $MX_2$. All of the compounds of hydrogen and a halogen have the formula HX, and these compounds are all colorless gases at room temperature. Compounds containing a halogen and one other element (except oxygen) are typically called **halides.**

FIGURE 5.10
Sodium metal reacts very readily with chlorine gas to form sodium chloride, which you can see rising as a cloud above the flask.

### The Noble Gases

The elements in Group 8A (He, Ne, Ar, Kr, Xe, and Rn) are called the **noble gases.** As the name implies, these are all gases at room temperature. The no-

**TABLE 5.1 Some Halogen Properties**

| Element | Period | Melting point (°C) | Boiling point (°C) | Atomic Radius (pm) |
|---|---|---|---|---|
| F | 2 | −219 | −188 | 57 |
| Cl | 3 | −101 | −35 | 97 |
| Br | 4 | −7.2 | 59 | 112 |
| I | 5 | 114 | 184 | 132 |

ble gases are extremely unreactive, and all of them are found in nature as uncombined atoms (that is, they are *monatomic*).

For many years it was believed that the noble gases could not undergo chemical reactions at all; thus they were called the *inert gases*. Some theoretical chemists had their doubts, and in 1962 the Canadian chemist Neil Bartlett succeeded in making a compound of xenon and the extremely reactive halogen fluorine. Since then, compounds of xenon and fluorine, xenon and oxygen, and krypton and fluorine have been prepared. This kind of compound is quite rare and usually interesting only to specialists and theoretical chemists.

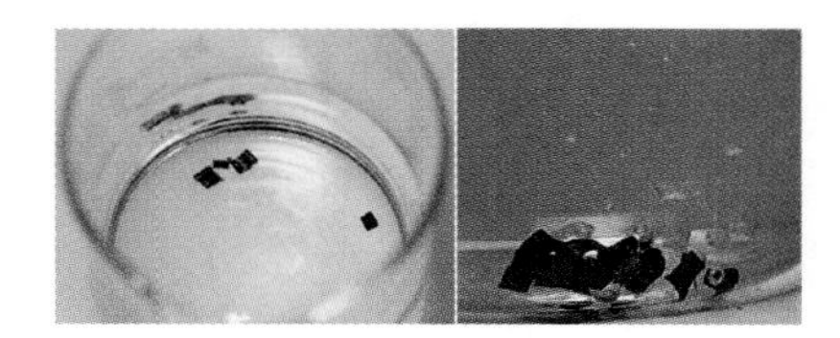

FIGURE 5.11
Magnesium doesn't react with water at room temperature, but it reacts vigorously when the water is boiling.

### Trends Across Periods

The properties of the elements in a period vary in a different way from those in a group. The element at the far left side of a period is an active alkali metal. As you move to the right, the next element you reach is a somewhat less reactive alkaline earth metal. Then, starting in the fourth period, are the metals in the B groups (see Section 5.3). In the third period and above you will then find more metals. Next you will find a semimetal or two, and the nonmetals follow. The nonmetals are more active the closer they are to the right side of the table until you reach the halogens. The halogen is the most reactive of the nonmetals in a period. The last and always the least reactive element in a period is a noble gas.

When metals react, they tend to lose electrons and form cations. Nonmetals typically gain electrons to form anions when they react with metals. When nonmetals react with each other they tend to share electrons. You will find a more complete discussion of these metal-nonmetal and nonmetal-nonmetal interactions in Chapter 10.

The formulas of the compounds that the elements in a period form also vary, usually in a systematic way. For example, the formulas of the compounds that contain one atom of a second-period element and chlorine are LiCl, $BeCl_2$, $BCl_3$, $CCl_4$, $NCl_3$, $Cl_2O$, and ClF. None of the other periods are quite this predictable, but the tendency remains.

FIGURE 5.12
These halogens, from left to right, are chlorine ($Cl_2$, a greenish-yellow gas), bromine ($Br_2$, a reddish-brown liquid), and iodine ($I_2$, a bluish-black solid). Fluorine ($F_2$) is not shown because it is so reactive and dangerous that special opaque containers must be used to store it.

### EXAMPLE 5.2 Classifying Elements by Group and Period

Using a periodic table, identify each of the following elements.

**a.** The halogen in the third period
**b.** The fourth-period element in Group 2B

**SOLUTIONS**

**a.** Chlorine (There is no halogen in the first row.)
**b.** Zinc

**EXERCISE 5.2**

Using a periodic table, identify each of the following elements.
**a.** The alkaline earth metal in the fifth period
**b.** The element in the second period and in Group 6A
**c.** The fifth-period element in Group 4B

### EXAMPLE 5.3 Chemical Formulas from the Periodic Table

Give the chemical formula for the compound containing each pair of elements.
**a.** Sodium and bromine
**b.** Calcium and iodine
**c.** Hydrogen and fluorine

**SOLUTIONS**

**a.** Sodium is in Group 1A (alkali metals) and bromine is in Group 7A (halogens). The general formula is MX, so the answer is NaBr.
**b.** Calcium is in Group 2A and iodine is in Group 7A, so the general formula is $MX_2$. The answer is $CaI_2$.
**c.** The general formula for the hydrogen halides is HX, so the compound is HF.

**EXERCISE 5.3**

Give the formula of the compound containing each pair of elements.
**a.** Barium and iodine
**b.** Fluorine and chlorine
**c.** Hydrogen and bromine
**d.** Potassium and fluorine

## 5.3 The Blocks of Elements

**GOAL:** To identify elements as main group elements, transition metals, lanthanides, or actinides.

As chemists discovered elements and filled in the periodic table, they found a pattern. There are only two elements in the first period in the top row of the table. Each of the next two rows (periods 2 and 3) has eight elements, and each of the two rows after that (periods 4 and 5) has eighteen. The last complete row (period 6) has 32 elements.

As you saw in Chapter 4, an atom has a small central nucleus that contains protons and (except for the most common isotope of hydrogen) neutrons and is surrounded by a cloud of electrons. Every atom of any given element contains the same number of protons (that is, has the same atomic number). In neutral atoms the number of electrons is equal to the number of protons.

A typical lithium atom has 3 protons, 4 neutrons, and 3 electrons; a typical cesium atom has 55 protons, 78 neutrons, and 55 electrons. From this viewpoint, lithium and cesium are very different. In chemical behavior, though, the two elements are remarkably similar. They react with hydrogen to form LiH and CsH. They react with chlorine to form LiCl and CsCl and with sulfur to form $Li_2S$ and $Cs_2S$. Each of these elements, when added to water, reacts rapidly and violently to produce a metal hydroxide compound (LiOH or CsOH) plus hydrogen gas ($H_2$).

How can two atoms that are so different react so similarly? The answer lies, as it almost always does in chemistry, with the electrons. Here is the explanation chemists have developed to account for the similarities.

Electrons in atoms behave as if they occupy **shells,** which you can imagine as layers around the nuclei. (In Chapter 6 you'll find out why electrons behave this way.) All of the elements in any given period of the periodic table have the same number of electron shells. Furthermore, the first-period elements have electrons in one shell, the second-period elements have electrons in two shells, and so on. *The number of occupied electron shells in an atom is the same as the number of its element's period* (see Figure 5.18).

Imagine building up an atom by adding one proton and one electron at a time (plus the appropriate number of neutrons). Table 5.2 shows the results for the first 18 elements. Hydrogen and helium, the only first-period ele-

**TABLE 5.2 Electron Counts in Shells for the First Eighteen Elements**

| Atomic Number | Element | First-shell electrons | Second-shell electrons | Third-shell electrons | |
|---|---|---|---|---|---|
| 1 | Hydrogen | 1 | | | |
| 2 | Helium | 2 | | | Noble gas |
| 3 | Lithium | 2 | 1 | | Alkali metal |
| 4 | Beryllium | 2 | 2 | | |
| 5 | Boron | 2 | 3 | | |
| 6 | Carbon | 2 | 4 | | |
| 7 | Nitrogen | 2 | 5 | | |
| 8 | Oxygen | 2 | 6 | | |
| 9 | Fluorine | 2 | 7 | | |
| 10 | Neon | 2 | 8 | | Noble gas |
| 11 | Sodium | 2 | 8 | 1 | Alkali metal |
| 12 | Magnesium | 2 | 8 | 2 | |
| 13 | Aluminum | 2 | 8 | 3 | |
| 14 | Silicon | 2 | 8 | 4 | |
| 15 | Phosphorus | 2 | 8 | 5 | |
| 16 | Sulfur | 2 | 8 | 6 | |
| 17 | Chlorine | 2 | 8 | 7 | |
| 18 | Argon | 2 | 8 | 8 | Noble gas |

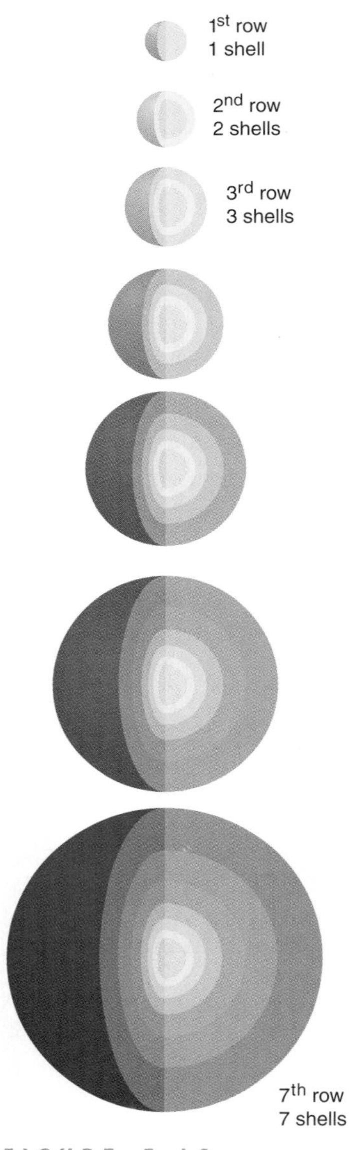


FIGURE 5.18
The number of occupied shells in an atom of a given element is the same as the number of the element's row, or period.

ments, have electrons in the first shell. The two electrons in helium fill the first shell, which has a capacity of only two electrons.

This is the first example of a general rule: The arrangement of electrons in a noble gas—the **electron configuration** of a noble gas—is particularly **stable.** This means that such configurations are easy to form and difficult to change once formed. Another way of saying this is that such configurations have *low potential energy.*

Now consider adding one more proton and one more electron (as well as two neutrons) to helium, making a lithium atom. Because the first electron shell is already filled, the third electron becomes the only electron in the second shell. In Table 5.2 you see that the noble gas neon (atomic number 10) has two electrons in the first shell and eight electrons—the maximum electron capacity—in the second shell. Neon is followed by sodium, the next alkali metal, in which the eleventh electron is the only electron in the third shell.

Every one of the alkali metals (Group 1A) has a similar electron configuration: there is exactly one electron in the outermost occupied shell, and all the inner shells are identical to those of the preceding noble gas. That is why the alkali metals are so similar chemically—they have the same number of electrons in their outer shells. This is an example of another general rule: elements with similar electron configurations react in similar ways.

The electrons in the outermost occupied shell of an atom are called the **valence electrons.** In the chemical reactions of main group elements (and in many reactions of the transition metals), these are the only electrons that take part. All members of a group in the periodic table have the same number of valence electrons, so they show similar chemical behavior.

For the main group elements (except helium), the number of valence electrons is equal to the group number. For example, all of the alkali metals (Group 1A) have one electron in their valence shell and all of the halogens (Group 7A) have seven valence electrons. Helium, a noble gas (Group 8A), is an exception since it takes only two electrons to fill the first shell.

### EXAMPLE 5.5 The Number of Shells and Valence Electrons

For each of the following elements, (1) how many occupied shells of electrons, and (2) how many valence electrons are there in a neutral atom?
**a.** Na **b.** Si **c.** C **d.** Sr

**SOLUTIONS**

**a.** Sodium is in the third period, and Group IA, so its neutral atom has three occupied shells of electrons with one valence electron.
**b.** Silicon is also in the third period, so it has three occupied electron shells. It's in Group 4A, so it has four valence electrons.
**c.** Carbon is in the second period, so it has two shells with electrons in them, and it's in Group 4A, so (like silicon) it has four valence electrons.

**d.** Strontium is in the fifth period, so it has five shells with electrons in them, and it's in Group 2A, so it has two valence electrons.

**EXERCISE 5.5**

For each of the following elements, tell (1) how many occupied shells of electrons and (2) how many valence electrons there are in a neutral atom.

**a.** Cl  **b.** Ra  **c.** As  **d.** O

## 5.5 Subshells and the Blocks of Elements

**GOAL:** To explain how the shell model relates to the layout of the periodic table.

Why do the periods at the bottom of the periodic table have more elements than the ones at the top? The explanation lies in the details of the structure of the electron shells.

The first shells are small and close to the nucleus and have relatively little capacity for electrons. The innermost shell can hold only two electrons, so there are only two elements in the first period. The second shell can hold a maximum of eight electrons. This gives us eight elements in the second period. To understand what happens after that, we need to take a closer look at the structure of the shells. The first shell is the smallest of all the shells. The electrons that occupy this shell are called, in general terms, *s electrons.* They are also called 1*s* electrons, where the number 1 is the shell number. This shell can hold either one 1*s* electron (as in a hydrogen atom) or two 1*s* electrons (as in a helium atom). These two electrons fill the first shell. Additional electrons must go in the second shell.

The second shell consists of two **subshells.** The first subshell of the second shell is similar to the 1*s* shell, though it extends further away from the nucleus. This subshell can accommodate two electrons, which are called *2s electrons.* The second subshell of the second shell can hold up to six electrons. Electrons in this subshell are called *2p electrons.*

For example, a carbon atom (in Group 4A) has four *valence* electrons: two in the 2*s* subshell and two in the 2*p* subshell. Oxygen (Group 6A) has six valence electrons, two in the 2*s* subshell and four in the 2*p* subshell. The completely filled second shell of a neon atom contains two 2*s* electrons and six 2*p* electrons.

Scientists often use a shorthand notation to write the electron configurations of the atoms. For a hydrogen atom the electron configuration is $1s^1$. The 1 stands for the first shell, the *s* tells you what subshell the electrons are in, and the superscript 1 tells you how many electrons are in that subshell. The electron configuration for oxygen is written $1s^2 2s^2 2p^4$, and that for neon is $1s^2 2s^2 2p^6$.

H; 1 electron in first shell.

He; 2 electrons in first shell.

Li; 2 electrons in the first shell and 1 electron in the second shell.

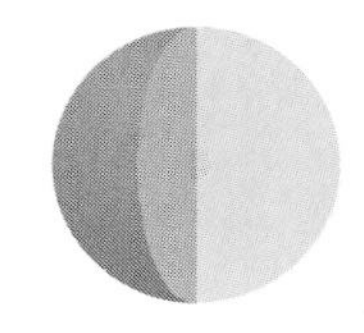

Ne; 2 electrons in the first shell and 8 electrons in the second shell.

FIGURE 5.19

Let's take a closer look at writing an electron configuration, using lithium as an example. Lithium has the atomic number 3, so we need to put three electrons into shells. We start with the innermost shell (the 1*s* shell) and fill it up with two electrons. Next, the last electron goes into the next shell or valence shell. The first one or two electrons to go in a shell always go to the first subshell. This means that the third electron in lithium is a 2*s* electron. There are no more electrons in a lithium atom, so its complete electron configuration is $1s^22s^1$. All of the alkali metals have a single *s* electron in their valence shells. Lithium contains a 2*s* electron, sodium a 3*s* electron, potassium a 4*s* electron, and so on.

An atom of beryllium (Group 2A, atomic number 4) contains two *s* electrons in its first shell and two more *s* electrons in its valence shell. The electron configuration is $1s^22s^2$. Atoms of all the other alkaline earth metals also have two *s* electrons in their valence shells. Boron (atomic number 5) contains two 2*s* electrons and one 2*p* electron in its valence shell. The complete electron configuration of boron is $1s^22s^22p^1$.

The electrons of the elements in the third period of the periodic table fill up the 3*s* and 3*p* subshells the same way the electrons of the elements in the second period fill the 2*s* and 2*p* subshells. The electron configurations of the elements in the first three periods are given in Table 5.3. You might compare this table with Table 5.2, which shows the numbers of electrons in the various shells for the first 18 elements.

The third electron shell contains three subshells. The first two subshells, 3*s* and 3*p*, are similar to the subshells of the second shell. Both subshells have diameters larger than the second-period subshells but smaller than the third subshell. This subshell is designated 3*d*, and it can hold up to ten electrons. You might think this subshell would fill right after the 3*p* subshell, but this isn't what happens.

**TABLE 5.3 Electron Configurations of the First Eighteen Elements**

| | | |
|---|---|---|
| H | $1s^1$ | |
| He | $1s^2$ | Noble gas |
| Li | $1s^22s^1$ | Alkali metal |
| Be | $1s^22s^2$ | |
| B | $1s^22s^22p^1$ | |
| C | $1s^22s^22p^2$ | |
| N | $1s^22s^22p^3$ | |
| O | $1s^22s^22p^4$ | |
| F | $1s^22s^22p^5$ | |
| Ne | $1s^22s^22p^6$ | Noble gas |
| Na | $1s^22s^22p^63s^1$ | Alkali metal |
| Mg | $1s^22s^22p^63s^2$ | |
| Al | $1s^22s^22p^63s^23p^1$ | |
| Si | $1s^22s^22p^63s^23p^2$ | |
| P | $1s^22s^22p^63s^23p^3$ | |
| S | $1s^22s^22p^63s^23p^4$ | |
| Cl | $1s^22s^22p^63s^23p^5$ | |
| Ar | $1s^22s^22p^63s^23p^6$ | Noble gas |

The filling of shells in elements of the fourth period of the periodic table is not quite so orderly as for the first three periods. The $4s$ subshell has a lower potential energy than the $3d$ subshell, and it fills first. (The reasons for this are quite subtle and are beyond the scope of this text.) The next element after argon is potassium, and its additional electron goes into the $4s$ subshell. The next element, calcium, contains two $4s$ electrons. The electron configurations of these two elements are written this way:

K $1s^22s^22p^63s^23p^64s^1$ Ca $1s^22s^22p^63s^23p^64s^2$

Since it is often cumbersome to write out complete electron configurations when atoms contain many electrons, scientists often use an abbreviated format for these electron configurations. As Table 5.3 shows, argon has a filled valence shell with a configuration of $1s^22s^22p^63s^23p^6$. The electron configuration for potassium contains this set of electrons plus its valence electron. Therefore, scientists write the electron configuration for potassium as $[Ar]4s^1$. The [Ar] in this electron configuration stands for $1s^22s^22p^63s^23p^6$, the electron configuration of argon. In this abbreviated notation the configuration for calcium is $[Ar]4s^2$, where [Ar] represents the inner electrons and $4s^2$ represents the valence electrons.

To the right of calcium in the periodic table is scandium. In this element all lower-energy subshells are full, so electrons start to add to the $3d$ subshell. Since all the transition elements have $d$ electrons in the valence shell, they are sometimes called *d-block elements.* The electron configuration of scandium is $[Ar]4s^23d^1$.

With the last transition metal in the fourth period, the $3d$ shell is filled. Zinc contains ten $3d$ electrons and has an electron configuration of $[Ar]4s^23d^{10}$. The next element is gallium, which has one $4p$ electron. The electron configuration of gallium is $[Ar]4s^23d^{10}4p^1$.

The fourth period contains 18 elements. The last element in this period is krypton, which has the electron configuration $[Ar]4s^23d^{10}4p^6$. The fifth period also has 18 elements, and ends with xenon ($[Kr]5s^24d^{10}5p^6$. These two periods contain ten elements more than periods 2 and 3, corresponding to the ten electrons each in the $3d$ and $4d$ orbitals.

**We didn't discuss the electron configurations of the transition metals. You can study the details of the filling process in a more advanced course.**

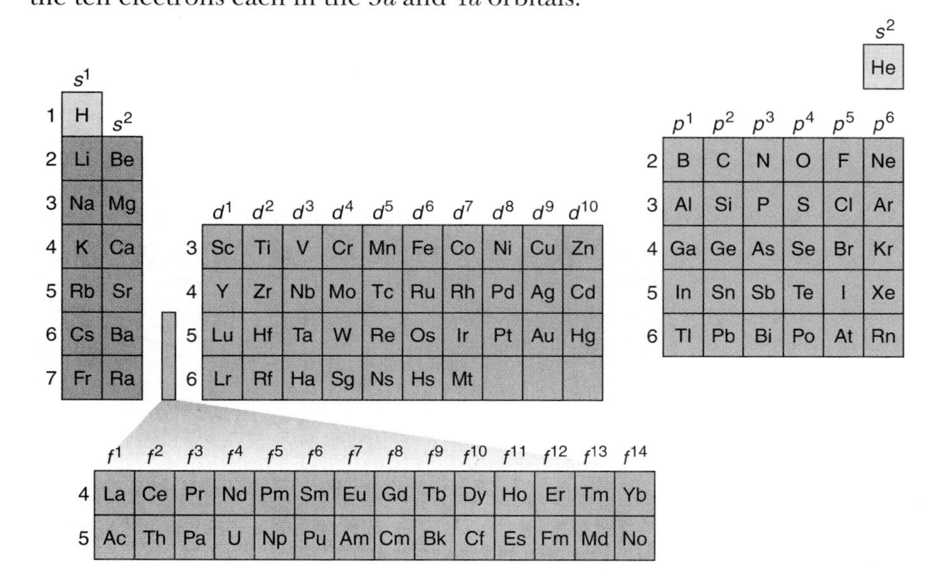


FIGURE 5.20
Elements can also be classified by what subshells are being filled in the valence shell. The transition metals add $d$ electrons, and the lanthanides and the actinides add $f$ electrons. This method subdivides the main group elements into an $s$ block and a $p$ block. Note that helium is an $s$-block element.

The sixth period introduces a fourth kind of subshell, with a capacity of 14 electrons. Electrons that occupy this type of subshell are called *f electrons.* Chemists sometimes refer to the lanthanides and actinides as *f-block elements.* The sixth period contains 32 elements; 14 more than the fifth period.

**STUDY ALERT**

**The electron classification scheme discussed in this section can be summarized as follows. The electrons that fill a particular kind of subshell have names. The electrons that fill the first subshell are called *s electrons.* The second subshell, the one that begins to fill with boron in the second period, contains *p electrons.* The electrons in the third (transition metal) subshell are called *d electrons,* and the ones in the fourth (lanthanide and actinide) subshell are called *f electrons.* These subshells define the blocks in the periodic table (see Figure 5.20).**

### EXAMPLE 5.6 Classifying Valence Electrons

How many valence electrons does an atom of each of the following elements have? What is the electron configuration of each?

**a.** C **b.** Al **c.** Cl **d.** Br

**SOLUTIONS**

**a.** Carbon is in Group 4A, so it has four valence electrons. Its electron configuration is $1s^22s^22p^2$ or $[He]2s^22p^2$.

**b.** Aluminum is in Group 3A, so it has three valence electrons. The electron configuration is $[Ne]3s^23p^1$.

**c.** Chlorine is in Group 7A, so it has seven valence electrons. Its electron configuration is $[Ne]3s^23p^5$.

**d.** Bromine is in Group 7A. It has seven valence electrons, as predicted by its group number. Because bromine is to the right of the fourth-period transition metals, its $3d$ subshell is filled. Bromine's electron configuration is $[Ar]4s^23d^{10}4p^5$.

**EXERCISE 5.6**

How many valence electrons does an atom of each of the following elements have? What is the electron configuration of each?

**a.** P **b.** O **c.** Cs

## 5.6 Some Periodic Trends

**GOAL:** To understand how variations in atoms' relative sizes, ability to attract electrons, and number of bonds formed are related to the elements' positions on the periodic table.

You can use the periodic table to predict many things about the elements. For example, as you saw in Section 5.2, you can use the table to predict the formulas of at least some of the compounds an element forms. (There is more on this topic later in this section and in Chapter 7). Properties are the most predictable for the main group elements, especially those near the top and sides of the periodic table.

The elements in the center and near the bottom of the periodic table—particularly the transition metals, the lanthanides, and the actinides—have more than their share of peculiarities. There are sometimes exceptions among them to the overall trends of the table. The properties of hydrogen are also hard to predict from the periodic table, mainly because hydrogen has no inner shell electrons at all.

In this section we look at three properties of elements that can be predicted most easily from the elements' positions on the periodic table. These are the relative sizes of atoms and ions, the ability of an atom to attract electrons to itself in a compound, and combining properties. This last property provides some simple rules for predicting the formulas of substances made from main group elements.

### Atomic and Ionic Radii

The measure chemists typically use for the size of an atom is the **atomic radius.** You might think of this as the radius of the atom's electron cloud. However, individual electrons are so small and move so fast that they're impossible to pin down exactly. Thus chemists can use the distance between nuclei in diatomic molecules to find the atomic radius. The atomic radius of hydrogen, for instance, is equal to half the distance between the hydrogen nuclei in $H_2$.

**This is one of several different methods chemists can use to measure atomic radii.**

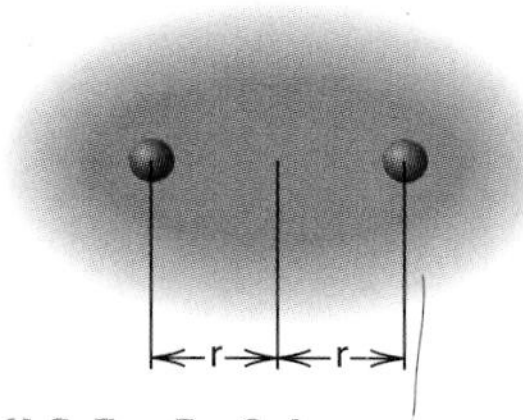


FIGURE 5.21
The atomic radius *(r)* is exactly one-half the distance between atomic nuclei in a homonuclear diatomic molecule.

You know that atoms are very small, and that it would take tens of millions of them to form a line only a centimeter long. Still, some atoms are larger than others. How do you predict which atoms are the bigger ones?

One of the rules should not surprise you. The atoms get larger as you go down a group within the periodic table. Potassium atoms are larger than sodium atoms, which are larger than lithium atoms, for instance. This makes sense because the outermost shells are larger farther down a group in the table.

Within a period of the table the largest atoms are the ones on the left. Lithium atoms are larger than beryllium atoms, which are larger than boron atoms, for instance. This rule might surprise you a little. A nitrogen atom is twice as massive as a lithium atom, but as you can see from Figure 5.22, the radius of the nitrogen atom is barely half that of the lithium atom. Why is the nitrogen atom so much smaller?

The answer comes from some aspects of electronic structure you already know about. The size of the outermost shell of electrons defines the size of an atom. The more strongly the outermost electrons are pulled toward the nucleus, the smaller the atom will be.

**STUDY ALERT**

**In the following discussion, keep Coulomb's Law (discussed in Chapter 4) firmly in mind. In words, the electromagnetic force is larger between greater charges and at smaller distances. The Coulomb's Law equation is $F = k(Q_1Q_2/r^2)$. For the purposes of this section, $F$ is the force of attraction between the nucleus and the valence electron, $Q_1$ is the effective nuclear charge, $Q_2$ is the charge on the electron, and $r$ is the atomic radius.**

Think about what attracts the outermost electron to a lithium atom, whose electron configuration is $1s^2 2s^1$. The valence electron (the $2s$ electron) has a negative charge, so (loosely speaking) it is held in place in the atom by electrostatic attraction to the positive charge of the nucleus. However, the two electrons of the inner shell (the two $1s$ electrons) screen the outer-shell electron from the nucleus. The two inner-shell electrons have a total charge of $-2$, and the nucleus has a total charge of $+3$. The outermost electron, then, is attracted to the lithium atom by a net charge of $+1$. This is called the **effective nuclear charge.**

A check of the Group 1A elements reveals that they all have an effective nuclear charge of $+1$. For example, the electron configuration of a potassium atom is $1s^2 2s^2 2p^6 3s^2 3p^6 4s^1$. The atomic number of potassium is 19, and the inner shells contains 18 electrons (two each in the $1s$, $2s$, and $3s$ subshells, and six each in the $2p$ and $3p$ subshells). The effective nuclear charge then is $(+19) + (-18) = +1$.

Next, let's think about a beryllium atom, with its $1s^2 2s^2$ electron configuration. The nucleus has a $+4$ charge, and the atom has two inner-shell elec-

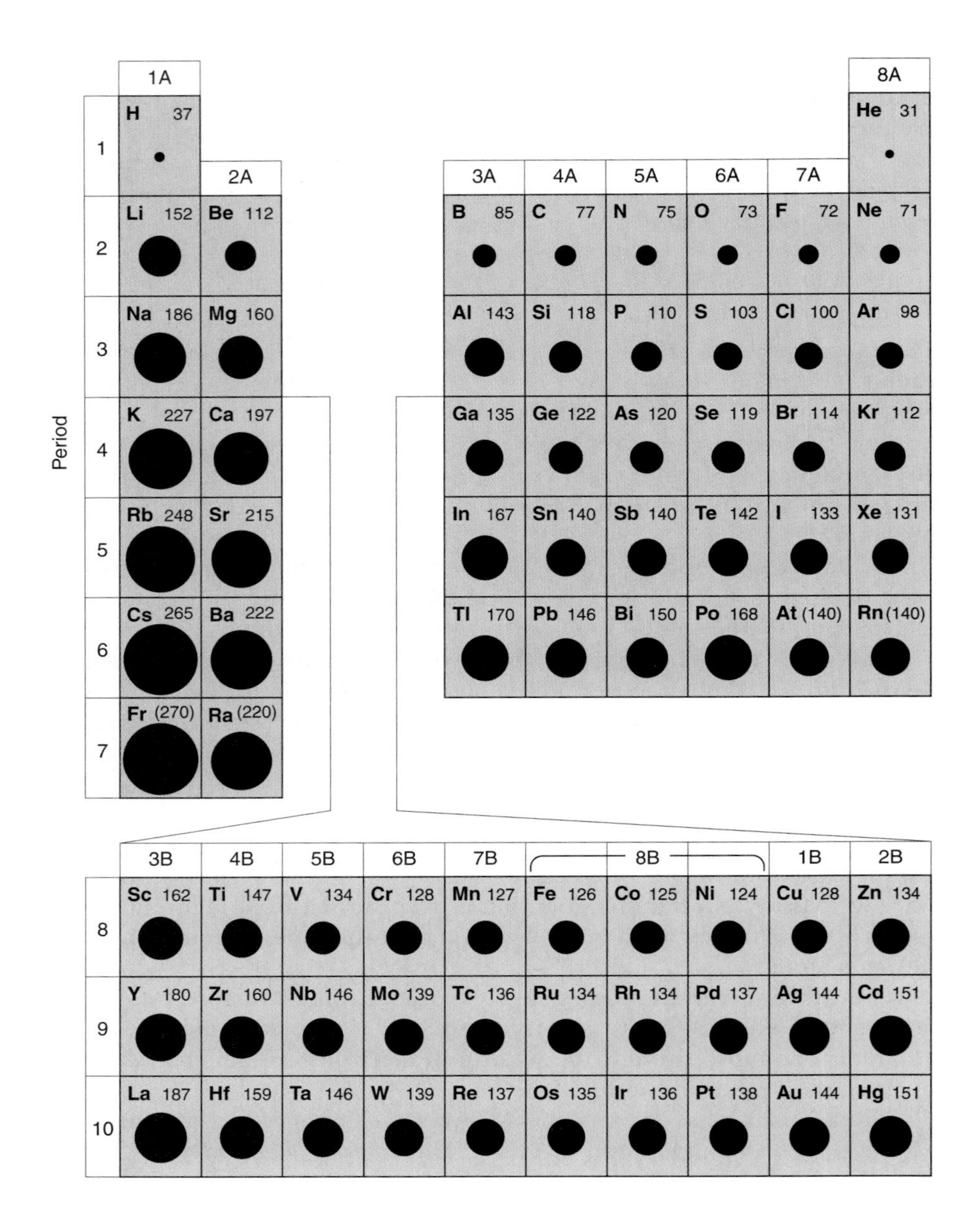


FIGURE 5.22
The atomic radii of the main group elements and the transition metals are given in picometers. The relative sizes of the atoms are drawn to scale.

trons (the two $1s$ electrons). The atom's effective nuclear charge, then, is $+2$. The atom has two outer-shell electrons, and each of them is attracted to the beryllium nucleus by that $+2$ charge.

Once again, by inspecting the periodic table, you can see that all the alkaline earth metals (Group 2A) have the same effective nuclear charge. An atom of magnesium, for example, has an electron configuration of $1s^2 2s^2 2p^6 3s^2$, and it has 12 protons in the nucleus. Since the magnesium atom has two valence electrons ($3s^2$) and ten inner-shell electrons ($1s^2 2s^2 2p^6$), the effective nuclear charge is $(+12) + (-10) = +2$.

This trend continues across the periodic table. **The effective nuclear charge of any main group element is equal to the group number.** For example, the effective nuclear charge of carbon (in Group 4A) is $+4$, and that of iodine (in Group 7A) is $+7$.

It is now relatively easy to understand why atoms get smaller toward the right side of the periodic table. The larger effective nuclear charges have a stronger force of attraction for the valence electrons (remember Coulomb's Law). Since the force pulling the electrons toward the nucleus is stronger, the valence electrons are held closer to the nucleus. This means that a beryllium atom will be smaller than a lithium atom and larger than a boron atom.

Atoms get larger toward the bottom of a group in the periodic table simply because the valence electrons are in a larger valence shell. For example, the valence electron of the lithium atom is the 2*s* electron, and the valence electron of the potassium atom is the 4*s* electron. The effective nuclear charges of these atoms are identical at + 1, but the 4*s* electron is farther from the nucleus than the 2*s* electron. Therefore, a potassium atom has a larger atomic radius than a lithium atom does.

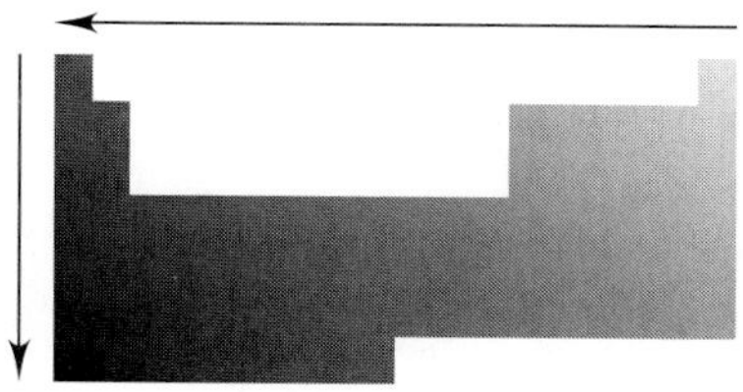

FIGURE 5.23
Trends in atomic radii

You will frequently encounter both cations and anions in the compounds you will study in this course. Table salt (NaCl), for example, contains $Na^+$ ions and $Cl^-$ ions.

How do the radii of ions compare with the radii of their parent atoms? The predictions are relatively easy to understand. A cation such as $Na^+$ contains fewer electrons than the parent atom. Since there are fewer electrons in the cation, the force of attraction of the nucleus for the remaining electrons is greater—the force of attraction per electron is larger. This means that all cations are smaller than their parent atoms.

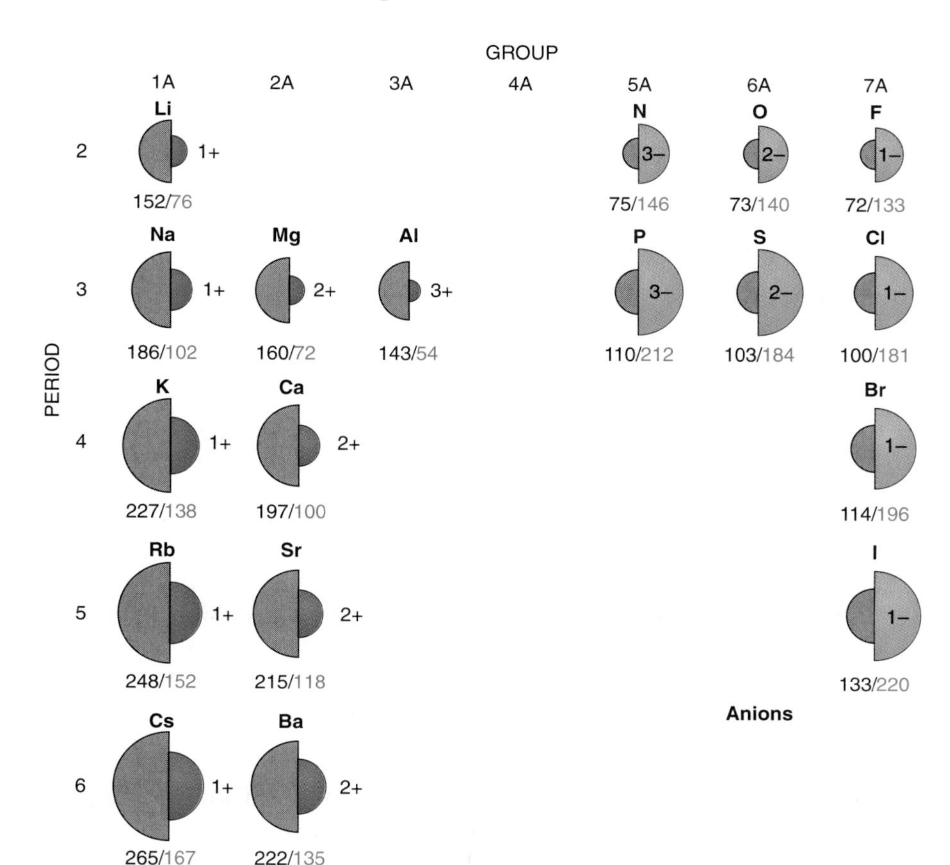


FIGURE 5.24
Cations are smaller than their parent atoms, and anions are larger. Radii are given in pm.

The pattern of electron loss by metal atoms is also important. Briefly, metal atoms have a tendency to lose all of their valence electrons. The ions formed by metals that behave this way contain one fewer occupied shell than the parent atoms, and this is another reason that the cations are smaller than the atoms. For instance, compare the electron configuration of Na ($1s^22s^22p^63s^1$) with $Na^+$ ($1s^22s^22p^6$).

Anions, in contrast, contain *more* electrons than their parent atoms. Therefore the force of attraction of the nucleus per electron decreases. This means that anions are always larger than the atoms they come from. The chloride ion, $Cl^-$, has a larger radius than a chlorine atom, and the hydride ion, $H^-$, has a larger radius than a hydrogen atom.

## EXAMPLE 5.7 Effective Nuclear Charges

**a.** What is the effective nuclear charge of a fluorine atom?
**b.** You can obtain the effective nuclear charge by subtracting the number of inner-shell electrons from the atomic number. What's an easier way?
**c.** Which of the elements in the fragment of the periodic table shown has the *largest* atomic radius?

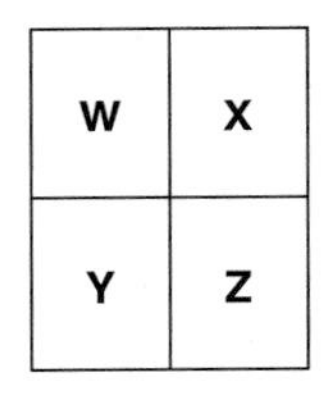


**d.** Which of the elements in the fragment of the periodic table in part (c) has the *smallest* atomic radius?
**e.** Among K, Ca, $K^+$, and $Ca^{2+}$, which has the *largest* radius?

**STUDY ALERT**

**Note the answer to part (d) of Example 5.7.**

**SOLUTIONS**

**a.** The atomic number of fluorine is 9, and it has two inner-shell electrons. Its effective nuclear charge is +7.
**b.** For the main group elements, the effective nuclear charge is equal to the group number.
**c.** Atoms lower down in the periodic table and toward the left side tend to be larger. Element Y has the largest atomic radius.
**d.** You learned how to find the *largest* atomic radius earlier in this section. This question asks for the *smallest* atomic radius. Read all questions carefully, and practice this kind of question as much as you can. If the largest atoms are toward the bottom and the left of the table, the smallest are to the top and the right. Element X has the smallest atomic radius.
**e.** The metal ions have one fewer occupied shell than the corresponding metals. For example, the electron configuration of $Ca^{2+}$ is [Ar], and that of Ca is $[Ar]4s^2$. Therefore the neutral atoms are larger. Potassium is to the left of calcium in the periodic table, so K has the largest atomic radius.

EXERCISE 5.7

**a.** Among S, Cl, Se, and Br, which has the *largest* atomic radius? Which has the *smallest?*

**b.** Among Cl, Br, $Cl^-$, and $Br^-$, which has the *largest* atomic radius? Which has the *smallest?*

### Electronegativity

A second property you can predict from the periodic table is **electronegativity.** This is the ability of the atoms of an element to attract electrons to themselves in a compound. Since electronegativity is a measure of an atom's ability to attract electrons, you will not be surprised to learn that it is related to the effective nuclear charge and the atomic radius.

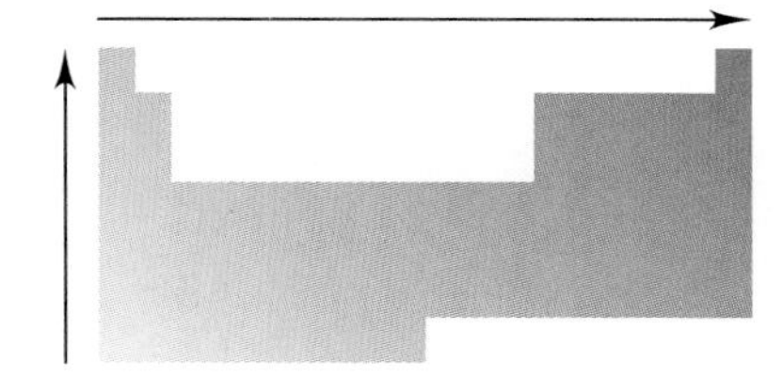

FIGURE 5.25
Trends in electronegativity

Atoms with large effective nuclear charges and small atomic radii attract electrons strongly (Coulomb's Law again)—and these atoms have high electronegativities. The reverse is also true. Large atoms with small effective nuclear charges have low electronegativities.

Knowledge of an element's electronegativity lets you predict the kinds of bonds its atoms can form. Atoms with relatively high electronegativities tend to form ions that have negative charges (they attract electrons strongly). Furthermore, when an atom shares electrons with an atom of another element, the more electronegative atom tends to pull electrons to itself. Both of these points will be covered in more depth in Chapters 7 and 10.

To find the more electronegative atoms in the periodic table, look toward the top and to the right. The top right of the table is where the small ele-

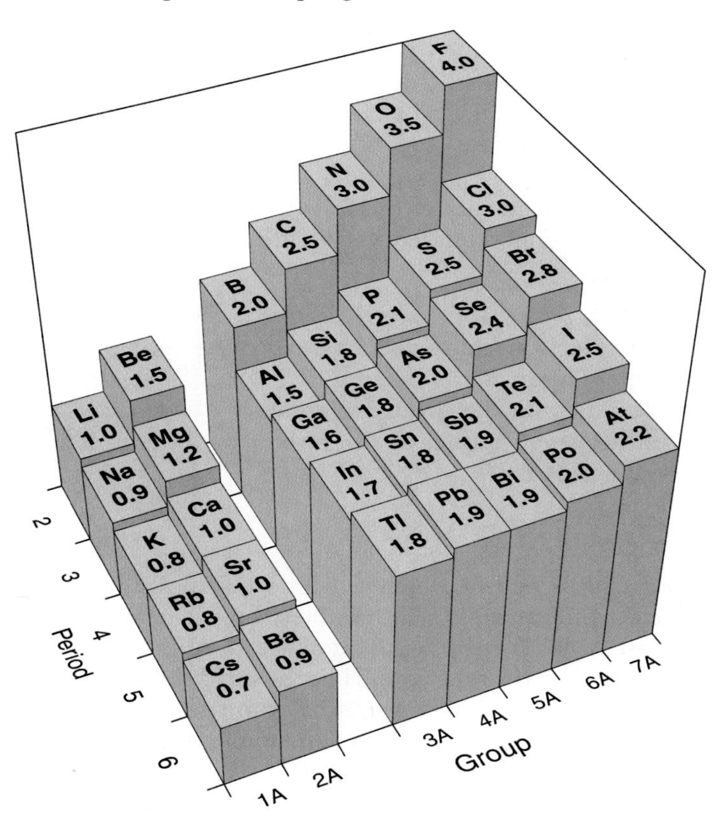


FIGURE 5.26
Electronegativities of the main group elements. The electronegativity of an element is a numerical measure of the ability of atoms of that element to attract electrons to themselves in compounds.

ments with large effective nuclear charges can be found. For example, fluorine is the smallest second-period element and has a effective nuclear charge of + 7.

You might think that helium is the most electronegative element. However, remember that electronegativity is the ability of an atom to attract electrons to itself *in compounds*. Helium does not form any compounds, so the concept of electronegativity doesn't apply to helium or to most of the noble gases. Fluorine is the most electronegative element.

The guidelines imply that the nonmetals have high electronegativities and the metals have low electronegativities. Because of this, the nonmetals are often spoken of as **electronegative** and the metals as **electropositive.**

Electronegativity is a relative term, of course. Oxygen is a very electronegative element, but in its few compounds with fluorine, the fluorine attracts the electrons more strongly. Electronegativities (like most other things you will study in chemistry) have numerical values. The electronegativities of the main group elements are shown in Figure 5.26.

### EXAMPLE 5.8 Electronegativity

**a.** Which of the elements in this fragment of the periodic table shown is the most electronegative?

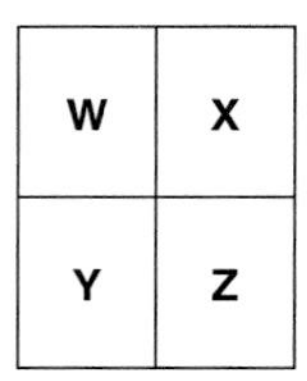


**b.** Which of the elements in the fragment of the periodic table in part (a) has the strongest tendency to form positive ions?

**SOLUTIONS**

**a.** The most electronegative elements are to the right and top of the periodic table. The answer is element X.

**b.** The element with the strongest tendency to form positive ions is the *least* electronegative one, or the one to the bottom left in the table. This is element Y.

**EXERCISE 5.8**

**a.** Among O, F, S, and Cl, which is the most electronegative? Which is the least electronegative? Which has the strongest tendency to form negative ions? Which has the largest atomic radius?

**b.** Among Na, Mg, K, and Ca, which is the most electronegative? Which is the most electropositive? Which has the strongest tendency to form positive ions? Which has the smallest atomic radius?

### Valence

A third property you can predict from the periodic table is the number of chemical bonds an atom will usually form. This number is called the **valence,** and it enables you to predict the charges on ions and the formulas of a wide range of compounds. You will use this information in Chapter 7, when you learn not only to predict formulas but also to name compounds from their formulas.

The rule that underlies valences is that *noble gas electron configurations are stable.* Atoms tend to gain, lose, or share electrons so as to have the same number of outer shell electrons as the nearest noble gas.

We'll use the elements that make up table salt, Na and Cl, as examples. An uncharged sodium atom has one valence electron, a 3*s* electron. When it gives up this electron, the resulting ion has the same electronic configuration that neon has ($1s^22s^22p^6$). Since the electron has a $-1$ charge, the sodium ion resulting from the removal of the electron has a charge of $+1$.

The uncharged chlorine atom has seven valence electrons. Adding another gives the anion the same electron configuration as the noble gas argon ($1s^22s^22p^63s^23p^6$) and a charge of $-1$. This makes good sense in terms of the concept of electronegativity. Sodium has a small effective nuclear charge and is relatively large, so it gives up its valence electron fairly easily. Chlorine has a large effective nuclear charge and attracts electrons strongly.

**STUDY ALERT**

**You use the same method to obtain the electron configurations of ions that you used to obtain the electron configurations of atoms—you fill shells and subshells with the number of available electrons. For example, Na atoms have 11 electrons and an electron configuration of $1s^22s^22p^63s^1$; $Na^+$ ions have 10 electrons and an electron configuration of $1s^22s^22p^6$.**

We have just explained what usually happens when metals bond to nonmetals. What happens when two nonmetals bond to each other, as in $CCl_4$ or $CO_2$? These compounds very seldom contain ions. Carbon has four valence electrons and tends to share them rather than give them up or acquire more. If carbon were to acquire four electrons, it would have a $-4$ charge. Similarly, carbon would have a $+4$ charge if it gave up four electrons. Coulomb's Law suggests that it would take a great deal of force to either add or remove four electrons from a carbon atom.

Instead, carbon tends to share its valence electrons with atoms bonded to it. (This concept is developed more fully in Chapter 10.) For example, in carbon tetrachloride ($CCl_4$), the central carbon atom shares four pairs of electrons (that is, eight electrons); this shared arrangement gives the carbon atom the electron configuration of neon.

You can think of each of the shared electron pairs as containing one electron from the carbon and one from a chlorine atom. Each chlorine atom shares two electrons with the central carbon atom and has six unshared electrons of its own. This gives the chlorine atoms the electron configuration of argon. Figure 5.27 illustrates this shared arrangement.

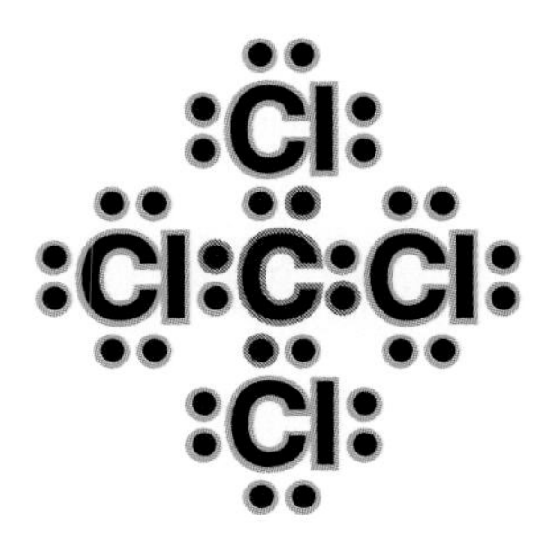


FIGURE 5.27
Carbon and chlorine share valence electrons in $CCl_4$. Each of the atoms acquires the electron configuration of a noble gas.

The valences of the main group elements follow a trend across a period. To summarize:

- The main group elements adopt a valence that gives them a noble gas electron configuration.
- The metals in Groups 1A, 2A, and 3A obtain a noble gas configuration of electrons by losing the electrons in their outer shell, forming cations. (Boron is not a metal, and shares electrons when it forms bonds.)
- The elements in groups 4A and 5A tend to share electrons. For example, carbon (Group 4A) shares electrons with four chlorine atoms to achieve an inert gas electron configuration (see Figure 5.27).

**The carbon and chlorine valence electrons are shown in different colors in Figure 5.27. In a $CCl_4$ molecule the electrons are indistinguishable from each other; they are all identical.**

FIGURE 5.28

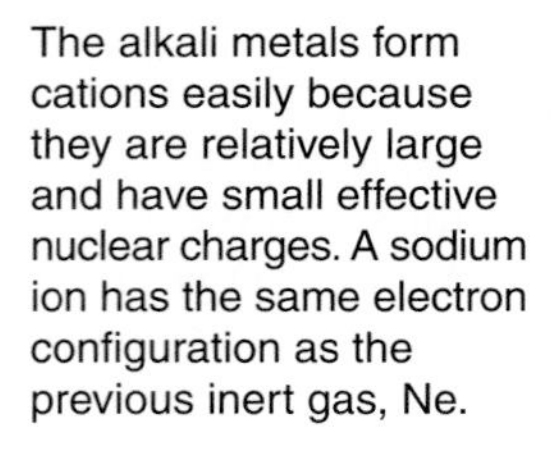


The alkali metals form cations easily because they are relatively large and have small effective nuclear charges. A sodium ion has the same electron configuration as the previous inert gas, Ne.

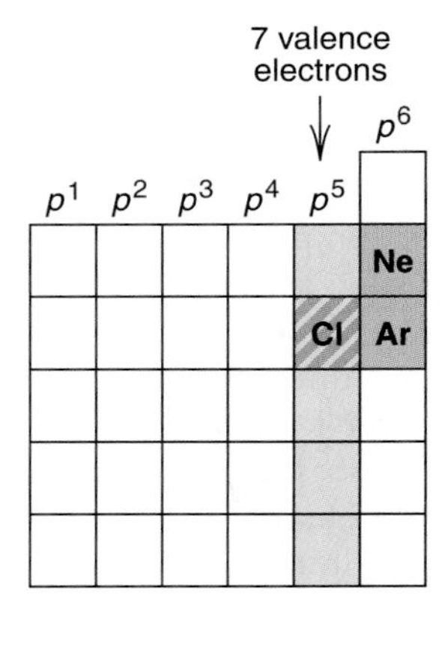


The halogens form anions relatively easily. They have large core charges and small atomic radii. A chloride ion has the same electron configuration as the nearest inert gas, Ar.

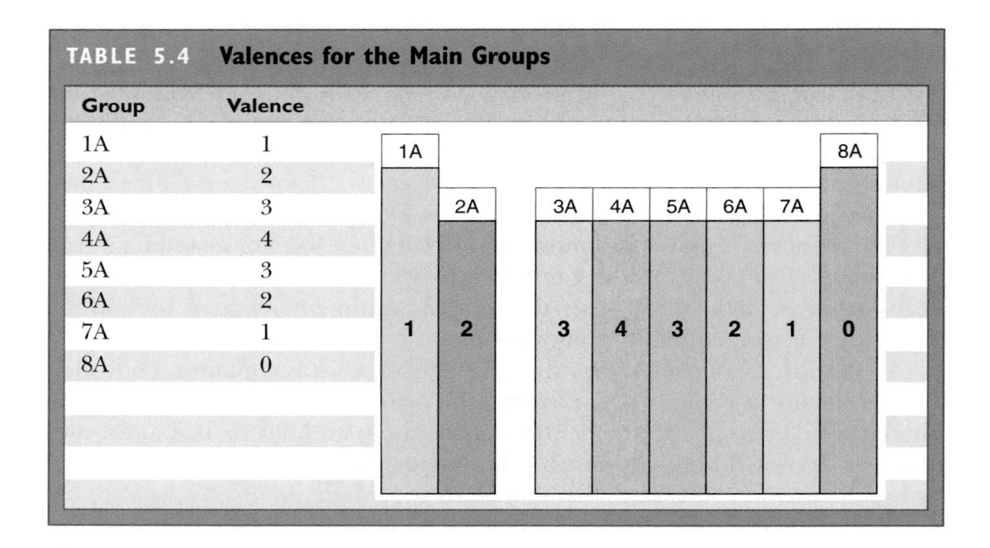

**TABLE 5.4 Valences for the Main Groups**

| Group | Valence |
|---|---|
| 1A | 1 |
| 2A | 2 |
| 3A | 3 |
| 4A | 4 |
| 5A | 3 |
| 6A | 2 |
| 7A | 1 |
| 8A | 0 |



- Group 6A and 7A elements tend to gain electrons and form anions when they form bonds with metals. For example, oxygen has six valence electrons ($2s^22p^4$), and gains two electrons to form $O^{2-}$ when bonded to metals such as Na and Ca. These elements tend to share electrons when bonded to other nonmetals.

You will find the most common valences of the main group elements given by group in Table 5.4. The valences of the elements near the top and sides of the periodic table are the most predictable.

We can predict a valence of the elements in each of the main groups in the periodic table. These group valences are given in Table 5.4. These valences are generally useful for the main group elements, especially those near the top of the periodic table.

**Some of the heavier main-group elements (for example, lead, tin, and arsenic) and most of the transition metals have more than one valence. See Chapter 7 for more information.**

### EXAMPLE 5.9 Valences

What is the expected valence of each of the following elements?
**a.** K **b.** Al **c.** N **d.** F

**SOLUTIONS**

**a.** Potassium is in Group 1A and has a valence of 1.
**b.** Aluminum is in Group 3A and has a valence of 3.
**c.** Nitrogen is in Group 5A and has a valence of 3.
**d.** Fluorine is in Group 7A and has a valence of 1.

**EXERCISE 5.9**

What is the expected valence of each of the following elements?
**a.** Ca **b.** B **c.** O **d.** Si

You can use the valences of the elements to predict the formulas of a large number of compounds that are composed of two elements. The general rule is that in any compound consisting of two elements, *the sums of the valences of the atoms of the two elements must be identical.* Here are some examples.

Water is $H_2O$. Hydrogen is in Group 1A and has a valence of 1. Oxygen is in Group 6A and has a valence of 2. Since the sum of the valences of the hydrogen atoms must equal the valence of the oxygen atom, you need two hydrogen atoms for each oxygen atom: $1 + 1 = 2$.

Methane is $CH_4$. Hydrogen has a valence of 1. Carbon is in Group 4A and has a valence of 4. Four hydrogen atoms are needed to balance the carbon atom: $1 + 1 + 1 + 1 = 4$.

The main use for valences, however, is to predict the formulas of compounds you don't know. For example, what is the formula of the compound that consists of sodium and sulfur? Sodium is in Group 1A and has a valence of 1. Sulfur is in Group 6A and has a valence of 2. The formula is $Na_2S$, because two sodium atoms balance one sulfur atom.

What is the formula of the compound that consists of oxygen and fluorine? Oxygen has a valence of 2, and fluorine has a valence of 1. This gives $OF_2$ as the formula.

Exceptions to the valence rules occur fairly frequently. For example, carbon makes two compounds with oxygen—CO and $CO_2$. Sulfur makes three compounds with fluorine—$SF_2$, $SF_4$, and $SF_6$. Among elements lower down in the periodic table, exceptions to the simple valence rules become more frequent. However, the fact that exceptions do occur does not mean that valences are useless. The trends are definitely there, and you can get a lot of mileage out of the valence rules.

### EXAMPLE 5.10 Predicting Formulas of Simple Compounds

What is the formula for a compound consisting of each of the following pairs of elements?

**a.** Sodium and iodine
**b.** Carbon and sulfur
**c.** Nitrogen and fluorine
**d.** Calcium and hydrogen

**SOLUTIONS**

**a.** Sodium (Group 1A) has a valence of 1; iodine (Group 7A) also has a valence of 1. The formula is NaI.
**b.** Carbon has a valence of 4. Sulfur has a valence of 2. The formula that balances the valences is $C_2$.
**c.** Nitrogen has a valence of 3. Fluorine has a valence of 1. The formula is $NF_3$.
**d.** Hydrogen has a valence of 1. Calcium is in Group 2A and has a valence of 2. The formula is $CaH_2$.

**EXERCISE 5.10**

What is the formula for a compound consisting of each of the following pairs of elements?

**a.** Calcium and iodine
**b.** Carbon and iodine
**c.** Hydrogen and sulfur
**d.** Magnesium and nitrogen

## 5.7 The Mole

**GOAL:** To calculate the molar masses of elements and compounds from atomic masses and the number of atoms from the number of moles, and vice versa.

**The use of hydrogen as a standard for determining the relative masses of the atoms was difficult for early scientists, and experimental problems led to inaccurate results. Chemists changed the standard to oxygen, which was assigned a mass of exactly 16, and this change allowed them to measure relative masses more accurately. In the middle of the twentieth century the standard was again changed, and now the relative masses of the elements are based on the most common isotope of carbon, which is assigned a mass of exactly 12.**

By the late eighteenth century, chemists had come to another very important conclusion. They had found that the ratio of the masses of the elements that make up any given compound is always the same. The atomic theory forms the basis of the explanation for this observation, and Dalton proposed that all molecules of any given compound contain the same numbers of atoms of each of the elements that make it up.

Chemists performed several important groups of experiments using sets of compounds that contained one element in common, such as a set of oxides. They compared the mass ratios for compounds in each set. By repeating such experiments many times with different sets of compounds, chemists were able to measure the relative atomic masses of most of the known elements. Hydrogen, the lightest element, was assigned a relative mass of 1. The chemists of the nineteenth century carefully measured the atomic masses of the other elements relative to hydrogen.

Initially, these chemists were confused about how many atoms of each element were present in some compounds, which caused errors in the values of the relative atomic masses. For instance, Dalton believed that the formula of water was HO. Because 8 g of oxygen reacts with 1 g of hydrogen to give 9 g of water, he thought that the atomic mass of oxygen was 8. The correction wasn't made for many years.

Studies of gases provided useful information. The French chemist Joseph Gay-Lussac (1778–1850) found that all gases expand by the same percentage when the temperature increases. A few years later, in 1809, he found that gases combine in proportions by volume that can be expressed as the ratios of small whole numbers. For instance, 1 L of oxygen gas reacts with 2 L of hydrogen gas to make water, with no gases left over at all.

Within a few years the Italian chemist Amadeo Avogadro (1776–1856) made a proposal based on Gay-Lussac's results. He suggested that a given number of atoms or molecules of *any* gas occupy the same volume at the same temperature and pressure, no matter what the gas is. He also proposed that the formula of water was $H_2O$, based on Gay-Lussac's result. It took some time (about 50 years!) before the scientific community accepted the entire proposal, but it cleared up a lot of the confusion about atomic masses.

Atomic masses played a major role in developing the periodic table, as you already know. These masses appear on almost all versions of the table. The atomic mass traditionally appears below the symbol of the element (see Figure 5.29 and the periodic table inside the front cover).

Once the relative masses of the atoms had been established the business of doing chemistry became much easier, because chemists could count atoms and molecules by weighing samples of elements. This is not as strange as it may first appear. There are twice as many paper clips in 200 g of clips as there are in 100 g. It is the same with atoms. There are twice as many copper atoms in 200 g of copper as there are in 100 g. It is not necessary to know how many paper clips or copper atoms each sample contains for this observation to be true.

Chemists extended this idea to compounds. For example, one molecule of carbon monoxide (CO) contains one atom of carbon and one atom of oxygen. Analysis of 28 g of CO shows that it contains 12 g of carbon and 16 g of oxygen. Since the number of atoms of carbon and the number of atoms of oxygen in the sample of CO are the same, there are exactly the same number of carbon atoms in 12 g of carbon as there are oxygen atoms in 16 g of oxygen.

When chemists first advanced this idea, they did not know the number of molecules of CO in 28 g. They knew neither the number of atoms in 12 g of carbon nor the number of atoms in 16 g of oxygen. They also did not know either the masses of any individual atoms or the mass of one CO molecule. They *did* know that the number of molecules in 28 g of CO and the number of atoms in both 12 g of carbon and 16 g of oxygen were exactly the same number.

This thinking allowed chemists to define a new quantity called the **mole** (abbreviated **mol**). The modern definition of a mole is the amount of a substance that contains as many formula units as there are atoms in exactly 12 g

MAIN–GROUP ELEMENTS (1A–2A) · TRANSITION ELEMENTS (3B–2B) · MAIN–GROUP ELEMENTS (3A–8A)

| Period | 1A | 2A | 3B | 4B | 5B | 6B | 7B | 8B | 8B | 8B | 1B | 2B | 3A | 4A | 5A | 6A | 7A | 8A |
|---|---|---|---|---|---|---|---|---|---|---|---|---|---|---|---|---|---|---|
| 1 | 1 **H** 1.008 | | | | | | | | | | | | | | | | | 2 **He** 4.003 |
| 2 | 3 **Li** 6.941 | 4 **Be** 9.012 | | | | | | | | | | | 5 **B** 10.81 | 6 **C** 12.01 | 7 **N** 14.01 | 8 **O** 16.00 | 9 **F** 19.00 | 10 **Ne** 20.18 |
| 3 | 11 **Na** 22.99 | 12 **Mg** 24.31 | | | | | | | | | | | 13 **Al** 26.98 | 14 **Si** 28.09 | 15 **P** 30.97 | 16 **S** 32.07 | 17 **Cl** 35.45 | 18 **Ar** 39.95 |
| 4 | 19 **K** 39.10 | 20 **Ca** 40.08 | 21 **Sc** 44.96 | 22 **Ti** 47.90 | 23 **V** 50.94 | 24 **Cr** 52.00 | 25 **Mn** 54.94 | 26 **Fe** 55.85 | 27 **Co** 58.93 | 28 **Ni** 58.70 | 29 **Cu** 63.55 | 30 **Zn** 65.39 | 31 **Ga** 69.72 | 32 **Ge** 72.59 | 33 **As** 74.92 | 34 **Se** 78.96 | 35 **Br** 79.90 | 36 **Kr** 83.80 |
| 5 | 37 **Rb** 85.47 | 38 **Sr** 87.62 | 39 **Y** 88.91 | 40 **Zr** 91.22 | 41 **Nb** 92.91 | 42 **Mo** 95.94 | 43 **Tc** (98) | 44 **Ru** 101.1 | 45 **Rh** 102.9 | 46 **Pd** 106.4 | 47 **Ag** 107.9 | 48 **Cd** 112.4 | 49 **In** 114.8 | 50 **Sn** 118.7 | 51 **Sb** 121.8 | 52 **Te** 127.6 | 53 **I** 126.9 | 54 **Xe** 131.3 |
| 6 | 55 **Cs** 132.9 | 56 **Ba** 137.3 | 71 **Lu** 175.0 | 72 **Hf** 178.5 | 73 **Ta** 180.9 | 74 **W** 183.9 | 75 **Re** 186.2 | 76 **Os** 190.2 | 77 **Ir** 192.2 | 78 **Pt** 195.1 | 79 **Au** 197.0 | 80 **Hg** 200.6 | 81 **Tl** 204.4 | 82 **Pb** 207.2 | 83 **Bi** 209.0 | 84 **Po** (209) | 85 **At** (210) | 86 **Rn** (222) |
| 7 | 87 **Fr** (223) | 88 **Ra** (226) | 103 **Lr** (260) | 104 **Rf** (Unq) | 105 **Ha** (Unp) | 106 **Sg** (Unh) | 107 **Ns** (Uns) | 108 **Hs** (Uno) | 109 **Mt** (Une) | 110 | 111 | 112 | | | | | | |

INNER TRANSITION ELEMENTS

| Period | | | | | | | | | | | | | | | |
|---|---|---|---|---|---|---|---|---|---|---|---|---|---|---|---|
| 6 | Lanthanides | 57 **La** 138.9 | 58 **Ce** 140.1 | 59 **Pr** 140.9 | 60 **Nd** 144.2 | 61 **Pm** (145) | 62 **Sm** 150.4 | 63 **Eu** 152.0 | 64 **Gd** 157.3 | 65 **Tb** 158.9 | 66 **Dy** 162.5 | 67 **Ho** 164.9 | 68 **Er** 167.3 | 69 **Tm** 168.9 | 70 **Yb** 173.0 |
| 7 | Actinides | 89 **Ac** (227) | 90 **Th** 232.0 | 91 **Pa** (231) | 92 **U** 238.0 | 93 **Np** (244) | 94 **Pu** (242) | 95 **Am** (243) | 96 **Cm** (247) | 97 **Bk** (247) | 98 **Cf** (251) | 99 **Es** (252) | 100 **Fm** (257) | 101 **Md** (258) | 102 **No** (259) |

FIGURE 5.29
The atomic mass appears below the element's symbol in the periodic table.

of the most common isotope of carbon. For instance, 1 mol of carbon atoms has a mass of 12.0 g, 1 mol of oxygen has a mass of 16.0 g, and 1 mol of calcium has a mass of 40.0 g. The mole is so useful as a unit that it was embraced by the scientific community fairly quickly. It is the standard SI unit for the amount of a substance.

**The molar mass of an element is sometimes called the *atomic weight*, and the molar mass of a compound is sometimes called the *molecular weight*, particularly in older textbooks and literature. These terms are imprecise, because there is a distinct difference between mass and weight (see Section 3.1).**

This basic idea is readily extended to molecules: 1 mol of $H_2O$ molecules contains 2 mol of hydrogen atoms and 1 mol of oxygen atoms, for instance. From the periodic table you can see that 1 mol of H atoms has a mass of 1 g, so 2 mol of H must have a mass of 2 g. One mole of oxygen atoms has a mass of 16 g. Adding these masses gives 18 g, the mass of 1 mol of water. Similar calculations show that 1 mol of hydrogen ($H_2$) has a mass of 2 g, and 1 mol of oxygen ($O_2$) has a mass of 32 g. The mass of 1 mol of a substance expressed in grams is called its **molar mass.** A molar mass has units of g $mol^{-1}$. Example 5.11 presents more calculations of molar mass.

FIGURE 5.30
There are twice as many paper clips in 200 g as in 100 g, and there are twice as many copper atoms in 200 g as in 100 g.

### EXAMPLE 5.11 Calculating Molar Mass

Calculate the molar mass of each compound.

**a.** Methane ($CH_4$), the principal component of natural gas
**b.** Octane ($C_8H_{18}$), a component of gasoline
**c.** Acetic acid ($CH_3CO_2H$), the main ingredient (other than water) in vinegar

**SOLUTIONS**

**a.** One mole of $CH_4$ contains 1 mol of carbon and 4 mol of hydrogen. Its molar mass is therefore $(1 \text{ mol C} \times 12 \text{ g mol}^{-1} \text{ C}) + (4 \text{ mol H} \times 1.0 \text{ g mol}^{-1} \text{ H}) = 12 \text{ g} + 4 \text{ g} = 16 \text{ g mol}^{-1} \text{ CH}_4$

**b.** There are 8 mol of carbon and 18 mol of hydrogen in 1 mol of octane.
$(8 \text{ mol C} \times 12 \text{ g mol}^{-1} \text{ C}) + (18 \text{ mol H} \times 1.0 \text{ g mol}^{-1} \text{ H}) =$
$114 \text{ g mol}^{-1}\ C_8H_{18}$

**c.** For acetic acid:
$(2 \text{ mol C} \times 12 \text{ g mol}^{-1} \text{ C}) + (4 \text{ mol H} \times 1.0 \text{ g mol}^{-1} \text{ H}) + (2 \text{ mol O} \times 16 \text{ g mol}^{-1} \text{ O}) = 60 \text{ g } CH_3CO_2H$

**EXERCISE 5.11**

Calculate the molar mass of each compound.

**a.** Ammonia, $NH_3$, used in household cleaners
**b.** Sodium carbonate, $Na_2CO_3$, used to manufacture glass, paper, and soap
**c.** Lycopene, $C_{40}H_{56}$, which gives the red color to tomatoes

**STUDY ALERT**
**The value of Avogadro's number ($6.02 \times 10^{23}$) is a "must know" for chemistry students.**

The connection between the masses of individual atoms and molecules and the mole was an unknown number, the number of particles in a mole. This was given the name **Avogadro's number.** A mole, then, contains Avogadro's number of particles. As the nineteenth century went on, chemists and physicists found ways to measure this number indirectly. Finally the masses of single atoms were measured directly using a mass spectrometer (see Chapter 3). All of these methods found that Avogadro's number has the enormous value of $6.02 \times 10^{23}$. There are $6.02 \times 10^{23}$ atoms of argon in 1 mol (39.95 g) of argon gas, $6.02 \times 10^{23}$ hydrogen molecules in 1 mol (2.016 g) of hydrogen gas, and the same number of molecules in 1 mol of ethyl alcohol ($C_2H_5OH$).

Keep in mind that Avogadro's number is a conversion factor:

$$1 \text{ mol of things} = 6.02 \times 10^{23} \text{ things}$$

To work with Avogadro's number, use the unit factor method you learned in Section 3.7. Example 5.12 works some typical problems involving Avogadro's number.

FIGURE 5.31
Each of these samples represents exactly 1 mol of an element. Each sample therefore contains the same number of atoms.

### EXAMPLE 5.12 Numbers of Atoms and Numbers of Moles

**a.** How many hydrogen *atoms* are there in 1 mol of hydrogen gas?
**b.** What is the mass of one hydrogen atom in grams?
**c.** How many moles of argon do $2.0 \times 10^{20}$ atoms of argon comprise?

**SOLUTIONS**

**a.** One mole of any pure substance (including hydrogen gas) contains $6.02 \times 10^{23}$ particles. In this case the particles are hydrogen molecules ($H_2$). Since each hydrogen molecule contains two hydrogen atoms, 1 mol of hydrogen molecules therefore contains

$$\frac{2 \text{ H atoms}}{\cancel{1 \text{ molecule } H_2}} \times \frac{6.02 \times 10^{23} \cancel{\text{ molecules } H_2}}{1 \text{ mol } H_2} = 12.0 \times 10^{23} \text{ atoms mol}^{-1}\ H_2$$

That is, $1.20 \times 10^{24}$ hydrogen atoms are present in a mole of $H_2$ gas.

FIGURE 5.32
Each of these samples represents exactly 1 mol of a compound. There are therefore exactly the same number of molecules in each sample.

**b.** One mole of hydrogen atoms has a mass of 1.00 g and contains $6.02 \times 10^{23}$ atoms. Using the unit factor method,

$$\frac{1.00\text{ g H}}{1\ \cancel{\text{mol H}}} \times \frac{1\ \cancel{\text{mol H}}}{6.02 \times 10^{23}\text{ atoms}} = 1.66 \times 10^{-24}\text{ g per H atom}$$

**c.** Here you have to convert atoms to moles. The basic use of a unit factor is the same.

$$2.0 \times 10^{20}\ \cancel{\text{atoms}} \times \frac{1\text{ mol Ar}}{6.02 \times 10^{23}\ \cancel{\text{atoms}}} = 3.3 \times 10^{-4}\text{ mol Ar}$$

**EXERCISE 5.12**

**a.** How many atoms total are there in 1 mol of ethyl alcohol ($C_2H_5OH$)?
**b.** How many atoms are there in 0.12 mol of hydrogen gas?

## EXPERIMENTAL EXERCISE 5.1

In the nineteenth century chemists determined the molar mass of an element by analyzing a series of pure substances containing the element. They had developed experimental techniques that allowed them to measure the molar mass of a compound and the mass of an element present in the compound. It was assumed that the part of the mass due to a particular element represented the sum of the masses of the atoms of that element present, and that the smallest mass obtained for an element for several compounds must be the molar mass of that element. Experimental results for an unknown element, symbolized by E, are given in the table below

| Compound | Molar Mass of Pure Substance (g) | Part of Molar Mass Due to Element E (g) |
|---|---|---|
| 1 | 28.05 | 28.05 |
| 2 | 44.13 | 28.01 |
| 3 | 30.00 | 14.02 |
| 4 | 17.05 | 14.01 |
| 5 | 63.75 | 14.01 |

**QUESTIONS**

**1.** What is the molar mass of E?
**2.** Which pure substance consists of only E atoms?
**3.** What is the formula for one molecule of E?
**4.** How many atoms are in 1 mol of E molecules?
**5.** How many atoms of E are there in one molecule of compound 2?
**6.** How many atoms of E are there in 1 mol of compound 5?
**7.** Which of the compounds has the formula $E_2O$?
**8.** What element is E?

## 5.8 Average Atomic Masses

**GOAL:** To define average atomic mass and to calculate it given the isotopic composition of a sample of an element.

As you learned in Section 4.9, samples of many of the naturally occurring elements consist of a mixture of isotopes. For example, the carbon atoms in your body consist mostly of atoms that contain six protons and six neutrons. However, about 1% of those carbon atoms contain six protons and seven neutrons. So what mass for carbon should appear on the periodic table? The answer is both practical and simple.

A mass reported on the periodic table is the **average atomic mass** of an element. This number is a weighted average of the masses of the isotopes of the element that are present under natural conditions. For example, bromine has an atomic number of 35, which means that the nucleus of each and every bromine atom contains 35 protons. In an ordinary sample of bromine, about half of the nuclei contain 44 neutrons and the rest contain 46 neutrons. The atomic mass of one isotope is 79 amu (atomic mass units), and the mass of the other is 81 amu. The average atomic mass (which is what you see in periodic tables) is very near 80 amu.

For chlorine the case is only slightly more complicated. About 75% of all chlorine atoms have a mass of about 35 amu, and the other 25% have a mass of about 37 amu. To find the weighted average, simply multiply the mass of each isotope by the fraction of the total it represents and add the products together. That is, for chlorine the average atomic mass is (0.75)(35 amu) + (0.25)(37 amu), or about 35.5 amu.

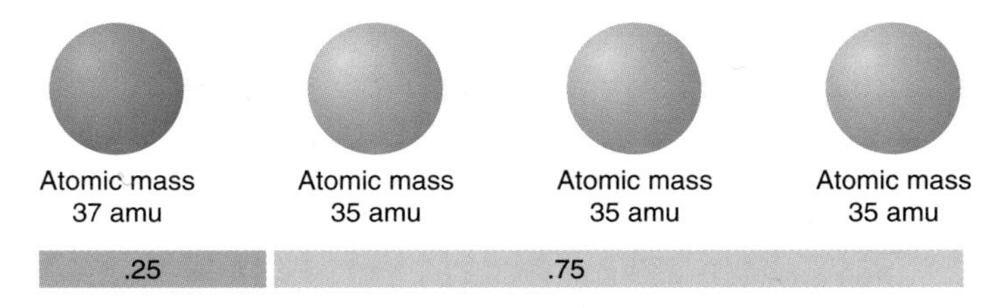


FIGURE 5.33
The average atomic mass of these four chlorine atoms is (0.75)(35 amu) + (0.25)(37 amu) = 35.5 amu. This percent composition is typical of all natural samples of chlorine.

### EXAMPLE 5.13 Average Atomic Masses

A sample of gallium consists of 60.0% of one isotope, having a mass of 68.9 amu, and 40.0% of another isotope, having a mass of 70.9 amu. What is the average atomic mass of this sample?

**If you wish to review percentages at this point, refer to the Tool Box on percentages in Chapter 2 on page 34.**

**SOLUTION**

The whole (100%) is equal to the sum of the parts. Suppose that you have a sample of 100 atoms of gallium: 60.0% of the 100 (that is, 0.600) have a mass of 68.9 amu, and the other 40.0% have a mass of 70.9 amu. The equation is

$$(0.600)(68.9\text{ amu}) + (0.400)(70.9\text{ amu}) = 69.7\text{ amu}$$

The average atomic mass of this sample of gallium is 69.7 amu.

**EXERCISE 5.13**

A sample of boron has 80% of one isotope, having a mass of 10.0 amu, and 20% of another isotope, having a mass of 11.0 amu. What is the average atomic mass of this sample?

CHAPTER BOX

## 5.1 MODERN METHODS FOR DETERMINING RELATIVE MASSES OF THE ELEMENTS

**Modern chemists** use mass spectrometers (see Section 4.9) to measure the masses and percent compositions of the isotopes of elements (among many other things). The best mass spectrometers are extremely sensitive, and chemists using them can measure atomic masses very accurately. For example, fluorine has an atomic mass of 18.998403 amu, and gold has an atomic mass of 196.966543 amu.

For most elements the isotopic composition of natural samples is constant throughout the earth. For a few elements, notably sulfur, the isotopic composition varies slightly. This is why the atomic masses of the elements reported on some periodic tables have different numbers of digits past the decimal point.

In many mass spectrometers, an electron beam converts atoms of samples into positive ions. This may seem odd, since electrons have negative charges, but it really does work. The incoming electron knocks an electron away from an atom of the sample and then departs.

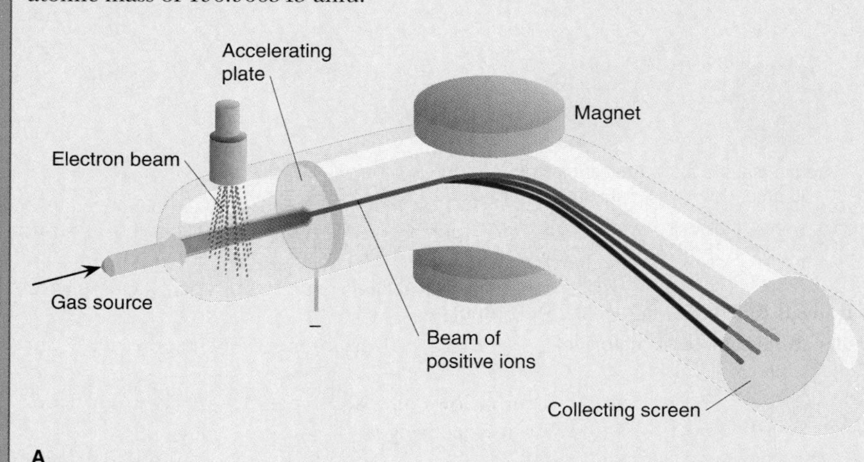


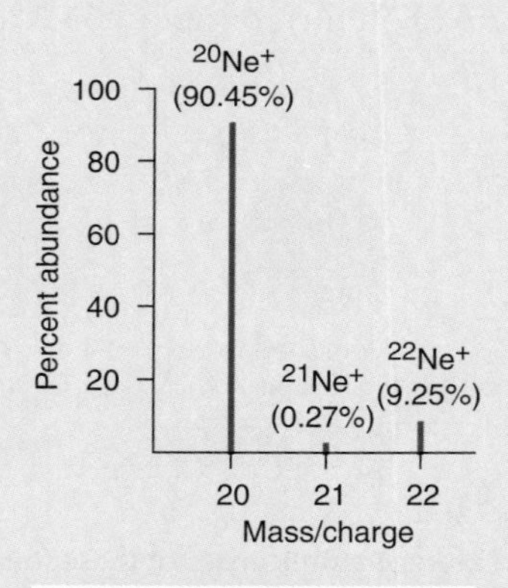


| | | |
|---|---|---|
| $^{20}Ne^+$ | = | 19.9924 amu |
| $^{21}Ne^+$ | = | 20.9938 amu |
| $^{22}Ne^+$ | = | 21.9914 amu |

B

**FIGURE 5.34**
(A) Schematic diagram of one type of mass spectrometer. The electron beam converts gas molecules into positive ions. The accelerating plate attracts the positive ions because it has a negative charge. The magnet deflects the beam of ions; more massive particles are not deflected as much as less massive ones are. (B) This mass spectrum of neon gives the abundance and mass of the individual isotopes.

## CHAPTER SUMMARY

1. Elements are typically classified as metals, nonmetals, or semimetals. Metals lie toward the left of the periodic table and nonmetals toward the right.
2. Semimetals have properties intermediate between those of metals and those of nonmetals.
3. The main group elements are designated as A groups in the periodic table used in this course. The transition metals are the B group elements. The lanthanides and actinides are located in a separate section below the main body of the periodic table.
4. An atom consists of a small central nucleus surrounded by a cloud of electrons arranged as shells and subshells.
5. Atoms with completely filled outer electron shells are particularly stable.
6. Valence electrons are the electrons in the outermost shell for all main group elements.
7. The valence electrons of main group elements are always in *s* and *p* subshells.
8. Atomic radii increase going down a group and decrease going from left to right across a period.
9. The effective nuclear charge of a main group element is equal to its group number.
10. Electronegative elements have large effective nuclear charges and small atomic radii.
11. The formulas of many simple compounds can be obtained by setting the sum of the valences of the atoms of the individual elements equal to each other.
12. The SI unit for measuring amount (the number of things) is the mole. One mole of anything is equal to Avogadro's number ($6.02 \times 10^{23}$) of that thing (electrons, atoms, molecules, etc.).
13. The atomic mass of an element reported on a periodic table is the average weighted mass of the naturally occurring isotopes of that element.

## PROBLEMS

### SECTION 5.1 Metals, Nonmetals, and Semimetals

1. Classify each of these as a metal, a nonmetal, or a semimetal.
   a. Mg (atomic number 12)
   b. As (atomic number 33)
   c. Sn (atomic number 50)
   d. U (atomic number 92)
   e. Br (atomic number 35)
   f. Xe (atomic number 54)
   g. K (atomic number 19)
   h. Mn (atomic number 25)
   i. O (atomic number 8)
   j. Cl (atomic number 17)
2. Classify each of these as a metal, a nonmetal, or a semimetal.
   a. Ne (atomic number 10)
   b. Si (atomic number 14)
   c. Al (atomic number 13)
   d. Pb (atomic number 82)
   e. I (atomic number 53)
   f. Hg (atomic number 80)
   g. B (atomic number 5)
   h. Fe (atomic number 26)
   i. Se (atomic number 34)
   j. Ar (atomic number 18)
3. Identify each element as a metal or nonmetal from the brief description of its properties at room temperature. Justify your answer.
   a. Liquid, does not conduct electricity, has low thermal conductivity
   b. Liquid, shiny silver color, excellent electrical conductor
   c. Dull yellow color, solid, powders easily when struck
   d. An unreactive gas
   e. Dull black solid, good electrical conductor, fractures easily under stress, poor thermal conductor
4. Identify each element as a metal or nonmetal from the brief description of its properties at room temperature. Justify your answer.
   a. Colorless, crystalline, very hard and brittle
   b. Red solid, poor thermal and electrical conductor
   c. Shiny silvery solid, used to make electrical wiring

d. Pale green, highly reactive gas
e. Shiny yellow solid, good electrical conductor

SECTION 5.2 **Columns and Rows: Groups and Periods**

5. Use the periodic table to identify each of these elements.
   a. 3rd period, Group 1A
   b. 2nd period, Group 3A
   c. 2nd period, Group 7A
   d. 3rd period, Group 6A
   e. 2nd period, Group 2A
   f. 2nd period, Group 1A
   g. 4th period, Group 3B
   h. 3rd period, Group 4A
   i. 1st period, Group 1A
   j. 6th period, Group 2A

6. Use the periodic table to identify each of these elements.
   a. 5th period, Group 6B
   b. 2nd period, Group 4A
   c. 1st period, Group 8A
   d. 6th period, Group 7A
   e. 5th period, Group 2B
   f. 3rd period, Group 5A
   g. 2nd period, Group 6A
   h. 5th period, Group 1B
   i. 5th period, Group 8A
   j. 2nd period, Group 3A

7. Refer to the periodic table to identify each of these elements.
   a. 5th-period alkali metal
   b. 2nd-period noble gas
   c. 3rd-period alkali metal
   d. 4th-period noble gas
   e. 6th-period noble gas
   f. 5th-period halogen
   g. 2nd-period alkali metal
   h. 3rd-period alkaline earth metal
   i. 2nd-period halogen
   j. 3rd-period noble gas

8. Refer to the periodic table to identify each of these elements.
   a. 4th-period halogen
   b. 2nd-period alkaline earth metal
   c. 5th-period noble gas
   d. 3rd-period halogen
   e. 4th-period alkali metal
   f. 4th-period alkaline earth metal
   g. 7th-period alkali metal
   h. 6th-period halogen
   i. 7th-period alkaline earth metal
   j. 6th-period alkaline earth metal

9. What is the chemical formula of the compound formed from each pair of elements?
   a. Potassium and bromine
   b. Calcium and chlorine
   c. Hydrogen and iodine
   d. Calcium and iodine

10. What is the chemical formula for the compound formed from each of the following pairs of elements?
   a. Barium and fluorine
   b. Strontium and chlorine
   c. Sodium and fluorine
   d. Magnesium and iodine

11. Identify each of these elements as an alkali metal, an alkaline earth metal, a halogen, or a noble gas from the description of its properties.
   a. Soft silvery solid that reacts violently with both $Cl_2$ and $H_2O$ at room temperature
   b. Gas that is extremely reactive at room temperature
   c. Solid that forms a compound with the formula MOH when reacted with water

12. Identify each of these elements as an alkali metal, an alkaline earth metal, a halogen, or a noble gas from the description of its properties.
   a. Unreactive toward most other substances
   b. Reacts with water to form a compound with the formula $M(OH)_2$
   c. Solid that reacts rapidly with $Cl_2$ and slowly with water at room temperature

SECTION 5.3 **The Blocks of Elements**

13. Classify each of these as either a main group element, a transition metal, a lanthanide, or an actinide:
   a. Sr (atomic number 38)
   b. Mo (atomic number 42)
   c. Pd (atomic number 46)
   d. Ce (atomic number 58)
   e. I (atomic number 53)
   f. Pu (atomic number 94)
   g. Os (atomic number 76)
   h. Ti (atomic number 22)
   i. S (atomic number 16)
   j. Cs (atomic number 55)

14. Classify each of these as either a main group element, a transition metal, a lanthanide, or an actinide:
   a. Au (atomic number 79)
   b. Pb (atomic number 82)
   c. C (atomic number 6)
   d. Cl (atomic number 17)
   e. Cu (atomic number 29)
   f. As (atomic number 33)
   g. Hg (atomic number 80)
   h. Xe (atomic number 54)
   i. Pr (atomic number 59)
   j. Pa (atomic number 91)

SECTION 5.4 **Atomic and Electronic Structure: The Shell Model**

15. How many electron shells are occupied in each of these atoms?
   a. H
   b. Na
   c. K
   d. C
   e. S
   f. Cl
   g. Kr
   h. F
   i. Br
   j. I

16. How many electron shells are occupied in each of these atoms?
   a. Ca
   b. Ti
   c. Ar
   d. Cs
   e. He
   f. O
   g. P
   h. Pb
   i. Cu
   j. As

17. Give the number of valence electrons in each of these neutral atoms.
   a. Be
   b. Al
   c. Si
   d. Se
   e. Br
   f. Kr
   g. He
   h. I
   i. S
   j. P

18. Give the number of valence electrons in each of these neutral atoms.
   a. In
   b. Sr
   c. H
   d. Pb
   e. As
   f. Xe
   g. Ba
   h. Mg
   i. Bi
   j. C

## SECTION 5.5 Subshells and the Blocks of Elements

19. Identify each of these elements from its electron configuration.
    a. $[He]2s^22p^2$
    b. $[Ne]3s^23p^5$
    c. $[Ne]3s^1$
    d. $[Ar]4s^2$
    e. $1s^2$
    f. $1s^22s^22p^6$
    g. $1s^22s^22p^63s^2$
    h. $[Ar]4s^23d^{10}4p^2$
    i. $1s^22s^1$
    j. $1s^22s^22p^1$

20. Identify each of these elements from its electron configuration.
    a. $[Kr]5s^1$
    b. $1s^22s^22p^5$
    c. $1s^22s^22p^63s^23p^4$
    d. $[Ne]3s^23p^5$
    e. $[Ar]4s^23d^{10}4p^1$
    f. $[Kr]5s^2$
    g. $[Ne]3s^23p^1$
    h. $[Ar]4s^23d^{10}4p^3$
    i. $1s^22s^2$
    j. $[He]2s^22p^4$

21. Without looking at a complete periodic table, write the electron configuration for each of the elements indicated by the letters (a) through (j) on the given portion of a periodic table.

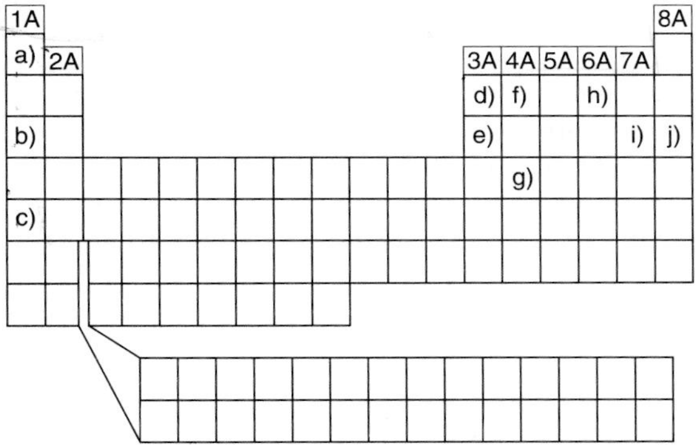

22. Without looking at a complete periodic table, write the electron configuration for each of the elements indicated by the letters (a) through (j) on the given portion of a periodic table.

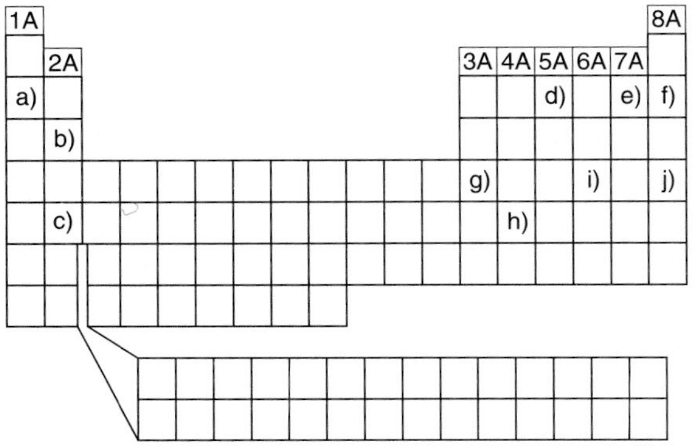

23. Write the electron configuration of each of these neutral atoms.
    a. H
    b. Li
    c. Cl
    d. Ar
    e. F
    f. O
    g. C
    h. B
    i. Be
    j. Ca

24. Write the electron configuration of each of these neutral atoms.
    a. K
    b. S
    c. Al
    d. He
    e. Mg
    f. P
    g. Si
    h. N
    i. Ne
    j. Na

## SECTION 5.6 Some Periodic Trends

25. Refer to the fragment of the periodic table shown below, in which W, X, Y, and Z represent main group elements in adjacent groups.

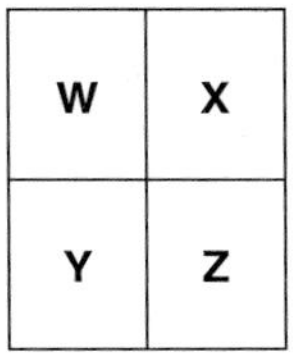

| W | X |
|---|---|
| Y | Z |

    a. Which element has the larger effective nuclear charge, W or X?
    b. Which element has the largest atomic radius?
    c. Which element has the highest electronegativity?
    d. If the valence of W is 3, what is the valence of X?
    e. If the valence of X is 0, what is the valence of W?
    f. If W reacts with Cl to form $WCl_4$, what is the formula of the compound of Y and Cl?

26. Refer to the figure in Problem 25.
    a. Which of the elements shown has the smaller effective nuclear charge, W or X?
    b. Which element has the smallest atomic radius?
    c. Which element has the smallest electronegativity?
    d. If the valence of W is 3, what is the valence of Y?
    e. If the valence of X is 5, what is the valence of Z?
    f. If W reacts with Cl to form $WCl_4$, what is the formula of the compound of X and Cl?

27. Write the electron configuration of each of these ions.
    a. $H^+$
    b. $Li^+$
    c. $Cl^-$
    d. $H^-$
    e. $F^-$
    f. $O^{2-}$
    g. $Ca^{2+}$
    h. $P^{3-}$

28. Write the electron configuration of each of these ions.
    a. $S^{2-}$
    b. $Al^{3+}$
    c. $Mg^{2+}$
    d. $I^-$
    e. $N^{3-}$
    f. $Na^+$
    g. $Br^-$
    h. $Sr^{2+}$

29. Identify each ion from its charge and electron configuration.

| Charge | Electron Configuration |
|---|---|
| a. +1 | $1s^22s^2$ |
| b. +1 | $1s^22s^22p^6$ |
| c. +2 | $1s^22s^22p^6$ |
| d. −1 | $1s^22s^22p^6$ |
| e. +3 | $1s^22s^22p^6$ |
| f. −2 | $1s^22s^22p^6$ |
| g. −3 | $1s^22s^22p^6$ |
| h. +1 | $1s^2$ |

30. Identify each ion from its charge and electron configuration.

| Charge | Electron Configuration |
|---|---|
| a. −1 | $[Ne]3s^23p^6$ |
| b. +2 | $[Ne]3s^23p^6$ |
| c. +2 | $1s^2$ |
| d. −2 | $[Ne]3s^23p^6$ |
| e. +3 | $[Ne]3s^23p^6$ |
| f. −3 | $[Ne]3s^23p^6$ |
| g. −1 | $1s^2$ |
| h. +1 | $[Ne]3s^23p^6$ |

31. What is the valence of each of these elements?
a. H b. Mg c. Si d. N e. F f. Ne g. I h. S i. Ba j. Ga

32. What is the valence of each of these elements?
a. Sr b. Cs c. Cl d. Cl e. C f. O g. Be h. Na i. Ar j. As

33. Give the formula of the compound formed from each of the following pairs of elements.
a. Sodium and iodine
b. Calcium and iodine
c. Magnesium and chlorine
d. Nitrogen and iodine
e. Fluorine and iodine
f. Lithium and bromine
g. Magnesium and sulfur
h. Aluminum and chlorine
i. Carbon and iodine
j. Oxygen and chlorine

34. Give the formula of the compound formed from each of the following pairs of elements.
a. Barium and oxygen
b. Gallium and bromine
c. Phosphorus and iodine
d. Aluminum and oxygen
e. Calcium and oxygen
f. Sodium and sulfur
g. Sulfur and chlorine
h. Cesium and phosphorus
i. Bromine and hydrogen
j. Hydrogen and sulfur

## SECTION 5.7 The Mole

35. How many atoms of carbon are in 1.0 mol of each of these compounds?
a. $C_2H_6O$ b. $CH_4$ c. $C_3H_8$ d. $Al_2(CO_3)_3$

36. How many atoms of carbon are in 1.0 mol of each of these compounds?
a. $C_{12}H_{22}O_{12}$ b. $C_8H_{18}$ c. $Na_2CO_3$ d. $C_{27}H_{46}O$

37. How many atoms of hydrogen are there in 3.0 mol of each of these compounds?
a. $CH_4$ b. $NaHCO_3$ c. $H_2SO_4$ d. $C_8H_{18}$

38. How many atoms of hydrogen are there in 7.0 mol of each of these compounds?
a. $C_2H_6O$ b. $H_3PO_4$ c. $C_6H_{12}O_6$ d. $Ca(OH)_2$

39. How many oxygen atoms are there in 2.5 mol of each of these compounds?
a. $H_2O$ b. $H_2O_2$ c. $CaCO_3$ d. $Al_2(CO_3)_3$

40. How many oxygen atoms are there in 4.7 mol of each of these compounds?
a. $HClO_4$ b. $HClO_3$ c. $CO_2$ d. $HNO_3$

41. What is the molar mass of each of these substances?
a. $N_2$ b. $H_2O_2$ c. KCl d. HClO e. $HClO_2$ f. $HClO_3$ g. $HClO_4$ h. $C_{12}H_{22}O_{11}$ i. $C_4H_{10}$ j. $CO_2$

42. What is the molar mass of each of these compounds?
a. $H_2SO_4$ b. HBr c. $Na_2CO_3$ d. $KHCO_3$ e. $Fe_2O_3$ f. $AlCl_3$ g. $Al_2S_3$ h. $PCl_5$ i. $NaI_3$ j. KCN

43. What is the mass of one *molecule* of each of the following?
a. $N_2$ b. $CH_4$ c. $O_2$ d. $H_2O$

44. What is the mass of one *molecule* of each of the following?
a. $NH_3$ b. $C_6H_{12}O_6$ c. $S_6$ d. Xe

45. a. How many moles of He are in $3.0 \times 10^{38}$ atoms of He?
b. How many moles of oxygen are in $9.0 \times 10^{23}$ molecules of $CO_2$?
c. How many moles of nitrogen are in $1.27 \times 10^{22}$ molecules of $N_2H_4$?
d. How many moles of carbon are in $1.5 \times 10^{24}$ molecules of $C_{40}H_{56}$?

46. a. How many moles of Al are in $3.0 \times 10^{19}$ atoms of Al?
    b. How many moles of hydrogen are in $4.78 \times 10^{12}$ molecules of $H_2O$?
    c. How many moles of nitrogen are in $9.9 \times 10^{22}$ molecules of $C_3N_3H_{11}$?
    d. How many moles of sulfur are in $7.6 \times 10^{24}$ molecules of $H_2S_2O_3$?

## SECTION 5.8 Average Atomic Masses

47. Naturally occurring samples of lithium consist of two isotopes: 92.58% of the atoms have a mass of 7.016 amu and the other 7.42% have a mass of 6.015 amu. What is the average atomic mass of naturally occurring lithium?

48. Naturally occurring samples of magnesium consist of three isotopes: 78.7% of the atoms have a mass of 23.985 amu, 10.2% have a mass of 24.986 amu, and 11.1% have a mass of 25.986 amu. What is the average atomic mass of naturally occurring magnesium?

49. Naturally occurring silver consists of two isotopes: 51.84% of the silver atoms have a mass of 106.905 amu and 48.16% of the silver atoms have a mass of 108.905 amu. What is the average atomic mass of naturally occurring silver?

50. Look at the mass spectroscopy data in Chapter Box 5.1, and calculate the average atomic mass of naturally occurring neon.

## ADDITIONAL PROBLEMS

51. How many grams of oxygen are there in 1 mol of each of these substances?
    a. $C_2H_4O_2$ b. $CO_2$ c. NO

52. How many grams of oxygen are there in 1 mol of each of these substances?
    a. $O_3$ b. $HNO_3$ c. $C_2H_2O_4$

53. How many grams of nitrogen are there in 2.00 mol of each of these compounds?
    a. $N_2H_4$ b. $NH_3$ c. $HNO_3$

54. How many grams of nitrogen are there in 3.00 mol of each of these compounds?
    a. $NO_2$ b. $KN_3$ c. $H_2NCH_2CH_2NH_2$

55. Give the valence of each element based on the number of its valence electrons.
    a. 1 valence electron
    b. 3 valence electrons
    c. 5 valence electrons
    d. 7 valence electrons

56. Give the valence of each element based on the number of its valence electrons.
    a. 2 valence electrons
    b. 4 valence electrons
    c. 6 valence electrons
    d. 8 valence electrons

57. What is the valence of the element E in each of these compounds?
    a. EH c. $EH_2$ e. $E_2O_3$
    b. $EH_3$ d. $EH_4$ f. $EO_2$

58. What is the valence of the element E in each of these compounds?
    a. ECl c. $ECl_4$ e. EO
    b. $ECl_2$ d. $ECl_3$ f. $E_2O$

### EXPERIMENTAL PROBLEM 5.1

In China (circa 700 AD) a street vendor selling pastries carried on a tray was accidentally bumped by a local merchant. The pastries went flying, and some were crushed by a passing cart. The contrite merchant offered to pay for the pastries. The street vendor insisted that there were 20 pastries on the tray, but the merchant was certain that the tray held no more than a dozen. A constable came by and gathered up all of the pastries, including the crushed pieces, and took everything to the local magistrate. The magistrate weighed 1 pastry and found that it had a mass of 14 g. All of the pastries plus the crushed pieces had a mass of 151 g.

**59.** If you were the magistrate, what decision would you make? (This is a very early example of counting by weighing).

## EXPERIMENTAL PROBLEM 5.2

As noted in Experimental Exercise 5.1, nineteenth-century chemists determined the molar mass of an element by analyzing a series of pure substances containing the element. They used experimental techniques that allowed them to measure the molar mass of a compound and the mass of the element present in the compound. The chemists assumed that the part of the mass due to a particular element represented the sum of the masses of the atoms of that element present, and that the smallest mass obtained for an element from a series of compounds must be the molar mass of that element. Experimental results for an unknown element, Y, are given in the table below.

| Compound | Molar Mass of Pure Substance (g) | Part of Molar Mass Due to Element Y (g) |
|---|---|---|
| 1 | 36.5 | 35.46 |
| 2 | 70.9 | 70.92 |
| 3 | 273.6 | 70.92 |
| 4 | 182.1 | 106.38 |
| 5 | 260.2 | 141.84 |
| 6 | 208.3 | 177.30 |

**QUESTIONS**

60. What is the molar mass of Y?
61. Which of the pure substances consists only of Y atoms?
62. What is the formula for one molecule of Y?
63. How many atoms of Y are in one molecule of compound 2?
64. How many atoms of Y are in one mole of compound 5?
65. Compound 6 contains only phosphorus and Y. What is the formula of compound 6?
66. What element is Y?

## SOLUTIONS TO EXERCISES

**Exercise 5.1**

The answer is (4). Ar (argon) lies to the right of the stairstep line in Figure 5.1.

**Exercise 5.2**

**a.** Strontium (Sr) **b.** Oxygen (O) **c.** Zirconium (Zr)

**Exercise 5.3**

**a.** The general formula for Group 2A halides is $MX_2$. The specific compound is $BaI_2$.
**b.** ClF, based on the general rule for compounds containing a second-period element and chlorine
**c.** HBr (the formula of all the hydrogen halides is HX)
**d.** Potassium is an alkali metal and F is a halogen, so KF is the answer.

**Exercise 5.4**

**a.** Carbon is a main group element (Group 4A).
**b.** Dysprosium is a lanthanide.
**c.** Potassium is a main group element and an alkali metal (Group 1A).
**d.** Chromium is a transition metal (Group 6B).
**e.** Californium is an actinide.
**f.** Bismuth is a main group element (Group 5A).

**Exercise 5.5**

**a.** (1) Chlorine is in the third period, so it has three occupied shells of electrons. (2) It's in Group 7A, so it has seven valence electrons.
**b.** (1) Radium is in the seventh period, so it has seven shells with electrons in them. (2) It's in Group 2A, so it has two valence electrons.

**c.** (1) Arsenic is in the fourth period, so it has four shells with electrons in them. (2) It's in Group 5A, so it has five valence electrons.
**d.** (1) Oxygen is in the second period, so it has two shells with electrons in them. (2) It's in Group 6A, so it has six valence electrons.

### Exercise 5.6

**a.** Phosphorus: Group 5A, 5 valence electrons, $[Ne]3s^2 3p^3$
**b.** Oxygen: Group 6A, 6 valence electrons, $1s^2 2s^2 2p^4$
**c.** Cesium: Group 1A, 1 valence electron, $[Xe]6s^1$ (Because Cs is to the left of the sixth-period transition metals, the $5d$ subshell is empty).

### Exercise 5.7

**a.** Se has the largest atomic radius and Cl has the smallest.
**b.** $Br^-$ has the largest atomic radius and Cl has the smallest.

### Exercise 5.8

**a.** From their positions on the periodic table, fluorine is the most electronegative, sulfur is the least electronegative, fluorine has the strongest tendency to form negative ions, and sulfur has the largest atomic radius.
**b.** Mg is the most electronegative; K is the most electropositive; K has the strongest tendency to form cations; Mg has the smallest atomic radius.

### Exercise 5.9

**a.** Calcium is in Group 2A and has a valence of 2.
**b.** Boron is in Group 3A and has a valence of 3.
**c.** Oxygen is in Group 6A and has a valence of 2.
**d.** Silicon is in Group 4A and has a valence of 4.

### Exercise 5.10

**a.** $CaI_2$
**b.** $CI_4$
**c.** $H_2S$
**d.** Magnesium has a valence of 2 and nitrogen has a valence of 3. To get these to add up equally $(2 + 2 + 2 = 3 + 3)$ you need three magnesium atoms and two nitrogen atoms: $Mg_3N_2$.

### Exercise 5.11

**a.** $(1 \text{ mol N} \times 14 \text{ g mol}^{-1} \text{ N}) + (3 \text{ mol H} \times 1.0 \text{ g mol}^{-1} \text{ H}) = 17 \text{ g}$
**b.** $(2 \text{ mol Na} \times 23.0 \text{ g mol}^{-1} \text{ Na}) + (1 \text{ mol} \times 12 \text{ g mol}^{-1} \text{ C}) + (3 \text{ mol O} \times 16 \text{ g mol}^{-1} \text{ O}) = 106 \text{ g}$
**c.** $(40 \text{ mol C} \times 12 \text{ g mol}^{-1}) + (56 \text{ mol H} \times 1.0 \text{ g mol}^{-1} \text{ H}) = 536 \text{ g}$

### Exercise 5.12

**a.** There are nine atoms in a molecule of ethyl alcohol, and there are $6.02 \times 10^{23}$ molecules in a mole of ethyl alcohol. Therefore 1 mol of ethyl alcohol contains

$$\frac{9 \text{ atoms}}{1 \text{ molecule } C_2H_6O} \times \frac{6.02 \times 10^{23} \text{ molecules } C_2H_6O}{1 \text{ mol } C_2H_6O} = 54.2 \times 10^{23} \text{ atoms mol}^{-1}$$

That is, there are $5.42 \times 10^{24}$ atoms in a mole of $C_2H_5OH$.

**b.** There are two atoms in a molecule of $H_2$, and there are $6.02 \times 10^{23}$ molecules in a mole of $H_2$. Therefore 0.12 mol of hydrogen molecules contains

$$0.12 \text{ mol } H_2 \times \frac{6.02 \times 10^{23} \text{ molecules } H_2}{1 \text{ mole } H_2} \times \frac{2 \text{ H atoms}}{1 \text{ molecule } H_2} = 1.4 \times 10^{23} \text{ H atoms}$$

That is, there are $1.4 \times 10^{23}$ H atoms in 0.12 mol of $H_2$.

### EXPERIMENTAL EXERCISE 5.1

1. The molar mass of E is 14.01 g, because that is the smallest mass for E obtained experimentally.
2. The molar mass of pure substance 1 and the molar mass due to element E are the same. Therefore pure substance 1 consists of only E atoms.
3. Pure substance 1 consists of two moles E atoms, so its formula is $E_2$.
4. Two atoms per molecule times $6.02 \times 10^{23}$ molecules per mole equals $1.20 \times 10^{24}$ atoms per mole of $E_2$.
5. Compound 2, like pure substance 1, contains two atoms of E per molecule.
6. Compound 5 contains 14.01 g of E per mole, or one atom of E per molecule. Therefore, 1 mol of this compound contains $6.02 \times 10^{23}$ E atoms.
7. The molar mass of $E_2O$ is 14.01 g + 14.01 g + 16.00 g = 44.02 g. This is the approximate molar mass of compound 2.
8. Like element E, nitrogen has an atomic mass of 14.01, and is the only element that does. Element E is nitrogen.

### Exercise 5.13

As before, the equation is

$$\text{Average atomic mass} = (0.80)(10.0 \text{ amu}) + (0.20)(11.0 \text{ amu}) = 10.2 \text{ amu}$$

The average atomic mass of this sample of boron is 10.2 amu.

6

# ATOMIC STRUCTURE IN MORE DETAIL



KEY TERMS

Chapters 4 and 5 examined some aspects of the structure of atoms. The classification of electrons as *s*, *p*, *d*, and *f* (Section 5.5) only hints at the level of detail that chemists and physicists use in dealing with atoms and molecules.

This chapter digs deeper into the topic of atomic structure. We begin with the subject of light, which is probably the most important tool scientists use to investigate the details of atomic and molecular structure. It is at the same time one of the best-understood and least-understood of natural phenomena. Scientists can measure the properties of light and predict what it will do as well as they can do anything. The same can be said about electrons, another major topic of this chapter.

When scientists look more closely at both light and electrons, their understanding seems to break down. Both light and electrons behave like waves in certain circumstances and like particles in other circumstances. What is something that acts this way really like? This chapter begins by describing the things about light that can be measured and predicted and ends by discussing the aspects nobody really understands.

FIGURE 6.1
Waves are easy to perceive in water. Sometimes both light and electrons behave as if they were waves rather than particles.

## 6.1 Waves

GOAL: To describe the properties of waves (wavelength, frequency, and velocity).

Light sometimes behaves as though it were waves and sometimes as though it were particles. Scientists have never observed light behaving like both a wave and a particle at the same time, and it looks as if they never will. This means we need to consider light's wave and particle properties separately. In this section we examine the properties of waves.

You probably are most familiar with waves in water (see Figure 6.1). Sound waves are another kind of wave you encounter often. Air molecules don't move up and down in sound waves the way molecules in water waves do. They vibrate toward and away from the source (see Figure 6.2). The mechanism of the ear translates the vibrations into something the brain can deal with.

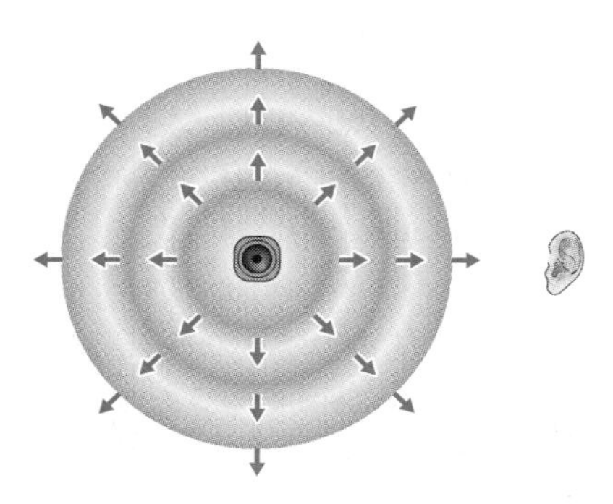

FIGURE 6.2
Sound also travels as waves. The atoms or molecules in the medium transmitting the sound move back and forth, creating zones of greater and lesser density.

Ordinary waves—simple ones, anyway—have four basic properties. Three of these properties are important when we consider light. **Wavelength** is the distance from the peak of one wave to the peak of the next (see Figure 6.3). The **frequency** is the number of waves that pass a given point per unit of time. A wave, like anything that moves, has a **velocity,** which is the distance the wave covers per unit of time.

The equation relating these three properties is

$$v = \nu\lambda$$

where $v$ is the velocity, $\nu$ (the Greek letter nu) is the frequency, and $\lambda$ (the Greek letter lambda) is the wavelength (see Figure 6.4).

One of the more interesting properties of light is that its velocity in a vacuum has a constant value: $3.00 \times 10^8$ m s$^{-1}$. Scientists universally use the letter $c$ to stand for the velocity of light, and thus the equation for the velocity of light is $c = \nu\lambda$. The constant velocity of light means that you can calculate its wavelength if you know its frequency, and vice versa.

Waves also have a height, or **amplitude.** For water waves the energy is related to the amplitude. Big waves have more energy than small waves. This is fairly obvious if you have ever been on a boat in choppy water. Big waves rock the boat more than small waves do because they have more energy.

However, and this is a very important point, the *total* energy of any light source has nothing to do with any measure of wave amplitude. The relative brightness of light bulbs has to do with the particle nature of light. An ordinary 75-watt bulb gives off more light particles than a 60-watt bulb does, for instance. We will discuss the energy of individual light particles later in this chapter, in Section 6.4.

These quantities used to describe waves have units, of course. Wavelengths are lengths, and scientists commonly express them in SI units of meters, with or without a prefix. Velocities are derived quantities and are usually given in meters per second ($m\ s^{-1}$). The units of frequency are cycles per second, meaning the number of wave crests or troughs (one cycle of the wave) that pass an observer each second:

$$1 \text{ cycle (wave) per second} = 1\ s^{-1} = 1 \text{ hertz (Hz)}$$

The hertz is not named for the car-rental company, but for the German physicist Heinrich Hertz (1857–1894).

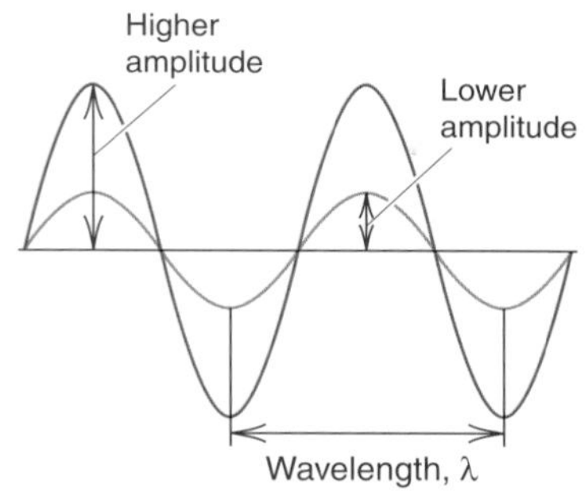


FIGURE 6.3
Wavelength is the distance between two wave troughs or crests; amplitude is the height of the wave. These two waves have the same wavelength and different amplitudes.

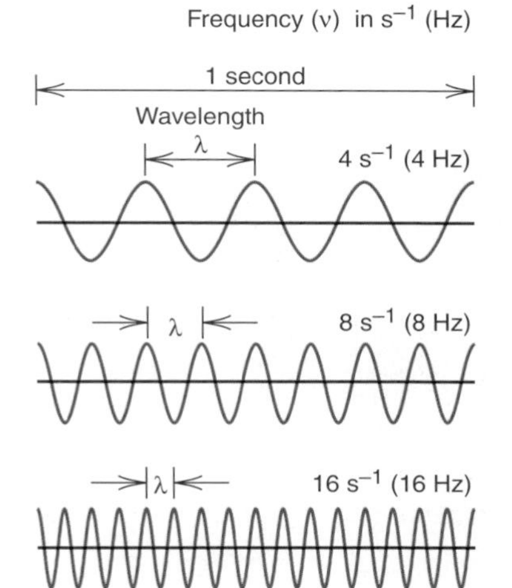


FIGURE 6.4
These three waves have different wavelengths and different frequencies. If the velocity of the waves is the same, shorter wavelengths mean higher frequencies.

### EXAMPLE 6.1 Velocity, Frequency, and Wavelength of Waves

**a.** Assuming that the two waves below have the same velocity, which one has the higher frequency? Explain your answer.

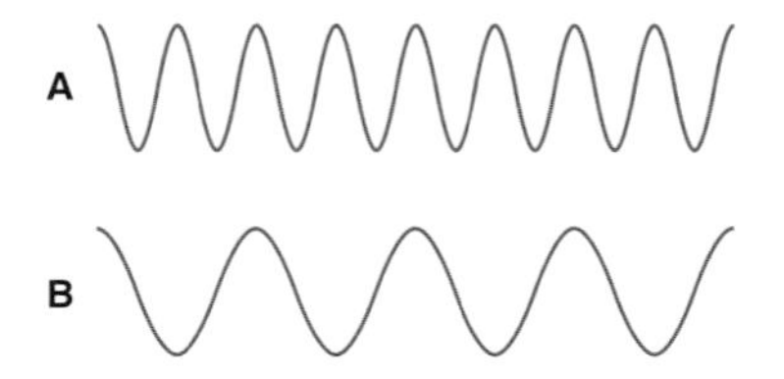


**b.** Suppose you just returned from a stroll along a lake shore. Some waves were breaking (as in Figure 6.5). The distance between wave tops was about 6 ft, and 9 waves broke in 15 s. What is the velocity of these waves in feet per second and in miles per hour?

**c.** The wavelength of a certain kind of radar waves is 0.10 m. The velocity of these waves is the same as the velocity of light, $3.00 \times 10^8\ m\ s^{-1}$. What is the frequency?

SOLUTIONS

**a.** The wave with the shorter wavelength (A) has the higher frequency. If the two waves have the same velocity, more full cycles of A will pass an observer in a given length of time.

FIGURE 6.5

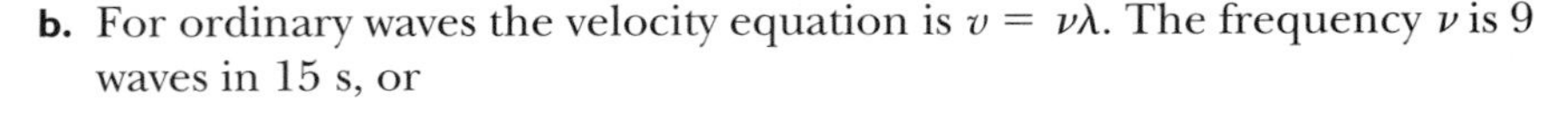

**b.** For ordinary waves the velocity equation is $v = \nu\lambda$. The frequency $\nu$ is 9 waves in 15 s, or

$$\nu = \frac{9 \text{ waves}}{15 \text{ s}} = 0.6 \text{ wave s}^{-1}$$

The wavelength, $\lambda$, is 6 ft. The velocity is then:

$$v = \nu\lambda = (0.6 \text{ s}^{-1})(6 \text{ ft}) = 3.6 \text{ ft s}^{-1}$$

The velocity of these waves is about 3.6 feet per second. Now the second part of this question is a unit-conversion problem. (If you need to review this, see Chapter 3.)

$$\frac{3.6 \cancel{\text{ft}}}{1 \cancel{\text{s}}} \times \frac{1 \text{ mi}}{5280 \cancel{\text{ft}}} \times \frac{60 \cancel{\text{s}}}{1 \cancel{\text{min}}} \times \frac{60 \cancel{\text{min}}}{1 \text{ h}} = 2.4 \text{ mi h}^{-1} = 2 \text{ mi h}^{-1}$$

The velocity of these waves is 2 miles per hour.

**c.** For light waves the velocity equation is $c = \nu\lambda$. You are given $c$ and $\lambda$, so you have:

$$3.00 \times 10^8 \text{ m s}^{-1} = \nu(0.10 \text{ m})$$

$$\frac{3.00 \times 10^8 \text{ m s}^{-1}}{0.10 \text{ m}} = \nu = 3.00 \times 10^9 \text{ s}^{-1}$$

FIGURE 6.6
Rainbows and reflections from a laser disk show the visible part of the electromagnetic spectrum.

**EXERCISE 6.1**

**a.** If the two waves shown below have the same frequency, which has the higher velocity? Explain your answer.

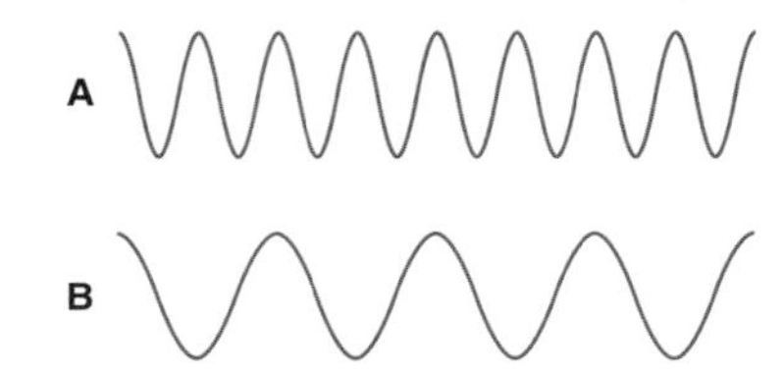


**b.** A speedboat passed you as you strolled along the lake shore. The tops of the waves it made were about 2.5 ft apart, and 12 of the waves broke on shore every 15 s. What is the velocity of these waves in feet per second and in kilometers per hour?

**c.** The frequency of a certain kind of radio waves is $9.8 \times 10^7 \text{ s}^{-1}$. The velocity of these waves is the same as the velocity of light, $3.00 \times 10^8 \text{ m s}^{-1}$. What is the wavelength of these waves in meters?

## 6.2 Light: The Electromagnetic Spectrum

**GOAL:** To list the families of electromagnetic radiation and give their properties.

After it rains, you can sometimes see a rainbow. You can also see the same familiar range of colors—red, orange, yellow, green, blue, violet—in the spray from a lawn sprinkler, in light that has passed through a glass prism, and in the reflection from a compact disc (see Figure 6.6). This range of colors is the visible spectrum. The existence of rainbows and the spectrum in general is due to the wave nature of light.

**The English astronomer Sir William Herschel (1738–1822) is probably better known as the discoverer of the planet Uranus than as the discoverer of infrared radiation.**

Certain types of waves, including radio and radar waves, have the properties of light but are invisible to the eye. The first important discovery about these waves was made by Sir William Herschel in 1800. He examined the heating effect of light of various colors (see Figure 6.7). When he used a prism to split sunlight into the colors that make it up, he found that the red portion heats objects strongly. Surprisingly, an object placed beyond the red end of the spectrum, where there was no light visible at all, became even hotter. Herschel proposed that a form of "light" invisible to the eye was responsible. Scientists now use the term **infrared radiation** for the invisible radiation that causes this heating.

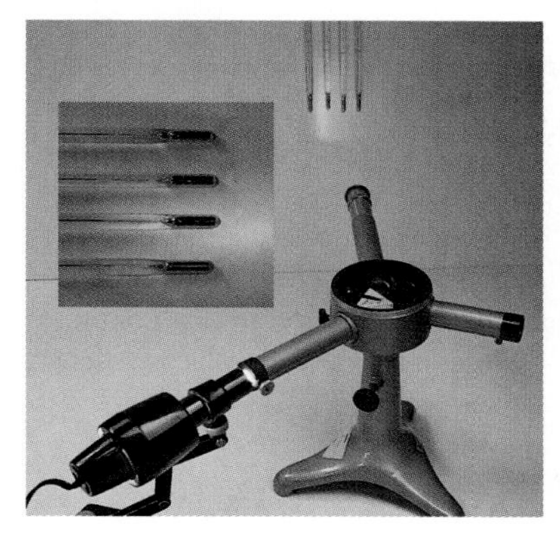

FIGURE 6.7
Herschel studied the heating ability of light of various colors. Here his experiment is repeated using modern apparatus.

Silver halides (AgCl, AgBr, and AgI) break down under the influence of light to give black grains of silver metal (among other things). This reaction is the basis of photography (see Figure 6.8). The German physicist Johann Ritter used this reaction to duplicate Herschel's discovery at the other end of the visible spectrum in 1801. Light from the invisible region of the spectrum beyond violet breaks AgCl down more effectively than any of the visible colors do. This "light" is now called **ultraviolet radiation.**

In the 1860s the great Scottish physicist James Clerk Maxwell proposed that all forms of invisible and visible light are variations of a single type of radiation. He proposed that this radiation travels through space as oscillating electric and magnetic fields oriented at right angles to each other (see Figure 6.9). Maxwell's idea was later verified, and scientists now refer to all of these varieties of "light" (visible and invisible) as **electromagnetic radiation.**

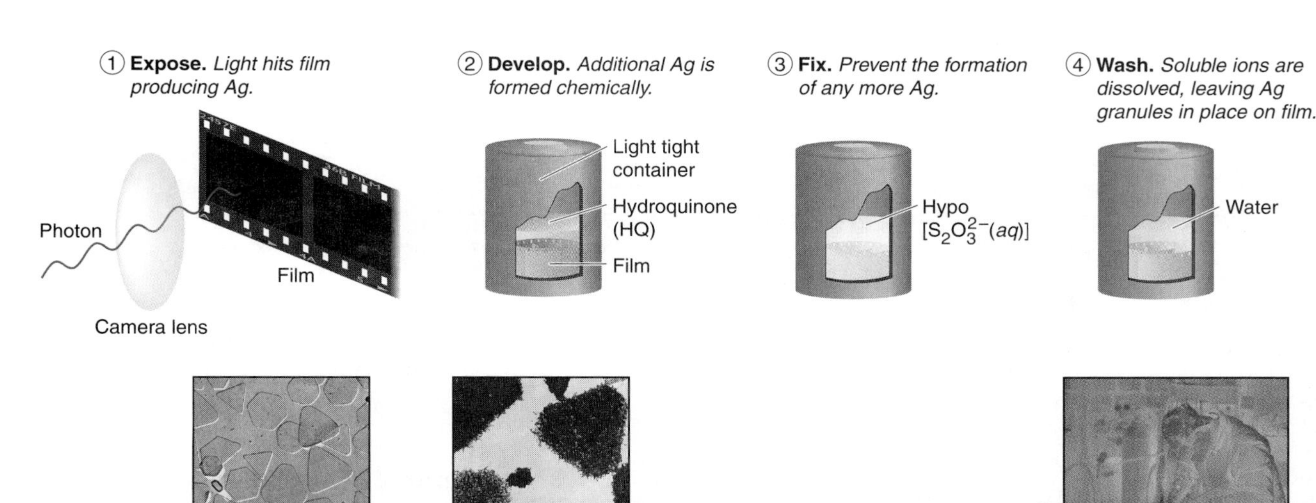
1 Expose. Light hits film producing Ag.
Photon
Camera lens
Film
2 Develop. Additional Ag is formed chemically.
Light tight container
Hydroquinone (HQ)
Film
3 Fix. Prevent the formation of any more Ag.
Hypo
[$S_2O_3^{2-}(aq)$]
4 Wash. Soluble ions are dissolved, leaving Ag granules in place on film.
Water
AgBr crystals before developing
AgBr crystals after developing
Negative

FIGURE 6.8
When you expose a photographic film, the visible light converts a small amount of silver halide (AgBr in this example) to silver metal. The developing process produces even more silver, and eventually yields a negative.

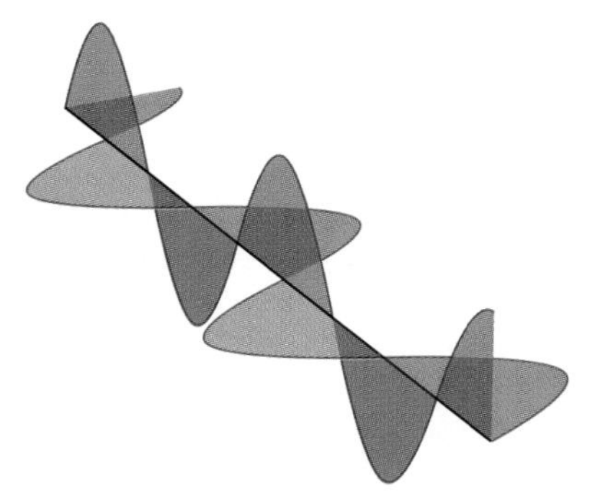

FIGURE 6.9
Electromagnetic radiation has electric and magnetic fields oscillating in perpendicular planes. Note that as the strength of the electric field increases (shown in RED), the strength of the magnetic field decreases proportionally (shown in BLUE). Consequently, the total energy of the radiation remains constant.

Many more varieties of electromagnetic radiation are now known, and they span a very wide range of frequencies and wavelengths.

Listed in order from low frequency to high and from long waves to short, here are the important families: **radio waves,** microwaves, infrared, visible, ultraviolet, X rays, and gamma rays. Figure 6.10 shows the relationship between frequency and wavelength for these families. For all of them, the velocity is $c = 3.00 \times 10^8$ m s$^{-1}$.

Broadcasts of AM radio stations use very low frequencies and very long wavelengths. The markings on your AM radio give the frequencies of the stations in kHz (kilohertz). This corresponds to wavelengths of several hundred meters (a couple or six football fields long). Television and FM radio broadcasts use somewhat higher frequencies. Another look at your radio shows that FM radio stations broadcast in the MHz (megahertz) range. These waves are a couple of meters long.

**Microwaves** are used in cooking and a host of other applications. They have higher frequencies (and shorter wavelengths) than radio waves. Microwave frequencies are about 1000 MHz, and microwaves have wavelengths in the millimeter range.

Herschel's discovery, infrared radiation (the prefix *infra* means "below"), is just below the visible spectrum in frequency. Red light has the lowest frequency of any visible light, and violet light has the highest. Ultraviolet ("beyond violet") light is known for its ability to produce suntan, sunburn, and sometimes skin cancer. Its frequencies are higher than those of visible light.

**X rays,** used in medicine, have even higher frequencies. The even more dangerous **gamma ($\gamma$) rays** are given off by some radioactive elements (see Chapter 17). Of all the types of electromagnetic radiation that scientists study, gamma rays have the highest frequencies and the shortest wavelengths.

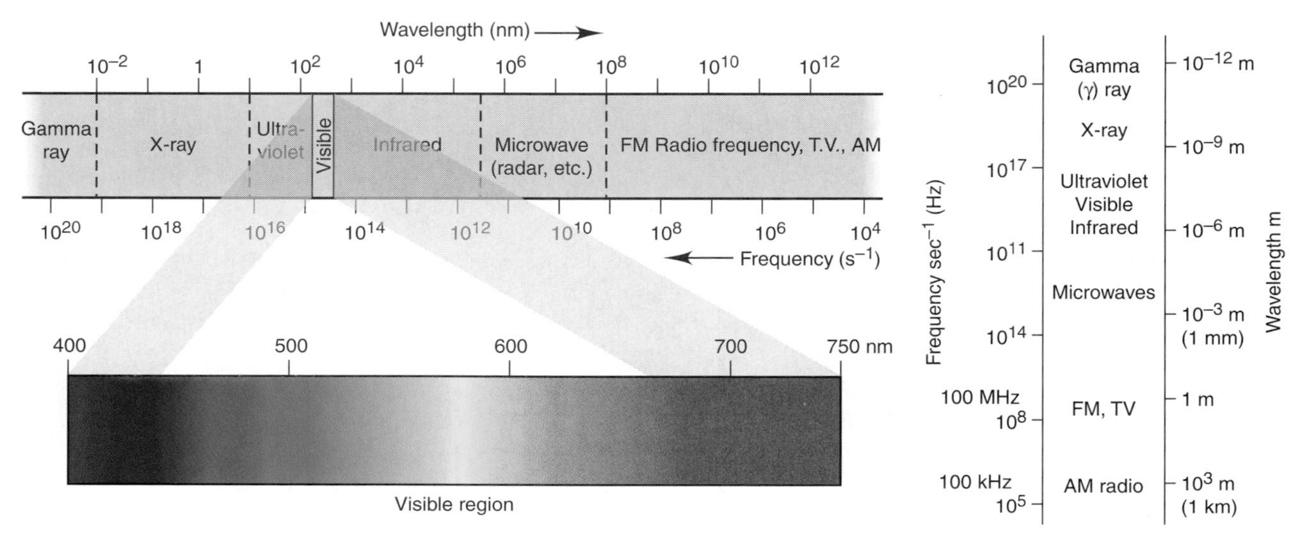


FIGURE 6.10
Various regions of electromagnetic radiation. These all have the same velocity in a vacuum, but they span a huge range of frequencies and wavelengths. The relatively narrow visible region of the spectrum is expanded to show the colors.

EXAMPLE 6.2

### Relative Wavelengths and Frequencies of Light

**a.** Which of these has the longest wavelengths? Which has the highest frequencies?
A. Infrared radiation C. Microwaves
B. Ultraviolet radiation D. Visible light

**b.** A popular FM music station in New Orleans broadcasts at 101.1 MHz. What is the wavelength of the radio waves broadcast by this station?

**STUDY ALERT**

**High frequencies go with short wavelengths (and vice versa).**

**SOLUTIONS**

**a.** You answer this kind of question using memorized information from earlier in the section. Of the types listed, microwaves have the longest wavelengths and ultraviolet radiation has the highest frequencies.

**b.** You solve this from the equation $c = \nu\lambda$. As always, $c = 3.00 \times 10^8$ m s$^{-1}$, and here $\nu = 101.1 \times 10^6$ s$^{-1}$.

$$3.00 \times 10^8 \text{ m s}^{-1} = (101.1 \times 10^6 \text{ s}^{-1})\lambda$$

$$\frac{3.00 \times 10^8 \text{ m s}^{-1}}{101.1 \times 10^6 \text{ s}^{-1}} = \lambda = 2.97 \text{ m}$$

**EXERCISE 6.2**

**a.** Which of the following has the shortest wavelengths? Which has the lowest frequencies?
A. Infrared radiation C. Microwaves
B. Ultraviolet radiation D. Visible light

**b.** An AM radio station in New Orleans broadcasts at 1280 kHz. What is the wavelength of the radio waves broadcast by this station?

Waves of light have one other important property—energy. This topic is discussed in Section 6.3.

CHAPTER BOX **6.1 NOT ALL RADIATION IS ELECTROMAGNETIC**

**The types** of radiation in Figure 6.10 are not the only ones. In the late nineteenth century physicists and chemists discovered many different phenomena that they called radiation. Some of these are electromagnetic, and some are not.

Radioactive elements give off three major kinds of radiation, designated by the Greek letters alpha ($\alpha$), beta ($\beta$), and gamma ($\gamma$). As you may remember from our discussion of Rutherford's experiment (Section 4.5), alpha rays (also called alpha particles) are helium nuclei, containing two protons and two neutrons. Beta rays are electrons with high energies. Neither alpha nor beta rays travel as fast as light in a vacuum (or anywhere else). Gamma rays travel at the velocity of light, and they are a form of electromagnetic radiation.

*Cosmic rays* rain down on the earth from space. Over 99% of them are particles, ranging from the protons and electrons you already know about to some stranger members of the physicists' particle zoo.

## 6.3 Light, Atoms, and Energy

**GOAL:** To summarize the results of early studies on light and its interactions with atoms and to explore the relationships among frequency, wavelength, and energy for a photon of light.

Section 6.2 considered light all by itself. This section will begin to examine what happens when light and atoms get together. Chemists and physicists learn more about both by studying these interactions.

Joseph von Fraunhofer (1787–1826) was a German optician noted for his glassworking ability. He was known all over Europe for the quality of the glassware and instruments he made. When he passed sunlight through his best glass prisms, the spectrum he saw was not continuous. There were certain frequencies missing, and these showed as dark lines (see Figure 6.11). Fraunhofer announced his discovery in 1814 and mapped the positions of several hundred of these lines. Nobody had any idea why sunlight would behave like that, so Fraunhofer's discoveries remained unexplained for half a century.

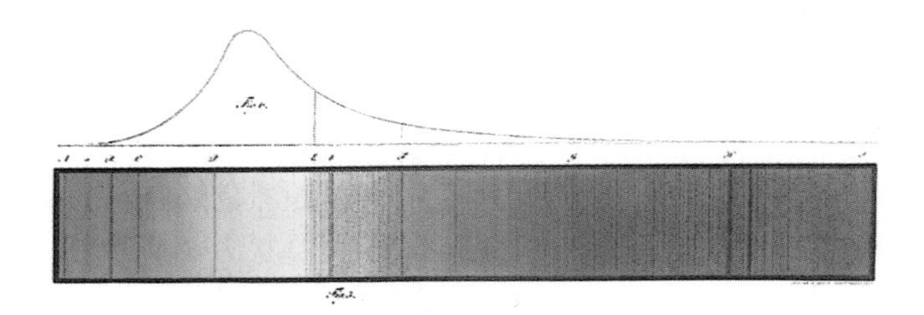

FIGURE 6.11
The dark lines Fraunhofer (shown in the margin) observed in the visible spectrum correspond to missing frequencies in sunlight.

In 1854 a young German physicist named Gustav Robert Kirchhoff was hired as a professor at Heidelberg. In the classroom he gave very thorough but very dull lectures. In the lab he teamed up with Robert Bunsen, another professor, to investigate the light that substances give off when they heat up. Bunsen developed a burner that gave a hot, faint-blue flame. Kirchhoff suggested using a prism to study the light that substances give off when heated in this flame.

When the team heated samples of elements to produce light and passed the light through a prism, they saw separated lines of pure colors rather than a continuous spectrum. Different elements produce lines of different colors—that is, light of different frequencies (see Figure 6.12). Each element's pattern of lines is unique, like a fingerprint, and can be used to identify that element. This discovery was very useful to chemists who were studying the composition of minerals. Some minerals gave off light that showed lines that did not correspond to those for any known element. Bunsen and Kirchhoff discovered the elements rubidium and cesium this way.

FIGURE 6.12
The faint blue of a Bunsen burner flame allows the colors that the heated elements give off to show up well. Strontium compounds give a crimson flame (left), while copper salts give green (center). You can see colors like these in fireworks displays.

Several scientists (including Kirchhoff) found that gaseous atoms absorb light. The frequencies are the same as the ones given off by hot, incandescent atoms of the same element (see Figure 6.13). This connected Kirchhoff's results with Fraunhofer's. Sodium atoms give off two distinctive bright

Sodium

FIGURE 6.13
Hot atoms of many elements give off light having distinct frequencies ("lines") in the spectrum. Physicists and astronomers have found the lines for helium, sodium, and many other elements in the solar spectrum. This is the spectrum for sodium.

lines in a burner flame, and Kirchhoff saw that these have the same frequencies as two of Fraunhofer's dark lines in sunlight.

Here, for the first time, was a way to find out what the sun and stars were made of. Astronomers found lines corresponding to many of the known elements in sunlight, along with lines of a mystery element. This element was called *helium* (from the Greek word for the sun). After helium was discovered in the sun, scientists looked for it on earth. The presence of helium-filled balloons in virtually every flower shop and grocery store is evidence of their success.

Kirchhoff's discoveries posed a new question: why do the elements produce lines; that is, why are some specific frequencies favored over the rest of the broad span of possibilities? One problem was that the lines emitted by the elements showed no obvious pattern. That perception began to change in 1885, when a math teacher at a Swiss girls' school began his research.

Johann Jakob Balmer studied the spectrum of hydrogen and saw a regular pattern (see Figure 6.14). He and his successors put together an equation that tells where the lines of the hydrogen spectrum fall:

$$\frac{1}{\lambda} = R_{\mathrm{H}}\left[\frac{1}{n^2} - \frac{1}{(n')^2}\right]$$

As usual, $\lambda$ is the wavelength; $R_{\mathrm{H}}$ is a constant equal to $1.097 \times 10^7\ \mathrm{m}^{-1}$, and $n$ and $n'$ are integers (simple whole numbers, 1, 2, 3, and so on). As long as $n'$ is larger than $n$, the equation predicts the wavelength of a line you can observe. These lines appear in the ultraviolet, visible, and infrared portions of the electromagnetic spectrum. They form a regular pattern. Although this equation worked, nobody yet knew why.

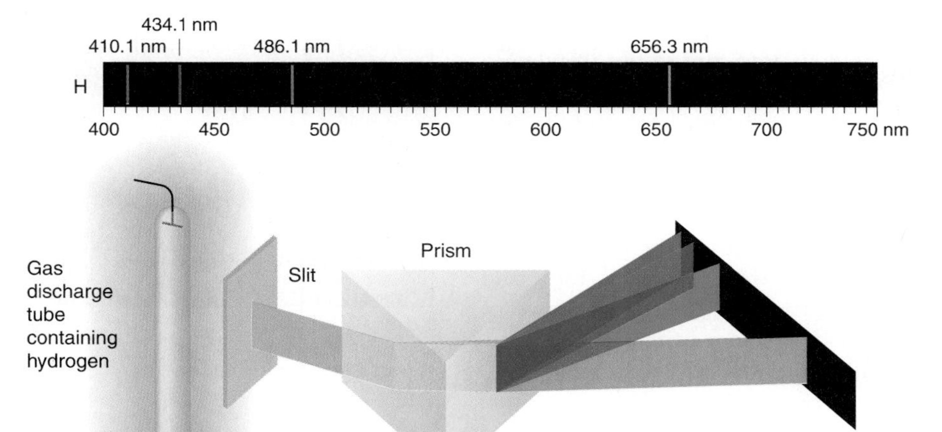


FIGURE 6.14
Four lines in the visible region of the hydrogen spectrum, discovered by Balmer.

In hindsight, all of these discoveries suggest that there was something fundamentally wrong with the way scientists of the time were interpreting the laws of physics and chemistry. Something was happening in these experiments with atoms and light that made no sense in terms of Dalton's view of the unbreakable atom. Arrhenius made his controversial proposals about the existence of ions at about the same time as Balmer's work appeared.

By the beginning of this century, chemists had largely accepted Arrhenius's ions. Physicists had more trouble with their part of the problem. Max Planck and Albert Einstein were the ones who got things moving. What follows is a condensed account of many years of effort.

The explanations that worked began with two hypotheses. First, light is made up of small units of energy—called **photons.** Second, the energy of a photon is a function of its frequency and nothing else. The equation for this energy is

$$E = h\nu$$

where $E$ is the energy, $\nu$ is the frequency of the light, and the constant $h$ is called Planck's constant and is equal to $6.63 \times 10^{-34}$ J s (joule-seconds). (The joule is the SI unit for energy, as you may remember from Section 4.1.)

Once again, light is acting in an odd way here. For ordinary waves the energy is a function of the amplitude, but light waves show no such effect.

## CHAPTER BOX 6.2 THE LUMINIFEROUS ETHER

**Light has** always been a tricky subject to deal with. When you think about the refraction of light through a prism from a mathematical standpoint, it's obvious that light is made up of waves. All the waves familiar to us propagate through something. Waves on a lake propagate through water, and sound waves propagate through the air (and through water and solids, too). Waves don't exist by themselves.

So physicists proposed that there was a substance called the *luminiferous* ("light-bearing") *ether,* an insubstantial fluid that pervaded the universe. Light was supposed to be a vibration of this ether. If this theory is true, there should be certain observable effects. For instance, if you are moving through this ether, light waves should seem to be shorter when you move toward their source and longer when you move away from them.

The American physicist Albert Abraham Michelson (1852–1931) devised an apparatus to test this theory. With financial help from Alexander Graham Bell, the inventor of the telephone, he split light waves, sent them in different directions, and combined them again. He found that the wavelength of light was the same in all directions. He started in 1881 and worked on his experiment for years, making it more careful and precise. His results were always the same.

There is no luminiferous ether. Light is a wave that does not need a medium in which to propagate.

FIGURE 6.15
Albert Abraham Michelson (1852–1931)

## EXAMPLE 6.3 Energy, Frequency, and Wavelength of Light

**a.** Consider the following types of electromagnetic radiation: radio waves, X rays, visible light, and ultraviolet radiation.
1. Which has the highest frequencies?
2. Which has the longest wavelengths?
3. Which has the highest energies?

**b.** One of the lines observed by Balmer was in the red region of the visible spectrum. This red light had a wavelength of 656 nanometers.
1. What is the energy, in joules, of this spectral line?
2. What is the energy, in kilojoules, of 1 mole of these photons?

### SOLUTIONS

**a.** 1. This is review from Section 6.2. X rays have the highest frequencies of this set.
2. High frequency corresponds to short wavelength; low frequency goes with long wavelength. Radio waves have the longest wavelengths of these types of radiation.
3. High frequency corresponds to high energy. X rays have the highest energies of this set.

**b.** 1. The equation is $E = h\nu$, where $h$ is Planck's constant ($6.63 \times 10^{-34}$ J s) and $\nu$ is the wavelength (656 nm $= 656 \times 10^{-9}$ m). It is important to remember to convert the wavelength to meters. To get the frequency, you use the relationship $c = \nu\lambda$ or

$$\nu = \frac{c}{\lambda} = \frac{3.00 \times 10^{8}\ \text{m s}^{-1}}{656 \times 10^{-9}\ \text{m}} = 4.57 \times 10^{14}\ \text{s}^{-1}$$

Then:

$$E = h\nu = (6.63 \times 10^{-34}\ \text{J s})(4.57 \times 10^{14}\ \text{s}^{-1}) = 3.03 \times 10^{-19}\ \text{J}$$

2. The answer $3.03 \times 10^{-19}$ J is the energy of one photon. Most information tables, however, report the energy per mole of photons. To get the molar energy, use Avogadro's number:

$$\frac{3.03 \times 10^{-19}\ \cancel{\text{J}}}{1\ \cancel{\text{photon}}} \times \frac{6.02 \times 10^{23}\ \cancel{\text{photons}}}{1\ \text{mol}} \times \frac{1\ \text{kJ}}{1000\ \cancel{\text{J}}} = 182\ \text{kJ mol}^{-1}$$

**STUDY ALERT**

**High frequency means high energy. Hold out your finger and move it up and down quickly. This represents a high frequency. It takes a lot of energy to move your finger this way.**

### EXERCISE 6.3

**a.** Consider the following types of electromagnetic radiation: radio waves, X rays, visible light, and ultraviolet radiation.
1. Which has the lowest frequencies?
2. Which has the shortest wavelengths?
3. Which has the lowest energies?

**b.** Another line observed in the spectrum of hydrogen is in the blue region of the visible spectrum. This blue light has a wavelength of 434 nm.
1. What is the energy of one of these photons, in joules?
2. What is the energy of 1 mol of these photons, in kilojoules?

## 6.4 Quantum Effects in Atoms

GOAL: To describe quantum effects involving the energy levels of electrons in atoms.

After Maxwell published his theory on electromagnetic radiation, the scientific community was satisfied that his wave theory explained all the observed phenomena involving light. However, difficulties in interpretation began to arise not too many years later, when physicists and chemists investigated the interactions between light and matter. Certain phenomena could only be explained by invoking the particle nature of light.

One of the most important of these experiments involved the *photoelectric effect.* Heinrich Hertz (for whom the units of frequency are named) discovered it while conducting experiments on radio waves. In exhibiting the photoelectric effect, certain metals give off electrons when light (usually ultraviolet light) shines on the surface (see Figure 6.16). These electrons are called *photoelectrons.* Physicists conducted experiments on a large number of metals, and their experimental results revealed several things:

- For each metal there is a unique threshold frequency (called $\nu_0$). Light with a frequency lower than $\nu_0$ does not cause the metal to give off any photoelectrons no matter how intense that light is.
- For any frequency above $\nu_0$ the number of electrons given off is proportional only to the intensity of the light.
- When the frequency of the light is greater than $\nu_0$, the kinetic energy of any photoelectrons the metal gives off is proportional to $\nu - \nu_0$.
- The kinetic energy of the individual photoelectrons is completely independent of the light intensity and depends only on $\nu$.

**The equation $E = h\nu$ originated from Einstein's efforts to interpret this experimental evidence.**

At the time this experiment was performed, existing theories predicted that shining more intense light should increase the kinetic energy of the electrons. After all, big waves increase the kinetic energy of a boat more than small waves do, by making it rock harder. Furthermore, the atomic theory in use at the time gave no reason why metals should even have a threshold frequency. Why did light of frequencies lower than $\nu_0$ fail to eject electrons from the metallic surface regardless of intensity?

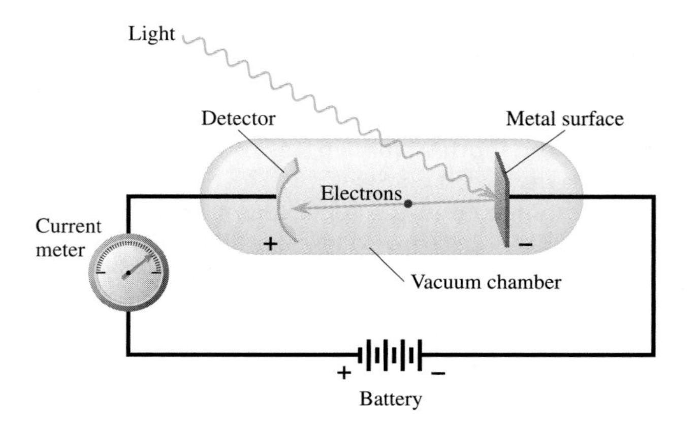


FIGURE 6.16
The photoelectric effect. This experiment is carried out in a vacuum. Light shines on the surface of a metal, causing the metal atoms to give off electrons. The battery makes the metallic surface into a negative electrode. When the electrons leave the metal, the negatively charged metallic surface repels them, and they travel down the tube toward the positive electrode. The number of electrons that traverse the tube is measured using an ammeter placed in the electric circuit.

Metals exhibiting the photoelectric effect do not release electrons spontaneously. Before an electron can escape a metal surface, it must gain at least a minimum amount of energy from a photon. If the photon has more energy than that, the metal emits an electron. For the photoelectric effect light behaves as if it exists in the form of particles. Yes, a photon of light does seem to have a frequency here, as a wave does, but it does not seem to have an amplitude. The energy of the light transfers as a single packet, and the amount of energy depends only on the frequency.

Furthermore, this experiment gave information about metal atoms. It strongly suggested that electrons in these atoms have certain energies, and that the atoms can only absorb photons that have more energy than the electrons do. When the energy of the photons is less than the energy of the electrons, the light has no effect on the electrons. Only when the light contains more energy than the electron can it knock the electron out of the metal atom. Different metals have different threshold energies, and the experiment suggests that the energy of the electrons in each metal atom is different. This was consistent with the results obtained by Balmer (Section 6.3) and others. Electrons in atoms have discrete, definite energies.

Chemists and physicists have developed a detailed model that accounts for these observations about atoms. Before discussing these details of what atoms are like, we need to look at one more key experiment. Light has obvious wave properties, and the fact that its energy comes in particle-like packets was something of a surprise. You already know that electrons are ordinarily classified as particles. Could these particles have wave properties?

**Einstein's complete equation was $E = h(\nu - \nu_0)$. That is, the energy of an ejected electron in the photoelectric effect is equal to Planck's constant times the difference between the light's frequency and the minimum necessary frequency.**

The French physicist Louis Victor de Broglie (1892–1976) proposed exactly that in 1924. (In fact, he proposed that *everything* has wave properties). Evidence to support his proposal was not long in coming. Two physicists, the American C. J. Davisson and the Briton G. P. Thomson (son of J. J. Thomson, who was the electron's discoverer) provided this evidence. These two experimenters independently found that moving electrons interfere with one another. Such interference is a property of waves. (For a more complete description of this phenomenon, see Section 6.7).

**According to de Broglie, all moving particles should have wave properties. However, the wavelength is inversely proportional to the mass of the particle. Any particle big enough to see, even through a microscope, is too big to have observable wave properties.**

The Danish physicist Niels Bohr (1885–1962) was the first to put forth a model of an atom with an equation for its energy levels. His equation worked for the hydrogen atom and no other kind. Many of his ideas found their way into the models chemists and physicists use today. Many aspects of these models are mathematically complex and will not concern us here. We will focus on a simplified description of the modern atomic model. This description has more depth than what you've seen up to now. Much of it involves quantum numbers, which we will discuss in more detail in the next section. Here are the basic ideas on which the theory is built:

- Electrons in atoms occupy **orbitals.** These are simply regions of space around a nucleus in which electrons can be found. (Bohr originally proposed that electrons occupy *orbits,* but this word implies that electrons are only particles, with no wave properties at all).
- The energy of an electron in an atom is proportional to its average distance from the nucleus. Electrons in orbitals farther from the nucleus have higher potential energies.
- Only a limited number of orbitals, each with a specific energy, are allowed in any atom. This means that the orbitals are *quantized.*

- When an atom absorbs light, an electron moves from a lower to a higher energy level (that is, from a lower-energy shell or subshell to one of higher energy). An atom emits light when an electron drops to a lower energy level (from a higher-energy orbital to a lower-energy one). The energy of the absorbed or emitted photon is equal to the energy difference between the two energy levels.

## EXAMPLE 6.4 Electrons in Orbitals

For the atom illustrated in the figure, the circles represent the average distance from the nucleus of electrons in the indicated energy levels. Which of the following electron transitions gives off a photon having the *highest* energy?

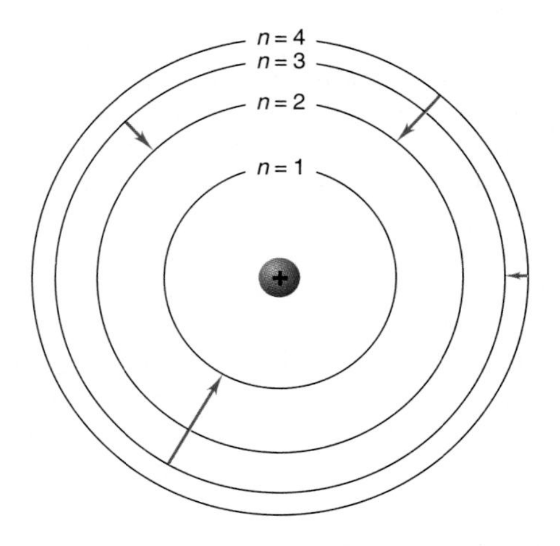


A. From level 3 to level 1
B. From level 4 to level 2
C. From level 4 to level 3
D. From level 3 to level 2

**SOLUTION**

The transition that gives off the highest-energy photon is the one that covers the largest average distance between energy levels. This is A: from energy level 3 to 1. Of course, the transition from energy level 4 to 1 gives off even more energy, but that isn't one of the listed answers. In fact, the energy of the photon (from the equation $E = h\nu = hc/\lambda$) is predicted correctly for a hydrogen atom by Balmer's equation (Section 6.3). In that equation, $n'$ is the number of the starting energy level for the transition and $n$ is the number of the final energy level. This prediction was one of the successes of Bohr's model of the atom. Unfortunately, it did not work well for atoms other than hydrogen.

**EXERCISE 6.4**

In the atom pictured in Example 6.4, for which transition between two energy levels is a photon having the *highest* energy absorbed?

You may have noticed that we haven't said anything about where the electrons in an atom are. We have merely referred to the "average distance" of an electron from the nucleus. One reason for this is theoretical. The best way to describe electrons in atoms is to consider only the wave nature of the electrons. Simply put, chemists and physicists think of electrons as waves confined to specific regions of space. This interpretation is difficult to visualize and highly mathematical in nature, and as far as is practical we will avoid it. The second reason was postulated in the late 1920s by the German physicist Werner Heisenberg (1901–1976).

Heisenberg knew that electrons displayed both particle and wave properties. On the one hand, a particle is localized in space and its position can be accurately known. You know exactly where your book is at this moment, don't you? On the other hand, a wave is spread out in space without a well-defined position. Where exactly is a wave as it approaches the beach? How can you picture something that has both wave and particle properties?

If electrons are particles, it is possible to find out exactly where they are; but if they are waves, this cannot be done. In thinking about this problem, Heisenberg came to the conclusion that there are limits on the accuracy with which an electron's position can be determined. This conclusion is known as the **Heisenberg Uncertainty Principle:** neither the position nor the momentum of an electron in an atom can be pinpointed at a particular instant in time. That's why scientists speak of orbitals instead of orbits, and average distance from the nucleus instead of actual distance. It is impossible to know exactly where any electron is at any given instant in time.

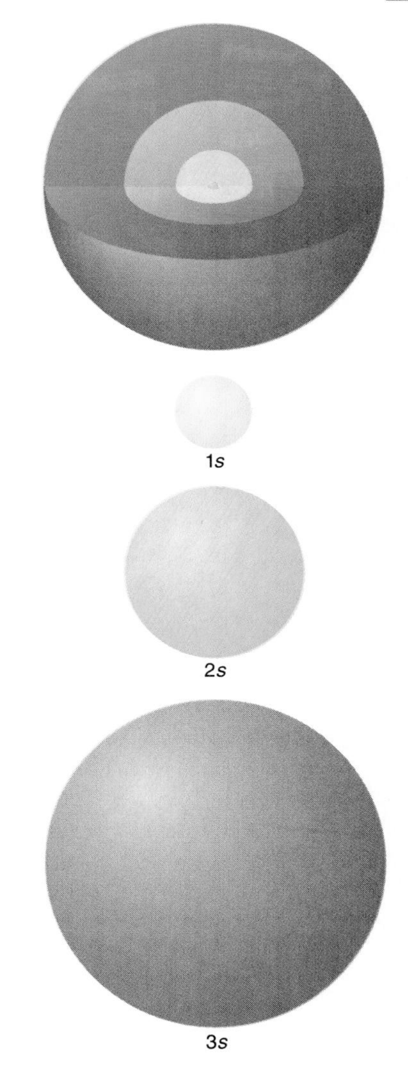


FIGURE 6.17
Atomic orbitals get larger as the value of *n* increases.

## 6.5 Quantum Numbers

**GOAL:** To introduce the three quantum numbers that specify the size, shape, and orientation in space of atomic orbitals.

Section 5.5 described in general terms the blocks of elements with *s*, *p*, *d*, and *f* electrons. This section develops this idea in more detail. Every electron in an atom, whether it is a valence electron or not, occupies an orbital—an *s*, *p*, *d*, or *f* orbital. You will recall that an orbital is a region of space around a nucleus in which an electron is found. Chemists use **quantum numbers** in the equations that describe these orbitals. You won't have to perform the calculations, but you will need to know something about what the numbers mean.

Scientists use the letter $n$ for the **principal quantum number.** Chemists use this number in calculating the energy of an electron. In fact, this is the same $n$ that Balmer used in his equation (see Section 6.3). The possible values of $n$ are the positive integers, starting at 1 and going up. The atomic orbital with $n = 1$ is the closest to the nucleus, and orbitals get progressively larger as $n$ increases, as shown in Figure 6.17.

**The value of *n* tells us which shell (which main energy level) the electron occupies.**

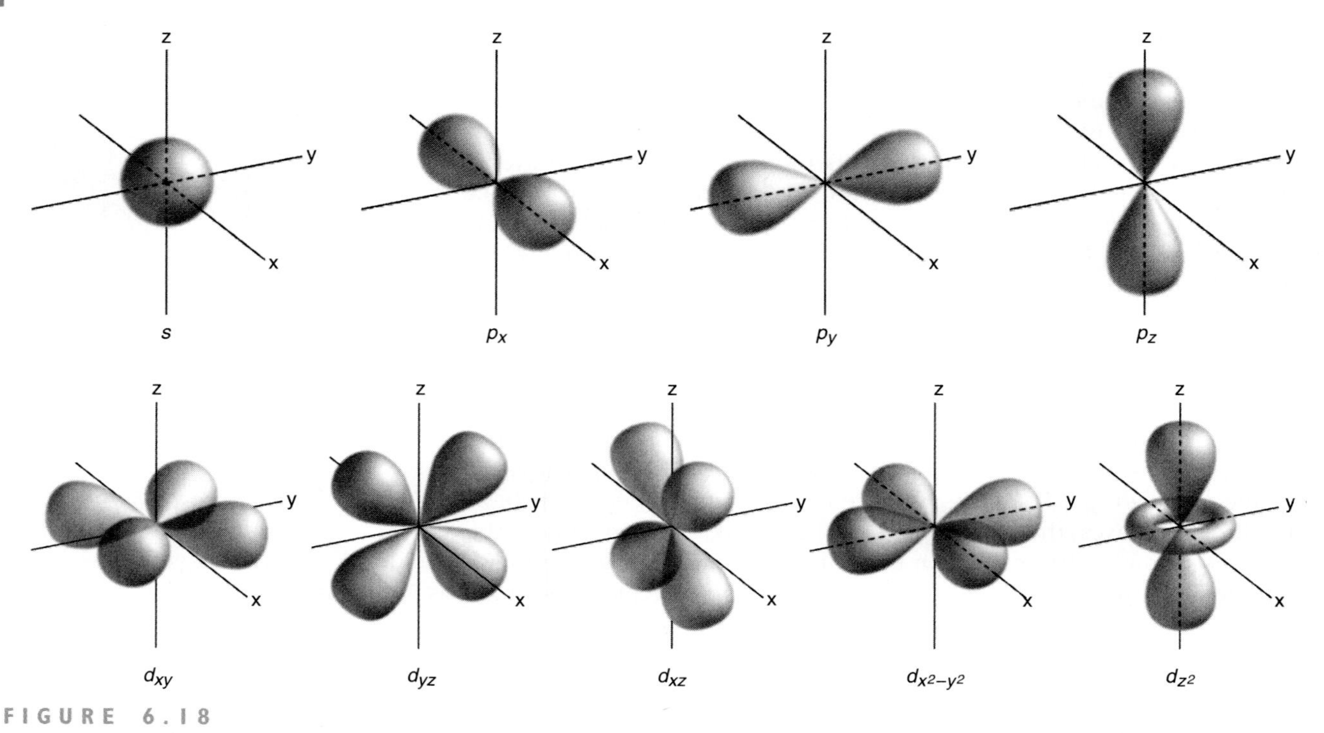


FIGURE 6.18
The angular quantum number $\ell$ specifies the shape of atomic orbitals. For *s* orbitals $\ell = 0$, for *p* orbitals $\ell = 1$, and for *d* orbitals $\ell = 2$.

If you want to know the value of $n$ for the orbital containing the valence electrons in any given element, look at the element's period number in the periodic table. The valence (and only) electron of a hydrogen atom is in the $n = 1$ shell. The valence electron of a lithium atom has $n = 2$, that of sodium has $n = 3$, and so on.

The **angular quantum number,** $\ell$, describes the shape of an orbital. If $\ell = 0$, the orbital is an *s* orbital, which is spherical. If $\ell = 1$, the orbital is a *p* orbital, usually drawn as a dumbbell shape. If $\ell = 2$, the orbital is a *d* orbital, and *f* orbitals have $\ell = 3$. Most of the *d* orbitals have a cloverleaf shape. See Figure 6.18 for simplified pictures of these orbitals.

Chemists very often use a shorthand designation for an orbital by giving the value of $n$ followed by the letter corresponding to the value of $\ell$. An orbital with $n = 3$ and $\ell = 1$, for instance, is called a 3*p* orbital.

**Remember that a subshell designation is also an orbital designation. The 3*s* subshell consists of one 3*s* orbital, for instance.**

EXAMPLE 6.5 **Subshell Designations**

What is the shorthand designation of each of the following subshells?

**a.** $n = 5$ and $\ell = 0$ **b.** $n = 2$ and $\ell = 1$ **c.** $n = 4$ and $\ell = 2$

SOLUTIONS

**a.** Since $\ell = 0$, this is an *s* subshell. Since $n = 5$, it is a 5*s* subshell.
**b.** 2*p* **c.** 4*d*

**EXERCISE 6.5**

What is the shorthand designation of each of the following subshells?
**a.** $n = 4$ and $\ell = 1$ **b.** $n = 3$ and $\ell = 2$ **c.** $n = 2$ and $\ell = 0$

As you can see, the values of $\ell$ are integers. The lowest possible value of $\ell$ is zero. The higher values of $\ell$ depend on the value of $n$. The highest possible value of $\ell$ is $n - 1$. All integer values of $\ell$ between 0 and $n - 1$ exist. For example, for $n = 1$, $\ell$ can have only a value of zero. For $n = 2$, $\ell$ can be either 0 (for the 2*s* orbital) or 1 (for the 2*p* orbital). The number of subshells that any given shell possesses equals the number of possible values for the quantum number $\ell$ for that shell. For example, the $n = 3$ shell has three subshells, because for $n = 3$ there are three different values for $\ell$ (2, 1, and 0).

**EXAMPLE 6.6** **Possible Subshell Types and the Value of *n***

**a.** What types of subshells are possible when $n = 3$? Explain your answer.
**b.** What is the lowest value of $n$ for which *f* subshells can exist? Explain your answer.

**SOLUTIONS**

**a.** What values of $\ell$ can there be if $n = 3$? The lowest value is 0 and the highest is $n - 1$, which is 2. The subshell types that can have $n = 3$ are *s* ($\ell = 0$), *p* ($\ell = 1$), and *d* ($\ell = 2$).
**b.** The *s*, *p*, *d*, and *f* subshells correspond to values of $\ell$ of 0, 1, 2, and 3, respectively. The highest allowed value of $\ell$ is $n - 1$, and if $n - 1 = 3$, then $n$ is 4. Only when $n$ is 4 or higher can *f* subshells exist. Figure 6.19 shows the combinations of $n$ and $\ell$ that are allowed for *s*, *p*, *d*, and *f* orbitals.

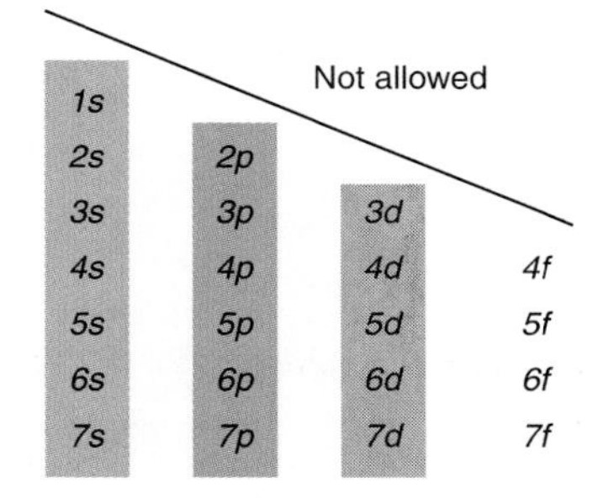


FIGURE 6.19
This shows the combinations of quantum numbers, and therefore the orbitals, that are allowed by the rules of quantum mechanics.

**EXERCISE 6.6**

**a.** What types of subshells are possible when $n = 2$? Explain your answer.
**b.** Can *d* subshells exist when $n = 4$? Explain your answer.

The **magnetic quantum number, $m_\ell$,** describes the orientation of an orbital in space. This quantum number can have any integer value from $-\ell$ to $+\ell$. This means that the lowest allowed value of $m_\ell$ is $-\ell$ and the highest is $+\ell$. If $n = 1$, for example, $\ell$ can be only 0, and therefore $m_\ell$ must be 0. If $\ell = 1$ (*p* subshells), $m_\ell$ can be $-1$, 0, or $+1$. Since $m_\ell$ has three values for *p* subshells ($-1$, 0, $+1$), there must be three different *p* orbitals oriented in three directions in space in a *p* subshell. This is true regardless of the value of $n$. There are three 2*p* orbitals and also three 5*p* orbitals. It is convenient to show the relative spatial orientation of the *p* orbitals on Cartesian coordinate axes (see Figure 6.20). In a *d* subshell, $\ell = 2$; so there are five values of $m_\ell$ (2, 1, 0, $-1$, $-2$) and five different *d* orbitals.

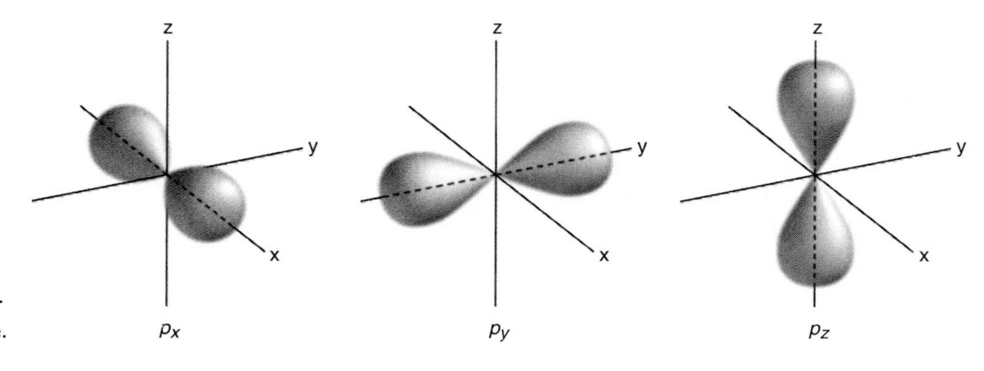


FIGURE 6.20
When $\ell = 1$ ($p$ subshells), $m_\ell$ can be $-1, 0$, or $+1$. This means that there are three $p$ orbitals in a subshell. The three are oriented along axes perpendicular to one another, as shown here.

At the most basic level, the quantum numbers $n$, $\ell$, and $m_\ell$ define an orbital. For example, all atoms have an orbital with $n = 1$, $\ell = 0$, and $m_\ell = 0$. There is only one orbital in each atom that has this set, or any other set, of these three quantum numbers.

### EXAMPLE 6.7 Counting and Designating Orbitals

How many atomic orbitals have $n = 3$, and what kinds are they?

**SOLUTION**

You can summarize the orbitals for which $n = 3$ by filling a table as shown in Figure 6.21: Adding up one $3s$ orbital, three $3p$ orbitals, and five $3d$ orbitals gives nine atomic orbitals that have $n = 3$.

| *n* | *l* | *m* | orbital |
|---|---|---|---|
| 3 | 0 | 0 | 3*s* |
| 3 | 1 | -1 | 3*p* |
| 3 | 1 | 0 | 3*p* |
| 3 | 1 | +1 | 3*p* |
| 3 | 2 | -2 | 3*d* |
| 3 | 2 | -1 | 3*d* |
| 3 | 2 | 0 | 3*d* |
| 3 | 2 | +1 | 3*d* |
| 3 | 2 | +2 | 3*d* |

FIGURE 6.21

**EXERCISE 6.7**

How many atomic orbitals in an atom have $n$ less than or equal to 2, and what kinds are they?

## 6.6 More on Electron Configurations

GOAL: To use the Pauli Exclusion Principle and the Aufbau Principle to determine the electron configurations of elements.

In previous sections you learned how to determine what type of orbital contains the valence electrons of an atom of a given element. By the end of this section you will be able to give a list of the orbitals in an atom that contain electrons, along with the number of electrons in each type of orbital. As you learned in Chapter 5, this kind of list is called an *electron configuration.*

First, however, there is one more quantum number and one more rule you need to be familiar with. Each electron behaves as if it were spinning on its axis, and there are only two allowed directions of spin for an electron in an atom. The **spin quantum number, $m_s$,** gives the direction of the spin. The allowed values of the spin quantum number are $+\frac{1}{2}$ and $-\frac{1}{2}$, which you can think of as the electron's spin axis pointing up or down. Chemists often represent electrons in orbitals with arrows, ↑ or ↓, that imply one value of $m_s$ or the other.

At the end of Section 6.5, you learned that no two *orbitals* in an atom may have the same three quantum numbers. A better, more complete statement of this rule, which is called the **Pauli Exclusion Principle,** says that no two *electrons* in an atom can have the same set of four quantum numbers. Since there are two values of $m_s$ for each orbital, this rule means that *no orbital can have more than two electrons in it.*

## EXAMPLE 6.8 The Electron Capacity of Orbitals

What is the maximum possible number of $2p$ electrons in an atom?

**SOLUTION**

First, you need to know the number of $2p$ orbitals. For $p$ orbitals $\ell = 1$ and $m_\ell$ can be $-1$, 0, or $+1$. This means that there are three $2p$ orbitals. Each orbital can hold two electrons, so the maximum capacity of the three $2p$ orbitals is six electrons. This is also true of the $3p$, $4p$, and $5p$ orbitals. There are three $p$ orbitals in any given set, so each set can accommodate a maximum of six electrons.

**EXERCISE 6.8**

What is the maximum possible number of $3d$ electrons in an atom?

**The rule that electrons add to atomic orbitals in order of increasing potential energy (that is, lowest energy first) is called the *Aufbau Principle* (from a German word for "build up"). Remember that, for each additional electron in the outer shell, there is another proton in the nucleus.**

The next rule for electron configurations is that electrons are added to atomic orbitals in the order of the orbitals' increasing potential energy. From Section 6.4 you can conclude that the first electron of an atom goes into the shell with $n = 1$, and so does the second one, because that shell has the lowest potential energy. That is, the first two electrons both go in the $1s$ orbital. What happens next? Does the third electron start filling the $2s$ orbital or a $2p$ orbital?

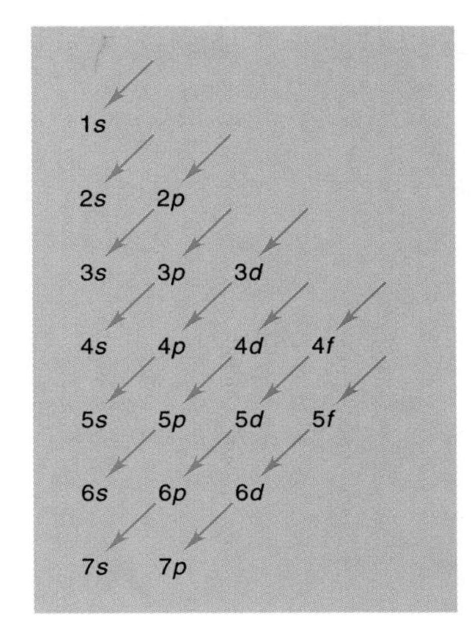


FIGURE 6.22
The arrows show the order of increasing energy for the allowed subshells. Start with the top arrow and use them to predict the order in which the subshells will fill as you write electron configurations: 1*s* first, then 2*s*, then 2*p*, 3*s*, 3*p*, 4*s*, 3*d*, 4*p*, 5*s*, 4*d*, 5*p*, 6*s*, 4*f*, 5*d*, and so on.

You might be able to guess from the periodic table, which contains two main group elements in what is called the $s$ block. Next there's a gap, and then the $p$ block begins. Yes, the $2s$ electrons come after the $1s$ electrons. The first element to have a $p$ electron is boron, with atomic number 5. A boron atom in its ground state (that is, its most stable state) has two $1s$ electrons, two $2s$ electrons, and one $2p$ electron.

This same sort of logic gives you the complete order in which the shells and subshells fill. Once the $2p$ orbitals are filled, the next element is sodium. This is an $s$-block element, so the $3s$ electrons are next. Aluminum begins the next block, a $p$ block, so the $3p$ electrons follow the $3s$. The next new block begins with potassium, so the $4s$ electrons are next. What happened to the $3d$ electrons? They start next, beginning with scandium. Then come the $4p$, followed by $5s$, $4d$, $5p$, and $6s$. The chart of Figure 6.22 can help you decide what order to use in filling the orbitals.

**The simple rules given here work reliably for the *s*- and *p*-block elements. They are less reliable for the *d*- and *f*-block elements. Some of the latter have electron configurations that are not as easy to predict.**

In Chapter 5 we wrote electron configurations based on the shell model of the atom. Now you know that the shells correspond to the principal quantum numbers ($n$) and subshells correspond to the angular quantum num-

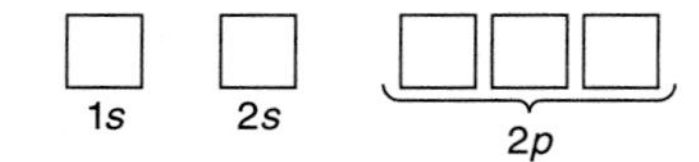


FIGURE 6.23
This diagram shows the 1s, 2s, and 2p orbitals represented as boxes.

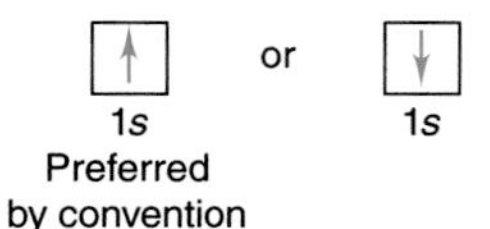


FIGURE 6.24
Real hydrogen atoms don't pay attention to scientific conventions, and either of these orbital diagrams is valid for the element.

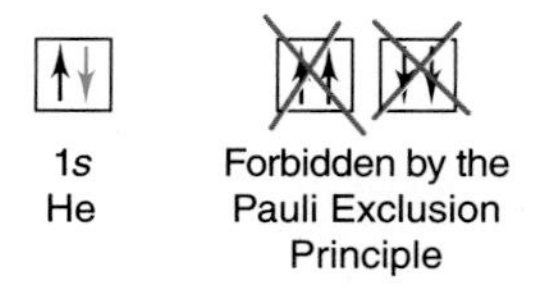


FIGURE 6.25

bers ($\ell$). You also know that the capacity of the second shell is eight electrons, and you know why. Two electrons can go into each orbital—that is, two in the 2*s* orbital and two in each of the three 2*p* orbitals, giving a total of eight.

Now let's take another look at electron configurations. It is sometimes useful to show electron configurations as orbital diagrams. To draw one of these diagrams, you first draw a box to represent each individual orbital and label these boxes (see Figure 6.23). Then you place arrows, ↑ or ↓, within the boxes to represent the electrons in the orbitals and the directions of their spins.

The electron configuration for hydrogen is $1s^1$. By convention, the first electron in an orbital is represented with its spin up ($m_s = +\frac{1}{2}$). The orbital diagram for hydrogen is shown in Figure 6.24.

For helium the electron configuration is $1s^2$. One important application of the Pauli Exclusion Principle is that the second electron placed in any orbital must have a spin opposite that of the first electron; that is, the electrons in an orbital are **paired** (have opposite spins). The orbital diagram for helium is shown in Figure 6.25.

The electron configuration for lithium is $1s^2 2s^1$, and that for beryllium is $1s^2 2s^2$. The electrons fill up the 2*s* orbital in the same way as they do the 1*s* orbital. You can see the orbital diagrams for these elements in Table 6.1.

In a boron atom (with an electron configuration of $1s^2 2s^2 2p^1$), there is an electron in a *p* orbital. By convention, this electron is placed in the first of the *p* orbital boxes, as shown in Figure 6.26.

For carbon ($1s^2 2s^2 2p^2$) two questions need to be answered. Where does the second *p* electron belong, and what is its spin? Coulomb's Law tells us

**TABLE 6.1 Orbital Diagrams for the First Ten Elements**

| Atomic Number | Element | Electron Configuration | Orbital Diagram 1s | 2s | 2p | | |
|---|---|---|---|---|---|---|---|
| 1 | H | $1s^1$ | ↑ | | | | |
| 2 | He | $1s^2$ | ↑↓ | | | | |
| 3 | Li | $1s^2 2s^1$ | ↑↓ | ↑ | | | |
| 4 | Be | $1s^2 2s^2$ | ↑↓ | ↑↓ | | | |
| 5 | B | $1s^2 2s^2 2p^1$ | ↑↓ | ↑↓ | ↑ | | |
| 6 | C | $1s^2 2s^2 2p^2$ | ↑↓ | ↑↓ | ↑ | ↑ | |
| 7 | N | $1s^2 2s^2 2p^3$ | ↑↓ | ↑↓ | ↑ | ↑ | ↑ |
| 8 | O | $1s^2 2s^2 2p^4$ | ↑↓ | ↑↓ | ↑↓ | ↑ | ↑ |
| 9 | F | $1s^2 2s^2 2p^5$ | ↑↓ | ↑↓ | ↑↓ | ↑↓ | ↑ |
| 10 | Ne | $1s^2 2s^2 2p^6$ | ↑↓ | ↑↓ | ↑↓ | ↑↓ | ↑↓ |

that electrons (all negatively charged) repel each other and that the closer together electrons are, the greater the repulsion. To minimize the repulsion, the last two electrons of carbon occupy two different *p* orbitals.

The answer to the second question is less obvious. The atom achieves the lowest potential energy when both electrons have the same spin (called **parallel spin**). To obtain the best box diagram for carbon, apply **Hund's Rule,** which is: In any set of orbitals with equal energies, the electron configuration with the lowest energy has the maximum number of electrons with parallel spin.

The orbital diagram for carbon is shown in Figure 6.27.

An electron configuration that follows the Aufbau Principle, the Pauli Exclusion Principle, and Hund's Rule (as do those shown in Table 6.1) has a lower potential energy than any other possible electron configuration and represents an atom in the **ground state.** Other electron configurations of higher potential energy represent **excited states.** Any electron configuration that violates the Pauli Exclusion Principle is not allowed—that is, it is *forbidden* (and never occurs). Compare the ground-state electron configuration for carbon in Table 6.1 with the two excited-state and one forbidden electron configurations in Figure 6.28.

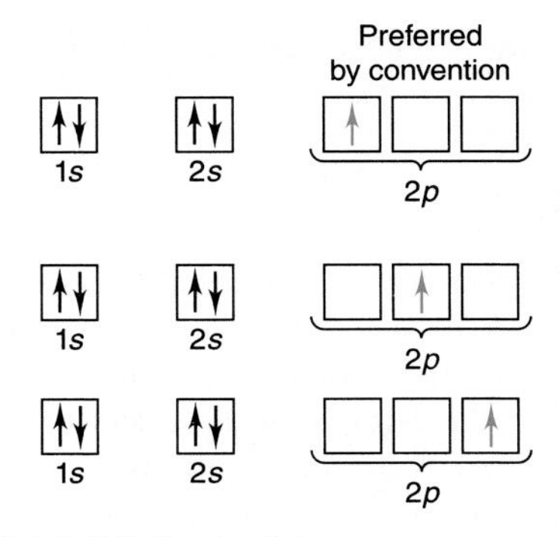


FIGURE 6.26
Because the three $2p$ orbitals are identical to each other in shape and energy, a single electron can occupy any one of them. By convention, chemists place the first electron in a subshell in the leftmost orbital.

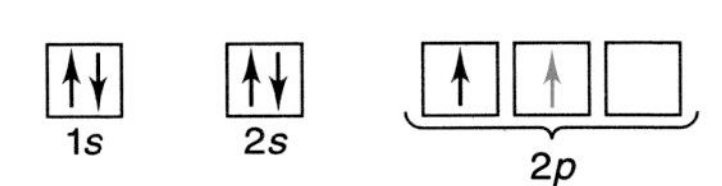


FIGURE 6.27
The two electrons that occupy the two $2p$ orbitals have parallel spins. Again, by convention, the boxes fill from left to right.

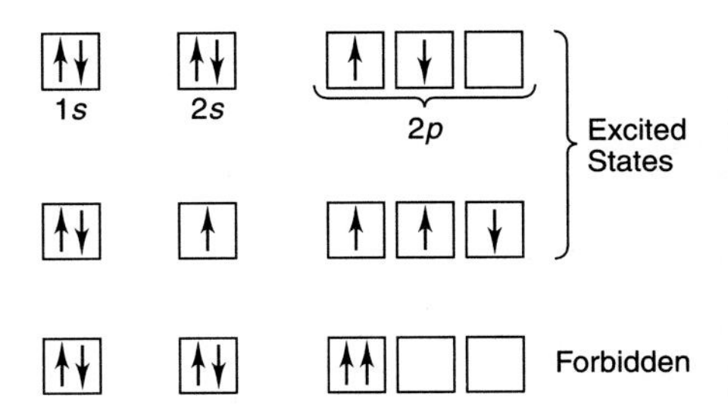


FIGURE 6.28
The top two electron configurations here are excited states. The top one violates Hund's Rule, and the middle one violates the guidelines in Figure 6.22. The bottom electron configuration is forbidden because it violates the Pauli Exclusion Principle. The two $2p$ electrons have the same four quantum numbers ($n = 2, \ell = 1, m_\ell = +1, m_s = +\frac{1}{2}$).

EXAMPLE 6.9

## Orbital Diagrams for Electron Configurations

Draw an orbital diagram showing the valence-shell electrons of each element.

**a.** Cl **b.** Kr **c.** K

SOLUTIONS

**a.** Chlorine is a third-period element and has the electron configuration $[Ne]3s^23p^5$. The box diagram for the valence-shell electrons is therefore:

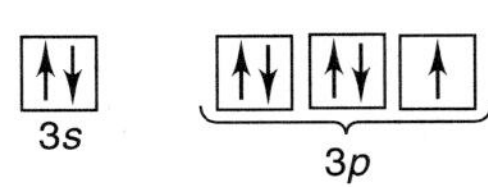


**b.** Krypton is the noble gas in the fourth period, and its valence shell contains eight electrons.

4s 4p

**c.** Potassium is the alkali metal in the fourth period, and its valence electron configuration is $[Ar]4s^1$.

EXERCISE 6.9

Draw an orbital diagram showing the valence-shell electrons of each element.

**a.** Mg **b.** S **c.** Al

### EXPERIMENTAL EXERCISE 6.1

When high-energy photons strike a sample of gaseous argon (atomic number 18), photoelectrons are ejected from the argon atoms. You can calculate the energy that an electron had in the argon atom by subtracting the kinetic energy of the photoelectron from the energy of the photon that ejected it. Experimental results for argon yield energy values of 309, 31.5, 24.1, 2.82, and 1.52 MJ $mol^{-1}$ for atomic electrons.

A. L. Chemist and a scientific team interpreted these results as follows. The atomic electrons with an energy of 309 MJ $mol^{-1}$ correspond to the 1*s* electrons of Ar. Those with energies of 31.5 and 24.1 MJ $mol^{-1}$ correspond to the 2*s* and 2*p* electrons, respectively. The 3*s* electrons and 3*p* electrons have atomic energies of 2.82 and 1.52 MJ $mol^{-1}$.

**QUESTIONS**

1. Is it reasonable that the lowest-energy orbital (1*s*) contains electrons that require the most energy to remove?
2. Other observations suggest that the energies of atomic electrons depend primarily on the principal quantum number *n*. Are the experimental results from argon consistent with this?
3. As noted above, some electrons in an argon atom have an energy of 309 MJ $mol^{-1}$. What is the kinetic energy of one of these electrons after being expelled from an argon atom by photons that have an energy of 350 MJ $mol^{-1}$?

### EXAMPLE 6.10 Ground, Excited, and Forbidden States

Identify each of these electron configurations as ground state, excited state, or forbidden.

**a.** $1s^2 3s^1$ **b.** $1s^3$ **c.** $1s^2 2s^2 3p^1$ **d.** $1s^2 2s^2 2p^2$

**SOLUTIONS**

**a.** Excited state, because the $n = 2$ level contains no electrons
**b.** Forbidden, because the 1*s* orbital can accommodate only two electrons
**c.** Excited state, because the $n = 2$ level contains no *p* electrons
**d.** Ground state

**EXERCISE 6.10**

Identify each electron configuration as ground state, excited state, or forbidden.

**a.** $1s^2 2s^1$ **b.** $2s^1$ **c.** $1s^2 2s^1 2p^6$ **d.** $1s^2 2s^3 2p^4$

## 6.7 The Nature of Light and Electrons: The Great Puzzle

**GOAL:** To discuss some puzzling aspects of electromagnetic radiation and electrons.

What are electrons really like? What is an atom really like? This and earlier chapters have viewed an atom as consisting of a tiny, heavy nucleus sur-

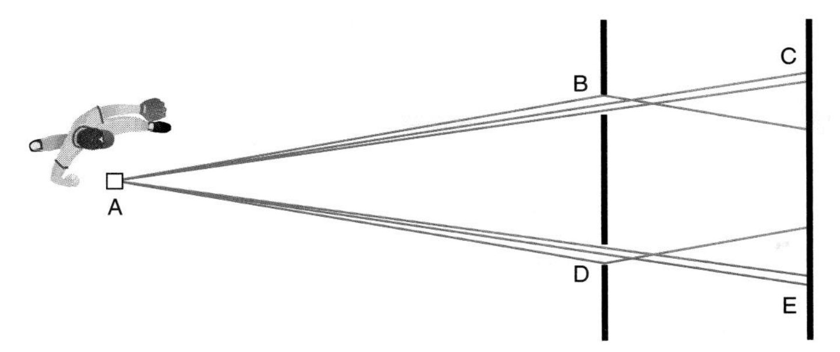


FIGURE 6.29

rounded by a fuzzy cloud of electrons. This is one of several mental models that scientists use when thinking about atoms. Each of these models has its uses.

This section looks at how an electron behaves when it's all by itself, not located in an atom. But before we do that, let's think about some models of how it *might* behave. As you know from the beginning of this chapter, there are basically two possibilities: the electron might act like a particle, or it might act like a wave.

First, let's consider how particles behave. Suppose you are standing at point A in the diagram in Figure 6.29 throwing baseballs in the general direction of the wall with two holes in it. Assume that you aren't aiming well at all. Most of the balls hit the wall and bounce back, but a few of them pass through the hole at point B. Some of these hit the edge of the hole and bounce a little sideways, but the overall result is that these balls will hit the far wall around point C. Other balls pass through the hole at point D (or bounce off an edge of the hole) and hit the far wall around point E. Particles have one more fairly obvious characteristic. When a ball hits the far wall, you'll hear a thump, and there won't be any noise between the thumps. Any particles will act this way.

Next, suppose the two walls are partly submerged in a swimming pool. (This is one way to think about what happens if electrons act like waves.) Suppose you are standing at point A in a pool, rhythmically bouncing up and down to make waves in the water. Most of the wave energy, like most of the baseballs, would bounce back off the wall, but some will make it through the holes. Waves are probably less familiar to you than particles, so the pattern might surprise you a little. First, the waves wash against the far wall continuously. The surface of the water moves up and down all the time. Water waves are not like baseballs, which are either hitting the wall or not.

When you measure the height of the waves at the far wall, the pattern you get isn't simple (see Figure 6.31). Where the wave tops from hole B reinforce the wave tops from hole D (as at point G), there is a high wave. Where a peak from B meets a trough from D (as at point H), the two waves cancel. At the far wall there are some areas where the waves reinforce one another, which means that the water is sloshing up and down strongly. At some points along the far wall, the waves cancel each other, giving very little motion.

These waves reinforce each other

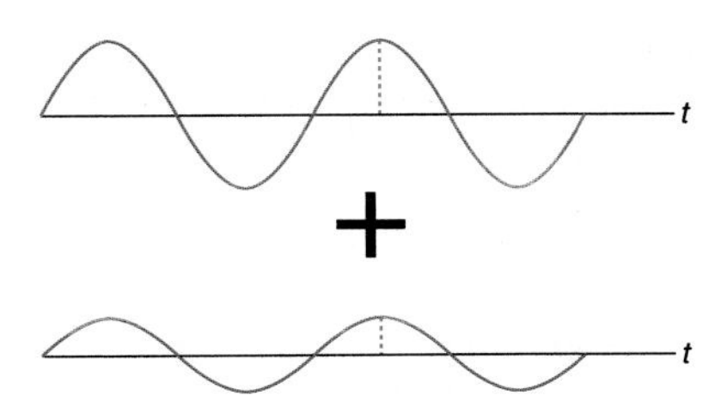


The waves add together to give a higher wave. This is called...

**Constructive interference**

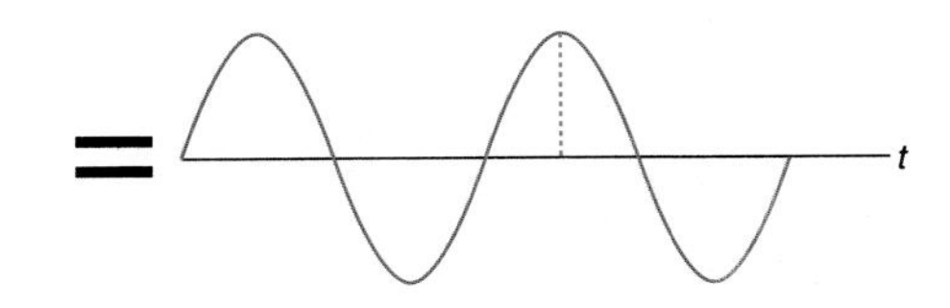


These waves are out of phase; they will interfere with each other

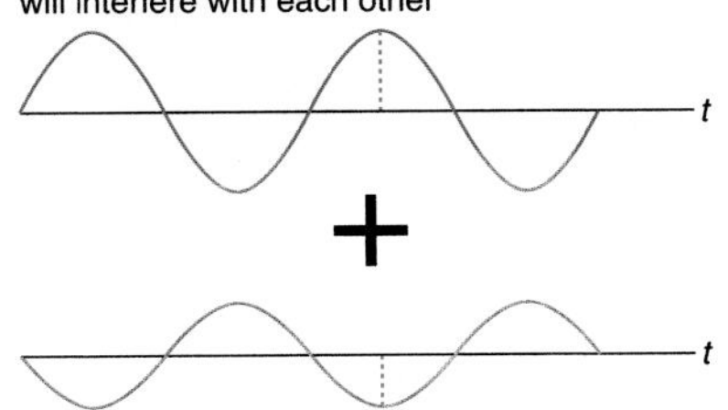


When this happens, the waves cancel each other out. This produces a dark area if light waves are interfering:

**Destructive interference**

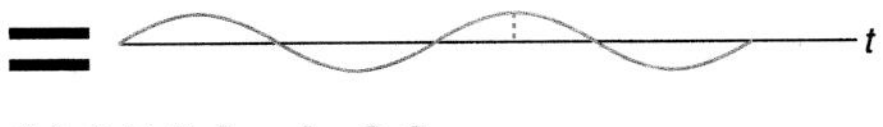


FIGURE 6.30

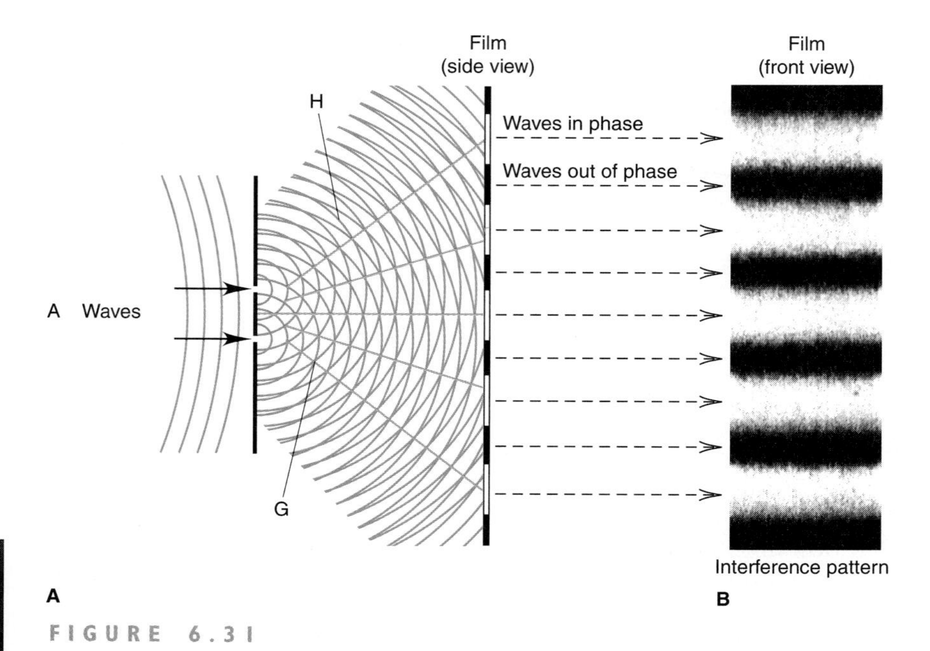


FIGURE 6.31

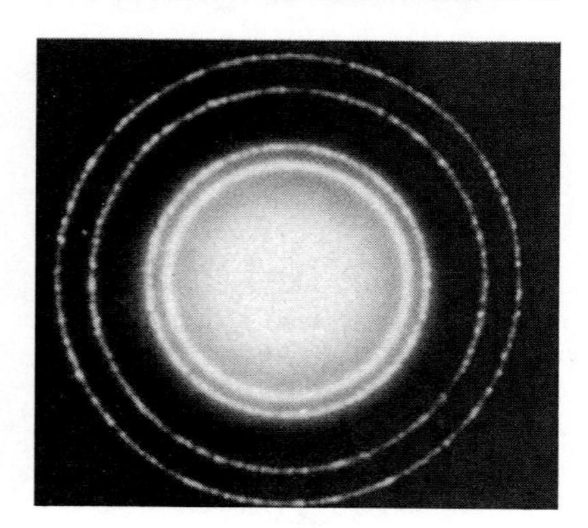

FIGURE 6.32
The spaces between atoms in a solid can provide "holes" for generating interference patterns. The photo on the top shows a pattern produced by X rays passing through aluminum foil. The pattern on the bottom comes from the same kind of interference with a beam of electrons. The patterns demonstrate the wave nature of both photons and electrons.

The resulting pattern along the wall is called an *interference pattern* (see Figure 6.31).

Now, what happens when you shoot electrons at a wall that has two very small holes? Do the electrons act like particles, or do they act like waves?

Since electrons are too small to see, all you can do to observe them is put something at point A that gives off electrons (a *source*) and something else at the far wall that tells you when and where electrons hit (a *detector*), and hope for the best.

If you set your source at point A to emit a continuous stream of electrons, the detector will register a continuous roar of electrons at the far wall. What happens when you slow down the source? If electrons are particles, the detector will register a hit every once in a while, with silence between. If electrons are waves, the roar of electrons hitting the detector will sink to a whisper, but should remain continuous. In a real physics lab this experiment gives an unambiguous answer. The electrons hit one at a time. Electrons are particles.

What happens when you shoot electrons at a wall with only one hole, let's say hole B in Figure 6.29? The detector will tell you that the electrons hit the far wall around point C. If you close hole B and open hole D, the electrons will hit the far wall around point E. Again the electrons seem to act like particles.

Now, open both holes, turn on the electron source, and see what happens. When you graph the locations where the electrons hit the far wall, you get the outline of an interference pattern. In other words, the electrons act like waves. If you slow the electron source down so that only one electron at a time is shooting out and use your detector to find the locations of the electron strikes over a period of a few hours, you still get an interference pattern. In other words, not only do electrons interfere with each other, one electron apparently can interfere with itself. Such an experiment is easier to do with photons, and the results are the same.

Can you think of a reasonable way to visualize an electron or a photon that does all of these things? If you can, you're probably all alone. Most physicists freely admit that nobody really understands what kind of thing could give these results.

## CHAPTER SUMMARY

1. Ordinary waves are described by their wavelength, frequency, amplitude, and velocity. Light waves are described by their wavelength, frequency, and velocity. The velocity of light waves in a vacuum is a constant, $c = 3.00 \times 10^8$ meters second$^{-1}$.
2. Visible light is one family of electromagnetic radiation. All electromagnetic radiation travels at the same velocity in a vacuum. Some important families, in order of increasing frequency, are radio waves, microwaves, infrared radiation, visible light, ultraviolet radiation, X rays, and gamma rays. The visible spectrum is only a very small part of the total electromagnetic spectrum.
3. Hot metal atoms emit photons of specific frequencies (and energies); cooler atoms absorb light photons having the same frequencies. Scientists can identify elements by either their emission or absorption spectra.
4. Light energy is quantized; that is, it comes in discrete energy units called photons. The energies of electrons in atomic orbitals are also quantized. Electrons and photons have properties of both particles and waves.
5. Because electrons have wave properties, it is impossible to determine their exact positions in atoms. This is one manifestation of Heisenberg's Uncertainty Principle.
6. The energy of an electron in an atom is related to the average distance of the electron from the nucleus. Electrons far from the nucleus have higher potential energies than electrons close to the nucleus.
7. Electrons in atoms are described by four quantum numbers: $n$ (principal quantum number), $\ell$ (angular quantum number), $m_\ell$ (magnetic quantum number), and $m_s$ (spin quantum number). The Pauli Exclusion Principle states that no two electrons in an atom can have the same set of four quantum numbers.
8. Orbital diagrams for atoms are drawn from electron configurations, which are in turn made using the Aufbau Principle, which states that electrons go into the subshells that have the lowest potential energy. In addition to the Aufbau Principle, orbital diagrams use the Pauli Exclusion Principle (no two electrons in an atom may have exactly the same values of the quantum numbers) and Hund's Rule (in any set of orbitals with equal energies, the electron configuration with the lowest energy has the maximum number of electrons with parallel spin).

## IMPORTANT EQUATIONS

$c = \nu\lambda$ $\qquad E = h\nu$

## PROBLEMS

### SECTION 6.1 Waves

1. If these two waves have the same velocity, which one has the lower energy? Explain your answer.

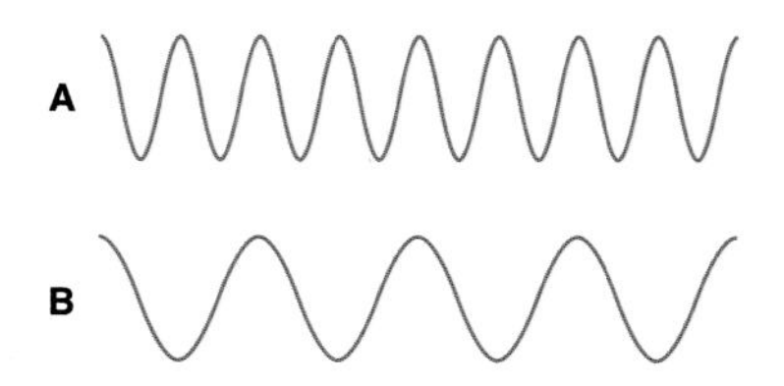


2. If the two waves shown in Problem 1 have the same energy, which has the higher velocity? Explain your answer.

3. Three of the variables that describe a wave are velocity, frequency, and wavelength. For each of these waves, calculate the missing variable from the two that are given.
   a. A wave in a moderate storm in the Atlantic, with a distance between wave crests of 80 m and a time between wave crests of 25 s
   b. A wave caused by a soft breeze on a pond, with a time between wave crests of 1.2 s and a wave velocity of 0.28 m $s^{-1}$
   c. A wave caused by an even softer breeze on a puddle, with a distance between wave crests of 3.3 cm and a wave velocity of 9.7 cm $s^{-1}$
   d. A sound wave in helium gas, with a velocity of 965 m $s^{-1}$ and a wavelength of 1.5 m
   e. A sound wave in oxygen gas, with a velocity of 316 m $s^{-1}$ and a wavelength of 1.5 m

4. For each of these waves, calculate the missing variable (velocity, frequency, or wavelength) from the two that are given.
   a. A high-pitched sound in dry air at 0°C with a velocity of $3.31 \times 10^2$ m $s^{-1}$ and a frequency of $1.5 \times 10^4$ $s^{-1}$
   b. A test vibration in a piece of copper with a wavelength of 21.2 cm and a velocity of $4.76 \times 10^3$ m $s^{-1}$
   c. A low-pitched sound heard by a scuba diver underwater, with a frequency of 135 $s^{-1}$ and a wavelength of 11.3 m
   d. A sound wave in neon gas, with a velocity of 435 m $s^{-1}$ and a wavelength of 0.77 m
   e. A sound wave in nitrogen gas, with a velocity of 334 m $s^{-1}$ and a wavelength of 0.77 m

5. For each of these problems, remember that the velocity of light is $3.00 \times 10^8$ m $s^{-1}$.
   a. The frequency of an extremely long radio wave is $2.9 \times 10^2$ Hz. What is the wavelength?
   b. An experiment uses infrared radiation with a wavelength of 2.3 $\mu$m (micrometers). What is the frequency of this radiation?

6. a. An experiment uses ultraviolet light with a wavelength of 230 nm (nanometers). What is the frequency of this radiation?
   b. Certain highly dangerous gamma rays have a frequency of $9.8 \times 10^{19}$ Hz. What is the wavelength of these rays?

### SECTION 6.2 Light: The Electromagnetic Spectrum

7. Which of these has the longest wavelength? Which has the highest frequency?
   A. X rays
   B. Radio waves
   C. Microwaves
   D. Gamma rays

8. Which of these has the shortest wavelength? Which has the lowest frequency?
   A. X rays
   B. Radio waves
   C. Microwaves
   D. Gamma rays

9. An FM radio station in New Orleans broadcasts at 101.1 MHz, and an AM station in the same city broadcasts at 1280 kHz. Which of the two stations broadcasts at the longer wavelength? Which waves have the higher frequency?

10. An FM radio station broadcasts at 95.5 MHz, and an AM station at 880 kHz. Which of the two broadcasts at the longer wavelength? Which waves have the higher frequency?

11. The shortest wavelength in the human range of vision is approximately $4 \times 10^{-7}$ m. What is the frequency of this light?

12. The longest wavelength in the human range of vision is approximately $7 \times 10^{-7}$ m. What is the frequency of this light?

13. Frequencies in the infrared spectrum are sometimes given in cycles per centimeter ($cm^{-1}$), rather than in cycles per second or hertz. For each of these frequencies, calculate the wavelength.
   a. 2960 $cm^{-1}$
   b. 675 $cm^{-1}$
   c. 3600 $cm^{-1}$
   d. 1690 $cm^{-1}$
   e. 1050 $cm^{-1}$

14. As in Problem 13, calculate the wavelength from each of these infrared frequencies.
   a. 1230 $cm^{-1}$
   b. 3050 $cm^{-1}$
   c. 870 $cm^{-1}$
   d. 3350 $cm^{-1}$
   e. 1710 $cm^{-1}$

15. Wavelengths of visible and ultraviolet light are usually given in nanometers (1 nm = $10^{-9}$ m). For each of these wavelengths, calculate the frequency in hertz. Reminder: the velocity of light is $3.00 \times 10^8$ m $s^{-1}$.
   a. 400 nm (short-wavelength visible light)

b. 176 nm (radiation absorbed by the hydrocarbon 1-pentene)
c. 308 nm (one wavelength absorbed by aluminum atoms)
d. 589.0 nm (bright light given off by hot sodium atoms)
e. 229 nm (a short wavelength absorbed by cadmium atoms)

16. As in Problem 15, calculate the frequency from the given wavelength.
a. 750 nm (long-wavelength visible light)
b. 451 nm (light absorbed by β-carotene, a yellow pigment found in carrots)
c. 663 nm (light absorbed by chlorophyll)
d. 408 nm (a wavelength strongly absorbed by $Sr^{2+}$ ions in air)
e. 176 nm (light absorbed by the hydrocarbon cyclobutane)

17. Which of these rays are *exclusively* electromagnetic radiation?
A. Cosmic C. Beta
B. Alpha D. Gamma

18. Which of these are *not* electromagnetic radiation?
A. X rays B. Gamma rays C. Beta rays

19. Calculate the frequency of electromagnetic radiation having each of these wavelengths.
a. 725 nm (visible, red)
b. $1 \times 10^{-9}$ m (X rays)
c. 7.2 pm (gamma rays)
d. 20 m (radio waves)
e. 490 nm (visible, green)

20. Calculate the frequency of electromagnetic radiation having each of these wavelengths.
a. $20 \times 10^{6}$ nm (microwaves)
b. $5 \times 10^{4}$ cm (infrared)
c. 600 nm (visible, yellow-orange)
d. $1.4 \times 10^{-7}$ m (ultraviolet)
e. 400 nm (visible, violet)

21. What is the wavelength of electromagnetic radiation with each of these frequencies?
a. $1.4 \times 10^{20}$ $s^{-1}$ (gamma radiation)
b. $4.3 \times 10^{14}$ $s^{-1}$ (visible)
c. 990 MHz (radio waves)
d. 1300 kHz (radio waves)
e. $6.7 \times 10^{9}$ $s^{-1}$ (microwaves)

22. What is the wavelength of electromagnetic radiation with each of these frequencies?
a. $5.6 \times 10^{13}$ Hz (infrared)
b. $8 \times 10^{18}$ $s^{-1}$ (X rays)
c. $7.5 \times 10^{14}$ Hz (visible)
d. 100 MHz (TV broadcast)
e. $9.7 \times 10^{15}$ $s^{-1}$ (ultraviolet)

## SECTION 6.3 Light, Atoms, and Energy

23. Which of the following types of electromagnetic radiation has the highest frequency? Which has the longest wavelength? Which has the highest energy?
A. Radio waves C. Visible light
B. X rays D. Ultraviolet radiation

24. Which of the following types of electromagnetic radiation has the lowest frequency? Which has the shortest wavelength? Which has the lowest energy?
A. Radio waves C. Visible light
B. X rays D. Ultraviolet radiation

25. Which of the following types of electromagnetic radiation has the highest frequency? Which has the longest wavelength? Which has the highest energy?
A. X rays C. Microwaves
B. Radio waves D. Gamma rays

26. Which of the following types of electromagnetic radiation has the lowest frequency? Which has the shortest wavelength? Which has the smallest energy?
A. X rays C. Microwaves
B. Radio waves D. Gamma rays

27. An FM radio station broadcasts at 92.9 MHz. What is the energy in joules of one of the photons broadcast by the station? What is the energy in kilojoules of 1 mol of these photons?

28. A radio station broadcasts at 880 kHz. What is the energy in joules of one of the photons broadcast by the station? What is the energy in kilojoules of 1 mol of these photons?

29. Certain gamma rays have a frequency of $9.8 \times 10^{19}$ Hz. What is the energy in joules of one photon of this radiation? What is the energy in kilojoules of 1 mol of these photons?

30. Certain gamma rays have a frequency of $1.12 \times 10^{20}$ Hz. What is the energy in joules of one of these photons? What is the energy in kilojoules of 1 mol of these photons?

31. Calculate the energy of a single photon of each of these frequencies.
a. $6.4 \times 10^{18}$ $s^{-1}$ d. 1250 kHz
b. $5.22 \times 10^{15}$ $s^{-1}$ e. $7.6 \times 10^{8}$ $s^{-1}$
c. 820 MHz

32. Calculate the energy of a single photon of each of these frequencies.
a. $9.1 \times 10^{10}$ $s^{-1}$ d. 125 kHz
b. $8.33 \times 10^{17}$ $s^{-1}$ e. $9.2 \times 10^{15}$ $s^{-1}$
c. $6.90 \times 10^{14}$ Hz

33. Compute the frequency of a photon having each of these energies.
a. $4.98 \times 10^{-19}$ J
b. $7.35 \times 10^{-17}$ J
c. $9.22 \times 10^{-22}$ J
d. $1.49 \times 10^{-26}$ J
e. $5.01 \times 10^{-16}$ J

34. Calculate the frequency of a photon having each of these energies.
a. $8.00 \times 10^{-20}$ J
b. $3.22 \times 10^{-14}$ J
c. $7.72 \times 10^{-18}$ J
d. $6.03 \times 10^{-15}$ J
e. $2.69 \times 10^{-21}$ J

35. Compute the energy of 1 mol of photons having each of these frequencies.
a. $3.70 \times 10^{20}$ $s^{-1}$
b. $6.99 \times 10^{18}$ $s^{-1}$
c. 455 MHz
d. 825 kHz
e. $9.721 \times 10^{10}$ $s^{-1}$

36. Compute the energy of 1 mol of photons having each of these frequencies.
a. $5.06 \times 10^{12}$ $s^{-1}$
b. $8.83 \times 10^{19}$ $s^{-1}$
c. $7.00 \times 10^{12}$ Hz
d. $3.94 \times 10^{22}$ $s^{-1}$
e. $9.2 \times 10^{15}$ $s^{-1}$

37. If 1 *mol* of photons has the energy given, what is the frequency of a single photon?
a. 320 MJ
b. 100 MJ
c. 125 kJ
d. 95.00 kJ
e. 25 kJ

38. If 1 *mol* of photons has the energy given, what is the frequency of a single photon?
a. 5.00 MJ
b. 2.87 J
c. 0.898 J
d. 6.25 MJ
e. $4.59 \times 10^{10}$ kJ

## SECTION 6.4 Quantum Effects in Atoms

39. Which electron transition in the atom pictured below gives off a photon having the lowest frequency? The lowest energy? The shortest wavelength? The circles represent the average distance from the nucleus of an electron in the energy level.

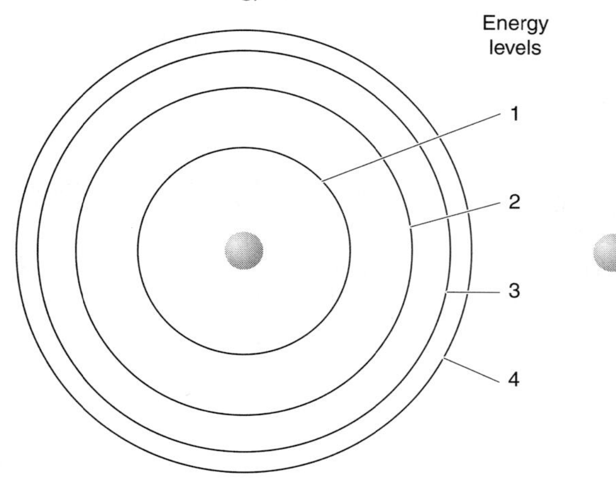


A. Level 3 to level 1
B. Level 4 to level 2
C. Level 4 to level 3
D. Level 3 to level 2

40. For the atom pictured in Problem 39, which of the following electron transitions gives off a photon having the highest frequency? The highest energy? The longest wavelength?
A. Level 3 to level 1
B. Level 4 to level 2
C. Level 4 to level 3
D. Level 3 to level 2

41. Each metal has a unique threshold energy for the photoelectric effect. The threshold energies of several metals are given below. What is the longest wavelength of light (in nm) that can cause the photoelectric effect in each metal?

| | Metal | Threshold energy (J) |
|---|---|---|
| a. | K | $3.68 \times 10^{-19}$ |
| b. | Na | $3.84 \times 10^{-19}$ |
| c. | Ag | $7.42 \times 10^{-19}$ |
| d. | Ba | $4.32 \times 10^{-19}$ |
| e. | Co | $8.00 \times 10^{-19}$ |

42. Here are the threshold energies of some metals. Again, what is the longest wavelength of light (in nm) that can cause the photoelectric effect in these metals?

| | Metal | Threshold energy (J) |
|---|---|---|
| a. | Ca | $4.69 \times 10^{-19}$ |
| b. | Cd | $6.75 \times 10^{-19}$ |
| c. | Cr | $7.20 \times 10^{-19}$ |
| d. | Fe | $7.70 \times 10^{-19}$ |
| e. | Mg | $5.56 \times 10^{-19}$ |

43. Which of the metals listed in Problems 41 and 42 exhibits the photoelectric effect when radiation with a wavelength of 500 nm (green light) shines on its surface?

44. Which of the metals listed in Problems 41 and 42 exhibits the photoelectric effect when radiation with a wavelength of 300 nm (blue light) shines on its surface?

## SECTION 6.5 Quantum Numbers

45. What is the shorthand designation for each of the following subshells?
a. $n = 5$ and $\ell = 4$
b. $n = 3$ and $\ell = 0$
c. $n = 3$ and $\ell = 1$
d. $n = 4$ and $\ell = 2$
e. $n = 1$ and $\ell = 0$
f. $n = 2$ and $\ell = 1$

46. What is the shorthand designation for each of the following subshells?
a. $n = 2$ and $\ell = 0$
b. $n = 3$ and $\ell = 2$
c. $n = 6$ and $\ell = 3$
d. $n = 4$ and $\ell = 1$
e. $n = 5$ and $\ell = 3$
f. $n = 7$ and $\ell = 0$

47. What are the values of $n$ and $\ell$ for each of these subshells?
    a. $5d$ e. $5f$
    b. $1s$ f. $3d$
    c. $5s$ g. $5p$
    d. $3s$

48. What are the values of $n$ and $\ell$ for each of these subshells?
    a. $4p$ e. $2p$
    b. $2s$ f. $4s$
    c. $4d$ g. $3p$
    d. $4f$

49. What types of subshells have $n = 4$? Explain your answer.

50. What types of subshells have $n = 5$? Explain your answer.

51. What is the lowest value of $n$ for which $p$ subshells can exist? Explain your answer.

52. Do $f$ subshells exist that have $n = 3$? Explain your answer.

53. How many orbitals can have $n = 4$, and what kinds are they?

54. How many orbitals can have $n = 5$, and what kinds are they?

55. Give reasonable values of $n$, $\ell$, and $m_\ell$ for each of these orbitals.

a

b

56. Give reasonable values of $n$, $\ell$, and $m_\ell$ for each of these orbitals.

a

b

57. Give all possible values of $m_\ell$ for orbitals with the shapes shown in Problem 55.

58. Give all possible values of $m_\ell$ for orbitals with the shapes shown in Problem 56.

59. How many $s$, $p$, $d$, and $f$ orbitals exist for each of these principal quantum numbers?
    a. 1 b. 3 c. 5

60. How many $s$, $p$, $d$, and $f$ orbitals exist for each of these principal quantum numbers?
    a. 2 b. 4 c. 6

## SECTION 6.6 More on Electron Configurations

61. What is the maximum number of electrons that each of these subshells can hold?
    a. $4f$ b. $5f$ c. $6f$

62. What is the maximum number of electrons that each of these subshells can hold?
    a. $1s$ b. $4s$ c. $1p$ d. $4p$

63. Draw an orbital diagram showing the valence shell electron configuration of each of these elements.
    a. Li d. P
    b. Mg e. Br
    c. Al

64. Draw an orbital diagram showing the valence shell electron configuration of each of these elements.
    a. Ar d. S
    b. Ca e. F
    c. Si

65. The ground-state electron configuration of many fourth-period transition metal ions is $[Ar]3d^x$, where $x$ is the number of electrons in the $3d$ subshell. Draw an orbital diagram showing the $3d$ electrons for each of these ions.

| Ion | Number of $3d$ electrons |
|---|---|
| a. $Fe^{2+}$ | 6 |
| b. $Ni^{2+}$ | 8 |
| c. $Ti^{3+}$ | 1 |
| d. $Zn^{2+}$ | 10 |
| e. $Cr^{3+}$ | 3 |

66. The ground-state electron configuration of many fourth-period transition metal ions is $[Ar]3d^x$, where $x$ is the number of electrons in the $3d$ subshell. Draw an orbital diagram showing the $3d$ electrons for each of the following ions.

| Ion | Number of $3d$ electrons |
|---|---|
| a. $Mn^{2+}$ | 5 |
| b. $V^{3+}$ | 2 |
| c. $Cr^{2+}$ | 4 |
| d. $Co^{2+}$ | 7 |
| e. $Cu^{2+}$ | 9 |

67. How many unpaired electrons are in the valence shell of each of these elements?
    a. Mg d. Si g. Ge
    b. Ca e. S h. Se
    c. Sr f. Ar

68. How many unpaired electrons are in the valence shell of each of these elements?
    a. Na d. Al g. As
    b. K e. P h. Br
    c. Rb f. Cl

69. Identify each of these electron configurations as ground state, excited state, or forbidden.

| | 1s | 2s | 2p | | |
|---|---|---|---|---|---|
| a. | ↑↓ | ↑↓ | ↑ | ↓ | |
| b. | ↑↓ | ↑↓ | ↑ | | ↑ |
| c. | ↑↓ | ↑↓ | | ↑ | ↑ |
| d. | ↑↓ | ↑ | ↑ | | |

70. Identify each of these electron configurations as ground state, excited state, or forbidden.

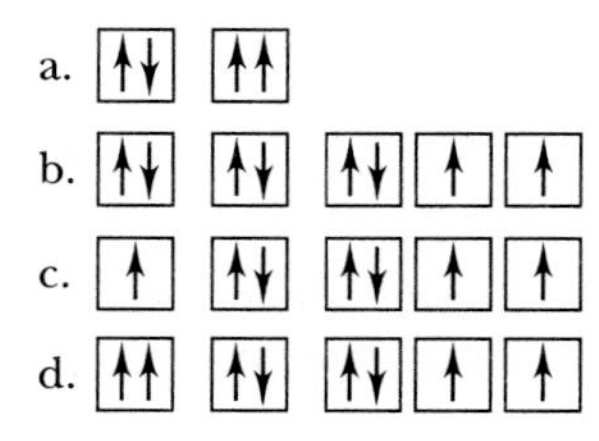

71. Identify each of these electron configurations as ground state, excited state, or forbidden.
    a. $[Ar]5s^1$
    b. $[Ar]4s^2$
    c. $[Ne]3s^23p^54s^1$
    d. $[Ne]3s^33p^3$
    e. $1s^22s^22p^4$

72. Identify each of these electron configurations as ground state, excited state, or forbidden.
    a. $1s^15s^1$
    b. $[Ne]3s^13p^5$
    c. $[Ar]3d^{12}4s^2$
    d. $1s^23d^1$
    e. $1s^22s^3$

**SECTION 6.7 The Nature of Light and Electrons: The Great Puzzle**

73. Draw a figure similar to Figure 6.31 that shows the destructive interference between two photons.

74. Draw a figure similar to Figure 6.31 that shows the constructive interference between two photons.

75. What experimental evidence supports each of these conclusions?
    a. Electrons have particle properties.
    b. Electrons have wave properties.

76. What experimental evidence supports each of these conclusions?
    a. Photons have particle properties.
    b. Photons have wave properties.

**ADDITIONAL PROBLEMS**

77. Different metals have different threshold energies for the photoelectric effect. Based on what you know about atomic structure, effective nuclear charges, and Coulomb's Law, predict which metal from each pair has the lower threshold energy. Explain your answers.
    a. Na or K
    b. K or Ca
    c. Ba or Pb

78. As you did in the previous problem, predict which metal from each pair has the lower threshold energy, and explain your answers.
    a. Sn or Pb
    b. Li or Cs
    c. Mg or Ba

## SOLUTIONS TO EXERCISES

### Exercise 6.1

**a.** The longer wave (B) has the higher velocity if the two waves have the same frequency. The longer waves must move faster if the same number of wave tops (or troughs) are to pass a given point in a given length of time.

**b.** The equation is $v = \nu\lambda$. The frequency $\nu$ is 12 waves in 15 s, or

$$\frac{12 \text{ waves}}{15 \text{ s}} = 0.80 \text{ s}^{-1}$$

The wavelength $\lambda$ is 2.5 ft. The velocity is then:

$$v = \nu\lambda = (0.80 \text{ s}^{-1})(2.5 \text{ ft}) = 2.0 \text{ ft s}^{-1}$$

The velocity of these waves is 2.0 feet per second. The next part of the question is a unit-conversion problem.

$$2.0 \text{ ft s}^{-1} = \frac{2.0 \cancel{\text{ft}}}{1 \cancel{\text{s}}} \times \frac{12 \cancel{\text{in}}}{1 \cancel{\text{ft}}} \times \frac{2.54 \cancel{\text{cm}}}{1 \cancel{\text{in}}} \times$$

$$\frac{1 \cancel{\text{m}}}{100 \cancel{\text{cm}}} \times \frac{1 \text{ km}}{1000 \cancel{\text{m}}} \times \frac{60 \cancel{\text{s}}}{1 \cancel{\text{min}}} \times \frac{60 \cancel{\text{min}}}{1 \text{ h}} = 2.2 \text{ km h}^{-1}$$

The velocity of these waves is 2.2 kilometers per hour.

**c.** The equation is $c = \nu\lambda$. You are given $c$ and $\nu$, so the equation is

$$3.00 \times 10^8 \text{ m s}^{-1} = (9.8 \times 10^7 \text{ s}^{-1})\lambda$$

$$\frac{3.00 \times 10^8 \text{ m s}^{-1}}{9.8 \times 10^7 \text{ s}^{-1}} = \lambda = 3.06 \text{ m} = 3.1 \text{ m}$$

### Exercise 6.2

**a.** Of the types of radiation given, ultraviolet radiation has the shortest wavelengths and microwaves have the lowest frequencies.

**b.** Use $c = \lambda\nu$, with $c = 3.00 \times 10^8$ m s$^{-1}$ and $\nu = 1280 \times 10^3$ s$^{-1}$.

$$\lambda = \frac{c}{\nu} = \frac{3.00 \times 10^8 \text{ m s}^{-1}}{1280 \times 10^3 \text{ s}^{-1}} = 234 \text{ m (about as long as two football fields)}$$

### Exercise 6.3

**a.** 1. Radio waves   2. X rays   3. Radio waves

**b.** 1. $\nu = \frac{c}{\lambda} = \frac{3.00 \times 10^8 \text{ m s}^{-1}}{434 \times 10^{-9} \text{ m}} = 6.91 \times 10^{14} \text{ s}^{-1}$

$$E = h\nu = (6.63 \times 10^{-34} \text{ J s})(6.91 \times 10^{14} \text{ s}^{-1})$$
$$= 4.58 \times 10^{-19} \text{ J}$$

2. $(4.58 \times 10^{-19} \text{ J photon}^{-1})(6.02 \times 10^{23} \text{ photon mol}^{-1})(1 \times 10^{-3} \text{ kJ J}^{-1}) = 276 \text{ kJ mol}^{-1}$

### Exercise 6.4

Transitions that absorb photons go from lower-energy levels to higher-energy levels. The largest energy difference is between level 1 and level 4.

### Exercise 6.5

**a.** $4p$   **b.** $3d$   **c.** $2s$

### Exercise 6.6

**a.** If $n = 2$, $\ell$ can be 0 or 1. That is, for $n = 2$, there can be $s$ and $p$ subshells.

**b.** The value of $\ell$ is 2 for $d$ subshells. If $n = 4$, $\ell$ can be 0, 1, 2, or 3. Yes, $d$ orbitals with $n = 4$ (called $4d$ orbitals) do exist.

### Exercise 6.7

The lowest possible value of $n$ is 1. If $n = 1$, then $\ell = 0$, and $m_\ell = 0$. This is the $1s$ orbital. If $n = 2$, $\ell$ can be either 0 or 1. If $\ell = 0$, $m_\ell$ must be 0. That's one orbital, the $2s$. If $\ell = 1$, $m_\ell$ can be $-1$, 0, or $+1$. That gives three orbitals, all $2p$. There are thus a total of five orbitals with $n$ less than or equal to 2: one (the $1s$) with $n = 1$ and four (one $2s$ and three $2p$) with $n = 2$.

### Exercise 6.8

For $d$ subshells $\ell = 2$ and $m_\ell$ can be $-2$, $-1$, 0, $+1$, or $+2$. This means that there are five $3d$ orbitals. Each orbital can hold two electrons, so the capacity of $3d$ subshells is 10 electrons.

### Exercise 6.9

a. Mg
[↑↓] $4s$

b. S
[↑↓] $4s$   [↑↓][↑][↑] $4p$

c. Al
[↑↓] $4s$   [↑][ ][ ] $4p$

### Exercise 6.10

**a.** Ground state
**b.** Excited state
**c.** Excited state
**d.** Forbidden

### EXPERIMENTAL EXERCISE 6.1

**1.** Yes. The $1s$ electrons are closest to the nucleus and have the lowest potential energy. A lot of energy must be added to remove them to the high-energy region far from the nucleus.

**2.** Yes, they are. There are three distinct energy groupings that differ by one order of magnitude (that is, one power of 10): There is a one-member group with an energy around 300 MJ mol$^{-1}$, a two-member group of energies around 20–30 MJ mol$^{-1}$, and a two-member group with energies around 2–3 MJ mol$^{-1}$. The energy difference between groups is much larger than the energy difference between group members. This supports the other evidence that atomic electron energies depend primarily on the value of $n$.

**3.** Since electron energy must be conserved, simply subtract the energy it takes to remove the electron from the total energy an argon atom absorbs. That is, 350 MJ mol$^{-1}$ − 309 MJ mol$^{-1}$ = 41 MJ mol$^{-1}$, or the kinetic energy of the $1s$ electron after expulsion.

# 7

# SIMPLE INORGANIC COMPOUNDS: NAMES AND PROPERTIES



**KEY TERMS**

This chapter introduces the rules of inorganic **nomenclature**—that is, the rules for naming inorganic compounds.

Yes, there are compounds that have more than one name. NaCl is both sodium chloride and table salt, for instance. Second and even third names can come from times before there were systems of nomenclature (as table salt does), from different languages, or from other nomenclature systems. Don't worry about this right now.

There is a guiding principle behind the rules of any workable nomenclature system: each name must describe one and only one compound, and each compound must have only one name. Being close, or getting most of the name right, is not good enough. This is one of those situations in which it is meaningless to think of a partial success. Just as a light is either on or off, a name is either correct or wrong.

Until the first part of the 1800s, scientists thought that some compounds came from living things only (organic compounds) and others came from nonliving materials only (inorganic compounds). Then a chemist found a reaction that made an inorganic salt into an organic compound. A flood of similar reactions followed. Now, to chemists, organic compounds are the ones made up primarily of carbon and hydrogen, and inorganic compounds are the rest. For more details about this distinction, see Chapter 18 (Organic Chemistry).

This chapter is closely based on Chapter 5. Before you can use nomenclature systems well, you must be able to find elements in their groups in the periodic table and identify them as metals or nonmetals.

## 7.1 Ions of Metallic Elements

GOAL: To name a cation formed from a metallic element when given the symbol of the cation and to write the symbol when given the name.

When metallic elements react, they very often form ions with positive charges (called cations). For the Group 1A and 2A metals the name of the cation is the name of the element plus the word *ion*. You may remember that the charge of the cation in these elements is the same as the group number, because these metal atoms lose the *s* electrons from their valence shells (see Section 5.6). For example, sodium is in Group 1A. The ion formed by sodium has a charge of $+1$ and the symbol $Na^+$. It is called simply the sodium ion (see Figure 7.1). Strontium is in Group 2A. Its ion is the strontium ion, whose symbol is $Sr^{2+}$.

STUDY ALERT

You *must* show the plus sign indicating a positive charge, though you omit this sign from almost every other positive number. Also, when an ion has a $+1$ or $-1$ charge, you write only the sign; the number 1 is assumed. Examples include $Na^+$ and $Cl^-$, which are *not* written $Na^{1+}$ and $Cl^{1-}$.

Few metals outside Groups 1A and 2A behave in this simple fashion. Almost all metals in other groups of the periodic table have more complex chemical behavior. The Group 3A element aluminum behaves simply, always forming a cation with a $+3$ charge. However, the cations formed from the Group 3A metals indium (In) and thallium (Tl) can have a charge of either $+1$ or $+3$. This ability to form two or more different cations is fairly common among metals.

Let's focus for the moment on the fifteen most common "well-behaved" metals, the ones that form only one cation when they react. A "must know" list is given in Table 7.1. You name all of these ions in exactly the same way: the name of the metal followed by the word *ion*.

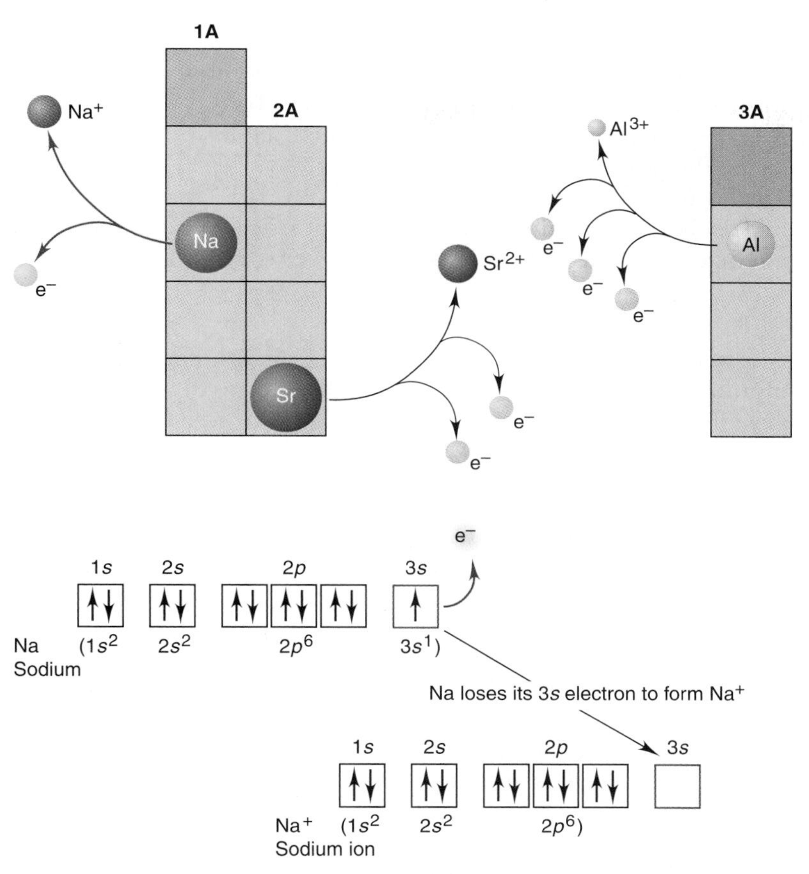


FIGURE 7.1
The Group 1A and 2A metals form ions by losing their s electrons from their valence shell. Group 1A metals such as sodium form ions with a charge of +1, and Group 2A metals such as strontium form ions with a +2 charge.

**TABLE 7.1 Metals That Form Only One Cation When They React**

| Metal | Cation Formed | Name |
|---|---|---|
| All Group 1A metals | $Li^+$, $Na^+$, $K^+$, $Rb^+$, $Cs^+$, $Fr^+$ | |
| All Group 2A metals | $Be^{2+}$, $Mg^{2+}$, $Ca^{2+}$, $Sr^{2+}$, $Ba^{2+}$, $Ra^{2+}$ | |
| Aluminum | $Al^{3+}$ | |
| Silver | $Ag^+$ | metal name ion |
| Zinc | $Zn^{2+}$ | |

## EXAMPLE 7.1 Names and Symbols of Simple Cations

**a.** What is the symbol for the barium ion?
**b.** What is the name of $K^+$?
**c.** What is the symbol for the zinc ion?

SOLUTIONS

**a.** The element barium is in Group 2A, and its symbol is Ba. The symbol for the barium ion combines these two facts: $Ba^{2+}$.
**b.** K is the symbol for the Group 1A element potassium. $K^+$ is therefore called the potassium ion.
**c.** Zinc is not a member of Group 1A or 2A. It is, however, one of the metals that forms only one cation when it reacts. It is listed in Table 7.1 as one of the "must know" ions. The answer is $Zn^{2+}$.

EXERCISE 7.1

**a.** Give the symbols for the magnesium ion, the lithium ion, and the aluminum ion.
**b.** Give the names of $Ca^{2+}$, $Li^+$, and $Ag^+$.

Most metals form more than one ion. Obviously, if these ions are to have unique names (the one basic rule for a nomenclature system), the name of the ion must include the charge. Several systems have been developed over the years, and the most widely used system is also among the simplest. To indicate the charge on a cation using this system, you write the charge of the metal as a Roman numeral placed in parentheses after the name of the metal.

**When reading the names of ions out loud, say the name of the metal and then the number corresponding to the Roman numeral: iron(III) is "iron-three."**

Iron, for example, exists in some compounds as $Fe^{2+}$ and in others as $Fe^{3+}$. The name of $Fe^{2+}$ is iron(II) ion, where the (II) indicates that the charge on the iron is $+2$. $Fe^{3+}$ is named the same way: the iron(III) ion. Note that there is no space between the end of the name of the metal and the left parenthesis, and that there *is* a space between the right parenthesis and the word *ion*. The names of these metal ions *must include the charge.*

FIGURE 7.2
The different cations formed by a metallic element have different chemical and physical properties. The purple solution on the left contains the $Cr^{3+}$ ion, and the orange solution on the right contains something that might be called the $Cr^{6+}$ ion. Because such ions have different properties, it's important to indicate the charge of the ion in the name.

**EXAMPLE 7.2 Names and Symbols of More Cations**

**a.** What is the symbol for the chromium(III) ion (see Figure 7.2)?
**b.** What is the name of $Mn^{2+}$?

SOLUTIONS

**a.** The symbol for the element chromium is Cr. The chromium(III) ion is therefore $Cr^{3+}$.
**b.** Mn is the symbol for the element manganese. Manganese is not listed in Table 7.1, so a Roman numeral is needed. $Mn^{2+}$ is the manganese(II) ion.

EXERCISE 7.2

**a.** Give the symbols for lead(II) ion, copper(I) ion, and titanium(II) ion.
**b.** Give the names for $Co^{3+}$, $Sn^{2+}$, and $Mn^{3+}$.

CHAPTER BOX **7.1 NOMENCLATURE STUDY HINTS**

**In terms** of concepts, this chapter is one of the easiest in this book. In terms of practical application, it is one of the more difficult, because it contains large amounts of material that you must both memorize and apply intelligently. There is no getting around this, but there are several things that you can and should do to help your memorization efforts. First, you must recognize that if you wait until the night before the test, you are dead, dead, dead! No matter what else you do, start early.

One of the easiest and best study methods is to write the names of the entries in the "must know" tables on cards, one card per entry. Put the name on one side of the card and the formula (or solubility or whatever) on the other. Carry a stack of these cards with you, and run through as many as you can at odd moments during the day. If you wait for a bus, use the cards. If you get to class early, use the cards until the instructor shows up. Use the cards!

Then make sure you work practice problems. Even if you know the names and formulas of the ions, you must apply rules to combine ion names to get the names of compounds. It is all too easy to get lost along the way, even though each and every one of the steps is simple. Practice!

## 7.2 Ions of Nonmetallic Elements

**GOAL:** To name ions formed from nonmetals, including oxyanions and other polyatomic ions.

When nonmetals react to form ions, the ions have negative charges (that is, they are anions). To name a **monatomic** anion (an ion formed from one atom), drop the end of the name of the element and add the suffix *-ide* (see Table 7.2). The charge on the anion is equal to the group number of the element minus 8 (hydrogen is the one fairly important exception to this rule).

**STUDY ALERT**
**Students misspell the name of the element F and its anion very often. There is no "flour" in fluorine.**

**TABLE 7.2 Names of Some Common Anions**

| Element | Name of Anion | Symbol |
|---|---|---|
| Hydr ogen | Hydr ide | $H^-$ |
| Nitr ogen | Nitr ide | $N^{3-}$ |
| Phosph orus | Phosph ide | $P^{3-}$ |
| Ox ygen | Ox ide | $O^{2-}$ |
| Sulf ur | Sulf ide | $S^{2-}$ |
| Selen ium | Selen ide | $Se^{2-}$ |
| Hal ogen | Hal ide | $X^-$ (X = F, Cl, Br, I) |

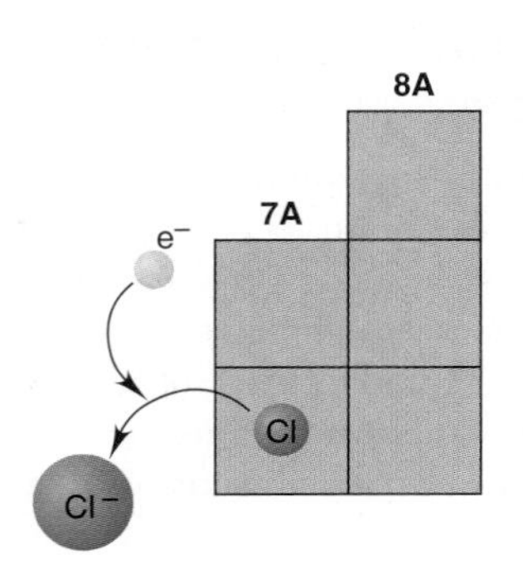


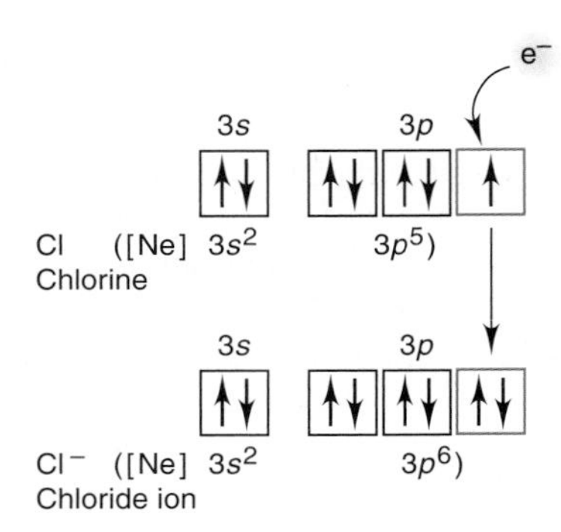

FIGURE 7.3
A chloride ion is formed when one electron adds to a $3p$ orbital of a chlorine atom, filling its valence shell.

Let's see how this applies to chlorine. Chlorine is a nonmetal and often forms an anion when it reacts, adding one electron to a $3p$ orbital to fill its valence shell (Figure 7.3). It is a member of Group 7A of the periodic table, so the charge on the ion is $7 - 8 = -1$. The symbol for this anion is $Cl^-$. To name the anion, change the ending *-ine* in the name of the element to *-ide* and add the word *ion:* The name is chloride ion.

Most periodic tables list hydrogen in Group 1A. If you blindly follow the guidelines, you can *incorrectly* conclude that the charge of the anion formed by hydrogen is $1 - 8 = -7$. Your chemical sense should indicate that this is unreasonable—you already know that hydrogen has a valence of 1. The electronic configuration of the anion is $1s^2$, the same as the configuration of helium. The charge of the anion formed by hydrogen is $-1$, and it is called the hydride ion.

**Collectively, the anions of the halogens are called the halide ions.**

In following the guidelines for naming anions, exactly how much of the name of an element do you drop as its ending? The names of all of the halogens end in *-ine,* which makes their case a simple one. However, the division between the main part of the name and the ending for hydrogen and for the nonmetals in Groups 5A and 6A is not always obvious. You will just have to memorize the names of these anions. Table 7.2 (p. 225) lists the names and symbols of some common anions.

## EXAMPLE 7.3 Names and Symbols for Monatomic Anions

**a.** What is the symbol for the sulfide ion?
**b.** What is the name of $Cl^-$?

**SOLUTIONS**

**a.** The sulfide ion is the monatomic ion of sulfur (S). Its charge is $6 - 8 = -2$. The symbol is therefore $S^{2-}$.
**b.** Cl is the symbol for chlorine. $Cl^-$ is the chloride ion.

**EXERCISE 7.3**

**a.** What are the symbols for the nitride, bromide, and oxide ions?
**b.** Name $I^-$, $P^{3-}$, and $O^{2-}$.

**TABLE 7.3 Polyatomic Ions and Their Names**

| Polyatomic Cations | | Polyatomic Anions Ending in -*ide* | | Polyatomic Anions Ending in -*ate* | | | |
|---|---|---|---|---|---|---|---|
| **Symbol** | **Name** | **Symbol** | **Name** | **Symbol** | **Name** | **Symbol** | **Name** |
| $NH_4^+$ | ammonium ion | $OH^-$ | hydroxide ion | $CO_3^{2-}$ | carbonate ion | $ClO_3^-$ | chlorate ion |
| $Hg_2^{2+}$ | mercurous ion | $CN^-$ | cyanide ion | $HCO_3^-$ | hydrogen carbonate ion | $BrO_3^-$ | bromate ion |
| | | | | | (often called bicarbonate ion) | $IO_3^-$ | iodate ion |
| | | | | $SO_4^{2-}$ | sulfate ion | $CH_3CO_2^-$ | acetate ion |
| | | | | $HSO_4^-$ | hydrogen sulfate ion | $CrO_4^{2-}$ | chromate ion |
| | | | | | (often called bisulfate ion) | $Cr_2O_7^{2-}$ | dichromate ion |
| | | | | $NO_3^-$ | nitrate ion | $MnO_4^-$ | permanganate ion |
| | | | | $PO_4^{3-}$ | phosphate ion | | |

A nonmetal often combines with another nonmetal or a metal to form an ion containing two or more atoms, called a **polyatomic ion.** A polyatomic ion is a chemical unit in its own right (see the discussion on $Al(OH)_3$ and $Fe_3(PO_4)_2$ in Section 2.7). It has specific chemical properties and its own name. Only a few polyatomic ions are named from a system; the rest must be memorized. Every polyatomic ion has its own characteristic charge. The charge of an ion is as important as the identity and numbers of its atoms, so act accordingly.

**A polyatomic ion that contains a metal almost always contains a nonmetal, too. The only important exception is the diatomic ion $Hg_2^{2+}$. The common name for this ion is mercurous ion, and chemists use this name almost exclusively.**

A list of polyatomic ions is given in Table 7.3. The list is divided into three sections: polyatomic cations, polyatomic anions that have an *-ide* suffix, and polyatomic anions that have an *-ate* suffix. When you memorize these names, you will find it useful to learn the groups separately.

**The formula for acetate ion in Table 7.3 is sometimes written as $C_2H_3O_2^-$. The acetate ion is better classified as an organic ion (see Chapter 18) rather than an inorganic one. It is included in this table because salts containing acetate ion (particularly sodium acetate) are common laboratory chemicals.**

The carbonate, sulfate, nitrate, phosphate, and chlorate ions all contain several oxygen atoms plus a single atom of another element. These polyatomic ions are collectively referred to as **oxyanions.** All except the carbonate ion are members of ion families in which only the number of oxygen atoms varies. For example, the oxyanions of chlorine are $ClO^-$, $ClO_2^-$, $ClO_3^-$, and $ClO_4^-$. The oxyanions of sulfur are $SO_3^{2-}$ and $SO_4^{2-}$. Note that the charge is the same for all of the members of each family; the only difference is the number of oxygen atoms. The carbonate ion is alone in its family.

To name the members of a particular group of oxyanions, you start with the ion whose name ends in *-ate* (see Table 7.3). This is the parent ion, from which all the other names are derived. What happens next depends on the number of oxygen atoms.

1. If a member has one fewer oxygen than the parent ion, replace the *-ate* suffix in the name with *-ite.* For example, for oxyanions of chlorine (see Figure 7.4), the parent, chlorate ion, has three oxygen atoms in its formula: $ClO_3^-$. The ion having one fewer oxygen than the chlorate ion is $ClO_2^-$, which is named *chlorite.*
2. If a member has two fewer oxygens than the parent ion, the name is prefixed with *hypo-* and *-ite* replaces *-ate.* The ion with two fewer oxygens than chlorate is $ClO^-$, which has the name *hypochlorite.* Note that

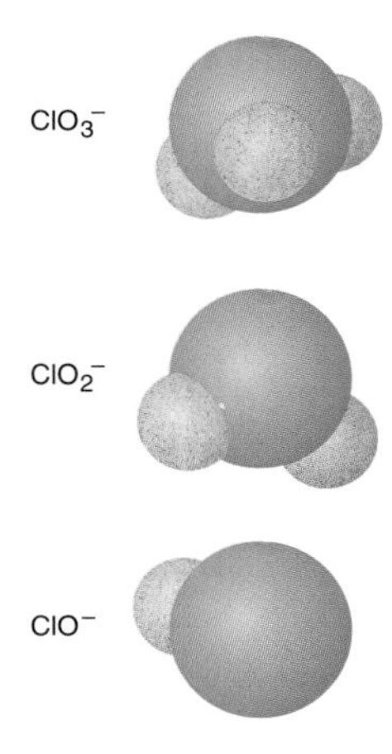

FIGURE 7.4

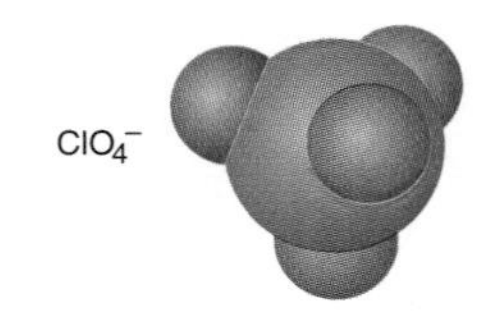


FIGURE 7.5

the *hypo-* prefix and the *-ite* suffix go together. *Hypo-* and *-ate* never occur in the same name.

3. If a member has one more oxygen than the parent ion, add the prefix *per-* to the name. $ClO_4^-$ is the ion with one more oxygen than chlorate, and it has the name *perchlorate.* The *per-* prefix occurs only with the *-ate* suffix.

Just because there are four possible names doesn't mean that all families of oxyanions have four members. You already know that carbonate is the only member of its family, for example.

To get the formula from the name, you follow the guidelines in reverse order. Consider the hypobromite ion. The *hypo-* prefix tells you that this ion has two fewer oxygens in it than bromate does. You know that bromate is $BrO_3^-$ because you have memorized Table 7.3 (you see how essential this is). You know that the charge on the hypobromite ion is $-1$ because all members of an oxyanion family have the same charge.

Because you have followed the guidelines, you can confidently conclude that the formula for hypobromite is $BrO^-$, even though it is unlikely that you ever saw it before. That's the real power of a good nomenclature system. It lets anyone get the correct formula from the name, and vice versa.

EXAMPLE 7.4 **Names and Formulas of Polyatomic Anions**

**a.** What is the formula of the sulfate ion?
**b.** What is the name of $HCO_3^-$?
**c.** What is the formula of the sulfite ion?
**d.** What is the name of $NO_2^-$?
**e.** What is the formula of the hypoiodite ion?

**SOLUTIONS**

**a.** Before you solve this set of problems, you have to memorize the formulas of the ions in Table 7.3. The sulfate ion is $SO_4^{2-}$.
**b.** You also have to be able to recognize and name the ions in Table 7.3 when you see them. $HCO_3^-$ is the hydrogen carbonate (or bicarbonate) ion.
**c.** The sulfite ion has the ending *ite,* so it must have one fewer oxygen atom than the sulf*ate* ion does ($4 - 1 = 3$). It also has the same charge ($-2$) as the sulfate ion. The formula of the sulfite ion must be $SO_3^{2-}$.
**d.** $NO_2^-$ looks like $NO_3^-$, the nitrate ion. It has the same charge and one fewer oxygen. The ending should be changed from *-ate* to *-ite,* making this the nitrite ion.
**e.** Not surprisingly, the oxyanions of bromine and iodine are named exactly like those of chlorine. The hypochlorite ion is $ClO^-$, so the hypoiodite ion is $IO^-$.

EXERCISE 7.4

**a.** What are the names of $BrO_4^-$, $CH_3CO_2^-$, and $IO_2^-$?
**b.** What are the formulas for the hydroxide, cyanide, periodate, and hypobromite ions?

## 7.3 Formulas of Simple Ionic Compounds

GOAL: To write the formulas of simple ionic compounds when given the names of the ions that make up the compounds.

In the previous two sections you learned how to name ions, but ions are only pieces of compounds. **Ionic compounds** form when cations and anions combine. The rule for determining the formula of any ionic compound is that **the total charge of the compound must be zero.** That is, the sum of the charges on the cations must be equal to (and, of course, opposite in sign from) the sum of the charges on the anions.

The simplest case occurs when the charges on the ions are equal and opposite. You simply write a formula that contains one ion of each kind, and the charges total to zero. The formula of the compound containing sodium ion ($Na^+$) and chloride ion ($Cl^-$) is NaCl (see Figure 7.6).

**When only one atom of an element is present in a compound, the subscript 1 isn't used (see Section 2.7). The formula for sodium chloride is written NaCl, *not* $Na_1Cl_1$.**

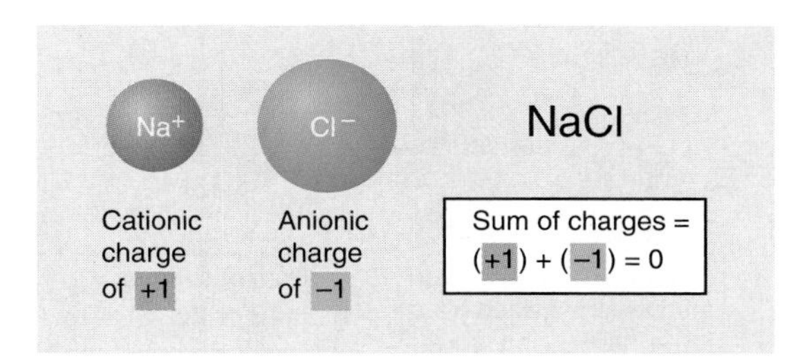


FIGURE 7.6

Calcium ion and sulfate ion have charges of +2 and −2, respectively, so their compound is $CaSO_4$. The formula of the compound made from aluminum ion (+3 charge) and phosphate ion (−3 charge) is $AlPO_4$.

### CHAPTER BOX 7.2 MISTAKES TO AVOID IN WRITING FORMULAS

**Three types** of errors are often made by students learning to write formulas. Please take note of these so that you can avoid making them.

1. *Failure to use subscripts.* For example, it's $Mg_3N_2$ and not 3Mg2N.
2. *Using incorrect symbols for the elements.* It's AgCl and not AGCl or AgCL.
3. *Incorrect use of parentheses.* It's $CaSO_4$ and not $Ca(SO_4)$ or $(Ca)(SO_4)$.

What if the charges on the ions are unequal? Consider the compound formed of $Ca^{2+}$ ions and $Cl^-$ ions. To make the total charge zero, you need twice as many $Cl^-$ ions as $Ca^{2+}$ ions, which gives the formula $CaCl_2$ (see Figure 7.7).

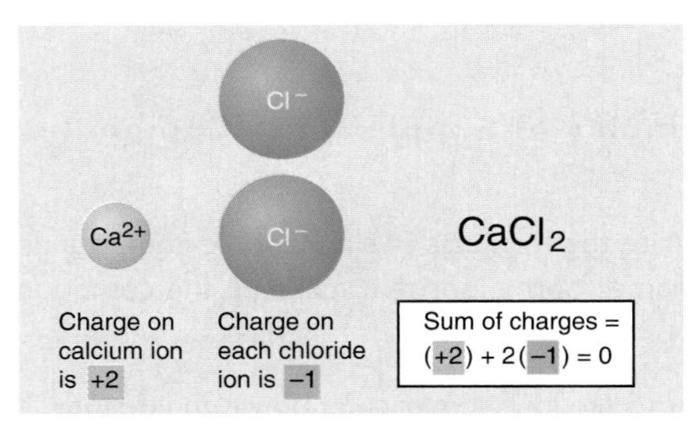


FIGURE 7.7

The formula for the compound formed between $Mg^{2+}$ and $N^{3-}$ is $Mg_3N_2$ (see Figure 7.8).

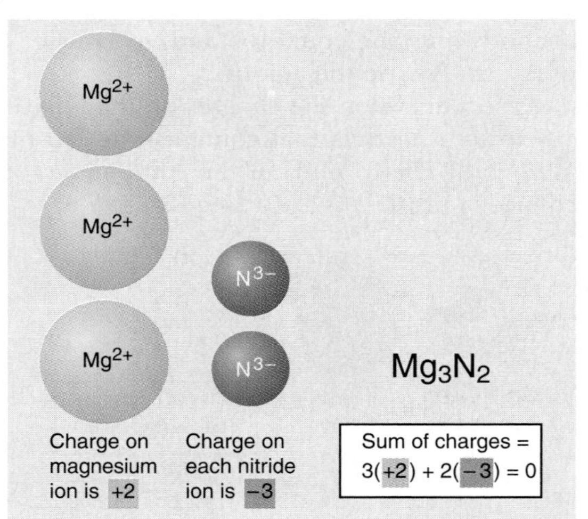


FIGURE 7.8

In Section 2.7 you learned that some formulas—for example, $Al(OH)_3$—contain parentheses. Parentheses are used to set off polyatomic ions in formulas for compounds that contain *more than one* polyatomic ion of a particular type. Some other examples of such formulas are $(NH_4)_2SO_4$, $Ca(CN)_2$, and $Fe(ClO_3)_3$. Don't use parentheses when only one polyatomic ion of a particular type is present, as in $AgNO_3$, $NH_4OH$, or $CaSO_4$.

EXAMPLE 7.5 **Predicting the Formulas of Ionic Compounds**

What is the formula of the ionic compound containing each pair of ions?

**a.** Ammonium ion and iodide ion
**b.** Potassium ion and sulfate ion
**c.** Mercurous ion and nitrate ion
**d.** Iron(III) ion and oxide ion
**e.** Calcium ion and phosphate ion

**SOLUTIONS**

**a.** The ammonium ion is $NH_4^+$, and the iodide ion (note the *-ide* ending) is the monatomic $I^-$ ion. The $+1$ and $-1$ charges balance, and the formula is $NH_4I$.

**b.** The potassium ion is $K^+$, and the sulfate ion is $SO_4^{2-}$. Two potassium ions must be present to yield a $+2$ charge to offset the $-2$ charge of the sulfate ion. This gives $K_2SO_4$ as the answer.

**c.** The mercurous ion is $Hg_2^{2+}$, and the nitrate ion is $NO_3^-$. Two nitrate ions are needed to offset the $+2$ charge of the mercurous ion. The correct formula is written $Hg_2(NO_3)_2$. While the formula $Hg_2N_2O_6$ contains all the necessary atoms, it does not make clear that the nitrate ion is present, so it is not correct (see Section 2.7 to review this concept). The formula $(HgNO_3)_2$ is also incorrect because $Hg_2^{2+}$ is polyatomic.

**d.** Iron(III) ion is $Fe^{3+}$, and oxide ion is $O^{2-}$. Two iron(III) ions have a total charge of $+6$, and three oxide ions have an offsetting charge of $-6$. This gives a formula of $Fe_2O_3$.

**e.** Calcium ion has a $+2$ charge, and the phosphate ion is $PO_4^{3-}$. Calcium phosphate is $Ca_3(PO_4)_2$.

**EXERCISE 7.5**

What is the formula of the ionic compound containing each pair of ions?

**a.** Sodium ion and hypochlorite ion
**b.** Lithium ion and sulfate ion
**c.** Ammonium ion and hydrogen sulfate ion
**d.** Calcium ion and carbonate ion
**e.** Cobalt(III) ion and sulfate ion
**f.** Tin(IV) ion and chloride ion

## 7.4 Naming Simple Ionic Compounds

**GOAL:** To name simple ionic compounds from their formulas.

How do you construct the names of simple ionic compounds? You have probably figured out the rule by now. The name of a compound is simply the combination of the names of the ions that make it up, with the name of the cation first. If you can recognize the ions that make up the compound, you can name the compound.

There's only one potential problem here that has any subtlety. If the cation comes from a transition metal or a metal from the center of the periodic table (such as Sn or Pb), you'll have to figure out what the charge on it is from the charge on the anion or anions. You do this by using the fact that the total charge of the cations plus the total charge of the anions is zero.

Look at $Fe_3(PO_4)_2$, for example (see Figure 7.9). Phosphate ion has a $-3$ charge, and there are two of them in the compound. The sum of the charges on the anions is $-6$. Therefore the sum of the charges on the cations must

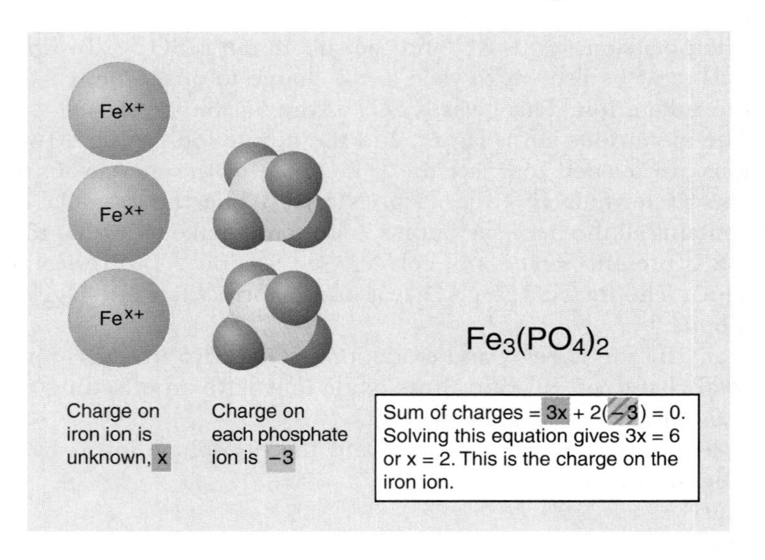
Fe^x+
Fe^x+
Fe^x+
Fe3(PO4)2
Charge on iron ion is unknown, x
Charge on each phosphate ion is −3
Sum of charges = 3x + 2(−3) = 0. Solving this equation gives 3x = 6 or x = 2. This is the charge on the iron ion.

FIGURE 7.9

equal $+6$, which means that the charge on each iron ion is $+2$. Such an iron ion is called iron(II), and the name of the compound is iron(II) phosphate.

If you have any difficulty in figuring out how to name ionic compounds you can use the chart in Figure 7.10 for naming cations and that in Figure 7.11 for help with anions. You will find that if you memorized the names in the earlier tables and if you name a few dozen compounds as practice, you will not need the charts.

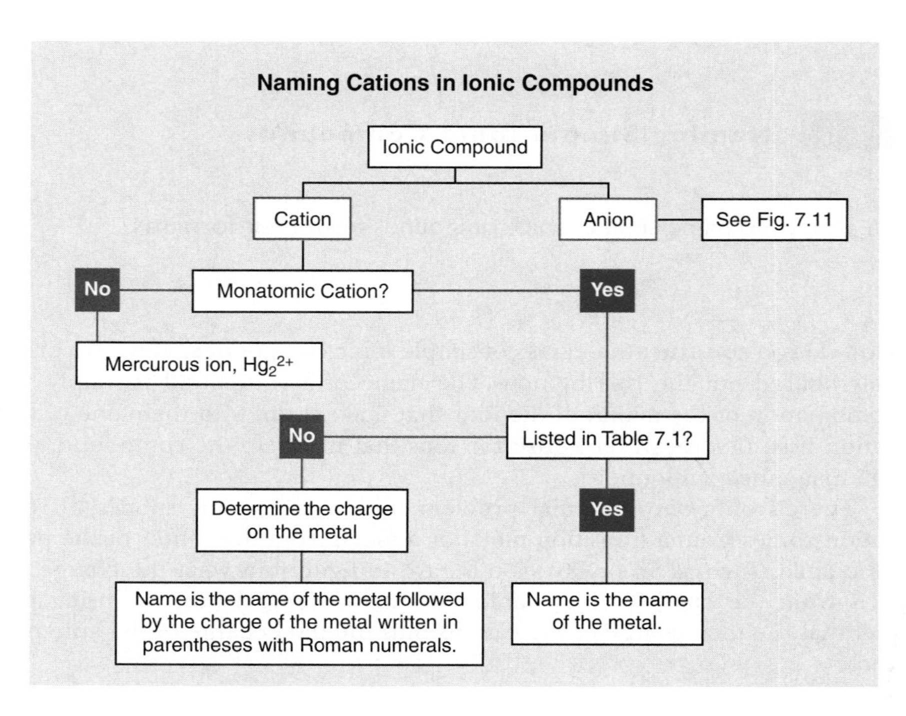
Naming Cations in Ionic Compounds
Ionic Compound
Cation
Anion
See Fig. 7.11
No
Monatomic Cation?
Yes
Mercurous ion, Hg2^2+
No
Listed in Table 7.1?
Yes
Determine the charge on the metal
Name is the name of the metal followed by the charge of the metal written in parentheses with Roman numerals.
Name is the name of the metal.

FIGURE 7.10

## CHAPTER BOX 7.3 THE ORIGINS OF SYSTEMATIC NOMENCLATURE

**In the** late 1700s a chemist named Guyton de Morveau was writing an article on chemistry for an encyclopedia. He went to read up on the history of chemistry and became confused. Since he lived in Paris, he did the sensible thing and went to Antoine Lavoisier for help.

Lavoisier decided that de Morveau's problem was rooted in the language of chemistry. Early scientist-discoverers had no systematic method for naming compounds. In fact, alchemists deliberately went out of their way to be obscure when naming substances, because they wanted to keep their recipes secret. There is really no point in waxing eloquent on the properties of pompholyx, algaroth, orpiment, or realgar if your readers have no idea what these substances are.

Guyton de Morveau and Lavoisier, with their coworkers Berthollet and Fourcroy, published their proposed systematic language for chemistry in 1787. Their names for compounds indicated what elements the compounds were made of. Pompholyx, for instance, became zinc oxide under the new system. By making it much easier for scientists to communicate with each other, the system was a huge step forward for the then-young science of chemistry.

Lavoisier at work.

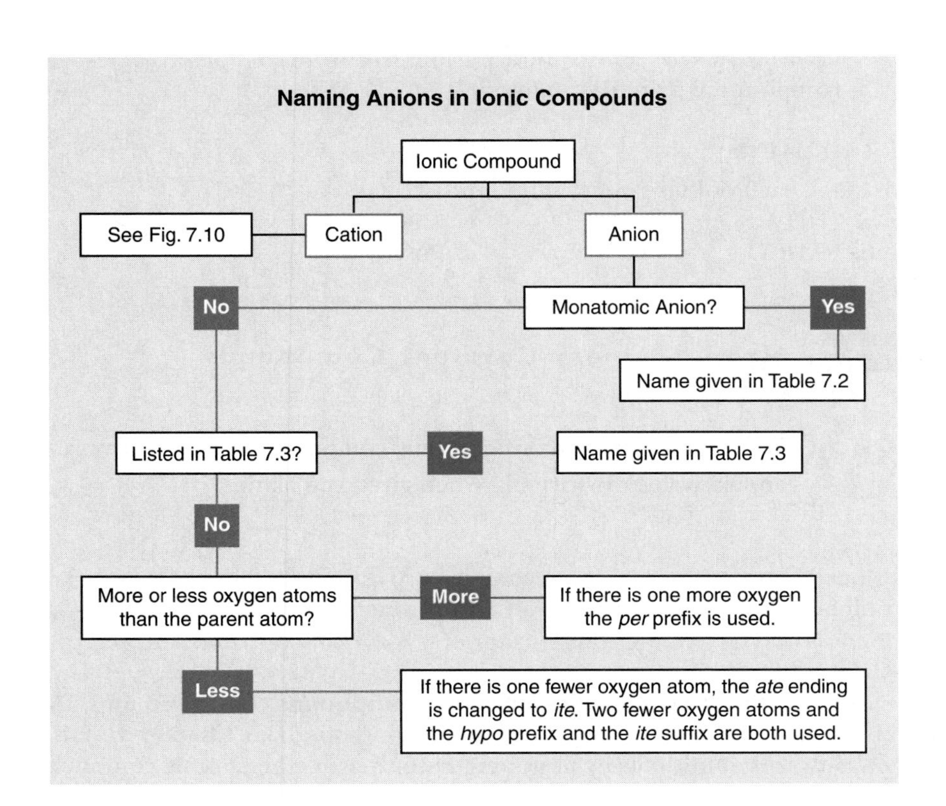
Naming Anions in Ionic Compounds
Ionic Compound
See Fig. 7.10
Cation
Anion
No
Monatomic Anion?
Yes
Name given in Table 7.2
Listed in Table 7.3?
Yes
Name given in Table 7.3
No
More or less oxygen atoms than the parent atom?
More
If there is one more oxygen the *per* prefix is used.
Less
If there is one fewer oxygen atom, the *ate* ending is changed to *ite*. Two fewer oxygen atoms and the *hypo* prefix and the *ite* suffix are both used.

FIGURE 7.11

### EXAMPLE 7.6 Naming Simple Ionic Compounds

Name each of these ionic compounds.

**a.** $K_2S$ **c.** $MnO_2$ **e.** $Fe(NO_3)_3$
**b.** $NH_4NO_3$ **d.** $Fe(NO_3)_2$ **f.** $Fe_2(SO_4)_3$

**SOLUTIONS**

**a.** The compound has potassium and sulfide ions, so the name is potassium sulfide.

**b.** This compound contains two polyatomic ions, and you must be able to find the boundary between them. They are ammonium ion ($NH_4^+$) and nitrate ion ($NO_3^-$), and the name is ammonium nitrate.

**c.** Manganese is not one of the metals that forms only one cation when it reacts (the ones listed in Table 7.1), so a Roman numeral must appear in the compound's name. The oxide ion has a $-2$ charge; two of them have a total charge of $-4$. There is only one manganese ion, so its charge must be $+4$. This compound's name is manganese(IV) oxide.

**d.** Iron also requires the use of a Roman numeral to identify its ions. The nitrate ion has a $-1$ charge, and there are two of them. Therefore in this compound the iron ion has a $+2$ charge to offset the $-2$ charge of the nitrates, and the compound is iron(II) nitrate.

**e.** This iron compound contains three nitrate ions and one iron ion. It's iron(III) nitrate. Note that both this and the compound in part d are iron nitrates, but they have different formulas, so they must have different names.

**f.** This compound contains two iron ions. The three sulfate ions have a total charge of $-6$, so the iron ions are iron(III). The name of this compound is iron(III) sulfate.

**EXERCISE 7.6**

Name each of these ionic compounds.

**a.** $(NH_4)_2SO_4$ **c.** $K_2SO_3$ **e.** $CaCr_2O_7$
**b.** $NaHCO_3$ **d.** NiS **f.** PbO

## 7.5 Naming Binary Covalent Compounds

**GOAL:** To name a two-element covalent compound from its formula and to write the formula when given the name.

**The convention for writing formulas of binary covalent compounds is to write the symbol for the element with the lower electronegativity first. Recall that the elements closer to the left side of the periodic table tend to be less electronegative. If two elements are in the same group, the more electronegative element is closer to the top. See Section 5.6 for a discussion of electronegativity.**

A **binary compound** is one that contains two different elements. Some familiar binary compounds are table salt (NaCl), methane (natural gas, $CH_4$), and ammonia ($NH_3$). Among these examples, NaCl contains ions and the other two do not.

Among other characteristics, **covalent compounds** contain no ions. (You will learn more about covalent and ionic compounds in Chapter 10.) Most covalent compounds consist of nonmetals only, while most ionic compounds contain at least one metal. Ionic compounds containing the ammonium ion,

## CHAPTER BOX 7.4 THE ALCHEMISTS' NAMES ARE STILL AROUND

**The names** the alchemists used for compounds existed long before the system that Lavoisier and his colleagues introduced. Some of these names were widely used outside alchemy and science, and many people still use them today.

Miners and mineralogists still use the old names. The most common lead ore, a mineral with the formula PbS, has been called galena rather than lead(II) sulfide for hundreds of years.

Sometimes it's necessary to have different names for the same chemical compound. Calcite and aragonite are minerals with different crystal structures, and limestone and marble are rocks with different properties and different geological histories. All of them have the basic formula $CaCO_3$.

Artists use the old names for some of their pigments. For example, vermilion is mercury(II) sulfide, verdigris is copper(II) acetate, and ocher is essentially iron(III) oxide.

FIGURE 7.12
The colors of these artist's pigments come from the metals they contain: white from titanium, brown from iron or manganese, and yellow, orange, and red from cadmium.

the only common nonmetallic cation, are an important exception. $NH_4Cl$, for example, is an ionic compound that contains $NH_4^+$ and $Cl^-$.

When you name binary covalent compounds from their formulas, you almost always name the elements in the order in which they are written from left to right. Use the full name of the first element, and given the second element the *-ide* suffix. For example, HCl is hydrogen chloride.

Certain prefixes indicate how many atoms of each element are present in binary covalent molecules; these are listed in Table 7.4. There are three important conventions for using the prefix *mono.* First, you never use this prefix for the first element you name. Second, when *mono* is combined with the word *oxide,* you omit one of the two consecutive letter o's. This is done to make pronunciation easier. For example, CO is carbon monoxide. It is not monocarbon monoxide, carbon monooxide, or monocarbon monooxide. Third, *mono* is used with the second element in a name only in situations where two elements make more than one binary compound. For example, carbon monoxide is correct because carbon forms two binary compounds with oxgen (CO and $CO_2$). Hydrogen chloride is correct for HCl because there is only one compound made between hydrogen and chlorine.

Here are some examples:

| Compound | Name |
|---|---|
| NO | Nitrogen monoxide |
| $NO_2$ | Nitrogen dioxide |
| $PCl_3$ | Phosphorus trichloride |
| $N_2F_4$ | Dinitrogen tetrafluoride |
| $S_2Cl_2$ | Disulfur dichloride |

Finally, almost all of the common covalent compounds that contain hydrogen are named in special ways. Some have common names you already know: $H_2O$ is water and $NH_3$ is ammonia, and $CH_4$ is methane. $H_2S$ is hydrogen sulfide, and the less common $SiH_4$ is called silane. HF, HCl, HBr, and HI

**STUDY ALERT**

**Do not confuse *binary* and *heteronuclear diatomic.* A binary compound consists of two different elements. A molecule of a binary compound can contain two or more atoms. For example, the compound $P_4O_{10}$ is a binary compound with 14 atoms in a molecule. Heteronuclear diatomic molecules contain exactly one atom each of two different elements. Thus *diatomic* means two atoms, but *binary* means two elements.**

**TABLE 7.4 Prefixes for Naming Binary Covalent Compounds**

| Prefix | Number of Atoms |
|---|---|
| mono- | 1 |
| di- | 2 |
| tri- | 3 |
| tetra- | 4 |
| penta- | 5 |
| hexa- | 6 |
| hepta- | 7 |
| octa- | 8 |
| nona- | 9 |
| deca- | 10 |

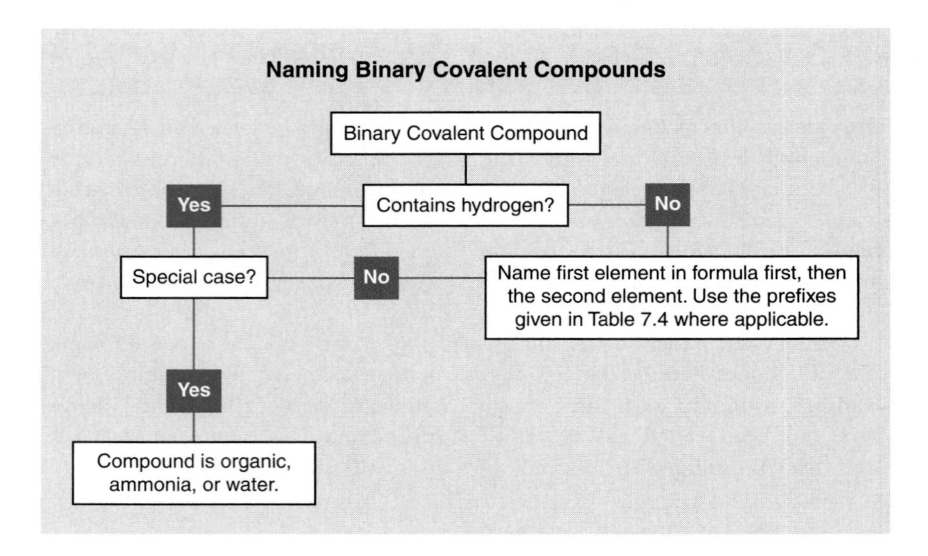


FIGURE 7.13

follow the general rule given earlier in this section and are named hydrogen fluoride, hydrogen chloride, hydrogen bromide, and hydrogen iodide, respectively.

Figure 7.13 summarizes the naming of binary covalent compounds.

### EXAMPLE 7.7 Naming Binary Covalent Compounds

What is the name of each of these compounds?

**a.** $PCl_5$ **b.** $N_2O_4$ **c.** $P_4O_{10}$ **d.** $N_2O$ **e.** $NH_3$

**SOLUTIONS**

**a.** P is phosphorus. Cl is chlorine, and you need to use the *-ide* ending. The prefix for five is *penta-*. This compound is phosphorus pentachloride.
**b.** Dinitrogen tetroxide
**c.** Tetraphosphorus decoxide
**d.** Dinitrogen monoxide (This is different from the compound in part b and must have a different name. Nitrogen oxide is wrong in both cases.)
**e.** Ammonia (You didn't say nitrogen trihydride, did you?)

**EXERCISE 7.7**

What is the name of each of these compounds?

**a.** $N_2O_5$ **b.** $NI_3$ **c.** $CO_2$ **d.** $CH_4$ **e.** HBr

When you are writing the formula of a compound from its name, you simply put down the symbols of the elements and the appropriate number of atoms in the order given. For example, nitrogen monoxide is NO, and sulfur tetrafluoride is $SF_4$.

EXAMPLE 7.8 **Formulas of Binary Covalent Compounds**

What is the formula of each of these compounds?
**a.** Carbon tetrachloride
**b.** Dinitrogen trioxide

**SOLUTIONS**

**a.** Carbon is C, and chlorine is Cl. The name indicates one carbon atom (no prefix) and four chlorine atoms (tetra). The formula is $CCl_4$.
**b.** There are two nitrogens and three oxygens: the formula is $N_2O_3$.

**EXERCISE 7.8**

What is the formula of each of these compounds?
**a.** Silicon dioxide
**b.** Nitrogen trifluoride
**c.** Tetraphosphorus decasulfide
**d.** Dinitrogen tetrachloride

## 7.6 Naming Acids

**GOAL:** To name an acid given its formula and to write the formula given the name.

For a variety of reasons, chemists use several different definitions of an acid. Chapters 8 and 15 cover some of these definitions and their uses. For the moment, you can think of an acid as a compound that forms between the cation $H^+$ and an anion.

Let's begin with the **oxyacids.** These are compounds that contain an oxyanion and enough hydrogens (think of them as $H^+$ ions) to balance the oxyanion's negative charge. To name most of these compounds, you start with the name of the oxyanion:

1. If the oxyanion name ends in the suffix *-ate,* replace the suffix with *-ic acid.*
2. If the oxyanion name ends in the suffix *-ite,* replace the suffix with *-ous acid.*

For example, $HNO_3$ consists of a nitrate group and one hydrogen. Take *nitrate,* remove *-ate,* and add *-ic acid.* This gives you *nitric acid,* the correct name of this oxyacid. The oxyacid $HNO_2$ is named similarly. The $NO_2^-$ group is nitrite, and you remove the *-ite* and add *-ous acid.* The name of $HNO_2$ is *nitrous acid.*

**It is convenient to think of acids as compounds containing hydrogen cations and their respective anions, because that way they are easier to name, and some of their chemical properties (Chapter 8) are easier to interpret. Acids, however, are covalent compounds that contain no ions.**

Unfortunately, there are exceptions to this pattern. $H_2SO_4$ (from the sulfate ion) is not sulfic acid but *sulfuric acid* (and $H_2SO_3$, named from the sulfite, is *sulfurous acid*). $H_3PO_4$ is *phosphoric acid.* The name *acetic acid* comes from the name of the acetate anion according to the system, but one of its hydrogens is usually written with the oxygens at the end of the formula: $CH_3CO_2H$. It's memorization time once again.

You don't mention the number of hydrogen atoms present when you name an acid. The number of hydrogens is extremely important to the chemistry of these compounds, and you have to know (or you have to be able to figure out) how many there are. You can deduce the number of hydrogen atoms from the name of the oxyacid by recalling that there is one hydrogen atom for every negative charge on the oxyanion for which the acid is named. For instance, you can figure out that phosphoric acid has three hydrogen atoms because the phosphate ion has a $-3$ charge ($PO_4^{3-}$).

### EXAMPLE 7.9 Names and Formulas of Oxyacids

**a.** What is the name of $HClO_4$?
**b.** What is the formula of sulfuric acid?
**c.** What is the name of $HNO_2$?
**d.** What is the formula of bromous acid?

**SOLUTIONS**

**a.** This acid contains $ClO_4^-$, the perchlorate ion. Remove *-ate* from the oxyanion name and add *-ic acid.* This gives you perchloric acid.
**b.** This is one of the names you have memorized. Remember that it comes from the sulfate ion, which is $SO_4^{2-}$. Add two hydrogens ($H^+$) to cancel the two negative charges on the sulfate ion, and you get $H_2SO_4$.
**c.** $NO_2^-$ is the nitrite ion. The acid is nitrous acid.
**d.** To find the name of the oxyanion, remove *-ous acid* and add *-ite,* yielding the bromite ion, $BrO_2^-$. One $H^+$ is needed to counteract the single negative charge, so the formula of bromous acid is $HBrO_2$.

**EXERCISE 7.9**

**a.** What are the names of HOCl, $HClO_2$, and $HClO_3$?
**b.** What are the formulas of phosphoric acid, bromic acid, and carbonic acid?

Finally, there are four binary compounds containing hydrogen and a halogen that act as acids. These are HF, HCl, HBr, and HI. In pure form these compounds are colorless gases, named as binary covalent compounds: hydrogen fluoride, hydrogen chloride, hydrogen bromide, and hydrogen iodide, respectively. The gases are uncommon, however.

These compounds are much more commonly manufactured and used as solutions in water. If you look for HCl in most laboratories, you will probably find only HCl dissolved in water. These water solutions have different names from the pure compounds. When dissolved in water, all four of the hydrogen

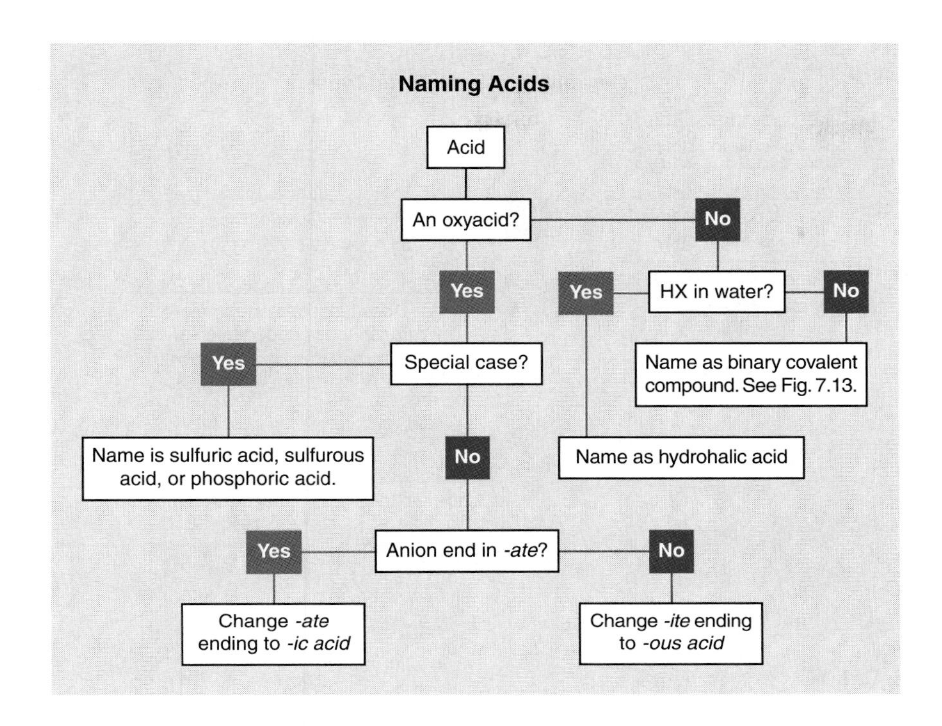


FIGURE 7.14

halides are named as *hydrohalic acids*—that is, as hydrofluoric acid, hydrochloric acid, hydrobromic acid, and hydroiodic acid.

Figure 7.14 is a chart summarizing the naming of acids.

## 7.7 Naming Unclassified Compounds

**GOAL:** To classify and name a compound when given the formula.

In the previous sections you learned the rules for naming ions, ionic compounds, binary covalent compounds, and acids. You know that it is necessary to memorize the entries in Tables 7.1, 7.2, 7.3, and 7.4. Unfortunately, even though you must memorize it, that information by itself is often not enough to name a compound properly. You must also know the rules for naming each of the types of compounds.

Furthermore, a formula by itself doesn't contain a label that tells you what type of compound it represents and what rule to use. The very first step in naming a compound from its formula alone is to classify the compound as ionic or covalent or as an acid. Only then will you know what rules to apply.

A general procedure for naming compounds is charted in Figure 7.15 on the next page. If you follow this chart, you will be able to name any of the compounds discussed in this chapter. Remember that you will not have the charts with you for your exams. You must practice naming compounds until you feel confident that there is no nomenclature question your instructor can ask that you cannot answer.

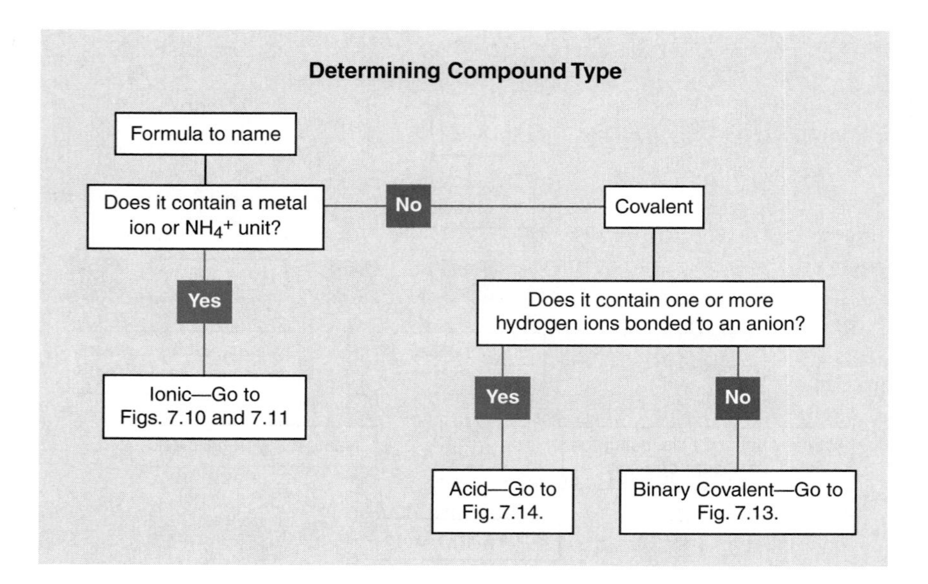


FIGURE 7.15

## EXAMPLE 7.10 Classifying and Naming Compounds

Classify and then name each compound.

**a.** $CaCl_2$ **b.** $FeCl_3$ **c.** $NCl_3$ **d.** $(NH_4)_2SO_4$ **e.** $HNO_3$

### SOLUTIONS

**a.** $CaCl_2$ contains a metal, calcium. All of the compounds in this chapter that contain metals are ionic (or can be named as ionic compounds). Calcium is in Group 2A, so the name does *not* need a Roman numeral. You just name the two ions in the compound: calcium chloride.

**b.** $FeCl_3$ contains a metal, so the compound is ionic. Iron's ions are distinguished using Roman numerals. The name is iron(III) chloride.

**c.** $NCl_3$ contains only nonmetals. Therefore, it's covalent, and you have to indicate that three chlorine atoms are present. The name is nitrogen trichloride.

**d.** $(NH_4)_2SO_4$ contains only polyatomic ions. To get this one right, you have to remember that it is one of the exceptions to the rule that all ionic compounds contain metals. This compound (and all the other exceptions you will see in this chapter) contains $NH_4^+$, the ammonium ion. This is ammonium sulfate, an ionic compound.

**e.** $HNO_3$ contains only nonmetals, so it's a covalent compound. Its formula begins with H, so it's an acid—nitric acid.

### EXERCISE 7.10

Classify and then name each compound.

**a.** $K_2S$ **b.** $CoCl_3$ **c.** $SF_6$ **d.** $Ba_3(PO_4)_2$ **e.** $HNO_2$

## 7.8 Classifying Compounds from Their Properties

**GOAL:** To classify compounds as ionic or covalent from their physical properties.

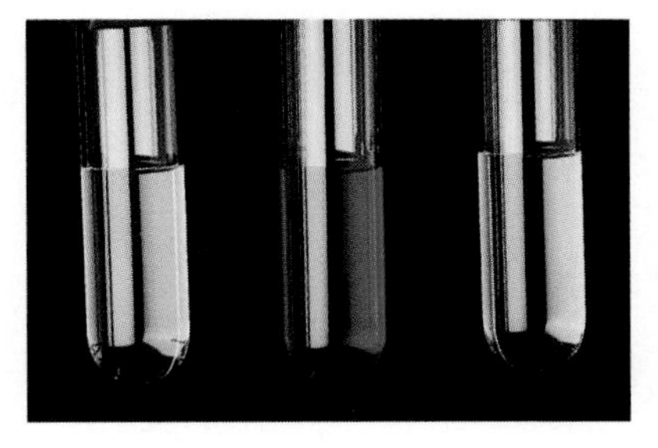

FIGURE 7.16
Many transition metal ions give colored solutions when dissolved in water. From left to right, these solutions contain iron(III) ion ($Fe^{3+}$), permanganate ion ($MnO_4^-$), and dichromate ion ($Cr_2O_7^{2-}$).

You now know how to name a wide variety of chemical compounds. What do these compounds actually look like?

All pure ionic compounds are solids under normal conditions. They melt only at very high temperatures (typically above 300°C), and typically are hard and brittle. Ionic compounds of main group elements are almost all white solids when pure. If they dissolve in water at all, they form clear, colorless solutions. Many ions of transition metals form colored solutions if they dissolve in water. Iron(III) ion is rusty red-brown, permanganate ion is purple, and dichromate ion is yellow-orange, for instance (see Figure 7.16).

Covalent compounds tend to have much lower melting and boiling temperatures than ionic compounds, though there are exceptions. Most of the binary covalent compounds discussed in Section 7.5 are gases (for example, HCl) or liquids. Like ionic compounds, some of them are soluble in water and some are not.

One exception to the rule that covalent compounds have low melting points is $SiO_2$, silicon dioxide (the mineral *quartz*). A crystal of this is a single giant network, and it takes an enormous amount of heat to turn it into a liquid. Diamonds are also giant covalent networks, and the melting point of a gem-quality diamond is extremely high (above 3000°C).

The oxyacids are all colorless liquids, though nitric acid develops a brown color if it stands. All the hydrohalic acids are colorless liquids, too. The way to recognize that you have an acid in the laboratory is to use its chemical properties. The simplest test is to place a drop of the liquid on a piece of blue litmus paper (see Figure 7.17). Acids turn this test paper red; other types of compounds leave its color unchanged. (We will discuss this more fully in Chapter 15.)

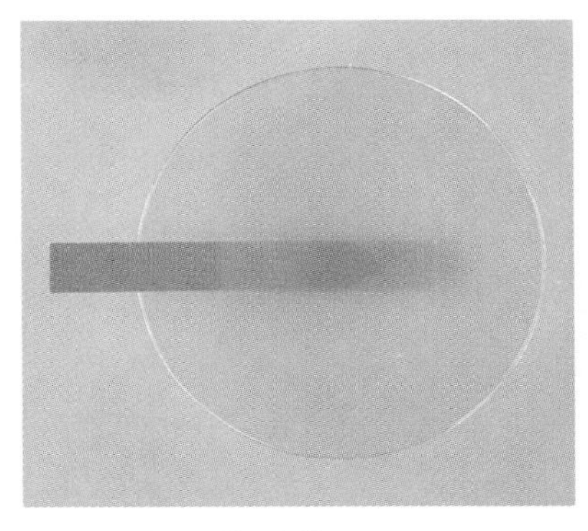

FIGURE 7.17
Acids turn blue litmus paper red. Substances that are not acids don't change the color of this paper.

### EXAMPLE 7.11 Properties of Ionic and Covalent Compounds

Classify each of these substances as either ionic or covalent. Explain your answers.

**a.** A colorless solid at room temperature that melts at 770°C
**b.** Candle wax
**c.** A compound that is a liquid at room temperature and has a boiling point of 85°C

**SOLUTIONS**

**a.** The high melting point suggests an ionic compound. (The compound described is KCl.)
**b.** Candle wax is a soft solid that melts easily; therefore it's covalent.
**c.** Compounds that are liquids at room temperature are typically covalent. The relatively low boiling point confirms this. (The compound described is the alcohol in alcoholic beverages, $C_2H_5OH$.)

**EXERCISE 7.11**

Classify each of these substances as either ionic or covalent. Explain your answers.

**a.** A snail shell
**b.** Cotton
**c.** A compound that is a solid at room temperature and has a melting point of 855°C

## 7.9 Predicting the Solubility of Ionic Compounds in Water

**GOAL:** To predict whether an ionic compound is soluble in water.

Most ionic compounds are insoluble in water. There are exceptions, and those are quite important. Before listing the ionic compounds that do dissolve in water, let's think in terms of what the ions are doing to see why most ionic compounds are insoluble.

By definition, ionic compounds are composed of positively charged cations and negatively charged anions. In an ionic crystal, each cation is completely surrounded by anions, and vice versa. The numbers and exact geometric arrangement of anions and cations depend on the nature of the crystal and are not important for the purposes of this discussion. The point is that each cation is firmly held in place by forces of attraction of many anions, and vice versa. For an ionic crystal to dissolve, a large amount of energy is needed to overcome these forces and pull the ions away from each other (see Figure 7.18). In most cases water molecules cannot provide enough energy to disrupt an ionic crystal. Water can supply the needed energy to dissolve only a few ionic compounds.

When ionic substances dissolve in water, the ionic crystal is destroyed and the ions are separated. For example, when NaCl dissolves in water, the solution contains $Na^+$ ions. There are *no* NaCl molecules in the solution. (See Figure 7.18.) The same thing is true for $CaCl_2$. The water solution contains only $Ca^{2+}$ and $Cl^-$ ions, each completely surrounded by water molecules, and there are no $CaCl_2$ molecules present. All of the ionic substances that dissolve in water separate into their component ions in this way.

**TABLE 7.5 Ionic Compounds That Dissolve in Water**

| |
|---|
| • Compounds containing the nitrate or acetate ion ($NO_3^-$ or $CH_3CO_2^-$). There are no important exceptions. |
| • Compounds containing a Group 1A metal cation ($Li^+$, $Na^+$, $K^+$, etc.). There are no important exceptions. |
| • Compounds containing the $NH_4^+$ ion. There are no important exceptions. |
| • Compounds containing the $Cl^-$ ion. This has some exceptions: AgCl, $PbCl_2$, and $Hg_2Cl_2$ are insoluble in water. |
| • Compounds containing the $SO_4^{2-}$ ion. This also has some exceptions: $PbSO_4$, $BaSO_4$, and $CaSO_4$ are insoluble in water. |

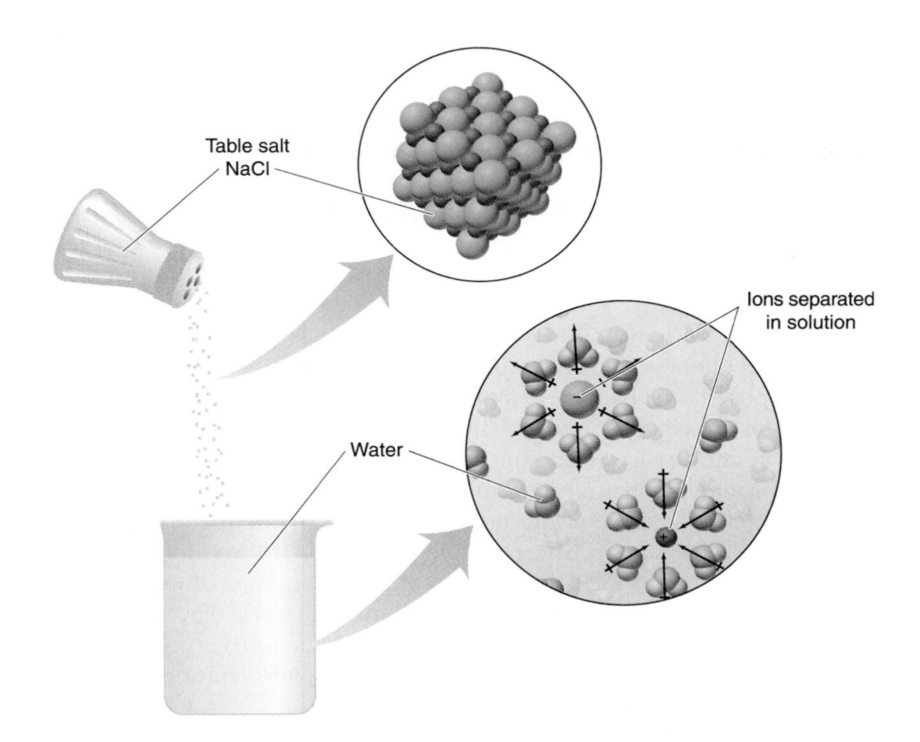


FIGURE 7.18
When ordinary table salt dissolves in water, the ionic crystal is destroyed. The sodium and chloride ions separate, with each ion completely surrounded by water molecules in solution. The energy needed to dissolve the crystal comes from the forces of attraction between the water molecules and the ions. (You will find a more detailed explanation of these forces in Chapter 12.)

A general rule of thumb is that all ionic compounds are insoluble in water except those listed in Table 7.5. It would be difficult to overemphasize the importance of this list. Because so few inorganic compounds dissolve in water, it is critical that you know which ones do.

## EXAMPLE 7.12 Predicting Solubility of Ionic Compounds in Water

Which of these ionic compounds are soluble in water?

**a.** $NaNO_3$ **b.** AgCl **c.** $CaCO_3$

**SOLUTIONS**

**a.** This compound is soluble, for two reasons: $Na^+$ is a Group 1A metal cation, and its compounds are soluble in water. Compounds containing the nitrate ion are also soluble.

**b.** Most chloride compounds are soluble, but AgCl is one of the exceptions listed in Table 7.5. It's insoluble in water.

**c.** Neither calcium ion nor carbonate ion is mentioned in the list in Table 7.5. Therefore, $CaCO_3$ is insoluble in water.

**EXERCISE 7.12**

Which of these ionic compounds are soluble in water?

**a.** $K_2S$ **c.** $PbCl_2$ **e.** NaOH

**b.** FeS **d.** $FeCl_2$ **f.** $Fe(OH)_2$

## CHAPTER SUMMARY

1. A nomenclature system is a set of rules for naming compounds and ions. Each compound must have a unique name, and each name must belong to one and only one compound.
2. Each metal from Groups 1A and 2A forms only one cation (positively charged ion). The positive charge on each of these ions is equal to the element's group number; examples are $Na^+$ and $Ca^{2+}$. Aluminum, zinc, and silver follow the same rule. The name of a metal ion is simply the name of the element followed by the word *ion.*
3. Most other metals can form more than one cation. Names for these ions are composed of the element's name followed by a Roman numeral indicating the charge followed by the word *ion.* Examples are the iron(II) ion, $Fe^{2+}$, and the iron(III) ion, $Fe^{3+}$.
4. Many nonmetals form monatomic (single-atom) anions (ions with negative charges) that have a charge equal to the element's group number minus 8. Examples are $Cl^-$, chloride ion, and $O^{2-}$, oxide ion. The hydride ion, $H^-$, is an exception to this rule. The names of nonmetal ions have the ending *-ide,* but are otherwise somewhat irregular.
5. Nonmetals can also form polyatomic (multiple-atom) ions, which may also include metal atoms. The names, formulas, and charges of common polyatomic ions must be memorized.
6. Polyatomic anions containing oxygen are called oxyanions. If the name, formula, and charge of one member of an oxyanion family is known, the others can be deduced. The parent ion of the family has an *-ate* suffix on its name. The member with one more oxygen has a *per-* prefix added to the name. With one fewer oxygen, *-ite* replaces *-ate.* With two fewer oxygens, the *-ite* ending is used and the prefix *hypo-* is added. An example of an oxyanion family is $ClO_4^-$, perchlorate ion; $ClO_3^-$, chlorate ion; $ClO_2^-$, chlorite ion; and $ClO^-$, hypochlorite ion.
7. The name of a simple ionic compound is formed by combining the names of the ions that make it up: for example, sodium chloride. The formulas of such compounds are obtained from the names by making sure that the total positive and total negative charges cancel each other out. All compounds have a total charge of zero.
8. Binary covalent compounds contain two different nonmetallic elements. Their names are formed from the names of the elements they contain with prefixes to indicate more than one of an ion as well as the usual monatomic ion suffix *-ide.* Examples are $N_2O$, dinitrogen monoxide, and $PCl_3$, phosphorus trichloride. Water ($H_2O$), ammonia ($NH_3$), and methane ($CH_4$) are special cases of binary covalent compounds whose names must be memorized.
9. Acids are covalent compounds that share a number of properties, including the ability to turn blue litmus paper red. Oxyacids combine oxyanions with $H^+$ ions. In their names the *-ate* ending of the oxyanion is replaced with *-ic acid,* and the *-ite* ending with *-ous acid,* though there are exceptions.
10. Ionic compounds are solids with high melting points. Covalent compounds are often low-melting solids, liquids, or gases, though there are exceptions.
11. Most ionic compounds are insoluble in water. The identities of many of those that do dissolve can be deduced from a few simple rules.

## PROBLEMS

### SECTION 7.1 Ions of Metallic Elements

1. What is the name of each of these ions?
   a. $Na^+$ d. $Mg^{2+}$
   b. $Rb^+$ e. $Zn^{2+}$
   c. $Ba^{2+}$

2. What is the name of each of these ions?
   a. $Cs^+$ d. $Ra^{2+}$
   b. $Li^+$ e. $Al^{3+}$
   c. $Sr^{2+}$

3. Write the symbol for each of these ions.
   a. Potassium ion d. Lithium ion
   b. Silver ion e. Beryllium ion
   c. Radium ion

4. Write the symbol for each of these ions.
   a. Rubidium ion d. Strontium ion
   b. Aluminum ion e. Sodium ion
   c. Calcium ion

5. Name each of the following ions.
   a. $Ti^{2+}$ d. $Fe^{3+}$
   b. $V^{3+}$ e. $Co^{2+}$
   c. $Ag^+$

6. Name each of the following ions.
   a. $Pb^{2+}$ d. $Au^{3+}$
   b. $Sn^{2+}$ e. $Bi^{3+}$
   c. $Tl^{3+}$

7. Name each of the following ions.
   a. $Cr^{3+}$ d. $Cu^{2+}$
   b. $Sn^{4+}$ e. $Co^{3+}$
   c. $Tl^+$

8. Name each of the following ions.
   a. $Bi^{5+}$ d. $Cr^{6+}$
   b. $Mn^{7+}$ e. $Pt^{2+}$
   c. $Ni^{2+}$

9. Give the symbol for each of these ions.
   a. Titanium(III) ion d. Mercury(II) ion
   b. Iron(II) ion e. Lead(IV) ion
   c. Cobalt(III) ion

10. Give the symbol for each of these ions.
    a. Manganese(VI) ion d. Titanium(IV) ion
    b. Chromium(II) ion e. Iron(III) ion
    c. Arsenic(III) ion

11. Give the symbol for each of these ions.
    a. Vanadium(II) ion d. Molybdenum(II) ion
    b. Manganese(III) ion e. Cobalt(II) ion
    c. Gold(I) ion

12. Give the symbol for each of these ions
    a. Iron(III) d. Nickel(II)
    b. Vanadium(V) e. Copper(I)
    c. Chromium(III)

### SECTION 7.2 Ions of Nonmetallic Elements

13. Name each of these ions.
    a. $Br^-$ d. $I^-$
    b. $N^{3-}$ e. $P^{3-}$
    c. $H^-$

14. Name each of these ions.
    a. $F^-$ d. $O^{2-}$
    b. $S^{2-}$ e. $Cl^-$
    c. $Se^{2-}$

15. What is the symbol for each of these ions?
    a. Iodide ion d. Bromide ion
    b. Phosphide ion e. Nitride ion
    c. Oxide ion

16. What is the symbol for each of these ions?
    a. Hydride ion d. Fluoride ion
    b. Selenide ion e. Sulfide ion
    c. Chloride ion

17. What is the formula of each of these ions?
    a. Acetate ion d. Bromite ion
    b. Hypochlorite ion e. Mercurous ion
    c. Hydrogen carbonate ion

18. What is the formula of each of these ions?
    a. Hydroxide ion d. Carbonate ion
    b. Chlorate ion e. Nitrate ion
    c. Periodate ion

19. What is the formula of each of these ions?
    a. Nitrite ion d. Phosphate ion
    b. Perchlorate ion e. Ammonium ion
    c. Hydrogen sulfate ion

20. What is the formula of each of these ions?
    a. Cyanide d. Phosphate
    b. Hydrogen carbonate e. Chromate
    c. Sulfate

21. What are the names of these ions?
    a. $MnO_4^-$ d. $Hg_2^{2+}$
    b. $SO_3^{2-}$ e. $OH^-$
    c. $NH_4^+$

22. What are the names of these ions?
    a. $BrO^-$ d. $NO_2^-$
    b. $Cr_2O_7^{2-}$ e. $IO_3^-$
    c. $HSO_4^-$

23. What is the name of each of these ions?
 a. $BrO_3^-$
 b. $CO_3^{2-}$
 c. $SO_4^{2-}$
 d. $PO_4^{3-}$
 e. $CN^-$

24. What is the name of each of these ions?
 a. $NO_3^-$
 b. $HCO_3^-$
 c. $IO_2^-$
 d. $CrO_4^{2-}$
 e. $CH_3CO_2^-$

## SECTION 7.3 Formulas of Simple Ionic Compounds

25. Give the formula of the compound consisting of each pair of ions.
 a. $Ca^{2+}$ and $CN^-$
 b. $Al^{3+}$ and $O^{2-}$
 c. $Na^+$ and $SO_4^{2-}$
 d. $Ba^{2+}$ and $CO_3^{2-}$
 e. $NH_4^+$ and $PO_4^{3-}$

26. Give the formula of the compound consisting of each pair of ions.
 a. $K^+$ and $Cr_2O_7^{2-}$
 b. $Ca^{2+}$ and $H^-$
 c. $Ag^+$ and $CrO_4^{2-}$
 d. $Sr^{2+}$ and $HSO_4^-$
 e. $Fe^{2+}$ and $P^{3-}$

27. Give the formula of the compound containing each pair of ions.
 a. Silver ion and chloride ion
 b. Aluminum ion and oxide ion
 c. Calcium ion and chloride ion
 d. Sodium ion and hydrogen carbonate ion
 e. Ammonium ion and sulfide ion

28. Give the formula of the compound containing each pair of ions.
 a. Lithium ion and perchlorate ion
 b. Mercury(II) ion and sulfate ion
 c. Copper(I) ion and bromide ion
 d. Sodium ion and perchlorate ion
 e. Iron(II) ion and bromide ion

29. Write the formula for the compound consisting of the two ions given.
 a. Iron(III) ion and dichromate ion
 b. Aluminum ion and chromate ion
 c. Zinc ion and iodate ion
 d. Sodium ion and hypochlorite ion
 e. Nickel(II) ion and nitrate ion

30. Write the formula for the compound consisting of the two ions given.
 a. Ammonium ion and sulfate ion
 b. Sodium ion and sulfite ion
 c. Lead(II) ion and carbonate ion
 d. Arsenic(V) ion and chloride ion
 e. Manganese(IV) ion and oxide ion

31. For each pair of ions, give the formula of the compound they form.
 a. Strontium ion and carbonate ion
 b. Mercury(II) ion and fluoride ion
 c. Barium ion and sulfate ion
 d. Vanadium(II) ion and sulfide ion
 e. Iron(III) ion and phosphate ion

32. For each pair of ions, give the formula of the compound they form.
 a. Tin(II) ion and chloride ion
 b. Potassium ion and hypochlorite ion
 c. Magnesium ion and acetate ion
 d. Ammonium ion and hydrogen carbonate ion
 e. Calcium ion and hypochlorite ion

33. Write the formula of the compound formed by each pair of ions.
 a. Sodium ion and perchlorate ion
 b. Aluminum ion and carbonate ion
 c. Zinc ion and phosphide ion
 d. Calcium ion and nitride ion
 e. Strontium ion and nitrate ion

34. Write the formula of the compound formed by each pair of ions.
 a. Iron(II) ion and phosphate ion
 b. Cobalt(III) ion and nitrate ion
 c. Copper(II) ion and nitrite ion
 d. Chromium(II) ion and sulfite ion
 e. Copper(I) ion and cyanide ion

35. Each of these formulas contains one or more errors. Find the error(s) and write the formula correctly.
 a. $Ag(NO_3)$
 b. $MG_3N_2$
 c. $NH_4CO_3$
 d. $FECl_3$
 e. CaP
 f. $Hg_2I_3$
 g. $NaCo_3$

36. Each of these formulas contains one or more errors. Find the error(s) and write the formula correctly.
 a. $(Hg_2)(NO_3)_2$
 b. $(NH_4)_3SO_4$
 c. $Na_2(SO_4)$
 d. AlO
 e. $NH_4Cr_2O_7$
 f. $Al(Cl)_3$
 g. $Mg_2P_3$

## SECTION 7.4 Naming Simple Ionic Compounds

37. Name each of these ionic compounds.
 a. $CaBr_2$
 b. $NH_4I$
 c. $CuBr_2$
 d. CaO
 e. $Sr(OH)_2$

38. Name each of these ionic compounds.
 a. $Ba_3P_2$
 b. $Hg_2Cl_2$
 c. $MnS_2$
 d. $Na_2Cr_2O_7$
 e. $Ag_2O$

39. Name each ionic compound.
    a. $Na_3PO_4$
    b. $CaSO_3$
    c. $Al_2S_3$
    d. $BaI_2$
    e. $PbBr_4$

40. Name each ionic compound.
    a. $Sn(OCl)_2$
    b. AuBr
    c. $Cr(NO_3)_3$
    d. AlP
    e. $Ni(OH)_2$

41. What is the name of each ionic compound?
    a. $MgCl_2$
    b. $(NH_4)_2CO_3$
    c. CuBr
    d. FeS
    e. $CaCr_2O_7$

42. What is the name of each ionic compound?
    a. $Li_3N$
    b. $HgCl_2$
    c. $NaNO_2$
    d. $PbBr_2$
    e. $Sn(OBr)_4$

43. Name each of these ionic compounds.
    a. $AuBr_3$
    b. $Cr(NO_3)_2$
    c. $(NH_4)_2SO_4$
    d. $KHSO_4$
    e. $Ca(HSO_4)_2$

44. Name each of these ionic compounds.
    a. $KMnO_4$
    b. CuCN
    c. $FeBr_3$
    d. $Al(OH)_3$
    e. LiI

45. For each ionic compound, give the correct name.
    a. FeSe
    b. $CH_3CO_2Na$
    c. $Fe(ClO_4)_2$
    d. NaOI
    e. $Ag_2CrO_4$

46. For each ionic compound, give the correct name.
    a. $KClO_3$
    b. $WCl_6$
    c. $CrO_3$
    d. $Mg(MnO_4)_2$
    e. $Li_2CO_3$

47. Name the following ionic compounds.
    a. $(CH_3CO_2)_2Hg$
    b. $Hg_2I_2$
    c. $PtCl_4$
    d. $TlHSO_4$
    e. $Al_2(CO_3)_3$

48. Name the following ionic compounds.
    a. AgI
    b. $Cu(NO_2)_2$
    c. $Zn(IO_3)_2$
    d. $Sn(ClO_4)_2$
    e. $PtSO_4$

## SECTION 7.5 Naming Binary Covalent Compounds

49. Name each of these covalent compounds.
    a. $CH_4$
    b. $NO_2$
    c. $SO_3$
    d. $BCl_3$
    e. HF
    f. $SF_6$

50. Name each of these covalent compounds.
    a. $PCl_3$
    b. $BCl_3$
    c. $P_4O_{10}$
    d. $N_2O$
    e. $Cl_2O_8$
    f. $S_2F_{10}$

51. Name the following covalent compounds.
    a. $P_2O_4$
    b. $P_2Se_3$
    c. $SeC_2$
    d. $P_4S_7$
    e. $SeO_3$
    f. $XeO_2$

52. Name the following covalent compounds.
    a. $NH_3$
    b. $N_2O_5$
    c. $SO_2$
    d. $BF_3$
    e. HI
    f. $SF_2$

53. For each covalent compound, give the correct name.
    a. $PCl_5$
    b. $BF_3$
    c. $S_2Cl_2$
    d. $H_2S$
    e. $Cl_2O_7$
    f. $SeBr_4$

54. For each covalent compound, give the correct name.
    a. $S_4N_2$
    b. $S_2O_2$
    c. $P_4Se_3$
    d. $Se_2Br_2$
    e. $SeO_2$
    f. $XeO_3$

55. Give the correct formulas for these covalent compounds.
    a. Carbon disulfide
    b. Oxygen difluoride
    c. Phosphorus pentachloride
    d. Disulfur dichloride
    e. Iodine trichloride

56. Give the correct formulas for these covalent compounds.
    a. Nitrogen trichloride
    b. Arsenic pentachloride
    c. Dinitrogen tetroxide
    d. Diphosphorus tetraiodide
    e. Triselenium dinitride

57. Write the formula for each of these covalent compounds.
    a. Boron nitride
    b. Sulfur hexafluoride
    c. Hydrogen sulfide
    d. Dinitrogen monoxide
    e. Iodine pentafluoride

58. Write the formula for each of these covalent compounds.
    a. Disulfur tetrafluoride
    b. Nitrogen triiodide
    c. Arsenic trifluoride
    d. Tetraphosphorus hexaoxide
    e. Diselenium dichloride

## SECTION 7.6 Naming Acids

59. Name each of these acids.
    a. $CH_3CO_2H$
    b. $H_2SO_4$
    c. HCl gas
    d. HCl dissolved in water
    e. $HNO_3$

60. Name each of these acids.
a. $H_3PO_3$
b. HBrO
c. $HBrO_2$
d. $HClO_4$
e. HF gas

61. Name each acid.
a. $HBrO_3$
b. $HBrO_4$
c. HI gas
d. HI dissolved in water
e. $H_2SO_3$

62. Name each acid.
a. $H_3PO_4$
b. $H_2CO_3$
c. $H_3PO_2$
d. $HNO_2$
e. HF dissolved in water

63. Give the formula of each of these acids.
a. Hydrogen fluoride
b. Hydrofluoric acid
c. Hypoiodous acid
d. Iodous acid
e. Sulfuric acid

64. Give the formula of each of these acids.
a. Periodic acid
b. Nitrous acid
c. Carbonic acid
d. Hydrochloric acid
e. Sulfurous acid

65. Write the correct formula for each acid.
a. Hypochlorous acid
b. Chlorous acid
c. Chloric acid
d. Phosphoric acid

66. Write the correct formula for each acid.
a. Perchloric acid
b. Nitric acid
c. Acetic acid
d. Hydrochloric acid

### SECTION 7.7 Naming Unclassified Compounds

67. Classify each of these compounds as either ionic or covalent. Identify the covalent compounds as either binary covalent compounds or oxyacids.
a. $CaCl_2$
b. $FeCl_3$
c. $NCl_3$
d. $(NH_4)_2SO_4$
e. $HNO_3$

68. Classify each of these compounds as either ionic or covalent. Identify the covalent compounds as either binary covalent compounds or oxyacids.
a. HOI
b. $NH_3$
c. $Cl_2O$
d. CaO
e. $KHSO_4$

69. Classify each compound as either ionic or covalent. Identify each covalent compound as either a binary covalent one or an oxyacid.
a. $K_2S$
b. $CoCl_3$
c. $SF_6$
d. $Ba_2(PO_4)_3$
e. $HNO_2$

70. Classify each compound as either ionic or covalent. Identify each covalent compound as either a binary covalent one or an oxyacid.
a. $CH_4$
b. $H_2O$
c. $HClO_4$
d. BaO
e. $KHCO_3$

71. Name each of these compounds.
a. $SiF_4$
b. $SnCl_2$
c. $PbCl_4$
d. HBr
e. $HBrO_3$

72. Name each of these compounds.
a. $HIO_4$
b. $NaNO_3$
c. KOI
d. $MgCl_2$
e. $PCl_5$

73. Name each compound.
a. $Ba(OH)_2$
b. $HClO_3$
c. $H_2O$
d. $NH_3$
e. $XeO_3$

74. Name each compound.
a. $CoBr_3$
b. NaH
c. $H_3PO_4$
d. CuOH
e. $Li_2Cr_2O_7$

75. For each of these compounds, give the correct name.
a. $SO_2$
b. $Mg(CN)_2$
c. HF
d. $KNO_2$
e. $FeCO_3$

76. For each of these compounds, give the correct name.
a. $Li_2SO_3$
b. $Ca(HCO_3)_2$
c. $KMnO_4$
d. BaO
e. $K_2Cr_2O_7$

77. What is the name of each of these compounds?
a. $FeSO_4$
b. HBr
c. $CH_4$
d. $Na_2S$
e. $XeF_2$

78. What is the name of each of these compounds?
a. $CoBr_2$
b. $SrH_2$
c. $H_2SO_4$
d. CuI
e. $Li_2CrO_4$

79. Give the formula of each compound.
a. Iodous acid
b. Manganese(II) nitrate
c. Titanium(IV) oxide
d. Sulfuric acid
e. Carbon monoxide

80. Give the formula of each compound.
a. Arsenic trichloride
b. Silver bromide
c. Calcium acetate
d. Sodium acetate
e. Potassium carbonate

81. Write the correct formula for each of these compounds.
a. Sulfur trioxide
b. Copper(I) oxide
c. Silver chromate
d. Hydrogen iodide
e. Dinitrogen tetrachloride

82. Write the correct formula for each of these compounds.
    a. Sodium selenide
    b. Titanium(IV) phosphide
    c. Strontium nitride
    d. Ammonium hydrogen carbonate
    e. Silver chloride

83. What is the formula of each of these compounds?
    a. Potassium sulfide
    b. Magnesium chloride
    c. Ammonium sulfide
    d. Potassium chlorate
    e. Cobalt(III) bromide

84. What is the formula of each of these compounds?
    a. Calcium nitrate
    b. Nitrogen triiodide
    c. Aluminum sulfate
    d. Magnesium sulfate
    e. Sodium chromate

85. For each of these compounds, write the correct formula.
    a. Tin(IV) oxide
    b. Cobalt(II) iodide
    c. Silver oxide
    d. Hydrogen chloride
    e. Dinitrogen monoxide

86. For each of these compounds, write the correct formula.
    a. Calcium selenide
    b. Sodium phosphide
    c. Iron(III) nitride
    d. Ammonium hydrogen sulfate
    e. Nitrogen dioxide

## SECTION 7.8 Classifying Compounds from Their Properties

87. Classify each of these substances as (essentially) ionic or covalent.
    a. Ice cubes
    b. Rocks
    c. A football
    d. Hair
    e. Skin

88. Classify each of these substances as (essentially) ionic or covalent.
    a. Plastic wrap
    b. Air
    c. Petrified wood
    d. Bread
    e. Baking soda

89. Classify each substance as (essentially) ionic or covalent.
    a. Gravel
    b. Water
    c. Paper
    d. Ammonia
    e. Bones

90. Classify each substance as (essentially) ionic or covalent.
    a. A bay leaf
    b. An orange rind
    c. A fossil
    d. Gum
    e. Wood

91. From the description of its properties at room temperature, classify each substance as either ionic or covalent.
    a. A highly water-soluble gas
    b. A gas that is insoluble in water
    c. A solid that contains no metals and is insoluble in water
    d. A solid containing both metals and nonmetals that is insoluble in water
    e. A liquid that turns blue litmus red

92. From the description of its properties at room temperature, classify each substance as either ionic or covalent.
    a. A liquid that has no effect on blue litmus
    b. A soft solid that melts in hot water
    c. Colorless crystals that don't melt in a frying pan at high heat
    d. A water-insoluble solid that burns easily
    e. A water-insoluble liquid

93. Classify each of the substances described as either ionic or covalent.
    a. Contains only $Pb^{2+}$ and $Cl^-$
    b. Contains only nitrogen and oxygen
    c. Has a melting point of 2000°C
    d. Has a melting point of 150°C
    e. Water-soluble and a liquid at room temperature

94. Classify each of the substances described as either ionic or covalent.
    a. Contains only nitrogen
    b. Boils at 350 K
    c. Melts at 350 K
    d. Freezes at 500°C
    e. Sublimes at room temperature

## SECTION 7.9 Predicting the Solubility of Ionic Compounds in Water

95. Classify each of these compounds as being either soluble or insoluble in water.
    a. $Fe(OH)_3$
    b. $Al_2O_3$
    c. $K_2Cr_2O_7$
    d. $(NH_4)_2SO_4$
    e. $Hg_2Cl_2$

96. Classify each of these compounds as being either soluble or insoluble in water.
    a. $BaSO_4$
    b. $Na_2O$
    c. $K_2CrO_4$
    d. $Ca(OH)_2$
    e. KOH

97. For each compound, decide whether it is soluble or insoluble in water.
    a. LiOH
    b. $Fe_2S_3$
    c. $Ag_2CrO_4$
    d. $NaHCO_3$
    e. $HgCl_2$

98. For each compound, decide whether it is soluble or insoluble in water.
    a. $CaCO_3$
    b. $KHSO_4$
    c. $(NH_4)_2SO_4$
    d. $Fe(OH)_3$
    e. $FeCl_2$

## SOLUTIONS TO EXERCISES

### Exercise 7.1

**a.** Magnesium is in Group 2A, so its ion has a charge of +2 and the symbol is $Mg^{2+}$. Lithium is in Group 1A; the symbol for its ion is $Li^{+}$. Aluminum is in Group 3A and forms only the $Al^{3+}$ ion (memorized).
**b.** $Ca^{2+}$, from calcium metal, is calcium ion. $Li^{+}$ is lithium ion, and $Ag^{+}$ is silver ion.

### Exercise 7.2

**a.** The symbol for lead is Pb, so the lead(II) ion is $Pb^{2+}$. Similarly, the copper(I) ion is $Cu^{+}$, and the titanium(II) ion is $Ti^{2+}$.
**b.** Co is the symbol for cobalt, so $Co^{3+}$ is the cobalt(III) ion. Similarly, $Sn^{2+}$ is the tin(II) ion, and $Mn^{3+}$ is the manganese(III) ion.

### Exercise 7.3

**a.** The nitride ion is formed from nitrogen, which is in Group 5A. This ion's charge is $5 - 8 = -3$, and its symbol is $N^{3-}$. Bromine is in Group 7A, so its ion has a charge of $-1$. Its symbol is $Br^{-}$. Since oxygen is in Group 6A, its ion has a charge of $-2$, and its symbol is $O^{2-}$.
**b.** $I^{-}$ is the ion of iodine, so its name is iodide ion. $P^{3-}$ is the monatomic anion of phosphorus and is called phosphide ion. $O^{2-}$ is the oxide ion.

### Exercise 7.4

**a.** The bromate ion is $BrO_3^{-}$. $BrO_4^{-}$ has one more oxygen atom (and the same charge), so it must be the perbromate ion. $CH_3CO_2^{-}$ is the acetate ion. $IO_2^{-}$ has one fewer oxygen than the iodate ion, $IO_3^{-}$, so it must be the iodite ion.
**b.** Hydroxide is $OH^{-}$. Cyanide is $CN^{-}$. Periodate (one more oxygen than iodate) and hypobromite (two fewer oxygen atoms than bromate) are $IO_4^{-}$ and $BrO^{-}$, respectively.

### Exercise 7.5

**a.** Sodium ion is $Na^{+}$, and hypochlorite ion is $ClO^{-}$. Sodium hypochlorite is $NaClO$.
**b.** Lithium ion is $Li^{+}$, and sulfate ion is $SO_4^{2-}$, so the formula is $Li_2SO_4$.
**c.** Ammonium is $NH_4^{+}$, and hydrogen sulfate is $HSO_4^{-}$, so the formula is $NH_4HSO_4$.
**d.** Calcium ion is $Ca^{2+}$, and carbonate ion is $CO_3^{2-}$, so the formula is $CaCO_3$.
**e.** Cobalt(III) ion is $Co^{3+}$, and sulfate ion is $SO_4^{2-}$. The formula is $Co_2(SO_4)_3$.
**f.** Tin(IV) ion is $Sn^{4+}$, and chloride ion is $Cl^{-}$. The formula is $SnCl_4$.

### Exercise 7.6

**a.** From the names of the two ions, this is ammonium sulfate.
**b.** This compound has two names, sodium hydrogen carbonate and sodium bicarbonate.
**c.** $SO_4^{2-}$ is the sulfate ion, so $SO_3^{2-}$ must be the sulfite ion. The name of the compound is potassium sulfite.
**d.** The sulfide ion is $S^{2-}$, so the nickel ion in this compound must be $Ni^{2+}$. The name of the compound is nickel(II) sulfide.
**e.** Calcium dichromate
**f.** Lead(II) oxide

### Exercise 7.7

**a.** Dinitrogen pentoxide
**b.** Nitrogen triiodide
**c.** Carbon dioxide
**d.** Methane
**e.** Hydrogen bromide

### Exercise 7.8

**a.** $SiO_2$
**b.** $NF_3$
**c.** $P_4S_{10}$
**d.** $N_2Cl_4$

### Exercise 7.9

**a.** HOCl (or HClO) is hypochlorous acid, $HClO_2$ is chlorous acid, and $HClO_3$ is chloric acid.
**b.** Phosphoric acid is $H_3PO_4$, bromic acid is $HBrO_3$, and carbonic acid is $H_2CO_3$.

### Exercise 7.10

**a.** This compound contains a metal, so it's ionic. Potassium is a Group 1A metal, so no Roman numeral is needed. The name is potassium sulfide.
**b.** Cobalt is a metal, so the compound is ionic. Since cobalt is a transition metal, the name has a Roman numeral. Since the three chloride ions each have a charge of $-1$, this cobalt ion must have a $+3$ charge. The name is cobalt(III) chloride.
**c.** This compound contains only nonmetals, so it's a binary covalent compound. The name is sulfur hexafluoride.
**d.** Barium is a metal from Group 2A, so this compound is ionic. Barium (like the rest of the Group 2A metals) doesn't need a Roman numeral to differentiate ions. The name is barium phosphate.
**e.** This compound contains a hydrogen bonded to an oxyanion, so it's an oxyacid; the name is nitrous acid.

**Exercise 7.11**

**a.** Snail shells are hard, brittle, and can be heated to high temperatures. Snail shells are ionic. (They're made up almost entirely of $CaCO_3$.)

**b.** Cotton is a soft, flexible solid that does not melt easily. Ionic compounds are brittle; therefore cotton is covalent.

**c.** Solids that have such high melting points are almost always ionic.

**Exercise 7.12**

**a.** This compound contains a Group 1A metal, so it is soluble in water.

**b.** Neither of these ions is listed in Table 7.5, so this compound is insoluble in water.

**c.** This is listed in Table 7.5 as an exception to the general solubility of chloride compounds. It is insoluble.

**d.** This chloride compound is not one of the listed exceptions, so it is soluble in water.

**e.** This compound contains a Group 1A metal, so it is soluble in water.

**f.** Neither of these ions is listed in Table 7.5, so the compound is insoluble in water.

8

# INTRODUCTION TO CHEMICAL REACTIONS



KEY TERMS

Unless you happen to be a chess enthusiast, the sentence "23.e6!! (Δ24. ef+−)" is gibberish. The chess community has developed a written language of its own, which chess players use to communicate their ideas exactly, even across barriers of language and culture. Chemists do the same thing when writing chemical equations to describe reactions. Letters and numbers are used in a very precise way to write a kind of sentence. Each of these chemical sentences contains a fair amount of explicit information and, as you will see in Chapter 9, much more implied information. Your major task for this chapter is to learn to read, write, and interpret this kind of sentence.

**A chemical *reaction* is a change you can observe in one way or another (not necessarily with your eyes). You can write a chemical *equation* to describe a reaction.**

Later in the chapter we will introduce various types of chemical reactions. The names and equations that describe the reaction types are important. They are a major part of the scientific language you will need to communicate effectively about chemistry with your instructor and your fellow students (see Section 1.8).

## INTRODUCTION TO CHEMICAL EQUATIONS

### 8.1 Information in Chemical Equations

GOAL: To translate chemical equations into English sentences and to distinguish between the reactants and the products in a chemical equation.

You can think of a chemical equation as a sentence written in a type of shorthand. Here is an equation that describes the burning of the main ingredient in natural gas:

$$CH_4 + 2\ O_2 \longrightarrow CO_2 + 2\ H_2O$$

Chemists have two distinct ways of translating this equation into an English sentence: "One molecule of methane reacts with two molecules of oxygen to give one molecule of carbon dioxide and two molecules of water" or "One mole of methane reacts with two moles of oxygen to give one mole of carbon dioxide and two moles of water." Both of these are illustrated in Figure 8.1. Note that the second translation uses the word *mole* rather than *molecule.*

FIGURE 8.1
You can interpret chemical equations in terms of either molecules or moles.

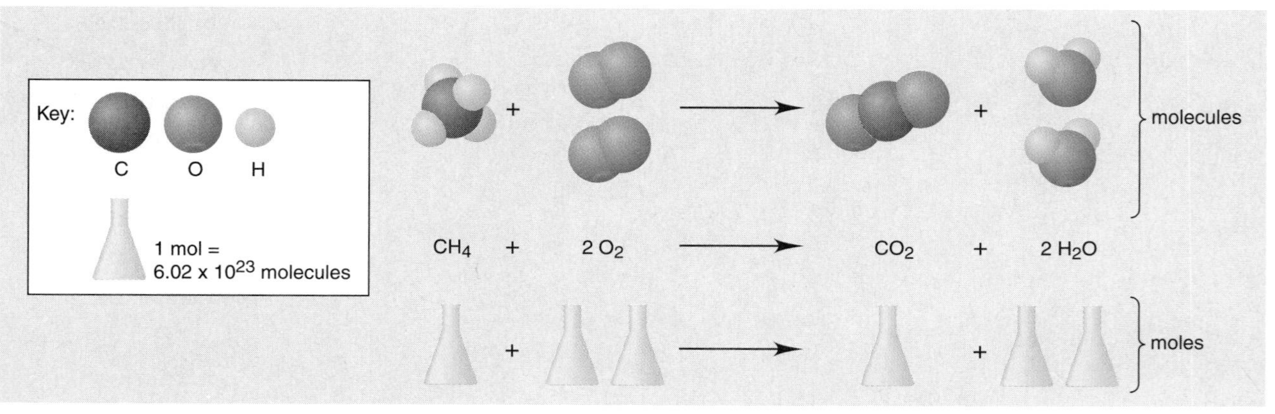

As you know from Chapter 5, a mole consists of Avogadro's number of molecules, and it is an amount that you can see. The best time to think in terms of atoms and molecules instead of moles is when you're balancing equations (discussed in Section 8.2). Atoms and molecules make more sense then. Generally, though, chemists think in terms of moles instead of molecules because they use these larger quantities in carrying out reactions. When adding an ounce of baking soda to a recipe, a chemist thinks of the amount as 0.2 mole instead of $1 \times 10^{23}$ molecules, mostly because it's more practical.

The chemicals a chemist starts with, the ones that react with each other, are called the **reactants.** In the equation in Figure 8.2 these are $CH_4$ and $O_2$. The reactants appear to the left of the arrow, →, on what is called the left side of the equation. The arrow means "produces," "gives," "forms," or "yields." The reaction changes the reactants into **products.** The products appear on the right side of the equation (that is, to the right of the arrow). In the example in Figure 8.2 these are $CO_2$ and $H_2O$.

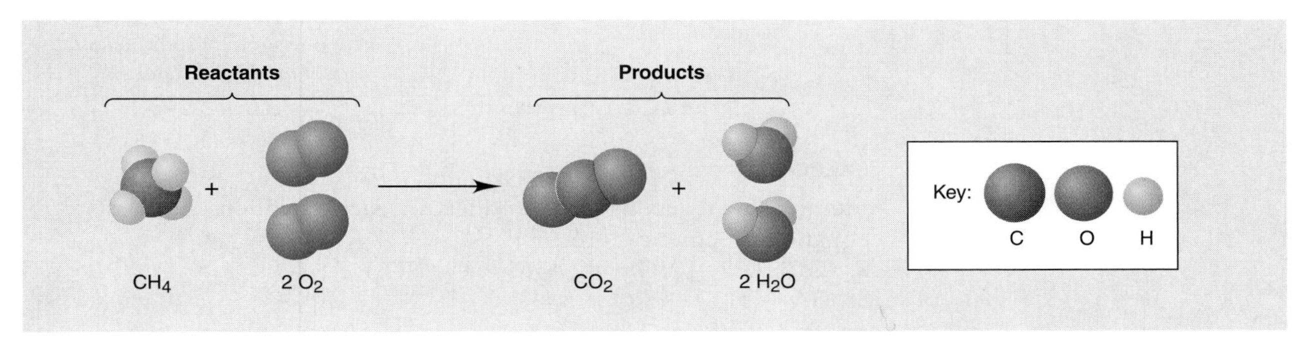


FIGURE 8.2

A number (called a *coefficient*) to the left of the formula of a reactant or a product indicates how many moles (or molecules) of that substance are consumed or formed. Thus, as Figure 8.3 shows, "2 $O_2$" means "two moles (or molecules) of $O_2$." (Recall that the subscript 2 in $O_2$ means that there are two oxygen atoms in each oxygen molecule.) Chemists usually don't write the coefficient when one mole of a substance is involved. The equation in Figure 8.2 tells you that there is *one* mole of $CH_4$ as a reactant and *one* mole of $CO_2$ as a product.

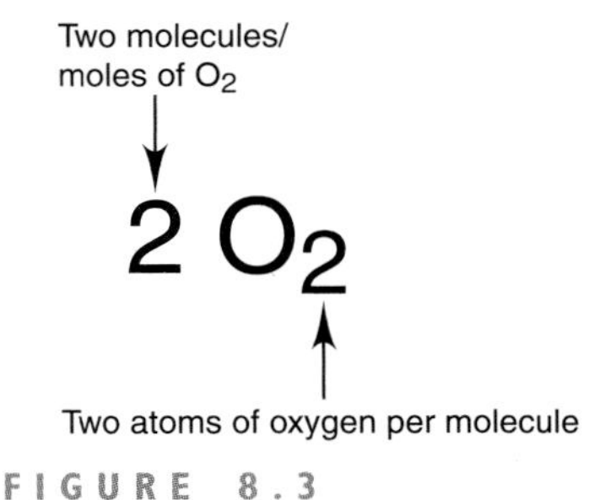


FIGURE 8.3

## EXAMPLE 8.1 Reading Chemical Equations

Identify the reactants and products in each equation, and restate the equation in sentence form.

**a.** $N_2 + 3\ H_2 \rightarrow 2\ NH_3$

**b.** $2\ Na + 2\ H_2O \rightarrow 2\ NaOH + H_2$

**c.** $H_2SO_4 + 2\ NaOH \rightarrow Na_2SO_4 + 2\ H_2O$

SOLUTIONS

| Equation | Description of the Reaction |
|---|---|
| a) $N_2 + 3\ H_2 \rightarrow 2\ NH_3$ | Reactants: $N_2$ and $H_2$<br>Product: $NH_3$<br>Sentence form: One mole of nitrogen reacts with three moles of hydrogen to give two moles of ammonia. |
| b) $2\ Na + 2\ H_2O \rightarrow 2\ NaOH + H_2$ | Reactants: Na and $H_2O$<br>Products: NaOH and $H_2$<br>Sentence form: Two moles of sodium react with two moles of water to give two moles of sodium hydroxide and one mole of hydrogen. |
| c) $H_2SO_4 + 2\ NaOH \rightarrow Na_2SO_4 + 2\ H_2O$ | Reactants: $H_2SO_4$ and NaOH<br>Products: $Na_2SO_4$ and $H_2O$<br>Sentence form: One mole of sulfuric acid reacts with two moles of sodium hydroxide to give one mole of sodium sulfate and two moles of water. |

**EXERCISE 8.1**

Identify the reactants and products in each equation, and restate the equation in sentence form.

**a.** $CaCl_2 + 2\ AgNO_3 \rightarrow 2\ AgCl + Ca(NO_3)_2$
**b.** $CH_4 + H_2O \rightarrow CO + 3\ H_2$
**c.** $Mg + 2\ HBr \rightarrow MgBr_2 + H_2$

## 8.2 Balanced and Unbalanced Equations

GOAL: To determine whether or not a given chemical equation is balanced.

For chemical equations to be useful, they must be **balanced equations**—that is, they should include the correct coefficients for all the reactants and products. A balanced equation tells you the ratios of the amounts of all of the reactants and products, and you will see how useful this is in Chapter 9. Unbalanced equations can mislead you, because they don't fully describe the chemical reactions they represent.

The scheme for balancing equations presented in this section rests on a scientific principle, the **Law of Conservation of Mass.** The atomic theory (Section 2.5) states that atoms are not transformed into different atoms by chemical reactions. Furthermore, detectable amounts of matter are neither created nor destroyed by chemical methods (Section 4.4). Thus, all atoms that are present in the reactants for a chemical reaction must also be present in the products after the reaction occurs. To balance mass in a chemical equation, you must balance the atom counts.

Let's look at the reaction between methane and oxygen (see Figure 8.4) in light of the Law of Conservation of Mass. In other words, let's count the atoms in this equation:

$$CH_4 + 2\ O_2 \longrightarrow CO_2 + 2\ H_2O$$

You can count a total of four oxygen atoms on the left side of the equation—two oxygen molecules that contain two oxygen atoms each. There are also four atoms of oxygen on the right side of the equation—two oxygen atoms in the one molecule of $CO_2$ and one oxygen atom in each of the two molecules of $H_2O$. Each side of the equation has one carbon atom and four hydrogen atoms. Since the two sides of the equation have the same numbers and kinds of atoms, the equation is balanced (see Figure 8.5).

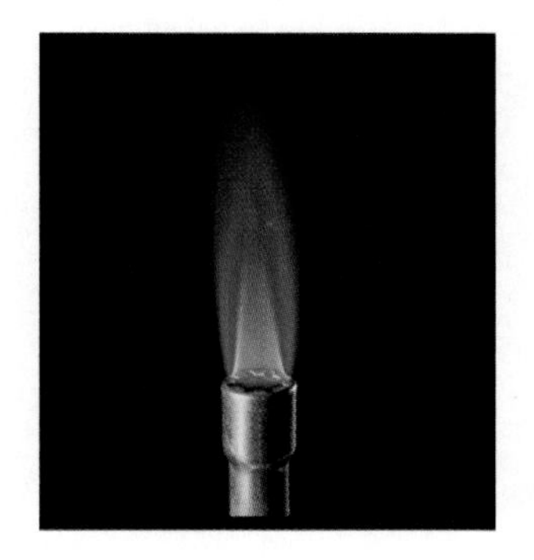

FIGURE 8.4
The reaction between methane and oxygen produces this flame.

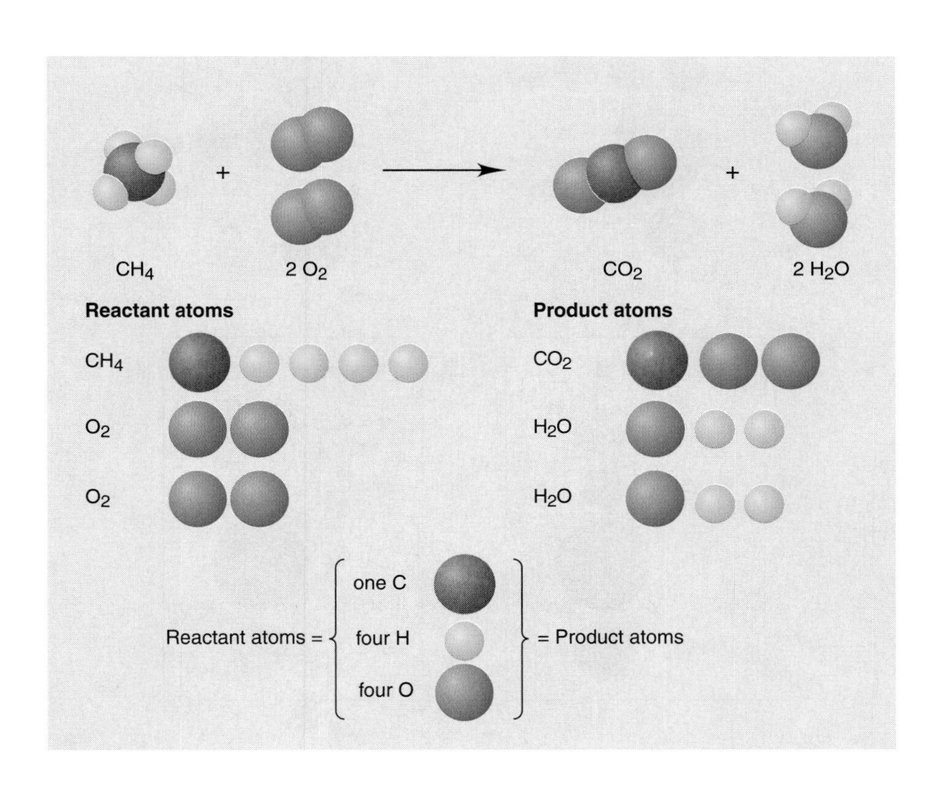


FIGURE 8.5
The equation for the reaction between methane and oxygen shows how the Law of the Conservation of Mass applies to chemical reactions. The two sides of the equation contain the same numbers and kinds of atoms.

It's useful to keep in mind that you are counting atoms, *not* molecules. For example, $CO_2$ does *not* contain a molecule of $O_2$. Figure 8.5 shows you that after an $O_2$ molecule reacts, it no longer exists as $O_2$. Chemical reactions destroy reactant *molecules* ($CH_4$ and $O_2$ in this case), but the *atoms* do not change.

EXAMPLE 8.2 **Inspecting Equations to Determine Whether They Are Balanced**

Determine whether each of these equations is balanced.

**a.** $H_2SO_4 + NaCl \rightarrow NaHSO_4 + HCl$

**b.** $NH_4Cl + Ca(OH)_2 \rightarrow NH_3 + 2\ H_2O + CaCl_2$

**SOLUTIONS**

**a.** Atoms of five elements are present: H, S, O, Na, and Cl. You must count the total number of each kind of atom on each side of the equation, as follows:

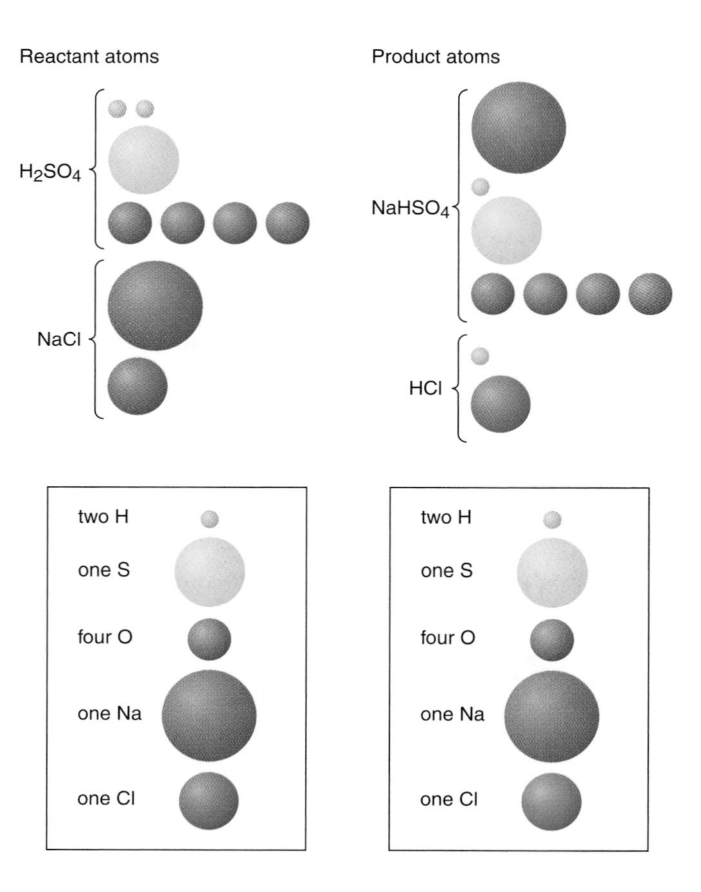


*Conclusion:* The equation is balanced.

**b.** There are five elements present: N, H, Cl, Ca, and O. When counting the oxygen and hydrogen atoms for the reactants, remember that the parentheses and the subscript 2 in $Ca(OH)_2$ mean that two OH groups are present.

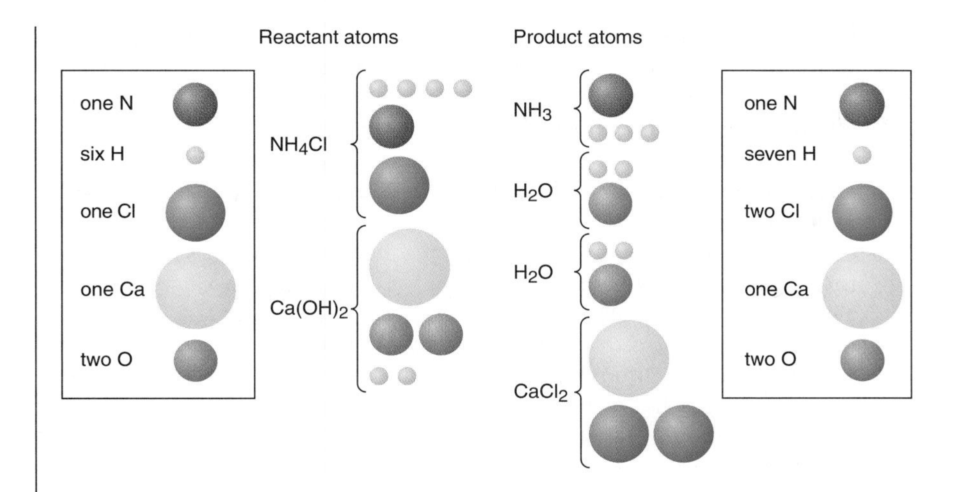


The nitrogen, calcium, and oxygen atoms are balanced, but the chlorine and hydrogen atoms are not. *Conclusion:* The equation is not balanced. An equation is not balanced unless *all* of the masses are conserved.

**EXERCISE 8.2**

Show that the equations in Exercise 8.1 are balanced.

## 8.3 Balancing Simple Chemical Equations

**GOAL:** To balance simple chemical equations from unbalanced equations given as either formulas or sentences.

Once you recognize that a chemical equation is unbalanced, the next step is to balance it. Learning to do this is very much like learning to ride a bicycle. It may seem tricky when you start, but after some practice you'll be saying, "Look, ma! No hands!" Of course, if you take a tumble showing off, then you need more practice.

Chemists call one common method **balancing equations by inspection.** As the name suggests, this is a trial-and-error approach. If you follow the series of steps below, you'll find that the technique is not as haphazard as it sounds.

1. *Write the complete unbalanced equation.* This means that first you write the correct chemical formulas for all the reactants and products. If the correct formulas aren't given in the statement of the problem, you get them from your memorized knowledge and the rules of nomenclature (Chapter 7). Once you have written the correct formulas for the reac-

**STUDY ALERT**
**Beginners in equation balancing sometimes change formulas to try to conserve mass. This is a major mistake. *Don't do it!***

tants and products, you *must not* change them. (If you change a subscript in a formula, you produce the formula for a compound that is either different or nonexistent. You are no longer describing the same reaction you started with.)

2. *Choose an element that appears in only one species on the left side and only one species on the right side.* You have now selected one reactant and one product. Pick the species that has the larger number of atoms of the element in it (if possible) and give it a coefficient of 1. Then balance the number of atoms of the element you originally chose by assigning a coefficient to the other species.
3. *Now that you have determined one coefficient, assign the others.* You do this one element at a time, balancing the numbers of atoms for all the elements present. If a fraction appears as a coefficient, leave it for the moment and complete the balancing.
4. *If the equation contains a coefficient that is a fraction, multiply* all *of the coefficients by that fraction's denominator so that they are all whole numbers.* If you interpret an equation in terms of molecules, a fraction of a molecule is meaningless. On the other hand, if you interpret the equation in terms of moles, a fraction of a mole is reasonable. Chemists, however, rarely use fractional coefficients in balanced equations. The circumstances in which this is useful are rare.
5. *Check* to make sure that the totals of each of the atoms on one side of the equation equal the totals of the same atoms on the other side of the equation.

**Remember that coefficients go in front of the formula containing the element you are balancing, and that you normally don't show a coefficient of 1 in the final balanced equation.**

Applying these guidelines to practical situations is fairly straightforward. The next example illustrates the general method.

## EXAMPLE 8.3a Balancing Chemical Equations

Balance the following equation in two ways—(a) with and (b) without the use of fractional coefficients.

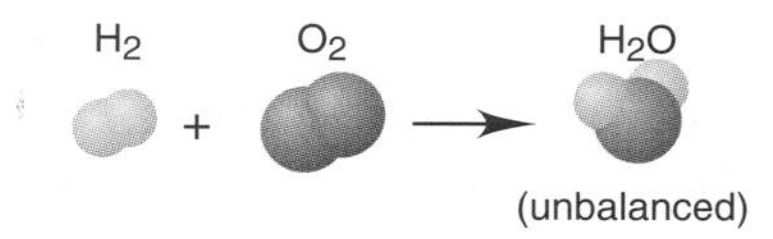


**SOLUTION**

In Step 1 you write down the formulas of the reactants and products. In this case they're given, so go on to step 2.

In Step 2 you pick an element and balance it. There are only two elements present, hydrogen and oxygen, and each one appears once on each side. Hydrogen is already balanced, so it's sensible to balance oxygen. There are two oxygen atoms on the left, and only one oxygen atom on the right. How do you balance oxygen? (Try it yourself before reading further.) One way is to double the number of water molecules on the right (another way is shown in Example 8.3b):

**You didn't change a subscript, did you? The equation $H_2 + O_2 \rightarrow H_2O_2$ is completely balanced, but it has nothing to do with the problem at hand. $H_2O_2$ is a compound named hydrogen peroxide, which is used in the household as a bleaching agent and an antiseptic. Hydrogen peroxide, however, is not water. We beseech, beg, and implore you not to make this type of blunder when balancing equations.**

$H_2$ + $O_2$ ⟶ 2 $H_2O$

(unbalanced)

Now oxygen is balanced, but hydrogen is not—the hydrogen atoms become unbalanced when the coefficient 2 is placed in front of $H_2O$. You must go on to the next step.

In Step 3 you balance the rest of the atoms. Here this means the hydrogen atoms. (Of course, you can remove the water molecule with its two extra hydrogens on the right, but that just puts you back to where you started, with the oxygens unbalanced.) On the left side you can change the number of $H_2$ molecules without changing the number of oxygen atoms. With two $H_2$ molecules on the reactant side, there will be four hydrogen atoms on the left and four on the right. The equation is then:

$$2\,H_2 + O_2 \longrightarrow 2\,H_2O \quad \text{(balanced)}$$

2 $H_2$ + $O_2$ ⟶ 2 $H_2O$

(balanced)

You go on to Step 4 only if there are fractional coefficients. Since there are none, check your answer as your last step (Step 5). There are four hydrogen atoms and two oxygen atoms on each side of the equation, so it really is balanced.

To see how Step 4 works, let's balance the same equation in a different way. Keep in mind that there are typically several possible routes to the correct answer—not only for balancing equations, but for many of the problems you will encounter in this book.

## EXAMPLE 8.3b Balancing Equations with Fractional Coefficients

We are still working on the unbalanced equation $H_2 + O_2 \rightarrow H_2O$.

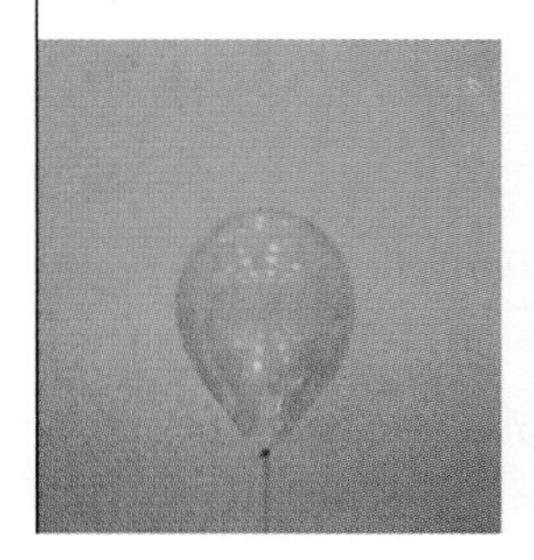
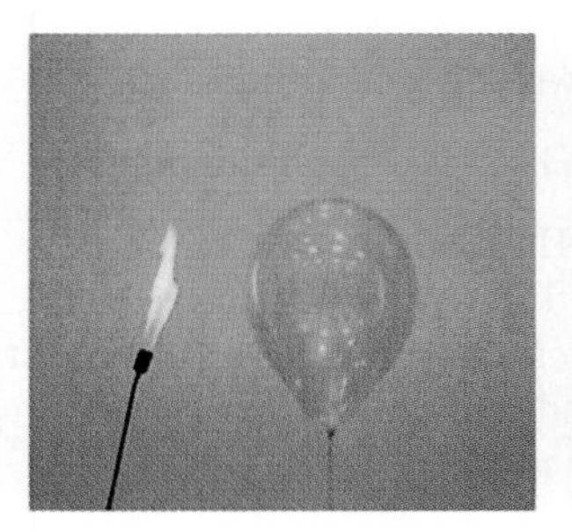
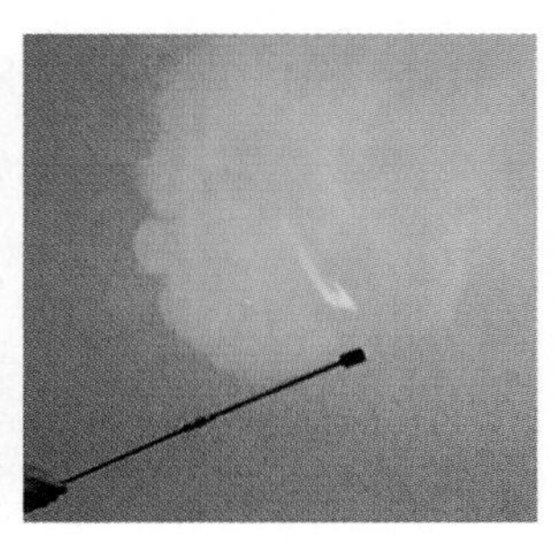

FIGURE 8.6
A balloon filled with a mixture of hydrogen and oxygen explodes when ignited. This is a classic demonstration of a chemical reaction.

SOLUTION

As you know, hydrogen is already balanced in the equation as given. In Example 8.3a, you balanced the oxygens by placing the coefficient 2 in front of the water, which unbalanced the hydrogen atoms. You can balance oxygen without disturbing hydrogen, like this:

$$H_2 + \tfrac{1}{2}\,O_2 \longrightarrow H_2O \quad \text{(balanced)}$$

Formally speaking, the equation is correctly balanced at this point and you could consider the problem finished. As a matter of convention, however, chemists seldom use fractional coefficients in balanced equations. Don't forget Step 4 when you balance equations.

This equation is balanced. There are two hydrogen atoms and one oxygen atom on the left, and the same numbers on the right. Unfortunately, there's a fractional coefficient ($\frac{1}{2}$) on the left—on to Step 4.

Step 4 requires that you multiply all the coefficients by the denominator of the fraction. The coefficient of $H_2$ becomes $2 \times 1 = 2$. The coefficient of $O_2$ becomes $\frac{1}{2} \times 2 = 1$. The coefficient of $H_2O$ becomes $2 \times 1 = 2$. The equation is then:

$$2\,H_2 + O_2 \longrightarrow 2\,H_2O \quad \text{(balanced)}$$

In Step 5, check the answer. There are once again four hydrogen atoms and two oxygen atoms on each side of the equation. This is the same as the answer you got in Example 8.3a.

EXERCISE 8.3

Write a balanced equation for each of these reactions.

**a.** $HBr + Na_2CO_3 \rightarrow H_2O + CO_2 + NaBr$

**b.** $C_2H_6 + O_2 \rightarrow CO_2 + H_2O$

A related type of problem requires you to write down the reactants and products from a sentence that describes a chemical reaction. You do this by applying your memorized knowledge and the rules of nomenclature.

## EXAMPLE 8.4 Writing and Balancing Equations from Written Statements

Write a balanced equation for this reaction: Iron metal reacts with hydrochloric acid to produce iron(II) chloride and hydrogen.

SOLUTION

In 1783 Jacques Charles used the reaction between an acid and a metal to make enough hydrogen gas to fill a large balloon (see Section 11.4).

In Step 1 you write down the reaction:

$$Fe + HCl \longrightarrow FeCl_2 + H_2 \quad \text{(unbalanced)}$$

The formula for iron(II) chloride comes from its name (Chapter 7), and you know by this time that hydrogen is diatomic.

In Step 2 you select and balance an element. The elements present are iron, hydrogen, and chlorine. At this point iron is balanced, but chlorine and hydrogen are not.

$$Fe + HCl \longrightarrow FeCl_2 + H_2$$

You can do Step 3 by balancing either the chlorine or the hydrogen. You balance chlorine first by placing the coefficient 2 in front of the HCl:

$$Fe + 2\ HCl \longrightarrow FeCl_2 + H_2$$

The iron and chlorine atoms are now balanced, which leaves only the hydrogen to be considered. An inspection shows that hydrogen becomes balanced when chlorine is. Since there are no fractional coefficients, Step 4 doesn't apply, and you have the correct answer. Since chemists don't normally write coefficients of 1, the balanced equation is

$$Fe + 2\ HCl \longrightarrow FeCl_2 + H_2 \quad \text{(balanced)}$$

A Step 5 check shows that the equation is balanced.

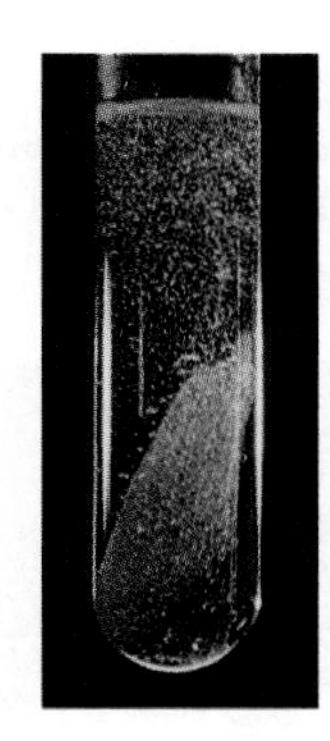

FIGURE 8.7
Iron metal reacts with hydrochloric acid to give iron(II) chloride and hydrogen gas.

**EXERCISE 8.4**

Write a balanced equation for each of these reactions.

**a.** Adding nitric acid to zinc metal produces zinc nitrate and hydrogen.
**b.** Phosphorus reacts with chlorine to produce phosphorus pentachloride.

## CLASSES OF CHEMICAL REACTIONS

The previous sections introduced the equations—the sentences—that chemists use to describe chemical reactions. Next we will venture into the almost limitless realm that the equations describe. What happens when two chemical substances mix? The rest of this chapter discusses just a few types of reactions.

### 8.4 Acid-Base Reactions

GOAL: To examine the properties of acids and bases and to write balanced equations for simple acid-base reactions.

Acids and bases are two of the most important classes of compounds in any kind of chemistry. They are so common that you encounter them in one form or another every day. The distinctive taste of vinegar comes from acetic acid, and the pleasant sourness of orange juice comes from citric acid. Many ant bites hurt because they contain formic acid. Proteins are made from amino acids, and the coding substance for life on earth, DNA, is a nucleic acid. Forests and lakes are sick or dying in many parts of the world because of acid rain. Most drain cleaners work because they contain a base, sodium hydroxide. Antacid tablets typically contain a mixture of bases.

To chemists, the fact that acids and bases react with one another is a key defining characteristic for both. These substances have several other properties you can use to identify them.

### Properties of Acids and Bases

**Never taste any unknown substance! Some early chemists did, and they died young as a result.**

**ACIDS TASTE SOUR.** The tastes of vinegar and lemon juice are classic examples of the sourness of acids (see Figure 8.8). The word *acid* comes from the Latin *acidus,* which means "sour."

FIGURE 8.8
These items have a sour taste because of their acid content. Lemon and lime juices contain citric acid; vinegar contains acetic acid.

**ACIDS REACT WITH MANY METALS TO GIVE HYDROGEN GAS.** A large number of metals (but not all of them) react with acids to produce hydrogen gas, $H_2$. Metals that do this include zinc, iron, aluminum, and those in Groups 1A and 2A, among others. Chemists have known about the reaction between hydrochloric acid and iron for more than two centuries (see Example 8.4):

$$2\ HCl + Fe \longrightarrow FeCl_2 + H_2$$

**ACIDS REACT WITH CARBONATES TO FORM CARBON DIOXIDE.** Strong acids react with carbonate salts to form carbon dioxide. Two $H^+$ ions from the acid react with $CO_3^{2-}$ to form $H_2O$ and $CO_2$. The third product of the reaction is the compound containing the metal cation and the acid anion. This reaction between nitric acid and sodium carbonate is typical:

$$2\ HNO_3 + Na_2CO_3 \longrightarrow 2\ NaNO_3 + H_2O + CO_2$$

Chemists use this reaction between acids and carbonates as a classic test to determine if an unknown substance is an acid. To perform this test, you place a couple of drops of an unknown solution on a piece of $CaCO_3$ (limestone or marble). The formation of $CO_2$ bubbles is easy to see and provides strong evidence that the solution is acidic (see Figure 8.9). The absence of bubbles also provides information—it shows that the test solution is not acidic. (It absolutely, positively does *not* show that the test solution is basic. Not being acidic doesn't mean that the solution is basic. It could also be neutral—neither acidic nor basic.)

FIGURE 8.9
Carbon dioxide evolves when hydrochloric acid reacts with calcium carbonate. This reaction is a positive chemical test for an acid. (Geologists sometimes use this reaction to test for rocks and minerals that contain the carbonate ion.)

**ACIDS CHANGE THE COLOR OF INDICATORS.** An *indicator* is a substance that has one color in acidic solutions and another color in basic solutions. The most famous acid-base indicator is called *litmus.* Acids turn blue litmus red. Litmus is a plant pigment, as are most of the indicators chemists have used for a long time. You can also use grape juice or tea as an acid-base indicator. Modern chemists have a wide range of indicators to choose from (more about them is found in Chapter 15).

Bases also have characteristic properties. To chemists the most important of these is the ability to neutralize acids. Here are some of the other typical properties of bases.

**Bases are frequently referred to as *alkalies.* An *alkaline* solution is one that is basic. These words are derived from the Arabic *alkali,* which means "ashes of a plant." Historically, plant ashes were a key source of bases. Our ancestors used wood ash and animal fat to manufacture soap. The wood ash provided the base necessary for the process.**

**BASES TASTE BITTER.** If you have ever eaten vegetables that were burned in cooking, you have an idea of the taste.

**BASES CHANGE THE COLOR OF INDICATORS.** Bases turn red litmus blue, for instance.

### Definitions of Acid and Base

Chemists use several definitions of acid and base, depending on the circumstances (for more details see Chapter 15). To chemists the two types of substances form an inseparable pair, because their natures are largely defined by the fact that they react with each other.

**TABLE 8.1 Some Common Laboratory Acids and Bases**

| Acids (water solutions) | | Bases | |
|---|---|---|---|
| Hydrochloric acid, | $HCl$ | Sodium hydroxide, | $NaOH$ |
| Hydrobromic acid, | $HBr$ | Potassium hydroxide, | $KOH$ |
| Hydroiodic acid, | $HI$ | | |
| Sulfuric acid, | $H_2SO_4$ | | |
| Nitric acid, | $HNO_3$ | | |
| Perchloric acid, | $HClO_4$ | | |

This section introduces the oldest and simplest definitions of acid and base, which were originally proposed by Svante Arrhenius. Around the end of the 19th century, Arrhenius was investigating how substances behave when dissolved in water (as were many other chemists of that time). This topic was important (and still is) not only because water is the essential medium of life on earth, but also because many industrial processes require it. The subject was even given a special name, *aqueous chemistry*. When a substance is dissolved in water, chemists call the result an *aqueous solution*.

**You may remember Arrhenius from reading about his proposals about the ionic nature of salts (Section 4.5).**

Arrhenius designed his definitions to describe the behavior of acids and bases in aqueous solutions. He said that an **acid** is any substance that produces hydrogen ions ($H^+$) when dissolved in water, and a **base** is any substance that produces hydroxide ions ($OH^-$) when dissolved in water. These definitions describe most of the common acids and bases you might use in the laboratory (see Table 8.1). In Chapter 15 you will learn other, more broadly useful definitions of acids and bases.

In Chapter 7 you thought of acids as compounds having an $H^+$ ion attached to an anion. You can use this characteristic to identify Arrhenius acids. For example, $HCl$ (thought of as a combination of $H^+$ and $Cl^-$) is a binary acid, and $HNO_3$ ($H^+$ and $NO_3^-$) is an oxyacid. You also know that some acids can produce more than one $H^+$ ion per molecule. For example, $H_2SO_4$ can produce two $H^+$ ions, and $H_3PO_4$ can produce three. Acids that yield more than one $H^+$ ion per molecule are called *polyprotic acids* (*poly-* is a prefix

## CHAPTER BOX 8.1 ARRHENIUS AND GLOBAL WARMING

**The Swedish** chemist Svante Arrhenius made two other important proposals. He was the first to suggest that compounds such as sodium chloride dissociate into individual ions when dissolved in water (Section 4.5). He also pointed out that the carbon dioxide gas in the atmosphere acts as a heat trap, allowing sunlight to reach the earth's surface but preventing the heat energy from radiating back out into space at night. This insulating capacity is the *greenhouse effect*. The controversial idea of global warming, that increases in atmospheric carbon dioxide and other heat-trapping gases from industry and land use are causing the earth's climate to heat up, goes back to this proposal by Arrhenius.

FIGURE 8.10
Svante Arrhenius (1859–1927)

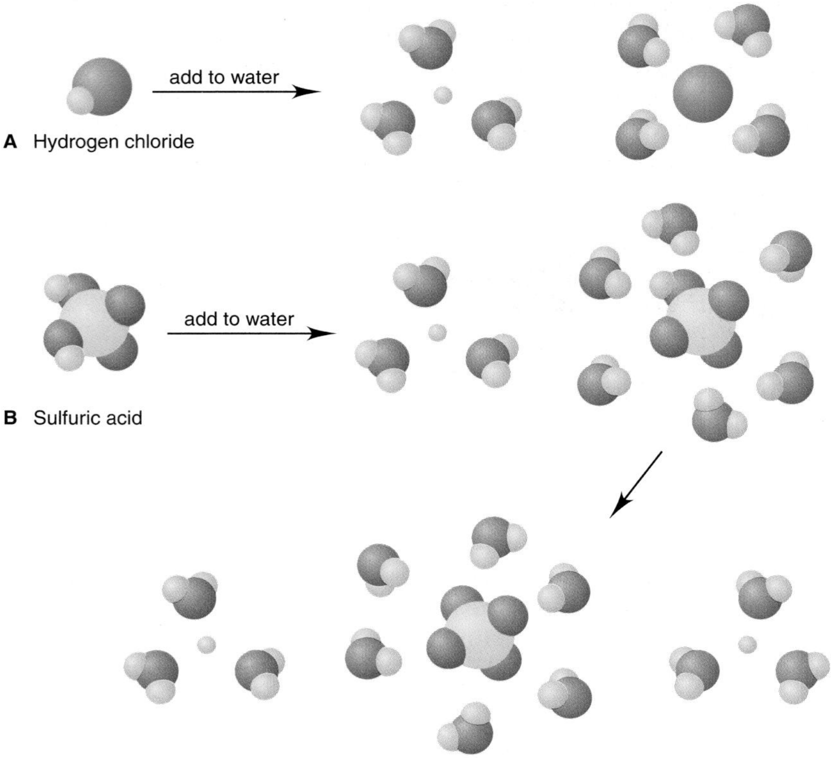

add to water

A Hydrogen chloride

add to water

B Sulfuric acid

FIGURE 8.11
(A) Hydrochloric acid is monoprotic. That is, when it dissolves in water, the covalent HCl molecules form $H^+$ and $Cl^-$ ions. (B) Sulfuric acid is diprotic. When $H_2SO_4$ molecules dissolve in water, they form $H^+$ and $HSO_4^-$ ions. Removing the second $H^+$ ion is more difficult, and the second reaction occurs only to a small extent.

meaning many, and *-protic* comes from the proton, $H^+$). The Greek prefixes *mono-, di-, tri-,* etc., are used for specific acids (see Figure 8.11). For instance, sulfuric acid ($H_2SO_4$) is *diprotic,* and phosphoric acid ($H_3PO_4$) is *triprotic.*

**Similar terms are applied to bases. For example, $Ba(OH)_2$ is *dibasic* because it contains two $OH^-$ ions.**

The common Arrhenius bases are water-soluble hydroxides. There are only a few of these bases—the hydroxides of the Group 1A metals and some slightly soluble hydroxides of Group 2A metals.

### Neutralization Reactions

When aqueous mixtures of Arrhenius acids and bases react, the $H^+$ ion (from the acid) and the $OH^-$ ion (from the base) combine to give water. The other two ions—an anion from the acid and a cation from the base—combine to form a compound called a *salt.* Chemists call this process **neutralization** because the solution no longer contains either an acid or a base when an equivalent amount of each has reacted. According to the Arrhenius definition, once the sources of $H^+$ and $OH^-$ have reacted, the acid and base no longer exist. This reaction is an example of a neutralization reaction:

**STUDY ALERT**

**Even though you can think of a water molecule, $H_2O$, as containing an $H^+$ ion bonded to an anion ($OH^-$), water is neither an acid nor a base in the Arrhenius sense. Rather, it is the product—the final "resting place"—of the reaction of the ions.**

$$HCl + NaOH \longrightarrow H_2O + NaCl$$

Note that the product is simply salt water, which is a solution that contains neither an acid nor a base. Such solutions are said to be **neutral solutions.** The anion of the salt ($Cl^-$) came from the acid, and the cation ($Na^+$) came from the base. Because NaCl is soluble in water (Table 7.5), the final product of the reaction is an aqueous solution that contains $Na^+$ and $Cl^-$ ions. If you evaporate this solution, solid NaCl will be left behind.

A typical acid-base reaction occurs when a person suffering from heartburn takes an antacid tablet. Heartburn is pain caused by stomach acid, which is HCl. Everyone has some acid in the stomach to help digest food. The stomach lining can usually tolerate the acidic conditions, but if some acid moves up into the esophagus, pain can be the result. The treatment for heartburn is to neutralize the acid. Aluminum hydroxide is one of the many bases that are used for this (see Figure 8.12).

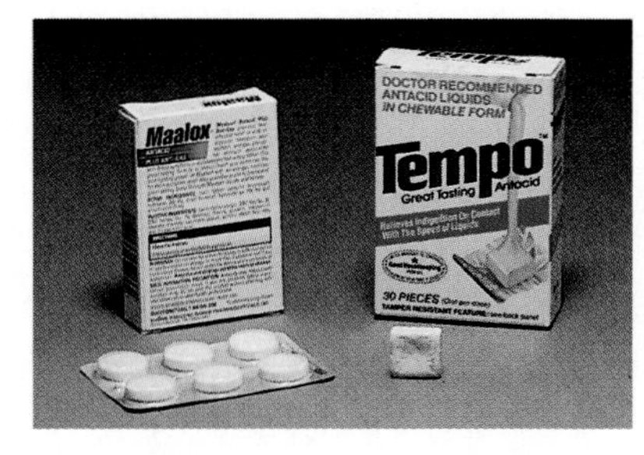


FIGURE 8.12
These antacids contain $Al(OH)_3$.

The Arrhenius theory tells you how to predict the products of this neutralization reaction. The $H^+$ from the hydrochloric acid reacts with the $OH^-$ from the aluminum hydroxide to form water; the aluminum ion and the chloride ion left over form aluminum chloride:

$$HCl + Al(OH)_3 \longrightarrow H_2O + AlCl_3 \quad \text{(unbalanced)}$$

The next example shows how to balance such equations.

## EXAMPLE 8.5 Balancing Equations for Acid-Base Reactions

Balance the equation for the neutralization reaction between HCl and $Al(OH)_3$.

**STUDY ALERT**
**Try to balance the equation yourself before reading further.**

**SOLUTION**

Step 1 calls for you to write the correct formulas, but this has already been done. In Step 2 you balance one of the elements. You can see that aluminum is balanced, but hydrogen, chlorine, and oxygen are not:

$$HCl + Al(OH)_3 \longrightarrow H_2O + AlCl_3$$

In Step 3 you balance the rest of the atoms. Since chlorine appears in only one formula on each side, it's a good idea to balance that next. You can do this by using three HCl molecules instead of one:

$$3\ HCl + Al(OH)_3 \longrightarrow H_2O + 1\ AlCl_3$$

Oxygen appears in only one molecule on each side, so you balance that next, by putting three $H_2O$ molecules on the right instead of one:

$$3\ HCl + Al(OH)_3 \longrightarrow 3\ H_2O + AlCl_3$$

Now aluminum, chlorine, and oxygen are balanced. How about hydrogen? Yes, that's balanced, too, with six atoms on each side of the equation (Step 5).

**EXERCISE 8.5**

Balance the equation for each of these neutralization equations.

**a.** $H_3PO_4 + NaOH \rightarrow Na_3PO_4 + H_2O$
**b.** Sulfuric acid reacts with potassium hydroxide to give potassium sulfate and water.
**c.** $NaHSO_4 + Na_3PO_4 \rightarrow Na_2SO_4 + NaH_2PO_4$
**d.** Sulfuric acid reacts with aluminum hydroxide to give aluminum sulfate and water.

FIGURE 8.13
$Ca(OH)_2$, a solid insoluble in water, reacts with an aqueous solution of nitric acid to give a clear solution.

What do you do if you are given only an acid and a base as reactants and asked to balance the resulting equation? As usual, you work in steps. First, you predict the formulas of the products, and then you can balance the equation. This is easier than it sounds. You know that one of the products is water. The other is the ionic compound made by combining the anion from the acid and the cation from the base. For example, consider the reaction between hydrobromic acid and sodium hydroxide. What are the products, and what is the balanced equation?

First, write the formulas for the reactants followed by the arrow that means "gives":

$$\mathbf{HBr + NaOH \longrightarrow ?}$$

The $H^+$ from the acid and the $OH^-$ from the base combine to give water:

$$HBr + NaOH \longrightarrow \mathbf{H_2O + ?}$$

For the other product, the anion from the acid ($Br^-$) and the cation from the base ($Na^+$) combine to form sodium bromide:

$$HBr + NaOH \longrightarrow H_2O + \mathbf{NaBr}$$

Is this equation balanced? Count the atoms, and you'll see it's a winner.

### EXAMPLE 8.6 Balancing Equations with Polyprotic and Polybasic Compounds

What are the products, and what is the balanced equation, for the reaction between $HNO_3$ and $Ca(OH)_2$?

**SOLUTION**

You can get the solutions to problems of this type by proceeding in an orderly way. First, write the reactants and the arrow:

$$\mathbf{HNO_3 + Ca(OH)_2 \longrightarrow ?}$$

Next, predict the products. One of the products is water, and the other is the salt formed between the calcium and nitrate ions. Use the rules and guidelines given in Chapter 7 to write the correct formula for the salt:

$$HNO_3 + Ca(OH)_2 \longrightarrow \mathbf{H_2O + Ca(NO_3)_2}$$

If you write the formula for calcium nitrate incorrectly (as $CaNO_3$, for example), you will never get a correct, balanced equation. A quick inspection shows that this equation is unbalanced.

You could balance this equation using the guidelines given in the previous section. However, paying attention to the *chemistry* as well as the arithmetic provides an important clue: $HNO_3$ produces only one $H^+$ ion, and $Ca(OH)_2$ produces two $OH^-$ ions. Clearly two $H^+$ ions are necessary to react with both of the $OH^-$ ions, and two molecules of water will be formed. To produce two $H^+$ ions, two $HNO_3$ molecules are needed.

$$\mathbf{2}\ HNO_3 + Ca(OH)_2 \longrightarrow \mathbf{2}\ H_2O + Ca(NO_3)_2$$

Is this equation balanced? When you count the atoms (Step 5) you will find that it is.

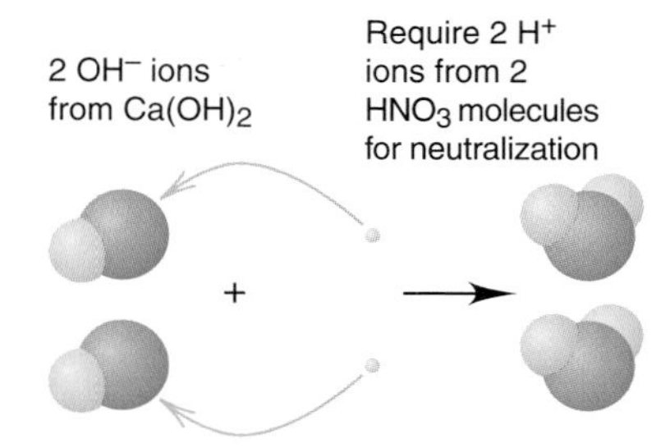


FIGURE 8.14
Looking closely at the chemistry can provide time-saving shortcuts when balancing equations. In this example, $Ca(OH)_2$ contains two $OH^-$ ions, which means that two $H^+$ ions are needed for neutralization. You need two $HNO_3$ molecules to supply these two $H^+$ ions.

**EXERCISE 8.6**

Predict the products and give a balanced equation for each of these reactions:

**a.** $HNO_3 + KOH \rightarrow$

**b.** $H_2SO_4 + Mg(OH)_2 \rightarrow$

**c.** The reaction of perchloric acid with sodium hydroxide

**d.** The reaction of sulfuric acid with potassium hydroxide

## CHAPTER BOX 8.2 SULFURIC ACID, THE INVISIBLE GIANT

**The U.S.** chemical industry makes and uses more sulfuric acid ($H_2SO_4$) than any other chemical, yet consumers almost never see any of it. Manufacturers use this exceptionally strong and corrosive acid to make other things. The fertilizer industry mixes sulfuric acid with phosphate rock in an acid-base reaction. This converts the rock from an inert substance to one that can be absorbed by plants. Various other industries use sulfuric acid in making plastics, dyes, fibers, other acids, and starting materials for a host of industrial processes. The growth of nations can be measured by looking at their consumption of sulfuric acid—developed countries use more sulfuric acid than less developed ones do.

Although sulfuric acid is very useful in industry, it has harmful effects, as well. Both the burning of coal and other fuels and the refining of copper and other metals add sulfuric acid to the atmosphere. It comprises most of the acid in acid rain. How can the acid be removed before it does any damage? One method involves adding a base to the gases in smokestacks. The acid-base reaction destroys the sulfuric acid before it reaches the atmosphere.

Do you want to see some sulfuric acid at a safe distance? The surface of the bright planet Venus had to be mapped by radar, because it is hidden by a blanket of permanent clouds, which are mostly sulfuric acid.

FIGURE 8.15
(Left) The clouds of Venus. In visible light Venus is a featureless ball. This photo was made with an ultraviolet light filter and is artificially colored and enhanced to show the cloud structure. (Right) The surface of Venus. Radar mounted on the spacecraft Magellan penetrated the clouds to produce this image.

## 8.5 Oxidation-Reduction Reactions: Combustion

**GOAL:** To introduce a type of simple oxidation-reduction reaction called combustion reactions and to write balanced equations for these reactions.

These original definitions of oxidation and reduction have been refined and extended over the years; see Chapter 16.

Oxidation and reduction, like neutralization, are important to society. These reactions produce most of the energy we use to cook, light our homes, and propel our cars. Batteries (and most power plants) work, iron rusts, and you convert your meals into the energy of life through these reactions. (We will discuss oxidation-reduction reactions in much more depth in Chapter 16.)

According to the oldest and simplest definition, an element or compound that gains oxygen in a reaction is **oxidized.** Similarly, an element or compound is **reduced** when it loses oxygen. By now you know enough about the conservation of mass to realize that if one substance gains oxygen atoms, another substance must lose them. Therefore, if something is oxidized, something else must be reduced at the same time. The reverse is also true.

The iron industry uses this reaction in making iron metal from iron ore (see Figure 8.16):

$$Fe_2O_3 + 3\ CO \longrightarrow 2\ Fe + 3\ CO_2$$

If you compare the reactants and the products, you will see that $Fe_2O_3$ loses its oxygen atoms, and therefore $Fe_2O_3$ is reduced. Carbon monoxide gains an oxygen atom to form $CO_2$, so carbon monoxide is oxidized.

How do these definitions apply to the following reaction, which occurs when charcoal burns in a backyard barbecue?

$$C + O_2 \longrightarrow CO_2$$

It is easy to see that carbon is oxidized, because it gains two oxygen atoms. What is reduced? The $O_2$ molecule is reduced because it is destroyed in the reaction with carbon (see Figure 8.17). Oxygen is almost always reduced when it reacts.

The reaction between carbon and oxygen is an example of a *combustion reaction,* an especially useful type of oxidation-reduction reaction. **Combustion** is the reaction that occurs when a substance burns in oxygen (or air). The substance being burned is oxidized, and oxygen is reduced by these reactions.

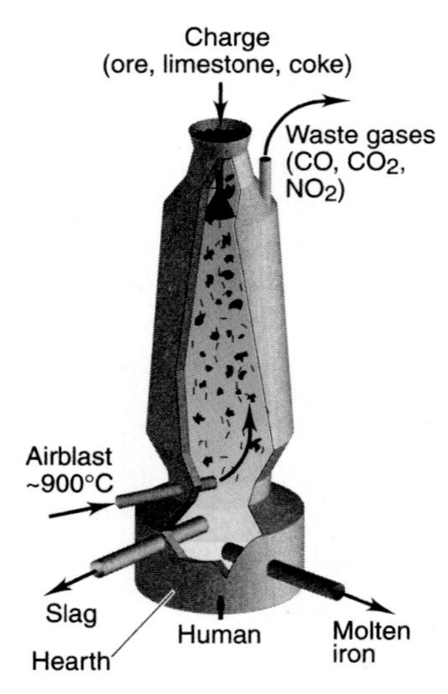


FIGURE 8.16
A blast furnace takes iron ore, limestone ($CaCO_3$), coke (carbon), and air and produces molten iron along with slag and various gases. First, oxygen in the airblast reacts with the carbon to make carbon monoxide (CO) and heat; these do the rest.

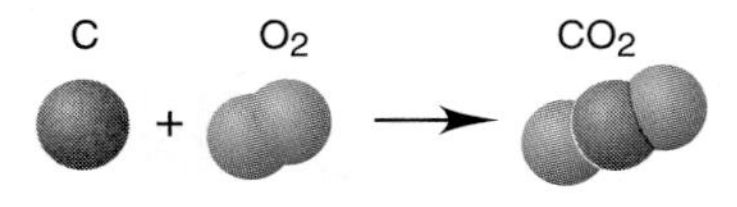


FIGURE 8.17
Carbon gains two oxygen atoms and is oxidized. All the atoms in $O_2$ are "lost," so it is reduced.

Humans have used combustion reactions for hundreds of thousands of years. We still burn wood in campfires, coal in power plants, natural gas in stoves, and gasoline in the engines of cars. We almost always use combustion reactions to produce heat. The fuel is typically a compound containing carbon and hydrogen or carbon, hydrogen, and oxygen (see Figure 8.18).

The first equation presented in this chapter showed the reaction of methane with oxygen. Methane is the main component of natural gas, and this reaction is a typical combustion reaction:

$$CH_4 + 2\,O_2 \longrightarrow CO_2 + 2\,H_2O$$

Note that the products are carbon dioxide and water. Combustion reactions almost always give these two products when the reactants contain only carbon and hydrogen and usually do so even when the carbon-containing reactant also includes oxygen. Once you know this, you can easily predict the products of a combustion reaction. For example, ethanol, $C_2H_5OH$, is sometimes mixed in gasoline. The equation for the combustion of ethanol is

$$C_2H_5OH + 3\,O_2 \longrightarrow 2\,CO_2 + 3\,H_2O$$

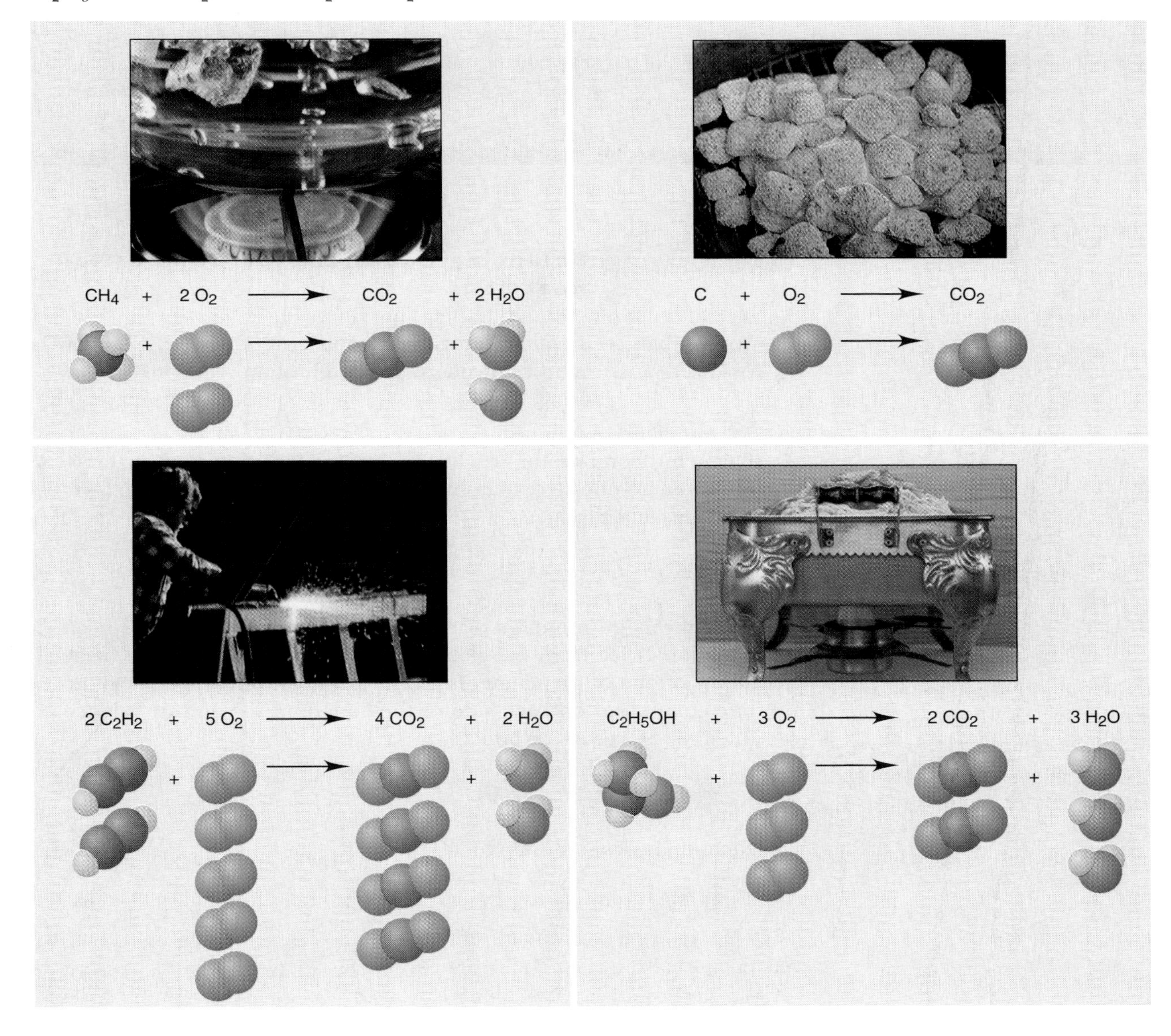


FIGURE 8.18
These are typical combustion reactions. Methane, $CH_4$, burns in a gas stove (upper left). Charcoal, pure carbon, burns in barbecues (upper right). A welder's torch burns acetylene, $C_2H_2$ (lower left), and food can be kept warm by burning a jelly-like substance that is primarily alcohol, $C_2H_5OH$ (lower right).

## CHAPTER BOX 8.3 MORE ABOUT ACID RAIN

**Most coal** contains small amounts of sulfur or compounds containing sulfur, or both. When coal burns, sulfur dioxide is a product:

$$S + O_2 \longrightarrow SO_2$$

Sulfur dioxide also forms as a product when refineries heat metal-containing sulfide ores in air:

$$Cu_2S + O_2 \longrightarrow 2\ Cu + SO_2$$

This process, called *roasting*, was a common early industrial method of obtaining pure metals from sulfide ores. The sulfur dioxide then reacts with more oxygen in the air to form sulfur trioxide.

$$2\ SO_2 + O_2 \longrightarrow 2\ SO_3$$

This sulfur trioxide reacts with water vapor to form sulfuric acid:

$$SO_3 + H_2O \longrightarrow H_2SO_4$$

Most of the sulfuric acid in the atmosphere comes from this reaction (a small amount comes from active volcanoes), and this is why unrestricted emissions of sulfur dioxide can devastate the environment (see Figure 8.19).

FIGURE 8.19
This spot near the metal smelters of Mt. Mitchell, North Carolina was once a beautiful forest. Now that emission controls are in use, the area is recovering.

### EXAMPLE 8.7 Balancing Equations for Combustion Reactions

Write a balanced equation for the combustion of octane, $C_8H_{18}$. What is oxidized, and what is reduced, in this combustion reaction?

**SOLUTION**

Balancing combustion reactions looks trickier than it actually is. As always, you go one step at a time. In Step 1 you write down the formulas of reactants and products:

$$C_8H_{18} + O_2 \longrightarrow CO_2 + H_2O$$

With the large numbers of carbon and hydrogen atoms in octane, the equation is far from balanced at this point. In Step 2 you balance the atoms of one of the elements. Both carbon and hydrogen appear in only one compound on each side of the equation, so you can balance either one first. Start with carbon:

$$C_8H_{18} + O_2 \longrightarrow 8\ CO_2 + H_2O$$

Next, do hydrogen (Step 3):

$$C_8H_{18} + O_2 \longrightarrow 8\ CO_2 + 9\ H_2O$$

Now you need to balance the oxygens (Step 3, continued). The products contain a total of 25 oxygen atoms, 16 in the eight $CO_2$ molecules and 9 in the nine $H_2O$ molecules. To balance oxygen you need $^{25}/_2$ oxygen molecules (because $^{25}/_2$ $O_2$ molecules contain 25 O atoms):

$$C_8H_{18} + {}^{25}/_2\,O_2 \longrightarrow 8\,CO_2 + 9\,H_2O$$

This version of the equation is balanced, with 8 carbon atoms, 18 hydrogen atoms, and 25 oxygen atoms on each side. However, it contains a fractional coefficient. In Step 4 you double all of the coefficients to eliminate the fraction. Don't forget to double *all* of the coefficients:

$$2\,C_8H_{18} + 25\,O_2 \longrightarrow 16\,CO_2 + 18\,H_2O$$

Step 5 check shows that this is the balanced equation. The atoms in octane gain oxygen, so octane is oxidized. Oxygen is reduced, as it usually is.

**EXERCISE 8.7**

**a.** For each of these reactions, predict the product or products and tell what species is oxidized.
   1. Sulfur burns in air.
   2. You burn a paraffin wax candle (composed of compounds containing carbon and hydrogen).
   3. You turn on a butane ($C_4H_{10}$) lighter.

**b.** Write a balanced equation for the combustion of each of these substances.
   1. Pentane, $C_5H_{12}$
   2. Methanol, $CH_3OH$
   3. Propane, $C_3H_8$

## 8.6 The Reactions of Active Metals and Water

**GOAL:** To write balanced equations for another type of simple oxidation-reduction reaction—the reactions of active metals with water.

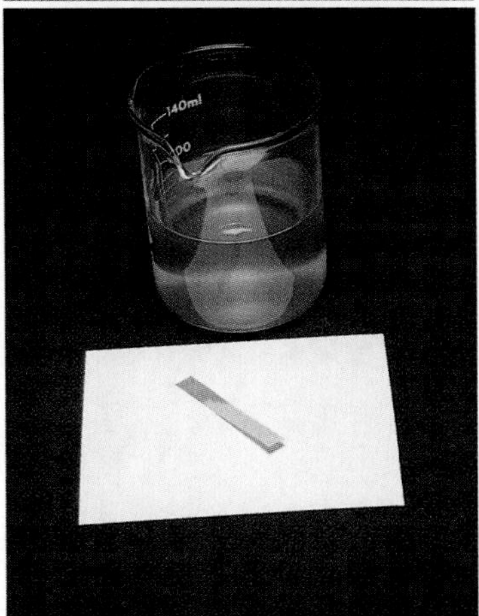

FIGURE 8.20
When sodium metal comes in contact with water, a vigorous reaction occurs (above). The gas bubbles are hydrogen, and the solution contains sodium hydroxide, which also forms in the reaction. The fact that the solution is basic can be confirmed by a litmus test (below).

In Section 8.4 you learned that many metals react with acids to form hydrogen gas. Some of the more reactive metals such as sodium react with ordinary water to form hydrogen gas (see Figure 8.20). All of the metals in Group 1A and the heavier metals from Group 2A (Ca, Sr, and Ba) react in this way. The products are the hydroxide of the metal and hydrogen gas. Metals that give this reaction when you mix them with water are called *active metals.*

The next example shows how to write the balanced equation for the reaction of sodium metal with water.

## EXPERIMENTAL EXERCISE

## 8.1

A.L. Chemist and a scientific team were asked to analyze a liquid sample that had a density of 2.0 g $mL^{-1}$. One member of the team added a few drops of the sample to some iron filings and reaction 1 occurred. One of the products of reaction 1 was the colorless gas 1, and the other product was iron(II) hydrogen sulfate (see Figure 8.21).

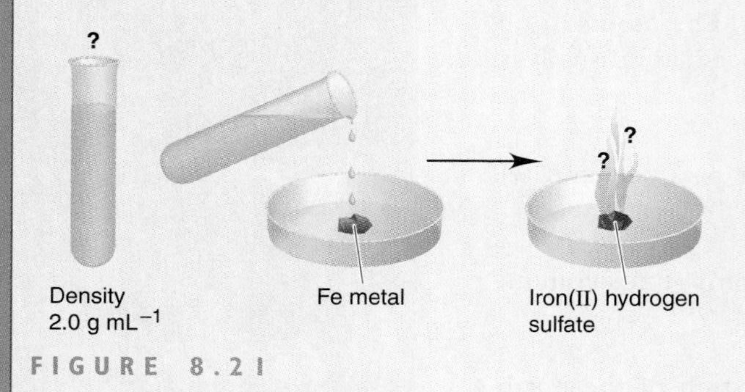


FIGURE 8.21

The team of chemists collected gas 1 and mixed it with air in a test chamber. When they lit a spark in this mixture of gases, a very vigorous reaction occurred. One team member noted that water condensed at the bottom of the test chamber (see Figure 8.22).

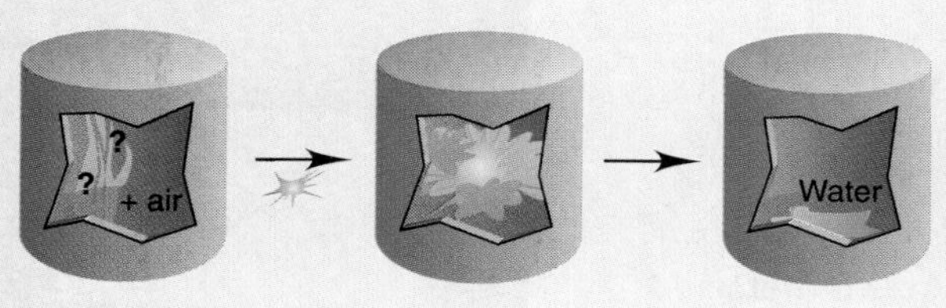


FIGURE 8.22

Another team member prepared solution A by dissolving 1.0 g of the original liquid sample in 100 mL of water. This chemist then dripped a portion of solution A on a piece of calcium carbonate. A chemical reaction (reaction 2) occurred, and the colorless gas 2 was given off. Gas 2 was also mixed with air in a test chamber, but a spark in this mixture of gases gave no reaction at all (see Figure 8.23).

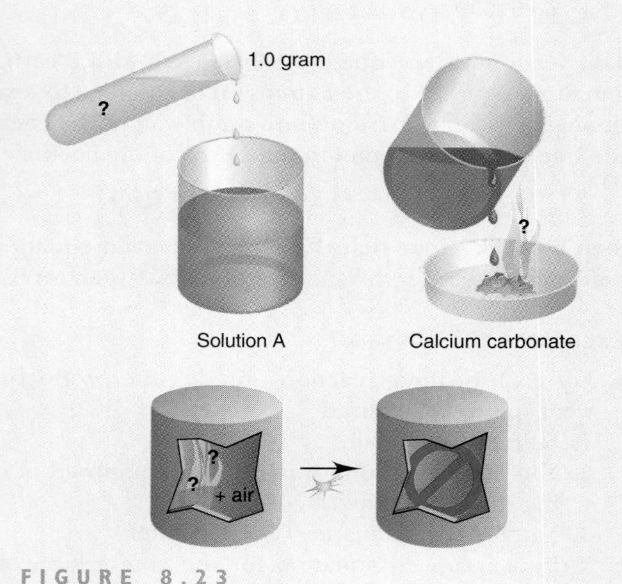


FIGURE 8.23

The team concluded that gas 1 was hydrogen and that the unknown sample was sulfuric acid.

### QUESTIONS

1. What volume of the original liquid sample was used to prepare solution A?
2. Why did the team conclude that the sample was sulfuric acid rather than any of the other common acids ($HNO_3$, HCl, etc.)?
3. Is the team's conclusion that gas 1 is hydrogen gas supported by experimental evidence?
4. What is the identity of gas 2?
5. Give a balanced equation for the reaction between the original liquid sample and iron.
6. What would you observe if the original sample were tested with red litmus paper?
7. What would you observe if solution A were tested with blue litmus paper?

### EXAMPLE 8.8 Balancing Equations for Reactions of Active Metals and Water

Predict the products of the reaction of sodium metal with water, and balance the equation.

**SOLUTION**

First, of course, you write the unbalanced equation with correct formulas. Sodium is a Group 1A metal, so its hydroxide is NaOH. The unbalanced equation is

$$Na + H_2O \longrightarrow NaOH + H_2$$

Second, you balance an element. Sodium and oxygen are balanced, but hydrogen is not. So the next step is to balance the H atoms:

$$Na + H_2O \longrightarrow NaOH + \tfrac{1}{2}\, H_2$$

Doubling all the coefficients to remove the fractional coefficient gives the final balanced equation:

$$2\ Na + 2\ H_2O \longrightarrow 2\ NaOH + H_2$$

**EXERCISE 8.8**

Write a balanced equation for the reaction of each of these active metals with water.

**a.** Ca **b.** Rb

## MORE INFORMATION FROM EQUATIONS

Now you know how to predict what happens when two chemical substances are mixed together, in some cases at least. You probably also know that there is more to chemical reactions than that. You can mix either hydrogen or methane with oxygen without getting a reaction. Gasoline (fortunately) doesn't burst spontaneously into flame as you pump it into your gas tank. Something else is needed, either a spark or another substance. Furthermore, combustion produces carbon dioxide and water, but these are almost always discarded; the usual reason for burning anything is to get heat or energy from the reaction. The next sections introduce these topics in more depth.

## 8.7 Reaction Conditions

**GOAL:** To write and interpret chemical equations in which the states of reactants and products, the reaction conditions, and the formulas of catalysts are given.

You know how to write and balance equations for several kinds of reactions, but there is much more information about reactions that is useful. Are the reactants and products solids, liquids, or gases? Are they dissolved in water? Is heat, light, high pressure, or anything else required to make the reaction work the way you want it to? Some reactions are speeded up by the presence of **catalysts,** which are substances that cause a reaction to go faster without being consumed in the reaction. All of these factors are collectively known as *reaction conditions.*

This section shows how to indicate the reaction conditions when writing equations. Whether or not you write the conditions explicitly depends on how much information you need to convey in a given situation.

You already know the physical states of many substances at room temperature and pressure, and you can figure out the states of many others by using common sense and your knowledge of the typical physical properties of ionic and covalent compounds. Chemists run most acid-base reactions, for instance, by dissolving the reactants in water and then mixing the solutions. Some of the products of these reactions are also water-soluble. To indicate that a compound is dissolved in water, chemists write ($aq$), for **aqueous,** immediately after the formula of the compound in question. Acid-base reactions are usually run at room temperature, and therefore water forms as a liquid, indicated by ($l$). For example, the reaction between hydrochloric acid and sodium hydroxide can be rewritten as follows:

$$HCl(aq) + NaOH(aq) \longrightarrow H_2O(l) + NaCl(aq)$$

Often one or more of the substances involved in a reaction is insoluble. Many hydroxides (used as reactants) and many salts (products) are insoluble. (You should review the solubility rules in Section 7.9 if you need them.) When a solid substance reacts or forms in a reaction, chemists indicate its solid nature by writing ($s$), for "solid," immediately after the formula. Insoluble gases are indicated with a ($g$) after the formula. The reaction between hydrochloric acid and aluminum hydroxide illustrates how soluble and insoluble substances are identified in a chemical equation:

**STUDY ALERT**

**HCl ($aq$) means that hydrogen chloride, HCl ($g$), is dissolved in water. You learned in Chapter 7 that an aqueous solution of hydrogen chloride is called hydrochloric acid. HCl($aq$) does *not* mean liquid hydrogen chloride; that is written as HCl($l$).**

$$HCl(aq) + Al(OH)_3(s) \longrightarrow 3\ H_2O(l) + AlCl_3(aq)$$

Hydrogen chloride gas, $HCl(g)$, dissolves in water to make hydrochloric acid, $HCl(aq)$. Water, of course, is a liquid. The chloride salt $AlCl_3$, like most metal chlorides, is soluble in water. $Al(OH)_3$ is one of the many insoluble hydroxides, and its solid state is indicated by ($s$).

Combustion reactions frequently take place in the gas phase. Many of the reactants are gases, including hydrogen and methane. Octane is a liquid at room temperature, but it vaporizes before it burns. The reactant oxygen and the product carbon dioxide are both gases. The other product of a combustion reaction is water, produced at the high temperatures of flames; water is also a gas under those circumstances. Here's how all this information is included in the equation for the combustion of octane:

**STUDY ALERT**

**There are no specific guidelines telling when an equation should include the states of reactants and products or the conditions under which reactions run. Whether or not these things are shown depends on how much information the reader needs to have about a particular chemical reaction. As a student, you must be able to include the conditions when asked to do so, and you must be able to interpret equations that contain conditions.**

$$2\ C_8H_{18}(g) + 25\ O_2(g) \longrightarrow 16\ CO_2(g) + 18\ H_2O(g)$$

When a reaction requires heat, the word *heat* is written over or under the arrow:

$$2\ C_8H_{18}(g) + 25\ O_2(g) \xrightarrow{\text{heat}} 16\ CO_2(g) + 18\ H_2O(g)$$

$$C(s) + O_2(g) \xrightarrow[\text{heat}]{} CO_2(g)$$

The second equation, for the combustion of carbon, translates as follows: "One mole of solid carbon reacts with one mole of oxygen gas when heated, to yield one mole of carbon dioxide gas." In general, chemists write any conditions that are necessary to make a reaction run efficiently, or at all, over or under the arrow. The preparation of ammonia from hydrogen and nitrogen is another example:

$$N_2(g) + 3\,H_2(g) \xrightarrow[\text{pressure}]{\text{heat}} 2\,NH_3(g)$$

Catalysts are substances that make reactions work more quickly and efficiently. A car's catalytic converter takes unburned and partly burned fuel that would otherwise become air pollution and combines it with oxygen to give $CO_2$ and $H_2O$ (see Figure 8.24). Some combustion reactions run more efficiently and at lower ignition temperatures with a catalyst present. With platinum metal as the catalyst, the equation for the combustion of methane is

$$CH_4(g) + 2\,O_2(g) \xrightarrow{\text{Pt}} CO_2(g) + 2\,H_2O(g)$$

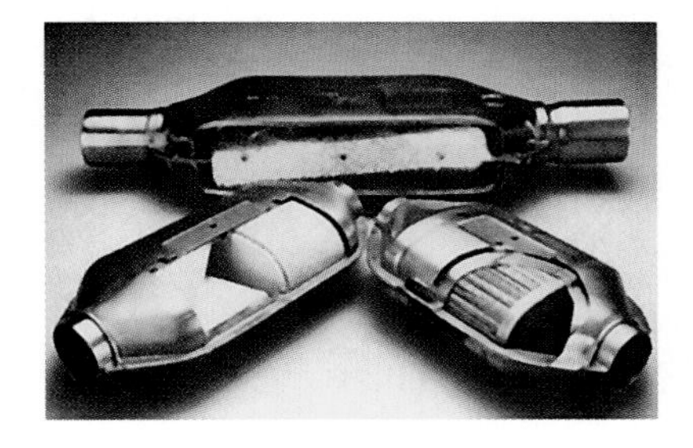

FIGURE 8.24
Catalytic converters are part of an automobile's exhaust system. In addition to converting partly burned fuel to $CO_2$ and $H_2O$, these converters also reduce oxides of nitrogen (NO and $NO_2$), converting them back to $N_2$ and $O_2$.

EXAMPLE 8.9 **Reading Chemical Equations That Contain Reaction Conditions**

Translate this equation for the combustion of methane into an English sentence.

$$CH_4(g) + 2\,O_2(g) \xrightarrow{\text{Pt}} CO_2(g) + 2\,H_2O(g)$$

**SOLUTION**

Translating the chemical equation for the combustion of methane in the presence of platinum into an English sentence requires a simple extension of what you have already learned. You simply include all conditions and states of matter in the sentence:

One mole of methane gas reacts with two moles of oxygen gas
in the presence of a platinum catalyst to yield
one mole of carbon dioxide gas and two moles of water vapor.

**EXERCISE 8.9**

**a.** Convert each of these equations into an English sentence.
   **1.** $2\,H_2O(l) \xrightarrow{\text{electric current}} 2\,H_2(g) + O_2(g)$
   **2.** $2\,SO_2(g) + O_2(g) \xrightarrow[\text{heat}]{V_2O_5} 2\,SO_3(g)$

**b.** For each of these reactions, write a balanced equation that includes the reaction conditions.
   **1.** One mole of nitrogen gas reacts with three moles of hydrogen gas at 450°C and 270 atmospheres to give two moles of ammonia gas.
   **2.** One mole of methane gas reacts with one mole of $H_2O$ gas in the presence of a nickel catalyst at 700°C to give one mole of carbon monoxide gas and three moles of hydrogen gas.
   **3.** The reaction of iron(III) hydroxide with aqueous nitric acid
   **4.** The reaction of solid cesium metal with water

## 8.8 Net Ionic Equations

**GOAL:** To write a balanced equation for a reaction in aqueous solution that shows only the species that actually take part in the reaction.

In previous sections you learned how to write balanced equations that provide information to readers about what happens in a reaction in terms of what an observer can see. This section considers some chemistry that is invisible to the eye. What actually happens in a flask when a reaction in aqueous solution takes place?

Remember from Chapter 7 that both hydroxides and salts are ionic compounds—made up of separate charged particles called *ions* that are held together by the attraction of opposite charges. A few ionic compounds dissolve in water because the water molecules can pull the ions apart (see Section 7.9). If you dissolve solid sodium chloride, $NaCl(s)$, in water, what you really have are the ions $Na^+(aq)$ and $Cl^-(aq)$.

When an Arrhenius acid dissolves in water, it reacts with a water molecule to form the hydronium ion, $H_3O^+$. For example, when $HCl(g)$ dissolves in water, this reaction occurs:

$$HCl(g) + H_2O(l) \longrightarrow H_3O^+(aq) + Cl^-(aq)$$

Acids that react in this manner are called *strong acids:* A **strong acid** reacts completely with water to form $H_3O^+(aq)$ and the appropriate aqueous anion. Only a few acids are classified as strong, and these are listed below.

**Strong Acids**

HX (X = Cl, Br, or I)
$HNO_3$
$HClO_4$
$H_2SO_4$ (one H only)

**STUDY ALERT**
**Acid strength (and weakness) is defined by how the individual substance behaves chemically when it *reacts* with water. All the acids you have studied react completely with base ($OH^-$) to *yield* water as one of the products. Examine closely the definitions of Arrhenius acid, strong acid, and weak acid. Be certain that the differences in the definitions are clear to you.**

Note that, although sulfuric acid is diprotic, only one of its acidic protons reacts with water. The reaction that occurs when pure sulfuric acid dissolves in water is

$$H_2SO_4(l) + H_2O(l) \longrightarrow HSO_4^-(aq) + H_3O^+(aq)$$

A **weak acid** is an acid that reacts only slightly with water. Acetic acid ($CH_3COOH$), the active ingredient in vinegar, is a common weak acid. In 0.1 M aqueous solutions of acetic acid, only about 1% of the acetic acid molecules react to form hydronium ions and acetate ions. The rest of the acetic acid (99%) remains in solution as discrete $CH_3CO_2H$ molecules. Other weak acids are HF, the ammonium ion ($NH_4^+$), and the hydrogen sulfate ion ($HSO_4^-$).

In summary, strong acids react completely with water, while weak acids react with water only slightly. You will find out more about the differences between strong and weak acids in Chapter 15.

The most common **strong bases** are water-soluble metal hydroxides—including the Group 1A metal hydroxides, for example, NaOH, and the few Group 2A hydroxides that are slightly soluble, such as $Ca(OH)_2$. These are strong bases because they completely ionize in water to give metal ions and hydroxide ions.

$$NaOH(s) \xrightarrow{H_2O} Na^+(aq) + OH^-(aq)$$

To find out about weak bases, see Chapter 15.

In view of all this, let's reexamine the reaction between hydrochloric acid and aqueous sodium hydroxide.

EXAMPLE 8.10

## Writing Equations for Reactions Involving Ionic Species

Rewrite the equation for the reaction between hydrochloric acid and aqueous sodium hydroxide to show only the ions that actually react. Remember that hydrogen chloride is $HCl(g)$ and hydrochloric acid is $HCl(aq)$. With all components shown, the equation is

$$HCl(aq) + NaOH(aq) \longrightarrow H_2O(l) + NaCl(aq)$$

SOLUTION

Your task is to rewrite this equation so that it indicates what actually happens *in solution* when the reaction occurs. The equation given above tells you what a chemist might mix to make the solution, but you need to focus on what takes place in the solution.

Since hydrochloric acid is a strong acid, $HCl(aq)$ actually consists of $H_3O^+(aq)$ and $Cl^-(aq)$. Sodium hydroxide is a water-soluble metal hydroxide, and $NaOH(aq)$ exists in solution as $Na^+(aq)$ and $OH^-(aq)$. Similarly, $NaCl(aq)$ is $Na^+(aq)$ and $Cl^-(aq)$. Only $H_2O(l)$ doesn't ionize. The complete ionic equation is cumbersome-looking.

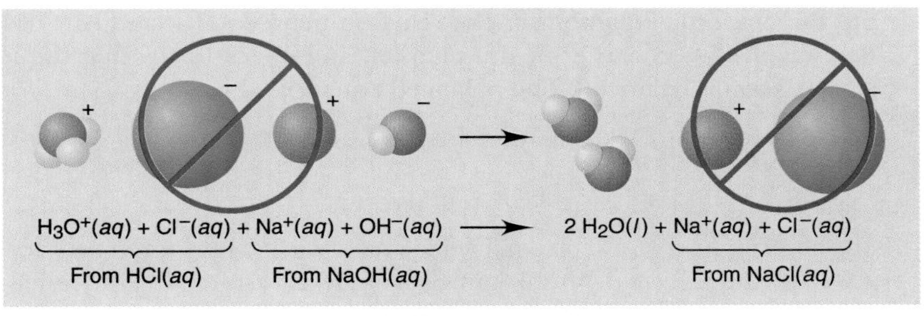


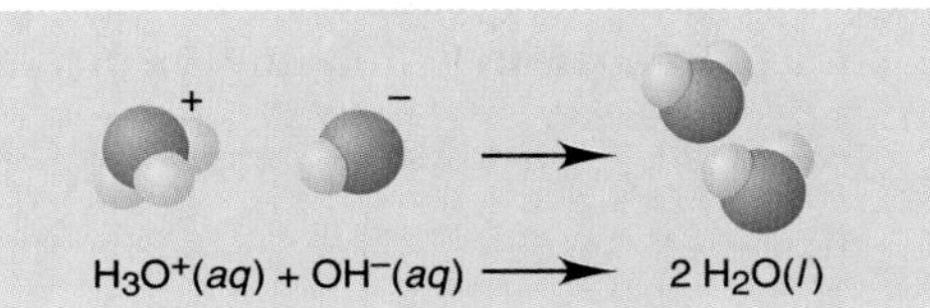


Note that there is one more $H_2O$ molecule among the products in the net ionic equation, as compared to the balanced equation given in the text above. The extra $H_2O$ is due to the fact that an $H_2O$ molecule went into forming $H_3O^+$ when $HCl(g)$ reacted with water.

In this equation, two species, $Na^+(aq)$ and $Cl^-(aq)$, don't change from one side of the equation to the other. You can cancel them out—subtract

Water does ionize to an extremely small extent (see Chapter 15). We are on safe turf to ignore it here.

them from both sides of the equation—to simplify it considerably.

$$H_3O^+(aq) + \cancel{Cl^-(aq)} + \cancel{Na^+(aq)} + OH^-(aq) \longrightarrow 2\,H_2O(l) + \cancel{Na^+(aq)} + \cancel{Cl^-(aq)}$$

The result is called a **net ionic equation,** because it shows only the ions involved in a chemical reaction:

$$H_3O^+(aq) + OH^-(aq) \longrightarrow 2\,H_2O(l)$$

***Net*** **has the same meaning here as it does in the phrase *net profit*—we have removed the things that have no effect.**

The formation of water is the only reaction that ever happens when strong acids react with hydroxide ions in water. The other ions present act as the sodium and chloride ions did in this example. No chemical change happens to them. They are called **spectator ions,** because in a sense they "watch" the reaction occur without participating.

**EXERCISE 8.10**

Write the net ionic equation and identify the spectator ions for the reaction that occurs when $HNO_3(aq)$ is added to $KOH(aq)$.

**STUDY ALERT**

**Although the charges on each side of an equation don't have to add up to zero, they do have to add up to the same number.**

The net ionic equation $H_3O^+(aq) + OH^-(aq) \rightarrow 2\,H_2O(l)$ demonstrates a vital principle—the **Conservation of Electrical Charge.** Charge, like mass and energy, is neither created nor destroyed in a chemical reaction. In this equation the reactants have a total charge of zero, consisting of one positive charge ($H_3O^+$) and one negative charge ($OH^-$). Since neither water molecule has a charge, the products also have a total charge of zero. In any chemical equation the total charges to the left of the arrow must equal the total charges to the right of the arrow.

**EXAMPLE 8.11** **Checking Net Ionic Equations for Charge Balance**

Write the net ionic equation for the reaction between HCl and $Al(OH)_3$, which was discussed earlier in this chapter, and check to see that the net ionic equation is balanced. The balanced equation is

$$3\,HCl(aq) + Al(OH)_3(s) \longrightarrow 3\,H_2O(l) + AlCl_3(aq)$$

**SOLUTION**

Hydrochloric acid is a strong acid and reacts with water to form $H_3O^+(aq)$ and $Cl^-(aq)$. Aluminum hydroxide is insoluble, so it remains as $Al(OH)_3(s)$. $AlCl_3$ is a soluble chloride salt that gives $Al^{3+}(aq)$ and $Cl^-(aq)$ in water. The complete ionic equation for this reaction is

$$3\,H_3O^+(aq) + 3\,Cl^-(aq) + Al(OH)_3(s) \longrightarrow 6\,H_2O(l) + Al^{3+}(aq) + 3\,Cl^-(aq)$$

Subtracting the spectator ions from both sides of this equation gives the net ionic equation:

$$3\,H_3O^+(aq) + Al(OH)_3(s) \longrightarrow 6\,H_2O(l) + Al^{3+}(aq)$$

It is a very useful practice to check the final result for charge balance. If the charges don't balance, there must be an error in your equation. Net ionic equations must balance with respect to both mass and charge. Note that the total charge on the left is +3, resulting from a +1 charge on each of the three $H_3O^+$ ions. The charge on the right is also +3, from the single $Al^{3+}$ ion. The charges on the two sides of the equation are equal, so electrical charge is conserved. The masses also balance; there are 12 hydrogen atoms, 6 oxygen atoms, and one aluminum atom on each side of the equation.

**EXERCISE 8.11**

Write each of these equations in net ionic form, showing the states of all species present.

**a.** $2\ HNO_3 + Ca(OH)_2 \xrightarrow{H_2O} 2\ H_2O + Ca(NO_3)_2$

**b.** $2\ Na + 2\ H_2O \longrightarrow 2\ NaOH + H_2$

**c.** $AgNO_3(aq) + NaCl(aq) \longrightarrow AgCl(s) + NaNO_3(aq)$

**d.** $(NH_4)_2S(aq) + Hg(NO_3)_2(aq) \longrightarrow 2\ NH_4NO_3(aq) + HgS(s)$

**e.** $Ba(NO_3)_2 + Na_2SO_4 \xrightarrow{H_2O} BaSO_4 + 2\ NaNO_3$

**f.** $Hg_2(NO_3)_2 + 2\ KCl \xrightarrow{H_2O} Hg_2Cl_2 + 2\ KNO_3$

## 8.9 Energy as Reactant and Product

**GOAL:** To write balanced equations that include energy as either a reactant or a product.

People usually use combustion reactions to produce heat, which is then used to generate power, move a vehicle, cook food, or simply provide warmth. Most balanced equations show only the mass and charge balance, but sometimes it's useful to show the amount of energy as a reactant or product.

Here, for one last time in this chapter, is the equation for the combustion of one mole of methane, with the amount of energy produced included:

$$CH_4(g) + 2\ O_2(g) \longrightarrow CO_2(g) + 2\ H_2O(g) + \mathbf{890\ kJ}$$

**When energy is included in an equation, the other reactants and products are usually interpreted in terms of moles rather than molecules.**

In other words, in addition to producing carbon dioxide and water, the combustion of one mole of methane produces 890 kJ of energy. Reactions that give energy as a product are **exothermic reactions.**

This reaction demonstrates the third fundamental conservation principle used by chemists, the **conservation of energy.** The potential energy of the reactants is converted to heat energy, which is written explicitly here as one of the products.

Some reactions consume energy rather than producing it. One example is the reaction of one mole of hydrogen with one mole of iodine to produce hydrogen iodide in the gas phase:

$$H_2(g) + I_2(g) + \mathbf{52\ kJ} \longrightarrow 2\ HI(g)$$

Reactions that use energy as if it were a reactant are called **endothermic reactions.**

## CHAPTER BOX 8.4 WHAT'S WRONG WITH $H^+$?

**Many chemists** write the formula for the $H^+$ ion as a reactant or product because it's a convenient piece of shorthand. However, nothing that even vaguely resembles the $H^+$ ion can exist in solution. The reason why this ion doesn't exist in solution involves electrostatic forces, as do so many reasons in chemistry.

Consider the smallest common cation, $Li^+$. Its nucleus contains three protons and four neutrons, and two electrons surround the nucleus. The radius of this lithium ion is about half the radius of the lithium atom. The two electrons that surround the nucleus help to maintain the size of the ion.

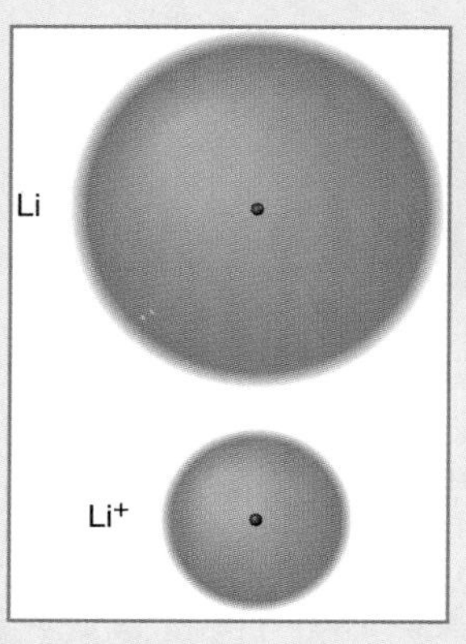


FIGURE 8.25
Lithium atoms have about twice the diameter of lithium ions. The three electrons (two 1*s*, one 2*s*) in the atom and the two electrons (1*s*) in the ion give these objects their size.

The species $H^+$ consists of a single proton and no electrons at all. Its nucleus is surrounded by nothing. The radius of $Li^+$ is about $10^{-10}$ m, but the radius of $H^+$ is about $10^{-15}$ m. The $H^+$ ion is about 100,000 times smaller than a hydrogen atom.

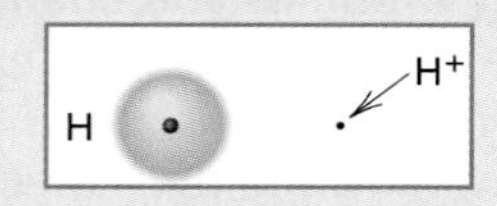


FIGURE 8.26
The $H^+$ ion is not drawn to scale because it would be too small to see. Its diameter is about one hundred thousand times smaller than the $Li^+$ ion in Figure 8.25.

Imagine what happens to an atom that has electrons as it approaches a lithium ion. The two electrons of $Li^+$ will begin to push the approaching atom's electrons away before the atom gets very close to the lithium nucleus. Although the three protons in the lithium nucleus have a larger charge than the two electrons, the electrons are much closer to the approaching atom. Like charges repel one another, and all electrons have negative charges.

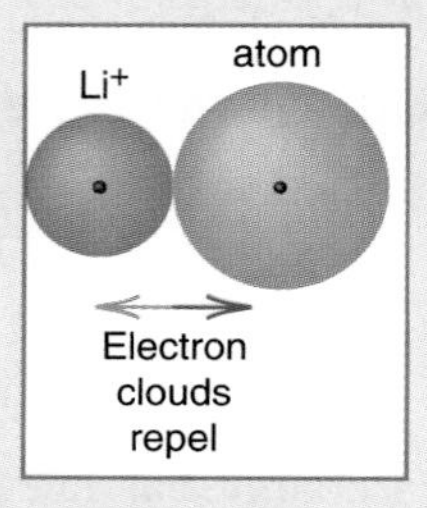


FIGURE 8.27
When another atom (or even an anion) comes close to a lithium ion, the 1*s* electrons in the $Li^+$ repel the valence electrons in the atom.

Now consider what happens when an atom approaches an $H^+$ ion. There are no electrons around $H^+$ to repel approaching electrons. The positive charge of the hydrogen nucleus is the only charge on $H^+$, and it can only attract electrons.

To appreciate the difference in the forces of attraction between approaching electrons and the nuclei of the two ions, think about the following situation. You are holding a magnet, and there is another magnet a mile away. How strongly will the magnets pull on each other? Now move the two magnets until they're half an inch apart. That's how great the difference is. So, if an $H^+$ ion appears somehow in a solution, it immediately fastens onto any electrons that happen to be near it. In aqueous solution, the most readily available electrons are the ones on the oxygen atom of $H_2O$.

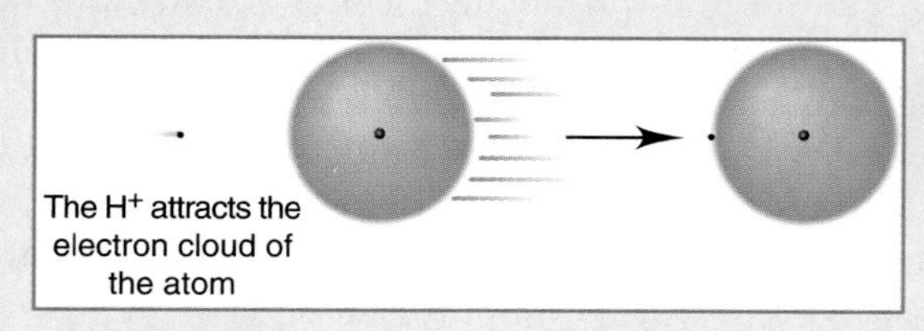


FIGURE 8.28
An $H^+$ ion has no electrons, so there are no repulsive forces between it and the electron cloud of an approaching atom. In water, $H^+$ attaches itself to an oxygen atom, forming $H_3O^+$.

The only time it is necessary to include the heat consumed or evolved in a reaction is when the heat itself is the focus of attention in the reaction. Except when chemists or physicists are analyzing heat, they write equations without showing how much (or even whether) energy is consumed or produced.

## EXAMPLE 8.12 Balancing Equations Including Heat

Solid potassium metal reacts with chlorine gas to give solid potassium chloride. The reaction is highly exothermic, and the amount of heat per mole of $Cl_2$ in the balanced equation is 873 kJ. Write a balanced equation for this reaction, including the heat.

FIGURE 8.29
Potassium reacts with chlorine to form $KCl(s)$.

**SOLUTION**

First, write the formulas of the reactants and products:

$$K(s) + Cl_2(g) \longrightarrow KCl(s)$$

Then, balance the atoms (you know how by now):

$$2\ K(s) + Cl_2(g) \longrightarrow 2\ KCl(s)$$

The reaction is exothermic. This means that heat is given off by the reaction; that is, heat is a product. Changing the equation to reflect this gives:

$$2\ K(s) + Cl_2(g) \longrightarrow 2\ KCl(s) + 873\ kJ$$

**EXERCISE 8.12**

Solid ammonium nitrate dissolves in water to give an aqueous solution of ions. Because dissolving ammonium nitrate in water at room temperature gives a very cold solution, this is the basis of a type of commercial ice pack. The heat involved is 26 kJ per mole of ammonium nitrate. Write an equation for this process, including the heat.

FIGURE 8.30
Ammonium nitrate absorbs heat when it dissolves in water, and this endothermic reaction forms the basis for chemical ice packs like the one on the right. Ammonium nitrate is also a widely used fertilizer. Mixtures of ammonium nitrate with diesel fuel are explosives, sometimes used by terrorists.

## CHAPTER BOX 8.5 CONSERVE ENERGY!

**Since it** is a scientific principle that energy *must* be conserved, why does the environmental movement insist that Americans *should* conserve energy more than they do? The reason for the apparent paradox once again lies in the fact that scientific usage of words is not exactly the same as popular usage.

When (for instance) methane burns in a gas stove, the potential energy in the mixture of methane and oxygen changes to heat. You use the heat to cook food. By the time you have finished eating, the energy has dissipated. It's all around your kitchen, or out the open window. It still exists (energy is conserved), but there's no way you can collect it and use it again. (That fact is another important scientific principle, the Second Law of Thermodynamics. For more information on this, see Chapter 4.)

In scientific terms, the call to conserve energy is a plea to save *potential* energy, to burn less gas and use less electricity, to convert less of it to heat. This is a worthwhile goal.

## CHAPTER SUMMARY

1. A chemical equation is a kind of sentence that identifies the reactants, the products, and sometimes other components of a chemical reaction. Equations can be translated into words in terms of either molecules or moles.
2. A balanced chemical equation provides the exact molar ratio of each reactant and product to the other reactants and products. By convention, no fractional coefficients appear in a balanced equation.
3. Acids have a typical sour taste, while bases taste bitter. Acids react with many metals to form hydrogen gas and a salt. They also react with carbonates to form carbon dioxide gas, water, and a salt. Acids turn the indicator blue litmus red, and bases turn red litmus blue.
4. According to Arrhenius's definitions, acids are substances that produce hydrogen ions ($H^+$) when dissolved in water, while bases produce hydroxide ions ($OH^-$) when dissolved in water. Also according to Arrhenius, acids and bases react to form salts and water. This process is called neutralization.
5. The combustion (reaction with oxygen) of compounds that contain carbon, hydrogen, and possibly oxygen gives heat, carbon dioxide, and water as reaction products. Combustion reactions are a type of oxidation-reduction reaction.
6. According to one simple definition, oxidation occurs when a substance gains oxygen atoms, and reduction occurs when a substance loses oxygen atoms.
7. Active metals react with water to produce hydrogen gas and a metal hydroxide.
8. The states of substances involved in reactions are indicated by (*s*) for solid, (*g*) for gas, (*l*) for liquid, and (*aq*) for aqueous solution. The appropriate symbol is placed immediately after the formula of the substance—for example, $H_2O(l)$ and $H_2O(g)$.
9. Reaction conditions such as temperature, pressure, or the presence of a catalyst are written over or under the arrow in a chemical equation. These are generally shown only when readers need the information to understand the key points of the reaction.

10. A net ionic equation shows only the species that undergo chemical changes in a reaction, with any spectator ions.
11. Endothermic reactions are reactions in which heat is absorbed (it can be thought of as a reactant). Exothermic reactions are reactions in which heat is given off (it can be thought of as a product).

## PROBLEMS

### SECTION 8.1 Information in Chemical Equations

1. Translate each of these sentences into a chemical equation.
   a. One mole of calcium hydroxide reacts with one mole of sulfuric acid to give one mole of calcium sulfate and two moles of water.
   b. One mole of magnesium metal reacts with two moles of hydrochloric acid to give one mole of magnesium chloride and one mole of hydrogen.
   c. One mole of sodium hydride reacts with one mole of hydrogen chloride to give one mole of sodium chloride and one mole of hydrogen.
   d. One mole of ammonia reacts with one mole of hydrogen bromide to give one mole of ammonium bromide.

2. Translate each of these sentences into a chemical equation.
   a. Two moles of aluminum metal react with three moles of sulfuric acid to give one mole of aluminum sulfate and three moles of hydrogen.
   b. Three moles of sodium hydroxide react with one mole of phosphoric acid to give one mole of sodium phosphate and three moles of water.
   c. Two moles of acetylene ($C_2H_2$) react with five moles of oxygen to give four moles of carbon dioxide and two moles of water.
   d. Two moles of copper(II) oxide react with one mole of carbon to give two moles of copper and one mole of carbon dioxide.

3. Translate each of these sentences into a chemical equation, and identify the reactants and products.
   a. One mole of iron(III) oxide reacts with three moles of carbon monoxide to give two moles of iron and three moles of carbon dioxide.
   b. Three moles of magnesium react with one mole of nitrogen to give one mole of magnesium nitride.
   c. Two moles of cesium react with two moles of water to give two moles of cesium hydroxide and one mole of hydrogen.
   d. Two moles of potassium chlorate give two moles of potassium chloride and three moles of oxygen.

4. Translate each of these sentences into a chemical equation, and identify the reactants and products.
   a. Two moles of silver nitrate react with one mole of potassium chromate to form one mole of silver chromate and two moles of potassium nitrate.
   b. One mole of aluminum oxide reacts with six moles of hydrogen bromide to yield two moles of aluminum bromide and three moles of water.
   c. One mole of barium reacts with one mole of sulfuric acid to produce one mole of barium sulfate and one mole of hydrogen.
   d. Two moles of mercury(II) oxide react to give one mole of oxygen gas and two moles of mercury.

5. For each reaction described, write an equation and identify the reactants and products.
   a. One molecule of nitrogen reacts with three molecules of hydrogen to give two molecules of ammonia.
   b. One molecule of nitrogen reacts with one molecule of oxygen to give two molecules of nitrogen monoxide.
   c. Two atoms of phosphorus react with three molecules of chlorine, yielding two molecules of phosphorus trichloride.
   d. One molecule of methane reacts with four molecules of chlorine to give one molecule of carbon tetrachloride and four molecules of hydrogen chloride.

6. For each reaction described, write an equation and identify the reactants and products.
   a. One molecule of methane reacts with four molecules of fluorine to form one molecule of carbon tetrafluoride and four molecules of hydrogen fluoride.
   b. Six molecules of hydrogen sulfide react with two molecules of sulfur trioxide to give six molecules of water and one $S_8$ molecule.
   c. Two molecules of sulfur dioxide react with one molecule of oxygen to produce two molecules of sulfur trioxide.
   d. Two atoms of sulfur react with one molecule of chlorine to give one molecule of disulfur dichloride.

7. Write each of these chemical equations as a sentence in English.
   a. $Ni + 2\ HCl \rightarrow NiCl_2 + H_2$
   b. $2\ NH_3 + H_2SO_4 \rightarrow (NH_4)_2SO_4$
   c. $HI + LiOH \rightarrow LiI + H_2O$
   d. $Cu + 2\ AgNO_3 \rightarrow 2\ Ag + Cu(NO_3)_2$
   e. $CO_2 + H_2O \rightarrow H_2CO_3$

8. Write each of these chemical equations as a sentence in English.
   a. $CaH_2 + 2\ H_2O \rightarrow Ca(OH)_2 + 2\ H_2$
   b. $Cu_2S + O_2 \rightarrow 2\ Cu + SO_2$
   c. $P_4O_{10} + 10\ C \rightarrow 4\ P + 10\ CO$
   d. $CuS + O_2 \rightarrow Cu + SO_2$
   e. $H_2SO_4 + FeS \rightarrow FeSO_4 + H_2S$

9. Translate each of these equations into an English sentence, and identify the reactants and products.
   a. $HOCl + NaOH \rightarrow NaOCl + H_2O$
   b. $2\ HNO_3 + Ba \rightarrow Ba(NO_3)_2 + H_2$
   c. $TiCl_4 + Ti \rightarrow 2\ TiCl_2$
   d. $Ba(NO_3)_2 + K_2SO_4 \rightarrow 2\ KNO_3 + BaSO_4$
   e. $2\ H_2S + O_2 \rightarrow 2\ S + 2\ H_2O$

10. Translate each of these equations into an English sentence, and identify the reactants and products.
    a. $2\ H_2S + 3\ O_2 \rightarrow 2\ H_2O + 2\ SO_2$
    b. $2\ P + 5\ Cl_2 \rightarrow 2\ PCl_5$
    c. $CaCO_3 \rightarrow CaO + CO_2$
    d. $3\ MnO_2 + 4\ Al \rightarrow 3\ Mn + 2\ Al_2O_3$
    e. $Fe_2O_3 + 3\ CO \rightarrow 2\ Fe + 3\ CO_2$

## SECTION 8.2 Balanced and Unbalanced Equations

11. Determine whether each of these equations is balanced.
    a. $AgNO_3 + LiCl \rightarrow AgCl + NaNO_3$
    b. $Hg_2(NO_3)_2 + NaCl \rightarrow Hg_2Cl_2 + NaNO_3$
    c. $AlCl_3 + NaOH \rightarrow Al(OH)_3 + NaCl$
    d. $AgNO_3 + Na_2CrO_4 \rightarrow Ag_2CrO_4 + NaNO_3$
    e. $K_2CO_3 + CaCl_2 \rightarrow CaCO_3 + 2\ KCl$

12. Determine whether each of these equations is balanced.
    a. $PbS + 3\ O_2 \rightarrow PbO + 2\ SO_2$
    b. $2\ PbO + C \rightarrow 2\ Pb + CO_2$
    c. $2\ CuFeS_2 + 3\ O_2 \rightarrow Cu_2S + 2\ FeO + 3\ SO_2$
    d. $Cu + 4\ HNO_3 \rightarrow Cu(NO_3)_2 + 2\ NO_2 + 2\ H_2O$
    e. $Al + 3\ HBr \rightarrow AlBr_3 + 2\ H_2$

13. Determine whether each of these equations is balanced.
    a. $(NH_4)_3PO_4 + 3\ CaCl_2 \rightarrow Ca_3(PO_4)_2 + 3\ NH_4Cl$
    b. $Pb(NO_3)_2 + 2\ KCl \rightarrow PbCl_2 + 2\ KNO_3$
    c. $Ba(ClO_4)_2 + Na_2SO_4 \rightarrow BaSO_4 + NaClO_4$
    d. $FeCl_3 + KOH \rightarrow Fe(OH)_3 + KCl$
    e. $K_2S + HgCl_2 \rightarrow 2\ KCl + HgS$

14. Determine whether each of these equations is balanced.
    a. $Fe + 2\ HCl \rightarrow FeCl_2 + H_2$
    b. $MgO + HCl \rightarrow MgCl_2 + H_2O$
    c. $Pb(OH)_2 + 2\ NaOH \rightarrow Na_2Pb(OH)_4$
    d. $2\ Pb(OH)_2 + 2\ HCl \rightarrow PbCl_2 + 2\ H_2O$
    e. $2\ Fe_2O_3 \rightarrow Fe_2O_3 + 3\ H_2O$

## SECTION 8.3 Balancing Simple Chemical Equations

15. Balance each of these equations.
    a. $F_2 + I_2 \rightarrow IF_3$
    b. $F_2 + I_2 \rightarrow IF_7$
    c. $CaCO_3 + HCl \rightarrow CaCl_2 + CO_2 + H_2O$
    d. $Fe + O_2 \rightarrow Fe_2O_3$
    e. $CuO + HCl \rightarrow CuCl_2 + H_2O$

16. Balance each of these equations.
    a. $CuCl_2 + NaOH \rightarrow Cu(OH)_2 + NaCl$
    b. $Al_2O_3 + HBr \rightarrow AlBr_3 + H_2O$
    c. $C + H_2O \rightarrow CO + H_2$
    d. $CO + O_2 \rightarrow CO_2$
    e. $Mg + CO_2 \rightarrow MgO + C$

17. Balance each of these equations.
    a. $CS_2 + Cl_2 \rightarrow CCl_4 + S_2Cl_2$
    b. $CH_4 + NH_3 \rightarrow HCN + H_2$
    c. $CaO + C \rightarrow CaC_2 + CO$
    d. $N_2 + H_2 \rightarrow NH_3$
    e. $N_2O_5 \rightarrow NO_2 + O_2$

18. Balance each of these equations.
    a. $SO_2 + O_2 \rightarrow SO_3$
    b. $Mg(OH)_2 + H_2SO_4 \rightarrow Mg(HSO_4)_2 + H_2O$
    c. $Fe + H_2O \rightarrow Fe_3O_4 + H_2$
    d. $Na_2CO_3 + C \rightarrow Na + CO$
    e. $NO + Cl_2 \rightarrow NOCl$

19. Balance each of these equations.
    a. $Cr(OH)_3 \rightarrow Cr_2O_3 + H_2O$
    b. $MnO_2 + HCl \rightarrow MnCl_2 + Cl_2 + H_2O$
    c. $SiCl_4 + H_2O \rightarrow Si(OH)_4 + HCl$
    d. $B + Cl_2 \rightarrow BCl_3$
    e. $P + NO \rightarrow P_4O_6 + N_2$

20. Balance each of these equations.
    a. $(NH_4)_2Cr_2O_7 \rightarrow N_2 + H_2O + Cr_2O_3$
    b. $NO + H_2 \rightarrow N_2 + H_2O$
    c. $CO + NO \rightarrow N_2 + CO_2$
    d. $P_4O_{10} + H_2O \rightarrow H_3PO_4$
    e. $KClO_3 \rightarrow KCl + O_2$

21. Write a balanced equation for each of these reactions.
    a. Sulfuric acid reacts with sodium cyanide to make hydrogen cyanide and sodium sulfate.
    b. Ammonia reacts with oxygen to give dinitrogen monoxide and water.

c. Magnesium nitride reacts with water to yield magnesium hydroxide and ammonia.
d. Iron(III) nitrate reacts with potassium iodide to give iron(II) nitrate, potassium nitrate, and iodine.
e. Calcium sulfide reacts with nitric acid to give calcium nitrate and hydrogen sulfide.

22. Write a balanced equation for each of these reactions.
a. Sodium sulfate reacts with carbon to yield sodium sulfide and carbon dioxide.
b. Sodium hydrogen carbonate decomposes to give sodium carbonate, water, and carbon dioxide.
c. Ammonium chloride reacts with calcium oxide to give calcium chloride, ammonia, and water.
d. Copper(I) oxide reacts with carbon monoxide to give copper and carbon dioxide.
e. Sodium bromide reacts with chlorine to yield sodium chloride and bromine.

23. Write and balance the equation for each of these reactions.
a. Tin(IV) oxide reacts with hydrogen to give tin and water.
b. Manganese(IV) oxide reacts with aluminum to give manganese and aluminum oxide.
c. Lead(II) sulfide reacts with oxygen to give lead(II) oxide and sulfur dioxide.
d. Aluminum reacts with hydrochloric acid to yield aluminum chloride and hydrogen gas.
e. Iron(III) hydroxide decomposes to form iron(III) oxide and water.

24. Write and balance the equation for each of these reactions.
a. Copper(II) chloride reacts with aluminum sulfide and water to give copper(II) sulfide, aluminum hydroxide, and hydrochloric acid.
b. Aluminum oxide reacts with chlorine and carbon to give aluminum chloride and carbon monoxide.
c. Zinc sulfide reacts with oxygen to yield zinc oxide and sulfur dioxide.
d. Phosphorus trihydride (commonly known as phosphine) reacts with oxygen to produce tetraphosphorus decoxide and water.
e. Calcium phosphate and sulfuric acid react to yield calcium sulfate and phosphoric acid.

25. Balance any of the equations in Problem 11 that were not balanced.

26. Balance any of the equations in Problem 12 that were not balanced.

27. Balance any of the equations in Problem 13 that were not balanced.

28. Balance any of the equations in Problem 14 that were not balanced.

## SECTION 8.4 Acid-Base Reactions

29. Balance each of these equations.
a. $HBr + KOH \rightarrow KBr + H_2O$
b. $HNO_3 + Ca(OH)_2 \rightarrow Ca(NO_3)_2 + H_2O$
c. $H_2SO_4 + Sr(OH)_2 \rightarrow SrSO_4 + H_2O$
d. $H_3PO_4 + Ba(OH)_2 \rightarrow Ba_3(PO_4)_2 + H_2O$
e. $HI + Al(OH)_3 \rightarrow AlI_3 + H_2O$

30. Balance each of these equations.
a. $H_2SO_4 + Fe(OH)_3 \rightarrow Fe_2(SO_4)_3 + H_2O$
b. $KHSO_4 + Ba(OH)_2 \rightarrow BaSO_4 + K_2SO_4 + H_2O$
c. $HI + CaCO_3 \rightarrow CaI_2 + H_2O + CO_2$
d. $H_2SO_4 + CuOH \rightarrow Cu_2SO_4 + H_2O$
e. $HBr + Mg(OH)_2 \rightarrow MgBr_2 + H_2O$

31. Balance each of these equations.
a. $HClO_4 + Sr(OH)_2 \rightarrow Sr(ClO_4)_2 + H_2O$
b. $H_3PO_4 + NaOH \rightarrow Na_3PO_4 + H_2O$
c. $HNO_3 + Mg(OH)_2 \rightarrow Mg(NO_3)_2 + H_2O$
d. $H_2SO_4 + Fe(OH)_3 \rightarrow Fe_2(SO_4)_3 + H_2O$
e. $HBr + K_2CO_3 \rightarrow KBr + H_2O + CO_2$

32. Balance each of these equations.
a. $HNO_3 + LiOH \rightarrow LiNO_3 + H_2O$
b. $HNO_3 + Ca(OH)_2 \rightarrow Ca(NO_3)_2 + H_2O$
c. $H_3PO_4 + Ca(OH)_2 \rightarrow Ca_3(PO_4)_2 + H_2O$
d. $H_2SO_4 + RbOH \rightarrow Rb_2SO_4 + H_2O$
e. $HF + Fe(OH)_3 \rightarrow FeF_3 + H_2O$

33. Write a balanced equation for each of the reactions described.
a. Hydrochloric acid reacts with potassium hydroxide to give potassium chloride and water.
b. Phosphoric acid reacts with sodium hydroxide to give sodium phosphate and water.
c. Sodium hydrogen sulfate reacts with sodium hydroxide to give water and sodium sulfate.
d. Lithium hydrogen carbonate reacts with lithium hydroxide to give lithium carbonate and water.
e. Hydrochloric acid reacts with calcium hydroxide to give calcium chloride and water.

34. Write a balanced equation for each of the reactions described.
a. Sulfuric acid reacts with strontium hydroxide to give strontium sulfate and water.
b. Perchloric acid reacts with iron(II) hydroxide to give iron(II) perchlorate and water.
c. Hydroiodic acid reacts with aluminum hydroxide, yielding aluminum iodide and water.
d. Hydrobromic acid reacts with zinc hydroxide to form zinc bromide and water.
e. Potassium hydrogen sulfate reacts with potassium hydroxide to give potassium sulfate and water.

35. Write and balance the equation for each of these reactions.
   a. Phosphoric acid reacts with barium hydroxide to give barium phosphate and water.
   b. Sulfuric acid reacts with potassium hydroxide yielding potassium sulfate and water.
   c. Hydrogen sulfide reacts with sodium hydroxide to give sodium sulfide and water.
   d. Acetic acid reacts with calcium hydroxide forming calcium acetate and water.
   e. Hydroiodic acid reacts with calcium carbonate to give calcium iodide, water, and carbon dioxide.

36. Write and balance the equation for each of these reactions.
   a. Hydrofluoric acid reacts with lithium hydroxide to give lithium fluoride and water.
   b. Phosphoric acid reacts with potassium hydroxide to give potassium phosphate and water.
   c. Sodium hydrogen carbonate reacts with sodium hydroxide to give sodium carbonate and water.
   d. Sulfuric acid reacts with barium hydroxide to give barium sulfate and water.
   e. Calcium hydroxide reacts with nitric acid to give calcium nitrate and water.

37. For each pair of reactants, predict the products and write the balanced equation.
   a. Hydrobromic acid and sodium hydroxide
   b. Nitric acid and barium hydroxide
   c. Sulfuric acid and calcium hydroxide
   d. Acetic acid and potassium hydroxide
   e. Potassium hydrogen sulfate and potassium hydroxide

38. For each pair of reactants, predict the products and write the balanced equation.
   a. Hydrofluoric acid and lithium hydroxide
   b. Sulfuric acid and aluminum hydroxide
   c. Nitric acid and magnesium hydroxide
   d. Calcium hydroxide and acetic acid
   e. Sodium bicarbonate and hydrochloric acid

39. Predict the products and balance the equation for each of these pairs of reactants.
   a. Sulfuric acid and potassium hydroxide
   b. Perchloric acid and aluminum hydroxide
   c. Hydrochloric acid and sodium carbonate
   d. Hydroiodic acid and zinc hydroxide
   e. Hydrobromic acid and zinc hydroxide

40. Predict the products and balance the equation for each of these pairs of reactants.
   a. Potassium hydrogen carbonate and potassium hydroxide
   b. Perchloric acid and potassium carbonate
   c. Periodic acid and strontium hydroxide
   d. Sulfuric acid and iron(III) hydroxide
   e. Perbromic acid and nickel(II) hydroxide

## SECTION 8.5 Oxidation-Reduction Reactions: Combustion

41. Identify the substance oxidized and the substance reduced in each of these reactions.
   a. $2\ C_3H_6 + 9\ O_2 \rightarrow 6\ CO_2 + 6\ H_2O$
   b. $Fe_2O_3 + 2\ Al \rightarrow 2\ Fe + Al_2O_3$
   c. $Cu_2O + CO \rightarrow 2\ Cu + CO_2$
   d. $2\ SO_2 + O_2 \rightarrow 2\ SO_3$
   e. $2\ PbO + C \rightarrow 2\ Pb + CO_2$

42. Identify the substance oxidized and the substance reduced in each of these reactions.
   a. $C_2H_4 + 3\ O_2 \rightarrow 2\ CO_2 + 2\ H_2O$
   b. $3\ MnO_2 + 4\ Al \rightarrow 3\ Mn + 2\ Al_2O_3$
   c. $Cu_2S + O_2 \rightarrow 2\ Cu + SO_2$
   d. $2\ CO + O_2 \rightarrow 2\ CO_2$
   e. $Si + O_2 \rightarrow SiO_2$

43. Write a balanced equation for the reaction of each of these substances with oxygen.
   a. $C_4H_{10}$ (butane)
   b. Aluminum metal
   c. $C_3H_8O$ (rubbing alcohol)
   d. $C_{10}H_{22}$ (decane)
   e. Iron metal

44. Write a balanced equation for the reaction of each of these substances with oxygen.
   a. Sulfur
   b. $C_4H_8$ (butene)
   c. $C_2H_6O_2$ (antifreeze)
   d. $C_6H_6$ (benzene)
   e. $C_7H_9$ (toluene)

45. Write and balance the equation for the reaction of each substance with oxygen.
   a. Calcium metal
   b. Phosphorus (forms $P_4O_{10}$)
   c. $C_{12}H_{22}O_{12}$ (table sugar)
   d. Lithium metal
   e. $C_6H_{14}$ (hexane)

46. Write and balance the equation for the reaction of each substance with oxygen.
   a. $C_6H_{12}$ (cyclohexane)
   b. Barium metal
   c. Carbon monoxide
   d. $C_5H_{12}$ (pentane)
   e. Zinc metal

47. For each substance, write a balanced equation for its reaction with oxygen.
   a. $CH_2O$ (formaldehyde)
   b. Ammonia (forms NO and water)
   c. $C_5H_{10}$ (pentene)
   d. Chromium metal (forms $CrO_3$)
   e. $C_8H_{18}$ (a component of gasoline)

48. For each substance, write a balanced equation for its reaction with oxygen.
   a. $N_2H_4$ (hydrazine; forms NO and water)
   b. Chlorine
   c. Magnesium metal
   d. $C_4H_{10}O$ (butanol)
   e. Manganese metal [forms manganese(II) oxide]

## SECTION 8.6 The Reactions of Active Metals and Water

49. Write a balanced equation for the reaction of each of these metals with water.
   a. Lithium
   b. Aluminum
   c. Barium

50. Write a balanced equation for the reaction of each of these metals with water.
   a. Cesium
   b. Potassium
   c. Strontium

## SECTION 8.7 Reaction Conditions

51. Write each of these equations as an English sentence.
   a. $4\ P(s) + 3\ O_2(g) \rightarrow P_4O_6(s)$
   b. $4\ P(s) + 5\ O_2(g) \rightarrow P_4O_{10}(s)$
   c. $Fe_2O_3(s) + 3\ CO(g) \xrightarrow{\text{heat}} 2\ Fe(l) + 3\ CO_2(g)$
   d. $2\ PbS(s) + 3\ O_2(g) \xrightarrow{\text{heat}} 2\ PbO(s) + 2\ SO_2(g)$

52. Write each of these equations as an English sentence.
   a. $NH_3(g) + HCl(g) \rightarrow NH_4Cl(s)$
   b. $H_2SO_4(l) + NaCl(s) \rightarrow HCl(g) + NaHSO_4(s)$
   c. $Xe(g) + 3\ F_2(g) \xrightarrow{\text{heat}} XeF_6(s)$
   d. $MgCl_2(l) \xrightarrow[\text{heat}]{\text{electric current}} Mg(s) + Cl_2(g)$

53. Translate each of these equations into an English sentence.
   a. $2\ H_2(g) + C(s) \xrightarrow{\text{heat}} CH_4(g)$
   b. $8\ Cu(s) + S_8(s) \xrightarrow{\text{heat}} 8\ CuS(s)$
   c. $H_2(g) + Cl_2(g) \xrightarrow{\text{light}} 2\ HCl(g)$
   d. $2\ KClO_3(s) \xrightarrow[\text{heat}]{MnO_2} 2\ KCl(s) + 3\ O_2(g)$

54. Translate each of these equations into an English sentence.
   a. $H_2(g) + S(s) \xrightarrow{\text{heat}} H_2S(g)$
   b. $4\ Ag(s) + O_2(g) \xrightarrow{\text{heat}} 2\ Ag_2O(s)$
   c. $2\ NO(g) \xrightarrow{Pt} N_2(g) + O_2(g)$
   d. $CH_4(g) + NH_3(g) \xrightarrow[Pt]{1200°C} HCN(g) + 3\ H_2(g)$

55. Write a balanced equation for each of these reactions.
   a. Sulfur trioxide gas is produced when sulfur dioxide gas is heated with oxygen gas in the presence of $V_2O_5$.
   b. Adding aqueous sulfuric acid to magnesium metal produces aqueous magnesium hydrogen sulfate and hydrogen gas.
   c. Heating nitrogen and hydrogen gases under pressure and in the presence of copper(II) oxide produces ammonia gas.
   d. Burning propane gas ($C_3H_8$) in air yields carbon dioxide and water, both as gases.

56. Write a balanced equation for each of these reactions.
   a. Adding hydrobromic acid to solid sodium carbonate produces aqueous sodium bromide, carbon dioxide, and water.
   b. Aqueous sulfuric acid forms when sulfur trioxide gas bubbles through water.
   c. Solid carbon and hydrogen gas form when methane is heated.
   d. Solid magnesium nitride forms when magnesium and nitrogen are heated.

57. Write a balanced equation for each of these reactions.
   a. Heating nitrogen monoxide gas in the presence of platinum metal produces nitrogen and oxygen.
   b. Heating nitrogen dioxide gas in the presence of rhodium metal produces nitrogen and oxygen.
   c. Heating lead(II) sulfide with oxygen gas produces lead(II) oxide and sulfur dioxide gas.
   d. Heating $CuFeS_2(s)$ with oxygen yields copper(I) sulfide, iron(II) oxide, and sulfur dioxide.

58. Write a balanced equation for each of these reactions.
   a. Hydrogen reacts with oxygen in the presence of platinum to form water.
   b. Hydrochloric acid reacts with iron metal to produce hydrogen and an aqueous solution of iron(II) chloride.
   c. Magnesium oxide is produced when magnesium is heated with oxygen.
   d. Sulfur dioxide gas is formed when melted sulfur is heated with oxygen gas.

59. Write and balance the equation for each of these reactions.
   a. The gases hydrogen sulfide and sulfur dioxide react to form water and elemental sulfur when heated.
   b. Hydrogen gas and calcium nitrate are formed when aqueous nitric acid is added to calcium metal.
   c. Heating magnesium metal with carbon dioxide gas forms carbon and magnesium oxide.
   d. Burning methanol ($CH_3OH$) in oxygen forms carbon dioxide and water.

60. Write and balance the equation for each of these reactions.
    a. Adding sulfuric acid to iodide forms hydrogen sulfate and hydrogen iodide.
    b. When an electric current passes through molten sodium chloride, liquid sodium metal and chlorine are produced.
    c. Heating tetraphosphorus decoxide with potassium oxide yields potassium phosphate.
    d. When carbon dioxide gas is bubbled through an aqueous solution of calcium hydroxide, calcium carbonate and water are produced.

61. Write a balanced equation for each of these reactions.
    a. Calcium nitrate forms when dinitrogen pentoxide gas reacts with calcium oxide.
    b. Heating aluminum hydroxide produces aluminum oxide and steam.
    c. At 600 K, iron(III) oxide reacts with hydrogen to produce iron(II) oxide and water.
    d. When aqueous sulfuric acid is poured over iron metal, hydrogen gas and an aqueous solution of iron(II) sulfate are the products.

62. Write a balanced equation for each of these reactions.
    a. Heating iron(III) hydroxide to 500°C produces iron(III) oxide and water.
    b. Bubbling hydrogen sulfide gas through an aqueous solution of copper(II) chloride produces hydrochloric acid and copper(II) sulfide.
    c. Heating copper(II) oxide with copper metal produces copper(I) oxide.
    d. Heating mercury(II) oxide yields oxygen and liquid mercury.

## SECTION 8.8 Net Ionic Equations

63. Balance each of these equations, and rewrite it as a net ionic equation.
    a. $HI(aq) + AgNO_3(s) \rightarrow AgI(s) + HNO_3(aq)$
    b. $Pb(NO_3)_2(aq) + NaCl(aq) \rightarrow PbCl_2(s) + NaNO_3(aq)$
    c. $HCl(aq) + CaCO_3(s) \rightarrow CaCl_2(aq) + CO_2(g) + H_2O(l)$
    d. $(NH_4)_2S(aq) + HCl(aq) \rightarrow NH_4Cl(aq) + H_2S(g)$

64. Balance each of these equations, and rewrite it as a net ionic equation.
    a. $Na_2CO_3(aq) + CaCl_2(aq) \rightarrow NaCl(aq) + CaCO_3(s)$
    b. $H_3PO_4(l) + NaOH(aq) \rightarrow Na_2HPO_4(aq) + H_2O(l)$
    c. $NaH_2PO_4(aq) + Na_3PO_4(aq) \rightarrow Na_2HPO_4(aq)$
    d. $NaHSO_4(aq) + NaHCO_3(aq) \rightarrow Na_2SO_4(aq) + CO_2 + H_2O$

65. Balance each equation, and rewrite it as a net ionic equation.
    a. $HNO_3(aq) + Mg(OH)_2(aq) \rightarrow Mg(NO_3)_2(aq) + H_2O(l)$
    b. $Hg_2(NO_3)_2(aq) + NaCl(aq) \rightarrow Hg_2Cl_2(s) + NaNO_3(aq)$
    c. $H_2SO_4(aq) + NaHCO_3(aq) \rightarrow Na_2SO_4(aq) + CO_2(g) + H_2O(l)$
    d. $(NH_4)_2S(aq) + NaOH(aq) \rightarrow Na_2S(aq) + NH_3(g) + H_2O(l)$

66. Balance each equation, and rewrite it as a net ionic equation.
    a. $Na_2SO_4(aq) + Ba(NO_3)_2(aq) \rightarrow NaNO_3(aq) + BaSO_4(s)$
    b. $CsHSO_4(aq) + CsOH(aq) \rightarrow Cs_2SO_4(aq) + H_2O(l)$
    c. $FeCl_3(aq) + KOH(aq) \rightarrow Fe(OH)_3(s) + KCl(aq)$
    d. $NaHCO_3(aq) + HCl(aq) \rightarrow NaCl(aq) + CO_2(g) + H_2O(l)$

67. Balance and rewrite each equation as a net ionic equation.
    a. $Pb(NO_3)_2(aq) + CaCl_2(aq) \rightarrow PbCl_2(s) + Ca(NO_3)_2(aq)$
    b. $HClO_4(aq) + Ca(OH)_2(aq) \rightarrow Ca(ClO_4)_2(aq) + H_2O(l)$
    c. $Na_2S(aq) + HCl(aq) \rightarrow H_2S(g) + NaCl(aq)$
    d. $Na_2S(aq) + NiCl_2(aq) \rightarrow NiS(s) + NaCl(aq)$

68. Balance and rewrite each equation as a net ionic equation.
    a. $AlCl_3(aq) + NaOH(aq) \rightarrow Al(OH)_3(s) + NaCl(aq)$
    b. $FeCl_2(aq) + AgNO_3(aq) \rightarrow AgCl(s) + Fe(NO_3)_2(aq)$
    c. $HI(aq) + KOH(aq) \rightarrow KI(aq) + H_2O(l)$
    d. $Pb(NO_3)_2(aq) + K_2Cr_2O_7(aq) \rightarrow PbCr_2O_7(s) + KNO_3(aq)$

69. Predict the products and write the net ionic equation for the reaction that occurs when aqueous solutions of each pair of compounds are mixed.
    a. Sodium sulfide and silver nitrate
    b. Sulfuric acid and potassium hydroxide
    c. Sulfuric acid and barium chloride
    d. Potassium hydroxide and copper(II) chloride
    e. Potassium carbonate and nitric acid

70. Predict the products and write the net ionic equation for the reaction that occurs when aqueous solutions of each pair of compounds are mixed.
    a. Zinc nitrate and potassium sulfide
    b. Aluminum chloride and sodium hydroxide
    c. Nickel(II) chloride and potassium hydroxide
    d. Nitric acid and sodium hydrogen carbonate
    e. Perchloric acid and calcium hydroxide

## SECTION 8.9 Energy as Reactant and Product

71. Translate each equation into an English sentence.
    a. $2\ MgO(s) + C(s) + 810\ kJ \rightarrow 2\ Mg(s) + CO_2(g)$
    b. $2\ Na(s) + Cl_2(g) \rightarrow 2\ NaCl(s) + 822\ kJ$
    c. $HCl(g) + H_2O(l) \rightarrow H_3O^+(aq) + Cl^-(aq) + 84\ kJ$
    d. $2\ C(s) + H_2(g) + 227\ kJ \rightarrow C_2H_2(g)$ (acetylene)

72. Translate each equation into an English sentence.
    a. $2\ C(s) + O_2(g) \rightarrow 2\ CO(g) + 221\ kJ$
    b. $N_2(g) + 2\ O_2(g) \rightarrow 2\ NO_2(g) + 66\ kJ$
    c. $3\ O_2(g) + 285\ kJ \rightarrow 2\ O_3(g)$ (ozone)
    d. $S(s) + O_2(g) \rightarrow SO_2(g) + 297\ kJ$

73. Translate each of these equations into English.
    a. $N_2(g) + O_2(g) + 181\ kJ \rightarrow 2\ NO(g)$
    b. $Ni(s) + Cl_2(g) \rightarrow NiCl_2(s) + 305\ kJ$
    c. $2\ S(s) + 3\ O_2(g) \rightarrow 2\ SO_3(g) + 791\ kJ$
    d. $C(s) + 2\ H_2(g) \rightarrow CH_4(g) + 75\ kJ$

74. Translate each of these equations into English.
    a. $2\ SO_2(g) + O_2(g) \rightarrow 2\ SO_3(g) + 198\ kJ$
    b. $SO_3(g) + H_2O(l) \rightarrow H_2SO_4(l) + 133\ kJ$
    c. $2\ NO(g) + O_2(g) \rightarrow 2\ NO_2(g) + 119\ kJ$
    d. $4\ B(s) + 3\ O_2(g) \rightarrow 2\ B_2O_3(s) + 2547\ kJ$

75. A sample of ice at 0°C can be heated and the temperature won't change. The solid water simply becomes liquid water. If you do the heating carefully, the temperature will remain at 0°C until all the ice has melted. The amount of heat absorbed per mole of ice is 46 kJ. Write an equation for this process, including the heat.

76. In what chemists call the *thermite reaction,* iron(III) oxide reacts with aluminum metal to yield iron metal and aluminum oxide. The reaction is highly exothermic, and the amount of heat per mole of iron(III) oxide is 852 kJ. Write a balanced equation for this reaction, including the heat.

## ADDITIONAL PROBLEMS

77. Balance each equation.
    a. $Cl_2(g) + O_2(g) \rightarrow Cl_2O(g)$
    b. $P_4(s) + Cl_2(g) \rightarrow PCl_3(l)$
    c. $H_3O^+(aq) + Na(s) \longrightarrow Na^+(aq) + OH^-(aq) + H_2(g)$
    d. $C_5H_{12}(l) + O_2(g) \xrightarrow{heat} CO_2(g) + H_2O(g)$
    e. $C_4H_{10}(g) + O_2(g) \xrightarrow{heat} CO_2(g) + H_2O(g)$

78. Balance each equation.
    a. $P_4(s) + Cl_2(g) \rightarrow PCl_5(l)$
    b. $NH_4^+(aq) + Mg(OH)_2(s) \rightarrow NH_3(aq) + H_2O(l) + Mg^{2+}(aq)$
    c. $C_7H_{16}(l) + O_2(g) \xrightarrow{heat} CO_2(g) + H_2O(g)$
    d. $C_4H_8(l) + O_2(g) \xrightarrow{heat} CO_2(g) + H_2O(g)$
    e. $Mg(s) + N_2(g) \xrightarrow{heat} Mg_3N_2(s)$

79. Write each of these equations as an English sentence.
    a. $H_2SO_4(aq) + NaBr(s) \rightarrow Na^+(aq) + HSO_4^-(aq) + HBr(g)$
    b. $S^{2-}(aq) + 2\ H_3O^+(aq) \rightarrow H_2S(g) + 2\ H_2O(l)$
    c. $C_2H_2(g)$ (acetylene) $+ 2\ H_2(g) \xrightarrow{Pt} C_2H_6(g)$ (ethane)
    d. $S(l) + NaCN(s) \xrightarrow{heat} NaSCN(s)$ (sodium thiosulfate)

80. Write each of these equations as an English sentence.
    a. $2\ HNO_3(aq) + CaO(s) \rightarrow Ca^{2+}(aq) + 2\ NO_3^-(aq) + H_2O(l)$
    b. $Ba^{2+}(aq) + SO_4^{2-}(aq) \rightarrow BaSO_4(s)$
    c. $2\ HBr(aq) + CO_3^{2-}(aq) \rightarrow H_2O(l) + CO_2(g) + 2\ Br^-(aq)$
    d. $CH_3CH_2OH(l) + O_2(g) \xrightarrow[heat]{Ag} CH_3COOH(l) + H_2O(g)$

## EXPERIMENTAL PROBLEM 8.1

A certain class of compounds called *cycloalkanes* has the general formula $(CH_2)_n$, where $n \geq 3$. A. L. Chemist and his team were asked for information about the combustion of these compounds, so team members went to the library to obtain information.

The combustion of cycloalkanes yields carbon dioxide, water, and heat. The general equation for the process is

$$(CH_2)_n + \frac{3n}{2} O_2(g) \longrightarrow n\ CO_2(g) + n\ H_2O(g) + \text{heat}$$

The heats of combustion for various cycloalkanes are as follows:

| *Name* | *n* | *Heat of combustion (kJ mol$^{-1}$)* |
|---|---|---|
| Cyclopropane | 3 | 2090 |
| Cyclobutane | 4 | 2743 |
| Cyclopentane | 5 | 3318 |
| Cyclohexane | 6 | 3949 |

### QUESTIONS

**81.** Which of the cycloalkanes listed has the smallest heat of combustion *per $CH_2$ unit?*

**82.** Which of the cycloalkanes listed has the largest heat of combustion *per $CH_2$ unit?*

**83.** What is the trend represented by this data?

**84.** All cycloalkanes for which $n \geq 15$ have heats of combustion of 658.2 kJ mol$^{-1}$ per $CH_2$ unit. Is this consistent with the trend identified in your answer to Question 83?

**85.** Cyclodecane ($n = 10$) has a heat of combustion of 663.2 kJ mol$^{-1}$ per $CH_2$ unit. How much heat evolves when 1 mol of cyclodecane burns?

**86.** Considering all the information given or calculated in this problem, explain why scientists review large data sets before stating trends or guidelines.

## SOLUTIONS TO EXERCISES

### Exercise 8.1

**a.** Reactants: $CaCl_2$ and $AgNO_3$. Products: AgCl and $Ca(NO_3)_2$. Sentence: One mole of calcium chloride reacts with two moles of silver nitrate to form two moles of silver chloride and one mole of calcium nitrate.

**b.** Reactants: $CH_4$ and $H_2O$. Products: CO and $H_2$. Sentence: One mole of methane reacts with one mole of water to yield one mole of carbon monoxide and three moles of hydrogen.

**c.** Reactants: Mg and HBr. Products: $MgBr_2$ and $H_2$. Sentence: One mole of magnesium reacts with two moles of hydrogen bromide, forming one mole of magnesium bromide and one mole of hydrogen.

### Exercise 8.2

**a.** Atoms present: Ca, Cl, Ag, N, O. Reactant side: one calcium and two chlorine atoms in $CaCl_2$ plus two silver, two nitrogen, and six oxygen atoms in 2 $AgNO_3$. Product side: two silver and two chlorine atoms in 2 AgCl plus one calcium, two nitrogen, and six oxygen atoms in $Ca(NO_3)_2$.

**b.** Atoms present: C, H, O. Reactant side: one carbon and four hydrogen atoms in $CH_4$ plus two hydrogens (giving a total of six hydrogen atoms) and one oxygen in water. Product side: one carbon and one oxygen in CO plus six hydrogens in 3 $H_2$.

**c.** Atoms present: Mg, H, Br. Reactant side: one magnesium atom from Mg plus two hydrogen and two bromine atoms from 2 HBr. Product side: One magnesium and two bromine atoms from $MgBr_2$ plus two hydrogen atoms from $H_2$.

### Exercise 8.3

**a.** Step 1 is already done. The correct formulas for reactants and products were given. Step 2: All the elements except oxygen appear only once on each side, so you can start with any atoms other than oxygen. The carbon, bromine, and oxygen atoms are balanced, so they can be ignored for now. The hydrogen atoms are not balanced, so starting with hydrogen makes sense:

$$\mathbf{2}\ HBr + Na_2CO_3 \longrightarrow H_2O + CO_2 + NaBr$$
(not completely balanced)

This balanced the hydrogen atoms and simultaneously unbalanced the bromine atoms. Step 3: Balance the rest of the atoms. Only sodium and bromine remain unbalanced, and starting with either one gives the same result:

$$2\ HBr + Na_2CO_3 \longrightarrow H_2O + CO_2 + \mathbf{2}\ NaBr \text{(balanced)}$$

An inspection shows that the equation is now balanced. Step 4 is unnecessary.

**b.** Step 1 is given.
Step 2: You can start with carbon since that appears only once on each side of the equation.

$$C_2H_6 + O_2 \longrightarrow \mathbf{2}\ CO_2 + H_2O \quad \text{(not completely balanced)}$$

Step 3: Balance the rest of the atoms. It's best to pick hydrogen first, because it occurs only once on each side of the unbalanced equation.

$$C_2H_6 + O_2 \longrightarrow \mathbf{2}\ CO_2 + \mathbf{3}\ H_2O \quad \text{(not completely balanced)}$$

Next balance the oxygen atoms. At the moment there are two on the reactant side and seven on the product side. To get seven oxygen atoms on the reactant side, assign a coefficient of 7/2 to $O_2$:

$$C_2H_6 + \tfrac{7}{2}O_2 \longrightarrow \mathbf{2}\ CO_2 + 3\ H_2O \quad \text{(balanced with a fractional coefficient)}$$

Step 4: Multiply all coefficients by the denominator of the fraction–in this case, 2.

$$2\ C_2H_6 + 7\ O_2 \longrightarrow 4\ CO_2 + 6\ H_2O \quad \text{(balanced)}$$

An inspection shows the equation is balanced.

### Exercise 8.4

**a.** Step 1:

$$HNO_3 + Zn \longrightarrow Zn(NO_3)_2 + H_2 \quad \text{(unbalanced)}$$

Step 2: Start with hydrogen, which appears once on each side.

$$\mathbf{2}\ HNO_3 + Zn \longrightarrow Zn(NO_3)_2 + H_2 \quad \text{(balanced)}$$

Steps 3 and 4 are unnecessary. The rest of the atoms become balanced with the hydrogens. A check shows the equation is balanced.

**b.** Step 1:

$$P + Cl_2 \longrightarrow PCl_5 \quad \text{(unbalanced)}$$

Steps 2 and 3: The phosphorus is balanced, so you can work on balancing the chlorine. There are two chlorine atoms on the left side of the equation and five on the right. These can be balanced by using a coefficient of 5/2 for the chlorine.

$$P + \tfrac{5}{2}Cl_2 \longrightarrow PCl_5$$

Step 4 removes the fractional coefficients and gives:

$$2\ P + 5\ Cl_2 \longrightarrow 2\ PCl_5$$

### Exercise 8.5

**a.** Step 1 is done. Step 2: Phosphorus is balanced but hydrogen, sodium, and oxygen are not. Since sodium appears only once on each side, it is a good place to begin.

$$H_3PO_4 + \mathbf{3}\ NaOH \longrightarrow Na_3PO_4 + H_2O$$

Step 3: Balance the rest of the atoms. Hydrogen is the next logical choice because it occurs in fewer compounds than oxygen does.

$$H_3PO_4 + 3\ NaOH \longrightarrow Na_3PO_4 + 3\ H_2O$$

Inspection shows that O is balanced, too. Step 4 is unnecessary, and a Step 5 check shows we are finished.

**b.** Step 1: Writing the correct formulas is essential. Sulfuric acid is the name you memorized for $H_2SO_4$, sulfate has a $-2$ charge and is written $SO_4^{2-}$, potassium is a Group 1A metal and has a $+1$ charge, and hydroxide is $OH^-$ with a $-1$ charge. This leads to the equation

$$H_2SO_4 + KOH \longrightarrow K_2SO_4 + H_2O$$

Step 2: Sulfur is balanced, but hydrogen, potassium, and oxygen are not. Potassium appears only once on each side, so it's a good place to start.

$$H_2SO_4 + \mathbf{2}\ KOH \longrightarrow K_2SO_4 + H_2O$$

Step 3: Balance the rest of the atoms. Inspection shows that hydrogen and oxygen remain unbalanced. These both can be balanced in one step.

$$H_2SO_4 + 2\ KOH \longrightarrow K_2SO_4 + 2\ H_2O$$

A Step 5 check shows that the equation is balanced.

**c.** The answer to Step 1 is given.
Step 2: Sulfur, phosphorus, and oxygen are balanced, but sodium and hydrogen are not. Hydrogen appears only once on each side, so it's a good place to start.

$$\mathbf{2}\ NaHSO_4 + Na_3PO_4 \longrightarrow Na_2SO_4 + NaH_2PO_4$$

This balances the hydrogens but unbalances sulfur and oxygen. Sodium remains unbalanced.

Step 3: Balance the remaining atoms. Since sulfur appears only once on each side, start there.

$$2\ NaHSO_4 + Na_3PO_4 \longrightarrow \mathbf{2}\ Na_2SO_4 + NaH_2PO_4$$

Inspection shows that the equation is balanced.

**d.** Step 1: Obtain the correct formulas. Aluminum is a metal that forms only one cation when it reacts, $Al^{3+}$ (see Table 7.1). Sulfuric acid is the name you memorized for $H_2SO_4$, sulfate is $SO_4^{2-}$, and hydroxide is $OH^-$. The equation is

$$H_2SO_4 + Al(OH)_3 \longrightarrow Al_2(SO_4)_3 + H_2O$$

Step 2: None of the atoms are balanced. Aluminum appears only once on each side, so you can start there.

$$H_2SO_4 + \mathbf{2}\ Al(OH)_3 \longrightarrow Al_2(SO_4)_3 + H_2O$$

Step 3: Balance the rest of the atoms. Sulfur appears once each side, so it's balanced next.

$$\mathbf{3}\ H_2SO_4 + 2\ Al(OH)_3 \longrightarrow Al_2(SO_4)_3 + H_2O$$

This leaves hydrogen and oxygen unbalanced. Pick either one to balance next, yielding

$$3\ H_2SO_4 + 2\ Al(OH)_3 \longrightarrow Al_2(SO_4)_3 + \mathbf{6}\ H_2O$$

All the atoms are balanced, and there are no fractional coefficients. You are finished with this problem.

**Exercise 8.6**

**a.** First, predict the products:

$$HNO_3 + KOH \longrightarrow \mathbf{H_2O + KNO_3}$$

The equation is balanced as written.

**b.** First, predict the products. Magnesium is a Group 2A metal, so its ion has a charge of $+2$. Sulfate has a charge of $-2$. Therefore, the products are water and $MgSO_4$:

$$H_2SO_4 + Mg(OH)_2 \longrightarrow \mathbf{H_2O + MgSO_4}$$

Your next step is to balance the equation. Note that magnesium hydroxide contains two hydroxide ions and sulfuric acid is diprotic. This suggests that the two $H^+$ ions necessary to balance both $OH^-$ ions come from one $H_2SO_4$ molecule and that two molecules of water are formed:

$$H_2SO_4 + Mg(OH)_2 \longrightarrow \mathbf{2}\ H_2O + MgSO_4$$

A check shows that the equation is now balanced.

**c.** The first step is to write the formulas of the reactants from the names, following the rules from Chapter 7. Chlorate, $ClO_3^-$, is in Table 7.3. The *per-* suffix means you have to add one more oxygen to get perchlorate, which is $ClO_4^-$. Perchloric acid is $HClO_4$.

$$HClO_4 + NaOH \longrightarrow ?$$

Next, you predict the products:

$$HClO_4 + NaOH \longrightarrow \mathbf{H_2O + NaClO_4}$$

The equation is now balanced. Is this the answer you got before you looked at this?

**d.** You know that the formula for sulfuric acid is $H_2SO_4$. Potassium is a Group 1A metal, so potassium hydroxide is KOH.

$$H_2SO_4 + KOH \longrightarrow ?$$

Next, predict the products. This requires identifying sulfuric acid as diprotic and recognizing that both hydrogen ions react.

$$H_2SO_4 + KOH \longrightarrow \mathbf{H_2O + K_2SO_4}$$

You can balance the equation most easily by thinking about the chemistry. Since $H_2SO_4$ gives two $H^+$ ions, it needs two $OH^-$ ions to react and form two water molecules. Including this in the equation gives

$$H_2SO_4 + \mathbf{2}\ KOH \longrightarrow \mathbf{2}\ H_2O + K_2SO_4$$

The equation is now balanced.

**Exercise 8.7**

**a. 1.** $SO_2$ is the product (Chapter Box 8.3 tells you this). Sulfur is oxidized, and oxygen is reduced.

**2.** $CO_2$ and $H_2O$ are the products. The wax is oxidized. Oxygen is reduced, as usual.

**3.** $CO_2$ and $H_2O$ are the products. The butane is oxidized and oxygen is reduced.

**b. 1.** In Step 1 you write down the formulas of the reactants and products.

$$C_5H_{12} + O_2 \longrightarrow CO_2 + H_2O$$

In Step 2 you balance one element. Carbon is a good place to start.

$$C_5H_{12} + O_2 \longrightarrow \mathbf{5}\ CO_2 + H_2O$$

In Step 3 you balance the rest of the atoms. It's best to start with hydrogen, since it occurs only once on each side of the equation.

$$C_5H_{12} + O_2 \longrightarrow 5\ CO_2 + \mathbf{6}\ H_2O$$

Next, balance oxygen. There are 16 oxygen atoms on the right hand side of this equation: 10 in 5 $CO_2$ and 6 in 6 $H_2O$.

$$C_5H_{12} + \mathbf{8}\ O_2 \longrightarrow 5\ CO_2 + 6\ H_2O$$

There are no fractional coefficients, so the equation is balanced.

**2.** In Step 1 you write the equation showing all reactants and products.

$$CH_3OH + O_2 \longrightarrow CO_2 + H_2O$$

In Step 2 you balance one element. You can start with either hydrogen or carbon since they both appear only once on each side. Start with hydrogen.

$$CH_3OH + O_2 \longrightarrow CO_2 + \mathbf{2}\ H_2O$$

In Step 3 you balance the rest of the atoms. Since carbon is already balanced, you need to look only at oxygen. There are three oxygen atoms on the left and four on the right. It would be an error to put the coefficient 2 in front of $CH_3OH$, because that unbalances the carbons and the hydrogens. You can work only with $O_2$ at this stage. There is one $O_2$, and you need one more oxygen atom, or $\frac{1}{2}O_2$. Since $1 + \frac{1}{2} = \frac{3}{2}$, you get

$$CH_3OH + \tfrac{3}{2}O_2 \longrightarrow CO_2 + 2\ H_2O$$

Step 4 removes the fractional coefficients:

$$2\ CH_3OH + 3\ O_2 \longrightarrow 2\ CO_2 + 4\ H_2O$$

3. In Step 1 you write the unbalanced equation:

$C_3H_8 + O_2 \longrightarrow CO_2 + H_2O$

In Step 2 you balance one of the elements, say, carbon.

$C_3H_8 + O_2 \longrightarrow \mathbf{3}\ CO_2 + H_2O$

In Step 3 you work on the rest of the atoms. Start with hydrogen first, since it occurs only once on each side:

$C_3H_8 + O_2 \longrightarrow 3\ CO_2 + \mathbf{4}\ H_2O$

Next, balance oxygen:

$C_3H_8 + \mathbf{5}\ O_2 \longrightarrow 3\ CO_2 + 4\ H_2O$

There are no fractional coefficients, so the problem is solved.

**Exercise 8.8**

**a.** Calcium is an active metal in Group 2A, so it has a $+2$ charge after it reacts. Two hydroxide ions are needed to balance this charge, and the formula for calcium hydroxide is $Ca(OH)_2$. The unbalanced equation is

$Ca + H_2O \longrightarrow Ca(OH)_2 + H_2$

Calcium is already balanced. Oxygen appears only once on each side, so balance that next:

$Ca + \mathbf{2}\ H_2O \longrightarrow Ca(OH)_2 + H_2$

This coefficient simultaneously balanced the hydrogens, so you're finished.

**b.** Rubidium is a Group 1A metal and acquires a charge of $+1$ when it reacts. The formula for rubidium hydroxide is therefore RbOH. The equation is

$Rb + H_2O \longrightarrow RbOH + H_2$

The rubidium atoms are already balanced, as are the oxygens. This leaves only the hydrogens unbalanced. There are two hydrogens on the right and three on the left.

$Rb + H_2O \longrightarrow RbOH + \frac{1}{2}H_2$

Doubling all coefficients to remove the fraction gives the final balanced equation:

$2\ Rb + 2\ H_2O \longrightarrow 2\ RbOH + H_2$

**Exercise 8.9**

**a. 1.** When an electric current is passed through two moles of liquid water, two moles of hydrogen gas and one mole of oxygen gas are produced.

**2.** Two moles of sulfur dioxide gas and one mole of oxygen gas react in the presence of $V_2O_5$ catalyst and heat to yield two moles of sulfur trioxide gas.

**b. 1.** $N_2(g) + 3\ H_2(g) \xrightarrow[270\ \text{atm}]{450°C} 2\ NH_3(g)$

**2.** $CH_4(g) + H_2O(g) \xrightarrow[Ni]{700°C} CO(g) + 3\ H_2(g)$

**3.** Here you must write the formulas of the reactants, predict the products and write their formulas, and balance the equation. The formulas come first. Iron(III) hydroxide is $Fe(OH)_3$, and it's one of the insoluble metal hydroxides. Aqueous nitric acid is $HNO_3(aq)$. This is an acid-base reaction, so the products are water and a salt composed of iron(III) ion and nitrate ion. Water is a liquid, $H_2O(l)$. You can deduce the formula for iron(III) nitrate from the fact that iron(III) has a $+3$ charge and nitrate has a $-1$ charge. All nitrate salts are water soluble, so this product is written as $Fe(NO_3)_3(aq)$. The equation without coefficients is

$$Fe(OH)_3(s) + HNO_3(aq) \longrightarrow H_2O(l) + Fe(NO_3)_3(aq) \quad \text{(unbalanced)}$$

To balance the equation, note that iron(III) hydroxide contains three hydroxide ions, and this requires three nitric acid molecules and the formation of three water molecules for complete neutralization:

$$Fe(OH)_3(s) + \mathbf{3}\ HNO_3(aq) \longrightarrow \mathbf{3}\ H_2O(l) + Fe(NO_3)_3(aq)$$

The equation is now balanced.

**4.** Again, this requires you to figure out the formulas of reactants and products, write the equation, and, finally, balance it. Cesium is a Group 1A metal and acquires a $+1$ charge when it reacts. Group 1A elements are active metals that form a metal hydroxide and hydrogen gas when they react with water. Cesium is a solid (as are all metals but mercury), water is a liquid, cesium hydroxide is a soluble Group 1A hydroxide, and hydrogen is a gas. All this gives

$$Cs(s) + H_2O(l) \longrightarrow CsOH(aq) + H_2(g) \quad \text{(unbalanced)}$$

You balance this equation as you did the one in Exercise 8.8b, with Cs substituting for Rb. The final result is

$$2\ Cs(s) + 2\ H_2O(l) \longrightarrow 2\ CsOH(aq) + H_2(g)$$

**Exercise 8.10**

Nitric acid is a strong acid, and potassium hydroxide is a water-soluble metal hydroxide, a strong base. Therefore the reactants in solution are actually $H_3O^+(aq)$ and $NO_3^-(aq)$ from the nitric acid and $K^+(aq)$ and $OH^-(aq)$ from $KOH(aq)$. The products of this acid-base reaction are water and the salt $KNO_3(aq)$. Since all nitrates are soluble, this product in solution consists of $K^+(aq)$ and $NO_3^-(aq)$. The overall reaction is therefore

$$H_3O^+(aq) + NO_3^-(aq) + K^+(aq) + OH^-(aq) \longrightarrow 2\,H_2O(l) + K^+(aq) + NO_3^-(aq)$$

Subtracting the ions that remain unchanged as the reaction occurs gives this result:

$$H_3O^+(aq) + OH^-(aq) \longrightarrow 2\,H_2O(l)$$

The spectator ions are $K^+(aq)$ and $NO_3^-(aq)$. The sum of the charges on each side of the equation equals zero, so the charges are balanced. The mass is too, so we are finished.

**Exercise 8.11**

**a.** $HNO_3$ is a strong acid. $Ca(OH)_2$ in an insoluble hydroxide. $Ca(NO_3)_2$ is a nitrate and is therefore soluble. The complete ionic equation is therefore

$$2\,H_3O^+(aq) + 2\,NO_3^-(aq) + Ca(OH)_2(s) \longrightarrow 4\,H_2O(l) + 2\,Ca^{2+}(aq) + NO_3^-(aq)$$

To obtain the net ionic equation, subtract the spectator ion $NO_3^-$. This gives

$$2\,H_3O^+(aq) + Ca(OH)_2(s) \longrightarrow 4\,H_2O(l) + Ca^{2+}(aq)$$

Checking for atom balance shows eight hydrogen, four oxygen, and one calcium atom on each side. A charge balance check shows that each side has a total charge of +2. All is well, and the equation is balanced.

**b.** Sodium is a metallic solid, water is a liquid, sodium hydroxide is a Group 1A water-soluble metal hydroxide, and hydrogen is a gas.

$$2\,Na(s) + 2\,H_2O(l) \longrightarrow 2\,Na^+(aq) + 2\,OH^-(aq) + H_2(g)$$

There are no spectator ions in this equation, so there is nothing to subtract out. A charge balance check shows a charge of zero on each side of the equation. Inspection verifies the mass balance, and the task is done.

**c.** Nitrate salts are soluble. NaCl is soluble, and AgCl is an exception to the guideline that chloride salts are soluble (refer to Table 7.5). Putting all this together yields

$$Ag^+(aq) + NO_3^-(aq) + Na^+(aq) + Cl^-(aq) \longrightarrow AgCl(s) + Na^+(aq) + NO_3^-(aq)$$

Subtracting the spectator ions (sodium and nitrate ions in this case) gives

$$Ag^+(aq) + Cl^-(aq) \longrightarrow AgCl(s)$$

A charge balance check shows a total charge of zero on both sides. Inspection verifies that mass balances.

**d.** Ammonium salts and nitrate salts are soluble, and HgS is not. The complete equation is lengthy:

$$2\,NH_4^+(aq) + S^{2-}(aq) + Hg^{2+}(aq) + 2\,NO_3^-(aq) \longrightarrow 2\,NH_4^+(aq) + 2\,NO_3^-(aq) + HgS(s)$$

Subtracting the spectator ions gives

$$Hg^{2+}(aq) + S^{2-}(aq) \longrightarrow HgS(s)$$

There is a total charge of zero on each side, and the masses are also balanced.

**e.** The reaction occurs in water, so you must make some predictions about the solubilities of the species involved. All nitrate salts and Group 1A salts are soluble. $BaSO_4$ is insoluble. This gives.

$$Ba^{2+}(aq) + 2\,NO_3^-(aq) + 2\,Na^+(aq) + SO_4^{2-}(aq) \longrightarrow BaSO_4(s) + 2\,Na^+(aq) + 2\,NO_3^-(aq)$$

When you remove the spectator ions, you get

$$Ba^{2+}(aq) + SO_4^{2-}(aq) \longrightarrow BaSO_4(s)$$

A check of the charge and mass balances demonstrates that all is well.

**f.** Again, this reaction takes place in water, and the solubility guidelines of Chapter 7 are needed. Nitrate salts are all soluble. Here it is also important to remember that $Hg_2^{2+}$ is a diatomic cation and that $Hg_2Cl_2$ is one of the insoluble chloride salts.

$$Hg_2^{2+}(aq) + 2\,NO_3^-(aq) + 2\,K^+(aq) + 2\,Cl^-(aq) \longrightarrow Hg_2Cl_2(s) + 2\,K^+(aq) + 2\,NO_3^-(aq)$$

The net ionic equation is

$$Hg_2^{2+}(aq) + 2\,Cl^-(aq) \longrightarrow Hg_2Cl_2(s)$$

A charge balance and mass balance check verifies this equation.

**Exercise 8.12**

$$NH_4NO_3(s) + 26\text{ kJ} \xrightarrow{H_2O} NH_4^+(aq) + NO_3^-(aq)$$

**Experimental Exercise 8.1**

**1.** This is a straightforward density problem, best solved using the unit factor method:

$$1.0\text{ g sample} \times \frac{1\text{ mL sample}}{2.0\text{ g sample}} = 0.50\text{ mL sample}$$

2. The formation of iron(II) hydrogen sulfate in reaction 1 suggests the presence of hydrogen sulfate, $HSO_4^-$, in the acid. The acid that contains hydrogen sulfate is $H_2SO_4$.
3. Yes, it is, for two reasons. First, when acids react with active metals, hydrogen gas is one of the products. Also, when hydrogen gas burns in atmospheric oxygen, water is a reaction product. These two points taken together provide strong support for the team's conclusions.
4. Acids react with carbonates to form carbon dioxide, which is the most likely candidate for the identity of gas 2. Carbon dioxide doesn't react with any of the components of air under these conditions.
5. The reactants are sulfuric acid and iron; the products are $H_2$ and $Fe(HSO_4)_2$. The balanced equation is

   $$2\ H_2SO_4 + Fe \longrightarrow Fe(HSO_4)_2 + H_2$$

6. The sample is an acid, so you would expect no color change with red litmus. You would observe that nothing happens—often a useful observation.
7. Since solution A is acidic, you would observe that the blue litmus turns red.

9

# FORMULAS AND REACTIONS: STOICHIOMETRY



KEY TERMS

FIGURE 9.1
Companies that manufacture ammonia mix $H_2(g)$ and $N_2(g)$ at high temperatures in the presence of a catalyst. If the goal is to prepare 400 tons of $NH_3(g)$, a manufacturer must carefully calculate the amounts of $N_2(g)$ and $H_2(g)$ needed to start. This calculation requires a balanced equation.

In Chapter 8 you learned to balance simple chemical equations. Since chemical changes are basically what chemists study, balancing equations is a fundamental skill in chemistry. For many of the most common problems in chemistry, however, the balanced equation is just a beginning. For example, if you are in the ammonia manufacturing business, you need a balanced equation before you can calculate how much hydrogen gas your company must buy to make 400 tons of ammonia for a customer. The balanced equation by itself, however, does not give the answer to this question. The equation is only one of the things you need in order to calculate the answer. You will learn the rest of the steps needed to solve this type of problem in this chapter.

This chapter also explores other implications of balanced equations. The major theme of this chapter is the study of **stoichiometry,** the area of chemistry that considers how much of one substance reacts with another and how much of a substance forms (or can form) when a given amount of another substance reacts.

## 9.1 Unit Factors from Balanced Equations

GOAL: To obtain unit factors from balanced equations and to use them in calculations.

For solving the type of problem faced by the ammonia manufacturer referred to above, you need fundamental skills from Chapter 8 and earlier. One necessary skill is the translation of equations into words. For instance, the equation $CH_4 + 2\ O_2 \rightarrow CO_2 + 2\ H_2O$ can be translated into the English sentence "One mole of methane reacts with two moles of oxygen to give one mole of carbon dioxide and two moles of water." The interpretation of a chemical equation you will use most often in this chapter is the one that uses moles. This allows you to construct some very useful unit factors.

**If you need review, this topic is covered in more depth in Section 3.7.**

You will also need to remember what unit factors are used for. So far you have used them primarily to convert a quantity from one unit to another. To convert a number of grams into kilograms, for instance, you multiply the number of grams by the unit factor 1 kg/1000 g. The units of grams cancel, leaving the answer in units of kilograms.

A balanced equation provides a different kind of unit factor, one that expresses how much of one substance reacts with—or is produced by—a specific amount of another substance. To construct these unit factors, you must start with a balanced equation; then you take the formula of one compound with its coefficient and divide it by the formula of any other compound with its coefficient. This kind of unit factor lets you convert a given number of moles of one compound into the number of moles of another compound involved in the reaction. Here is a balanced equation, followed by four of the twelve unit factors you can get from it:

$$CH_4 + 2\,O_2 \longrightarrow CO_2 + 2\,H_2O$$

$$\frac{1\text{ mol }CH_4}{2\text{ mol }O_2} \quad \frac{1\text{ mol }CH_4}{1\text{ mol }CO_2} \quad \frac{1\text{ mol }CH_4}{2\text{ mol }H_2O} \quad \frac{2\text{ mol }O_2}{1\text{ mol }CH_4}$$

These unit factors can be translated into words. The first one can be translated this way: "For each mole of $CH_4$ you have, you need 2 mol of $O_2$ for a complete reaction." If you were designing a burner for natural gas, you might need this kind of information.

### EXAMPLE 9.1 Obtaining Unit Factors from Balanced Equations

**a.** Construct two unit factors from this balanced equation:

$$2\,C_4H_8 \xrightarrow{H_2SO_4} C_8H_{16}$$

**b.** Construct six unit factors from this balanced equation:

$$2\,H_2 + O_2 \longrightarrow 2\,H_2O$$

**SOLUTIONS**

**a.** You simply divide one of the formulas with its coefficient by the other and then reverse the process, to obtain two unit factors:

$$2\,C_4H_8 \longrightarrow C_8H_{16}$$

$$\frac{2\text{ mol }C_4H_8}{1\text{ mol }C_8H_{16}} \quad \frac{1\text{ mol }C_8H_{16}}{2\text{ mol }C_4H_8}$$

Sulfuric acid is a catalyst here. Since catalysts speed up reactions without being consumed, one molecule of a catalyst working over and over is enough (in theory) to keep a reaction going as long as the other reactants last. You can't construct unit factors for catalysts.

**b.** This sort of problem requires you to be systematic. You can start with hydrogen. There will be two factors that link hydrogen and oxygen and two other factors that link hydrogen and water:

$$2\,H_2 + O_2 \longrightarrow 2\,H_2O$$

$$\frac{2\text{ mol }H_2}{1\text{ mol }O_2} \quad \frac{1\text{ mol }O_2}{2\text{ mol }H_2} \quad \frac{2\text{ mol }H_2}{2\text{ mol }H_2O} \quad \frac{2\text{ mol }H_2O}{2\text{ mol }H_2}$$

Finally, there will be two factors that link oxygen and water:

$$\frac{1\text{ mol }O_2}{2\text{ mol }H_2O} \quad \frac{2\text{ mol }H_2O}{1\text{ mol }O_2}$$

You now have the requested six unit factors, which is all you can get from this equation.

**EXERCISE 9.1**

**a.** Construct two unit factors from this balanced equation:

$$2\,NO_2 \longrightarrow N_2O_4$$

**STUDY ALERT**

**You may have noticed that the number of unit factors that can be made from a given chemical equation can be quite large. There are so many possible unit factors that there's no way anyone could ever memorize them all, and they change with every equation. You have to know how to construct them.**

**b.** Construct twelve unit factors from this balanced equation:

$$HCl + NaOH \longrightarrow NaCl + H_2O$$

**c.** Balance the following equation, and then construct four unit factors from it. (Twelve are possible.)

$$NaOH + H_2SO_4 \longrightarrow Na_2SO_4 + H_2O \quad \text{(unbalanced)}$$

## 9.2 Using Unit Factors to Solve Mole-Mole Problems

**GOAL:** To calculate the number of moles of a species needed or produced when given the number of moles of another species present in a reaction.

The unit factors of Section 9.1 let you convert a given number of moles of one substance to the corresponding number of moles of any other substance in a balanced equation. This conversion is vital for anyone studying, planning, or analyzing a reaction.

Consider the reaction that Joseph Priestley (1733–1804) used for his 1774 discovery of oxygen (see Figure 9.2):

$$HgO \xrightarrow{\text{heat}} Hg + O_2 \quad \text{(not balanced)}$$

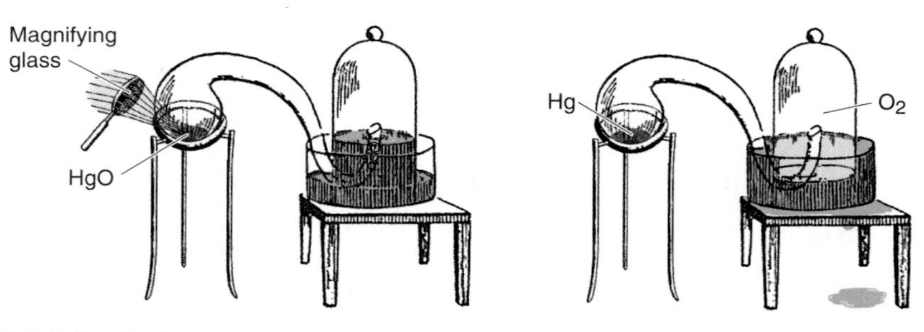


FIGURE 9.2
Joseph Priestley prepared oxygen by heating a sample of bright red mercury(II) oxide on a sunny day using a magnifying glass. When gases are reaction products, they are often collected by bubbling them through water into an inverted flask or cylinder. (You may remember from Chapter 1 that Lavoisier used this same reaction to investigate other chemical phenomena. See Experimental Problem 1.1 and Figure 1.7 for details.)

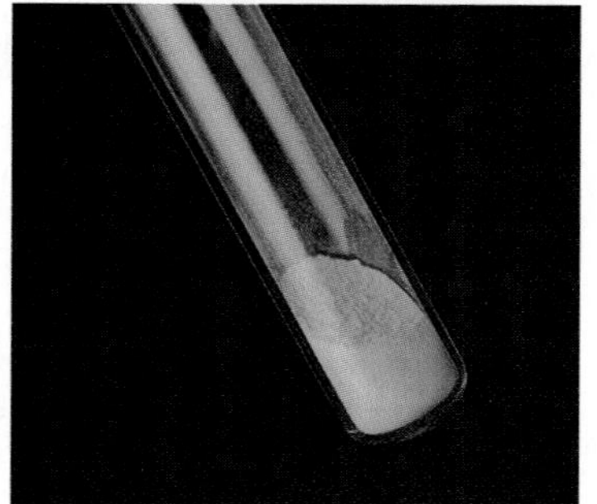

FIGURE 9.3
When mercury(II) oxide (a naturally occurring mineral known as cinnabar) is strongly heated, oxygen gas and mercury metal form.

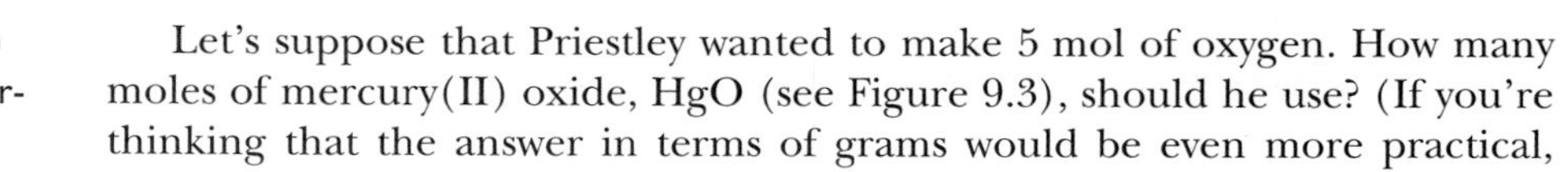

Let's suppose that Priestley wanted to make 5 mol of oxygen. How many moles of mercury(II) oxide, HgO (see Figure 9.3), should he use? (If you're thinking that the answer in terms of grams would be even more practical, you're right. We'll get to that later.) The first step is to balance the equation:

$$2\ \mathrm{HgO} \xrightarrow{\text{heat}} 2\ \mathrm{Hg} + \mathrm{O_2} \quad \text{(balanced)}$$

The second step is to find the appropriate unit factor. As usual, you want the unit factor that allows you to cancel out the units you don't need, leaving the units you do need. You were given a number of moles of $O_2$, and you need the number of moles of HgO. The calculation should be no surprise:

$$5\ \cancel{\mathrm{mol\ O_2}}\left(\frac{2\ \mathrm{mol\ HgO}}{1\ \cancel{\mathrm{mol\ O_2}}}\right) = 10\ \mathrm{mol\ HgO}$$

For Priestley to end up with 5 moles of oxygen, he should begin with 10 moles of HgO.

### EXAMPLE 9.2 Calculating Moles of One Species from Moles of Another Species

**a.** If Priestley made 5 mol of oxygen from 10 mol of mercury(II) oxide, how many moles of mercury metal would he produce at the same time?

**b.** Next, let's suppose that Priestley made oxygen from potassium chlorate instead of mercury(II) oxide. The equation for this reaction is

$$\mathrm{KClO_3} \longrightarrow \mathrm{KCl} + \mathrm{O_2} \quad \text{(not balanced)}$$

How many moles of $KClO_3$ are needed to make 5.0 mol of $O_2$?

**SOLUTIONS**

**a.** The balanced equation is $2\ \mathrm{HgO} \rightarrow 2\ \mathrm{Hg} + \mathrm{O_2}$. The equation gives a unit factor you can use to calculate moles of mercury:

$$5\ \cancel{\mathrm{mol\ O_2}}\left(\frac{2\ \mathrm{mol\ Hg}}{1\ \cancel{\mathrm{mol\ O_2}}}\right) = 10\ \mathrm{mol\ Hg}$$

This reaction makes 10 mol of mercury metal in addition to 5 mol of oxygen.

**b.** First, you must balance the equation:

$$2\ \mathrm{KClO_3} \longrightarrow 2\ \mathrm{KCl} + 3\ \mathrm{O_2} \quad \text{(balanced)}$$

Then you select the unit factor you need and use it:

$$5.0\ \cancel{\mathrm{mol\ O_2}}\left(\frac{2\ \mathrm{mol\ KClO_3}}{3\ \cancel{\mathrm{mol\ O_2}}}\right) = \frac{10}{3}\ \mathrm{mol\ KClO_3} = 3.3\ \mathrm{mol\ KClO_3}$$

To make 5.0 mol of oxygen, Priestley would need 3.3 mol of potassium chlorate.

**EXERCISE 9.2**

**a.** How many moles of NaOH are needed to react with phosphoric acid to yield 0.0013 mol of $Na_3PO_4$?

**b.** How many moles of chlorine are needed to make 3.8 mol of $AlCl_3$ from the reaction $\mathrm{Al} + \mathrm{Cl_2} \rightarrow \mathrm{AlCl_3}$ (equation not balanced)?

## 9.3 More Complicated Mole-Mole Problems

GOAL: To calculate how many moles of a reactant are used up, produced, or remain after a reaction goes partially to completion, or when one reactant is used in excess.

Section 9.2 contains an unrealistic assumption—the idea that chemical reactions work perfectly and that every mole of reactant yields the maximum possible amount of product. In the real world of the laboratory, it doesn't often work that way. Many reactions start well but then grind to a halt, with some of the reactants still left unconverted into products. (For more information about this, see Chapter 14.) You can push some of these reluctant reactions closer to the desired end by adding more of one reactant (usually the less expensive one) than is strictly necessary. Chemists describe this addition as using an **excess** of that reactant.

Here's a problem of this sort. A chemist mixed 5 mol of hydrogen with 3 mol of iodine in an attempt to make hydrogen iodide. When the reaction stopped, there was 1 mol of iodine left. How much unreacted $H_2$ was left at the end, and how much HI was made? To solve this kind of problem, you first balance the equation. Then you make up and use a table. For this example the balanced equation is a simple one:

$$H_2 + I_2 \longrightarrow 2\ HI$$

Set up the table by listing the substances in the balanced equation across the top:

| Moles: | $H_2$ | $I_2$ | HI |
|---|---|---|---|
| At the beginning | | | |
| Change | | | |
| At the end | | | |

Start to fill in the table by using the information you were given. There were 5 mol of $I_2$ and 3 mol of $H_2$ at the beginning. You can also say that there was no HI at the beginning, since no product can be formed until the reaction starts. There was 1 mol of $I_2$ left at the end.

| Moles: | $H_2$ | $I_2$ | HI |
|---|---|---|---|
| At the beginning | 5 | 3 | 0 |
| Change | | | |
| At the end | | 1 | |

Now you can start calculating. If there were 3 mol of $I_2$ at the beginning and only one in the end, two were used up (3 mol $I_2$ − 1 mol $I_2$ = 2 mol $I_2$). This is the change in the number of moles of $I_2$. A negative sign shows that the number of moles of iodine decreased:

| Moles: | $H_2$ + | $I_2$ ⟶ | 2 HI |
|---|---|---|---|
| Beginning | 5 | 3 | 0 |
| Change | | −2 | |
| End | | 1 | |

Since you know that 2 mol of $I_2$ were used up, you can use the unit factor method to calculate both the number of moles of $H_2$ that were used up and the number of moles of HI that were formed:

$$2\ \cancel{\text{mol } I_2}\left(\frac{1\text{ mol } H_2}{1\ \cancel{\text{mol } I_2}}\right) = 2\text{ mol } H_2 \text{ used up}$$

$$2\ \cancel{\text{mol } I_2}\left(\frac{2\text{ mol HI}}{1\ \cancel{\text{mol } I_2}}\right) = 4\text{ mol HI formed}$$

Now you can fill in two more numbers in the table. Results of this kind of calculation always go in the "Change" row. The change in $H_2$ is negative (it's being used up), and the change in HI is positive:

| Moles: | $H_2$ + | $I_2 \longrightarrow$ | 2 HI |
|---|---|---|---|
| Beginning | 5 | 3 | 0 |
| Change | − 2 | − 2 | + 4 |
| End | | 1 | |

The last two numbers, the answers to the two parts of the problem, come from simple arithmetic again. You started with 5 mol of $H_2$ and used 2 mol, leaving 3 mol. You started with no HI and made 4 mol, and that's the other answer you need.

| Moles: | $H_2$ + | $I_2 \longrightarrow$ | 2 HI |
|---|---|---|---|
| Beginning | 5 | 3 | 0 |
| Change | − 2 | − 2 | + 4 |
| End | 3 | 1 | 4 |

EXAMPLE 9.3 **How Much Did You Make, and How Much Is Left?**

A team of chemists took 1.2 mol of pure $KClO_3$ and heated it until they obtained 0.9 mol of oxygen. How many moles of $KClO_3$ did they have left? (Reminder: The unbalanced equation is $KClO_3 \rightarrow KCl + O_2$. To create useful unit factors, though, you must have a *balanced* equation.)

**SOLUTION**

First, you must write the balanced equation:

This equation was balanced in Example 9.2.

$$2\ KClO_3 \longrightarrow 2\ KCl + 3\ O_2$$

Now it's time to construct a table containing the given information (You can ignore KCl since it is not part of this particular problem.)

| Moles: | $2\ KClO_3 \longrightarrow$ | 2 KCl + | $3\ O_2$ |
|---|---|---|---|
| Beginning | 1.2 | | 0 |
| Change | | | |
| End | | | 0.9 |

The change in the amount of $O_2$ is 0.9 mol:

| Moles: | $2\ KClO_3$ | $\longrightarrow$ | $2\ KCl$ | + | $3\ O_2$ |
|---|---|---|---|---|---|
| Beginning | 1.2 | | | | 0 |
| Change | | | | | + 0.9 |
| End | | | | | 0.9 |

Now calculate the change in the amount of $KClO_3$ by using the balanced equation and the unit factor method:

$$0.9\ \cancel{\text{mol } O_2}\left(\frac{2\text{ mol } KClO_3}{3\ \cancel{\text{mol } O_2}}\right) = 0.6\text{ mol } KClO_3 \text{ used up}$$

The amount of $KClO_3$ used goes in the table:

| Moles: | $2\ KClO_3$ | $\longrightarrow$ | $2\ KCl$ | + | $3\ O_2$ |
|---|---|---|---|---|---|
| Beginning | 1.2 | | | | 0 |
| Change | − 0.6 | | | | + 0.9 |
| End | | | | | 0.9 |

The amount of $KClO_3$ left over is 0.6 mol (1.2 mol $KClO_3$ − 0.6 mol $KClO_3$ = 0.6 mol $KClO_3$).

**EXERCISE 9.3**

The same team of chemists heated 1.2 mol of HgO until they had 0.4 mol of $O_2$. How many moles of HgO did they have left? How many moles of Hg did they make?

## 9.4 Converting between Moles and Grams

**GOAL:** To calculate how many grams of a substance you have when you are given the number of moles, and vice versa.

The most important practical problems in stoichiometry are ones in which you need to know how much (in units of mass) of a reaction product you can expect to make, knowing only how much reactant (also in units of mass) you have. You cannot find out the number of moles of a substance directly, by weighing or by any other means. The problems in Sections 9.2 and 9.3 are the middle steps of this kind of problem. To complete the ends, you have to know how to convert the number of grams of a substance into a number of moles, and vice versa. That's what this section is concerned with.

The key to doing this type of problem is the fact that a mole of any element has a known mass. Chemists have been measuring these masses for almost two centuries with increasing accuracy and precision. You can find molar masses in lists (such as the one on the end papers of this book) and on most periodic tables. The mass of a mole of a compound is the sum of the masses of the moles of the atoms that make it up.

For example, what is the mass of 1 mol of NaCl? Well, the mass of 1 mol of Na is 22.99 g, and the mass of 1 mol of Cl is 35.45 g. Since 1 mol of NaCl contains 1 mol of Na atoms and 1 mol of Cl atoms, the mass of 1 mol of NaCl is the sum of these two masses, which is 58.44 g. This information can be made into two useful unit factors:

$$\frac{58.44 \text{ g NaCl}}{1 \text{ mol NaCl}} \quad \text{and} \quad \frac{1 \text{ mol NaCl}}{58.44 \text{ g NaCl}}$$

What is the mass of 0.50 mol of NaCl? It's:

$$0.50 \text{ mol NaCl}\left(\frac{58.44 \text{ g NaCl}}{1 \text{ mol NaCl}}\right) = 29 \text{ g NaCl}$$

The number 0.50 has two significant figures, and your answer should too (Section 3.3).

**Of course 1 mol of NaCl contains 1 mol of $Na^+$ ions and 1 mol of $Cl^-$ ions. You can ignore this when computing the molar mass of NaCl. (Why?)**

### EXAMPLE 9.4 Determining the Mass of a Given Number of Moles of a Compound

What is the mass of each of the following?

**a.** 1.0 mol of NaOH
**b.** 0.50 mol of $Na_2O$
**c.** 1.37 mol of $Ca(OH)_2$

**SOLUTIONS**

**a.** One mole of NaOH contains 1.0 mol each of sodium, oxygen, and hydrogen atoms. From the table inside the front cover of this book, the molar masses of Na, O, and H in grams per mole are 22.99, 16.00, and 1.008 g mol$^{-1}$, respectively. A calculator gives the sum of these molar masses as 39.998, which should be rounded off to 40.00 g mol$^{-1}$ for NaOH.

**b.** The molar masses of Na and O are 22.99 and 16.00 g mol$^{-1}$, respectively. The formula $Na_2O$ contains two Na atoms and one O atom, so the molar mass of $Na_2O$ is

$$2(22.99 \text{ g mol}^{-1} \text{ Na}) + 16.00 \text{ g mol}^{-1} \text{ O} = 61.98 \text{ g mol}^{-1}$$

The mass of 0.50 mol is

$$0.50 \cancel{\text{mol Na}_2\text{O}}\left(\frac{61.98 \text{ g Na}_2\text{O}}{1 \cancel{\text{mol Na}_2\text{O}}}\right) = 30.99 \text{ g Na}_2\text{O}$$

which you round off to 31 g $Na_2O$.

**c.** The molar masses of Ca, O, and H are 40.08, 16.00, and 1.008 g mol$^{-1}$, respectively. The formula $Ca(OH)_2$ contains one Ca atom, two O atoms, and two H atoms. The molar mass of $Ca(OH)_2$ is

$$40.08 \text{ g mol}^{-1} \text{ Ca} + 2(16.00 \text{ g mol}^{-1} \text{ O}) + 2(1.008 \text{ g mol}^{-1} \text{ H}) = 74.096 \text{ or } 74.10 \text{ g mol}^{-1} \text{ Ca(OH)}_2$$

The mass of 1.37 mol of $Ca(OH)_2$ is

$$1.37 \cancel{\text{mol Ca(OH)}_2}\left(\frac{74.10 \text{ g Ca(OH)}_2}{1 \cancel{\text{mol Ca(OH)}_2}}\right) = 101.517 \text{ g}$$

which you round off to 102 g $Ca(OH)_2$.

FIGURE 9.4
NaOH is the active ingredient in several brands of drain cleaners. Follow the label instructions—handle with care!

**EXERCISE 9.4**

What is the mass of each of the following?

**a.** 1 mol of $CaCl_2$

**b.** 0.500 mol of $H_2SO_4$

**c.** 1.37 mol of $Ca_3(PO_4)_2$

Alternatively, suppose that you have, for example, 100.0 g of NaCl. How many moles of NaCl is this? The best way to get the answer is to use the unit factor method:

$$100.0 \, \cancel{\text{g NaCl}} \left( \frac{1 \text{ mol NaCl}}{58.44 \, \cancel{\text{g NaCl}}} \right) = 1.711 \text{ mol}$$

**EXAMPLE 9.5** **Determining the Number of Moles from the Mass of a Compound**

**a.** You have 0.14 g of NaOH. How many moles is this?

**b.** You have 12.7 g of $Al(OH)_3$. How many moles is this?

**SOLUTIONS**

**a.** The molar mass of NaOH is

$$22.99 \text{ g mol}^{-1} + 16.00 \text{ g mol}^{-1} + 1.008 \text{ g mol}^{-1} = 40.00 \text{ g mol}^{-1}$$

If you have 0.14 g of NaOH, you have:

$$0.14 \, \cancel{\text{g NaOH}} \left( \frac{1 \text{ mol NaOH}}{40.00 \, \cancel{\text{g NaOH}}} \right) = 0.0035 \text{ mol NaOH}$$

$$= 3.5 \times 10^{-3} \text{ mol NaOH}$$

**b.** The molar mass of $Al(OH)_3$ is

$$26.98 \text{ g mol}^{-1} + 3(16.00 \text{ g mol}^{-1}) + 3(1.008 \text{ g mol}^{-1})$$
$$= 78.004 \text{ g mol}^{-1} \text{ (which rounds to 78.00 g mol}^{-1}\text{)}$$

If you have 12.7 g of $Al(OH)_3$, you have

$$12.7 \, \cancel{\text{g Al(OH)}_3} \left( \frac{1 \text{ mol Al(OH)}_3}{78.004 \, \cancel{\text{g Al(OH)}_3}} \right) = 0.163 \text{ mol Al(OH)}_3$$

(after rounding off)

FIGURE 9.5
This is a ruby, a naturally occurring sample of aluminum oxide. The color comes from a chromium oxide impurity.

**EXERCISE 9.5**

**a.** You have 1.37 g of $CaCO_3$. How many moles is this?

**b.** You have 0.0012 g of $Al_2O_3$. How many moles is this?

## CHAPTER BOX 9.1 MOLECULES AND MOLES

**It may** seem strange that this chapter does not consider the masses of single atoms and molecules. You can, for instance, calculate the mass of a mole of a compound by multiplying the mass of a molecule by Avogadro's number. This works, but it's not necessary or even especially useful to do so. One mole of *any* substance contains Avogadro's number of atoms, molecules, or formula units, whatever is appropriate. As Figure 9.6 indicates, chemists worked with moles long before they had any idea what the value of Avogadro's number was (see Section 5.7).

FIGURE 9.6
Scientists knew that 1 mol of iron atoms (55.85 g) is more massive than 1 mol of sulfur atoms (32.07 g) long before they knew the value of Avogadro's number.

The first step was taken by Joseph Proust (1754–1826), who stated the Law of Definite Proportions: Each chemical compound has a fixed and definite composition in terms of mass. The idea that compounds have definite formulas is not far beyond this.

John Dalton decided that only a few simple combinations of any two elements should be possible. Since 1 g of hydrogen reacts with 8 g of oxygen (using modern units) to make water (leaving neither hydrogen nor oxygen left over), he decided that the ratio of the atomic weights of H and O was 1:8 and that the formula of water was HO.

Dalton went further. He knew that some pairs of elements form more than one compound. He studied the oxides of carbon and found that 12 g of carbon could react with either 16 or 32 g of oxygen to form two different compounds. The formulas of these two compounds, he decided, were most likely CO and $CO_2$. Turning to the oxides of nitrogen, he found that 14 g of nitrogen could combine with 8, 16, or 32 g of oxygen. He speculated that the formulas of these oxides might be NO, $NO_2$, and $NO_4$, but based on the mass ratios were more likely to be $N_2O$, NO, and $NO_2$.

The British chemist Thomas Thomson (1773–1852) was convinced. He applied Dalton's ideas to a wide variety of compounds and found that they worked. Thomson was the author of a widely used textbook, and he put his own and Dalton's results in his third edition, published in 1807. Thomson's work, along with similar research by the Swedish chemist Jons Jakob Berzelius (1779–1848), really established Dalton's ideas.

The confusion over which of the possible ratios for many formulas was correct (HO or $H_2O$ for water, for instance) lasted half a century.

## 9.5 Solving Gram-Gram Problems

**GOAL:** To use the number of grams of one reactant to calculate either the number of grams of a product or the number of grams of another reactant.

In this section you will learn to solve a type of problem that requires you to combine several of the procedures you have learned already. A problem of this kind faced by a working chemist might go like this: When 6.0 g of hydrogen reacts with an excess of oxygen, how many grams of water will be produced if all of the hydrogen is consumed? (The reaction itself is shown in Figure 9.11 on page 311.)

This kind of problem is solved in three basic steps (see Figure 9.7), and you already know how to do all of them. The first and third portions come from Section 9.4 on converting between moles and grams. The middle part comes from Section 9.2, which covered mole-mole problems. Here's an overview of how the procedure works.

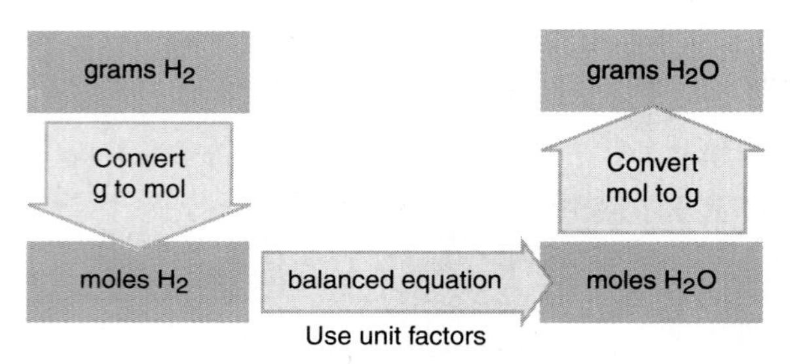


FIGURE 9.7

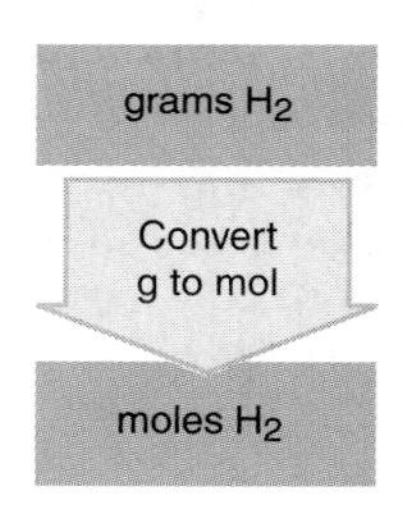


FIGURE 9.8

Write the equation first: $2\ H_2 + O_2 \longrightarrow 2\ H_2O$. Then you convert the number of grams of hydrogen to the number of moles (see Figure 9.8). Since hydrogen is $H_2$, its molar mass is $2(1.008\ \text{g mol}^{-1}) = 2.016\ \text{g mol}^{-1}$. The calculation of the number of moles, using a unit factor, goes like this:

$$6.0\ \cancel{\text{g H}_2}\left(\frac{1\ \text{mol H}_2}{2.016\ \cancel{\text{g H}_2}}\right) = 3.0\ \text{mol H}_2$$

The units of grams cancel, leaving the units of moles.

In the second step of the procedure (see Figure 9.9), you need the balanced equation. For the reaction of $H_2$ with $O_2$ to give $H_2O$, this is $2\ H_2 + O_2 \rightarrow 2\ H_2O$. You can now use a unit factor from the equation to convert 3.0 mol of $H_2$ into the number of moles of water that can be made from the $H_2$:

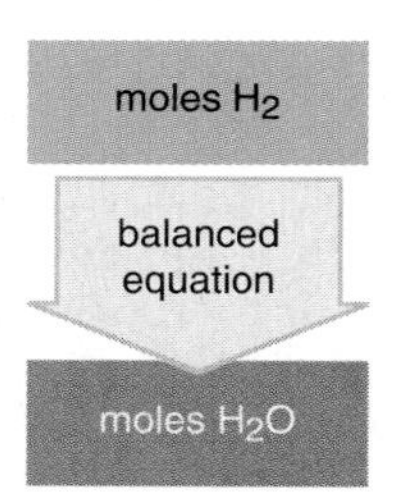


FIGURE 9.9

$$3.0\ \cancel{\text{mol H}_2}\left(\frac{2\ \text{mol H}_2\text{O}}{2\ \cancel{\text{mol H}_2}}\right) = 3.0\ \text{mol H}_2\text{O}$$

Next, you convert 3.0 mol of $H_2O$ to grams (see Figure 9.10). The molar mass of $H_2O$ is $2(1.008\ \text{g mol}^{-1}) + 16.00\ \text{g mol}^{-1} = 18.02\ \text{g mol}^{-1}$. The number of grams of water in 3.0 mol is

$$3.0\ \text{mol H}_2\text{O}\left(\frac{18.02\ \text{g H}_2\text{O}}{1\ \text{mol H}_2\text{O}}\right) = 54\ \text{g H}_2\text{O}$$

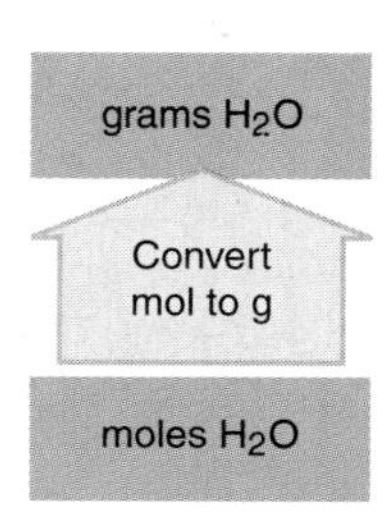


FIGURE 9.10

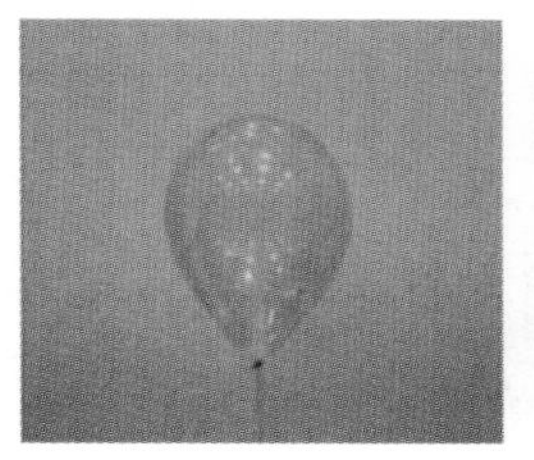
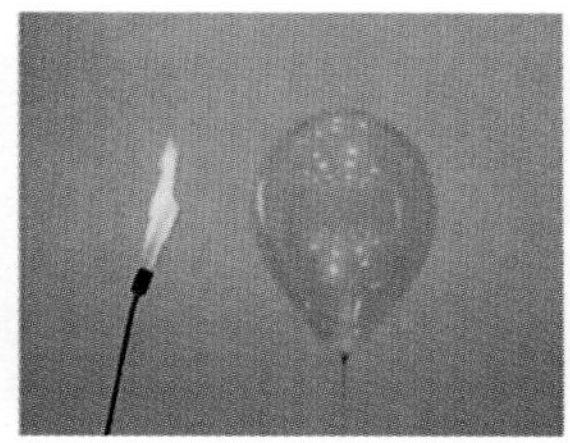
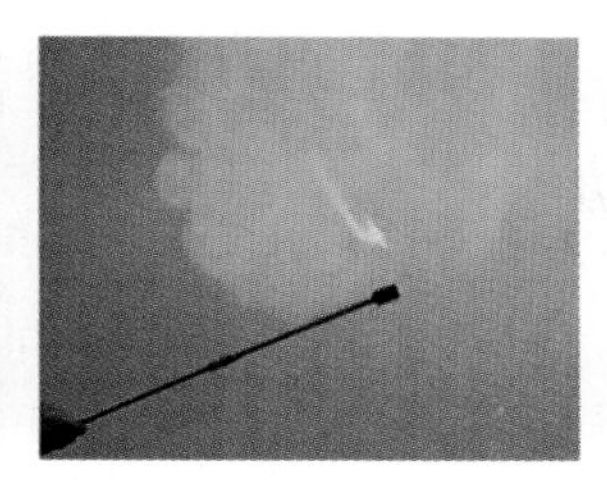
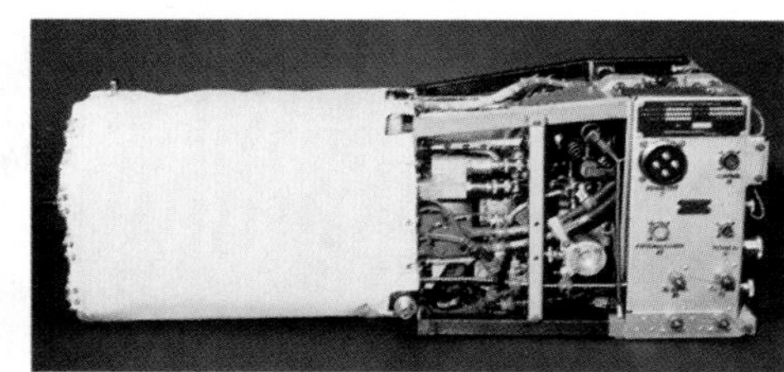

FIGURE 9.11
Exploding balloons—in which hydrogen reacts with oxygen from air in the reaction $2\ H_2 + O_2 \rightarrow 2\ H_2O$—are staples of chemical demonstrations (left). The same reaction in the more controlled environment of a fuel cell (right) may someday provide energy without significant pollution, although the method is still too expensive for widespread use.

Does this procedure seem cumbersome? To make it shorter, you can combine the three steps into one:

$$6.0\ \cancel{\text{g H}_2}\left(\frac{1\ \cancel{\text{mol H}_2}}{2.016\ \cancel{\text{g H}_2}}\right)\left(\frac{2\ \cancel{\text{mol H}_2\text{O}}}{2\ \cancel{\text{mol H}_2}}\right)\left(\frac{18.02\ \text{g H}_2\text{O}}{1\ \cancel{\text{mol H}_2\text{O}}}\right) = 54\ \text{g of H}_2\text{O}$$

The units cancel out, as they typically do with the unit factor method.

**STUDY ALERT**

**If the units don't cancel out, you've done something wrong. Making sure that the units cancel out is one of the most important ways of checking whether your answer is reasonable. Perhaps the most common error made by students learning to work chemistry problems is the failure to put complete units with numbers they write down. The units are a necessary and integral part of the number—you *must* use them (please!).**

### EXAMPLE 9.6 Solving Gram-Gram Problems

**a.** How many grams of hydrogen gas form when 5.0 g of sodium metal react with excess water?

**b.** You need to make 20 g of silver chloride from silver nitrate and sodium chloride. How much silver nitrate do you require? (Note: Silver nitrate is expensive, as anything containing silver is. You'd better not buy too much.)

**SOLUTIONS**

**a.** To construct the three required unit factors, you need three pieces of information. These are the molar mass of Na ($22.99\ \text{g mol}^{-1}$), the balanced equation ($2\ \text{Na} + 2\ \text{H}_2\text{O} \rightarrow 2\ \text{NaOH} + \text{H}_2$), and the molar mass of $H_2$ ($2 \times 1.008\ \text{g mol}^{-1}\ \text{H} = 2.016\ \text{g mol}^{-1}\ \text{H}_2$). These allow you to construct the three unit factors:

$$5.0\ \cancel{\text{g Na}}\left(\frac{1\ \cancel{\text{mol Na}}}{22.99\ \cancel{\text{g Na}}}\right)\left(\frac{1\ \cancel{\text{mol H}_2}}{2\ \cancel{\text{mol Na}}}\right)\left(\frac{2.016\ \text{g H}_2}{1\ \cancel{\text{mol H}_2}}\right) = 0.22\ \text{g H}_2$$

You get only 0.22 g of $H_2$ from 5.0 g of sodium. This doesn't seem like much, but hydrogen gas has a much smaller molar mass than sodium metal does and 2 mol Na yields only 1 mol $H_2$ according to the balanced equation.

**b.** Again you need three pieces of information to construct the three required unit factors. The molar mass of silver chloride is

$$107.9\ \text{g mol}^{-1}\ \text{Ag} + 35.45\ \text{g mol}^{-1}\ \text{Cl} = 143.4\ \text{g mol}^{-1}\ \text{AgCl}$$

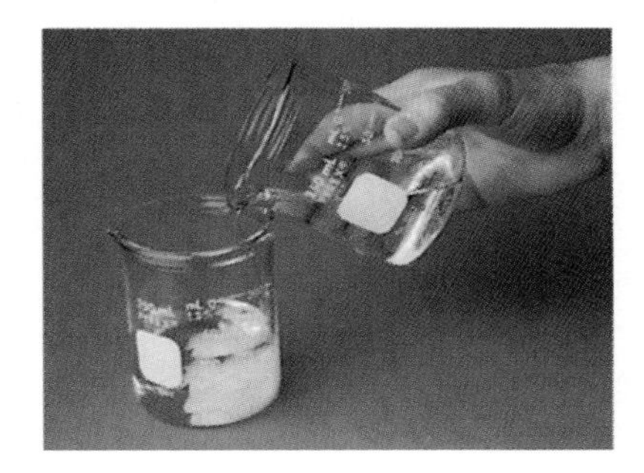

FIGURE 9.12
To make NaCl react with $AgNO_3$ in the laboratory, you dissolve each compound (both are white solids) separately in water. When you mix the two solutions, insoluble white AgCl forms. The other product, $NaNO_3$, stays in solution. You can isolate the AgCl by filtration.

The balanced equation is

$$AgNO_3 + NaCl \longrightarrow AgCl + NaNO_3$$

The molar mass of silver nitrate is

$$107.9 \text{ g mol}^{-1} \text{ Ag} + 14.01 \text{ g mol}^{-1} \text{ N} + 3(16.00 \text{ g mol}^{-1} \text{ O}) = 169.9 \text{ g mol}^{-1} \text{ AgNO}_3$$

You then set up the equation and solve it:

$$20 \cancel{\text{g AgCl}} \left(\frac{1 \cancel{\text{mol AgCl}}}{143.4 \cancel{\text{g AgCl}}}\right)\left(\frac{1 \cancel{\text{mol AgNO}_3}}{1 \cancel{\text{mol AgCl}}}\right)\left(\frac{169.9 \text{ g AgNO}_3}{1 \cancel{\text{mol AgNO}_3}}\right) = 24 \text{ g AgNO}_3$$

You need to start with slightly more $AgNO_3$ than the AgCl you will produce, which makes sense because $AgNO_3$ has a larger molar mass than AgCl (167 g mol$^{-1}$ compared to 143 g mol$^{-1}$) and the balanced equation states that 1 mol of silver nitrate yields 1 mol of silver chloride.

**EXERCISE 9.6**

**a.** How many grams of $Na_2SO_4$ will be produced by the reaction of 9.2 g of NaOH with excess $H_2SO_4$?

**b.** How many grams of $CO_2$ gas will you get from the reaction of 1.00 g of $Na_2CO_3$ with excess acetic acid ($CH_3CO_2H$)? The balanced equation is

$$Na_2CO_3 + 2\ CH_3CO_2H \longrightarrow H_2O + CO_2 + 2\ CH_3CO_2Na$$

## 9.6 The Limiting Reactant: Mole-Mole Problems

**GOAL:** To determine which reactant will be used up first and to calculate both the amount of the other reactant remaining and the yield of the products in moles, given the numbers of moles for all reactants.

Suppose you are a chemist planning a reaction, and you want to know how much product to expect. Most of the problems in Section 9.5 mentioned "an excess" of one reactant. If one reactant is in excess, you know that the other reactant (or one of the other reactants) is the one that will be used up first. But suppose all you know at the beginning of a reaction is the number of moles of each of the reactants. What do you do then?

In such a case at least one of the reactants is probably in excess, and there will be some of it left over at the end of the reaction. Consider that the reactant that is not in excess will be completely used up during the reaction. The reactant that is used up is called the **limiting reactant,** because it places a limit on how much product you can get from the reaction. (Occasionally you will have just enough of each reactant so that they are all used up.)

As one possible problem of this type, let's consider the reaction of $H_2$ and $Cl_2$ to give HCl. Suppose you start with 5 mol of $H_2$ and 1 mol of $Cl_2$ and allow them to react completely. How many moles of HCl will you make?

To see how to figure out the answer, let's consider a related problem involving something more visible than molecules. Suppose you have five plastic lids and one plastic bowl on which they fit. How many different kinds of food can you store? The result is shown in Figure 9.13.

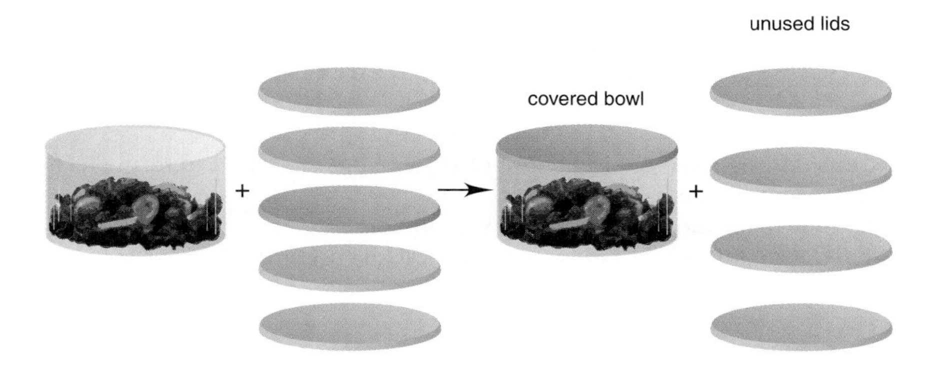


**FIGURE 9.13**
If you have five lids but only one bowl bottom, you will be able to use only one of the lids. The rest are not going to be any help in storing food.

What does this have to do with the reaction of $H_2$ and $Cl_2$? To answer this question, let's consider the balanced equation:

$$H_2 + Cl_2 \longrightarrow 2\ HCl$$

Remember that this equation can be translated into English in two ways. The most important one this time is "One molecule of hydrogen ($H_2$) plus one molecule of chlorine ($Cl_2$) gives two molecules of HCl." One molecule of $Cl_2$ can react with only one molecule of $H_2$. Once that happens, the $Cl_2$ is all gone, and no more HCl can form regardless of how many extra molecules of $H_2$ are present (see Figure 9.14). The chlorine limits the amount of product that can be formed, so chlorine is the limiting reactant.

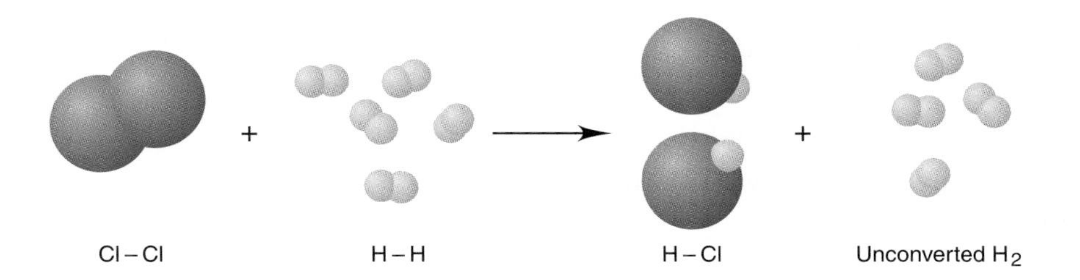


**FIGURE 9.14**
Once this reaction uses up the $Cl_2$, no more HCl can form.

If you start with 5 mol of $H_2$ and 1 mol of $Cl_2$, what will you end up with? Once you identify the limiting reactant, you can find the amount of HCl you can make by using the familiar unit factor method. In this case it is easy to see that chlorine is the limiting reactant, because the balanced equation tells you directly that 1 mol of $Cl_2$ reacts with only 1 mol of $H_2$. All the extra $H_2$ in the universe won't help make any more HCl if only 1 mol of $Cl_2$ is present.

$$1\ \cancel{\text{mol Cl}_2}\left(\frac{2\ \text{mol HCl}}{1\ \cancel{\text{mol Cl}_2}}\right) = 2\ \text{mol HCl}$$

You will make 2 mol of HCl and have 4 mol of $H_2$ left over (you started with 5 mol of $H_2$ minus the 1 mole of $H_2$ that reacted with the 1 mol of $Cl_2$—see Figure 9.14).

When the coefficients of the reactants are not both 1, this kind of problem is slightly more complicated. One method is to *identify the limiting reactant by dividing the number of moles of each reactant by its coefficient in the balanced equation. The limiting reactant is the one that gives the smallest result.* Example 9.7 illustrates the technique.

### EXAMPLE 9.7 Solving Limiting Reactant Mole-Mole Problems

Consider the reaction of 3.0 mol of $H_2$ with 2.0 mol of $O_2$ to form $H_2O$.

**a.** What is the limiting reactant?
**b.** How many moles of water will you obtain?
**c.** How many moles of the reactant originally present in excess will be left over?

**SOLUTIONS**

**a.** First, of course, it's necessary to have a balanced equation. The equation for this familiar reaction is

$$2\ H_2 + O_2 \longrightarrow 2\ H_2O$$

Next, determine the limiting reactant by dividing the starting amount of each reactant by its reaction coefficient:

$$\frac{3.0 \text{ mol } H_2}{2} = 1.5 \qquad \frac{2.0 \text{ mol } O_2}{1} = 2.0$$

The limiting reactant is $H_2$, since it gives the smallest number in the calculations.

**b.** You find out how much $H_2O$ forms by using the unit factor method. Since $H_2$ limits the amount of $H_2O$ that can form, you calculate the moles of $H_2O$ produced from the amount of $H_2$ present at the start. Here's the equation:

$$3.0 \cancel{\text{mol } H_2} \left(\frac{2 \text{ mol } H_2O}{2 \cancel{\text{mol } H_2}}\right) = 3.0 \text{ mol } H_2O$$

The amount of water formed is 3.0 mol.

**c.** How much of the reactant in excess, $O_2$, is left over? As usual, you calculate the change in the amount of $O_2$ by using the unit factor method:

$$3.0 \cancel{\text{mol } H_2} \left(\frac{1 \text{ mol } O_2}{2 \cancel{\text{mol } H_2}}\right) = 1.5 \text{ mol } O_2$$

Since 1.5 mol of $O_2$ was used up in the reaction and you had 2.0 mol of $O_2$ in the beginning, 0.5 mol of $O_2$ remains unreacted. This can be summarized by using a table similar to the ones in Section 9.3.

| Moles: | $2\ H_2$ | $+\ O_2 \longrightarrow$ | $2\ H_2O$ |
|---|---|---|---|
| Beginning | 3.0 | 2.0 | 0 |
| Change | −3.0 | −1.5 | +3.0 |
| End | 0 | 0.5 | 3.0 |

The amount of $O_2$ left over is 0.5 mol.

EXERCISE 9.7

Suppose you start with 0.50 mol of $CH_4$ and 3.0 mol of $Cl_2$ and run this reaction:

$CH_4 + Cl_2 \longrightarrow CCl_4 + HCl$ (not balanced)

**a.** What is the limiting reactant?
**b.** How many moles of $CCl_4$ will you obtain?

## 9.7 The Limiting Reactant: Gram-Gram Problems

GOAL: To identify the reactant that will be used up first, given the masses of all reactants in a reaction, and to use that knowledge to calculate the yields of the products in grams.

To solve the problems in this section, you once again need to combine strategies from two previous sections. The problems here are both gram-gram problems (like those in Section 9.5) and limiting reactant problems (like those in Section 9.6). As in Section 9.6, one of the reactants is used up first, leaving some amount of another reactant left over. As in Section 9.5, the quantities of reactants are given in grams instead of moles.

Like almost all strategies for solving numerical problems, the strategy for solving gram-gram limiting reactant problems involves several steps. First, you convert the masses of the reactants into moles. Then you solve the corresponding mole-mole limiting reactant problem. Finally you convert the answer back to units of mass (see Figure 9.15).

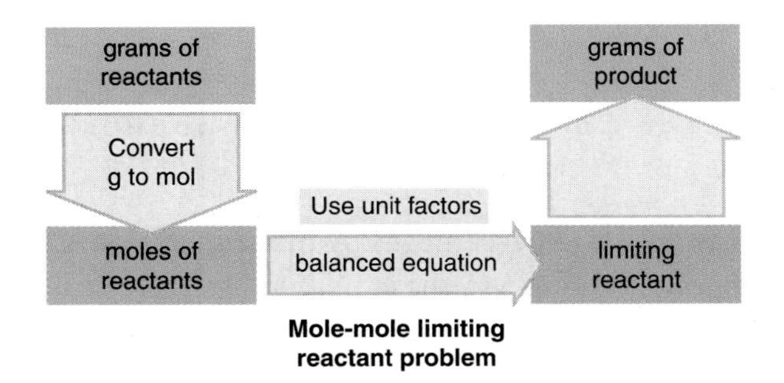


FIGURE 9.15

### EXAMPLE 9.8 Solving Limiting Reactant Gram-Gram Problems

You start the reaction that forms water from its elements with 20 g of $H_2$ and 20 g of $O_2$.
**a.** What is the limiting reactant?
**b.** What mass of $H_2O$ forms from the reaction?

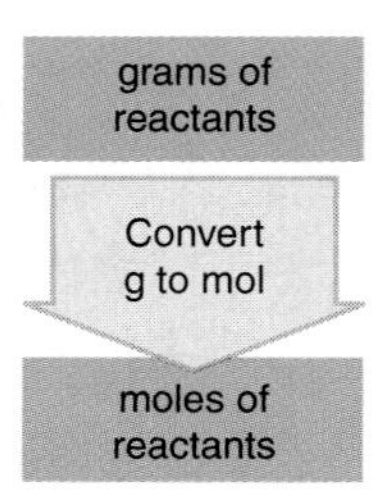


FIGURE 9.16

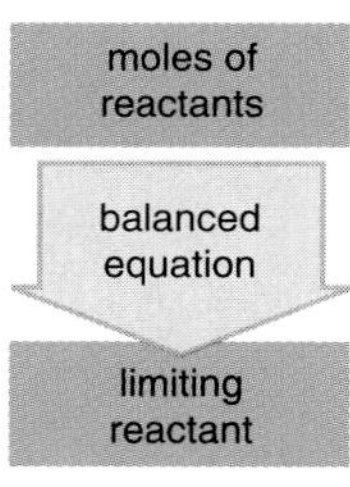


Mole-mole limiting reactant problem

FIGURE 9.17

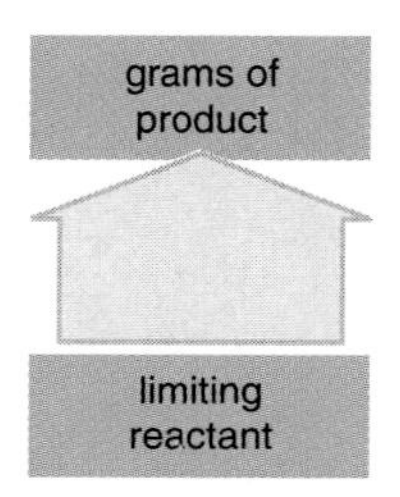


FIGURE 9.18

**SOLUTIONS**

First, you need the balanced equation: $2\ H_2 + O_2 \rightarrow 2\ H_2O$. Next, convert the given masses of reactants to moles:

$$20\ \cancel{g\ H_2}\left(\frac{1\ mol\ H_2}{2.016\ \cancel{g\ H_2}}\right) = 10\ mol\ H_2$$

$$20\ \cancel{g\ O_2}\left(\frac{1\ mol\ O_2}{32.00\ \cancel{g\ O_2}}\right) = 0.63\ mol\ O_2$$

Now identify the limiting reactant:

$$\frac{20\ mol\ H_2}{2} = 10 \qquad \frac{0.63\ mol\ O_2}{1} = 0.63$$

Oxygen is the limiting reactant.

Next, the amount of water produced is determined by the amount of oxygen consumed. This means you use the unit factor (from the balanced equation) that converts moles of oxygen to moles of water and the unit factor (based on the molar mass) that converts moles of water to grams of water:

$$20\ \cancel{g\ O_2}\left(\frac{1\ \cancel{mol\ O_2}}{32.00\ \cancel{g\ O_2}}\right)\left(\frac{2\ \cancel{mol\ H_2O}}{1\ \cancel{mol\ O_2}}\right)\left(\frac{18.02\ g\ H_2O}{1\ \cancel{mol\ H_2O}}\right) = 23\ g\ H_2O$$

The mass of water formed is 23 g.

**EXERCISE 9.8**

Suppose you react 20 g of aluminum metal with 25 g of chlorine gas.

**a.** What is the limiting reactant?

**b.** What mass of aluminum chloride will you get?

## 9.8 Percent Composition of Compounds

**GOAL:** To calculate the percentages by mass of elements present in a compound, and to use these percentages to calculate the ratios by mass of those elements present in a sample of the compound.

**Keep in mind that *percentage* means "percentage by mass" unless you are told otherwise.**

It is often important, in practical terms, to relate the masses of elements in a sample of a compound to the formula of the compound. How much copper can you get from a certain sample of copper ore? Or perhaps you have found out (by experiment) the percentage of carbon in a newly prepared compound. Does this percentage match what you expect from the proposed formula of the compound? Example 9.9 gives the technique for solving this type of problem.

## EXAMPLE 9.9 Calculating Percentage by Mass

**a.** What is the percentage by mass of copper in $Cu_2O$?
**b.** A sample of copper ore contains 32.1% copper. What mass of copper is present in 1000 kg of this ore?
**c.** What mass of iron is present in 200 kg of $Fe_3O_4$?

FIGURE 9.19
In mineral form, $Cu_2O$ is called *cuprite*.

**SOLUTIONS**

**a.** The formula for finding a percentage is

$$\frac{\text{Part}}{\text{Whole}} \times 100\% = \text{Percentage}$$

There are two atoms of copper and one atom of oxygen in the formula of $Cu_2O$. The "Part" is the mass of the Cu atoms, and the "Whole" is the total mass of the $Cu_2O$. From the periodic table, the atomic mass of copper is 63.55 and that of oxygen is 16.00. For simplicity, assume you have 1 mol of $Cu_2O$, that contains 1 mol of O atoms and 2 mol of Cu atoms. The percentage by mass of copper is then:

$$\frac{\text{Part}}{\text{Whole}} \times 100\% = \frac{2(\text{molar mass Cu})}{\text{molar mass Cu}_2\text{O}} \times 100\% = \%\ \text{Cu}$$

$$\frac{\text{Part}}{\text{Whole}} \times 100\% = \frac{2(63.55\text{ g Cu})}{2(63.55\text{ g Cu}) + 1(16.00\text{ g O})} \times 100\% = 88.82\%$$

**b.** This time you know the percentage (32.1% copper) and the whole (1000 kg ore), and want to determine the part (grams of copper). Rearranging the percentage equation gives:

$$\text{Part} = \frac{(\text{Percentage})(\text{Whole})}{100\%} = \text{mass of Cu}$$

$$\text{Part} = \frac{(32.1\%)(1000\text{ kg})}{(100\%)} = 321\text{ kg}$$

That is, 321 kg of copper are present in 1000 kg of this ore.

**c.** This problem takes one more step to solve. First, you have to find the percentage by mass of iron in $Fe_3O_4$, as you did for copper in part a:

$$\frac{\text{Part}}{\text{Whole}} \times 100\% = \frac{3(\text{molar mass Fe})}{\text{molar mass Fe}_3\text{O}_4} \times 100\% = \%\text{ of Fe}$$

$$\frac{\text{Part}}{\text{Whole}} \times 100\% = \frac{3(55.85\text{ g Fe})}{3(55.85\text{ g Fe}) + 4(16.00\text{ g O})} \times 100\% = 72.36\%$$

Then you use this percentage to find the "Part" of the 200-kg sample that is iron, as you did for copper in part b:

$$\text{Part} = \frac{(\text{Percentage})\ (\text{Whole})}{100\%} = \text{mass of Fe}$$

$$\text{Part} = \frac{(72.36\%)(200\text{ kg})}{100\%} = 145\text{ kg}$$

There are 145 kg of iron in a 200-kg sample of $Fe_3O_4$.

**EXERCISE 9.9**

**a.** What is the percentage of copper in $CuSO_4$?
**b.** What is the percentage by mass of each element in $C_2H_6O$?
**c.** What mass of iron is present in 100 tons of $Fe_2O_3$?
**d.** What mass of CaO can you get from 100,000 kg of $CaCO_3$?

## 9.9 Empirical Formulas of Compounds

**GOAL:** To calculate the empirical formula of a compound, given either its composition by mass or its percent composition.

If you know the name of a compound, getting the formula is straightforward. If you know what elements make up a compound, you can sometimes figure out what the formula must be. These problems were covered earlier, primarily in Chapter 7. But what if you don't know the name of a compound, and the elements present in it can form more than one substance? This is a common practical difficulty. Many thousands of known compounds are made up of only carbon and hydrogen, for instance. How do chemists know the formulas of the compounds they use? This section provides an answer to that question.

Today chemists have many sophisticated tools for learning the formula of an unknown compound. One kind of information that even early chemists had for many substances was the composition by mass. From this and the atomic masses of the elements, they could calculate a type of formula called an **empirical formula.** This is the simplest formula for a compound, one that gives the numbers of atoms present in their *smallest* whole-number ratio.

**The word *empirical* means "derived from experience or experiment," so that definition fits these formulas, too.**

Suppose you have determined that a compound contains 71.4% calcium and 28.6% oxygen. To find the empirical formula from this percent composition data, follow the series of simple steps shown in Figure 9.20.

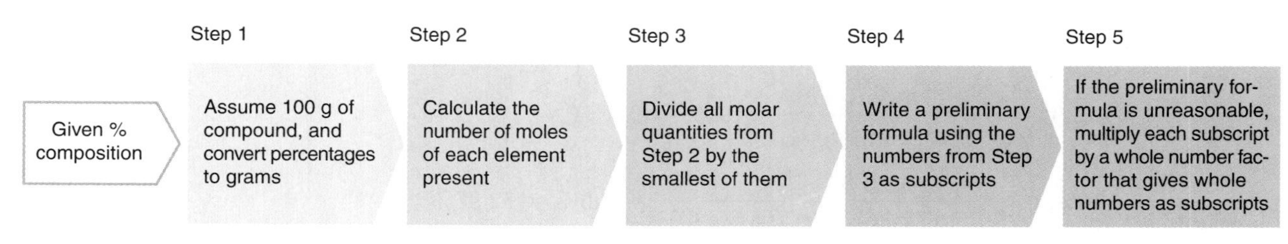
Step 1
Step 2
Step 3
Step 4
Step 5
Given % composition
Assume 100 g of compound, and convert percentages to grams
Calculate the number of moles of each element present
Divide all molar quantities from Step 2 by the smallest of them
Write a preliminary formula using the numbers from Step 3 as subscripts
If the preliminary formula is unreasonable, multiply each subscript by a whole number factor that gives whole numbers as subscripts

FIGURE 9.20

**Step 1** Assume you have exactly 100 g of your compound, and convert the percentages to grams. In this case in 100 g of the calcium-oxygen compound, 71.4 g is calcium and 28.6 g is oxygen. (The reason for assuming 100 g of sample is to make this conversion simple.)

**Step 2** Calculate the number of moles of each element from the number of grams, using unit factors. For your sample:

$$71.4 \text{ g Ca} \left(\frac{1 \text{ mol Ca}}{40.08 \text{ g Ca}}\right) = 1.78 \text{ mol Ca}$$

$$28.6 \text{ g O} \left(\frac{1 \text{ mol O}}{16.00 \text{ g O}}\right) = 1.79 \text{ mol O}$$

**Step 3** Divide all numbers from Step 2 by the smallest number obtained in the calculations. In this case:

$$\frac{1.79}{1.78} = 1.01 \qquad \frac{1.78}{1.78} = 1$$

It is probably obvious to you that the number of moles of calcium is equal to the number of moles of oxygen in this problem, at least to the limits of experimental error (5% error is typically allowed in these calculations), but let's do Step 4 anyway.

**Step 4** Write down a preliminary formula using the numbers from Step 3 as the subscripts. If the preliminary formula makes chemical sense, it is the empirical formula and the problem is completed. Step 3 gives a preliminary formula of $Ca_{1.00}O_{1.01}$, which does not make sense chemically. However, the number 1.01 is within 5% of 1.00, so the formula is really $Ca_1O_1$, or CaO. This is the formula of a real compound, calcium oxide (see Figure 9.21).

FIGURE 9.21
CaO, called *lime,* is an ingredient in many formulas for cement and mortar. Reinforced cement now is the basic material for much of our highway system.

Suppose, however, that the preliminary formula in Step 4 is unreasonable—for example, $X_{1.5}Y_1$. It is impossible to have 1.5 atoms of anything in a chemical formula, so you immediately recognize that something is wrong. When this happens, you need one more step.

**Step 5** If the preliminary formula from Step 4 is impossible, multiply each subscript in the preliminary formula by a whole number, beginning with 2 (then, if necessary, try 3, then 4, etc.) until you get a formula that contains only whole numbers as subscripts. This will be the answer. For $X_{1.5}Y_1$, multiplying each subscript by 2 gives $2 \times 1.5 = 3$ for X, and $2 \times 1 = 2$ for Y, which yields $X_3Y_2$, a sensible empirical formula.

Example 9.10 provides another example illustrating this procedure.

EXAMPLE 9.10 **Finding the Empirical Formula from the Percent Composition**

What is the empirical formula of a compound that consists of 88.8% carbon and 11.1% hydrogen?

SOLUTION

**Step 1** In a 100-g sample of this compound, there are 88.8 g of carbon and 11.1 g of hydrogen.

**Step 2** Calculate the number of moles of carbon and hydrogen:

$$88.8 \text{ g C} \left(\frac{1 \text{ mol C}}{12.01 \text{ g C}}\right) = 7.39 \text{ mol C}$$

$$11.1 \text{ g H} \left(\frac{1 \text{ mol H}}{1.008 \text{ g H}}\right) = 11.0 \text{ mol H}$$

**Step 3** Next, divide each of these two numbers by the smaller. This gives you 1.00 for C and 11.0/7.39 = 1.49 for H.

**Step 4** Write down the preliminary formula $C_{1.00}H_{1.49}$, which is impossible. Note that 1.49 is *not* within 5% of either 1.00 or 2.00, so the formula can be neither $C_1H_1$ or $C_1H_2$. Go to Step 5.

**Step 5** Multiplying each subscript in $C_{1.00}H_{1.49}$ by 2 gives 2.00 for C and 2.98 for H. The 2.98 is within 5% of 3.00. This gives you the empirical formula $C_2H_3$, which is the answer.

For an outline of this, see Figure 9.22.

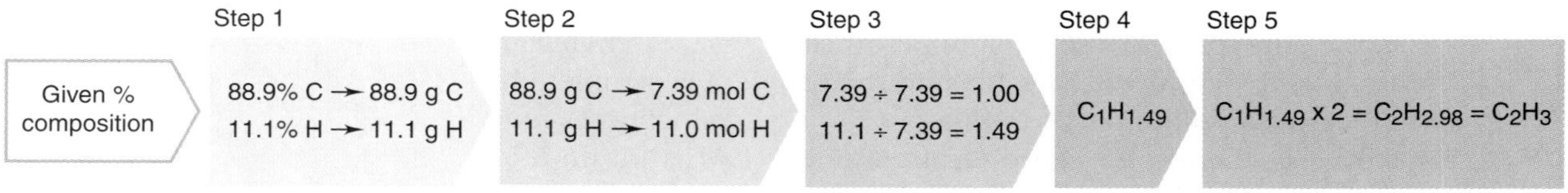

| | Step 1 | Step 2 | Step 3 | Step 4 | Step 5 |
|---|---|---|---|---|---|
| Given % composition | 88.9% C → 88.9 g C<br>11.1% H → 11.1 g H | 88.9 g C → 7.39 mol C<br>11.1 g H → 11.0 mol H | 7.39 ÷ 7.39 = 1.00<br>11.1 ÷ 7.39 = 1.49 | $C_1H_{1.49}$ | $C_1H_{1.49} \times 2 = C_2H_{2.98} = C_2H_3$ |

FIGURE 9.22

**EXERCISE 9.10**

What is the empirical formula of a compound that consists of 40.0% carbon, 6.67% hydrogen, and 53.3% oxygen?

## 9.10 Molecular Formulas from Empirical Formulas

GOAL: To calculate molecular formulas, given empirical formulas and molar masses, and vice versa.

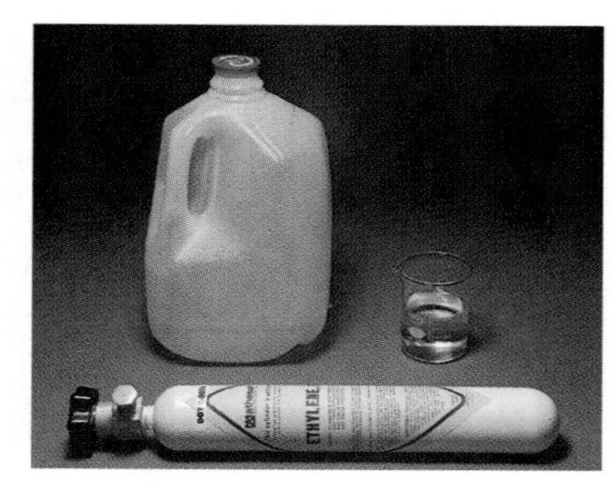


FIGURE 9.23
Ethylene ($C_2H_4$, a gas), cyclohexane ($C_6H_{12}$, a liquid), and polyethylene (a solid of high molecular mass; the milk jug is made from it) all have the empirical formula $CH_2$.

The major advantage of empirical formulas is that you can calculate them from simple percent composition data. They also have one major disadvantage: You can't be sure the empirical formula gives the exact numbers of atoms that are really present in a molecule of the substance. The actual number of each element in a molecule of a compound is called the **molecular formula.**

For example, carbon's valence of 4 tells you that a compound with the empirical formula $CH_2$ almost certainly has a different molecular formula. The molecular formula may be $C_2H_4$, or $C_3H_6$, or $C_4H_8$, or any other multiple of $CH_2$ (see Figure 9.23). The empirical formula gives the smallest whole-number ratio of the atoms present, *not* the actual numbers of atoms (although sometimes the two are the same). It's a least–common-denominator formula.

To find the empirical formula from the molecular formula, divide all of the subscripts by their largest common factor. The compounds acetylene ($C_2H_2$) and benzene ($C_6H_6$) both have the empirical formula CH. The simple sugar glucose ($C_6H_{12}O_6$) has the empirical formula $CH_2O$. However, the molecular formula of table sugar (sucrose; see Figure 9.24) is $C_{12}H_{22}O_{11}$, the same as its empirical formula.

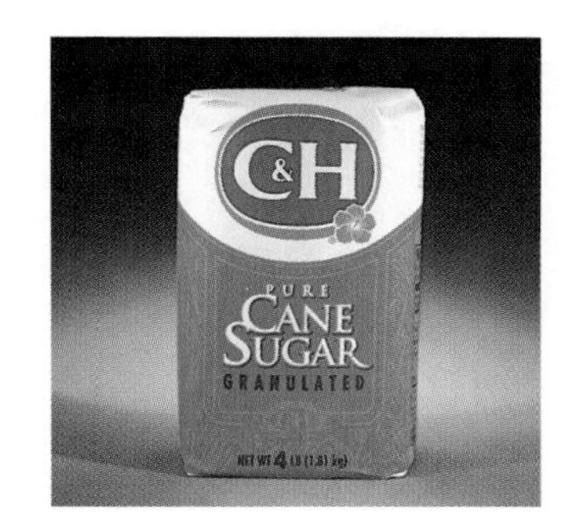


FIGURE 9.24
Table sugar, or sucrose, is $C_{12}H_{22}O_{11}$.

EXAMPLE 9.11

### Obtaining the Empirical Formula from the Molecular Formula

A compound commonly used in mothproofing is *p*-dichlorobenzene, which has the molecular formula $C_6H_4Cl_2$. What is its empirical formula?

**SOLUTION**

All the subscripts in the molecular formula can be divided by 2. This gives the empirical formula $C_3H_2Cl$.

**EXERCISE 9.11**

For each molecular formula, give the empirical formula.

**a.** $C_4H_8$ **b.** $P_4O_{10}$ **c.** $C_2H_6O$ **d.** $B_2H_6$

To find the molecular formula from the empirical formula, you need one more piece of information—the molar mass of the compound. There are several ways to find this quantity by experiment. (You will learn one classical method when you study gases in Chapter 11.) To make the conversion, you add the atomic masses of the elements in the empirical formula, this gives the empirical formula mass. Then you divide the compound's molar mass by the empirical formula mass to get a multiplication factor. Finally, you multiply each of the subscripts in the empirical formula by this multiplication factor.

EXAMPLE 9.12

### Finding the Molecular Formula from the Empirical Formula and Molecular Mass

A compound has the empirical formula CH and a molecular mass of 78. What is the molecular formula of this compound (Figure 9.25)?

**SOLUTION**

The mass for the empirical formula CH is 12 + 1 = 13. The molecular mass of the compound is given as 78. Dividing gives 78/13 = 6. Multiply each of the (understood) subscripts 1 in the empirical formula by 6. The molecular formula is $C_6H_6$.

| Add masses of atoms in empirical formula | Divide mass of compound by mass of empirical formula | Use result of Step 2 to multiply each subscript of empirical formula |
|---|---|---|
| CH<br>12 + 1 = 13 | 78 ÷ 13 = 6 | $C_{1x6}H_{1x6}$<br>$C_6H_6$ |

FIGURE 9.25

**EXERCISE 9.12**

A starting material used in making polyester fibers is *p*-xylene, which has the empirical formula $C_4H_5$ and a molecular mass of 106. What is the molecular formula of *p*-xylene?

FIGURE 9.26
The plastic used in this soft drink bottle is a polyester that's made from *p*-xylene, in a process of several steps. Once used, the polyester can be broken down, purified, and remade. This is one of the most effective recycling processes there is—you should definitely recycle such bottles!

## EXPERIMENTAL EXERCISE 9.1

You are a member of A. L. Chemist's team, given the task of determining the empirical formula and molecular formula of an inorganic salt that contains only iron and chlorine. A 1000-mg sample of the salt dissolved completely in water to give solution 1. You added $AgNO_3(aq)$ to solution 1, forming a precipitate and solution 2. All of the chlorine in solution 1 precipitated as AgCl(*s*) because you added an excess of $AgNO_3(aq)$. After drying it carefully, you weighed the AgCl(*s*); it had a mass of 2260 mg. Your team then calculated the mass of chlorine present in the original sample by using the unit factor method:

$$2260\ \cancel{\text{mg AgCl}}\left(\frac{35.45\ \text{mg Cl}}{143.32\ \cancel{\text{mg AgCl}}}\right) = 559.0\ \text{mg Cl}$$

**QUESTIONS**

1. Two members of your team stated that the unknown sample contained 55.90% chlorine. Show that they are correct.
2. If the sample is 55.90% chlorine, what is the percentage of iron in the sample?
3. What is the empirical formula of the salt?
4. In a separate experiment a teammate determined that the molar mass of the salt is 126.8 g. What is the molecular formula?
5. What ions are present in solution 1?
6. What is the net ionic equation for the reaction that occurred when $AgNO_3(aq)$ was added to solution 1?
7. What ions are present in solution 2?

## CHAPTER SUMMARY

1. The unit factor method is very useful for solving most of the problem types in this chapter. Keep in mind that the magnitudes and the units are equally important in unit factors. Without the magnitude you can't get a numerical answer. Without the unit you have no way of knowing how to combine the numbers to get the correct answer.
2. Chemical equations must always be balanced before you can use them in calculations.

3. Given the number of moles of one reactant or product, you can calculate the number of moles of any other substance in a reaction by using the unit factor method. These calculations are called mole-mole problems.
4. Given the number of grams of one reactant or product, you can calculate the number of grams of any other substance in a reaction by using the unit factor method. First, you convert grams to moles; then you solve a mole-mole problem; finally, you convert moles to grams. These calculations are called gram-gram problems.
5. A limiting reactant is the one that limits the amount of product that can be formed.
6. An empirical formula gives the simplest whole-number ratio for the atoms in a compound.
7. Molecular formulas can be obtained from empirical formulas by dividing the molar mass by the empirical formula mass, then multiplying all the subscript numbers in the empirical formula by the resulting quantity.

## PROBLEMS

### SECTION 9.1 Unit Factors from Balanced Equations

1. Translate each of these unit factors derived from the balanced equation $Fe_2O_3 + 2\ Al \rightarrow 2\ Fe + Al_2O_3$ into an English sentence.

a. $\frac{1\ mol\ Fe_2O_3}{2\ mol\ Fe}$

b. $\frac{2\ mol\ Fe}{1\ mol\ Al_2O_3}$

c. $\frac{2\ mol\ Al}{2\ mol\ Fe}$

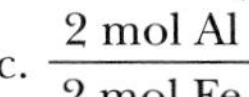

d. $\frac{1\ mol\ Fe_2O_3}{2\ mol\ Al}$

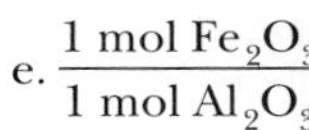

e. $\frac{1\ mol\ Fe_2O_3}{1\ mol\ Al_2O_3}$

FIGURE 9.27
The reaction between Al and $Fe_2O_3$, called the *thermite reaction,* is so exothermic that the iron metal produced is molten.

2. Translate each of these unit factors derived from the balanced equation $3\ H_2SO_4(l) + Na_3PO_4(s) \rightarrow 3\ NaHSO_4(s) + H_3PO_4(l)$ into an English sentence.

a. $\frac{3\ mol\ H_2SO_4}{1\ mol\ Na_3PO_4}$

b. $\frac{1\ mol\ Na_3PO_4}{3\ mol\ NaHSO_4}$

c. $\frac{1\ mol\ H_3PO_4}{1\ mol\ Na_3PO_4}$

d. $\frac{3\ mol\ H_2SO_4}{3\ mol\ NaHSO_4}$

e. $\frac{3\ mol\ NaHSO_4}{1\ mol\ Na_3PO_4}$

3. Construct all possible unit factors from each of the following equations. If an equation is not balanced as shown, balance it first.
a. $NO + NO_2 \rightarrow N_2O_3$
b. $Al + HCl \rightarrow AlCl_3 + H_2$
c. $C_2H_6 + O_2 \xrightarrow[heat]{Pt} CO_2 + H_2O$

4. Construct all possible unit factors from each of the following equations. If an equation is not balanced as shown, balance it first.
a. $Al + O_2 \rightarrow Al_2O_3$
b. $Ca(OH)_2 + HNO_3 \rightarrow Ca(NO_3)_2 + H_2O$
c. $Pb(OH_2)(s) \xrightarrow{heat} PbO(s) + H_2O(g)$

5. Write all possible unit factors for each equation. Remember to balance any equation that is not already balanced.
a. $Na + H_2O \rightarrow NaOH + H_2$
b. $C_3H_6 + O_2 \xrightarrow{heat} CO_2 + H_2O$
c. $NaOH + H_2SO_4 \rightarrow Na_2SO_4 + H_2O$

6. Write all possible unit factors for each equation. Remember to balance any equation that is not already balanced.
a. $CaC_2 + H_2O \rightarrow Ca(OH)_2 + C_2H_2$
b. $Mg_3N_2 + H_2O \rightarrow Mg(OH)_2 + NH_3$
c. $C_6H_6 + HNO_3 \xrightarrow{H_2SO_4} C_6H_4(NO_2)_2 + H_2O$

7. Write all possible unit factors for each equation. Remember to balance any equation that is not already balanced.
a. $2\ NO_2 \rightarrow N_2O_4$
b. $C_6H_6 + O_2 \xrightarrow[Pd]{heat} CO_2 + H_2O$
c. $NO_2 \xrightarrow{Rb} N_2 + O_2$

8. Write all possible unit factors for each equation. Remember to balance any equation that is not already balanced.
   a. $SO_2 + O_2 \xrightarrow{V_2O_5} SO_3$
   b. $K_3PO_4 + HNO_3 \rightarrow KH_2PO_4 + KNO_3$
   c. $HBr + Fe \rightarrow FeBr_2 + H_2$
9. For each of these equations, balance the equation if necessary and write all the unit factors.
   a. $H_2SO_4 + NaI \rightarrow HI + NaHSO_4$
   b. $P + H_2 \rightarrow PH_3$
   c. $CH_4 + O_2 \rightarrow CO_2 + H_2O$
10. For each of these equations, balance the equation if necessary and write all the unit factors.
   a. $Al_2O_3 + CH_3CO_2H \rightarrow Al(O_2CCH_3)_3 + H_2O$
   b. $Fe(OH)_3 + H_2SO_4 \rightarrow Fe_2(SO_4)_3 + H_2O$
   c. $C_4H_6 + H_2 \xrightarrow{Pd} C_4H_{10}$

## SECTION 9.2 Using Unit Factors to Solve Mole-Mole Problems

11. For each of these balanced equations, how many moles of the *first* reactant are required to produce 0.5 mol of product?
   a. $4\ Fe + 3\ O_2 \rightarrow 2\ Fe_2O_3$
   b. $3\ H_2 + 2\ N_2 \rightarrow 2\ NH_3$
   c. $SO_3 + H_2O \rightarrow H_2SO_4$
12. For each of these balanced equations, how many moles of the *first* reactant are required to produce 0.25 mol of product?
   a. $2\ CuCl + Cl_2 \rightarrow 2\ CuCl_2$
   b. $Cu + CuCl_2 \rightarrow 2\ CuCl$
   c. $5\ O_2 + 4\ P \rightarrow P_4O_{10}$
13. How many moles of the *first* reactant in each of these balanced equations are required to produce 2.5 mol of product?
   a. $4\ P + 3\ O_2 \rightarrow P_4O_6$
   b. $3\ Br_2 + 2\ Fe \rightarrow 2\ FeBr_3$
   c. $4\ HCl + C_8H_8 \rightarrow C_8H_{12}Cl_4$
14. How many moles of the *first* reactant in each of these balanced equations are required to produce 4.2 mol of the *first* product?
   a. $2\ H_2 + O_2 \rightarrow 2\ H_2O$
   b. $2\ H_2O + CaH_2 \rightarrow Ca(OH)_2 + H_2$
   c. $10\ O_2 + C_8H_8 \rightarrow 8\ CO_2 + 4\ H_2O$
15. For each balanced equation, determine how many moles of the *first* reactant are required to produce 0.33 mol of the *first* product.
   a. $6\ HCl + Al_2(CO_3)_3 \rightarrow 2\ AlCl_3 + 3\ CO_2 + 3\ H_2O$
   b. $6\ H_2O + P_4O_{10} \rightarrow 4\ H_3PO_4$
   c. $2\ NH_4NO_3 \rightarrow 2\ N_2 + O_2 + 4\ H_2O$
16. For each balanced equation, determine how many moles of the *first* reactant are required to produce 0.050 mol of the *first* product.
   a. $2\ NI_3 \rightarrow N_2 + 3\ I_2$
   b. $Al_2S_3 + 6\ H_2O \rightarrow 2\ Al(OH)_3 + 3\ H_2S$
   c. $6\ HCl + Fe_2S_3 \rightarrow 2\ FeCl_3 + 3\ H_2S$
17. Determine how many moles of the *first* reactant in each balanced equation are required to produce 1.05 mol of the *first* product.
   a. $2\ H_2O_2 \rightarrow 2\ H_2O + O_2$
   b. $C_{12}H_{24} + 18\ O_2 \rightarrow 12\ CO_2 + 12\ H_2O$
   c. $9\ O_2 + 2\ C_4H_2 \rightarrow 8\ CO_2 + 2\ H_2O$
18. Determine how many moles of the *first* reactant in each balanced equation are required to produce 2.7 mol of the *first* product.
   a. $Ag_2O + 2\ HBr \rightarrow 2\ AgBr + H_2O$
   b. $C_6H_8 + 8\ O_2 \rightarrow 6\ CO_2 + 4\ H_2O$
   c. $21\ O_2 + 2\ C_8H_{10} \rightarrow 16\ CO_2 + 10\ H_2O$
19. Solve each of these problems, writing and/or balancing the equation first if necessary.
   a. How many moles of AgCl form if 2.0 mol of $AgNO_3$ reacts with an excess of NaCl? The equation is

      $$AgNO_3 + NaCl \longrightarrow AgCl + NaNO_3$$

   b. How many moles of $Na_2SO_4$ are made from the reaction of 1.58 mol of $H_2SO_4$ with excess NaOH?
   c. How many moles of hydrogen gas are prepared from the reaction of 0.25 mol of zinc metal with excess hydrochloric acid? The equation is

      $$Zn(s) + HCl(aq) \longrightarrow ZnCl_2(aq) + H_2(g)$$

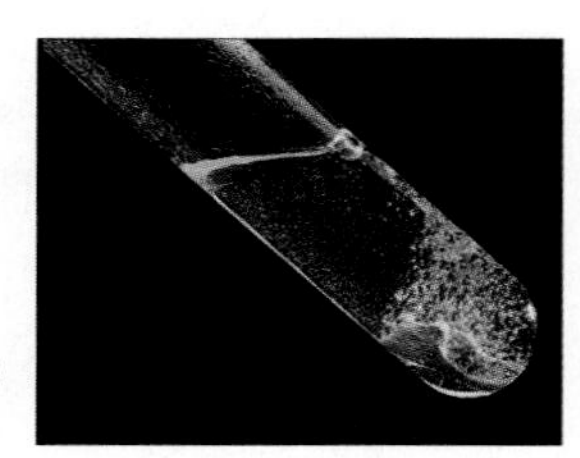

FIGURE 9.28
Zinc is one of many metals that react with acids to form hydrogen gas.

20. Solve each of these problems, writing and/or balancing the equation first if necessary.
   a. How many moles of $Cl_2$ are needed to react completely with 3.1 mol of Al to form $AlCl_3$?
   b. How many moles of oxygen are needed for the complete combustion of 12.5 mol of ethylene ($C_2H_4$)?
   c. How many moles of $PbCl_2$ are produced if 4 mol of $Pb(NO_3)_2$ reacts with an excess of NaCl? The equation is

      $$Pb(NO_3)_2 + NaCl \longrightarrow PbCl_2 + NaNO_3$$

21. Solve each of the following problems. If necessary, first balance the equation.
   a. How many moles of $CaCO_3$ are formed in the reaction of 0.23 mol of $Na_2CO_3$ with excess $CaCl_2$? The equation is

   $$Na_2CO_3 + CaCl_2 \longrightarrow NaCl + CaCO_3$$

   b. How many moles of hydrogen gas are prepared from the reaction of 0.66 mol of aluminum metal with excess hydrochloric acid? The equation is

   $$Al(s) + HCl(aq) \longrightarrow AlCl_3(aq) + H_2(g)$$

   c. How many moles of $Br_2$ are needed to react completely with 0.50 mol of Ca to form $CaBr_2$? The equation is

   $$Ca + Br_2 \longrightarrow CaBr_2$$

22. Solve each of the following problems. If necessary, first balance the equation.
   a. How many moles of oxygen are needed to react completely with 7.5 mol of acetylene ($C_2H_2$)? The equation is

   $$C_2H_2 + O_2 \longrightarrow CO_2 + H_2O$$

FIGURE 9.29
Acetylene burns with a flame hot enough to melt most metals. It has been used for welding for about 75 years.

   b. How many moles of hydrochloric acid are needed to react completely with 2.7 mol of calcium carbonate? The equation is

   $$CaCO_3(s) + HCl(aq) \longrightarrow CaCl_2(aq) + CO_2(g) + H_2O(l)$$

   c. How many moles of AgBr form when 0.074 mol of $AgNO_3$ reacts with excess $CaBr_2$? The equation is

   $$AgNO_3 + CaBr_2 \longrightarrow Ca(NO_3)_2 + AgBr$$

23. How many moles of $CO_2$ form when each of these samples reacts with a large excess of $O_2$?
   a. $2.5 \times 10^3$ mol C
   b. $3.01 \times 10^4$ mol $C_3H_8$
   c. $3.1 \times 10^{-4}$ mol $C_2H_2$

24. How many moles of $CO_2$ form when each of these samples reacts with a large excess of $O_2$?
   a. $1.2 \times 10^5$ mmol CO
   b. $2.01 \times 10^{-2}$ mmol $C_2H_6$
   c. $6.7 \times 10^2$ mol $CH_4$

## SECTION 9.3 More Complicated Mole-Mole Problems

25. Solve each problem. Don't forget to start with a balanced equation.
   a. A mixture containing 10 mol $N_2$ and 2.5 mol $O_2$ produced 1.25 mol NO as the only reaction product. How many moles of $N_2$ and $O_2$ remain unreacted?
   b. A mixture containing 5 mol $H_2$ and 25 mol $CO_2$ produced 3.5 mol CO and some $H_2O$. How many moles of water were formed and how many moles of $H_2$ remain unreacted?

26. Solve each problem. Don't forget to start with a balanced equation.
   a. A mixture containing 7 mmol C and 5 mmol of $CO_2$ produced 8 mmol CO as the only product. How many moles of C and $CO_2$ remain unreacted?
   b. A mixture containing 20 mmol $H_2$ and 35 mmol $I_2$ produced 30 mmol of HI as the only reaction product. How many moles of $H_2$ and $I_2$ remain unreacted?

27. Solve each of these problems. Don't forget to balance the equation.
   a. A mixture of 50 mmol of $H_2$ and 150 mmol $O_2$ reacted to form 50 mmol of $H_2O$ as the only product. How many moles of $H_2$ and $O_2$ remain unreacted?
   b. A mixture of 1000 mol $H_2$ and 5000 mol $N_2$ reacted to form 540 mol $NH_3$ as the only product. How many moles of $H_2$ and $N_2$ remain unreacted?

28. Solve each of these problems. Don't forget to balance the equation.
   a. A mixture of 1.2 mol of $H_2$ and 1.0 mol of $N_2$ reacted, giving 0.40 mol of $NH_3$. How many moles of $H_2$ and $N_2$ remained unreacted?
   b. A mixture of 0.50 mol $CH_3CO_2H$ and 0.60 mol $CH_3OH$ reacted to give 0.35 mol of $CH_3CO_2CH_3$ and $H_2O$. How much $CH_3CO_2H$ and $CH_3OH$ remained?

29. A team of chemists burned pure carbon in an atmosphere that did not contain enough oxygen to react with all of the carbon. The equation is $C + O_2 \rightarrow CO$ (unbalanced). If the chemists started with 4.0 mol of carbon and 1.5 mol of oxygen, how many moles of carbon were left over? How many moles of carbon monoxide were produced? (Assume all of the oxygen was consumed.)

30. A team of chemists put 1.0 mol of chlorine gas in a container with 1.5 mol of aluminum metal. Assuming all the $Cl_2$ reacted, how many moles of $AlCl_3$ were prepared? How many moles of Al were left unreacted?

31. A team of chemists added 0.50 mol of liquid bromine to 1.2 mole of zinc metal. The equation for the reaction is $Br_2 + Zn \rightarrow ZnBr_2$. Later the chemists removed the $Br_2$ that hadn't reacted and found they had 0.20 mol of it. How many moles of Zn were left? How many moles of $ZnBr_2$ did the reaction make?

32. A team of chemists heated some iron metal with oxygen gas and a catalyst. The equation for the reaction is $Fe + O_2 \rightarrow FeO$ (unbalanced). If the chemists started with 2.0 mol of iron and 0.8 mol of oxygen, how many moles of iron were left? How many moles of this iron oxide were produced? (Assume all of the oxygen was consumed.)

33. Some chemists put 0.70 mol of chlorine gas in a container with 1.1 mol of magnesium metal. The equation for the reaction is $Mg + Cl_2 \rightarrow MgCl_2$. The reaction ran hot, so they removed the gas and found they had 0.20 mol of $Cl_2$ left over. How many moles of $MgCl_2$ did the reaction produce, and how many moles of Mg metal remained after the experiment?

34. Some chemists mixed 0.50 mol of sulfur with 1.2 mol of lithium metal. The equation for the reaction is $Li + S \rightarrow Li_2S$. Later they removed the sulfur that hadn't reacted and found they had 0.30 mol of it left. How many moles of Li remained? How many moles of $Li_2S$ did the reaction make?

35. Calcium hydride ($CaH_2$) reacts with hydrochloric acid to form calcium chloride and hydrogen gas. How many moles of $CaH_2$ are left and how many moles of $H_2$ form when 1.5 mol of HCl reacts with 12 mol of $CaH_2$?

36. A reaction mixture containing 2.00 mol of $SO_2$ and 5.00 mol of $O_2$ produced a product mixture containing 1.25 mol of $SO_3$. How many moles of $SO_2$ reacted, and how many moles of oxygen were left over?

37. A mixture containing 25 mol $CH_4$ and 50 mol $NH_3$ produced 70 mol $H_2$ and some HCN. How many moles of $CH_4$ and $NH_3$ remained unreacted, and how many moles of HCN were formed?

38. A mixture containing 20 mol CaO and 100 mol C reacted to form 15 mol of $CaC_2$ and some CO. How many moles of CaO and C remained unreacted, and how many moles of CO were formed?

## SECTION 9.4 Converting Between Moles and Grams

39. What is the mass, in grams, of each of the following?
    a. 3.00 mol KOH
    b. 0.030 mol $MgSO_4$
    c. 2.00 mol ZnO
    d. 0.50 mol $K_2CO_3$

40. What is the mass, in grams, of each of the following?
    a. 2.5 mol $AlCl_3$
    b. 5.25 mol $Fe(NO_3)_3$
    c. 2.1 mol $Al(NO_3)_3$
    d. 0.295 mol $C_8H_{16}$

41. Find the mass, in grams, of each sample.
    a. $7.5 \times 10^{-2}$ mol $Ca(NO_3)_2$
    b. $20 \times 10^{-3}$ mol $CaF_2$
    c. $6.78 \times 10^{-3}$ mol $C_4H_8$
    d. $4.0 \times 10^{-3}$ mol AgCN

42. Find the mass, in grams, of each sample.
    a. $12.1 \times 10^{-3}$ mol $Ca_3(PO_4)_2$
    b. $1.32 \times 10^{-2}$ mol $CaC_2$
    c. $7.22 \times 10^{-3}$ mole $Mg(ClO_4)_2$
    d. $5.00 \times 10^{-3}$ mol $ZnBr_2$

43. What is the mass, in grams, of each of these samples?
    a. 24.55 mmol $H_2SO_4$
    b. $8.25 \times 10^{-6}$ mol $Ca(ClO_3)_3$
    c. 55 kmol $Na_2HPO_4$
    d. 20 mmol $Ag_2S$

44. What is the mass, in grams, of each of these samples?
    a. $4.01 \times 10^{-5}$ mol $Fe_2O_3$
    b. 5.024 mol $Fe_3O_4$
    c. 9.880 mol $HClO_4$
    d. 3.11 mmol $TiCl_3$

45. Determine the mass, in grams, of each sample.
    a. $7.89 \times 10^3$ mol $NH_3$
    b. 48.112 mol $KBrO_3$
    c. 4.4 mmol CO
    d. 6.00 mol $NaHCO_3$

46. Determine the mass, in grams, of each sample.
    a. 2.3 mol $CH_3CO_2H$
    b. 4.77 mmol $C_6H_{12}O_6$
    c. $1.273 \times 10^2$ mol $KMnO_4$
    d. 7.5 mol $K_2SO_4$

47. How many moles of the substance are present in each of the following samples?
    a. 12.1 g Al
    b. 20.0 mg NaCl
    c. $5.0 \times 10^{-2}$ g $SiO_2$
    d. 2.2 kg $Fe_2(CO_3)_3$

48. How many moles of each compound are present in these samples?
    a. $7.31 \times 10^{-4}$ g $AgClO_4$
    b. 0.45 g $C_7H_7O_2$
    c. 60 mg $ZnBr_2$
    d. 12.6 $\mu$g $C_{40}H_{56}$

49. How many moles of each compound are present in these samples?
   a. 35.7 g $Ca(H_2PO_4)_2$
   b. $12 \times 10^7$ g $CO_2$
   c. 1.00 kg $H_2O$
   d. $1 \times 10^{-5}$ g PbS

50. How many moles of each compound are present in these samples?
   a. $4.33 \times 10^{-4}$ g $Ni(OH)_2$
   b. 246.11 g $Mg(HSO_4)_2$
   c. 67.99 mg $K_2CO_3$
   d. 300 kg $O_2$

51. Convert each of these masses to moles.
   a. 47.66 μg $C_6H_{12}O_6$
   b. 30 g $C_{12}H_{22}O_{11}$
   c. 525 kg $(NH_4)_2CO_3$
   d. $1.27 \times 10^6$ g $CaCO_3$

52. Convert each of these masses to moles.
   a. 1.75 g $Hg_2Cl_2$
   b. 12.1 mg $K_2Cr_2O_7$
   c. $5.4 \times 10^4$ kg $Ca_3(PO_4)_2$
   d. 5.278 g $NaHCO_3$

53. For each sample, determine the number of moles of the substance present.
   a. 8.2 g $C_8H_{12}$
   b. 9.75 kg $Na_2CrO_4$
   c. $2.19 \times 10^{-4}$ g $Al(NO_3)_3$
   d. 2.0 g NaOH

54. For each sample, determine the number of moles of the substance present.
   a. 9.8 g $Al_2(CO_3)_3$
   b. $1.23 \times 10^3$ kg $NH_4NO_3$
   c. 12.2 mg $CaSO_4$
   d. $8.92 \times 10^{-4}$ g AuCN

55. For each sample, determine the number of moles of the substance present.
   a. 0.035 g of magnesium metal
   b. 20 kg of road-clearing salt ($CaCl_2$)
   c. $7.5 \times 10^{-2}$ g of $AsCl_3$
   d. 22.9 g of $Fe_3(PO_4)_2$

56. For each sample, determine the number of moles of the substance present.
   a. $5.723 \times 10^{-3}$ g of gold(I) iodide
   b. 25 g $Ni(CO)_4$
   c. 100 g $Mo_2(CO)_6(C_5H_5)_2$
   d. $8.33 \times 10^5$ g $PCl_5$

57. Calculate the number of moles for each sample.
   a. 10.0 mL of toluene, $C_7H_8$ (the density of toluene is 0.867 g $mL^{-1}$)
   b. 235 mL of methyl salicylate, $C_8H_8O_3$ (density = 1.174 g $mL^{-1}$)

58. Calculate the number of moles for each sample.
   a. 50 L $NH_3(g)$ (density = 0.759 g $L^{-1}$)
   b. 250 mL $O_2$ (density = 1.43 g $L^{-1}$)

59. How many moles of each substance are present in the sample described?
   a. 450 mL benzene, $C_6H_6$ (density = 0.880 g $mL^{-1}$)
   b. 6.7 mL of dimethyl phthalate, $C_{10}H_{10}O_4$ (density = 1.19 g $mL^{-1}$)

60. How many moles of each substance are present in the sample described?
   a. 780 mL of ethanol, $CH_3CH_2OH$ (density = 0.789 g $mL^{-1}$)
   b. $2.00 \times 10^3$ L HBr gas (density = 3.612 g $L^{-1}$)

61. For each sample described, determine how many moles of the substance are present.
   a. 50.2 mL $Br_2$ (density = 3.12 g $mL^{-1}$)
   b. 1.000 L $H_2O$ (density = 1.000 g $mL^{-1}$)

62. For each sample described, determine how many moles of the substance are present.
   a. 2.5 mL benzaldehyde, $C_7H_6O$ (density = 1.04 g $mL^{-1}$)
   b. 28.5 L octane, $C_8H_{18}$ (density = 0.703 g $mL^{-1}$)

## SECTION 9.5 Solving Gram-Gram Problems

Don't forget to balance the equation when necessary.

63. How many grams of $H_2$ can form when the given amount of each metal reacts with an excess of HCl(*aq*) to form $H_2$ and a metal chloride salt?
   a. 12.00 kg Fe (forms $FeCl_2$)
   b. 10.0 g Mg

64. How many grams of $H_2$ can form when the given amount of each metal reacts with an excess of HCl(*aq*) to form $H_2$ and a metal chloride salt?
   a. 25.25 g Ni (forms $NiCl_2$)
   b. 55.0 g Al

65. How many grams of $H_2$ can form when the given amount of each metal reacts with an excess of HCl(*aq*) to form $H_2$ and a metal chloride salt?
   a. 155.0 mg Pb (forms $PbCl_2$)
   b. 5.7 kg Sr

66. How many grams of $H_2$ can form when the given amount of each metal reacts with an excess of HCl(*aq*) to form $H_2$ and a metal chloride salt?
   a. 75 g Fe (forms $FeCl_2$)
   b. 40 g Zn

67. Sodium chloride reacts with sulfuric acid to form sodium hydrogen sulfate and hydrogen chloride gas. How many grams of sodium hydrogen sulfate can be prepared from 100 g of sodium chloride using this reaction?

68. Heating solid sodium sulfate with carbon produces sodium sulfide and carbon dioxide. How many grams of sodium sulfate are needed to prepare 50.00 g of sodium sulfide using this reaction?

69. When a mixture containing solid sodium sulfide and solid calcium carbonate is heated, calcium sulfide and sodium carbonate form. How many grams of sodium carbonate can be prepared from 175 g of sodium sulfide using this reaction?

70. Sodium chloride reacts with ammonium hydrogen carbonate to form sodium hydrogen carbonate and ammonium chloride. How many grams of sodium chloride are needed to prepare 500 g of sodium hydrogen carbonate using this reaction?

71. Ammonium chloride reacts with calcium oxide to produce ammonium gas, water, and calcium chloride. How many grams of ammonia can be produced from 600 g of ammonium chloride using this reaction?

72. Heating a mixture of $NH_3(g)$, $CO_2(g)$, and $H_2O(g)$ produces $NH_4HCO_3(s)$ as the only reaction product. How many grams of $NH_3$ are needed to make 1.00 kg of $NH_4HCO_3$ using this reaction?

73. A mixture of $Na_2CO_3(aq)$, $H_2O(l)$, and $CO_2(g)$ produces $NaHCO_3(aq)$ as the only product. How many grams of $Na_2CO_3$ are needed to produce 750 g of $NaHCO_3$ using this reaction?

74. Heating $Na_2SO_3(aq)$ and $S(s)$ produces $Na_2S_2O_3(aq)$. How many grams of sulfur are needed to react completely with 650 g of $Na_2SO_3$ in this reaction?

75. KOH reacts with $Cl_2$ to form KCl, KOCl, and $H_2O$. How much KOCl can be made from 1.00 g of KOH using this reaction?

76. Heating $KOCl(aq)$ produces $KClO_3(aq)$ and $KCl(aq)$. How many grams of $KClO_3$ can be prepared from 705 g of KOCl using this reaction?

77. How many grams of hydrogen gas are produced from the reaction of 2.3 g of calcium metal with excess water?

78. How much barium nitrate is required to make 2.50 g of barium sulfate from barium nitrate and sodium sulfate?

79. How many grams of carbon dioxide gas form if 1.0 kg of butane ($C_4H_{10}$) burns completely in an excess of oxygen?

80. How much copper will you get if you react 500 kg of $Cu_2S$ with oxygen? The equation for the reaction is

$$Cu_2S(l) + O_2(g) \longrightarrow Cu(l) + SO_2(g) \quad \text{(unbalanced)}$$

## SECTION 9.6 The Limiting Reactant: Mole-Mole Problems

81. Heating a mixture of $As_2O_3$ and C produces elemental As and $CO_2$. How much As can be prepared by heating 10 mol $As_2O_3$ with 12 mol C?

82. Heating sand ($SiO_2$) and coke (pure C) in an electric furnace produces carborundum (CSi) and carbon monoxide. How many moles of carborundum are produced by heating 10 mol $SiO_2$ with 25 mol C?

83. How many moles of $SiF_4$ are formed if 5 mol of $SiO_2$ is mixed with 8.00 mol of HF? The equation for the reaction is:

$$SiO_2 + HF \longrightarrow SiF_4 + H_2O \quad \text{(unbalanced)}$$

84. Strong heating of a mixture of $Na_2CO_3(s)$ and $C(s)$ produces $Na(l)$ and $CO(g)$. How many moles of sodium metal can be prepared by heating 10 mol $Na_2CO_3$ with 10 mol C?

85. Benzene ($C_6H_6$) reacts with hydrogen in the presence of a platinum catalyst to form cyclohexane ($C_6H_{12}$). How many moles of $C_6H_{12}$ can be prepared from 0.50 mol $C_6H_6$ and 1.2 mol $H_2$?

86. Heating a mixture of zinc sulfide and oxygen produces zinc oxide and sulfur dioxide. How many moles of ZnO can be prepared from 2.5 mol ZnS and 3.5 mol $O_2$?

87. How many moles of $FeBr_3$ can be prepared by mixing 5.11 mol Fe with 7.25 mol $Br_2$?

88. Heating a mixture of $FeS_2$ (pyrite) with oxygen produces $Fe_2O_3$ and gaseous $SO_2$. How many moles of $SO_2$ are formed by heating 1.25 mol of $FeS_2$ with 6.33 mol $O_2$?

FIGURE 9.30
The mineral pyrite is a naturally occurring compound commonly known as "fool's gold."

89. Mixing $FeCl_3(aq)$ with $KCN(aq)$ produces $K_3Fe(CN)_6$ and KCl. How many moles of KCl form from a mixture containing 2.5 mol $FeCl_3$ and 18 mol KCN?

90. How many moles of ammonia can be prepared from 20 mol $H_2$ and 15 mol $N_2$?

91. A reaction mixture contains 2.0 mol of calcium metal and 3.0 mol of water. What is the limiting reactant, how many moles of the reactant in excess will be left over, and how many moles of the product $Ca(OH)_2$ will be formed? The equation is

$$Ca + 2\,H_2O \longrightarrow Ca(OH)_2 + H_2$$

92. A reaction mixture contains 0.50 mol of sodium metal and 0.25 mol of $H_2O$. What is the limiting reactant, how many moles of the reactant in excess will be left over, and how many moles of hydrogen gas will be formed? The equation is

$$2\ Na + 2\ H_2O \longrightarrow 2\ NaOH + H_2$$

## SECTION 9.7 The Limiting Reactant: Gram-Gram Problems

93. Each of these reactions was carried out by using 12.5 g of each of the two reactants. All of the equations are balanced. In each case what is the limiting reactant, and how many grams of product are formed?
   a. $2\ FeBr_2 + Br_2 \rightarrow 2\ FeBr_3$
   b. $2\ CuCl + Cl_2 \rightarrow 2\ CuCl_2$
   c. $2\ Hg + O_2 \rightarrow 2\ HgO$
   d. $N_2 + 3\ H_2 \rightarrow 2\ NH_3$
   e. $2\ CO + O_2 \rightarrow 2\ CO_2$

94. Each of these reactions was carried out by using 7.5 g of each of the two reactants. All of the equations are balanced. In each case what is the limiting reactant, and how many grams of product are formed?
   a. $2\ NO + O_2 \rightarrow 2\ NO_2$
   b. $2\ Mg + O_2 \rightarrow 2\ MgO$
   c. $2\ CuBr + Br_2 \rightarrow 2\ CuBr_2$
   d. $Mg + S \rightarrow MgS$
   e. $Ca + Br_2 \rightarrow CaBr_2$

95. You can make $AlBr_3$ by reacting Al metal with $Br_2$. If your reaction mixture contains 1.00 g of Al and 2.00 g of $Br_2$, what is the limiting reactant and how many grams of $AlBr_3$ will form?

96. Heating a mixture of $H_2S$ and $SO_2$ produces water and sulfur. What is the limiting reactant and how many grams of sulfur are produced from a reaction mixture containing 150 g $H_2S$ and 75 g $SO_2$?

97. Heating $Cr_2O_3$ with Al produces Cr and $Al_2O_3$. What is the limiting reactant and how many grams of chromium are formed from a reaction mixture that contains 75 g Al and 125 g $Cr_2O_3$?

98. A mixture containing $MnO_2$ and HCl produces $MnCl_2$, $Cl_2$ and water when heated. What is the limiting reactant and how many grams of chlorine gas can be prepared from a reaction mixture containing 160 g $MnO_2$ and 200 g HCl?

99. In the first quarter of this century, a common laboratory preparation of $Br_2$ used this reaction:

$$2\ NaBr + 2\ H_2SO_4 + MnO_2 \longrightarrow Na_2SO_4 + MnSO_4 + 2\ H_2O + Br_2 \quad \text{(balanced)}$$

What is the limiting reactant and how much $Br_2$ is formed from a reaction mixture containing 100 g NaBr, 100 g $MnO_2$, and an excess of $H_2SO_4$?

100. In the middle of the nineteenth century, a common lab preparation of hydroiodic acid used this reaction:

$$H_2S(aq) + I_2(aq) \longrightarrow 2\ HI(aq) + S(s) \quad \text{(balanced)}$$

What is the limiting reactant and how many grams of HI are formed if the reaction mixture contains 50 g $H_2S$ and 100 g $I_2$?

101. A century ago dental hygiene was a fledgling concept, and toothpaste was unheard of. Practitioners used a tooth powder, commonly known at the time as precipitated chalk (the modern name is calcium carbonate). The balanced equation is:

$$Na_2CO_3(aq) + CaCl_2(aq) \longrightarrow CaCO_3(s) + 2\ NaCl(aq)$$

What is the limiting reactant and how many grams of calcium carbonate can be prepared from a reaction mixture containing 50 g $Na_2CO_3$ and 75 g $CaCl_2$?

102. Early chemists bubbled samples of unknown gases through a mixture called lime water ($Ca(OH)_2(aq)$) to test for the presence of $CO_2$. Carbon dioxide reacts with lime water to produce $CaCO_3$. The balanced equation is

$$Ca(OH)_2(aq) + CO_2(g) \xrightarrow{H_2O} CaCO_3(s) + H_2O(l)$$

What is the limiting reactant and how many grams of calcium carbonate result from bubbling 0.050 g $CO_2$ through lime water that contains 0.25 g $Ca(OH)_2$?

103. When heated, nitrogen and oxygen can react to form either NO or $NO_2$ depending on the relative amounts of the two reactants that are present. If 77.84 g of $N_2$ is mixed with 85 g of $O_2$, what is the limiting reactant and what product is formed?

104. When heated, carbon and oxygen can react to form either CO or $CO_2$ depending on the relative amounts of the two reactants that are present. If 198 g of C is mixed with 250 g of $O_2$, what is the limiting reactant and what product forms?

105. A reaction mixture contains 5.0 g of $AgNO_3$ and 25.0 g of $CaCl_2$ dissolved in water. What is the limiting reactant, and how many grams of AgCl are formed? The balanced equation for the reaction is

$$2\ AgNO_3 + CaCl_2 \longrightarrow 2\ AgCl + Ca(NO_3)_2$$

106. A reaction mixture contains 20 g of $Pb(NO_3)_2$ and 20 g of NaCl. What is the limiting reactant, and how many grams of $PbCl_2$ are formed? The balanced equation for the reaction is

$$Pb(NO_3)_2 + 2\ NaCl \longrightarrow PbCl_2 + 2\ NaNO_3$$

107. A reaction mixture contains 12 g of acetylene ($C_2H_2$) and 56 g of oxygen. If you set off a spark in this mixture and a combustion reaction occurs, how many grams of $H_2O$ are formed? What is the limiting reactant?

108. A reaction mixture contains 12.0 g of $AgNO_3$ and 15.0 g of $AlCl_3$ dissolved in water. What is the limiting reactant, and how many grams of AgCl are formed? The balanced equation for the reaction is

$$3\ AgNO_3 + AlCl_3 \longrightarrow 3\ AgCl + Al(NO_3)_3$$

109. A reaction mixture contains 1.25 g of $Pb(NO_3)_2$ and 2.00 g of NaCl. What is the limiting reactant, and how many grams of $PbCl_2$ are formed? The balanced equation for the reaction is

$$Pb(NO_3)_2 + 2\ NaCl \longrightarrow PbCl_2 + 2\ NaNO_3$$

110. A reaction mixture contains 50 g of $Pb(NO_3)_2$ and 10 g of $MgCl_2$. What is the limiting reactant, and how many grams of $PbCl_2$ are formed? The balanced equation for the reaction is

$$Pb(NO_3)_2 + MgCl_2 \longrightarrow PbCl_2 + Mg(NO_3)_2$$

## SECTION 9.8 Percent Composition of Compounds

111. What is the mass percent of chlorine in each of these compounds?
 a. AgCl
 b. $PbCl_2$
 c. HCl
 d. $Cl_2O$
 e. $C_6H_{11}Cl$

112. What is the mass percent of chlorine in each of these compounds?
 a. $C_6H_5Cl$
 b. $ZnCl_2$
 c. $FeCl_2$
 d. $FeCl_3$
 e. $CH_2Cl_2$

113. What is the percentage by mass of the given element in each of these compounds?
 a. Iron in $FeBr_2$
 b. Sodium in $Na_2SO_4$
 c. Sodium in $Na_2SO_3$
 d. Nitrogen in $Zn(NO_3)_2$
 e. Oxygen in $Ca(OCl)_2$

114. What is the percentage by mass of the given element in each of these compounds?
 a. Potassium in $KHCO_3$
 b. Nitrogen in $N_2O_4$
 c. Carbon in $C_{12}H_{24}$
 d. Nitrogen in $NH_4Cl$
 e. Phosphorus in $P_4O_{10}$

115. Determine the percentage by mass of the given element in each of these compounds.
 a. Sulfur in ZnS
 b. Hydrogen in NaH
 c. Oxygen in $KClO_4$
 d. Nitrogen in $NH_4CN$
 e. Silver in AgCl

116. Determine the mass percentage of the given element in each of these compounds.
 a. Chromium in $K_2Cr_2O_7$
 b. Carbon in $CO_2$
 c. Hydrogen in $H_2O$
 d. Oxygen in $Sr(OH)_2$
 e. Uranium in $UF_6$

117. What is the percentage by mass of each element in each of these compounds?
 a. $C_6H_6$
 b. $CrO_3$
 c. $C_7H_8$
 d. KI
 e. KCN

118. What is the percentage by mass of each element in each of these compounds?
 a. $CaH_2$
 b. $Ca(HCO_3)_2$
 c. $Fe_3O_4$
 d. $Fe_2(SO_4)_3$
 e. $Ni(NO_3)_2$

119. What is the percentage by mass of each element in each of these compounds?
 a. $C_2H_6O$
 b. $C_8H_{18}$
 c. $C_6H_5Br$
 d. $KMnO_4$
 e. $NH_4NO_3$

120. What is the percentage by mass of each element in each of these compounds?
 a. $Ca_3(PO_4)_2$
 b. $NiSO_4$
 c. $Ag_2CrO_4$
 d. $Ca(OCl)_2$
 e. $Al(BrO_3)_3$

121. Compute the mass of the given element in each sample.
 a. Gold in a 150-g rock sample that is 0.20% gold
 b. Nickel in a 75-kg sample of steel that is 18.2% nickel
 c. Oxygen in 10 g of a compound that is 14% oxygen
 d. Barium in 14.66 mg of a compound that is 58.86% barium
 e. Silver in 1.245 g of a compound that is 65.04% silver

122. Compute the mass of the given element in each sample.
 a. Chlorine in 14.225 mg of a compound that is 24.73% chlorine
 b. Platinum in 2000 kg of an ore sample that is 2.4% platinum
 c. Magnesium in a 5.00-g steel sample that is 4.55% magnesium
 d. Tin in a 125-kg bronze sculpture that is 4.11% tin
 e. Gold in a 500-g quartz sample containing 0.15% gold

123. Determine the mass of the element that is present in each of the samples described.
 a. Tin in $5.25 \times 10^3$ kg of tin ore that is 22% tin
 b. Chromium in a 1500-kg sample of steel that is 16.5% chromium
 c. Copper in a 2.00-g sample of an alloy that is 32% copper
 d. Manganese in a 22.56-mg sample of bronze that is 0.331% manganese
 e. Bromine in 8.99 g of a compound that is 42.5% bromine

124. Determine the mass of the element that is present in each of the samples described.
   a. Phosphorus in 23.66 g of a compound that is 14.87% phosphorus
   b. Sulfur in 40.76 mg of a compound that is 2.01% sulfur
   c. Silver in $5.33 \times 10^6$ kg of ore containing 0.11% silver
   d. Iron in a 14.77-g sample of steel that is 92% iron
   e. Titanium in a 769-mg sample of an alloy that is 0.444% titanium

125. What mass of the given element is present in each sample?
   a. Iron in 75 g of FeO
   b. Nitrogen in 100 mg of NO
   c. Carbon in 200 μg of $CO_2$
   d. Chlorine in 577.6 mg of AgCl
   e. Barium in 1.2229 g of $BaSO_4$

126. What mass of the given element is present in each sample?
   a. Silver in 750 mg of $Ag_2O$
   b. Chromium in 75 g of $K_2Cr_2O_7$
   c. Oxygen in 87.7 g of $Ca(HCO_3)_2$
   d. Zinc in 99.22 g of ZnS
   e. Sulfur in 46 g of $S_2Cl_2$

127. Calculate the mass of the element that is present in each of these samples.
   a. Iron in 124 g of $Fe_3O_4$
   b. Aluminum in 45 g of $AlBr_3$
   c. Carbon in 76 g of $C_2H_2$
   d. Phosphorus in 55 kg of $Ca_3(PO_4)_2$
   e. Calcium in 900 kg of $CaSO_4$

128. Calculate the mass of the element that is present in each of these samples.
   a. Nitrogen in 96.22 mg of $NH_4NO_3$
   b. Hydrogen in 20 mg of $(NH_4)_2SO_4$
   c. Copper in $5.33 \times 10^4$ kg of $Cu_2S$
   d. Bromine in 55 g of $PBr_3$
   e. Arsenic in 30 mg of $AsH_3$

129. What mass of each compound can be obtained (in theory) from the given amount of starting material?
   a. CaO from heating 20 mg of $CaCO_3(s)$
   b. PbO from heating 75 g of $Pb(OH)_2(s)$
   c. MgO from heating 99 g of $MgCO_3$

130. What mass of each compound can be obtained (in theory) from the given amount of starting material?
   a. CuO from heating 187 g of $Cu(OH)_2$
   b. $H_2O$ from heating 99 g of $Ca(OH)_2$
   c. $CO_2$ from heating 1250 kg of $CaCO_3$

## SECTION 9.9 Empirical Formulas of Compounds

131. Find the empirical formula of each compound from its percent composition.
   a. Aluminum, 52.9%; oxygen, 47.1%
   b. Copper, 64.2%; chlorine, 35.8%
   c. Carbon, 85.7%; hydrogen, 14.3%

132. Find the empirical formula of each compound from its percent composition.
   a. Carbon, 92.3%; hydrogen, 7.7%
   b. Nitrogen, 46.68%; oxygen 53.32%
   c. Copper, 47.3%; chlorine, 52.7%

133. Determine the empirical formula of each compound from the given percent composition.
   a. Hydrogen, 5.9%, oxygen, 94.1%
   b. Carbon, 75.0%; hydrogen, 25.0%
   c. Carbon, 54.5%; hydrogen, 9.1%; oxygen, 36.4%

134. Determine the empirical formula of each compound from the given percent composition.
   a. Nitrogen, 30.45%; oxygen, 69.55%
   b. Nitrogen, 63.65%; oxygen, 36.35%
   c. Phosphorus, 56.34%; oxygen, 43.66%

135. Use the percent composition data to calculate the empirical formula of each of these compounds.
   a. Iron, 63.53%; sulfur, 36.47%
   b. Nitrogen, 36.86%; oxygen, 63.14%
   c. Iron, 46.55%; sulfur, 53.45%

136. Use the percent composition data to calculate the empirical formula of each of these compounds.
   a. Lead, 59.38%; chlorine, 40.62%
   b. Chromium, 24.55%; bromine, 75.45%
   c. Phosphorus, 7.19%; bromine, 92.81%

137. What is the empirical formula of each of these compounds?
   a. Chlorine, 38.35%; fluorine, 61.65%
   b. Iodine, 57.19%; fluorine, 42.81%
   c. Carbon, 42.86%; oxygen, 57.14%

138. What is the empirical formula of each of these compounds?
   a. Hydrogen, 11.1%; oxygen, 88.9%
   b. Calcium, 62.5%; carbon, 37.5%
   c. Carbon, 27.27%; oxygen 72.73%

139. Calculate each compound's empirical formula from the given percent composition.
   a. Sodium (28.39%), chromium (32.10%), and oxygen
   b. Potassium (43.18%), chlorine (39.15%), and oxygen
   c. Sodium (17.04%), bromine (59.23%), and oxygen

140. Calculate each compound's empirical formula from the given percent composition.
   a. Sodium (32.37%), sulfur (22.57%), and oxygen
   b. Carbon (40.00%), hydrogen (6.73%), and oxygen
   c. Carbon (52.17%), hydrogen (13.0%), and oxygen

## SECTION 9.10 Molecular Formulas from Empirical Formulas

141. What is the empirical formula of each compound?
a. Deoxyribose, $C_5H_{10}O_4$, the D in DNA
b. Adipic acid, $C_6H_{10}O_4$, used to make nylon
c. Ethylene glycol, $C_2H_6O_2$, used in antifreeze
d. Hydrogen peroxide, $H_2O_2$, used as an antiseptic
e. Hydrazine, $N_2H_4$, used as a rocket fuel

142. What is the empirical formula of each compound?
a. Ribose, $C_5H_{10}O_5$, the R in RNA
b. Hexamethylenediamine, $C_6H_{16}N_2$, used to make nylon
c. Alanine, $C_3H_7NO_2$, an amino acid
d. Styrene, $C_8H_8$, used to make a type of plastic
e. Sulfur, $S_8$, used to make sulfuric acid

143. Find the empirical formula of each of these compounds.
a. $Na_2O_2$
b. $P_4O_{10}$
c. $C_4H_8$
d. $K_2Cr_2O_7$
e. $C_2H_4O_2$

144. Find the empirical formula of each of these compounds.
a. $S_2Cl_2$
b. $P_4O_6$
c. $C_4H_{10}O_2$
d. $P_4S_3$
e. $N_2O_4$

145. Calculate the molecular formula for each of these compounds from the empirical formula and molar mass.

| | Empirical formula | Molar mass (g) |
|---|---|---|
| a. | CH | 52 |
| b. | $C_2H_3$ | 81 |
| c. | $CaC_2$ | 64 |
| d. | $C_2HNO_2$ | 213 |
| e. | NO | 30 |

146. Calculate the molecular formula for each of these compounds from the empirical formula and molar mass.

| | Empirical formula | Molar mass (g) |
|---|---|---|
| a. | As | 300 |
| b. | HO | 34 |
| c. | $CH_2O$ | 90 |
| d. | $CH_2$ | 70 |
| e. | BN | 74.4 |

147. Use the empirical formula and the molar mass to calculate the molecular formula for each of these compounds.

| | Empirical formula | Molar mass (g) |
|---|---|---|
| a. | $NH_2$ | 32 |
| b. | CO | 28 |
| c. | $CH_2$ | 112 |
| d. | CH | 104 |
| e. | P | 124 |

148. Use the empirical formula and the molar mass to calculate the molecular formula for each of these compounds.

| | Empirical formula | Molar mass (g) |
|---|---|---|
| a. | $PNCl_2$ | 348 |
| b. | $BH_3$ | 27.6 |
| c. | $AlBr_3$ | 533.4 |
| d. | $C_7H_{16}$ | 100 |
| e. | $Mn(CO)_5$ | 390 |

## ADDITIONAL PROBLEMS

149. a. How many grams of $H_2O$ can be formed from 75 g of $H_2$?
b. How many grams of $AlCl_3$ can be formed from 125 g of Al?
c. How many grams of Cu can be obtained from 1.00 kg of CuO?
d. How many grams of C are needed to prepare 1.44 g of $CH_4$?
e. How many grams of Fe can be obtained from 12.22 g of $Fe_2O_3$?

150. a. How many grams of $O_2$ are needed to prepare 14 g of $H_2O$?
b. How many grams of $Cl_2$ are needed to prepare 68 g of $FeCl_3$?
c. How many grams of Zn can be obtained from 125 g of ZnO?
d. How many grams of C are needed to prepare 25.4 g of $CBr_4$?
e. How many grams of Fe can be obtained from 44.5 g of $Fe_3O_4$?

151. What mass of each compound can be obtained (in theory) from the given amount of starting material?
a. $C_2H_4$ from heating $5.00 \times 10^3$ kg of $C_4H_8$
b. $CO_2$ from burning 76 g of $C_8H_{18}$
c. $H_2O$ from burning 10 L of $C_8H_{18}$ (density = 0.70 g $mL^{-1}$)

152. What mass of each compound can be obtained (in theory) from the given amount of starting material?
a. $N_2$ from continuously sparking 65 g of $NH_3(g)$
b. $Cl_2$ by passing an electric current through 500 g of $NaCl(l)$
c. $CO_2$ from burning 10 L of $C_8H_{18}$ (density = 0.70 g $mL^{-1}$)

153. Which of the samples in each pair contains the larger mass of carbon?
a. 2 mol HCN or 0.5 mol $C_5H_{10}$
b. 5 mol $CH_4$ or 40 g $C_2H_6$
c. 1 mmol of $C_6H_6$ or 10 mg of $C_{12}H_{26}$
d. 1 mol $Ca(HCO_3)_2$ or 1 mol $CO_2$
e. 20 g $CH_4$ or 40 g $Na_2CO_3$

## EXPERIMENTAL PROBLEM 9.1

Compounds containing arsenic are quite poisonous, and during the first half of this century some pesticides and rat poisons contained $As_2O_3$ as the active agent. Arsenic poisoning from these agents resulted in several deaths, both accidental and not so accidental, during the last couple of centuries. Now $As_2O_3$ is almost never used in pesticides.

During the last half of the nineteenth century a sensitive analysis called the Marsh test was developed to detect the presence of arsenic. This test is sensitive enough to detect fractions of a milligram of arsenic. The Marsh test used two reactions of arsenic:

Reaction 1—Most arsenic compounds (such as $As_2O_3$) react with $H_2$ to form arsine, $AsH_3(g)$.

Reaction 2—Arsine decomposes to As(*s*) and $H_2O(g)$ when heated with a limited supply of oxygen.

A typical apparatus used in the Marsh test is shown in Figure 9.31.

Once the apparatus is set up, the reaction between Zn(*s*) and $H_2SO_4(aq)$ produces $H_2(g)$. The arsenic-containing compound is then added immediately through the thistle tube, and it forms $AsH_3(g)$. The arsine sweeps along with the hydrogen through a drying tube that removes $H_2O(g)$ from the mixture of gases. The arsine is then decomposed, and arsenic deposits as a brown-black solid. After the hydrogen gas has passed through the system, it is burned, primarily for safety reasons.

### QUESTIONS

**154.** Write a balanced equation for Reaction 1 using $As_2O_3$ as the arsenic-containing compound.

**155.** Write a balanced equation for Reaction 2.

**156.** Where does the oxygen that reacts with $AsH_3(g)$ in Reaction 2 come from?

**157.** Why is the arsenic-containing compound added *immediately* after the reaction between Zn(*s*) and $H_2SO_4(aq)$ begins?

**158.** One of the exhaust gases doesn't burn and contains oxygen. What is it?

**159.** This test can detect 0.2 mg of As. How much $As_2O_3$ must be present in a sample for this amount of arsenic to be produced?

**160.** What is the limiting reactant in Reaction 2?

**161.** Arsine has a garliclike odor and is extremely poisonous. One reason nineteenth-century scientists burned the exhaust gases was to destroy any arsine that didn't decompose. What is the arsenic-containing product when $AsH_3$ burns in the excess $O_2$ present in the atmosphere?

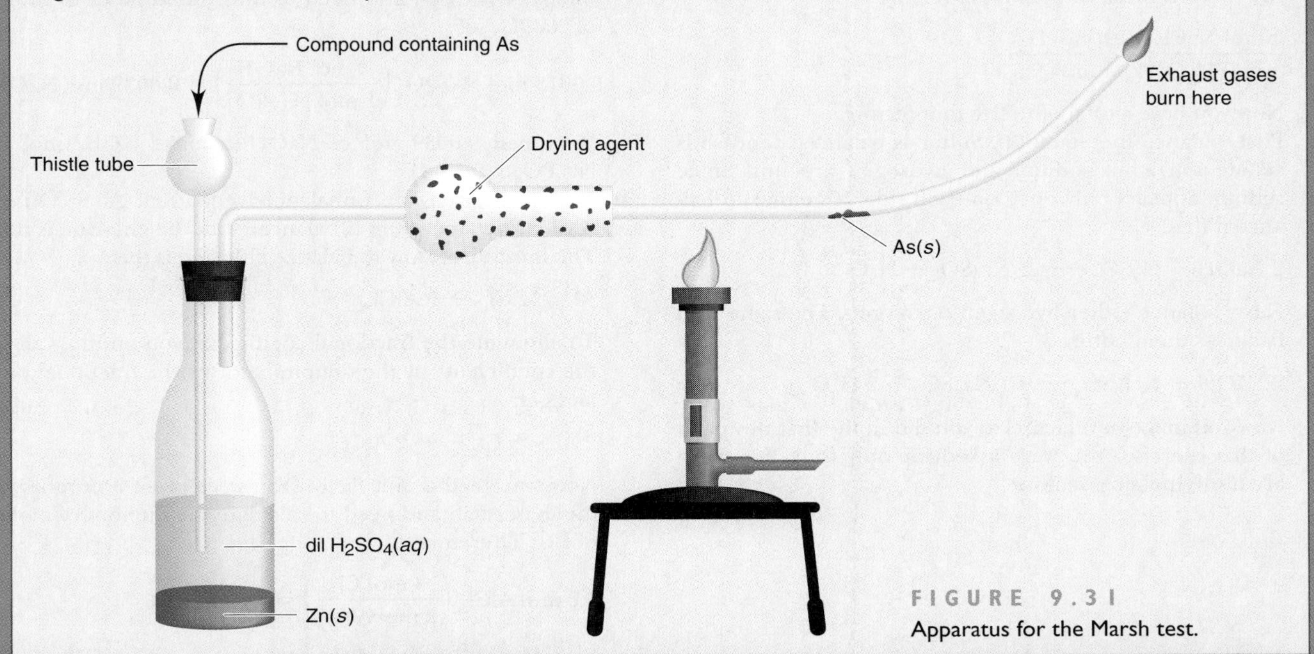


FIGURE 9.31
Apparatus for the Marsh test.

## SOLUTIONS TO EXERCISES

### Exercise 9.1

**a.** In the equation $2\ NO_2 \rightarrow N_2O_4$, there are two formulas with coefficients, $2\ NO_2$ and (1) $N_2O_4$. You divide the first by the second, and then you divide the second by the first:

$$\frac{2\ \text{mol}\ NO_2}{1\ \text{mol}\ N_2O_4} \qquad \frac{1\ \text{mol}\ N_2O_4}{2\ \text{mol}\ NO_2}$$

**b.** There are four formulas in the equation $HCl + NaOH \rightarrow NaCl + H_2O$, and all of them have a coefficient of 1. You're told that there are twelve unit factors. Here's one way to solve this systematically. Take the first formula in the equation, HCl, and divide it by the others in succession; then invert all the unit factors you got that way:

$$\frac{1\ \text{mol HCl}}{1\ \text{mol NaOH}} \qquad \frac{1\ \text{mol HCl}}{1\ \text{mol NaCl}} \qquad \frac{1\ \text{mol HCl}}{1\ \text{mol}\ H_2O}$$

$$\frac{1\ \text{mol NaOH}}{1\ \text{mol HCl}} \qquad \frac{1\ \text{mol NaCl}}{1\ \text{mol HCl}} \qquad \frac{1\ \text{mol}\ H_2O}{1\ \text{mol HCl}}$$

That takes care of HCl. Next, take the second formula, NaOH, and combine it with the third and fourth formulas; then invert the resulting factors:

$$\frac{1\ \text{mol NaOH}}{1\ \text{mol NaCl}} \qquad \frac{1\ \text{mol NaOH}}{1\ \text{mol}\ H_2O}$$

$$\frac{1\ \text{mol NaCl}}{1\ \text{mol NaOH}} \qquad \frac{1\ \text{mol}\ H_2O}{1\ \text{mol NaOH}}$$

That takes care of NaOH. The last two factors come from the two products, NaCl and $H_2O$:

$$\frac{1\ \text{mol NaCl}}{1\ \text{mol}\ H_2O} \qquad \frac{1\ \text{mol}\ H_2O}{1\ \text{mol NaCl}}$$

Now you have a total of twelve unit factors.

**c.** First, balance the equation. Sulfur is balanced (and so is sulfate ion), but sodium and hydrogen are not. Since sodium appears only once on each side, it's easier to balance it first:

$$2\ NaOH + H_2SO_4 \longrightarrow 1\ Na_2SO_4 + H_2O$$

Next, balance either hydrogen or oxygen. The same step balances them both:

$$2\ NaOH + 1\ H_2SO_4 \longrightarrow 1\ Na_2SO_4 + 2\ H_2O$$

Now you make unit factors as you did in the first two parts of this exercise. You were asked for only four, but there are twelve possible factors:

$$\frac{2\ \text{mol NaOH}}{1\ \text{mol}\ H_2SO_4} \qquad \frac{2\ \text{mol NaOH}}{1\ \text{mol}\ Na_2SO_4} \qquad \frac{2\ \text{mol NaOH}}{2\ \text{mol}\ H_2O}$$

$$\frac{1\ \text{mol}\ H_2SO_4}{2\ \text{mol NaOH}} \qquad \frac{1\ \text{mol}\ Na_2SO_4}{2\ \text{mol NaOH}} \qquad \frac{2\ \text{mol}\ H_2O}{2\ \text{mol NaOH}}$$

$$\frac{1\ \text{mol}\ H_2SO_4}{1\ \text{mol}\ Na_2SO_4} \qquad \frac{1\ \text{mol}\ H_2SO_4}{2\ \text{mol}\ H_2O} \qquad \frac{1\ \text{mol}\ Na_2SO_4}{1\ \text{mol}\ H_2SO_4}$$

$$\frac{2\ \text{mol}\ H_2O}{1\ \text{mol}\ H_2SO_4} \qquad \frac{1\ \text{mol}\ Na_2SO_4}{2\ \text{mol}\ H_2O} \qquad \frac{2\ \text{mol}\ H_2O}{1\ \text{mol}\ Na_2SO_4}$$

You should be able to find your four answers among these.

### Exercise 9.2

**a.** First, you need an equation. From the wording of the problem, you know that NaOH reacts with phosphoric acid (which is $H_3PO_4$, as you know from Chapter 7) to give $Na_3PO_4$. From your knowledge of acid-base reactions (Section 8.4), you know that water is the other product:

$$NaOH + H_3PO_4 \longrightarrow Na_3PO_4 + H_2O$$

Balancing comes next. Phosphorus is balanced (and so is phosphate). Sodium is balanced next:

$$3\ NaOH + 1\ H_3PO_4 \longrightarrow 1\ Na_3PO_4 + H_2O$$

Hydrogen and oxygen are balanced last, giving the balanced equation:

$$3\ NaOH + 1\ H_3PO_4 \longrightarrow 1\ Na_3PO_4 + 3\ H_2O$$

Now you need a unit factor. You were given a number of moles of $Na_3PO_4$ and need to find the number of moles of NaOH, so:

$$0.0013\ \cancel{\text{mol}\ Na_3PO_4}\left(\frac{3\ \text{mol NaOH}}{1\ \cancel{\text{mol}\ Na_3PO_4}}\right) = 0.0039\ \text{mol NaOH}$$

You need 0.0039 mol of NaOH to make 0.0013 mol of $Na_3PO_4$.

**b.** You were given the unbalanced equation: $Al + Cl_2 \rightarrow AlCl_3$. The aluminum is balanced, but the chlorine is not. The most direct way to balance chlorine is this:

$$Al + \tfrac{3}{2}\ Cl_2 \longrightarrow AlCl_3$$

To eliminate the fractional coefficient, you multiply all of the coefficients by the denominator of the fractional one and get:

$$2\ Al + 3\ Cl_2 \longrightarrow 2\ AlCl_3$$

Now you need a unit factor. You were given a number of moles of $AlCl_3$ and need to calculate the number of moles of $Cl_2$. Therefore:

$$3.8\ \cancel{\text{mol}\ AlCl_3}\left(\frac{3\ \text{mol}\ Cl_2}{2\ \cancel{\text{mol}\ AlCl_3}}\right) = 5.7\ \text{mol}\ Cl_2$$

You need 5.7 mol of $Cl_2$ to make 3.8 mol of $AlCl_3$ from the reaction given.

### Exercise 9.3

First, you need a balanced equation. This one has been used several times in this chapter:

$$2\ HgO \longrightarrow 2\ Hg + O_2$$

Next, you need a table. Here it is, with the amounts given filled in:

| Moles: | $2\ HgO \longrightarrow$ | $2\ Hg\ +$ | $O_2$ |
|---|---|---|---|
| Beginning | 1.2 | 0 | 0 |
| Change | | | |
| End | | | 0.4 |

The change in the amount of $O_2$ is $+0.4$ mol. From this you can calculate the changes in the amounts of Hg and HgO using the unit factor method:

$$0.4\ \cancel{mol\ O_2}\left(\frac{2\ mol\ Hg}{1\ \cancel{mol\ O_2}}\right) = 0.8\ mol\ Hg$$

$$0.4\ \cancel{mol\ O_2}\left(\frac{2\ mol\ HgO}{1\ \cancel{mol\ O_2}}\right) = 0.8\ mol\ HgO$$

Now you go back to the table and fill in the changes—the amounts of Hg and $O_2$ made and the amount of HgO used up:

| Moles: | $2\ HgO \longrightarrow$ | $2\ Hg\ +$ | $O_2$ |
|---|---|---|---|
| Beginning | 1.2 0 | 0 | 0 |
| Change | **− 0.8** | **+ 0.8** | **+ 0.4** |
| End | 0.4 | 0.8 | 0.4 |

As usual, the final step is simple. The amount of HgO left over is 0.4 mol (by subtraction), and the amount of Hg made is 0.8 mol (by addition).

### Exercise 9.4

**a.** Molar masses are 40.08 g for Ca and 35.45 g for Cl. $CaCl_2$ contains one Ca and two Cl atoms. The molar mass of $CaCl_2$ is 110.98 g $mol^{-1}$.

**b.** Molar masses are H, 1.008 g; S, 32.06 g; O, 16.00 g. One mole of $H_2SO_4$ contains 2 mol H, 1 mol S, and 4 mol O. The molar mass of $H_2SO_4$ is 98.076 g, which should be rounded off to 98.08 g $mol^{-1}$. The mass of 0.500 mol is

$$(0.500\ mol\ H_2SO_4)(98.076\ g\ mol^{-1}\ H_2SO_4) = 49.038\ g\ H_2SO_4$$

This is rounded off to 49.0 g $H_2SO_4$.

**c.** Molar masses are Ca, 40.08 g; P, 30.97 g; O, 16.00 g. $Ca_3(PO_4)_2$ contains three Ca, two P, and eight O atoms. The molar mass of $Ca_3(PO_4)_2$ is 310.18 g $mol^{-1}$. The mass of 1.37 mol is

$$[1.37\ mol\ Ca_3(PO_4)_2][310.18\ g\ mol^{-1}\ Ca_3(PO_4)_2] = 425\ g\ Ca_3(PO_4)_2 \quad \text{(rounded off)}$$

### Exercise 9.5

**a.** The molar mass of $CaCO_3$ is 40.08 g $mol^{-1}$ + 12.01 g $mol^{-1}$ + 3(16.00) g $mol^{-1}$ = 100.09 g $mol^{-1}$. The unit factor method gives:

$$1.37\ \cancel{g\ CaCO_3}\left(\frac{1\ mol\ CaCO_3}{100.09\ \cancel{g\ CaCO_3}}\right) = 1.37 \times 10^{-2}\ mol\ CaCO_3 \text{ (after rounding off)}$$

**b.** The molar mass of $Al_2O_3$ is 2(26.98 g $mol^{-1}$) + 3(16.00 g $mol^{-1}$) = 101.96 g $mol^{-1}$. The unit factor method gives:

$$0.0012\ \cancel{g\ Al_2O_3}\left(\frac{1\ mol\ Al_2O_3}{101.96\ \cancel{g\ Al_2O_3}}\right) = 1.2 \times 10^{-5}\ mol\ \ Al_2O_3$$

### Exercise 9.6

**a.** First, you need the molar mass of $Na_2SO_4$:

$$2(22.99\ g\ mol^{-1}\ Na) + 32.06\ g\ mol^{-1}\ S + 4(16.00\ g\ mol^{-1}\ O) = 142.04\ g\ mol^{-1}\ Na_2SO_4$$

The balanced equation is

$$2\ NaOH + H_2SO_4 \longrightarrow Na_2SO_4 + 2\ H_2O$$

The molar mass of NaOH is

$$22.99\ g\ mol^{-1}\ Na + 16.00\ g\ mol^{-1}\ O + 1.008\ g\ mol^{-1}\ H = 40.00\ g\ mol^{-1}\ NaOH$$

Now you set up a unit factor equation and solve it:

$$9.2\ \cancel{g\ NaOH}\left(\frac{1\ \cancel{mol\ NaOH}}{40.00\ \cancel{g\ NaOH}}\right)\left(\frac{1\ \cancel{mol\ Na_2SO_4}}{2\ \cancel{mol\ NaOH}}\right)\left(\frac{142.04\ g\ Na_2SO_4}{1\ \cancel{mol\ Na_2SO_4}}\right) = 16\ g\ Na_2SO_4$$

**b.** First, you need the molar mass of $Na_2CO_3$:

$$2(22.99\ g\ mol^{-1}\ Na) + 12.01\ g\ mol^{-1}\ C + 3(16.00\ g\ mol^{-1}\ O) = 105.99\ g\ mol^{-1}\ Na_2CO_3$$

The balanced equation was given, and the molar mass of $CO_2$ is

$$12.01\ g\ mol^{-1}\ C + 2(16.00\ g\ mol^{-1}\ O) = 44.01\ g\ mol^{-1}\ CO_2$$

The unit factor equation is

$$1.00\ \cancel{g\ Na_2CO_3}\left(\frac{1\ \cancel{mol\ Na_2CO_3}}{105.99\ \cancel{g\ Na_2CO_3}}\right)\left(\frac{1\ \cancel{mol\ CO_2}}{1\ \cancel{mol\ Na_2CO_3}}\right)\left(\frac{44.01\ g\ CO_2}{1\ \cancel{mol\ CO_2}}\right) = 0.415\ g\ CO_2$$

Since two Na atoms and an O atom are "lost" in forming a molecule of $CO_2$ from $Na_2CO_3$, the number of grams of $CO_2$ you get is less than the number of grams of $Na_2CO_3$ you started with. You should get into the habit of doing this kind of simple analysis to make certain your answers make sense. Common sense can go a long way in keeping you out of academic hot water.

**Exercise 9.7**

**a.** You start by balancing the equation. In the equation as given, carbon is balanced and chlorine appears in three places, so you start by balancing hydrogen:

$$CH_4 + Cl_2 \longrightarrow CCl_4 + 4\,HCl$$

Once this is done, chlorine is easy to balance, and balancing it gives you the balanced equation:

$$CH_4 + 4\,Cl_2 \longrightarrow CCl_4 + 4\,HCl$$

Next, you find the limiting reactant by dividing the amount of each starting material by its coefficient in the balanced equation:

$$\frac{0.5 \text{ mol } CH_4}{1} = 0.5 \qquad \frac{3.0 \text{ mol } Cl_2}{4} = 0.75$$

Thus $CH_4$ is limiting.

**b.** Calculation of the number of moles of $CCl_4$ formed is done by the usual unit factor method:

$$0.50 \cancel{\text{ mol } CH_4}\left(\frac{1 \text{ mol } CCl_4}{1 \cancel{\text{ mol } CH_4}}\right) = 0.50 \text{ mol } CCl_4$$

This reaction forms 0.50 mol of $CCl_4$.

**Exercise 9.8**

**a.** First, you need a balanced equation for the reaction:

$$2\,Al + 3\,Cl_2 \longrightarrow 2\,AlCl_3$$

Find the numbers of moles of the reactants:

$$20 \cancel{\text{ g Al}}\left(\frac{1 \text{ mol Al}}{26.98 \cancel{\text{ g Al}}}\right) = 0.74 \text{ mol Al}$$

$$25 \cancel{\text{ g } Cl_2}\left(\frac{1 \text{ mol } Cl_2}{70.90 \cancel{\text{ g } Cl_2}}\right) = 0.35 \text{ mol } Cl_2$$

Identify the limiting reactant:

$$\frac{0.74 \text{ mol Al}}{2} = 0.37 \qquad \frac{0.35 \text{ mol } Cl_2}{3} = 0.12$$

Chlorine is limiting.

**b.** To obtain the mass of $AlCl_3$ formed, construct the appropriate unit factors from the balanced equation and the molar masses, arrange them in an equation, and calculate:

$$25 \cancel{\text{ g } Cl_2}\left(\frac{1 \cancel{\text{ mol } Cl_2}}{70.90 \cancel{\text{ g } Cl_2}}\right)\left(\frac{2 \cancel{\text{ mol } AlCl_3}}{3 \cancel{\text{ mol } Cl_2}}\right)\left(\frac{133.33 \text{ g } AlCl_3}{1 \cancel{\text{ mol } AlCl_3}}\right) = 31 \text{ g } AlCl_3$$

**Exercise 9.9**

**a.** A mol of $CuSO_4$ contains one mol of copper, one mol of sulfur, and four mol of oxygen. The "Part" is the mass of the copper, and the "Whole" is the mass of the $CuSO_4$. From the periodic table, the atomic mass of copper is 63.55, that of sulfur is 32.07, and that of oxygen is 16.00. For 1 mol of $CuSO_4$, then:

$$\frac{\text{Part}}{\text{Whole}} \times 100\% = \frac{1(63.55 \text{ g Cu})}{1(63.55 \text{ g Cu}) + 1(32.07 \text{ g S}) + 4(16.00 \text{ g O})} \times 100\% = 39.81\% \text{ Cu}$$

**b.** The equation for the percentage of carbon in $C_2H_6O$ is

$$\frac{\text{Part}}{\text{Whole}} \times 100\% = \frac{2(12.01 \text{ g C})}{2(12.01 \text{ g C}) + 6(1.008 \text{ g H}) + 1(16.00 \text{ g O})} \times 100\% = 52.14\% \text{ C}$$

The equation for the percentage of hydrogen is

$$\frac{\text{Part}}{\text{Whole}} \times 100\% = \frac{6(1.008 \text{ g H})}{2(12.01 \text{ g C}) + 6(1.008 \text{ g H}) + 1(16.00 \text{ g O})} \times 100\% = 13.13\% \text{ H}$$

The equation for the percentage of oxygen is:

$$\frac{\text{Part}}{\text{Whole}} \times 100\% = \frac{1(16.00 \text{ g O})}{2(12.01 \text{ g C}) + 6(1.008 \text{ g H}) + 1(16.00 \text{ g O})} \times 100\% = 34.73\% \text{ O}$$

**c.** The atomic masses are $55.85 \text{ g mol}^{-1}$ for iron and $16.00 \text{ g mol}^{-1}$ for oxygen. The equation to find the percentage of iron is

$$\frac{\text{Part}}{\text{Whole}} \times 100\% = \frac{2(55.85 \text{ g Fe})}{2(55.85 \text{ g Fe}) + 3(16.00 \text{ g O})} \times 100\% = 69.94\% \text{ Fe}$$

Then (100 tons) (0.6994) = 69.9 tons of iron in 100 tons of $Fe_2O_3$.

**d.** The atomic masses are $40.08 \text{ g mol}^{-1}$ for calcium, $12.01 \text{ g mol}^{-1}$ for carbon, and $16.00 \text{ g mol}^{-1}$ for oxygen. The equation to find the percentage of CaO is

$$\frac{\text{Part}}{\text{Whole}} \times 100\% = \frac{1(40.08) + 1(16.00)}{1(40.08) + 1(12.01) = 3(16.00)} \times 100\% = 56.03\%$$

Then (100,000 kg) (0.5603) = 56,030 kg of CaO (in theory) from 100,000 kg of $CaCO_3$.

**Exercise 9.10**

This compound has three elements, but the procedure is the same:

$$40.0\ \cancel{g\ C}\left(\frac{1\ \text{mol C}}{12.01\ \cancel{g\ C}}\right) = 3.33\ \text{mol C}$$

$$6.67\ \cancel{g\ H}\left(\frac{1\ \text{mol H}}{1.008\ \cancel{g\ H}}\right) = 6.62\ \text{mol H}$$

$$53.3\ \cancel{g\ O}\left(\frac{1\ \text{mol O}}{16.00\ \cancel{g\ O}}\right) = 3.33\ \text{mol O}$$

Dividing each of these numbers by the smallest gives 1, $6.62/3.33 = 1.99$, and 1. The formula is $C_1H_{1.99}O_1$. The number 1.99 is within 5% of 2, so rounding off that subscript gives you the formula $CH_2O$, which is the answer.

**Exercise 9.11**

**a.** Both subscripts are divisible by 4, so the empirical formula is $CH_2$.
**b.** Both subscripts are divisible by 2, giving $P_2O_5$.
**c.** The last subscript is 1, which cannot be made any smaller. The empirical formula is the same as the molecular formula, $C_2H_6O$.
**d.** Dividing the subscripts by 2 gives the empirical formula, $BH_3$.

**Exercise 9.12**

The formula $C_4H_5$ has a mass of $(12 \times 4) + (5 \times 1) = 53$. Dividing the molecular mass by this number gives $106/53 = 2$. The molecular formula of *p*-xylene is $C_{2\times4}H_{2\times5}$, or $C_8H_{10}$.

**Experimental Exercise 9.1**

**1.** $$\text{Percentage} = \frac{\text{Part}}{\text{Whole}} \times 100\%$$

$$= \frac{559.0\ \text{mg Cl}}{1000\ \text{mg sample}} \times 100\% = 55.90\%\ \text{Cl}$$

**2.** $100\% = \%\ \text{Cl} + \%\ \text{Fe}$

$100\% = 55.90\%\ \text{Cl} + \%\ \text{Fe}$

$\%\ \text{Fe} = (100.00 - 55.90)\% = 44.10\%\ \text{Fe}$

**3.** A 100-g sample contains 55.90 g of chlorine and 44.10 g of iron. First, convert masses to moles:

$$\text{mol Cl} = 55.90\ \cancel{g\ Cl}\left(\frac{1.000\ \text{mol Cl}}{35.45\ \cancel{g\ Cl}}\right) = 1.577\ \text{mol Cl}$$

$$\text{mol Fe} = 44.10\ \cancel{g\ Fe}\left(\frac{1\ \text{mol Fe}}{55.85\ \cancel{g\ Fe}}\right) = 0.7896\ \text{mol Fe}$$

Next, divide each number of moles by the smaller number:

$$\frac{1.577}{0.7896} = 1.997 \approx 2 \qquad \frac{0.7896}{0.7896} = 1.000$$

This means that there are 2 chlorine atoms for every iron atom and that $FeCl_2$ is the empirical formula.

**4.** The empirical formula mass of $FeCl_2$ is $55.85 + (2 \times 35.45) = 126.75$. This is close enough to the experimentally determined molar mass of 126.8 g to say that the empirical and molecular formulas are identical. The molecular formula is $FeCl_2$.
**5.** Solution 1 contains $Fe^{2+}$ ions and $Cl^-$ ions.
**6.** $Ag^+(aq) + Cl^-(aq) \rightarrow AgCl(s)$
**7.** Solution 2 contains $Fe^{2+}$ ions, $NO_3^-$ ions, and $Ag^+$ ions. You added an excess of $AgNO_3(aq)$, which means that essentially all of the $Cl^-$ ions but not all of the $Ag^+$ ions reacted.

# CHEMICAL BONDING AND MOLECULAR STRUCTURE



**KEY TERMS**

**Angular structure, 360**

**Bent structure, 360**

**Bond, 340**

**Bond angle, 356**

**Bond length, 344**

**Covalent bond, 344**

**Dipole moment, 362**

**Dissociates, 341**

**Double bond, 351**

**Ionic bond, 341**

**Lewis diagrams, 347**

**Linear structure, 356**

**Metallic bond, 345**

**Octahedral structure, 367**

**Octet rule, 347**

**Polar covalent bond, 344**

**Polarity, 361**

**Resonance hybrid, 355**

**Resonance structure, 354**

**Single bond, 351**

**Tetrahedral structure, 358**

**Trigonal bipyramid, 366**

**Trigonal planar structure, 357**

**Trigonal pyramid, 359**

**Triple bond, 351**

**VSEPR theory, 356**

In previous chapters, especially Chapter 2, you encountered the idea that the arrangement of atoms in molecules has something to do with how the molecules behave. This chapter explores in more detail how molecules form and the shapes they have. If atoms did not somehow attract one another and form molecular structures, the universe would consist of a collection of atoms forever moving randomly in the vastness of space—a rather boring place from a chemist's (or any other) point of view.

Atoms do attract one another, and so do molecules. These attractive forces create the wide range of physical and chemical properties you observe every day. Chapter 12 will discuss the forces of attraction that molecules have for one another—*intermolecular* forces of attraction. This chapter takes a closer look at the forces of attraction between the atoms that make up chemical compounds—which can be called *intramolecular* forces.

When two or more atoms have a strong attraction for each other, chemists say that they form a **bond.** This chapter begins with a description of what bonds are, how they work, and how bonding influences the properties that all substances have. It also explores how atoms in molecules fit together, the shapes that molecules take, and (to a limited extent) how molecules behave.

## 10.1 Types of Bonding: Ionic, Covalent, and Metallic

**GOAL:** To describe the forces that hold atoms together in compounds.

FIGURE 10.1
Copper (left) can react with chlorine (left center) to form either CuCl (right center) or $CuCl_2$ (right). Which compound forms depends on which reactant is present in excess. The physical and chemical properties of these four substances are very different.

***Don't even think about repeating Benjamin Franklin's kite experiment.*** **Although the cornerstone of science is the ability to reproduce experiments, the first two people who tried to repeat Franklin's experiment were killed when the electric current from a lightning bolt traveled down their kite strings and passed through their bodies. Franklin was very lucky that his brilliant career did not end prematurely.**

Once scientists came to accept the idea that matter consists of very small particles and is not some sort of continuous, infinitely divisible fluid, they began to speculate (without having much evidence) on what these tiny particles were like. Are they small, hard spheres like tiny billiard balls? Do atoms of different elements have different shapes? Do atoms group together in molecules because they have something like little hooks on them, or are they stuck together with something like glue?

Early chemists knew that elements keep their identities in some way when they form compounds. The ancients refined copper from copper ores and observed that the metal turned green with age and returned to a state much like that of the original ore. Why are copper (a reddish metal) and chlorine (a greenish yellow gas) so different from copper(I) chloride (a white solid that is insoluble in water) and copper(II) chloride (a blue solid that is water-soluble)? (See Figure 10.1).

In the 18th century many experimenters began playing with static electricity—that's what gives you a shock and sometimes produces a visible spark when you shuffle across a carpet and then touch a metallic object on a dry day. Benjamin Franklin's famous kite experiment showed that these tiny sparks and huge bolts of lightning are two manifestations of the same thing. Franklin thought about his experiments and experiences and proposed that there are two kinds of electricity—one positive and the other negative. We have already discussed this at some length beginning in Section 4.4.

The Italian physicist Count Alessandro Volta (1745–1827) made a key discovery. Following up on some earlier experiments, he made a pile of alternating disks of copper and zinc separated by pieces of wet cardboard (see Figure 10.2). He found that an electrical current ran between the copper disk at one end and the zinc disk at the other. This *Voltaic pile* is the ancestor of every battery in use today.

Word of Volta's experiments got around. In 1800, soon after they heard about Volta's work, the English chemists William Nicholson (1753–1815) and Anthony Carlisle (1768–1840) made their own Voltaic pile (Section 4.4). They ran wires from their battery into a container of water and produced the elements hydrogen and oxygen. This experiment showed that there must be a link between electricity and chemistry, but the nature of that link remained a puzzle for a long time.

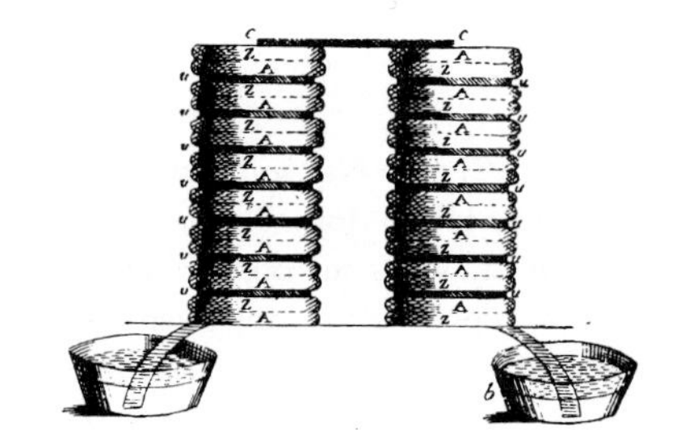

FIGURE 10.2
This crude battery, called a Voltaic pile, generates an electric current. One of the first experiments done with a Voltaic pile was to pass a current through water. This process formed $H_2(g)$ and $O_2(g)$ and established that a link exists between electricity and chemistry.

### Ionic Bonding

Svante Arrhenius knew that pure water is a poor conductor of electric current. He also knew that some substances (including table salt, or sodium chloride) dissolve in water, and that the resulting solutions conduct electric current very well. He studied this phenomenon and proposed that when sodium chloride dissolves, it **dissociates** into separate sodium and chloride ions that have opposite charges. Because these ions move independently of each other in solution, electric current can flow. (See Figure 10.3).

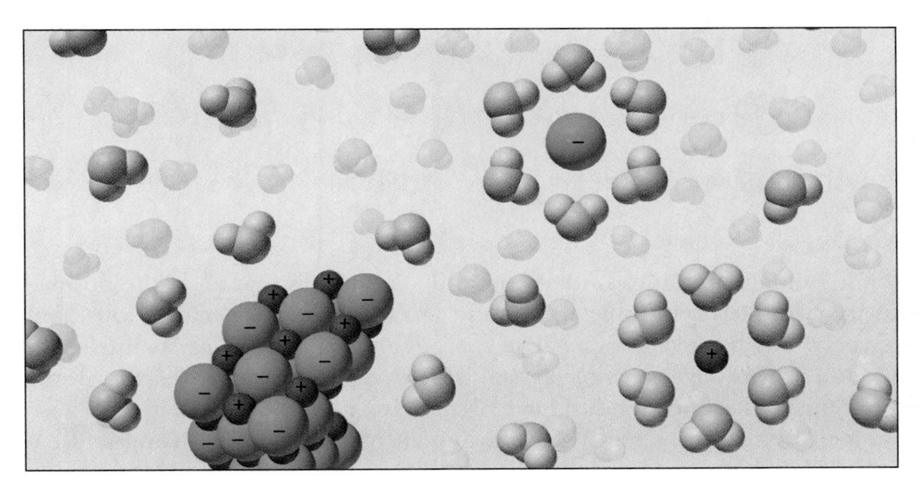

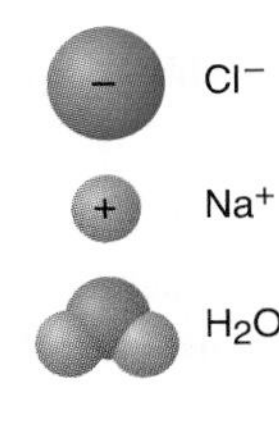

FIGURE 10.3
When solid NaCl dissolves in water, it dissociates into $Na^+$ and $Cl^-$ ions. In solution these ions move independently and can conduct an electric current.

In solid sodium chloride the sodium and chloride ions attract each other, forming what chemists call an **ionic bond.** The electrostatic force—the attraction of unlike charges or the repulsion of like charges—operates in all directions. If a sodium ion has a chloride ion on one side, it can still attract another chloride ion on the other side. In fact, in a salt crystal each sodium

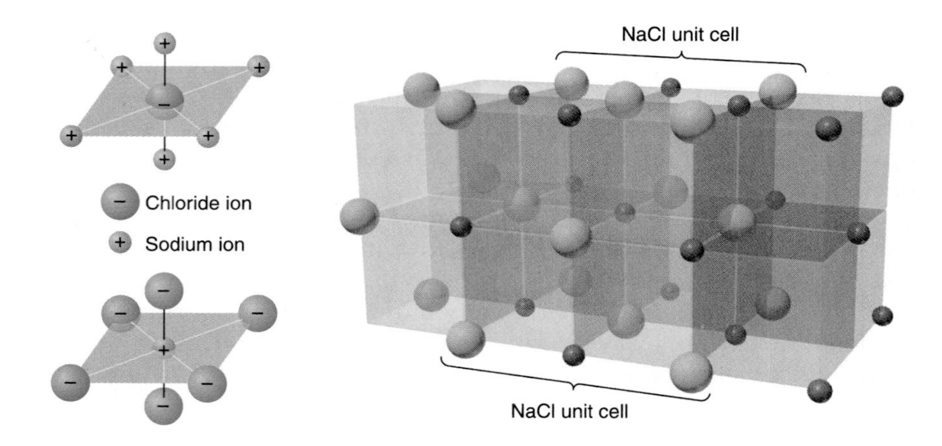


FIGURE 10.4
In solid NaCl each $Na^+$ ion is surrounded by six $Cl^-$ ions (lower left), and each $Cl^-$ ion is surrounded by six $Na^+$ ions (upper left). In any given salt crystal, no matter how small, the sodium and chloride ions stack on top of each other in a seemingly endless three-dimensional array. At right is part of such an array. Each of the lines connecting atoms represents an ionic bond.

ion is surrounded by six chloride ions and each chloride ion is surrounded by six sodium ions (see Figure 10.4).

Some other compounds have this type of crystalline structure—called the NaCl structure—but most ionic compounds have different numbers and arrangements of ions. In cesium chloride (CsCl), for instance, each $Cs^+$ ion is surrounded by eight $Cl^-$ ions, and vice versa. The fact that the $Cs^+$ ion (radius = 167 pm) is larger than the $Na^+$ ion (radius = 97 pm) makes the difference here. This means that there is more space available for chloride ions around each cation in CsCl than there is in NaCl. (See Figure 10.5.)

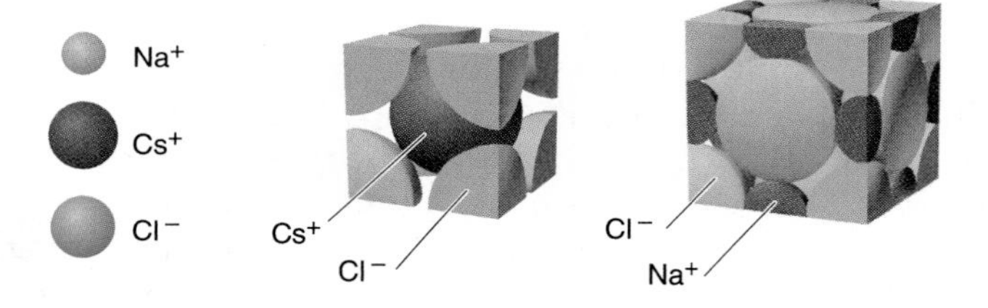


FIGURE 10.5
Because the $Cs^+$ ion is larger than the $Na^+$ ion, more $Cl^-$ ions can pack around $Cs^+$ (eight) in cesium chloride than around $Na^+$ (six) in sodium chloride.

**Here are some examples of the high melting points typical of ionic compounds:**

| Compound | Melting Point(°C) |
|---|---|
| NaCl | 801 |
| $Na_2CO_3$ | 851 |
| $CaSO_4$ | 1000 |
| $Al_2O_3$ | 2072 |

The fundamental idea underlying ionic bonding is that all of the positive ions in a sample of an ionic compound attract all of the negative ions, and vice versa. The positive and negative charges alternate in the crystalline structure, and in this way the whole crystal is held together by electrostatic attraction. Ionic compounds tend to have high melting and boiling points because the electrostatic forces that hold the ions in place are fairly strong and extend in all directions.

### Covalent Bonding

The forces of attraction in molecules such as hydrogen ($H_2$) were harder for pioneering chemists to understand. The two hydrogen *atoms* in a hydrogen *molecule* are even harder to break apart than the two ions in NaCl. On the other hand, the attraction of one $H_2$ molecule for another is about as weak as an intermolecular interaction can get. As you know, hydrogen is a gas at room temperature. The temperature has to drop to about 14 K (−259°C) before $H_2$ molecules will attract one another enough to form a liquid. Almost any movement at all by the $H_2$ molecules in this liquid is enough to release them into the gas phase.

Clearly two different forces exist. One of these forces works on the atomic level—holding two hydrogen atoms together to form an $H_2$ molecule. The other force works at the molecular level—holding $H_2$ molecules together. Obviously the force between the molecules (the intermolecular force) is much weaker than the force within a molecule (the intramolecular force) (see Figure 10.6).

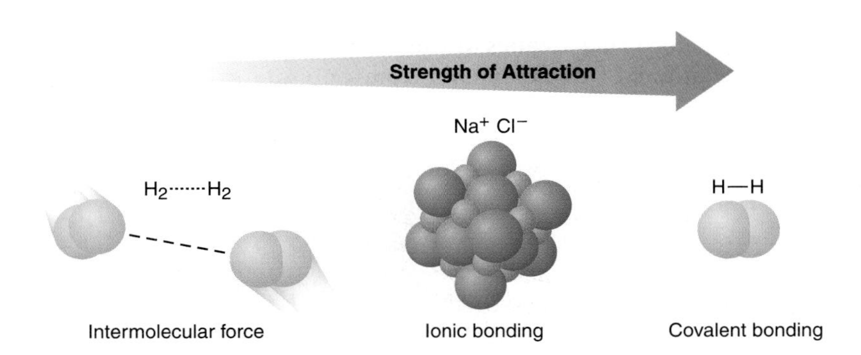


FIGURE 10.6
The strength of the attractive force increases from intermolecular attractions to covalent bonds.

Chemists in the late nineteenth century knew that most molecules made up of atoms of nonmetals behaved somewhat like $H_2$ molecules, but they didn't have an explanation of what underlies this behavior. Chemists had to wait until Rutherford's work clarified the structure of the atom before they could begin to understand what the structures of molecules were really like.

To get some idea of the problem, consider a hydrogen atom. At its center is a proton with a positive charge. This proton is very small relative to the atom as a whole and has most of the atom's mass. It is surrounded by nearly empty space that contains an electron. The electron has only about $\frac{1}{2000}$ of the mass of the proton, but its negative charge has a magnitude equal to that of the proton's positive charge.

The positively charged proton has a very strong attraction for the negatively charged electron, and this electrostatic force of attraction holds the electron and the proton together in the hydrogen atom. Although the attractive force extends to an infinite distance from the proton, its strength falls off very rapidly with increasing distance (Coulomb's Law tells you that the attractive force is inversely proportional to the *square* of the distance).

Now consider what happens when two hydrogen atoms approach one another. The proton of each hydrogen atom attracts the electron of the other, as illustrated in Figure 10.7. These attractive forces bring the atoms closer together, which increases the attractive force of each proton for the other atom's electron (the shorter the distance between charges, the stronger the force). At some point the repulsive forces between the two protons and between the two electrons begin to have an effect. When the proton-electron attraction balances the proton-proton and electron-electron repulsions, the atoms cannot move any closer together.

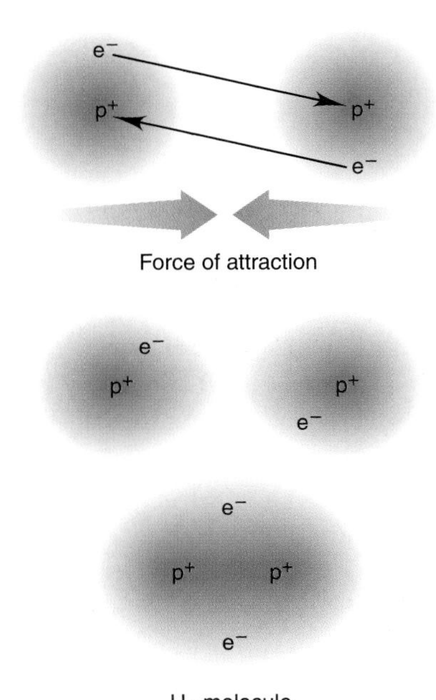


FIGURE 10.7
Formation of an $H_2$ molecule from two hydrogen atoms. When the atoms are far apart (top), the force between them is tiny. As they approach one another (center), the proton of each atom attracts the electron of the other. The distance at which the attractive and repulsive forces balance each other (bottom) is called the *bond length.*

This distance at which the forces are balanced is known as the **bond length.** In the $H_2$ molecule it is equal to 74 pm. At this distance the two protons have equal attractions for the electrons. In other words, the individual hydrogen atoms share the electrons, which are called an *electron pair.* This kind of bond, in which electron pairs are shared, is called a **covalent bond.**

The two atoms in an $H_2$ molecule share the electrons equally. The opposite situation exists in sodium chloride: Both sodium atoms and chlorine atoms have electrons that might be shared, but no sharing takes place in NaCl. Each sodium atom gives up its electron to a chlorine atom to form the $Na^+$ and $Cl^-$ ions.

Whether two atoms form an ionic bond or a covalent bond can be predicted from the electronegativities of the element(s). In NaCl an ionic bond forms between atoms of two elements from opposite sides of the periodic table, atoms that have very different electronegativities. In $H_2$ the nonpolar covalent bond forms between atoms that are identical and that therefore have the same electronegativity. (See Figure 10.8.)

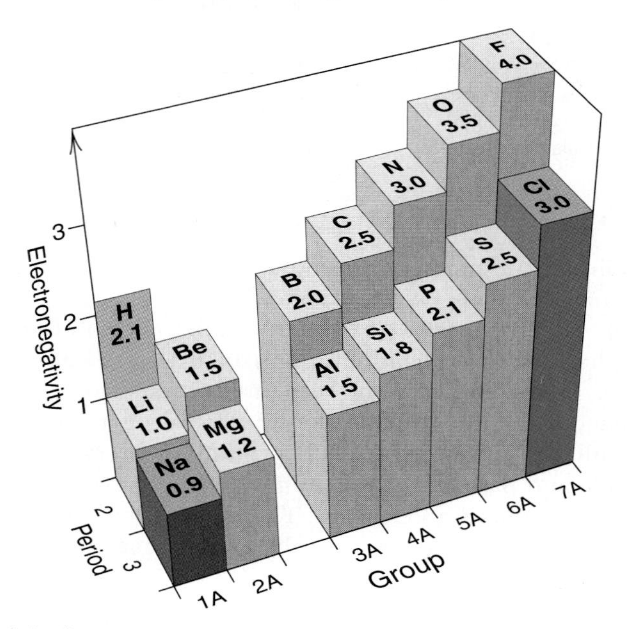


FIGURE 10.8

Bonds between atoms with large electronegativity differences (as in NaCl) tend to be ionic. When the difference in electronegativity is zero (as in $H_2$), purely covalent bonds form. Intermediate cases, in which polar covalent bonds form, do exist (HCl is an example).

There are intermediate cases, of course. Hydrogen chloride (HCl) is one of them. Hydrogen and chlorine are far apart on the periodic table and have considerably different electronegativities. The difference is not as great, however, as the difference between sodium and chlorine. In this intermediate case the hydrogen and chlorine atoms share a pair of electrons—but *not* equally. The more electronegative chlorine atom pulls the shared electrons toward itself. This kind of bond is called a **polar covalent bond,** because one end of the molecule (toward the more electronegative atom) is slightly negative and the other end of the molecule is slightly positive (see Figure 10.9).

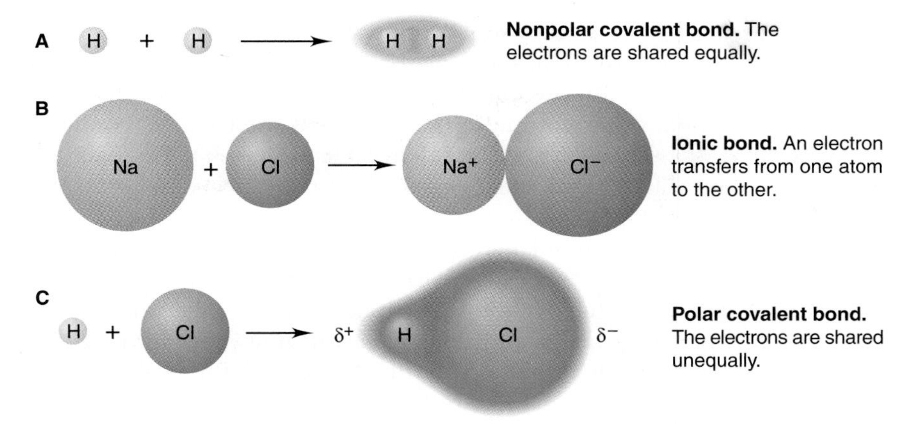


FIGURE 10.9
The amount of electron sharing varies for the three kinds of bonds shown: (A) nonpolar covalent, (B) ionic, and (C) polar covalent.

### Metallic Bonding

Samples of metals and metal alloys behave very differently from samples of ionic and covalent compounds. Most metals have high melting points, and all of them have high boiling points. From these facts you can conclude that some kind of fairly strong force holds all of the metal atoms in a sample together, just as ionic bonding holds together all the ions in a crystal.

However, the bonding in metals can't very well be ionic. A piece of copper, for instance, is made up of only one kind of atom; all copper atoms have the same electronegativity. Also, metals are malleable and ductile, while solid salts (and solid covalent compounds, too) are brittle. From this sort of evidence you should conclude that the bond holding metal atoms together must be somehow different.

This different type of bond is the **metallic bond,** the last of the types of chemical bonds to be understood by chemists. The nucleus of a metal atom holds the inner-shell electrons strongly, but the valence electrons (the few in the outermost shell) only weakly. These valence electrons form a "sea" of electrons, free to move from atom to atom throughout a sample of a metal. This free flow of electrons means that samples of most metals bend or stretch under pressure but are hard to break apart. This electron flow is also the reason why solid metals are good conductors of electricity.

Let's look specifically at the metallic bonding in a sample of sodium, which has the electron configuration $1s^22s^22p^63s^1$. The single $3s$ electron is only weakly held by the sodium atom (this is why the $Na^+$ ion forms so easily). In a chunk of sodium metal, the $3s$ orbital of one atom can overlap with the $3s$ orbitals of all the neighboring atoms (see Figure 10.10).

**It takes a lot of energy to separate atoms from a sample of a metal and release them into the gas phase. Here are the boiling points of some typical metals:**

| Metal | Boiling Point (°C) |
|---|---|
| Mg | 1090 |
| Pb | 1750 |
| Sn | 2603 |
| Fe | 2862 |
| Ni | 2914 |

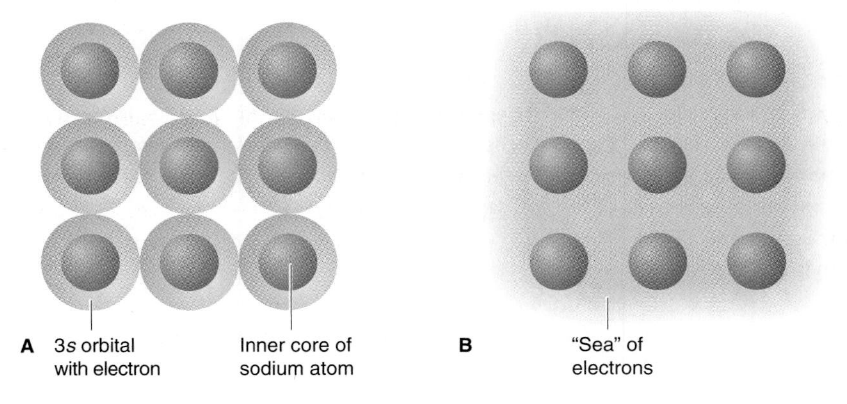


FIGURE 10.10
(A) In sodium metal the 3s orbital on each sodium atom overlaps with the 3s orbitals of neighboring sodium atoms. (B) This overlap produces a "sea" of loosely held electrons, none of which is associated with any individual atom. Instead they are associated with the sample as a whole.

In fact, because of this extensive overlapping, the 3*s* electrons are no longer associated with individual sodium atoms. Instead, these loosely bound electrons are associated with the entire structure. This type of metallic bonding occurs in all metals. Aluminum, for example, has the electron configuration $1s^22s^22p^63s^23p^1$. There are three valence electrons ($3s^23p^1$). Each aluminum atom contributes three electrons to the sea of electrons in forming the metallic bond.

It is relatively easy to pound metals into thin sheets (that is, metals are malleable), because it is relatively easy to change the arrangement of the atoms without disrupting the bonding. Since the valence electrons don't associate with any particular atom and since electrons move rapidly, pounding a metal sample flat has little influence on the electrostatic forces that hold its atoms together. On the other hand, crystals of ionic compounds are brittle. They shatter under stress because displacement of ions in the crystals moves similar charges near each other and the repulsions crack the crystal.

**Section 12.2 and Figure 12.14 consider this topic more fully.**

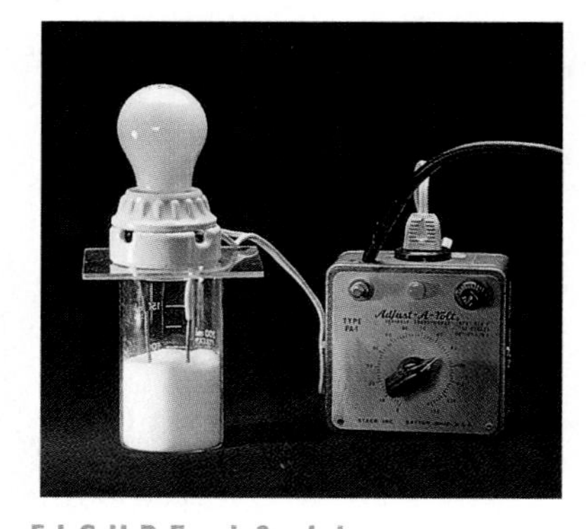

FIGURE 10.11
The beaker contains NaCl(s). Will the bulb light up when the current is turned on? That is, will the NaCl(s) conduct electric current?

### EXAMPLE 10.1 Predicting Electrical Conductivity

Ionic compounds in solution and solid metals are good conductors of electricity. Are solid samples of ionic compounds good conductors? Explain your answer.

**SOLUTION**

For a sample to conduct electricity, something that has a charge must move. In a sample of a solid metal, the valence-shell electrons are free to move. In solutions of ionic compounds, the ions are separate and free to move, carrying the current. In a sample of a solid ionic compound, the ions are locked in place in the crystal structure. Solid samples of ionic compounds are poor conductors of electricity.

**EXERCISE 10.1**

Are solid samples of covalent compounds good conductors of electricity? Explain your answer.

### EXPERIMENTAL EXERCISE 10.1

A. L. Chemist gave a blindfolded student samples of two solid substances, one of which was a metal and the other a nonmetal. The two samples had the same size and shape and approximately the same density, and both had smooth polished surfaces. A. L. asked the student, "Which of these is a metal?" The student held each sample for a few seconds and then answered the question correctly.

Most seemingly miraculous blindfold tricks involve peeking down the side of the nose under the blindfold. Assuming that there was no cheating, how did the student know which of the two samples was a metal?

## 10.2 Lewis Diagrams for Simple Compounds

GOAL: To draw Lewis diagrams for compounds composed of hydrogen atoms and atoms that obey the octet rule.

FIGURE 10.12
Gilbert N. Lewis (1875–1946) was a brilliant chemist and an inspiring teacher. He is known for his contributions in the areas of acid-base behavior and chemical bonding. He formulated the idea of Lewis diagrams in 1902 while teaching basic college chemistry.

This section introduces **Lewis diagrams,** which were an invention of Gilbert N. Lewis (Figure 10.12). These diagrams have two purposes. First, they show how the atoms in a compound or ion connect to one another. Second, they account for all the valence electrons of all the atoms in a compound or ion.

All of the rules for drawing Lewis diagrams to reflect the structures of real compounds are simple. As often happens in the sciences, however, you have to apply several of these rules at the same time. The key to understanding this section (and to answering questions about this material correctly on tests) is to practice.

Compounds come in two broad classifications—ionic and covalent. In Chapter 7 you found you had to decide whether a compound was ionic or covalent before you could name it. You'll have to know how to tell the difference in this chapter as well. The Lewis diagrams of ionic and covalent compounds are substantially different from one another.

**If you need a review, this material is in Section 7.7.**

### The Octet Rule

Section 5.4 noted the stability of atoms or ions that have the same electron configurations as the noble gases. For all of the main group elements except hydrogen, helium, lithium, and beryllium, this involves a valence shell that contains two *s* and six *p* electrons. A main group element that has a valence-shell electron configuration of $ns^2np^6$ is said to contain an *octet* of electrons. Guidelines for drawing Lewis diagrams include the following statement, known as the **octet rule:**

> A molecular structure is most stable when each atom in it contains an octet of electrons in its valence shell.

The octet rule is somewhat limited in scope, because many elements don't obey it. The most obvious exceptions are molecules containing hydrogen, which can accommodate two electrons at most in its valence shell. In fact, of the main group elements only carbon, oxygen, nitrogen, fluorine, and the alkali and alkaline earth metals obey the octet rule in virtually all of their compounds. Elements in the third row of the periodic table and beyond can use *d* orbitals to hold more than eight electrons in their valence shells (this behavior is discussed in more detail in Section 10.7).

**The hydride ion does not have an octet. It has only two electrons and is drawn like this:**

$H:^-$

In spite of these problems, the octet rule is extremely useful. Most known compounds and ions contain carbon, oxygen, or nitrogen atoms, and these elements almost always obey the octet rule. Furthermore, the octet rule is the starting point for drawing almost any Lewis diagram—even those that contain atoms that do not adhere to the rule (once again, see Section 10.7).

### Binary Ionic Compounds

The simplest structures to diagram are the binary ionic compounds, which consist of a metal and a nonmetal. You must draw the structures of the cation and anion that make up an ionic compound separately. Remember that metals are electropositive and form cations easily, and that nonmetals are electronegative and form anions easily.

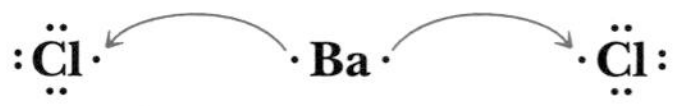

$$:\ddot{\underset{..}{Cl}}\cdot \leftarrow \cdot Ba \cdot \rightarrow \cdot \ddot{\underset{..}{Cl}}:$$

The rules for drawing Lewis diagrams for binary ionic compounds from their formulas are very simple.

1. For each of the ions, write the symbol of the element, leaving some space between symbols.
2. Add the positive charge to the symbol for the cation in the usual way.
3. Place four pairs of dots (to indicate an octet of electrons) around the symbol for the anion. Then indicate the negative charge on the ion in the usual way.
4. Check your answer. For compounds consisting of main group elements, the total number of electrons you show in your diagram must be equal to the sum of the group numbers of the atoms in the formula.

**STUDY ALERT**

**Checking your answer is absolutely essential. Be sure to do this every single time you draw a Lewis diagram.**

### EXAMPLE 10.2 Drawing Lewis Diagrams for Binary Ionic Compounds

Draw the Lewis diagram for each of these binary ionic compounds.

**a.** $BaCl_2$
**b.** Sodium oxide

**SOLUTIONS**

**a.** First, write the Ba symbol and two Cl symbols separately:

Cl Ba Cl

Barium is in Group 2A, so it has two valence electrons that it loses to form the $Ba^{2+}$ ion. Its Lewis diagram is just like its formula, $Ba^{2+}$. Next, surround each Cl symbol with four pairs of electrons (each electron is shown as a dot):

$$:\ddot{\underset{..}{Cl}}:^- \quad Ba^{2+} \quad :\ddot{\underset{..}{Cl}}:^-$$

**The cation has given up its valence electrons. Since a Lewis diagram shows only valence electrons, there are no dots around $Ba^{2+}$.**

Check: The formula indicates one barium atom and two chlorine atoms. The sum of the group numbers is $2 + 7 + 7 = 16$. Each of the two chloride ions has 8 electrons, also giving 16.

**b.** First, you have to know the formula of sodium oxide. Sodium is in Group 1A, so the sodium ion is $Na^+$. Oxygen is in Group 6A, so the oxide ion is $O^{2-}$. The formula of sodium oxide is therefore $Na_2O$. Its Lewis diagram is

$$Na^+ \quad :\ddot{\underset{..}{O}}:^{2-} \quad Na^+$$

Check: Add up the group numbers of the two sodium atoms and the oxygen atom: $1 + 1 + 6 = 8$. The oxide ion has 8 electrons, so this diagram is OK.

**EXERCISE 10.2**

Draw the Lewis diagram for each binary ionic compound.

**a.** $K_3N$
**b.** Calcium sulfide

### Covalent Compounds

In covalent compounds the atoms do not form separate ions but are held together by the sharing of electrons. Chemists call a shared pair of electrons a *bond* and symbolize it by drawing a line between the two atoms. A shared pair of electrons is called a *bonding pair.* Here are the guidelines for drawing Lewis diagrams for covalent compounds:

1. Draw a skeleton for the compound. Usually the atom with the smallest subscript in the formula is in the center (there are a few exceptions). Hydrogen can never be the central atom because it cannot make more than one bond (see below). Connect each of the other atoms to the central atom with a line.
2. Add pairs of electrons (shown as dots) to each of the atoms except hydrogen until each is surrounded by a total of eight electrons. Each bond counts as two electrons for *both* of the bonded atoms. Pairs of electrons that are associated with only one atom are called lone pairs or *nonbonding pairs,* to distinguish them from the shared electrons in bonds.
3. As always, hydrogen can have only two electrons, and any hydrogen atom in a molecule has one and only one bond to it (and no nonbonding pairs). The Lewis diagram of the hydrogen molecule (see Section 10.1) is this:

   **H—H**

4. Check to make sure that the number of electrons shown in the structure is equal to the sum of the group numbers of the atoms in the molecule.

### EXAMPLE 10.3 Drawing Lewis Diagrams for Covalent Compounds with Single Bonds

Draw the Lewis diagram for each of these covalent compounds.

**a.** Water
**b.** Nitrogen trichloride

FIGURE 10.13
Shape of $H_2O$ molecule.

**Lewis diagrams don't attempt to show the actual shapes of molecules. For example, the water molecule is V-shaped, not linear. Sections 10.4 and 10.5 are concerned with molecular shapes.**

SOLUTIONS

**a.** Water is, of course, $H_2O$. You begin by drawing the skeleton. Put the oxygen in the center (hydrogen is never the central atom) and bond one hydrogen to it on each side, like this:

$$\mathbf{H{-}O{-}H}$$

Each of the two hydrogens has a single bond (as it should), and the oxygen needs four more electrons (it has four associated with it already, in its two bonds). Add two pairs of dots to the oxygen, giving this diagram:

$$\mathbf{H{-}\underset{\cdot\cdot}{\overset{\cdot\cdot}{O}}{-}H}$$

Hydrogen is in Group 1, and oxygen is in Group 6A. The sum of the group numbers of the three atoms in the molecule is $1 + 1 + 6 = 8$. There are two bonds and two nonbonding pairs of electrons in a water molecule, for a total of 8 electrons.

**b.** Nitrogen trichloride is $NCl_3$. First, draw the skeleton. Put the nitrogen atom in the middle and bond the three chlorine atoms to it, like this:

$$\begin{array}{c}\mathbf{Cl{-}N{-}Cl}\\ | \\ \mathbf{Cl}\end{array}$$

The nitrogen atom has six electrons (in three bonds) and needs eight electrons. Each of the chlorine atoms has two electrons (in one bond) and needs six more electrons (three pairs). Here's what you get when you put in all pairs:

$$\begin{array}{c}:\underset{\cdot\cdot}{\overset{\cdot\cdot}{\mathbf{Cl}}}{-}\overset{\cdot\cdot}{\mathbf{N}}{-}\underset{\cdot\cdot}{\overset{\cdot\cdot}{\mathbf{Cl}}}:\\ | \\ :\underset{\cdot\cdot}{\mathbf{Cl}}:\end{array}$$

Now, the all-important check: Does the sum of the group numbers match the total number of electrons in the Lewis diagram? The sum of the group numbers is $5 + 7 + 7 + 7 = 26$. The structure contains three bonds (six electrons) and ten nonbonding pairs (20 electrons) for a total of 26 electrons. The diagram is correct.

EXERCISE 10.3

Draw the Lewis diagram for each of these covalent compounds.

**a.** Ammonia

**b.** The useful solvent dichloromethane, $CH_2Cl_2$

## 10.3 Multiple Bonds, Polyatomic Ions, and Resonance

GOAL: To draw Lewis diagrams for molecules containing multiple bonds and for polyatomic ions.

### Multiple Bonds

The procedures described in Section 10.2 worked well. If you followed the guidelines, you got the right answers. However, Lewis diagrams have their complications, and sometimes the guidelines become only hints and suggestions. So far all the bonds you have seen have been drawn as single lines representing **single bonds** between two atoms. Even small and simple compounds can contain **double bonds** (two lines that represent four electrons) or **triple bonds** (three lines indicating a total of six electrons).

If you are trying to draw a Lewis diagram for a compound with a double or triple bond, you will reach an impasse if you follow the rules from Section 10.2. When you count the electrons in your structure, you will get a number that is larger than the sum of the group numbers (which means the diagram shows too many electrons). To reduce the total number of electrons, you first remove two nonbonding pairs, one from each of two atoms that are already bonded together. Then draw a new bond between the two atoms, turning a single bond into a double bond or a double bond into a triple bond. Example 10.4 illustrates this process.

### EXAMPLE 10.4 Drawing Lewis Diagrams for Molecules with Multiple Bonds

Draw the Lewis diagram for nitrogen ($N_2$), the most abundant gas in the earth's atmosphere.

**SOLUTION**

This looks easy. First, you connect the two nitrogen atoms with a bond. Then fill in three pairs of electrons around each nitrogen atom. Since nitrogen is in Group 5A and the formula is $N_2$, the diagram should show 10 electrons. However, there are 14 (oops!). Since the paragraphs above suggest that a multiple bond is the answer to this problem, try removing one nonbonding pair from each nitrogen atom and changing the single bond to a double bond. This lowers the electron count to 12, which is closer to 10 but still not correct. Remove one more pair of nonbonding electrons from each nitrogen atom and connect the two atoms with a triple bond. This structure has ten electrons and is the correct one.

| First attempt | Second attempt | Third attempt |
|---|---|---|
| :N̤̈—N̤̈: | :N̈=N̈: | :N≡N: |
| There are 14 electrons, too many. | Now there are 12 electrons, still too many. | The correct structure, with 10 electrons. |

**EXERCISE 10.4**

Draw the Lewis diagram for carbon monoxide (CO), a poisonous gas produced by the incomplete combustion of carbon-containing fuels.

**There are a few exceptions to these guidelines, as you observed when you worked on Exercise 10.4.**

The atoms that most commonly form multiple bonds are C, N, and O atoms. (These are not the only ones, but they form multiple bonds far more frequently than atoms of any other element do). One thing to keep in mind is that carbon, nitrogen, and oxygen atoms almost always obey the octet rule. Therefore, in neutral molecules, carbon (in Group 4) typically makes four bonds and has no nonbonding electron pairs. Nitrogen (Group 5) makes three bonds and has one nonbonding pair, and oxygen (Group 6) makes two bonds and has two nonbonding pairs. This means that only a few arrangements in Lewis diagrams are reasonable for these atoms. These possibilities are as follows:

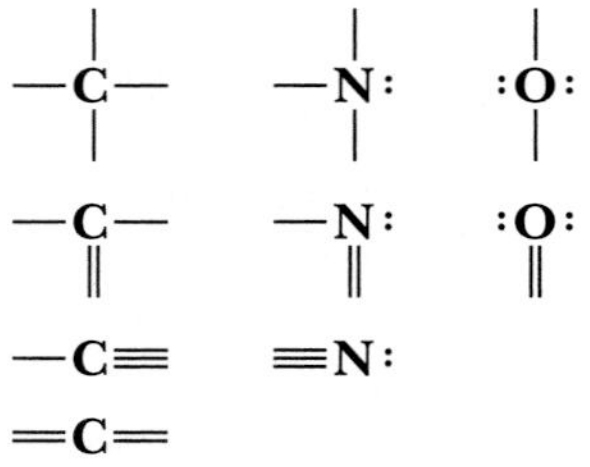

```
   |        |       |
 —C—      —N:     :O:
   |        |       |

 —C—      —N:     :O:
   ‖        ‖       ‖

 —C≡      ≡N:

 =C=
```

Note how much simpler Example 10.4 would have been if you had known that nitrogen typically makes three bonds.

## EXAMPLE 10.5 Drawing More Lewis Diagrams with Multiple Bonds

Draw a Lewis diagram for $H_2CO$ (formaldehyde, an important but somewhat toxic preservative and industrial chemical).

**SOLUTION**

The formula gives a strong hint as to how to arrange the skeleton. The two hydrogens should be attached to the carbon atom:

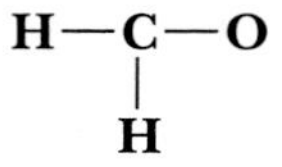

```
H—C—O
  |
  H
```

Carbon normally forms four bonds and oxygen forms two. The most likely structure, then, is one with a double bond between the carbon and oxygen atoms. If you then add two nonbonding pairs to the oxygen, you get this structure:

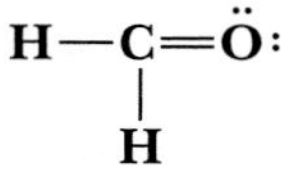

```
     ..
H—C=O:
  |
  H
```

Adding up the group numbers (2 for the two hydrogens, 4 for the carbon, and 6 for the oxygen) gives 12. A count of the electrons in the diagram also gives 12.

**EXERCISE 10.5**

Draw the Lewis diagram for $CO_2$.

### Polyatomic Ions

To draw Lewis diagrams for polyatomic ions, species that contain both covalent bonds and charges, you perform the first steps as you did in Section 10.2—that is, you draw the skeleton and satisfy the octet rule. The first complication comes when you determine the number of electrons the species should have.

To get the correct number, you need to add up the group numbers (as usual) and then *add one more electron for each negative charge on the ion.* Since an electron has a single negative charge, each negative charge on the ion means that there is an extra electron. For each positive charge on the ion, you subtract an electron (take away a negative charge). The final step for Lewis diagrams for polyatomic ions is to put brackets around the structure and write the charge to the upper right.

**The only common polyatomic cation is the ammonium ion, $NH_4^+$.**

In polyatomic anions the extra electron typically is associated with the more electronegative atom. When oxygen or nitrogen is involved, new bonding possibilities arise. Oxygen frequently has one bond and three nonbonding electron pairs (in anions). Sometimes nitrogen has two bonds and two nonbonding electron pairs (in anions), and sometimes it has four bonds and no pairs (in cations). Here is how these atoms often look in Lewis diagrams for polyatomic ions:

$$-\ddot{\underset{..}{O}}:^{-} \qquad -\ddot{\underset{..}{N}}-^{-} \qquad -\overset{|}{\underset{|}{N}}-^{+}$$

**EXAMPLE 10.6** **Drawing Lewis Diagrams for Polyatomic Ions**

Draw the Lewis diagram for each of these ions.

**a.** Hydroxide ion, $OH^-$

**b.** Cyanide ion, $CN^-$

**SOLUTIONS**

**a.** The only way you can draw the skeleton of this ion is with a single bond from the H to the O. Then you fill in three pairs of electrons around the O atom to make it obey the octet rule. You get this structure:

$$:\ddot{\underset{..}{O}}-H$$

The total number of electrons for the ion should be the sum of the group numbers plus one more for the negative charge:

1 from hydrogen + 6 from oxygen + 1 from the extra electron = 8

Counting electrons in the structure gives 8 electrons, as it should. Now you put brackets around the structure and indicate the charge:

$$\left[:\ddot{\underset{..}{O}}-H\right]^{-}$$

**b.** Draw a single bond from C to N and put three pairs of electrons around both C and N atoms:

$$:\ddot{\underset{..}{C}}-\ddot{\underset{..}{N}}:$$

This diagram has 14 electrons. Carbon is in Group 4A, and nitrogen is in Group 5A, giving 9 valence electrons; adding 1 electron for the negative charge yields 10 electrons. Thus 14 electrons is 4 too many. If you make two more bonds between C and N (see Example 10.4), you get a Lewis diagram that contains a triple bond and 10 total electrons:

$$[:C\equiv N:]^-$$

This is the correct answer.

**EXERCISE 10.6**

Draw the Lewis diagram for each of these ions:

**a.** Ammonium ion, $NH_4^+$

**b.** Amide ion, $NH_2^-$

### Equivalent Lewis Diagrams and Resonance

Occasionally when drawing Lewis diagrams, you will be faced with a situation where you must draw one or more multiple bonds to get a correct diagram, and it is not clear where the extra bond (or bonds) should go. The formate ion, $HCO_2^-$, illustrates this point. The Lewis diagram must show 12 valence electrons for the two oxygen atoms (Group 6A), 4 for the carbon (Group 4A), 1 for the hydrogen, and 1 for the negative charge, for a total of 18.

Using the usual procedure, you make carbon the central atom and give it and the two oxygens an octet of electrons, like this:

$$\begin{array}{c} \phantom{:O-}H\phantom{-O:} \\ \phantom{:O-}|\phantom{-O:} \\ :\ddot{\underset{..}{O}}-\underset{..}{C}-\ddot{\underset{..}{O}}: \end{array}$$

This structure has 20 electrons, which is too many. Two electrons can be "removed" by putting a double bond between the carbon atom and one of the oxygen atoms, but which one? There are two choices:

$$\left[\begin{array}{c} H \\ | \\ :\ddot{O}=C-\ddot{\underset{..}{O}}: \end{array}\right]^- \quad \left[\begin{array}{c} H \\ | \\ :\ddot{\underset{..}{O}}-C=\ddot{O}: \end{array}\right]^-$$

There is no rule that enables you to choose between these two. In both of these Lewis diagrams, the carbon atom forms a single bond to one of the oxygen atoms and a double bond to the other. The two diagrams are, in fact, both correct. They are said to be *equivalent.* The only difference between the diagrams is the position of the double bond.

Equivalent Lewis diagrams are called **resonance structures.** Chemists indicate that structures have this relationship by connecting them with a double-headed arrow:

$$\left[\begin{array}{c}\text{H}\\ |\\ :\ddot{\text{O}}{=}\text{C}{-}\underset{\cdot\cdot}{\ddot{\text{O}}}:\end{array}\right]^{-} \longleftrightarrow \left[\begin{array}{c}\text{H}\\ |\\ :\underset{\cdot\cdot}{\ddot{\text{O}}}{-}\text{C}{=}\ddot{\text{O}}:\end{array}\right]^{-}$$

At this point you might be wondering what the real structure is and what the ion looks like. There is some experimental evidence. The length of an ordinary carbon-oxygen single bond is 136 pm, and the length of an ordinary carbon-oxygen double bond is 123 pm. In $HCO_2^-$ the two carbon-oxygen bonds have the intermediate length of 127 pm. This measurement strongly suggests that each of the two resonance structures only approximates the actual electron distribution in the ion and that the real structure is something halfway between the two.

This halfway structure, an average of all the individual resonance structures, is called the **resonance hybrid.** The resonance structures are the best you can do if you follow the guidelines for drawing Lewis diagrams.

It is important to understand that the electrons do not flip back and forth, as the resonance structures seem to suggest. The $HCO_2^-$ anion is a composite of the two equivalent resonance structures shown above, and its two oxygen atoms are identical. Whenever you can draw two or more equivalent Lewis diagrams with the atoms in fixed positions, resonance is an explanation of the real structure. It occurs quite frequently for oxyanions and certain organic molecules called aromatic compounds (discussed in Chapter 18).

### EXAMPLE 10.7 Drawing Resonance Structures

Draw resonance structures for the carbonate ion, $CO_3^{2-}$.

**SOLUTION**

First, you have to draw the ion's skeleton. The guidelines for Lewis diagrams suggest that you should put the carbon in the center with the oxygens around it, and then place electron pairs:

$$\begin{array}{c}:\ddot{\text{O}}:\\ |\\ :\underset{\cdot\cdot}{\ddot{\text{O}}}{-}\underset{\cdot\cdot}{\text{C}}{-}\underset{\cdot\cdot}{\ddot{\text{O}}}:\end{array}$$

The sum of the group numbers plus 2 equals 24 electrons. The structure with no double bonds has 26 electrons, so you need to make a double bond from the carbon to one of the oxygen atoms. Since the three possible structures are equivalent, resonance is important. The three resonance structures look like this:

$$\left[\begin{array}{c}:\ddot{\text{O}}:\\ |\\ :\ddot{\text{O}}{=}\text{C}{-}\underset{\cdot\cdot}{\ddot{\text{O}}}:\end{array}\right]^{2-} \longleftrightarrow \left[\begin{array}{c}:\text{O}:\\ \|\\ :\underset{\cdot\cdot}{\ddot{\text{O}}}{-}\text{C}{-}\underset{\cdot\cdot}{\ddot{\text{O}}}:\end{array}\right]^{2-} \longleftrightarrow \left[\begin{array}{c}:\ddot{\text{O}}:\\ |\\ :\underset{\cdot\cdot}{\ddot{\text{O}}}{-}\text{C}{=}\ddot{\text{O}}:\end{array}\right]^{2-}$$

**EXERCISE 10.7**

Draw resonance structures for the nitrate ion, $NO_3^-$.

## 10.4 Lewis Diagrams and the Arrangement of Electrons: VSEPR Theory

GOAL: To use the knowledge that electron pairs repel one another to predict the shapes of molecules.

Now that you know how to draw Lewis diagrams, this section discusses how the atoms in a molecule are arranged in space. You can't necessarily determine the spatial arrangement directly from the Lewis diagram. These diagrams tell what is connected to what, but don't usually give the exact geometric arrangement of the atoms. This geometry is important, because (as you already know) the properties of substances are related to their molecular structures.

Section 10.2 mentioned that the three atoms in a water molecule assume a V-shaped geometry rather than falling into a straight line. How are the atoms in the carbonate ion arranged? Are they in a T shape, as the Lewis diagrams in Example 10.7 seem to indicate, or is the answer something else? In this section you will learn how to predict the geometry of compounds and ions using Lewis diagrams as a starting point. As usual, plenty of practice will allow you to get the ideas down.

You can make a good (though not always perfect) prediction of the arrangement of the atoms around a central atom from one basic idea. Electrons have negative charges, so they all repel one another. In molecules electrons cluster around the atoms, grouping together to form nonbonding pairs and single, double, and triple bonds. This means that these electrons can't just fly away from each other. The best they can do is arrange themselves around the central atom in such a way that they are as far apart as possible; this minimizes the forces that push them apart. This minimizing of repulsive forces is the basic principle underlying **Valence Shell Electron Pair Repulsion (VSEPR) theory,** the topic of this section.

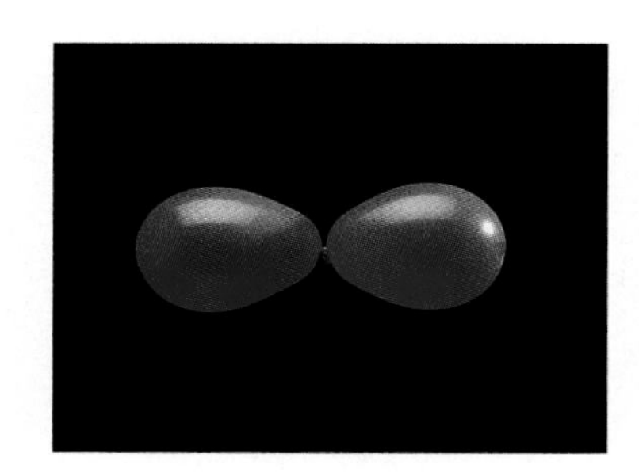

FIGURE 10.14
Here balloons represent the electrons in chemical bonds. Like electrons, balloons tied together push at each other and end up as far apart as possible. Here two balloons repel each other to give a linear structure.

For the relatively simple compounds we have looked at so far, there are three basic arrangements of electrons around a central atom. The first one is illustrated by carbon dioxide, $CO_2$. Its Lewis diagram looks like this:

$$\ddot{\underset{..}{O}}=C=\ddot{\underset{..}{O}}$$

The electrons associated with the carbon atom are in double bonds, which you can think of as two sets of four electrons each. How can the electrons in these sets get as far apart as possible? The answer is, by moving the atoms into a straight line with the carbon in the middle. This arrangement of atoms is called a **linear structure.** (See Figure 10.14.) Note that when you name molecular geometries, you pretend that you see only the nuclei of the atoms. The locations of the nuclei depend on the arrangement of the electrons, of course, but once you know where the nuclei fall you ignore the electrons in giving the structure its name.

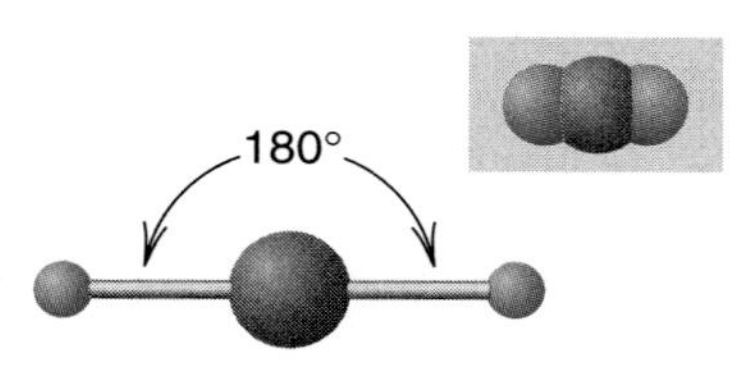


FIGURE 10.15
The bond angle in a linear molecule such as $CO_2$ is 180°.

Now that you know that the $CO_2$ molecule has a linear structure, you can predict another of its properties—the **bond angle.** This is the angle between two lines drawn from the central carbon atom to the oxygen atoms. In carbon dioxide and all other linear molecules, this angle is 180°, as shown in Figure 10.15.

Next, let's think about what happens if there are three sets of electrons around the central atom. Formaldehyde is a molecule of this kind. Here is its Lewis diagram:

H—C=Ö:
|
H

The carbon atom has three sets of electrons around it, in two single bonds and a double bond. How can three sets of electrons get as far apart as possible while remaining associated with the central atom? They form a flat triangular arrangement with the carbon atom in the middle. This is called a **trigonal planar structure.** (See Figure 10.16.) Any movement of the electron sets around the central carbon will bring them closer together.

The bond angles in an ideal trigonal planar structure are 120°, one-third of a full circle (see Figure 10.17). There are not many perfect trigonal planar structures, because it's not possible to divide an octet (eight electrons) evenly into three localized bonds. The very reactive species $BH_3$ and $CH_3^+$, which have only six valence electrons each, form exact trigonal planar structures. The H—C—H bond angle in formaldehyde, however, is 118°, and each H—C—O angle is 121°. This variation is easy to understand. The four electrons of the carbon-oxygen double bond take up more space and exert a greater repulsive force than do the two electrons in each of the carbon-hydrogen single bonds. This greater repulsion forces the hydrogen atoms closer together. (See Figure 10.18).

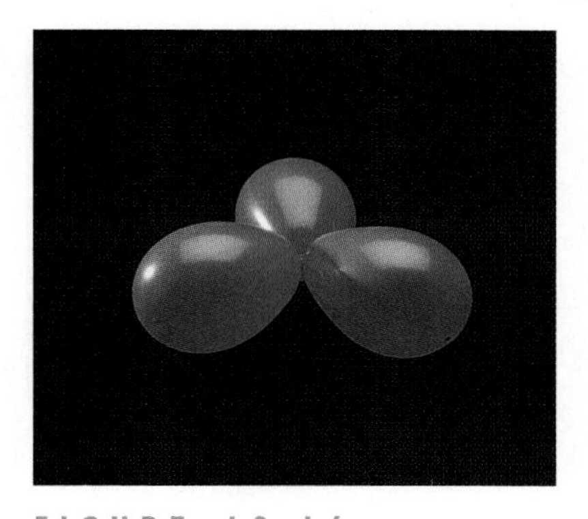

FIGURE 10.16
Three balloons tied together fall into a trigonal planar geometry.

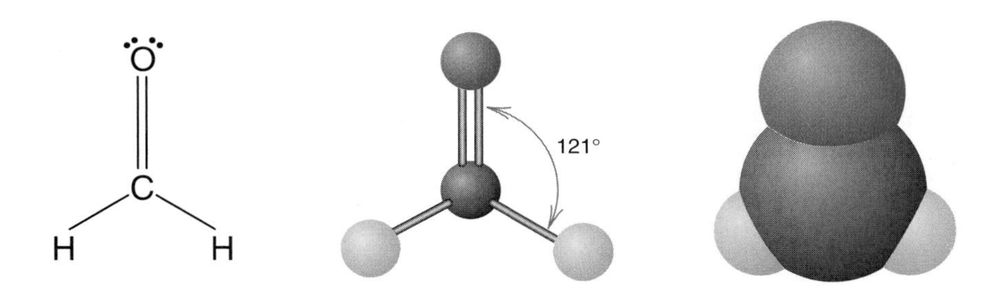


FIGURE 10.17

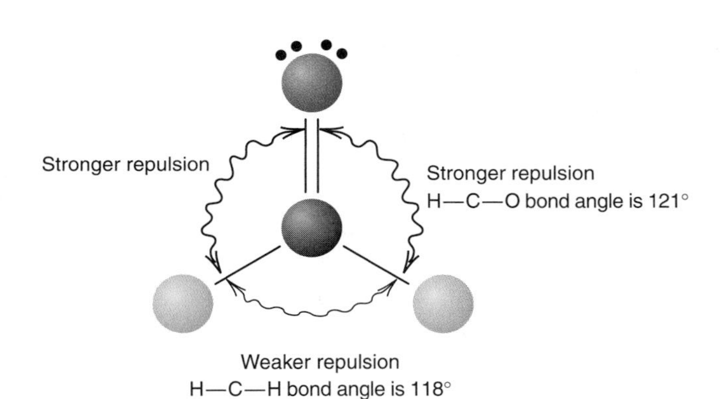


FIGURE 10.18
In the formaldehyde molecule the four electrons in the carbon-oxygen double bond repel the two electrons in each carbon-hydrogen single bond more strongly than the electrons in the single bonds repel each other. This increases the H—C—O bond angle and reduces the H—C—H bond angle. None of the bond angles in formaldehyde is the ideal 120° of a perfect trigonal planar molecule.

Another arrangement of electrons around a central atom involves four sets of electrons. Methane is one of many molecules with this arrangement. The Lewis diagram for methane is drawn as a square arrangement on two-dimensional paper:

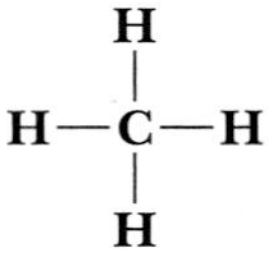


How do four sets of electrons get as far apart as possible? The necessary shape is called a *tetrahedron,* or a **tetrahedral structure.** Figure 10.19 shows several possible ways of looking at this arrangement. It isn't as easy to calculate the bond angles for this kind of structure as it is for the linear and trigonal planar structures. All of the H—C—H bond angles in methane are 109.5°.

Because chemists frequently find it necessary to show three-dimensional molecular geometries on the printed page, they use various conventions in drawing structures. In one method the central atom is assumed to lie in the plane of the paper. A bond shown as a solid wedge indicates that the atom on the thick end of the wedge is above the paper, and a dashed wedge points toward an atom that is behind the paper. A line indicates a bond that lies in the plane of the page. Figure 10.20 shows this.

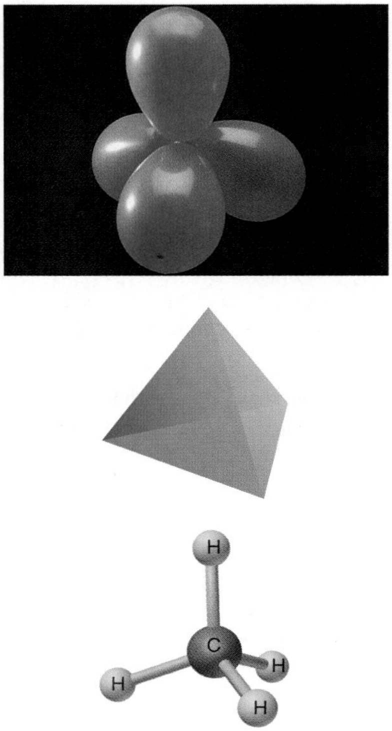


FIGURE 10.19
These objects are all tetrahedral. If you tie four balloons to a common center (as shown at *top*), they will fall into a tetrahedral arrangement. The basic geometric form is in the *center*, a sort of pyramid with four equilateral triangles as its sides and bottom. On the *bottom* is a molecular model put together to show the structure of methane.

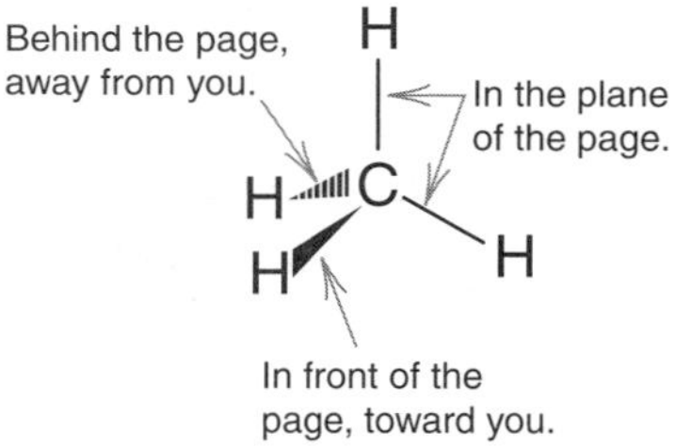


FIGURE 10.20
Representation of a three-dimensional methane molecule on a two-dimensional page.

**EXAMPLE 10.8** **Indicating Molecular Shapes on Paper**

Which of the following structures is a tetrahedron?

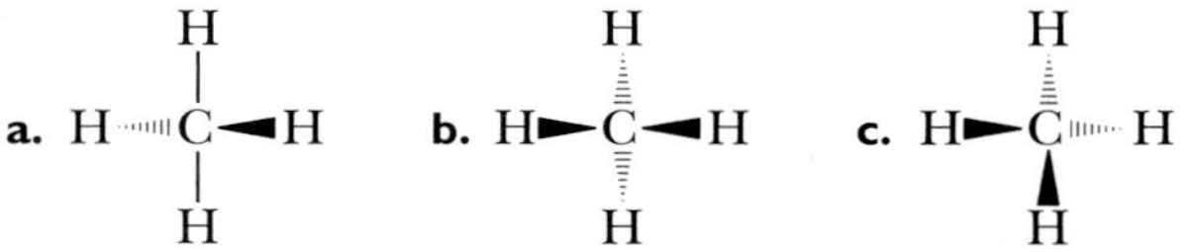


**SOLUTION**

Structure B is a tetrahedron. A and C are not (look carefully or build models); they are both flat (but not in the plane of the paper).

**EXERCISE 10.8**

Which of the following structures is a tetrahedron?

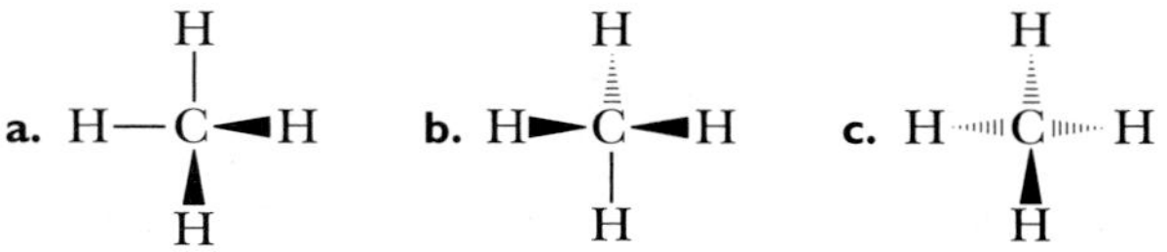

## 10.5 The Arrangement of Electrons and the Shapes of Molecules

**GOAL:** To name some other shapes of molecules and to estimate their bond angles.

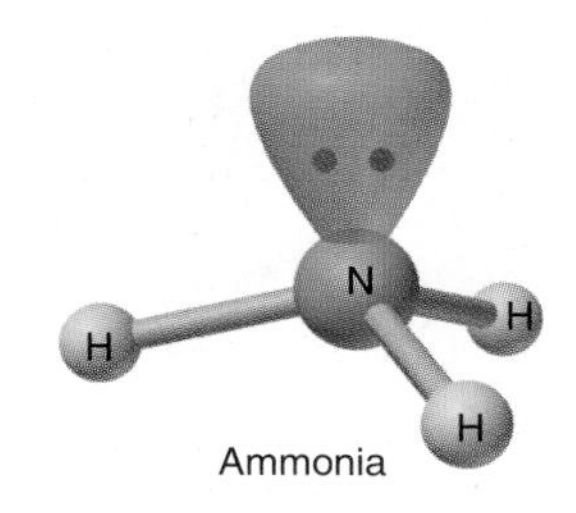


FIGURE 10.21
This picture of the $NH_3$ molecule shows that the arrangement of the four electron pairs (three bonding and one nonbonding) surrounding the nitrogen atom is tetrahedral. However, the nonbonding electrons occupy a larger volume of space than this picture suggests, and according to the Heisenberg Uncertainty Principle their exact location cannot be determined. Chemists ignore nonbonding electron pairs when naming molecular structures.

Once again, central atoms that obey the octet rule can have two, three, or four sets of electrons around them with two, three, or four atoms attached, respectively. Each of these possibilities results in an arrangement of atoms—the linear, trigonal planar, or tetrahedral structure, respectively. However, this simple scheme doesn't work for all molecules. Let's look at the structure of ammonia, $NH_3$, as an example. Here is the Lewis diagram:

$$\begin{array}{c} H-\ddot{N}-H \\ | \\ H \end{array}$$

There are four sets of electrons around the central nitrogen atom, so the molecule's shape should be tetrahedral. In one sense it really is a tetrahedron, with hydrogen atoms at three of the four corners and a pair of electrons at the other (see Figure 10.21). From Chapter 6, however, you may remember that the locations of electrons are impossible to pin down exactly. Heisenberg's Uncertainty Principle applies in full force to the smallest particles, particularly electrons.

Chemists may not be able to pin down electrons, but they have several ways to locate nuclei with considerable accuracy. Because of this, they describe molecular structures with names that tell only where the nuclei are. They do not view the structure of the ammonia molecule as a tetrahedron with a nitrogen atom in the center, hydrogen atoms at three corners, and a pair of electrons at the other. Instead the molecular structure is a pyramid with the nitrogen at the top, a **trigonal pyramid.** (See Figure 10.22.)

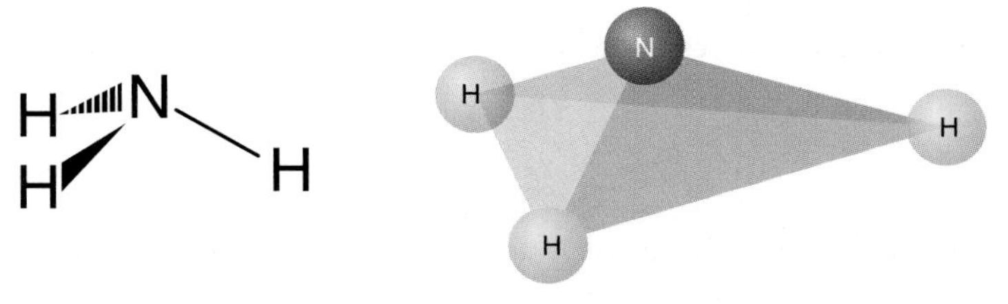


FIGURE 10.22
The structure of the $NH_3$ molecule is a trigonal pyramid.

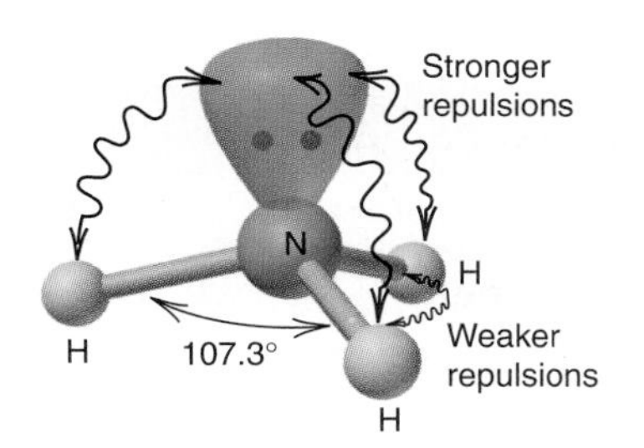


FIGURE 10.23
The repulsions between the nonbonding pair and the bonding pairs in ammonia are stronger than the repulsions between the bonding pairs. Because of this the H—N—H bond angle is 107.3° rather than the 109.5° in a perfect tetrahedron.

The bond angles in an ideal trigonal pyramid are 109.5°, as they are in a tetrahedron. The actual H—N—H bond angles in ammonia are smaller—107.3°. The nonbonding pair of electrons apparently takes up more space than the bonding pairs do. This seems reasonable. Bonding electron pairs are held in place by the force of two nuclei, one on each end of the bond. Nonbonding electrons are held by only one nucleus (see Figure 10.23).

You run into the same situation when you try to predict the structure of the water molecule. Here is the Lewis diagram for water:

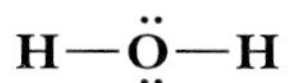

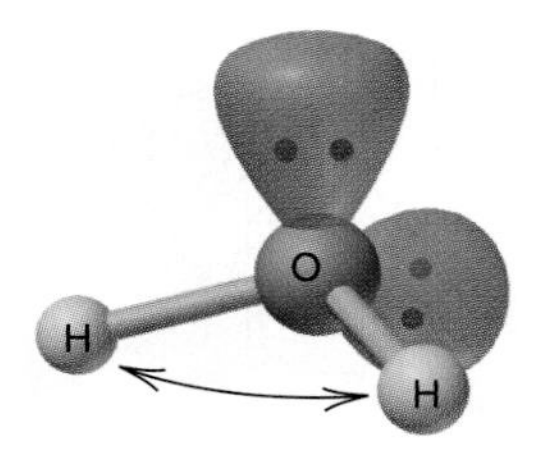


Predicted angle: 109.5°
Actual angle: 104.5°

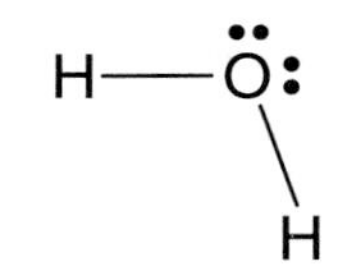


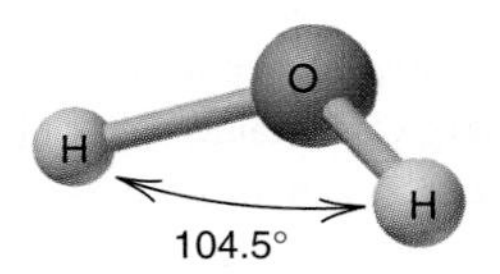


FIGURE 10.24
The four electron pairs (two bonding and two nonbonding) that surround the oxygen atom in a water molecule fall into a tetrahedral arrangement. As always, you ignore the nonbonding electron pairs when you identify the geometry of the structure. The water molecule has a bent, or angular, structure.

The two oxygen-hydrogen single bonds and the two pairs of nonbonding electrons on oxygen give the molecule four electron sets, so its structure might be considered tetrahedral. Considering only the locations of the nuclei, however, results in an **angular**, or **bent, structure.** The two nonbonding pairs of electrons force the hydrogens closer together than they would be in a tetrahedral structure, giving a bond angle of 104.5° instead of 109.5° (see Figure 10.24).

Another molecule whose shape isn't obvious from the Lewis diagram is nitrosyl chloride, which has the formula ClNO. Its Lewis diagram looks like this:

$$:\ddot{\underset{\cdot\cdot}{\mathrm{Cl}}}-\ddot{\mathrm{N}}=\ddot{\mathrm{O}}:$$

With three sets of electrons around the central atom (one single bond, one double bond, and one nonbonding pair), the molecular structure should be trigonal planar. However, one of the corners contains a pair of electrons rather than an atom. Thus chemists term this structure bent, or angular, like the water molecule's structure. The ideal Cl—N—O bond angle is 120°, but the actual bond angle is less than this (see Figure 10.25).

For a summary of structure names and bond angles, along with some example molecules, see Table 10.1.

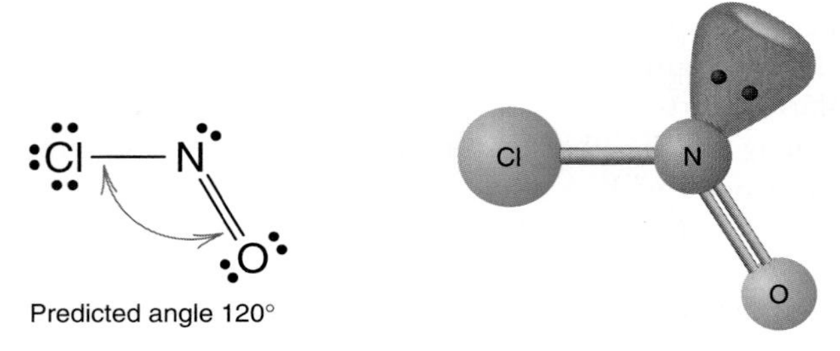


FIGURE 10.25
In the nitrosyl chloride molecule the repulsions between the nonbonding electron pair on nitrogen and the bonding electron pairs are greater than the repulsions between the bonding pairs. This repulsion compresses the bond angle, making it smaller than the predicted 120°.

**TABLE 10.1 Molecular Structure Names**

| Number of Sets of Electron Pairs | Predicted Bond Angle | Number of Atoms Attached to Central Atom | Molecular Structure Name | Example |
|---|---|---|---|---|
| 4 | 109.5° | 4 | Tetrahedral | $CH_4$ |
| 4 | 109.5° | 3 | Trigonal pyramid | $NH_3$ |
| 4 | 109.5° | 2 | Bent, or angular | $H_2O$ |
| 3 | 120° | 3 | Trigonal planar | $H_2CO$ |
| 3 | 120° | 2 | Bent, or angular | ClNO |
| 2 | 180° | 2 | Linear | $CO_2$ |

EXAMPLE 10.9 **Predicting Molecular Structures and Bond Angles**

**a.** What is the shape of $CH_2Cl_2$? What is the approximate Cl—C—Cl bond angle?
**b.** What is the shape of the $NO_2^+$ ion? What is the O—N—O bond angle?

**SOLUTIONS**

**a.** The Lewis diagram of $CH_2Cl_2$ is

```
      H
      |
:Cl—C—Cl:
      |
      H
```

Four atoms are attached to the central carbon atom, and there are no nonbonding electron pairs on the carbon, so $CH_2Cl_2$ is tetrahedral, like methane. The approximate bond angle (from Table 10.1) is 109.5°.

**b.** The Lewis diagram of the $NO_2^+$ ion is shown below. It's a linear species, so the bond angle is 180°.

$$[\ddot{O}{=}N{=}\ddot{O}]^+$$

**EXERCISE 10.9**

**a.** What is the shape of $Cl_2CS$? What is the approximate Cl—C—Cl bond angle?
**b.** What is the shape of the $NCl_2^-$ ion? What is the approximate bond angle?

## 10.6 Polar Molecules and Dipole Moments

**GOAL:** To predict the polarity of bonds from relative electronegativities and the polarity of molecules from the polarity of bonds.

One of the most useful properties you can predict from the periodic table is electronegativity, the ability of an atom to attract electrons to itself from another atom. However, the electronegativity of an atom of an element is not very useful by itself. It's the difference between the electronegativities of two atoms of different elements that are bonded to one another that is important and useful.

From the difference between the electronegativities of two atoms, you can figure out the **polarity** of the bond between them. In other words, you can determine which of the atoms in the bond pulls the electrons to itself more strongly, and to what extent. A completely polar bond is ionic. A covalent bond between two atoms of the same element is not polar at all, because the atoms share the electrons equally. As you already know (Section 10.1), polar covalent bonds between atoms of different elements fall somewhere between these two extremes.

In polar bonds the bonding electrons are not shared equally by the two atoms that form the bond. The more electronegative atom acquires a partial negative charge, and the less electronegative atom has a partial positive charge. When the center of positive charge and the center of negative charge are separated this way, you have a dipole. The magnitude and direction of the separation is called the **dipole moment.** This property has both a magnitude (defined as the amount of charge displaced times the distance between the positive and negative ends of the bond) and a direction (that is, it is a vector quantity).

On Lewis diagrams chemists indicate the polarity of a bond in either of two ways. One way is to draw an arrow with a crossbar at the more positive end of the bond (forming a plus sign) and the arrowhead at the more negative end. Alternatively, the lowercase Greek letter delta (δ, meaning "partial") followed by a plus or minus sign is placed next to the appropriate atom. This Lewis diagram for hydrogen chloride shows both of these methods of indicating the dipole moment:

$$\overset{\delta+}{H}-\overset{\delta-}{\ddot{\underset{..}{Cl}}}:$$
$$+\!\!\longrightarrow$$

A molecule is polar if, like HCl, it has a dipole moment. Nonpolar molecules such as $H_2$ have a dipole moment of zero. (See Figure 10.9.)

The dipole moment of a molecule is a property that can be measured in a laboratory. For diatomic molecules such as HCl, the dipole moment of the molecule is the same as the dipole moment of the one and only bond. For molecules with more than one bond, the dipole moment is the *sum* of the dipole moment *vectors* of all the bonds. The dipole moment of a more complicated molecule can be predicted in four reasonably easy steps:

1. Draw the Lewis diagram for the molecule.
2. Determine the molecule's three-dimensional structure (Table 10.1).
3. Draw the dipole moment arrows for all individual bonds in the molecule that connect different elements.
4. Add up the dipole moment arrows of the bonds to get the dipole moment arrow for the molecule. Arrows of equal length that go in opposite directions cancel each other; arrows that go in the same direction reinforce one another.

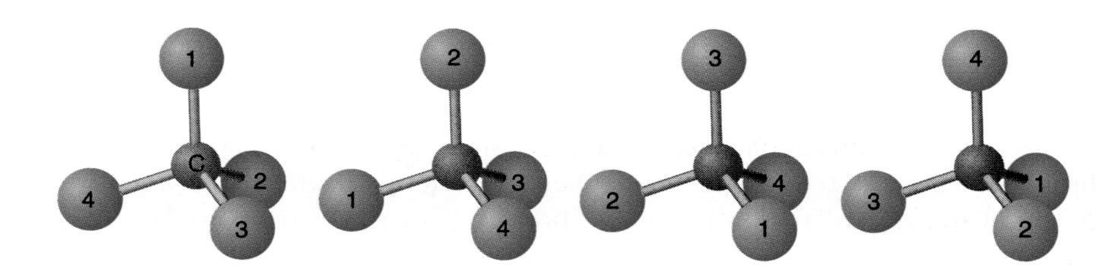


FIGURE 10.26
In carbon tetrachloride, $CCl_4$, the outer atoms are interchangeable, so the molecule cannot be polar.

If the outer atoms of the molecule are completely interchangeable, then the sum of the dipole moments of the individual bonds must equal zero, and the molecule *cannot* be polar. The tetrahedral $CCl_4$ molecule is an example. It makes no difference at all which of the four chlorine atoms is at the top of the tetrahedron (see Figure 10.26). Since all the chlorine atoms are interchangeable, $CCl_4$ is a nonpolar molecule, despite the fact that each of the four C—Cl bonds is polar.

### EXAMPLE 10.10 Predicting Dipole Moments

Indicate the direction of the dipole moment for each of these molecules.

**a.** Carbon dioxide

**b.** $CH_2Cl_2$

**SOLUTIONS**

**a.** The first two steps are to draw a Lewis diagram and to determine the molecular structure. As you learned earlier in this chapter, carbon dioxide has the following Lewis structure and is a linear molecule:

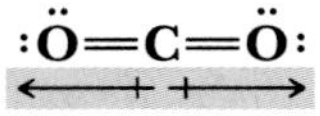

:Ö=C=Ö:

It is easy to see from this diagram that the two individual dipole moments are equal in magnitude and opposite in direction. When the two dipole moments arrows (vectors) are added, they cancel each other.

**b.** First, draw the Lewis diagram and determine the structure.

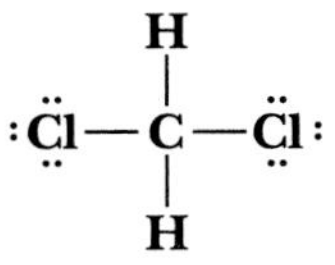

H — :C̈l—C—C̈l: — H Tetrahedral

Because Lewis diagrams do not necessarily show molecular structures, you must redraw the diagram to indicate the tetrahedral arrangement of the atoms. Since the outside atoms are not interchangeable, the dipole moments do not cancel out and the molecule is polar.

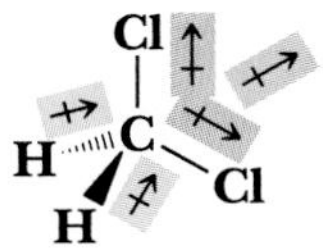

Cl, H, C, Cl, H

**EXERCISE 10.10**

Indicate the direction of the dipole moment for each molecule.

**a.** HCN

**b.** Water

EXPERIMENTAL EXERCISE

10.2

Almost a century ago a chemist speculated that $CH_2Cl_2$ might have a square planar structure. That is, its molecules might be in the shape of a flat square with the carbon atom at the center and the other four atoms at the corners. The chemist distilled a large amount of $CH_2Cl_2$ very carefully and isolated only one type of $CH_2Cl_2$. The molecules in this sample had a dipole moment not equal to zero.

**QUESTIONS**

1. How many different square planar structures of $CH_2Cl_2$ are possible? Draw the structures.
2. Show the direction of the dipole moment you would predict for each of the structures you drew.
3. Are the experimental results consistent with a square planar structure?

## 10.7 Beyond the Octet Rule

GOAL: To draw Lewis diagrams for molecules in which an atom violates the octet rule.

As you have seen, you can use the octet rule to draw diagrams of the molecular structures of a large number of compounds. However, the rule is obeyed consistently only by the elements carbon, nitrogen, oxygen, fluorine, and the alkali and alkaline earth metals. All other elements can disobey the octet rule at least in some of their compounds.

When the octet rule is violated, the central atom in a molecule is almost always the one involved. Central atoms most often violate the rule when they bond to a strongly electronegative element (oxygen, nitrogen, or one of the halogens, particularly fluorine or chlorine). When the central atom has more than an octet of electrons, it typically has *d* orbitals in its valence shell, which it can use to make bonds. The valence shells of carbon, nitrogen, oxygen, and fluorine atoms have no *d* orbitals, and consequently these atoms cannot make more than four bonds. There are only four valence-shell orbitals available in these atoms—one *s* orbital and three *p* orbitals.

Consider the compound phosphorus pentachloride, $PCl_5$. It isn't possible to draw a reasonable structure for $PCl_5$ in which all atoms obey the octet rule; you can test this for yourself if you want. Phosphorus is a third-period element and a member of Group 5A, and its valence-shell electron configuration can be written as $3s^23p^33d^0$. Since the valence shell contains *d* orbitals, they can be used in bonding. The best Lewis diagram that can be drawn for $PCl_5$ is this:

:C̈l: :C̈l: :C̈l—P :C̈l: :C̈l: (Lewis diagram of $PCl_5$)

This structure has the correct number of valence electrons (40), but the phosphorus atom is surrounded by 10 (not 8) electrons. Like many elements in the lower portions of the periodic table, phosphorus can *expand its octet* when necessary.

EXAMPLE 10.11 **Drawing Lewis Diagrams When Atoms Violate the Octet Rule**

Draw the Lewis diagram for each compound.

**a.** Iodine pentafluoride, $IF_5$ **b.** Sulfur trioxide, $SO_3$

SOLUTIONS

**a.** First, draw the skeleton of the molecule. Attach the five fluorine atoms to the central iodine. Then apply the octet rule (as best you can) by placing three pairs of electrons around each of the five fluorine atoms. (Remember, typically only the central atom violates the rule.) This gives you a total of 40 electrons in your diagram, 30 around the fluorine atoms and 10 in the bonds. Next, find the number of electrons there should be by adding up the group numbers. There are six atoms, all in Group 7A, so there should be 42 electrons. To make your structure fit this requirement, place a nonbonding pair of electrons on the iodine atom. This gives you the following structure, which is the correct one:

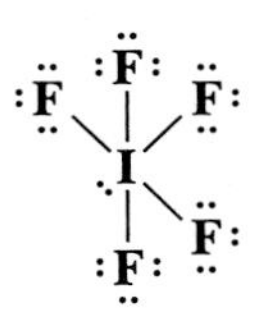

:F: F: F: I: F: :F:

**b.** Sulfur trioxide can be a little confusing, because it is possible to draw a structure for it that obeys the octet rule. (Try it. There are three resonance structures of $SO_3$ that obey the octet rule; each contains two oxygen atoms with single bonds to sulfur and one with a double bond.) First, you draw the skeleton, surrounding the central sulfur atom with the three oxygens. Knowing that oxygen almost always makes two bonds in neutral molecules, you draw a double bond from each of the oxygens to the central sulfur. Then you check the number of valence electrons (four atoms, all in Group 6A). The diagram shows 24 electrons, and it should. Here it is:

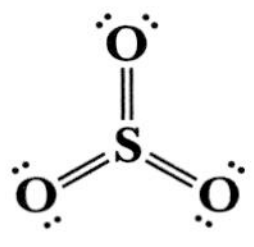

O O S O

EXERCISE 10.11

Draw the Lewis diagram for each compound.

**a.** Sulfuric acid, $H_2SO_4$ [Hint: Sulfuric acid is an oxyacid, so another way to write the formula is $(HO)_2SO_2$.]

**b.** Sulfur tetrafluoride, $SF_4$

FIGURE 10.27
If you attach five balloons to a common center, they will form a trigonal bipyramid.

## 10.8 Molecules That Violate the Octet Rule

GOAL: To predict the Lewis diagrams and shapes of molecules whose central atoms have five or six sets of electrons around them.

In Sections 10.4 and 10.5 you studied the shapes of molecules in which the central atom was surrounded by two, three, or four sets of electrons. The linear, trigonal planar, tetrahedral, etc., structures obviously can't describe molecules that have five or six atoms around their central atom. How do you draw Lewis diagrams for molecules such as $PCl_5$ and $SF_6$, and what shapes do such molecules have?

You can get some idea of the shape of a molecule with five atoms around a central atom. One way is to tie five balloons to a common center (see Figure 10.27). The shape the five balloons form is called a **trigonal bipyramid.** You can also take two tetrahedrons and hold them so that they share one side to make this shape (see Figure 10.28).

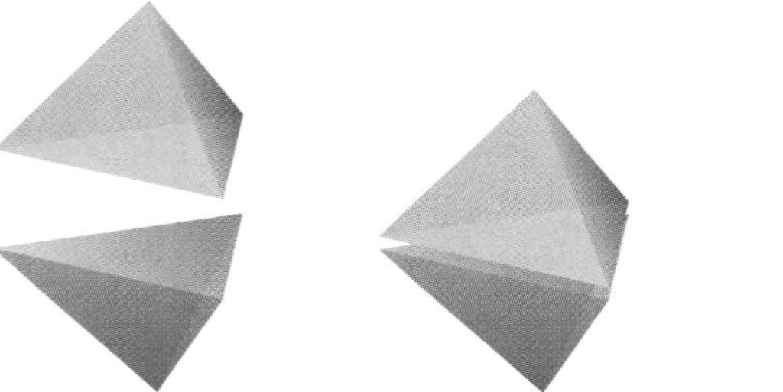

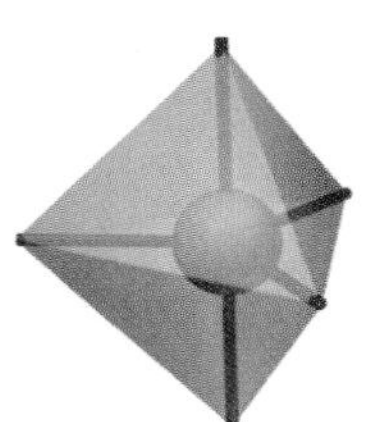

FIGURE 10.28
You can make a trigonal bipyramid by placing two tetrahedrons face to face, or by assembling six equilateral triangles.

The trigonal bipyramid is unusual because it contains two different bond angles. In a linear structure, the bond angle is always 180°. In a trigonal planar structure, the angles are all 120°, and in a tetrahedron they are all 109.5°. The trigonal bipyramid has some bond angles of 120° and others of 90°, depending on what atoms are involved. You can think of this structure as combining a linear structure (the atoms at the top and bottom of Figure 10.29) and a trigonal planar structure (the other three outer atoms). Phosphorus pentachloride, $PCl_5$, has this molecular structure.

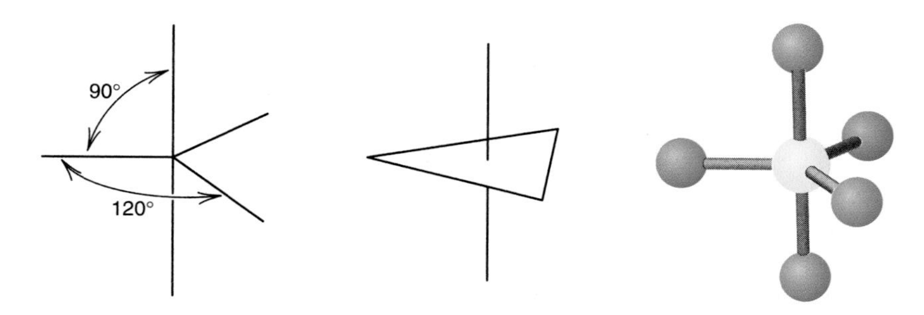


FIGURE 10.29
At the left are shown the bond angles in a trigonal bipyramid. The structure in the center shows another way to represent a trigonal bipyramid on a flat sheet of paper. The model at the right shows $PCl_5$, which has this structure.

If there are six atoms connected to the central atom, the molecular shape is somewhat simpler. Six balloons tied to a central point fall into this shape readily (see Figure 10.30). It's called an *octahedron,* or an **octahedral structure.** The bond angles in octahedral molecules are all 90°. As Figure 10.31 shows, sulfur hexafluoride ($SF_6$) has an octahedral molecular structure.

### EXAMPLE 10.12 More Lewis Diagrams and Molecular Shapes

Draw the Lewis diagram for the ion $PF_6^-$ and identify its structure.

**SOLUTION**

This is a polyatomic ion. The skeleton has a phosphorus atom at the center, with single bonds to each of the six fluorine atoms. If you place six nonbonding electrons around each of the fluorine atoms, all of them will obey the octet rule; the central phosphorus atom violates it. Your electron count for the ion includes 42 from the six F atoms, 5 from the P atom, and 1 for the negative charge, for a total of 48. The Lewis diagram is:

```
        ..
  [    :F:     ]-
   ..   |   ..
  :F.   |   .F:
   .. \ | /  ..
        P
   .. / | \  ..
  :F    |    F:
   ..  :F:   ..
        ..     ]
```

This ion has an octahedral structure.

**EXERCISE 10.12**

Draw the Lewis diagram for $IF_3$.

FIGURE 10.30
If you put six balloons together, you will get an octahedron like this.

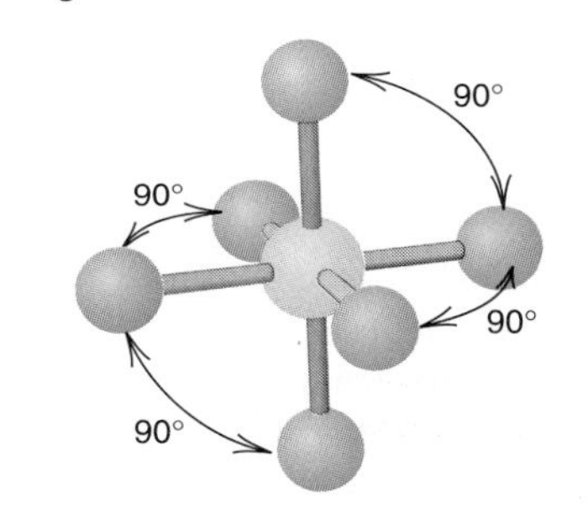


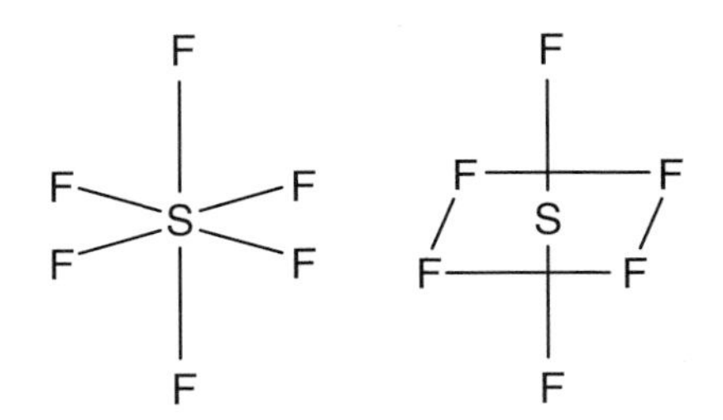


FIGURE 10.31
For sulfur hexafluoride, all the bond angles in the octahedral structure are 90°, and all the corners are equivalent. The two $SF_6$ structures shown here illustrate two of the ways chemists draw octahedral structures on flat paper.

## CHAPTER SUMMARY

1. Three types of bonding can be found in common compounds and elements: ionic, covalent, and metallic.
2. A sample of an ionic compound is a three-dimensional array of cations interspersed with anions. The strength of this electrostatic attraction explains why all known ionic compounds are solids at room temperature.
3. Covalent bonds require the sharing of one pair of electrons (single bond), two pairs (double bond), or three pairs (triple bond). The individual bonds are quite strong—typically stronger than ionic bonds.
4. Samples of metals are also three-dimensional arrays. The valence shell of each individual metal atom overlaps the valence shells of all neighboring metal atoms. This overlapping gives a sea of loosely held electrons. Metals are good electrical conductors because the electrons are free to move from atom to atom throughout the entire sample of the metal.
5. Lewis diagrams are representations of how atoms are bonded to each other and of the nonbonding electron pairs (if any) associated with each atom. These diagrams use the symbols of the elements to show nuclei

and inner electrons, pairs of dots to show nonbonding pairs of electrons, and lines to show bonds (bonding electron pairs). Simple Lewis diagrams provide no direct information about the three-dimensional aspects of molecular structures, but they can be used in predicting these structures.

6. In Lewis diagrams for polyatomic ions it is necessary to add one electron to the total electron count for each negative charge present and to subtract one electron for each positive charge.
7. VSEPR theory says that the positioning of electron pairs around a central atom is determined by the fact that electrons repel each other. All electron sets (single bonds, double bonds, triple bonds, nonbonding pairs) will be as far apart as possible.
8. The structure assigned to a molecule may not be the same as the geometric arrangement of electrons. Only the locations of nuclei are considered when naming molecular structures. (See Table 10.1 for a summary.)
9. All bonds involving atoms of different elements are polar. Whether or not a molecule is polar depends on its geometry. For example, $CO_2$ is linear and nonpolar, while $H_2O$ is bent and polar.
10. The octet rule is strictly obeyed only by carbon, nitrogen, oxygen, and fluorine and many monatomic ions such as $Na^+$. Lewis diagrams involving any other atom as the central atom may require expanding its valence shell.
11. When a central atom makes five bonds and has no nonbonding electrons associated with it, the molecular structure is usually a trigonal bipyramid. When a central atom makes six bonds and has no nonbonding electrons associated with it, the molecular structure is usually octahedral.

## PROBLEMS

### SECTION 10.1 Types of Bonding: Ionic, Covalent, and Metallic

1. Explain why NaBr(*l*) conducts electricity and NaBr(*s*) does not.

2. Would you expect aqueous solutions of covalent compounds to conduct electricity? Explain your answer.

3. What kind of bonding (ionic, covalent, or metallic) is most likely to be present in each of these samples?
   a. A liquid that is transparent at room temperature
   b. A ductile solid
   c. A liquid that conducts electricity at room temperature
   d. A compound that is a liquid at room temperature and doesn't conduct electricity
   e. A purple solid that sublimes easily

4. What kind of bonding (ionic, covalent, or metallic) is most likely to be present in each of these samples?
   a. A malleable solid
   b. A brittle colorless solid with high melting point
   c. A room temperature gas
   d. A deep red liquid at room temperature
   e. An electrical insulator

5. Lead metal reacts with excess chlorine to form $PbCl_4$, which has a melting point of $-15°C$. $PbCl_4$ is unstable and, if heated to 50°C, decomposes to give $PbCl_2(s)$ and $Cl_2(g)$. What type of bonding is present in $PbCl_4$, ionic or covalent? Explain your answer.

6. $PbCl_2$ is insoluble in cold water but dissolves slightly in hot water to give a solution that conducts electricity. The melting point of $PbCl_2$ is 501°C. What type of bonding is present in $PbCl_2$, ionic or covalent? Explain your answer.

### SECTION 10.2 Lewis Diagrams for Simple Compounds

7. Draw the Lewis diagram for each of these ionic compounds.
 a. NaCl  b. $Li_2O$  c. $AlF_3$

8. Draw the Lewis diagram for each of these ionic compounds.
 a. LiI  b. $BaF_2$  c. $K_3N$

9. Draw Lewis diagrams for these ionic compounds.
 a. NaI  b. $CaCl_2$  c. $Li_3P$

10. Draw Lewis diagrams for these ionic compounds.
 a. KBr  b. $Na_2S$  c. $AlI_3$

11. Draw the Lewis diagram for each of these compounds.
 a. $OF_2$  c. $CH_3Cl$
 b. $Cl_2$  d. $H_2S$

12. Draw the Lewis diagram for each of these compounds.
 a. $CH_4$  c. $Cl_2O$
 b. $Br_2$  d. $PCl_3$

13. Draw the Lewis diagram for each of these compounds.
 a. $SiH_4$  c. $H_3C—OH$
 b. $NF_3$  d. $H_2N—NH_2$

14. Draw the Lewis diagram for each of these compounds.
 a. $SiF_4$  c. $AsH_3$
 b. $CH_3—CH_3$  d. $H_2N—OH$

### SECTION 10.3 Multiple Bonds, Polyatomic Ions, and Resonance

15. Draw the Lewis diagram for each of these compounds.
 a. $N_2O$ (nitrogen in the middle)
 b. CO
 c. SCO
 d. $H_2CNH$ (atoms in that order)

16. Draw the Lewis diagram for each of these compounds.
 a. $CS_2$
 b. HCCH (skeleton as listed here)
 c. $H_2CCH_2$ (skeleton as listed here)
 d. $Cl_2CO$

17. Draw a Lewis diagram for each of these ions.
 a. $NO_2^+$  c. $SCN^-$
 b. $OCl^-$  d. $O_2^{2-}$

18. Draw a Lewis diagram for each of these ions.
 a. $OCN^-$  c. $HCO^+$
 b. $C_2^{2-}$  d. $CNO^-$

19. Draw two resonance structures for the bicarbonate (hydrogen carbonate) ion, $HCO_3^-$ (the skeleton is $HOCO_2^-$).

20. Draw two resonance structures for the nitrite ion, $NO_2^-$.

21. Draw two resonance structures for sulfur dioxide, $SO_2$.

22. Ozone, $O_3$, is one of the few compounds for which the guidelines that all oxygen atoms make two bonds in neutral compounds doesn't hold. The fact that the rule is violated by at least one of the oxygen atoms in ozone is one reason why ozone is an unstable, highly reactive species. Draw two resonance structures for ozone.

### SECTION 10.4 Lewis Diagrams and the Arrangement of Electrons: VSEPR Theory

23. Which of the following structures is *not* a tetrahedron?

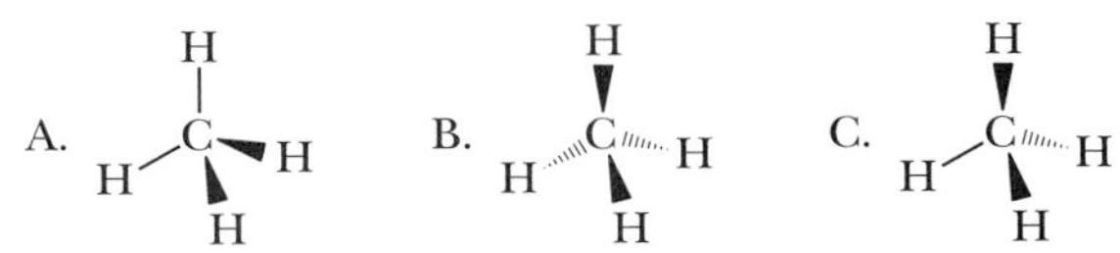

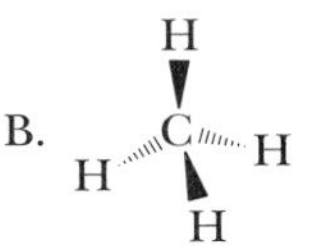
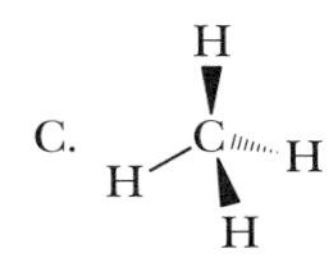

24. Which of the following structures is *not* a tetrahedron?

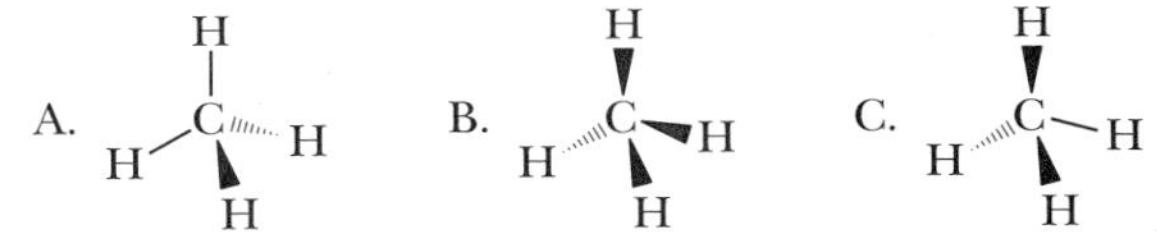

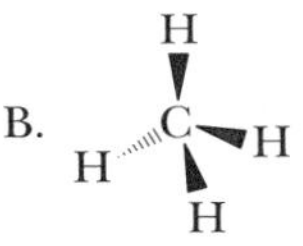

25. For each of these compounds, apply VSEPR theory to predict whether the actual bond angle is equal to, smaller than, or larger than the ideal angle predicted by Table 10.1, and explain your answer.
 a. $CS_2$ (S—C—S angle)
 b. $Cl_2C{=}CCl_2$ (Cl—C—C angle)
 c. $Cl_2C{=}CCl_2$ (Cl—C—Cl angle)

26. For each of these compounds, apply VSEPR theory to predict whether the actual bond angle is equal to, smaller than, or larger than the ideal angle predicted by Table 10.1, and explain your answer.
 a. H—C≡C—H (H—C—C angle)
 b. $CH_2I_2$ (H—C—H angle)
 c. $CH_2I_2$ (I—C—I angle)

27. For each of these compounds, apply VSEPR theory to predict whether the actual bond angle is equal to, smaller than, or larger than the ideal angle predicted by Table 10.1, and explain your answer.
 a. $F_2C{=}O$ (F—C—F angle)
 b. $F_2C{=}O$ (F—C—O angle)
 c. $CF_4$

28. For each of these compounds, apply VSEPR theory to predict whether the actual bond angle is equal to, smaller than, or larger than the ideal angle predicted by Table 10.1, and explain your answer.
 a. $H_2N—NH_2$ (H—N—N angle)
 b. $H_2N—NH_2$ (H—N—H angle)
 c. N≡C—C≡N (N—C—C angle)

29. The text mentioned that the (unstable) cation $CH_3^+$ has a trigonal planar structure, with an H—C—H bond angle of exactly 120°. Draw the Lewis diagram for $CH_3^+$ and show that this must be the case.

30. The text mentioned that the (highly unstable) molecule $BH_3$ has a trigonal planar structure, with an H—B—H bond angle of exactly 120°. Draw the Lewis diagram for $BH_3$ and show that this must be the case.

## SECTION 10.5 The Arrangement of Electrons and the Shapes of Molecules

31. What is the shape and approximate bond angle of each of these molecules?
    a. $Cl_2O$ b. $SiH_4$ c. $NCl_3$

32. What is the shape and approximate bond angle of each of these molecules?
    a. $CS_2$ b. $N_2O$ c. ClCN

33. Predict the shape and approximate bond angle of each of these molecules.
    a. $Cl_2S$ b. HOCl c. SCO

34. Predict the shape and approximate bond angle of each of these molecules.
    a. $PCl_3$ b. $SO_2$ c. $Cl_2CO$

35. Predict the shape and approximate bond angle for each of these ions.
    a. $BF_4^-$ b. $OCN^-$ c. $HCO_2^-$

36. Predict the shape and approximate bond angle for each of these ions.
    a. $NCS^-$ b. $NO_3^-$ c. $NO_2^-$

37. What is the shape and approximate bond angle of each of these ions?
    a. $HCO^+$ b. $CH_3^-$ c. $BrO_2^-$

38. What is the shape and approximate bond angle of each of these ions?
    a. $H_3O^+$ b. $H_2F^+$ c. $CO_3^{2-}$

39. Elemental sulfur has the formula $S_8$. Which of the following is a reasonable structure for the $S_8$ molecule? Explain your answer.

A. :S̈—S̈—S̈: (eight-membered ring of S atoms)

B. 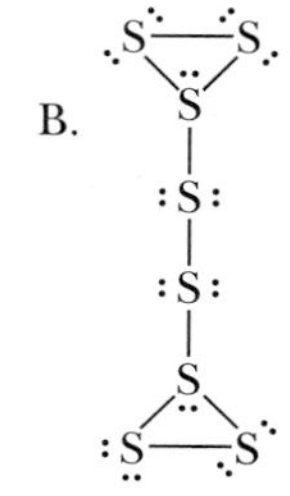

C. :S̈—S—S—S̈: with S̈ above and :S̈: below each central S

D. :S̈—S̈—S̈—S̈—S̈—S̈—S̈—S̈:

40. Elemental phosphorus has the formula $P_4$. Which of the following is a reasonable structure for the $P_4$ molecule? Explain your answer.

A. 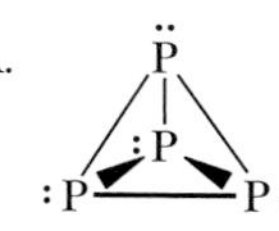

B. :P̈: / :P—P̈: / :P̈:

C. 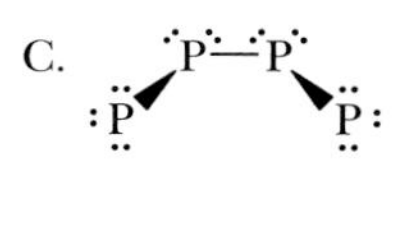

D. :P̈—P̈—P̈—P̈:

## SECTION 10.6 Polar Molecules and Dipole Moments

41. Draw the molecular structure for each of these compounds, showing the approximate bond angles, and use the arrow symbol to indicate the direction of the dipole moment for each molecule.
    a. BrF b. $H_2CO$ c. $CH_3Cl$

42. Draw the molecular structure for each of these compounds, showing the approximate bond angles, and use the arrow symbol to indicate the direction of the dipole moment for each molecule.
    a. CO b. $OF_2$ c. $CH_2F_2$

43. For each of these molecules, draw the structure, showing the approximate bond angles, and indicate the direction of the dipole moment using the arrow symbol.
    a. $H_2S$ b. HBr c. $SF_2$

44. For each of these molecules, draw the structure, showing the approximate bond angles, and indicate the direction of the dipole moment using the arrow symbol.
    a. $H_3SiCl$ b. SCO c. $HCCl_3$

45. What is the direction of the dipole moment of this molecule?

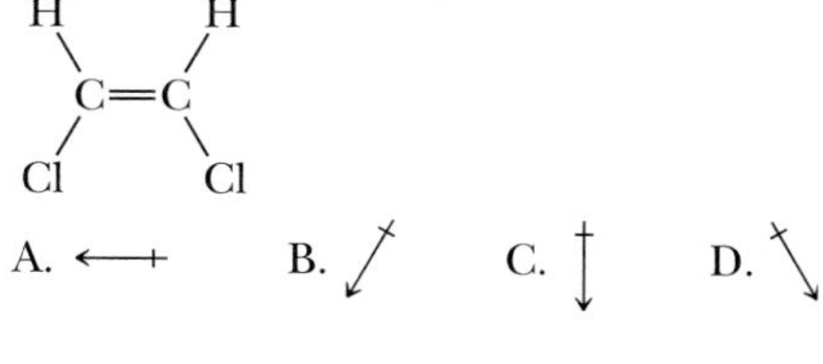

46. What is the direction of the dipole moment of this molecule?

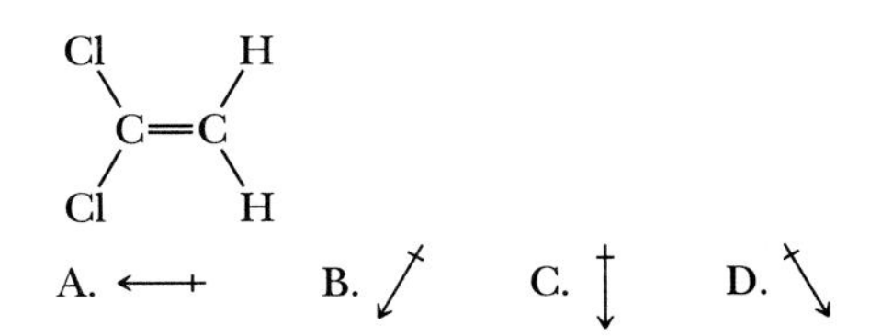


47. Which of these arrows indicates the direction of the dipole moment for the molecule shown?

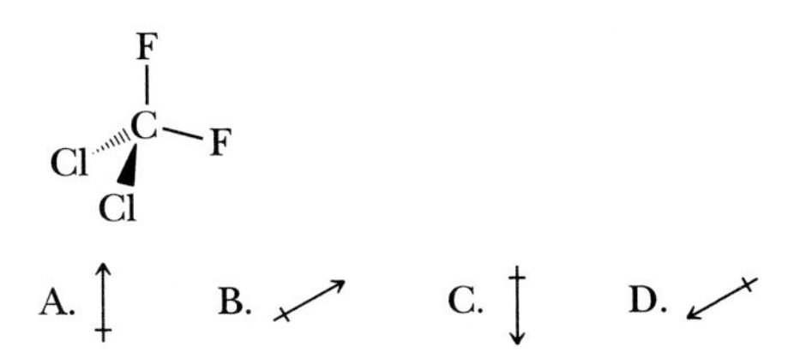


48. Which of these arrows indicates the direction of the dipole moment for the molecule shown?

H
C — Cl
Cl
Cl

A. B. C. D.

SECTION 10.7 **Beyond the Octet Rule**

49. Draw the Lewis diagram for each of the following.
a. $BrF_3$ c. $IF_4^+$
b. $POCl_3$ d. $ClO_4^-$

50. Draw the Lewis diagram for each of the following.
a. $I_3^-$ c. $SeF_4$
b. $IF_7$ d. $PF_5$

51. Draw the Lewis diagrams for these polyatomic ions.
a. $BrF_4^-$ c. $ICl_2^-$
b. $SF_5^-$ d. $FSO_3^-$ (sulfur is the central atom)

52. Draw the Lewis diagrams for these compounds or ions.
a. $SO_2F_2$ c. $XeF_6$
b. $SO_3^{2-}$ d. $S_2^{2-}$

SECTION 10.8 **Molecules That Violate the Octet Rule**

53. Draw a Lewis diagram for each species and name the shape of its structure.
a. $IF_6^+$ b. $PCl_5$ c. $ClO_3^-$

54. Draw a Lewis diagram for each species and name the shape of its structure.
a. $SO_2$ b. $XeO_3$ c. $SO_4^{2-}$

55. From the diagrams you drew in Problems 49 (b) and (d), name the shapes of the two species.

56. From the diagram you drew in Problem 50 (d), name the shape of the molecule.

57. From the diagram you drew in Problem 51 (d), name the shape of the ion.

58. From the diagram you drew in Problems 52 (a) and (b), name the shapes of the species.

**ADDITIONAL PROBLEMS**

59. The Lewis diagram for $SO_3$ can be drawn as either three resonance structures that obey the octet rule or a single diagram that does not. Draw all four diagrams.

60. The Lewis diagram for the phosphate ion, $PO_4^{3-}$, can be drawn as either a single diagram that obeys the octet rule or four resonance structures that do not. Draw all five diagrams.

61. One Lewis diagram that can be drawn for the perchlorate ion, $ClO_4^-$, obeys the octet rule but violates the guidelines that each oxygen atom should have two bonds to it. Another Lewis diagram can be drawn that violates the octet rule for the chlorine atom but complies with the rule that each oxygen atom should have two bonds to it. Draw both of these diagrams.

62. For the perchlorate ion Lewis diagrams can also be drawn that contain one, two, or three double bonds from chlorine to oxygen. (All of these violate the octet rule). There are a total of 14 such diagrams. Draw them all.

63. Determine whether any of these molecules has a dipole moment of zero.
a. $SF_6$ b. $SF_2$ c. $NF_3$

64. Determine whether any of these molecules has a dipole moment of zero.
a. $O_2$ b. $BeH_2$ c. $PF_5$

65. Draw a Lewis diagram for diazine, $N_2H_2$. Do you expect the molecule to be linear? Explain your answer.

66. Two different structures are possible for diazine, $N_2H_2$, and both can be represented by the same Lewis diagram. One of the structures is polar, and one is nonpolar. Draw both structures, indicating the geometry.

67. Draw a Lewis diagram of a molecule that has the given characteristics.
    a. Composed of C and H, nonpolar, a gas at room temperature
    b. Composed of N and H, polar, gas at room temperature
    c. Contains iodine only, nonpolar, solid at room temperature
    d. Composed of Al and I, an electrical insulator at room temperature

68. Draw a Lewis diagram of a molecule that has the given characteristics.
    a. Composed of S and F, nonpolar, gas at room temperature
    b. Composed of Mg and O, high boiling point, solid at room temperature
    c. Composed of H and O, polar, a liquid at room temperature
    d. Contains N only, nonpolar, a gas at room temperature

## SOLUTIONS TO EXERCISES

### Exercise 10.1

In all covalent solids the molecules are stuck in a crystal structure and cannot move around. Electrons in covalent bonds cannot flow from one molecule to another. For this reason, solid covalent compounds are poor conductors of electricity. (As is typical in chemistry, there are some exceptions. Most of these are giant covalent molecules in which electrons can flow from one end to the other. Graphite is probably the simplest example.)

### Exercise 10.2

**a.** Potassium is in Group 1A, so its ion is $K^+$. Nitrogen is in Group 5A, so its ion (the nitride ion) has a charge of $5 - 8 = -3$. The diagram is therefore:

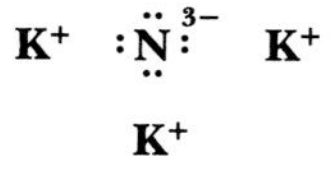

$K^+$ $:\ddot{\underset{..}{N}}:^{3-}$ $K^+$

$K^+$

**b.** Calcium is in Group 2A, so the charge on its ion is $+2$. Sulfur is in Group 6A, so its charge is $-2$. The correct diagram is

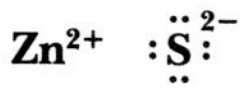

$Zn^{2+}$ $:\ddot{\underset{..}{S}}:^{2-}$

### Exercise 10.3

**a.** First, place the nitrogen atom in the center and bond the three hydrogens to it. The hydrogens with their single bonds are fine. The nitrogen has three bonds to it (six electrons) and needs two more electrons to obey the octet rule. So you add a nonbonding pair and get this structure:

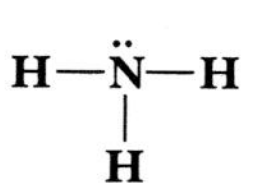

```
   ..
H—N—H
  |
  H
```

Group numbers: H, 1; N, 5. Sum of group numbers: $1 + 1 + 1 + 5 = 8$. The diagram shows eight electrons, six in the three bonds and two nonbonding electrons.

**b.** First, place the carbon atom in the center and bond the two hydrogens and the two chlorines to it. The hydrogens with their single bonds are fine. The carbon has four bonds (eight electrons) and obeys the octet rule. Each chlorine has one bond (two electrons) and needs six more electrons (three nonbonding pairs). The result is

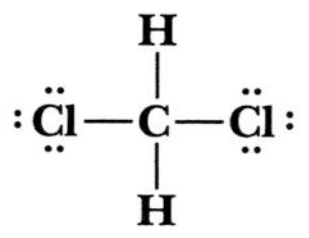

```
       H
  ..   |   ..
 :Cl—C—Cl:
  ..   |   ..
       H
```

The sum of the group numbers is $1 + 1 + 4 + 7 + 7 = 20$. The electrons shown in the diagram are in four bonds (8 electrons) and six nonbonding pairs (12 electrons), which gives the same total of 20 electrons.

### Exercise 10.4

The first attempt at a Lewis diagram gives you a total of 14 electrons. Since carbon is in Group 4A and oxygen is in Group 6A, you must end up with only 10 electrons in the final Lewis diagram. You proceed exactly as in Example 10.4.

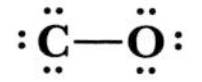

| $:\ddot{\underset{..}{C}}—\ddot{\underset{..}{O}}:$ | $:\ddot{C}=\ddot{O}:$ | $:C\equiv O:$ |
|---|---|---|
| 14 electrons | 12 electrons | 10 electrons (correct answer) |

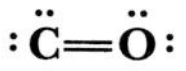

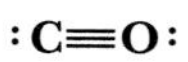

### Exercise 10.5

Carbon goes in the center, and each oxygen gets two bonds. This gives the following Lewis diagram:

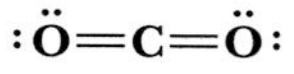

$:\ddot{O}=C=\ddot{O}:$

### Exercise 10.6

**a.** One bond links the central nitrogen atom to each of the four hydrogens (there's no other way to draw the skeleton). There are no nonbonding pairs, and the diagram looks like this:

```
[    H    ]+
     |
[ H—N—H  ]
     |
[    H    ]
```

The sum of the group numbers minus 1 for the positive charge gives 8 electrons. The structure diagrammed does in fact have 8 electrons. Note that nitrogen has four bonds and no pairs in this polyatomic cation.

**b.** A central nitrogen atom is bonded to two hydrogen atoms. Two nonbonding pairs go on the nitrogen, giving it an octet of electrons and a Lewis diagram with a total of 8 electrons. This matches the electron count: Nitrogen is in Group 5A, and hydrogen in Group 1A; 7 electrons plus 1 electron for the negative charge gives a total of 8 electrons.

```
[ :N̈—H ]−
   |
[  H    ]
```

### Exercise 10.7

The nitrate ion has a central nitrogen surrounded by three oxygens. To follow the octet rule and have the correct number of electrons (24), the structure needs a double bond, and there are three equivalent places to put it. The resonance structures look like this:

```
[    :Ö:     ]−       [     Ö:     ]−       [    :Ö:     ]−
      |          ⟷          ‖          ⟷          |
[ :Ö=N—Ö:    ]        [ :Ö—N—Ö:    ]        [ :Ö—N=Ö:    ]
```

### Exercise 10.8

Structure B is a tetrahedron, slightly distorted. A and C can be described as folded structures.

### Exercise 10.9

**a.** The Lewis diagram for $Cl_2CS$ is given below. The molecule has a trigonal planar structure and a Cl—C—Cl bond angle of approximately 120°.

```
       :S:
        ‖
  :C̈l—C—C̈l:
```

**b.** The Lewis diagram for $NCl_2^-$ is given below. The molecular structure is bent, or angular, and the bond angle is approximately 109.5°.

```
[ :C̈l—N̈—C̈l: ]−
```

### Exercise 10.10

**a.** The HCN molecule is linear, and the terminal atoms are not interchangeable. Nitrogen is the most electronegative of the three atoms, so the answer is:

```
H—C≡N:
+————→
```

**b.** The water molecule has a bent structure, and oxygen is the more electronegative of the two elements. The Lewis diagram must be drawn with the geometry indicated correctly to find the direction of the dipole moment. First, indicate the dipoles on the bonds of the Lewis diagram this way:

```
+——→
H—Ö:  ↖
   \    \
    H    +
```

Next, add these two vectors. You might imagine what would happen if the two arrows collided and stuck together—they'd go off to the upper right, like this:

```
       ↗
      /
H—Ö:+
   \
    H
```

**Exercise 10.11**

**a.** Acids (particularly oxyacids) frequently have hydrogen atoms attached to oxygens, as the hint suggested. In sulfuric acid the four oxygens, in turn, surround the central sulfur atom. Oxygen usually forms two bonds, so you should place double bonds between the oxygen atoms that have no hydrogens and the sulfur atom. Finally, add nonbonding electron pairs to the oxygens (two pairs each) so that they obey the octet rule. The diagram is

```
        :O:
         ‖
  H—Ö—S—Ö—H
         ‖
        :O:
```

The electron count is 32, which is equal to the sum of the group numbers.

**b.** By now you have established your routine. First, you place the four fluorine atoms around the central sulfur, with a single bond to each fluorine. Next, you put three nonbonding electron pairs around each fluorine atom, so that they all obey the octet rule. Finally, you count the electrons. You have 32, and the sum of the group numbers is 34, so you add a pair of electrons to the central sulfur. This gives you the following diagram, which is the correct answer:

```
   :F:
    |
   :S—F:
   / \
:F:   :F:
```

**Exercise 10.12**

All four atoms are in Group 7, so the diagram should show 28 electrons. Iodine is the central atom and expands its valence shell. Fluorine, as always, obeys the octet rule. The diagram is:

```
        F:
       /
:F—I
       \
        F:
```

**Experimental Exercise 10.1**

Unless the room was very hot, the samples would be noticeably cooler than the student's hand. Metals are good conductors of heat, and nonmetals are not. The part of the nonmetal sample touching the student's hand would warm up quickly, and the heat would flow only slowly to the rest of the sample. The heat in the part of the metal touching the student's hand would flow to the rest of the metal sample quite rapidly, meaning that the part of the metal touching the hand would feel colder longer.

**Experimental Exercise 10.2**

**1.** $CH_2Cl_2$ has two possible square planar structures:

```
   :Cl:              :Cl:
    |                 |
 H—C—Cl:           H—C—H
    |                 |
    H                :Cl:
```

**2.** The first structure has a dipole moment in the direction shown below. The other structure has a dipole moment of zero.

```
   :Cl:  ↗
    |
 H—C—Cl:
    |
    H
```

**3.** The fact that the chemist was able to isolate only one $CH_2Cl_2$ structure *suggests* that the compound is tetrahedral rather than square planar. If the square planar structure with the dipole moment is much more stable than the one without, the molecule might be square planar. More information is needed (and had been available for many years even then).

11

# GASES AND THE ATMOSPHERE



KEY TERMS

At the beginning of its geological history, the earth was warmer than it is now. Thousands of erupting volcanos spewed forth molten rock along with a mixture of gases that helped form the atmosphere. The most abundant gases were probably carbon dioxide, nitrogen, and water vapor; there was almost no oxygen. Over millions of years much of the water vapor cooled and condensed, falling as rain and forming the oceans. After more millions of years those oceans brought forth life.

FIGURE 11.1
World-wide destruction of forests, along with the burning of fossil fuels in industrial nations, is returning carbon dioxide to the atmosphere at a rapid pace.

Early living organisms were responsible for one of the most dramatic geological changes in the history of the earth. They developed *photosynthesis,* the process that converts sunlight, water, and carbon dioxide into sugars and gaseous oxygen. Photosynthesis was (and still is) so effective that the carbon dioxide content of the atmosphere dropped to 0.03% by volume, and the oxygen content rose to 21%.

New forms of life arose that could take advantage of the higher concentration of oxygen in the atmosphere. They absorbed the oxygen from water at first (but later breathed air) and used it to oxidize the sugars from plants they ate. This new type of metabolism provided more energy faster than the photosynthetic method, and it allowed animals to move around in search of food sources. The metabolism of all animals—including humans—works this way (see Figure 11.2). As you read this book, your lungs are pumping oxygen in and out without any conscious thought on your part.

FIGURE 11.2
Here biologists are measuring how fast an elephant uses oxygen during exercise, to find out at what rate it uses energy (metabolizes). Equipment on the cart measures the oxygen content of the air that the elephant exhales through the hose. The study showed that elephants use less energy per kilogram of body mass than any other land animal.

Air is much more than the source of oxygen for the metabolism of living things. The atmosphere traps and releases energy from the sun, helping to smooth temperature fluctuations. Variations of air pressure and temperature across land and sea are responsible for weather. Ozone in the upper atmosphere blocks most of the solar ultraviolet rays, which would otherwise cause cancers and mutations. Human and natural processes, from cars and factories to volcanos and the digestive tracts of animals, release huge amounts of pollutants into the air. Some of these compounds react further, including the ones that cause smog and damage the ozone layer (see Figure 11.3).

The gases that make up the air share certain characteristics with all other gases. As you saw in Chapter 2, gases are easy to compress, and they expand to fill their containers. Gases consist of individual atoms or molecules that zip around in relatively large volumes of otherwise empty space. The intermolecular forces of attraction between gas molecules are usually very small in comparison with their kinetic energy. This allows scientists to visualize these molecules as tiny balls bouncing off one another and the walls of the container and thus to expect all gases to exhibit regular and predictable physical properties. This mental picture has enabled scientists to construct a theory, complete with equations, that predicts how gas samples will react to changes in volume, pressure, and temperature. No successful theory of this kind exists for solids or liquids, because intermolecular forces in the condensed phases are much stronger and more complex than those in gases.

FIGURE 11.3
Automobile exhaust introduces NO into the atmosphere as a pollutant. The NO slowly reacts with atmospheric oxygen to form $NO_2$, which occasionally reaches concentrations high enough to be visible as a brown layer hanging over a city.

## 11.1 The Composition of the Atmosphere

GOAL: To identify the elements and compounds that make up the earth's atmosphere and describe their behavior.

Gases are all around you, and many of them are vital to your life. In your body the oxygen you breathe reacts with certain nutrients—carbohydrates, fats, and sometimes proteins—to give you the energy you need to live and move. Carbon dioxide and water are the other products of this process. Plants take up the carbon dioxide and water vapor you exhale and use the energy of sunlight to convert these gases back to oxygen and nutrients.

Other gases in the atmosphere have different roles. For instance, certain microorganisms (called *nitrogen-fixing bacteria*) take up nitrogen gas and convert it to nutrients that plants can use in making proteins, nucleic acids (DNA and RNA), and other compounds necessary for life. When the bacteria work too slowly, farmers provide nitrogen compounds in fertilizers to keep plant processes going (see Figure 11.5).

FIGURE 11.5
Atmospheric nitrogen is relatively inert, and plants cannot use it directly. Many fertilizers contain ammonium nitrate, $NH_4NO_3$, which puts nitrogen into the soil in a form that plants can use. Agricultural machinery is designed to spread fertilizer evenly over the ground.

FIGURE 11.4
Earth's blanket of atmosphere seems thin when viewed from close planetary orbit.

**TABLE 11.1 The Composition of Dry Air**

| Gas | Formula | Percent by Volume |
|---|---|---|
| Nitrogen | $N_2$ | 78.08% |
| Oxygen | $O_2$ | 20.95 |
| Argon | Ar | 0.93 |
| Carbon dioxide | $CO_2$ | 0.033 |
| Neon | Ne | 0.0018 |
| Helium | He | 0.00052 |
| Methane | $CH_4$ | 0.0002 |
| Krypton | Kr | 0.0001 |
| Hydrogen | $H_2$ | 0.00005 |

**Covalent compounds can also be liquids (for example, water and the compounds that make up gasoline) or solids (diamonds and candle wax) at room temperature. See Section 10.1.**

The list of elements that are gases at room temperature includes many of the nonmetallic elements: He, Ne, Ar, Kr, Xe, Rn, $H_2$, $N_2$, $O_2$, $F_2$, and $Cl_2$. Almost all the other substances that are gases at room temperature are relatively small covalent molecules composed of nonmetals. The gaseous compounds $NH_3$, HI, $CH_4$, $SO_2$, $SF_2$, and HCN are examples.

Nitrogen ($N_2$) is the most abundant gas on earth. It makes up about 78% by volume of the atmosphere at sea level. Oxygen comprises almost 21% of the atmosphere. The noble gases, the elements in Group 8A of the periodic table, make up most of the remaining 1% or so. Argon is by far the most common of these rather rare elements. The percent composition of the atmosphere by volume is given in Table 11.1.

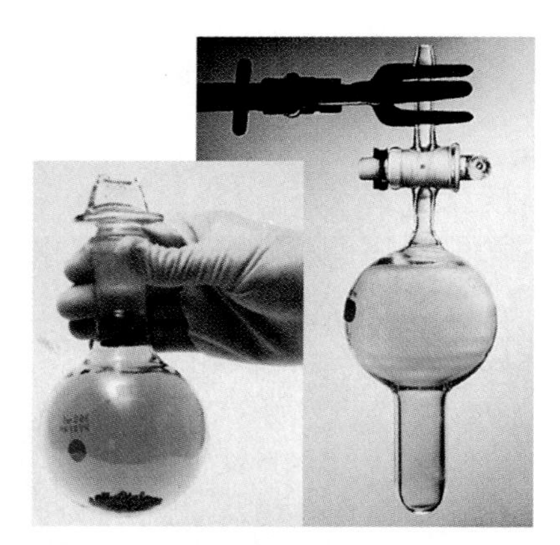

FIGURE 11.6
Most gases are colorless, including all the components of the atmosphere. The halogens have colors in the gas phase. For instance, heating solid iodine gives purple iodine vapor (left). Another colored gas is the atmospheric pollutant $NO_2$, which is brown (right).

A variety of gases make up the very small remaining portion of the atmosphere. Carbon dioxide ($CO_2$) is vital for the life of plants, and therefore for human life. However, scientists worry that human activities such as burning fossil fuels and clearing land for agriculture, along with animal metabolism and the natural decay of plant material, have caused an increase in the amount of $CO_2$ in the atmosphere.

The extra $CO_2$ absorbs infrared radiation that would normally carry energy off into space. (This is called the *greenhouse effect* and is illustrated in Figure 11.7. See Chapter Box 8.1 for the origin of the idea of global warming.) It seems likely that the earth's atmosphere is heating up, although how significant the warming is remains a hot topic of debate in the scientific community. Many processes, both natural and artificial, give off methane, which is another *greenhouse gas.*

Air pollution is a mixture of gases (and some liquid droplets and solid particles) that comes largely from human activities. Nitrogen oxides (there are several of them) are the products when $N_2$ reacts with $O_2$ at high temperatures. They can form in automobile engines and fossil fuel power plants and are also created by bolts of lightning. Coal and some heavy heating oils contain sulfur, which burns to give $SO_2$. This compound is a major cause of acid rain (see Chapter Box 8.3).

Hydrocarbons (compounds made up of only carbon and hydrogen) are a major component of fuels. They form a variety of gases when they burn incompletely. When light shines on a mixture of these compounds, *photochemical smog* forms. One of the compounds in this smog is ozone, which is as uncomfortable to humans at sea level as it is important in the upper atmosphere.

Ozone ($O_3$) forms in the upper atmosphere from reactions involving oxy-

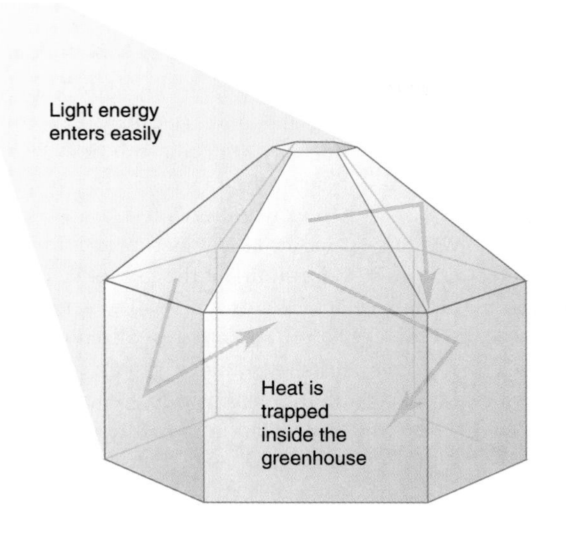


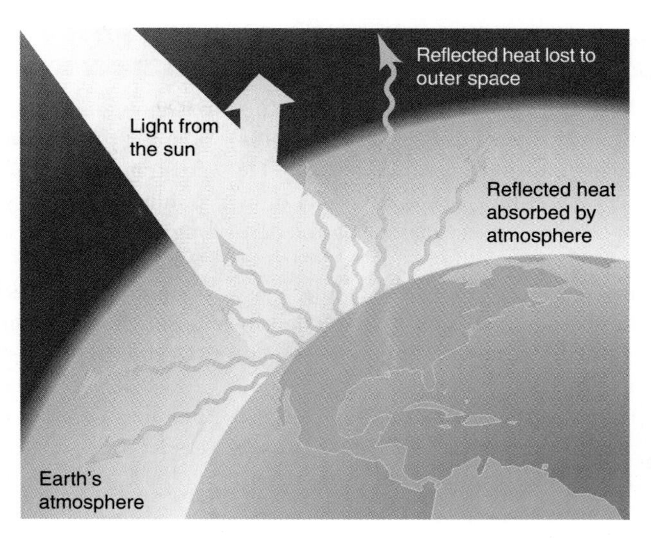


FIGURE 11.7
The earth is warmed by what is called the *greenhouse effect.* Certain gases in the atmosphere, including $CO_2$, have the same effect as the windowpanes in a greenhouse—they let sunlight in and prevent the resulting heat from leaving.

gen ($O_2$) and high-energy radiation. At sea level it's a pollutant, but in the stratosphere it absorbs dangerous ultraviolet radiation and keeps much of it from reaching the earth's surface. Because of effects like this, the upper atmosphere is much different from the lower atmosphere. At the upper boundary of the atmosphere, where it's hard to tell the difference between atmosphere and the vacuum of outer space, hydrogen is the most abundant gas.

Gases called *freons* (Freon-11 has the formula $CCl_3F$, and Freon-12 is $CCl_2F_2$) are the working compounds in refrigerators and air conditioners. These compounds have other uses, as solvents and as propellants in aerosol spray cans. Freons are neither toxic nor flammable, nor corrosive, which makes them all the more useful. Recently scientists have found that molecules of these gases last long enough to drift to the upper atmosphere, where a series of reactions involving high-energy radiation and chlorine from these compounds destroys the protective ozone (see Figure 11.8). Because of this effect, freons have been banned from spray cans in most countries, and they are gradually being replaced with other compounds in refrigerators and air conditioners.

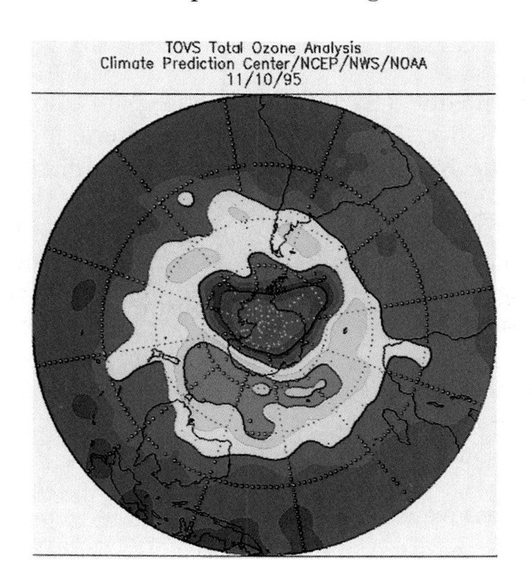


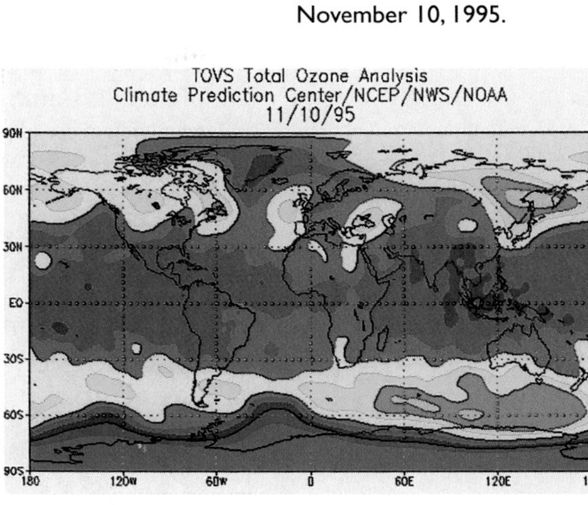


FIGURE 11.8
Two views of the famous ozone hole over Antarctica constructed from satellite camera views of the earth on November 10, 1995.

## CHAPTER BOX 11.1 THE DISCOVERY OF ARGON

**Near the** end of the 1800s, the English physicist Lord Rayleigh (1842–1919) was studying atomic masses. He very carefully measured the atomic mass of oxygen by measuring the density of the gas and found that it was not exactly 16 times the atomic mass of hydrogen. He got the same density (and the same atomic mass) no matter where the sample of oxygen came from. He found that oxygen from the air has the same atomic mass as oxygen made from mercury(II) oxide, for instance.

Rayleigh then performed a similar experiment with nitrogen and ran into a problem. Nitrogen gas from the air had a slightly higher density, and therefore a slightly higher atomic mass, than nitrogen formed from compounds. He thought that the nitrogen collected from the air might be contaminated, but he couldn't identify the contamination. He got so frustrated he wrote a letter that was published in the journal *Nature,* asking for suggestions.

The Scottish chemist William Ramsay (1852–1916) took up the problem. He remembered that Henry Cavendish (1731–1810), a great scientist of a century earlier, had combined nitrogen from the air with oxygen. Cavendish found that he had a bubble of gas left over. Ramsay thought that this bubble might be Rayleigh's contaminant. He used a different method to isolate such a bubble; he reacted nitrogen isolated from the air with magnesium. There was some gas left over, as he expected. Ramsay and Rayleigh heated this gas and studied the lines in the spectrum of light that it gave off. The two scientists found that they had discovered a new element. Because they couldn't find anything that the new gas would react with, they called it *argon,* from a Greek word for "inert."

Argon fell between chlorine and potassium in the periodic table. No periodic group was known in that location, so Ramsay realized that he had discovered a new group. He set out to fill it in, and within a few years he had isolated a sample of helium, which he identified as a substance that had been observed in the spectrum of the sun a generation earlier (see Section 6.3). He distilled a large sample of argon that he had isolated from nitrogen and found elements he called neon, krypton, and xenon in it. Radon was discovered during studies of radioactivity soon afterward.

### EXAMPLE 11.1 Atmospheric Gases

The ability of gases in the atmosphere to absorb radiation has many consequences for the environment. Which of the following gases is *least* important as an absorber of radiation?

A. Methane B. Carbon dioxide C. Oxygen D. Ozone

**SOLUTION**

Methane (A) and carbon dioxide (B) are greenhouse gases, and ozone (D) absorbs ultraviolet radiation in the upper atmosphere, protecting plants and animals from the hazards of this high-energy radiation. The answer is oxygen (C).

**EXERCISE 11.1**

What atmospheric gas is essential for animal metabolism?

## 11.2 Pressure

**GOAL:** To explain what pressure is and know its units.

"Nature abhors a vacuum," the ancients thought. By 1600 European engineers were making use of this concept to pump water. Their vacuum pumps had a limit, however. No matter what they tried, the pumps could raise water no further than about 33 feet. Galileo (1564–1642) looked into the problem but only concluded that there seemed to be some limit to the vacuum force. A promising young scientist named Evangelista Torricelli (1608–1647) served as Galileo's secretary for the last 3 months of the older man's life. Galileo recommended that Torricelli take up the problem.

**The phrase "Nature abhors a vacuum" is attributed to Baruch Spinoza, who lived in the 1600s. The idea was widespread before Spinoza came along, but this venerable wordsmith memorably summarized centuries of observations.**

Torricelli suspected that the effect might be a simple mechanical one involving air pressure. His reasoning was quite simple and easy to test. Torricelli thought that when the piston of a pump is pulled above the surface of the water, the air pressure over the water inside the pump becomes less (see Figure 11.9). The air pressure outside the pump stays constant, and the difference in air pressure inside and outside the pump forces the water to rise. If the normal pressure of the air in the atmosphere was only enough to support 33 feet of water, all the pumping in the world could force the water no higher.

**Torricelli's experiment (see below) is another one that you shouldn't try to duplicate, although it isn't quite as dangerous as the Franklin kite experiment. Mercury is the only metal that is liquid at room temperature, and it is much more volatile than other metals. There will be gaseous mercury atoms above any pool of mercury. Breathing these vapors can be harmful, particularly if you do it over a long period of time. Many early chemists and photographers (who used mercury vapors in the developing process) died prematurely from mercury poisoning.**

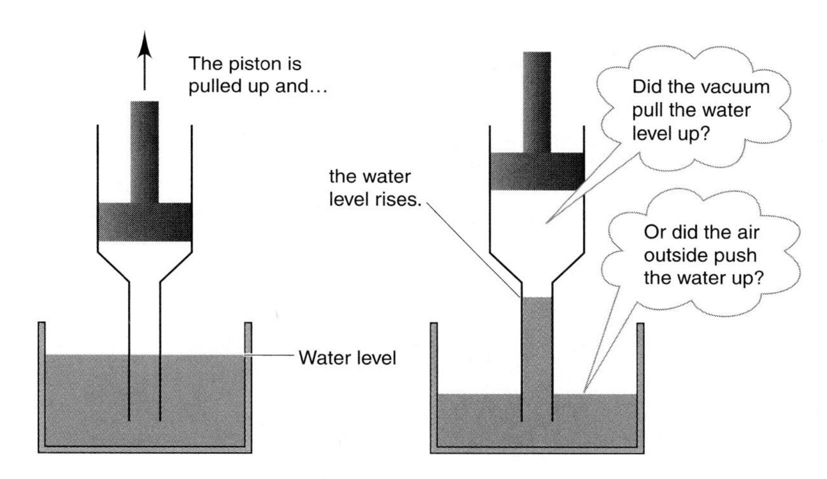
The piston is pulled up and...
the water level rises.
Water level
Did the vacuum pull the water level up?
Or did the air outside push the water up?

FIGURE 11.9
Torricelli contemplated the water pump and asked how it worked.

Torricelli tested his idea using mercury, which is much more dense than water and was easier for him to handle. He filled a 4-foot tube with mercury, put a stopper in the top, inverted the tube, and put the lower end in an open dish containing more mercury (see Figure 11.10). Then he unstoppered that

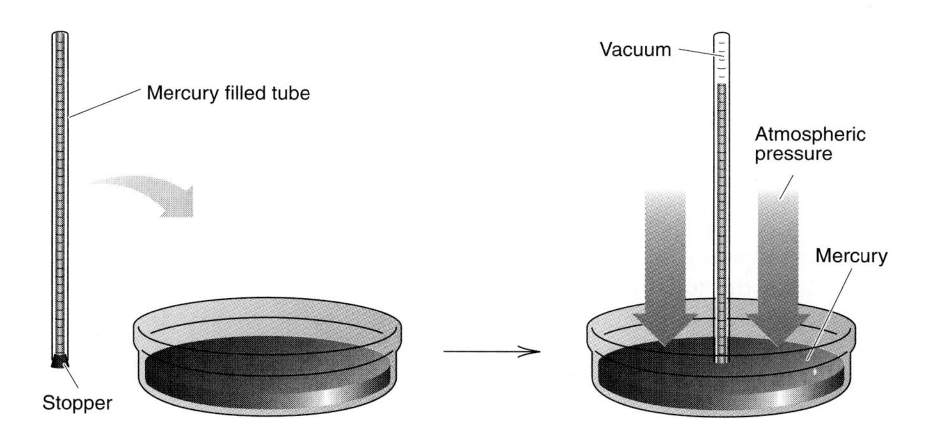
Mercury filled tube
Stopper
Vacuum
Atmospheric pressure
Mercury

FIGURE 11.10
Torricelli's experiment demonstrated that air pressure pushes the water up in a water pump (Figure 11.9) and that there are fixed limits to how much "push" is available. Torricelli used mercury in his experiments because it is much denser than water, allowing his mercury column to be 1/14 as long as a water column would have to be. A modern laboratory barometer, the kind scientists use to measure atmospheric pressure, works just like this. At sea level and at 0°C, the average pressure of the atmosphere supports a column of mercury 760 mm in height.

end. The mercury flowed out the bottom of the tube until about 30 inches remained. The height of the tube (as long as it was more than 30 inches) and the width of the tube (as long as it wasn't extremely narrow) made no difference. The pressure of the air was the only important variable. If you think a device like this could be useful in routinely measuring air pressure, you are right. Called a barometer, it is still in use today.

What is pressure? This is another of those words whose meaning is slightly different in science and in everyday life. A student studying for an exam, a basketball player at the foul line near the end of a close game, or a textbook author facing a deadline are said to be under pressure. This is somewhat parallel to scientists' definition, but not exactly the same. **Pressure** to a scientist is defined as *force per unit of area.* If you push an ice pick against a wall, the *force* is the same on the wall and on your hand (see Figure 11.11). The tip of the ice pick has a much smaller area than the handle, so it applies much more *pressure* to the wall.

**The difference between force and pressure is akin to the difference between mass and density. A ton of feathers has the same mass as a ton of lead, but the feathers occupy more volume and thus a ton of feathers has a lower density than a ton of lead.**

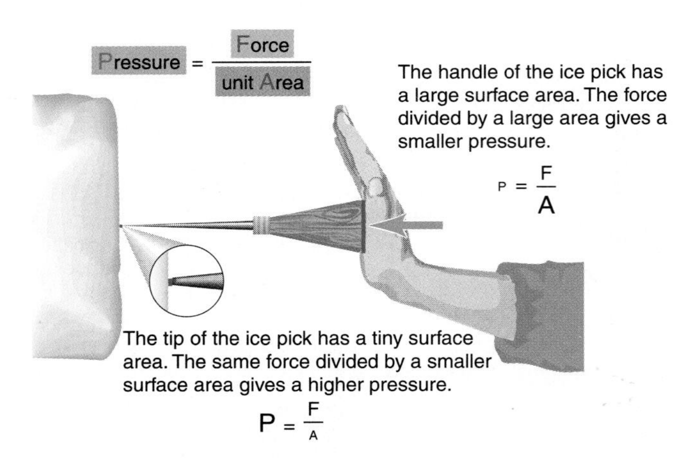

Pressure = Force / unit Area
The handle of the ice pick has a large surface area. The force divided by a large area gives a smaller pressure.
P = F / A
The tip of the ice pick has a tiny surface area. The same force divided by a smaller surface area gives a higher pressure.
P = F / A


FIGURE 11.11
The relationship of force to pressure.

This is why you want your kitchen knives to be as sharp as possible. A sharp knife has a smaller surface area along the edge of the blade in contact with the food. Therefore, a small force applies a large pressure. When kitchen knives are dull, you have to push harder to cut the food; this can result in the knife slipping and you cutting yourself. If you want to avoid kitchen accidents of this kind, sharpen all the knives you normally use.

## EXAMPLE 11.2 Force and Pressure

You are pushing on the handle of an ice pick with a force of 5 lb. The handle of the ice pick has an area of 1 $in^2$. What is the pressure exerted by your hand on the handle of the ice pick? (The pressure exerted by the handle on your hand is the same.)

**SOLUTION**

Pressure is force per unit of area. The force is 5 lb, and the area of the handle is 1 in$^2$. The pressure is

$$\frac{5 \text{ lb}}{1 \text{ in}^2} = 5 \text{ lb in}^{-2}, \text{ or 5 pounds per square inch}$$

**EXERCISE 11.2**

The tip of the ice pick in Example 11.2 has an area of 0.001 in$^2$.

**a.** What pressure does the tip exert on the wall you're pushing against?
**b.** Why do ice picks (and knives) become blunt so easily?

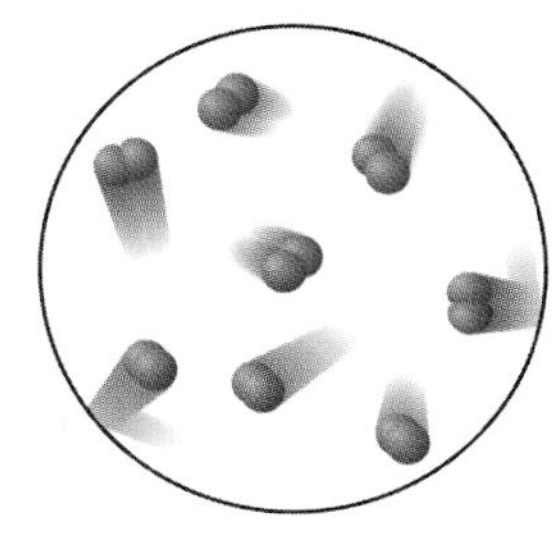

FIGURE 11.12
Gas molecules are constantly in motion, colliding with each other and with the sides of their container. They exert pressure by colliding with the walls of the container. The more gas molecules there are, the more collisions there are and the greater the pressure.

So how does all of this relate to the pressure of a gas? The answer is remarkably simple. Gas molecules are in constant motion and are widely separated. They collide with each other and with the sides of the container holding them (see Figure 11.12). Think of what happens if you punch a wall. You exert pressure on the wall equal to the force of the punch divided by the area of your hand. (You know you exert pressure because your hand hurts like blazes.) The harder you punch, the more pressure there is.

Molecules act this way when they hit the sides of the container. Molecules don't have much mass, but there are many of them: A mere 2 g of hydrogen gas contains $6 \times 10^{23}$ molecules, after all. The pressure that a gas exerts on the walls of a container is the sum of the forces of all the molecules hitting it, divided by the area of the walls.

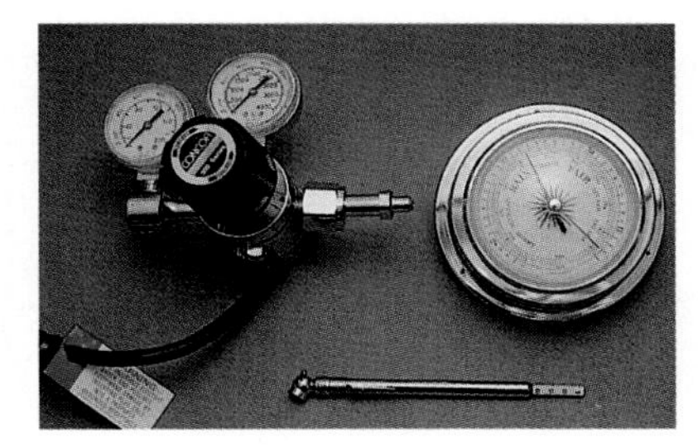

FIGURE 11.13
Pressure gauges connected to pressure reduction valves (left) usually measure the pressure in gas storage tanks in psi, atm, Pa, or kPa. Barometers (top, right) measure atmospheric pressure in inches of mercury, mm Hg, or torr. Tire gauges (bottom, right) usually measure tire pressure in psi.

There are, as usual in this nonstandardized world, many units for pressure (see Figure 11.13). Industrial engineers, mechanics, and scuba divers often use pounds per square inch (abbreviated psi), but scientists seldom use this English-system unit. Torricelli's mercury barometer has been used for centuries, and TV weather reporters in the United States give air pressures in inches of mercury.

Chemists and other scientists often use the height of mercury in millimeters (that is, in mm Hg) as a unit of pressure. The unit mm Hg is sometimes called the **torr** in Torricelli's honor. The atmospheric pressure at sea level averages about 760 mm Hg, and this is defined as **1 atmosphere (atm)** in another system of units often used by scientists. Finally, the official SI unit of pressure is the **pascal (Pa).** These units are related as follows:

$$760 \text{ mm Hg} = 760 \text{ torr} = 1 \text{ atm} = 1.01325 \times 10^5 \text{ Pa} = 14.7 \text{ psi}$$

**Pascals are derived SI units; 1 Pa = 1 kg m$^{-1}$ s$^{-2}$. It is the standard SI unit of pressure. However, because the pressures normally encountered in human experience are typically hundreds of kilopascals (kPa), even chemists seldom use the unit.**

The pascal is much too small to be convenient; after all, it takes over 100,000 of them to equal 1 atmosphere. Most scientists, including the authors of this book, use either *atmospheres* or *millimeters of mercury* (torr) as units to describe the pressures of gases instead of the standard SI units.

**EXAMPLE 11.3** **Units of Pressure**

**a.** What is the value of 760 mm Hg in inches of mercury?
**b.** A tire is rated for a maximum inflation pressure of 35 psig, which is psi in excess of atmospheric pressure. What is the total pressure in atmospheres and in torr?

**SOLUTION**

**a.** This is a unit conversion problem, which you know how to do—you use the unit factor method. The conversion factors you need are 1 cm = 10 mm (or you might have to use 1 m = $10^3$ mm and 1 m = $10^2$ cm) and 1 cm = 2.54 in:

$$760 \cancel{\text{mm}} \left(\frac{1 \cancel{\text{cm}}}{10 \cancel{\text{mm}}}\right)\left(\frac{1 \text{ in}}{2.54 \cancel{\text{cm}}}\right) = 29.9 \text{ in}$$

Therefore, 760 mm Hg is equivalent to 29.9 inches of mercury. This seems reasonable, since the barometric pressures you hear on weather reports range from around 29 to just over 30 inches.

**b.** First of all, the total pressure is (35 + 14.7) psi = 49.7 psi. This is now another unit conversion problem. The conversion factors are given above.

$$49.7 \cancel{\text{psi}} \left(\frac{1 \text{ atm}}{14.7 \cancel{\text{psi}}}\right) = 3.4 \text{ atm}$$

$$49.7 \cancel{\text{psi}} \left(\frac{760 \text{ torr}}{14.7 \cancel{\text{psi}}}\right) = 2.6 \times 10^3 \text{ torr}$$

**EXERCISE 11.3**

The pressure of a gas was reported as 730 mm Hg. What is this pressure in atmospheres?

Final levels of mercury

Height of mercury equivalent to one atmosphere of pressure

Initial level of mercury

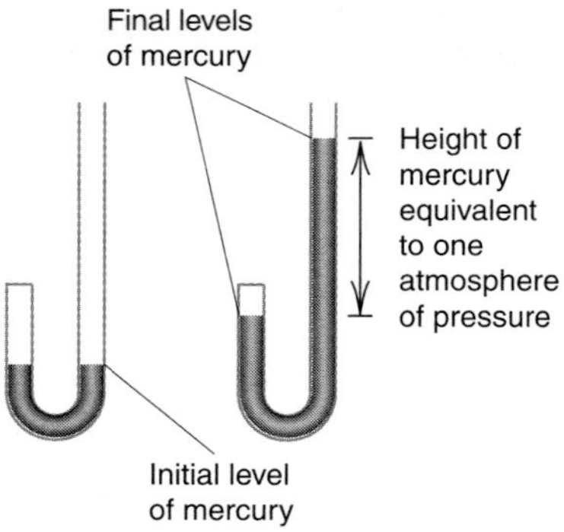

FIGURE 11.14
In Boyle's experiment with the "spring of air," he first trapped a small sample of air in the closed portion of his J-tube (on the left). Since the level of mercury is the same on both sides of the tube, the pressure of the trapped sample is 1 atm. When Boyle added enough mercury to raise the level on the right side of the J-tube to 760 mm above the left side, the pressure on the trapped air sample on the left became 2 atm. The volume of the trapped air on the left decreased to half its original value.

## 11.3 Boyle's Law: Pressure and Volume

**GOAL:** To understand the relationship between pressure and volume of a sample of gas at constant temperature, and to use this knowledge to perform calculations.

Robert Boyle was an early English scientist who studied gases (and many other things). His most famous experiment involved what he called the "spring of air." He used a J-shaped tube with the short end closed, like the

## CHAPTER BOX 11.2 ROBERT BOYLE

**Robert Boyle** (1627–1691) was considered something of an iconoclast in his day. His interest in experimentation was unusual and considered an oddity (see Figure A). Baruch Spinoza, the Dutch philosopher quoted at the beginning of Section 11.2, corresponded frequently with Boyle and tried to convince him that reason, rather than experiment, was the path to the truth.

When Boyle was 30, he redesigned, and greatly improved, the vacuum pump. For some time the vacuum produced by this kind of pump was called a *Boylean vacuum.* Boyle used the pump to evacuate a large cylinder and was the first to verify (by experiment!) Galileo's hypothesis that all objects fall at constant velocity in a vacuum.

Boyle was never able to offer a physical interpretation of his famous law, primarily because he did not know about the molecular nature of gases. Boyle's "spring of air" experiment was done in 1662. At that point in the history of science, theories on the nature of matter were in the embryo stage; most of the ones in use originated with the ancient Greeks. The writings of Democritus suggested that matter might consist of tiny, indivisible units called atoms, and Aristotle proposed that matter was a continuous fluid. If gases consist of a continuous fluid, as Aristotle thought, how could they be compressible? On the other hand, if gases consisted of atoms with a lot of space between, they could be compressed easily. Boyle and his friends were convinced that matter consisted of atoms, but they had no way to make use of this fact, so the idea languished. It wasn't until Dalton proposed his atomic theory in 1803, and his theory on the partial pressure of gases (Section 11.8) in 1801, that scientists began to understand exactly why gases behave the way they do.

FIGURE A
Robert Boyle (1627–1691) (right) was taught by private tutors, so he avoided the slavish devotion to Aristotle rampant in the universities at the time. He studied Galileo's work in Italy soon after the great Italian's death, developed an improved vacuum pump, and used it to experiment with gases. He was a strong upholder of the importance of experiments at a time when few philosophers supported this position.

one shown in Figure 11.14. With any sample of gas, he found that the pressure on the gas sample multiplied by the volume occupied by the sample remained constant. In equation form, this is

Boyle's Law $$P_1V_1 = P_2V_2 \quad \textbf{(at constant temperature)}$$

where $P_1$ and $V_1$ are the pressure and the volume of the sample in one experiment, and $P_2$ and $V_2$ are the pressure and the volume of the same sample in a different experiment. (The equation is valid only when both experiments are carried out at the same temperature.) This equation is now called **Boyle's Law.**

**Boyle's Law was the first application of mathematics to chemistry.**

Two key conclusions emerged from Boyle's experiments. The first is the equation above. The second is more subtle. Boyle's Law describes the behavior of gas samples regardless of what the gas is. In other words, it works for nitrogen, carbon dioxide, argon, or any mixture of gases (including air). You will see that you can predict many of the properties of a gas sample without knowing what atoms or molecules make it up.

It is easy enough to describe what happened to atoms and molecules as the volume changed in Boyle's experiment. A gas in a piston chamber con-

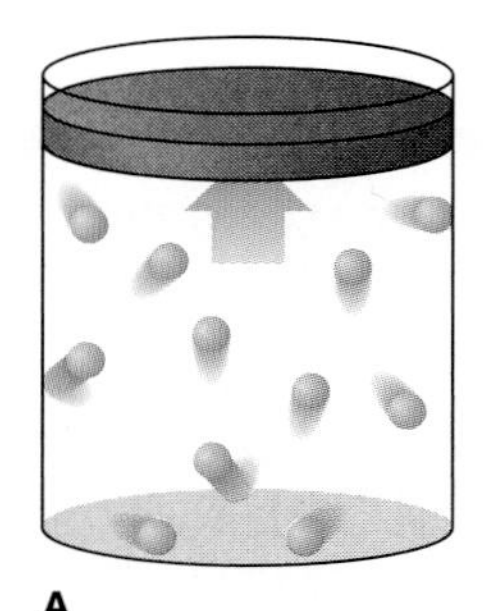

A

B

FIGURE 11.15
When the volume decreases, the average time between collisions of the molecules with the wall is shorter, and the surface area with which to collide is smaller.

tains a certain number of molecules striking the sides of the container (see Figure 11.15A). If you push the piston in to reduce the volume, the molecules have a shorter flight path and a smaller wall area to strike (see Figure 11.15B). The molecules hit the walls with the same force but twice as often, causing the pressure to double.

## EXAMPLE 11.4 Pressure and Volume (Boyle's Law)

A scuba diver filled a balloon with air at the surface of the ocean, tied a weight to it, and pulled it down to a depth of 20 m, where the pressure is 3.0 atm. If the volume of the filled balloon at the surface was 11 L and the atmospheric pressure at the surface was 1.0 atm, what is the final volume?

**SOLUTION**

Since pressure and volume are involved, this is a Boyle's Law problem, so you use the equation $P_1V_1 = P_2V_2$. At the surface the pressure is 1.0 atm (you have to know this), and the volume (given) is 11 L. These quantities are the values of $P_1$ and $V_1$; $P_2$ is 3.0 atm, and $V_2$ is the desired answer.

$$P_1V_1 = P_2V_2$$

$$(1.0\text{ atm})(11\text{ L}) = (3.0\text{ atm})V_2$$

$$\frac{(1.0\,\cancel{\text{atm}})(11\text{ L})}{3.0\,\cancel{\text{atm}}} = V_2 = \frac{11\text{ L}}{3.0} = 3.7\text{ L (rounded off)}$$

The volume of the balloon at a depth of 20 m is 3.7 L. Don't forget significant figures!

A second method that some students prefer begins with a table.

| | |
|---|---|
| $P_1 = 1.0$ atm | $P_2 = 3.0$ atm |
| $V_1 = 11$ L | $V_2 = ?$ |

You are finding a change in volume from $V_1$ to $V_2$. Since the pressure increases from $P_1$ to $P_2$, the volume must decrease from $V_1$ to $V_2$. Start with $V_1$ and multiply by a fraction made up of the two Ps in order to decrease the V.

$$11\text{ L} \times \frac{1.0\,\cancel{\text{atm}}}{3.0\,\cancel{\text{atm}}} = 3.7\text{ L}$$

**EXERCISE 11.4**

A scuba diver partly filled a balloon with air at a depth of 30 m (where the pressure is 4.0 atm). The volume of the balloon at 30 m was 5.3 L. The diver released the balloon, and it floated to the surface.

**a.** What is the volume of the balloon at the surface?

**b.** Why didn't the diver fill the balloon completely at the depth of 30 m?

## 11.4 Charles' Law: Volume and Temperature

**GOAL:** To understand the relationship between volume and temperature of a sample of gas at constant pressure and to use this knowledge to perform calculations.

Fifteen years after Boyle published his results, the French physicist Edmé Mariotte (1620–1684) discovered the same relationship without knowing of the Englishman's work. Mariotte went further than Boyle did. He noticed that air expands when the temperature rises and contracts when the temperature falls. A generation later another French physicist, Guillaume Amontons (1663–1705), took this work further. He found that the volume of air samples changed in a regular way when the temperature rose or fell. Through studies of different gases he showed that the identity of the gas doesn't matter, just as in Boyle's experiment. Amontons's results didn't give an equation as satisfying as Boyle's, so they were ignored for some time.

Almost a century later Jacques Alexandre César Charles was involved in hot-air ballooning, a craze begun by his fellow Frenchmen the Mongolfier brothers (see Figure 11.16). He worked with heated gases and rediscovered the relationship Amontons had found. Modern chemists state **Charles' Law** this way: The volume occupied by a sample of gas is directly proportional to its absolute temperature if the pressure is held constant.

**Yet another French physicist, Joseph Louis Gay-Lussac, independently discovered this law a few years after Charles did his work. Because Gay-Lussac published his work and Charles did not, the law is sometimes called Gay-Lussac's Law.**

FIGURE 11.16
Balloon flight has fascinated many people, from the Mongolfier era, which began in 1783 (left) until today (right).

The equation for Charles' Law used today is this:

Charles' Law $$\frac{V_1}{T_1} = \frac{V_2}{T_2} \quad \textbf{(at constant pressure)}$$

where $V_1$ and $T_1$ are the volume and *absolute* temperature of a sample of gas at one time, and $V_2$ and $T_2$ are the volume and absolute temperature at another time. You should take careful note of an important feature of this

CHAPTER BOX

## 11.3 THE ABSOLUTE TEMPERATURE SCALE

It is interesting to note that both Charles and Gay-Lussac were hot-air balloon enthusiasts. Gay-Lussac set an altitude record at 7 km (23,000 ft). The relationship between the volume of a gas and its temperature is very important to a hot-air balloonist. As the volume of the balloon increases, the amount of displaced air increases and the balloon becomes more buoyant. The reverse is also true.

Charles realized that hydrogen would be a far more efficient gas than air for use in balloons because it was so much less dense (and therefore more buoyant). He was the first to construct hydrogen balloons. However, hydrogen is dangerous to use because it can react violently with oxygen, as the destruction of the German airship *Hindenburg* in the 1930s clearly showed (see Figure A).

FIGURE A

The explosion of the dirigible *Hindenburg* in 1937 clearly demonstrated the hazards of using hydrogen in lighter-than-air craft. Modern blimps use helium, which doesn't react with anything at all and has a density not much greater than that of hydrogen. The *Hindenburg* was originally designed to use helium, but Germany had no domestic source of that gas.

Suppose you perform an experiment with an apparatus like that shown in Figure 11.17. Initially you use ice water, giving a temperature of 0°C, and adjust the gas volume to 1 L. If you warm the water in the bath successively to temperatures of 10°C, 20°C, 30°C, and so on, you will find that the volume of the gas increases in even steps. For every 1°C increase in temperature, the volume of the gas sample that was 1 L at 0°C will increase by 3.66 mL. The volume of the gas at 100°C will be 1.366 L. This is illustrated in Figure B.

Now suppose, instead of increasing the temperature, you reduce it. For each 1°C decrease in temperature, the volume of the gas sample will decrease by 3.66 mL (see Figure C). If no other factors interfere, the volume of the gas will be reduced to zero at a temperature of −273°C. At even lower temperatures the volume of the gas would become negative. Since a negative value for a volume is clearly absurd, scientists concluded that −273°C is the lowest temperature that can be reached. This temperature is called **absolute zero.**

In reality, of course, all gases condense to liquids before they reach −273°C. No gas can be cooled to the point at which it has zero volume. On the other hand, the idea that there exists a lowest possible temperature is an extremely important scientific concept. The British scientist Lord Kelvin was the first to suggest that scientists use an absolute temperature scale. Instead of setting zero on a temperature scale at the freezing point of water, the absolute temperature scale would have its zero point at this lowest possible temperature. Modern accurate measurements place absolute zero at −273.15°C.

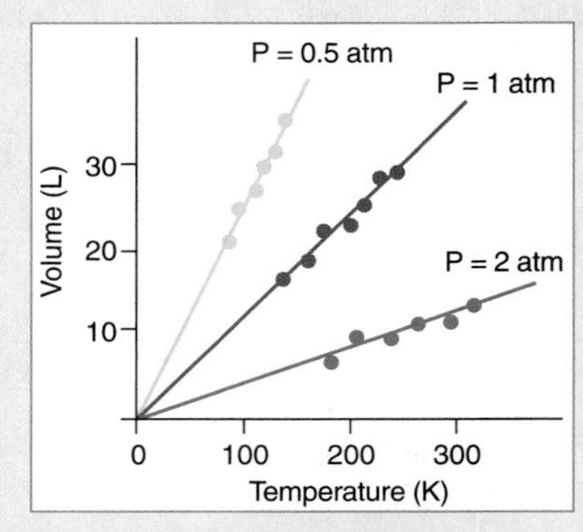


FIGURE C

Regardless of the initial pressure or volume of gas used, the volume of the gas steadily approaches zero as the temperature drops.

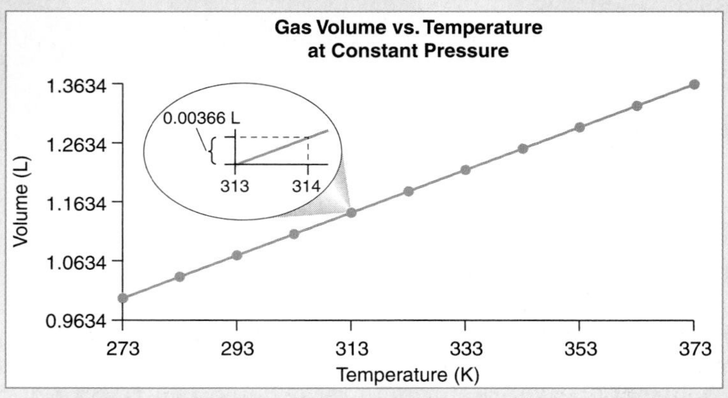


FIGURE B

Plot of volume occupied by a gas versus temperature at constant pressure.

equation. It is valid *only* if you use the absolute temperature scale—the Kelvin scale. It doesn't work if you use the Celsius (Centigrade) or Fahrenheit scales.

How could you test Charles' Law? One way might be with Boyle's J-tube apparatus or with a similar U-tube (see Figure 11.17). To keep the pressure constant you must add or remove mercury to maintain equal levels on the two sides of the U-tube. You submerge the tube in stirred water to keep the temperature uniform and heat or cool the water to change the temperature.

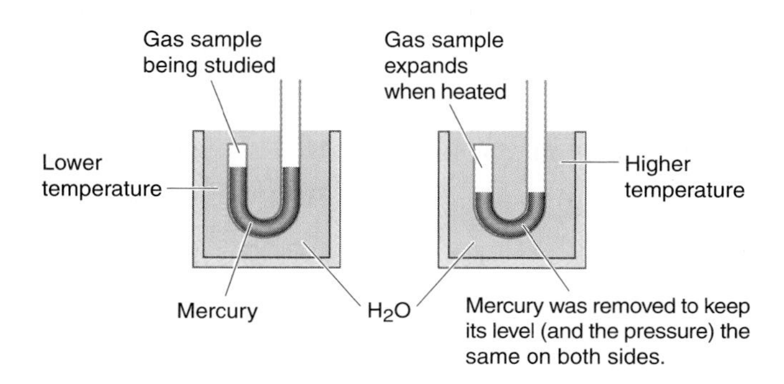


FIGURE 11.17
Apparatus for testing Charles' Law.

EXAMPLE 11.5 **Gas Temperature and Volume (Charles' Law)**

**a.** A sample of sulfur dioxide gas had a volume of 2.3 L at 300 K. What was the volume of this sample after it was heated to 350 K? The pressure was kept at 1 atm.

**b.** A sample of oxygen had a volume of 1.7 L at 100°C. What was the volume of this sample after it was cooled to 0°C? The pressure was kept at 1 atm.

**SOLUTIONS**

**a.** The problem involves volume and temperature of a gas, so Charles' Law is the one to use. The equation is $V_1/T_1 = V_2/T_2$. Since the sample began at 2.3 L and 300 K, those quantities are $V_1$ and $T_1$; 350 K is $T_2$. The solution is then:

$$\frac{2.3\ \text{L}}{300\ \text{K}} = \frac{V_2}{350\ \text{K}}$$

$$\frac{(2.3\ \text{L})(350\ \not{\text{K}})}{300\ \not{\text{K}}} = V_2 = 2.7\ \text{L}$$

The final volume of this sample is 2.7 L. Since the sample was heated, it's reasonable that $V_2$ is a little more than $V_1$, which was 2.3 L.

**b.** The identity of the gas sample doesn't matter here. What does matter is that the temperatures are given in the Celsius scale and have to be converted to kelvins. Here $V_1$ is 1.7 L, $T_1$ is 373 K, and $T_2$ is 273 K. The calculation is done as before:

$$\frac{1.7\ \text{L}}{373\ \text{K}} = \frac{V_2}{273\ \text{K}}$$

$$V_2 = \frac{(1.7\ \text{L})(273\ \cancel{\text{K}})}{373\ \cancel{\text{K}}} = 1.2\ \text{L}$$

The final volume of this gas sample is 1.2 L. Since the sample was cooled, it's reasonable that $V_2$ is a little less than $V_1$.

**EXERCISE 11.5**

**a.** A sample of oxygen gas had a volume of 5.0 L at 250 K. What was the volume of this sample after it was heated to 350 K? The pressure was kept at 1 atm.

**b.** A sample of helium gas had a volume of 5.0 L at 0°C. What was the volume of this sample after it was cooled to −100°C? The pressure was kept at 1 atm.

Why does the volume of a gas increase when the temperature increases at constant pressure? Scientists learned the answer to this question long after Boyle, Mariotte, and Amontons had died. When the temperature of a gas increases, its atoms or molecules move faster and strike the sides of their container with more force. To keep the pressure the same, the container must become larger—that is, the volume must increase. This means that fewer particles will strike any given area of the walls of the container. The greater amount of force per strike will be balanced by fewer strikes per unit area, keeping the pressure the same (see Figure 11.18).

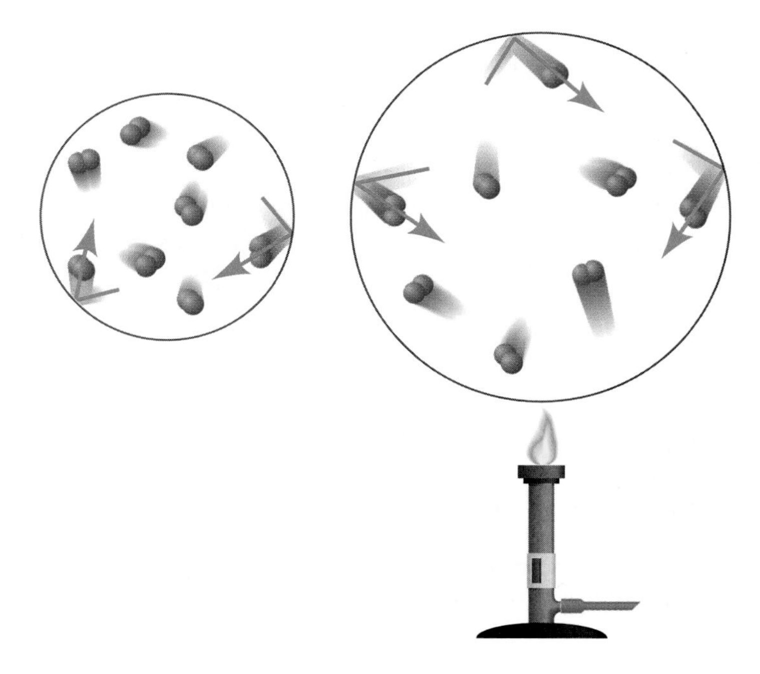

FIGURE 11.18
Heating a gas makes its molecules move faster, and they strike the walls of the container with greater force. For the pressure to remain constant, the volume must increase. The greater force of each collision with the walls of the container is counterbalanced by the smaller number of collisions within a given area.

## 11.5 Pressure, Temperature, and Volume Combined

**GOAL:** To understand and visualize the relationship between pressure and temperature, to perform calculations using this, and to perform calculations on samples when P, V, and T all change.

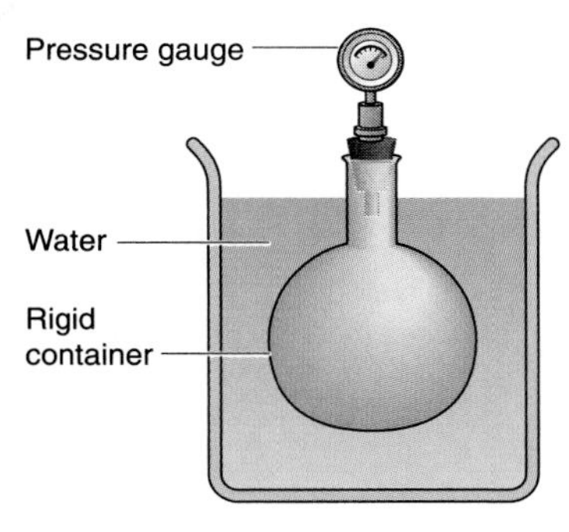


FIGURE 11.19
Apparatus for testing the relationship between pressure and temperature at constant volume. Because the container is rigid, the volume of the gas inside doesn't change significantly as the water is heated or cooled. As the temperature of the water changes, the pressure will change, which can be observed by reading the pressure gauge.

Guillaume Amontons (see Section 11.4) investigated the relationship between the pressure and the temperature of a gas as well as that between temperature and volume. These experiments required an apparatus having a constant volume, along with the same kind of water bath for regulating the temperature that was used in investigating Charles' Law (see Figure 11.19). The result of these experiments is that pressure and absolute temperature are directly proportional. That is, if a gas sample is heated in a rigid container, the pressure increases. The equation is:

$$\frac{P_1}{T_1} = \frac{P_2}{T_2}$$

where $P_1$ and $T_1$ are the initial pressure and absolute temperature of a sample of gas, and $P_2$ and $T_2$ are the final pressure and absolute temperature of the same sample. Once again, this equation is valid only if you use Kelvin temperatures rather than Celsius or Fahrenheit. Zero on the temperature scale has to be absolute zero. Theoretically, a gas at absolute zero exerts no pressure.

### EXAMPLE 11.6 Visualizing Changes in Gas Samples

**a.** A gas sample has a pressure of 700 torr at 0°C. What is its pressure at 100°C if the volume of the sample remains unchanged?

**b.** The atoms in a sample of argon gas in a rigid container are shown on the left below. The sample was cooled from 100°C to 10°C. Which of the four pictures on the right best represents the sample after cooling? Describe the major difference between the warmer and cooler samples.

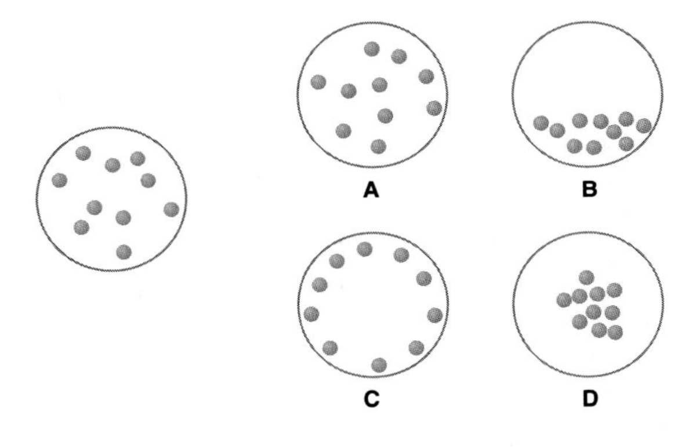

**SOLUTIONS**

**a.** This calculation uses the pressure-temperature equation given above. It is, of course, first necessary to convert the temperatures to the Kelvin scale: $P_1$ = 700 torr, $T_1$ = 273 K, $T_2$ = 373 K.

$$\frac{700 \text{ torr}}{273 \text{ K}} = \frac{P_2}{373 \text{ K}}$$

$$\frac{(700 \text{ torr})(373 \cancel{\text{K}})}{273 \cancel{\text{K}}} = P_2 = 956 \text{ torr}$$

**b.** **A**, best represents the sample after cooling. The atoms are moving more slowly in the cooler sample, on the average, but they are still spread throughout the volume of the container, as they are in all gas samples.

**EXERCISE 11.6**

A gas sample in a rigid container has a pressure of 1.0 atm and a temperature of 150°C. The sample cools to 25°C. What is its new pressure?

Is there an equation that expresses what happens if the pressure, volume, and temperature of a gas sample can all vary? Yes, there is. The combined equation should come as no surprise:

Combined Gas Law $$\frac{P_1V_1}{T_1} = \frac{P_2V_2}{T_2}$$

Here $P_1$, $V_1$, and $T_1$ are the initial pressure, volume, and absolute temperature of a sample of gas, and $P_2$, $V_2$, and $T_2$ are the final pressure, volume, and absolute temperature of the same sample. You can see that this equation contains all three of the gas laws we have discussed so far. For instance, if the temperature remains constant (that is, if $T_1 = T_2$), then the equation reduces to $P_1V_1 = P_2V_2$, which is Boyle's Law.

**EXAMPLE 11.7** **Calculating Pressure From Volume and Temperature**

A chemist placed an air sample that originally had a volume of 100 L, a pressure of 1.0 atm, and a temperature of 25°C in a 5.00-L container at 0°C. What is the pressure of the air sample in the new container?

**SOLUTION**

First, the temperatures must be converted to the Kelvin scale: 25°C = 298 K and 0°C = 273 K. Then you use the combined gas law:

$$\frac{P_1V_1}{T_1} = \frac{P_2V_2}{T_2}$$

$$\frac{(1.0 \text{ atm})(100 \text{ L})}{298 \text{ K}} = \frac{(P_2)(5.00 \text{ L})}{273 \text{ K}}$$

or

$$P_2 = \frac{(1.0 \text{ atm})(100 \text{ L})(273 \text{ K})}{(298 \text{ K})(5.00 \text{ L})} = 18 \text{ atm}$$

**EXERCISE 11.7**

A 100-L sample of air at 1.0 atm and 300 K was subjected to a pressure of 700 mm Hg at 273 K. What is the final volume of the air sample?

## 11.6 The Molar Volume of Gases

**GOAL:** To show that equal numbers of moles of all gases occupy the same volume at a given temperature and pressure.

Physicists and chemists understood the behavior of gases fairly well by 1800. The fact that all gases behave so similarly was very striking. Progress toward an explanation in molecular terms began from two starting points. One was the law of combining volumes, and the other was Gay-Lussac's proposal (Section 5.7) that when two gases react with each other, the volumes of gases that react have a ratio of small whole numbers.

Amadeo Avogadro was the one who saw the connection between combining volumes, atomic theory, and the general properties of gases. It made perfectly good sense to Avogadro that elements were composed of atoms and that atoms combined as units to form compounds. One of the reasons he liked Dalton's atomic theory so much is that he could use it to interpret Gay-Lussac's Law. The reasoning went like this. For the reaction between the gases carbon monoxide and oxygen, the balanced equation is

$$2\ CO(g) + O_2(g) \longrightarrow 2\ CO_2(g)$$

Chemists at the time knew that 2 L of $CO(g)$ react with 1 L of $O_2(g)$ to form 2 L of $CO_2(g)$; this is the law of combining volumes (see Figure 11.20). If in fact two *molecules* of $CO(g)$ react with one *molecule* of $O_2(g)$ to form two *molecules* of $CO_2(g)$, then it *must* be true that one volume of $CO(g)$ has the same number of molecules as one volume of $CO_2(g)$ and one volume of $O_2(g)$.

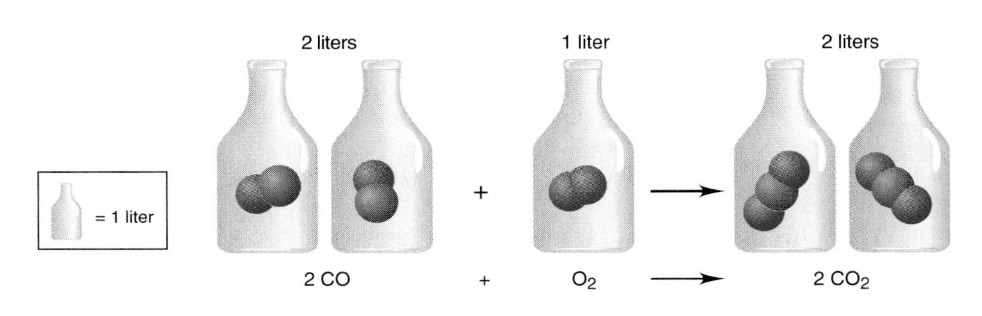


FIGURE 11.20
Two liters of carbon monoxide react with one liter of oxygen to produce two liters of carbon dioxide. The reactions of gases always behave this way; the ratio of the volumes of the gaseous reactants to each other and to the volumes of the gaseous products are ratios of simple whole numbers.

Avogadro published this idea in 1811. The modern wording goes like this: *Equal volumes of any gases under the same conditions of temperature and pressure contain the same number of particles.* Called **Avogadro's Hypothesis** in his honor, this idea withstood the test of time extremely well and is universally adopted today. In equation form Avogadro's Hypothesis is

$$\frac{V_1}{n_1} = \frac{V_2}{n_2}$$

where $V$ is a volume and $n$ is a number of moles of a gas.

**STUDY ALERT**

**Standard conditions of temperature and pressure (STP) are 273 K (0°C) and 1 atm. The molar volume of an ideal gas at STP is 22.4 L. For most instructors these are "must know" numbers.**

The molar mass can be used to find out how much volume a mole of gas occupies under certain circumstances. It's easy enough to adjust the pressure of a gas sample to 1 atm and the temperature to 0°C. These are called *standard conditions,* or **standard temperature and pressure (STP).** The volume of a mole of an ideal gas at STP, called the **standard molar volume,** is 22.414 L.

### EXAMPLE 11.8 Finding the Volume at Nonstandard Conditions

What is the volume of 0.500 mol of helium gas at 100°C and 1 atm?

**SOLUTION**

You know the volume of a mole of any gas at STP (22.414 L), and you know what STP means (1 atm pressure and 0°C). Since only the volume and the temperature (not the pressure) change, this can be treated as a Charles' Law problem, so you use:

$$\frac{V_1}{T_1} = \frac{V_2}{T_2}$$

$T_1 = 273$ K and $T_2 = 373$ K, but what about $V_1$? You can get $V_1$ by using the unit factor method; the molar volume at STP is the conversion factor.

$$0.500 \text{ \sout{mol He}} \left(\frac{22.414 \text{ L He}}{1 \text{ \sout{mol He}}}\right) = 11.2 \text{ L He} = V_1$$

$$\frac{11.2 \text{ L He}}{273 \text{ K}} = \frac{V_2}{373 \text{ K}}$$

$$V_2 = \frac{(11.2 \text{ L He})(373 \text{ \sout{K}})}{273 \text{ \sout{K}}} = 15.3 \text{ L He}$$

**EXERCISE 11.8**

What is the volume of 0.250 mol of argon gas at 400 K and 2.00 atm?

## 11.7 The Ideal Gas Law

**GOAL:** To use the ideal gas law to solve problems involving gases.

So far we've discussed gases in terms of four variables, pressure ($P$), volume ($V$), temperature ($T$), and amount ($n$, the number of moles in the sample). The Combined Gas Law tells us that $PV/T$ has a fixed value for any gas. That is, for any gas:

$$\frac{PV}{T} = k$$

where $k$ is a constant. From Avogadro's Hypothesis we know that at a fixed temperature and pressure, the volume of a gas sample is proportional to the number of moles of gas present. This means that the value of the constant $k$ depends on how many moles of gas there are. In equation form, $k = nR$, where $n$ is the number of moles of gas and $R$ is a different constant. Thus, for any gas:

$$\frac{PV}{T} = nR$$

This is called the **ideal gas equation** (for reasons we'll talk about later). Chemists usually write it in this form:

$$PV = nRT$$

where $P$ is the pressure, $V$ is the volume, $n$ is the number of moles of gas, and $T$ is the absolute temperature; $R$ is the *ideal gas constant,* which is, expressed in the units most often used by chemists:

$$R = 0.08206 \frac{\text{L atm}}{\text{mol K}} = 0.08206 \text{ L atm mol}^{-1} \text{ K}^{-1}$$

With the ideal gas law you can use the measured volume, pressure, and temperature of a sample of gas to find out how many moles it contains. Because $R$ is almost always given in units of L atm $\text{mol}^{-1}$ $\text{K}^{-1}$, you must use liters for volume, atmospheres for pressure, and kelvins for temperatures in all of your calculations with this value of $R$.

**Remember: You have to use absolute (Kelvin) temperatures to get the right answers with the equations in this chapter.**

Volume (L) — Temperature (K)

$$PV = nRT$$

Pressure (atm) — Number of moles — Gas constant (L atm $\text{mol}^{-1}\text{K}^{-1}$)

## EXAMPLE 11.9 Calculating Moles with the Ideal Gas Equation

**a.** A large gas cylinder has a volume of 125 L. How many moles of nitrogen gas does this cylinder hold at room temperature (20°C) and a pressure of 195 atm?

**b.** How many kg of nitrogen does this cylinder contain?

**SOLUTIONS**

**a.** The absolute temperature is 273 + 20 = 293 K. The ideal gas equation gives:

$$PV = nRT$$
$$(195\ \text{atm})(125\ \text{L}) = n(0.0821\ \text{L atm mol}^{-1}\ \text{K}^{-1})(293\ \text{K})$$
$$n = 1.01 \times 10^3\ \text{mol}$$

There are $1.01 \times 10^3$ mol of nitrogen gas present in the cylinder.

**b.** The formula of nitrogen gas is $N_2$, and its molar mass is 28.0 g $\text{mol}^{-1}$.

$$1.01 \times 10^3\ \text{mol}\left(\frac{28.0\ \text{g}}{1\ \text{mol}}\right) = 2.84 \times 10^4\ \text{g} = 28.4\ \text{kg}$$

This looks like a large amount, but it's only about 60 pounds. Many adults could pick this up and carry it. The metal cylinder itself weighs much more.

**EXERCISE 11.9**

A reaction produces pure carbon dioxide gas, and a chemist withdraws exactly 2.00 mL at STP from the reaction flask with a hypodermic syringe. How many moles of $CO_2$ were collected?

You can calculate the molar mass of a gas by using the Ideal Gas Law if you have enough information. You know that the molar mass has units of g $\text{mol}^{-1}$. The number of moles is a quantity you can get from the ideal gas equation, and the mass of the gas is usually given (because it's fairly easy to determine experimentally).

### EXAMPLE 11.10 Determining Molar Mass from the Ideal Gas Equation

A 13.3-g sample of a gas occupied a volume of 7.42 L at 20°C and a pressure of 740 mm Hg. What is the molar mass of this gas?

**SOLUTION**

How can you get an answer? You know that molar mass is grams divided by moles. You have the number of grams (13.3), and you need to know the number of moles. You can use the ideal gas equation to calculate the number of moles. Before you can do that, you have to put everything in the appropriate units: 20°C = 293 K, and 740 mm Hg = 0.974 atm.

$$PV = nRT$$
$$(0.974\ \text{atm})(7.42\ \text{L}) = n(0.0821\ \text{L atm mol}^{-1}\ \text{K}^{-1})(293\ \text{K})$$
$$7.227\ \text{mol} = n(24.055)$$
$$n = 0.300\ \text{mol}$$

Therefore, 0.300 mol of the gas has a mass of 13.3 g, and the molar mass (in g $mol^{-1}$) is

$$\frac{13.3 \text{ g}}{0.300 \text{ mol}} = 44.3 \text{ g mol}^{-1}$$

**EXERCISE 11.10**

A gas sample is either $CO_2$ or CO. To identify the gas, a chemist determined that 22 g of the gas occupies 12.2 L at 1.00 atm and 300 K. What is the gas, $CO_2$ or CO?

Another common use of the ideal gas equation is to compute the densities of gases under various conditions. For example, in Exercise 11.10, 22 g of an unknown gas occupies a volume of 12.2 L; so at 1 atm and 300 K, this gas has a density of 22 g/12.2 L = 1.8 g $L^{-1}$. In most gas density problems, you are given the mass and you calculate the volume from the ideal gas equation. Example 11.11 presents a practical application.

**The density of a gas usually has units of g/L, or g $L^{-1}$.**

## EXAMPLE 11.11 Calculating Gas Density

What is the density of methane, $CH_4$, at 300 K and 1.00 atm?

**SOLUTION**

You can pick any quantity of methane and use the ideal gas equation to compute its volume. Like most practicing scientists, however, you want to avoid unnecessary work. If you find the molar volume, it reduces the number of steps in the calculation. You get the molar volume from the ideal gas equation, $PV = nRT$:

$$V(1.00 \text{ atm}) = (1 \text{ mol})(0.0821 \text{ L atm mol}^{-1} \text{ K}^{-1})(300 \text{ K})$$

$$V = 24.6 \text{ L}$$

(Note that conditions are not at STP, 1 atm, and 273 K, so 22.4 L is not the correct molar volume.) Since 1 mol of methane has a mass of 16 g, the density of methane under these conditions is

$$\frac{16 \text{ g}}{24.6 \text{ L}} = 0.65 \text{ g L}^{-1}$$

**EXERCISE 11.11**

What is the density of the Freon $CCl_2F_2$ at STP?

Of course, you can calculate any of the four variables in the ideal gas equation ($P$, $V$, $n$, or $T$) if the other three are known. This kind of calculation is common, and it comes in many disguises.

### EXAMPLE 11.12 Ideal Gas Law Calculations

When Charles constructed the first hydrogen-filled balloon, he prepared hydrogen gas from the reaction between sulfuric acid and iron:

$$2\ H_2SO_4(aq) + Fe(s) \longrightarrow Fe(HSO_4)_2(aq) + H_2(g)$$

What volume of hydrogen gas is produced from excess sulfuric acid and 50 mol of iron at 750 mm Hg and 25°C?

**SOLUTION**

This kind of problem appears complicated, but it involves nothing more than a series of steps you already know how to carry out. The first task is developing a road map to the solution. You want to calculate the volume of a gas, and you are given the temperature and the pressure. To use the ideal gas equation, you also need to know the number of moles of hydrogen, so you start there.

Since $H_2SO_4(aq)$ is in excess, $Fe(s)$ is the limiting reactant. The balanced equation shows that 1 mol $Fe(s)$ yields 1 mol $H_2(g)$, and therefore the reaction produces 50 mol $H_2(g)$.

Next, you need to get everything in the correct units:

$$P = 750\ \text{mm Hg}\left(\frac{1\ \text{atm}}{760\ \text{mm Hg}}\right) = 0.987\ \text{atm}$$

$$T = 25°\text{C} = 25 + 273 = 298\ \text{K}$$

Now you can solve the ideal gas equation $PV = nRT$ for the volume of hydrogen gas. Remember that the equations of this chapter work only for gases.

$$(0.987\ \cancel{\text{atm}})\,V = (50\ \cancel{\text{mol}})\,(0.0821\ \text{L}\ \cancel{\text{atm mol}^{-1}\ \text{K}^{-1}})\,(298\ \cancel{\text{K}})$$

$$V = 1.2 \times 10^3\ \text{L}$$

Note that this is just a little over 1 $m^3$, not nearly enough hydrogen gas to fill a balloon that carries a human passenger. Charles used much more iron than this when he filled his balloon.

**EXERCISE 11.12**

If Charles' balloon had a volume of 1000 $m^3$, how many kilograms of iron would he need to fill it with hydrogen gas at 300 K and 1.00 atm if he used the reaction given in Example 11.12? (Note: 1 $m^3$ = 1000 L.)

## 11.8 When More Than One Gas Is Present: Partial Pressures

**GOAL:** To state Dalton's Law of partial pressures and use the concept to solve problems involving gaseous mixtures.

With the exception of molar mass calculations, the equations of the previous sections can be used for either pure gases or mixtures of gases, such as air. In all the experiments discussed so far in this chapter, the gases behave exactly the same way, regardless of their identity. Pure gases follow Boyle's Law, and so do gaseous mixtures. In fact, Boyle used air, a mixture of gases, in his experiments, and so did Charles and Gay-Lussac. The ideal gas equation contains nothing that identifies a particular gas, only the number of moles (and therefore the number of molecules) of gas present. This tells you that the molecules of each substance in a mixture of gases act independently of all other molecules in the mixture.

## EXAMPLE 11.13 Microscopic Appearance of a Gas Mixture

Which of the following *best* represents a container of air? The molecules shown in red represent oxygen, and those in blue represent nitrogen .

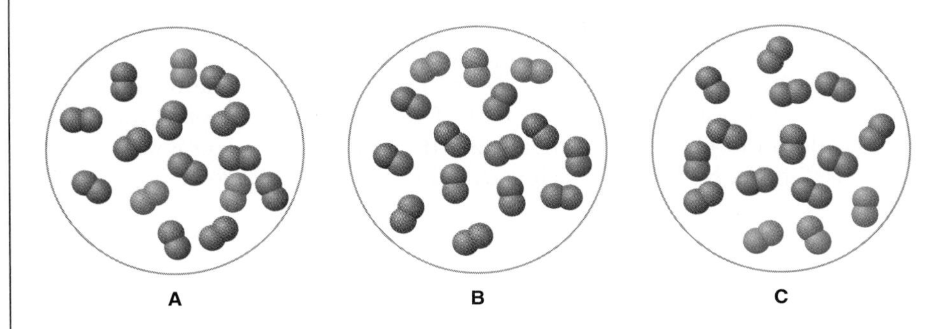


### SOLUTION

Diagram A shows the molecules distributed randomly throughout the container and is therefore the best representation. You already know that gases expand to fill their containers and that any sample of gas is mostly empty space. Any enclosed mixture of gases will quickly become completely homogeneous at any level that can be perceived.

You might reason that oxygen molecules are heavier than nitrogen molecules (32 g $mol^{-1}$ versus 28 g $mol^{-1}$) and that oxygen molecules therefore should tend to settle toward the bottom of the container, as in diagram C. However, as we will discuss in the next two sections, gas molecules are in constant random motion because of their thermal energy. Near room temperature, the average speed of gas molecules is too high (around 400 m $s^{-1}$, or 900 mi $h^{-1}$) for any particular group of molecules to settle to the bottom.

### EXERCISE 11.13

Most people assume that the atmosphere is homogeneous.

**a.** List evidence that demonstrates that this isn't true.

**b.** Why do people assume that the atmosphere is homogeneous?

**Yes, this is the same John Dalton who first successfully applied the atomic theory to chemistry. He discovered the Law of Partial Pressures while studying the weather.**

When a sample consists of a mixture of gases, each individual gas exerts a pressure called its **partial pressure.** John Dalton discovered that the total pressure of the mixture is equal to the sum of the partial pressures of the individual gases. This statement, now known as **Dalton's Law of Partial Pressures,** is expressed by the equation

$$P_{\text{total}} = P_1 + P_2 + P_3 + P_4 \cdots$$

**Every liquid has a unique vapor pressure at any given temperature. Solids do, too, but the vapor pressures of solids are almost always so small that they are ignored.**

Dalton wanted to know how much water a given volume of air could absorb at different temperatures. He found that the temperature was all that mattered. The total pressure and the identities and amounts of the other gases (as long as they did not react with water) had nothing to do with the amount of water the air could hold. The vapor pressure of water—the partial pressure of water vapor in contact with liquid water—depends only on temperature. (See Table 11.2.)

You have heard TV meteorologists give the relative humidity as part of the weather report. The **relative humidity** is the ratio of the partial pressure of the water vapor in the atmosphere to the pressure of the amount of water vapor that would saturate the atmosphere at the same temperature. This ratio is usually given as a percentage. For example, if the relative humidity is 50%, the atmosphere is holding half of the total amount of water it could hold. (It absolutely does not mean that the atmosphere is 50% water vapor.) One use of relative humidity is illustrated in Example 11.14.

**TABLE 11.2 Vapor Pressure of Water at Various Temperatures**

| Temperature (°C) | Vapor pressure (mm Hg) |
|---|---|
| 0 | 4.579 |
| 10 | 9.209 |
| 20 | 17.535 |
| 30 | 31.824 |
| 40 | 55.324 |
| 50 | 92.51 |
| 60 | 149.38 |
| 70 | 233.7 |
| 80 | 355.1 |
| 90 | 525.76 |
| 100 | 760.00 |
| 110 | 1074.56 |
| 120 | 1489.14 |

### EXAMPLE 11.14 Partial Pressures and the Percentage of Water in the Atmosphere

It's an ordinary late fall day along the coast of Antarctica (Figure 11.21), with a temperature of −50°C and a relative humidity of 50%. The vapor pressure of ice is 0.030 mm Hg at that temperature. What is the percentage of water in the atmosphere? Assume a total pressure of 1.0 atm.

**SOLUTION**

The relative humidity of 50% means that the partial pressure of the water in the atmosphere is exactly half the vapor pressure of pure water at the given temperature, which is −50°C. The partial pressure of the water vapor actually present in the atmosphere is then

$$(0.030 \text{ mm Hg})(0.50) = 0.015 \text{ mm Hg}$$

The total pressure is 1.0 atm = 760 mm Hg. The percentage of water in the air is

$$\frac{0.015 \text{ mm Hg}}{760 \text{ mm Hg}} \times 100\% = 0.000020 \times 100\% = 0.0020\%$$

There isn't much water in the air under these circumstances.

FIGURE 11.21
Antarctica

**EXERCISE 11.14**

It's a hot, rainy summer day in Houston (Figure 11.22), with a temperature of 37°C and a relative humidity of 100%. The vapor pressure of water is 47.1 mm Hg at that temperature. What is the percentage of water in the air? Assume a total pressure of 1 atm.

FIGURE 11.22
Houston, Texas

The point of Dalton's Law of Partial Pressures is that the pressure exerted by each gas in a mixture depends only on the amount (that is, the number of moles) of that gas present. This becomes obvious if we substitute $nRT/V$ for the pressure of each of the gases in Dalton's equation. Since $V$, $T$, and $R$ are identical for all gases present, they factor out:

$$P_{total} = P_1 + P_2 + P_3 + P_4 + \cdots$$

$$P_{total} = \frac{n_1RT}{V} + \frac{n_2RT}{V} + \frac{n_3RT}{V} + \frac{n_4RT}{V} + \cdots$$

$$= \frac{(n_1 + n_2 + n_3 + n_4 + \cdots)RT}{V}$$

This leaves us with the conclusion that the partial pressure of each gas in a mixture depends only on the number of moles of that gas. For example, if the atmosphere is 21% oxygen, 78% nitrogen, and 1% argon (mol percent), the partial pressure of each gas is 0.21 atm for oxygen, 0.78 atm for nitrogen and 0.01 atm for argon. This kind of analysis is useful for solving certain problems, as illustrated by Example 11.15.

**EXAMPLE 11.15** **Using Dalton's Law of Partial Pressures**

Helium (1.0 L) and nitrogen (2.0 L), both at 1.0 atm, were placed in an evacuated 10-L flask. What is the partial pressure of each gas in the flask, and what is the total pressure, assuming the temperature is held constant?

**SOLUTION**

For this type of problem there are many roads to the answer. One way is to calculate the partial pressure of each gas from Boyle's Law and then get the total pressure from Dalton's Law of Partial Pressures:

$$P_{He} = \frac{V_1P_1}{V_2} = \frac{(1.0\text{ L})(1.0\text{ atm})}{10\text{ L}} = 0.10\text{ atm}$$

$$P_{N_2} = \frac{V_1P_1}{V_2} = \frac{(2.0\text{ L})(1.0\text{ atm})}{10\text{ L}} = 0.20\text{ atm}$$

$$P = 0.10\text{ atm} + 0.20\text{ atm} = 0.30\text{ atm}$$

Another approach is to recognize that you start off with a total of 3 L of gas. According to Avogadro's Hypothesis, ⅓ of the gas particles are

He atoms and ⅔ are $N_2$ molecules. The volume increases to 10 L after mixing, so Boyle's Law indicates that the total pressure is (3.0 L)(1.0 atm)/(10 L) = 0.30 atm, the same result you calculated above. Because the partial pressure of each gas is directly proportional to the number of molecules of that gas present, the partial pressure of He is (⅓)(0.30 atm) = 0.10 atm and the partial pressure of $N_2$ is (⅔)(0.30 atm) = 0.20 atm.

**EXERCISE 11.15**

A 12-L flask immersed in a water bath at 27°C contains 2.0 g of $H_2$, 4.0 g of $CH_4$, and 2.0 g of $NH_3$. What are the partial pressures of these three gases? What is the total pressure?

## 11.9 Molecules in Motion: Diffusion and Effusion

**GOAL:** To describe the processes of diffusion and effusion and explain how molecular mass affects the rate of these processes.

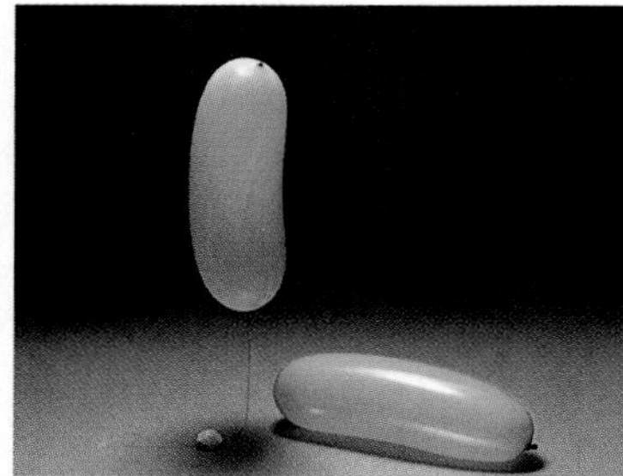

FIGURE 11.23
Helium, with an atomic mass of 4 amu, effuses through the walls of balloons much more rapidly than air does. This is why helium balloons lose their buoyancy so quickly.

By the 1820s chemists and physicists had established the fact that gases (and all other substances) are made of atoms and molecules. The fact that hydrogen moves out of a flask that has a hairline crack faster than air moves into the same flask caught the eye of a young Scottish chemist named Thomas Graham (1805–1869). Gas molecules move very rapidly, so Graham slowed them down by forcing them to work their way through such barriers as plaster of Paris plugs, very narrow tubes, and tiny holes in platinum sheets. In other words, Graham studied **diffusion**—the rate at which gases mix—and **effusion**—the rate at which a gas escapes under pressure through a tiny hole. (See Figure 11.23.)

Figure 11.24 shows visible evidence of variation in diffusion rates: HCl(*g*) diffused from one end of the tube (the left end) and $NH_3$(*g*) diffused from the other, forming a white cloud of solid $NH_4Cl$ particles where the two gases

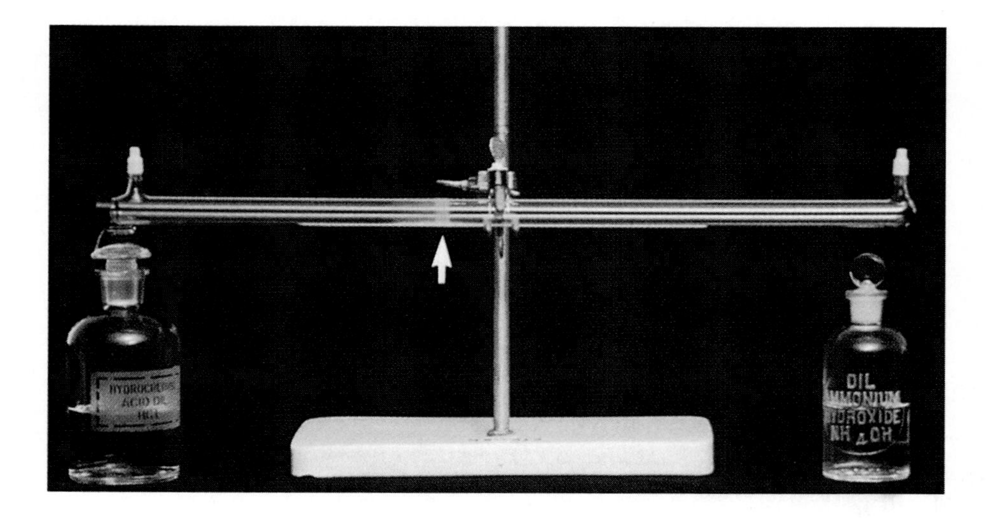


FIGURE 11.24
Apparatus for showing the relative rates of diffusion of $NH_3$ and HCl. The white cloud of $NH_4Cl$ forms closer to the HCl end, demonstrating that HCl diffuses more slowly.

## EXPERIMENTAL EXERCISE 11.1

The scientific team of A. L. Chemist was given a sample of gas to analyze. They were told by an unreliable source that the sample, labeled gas 1, was a mixture of $O_2$ and $N_2$, and their task was to find out the actual composition of the mixture.

The team performed three experiments. First, they mixed a sample of gas 1 with excess C(*s*) in a sealed container and strongly heated it (see Figure). All the oxygen in the mixture reacted with the C(*s*). The team called the mixture of gases collected at the end of this reaction gas 2.

In the second experiment the team passed a 10.00-L sample of gas 2 over CaO(*s*). The CaO absorbed all the $CO_2$(*g*), forming $CaCO_3$ and leaving 8.00 L of gas 3. (The volumes of both gas samples were measured at STP.) The team concluded that gas 1 contains 20% by volume $O_2$.

In the third experiment the team strongly heated the 8.00-L sample of gas 3 in the presence of excess Mg(*s*). All of the nitrogen reacted to form $Mg_3N_2$(*s*), leaving 0.10 L of gas 4. Gas 4 had a density of 1.79 g $L^{-1}$. Again all gas volumes and the density were measured at STP. The team concluded that gas 1 was simply a sample of air and was a mixture of $O_2$, $N_2$, and Ar.

**QUESTIONS**

1. Why is the right-hand vent left open when the reaction vessel in the first experiment is charged with gas 1?
2. Why are both vents closed in the first experiment, in which gas 1 is heated with C(*s*)?
3. Why is the left-hand vent closed when gas 2 is removed from the reaction chamber in the first experiment?
4. How many liters of gas 1 are necessary to make 10 L of gas 2?
5. How many moles of $CO_2$(*g*) form when 50 L of gas 1 measured at STP reacts with excess C(*s*)?
6. Justify the team's conclusion that gas 4 is argon.
7. What is the limiting reactant in the reaction between Mg and $N_2$?
8. Which of the following amounts of CaO should be used in the second experiment?
   A. 1.0 g   B. 10 g   C. 100 g   D. 1.0 kg
9. What mass of $Mg_3N_2$ forms in the third experiment?

**First Experiment**

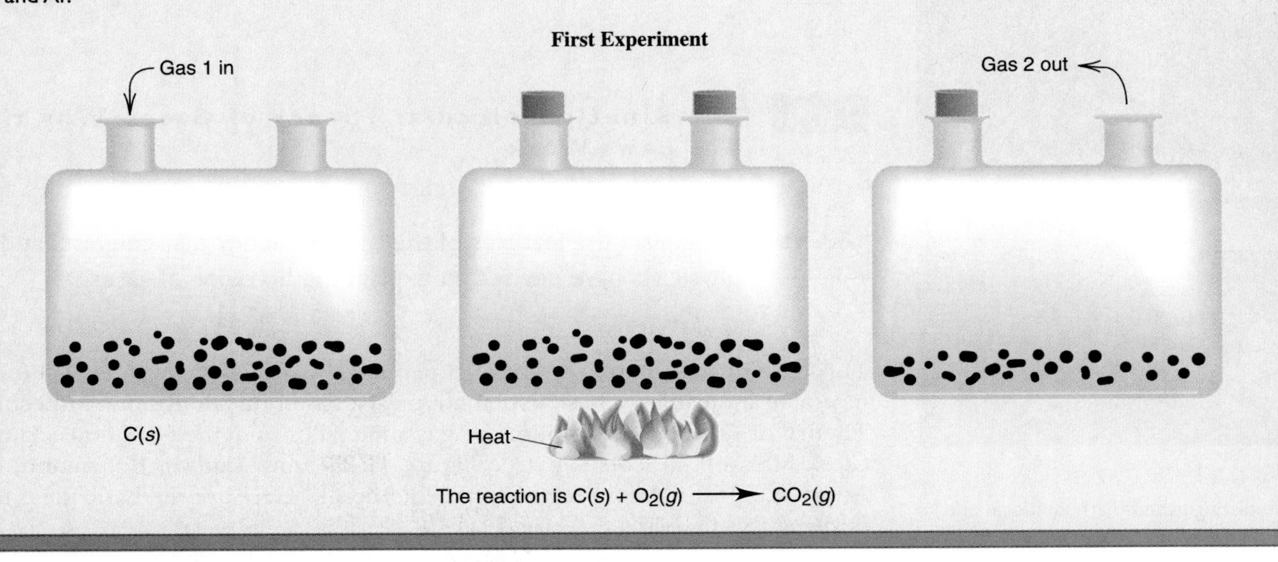


met and reacted. The cloud of particles forms closer to the HCl end than to the $NH_3$ end. In other words, the $NH_3$ diffuses faster than the HCl does.

Graham, of course, measured the distances the molecules traveled. From such experiments he found that the rates of diffusion and effusion are inversely proportional to the square root of the density of the gas. In more modern terms, the rates of diffusion and effusion are inversely proportional to the square root of the molar mass of the gas:

$$\text{Rate of diffusion} = \frac{\text{constant}}{\sqrt{\text{Density of the gas}}} = \frac{\text{constant}}{\sqrt{\text{Molar mass}}}$$

The value of the constant (and the rate of diffusion) changes with temperature.

### EXAMPLE 11.16 Relative Rates of Diffusion

Which of these gases diffuses most rapidly? Explain your answer.
A. Oxygen B. Nitrogen C. Neon D. Argon

**SOLUTION**

The gas that diffuses most rapidly is the one with the smallest molar mass. The molar masses are oxygen ($O_2$), 32 g mol$^{-1}$; nitrogen ($N_2$), 28 g mol$^{-1}$; neon (Ne), 20 g mol$^{-1}$; and argon (Ar), 40 g mol$^{-1}$. Neon (C) is the one that diffuses most rapidly.

**EXERCISE 11.16**

Which of these gases diffuses most slowly? Explain your answer.
A. Oxygen B. Nitrogen C. Neon D. Argon

## 11.10 The Kinetic Molecular Theory of Gases: Why the Gas Laws Work

**GOAL:** To explain the features of the overall theory that chemists and physicists have devised to explain the behavior of gases.

FIGURE 11.25
The distinguished mathematician and physicist James Clerk Maxwell (1831–1879) was renowned for putting ideas in rigorous mathematical terms. He took ideas about electricity and magnetism and put them in the form still used today and known as Maxwell's Equations.

Throughout this chapter we have explained the gas laws in terms of the behavior of molecules. These explanations are based on the Kinetic Molecular Theory of Gases, a description of gas molecules and their motions. James Clerk Maxwell, a Scotsman (see Figure 11.25), and Ludwig Boltzmann, an Austrian, put the theory together independently. Here are the basic ideas underlying this theory:

1. Any sample of a gas is composed of many particles that act like tiny hard objects.
2. The particles are much smaller than the spaces between them. Most of any gas sample is empty space.
3. The particles move in straight lines until they collide with each other or the walls of the container. They are in constant, random motion.

4. There is no force of attraction between gas molecules or between the molecules and the wall of the container.
5. None of the energy of a gas particle is lost when it collides with another molecule or with the wall of the container. That is, the collisions are *elastic*. One particle may gain kinetic energy while another loses it, but the total energy remains constant. (Bouncing balls are different. They eventually give up their kinetic energy as heat and roll to a stop.)
6. The average kinetic energy of a sample of gas particles depends on the absolute temperature of the gas, and nothing else.

The sixth statement explains the mathematical form of Graham's results on rates of diffusion. The equation for the kinetic energy of a particle is

**Kinetic energy of a moving object = KE =** $\frac{1}{2}mv^2$

where $m$ is the mass of the particle and $v$ is its velocity. If you rearrange this equation, you get:

$$\frac{2(\mathrm{KE})}{m} = v^2$$

$$\text{or } v = \sqrt{2(\mathrm{KE})/m}$$

In other words, the velocity of a gas particle is inversely proportional to the square root of the mass of the particle.

Graham's Law is probably the most difficult of the gas laws to explain using the Kinetic Molecular Theory of Gases. Example 11.17 offers some easier explanations.

### EXAMPLE 11.17 Explaining Gas Behavior Using the Kinetic Molecular Theory of Gases

**a.** You add more gas to a rigid container (that is, volume is constant) at room temperature. The pressure increases. Explain the reason for this behavior using the Kinetic Molecular Theory of Gases.

**b.** You seal the needle of a syringe and pull the plunger out, increasing the volume of a gas sample inside. This is all done at room temperature. What happens to the pressure inside the syringe? Explain the reasoning behind your prediction.

**SOLUTIONS**

**a.** With more gas particles inside the container, more of them will hit any unit area of the walls during a given time period. With more collisions and the same average energy per collision (because the temperature is constant), the force on the walls and therefore the pressure must increase.

**b.** The temperature is constant, so the average energy of the particles (and the force of their collisions with the walls of the syringe) is also constant. The area of the walls of the syringe and the distance molecules must travel along its barrel are both larger now, so the pressure (force per unit area) must decrease.

**EXERCISE 11.17**

a. You heat a rigid, sealed container of gas. What happens to the pressure inside the container? Explain the reasoning behind your prediction.
b. You have a round, sealed, metal container of gas in a cold water bath. You hit the container with a hammer and put a lot of dents in it. The pressure inside the container increases though the temperature remains constant. Explain why this occurs using the Kinetic Molecular Theory of Gases.

## CHAPTER BOX 11.4 THE MYTH OF THE IDEAL GAS

**An** *ideal gas* is a mythical beast, defined as a gas that obeys the Ideal Gas Law at all temperatures and pressures. The Ideal Gas Law gives an approximation (and a very good one) for the behavior of real gases under the conditions you are likely to find in the laboratory, factory, or home.

For real gases, deviations from ideal behavior occur at the extremes of temperature and pressure. For instance, at very low temperatures, gas molecules have small kinetic energies and low velocities. When the molecules collide, they have so little kinetic energy that intermolecular forces of attraction can take hold. Under these conditions, there can be a significant force of attraction between molecules. If some pairs of molecules stick together for an instant after they collide, the effect is to make the pressure lower than that predicted by the Ideal Gas Law. When the average kinetic energy becomes low enough, the intermolecular forces take over and the gas condenses into a liquid. All real gases condense to liquids at temperatures above absolute zero, and therefore all real gases must have intermolecular forces.

At relatively high densities (caused by low temperatures, high pressures, or both), the average distance between gas molecules decreases, and the molecules themselves occupy a significant fraction of the total volume. This deviation causes the volume occupied by real gas samples to be larger than predicted by the Ideal Gas Law.

A Dutch physicist, Johannes van der Waals, studied these deviations and developed an equation to describe the behavior of real gases. It is similar to the ideal gas equation but contains additional terms to account for the nonideal conditions.

## EXPERIMENTAL EXERCISE

## 11.2

A. L. Chemist's usual team measured the speed of gas molecules using an apparatus like the one in Figure A. They heated a gas sample to a specific temperature in the oven and then allowed the gas molecules to effuse through the exit port. The effusing molecules encounter a series of rotating disks. Each disk has a slit that is offset from the slit in the preceding disk by the same angle. As the disks rotate, each slit arrives successively at the same position relative to the stream of gas molecules. If the gas molecules travel the distance between disks in the same amount of time as it takes for the next slit to get into position, the molecules pass through all the slits and strike the detector. Molecules traveling at any other speed will not arrive at each disk at the same time as the slit lines up and will not pass through.

The team calculated the speed of the molecules arriving at the detector from the speed of rotation of the disks. Then the team varied the speed of disk rotation and measured the number of molecules arriving at the detector for each speed.

The team measured the speed of $CO_2$, $CH_4$, and $H_2$ molecules at 300 K and the speed of $N_2$ molecules at 273 K, 1273 K, and 2273 K. Their results appear in Figure B. The team offered these three general conclusions:

1. The molecules of a gas have a distribution of speeds. The most probable speed is the one at the peak of each curve.
2. Molecular speed depends on the molecular mass. Fast-moving molecules have a smaller molecular mass than slow-moving molecules.
3. Molecular speed depends on temperature. Molecules at high temperatures have higher speeds on the average than molecules at low temperatures.

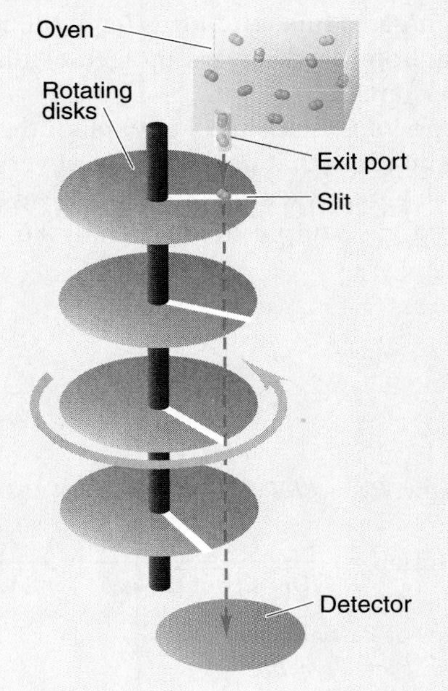


FIGURE A

### QUESTIONS

**1.** What would be an appropriate diameter for the exit port in Figure A?
A. 0.1 cm B. 1.0 cm C. 10 cm D. 100 cm

**2.** Name at least one variable, other than the speed at which the disks rotate, that the scientists must know to calculate the speed of the molecules.

**3.** What is the most probable speed for $CO_2$ at 300 K?

**4.** Draw a plot similar to those shown in Figure B for $NH_3$ at 300 K.

**5.** Draw a plot similar to those shown in Figure B for $N_2$ at 3273 K.

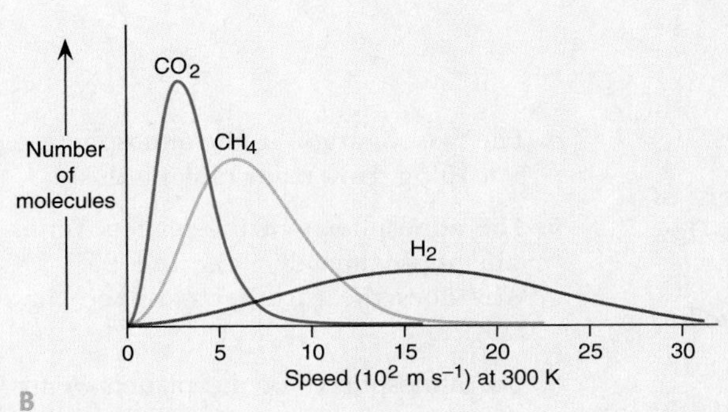


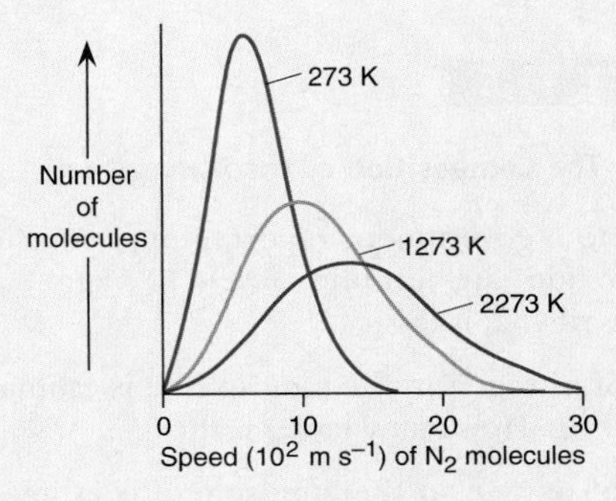


FIGURE B

## CHAPTER SUMMARY

1. The atmosphere is a mixture of gases consisting primarily of nitrogen (78 mol %), oxygen (20%), and argon (1%). Carbon dioxide and water vapor are important minor constituents. The atmosphere is homogeneous if relatively small volumes are considered.
2. Pressure is force per unit of area. Gas pressure results from the force of collisions between gas molecules and the sides of the container.
3. Boyle's Law states that the pressure of a sample of any gas times its volume remains the same as long as the temperature is constant.
4. Charles' Law states that there is a direct relationship between the volume of a sample of gas and its absolute temperature at constant pressure.
5. Standard conditions of temperature and pressure (STP) are 273 K and 1 atm. The molar volume of an ideal gas at STP is 22.4 L.
6. The ideal gas equation is $PV = nRT$, where the gas constant, $R$, has a value of 0.0821 L atm $mol^{-1}$ $K^{-1}$. This equation can be used to calculate any one of the four variables if the other three are known.
7. Diffusion is the rate at which gases mix, and effusion is the rate at which gases escape through a tiny hole. Heavier gas molecules diffuse and effuse more slowly than lighter ones.
8. The Kinetic Molecular Theory of Gases is used to explain the behavior of gas samples. It is based on the ideas that gases consist of very small, hard particles bouncing around in a near vacuum and that the average kinetic energy of the molecules in a gas sample depends only on the absolute temperature.

## IMPORTANT EQUATIONS

Boyle's Law: $P_1V_1 = P_2V_2$ (for a sample at constant temperature)

Charles' Law: $V_1/T_1 = V_2/T_2$ (for a sample at constant pressure)

Combined Gas Law: $P_1V_1/T_1 = P_2V_2/T_2$ (for a given sample)

Ideal Gas Law: $PV = nRT$

$$\text{Rate of diffusion} = \frac{\text{constant}}{\sqrt{\text{Density of the gas}}} = \frac{\text{constant}}{\sqrt{\text{Molar mass}}}$$

Dalton's Law of Partial Pressures:
$P_{TOT} = P_1 + P_2 + P_3 + P_4 + \cdots$

## PROBLEMS

### SECTION 11.1 The Composition of the Atmosphere

1. According to a government report, global emissions of carbon dioxide are around $6.0 \times 10^{12}$ kg $y^{-1}$. How many moles of $CO_2$ is this?

2. The mass of nitrogen in the atmosphere is estimated to be $4.0 \times 10^{21}$ g. How many moles is this?

3. The mass of oxygen in the atmosphere is estimated to be $1.2 \times 10^{21}$ g. How many moles is this?

4. The mass of argon in the atmosphere is estimated to be $5 \times 10^{19}$ g. How many moles is this?

5. The atmospheres of the planets Venus and Mars contain approximately 96% and 95% $CO_2$, respectively. Why does the earth's atmosphere have so much less $CO_2$?

6. The atmospheres of the planets Venus and Mars contain almost no detectable $O_2$. Why does the earth have so much more $O_2$ in its atmosphere?

7. The average temperature of the moon is slightly lower than the average temperature of the earth. What effect is responsible for this?

8. The planet Venus is hot, much hotter than expected for its distance from the sun. What effect is responsible for this? (Hint: See Problem 5.)

9. Which of the following gases produces environmental effects by absorbing low-energy radiation as it leaves the earth?
   A. Nitrogen  C. Ozone
   B. Methane  D. Argon

10. Which of the following gases produces environmental effects by absorbing high-energy radiation as it approaches the earth?
    A. Methane  C. Oxygen
    B. Nitrogen  D. Ozone

**SECTION 11.2 Pressure**

11. Suppose some gas molecules occupy a closed container. Which of these actions increases the pressure inside the container?
    A. Making the molecules move faster
    B. Expanding the size of the container
    C. Removing molecules from the container
    D. Changing the shape of the container

12. Suppose some gas molecules occupy a closed container. Which of these actions decreases the pressure inside the container?
    A. Adding more molecules to the container
    B. Changing the shape of the container
    C. Slowing down the molecules
    D. Decreasing the size of the container

13. A. L. Chemist read a barometer on a hot, clear summer day in New Orleans. The pressure was 30.15 in Hg. What is this pressure in mm Hg and in atmospheres?

14. An airplane recorded a barometric pressure of 26.91 in Hg in the eye of a hurricane. What is this pressure in mm Hg and in atmospheres?

15. The pressure in a filled scuba tank is 3000 psi. What is this pressure in atmospheres and in mm Hg?

16. An open-water scuba diver saw that the pressure gauge of his tank read 500 psi and decided to go to the surface before his air ran out. What is this pressure in mm Hg and in atmospheres?

17. It was a cold day on Mars, and the barometric pressure was 0.006 atm. What is this pressure in mm Hg?

18. It's a hot day on the planet Venus (but all days on Venus are hot). The barometric pressure is 90 atm. What is this pressure in mm Hg?

19. An open-end manometer is a device used to measure the pressure of a gas by comparing it to the atmospheric pressure. A U-shaped tube containing mercury connects to a flask containing the gas sample, as shown. To start, the mercury levels on the sides of the U-tube are equal. The stopcock is opened, and the pressure of the gas is measured. Redraw this apparatus, showing the mercury levels on both sides of the U-tube if the gas sample has a pressure smaller than the atmospheric pressure.

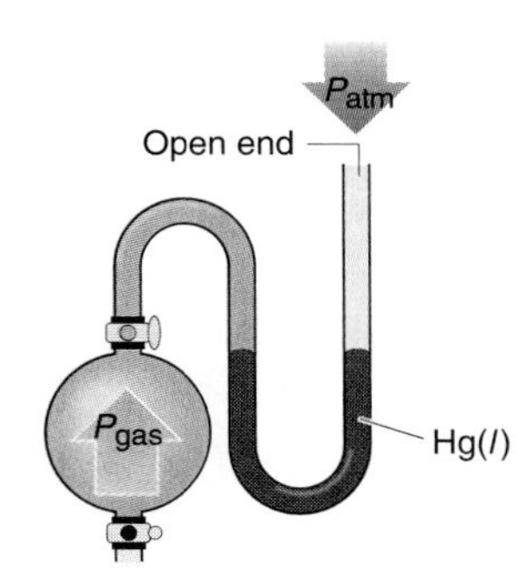


20. What is the pressure of the gas in the flask shown if atmospheric pressure is 740 mm Hg?

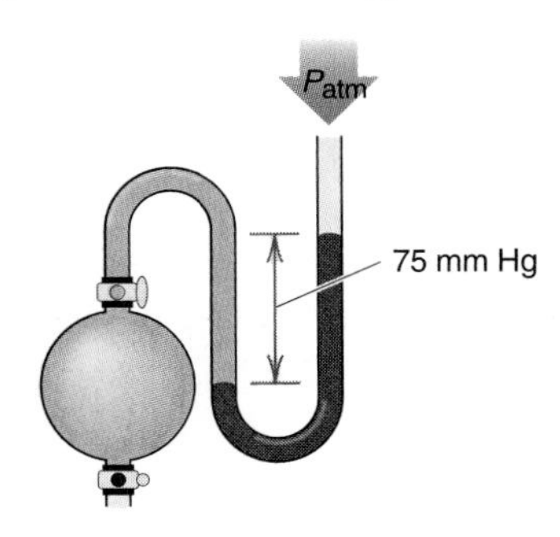


21. Newton defined force as mass times acceleration. The SI units for force are kg m $s^{-2}$ (called *newtons* in his honor). Show that the SI unit for pressure, the pascal, is equivalent to kg $m^{-1}$ $s^{-2}$.

22. If force is equal to mass times acceleration and the SI units for force are kg m $s^{-2}$, what are the SI units for acceleration?

23. Perform each of these conversions.
    a. 722 mm Hg to atm  c. 2.60 atm to kPa
    b. 27 psi to atm  d. 800 kPa to mm Hg

24. Perform each of these conversions.
    a. 1.45 atm to torr  c. 7900 kPa to atm
    b. 0.998 atm to psi  d. 740 torr to kPa

25. Carry out each of these conversions.
    a. 420 mm Hg to atm  c. $3.2 \times 10^{-2}$ atm to kPa
    b. 5.43 psi to atm  d. $1.14 \times 10^{5}$ Pa to mm Hg

26. Carry out each of these conversions.
    a. 0.87 atm to torr
    b. 45 atm to psi
    c. $8.1 \times 10^6$ Pa to atm
    d. $1.21 \times 10^3$ torr to kPa

27. Convert each quantity as indicated.
    a. $1.134 \times 10^3$ torr to atm
    b. $4.5 \times 10^3$ psi to atm
    c. 4.66 psi to Pa
    d. $1.92 \times 10^4$ kPa to mm Hg

28. Convert each quantity as indicated.
    a. $7.21 \times 10^{-1}$ atm to mm Hg
    b. $1.0 \times 10^6$ atm to psi
    c. 1.2 Pa to psi
    d. 420 torr to kPa

29. What is the pressure in psi if a force of 5.0 lb is exerted evenly over an area of 1.0 $ft^2$? What is this pressure in atmospheres?

30. What is the pressure in psi if a force of 11.5 lb is exerted evenly over an area of 7.3 $ft^2$? What is this pressure in mm Hg?

31. What is the pressure in psi if a force of $1.27 \times 10^3$ lb is exerted evenly over an area of 5.7 $ft^2$? What is this pressure in torr?

32. What is the pressure in psi if a force of $1.07 \times 10^{-4}$ pounds is exerted evenly over an area of 1.0 $ft^2$? What is this pressure in pascals?

33. What is the pressure in psi if a force of 540 lb is exerted evenly over an area of $1.2 \times 10^{-3}$ $ft^2$? What is this pressure in mm Hg?

34. What is the pressure in psi if a force of 12 lb is exerted evenly over an area of $1.0 \times 10^3$ $ft^2$? What is this pressure in atmospheres?

### SECTION 11.3 Boyle's Law: Pressure and Volume

35. Calculate the final pressure of each of these gases. The temperature remains constant at 25°C in all cases.
    a. A 25.0-mL sample of $SO_2$ at 700 torr expands to 2.00 L.
    b. A 15-mL sample of Kr at 0.554 atm expands to 23 mL.
    c. A 75.5-L sample of $PH_3$ at 850 mm Hg is compressed to 22.4 L.

36. Calculate the final pressure of each of these gases. The temperature remains constant at 25°C in all cases.
    a. A 20.0-mL sample of $N_2O$ at 0.0751 atm is compressed to 13 mL.
    b. A 23.6-L sample of $NO_2$ at 975 mm Hg expands to 125 L.
    c. A 68.6-mL sample of $H_2S$ at 4000 kPa is compressed to 25.6 mL.

37. Calculate the final pressure of each gas, assuming that the temperature remains constant at 25°C.
    a. 2.02 $cm^3$ of CO at 724 torr expands to 25.00 mL.
    b. 57.66 mL of $H_2$ at 2.00 atm is compressed to 5.11 mL.
    c. 1200 L of He at 1.00 atm expands to 5200 L.

38. Calculate the final pressure of each gas, assuming that the temperature remains constant at 25°C.
    a. 50.0 L of $AsH_3$ at 14.2 atm expands to 1000 L.
    b. 127 mL of $Cl_2O$ at 125 atm expands to 10.0 L.
    c. 12.0 L of $B_2H_6$ at 1.0 atm is compressed to 250 mL.

39. Calculate the final volume for each of these gases. The temperature remains constant at 125°C in all cases.
    a. 75.66 mL of $H_2O$ at 755 mm Hg is placed under 2.000 atm of pressure.
    b. 44.6 mL of $C_2H_6O$ at 2.00 atm is placed under 0.0118 atm of pressure.
    c. 124 mL of $SCl_2$ at 699 torr is placed under a pressure of 788 torr.

40. Calculate the final volume for each of these gases. The temperature remains constant at 125°C in all cases.
    a. $2.000 \times 10^3$ L of $NH_3$ at 1.00 atm is placed under a pressure of 500 torr.
    b. 2.555 atm of pressure is applied to 20.56 mL of $H_2$ at 1.000 atm.
    c. 50.0 atm of pressure is applied to $7.70 \times 10^3$ L of He at 0.977 atm.

41. Calculate the final volume of each gas, assuming that the temperature remains constant at 20°C.
    a. $2.00 \times 10^4$ kPa of pressure is applied to 30.4 mL of $CO_2$ at 1.00 atm.
    b. 23.2 mL of HCl at 822 torr is placed under 2.5 atm of pressure.
    c. $4.1 \times 10^5$ L of air at 1.00 atm is pumped into a tank until the pressure reaches 45.0 atm.

42. Calculate the final volume of each gas, assuming that the temperature remains constant at 20°C.
    a. $7.30 \times 10^5$ L of $N_2$ at 2.00 atm is placed under 25.0 atm of pressure.
    b. 4.00 mL of air at 760 torr is placed under a pressure of 200 torr.
    c. 8.25 mL of HI at 12.1 atm expands until the pressure is 1.0 atm.

43. A balloon has a volume of 120 L at sea level ($P = 1.0$ atm). What will the volume of this balloon be after a diver pulls it down to a depth of 30 m ($P = 4.0$ atm)?

44. A flexible container of air used by divers to raise salvaged items has a volume of 35 L at a depth of 40 m ($P = 4.5$ atm). What will the volume of this container be after it floats up to a depth of 10 m ($P = 2.0$ atm)?

45. An astronaut took a 1.0-L container of air that had a pressure of 5 atm and released the air into a large, very light, reflective space balloon. After inflation the balloon had a diameter larger than the length of a football field and a volume of $1.2 \times 10^6$ m$^3$. If the temperature of the cylinder and the balloon are the same, what is the final pressure in the balloon?

46. Some astronauts are building a space station. They seal up a portion of the station having a volume of 850 m$^3$ and begin to add oxygen to it. Their oxygen supplies are stored at a pressure of 200 atm. What volume of oxygen from the supply will be needed to bring the pressure inside the station to 0.20 atm?

47. Two people selling novelty balloons in a mall have a large cylinder of helium gas. If the cylinder has a volume of 150 L and a pressure of 120 atm, how many 16-L balloons can the salespeople fill with helium at a pressure of 1 atm? (Hint: Can they get *all* of the helium out of the tank?)

48. Two contractors have been paid to fill 20,000 balloons, each with a volume of 1.50 L, with helium for release at a sports event. (Yes, this is littering, but never mind that for the moment.) What volume of helium at 120 atm do they need to have on hand?

49. Explain, in molecular terms, why the pressure of a gas sample increases when the volume decreases at constant temperature.

50. Explain, in molecular terms, why the pressure of a gas sample decreases when the volume increases at constant temperature.

51. Redraw the figure below to show what happens if the 1.00-kg mass is changed to (a) a 2.00-kg mass and (b) a 0.500-kg mass.

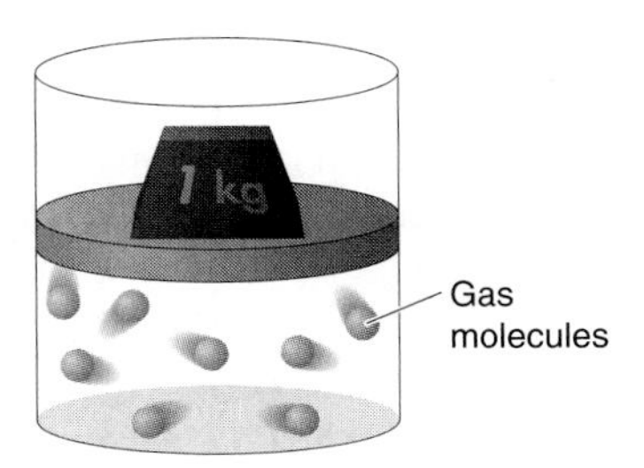


52. Redraw the figure below to show what happens after the stopcock is opened.

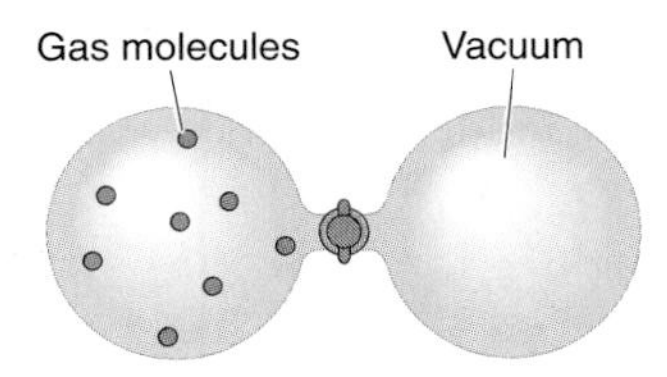


### SECTION 11.4 Charles' Law: Volume and Temperature

53. Calculate the final volume of each of these gases. The pressure remains constant at 1 atm.
    a. 27.8 mL of HI at 24.5°C is heated to 400°C.
    b. 23.3 L of $CH_4$ at 300 K is cooled to 150 K.
    c. 0.0754 L of $SO_2$ at 40°C is heated to 75°C.

54. Calculate the final volume of each of these gases. The pressure remains constant at 1 atm.
    a. 95.5 m$^3$ of $NH_3$ at 155°C is cooled to 15.0°C.
    b. 2.04 mL of $H_2$ at 100 K is heated to 550 K.
    c. 77.88 mL of $C_3H_8$ at 200°C is cooled to 20.0°C.

55. Calculate the final volume for each gas, assuming that the pressure remains constant at 1 atm.
    a. $4.99 \times 10^3$ L of $CH_4$ at 18.5°C is heated to 22.6°C.
    b. $7.55 \times 10^{-5}$ L of He at $-200$°C is heated to 100°C.
    c. $1.2 \times 10^3$ L of air at 100 K is heated to 150 K.

56. Calculate the final volume of each gas, assuming that the pressure remains constant at 1 atm.
    a. 44.6 mL of $H_2O$ at 100°C is heated to 200°C.
    b. 75.8 mL of $N_2$ at 1000 K is cooled to 300 K.
    c. $5.0 \times 10^5$ L of He at 50°C cools to 20°C.

57. Why does the volume of a gas sample increase when the temperature increases at constant pressure? Explain in terms of the behavior of molecules in the sample.

58. Why does the pressure of a gas sample increase when the temperature increases in a rigid container (constant volume)? Explain in terms of the behavior of molecules in the sample.

59. A sample of $CH_4(g)$ had a volume of 25.0 L at $-75$°C. What was the volume of this sample after it was heated to $-25$°C? The pressure was kept at 1 atm.

60. A sample of butane gas had a measured volume of 14.6 mL at 0°C. What was the volume of this sample after it was heated to 50°C? The pressure was kept at 1 atm.

61. A 12.0-L sample of $CO_2(g)$ had a temperature of 300 K. What was the volume of this sample after it was heated to 325 K? The pressure was constant at 1 atm.

62. A chemist worked with a 0.25-L sample of $C_2H_2(g)$ that had a temperature of 290 K. What was the volume of the same sample at 373 K? The pressure was constant at 1 atm.

63. A sample of argon had a volume of 155 mL at 0°C. What was the volume of this sample after it was cooled to $-40$°C? The pressure was kept at 1 atm throughout the experiment.

64. A 350-mL sample of argon was stored at 20°C. What was the volume of this sample after it was heated to 100°C? The pressure was kept at 1 atm throughout the experiment.

65. Atoms of argon gas in a flexible balloon are represented in the figure below. The balloon was heated from 20°C to 50°C. Redraw the picture to show what the balloon looks like after heating. Describe the major difference between your drawing and the original.

66. Atoms of argon gas in a flexible balloon are shown in the figure in Problem 65. Redraw the picture to show what the balloon looks like after cooling from 20°C to − 50°C. Describe the major difference between your drawing and the original.

### SECTION 11.5 Pressure, Temperature, and Volume Combined

67. Calculate the final volume of each of these gases.
    a. 33.7 mL of $SO_2$ at 588 torr and 55.3°C is placed under a pressure of 800 torr at 125°C.
    b. 12.6 L of $O_2$ at 2.55 atm and 300 K is placed under a pressure of 755 mm Hg at 350 K.
    c. 98.44 mL $PH_3$ at 770 torr and 40°C is placed under a pressure of 0.222 atm at 125°C.

68. Calculate the final volume of each of these gases.
    a. 722 $m^3$ of $O_2$ at 1.10 atm and 250 K is placed under a pressure of 5.00 atm at 300 K.
    b. $2.98 \times 10^{-4}$ L of $SF_2$ at 850 torr and 20°C is placed under a pressure of 700 torr at 100°C.
    c. 0.561 mL of Ar at 1.00 atm and 273 K is placed under a pressure of 5.00 atm at 200 K.

69. Calculate the final volume for each gas.
    a. 99.2 mL of $C_2H_2$ at 770 mm Hg and 300 K is placed under a pressure of 0.500 atm at 400 K.
    b. 47.9 mL of $SO_2$ at 2.0 atm and 35.5°C is placed under a pressure of 255 torr at 500 K.
    c. 50 L of He at 5.0 atm and − 150°C is placed under a pressure of 25 atm at 18.5°C.

70. Calculate the final volume for each gas.
    a. 10.0 mL of Ar at 13 atm and 1000 K is placed under a pressure of 25 atm at 300 K.
    b. 45.5 mL of $NH_3$ at 700 mm Hg and 20°C is placed under a pressure of 350 mm Hg at 60°C.
    c. 25.2 L of $O_2$ at 2.3 atm and − 45°C is placed under a pressure of 2.0 atm at − 100°C.

71. What is the final pressure of each of these gases?
    a. 35.7 mL of $AsH_3$ at 450 torr and 12.2°C is expanded to 2.0 L at 35.2°C.
    b. 48 $m^3$ of $N_2$ at 20 atm and 0°C is expanded to 100 $m^3$ at 100°C.
    c. 12.5 mL of $H_2S$ at 1.22 atm and − 5.3°C is compressed to 0.55 mL at 20°C.

72. What is the final pressure of each of these gases?
    a. 35.3 L of $CO_2$ at 5.0 atm and 11.2°C is expanded to 50.0 L at 25°C.
    b. 44.9 mL of Ar at 35 atm and 25°C is expanded to 2.2 L at − 100°C.
    c. 144 mL of CO at 0.011 atm and − 10°C is expanded to 155 mL at − 14°C.

73. Find the final pressure for each gas.
    a. 1.0 L of $CH_4$ at 3.5 atm and 50°C is compressed to 5.0 mL at 20°C.
    b. 2.3 mL of $C_3H_8$ at 1.5 atm and 80°C is compressed to 1.7 mL at 25°C.
    c. 7.5 mL of ClNO at 10.5 atm and 300 K expands to 25.0 mL at 320 K.

74. Find the final pressure for each gas.
    a. $1.25 \times 10^3$ L of $O_2$ at 90 atm and 30°C expands to $5.0 \times 10^5$ L at 20°C.
    b. 8.0 mL of $N_2O$ at 15 atm and 30°C expands to 125 mL at − 5°C.
    c. 35.7 mL of $C_2H_2$ at 700 torr and 32°C is compressed to 10.0 mL at 50°C.

75. A dive shop owner wants to fill a 25-L scuba tank with air at 200 atm and 30°C. If the air in the shop is at 1.0 atm and 20°C, how much air (in liters) is needed to fill the tank?

76. In a Navy diving experiment, a 10-L tank was filled with a helium-oxygen mixture at 450 atm and 0°C. If the gas mixture is at 1.0 atm and 25°C, how much of it (in liters) is needed to fill the tank?

77. An astronaut uses a 1.0-L container of air at a pressure of 5.0 atm and a temperature of 300 K to inflate a large, reflective space balloon. After inflation the balloon has a volume of $1.2 \times 10^6$ $m^3$ and a temperature (in the earth's shadow) of 100 K. What is the final pressure in the balloon?

78. Astronauts sealed a portion of an evacuated space station having a volume of 1350 $m^3$ and began to add oxygen to it. Their oxygen supplies were stored at a pressure of 120 atm and a temperature of 100 K. What volume of oxygen from the supply was needed to bring the pressure inside the station to 0.20 atm at 20°C?

79. A sealed rigid flask containing $N_2(g)$ at 0.977 atm and 15°C is placed in an oven. What is the temperature of the oven if, after 45 min, the pressure inside the flask is 3.99 atm?

80. A sealed bulb contained $He(g)$ at 1 atm and 0°C. An explorer took this bulb to the South Pole and found that the pressure inside the bulb was 0.85 atm. What was the temperature on that particular day?

## SECTION 11.6 The Molar Volume of Gases

81. What is the volume occupied by each of these gas samples at STP?
 a. 5.0 mol $O_2$
 b. 2.0 mmol Ar
 c. 14 mmol CO

82. What is the volume occupied by each of these gas samples at STP?
 a. 1.0 mol He
 b. 25 mol $N_2$
 c. 1.25 mmol $SO_2$

83. What is the volume occupied by each of these gas samples?
 a. 7.3 mol of $CO_2$ at 300 K and 1.0 atm
 b. 0.85 mol of $SO_2$ at 273 K and 3.5 atm
 c. 13.9 mol of Ne at 50°C and 2.7 atm

84. What is the volume occupied by each of these gas samples?
 a. 1.8 mol of $N_2$ at − 40°C and 2.12 atm
 b. 35.2 mmol of $H_2S$ at 50°C and a pressure of 760 mm Hg
 c. 0.355 mmol of He at 1000 K and a pressure of 800 torr

85. Calculate the volume occupied by each gas sample.
 a. 1.0 mol Ar at 0°C and 700 mm Hg
 b. 14 mol $N_2O$ at 200°C and 2.0 atm
 c. 4.0 kmol $O_2$ at 15°C and 25 atm

86. Calculate the volume occupied by each gas sample.
 a. 12 mol CO at − 10°C and 850 torr
 b. 1.75 mmol $B_2H_6$ at 21°C and 12 atm
 c. $7.2 \times 10^{-5}$ mol Rn at 10°C and 1.4 atm

87. A chemist mixed a 1.0-L sample of methane gas with an excess of chlorine gas and exposed the mixture to a bright light. At the end of the experiment, too little $CH_4(g)$ was left to measure easily. What volume (in liters) of $HCl(g)$ was produced?

88. Methane reacts with oxygen to produce carbon dioxide and water, all in the gas phase. How many liters of $H_2O(g)$ can be produced from 1.75 L of $CH_4(g)$ and excess $O_2(g)$ if the temperature and the pressure are the same both before and after the reaction?

89. Arsenic reacts with hydrogen to form $AsH_3(g)$. How many milliliters of $H_2(g)$ at STP are required to make 50.0 mL of $AsH_3(g)$?

90. What volume of $Cl_2$ gas at STP must be placed over excess Al metal to produce 500 g of $AlCl_3$?

91. For the first third of this century (before batteries became inexpensive and commonplace), coal miners used acetylene ($C_2H_2$) lamps attached to their hard hats as light sources in the mines. The acetylene was produced by reacting calcium carbide ($CaC_2$) with water:

$$CaC_2(s) + 2\,H_2O(l) \longrightarrow Ca(OH)_2(aq) + C_2H_2(g)$$

What volume of acetylene can be produced at STP from 125 g of $CaC_2$?

92. With the reaction given in Problem 91, how many grams of $CaC_2$ are needed to produce 10.0 L of $C_2H_2$ at STP?

## SECTION 11.7 The Ideal Gas Law

93. How many moles of gas are present in each of these samples?
 a. 30 mL of $H_2$ at 750 torr and 20°C
 b. 40.6 L of CO at 1.44 atm and − 20°C
 c. 125 mL of Ar at 25 atm and − 50°C

94. How many moles of gas are present in each of these samples?
 a. 1.22 L of $H_2S$ at 0.887 atm and 75°C
 b. 25 L of $C_2H_6$ at 25 torr and 0°C
 c. 420 mL of $NH_3$ at 0.0550 atm and 400 K

95. How many moles of gas are present in each sample?
 a. 25.5 mL of $CO_2$ at 10.0 atm and − 25°C
 b. $5.0 \times 10^4$ L of $O_2$ at 15 atm and 0°C
 c. 600 L of Ar at 50 atm and 750 K

96. How many moles of gas are present in each sample?
 a. 44.9 mL of $SO_2$ at 25 mm Hg and 20°C
 b. 12.37 L of $H_2$ at 14.9 psi and 300 K
 c. 8.7 mL of $H_2O$ at 1.7 atm and 400 K

97. What volume is occupied by each of these gas samples?
 a. 25.5 kg $NH_3$ at 50°C and 250 torr
 b. 50 mg $N_2O$ at 0°C and 750 mm Hg
 c. 102 g CO at STP

98. What volume is occupied by each of these gas samples?
 a. 88.5 mg $H_2S$ at 30°C and 700 mm Hg
 b. 12.5 g $SO_2$ at − 5°C and 0.229 atm
 c. 80.3 g $O_2$ at 15°C and 20 atm

99. What volume is occupied by each gas sample?
 a. 44.4 g of $AsH_3$ at 50°C and 750 torr
 b. 89.6 g of $H_2$ at STP
 c. 200 mg of Ar at 70°C and 2.00 atm

100. What volume is occupied by each gas sample?
 a. 75 kg of $CO_2$ at 300 K and 5.11 atm
 b. 11 g of $SO_2$ at STP
 c. $7.8 \times 10^3$ g of $N_2$ at 90 atm and 25°C

101. What is the pressure of each of these gas samples?
 a. 95 mg of Ne with a volume of 250 mL at 300 K
 b. 5.44 kg of $NH_3$ with a volume of 48.8 L at 400 K
 c. 750 g of $Cl_2$ with a volume of 35.6 L at 325 K

102. What is the pressure of each of these gas samples?
 a. 66.2 mg $F_2$ with a volume of 5.00 L at 298 K
 b. 0.115 g of CO with a volume of 26.8 mL at − 20°C
 c. 76.0 kg of $N_2$ with a volume of $5.00 \times 10^3$ L at 18°C

103. What is the pressure of each gas sample?
 a. 722 g of $N_2O$ with a volume of 25.5 L at 0°C
 b. 1.00 kg of $H_2O$ with a volume of 200 L at 250°C
 c. 87.9 g of Ar with a volume of 0.552 L at − 25°C

104. What is the pressure of each gas sample?
 a. 456 mg of $H_2S$ with a volume of 20.0 mL at 10°C
 b. 1.2 mg of Rn with a volume of 120 mL at 0°C
 c. 45 g of NO at 300 K with a volume of 50 L

105. a. A balloon has a volume of 7500 $m^3$. How many moles of helium gas does this balloon hold on a warm day (27°C) at a pressure of 760 mm Hg?
 b. How many grams of helium does the balloon contain?

106. a. A large balloon has a volume of 5000 $m^3$ at a high altitude. How many moles of helium does this balloon hold if the temperature is − 40°C and the pressure is 440 mm Hg?
 b. How many grams of helium does the balloon contain?

107. A small gas cylinder contains 8.0 g of $H_2(g)$ at a pressure of 120 atm and room temperature (20°C). What is the volume of this cylinder?

108. A gas cylinder contains 125 g of $N_2(g)$ at a pressure of 180 atm in a warm room (temperature = 25°C). What is the volume of this cylinder?

109. A 12-L cylinder contains 3.5 mol of oxygen gas at a temperature of 0°C. What is the pressure in this cylinder?

110. A 110-L cylinder contains 8.5 mol of argon gas at a temperature of 40°C. What is the pressure in this cylinder?

111. A 10.8-g sample of a gas occupies a volume of 11.1 L at 27°C and a pressure of 1.00 atm. What is the molar mass of this gas?

112. A 6.02-g sample of a gas occupies a volume of 3.70 L at 37°C and a pressure of 0.94 atm. What is the molar mass of this gas?

113. What is the density of each of these gas samples in $g\ L^{-1}$?
 a. $CF_4$ at STP
 b. $H_2S$ at 15.5°C and 750 torr
 c. $PCl_3$ at 125°C and 0.986 atm

114. What is the density of each of these gas samples in $g\ L^{-1}$?
 a. $H_2CCl_2$ at 75.0°C and 760 torr
 b. $SO_2$ at STP
 c. $H_2O$ at 450 K and 1.11 atm

115. Calculate the density of each gas sample in $g\ L^{-1}$.
 a. $CH_4$ at − 35.5°C and 700 mm Hg
 b. He at STP
 c. $CO_2$ at 325 K and 1.66 atm

116. Calculate the density of each gas sample in $g\ L^{-1}$.
 a. $SO_3$ at 300 K and 200 torr
 b. Rn at STP
 c. $C_2H_6$ at 25°C and 21 atm

117. What is the molar mass of each of these gases?
 a. A 5.27-g sample has a volume of 3.47 L at STP.
 b. A 3.53-g sample has a volume of 1.61 L at 298 K and 0.992 atm.
 c. A 3.11-g sample has a volume of 2.73 L at 300 K and 1.0 atm.

118. What is the molar mass of each of these gases?
   a. A 908-mg sample has a volume of 318 mL at STP.
   b. A 7.61-g sample has a volume of 12.2 L at 25°C and 0.95 atm.
   c. A 3.58-g sample has a volume of 1.97 L at 0°C and 1.02 atm.

119. Calculate the molar mass of each gas.
   a. 41.3 mg has a volume of 30.8 mL at STP.
   b. 555 mg has a volume of 213 mL at 0.822 atm and 273 K.
   c. 10.5 g has a volume of 7.14 L at 400 K and 1.15 atm.

120. Calculate the molar mass of each gas.
   a. 3.79 g has a volume of 4.61 L at 300 K and 1.00 atm.
   b. 42.8 mg has a volume of 60.0 mL at STP.
   c. 1.98 g has a volume of 1.17 L at 47°C and 1.01 atm.

121. A gas has an empirical formula of $CH_2$. What is its molecular formula if a 2.00-g sample has a volume of 1.15 L at 310 K and 1.05 atm?

122. A gas has an empirical formula of CH. What is its molecular formula if a 2.00-g sample has a volume of 1.72 L at STP?

123. Iron reacts with chlorine gas to form either $FeCl_2$ or $FeCl_3$ depending on which element is present in excess. What product forms if 25.0 g of Fe(*s*) reacts with 20.0 L of $Cl_2(g)$ at 298 K and 1.00 atm? How many moles of product are formed?

124. Calcium hydride ($CaH_2$) reacts vigorously with cold water to form $H_2(g)$ and $Ca(OH)_2(aq)$. What volume of $H_2(g)$ forms at STP from 50.0 g of $CaH_2$ and excess $H_2O$?

## SECTION 11.8 When More Than One Gas Is Present: Partial Pressures

125. It's a cool and pleasant fall day in New England, with a temperature of 17°C and a relative humidity of 40%. The vapor pressure of water is 14.5 mm Hg at that temperature; the total pressure is 1.0 atm. What is the percentage of water in the air?

126. It's a snowy winter day at a ski resort near Vancouver, with a temperature of −5°C and a relative humidity of 80%. The vapor pressure of ice is 3.01 mm Hg at that temperature; the total pressure is 1.0 atm. What is the percentage of water in the air?

127. Oxygen (25 L) and nitrogen (100 L), both at 1.0 atm, are pumped into an evacuated gas cylinder with a volume of 5.0 L. The temperature is 27°C. What is the partial pressure of each gas in the cylinder, and what is the total pressure?

128. Carbon monoxide (0.35 L) and hydrogen (1.10 L), both at 1.0 atm pressure, are carefully added to an evacuated flask with a volume of 11.5 L. The temperature is 20°C. What is the partial pressure of each gas in the flask, and what is the total pressure?

129. The figure below shows small portions of three identical flasks containing a mixture of gases at 20°C. Assume that the portion shown is representative of the entire flask in each case.

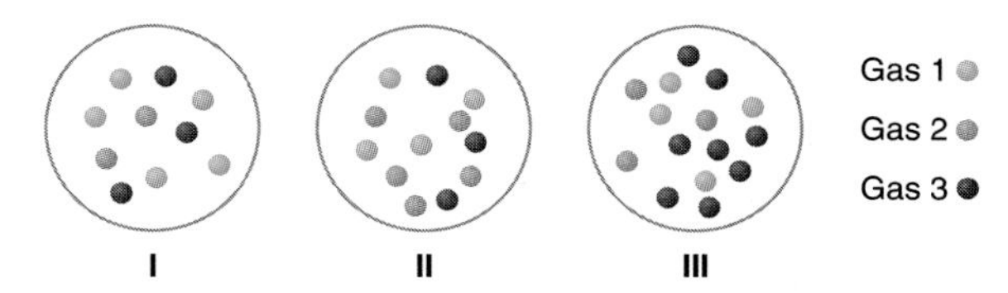


   a. Which of the flasks has the highest partial pressure of gas 1?
   b. If the total pressure in flask I is 2.0 atm, what is the partial pressure of gas 2?
   c. Which flask has the highest total pressure?
   d. If 2.0 mol of gas 2 reacts with 1.0 mol of gas 3 to produce 1.0 mol of a new gas, and if the pressure in flask I is initially 2.5 atm, what will the total pressure be once the reaction is over?

130. Refer to the figure in Problem 129, and again assume that the portions shown are representative of the entire contents of the flasks.
   a. Which of the flasks has the highest partial pressure of gas 3?
   b. If the total pressure in flask III is 3.0 atm, what is the partial pressure of gas 3?
   c. Which flask has the lowest total pressure?
   d. If 2.0 mol of gas 2 reacts with 1.0 mol of gas 3 to produce 1.0 mol of a new gas, and if the pressure in flask II is initially 2.5 atm, what will the total pressure be once the reaction is over?

## SECTION 11.9 Molecules in Motion: Diffusion and Effusion

131. Which of the following gases diffuses most rapidly?
   a. Helium  c. Fluorine
   b. Xenon  d. Chlorine

132. Which of the following gases diffuses most slowly?
   a. Helium  c. Fluorine
   b. Xenon  d. Chlorine

133. What is the ratio of the diffusion rate of fluorine to the diffusion rate of chlorine?

134. What is the ratio of the diffusion rate of oxygen to the diffusion rate of nitrogen?

135. What is the ratio of the diffusion rate of CO to the diffusion rate of $CO_2$?

136. What is the ratio of the diffusion rate of $Br_2$ to the diffusion rate of $NH_3$?

137. What is the ratio of the diffusion rate of $H_2O(g)$ to the diffusion rate of $H_2S$?

138. What is the ratio of the diffusion rate of HCl to the diffusion rate of HI?

139. The experiment shown in Figure 11.24 is illustrated below. Redraw this figure to show the relative position of the white cloud if HI is used instead of HCl.

140. Redraw the figure in Problem 139 to show the relative position of the white cloud if $AsH_3$ is used instead of $NH_3$.

### SECTION 11.10 The Kinetic Molecular Theory of Gases: Why the Gas Laws Work

141. You add more gas to a flexible container. The volume of the container expands while the temperature and pressure remain the same. Explain the reason for this behavior using the Kinetic Molecular Theory of Gases.

142. You have a syringe containing a gas sample. You heat the syringe, and the plunger moves farther out of the tube. In other words, the volume of the gas increases when the temperature increases but the pressure remains the same. Explain the reason for this behavior using the Kinetic Molecular Theory of Gases.

143. At high pressures most gases do not obey the Ideal Gas Law very well. Explain why.

144. At low pressures and relatively high temperatures most gases obey the Ideal Gas Law very well. Explain why.

145. The atomic radius of argon is approximately 100 pm. Use the equation for the volume of a sphere, $V = \frac{4}{3}\pi r^3$, to calculate the volume of 1 mol of Ar atoms. How does this compare with the volume occupied by 1 mol of Ar gas at STP (1000 L = 1 $m^3$)?

146. Air at room temperature and 1 atm of pressure has a density about $\frac{1}{800}$ of the density of water. Air under a pressure of 800 atm is considerably less dense than water, because of deviations from the Ideal Gas Law. Explain, in terms of molecular behavior, why air has a lower density at 800 atm than the Ideal Gas Law predicts.

### ADDITIONAL PROBLEMS

147. When a metal can is connected to a vacuum pump, the can collapses. Explain why.

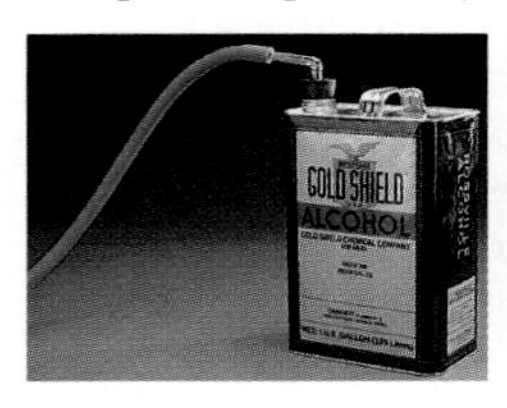

148. If you place about 50 mL of water in a can similar to the one shown in Problem 147, heat the can until the water boils, and then remove the heat and tightly stopper the can, it will collapse. Explain why.

149. Which has the greater density, dry air at 1.0 atm and 20°C or air with a relative humidity of 100% under the same conditions? Explain your answer.

150. Which has the greater density at STP, pure oxygen or a 50-50 mixture (by volume) of oxygen and nitrogen?

151. A volatile compound has the following percent composition by mass: 54.5% C, 9.1% H, 36.4% O. At 100°C and 1.00 atm, a 2.00-g sample of this gas occupies a volume of 1.39 L.
   a. What is the molar mass of the compound?
   b. What are its empirical and molecular formulas?

152. A volatile compound has the following percent composition by mass: 85.7% C and 14.3% H. At 100°C and 1.00 atm, a 2.00-g sample of this gas occupies a volume of 875 mL.
   a. What is the molar mass of the compound?
   b. What are its empirical and molecular formulas?

153. One method of preparation of HCN is to heat a mixture of $CH_4$ and $NH_3$ to 1500 K in the presence of a catalyst:

$$CH_4(g) + NH_3(g) \longrightarrow HCN(g) + 3\ H_2(g)$$

A reaction mixture containing 50.0 mol $CH_4$ and 50.0 mol $NH_3$ is placed in a reaction chamber having a volume of 500 L and heated to 1500 K. The catalyst has not yet been added, so no reaction occurs.
   a. What is the total pressure inside the reaction chamber?
   b. What is the partial pressure of each gas?

154. The catalyst is added to the reaction mixture in Problem 153, and the reaction occurs.
   a. If no detectible amounts of $CH_4$ or $NH_3$ remain, what is the total pressure inside the reaction chamber?
   b. What is the partial pressure of $HCN(g)$?
   c. The reaction chamber cools to 10°C. The $HCN(g)$ condenses to a colorless liquid and is removed. What is the pressure of $H_2(g)$ in the reaction chamber?

# EXPERIMENTAL PROBLEM

## 11.1

In the 1800s the French chemist A. Dumas developed a method for determining the molar mass of compounds that evaporate easily. Here's how this method works. First, you weigh a flask of known volume and then place a small sample of a volatile liquid in the flask (see Figure A). Next, you immerse the flask in a bath of boiling water and heat the sample until all the liquid evaporates. The temperature of the water (equal to the temperature of the gas) was 100°C (see Figure B). Then you remove the flask from the water bath and cool it down. The gas remaining in the flask condenses back into a liquid (see Figure C). The mass of liquid in the cooled flask is equal to the mass of gas that remained in the flask after heating. This mass is determined by reweighing the flask with the liquid and subtracting the original mass of the flask.

A. L. Chemist joined a team of scientists that used this technique to determine the molar mass of an unknown sample. They obtained this raw data:

$$T_{gas} = 100°C\ (373\ K)$$
$$P = 752\ mm\ Hg$$
$$\text{Volume of flask} = 527\ mL$$
$$\text{Mass of flask} = 126.37\ g$$
$$\text{Mass of flask + condensed sample} = 127.81\ g$$

The team then calculated the number of moles of gas present and the molar mass of the gas:

$$n = \frac{PV}{RT} = \frac{(0.989\ atm)(0.527\ L)}{(0.0821\ L\ atm\ mol^{-1}\ K^{-1})(373\ K)}$$
$$= 0.0170\ mol$$

$$\text{Molar mass} = \frac{127.81\ g - 126.37\ g}{1.70 \times 10^{-2}\ mol} = 84.7\ g\ mol^{-1}$$

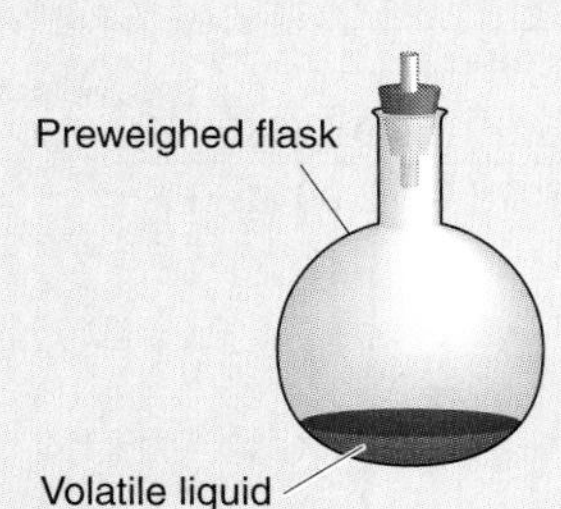


FIGURE A

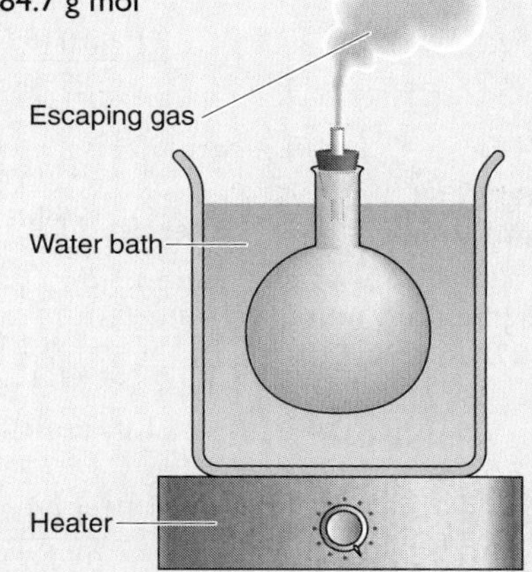


FIGURE B

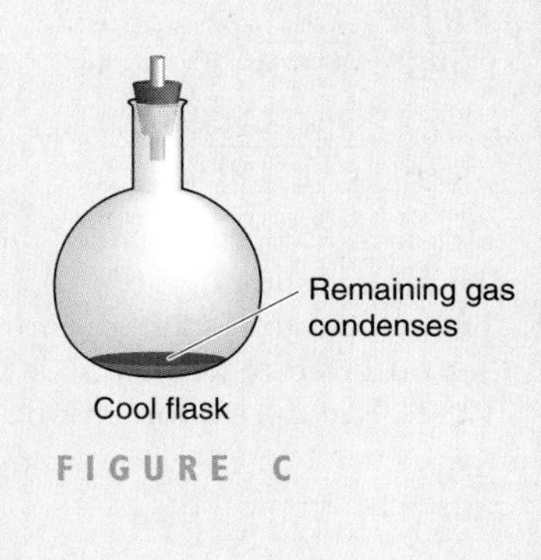


FIGURE C

### QUESTIONS

**155.** Why is it necessary to evaporate *all* of the liquid in step 2?

**156.** Will the flask in step 3 contain more sample if $P_{atm}$ = 765 mm Hg? Justify your answer.

**157.** Why can't this technique be used to determine the molar mass of NaCl?

**158.** What is the molecular formula of the sample if its empirical formula is $CH_2Cl_2$?

**159.** What is the pressure inside the flask in step 1? What gas is in the flask?

**160.** At what step is the pressure inside the flask greater than $P_{atm}$?

**161.** Why is 373 K used as the temperature of the gas in the calculations, even though the flask is cooled to room temperature in step 3?

**162.** The team used 5.00 mL of liquid sample in step 1. Would the results be different if the team used 8.00 mL? Explain your answer.

## SOLUTIONS TO EXERCISES

### Exercise 11.1

Oxygen reacts with carbohydrates and fats to produce carbon dioxide, water, and the energy of life.

### Exercise 11.2

**a.** The pressure on the wall is 5 lb/0.001 in$^2$, or $5 \times 10^3$ lb in$^{-2}$.

**b.** When you use an ice pick or a knife, the pressure on the working area (the tip of the ice pick or the edge of the knife blade) is very high, so the working area erodes or deforms easily.

### Exercise 11.3

This is another unit conversion problem, to be solved using the unit factor method. The conversion factor you need is 1 atm = 760 mm Hg.

$$730\ \cancel{\text{mm Hg}}\left(\frac{1\text{ atm}}{760\ \cancel{\text{mm Hg}}}\right) = 0.961\text{ atm}$$

### Exercise 11.4

**a.** This is a problem involving the equation $P_1V_1 = P_2V_2$. The pressure at 30 m ($P_1$) is 4.0 atm, and the volume ($V_1$) is 5.3 L. The pressure at the surface ($P_2$) is 1.0 atm, and $V_2$ is the unknown. The equation is

$$P_1V_1 = P_2V_2$$

$$(4.0\text{ atm})(5.3\text{ L}) = (1.0\text{ atm})V_2$$

$$V_2 = \frac{(4.0\ \cancel{\text{atm}})(5.3\text{ L})}{1.0\ \cancel{\text{atm}}} = 21\text{ L}$$

The volume of this balloon at the surface is 21 L.

**b.** The diver didn't fill the balloon completely at 30 m because it would have exploded after expanding on its way to the surface. Divers don't inhale deeply at depth and hold their breath on the way to the surface because the same thing can happen to their lungs.

### Exercise 11.5

**a.**
$$\frac{5.0\text{ L}}{250\text{ K}} = \frac{V_2}{350\text{ K}}$$

$$V_2 = \frac{(5.0\text{ L})(350\ \cancel{\text{K}})}{250\ \cancel{\text{K}}} = 7.0\text{ L}$$

This answer is reasonable; the temperature went up, so the volume should, too.

**b.** The temperatures are given in Celsius degrees and must be converted to kelvins: $V_1$ is 5.0 L, $T_1$ is 273 K, and $T_2$ is 173 K. The calculation is done as before:

$$\frac{5.0\text{ L}}{273\text{ K}} = \frac{V_2}{173\text{ K}}$$

$$V_2 = \frac{(5.0\text{ L})(173\ \cancel{\text{K}})}{273\ \cancel{\text{K}}} = 3.2\text{ L}$$

### Exercise 11.6

Here $P_1 = 1$ atm, $T_1 = 423$ K, and $T_2 = 298$ K. You solve for $P_2$.

$$\frac{1.0\text{ atm}}{423\text{ K}} = \frac{P_2}{298\text{ K}}$$

$$P_2 = \frac{(1.0\text{ atm})(298\ \cancel{\text{K}})}{423\ \cancel{\text{K}}} = 0.70\text{ atm}$$

### Exercise 11.7

Here the trick is to get all the pressure units the same so that they cancel: 1.00 atm = 760 mm Hg = $P_1$.

$$\frac{P_1V_1}{T_1} = \frac{P_2V_2}{T_2}$$

$$\frac{(760\text{ mm Hg})(100\text{ L})}{300\text{ K}} = \frac{(700\text{ mm Hg})(V_2)}{273\text{ K}}$$

$$V_2 = \frac{(760\ \cancel{\text{mm Hg}})(100\ \cancel{\text{L}})(273\ \cancel{\text{K}})}{(300\ \cancel{\text{K}})(700\ \cancel{\text{mm Hg}})} = 99\text{ L}$$

### Exercise 11.8

Here the temperature and the pressure change and the volume is the unknown factor. This is a Combined Gas Law problem. So you use:

$$\frac{P_1V_1}{T_1} = \frac{P_2V_2}{T_2}$$

If you assume that the gas starts at STP, $T_1 = 273$ K and $P_1 = 1$ atm. You obtain $V_1$ using the molar volume at STP as a conversion factor:

$$0.0250\ \cancel{\text{mol Ar}}\left(\frac{22.4\text{ L}}{1\ \cancel{\text{mol Ar}}}\right) = 5.60\text{ L} = V_1$$

Next, $P_2 = 2.00$ atm and $T_2 = 400$ K, and you can plug the values into the Combined Gas Law equation:

$$\frac{(1\text{ atm})(5.60\text{ L})}{273\text{ K}} = \frac{(2.00\text{ atm})(V_2)}{400\text{ K}}$$

$$V_2 = \frac{(1\ \cancel{\text{atm}})(5.60\text{ L})(400\ \cancel{\text{K}})}{(273\ \cancel{\text{K}})(2.00\ \cancel{\text{atm}})} = 4.10\text{ L}$$

### Exercise 11.9

Here $P = 1$ atm, $T = 273$ K, and $V = 2.00$ mL = $2.00 \times 10^{-3}$ L.

$$PV = nRT$$

$$(1.00\ \cancel{\text{atm}})(2.00 \times 10^{-3}\ \cancel{\text{L}}) = n(0.0821\ \cancel{\text{L}}\ \cancel{\text{atm}}\text{ mol}^{-1}\cancel{\text{K}}^{-1})(273\ \cancel{\text{K}})$$

$$n = \frac{(1.00)(2.00 \times 10^{-3})}{(0.0821\text{ mol}^{-1})(273)} = 8.92 \times 10^{-5}\text{ mol }CO_2$$

**Exercise 11.10**

The molar mass of $CO_2$ is 44 g $mol^{-1}$ and that of CO is 28 g $mol^{-1}$. To find the number of moles of the sample gas, you use the ideal gas equation:

$$PV = nRT$$
$$(1.00\ \cancel{atm})(12.2\ \cancel{L}) = n(0.0821\ \cancel{L}\ \cancel{atm}\ mol^{-1}\ \cancel{K}^{-1})(300\ \cancel{K})$$
$$n = 0.495\ mol$$

The molar mass is 22 g/0.495 mol = 44 g $mol^{-1}$, and therefore the gas is $CO_2$.

**Exercise 11.11**

This one is relatively easy, since you don't need to calculate the volume at STP—it is 22.4 L. The molar mass of this freon is

$$12.0\ g\ mol^{-1} + 2 \times 35.45\ g\ mol^{-1} + 2 \times 19.0\ g\ mol^{-1} = 121\ g\ mol^{-1}$$

The density is therefore:

$$\frac{121\ g}{22.4\ L} = 5.40\ g\ L^{-1}$$

**Exercise 11.12**

Here you need to calculate the mass of iron. To do this, you need the number of moles of iron, which you get from the balanced equation:

$$1\ mol\ Fe(s) \approx 1\ mol\ H_2(g)$$

You can obtain the number of moles of $H_2(g)$ from the ideal gas equation. First, however, you must express the volume in liters:

$$V = 1000\ m^3 = 1000\ m^3 \left(\frac{1000\ L}{1\ m^{-3}}\right) = 1.0 \times 10^6\ L$$

Then you use the ideal gas equation, $PV = nRT$ to get the number of moles of $H_2(g)$:

$$(1.00\ \cancel{atm})(1.0 \times 10^6\ \cancel{L}) = n(0.0821\ \cancel{L}\ \cancel{atm}\ mol^{-1}\ \cancel{K}^{-1})(300\ \cancel{K})$$
$$n = 4.06 \times 10^4\ mol\ of\ H_2(g)$$

This is equivalent to $4.06 \times 10^4$ mol of Fe($s$) (from the balanced equation). Next, convert moles of Fe into grams and then kilograms:

$$(4.06 \times 10^4\ mol\ Fe)(55.85\ g\ mol^{-1}) = 2.27 \times 10^6\ g$$
$$2.27 \times 10^6\ g \left(\frac{1\ kg}{1000\ g}\right) = 2.27 \times 10^3\ kg,$$

(or 2.27 metric tons of iron)

**Exercise 11.13**

**a.** The atmosphere cannot be homogeneous on a planetary scale. Atmospheric pressure in New Orleans (5 ft below sea level) is measurably higher than atmospheric pressure in Denver (1 mi above sea level). It can be raining and cool in London, clear and hot in Manila, and snowing in New York all at the same time. This is not the behavior of a homogeneous mixture.

**b.** The assumption is often made because the atmosphere in any small space, such as a classroom or laboratory, is in fact homogeneous.

**Exercise 11.14**

The percentage of water in the air is:

$$\frac{47.1\ mm\ Hg}{760\ mm\ Hg} \times 100\% = 0.0620 \times 100\% = 6.20\%$$

There's quite a difference between Antarctica and Houston, isn't there?

**Exercise 11.15**

According to Dalton, the three gases act as if they are independent. The partial pressures can be found by using $PV = nRT$, for each gas individually. In each case you first have to find the number of moles present. Of course, 27°C = 300 K. For $H_2$:

$$n = \frac{2.0\ g}{2.0\ g\ mol^{-1}} = 1.0\ mol$$

Then:

$$P_{H_2} = \frac{nRT}{V} = \frac{(1.0\ mol)(0.0821\ L\ atm\ mol^{-1}\ K^{-1})(300\ K)}{12\ L} = 2.1\ atm$$

For $CH_4$:

$$n = \frac{4.0\ g}{16.0\ g\ mol^{-1}} = 0.25\ mol$$

Then:

$$P_{CH_4} = \frac{nRT}{V} = \frac{(0.25\ mol)(0.0821\ L\ atm\ mol^{-1}\ K^{-1})(300\ K)}{12\ L} = 0.51\ atm$$

For $NH_3$:

$$n = \frac{2.0\text{ g}}{17.0\text{ g mol}^{-1}} = 0.12\text{ mol}$$

Then:

$$P_{NH_3} = \frac{nRT}{V}$$

$$= \frac{(0.12\text{ mol})(0.0821\text{ L atm mol}^{-1}\text{ K}^{-1})(300\text{ K})}{12\text{ L}}$$

$$= 0.25\text{ atm}$$

The total pressure is the sum of the partial pressures:

$$2.1\text{ atm} + 0.51\text{ atm} + 0.25\text{ atm} = 2.9\text{ atm}$$

### Exercise 11.16

The gas that diffuses most slowly is the one with the largest atomic or molecular mass. For this group of gases, the answer is argon (D).

### Exercise 11.17

**a.** The number of particles is constant, but their average kinetic energy increases. The molecules hit the sides of the container harder and more often, increasing the pressure.

**b.** The temperature and the number of particles remains constant, and so does the surface area of the inside of the container. The dents decrease the volume, and the distance between the walls decreases at the dents, so the molecules take less time on the average between collisions with those walls. The number of collisions increases, and therefore the pressure increases.

### EXPERIMENTAL EXERCISE 11.1

1. The reaction vessel contains air. It is necessary to allow atmospheric air to escape from the vessel in order to be certain that gas 1 completely fills the chamber. Remember, at this point in the experiment the team doesn't know the composition of gas 1.
2. It is necessary to keep atmospheric oxygen out of the sample.
3. The team doesn't want air from the atmosphere to contaminate gas 2 either.
4. A mole of $O_2$ gives a mole of $CO_2$ in the reaction with C(*s*). Neither $N_2$ nor Ar reacts. Therefore, if the amount of gas 2 isolated is 10 L, then 10 L of gas 1 is needed to begin with.
5. You know that 10 L of gas 1 produced 2.0 L of $CO_2$. This gives you one of the unit factors necessary to solve this problem. You also know that 1 mol of an ideal gas has a volume of 22.4 L at STP.

$$50\text{ L }\cancel{\text{gas 1}}\left(\frac{2.0\ \cancel{\text{L CO}_2}}{10.0\text{ L }\cancel{\text{gas 1}}}\right)\left(\frac{1\text{ mol CO}_2}{22.4\ \cancel{\text{L CO}_2}}\right) = 0.44\text{ mol CO}_2$$

6. Gas 4 has a density of 1.79 g $L^{-1}$ at STP. To get the molar mass, you find the mass of 22.4 L of the gas.

$$\left(\frac{1.79\text{ g gas 4}}{1\text{ L gas 4}}\right)\left(\frac{22.4\text{ L gas 4}}{1\text{ mol gas 4}}\right) = 40\text{ g mol}^{-1},$$

equal to the molar mass of Ar

7. By definition, *excess* and *limiting* are opposites. Since excess Mg was used, the limiting reactant must be $N_2$.
8. All gas volumes are 10 L or less and all are measured at STP. This means that there is always less than 0.5 mol of gas (10/22.4). Clearly, 100 g of CaO (56 g = 1 mol) is more than enough.
9. First, you need the balanced equation:

$$3\text{ Mg} + N_2 \longrightarrow Mg_3N_2$$

Since $N_2$ is the limiting reactant (see Question 7), you need to find the number of moles of $N_2$ to get the answer. Since all measurements were done at STP, you know 1 mol occupies 22.4 L. The amount of $N_2$ present is 8.0 L − 0.1 L = 7.9 L, and the molar mass of $Mg_3N_2$ is

$$(3 \times 24.3\text{ g mol}^{-1}) + (2 \times 14.0\text{ g mol}^{-1}) = 100.9\text{ g mol}^{-1}$$

$$7.9\ \cancel{\text{L N}_2}\left(\frac{1\ \cancel{\text{mol N}_2}}{22.4\ \cancel{\text{L N}_2}}\right)\left(\frac{1\ \cancel{\text{mol Mg}_3\text{N}_2}}{1\ \cancel{\text{mol N}_2}}\right)\left(\frac{100.9\text{ g Mg}_3\text{N}_2}{1\ \cancel{\text{mol Mg}_3\text{N}_2}}\right) = 35.6\text{ g Mg}_3\text{N}_2$$

### EXPERIMENTAL EXERCISE 11.2

**1.** If the gas is to effuse and not flow through the exit port, the opening should be pinhead-sized—answer A.

**2.** There are several possibilities, including the distance between the disks and the total distance the molecules travel.

**3.** 250 m $s^{-1}$ (read from Figure B).

**4.** Methane has a molar mass of 16 g $mol^{-1}$, and ammonia has a molar mass of 17 g $mol^{-1}$. Consequently, your plot should look very much like the plot for $CH_4$, with the most probable speed at a slightly lower value.

**5.** The curve should be similar to that for $N_2$ in Figure B, but should be flatter than for $N_2$ at 2273 K and the most probable value of the speed should be higher. A reasonable guess would be to put the most probable value just below 2000 m $s^{-1}$.

12

# LIQUIDS AND SOLIDS



## KEY TERMS

Unlike gases, in which the molecules bounce around in large volumes of empty space, liquids and solids have atoms, molecules, or ions that are in continuous contact with one another (see Figure 12.1). This is why neither liquids nor solids expand to fill their containers and why liquids and solids are not easily compressible. In this chapter we will consider in more detail the way liquids and solids behave and why.

You encounter liquids and solids constantly in daily life. Water is the most important liquid on earth, since all plants and animals need water for survival. Water has some unusual and unexpected properties in comparison with other liquid substances. These properties are described in detail later in this chapter.

Inspect the solids around you at the moment and think about their characteristics. There are many possible subclasses of solids. Plastic bags, rubber bands, and paper are flexible. Metals are malleable and ductile, but glass and diamonds are quite brittle. Table salt dissolves in water, but most other ionic compounds (rocks, sand, and soil, for instance) do not. The physical properties of a solid are intimately related to its structure. This chapter takes a closer look at the structure of solids.

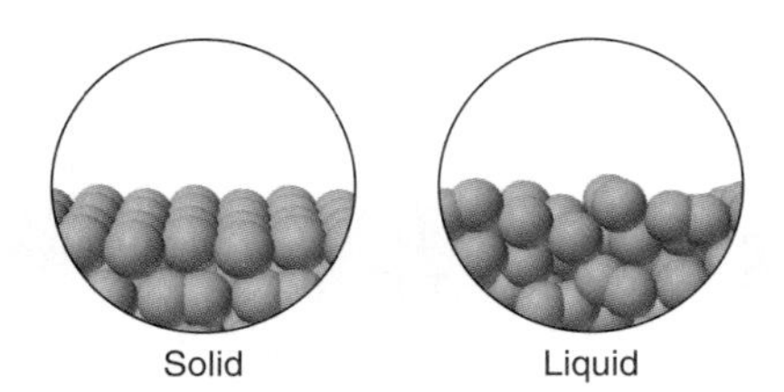


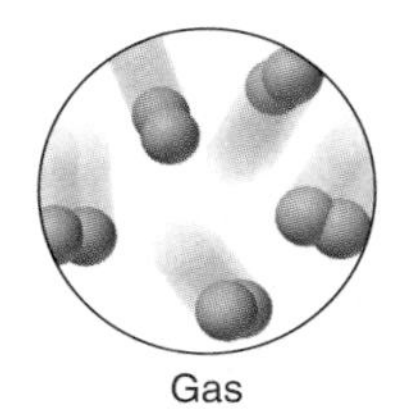


FIGURE 12.1
Relative spacing of molecules in a solid, a liquid, and a gas.

## 12.1 Solids and Liquids Classified

GOAL: To describe the various forms that solids and liquids can take.

From earlier chapters, you are already familiar with some of the classifications of solids and liquids. The type of bonding in a compound can often tell you whether the compound is a solid at room temperature. Ionic compounds are brittle solids. Almost all metals are solids, too, but they're ductile and malleable rather than brittle.

**Mercury is the most common liquid metal. Cesium (melting point = 28°C) and gallium (melting point = 30°C) are liquids on a hot day.**

The physical properties of covalent compounds show enormous variety. Some are gases that are very difficult to liquefy or solidify. Hydrogen is the most extreme case, with a melting point of −259°C. Only the noble gas helium (which engages in no bonding at all) has a lower melting point. Other covalent compounds are among the substances with the highest melting points. Diamond, which melts above 3500°C, is an example of this kind of substance. A diamond is a single giant molecule, made up of carbon atoms bonded to each other (see Chapter Box 12.1). Such solids are sometimes called **network solids**.

A distinction is also made between crystalline and amorphous solids (see Figure 12.3). Before scientists and scholars knew anything about the structures of molecules and ions, they called beautiful natural objects with outwardly symmetrical forms *crystals*. Early in this century, scientists discovered that they could use X-rays to determine the arrangement of atoms in crystals. Since then, solids have been classified as **crystalline solids** when there is a long-range repeating pattern in their atomic, molecular, or ionic structure and as **amorphous solids** (from the Latin, meaning "without form") when there is not.

FIGURE 12.2
Ordinary body temperature, 37°C, is higher than the melting point of gallium, 29.8°C. As you can see, a sample of gallium will melt in a person's hand.

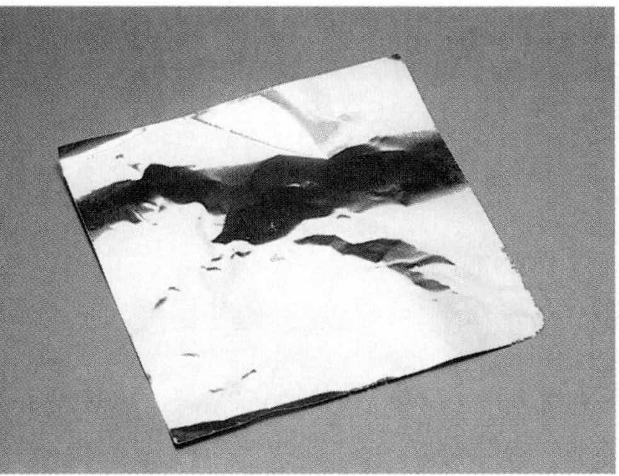
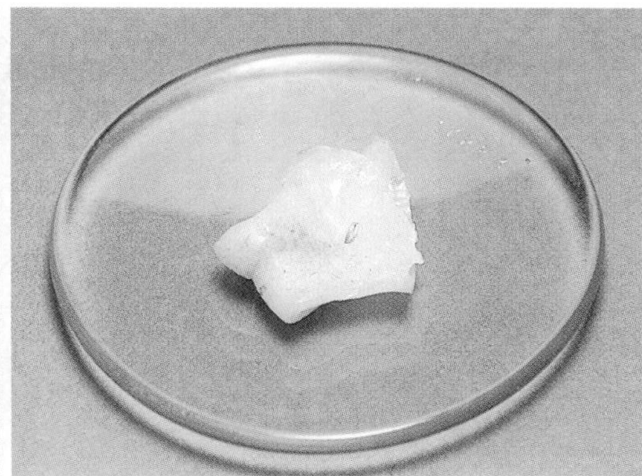

FIGURE 12.3
Solids take on a wide variety of outward forms: left, beryl; center, aluminum metal; right, PVC plastic. Beryl and aluminum are crystalline; PVC is amorphous.

Sulfur is an element that has both crystalline and amorphous forms. At room temperature the stable form of sulfur is crystalline *orthorhombic sulfur.* When solid sulfur crystals form from liquid sulfur at 119.0°C, a slightly different crystalline form called *monoclinic sulfur* results (see Figure 12.4). If liquid sulfur is heated to a temperature near 200°C, the liquid becomes very thick and syrupy and changes to a dark-red color. If this dark-red liquid is cooled rapidly (by being poured into water, for instance), a rubbery solid forms (see Figure 12.6). This is amorphous sulfur.

**The use of the word *crystal* to refer to a type of glass arose from the original definition of the word. Crystal glass has the symmetrical outward shape required by the older definition, but not the molecular order demanded by the new one.**

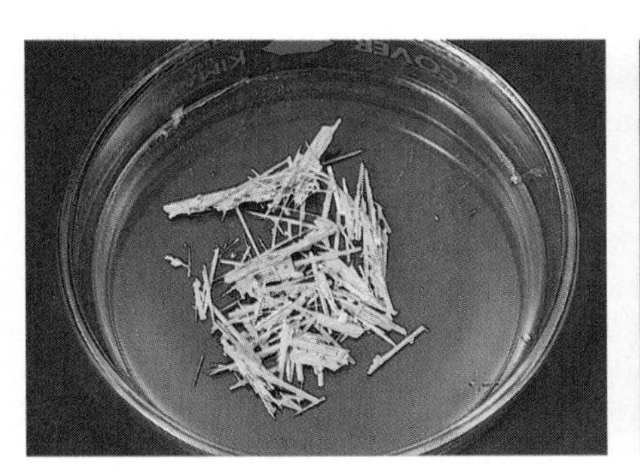

FIGURE 12.4
Orthorhombic sulfur (left) is stable at room temperature. At temperatures below 119°C, monoclinic sulfur (right) slowly converts to orthorhombic sulfur. Natural samples of monoclinic sulfur can be found near volcanos, where the temperature is above 119°C.

Both orthorhombic sulfur and monoclinic sulfur have molecules with the formula $S_8$. This molecule is a ring of eight sulfur atoms, with each atom bonded to two adjacent ones (see Figure 12.5). At higher temperatures the ring breaks open to form a chain of sulfur atoms. Although the six atoms in the middle of the chain are bonded to two adjacent atoms, the two atoms at

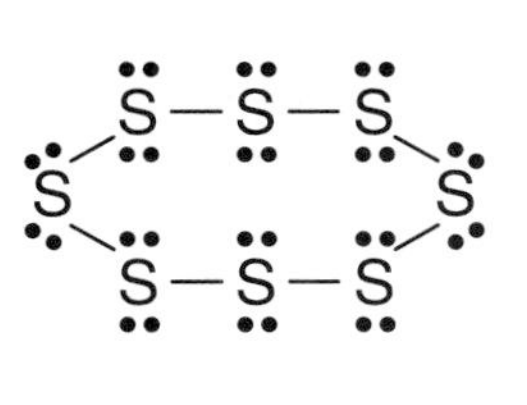

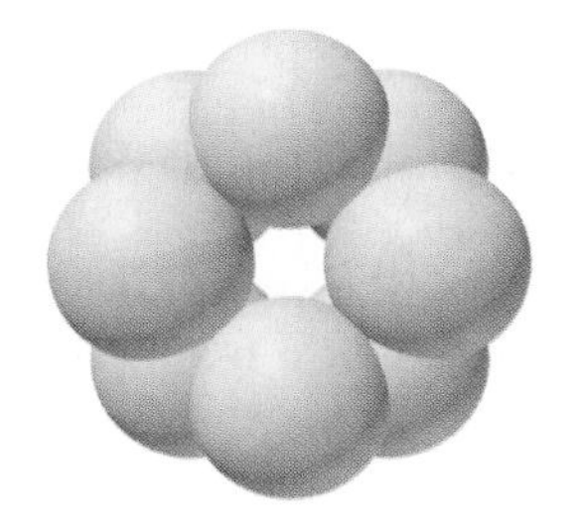
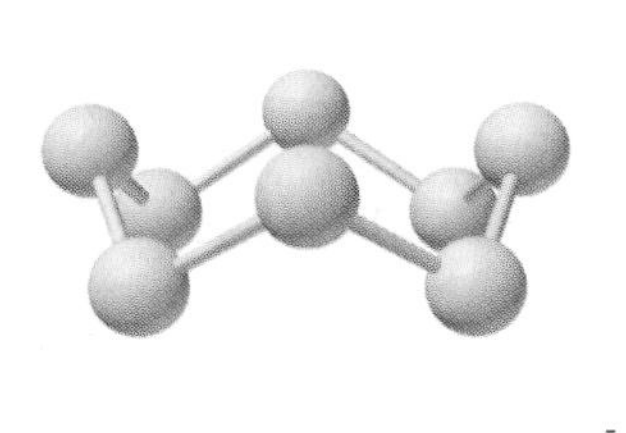

FIGURE 12.5
The $S_8$ molecules that comprise both orthorhombic and monoclinic sulfur have the form of a ring. Left to right: The Lewis diagram, a space-filling model, and a ball-and-stick model of this molecule.

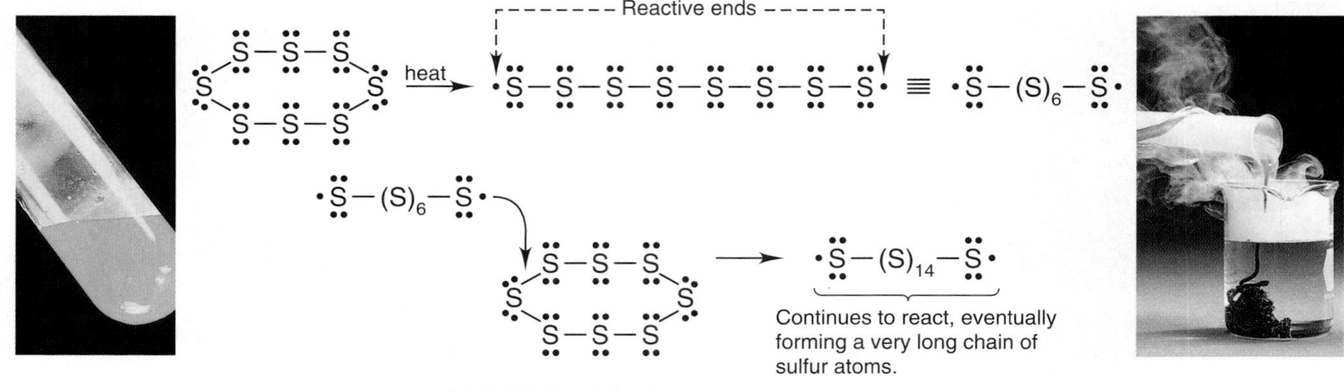
Reactive ends
heat
Continues to react, eventually forming a very long chain of sulfur atoms.

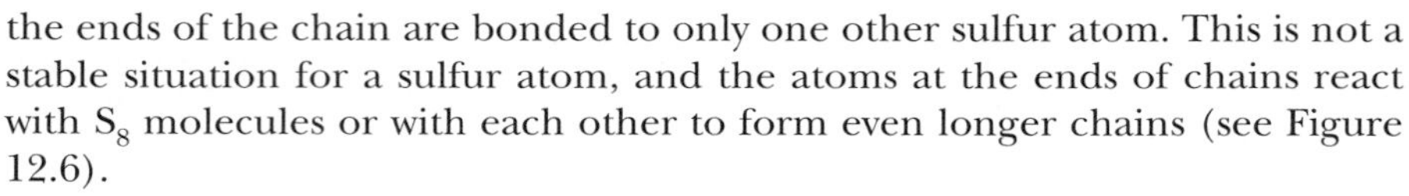

FIGURE 12.6

The photo on the left shows liquid sulfur that has just melted. The $S_8$ rings open when the sulfur is heated further. Ends of the open rings break open other rings to make long chains of sulfur molecules. Near 200°C liquid sulfur becomes dark red and flows very slowly. The photo on the right shows this liquid becoming rubbery amorphous sulfur as it cools in water. After several days the rubbery amorphous sulfur converts back to its original crystalline form.

the ends of the chain are bonded to only one other sulfur atom. This is not a stable situation for a sulfur atom, and the atoms at the ends of chains react with $S_8$ molecules or with each other to form even longer chains (see Figure 12.6).

When liquid sulfur cools slowly, the $S_8$ molecules fall into a crystal structure that has long-range order. When the high-temperature syrupy liquid cools rapidly, the chains of sulfur atoms fall into an irregular mat (see Figure 12.6).

Crystalline substances have sharp melting points. Orthorhombic sulfur (see Figure 12.7) melts at 112.8°C, for instance. When a crystalline solid melts, the orderly arrangement of molecules changes into the more random arrangement of the liquid state. This change can happen only if the molecules or atoms in the solid have enough kinetic energy to move past each other. When a chemist heats orthorhombic sulfur, the orderly crystalline arrangement changes abruptly to a random one, and the chemist observes a narrow melting point range.

FIGURE 12.7

(Top) This shows the arrangement of the $S_8$ molecules in crystals of orthorhombic sulfur. (Bottom) The long chains of covalently bonded sulfur atoms in amorphous sulfur make a rubbery tangle.

Amorphous solids, including amorphous sulfur, behave differently from crystalline solids. They simply soften gradually and become more fluid. Amorphous solids already have a random ordering of molecules somewhat like that of a liquid state. As a sample of such a solid is heated, the molecular motion within it gradually increases. Because the molecules were randomly arranged to begin with, there is no sharp transition between orderly solid and disordered liquid. Of course, this property can be useful. Glass is amorphous, without molecular order even when it is solid. The fact that glass does not melt sharply allows artists to mold and blow it into a variety of shapes (see Figure 12.8).

Does this mean that any irregular lump of a substance or anything that has an outward symmetry imposed by a manufacturer is amorphous? Not at

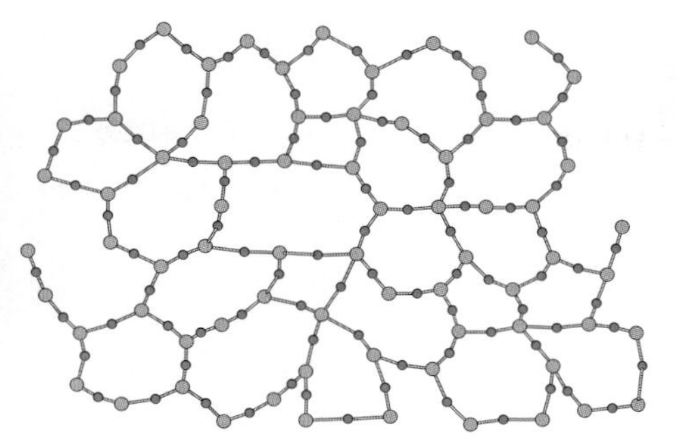

FIGURE 12.8
Glassblowers (left) exploit the wide melting range of glass in their work. Glass has no long-range molecular order (as illustrated by the simplified two-dimensional representation of $SiO_2$ glass, center), and this helps glassblowers impose forms on the glass (right).

all. Many materials are what is called **polycrystalline**—that is, they consist of many small crystals. You can usually make out the individual crystals in a lump of granite, for example, if you look closely (see Figure 12.9). You would need a microscope and other special techniques to see the crystals in some materials.

FIGURE 12.9
Polarized light micrograph of granite.

Finally, other elements besides sulfur, as well as many compounds, exist in several forms. The different forms are called **allotropes**. Diamonds and graphite (the black stuff in your "lead" pencil) are two forms of pure carbon. Graphite and diamonds are so different in appearance and physical properties that it took scientists a relatively long time to figure out that they were in fact allotropes of carbon instead of two different substances.

Diamonds are colorless, very hard crystals with a density of 3.35 g $cm^{-3}$. Graphite is black, so soft that can be used as a lubricant, and has a density of 2.25 g $cm^{-3}$. Diamonds are chemically inert—they will not even react with oxygen unless the temperature is above 800°C. Graphite reacts with several acids at room temperature. It is not too surprising that two centuries ago scientists doubted that graphite and diamonds were both pure carbon.

Think about water, motor oil, molasses, and gasoline—and you will get an idea of the properties of a wide range of ordinary liquids. All liquids flow, for instance, but some liquids flow more slowly than others. Slow-flowing liquids are said to have high **viscosities**. Molasses in January and amorphous sulfur both have very high viscosities, for example. Even though glass is technically a solid (an amorphous one), it can be thought of as a liquid with an extremely high viscosity (see Figure 12.10).

FIGURE 12.10
The arrangement of the atoms in glass is so disorderly that you can think of it as an extremely viscous liquid rather than a solid. Very old glass windows in European cathedrals are thicker at the bottom than at the top. Some of the glass has flowed downward over the centuries.

Chemists in the petroleum industry explore the properties of lubricating oils (among many other substances). They try to find compounds and mixtures that are fluid enough at low temperatures to lubricate engines on cold mornings and viscous enough at high temperatures to lubricate engines that are close to overheating. Gasoline typically has a lower viscosity and motor oils have a higher viscosity than water does.

A few substances have more complex behavior. Some compounds have what amounts to two melting points. They first lose some (but not all) of the order they have as crystalline solids and then lose the rest at a higher temper-

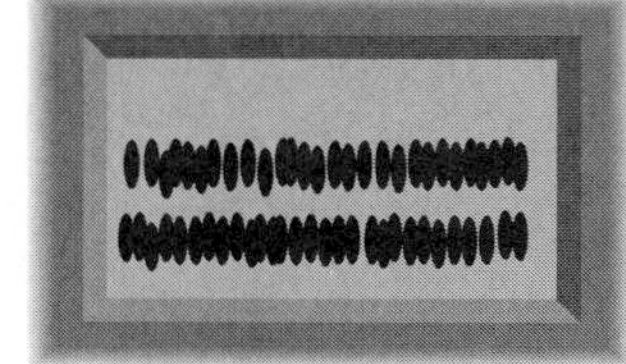

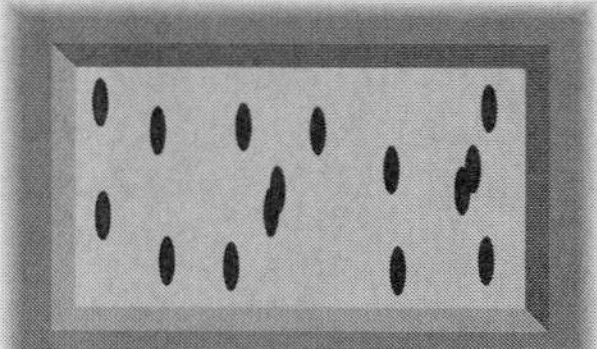

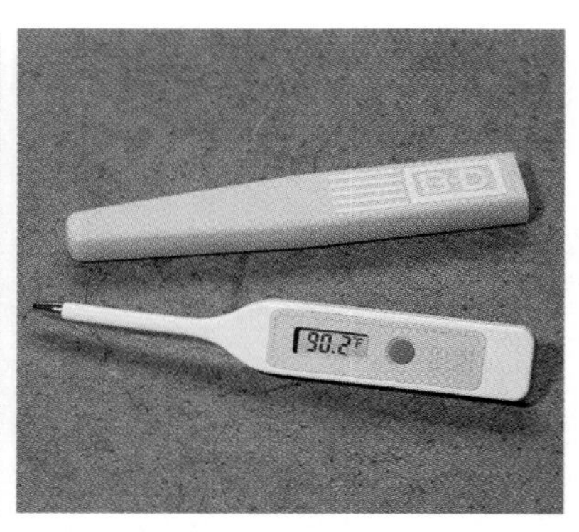


FIGURE 12.11
All the molecules of *smectic* liquid crystals (left) point in the same direction, and they also form layers perpendicular to their alignment. *Nematic* liquid crystals (right) are somewhat less ordered. All the molecules point in the same direction, but they don't form layers. A third form, *cholesteric* liquid crystals, has a more complex structure.

FIGURE 12.12
The lack of viscosity of superfluid helium gives rise to some unusual properties. For example, placing a capillary tube in the liquid gives rise to the cascading fountain you see here.

ature. These compounds in their partially ordered state are called *liquid crystals* (see Figure 12.11). Such materials can often change their properties under the influence of small changes in temperature, pressure, or electric field. This kind of ability to change properties underlies the liquid crystal displays you see on many calculators and digital watches.

Many solid substances have allotropes. As far as scientists know, however, only two substances have more than one liquid phase, hydrogen and helium. The boiling point of ordinary liquid helium is $-269°C$, just over 4 K. At even lower temperatures helium exists in a rare form called a *superfluid*. Among other odd properties, this liquid phase of helium has a viscosity of zero (see Figure 12.12).

### EXAMPLE 12.1 The Melting Points of Solids

A scientist determined the melting behavior of two solids by slowly and steadily adding heat to a sample of each. One of these graphs represents the results obtained for a crystalline solid, and the other those for an amorphous solid. Which is which? Explain your answer.

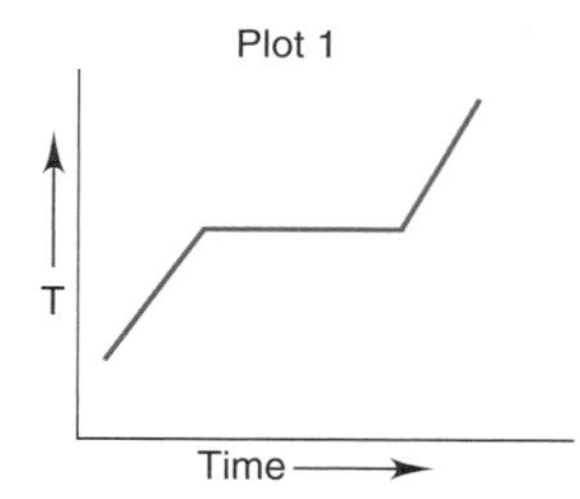


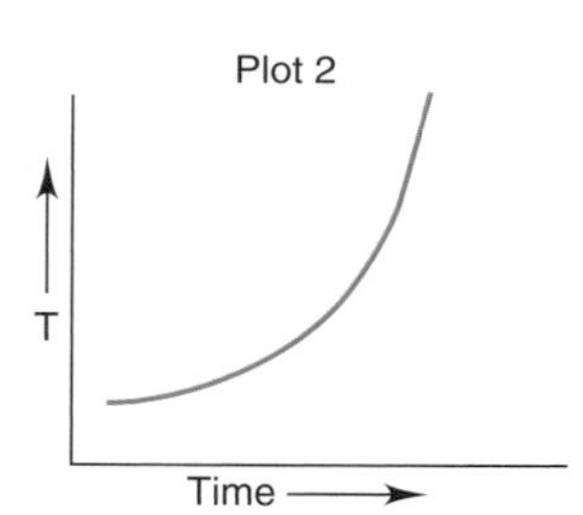


**SOLUTION**

Crystalline solids have sharp melting points, and amorphous solids do not. Therefore, Plot 1 shows the results for the crystalline solid and Plot 2 for the amorphous solid.

**EXERCISE 12.1**

The illustration below shows one type of melting point apparatus. To use it, you place a sample of a solid on the sample plate on top of the heating block. The heat control knob controls how fast the heating block warms up. Explain why it is necessary to heat the sample *slowly* in order to obtain an accurate melting point.

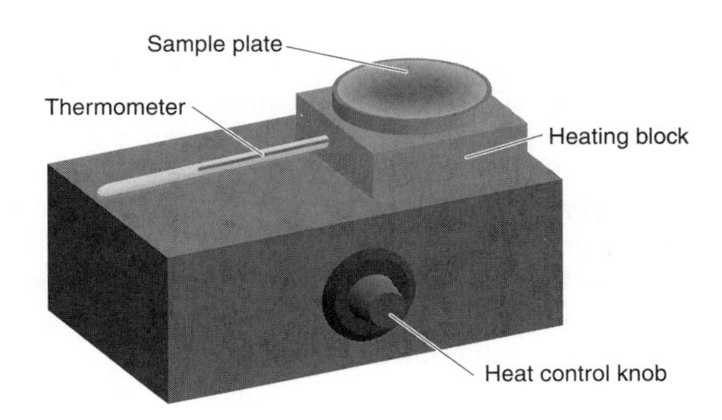


## 12.2 Strong Forces: Bonding in Ionic, Network, and Metallic Solids

**GOAL:** To describe the forces that hold ionic, network, and metallic solids together.

You are already familiar with the relationship between the forces holding molecules together and the physical properties of compounds. In this section we will look at this relationship in more detail, starting with the forces that hold the higher-melting solids together. Pieces of copper, salt crystals, and quartz crystals all have relatively high melting points. However, the copper is ductile while the others are brittle. Why?

The forces that hold a solid together are different for different substances. A network solid such as quartz, for instance, is linked throughout by covalent bonds (see Figure 12.13). Ionic compounds such as table salt are held together by the electrostatic attraction of the positive ions for the negative ions, a force called *ionic bonding*. The atoms of a metal such as copper are held together by the sea of electrons that constitutes the *metallic bond*. (You first heard about this in Section 10.1.) These forces are all very strong, which is why copper, salt, and quartz all have high melting and boiling points. On the other hand, they exhibit tremendous differences in ductility and brittleness.

Consider the metallic bond. The atoms held together share valence-shell electrons that are not localized to any particular nucleus. When the nuclei slide past one another, the electrons can shift very easily to accommodate the change. This is why metals are ductile and malleable.

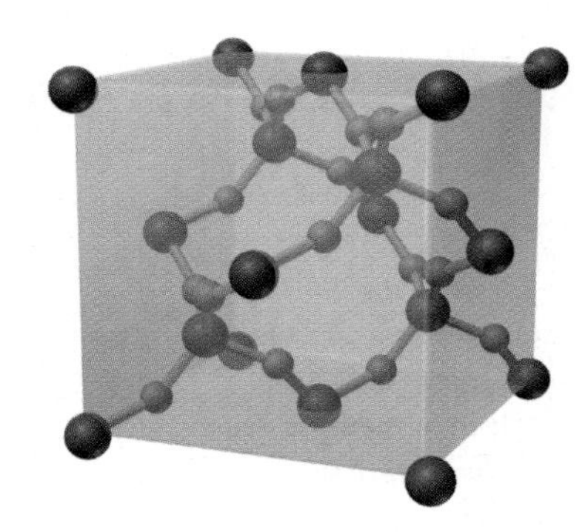

FIGURE 12.13
The minerals quartz, cristobalite (structure shown here), and tridymite are all composed of silica, $SiO_2$. They are all three-dimensional network solids with high melting points.

**The ancient Chinese clearly recognized the differences between substances that had these bonding types, though they had no theory of bonding. Three of their five elements were earth (ionic), wood (covalent), and metal (metallic of course). Their other two elements were water and fire.**

**Refer to Figure 10.10 if you need to review the bonding in metallic solids.**

Network solids behave differently. The covalent bonds holding their atoms together have very specific lengths and directions. These bonds must be broken to change the shape of a sample of a covalent network solid, and this bond breaking requires a lot of energy. Network solids, in other words, are much less malleable than metals are.

Some network solids consist of long chains that form only two-dimensional networks. Sometimes they can stretch (like rubber), but they rebound to their original shape. Network solids that form strong three-dimensional networks of bonds (like diamonds and quartz) are usually very brittle.

Let's take a closer look at diamonds and graphite, the two allotropic forms of carbon. You already know that diamonds are unreactive, hard, and have a very high melting point. Graphite is less dense than diamond, more reactive, and quite soft. The differences in properties arise from the structures. Diamond forms a three-dimensional array of bonds, while graphite forms a two-dimensional array of sheets (see Chapter Box 12.1).

The electrostatic force that holds ions together in ionic compounds is also three-dimensional in nature, but it is not directional (as covalent bonds are). So why aren't ionic compounds malleable? If the ions in an ionic crystal start to move past one another, sooner or later like charges will be next to one another. They repel one another, and the crystal breaks apart (see Figure 12.14). Ionic crystals are brittle, as you know from your own experience. Salt crystals and most rocks are made of ionic compounds.

**A polyatomic ion such as $SO_4^{2-}$ is held together by covalent bonds. Ionic bonding holds this ion in a crystalline structure with positive ions.**

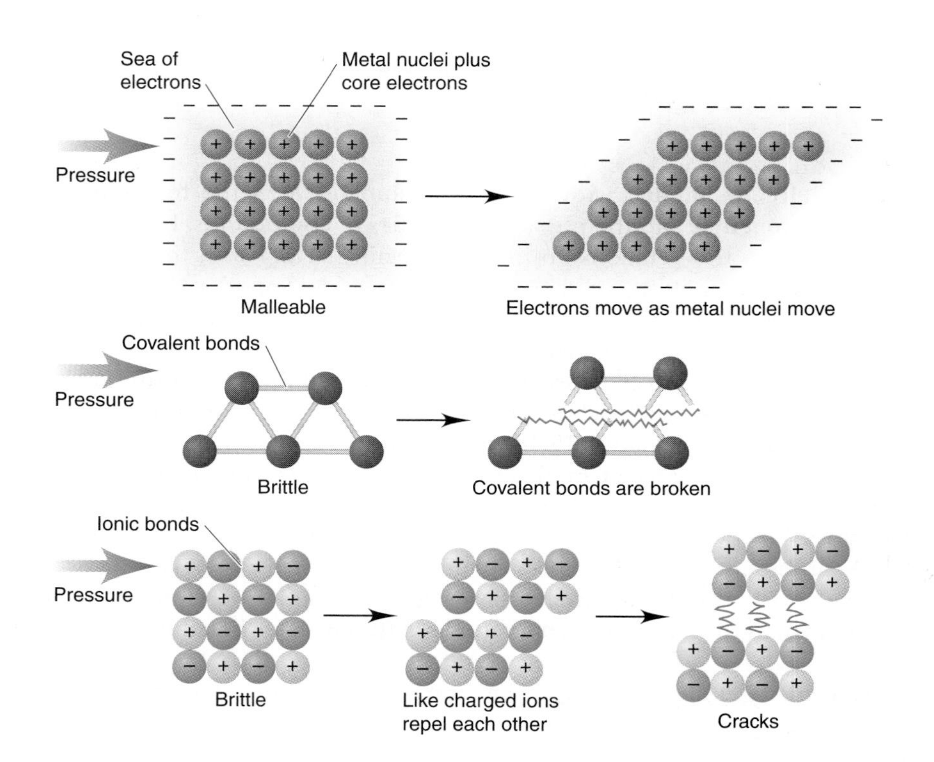


FIGURE 12.14
Structures of malleable and brittle substances.

(Top) In metals, a sea of valence-shell electrons hold the atoms (nuclei and core electrons) together. As a piece of metal deforms, the atoms move past one another, and the valence-shell electrons shift so that the structure holds together. This makes metals malleable and ductile.

(Center) In a three-dimensional network solid such as quartz or diamond, any attempt to move the atoms out of their original locations breaks the covalent bonds. Such substances are therefore brittle.

(Below) Deforming a crystal of an ionic compound brings ions with like charges next to one another. The crystal then breaks. These compounds are brittle.

## CHAPTER BOX 12.1 DIAMONDS AND DUST

**The most** important structural feature of diamond is its three-dimensional framework of covalent bonds. Because of this network, covalent bonds must be broken before the carbon atoms can move around, and this bond breaking requires considerable energy. Diamond won't melt below 3500°C and is chemically inert at ordinary temperatures. Diamond cutters use natural flaws in the crystal structure to decide where to split a diamond. Finding such a flaw can require days of study using a microscope, and diamond cutting is a task that only highly trained professionals are allowed to perform.

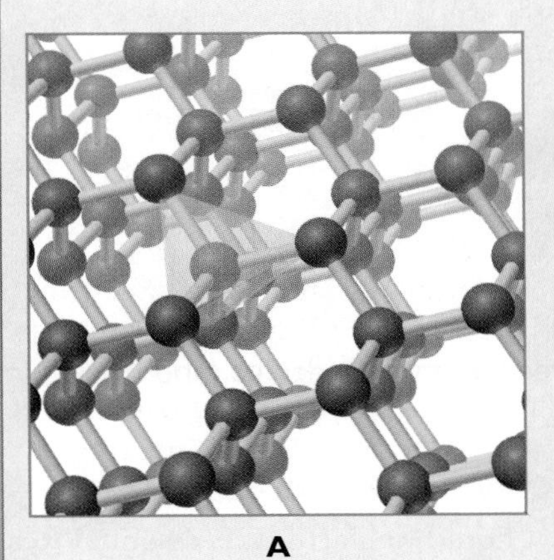

A B C

Diamonds. (A) The framework of carbon atoms in a diamond. (B) A diamond cutter at work. (C) The fabulous Hope Diamond.

The structure of graphite consists of flat sheets of carbon atoms stacked one on top of another, as shown below. Graphite is soft because these sheets can slide past one another easily, without breaking any of the covalent bonds within the sheets themselves. This structural feature makes graphite an excellent lubricant, and it also means that flakes of graphite can slide easily off the tip of a pencil onto a sheet of paper. Graphite samples are also much more reactive than diamonds.

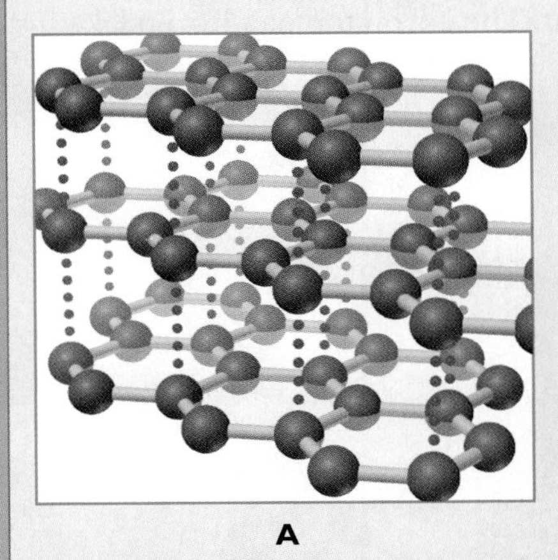

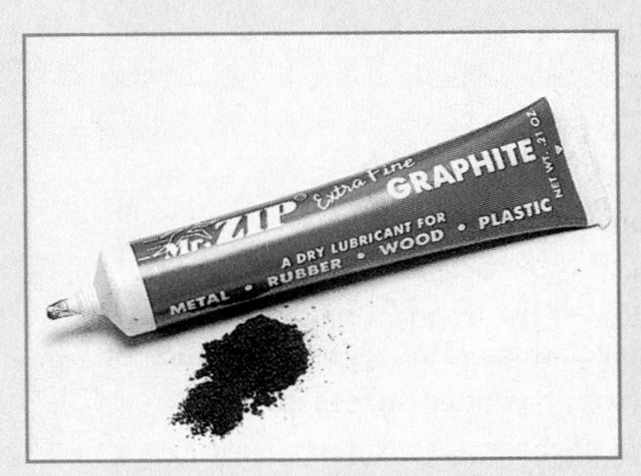


A B C

Graphite. (A) The framework of carbon atoms in graphite. (B) Pencils are the most familiar use of graphite. (C) Graphite is also a useful lubricant.

### EXAMPLE 12.2 Intermolecular Forces and Physical Properties

Is plastic wrap a three-dimensional network solid? Explain your answer.

**SOLUTION**

Because plastic wrap is flexible and because it stretches easily, you can conclude that it is not a three-dimensional network solid. (In fact it forms long chains; the molecules are so large that they are entangled.)

**EXERCISE 12.2**

Is wood a three-dimensional network solid? Explain your answer.

## 12.3 Hydrogen Bonding

**GOAL:** To describe the forces between water molecules (and molecules of some related substances).

Water is far and away the most important liquid on earth. It is essential to all plant and animal life. It is also the substance that keeps the temperature of almost all inhabited portions of the planet somewhere between $-20°C$ and 40°C. About 75% of the earth's surface area is covered by water. The oceans absorb heat from the sun during daylight hours, and this absorption prevents the temperature from rising too much. Then the water radiates this heat back into the atmosphere at night, and this keeps the atmosphere from getting too cold. Water holds heat very well relative to its small molecular mass.

Furthermore, solid water (ice) is less dense than liquid water—simply put, ice floats on water (see Figure 12.15). This property is rare. Almost all solids are more dense than their liquids. Life on earth would be very different, if it existed at all, if ice did not float. The ice on top of a frozen lake insu-

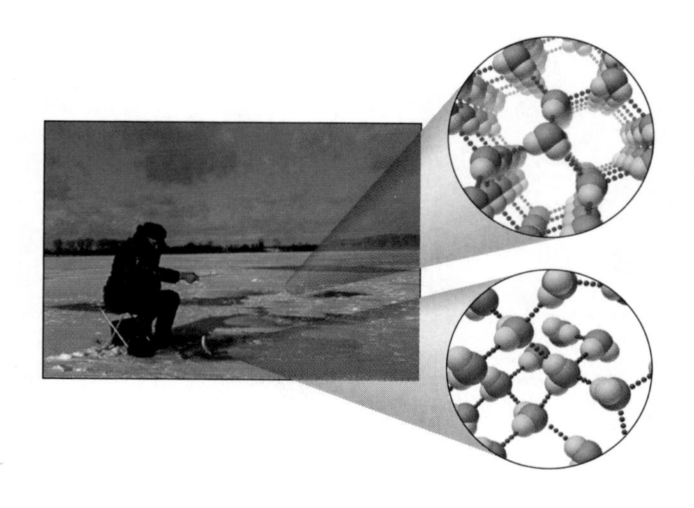

FIGURE 12.15
Water and ice. The crystal structure of ice (top) has open spaces in it, but some of these spaces don't exist in liquid water (bottom). This means that ice forms on the surface of a lake and doesn't sink to the bottom. Since the lake doesn't freeze solid, life goes on beneath the frozen surface.

lates the water below, so plant and animal life can continue beneath the surface of the ice. If ice sank, lakes in northern lands all around the globe would freeze solid every winter, killing the life in them. In spring only the top layer of water would thaw in the sunlight, leaving intact the great mass of ice below.

Water also has an extremely high boiling point, 100°C, for a molecule of its small size. If you think about the boiling process—the conversion of a liquid into a gas—you will realize that all that must happen for a liquid to boil is that its molecules must attain enough kinetic energy (through the input of heat) to allow them to escape the liquid phase. Since larger, more massive molecules are more difficult to move, you might conclude that large molecules have higher boiling points than do small molecules. This conclusion is only partially correct. Table 12.1 gives the boiling points of the compounds that hydrogen forms with Group 5A and 6A elements.

**TABLE 12.1 Boiling Points of Hydrogen Compounds of Group 5A and 6A Elements**

| Group 5A Compound | Molar Mass (g) | Boiling Point (°C) | Group 6A Compound | Molar Mass (g) | Boiling Point (°C) |
|---|---|---|---|---|---|
| $NH_3$ | 17 | −33 | $H_2O$ | 18 | 100 |
| $PH_3$ | 34 | −88 | $H_2S$ | 34 | −61 |
| $AsH_3$ | 78 | −55 | $H_2Se$ | 81 | −42 |
| $SbH_3$ | 125 | −17 | $H_2Te$ | 130 | −2 |

Note the trend in the lower part of this table (also see Figure 12.16). From $PH_3$ to $SbH_3$ and from $H_2S$ to $H_2Te$, both the molar masses and the boiling points steadily increase. So why do $NH_3$ and $H_2O$ have such high boiling points? According to the trend in the lower portion of the table, the boiling point of $NH_3$ should be lower than that of $PH_3$ (certainly not 54°C higher!). Also according to the trend, the boiling point of water should be well below − 61°C. Clearly, something other than molar mass influences the boiling points of these compounds.

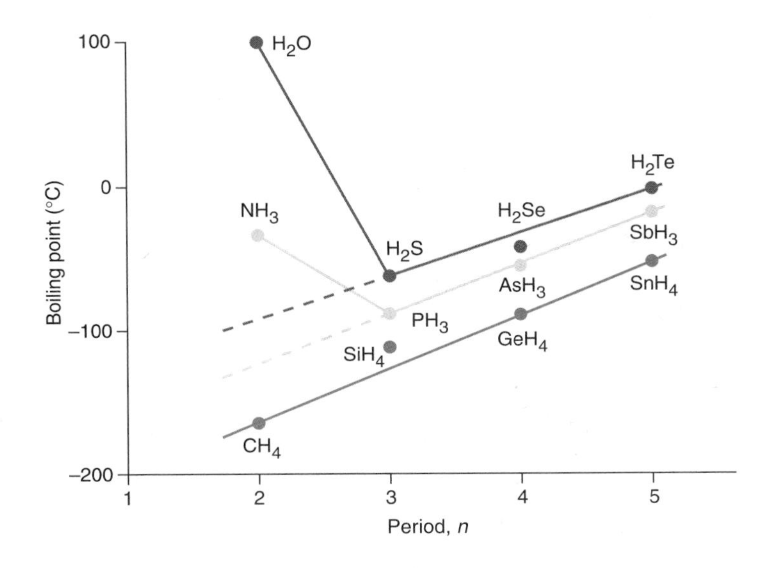


FIGURE 12.16 Boiling points of Group 4A, 5A, and 6A hydrides. The boiling points of the Group 4A hydrides are included here for purposes of comparison. Note that most of these compounds follow a general trend; those with more massive molecules have higher boiling points than those with less massive molecules. For example, $CH_4$ (16 g mol$^{-1}$) has a lower boiling point than $SiH_4$ (32 g mol$^{-1}$).

**TABLE 12.2 Covalent Bond Lengths for the Group 6A Hydrides**

| | Bond Length (pm) |
|---|---|
| H—O | 96 |
| H—S | 134 |
| H—Se | 146 |
| H—Te | 170 |

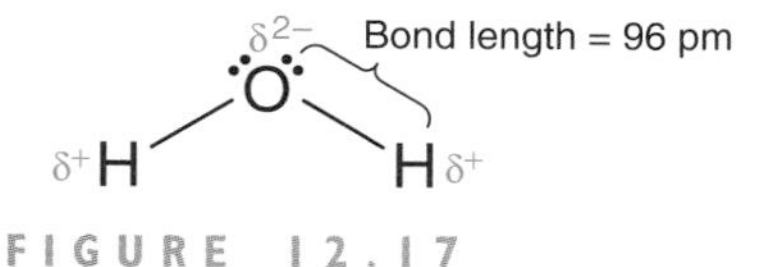


FIGURE 12.17
The polarity of the water molecule.

FIGURE 12.18
One water molecule is held to another by *hydrogen bonds,* which are different from covalent bonds. Hydrogen bonds are shown as dotted lines, as in this figure. Hydrogen bonds in water have about 4% as much strength as covalent O—H bonds. For any water molecule, a hydrogen bond can form to each of the electron pairs on the oxygen, and both hydrogen atoms can also form a hydrogen bond. Consequently, a single water molecule can be involved in a total of four hydrogen bonds (right).

Section 12.2 considered the strong forces that hold the atoms of metals, ionic compounds, and network solids together. These bonding forces involve attractions between atoms or ions. There must also be some attractions between molecules. All gases, even helium (which makes no compounds at all), can condense into liquids. Thus, intermolecular forces—forces of attraction between molecules—must exist. These forces are weaker than bonding forces, but do have measurable strength.

Let's begin by considering the relationships between the hydrogen and oxygen atoms in an ice cube. Each hydrogen atom in a water molecule is quite close to the oxygen atom, held there by a covalent bond (see Table 12.2). This covalent bond is polar, because oxygen is more electronegative than hydrogen (see Figure 12.17). For each oxygen atom there is a third (and frequently a fourth) hydrogen atom nearby, one that is part of another water molecule. This hydrogen atom is farther from the oxygen atom than the first two, too far away for a true covalent bond to exist. On the other hand, it is much closer to the oxygen atom than is any fifth hydrogen atom to the carbon atom in $CH_4$ or any third hydrogen atom to the sulfur atom in $H_2S$.

All hydrogen atoms in $H_2O$ have partial positive charges. Each one strongly attracts a nonbonding pair of electrons on an oxygen atom in another water molecule. At temperatures below 0°C the water molecules are locked together by this force, which is called **hydrogen bonding** (see Figure 12.18).

Next, let's consider what happens when an ice cube warms up. Hydrogen bonds are strong enough to hold molecules together at lower temperatures. As you heat ice, the water molecules move more and more vigorously as they gain kinetic energy. They are locked in place and can't move around freely the way gas molecules do, but they vibrate more and more strongly.

Finally, they break loose and begin to flow around one another. The hydrogen bonds still exist, but they break and form continuously as the molecules move. In other words, the ice cube melts and forms liquid water. If you keep heating the water, the molecules move more and more.

At last the molecules move so energetically that they knock each other away from the mass, and the water boils. For this to happen the hydrogen bonds holding the water molecules together have to break. Water has such a high boiling point because of the large amount of energy required to break the hydrogen bonds. Ammonia, $NH_3$, is also capable of hydrogen bonding, and it too has an unusually high boiling point (see Table 12.1).

A hydrogen bond has two ends, and each one has particular characteristics. At one end is a hydrogen atom, and at the other end is a pair of electrons on the strongly electronegative atom to which the hydrogen is bonded. These are necessary conditions. Hydrogen sulfide, $H_2S$, doesn't form hydrogen bonds to any significant extent because the sulfur atom is not electronegative enough. In practical terms, hydrogen bonding is important in covalent compounds that have O—H, N—H, and H—F bonds. These compounds have higher boiling points than similar compounds that are incapable of forming hydrogen bonds.

Hydrogen bonding can occur between two molecules of the same type—for example, two water molecules. It can also occur between two different molecules—such as a molecule of ammonia and a molecule of water. Finally, hydrogen bonding can occur between a molecule that is capable of hydrogen bonding and another that contains oxygen or nitrogen but no O—H or N—H covalent bond—for example, between a molecule of water and one of acetone, $(CH_3)_2C{=}O$.

Acetone is incapable of hydrogen bonding by itself because it contains no O—H bonds. However, the nonbonding electron pairs on the oxygen atom in acetone form hydrogen bonds with the hydrogen atoms in water molecules. This hydrogen bonding explains why acetone, unlike most organic liquids, dissolves easily in water. Figure 12.19 illustrates cases in which hydrogen bonding can occur.

**Hydrogen bonding is the strongest of the intermolecular forces. Hydrogen bonding is strong only for molecules that contain a bond between a hydrogen atom and an oxygen, nitrogen, or fluorine atom. Weaker hydrogen bonds occur with other molecules. Modern chemists speak of hydrogen bonding in molecules such as HCl and $H_2S$. These bonds are much weaker than those formed between molecules containing OH and NH groups, but they do exist.**

**Hydrogen bonding is also important to the sense of taste. It occurs between molecules in the food and receptor molecules in the tongue. The sour taste of acids (which contain O—H bonds) and the sweet taste of sugar (which contains many O—H bonds) are caused by hydrogen bonding.**

$H_3C-\ddot{O}\cdots H-\ddot{O}-CH_3$ (each O also bonded to H)
Hydrogen bonding between two molecules of the same compound
$H_3N\cdots\cdots H-\ddot{O}-H$
$H_2\ddot{N}-H\cdots\cdots \ddot{O}H_2$
Hydrogen bonding between two different molecules, both capable of hydrogen bonding with others like themselves
$(CH_3)_2C{=}\ddot{O}\cdots\cdots H-\ddot{O}-H$
Hydrogen bonding between two different molecules, one of which—in this case, acetone—is incapable of hydrogen bonding on its own

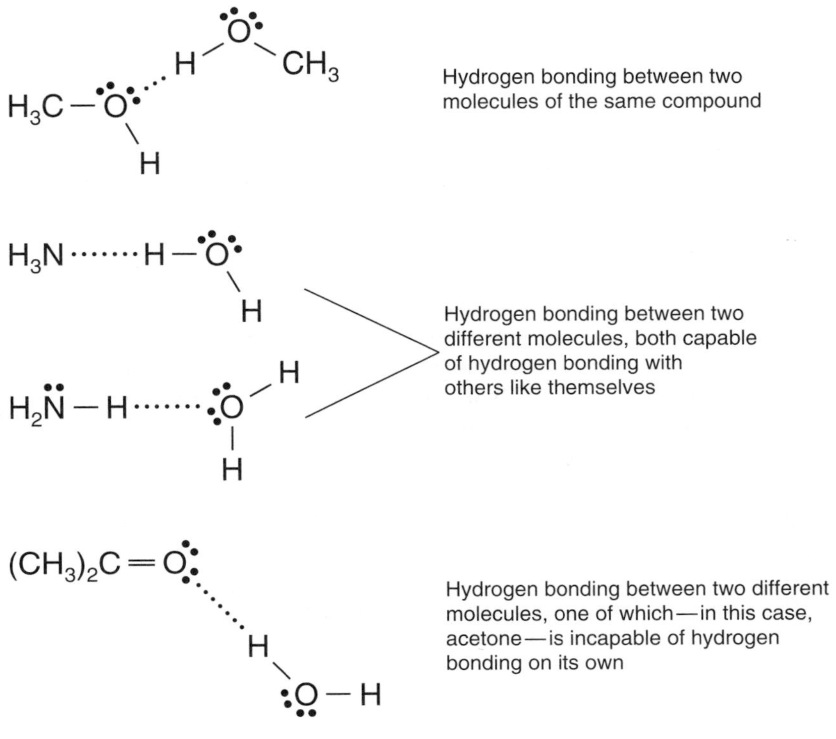

FIGURE 12.19
In what situations does hydrogen bonding occur?

## EXAMPLE 12.3 Relative Boiling Points

Which of these compounds has the highest boiling point? Which has the lowest? Explain your answers.

**A.** $CH_3CH_2OH$ **B.** $NH_4Cl$ **C.** $CH_3OCH_3$

SOLUTIONS

First, you need to figure out what kinds of forces hold each formula unit together. Compound A (ethanol) is a covalent compound that contains an OH group. Hydrogen bonding occurs between its molecules. Compound B contains an $NH_4$ unit, which is $NH_4^+$, the ammonium ion; B is an ionic compound. Compound C contains hydrogen atoms and an oxygen atom, but all the hydrogens are attached to the carbon atoms. (How do you know? You have to be able to draw the skeleton for a Lewis diagram.) Compound C is covalent, and there is no hydrogen bonding between its molecules. The highest boiling point is that of the ionic compound, B, and the lowest is that of C. (Note: At room temperature, A is a liquid, B is a solid, and C is a gas.)

EXERCISE 12.3

Which of these molecules are capable of hydrogen bonding with other molecules of the same compound? Explain the reasoning behind your answers.

**a.** $HN(CH_3)_2$
**b.** $CH_3CH_2CH_3$
**c.** $CH_3F$
**d.** $CH_3CH_2—O—CH_2CH_3$
**e.** $HOCH_2CH_3$
**f.** $OF_2$
**g.** $CH_3CH_2NH_2$
**h.** $CH_3CH_2CH_2OH$

FIGURE 12.20
The open structure of the molecules in ice gives it a density less than that of liquid water. Ice floats because of the hydrogen bonding in its structure.

Finally, let's return to the observation that ice floats. What is different about solid water that makes it less dense, not more, than liquid water? Again, the answer involves hydrogen bonding. In the crystal structure of ice, each oxygen atom is surrounded by four hydrogen atoms—two in the same water molecule, which are covalently bonded to the oxygen, and two in other water molecules, which are hydrogen bonded to the oxygen (see Figure 12.18). This bonding pattern results in a three-dimensional network in which each oxygen atom is at a corner of a six-sided ring, forming an open structure (see Figure 12.15). The hydrogen bonding actually forces the water molecules to stay farther apart than they are in the liquid phase. The increase in separation is small, but it gives ice a density of 0.917 g $mL^{-1}$ at 0°C, compared to 1.000 g $mL^{-1}$ for water.

## 12.4 Other Intermolecular Forces: Van der Waals Forces

GOAL: To describe the forces that hold the molecules of covalent compounds together.

Most covalent molecules are incapable of hydrogen bonding. Some are solids (such as candle wax and sulfur), some are liquids (such as gasoline and bromine), and some are gases (such as the components of the atmosphere). What forces attract one covalent molecule to another when there is no hydrogen bonding? There must be something, because even helium exists in the liquid phase if the kinetic energy of the atoms (that is, the temperature) is low enough. If there were no intermolecular forces of attraction at all, every substance would be a gas no matter what the temperature (see Figure 12.21).

FIGURE 12.21
Under high pressures or low temperatures, gases can become liquids and solids. Carbon dioxide in the high-pressure cylinder of a fire extinguisher is a liquid. Dry ice is the well-known solid form of carbon dioxide. As dry ice warms to room temperature, it sublimes (that is, it goes directly from the solid to the gas phase).

The forces of attraction between covalent molecules are electrostatic in nature. They work in a variety of closely related ways. As a group, they are named *van der Waals forces* after the Dutch physicist who investigated why the Ideal Gas Law doesn't hold exactly for real gases (Chapter Box 11.4). Depending on the types of molecules involved, these forces are called dipole-dipole forces, dipole–induced dipole forces, and induced dipole–induced dipole forces. Let's look at each type in more detail.

**Dipole-dipole forces** are electrostatic forces between polar molecules. They are very similar to hydrogen bonds, but have a different name because hydrogen bonds are so much stronger. Consider the molecule HCl, hydrogen chloride. Chlorine is more electronegative than hydrogen, so the chlorine end of the molecule has a partial negative charge and the hydrogen end has a partial positive charge. The HCl molecule is said to have a *permanent dipole.* The ends of two different HCl molecules attract each other (see Figure 12.22).

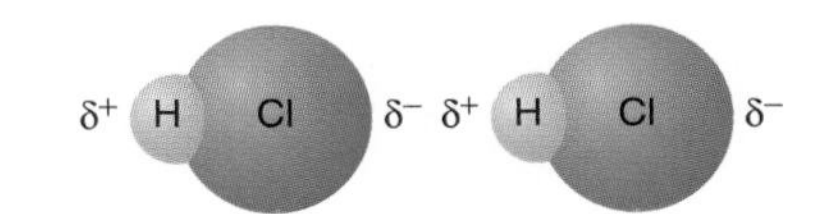


FIGURE 12.22
Opposite charges attract, so the negative end of one HCl molecule attracts the positive end of another HCl molecule. This dipole-dipole force is not very strong (about one-sixth as strong as the hydrogen bonding in water), so HCl has a low boiling point, about − 85°C. Compare this to the boiling point of HF (20°C), whose molecules hydrogen bond.

Dipole-dipole forces are considerably weaker than hydrogen bonds, so HCl is a gas at (and well below) room temperature. The boiling point of hydrogen fluoride, HF, which forms hydrogen bonds, is 20°C. The boiling point of HCl is very much lower, at − 85°C.

All polar compounds exhibit dipole-dipole forces, and they tend to have higher boiling points than do nonpolar compounds of similar masses. For example, both propane and acetaldehyde have molar masses of 44 g $mol^{-1}$. Propane is nonpolar and acetaldehyde is polar. The boiling point of propane is 68°C lower than the boiling point of acetaldehyde.

**Propane**
nonpolar, bp = −48°C

**Acetaldehyde**
polar, bp = 20°C

**Dipole–induced dipole forces** are also electrostatic in nature but occur between a molecule that has a permanent dipole and a molecule or atom that is nonpolar. Imagine an argon atom as it approaches the negative (chlorine) end of an HCl molecule. The isolated argon atom is electrically neutral overall and completely nonpolar. When the negative end of a polar molecule approaches an argon atom, it repels the electron cloud of the argon atom, pushing it away. The electron cloud around the chlorine atom attracts the argon nucleus, and since the argon electron cloud has been pushed off center, the forces do not balance. The polar molecule and the argon atom attract one another, although weakly.

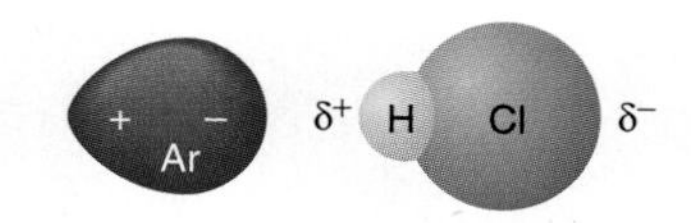


FIGURE 12.23
The positive end of an HCl molecule attracts the electron cloud of the Ar atom and polarizes it, inducing a dipole in the neutral atom.

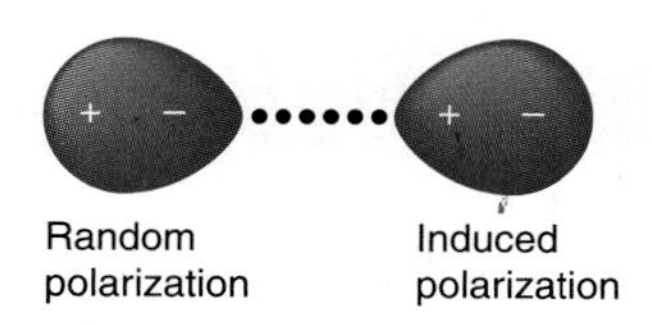


FIGURE 12.24
When one radon atom approaches another, the electrons can rearrange themselves so that the unlike charges attract each other more strongly than the like charges repel. The resulting *induced dipole–induced dipole force* is not very strong, as indicated by the boiling point of radon, − 62°C.

The argon atom is said to be **polarizable**. In other words, an approaching object with an electric charge can separate its charges and make it somewhat polar. The polarizability of an atom or molecule depends (in part) on how many electrons it contains. Large atoms and molecules are more polarizable than small ones. The force of attraction involved is even weaker than a dipole-dipole force.

The final type of electrostatic force is **induced dipole–induced dipole forces**. Visualize two nonpolar molecules as they approach each other. A random momentary polarization of one molecule can attract or repel the electrons in the other molecule. As a result, there is some attraction between the two molecules. In the smallest atoms and molecules, such as helium and hydrogen, this force of attraction is very, very weak. When the atoms or molecules are large and polarizable, the force of attraction can be considerable. Iodine, $I_2$, has a molecular mass of 254 amu and 106 electrons. This molecule is nonpolar, but iodine is a solid with a melting point of 113.5°C. The boiling point of radon (86 electrons, atomic mass of 222 amu) is higher than the boiling point of hydrogen chloride, at − 62°C (see Figure 12.24).

The intermolecular forces we have discussed so far vary widely in strength. Hydrogen bonding is the strongest, followed in order by dipole-dipole, dipole–induced dipole, and induced dipole–induced dipole forces. This order can change under certain circumstances. To illustrate this, let's consider the following molecule:

$$CH_3CH_2CH_2CH_2CH_2CH_2CH_2CH_2CH_2CH_2CH_2CH_2CH_2CH_2CH_2CH_2\underset{\displaystyle OH}{\underset{|}{C}}{=}O$$

The molecule contains an O—H bond, so it is capable of hydrogen bonding. The molecule is polar, with the oxygen-containing end having a partial negative charge. However, the strongest intermolecular forces between this molecule and others just like it are the induced dipole–induced dipole forces. The reasoning goes like this: The part of the molecule that participates in hydrogen bonding is very small compared to the whole molecule. Most of the molecule consists of nonpolar $CH_2$ units. Because there are so many $CH_2$ units with so much surface area, the sum of all the induced dipole–induced dipole interactions with another molecule of the same compound is stronger than the hydrogen bonding.

Keep in mind that *all* molecules undergo induced dipole–induced dipole interactions—even polar molecules. These forces are quite weak and are overshadowed by hydrogen bonding and dipole-dipole interactions for small molecules such as $H_2O$ and HCl. For large molecules they are almost always a major force, simply because the accumulation of lot of little things can be larger than one big thing. (Which would you rather have, one $100 bill or 30,000 pennies?)

### EXAMPLE 12.4 Identifying Intermolecular Forces

List all the types of intermolecular forces present in a sample of each of these compounds.

**a.** $CH_3CH_2CH_3$ **b.** $CH_3CH_2Cl$ **c.** $Br_2$ **d.** $CH_3OH$

**SOLUTIONS**

All of them exhibit induced dipole–induced dipole forces, (b) and (d) show dipole-dipole forces, and (d) is capable of hydrogen bonding.

**EXERCISE 12.4**

List all the types of intermolecular forces present in a sample of each of these compounds.

**a.** $NH_3$ **b.** $(CH_3)_2C{=}O$ **c.** NaBr **d.** $CH_4$

## 12.5 Crystal Structures

**GOAL:** To describe how atoms pack together to form crystals.

All elements have at least one solid, crystalline state. Forming this solid is most difficult for helium, requiring a temperature of less than 1.0 K and a pressure of 26 atm, but it can be done. Crystal structures of compounds, both ionic and covalent, can be quite complex. However, most metals and all the noble gases adopt relatively simple crystal structures in the solid state. How atoms arrange themselves in these structures is the topic of this section.

Suppose you have a flat tray, and you put enough marbles on it to cover its bottom with one layer and a few more. Then you shake the tray gently. Some of the marbles may fall into chaotic piles, but many of them will arrange themselves into patterns. If you look at the marbles toward the middle of many of these patterns, you will see that each one has six others touching it, forming a hexagon around the central marble (see Figure 12.25A). This kind of arrangement is the most efficient and space-conserving way to form a single layer of spherical objects on a flat surface.

Next, suppose that you begin adding marbles on top of the first layer. If you just drop a marble on at random, it will fall into a spot over a hole between three of the first-layer marbles (see Figure 12.25B). If you keep adding marbles, you will notice that only half of these holes are being covered. Each marble in the first layer is surrounded by six other marbles and six holes. Only three of these holes can be covered by marbles in the next layer (see Figure 12.25C and D).

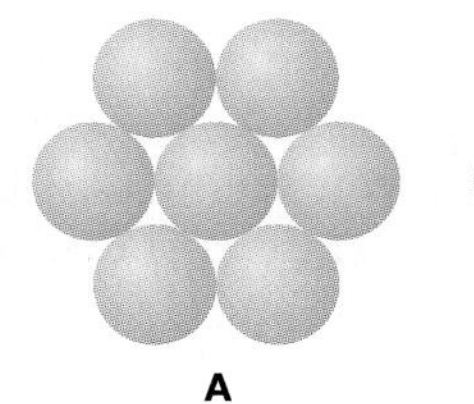
A

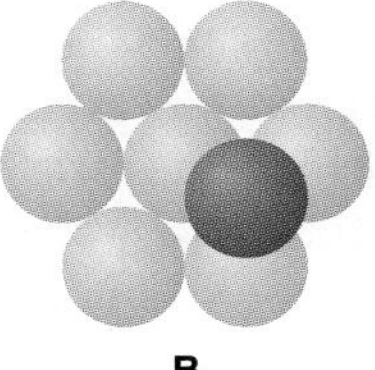
B

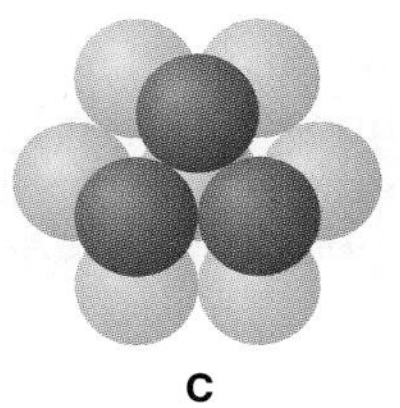
C

D

FIGURE 12.25
Packing spheres in a hexagonal array (A) is the most space-efficient way to cover a surface. When you add a sphere to start a second layer (B), the sphere sits above a hole between three of the bottom-layer spheres. There are two different ways to stack three spheres on the first seven, as shown in (C) and (D). This becomes important when you add a third layer.

## EXPERIMENTAL EXERCISE

### 12.1

A. L. Chemist analyzed a white crystalline solid, solid A. It had an empirical formula of $CH_2O$, a molar mass of 180 g, and a sharp melting point between 145°C and 147°C. A 10-g sample dissolved in 10 minutes in 100 mL of cold water, and a second 10-g sample dissolved within a few seconds in 100 mL of boiling water.

When a large sample of solid A was melted and quickly refrozen, a different substance formed, which was designated solid B. This solid was a single lump of a transparent substance with a glassy surface. It had a very wide melting point range that varied depending on how quickly the molten solid A was refrozen. A 10-g sample dissolved in 20 minutes in hot water. When the water was evaporated slowly, solid A formed and precipitated out of solution.

A. L. Chemist concluded that solid A and solid B were allotropic forms of the same substance and that solid B was an amorphous solid.

**QUESTIONS**

1. What is the molecular formula of solid A?
2. What type of bonding is present in solid A, ionic, covalent, or metallic?
3. What evidence supports the conclusion that solid B and solid A are allotropes?
4. What evidence supports the conclusion that solid B is amorphous?
5. Why did A. L. Chemist suspect that solid A was water-soluble?

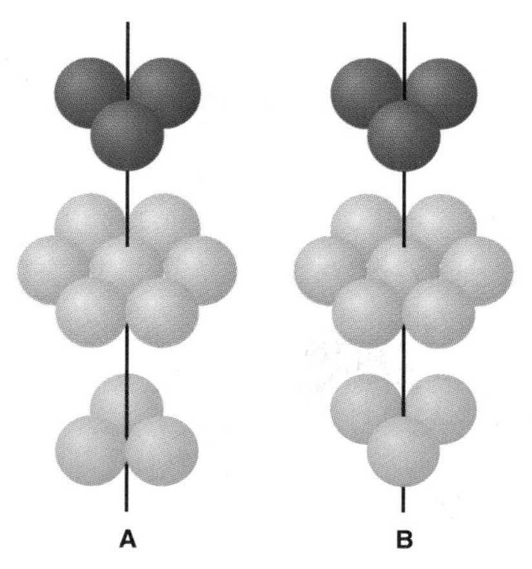


FIGURE 12.26
When you begin a third layer of spheres, you can put the spheres over the holes left by both the first and second layers (A), as in the crystal structure of copper, or you can put them directly above the spheres of the first layer (B), as in the crystal structures of magnesium and zinc.

Finally, you begin a third layer. If you have just three marbles in the second layer, there's only one way to put a marble in the third. This puts this third-layer marble directly above the central marble in a seven-marble group in the first layer. If you add more marbles, you can get a pattern like that in Figure 12.26B.

There is another way to place the third-layer marbles, though you must have more than six marbles on the second layer to start doing it. In this structure you place the third-layer marbles so they cover the first-level holes that weren't covered by the second-layer marbles. Figure 12.26A shows this structure.

Are atoms arranged this way in crystals? Yes, some of them are. The pattern in which the marbles in the third layer are directly above those in the first is called the **hexagonal close-packed structure**. Atoms in crystals of magnesium, zinc, and cobalt (among others) form this structure. The pattern with the third-layer marbles above first-layer holes is the **cubic close-packed structure**. Copper, aluminum, silver, gold, many other metals, and most of the noble gases crystallize with this structure.

Scientists call these structures close-packed because the available volume of space is filled as completely as possible. About 74% of the space is filled by marbles or atoms, while the rest is holes. Each sphere touches six atoms on its own plane, three in the plane above, and three in the plane below. If the spheres are atoms, this means that the *coordination number* of the atoms (the total number of neighbors touched by each atom) is 12.

There are other ways of packing marbles on a tray, as you might find if you experiment. Suppose, for instance, that your tray has square corners. If you add marbles while the tray is tilted, some of them will fall into a square arrangement. A second layer of marbles will fill all the holes, and the third layer repeats the first. This pattern is called a **body-centered cubic structure.** In a real crystal structure of this type, the atoms in any given layer have more separation than they do in the close-packed structures (see Figure 12.27). The four atoms above and the four below the holes in adjacent layers do touch a common central atom, giving a coordination number of 8. With this packing the spheres fill about 68% of the available volume. The alkali metals (sodium, potassium, etc.), some of the alkaline earth metals (calcium, strontium, and barium), and some of the transition metals (including iron) crystallize in this way.

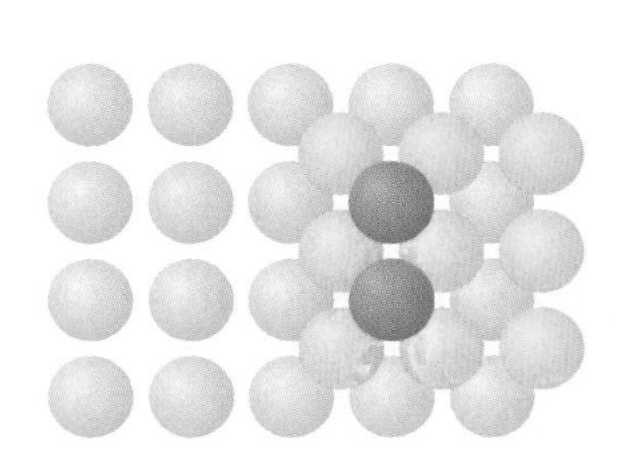

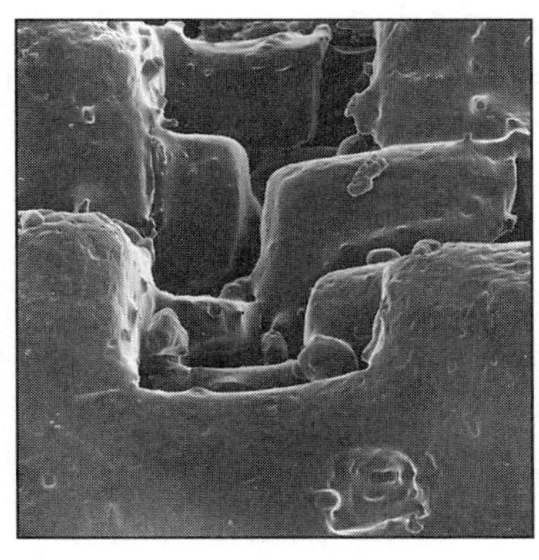

FIGURE 12.27
The atoms in layers of the body-centered cubic structure (left) form squares in which the atoms don't touch. Atoms in the layers above and below do touch atoms in a layer, giving a coordination number of eight. Crystals of sodium (right) have this structure.

Finally, suppose that you give your tray with square corners a very strong tilt. The marbles of all the layers might fall in a square pattern, with the atoms in all the layers directly in line with each other (see Figure 12.28). This pattern is called a **simple cubic structure**. It is a relatively inefficient way to fill space, because the spheres take up only 52% of the total volume. Of the elements, only radioactive polonium crystallizes with this structure.

What you have just read is an introduction to **crystallography**, the study of crystal structures. As you can imagine, the crystal structures of compounds are even more varied than those of elements. With atoms or ions of two or more sizes, and molecules or ions of irregular shape, more types of crystal structures are possible. In crystallography the experimental tools are powerful, and the mathematics (especially the geometry) is elegant. If this appeals to you, perhaps you will want to learn more about this subject in a more advanced course.

FIGURE 12.28
The atoms in the simple cubic structure stack directly on top of one another, leaving considerable empty space within the crystal structure.

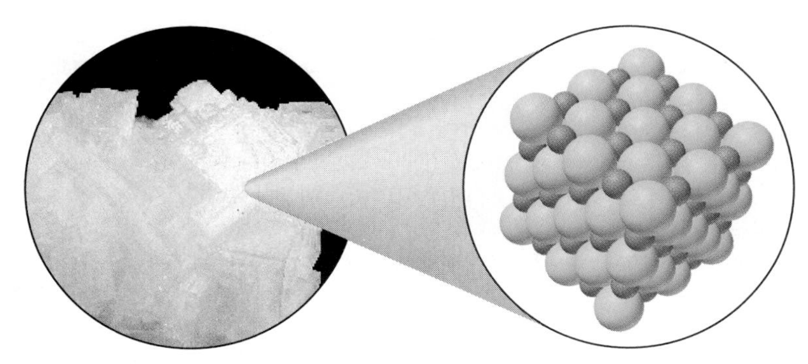

FIGURE 12.29
Sodium chloride (table salt) forms crystals shaped like cubes (left). This is a reflection of its body-centered cubic crystal structure (right).

### EXAMPLE 12.5 Crystal Structures

Imagine that atoms of a single element are packed in a rigid box in a regular pattern. Which of these crystal structures allows the smallest number of atoms to be packed in the box?

**A.** Cubic close-packed
**B.** Hexagonal close-packed
**C.** Simple cubic
**D.** Body-centered cubic

**SOLUTION**

The simple cubic structure fills the smallest percentage of the available space with atoms, so it must allow the smallest number of atoms in a given volume, if all the atoms are equal in size.

**EXERCISE 12.5**

For which of these elements do the atoms *not* pack into one of the four crystal structures (cubic close-packed, hexagonal close-packed, simple cubic, body-centered cubic) studied in this section?

**A.** Sodium **B.** Sulfur **C.** Iron **D.** Aluminum

## 12.6 Changes of State

**GOAL:** To examine the changes of state that samples of pure substances undergo as they melt, freeze, vaporize, and condense.

Up to this point we have looked at solids, liquids, and gases separately. The time has come to examine what happens when a substance changes from one of these phases to another. Such transitions are commonly referred to as *changes of state* (see Figure 12.30).

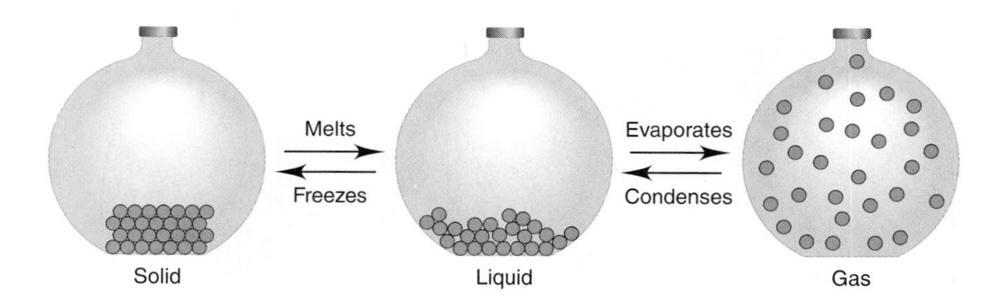


FIGURE 12.30
Changes in state, or phase changes.

Suppose you stuck a thermometer in a glass of ice water and wrote down times and temperatures of the mixture at regular intervals until all the ice melted. Then you continued to record temperatures until the cold water warmed up to room temperature. A plot of data obtained in this way looks like Figure 12.31. Although heat is being absorbed from the surroundings (you know this because the ice melts), the temperature of the ice-water mixture stays constant at 0°C until the last of the ice melts. The water warms up to room temperature only after there is no ice left in the glass.

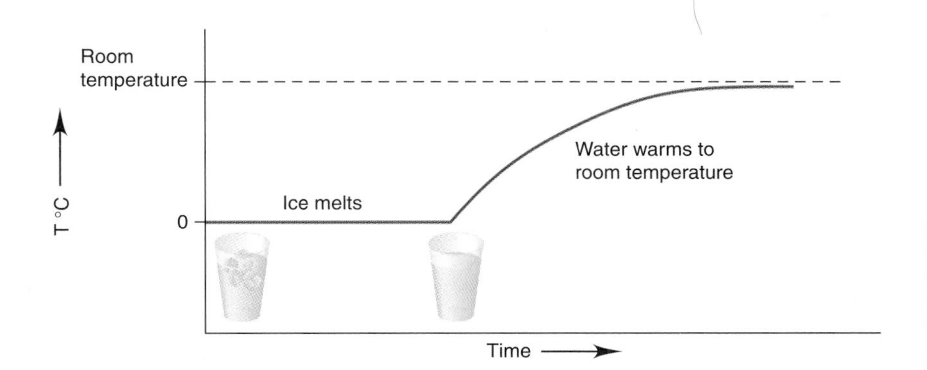


FIGURE 12.31
The temperature stays at 0°C for ice-water mixtures until all the ice melts. The water then warms up to room temperature.

What is happening at the molecular level? The Second Law of Thermodynamics (Section 4.2) says that heat flows from warmer objects (such as the air in the room) to colder ones (the ice-water mixture) and that systems tend to become disordered. The ice and the water in the glass are at the same temperature. However, the ice is a crystalline solid, and its water molecules are arrayed in an orderly pattern. As heat flows into the glass, the water molecules in the ice crystals absorb it, and some of the hydrogen bonds are broken, raising the potential energy and forming a less ordered pattern.

Because the potential energy of the molecules must increase in order to break up the internal structure of a crystalline solid, melting is an endothermic process (heat is absorbed). Exactly how much heat it takes to melt a particular crystal depends on the mass of the crystal and the strength of its intermolecular forces. The **molar heat of fusion** is defined as the amount of heat it takes to melt 1 mol of a substance at constant pressure. Table 12.3 gives the molar heats of fusion of some common substances.

**TABLE 12.3 Molar Heats of Fusion**

| Substance | Melting Point (°C) | Heat of Fusion (kJ $mol^{-1}$) |
|---|---|---|
| Hydrogen ($H_2$) | −259 | 0.9 |
| Methane ($CH_4$) | −182 | 0.94 |
| Ethanol ($CH_3CH_2OH$) | −115 | 5.01 |
| Water ($H_2O$) | 0 | 6.0 |
| Silver (Ag) | 961 | 11.3 |
| Sodium Chloride (NaCl) | 801 | 27.2 |

## EXAMPLE 12.6 Heat of Fusion

How much heat does it take to melt 50.0 g of ice at 0°C?

**SOLUTION**

First you convert grams of water into moles and then use the molar heat of fusion of water (6.0 kJ $mol^{-1}$) as a conversion factor to calculate the required amount of heat:

$$50.0 \text{ g } H_2O \left(\frac{1 \text{ mol } H_2O}{18 \text{ g } H_2O}\right)\left(\frac{6.0 \text{ kJ}}{1 \text{ mol } H_2O}\right) = 17 \text{ kJ}$$

**EXERCISE 12.6**

How much heat does it take to melt 50.0 g of ethanol at − 115°C?

Next, suppose you are heating a pot of water on a stove. As you did for the glass of ice water, you measure the temperature of the water periodically as it heats up and then boils away. A plot of the data obtained from this experiment (see Figure 12.32) has some similarities to the one you obtained for the ice-water mixture. As the stove adds heat to the warm water, the temperature increases, as you would expect. When the water starts to boil, the temperature stops increasing. As long as liquid water remains in the pot, the temperature of the water-steam mixture remains constant at 100°C. Finally, when all the liquid water is gone, the temperature of the gaseous water can start to increase again.

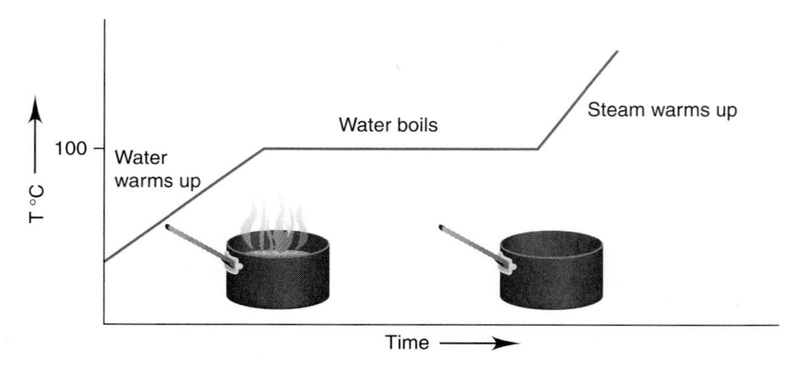


FIGURE 12.32
The temperature of a water and steam mixture remains at 100°C as long as the water is boiling. After all the water is gone, the steam will heat up.

Evaporation, like melting, is an endothermic process. The water absorbs heat, and some of the water molecules gain enough kinetic energy to enter the gas phase. Temperature is a manifestation of kinetic energy, and as long as the molecules with high kinetic energies leave the liquid phase, the temperature of the liquid will remain constant regardless of how fast you heat it. The amount of heat it takes to evaporate 1 mol of liquid at 1 atm is called the **molar heat of vaporization**. For values of the heats of vaporization of some common substances, see Table 12.4.

**TABLE 12.4 Molar Heats of Vaporization**

| Substance | Boiling Point (°C) | Heat of Vaporization (kJ $mol^{-1}$) |
|---|---|---|
| Hydrogen ($H_2$) | −253 | 0.12 |
| Methane ($CH_4$) | −164 | 10.4 |
| Ethanol ($CH_3CH_2OH$) | 78.5 | 38.6 |
| Water ($H_2O$) | 100 | 40.7 |
| Sodium Chloride (NaCl) | 1465 | 207 |

### EXAMPLE 12.7 Changes of State

Draw a graph similar to those in Figures 12.31 and 12.32 showing what happens to the temperature of the water if you take a pail of ice and water at 0°C, heat it until all the water evaporates, and continue heating for a short time.

**SOLUTION**

This is a combination of the two processes illustrated in Figures 12.31 and 12.32, so your graph should contain the essential elements of each.

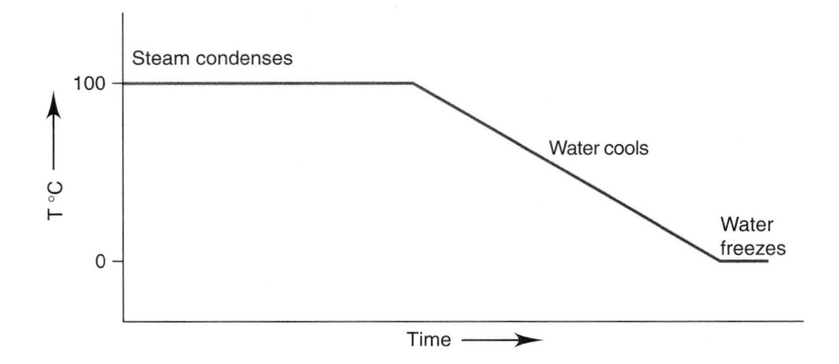


**EXERCISE 12.7**

Draw a graph similar to those in Figures 12.31 and 12.32 showing what happens to the temperature of the water if you take a flask filled with gaseous water at 100°C and condense it to ice at 0°C.

## CHAPTER SUMMARY

1. In both liquids and solids, molecules are in contact with one another. In liquids the molecules can move past one another (though they move slowly in viscous liquids); in solids they cannot.
2. Crystalline solids have regular, ordered arrays of atoms, or molecules. In amorphous solids the molecules or atoms are randomly arranged.
3. The properties of solids—ionic, metallic, and network—depend on the type of bonding present. Each type of solid has unique properties.
4. Significant hydrogen bonding occurs between hydrogen atoms covalently bonded to O, N, and F atoms and nonbonding pairs of electrons on O, N, and F atoms in other molecules. Hydrogen bonding is the strongest of the intermolecular forces.
5. Polar molecules have dipole-dipole interactions with each other. These forces are stronger when the molecules are more polarizable and when the charge separation in the dipole is large.
6. The dipoles in polar molecules can separate the centers of positive and negative charges in otherwise nonpolar molecules. These dipole-induced dipole interactions occur in mixtures of polar and nonpolar molecules.
7. Induced dipole–induced dipole interactions are possible for all molecules. Individually, these are the weakest of the intermolecular forces, but for larger molecules they can total up to be stronger than other intermolecular forces.
8. Atoms in crystals fall into regular, repeating patterns. The two patterns in which the maximum amount of available space is filled are called the hexagonal and the cubic close-packed structures. Others, less effective at filling the available space, are the body-centered cubic and the simple cubic structures. Elements crystallize with these structures. Ionic and covalent compounds have additional types of crystal structures.
9. As samples of pure substances melt or vaporize, they absorb heat without changing temperature; both of these processes are endothermic. The amount of heat required to melt a mole of a substance is called the molar heat of fusion. Similarly, the amount of heat required to vaporize a mole of a substance is called the molar heat of vaporization.

## PROBLEMS

### SECTION 12.1 Solids and Liquids Classified

1. Which of these is often used in calculator displays?
   A. Superfluids  C. Allotropes
   B. Liquid crystals  D. Amorphous solids

2. Glass is an example of which of the following?
   A. Superfluid  C. Allotrope
   B. Liquid crystal  D. Amorphous solid

3. Diamonds and graphite are examples of which of the following?
   A. Superfluids  C. Allotropes
   B. Liquid crystals  D. Amorphous solids

4. Which of the following has no known examples at room temperature?
   A. Superfluids  C. Allotropes
   B. Liquid crystals  D. Amorphous solids

5. Give two examples of amorphous solids not mentioned in the text.

6. Give two examples of network solids not mentioned in the text.

7. Name two elements that have allotropic forms that can exist as solids at room temperature.

8. Name one element that is known to exist as a superfluid near absolute zero.

9. You are given a clear, colorless block of a solid material and are told that it is either an NaCl crystal (an ionic solid), a block of quartz (a network solid), or a block of clear plastic (an amorphous solid). How would you tell which type of solid the sample is? Describe an experimental procedure.

10. If you heat a sample too rapidly during a melting-point experiment (see Exercise 12.1), would the measured melting point be higher or lower than its actual value? Justify your answer.

### SECTION 12.2 Strong Forces: Bonding Ionic, Network, and Metallic Solids

11. Name one general type of network solid that is not brittle.

12. Name five substances that consist of small covalent molecules with no extended structure.

13. For which of these substances do covalent bonds have to be broken to melt the substance?
A. C  B. $Na_2SO_4$  C. Na  D. $CH_3CH_3$

14. For which of these substances do ionic bonds have to be broken to melt the substance?
A. C  B. $Na_2SO_4$  C. Na  D. $CH_3CH_3$

15. The rubbery sulfur shown in the photo in Figure 12.6, which has the structure shown in Figure 12.7, can be classified as a one-, two-, or three-dimensional network solid. Which is correct? Explain your answer.

16. Name a two-dimensional and a three-dimensional network solid. Do not use sulfur as one of your examples. Describe the physical properties that arise from the number of dimensions.

17. Which of these elements forms a network covalent structure in the solid state?
A. Carbon  C. Sodium
B. Hydrogen  D. Fluorine

18. Which of these elements does not form an extended structure in the solid state?
A. Carbon  C. Sodium
B. Iron  D. Fluorine

19. Which of these solids does *not* have brittleness as a property?
A. Hydrogen  C. Potassium
B. Chlorine  D. Carbon

20. For which of these solids is it necessary to break covalent bonds to melt the solid?
A. Iodine  C. Sulfur (crystalline)
B. Calcium  D. Carbon (graphite)

21. Draw a picture similar to one of the ones in Figure 12.14 to illustrate what happens when a chunk of gold metal is pounded flat.

22. Draw a picture similar to one of the ones in Figure 12.14 to illustrate what happens when a chunk of sodium chloride is hit with a hammer.

### SECTION 12.3 Hydrogen Bonding

23. Which of these compounds has the highest boiling point? Explain your answer.
A. $NH_4F$  C. $CH_3CH_3$
B. $CH_3OH$  D. $CH_3NH_2$

24. Which of these compounds has the lowest boiling point? Explain your answer.
A. $NH_4F$  C. $CH_3CH_3$
B. $CH_3OH$  D. $CH_3NH_2$

25. For each of these compounds, indicate whether the molecules can form hydrogen bonds with other molecules of the same type. Explain your answer in each case.
a. $HCF_3$  c. HF
b. $H_2C{=}O$  d. $CH_3{-}O{-}CH_2CH_3$

26. For each of these compounds, indicate whether the molecules can form hydrogen bonds with other molecules of the same type. Explain your answer in each case.
a. $CH_3CH_2OH$  c. $O_3$
b. $H_2N{-}NH_2$  d. $CH_3CH_2CH_2CH_3$

27. Indicate whether each of these compounds can form hydrogen bonds with molecules of water. Explain or draw structures to justify your answer in each case.
a. $NCl_3$  b. HI  c. $HCCl_3$  d. $NH_4Cl$

28. Indicate whether each of these molecules or ions can form hydrogen bonds with water molecules. Explain or draw structures to justify your answer in each case.
a. $H_2C{=}O$  c. $H_3O^+$
b. $(CH_3CH_2)_2O$  d. $CH_3CH_2OH$

29. Explain the trend in boiling points for this series of compounds.

| Formula | Molar Mass (g mol$^{-1}$) | Boiling point (°C) |
|---|---|---|
| $CH_4$ | 16 | – 162 |
| $CH_3CH_3$ | 30 | – 88.5 |
| $CH_3CH_2CH_3$ | 44 | – 42 |

30. Explain the trend in boiling points for this series of compounds.

| Formula | Molar mass (g $mol^{-1}$) | Boiling point (°C) |
|---|---|---|
| HF | 20 | 20 |
| HCl | 36 | −85 |
| HBr | 61 | −67 |

## SECTION 12.4 Other Intermolecular Forces: Van der Waals Forces

31. List all the intermolecular forces present in a sample of each compound.
a. $NH_4Br$
b. $H_2C{=}O$
c. $CH_3SH$
d. $CH_3CH_3$

32. List all the intermolecular forces present in a sample of each compound.
a. $CH_3CO_2H$
b. HBr
c. $CH_2Cl_2$
d. $H_2C{=}CH_2$

33. Draw a picture similar to Figure 12.23 or 12.24, illustrating what you think happens when a cation approaches a polar molecule.

34. Draw a picture similar to Figure 12.23 or 12.24, illustrating what you think happens when an anion approaches a polar molecule.

35. Draw a picture similar to Figure 12.23 or 12.24, illustrating what you think happens when a cation approaches an atom of a noble gas.

36. Draw a picture similar to Figure 12.23 or 12.24, illustrating what you think happens when an anion approaches an atom of a noble gas.

37. Based on your drawings for Problems 33 and 34, do *ion-dipole forces* exist in those two situations?

38. Based on your drawings for Problems 35 and 36, do *ion–induced dipole forces* exist in those two situations?

39. Give an example of a compound whose molecules do not form hydrogen bonds among themselves but will form hydrogen bonds with molecules of $CH_3OH$.

40. Give an example of a compound whose molecules are capable of hydrogen bonding among themselves but have induced dipole–induced dipole forces as the strongest interaction between them.

## SECTION 12.5 Crystal Structures

41. Atoms of a single element fall into a regular pattern in a rigid box. Which of these crystal structures allows the smallest number of atoms to be packed along one edge of the box?
A. Cubic close-packed
B. Hexagonal close-packed
C. Simple cubic
D. Body-centered cubic

42. X-ray crystallography is a tool used by chemists and physicists to study crystal structures. For which of these substances is X-ray crystallography most useful?
A. Polyethylene
B. Sodium metal
C. Glass
D. Amorphous sulfur

43. Suppose you are a two-dimensional being living in a two-dimensional universe, with an interest in the two-dimensional crystal structures that you find in your world. Draw a diagram that shows close packing of circles in a single two-dimensional layer.

44. Suppose you are a two-dimensional being with an interest in two-dimensional crystal structures. Draw a diagram that shows square packing of circles in a single two-dimensional layer.

## SECTION 12.6 Changes of State

45. Calculate the amount of heat that is absorbed when each sample melts at its normal melting point. The molar heats of fusion are given in Table 12.3.
a. 2.0 mol of hydrogen
b. 5.9 mol of methane
c. $1.97 \times 10^{-3}$ mol of ethanol
d. $2.3 \times 10^{2}$ mol of water

46. Calculate the amount of heat that is absorbed when each sample melts at its normal melting point.
a. 0.66 mol of silver metal
b. 9.5 mol of $H_2O$
c. $1.3 \times 10^{4}$ mol of NaCl
d. $4.29 \times 10^{-3}$ mol of methane

47. Calculate the amount of heat that is absorbed when each of these samples melts at its normal melting point. The molar heats of fusion are given in Table 12.3.
a. 4.0 mol of $H_2O$
b. 7.1 mol of $CH_3CH_2OH$
c. $2.55 \times 10^{4}$ mol of Ag
d. $9.2 \times 10^{-6}$ mol of $H_2$

48. Calculate the amount of heat that is absorbed when each of these samples melts at its normal melting point.
a. 0.39 mol of NaCl
b. $4.32 \times 10^{2}$ mol of hydrogen
c. $6.1 \times 10^{-3}$ mol of silver metal
d. 1.4 mol of $H_2O$ (an ice cube)

49. How much heat is required to melt each sample at its normal melting point?
    a. 1.75 g of Ag metal
    b. 7.7 kg of water
    c. $1.1 \times 10^3$ kg of sodium chloride
    d. $4.6 \times 10^{-3}$ g of $CH_4$

50. How much heat is required to melt each sample at its normal melting point?
    a. $2.78 \times 10^9$ kg of $H_2$
    b. 9.2 g of methane
    c. $1.97 \times 10^{-3}$ g of $CH_3CH_2OH$
    d. $2.3 \times 10^2$ kg of $H_2O$

51. What amount of heat is required to melt each of these samples at its normal melting point?
    a. 125 mg of sodium chloride
    b. 9.443 g of ethanol
    c. 28 g of Ag metal (1 ounce)
    d. $5.09 \times 10^7$ kg of water

52. What amount of heat is required to melt each of these samples at its normal melting point?
    a. 1.7 mg of $CH_3CH_2OH$
    b. 50 kg of $H_2O$
    c. $1.1 \times 10^{-5}$ kg of hydrogen
    d. $2.7 \times 10^3$ g of silver metal

53. Calculate the amount of heat that is required to vaporize each sample at its normal boiling point. The molar heats of vaporization are given in Table 12.4.
    a. 12.7 mol of hydrogen
    b. 0.173 mol of methane
    c. $1.2 \times 10^{-2}$ mol of ethanol
    d. $4.19 \times 10^7$ mol of water

54. Calculate the amount of heat that is required to vaporize each sample at its normal boiling point.
    a. 0.222 mol of $CH_4$
    b. 8.3 mol of $H_2O$
    c. $7.9 \times 10^3$ mol of NaCl
    d. $6.8 \times 10^{-5}$ mol of table salt

55. Calculate the amount of heat that is required to vaporize each of these samples at its normal boiling point. The molar heats of vaporization are given in Table 12.4.
    a. 117 mol of ethanol
    b. 12 mmol of water
    c. $9.9 \times 10^5$ mol of hydrogen
    d. $1.2 \times 10^2$ mol of methane

56. Calculate the amount of heat that is required to vaporize each of these samples at its normal boiling point. The molar heats of vaporization are given in Table 12.4.
    a. 22 mol of NaCl
    b. 2 mmol of $CH_3CH_2OH$
    c. $3.3 \times 10^3$ mol of $CH_4$
    d. 1.7 mmol of $H_2O$

57. How much heat is required to vaporize each sample at its normal boiling point? The molar heats of vaporization are given in Table 12.4.
    a. 7.23 g of ethanol
    b. $1.4 \times 10^{24}$ g of $H_2O$ (all the world's oceans)
    c. $8.81 \times 10^3$ kg of sodium chloride
    d. $3 \times 10^{-3}$ g of $CH_4$

58. How much heat is required to vaporize each sample at its normal boiling point? Molar heats of vaporization are given in Table 12.4.
    a. 1.8 kg of $H_2$
    b. $1.2 \times 10^8$ kg of $CH_4$ (the liquid natural gas in a full tanker)
    c. $1.97 \times 10^4$ g of $CH_3CH_2OH$
    d. 7.7 kg of water

59. What amount of heat is needed to vaporize each of these samples at its normal boiling point?
    a. 3.1 mg of sodium chloride
    b. 45 kg of $CH_4$
    c. $7.50 \times 10^4$ g of ethanol
    d. $2 \times 10^5$ g of $H_2O$ (a bathtub full)

60. What amount of heat is needed to vaporize each of these samples at its normal boiling point?
    a. 9 mg of $CH_3CH_2OH$
    b. 300 g of water (a glassful)
    c. $2.12 \times 10^2$ g of $H_2$
    d. $9.95 \times 10^4$ g of methane

61. In hot and dry regions of the world, some people use machines called *evaporative coolers* to keep their homes comfortable. Water is used in these coolers. Explain how such a machine might work.

62. Perspiration is the means your body uses to keep you from overheating. Explain how the evaporation of water from your skin keeps you cool on a hot day.

63. The molar heats of fusion of several substances are given in Table 12.3. What is the relationship between the strength of the intermolecular forces holding a crystal together and the magnitude of the molar heat of fusion?

64. The molar heats of vaporization of several substances are given in Table 12.4. What is the relationship between the magnitude of the molar heat of vaporization and the strength of the intermolecular forces present in a liquid?

## ADDITIONAL PROBLEMS

65. Although carbon has a smaller molar mass than methane does, it has a much higher melting point. Explain why.

66. Explain why most network solids (rocks, gemstones, plastics, etc.) are insoluble in water and most other solvents.

## EXPERIMENTAL PROBLEM 12.1

During a study of gases, A. L. Chemist hypothesized that some, but not all, acetic acid molecules hydrogen bond in the gas phase to form dimers. The structures of acetic acid (molar mass = 60 g $mol^{-1}$) and the proposed dimer are as follows:

$H_3C-C(=O)-O-H$ | $H_3C-C(=O\cdots H-O)(-O-H\cdots O=)C-CH_3$

**Acetic acid** | **Acetic acid dimer**

Two experiments were performed, and these data were obtained.

Experiment 1: A 1.100-g sample of $CH_3CO_2H(g)$ occupied a volume of 637 mL at 110°C and 454 torr.

Experiment 2: A 0.810-g sample of $CH_3CO_2H(g)$ occupied a volume of 644 mL at 156°C and 458 torr.

**QUESTIONS**

**67.** Calculate the molar mass of acetic acid from the results of Experiment 1.

68. Does this result support A. L. Chemist's hypothesis?

**69.** Compute the molar mass of acetic acid from the results of Experiment 2.

70. A. L. Chemist concluded that since hydrogen bonds are weak and more easily broken than covalent bonds, these experimental results support his hypothesis. Do you agree? Explain your answer.

## SOLUTIONS TO EXERCISES

### Exercise 12.1

If you heat the sample too rapidly, the heating block (where the thermometer is located) gets warmer at a faster rate than the sample plate. It is necessary to heat slowly enough that the plate and the block are at the same temperature all the time.

### Exercise 12.2

Wood seems brittle in large blocks, but when cut thin, it's flexible in all directions. Paper, which is usually made from wood, is very flexible. Wood (or cellulose, its major constituent) is a two-dimensional network solid.

### Exercise 12.3

Molecules a, e, g, and h can hydrogen bond. $HN(CH_3)_2$ (a) contains an N—H bond; $HOCH_2CH_3$ (e), an O—H bond; $CH_3CH_2NH_2$ (g), two N—H bonds; and $CH_3CH_2CH_2OH$ (h), one O—H bond.

### Exercise 12.4

All four molecules have induced dipole–induced dipole forces. (a) $NH_3$ is capable of hydrogen bonding and has dipole-dipole forces. (b) $(CH_3)_2C{=}O$ has dipole-dipole forces. (c) NaBr has ionic bonds.

### Exercise 12.5

Sulfur forms $S_8$ molecules that are irregular in shape. The other three have metallic bonding, and their atoms will pack into one of the crystal structures.

### Exercise 12.6

$$50.0 \text{ g ethanol}\left(\frac{1 \text{ mol ethanol}}{46 \text{ g ethanol}}\right)\left(\frac{5.01 \text{ kJ}}{1 \text{ mol ethanol}}\right) = 5.49 \text{ kJ}$$

### Exercise 12.7

Your graph should look something like this:

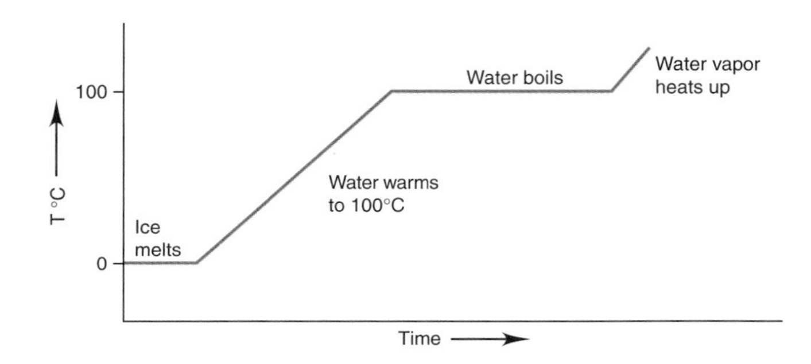

### Experimental Exercise 12.1

1. The empirical formula is $CH_2O$, so the empirical formula mass is 12 g [carbon] + 16 g [oxygen] + (2)(1 g) [hydrogen] = 30 g. The molar mass is 180 g. Since (180 g)/(30 g) = 6, the molecular formula is $C_6H_{12}O_6$.

2. A compound consisting of carbon, hydrogen, and oxygen with a melting point below 300°C must have covalent bonds.

3. The two solids are easily interconverted. Solid A converts to solid B at moderately high temperatures (146°C), and solid B reconverts to solid A in aqueous solution at 100°C. Note that this is not proof but evidence that supports the conclusion.

4. The glassy surface of the sample is certainly a clue. Better evidence is the wide melting range of solid B and the fact that the melting-point range depends on experimental conditions.

5. The fact that large numbers of oxygen atoms are present in the molecule is suggestive. Note that it really doesn't matter whether the oxygen was bonded as —O—, —OH, =O, or any combination of these; any of these can form hydrogen bonds with water, increasing the solubility.

13

# SOLUTIONS



**KEY TERMS**

Homogeneous mixtures are called solutions. In most of the solutions you are familiar with, water is the principal component. Vinegar is a solution, a homogeneous mixture, of 95% water and 5% acetic acid. Liquid bleach is a solution of water and sodium hypochlorite, and household ammonia is a solution of ammonia and water. Of course, not all solutions are liquids, and liquid solutions don't necessarily contain water. The coins you carry are made from solid solutions of two or more metals, called alloys. (Dimes and quarters, with their copper-colored edges and silvery faces, are obviously not homogeneous.) The gasoline you put in your car is a homogeneous mixture of many compounds, almost all of which contain only carbon and hydrogen. In this chapter we will look primarily at water-based (aqueous) solutions, because they are so common in daily life.

Liquid solutions can be made from two or more liquids, from solids and liquids, or from gases and liquids. All of these possibilities are reasonably common. Vinegar, bleach, and household ammonia are examples that illustrate this (see Figure 13.1).

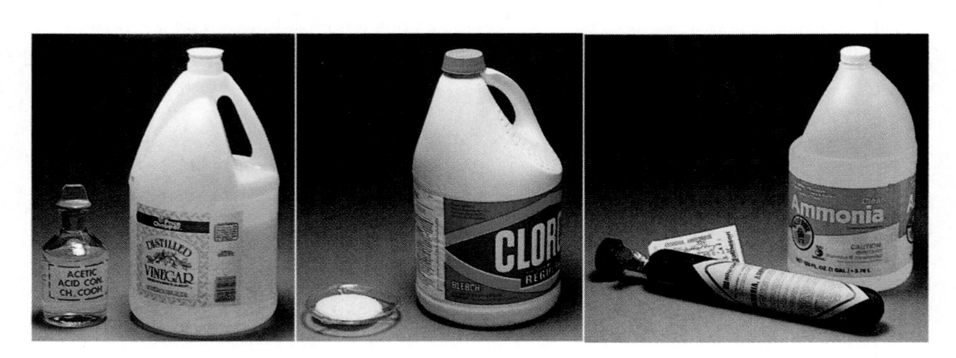


FIGURE 13.1
Solutions can be made with a liquid and either a solid, a liquid, or a gas. Three common household items—vinegar, bleach, and ammonia—illustrate this. Vinegar is water plus a liquid (acetic acid), bleach is a water plus a solid (sodium hypochlorite), and household ammonia is water plus a gas (ammonia, $NH_3$).

It is almost impossible to tell whether a liquid is a solution or a pure substance simply by looking at it. Three of the flasks shown in Figure 13.2 contain colorless liquids, which look very much alike. The solution of ethanol and water in flask 2 is essentially impossible to distinguish by sight from the pure ethanol in flask 3 and the pure water in flask 1. Flask 4 has a boundary between two liquid phases. It contains a heterogeneous mixture, not a solution.

FIGURE 13.2
The first test tube (on the left) contains pure water, the third contains ethanol, and the second contains a mixture of the two. Since these three samples are all colorless, homogeneous liquids, they are almost impossible to tell apart just by looking. The fourth test tube (on the right) contains a heterogeneous mixture of water and octane (a component of gasoline) and has a visible boundary between the two layers.

## 13.1 Solvents and Solutes

GOAL: To learn the terminology for describing solutions.

Before beginning to discuss **solutions,** let's review some important terminology, some of which you have seen before. When two substances form a solution, one of them **dissolves** in the other. When a solid or a gas dissolves in a liquid to make a solution, the liquid is the **solvent** and the other substance is the **solute.** If two liquids form a solution, chemists usually consider the solvent to be the one present in the larger amount. Solutions containing water as the solvent are **aqueous solutions.**

When one substance dissolves in another to form a solution, the process usually involves a physical rather than a chemical change. That is, you can often separate the solvent and solute by evaporating the solvent or distilling the mixture. However, some solutions form because of chemical changes. For example, sodium reacts with water and gives off hydrogen gas:

$$2\,Na(s) + 2\,H_2O(l) \longrightarrow 2\,NaOH(aq) + H_2(g)$$

The other product of the reaction is an aqueous solution of sodium hydroxide, $NaOH(aq)$. You shouldn't say that "metallic sodium dissolves in water," because the metal no longer exists after the process is complete. There is no way to get the sodium metal back through distillation.

Many solutions are prepared by chemical reactions. Alcoholic beverages are an example. All of these beverages are primarily solutions of ethanol, $CH_3CH_2OH$, in water. Federal regulations require that all alcohol for human consumption be produced by fermentation. This process starts with sugars that occur naturally in grapes and grain and uses yeast to convert the sugars to ethanol. To make whiskey, wine, or beer, distillers and brewers begin with aqueous solutions of sugars to which some yeast has been added and end up (after a very complex series of chemical reactions) with the potentially dangerous and intoxicating beverages many people drink (see Figure 13.3).

FIGURE 13.3
Brewers use large vats for making beer, carefully controlling the temperature and pressure of the reaction mixture.

Manufacturers do not prepare the bleach you buy in the grocery store by adding solid sodium hypochlorite, NaOCl, to water. They prepare it in solution by passing an electric current through a cold aqueous solution of NaCl in a process known as electrolysis. The overall reaction is

$$NaCl(aq) + H_2O(l) \xrightarrow[0°C]{\text{electric current}} NaOCl(aq) + H_2(g)$$

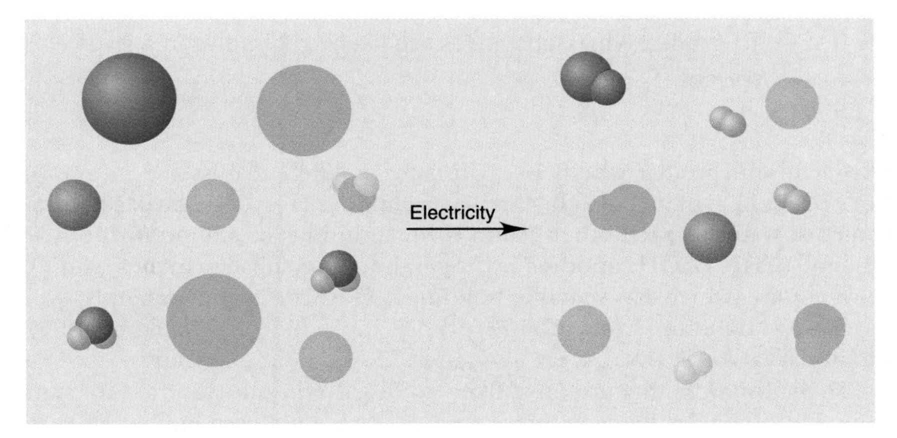


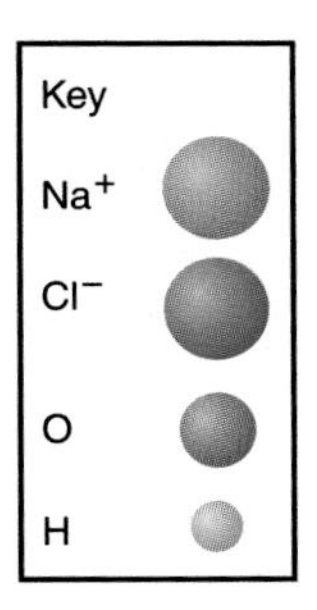


Companies prepare bleach this way, by passing electricity through aqueous solutions of NaCl. This is an example of an electrolysis reaction.

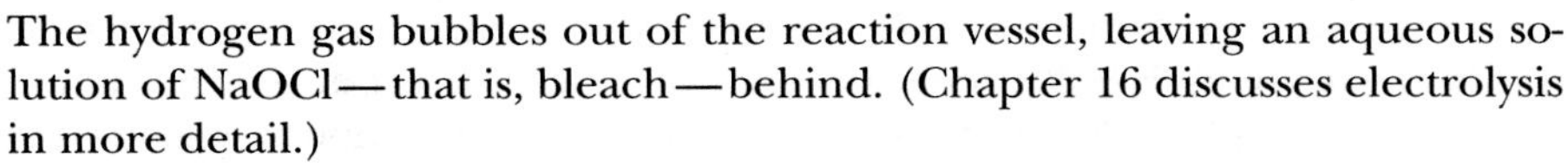

The hydrogen gas bubbles out of the reaction vessel, leaving an aqueous solution of NaOCl—that is, bleach—behind. (Chapter 16 discusses electrolysis in more detail.)

Chemists prepare most common solutions using simpler methods. To sweeten coffee or tea, for instance, you add sugar or artificial sweetener to the beverage and stir until you have a solution.

**STUDY ALERT**

**One error that some beginning students make is to say that sodium chloride (or whatever) "melts" in water when they mean "dissolves" in water.**

### EXAMPLE 13.1 The Terminology of Solutions

For each of these homogeneous mixtures, identify the solvent and the solute.

**a.** 10 g of NaCl(*s*) mixed with 100 mL of water
**b.** 100 mL of ethanol, $CH_3CH_2OH$, mixed with 20 mL of water
**c.** 100 mL of $CO_2(g)$ at 1 atm mixed under high pressure with 100 mL of $H_2O(l)$

**SOLUTIONS**

**a.** Water is the solvent here, and the solid, NaCl, is the solute.
**b.** The solvent is the substance present in a larger amount. Ethanol is the solvent, and water is the solute.
**c.** Carbon dioxide is a gas, so $H_2O$ is the solvent and $CO_2$ is the solute.

**EXERCISE 13.1**

For each of these homogeneous mixtures, identify the solvent and the solute.

**a.** Household ammonia
**b.** Household bleach
**c.** 100 mL of acetic acid, $CH_3CO_2H$, mixed with 10 mL of water

## 13.2 Principles of Solubility

**GOAL:** To predict what substances are likely to dissolve in a given solvent.

How can you predict whether a substance is likely to dissolve in another? Let's begin by looking at some familiar examples. Hydrogen bonds hold molecules of water to each other. Water forms solutions in any proportions with ethanol ($CH_3CH_2OH$, another hydrogen-bonding substance) but won't dissolve a coin (which has metallic bonding), most rocks and minerals (which have ionic bonding), or gasoline (whose molecules are held together by van der Waals forces).

The substances that are most likely to dissolve in one another are similar to one another. To see why this is true, let's take a closer look at what happens when molecules of a solute come into contact with molecules of a liquid solvent. A solvent is a liquid (and not a gas) because of the intermolecular forces of attraction between its molecules. These attractive forces constitute the glue that keeps the molecules from entering the gas phase.

When you mix two substances to make a solution, the solute molecules must insert themselves between solvent molecules. This process disturbs the attractive forces that already exist, both those between solvent molecules and those between solute molecules. For this to occur, the disrupted forces must be replaced by intermolecular forces that are at least as strong. If the forces between solvent molecules are very different from those between solute molecules, the forces between solvent and solute molecules are likely to be weak, and the solute is unlikely to dissolve very much.

The reasons why all gases make solutions with other gases if they do not react chemically are that the intermolecular forces between gas molecules are negligible and the gas molecules are far apart.

Sometimes this type of analysis can get a little tricky. Acetone, $(CH_3)_2C{=}O$, is like ethanol in that it dissolves in water in any proportion. This is true even though acetone molecules cannot form hydrogen bonds by themselves; the forces between one acetone molecule and another are dipole-dipole forces. However, acetone molecules can form hydrogen bonds with water molecules (see Figure 12.19). Furthermore, both acetone molecules and water molecules are polar, so they can also interact through dipole-dipole forces. The solubility of acetone in water is not so surprising after all.

The general guideline for predicting what substances will dissolve in what solvents is this: "Like dissolves like." That is, if the molecules of a substance are held together by a certain type of intermolecular force, that substance is most likely to dissolve in a solvent whose molecules are held together by the same type of force.

**STUDY ALERT**

**Remember from Chapter 12 that van der Waals forces are the strongest intermolecular forces (in the aggregate) for most larger organic molecules. You will find relatively few large molecules that are soluble in water.**

Water forms mixtures in any proportions with several of the organic compounds called *alcohols,* including methanol ($CH_3OH$), ethanol ($CH_3CH_2OH$), and 1-propanol ($CH_3CH_2CH_2OH$). However, 1-butanol ($CH_3CH_2CH_2CH_2OH$), which has slightly larger molecules than the three alcohols just mentioned, dissolves only slightly. The part of the 1-butanol molecule composed of only carbon and hydrogen attracts other butanol molecules by van der Waals forces and does not form hydrogen bonds (see Figure 13.4). Larger alcohols are even less soluble in water.

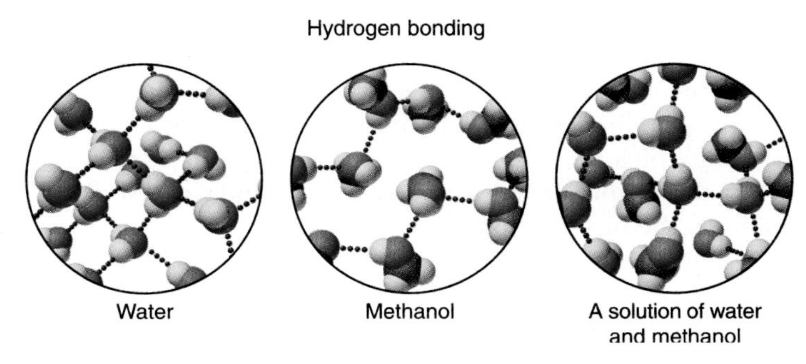


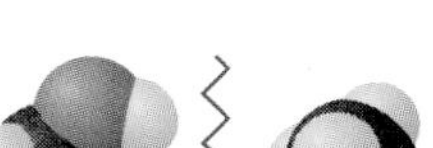


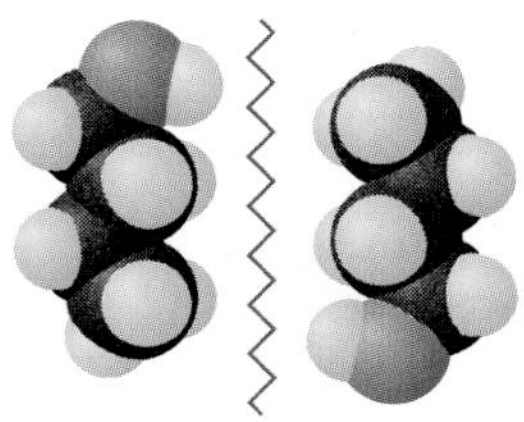

FIGURE 13.4
In both methanol and water (left), hydrogen bonding is the strongest intermolecular force. When methanol dissolves in water, hydrogen bonds between water molecules are replaced by hydrogen bonds between methanol molecules and water molecules. Methanol is soluble in water in all proportions. In 1-butanol (right), the most important intermolecular forces are the van der Waals forces; 1-butanol is only slightly soluble in water.

Suppose you need to dissolve a metal, not by using a chemical reaction, but in such a way that the metal retains its chemical properties. What solvent could you use? For hundreds of years chemists have used mercury, the only

## CHAPTER BOX 13.1 CRYOLITE AND THE MANUFACTURE OF ALUMINUM

**If you** were to try to make aluminum metal from its oxide, $Al_2O_3$, you would need to have the oxide in a phase in which the ions are mobile. You could try melting the compound directly, but the melting point of $Al_2O_3$ is 2072°C, which is hot enough to melt almost any ordinary laboratory apparatus. This poses a major practical problem for your manufacturing process. You could dissolve the $Al_2O_3$ in strong aqueous acid, but water interferes with the aluminum-making process (hydrogen is the product instead). What other solvent might you use?

The American chemist Charles Hall (1863–1914) made the necessary discovery in 1886 while he was still an undergraduate at Oberlin College. He used the mineral cryolite (an ionic compound with the formula $Na_3AlF_6$), which forms a mixture with $Al_2O_3$ that melts at 1000°C. This may not sound like much of an improvement, but it made the large-scale production of aluminum metal economically feasible.

Hall was inspired to begin his research by a college professor who told him that he could get rich if he invented an inexpensive method of aluminum manufacture. (Aluminum was selling for about $5.50 an ounce—a week's pay—at the time). Hall worked in a woodshed with borrowed equipment. It took him a little more than a year to show that cryolite was a suitable solvent for $Al_2O_3$. He used an iron frying pan for his reaction flask and a forge as his heat source. He ended up incredibly wealthy. If you want to do the same thing, figure out an inexpensive way to manufacture ammonia—you won't be able to spend all the money you will end up with.

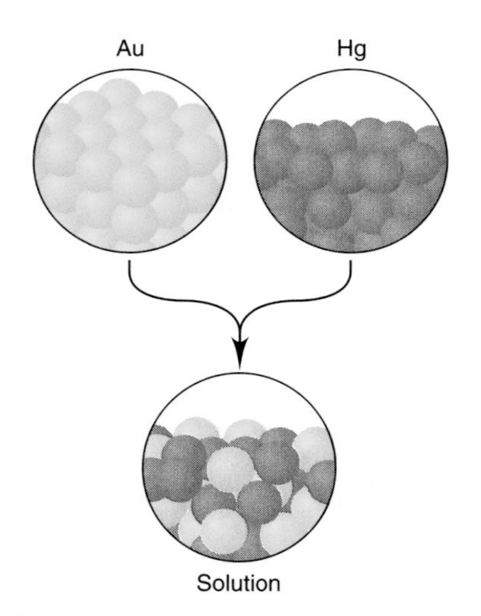


FIGURE 13.5
Metallic bonding is the primary force between mercury atoms and also between gold atoms. Gold readily dissolves in Hg($l$).

metal that is a liquid at room temperature. Mercury dissolves almost all of the other metals. It dissolves them so well, in fact, that you must be very careful with rings and other jewelry if you break a mercury thermometer in the lab. Those little droplets of mercury will dissolve the gold in your ring in nothing flat (see Figure 13.5). (Mercury is also poisonous and can be absorbed through your skin.)

Iron doesn't dissolve in mercury, though it's the only metallic element that doesn't. This illustrates that the "like dissolves like" guideline isn't perfect. It's a good guideline, not a hard-and-fast physical law.

The solubility of ionic compounds is more difficult to predict than the simple rule of thumb presented in this section would suggest. We'll examine such solubility in the next section.

### EXAMPLE 13.2 Predicting Solubilities Based on Intermolecular Forces

**a.** Which of these compounds is the most soluble in water?
A. $CaF_2$ B. $CH_3OH$ C. $CH_3Cl$ D. $CH_4$

**b.** Which of these compounds is the most soluble in liquid argon (Ar)?
A. NaCl B. $CH_3OH$ C. $CH_4$ D. $CaSO_4$

SOLUTIONS

**a.** First, classify the solvent: The strongest intermolecular forces in water are hydrogen bonds. Then classify the possible answers: $CaF_2$ is composed of a metal and a nonmetal and is ionic. $CH_3OH$, with its OH group, is a hydrogen-bonding substance. This may be the answer, but continue to make sure. $CH_3Cl$ has dipole-dipole forces, and $CH_4$ is a nonpolar molecule that exhibits only van der Waals forces. Compound B is the answer; $CH_3OH$ is the most soluble of these compounds in water.

**b.** The atoms in liquid argon are held to one another by van der Waals forces. Of the four substances listed, the most soluble in Ar is $CH_4$, another substance with van der Waals forces. (NaCl and $CaSO_4$ are ionic, and $CH_3OH$ has hydrogen bonding.)

**EXERCISE 13.2**

**a.** Which of these compounds is the least soluble in water?
A. $CaF_2$ B. NaCl C. $CH_3OH$ D. $CH_3CH_2OH$

**b.** Which of these compounds is the most soluble in methanol ($CH_3OH$)?
A. $H_2O$ B. HgS C. CaO D. $(NH_4)_2S$

## 13.3 Predicting the Solubility of Ionic Compounds in Water

**GOAL:** To predict whether a given ionic compound is soluble in water.

Relatively few substances will form a solution in a solvent that has intermolecular forces different from those in the substance itself. The ionic compounds that dissolve in water are the most important of these exceptions, as far as living organisms are concerned. Many of these compounds are necessary for the proper functioning of the human body, for instance. If you removed all cells, viruses, and other particles from blood, you would be left with a liquid that is a weak solution of ionic compounds in water. The compound present in the highest concentration would be sodium chloride.

The "like dissolves like" guideline applies most of the time, and almost all ionic compounds are insoluble in water. Table 13.1 lists the exceptions, the ionic compounds that are water-soluble (see also Section 7.9). Let's think about what ions do as they dissolve (or don't), to see why most ionic compounds are insoluble in water.

**TABLE 13.1 Ionic Compounds that Dissolve in Water**

- Salts containing the nitrate ($NO_3^-$) and acetate ($CH_3CO_2^-$) ions; no important exceptions.
- Salts containing Group 1A metal cations ($Li^+$, $Na^+$, $K^+$, etc.); no important exceptions.
- Salts containing the $NH_4^+$ cation; no important exceptions.
- Most chloride salts, with some important exceptions: AgCl, $PbCl_2$, and $Hg_2Cl_2$ are insoluble.
- Most sulfate salts, with some important exceptions: $PbSO_4$, $BaSO_4$, and $CaSO_4$ are insoluble.

By definition, ionic compounds are composed of cations (positively charged) and anions (negatively charged). In an ionic crystal each cation is surrounded by anions, and each anion is surrounded by cations. The cations are firmly held in place by the electrostatic forces of attraction of many anions, and vice versa. To dissolve an ionic crystal, something must supply a large amount of energy to overcome the electrostatic forces of attraction and pull the ions away from the crystal. Water, being polar, can attract cations toward its oxygen atoms, or anions toward its hydrogen atoms. In most cases, however, the binding between water molecules and ions cannot provide enough energy to disrupt the structure of an ionic crystal. For a few ionic compounds, though, it can and does supply enough energy.

For ionic substances that do dissolve in water, the crystal structure ceases to exist and the ions separate. For example, when NaCl dissolves in water, the result is $Na^+$ ions and $Cl^-$ ions in aqueous solution. The same thing is true when $CaCl_2$ dissolves in water. The solution contains $Ca^{2+}(aq)$ and $Cl^-(aq)$, and there are no $CaCl_2$ units present. All of the ionic substances that dissolve in water separate into ions (as shown in simplified form in Figure 13.6).

The energy needed to disrupt the crystal and provide the necessary solute-solvent interactions comes from a type of intermolecular force we have not yet discussed in any detail. These interactions are **ion-dipole forces.** The negative end of a polar molecule stabilizes the cation, and the positive end stabilizes the anion as shown in Figure 13.6. You might think of the ion-dipole forces as serving three functions: They provide the energy necessary to disrupt the ionic crystal, they stabilize the ions in solution, and they constitute attractive forces between solute and solvent molecules to replace the broken hydrogen bonds between solvent molecules. All of these things must happen if an ionic compound is to dissolve in water.

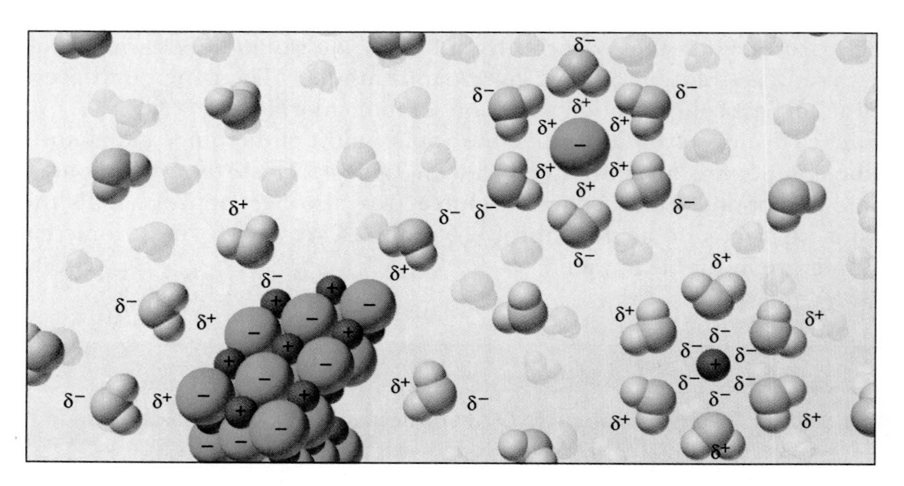


FIGURE 13.6

Electrostatic forces of attraction between the negative end of a polar water molecule and the positive charge of the cation hold $Na^+$ ions in solution. In dilute solutions of NaCl, each sodium ion is surrounded by 8 to 12 water molecules (for clarity, fewer than this are shown here). The $Cl^-$ ions are stabilized in a similar way. From the point of view of the solvent, the hydrogen bonds that are broken to accommodate the dissociation of NaCl are replaced with ion-dipole forces, which have approximately the same strength.

So few ionic compounds dissolve in water because the energy needed to break down the crystal structure is almost always much larger than the ion-dipole forces can supply. The electrostatic forces holding ionic crystals together are very strong, particularly when the ions are small, the charges are large, or both. A rule of thumb is that all ionic compounds are insoluble in water except those listed in Table 13.1. It would be difficult to overemphasize the importance of this list. Because so few inorganic compounds dissolve in water, you will need to know which ones they are.

### EXAMPLE 13.3 Predicting Solubility in Water

For each compound, predict whether it is soluble in water.

**a.** $NaNO_3$ **b.** AgCl **c.** $CaCO_3$
**d.** $Ca(NO_3)_2$ **e.** $BaSO_4$ **f.** $Al(OH)_3$

**SOLUTIONS**

**a.** $NaNO_3$ is water soluble, and there are two ways to know this: Na is a Group 1A metal, so its salts are soluble in water. Salts containing the nitrate ion are also soluble.
**b.** Most chloride salts are water soluble, but AgCl is one of the exceptions listed in Table 13.1. It's insoluble.
**c.** Neither calcium ion nor carbonate ion is mentioned in Table 13.1 as ions that are water soluble. Therefore, $CaCO_3$ is insoluble.
**d.** $Ca(NO_3)_2$ contains the nitrate ion, $NO_3^-$. The general rule is that nitrate salts are soluble in water, so you can conclude that this compound is.
**e.** Most sulfate salts are soluble in water, but this is one of the exceptions listed in Table 13.1, which you have to memorize. $BaSO_4$ is insoluble in water.
**f.** Table 13.1 mentions neither the aluminum ion nor the hydroxide ion. Therefore, $Al(OH)_3$ obeys the general rule that ionic compounds are insoluble in water.

**EXERCISE 13.3**

For each compound, predict whether it is soluble in water.

**a.** $K_2S$ **b.** FeS **c.** $PbCl_2$ **d.** $FeCl_2$
**e.** NaOH **f.** $Fe(OH)_2$ **g.** HgS

## 13.4 Saturated and Unsaturated Solutions

**GOAL:** To use concentration and solubility data to predict whether or not a solution is saturated.

Suppose you add a small quantity of solid sodium chloride (table salt) to a flask of water, and shake the flask until the salt dissolves completely. If you

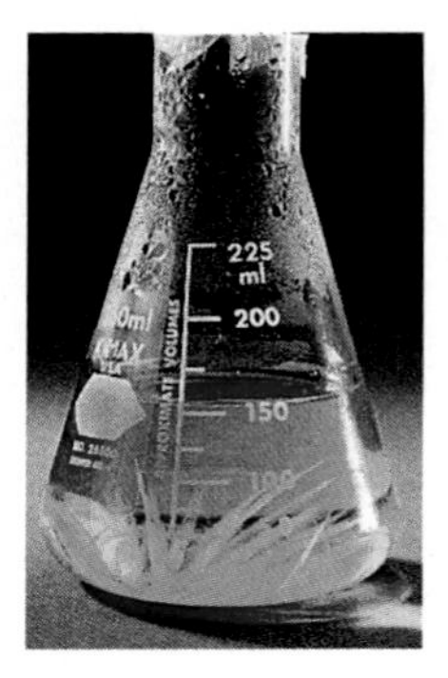


FIGURE 13.7
As a solution cools, it can become saturated again and the excess solute appears as crystals in the solution.

keep adding small quantities of salt, it will dissolve more and more slowly as you shake. Finally, some of the salt won't dissolve no matter how much you shake the flask. You can add more salt to the flask, but no more will dissolve unless you change the temperature. The salt solution is now a **saturated solution.** When the solution has some sodium chloride in it but can still dissolve more, it is said to be an **unsaturated solution.** The **solubility** of a compound is the amount that dissolves in a given amount of solvent to form a saturated solution at a particular temperature. Chemists often report solubility data in units of grams of solute per 100 mL of solvent. They also report the temperature, since the solubility depends on the temperature of the solution.

Standard reference books, such as the *CRC Handbook of Chemistry and Physics,* give the solubilities of many compounds in water. For example, the solubility of sodium chloride in water is 35 g per 100 mL at 0°C and 39 g per 100 mL at 100°C. Sodium chloride is unusual in that it is almost as soluble in cold water as it is in hot water. The solubilities of most compounds change with temperature more than that of sodium chloride does. Sodium carbonate ($Na_2CO_3$), for example, has a solubility of 7 g per 100 mL at 0°C and 46 g per 100 mL at 100°C. Most water-soluble solids are more soluble at high temperatures than at low temperatures, although there are exceptions.

Gases, unlike solids, are more soluble in liquids at lower temperatures than at higher temperatures. For example, ammonia gas is more than 10 times as soluble in water at 0°C as at 100°C. The solubilities of oxygen gas are 4.89 mL per 100 mL of water at 0°C and 2.46 mL per 100 mL of water at 100°C. The lower solubility of oxygen in warm water can be a problem for fish (see Figure 13.8).

FIGURE 13.8
The *nishiki-goi* (Japanese colored carp) in garden ponds like this can die during the hottest days of summer. The heat causes the oxygen concentration to drop to levels too low to sustain the lives of the fish.

## EXAMPLE 13.4 Saturated and Unsaturated Solutions

The solubility of calcium acetate is 37 g per 100 mL of water at 0°C.

**a.** How much calcium acetate will dissolve, and how much will remain undissolved, when 100 mL of water is added to 50 g of calcium acetate at 0°C?

**b.** How much calcium acetate will dissolve, and how much will remain undissolved, when 10 mL of water is added to 10 g of calcium acetate at 0°C?

**SOLUTIONS**

**a.** The amount of calcium acetate being added, 50 g, is larger than the maximum amount that 100 mL of water can dissolve at 0°C, which is 37 g. The solution will be saturated when 37 g has dissolved, and the remaining 13 g will not dissolve.

**b.** This answer can be found by using the solubility as a unit factor:

$$10\ \cancel{\text{mL water}} \times \frac{37\ \text{g calcium acetate}}{100\ \cancel{\text{mL water}}} = 3.7\ \text{g calcium acetate}$$

The solution will become saturated when 3.7 g has dissolved, and the remaining 6.3 g will remain undissolved.

EXERCISE 13.4

The solubility of calcium chloride is 74 g per 100 mL of water at 20°C.

**a.** How much calcium chloride will dissolve, and how much will remain undissolved, when 100 mL of water is added to 50 g of $CaCl_2$ at 20°C?

**b.** How much calcium chloride will dissolve, and how much will remain undissolved, when 15 mL of water is added to 20 g of $CaCl_2$ at 20°C?

## CHAPTER BOX 13.2 WATER-SOLUBLE ROCKS

**There are** a few exceptions to the guideline that rocks and minerals do not dissolve in water. The solubility of calcium sulfate ($CaSO_4$), the main component of the mineral *gypsum,* is 0.21 g per 100 mL of water at 30°C. Since a saturated calcium sulfate solution is too dilute to be useful in the laboratory, Table 13.1 lists this compound as insoluble in water. However, the slow and steady dissolving power of rain over thousands of years will have a significant effect. Exposed layers of gypsum on mountainsides slowly dissolve in rainwater, as the water steadily carries tiny amounts of calcium sulfate away. If the water flows into a lake with no outlet, new gypsum beds will form. This is how the White Sands in New Mexico formed and continue to form (top photo at right). As the water in such a lake evaporates, gypsum particles blow away in the wind.

Limestone ($CaCO_3$) is very insoluble in water, but it can dissolve by way of a chemical reaction with acidic water. Carbon dioxide in air and soil reacts with water to form carbonic acid ($H_2CO_3$), a weak acid. The carbonic acid in turn reacts with the limestone by this overall process:

$$H_2CO_3(aq) + CaCO_3(s) \rightleftharpoons Ca^{2+}(aq) + 2\,HCO_3^-(aq)$$

This process has created most of the world's great caves, among other things. The stalactites and stalagmites that decorate caves form when the carbon dioxide evaporates and the calcium carbonate redeposits (as in the photo at right, bottom). You will learn more details of this process when we look at the topic of chemical equilibrium in Chapter 14.

(Top) Dunes of gypsum (calcium sulfate) sand in White Sands National Monument, New Mexico. (Bottom) Calcite (calcium carbonate) cave formations in Carlsbad Caverns in New Mexico.

## 13.5 Precipitation Reactions

GOAL: To write equations for reactions that produce water-insoluble ionic products from water-soluble ionic starting materials.

Chapter 8 introduced several types of chemical reactions. This section introduces another type, **precipitation reactions.** To run such a reaction, chemists prepare aqueous solutions of two water-soluble compounds and then mix the solutions. If one of the products is insoluble, crystals of this product appear in the solution and fall to the bottom. The solid product is called a **precipitate.**

For example, consider what would happen if you mix a solution of sodium carbonate ($Na_2CO_3$) with a solution of calcium chloride ($CaCl_2$). $Na_2CO_3$ and $CaCl_2$ are both soluble in water. This means that the sodium carbonate solution contains $Na^+(aq)$ ions and $CO_3^{2-}(aq)$ ions, and the calcium chloride solution contains $Ca^{2+}(aq)$ ions and $Cl^-(aq)$ ions. At the instant you mix these two solutions before anything (any reaction) happens, the reaction mixture contains all four ions (see Figure 13.9).

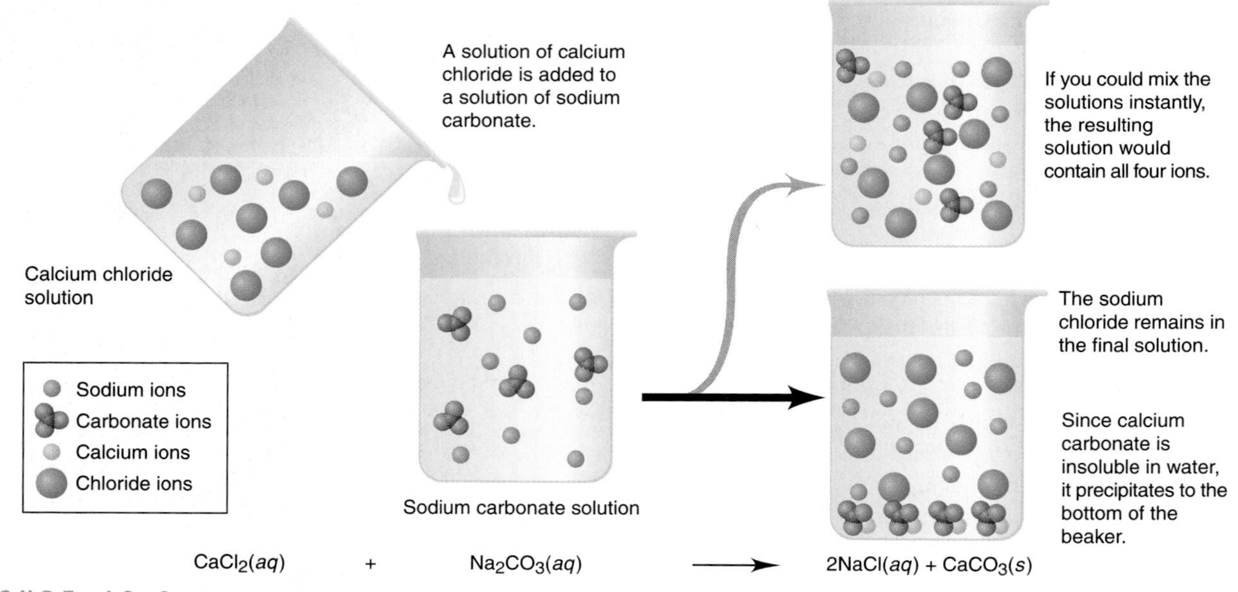
A solution of calcium chloride is added to a solution of sodium carbonate.
If you could mix the solutions instantly, the resulting solution would contain all four ions.
Calcium chloride solution
The sodium chloride remains in the final solution.
Sodium ions
Carbonate ions
Calcium ions
Chloride ions
Since calcium carbonate is insoluble in water, it precipitates to the bottom of the beaker.
Sodium carbonate solution
$CaCl_2(aq)$ + $Na_2CO_3(aq)$ → $2NaCl(aq) + CaCO_3(s)$

FIGURE 13.9
Depiction of what happens during the reaction: $CaCl_2(aq) + Na_2CO_3(aq) \rightarrow 2\,NaCl(aq) + CaCO_3(s)$

You can predict the possible products of this sort of reaction by reshuffling the ions in the reactants. That is, you put the cation from one starting material with the anion from the other, and vice versa. For example, the ionic compounds $CaCO_3$ and NaCl are the possible products of the reaction between $CaCl_2$ and $Na_2CO_3$. Sodium chloride is soluble in water, but calcium carbonate is not. As soon as the two reactants mix, insoluble calcium carbonate crystals appear and eventually settle to the bottom of the flask. This precipitate can be separated from the final solution (see Figure 13.10). The equation for the overall reaction can be written this way:

$$Na_2CO_3(aq) + CaCl_2(aq) \longrightarrow 2\ NaCl(aq) + CaCO_3(s)$$

Two moles of NaCl are formed for each mole of $Na_2CO_3$ used. From the balanced equation you can write a *net ionic equation,* as you did for some of the reactions in Chapter 8. There are two ways to do this, a long way and a short way.

To use the longer method, first you separate all soluble salts into the ions that make them up and add "$(aq)$" to each:

$$2\ Na^+(aq) + CO_3^{2-}(aq) + Ca^{2+}(aq) + 2\ Cl^-(aq) \longrightarrow 2\ Na^+(aq) + 2\ Cl^-(aq) + CaCO_3(s)$$

Then you cancel out all ions that are the same on the left and the right sides of the arrow (the spectator ions):

$$\cancel{2\ Na^+(aq)} + CO_3^{2-}(aq) + Ca^{2+}(aq) + \cancel{2\ Cl^-(aq)} \longrightarrow \cancel{2\ Na^+(aq)} + \cancel{2\ Cl^-(aq)} + CaCO_3(s)$$

This yields the net ionic equation:

$$Ca^{2+}(aq) + CO_3^{2-}(aq) \longrightarrow CaCO_3(s)$$

To use the shorter method of solving the same problem, you start with the formula of the solid product on the right side of the equation, like this:

$$\longrightarrow CaCO_3(s)$$

The calcium ion and the carbonate ion make up this product and must be the starting materials. You write the formulas of these ions followed by "$(aq)$," and this gives you the same net ionic equation:

$$Ca^{2+}(aq) + CO_3^{2-}(aq) \longrightarrow CaCO_3(s)$$

To use this method, you must know the correct formulas (particularly the charges) of the ions. If you write the charges of the calcium and carbonate ions as $+1$ and $-1$, for instance, your equation will look okay but will be wrong. If you use the long method, you can probably tell that the charge on the carbonate ion is $-2$ from the fact that two sodium ions with $+1$ charges associate with it.

**Ionic compounds must have an overall charge of zero, so anions must combine with cations.**

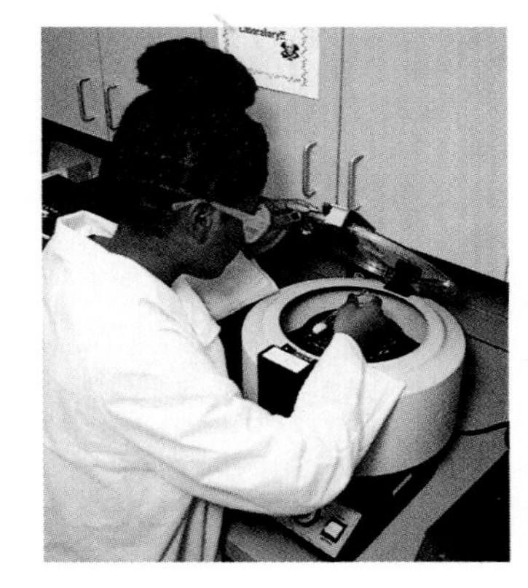

FIGURE 13.10
Chemists can remove precipitates from reaction mixtures by using a centrifuge (shown here) or by filtration (Figure 2.12).

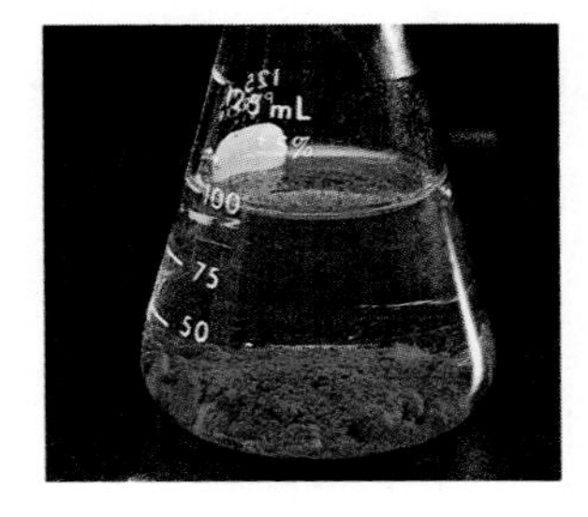


FIGURE 13.11
The photograph on top shows aqueous solutions of sodium chromate (yellow) and silver nitrate (colorless). When the two solutions are mixed, an insoluble precipitate, silver chromate (red), forms as shown in the photograph at the bottom. Reactions of this kind are called *precipitation reactions*.

### EXAMPLE 13.5 Predicting the Products of Precipitation Reactions

**a.** Give the balanced equation for the reaction that occurs when an aqueous solution of $AgNO_3$ is mixed with an aqueous solution of $Na_2CrO_4$ in Figure 13.11.

**b.** Give the net ionic equation for the same reaction.

SOLUTIONS

a. The formulas of the reactants are given, so write them down. Then write the formulas of the products you get if you reshuffle the ions. Don't forget that the compounds you propose as products have to contain both an anion and a cation.

$$AgNO_3 + Na_2CrO_4 \longrightarrow Ag_2CrO_4 + NaNO_3$$

Next, consider the solubilities of the products. From Table 13.1, $NaNO_3$ is soluble in water because it consists of $Na^+$ and $NO_3^-$, but $Ag_2CrO_4$ is insoluble. The equation therefore is

$$AgNO_3(aq) + Na_2CrO_4(aq) \longrightarrow Ag_2CrO_4(s) + NaNO_3(aq)$$

Is this equation balanced? (You should always check.) No, it isn't, so there is more work to be done to get this balanced equation, which is the answer:

$$2\ AgNO_3(aq) + Na_2CrO_4(aq) \longrightarrow Ag_2CrO_4(s) + 2\ NaNO_3(aq)$$

b. To use the long way, start with the answer to part (a), separate the soluble compounds into ions, and leave the precipitate, marked "(*s*)," alone. This gives:

$$2\ Ag^+(aq) + 2\ NO_3^-(aq) + 2\ Na^+(aq) + CrO_4^{2-}(aq) \longrightarrow Ag_2CrO_4(s) + 2\ Na^+(aq) + 2\ NO_3^-(aq)$$

Next, remove the ions that appear on both sides of the equation:

$$2\ Ag^+(aq) + \cancel{2\ NO_3^-(aq)} + \cancel{2\ Na^+(aq)} + CrO_4^{2-}(aq) \longrightarrow Ag_2CrO_4(s) + \cancel{2\ Na^+(aq)} + \cancel{2\ NO_3^-(aq)}$$

This yields the net ionic equation:

$$2\ Ag^+(aq) + CrO_4^{2-}(aq) \longrightarrow Ag_2CrO_4(s)$$

You get the same equation if you start with the formula of the solid product and write its component ions as the reactants.

EXERCISE 13.5

a. Predict the products of the reaction that occurs when you mix aqueous mercury(II) nitrate with aqueous ammonium sulfide.
b. Give the net ionic equation for the same reaction.

## 13.6 Electrolytes: Dissociation in Solution

GOAL: To predict whether an aqueous solution of a compound will conduct electricity.

As soon as chemists learned how to produce electric currents, they investigated the responses of substances and solutions to electricity. The ability of a substance to conduct electricity is a fundamental property. Metallic elements conduct electric currents, and nonmetals (with a few exceptions, of which graphite is most important) do not. Solid and liquid samples of almost all compounds—whether ionic or covalent, organic or inorganic—do not conduct electricity well. Pure water is a poor conductor.

Furthermore, dissolving ethyl alcohol or sugar in the water does not increase the electrical conductivity. Ethyl alcohol and sugar are **nonelectrolytes.** On the other hand, the aqueous solutions of some compounds conduct electricity very well. These compounds are called **electrolytes.** Sodium chloride is one. Solutions of sodium chloride in water are as good as metal wires at conducting electricity (see Figure 13.12). If a sample of a substance is to conduct an electric current, charges must somehow be able to move from place to place within the sample.

See the disussion of metallic bonding in Section 10.1 for the explanation of the electrical conductivity of metals.

FIGURE 13.12
In a standard test of electrical conductivity, the ends of two wires are placed in the solution being tested. The wires are attached to a source of electricity and a light bulb. If the solution conducts electricity, the bulb lights up. If it doesn't, the bulb stays off. Pure water (left) is a poor conductor, but salt water (right) conducts electricity well.

Why do certain aqueous solutions conduct electricity? When ionic compounds dissolve in water, the ions separate and water molecules surround them. This process is called **dissociation.** Each separate ion has a charge and can move through the solution toward the wire that has a charge opposite to its own (see Figure 13.13). When the ions move, an electric current flows through the solution because charges are moving.

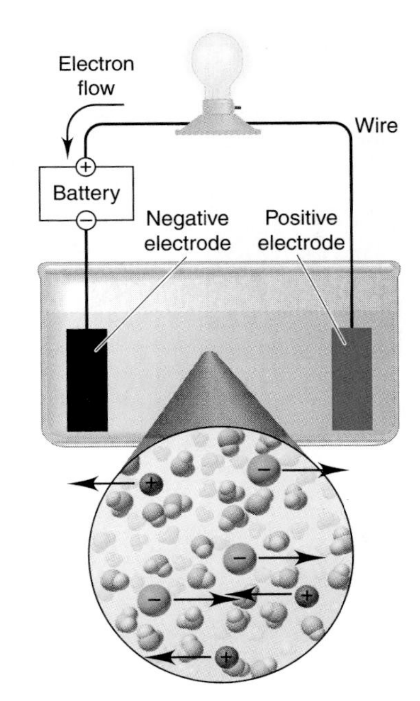


FIGURE 13.13
What happens in a solution during a standard test of electrical conductivity: The current in the wires consists of electrons. The current in the solution consists of moving ions.

## EXAMPLE 13.6 Electrolytes

Which of these compounds is an electrolyte—that is, which one dissolves in water to give a solution that conducts electricity?

**A.** $CH_3OH$ **B.** $Al_2O_3$ **C.** $KNO_3$

SOLUTION

Compound A is made up entirely of nonmetals and is covalent. Its aqueous solution does not conduct electricity. Compound B is composed of a metal and a nonmetal and is therefore ionic. However, it doesn't dissolve in water, so it can't be an electrolyte. Compound C is both ionic and water-soluble; it is the electrolyte of this group.

EXERCISE 13.6

Which of these compounds is an electrolyte?

**A.** HgS **B.** $NH_4Cl$ **C.** $C_2H_6$

## 13.7 Strong and Weak Acids and Bases

**GOAL:** To distinguish between strong and weak acids and bases on the basis of their ability to conduct electricity and their behavior in solution.

| TABLE 13.2 Strong Acids |
|---|
| HCl |
| HBr |
| HI |
| $HNO_3$ |
| $H_2SO_4$ |
| $HClO_4$ |

Salts are not the only electrolytes; strong acids and bases are also electrolytes. Strong bases are water-soluble ionic compounds, primarily the water-soluble hydroxides. As for strong acids, they react with water to form a solution containing the hydronium ion ($H_3O^+$) and the anion of the acid (see Section 8.8). For example, when HCl(*g*) dissolves in water, the ions $H_3O^+(aq)$ and $Cl^-(aq)$ form. Solutions of strong acids and bases conduct electricity just as NaCl does. A list of strong acids is given in Table 13.2.

Acetic acid ($CH_3CO_2H$) is a weak acid that dissolves in water (as you should expect, since it is a small polar molecule that is capable of hydrogen bonding). The light bulb glows dimly when an acetic acid solution is tested for electrical conductivity. Solutions of ammonia ($NH_3$) in water, which are weakly basic, also produce a dim light in the electrical conductivity test. Salts, strong acids, and strong bases are **strong electrolytes,** whereas weak acids and bases are **weak electrolytes** (see Figure 13.14).

**Strong electrolytes** Water-soluble salts
Strong acids
Strong bases

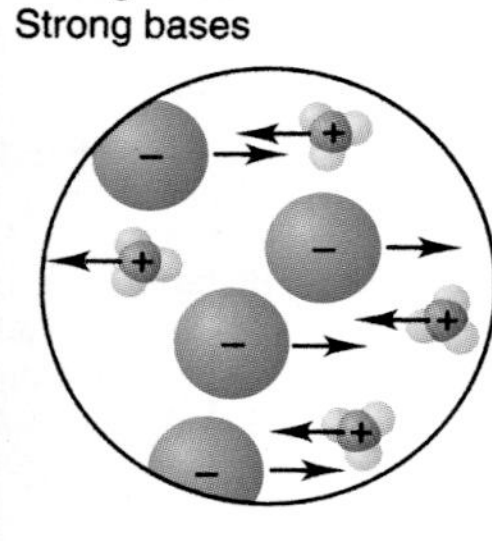


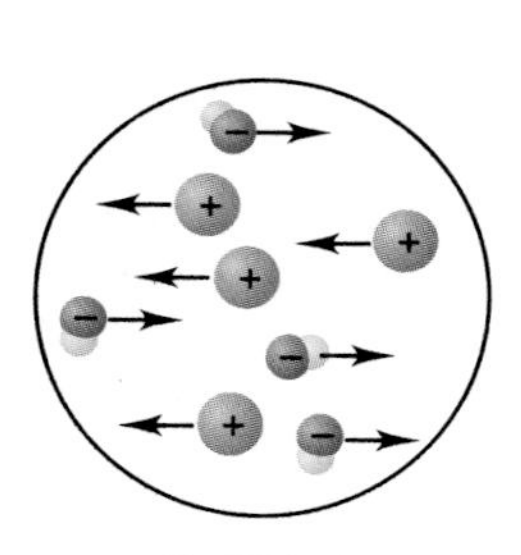


**Weak electrolytes** Weak acids
Weak bases

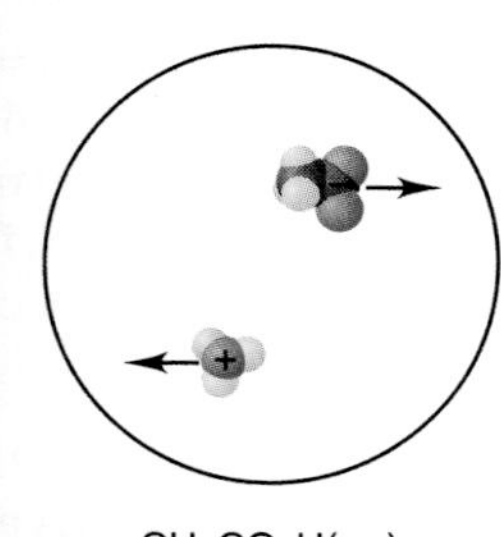


FIGURE 13.14
Strong and weak electrolytes in solution. HCl reacts completely with water to form a solution containing only $H_3O^+$ and $Cl^-$ ions. NaOH, like other water-soluble ionic compounds, completely dissociates to give a solution containing only $Na^+$ and $OH^-$ ions. Both HCl and NaOH are strong electrolytes. Acetic acid reacts only slightly with water, yielding a solution containing relatively few acetate and hydronium ions. Because the ion concentration is low, the acetic acid solution conducts electricity only weakly, and the light in the test apparatus does not shine as brightly as it does with solutions of HCl and NaOH.

The acid-base properties of compounds parallel their properties as electrolytes. Concentrated aqueous solutions of hydrogen chloride are very corrosive, but similar solutions of acetic acid are much less so. Vinegar is a 5% solution of $CH_3CO_2H$ in water, and cooks use it in many recipes, in salad dressing for example. A 5% solution of HCl is a potentially dangerous substance that chemists and other workers must handle with care. As you might expect, these differences in behavior come from differences in the behavior of the molecules in solution. Both HCl and $CH_3CO_2H$ derive their acidic properties from the reaction with water to form the $H_3O^+$ ions:

$$HCl(g) + H_2O(l) \longrightarrow H_3O^+(aq) + Cl^-(aq)$$
$$CH_3CO_2H(aq) + H_2O(l) \rightleftharpoons H_3O^+(aq) + CH_3CO_2^-(aq)$$

When hydrogen chloride dissolves in water, the reaction goes virtually to completion, yielding a high concentration of ions, including $H_3O^+$. The resulting solution conducts electricity well. When acetic acid dissolves in water, only a small percentage of its molecules react. The resulting solution has a relatively low concentration of ions, meaning a low electrical conductivity. The low concentration of $H_3O^+$ in particular gives it only weak acidic properties.

### EXAMPLE 13.7 Strong and Weak Electrolytes

Which of these compounds reacts strongly with water to produce ions in aqueous solution?

**A.** $CH_3CO_2H$ **B.** $KNO_3$ **C.** HBr

**SOLUTION**

Acetic acid (A) is not listed in Table 13.2 and therefore is a weak acid that does not react strongly with water. Potassium nitrate (B) is a strong electrolyte, because it is a water-soluble ionic compound; it doesn't react with water to produce ions. Hydrogen bromide (C) is a strong acid (listed in Table 13.2) and is therefore the answer.

**EXERCISE 13.7**

Which of these compounds reacts strongly with water to produce ions in aqueous solution?

**A.** $NH_3$ **B.** $H_2SO_4$ **C.** $Na_2CO_3$

### EXAMPLE 13.8 Identifying Electrolytes

Classify each compound as a strong electrolyte, a weak electrolyte, or a nonelectrolyte.

**a.** $CH_3I$ **b.** HI **c.** HOCl **d.** $CaF_2$ **e.** $CF_4$

**SOLUTIONS**

**a.** $CH_3I$ is a covalent compound that doesn't dissolve in water; it is a nonelectrolyte.
**b.** HI reacts completely with water to form $H_3O^+$ and $I^-$ ions; it is a strong electrolyte.
**c.** HOCl is an oxyacid that is not listed in Table 13.2; it is therefore a weak electrolyte.
**d.** $CaF_2$ is insoluble in water and is a nonelectrolyte.
**e.** $CF_4$ is a water-insoluble covalent compound, a nonelectrolyte.

**EXERCISE 13.8**

Classify each compound as a strong electrolyte, a weak electrolyte, or a nonelectrolyte.

**a.** $HNO_3$ **b.** $K_2SO_4$ **c.** $CH_3OH$ **d.** $CH_3CO_2H$ **e.** KBr

## 13.8 Concentration of Solutions: Molarity

**GOAL:** To determine the concentrations of solutions in units of moles of solute per liter of solution.

Making up solutions and calculating the amounts of substances present in them are basic skills for chemists, and they are also important for people in many related professions. There are several widely used ways to express the amount of a substance in a given amount of solution. The most important of these for chemists is **molarity,** which is the number of moles of a solute per liter of solution. The abbreviation for molarity is *M.* That is, if a solution contains 0.20 mol of NaCl per 1 L of solution, the concentration is written as 0.20 *M* NaCl.

$$M = \frac{\text{mol solute}}{\text{L of solution}}$$

Why are concentrations so useful for chemists? If chemists know how many moles of a substance are present in a given amount of solution, they can figure out how much of another substance is needed to react with it and how much of a third substance will be obtained as the reaction product. These calculations are always important for chemists. You had some indication of this in Chapter 9.

### EXAMPLE 13.9 Calculating Molarity

Calculate the molarity of each of these solutions.

**a.** A solution containing 7.1 mol of NaCl dissolved in enough water to give 5.2 L of solution

**b.** A solution containing 12.9 g of sodium carbonate ($Na_2CO_3$) dissolved in enough water to give 125 mL of solution

**SOLUTIONS**

**a.** Molarity is defined as moles of solute per liter of solution, and you have the number of moles and the number of liters. This is just as straightforward as it seems to be:

$$\text{Molarity} = \frac{\text{Moles of solute}}{\text{Liters of solution}} = \frac{7.1 \text{ mol NaCl}}{5.2 \text{ L}} = 1.4\ M\,\text{NaCl}$$

**b.** This time you have neither the number of moles of solute nor the number of liters of solution, but you do have enough information to calculate both. To find the number of moles of $Na_2CO_3$, divide the number of grams (given) by the molar mass (which you know how to find):

$$\frac{12.9 \text{ g } Na_2CO_3}{(2 \times 23.0 \text{ g mol}^{-1} \text{ Na}) + (1 \times 12.0 \text{ g mol}^{-1} \text{ C}) + (3 \times 16.0 \text{ g mol}^{-1} \text{ O})} = 0.122 \text{ mol } Na_2CO_3$$

The volume given is 125 mL, which you convert to 0.125 L using the conversion factor 1 L = 1000 mL. Then you use the equation that defines molarity:

$$\text{Molarity} = \frac{\text{Moles of solute}}{\text{Liters of solution}} = \frac{0.122 \text{ mol } Na_2CO_3}{0.125 \text{ L solution}}$$

$$= 0.976\ M\ Na_2CO_3$$

**EXERCISE 13.9**

Calculate the molarity of each of these solutions.

**a.** A solution containing 0.32 mol of $CH_3OH$ dissolved in enough water to give 1.57 L of solution

**b.** A solution containing 77.5 g of calcium chloride ($CaCl_2$) dissolved in enough water to give 885 mL of solution

## 13.9 Preparing Solutions

**GOAL:** To learn how to prepare a solution of a given molarity.

Now that you know how to calculate molarity, the next step is to learn procedures for making up solutions. Given bottles of solute and solvent and the need for a solution of a specific concentration, how do you go about making it? The first thing you must do is pay strict attention to the definition of molarity, which is moles of solute per liter of *solution*. This means you *cannot* prepare a 1.0 $M$ solution of anything by weighing out the mass of 1 mol of the substance and then adding a liter of water. If you did that, you would end up with 1 mol of solute per liter of *solvent*, and that is *not* the definition of molarity.

To prepare a 1.00 $M$ solution, you need to weigh out the mass of 1.00 mol of the solute and add only enough water to bring the total volume of solution to 1.00 L. This may sound tricky, but it is not. Chemists use special flasks, called *volumetric flasks*, when preparing solutions (see Figure 13.15). These flasks come in a variety of convenient sizes.

FIGURE 13.15
Chemists use volumetric flasks like this one (top) to prepare exact amounts of solutions of specific molar concentrations. Such flasks come in a variety of sizes. To obtain a specific volume of solution, the chemist adds solvent up to the mark on the neck of the flask (bottom).

A

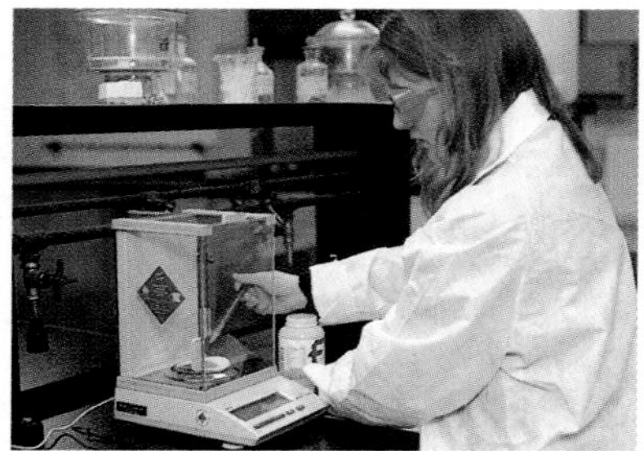

B

C

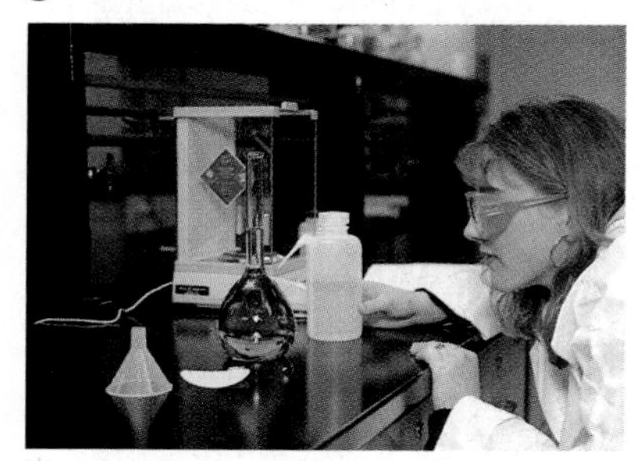

D

FIGURE 13.16
Making up a solution. (A) First, calculate the number of moles and the number of grams of substance you need. (B) Then, weigh out this amount. (C) Add the solute to about half of the solvent and swirl to dissolve. (D) Add the rest of the solvent, up to the mark on the flask, and swirl the flask again to give a homogeneous mixture.

Suppose you need 500 mL of 1.50 *M* NaCl, for instance. Here is a four-step method for making up a given amount of solution of a specific molarity (see Figure 13.16):

1. Calculate the number of grams of solute you need. First, molarity (moles solute per liter solution) times the amount of solution (in liters) equals the number of moles of solute. For the NaCl solution:
   In general, moles of solute = V (in liters) $\times$ *M*

   $$1.50 \text{ mol L}^{-1} \times 0.500 \text{ L} = 0.750 \text{ mol NaCl}$$

   The number of moles times the molar mass gives the number of grams. In this case:

   $$0.750 \text{ mol NaCl} \times 58.45 \text{ g mol}^{-1} \text{ NaCl} = 43.8 \text{ g NaCl}$$

2. Weigh out the calculated number of grams of solute.
3. Add about half of the needed solvent to a volumetric flask (a 500-mL one in this case), add the solute to the flask, and swirl the flask to dissolve the solute.
4. Add more solvent until the liquid level comes exactly to the mark on the neck of the flask (see Figure 13.15). Mix the solution so that it will be uniform throughout.

## EXAMPLE 13.10 Preparing a Solution of a Given Molarity

Outline the procedure for making up 250 mL of 0.50 *M* $CaCl_2$ in water.

**SOLUTION**

Use the four-step procedure:

**1.** You need 0.250 L of 0.500 *M* $CaCl_2$. This means you need:

$$(0.250 \text{ L})(0.50 \text{ mol L}^{-1}) = 0.125 \text{ mol CaCl}_2$$

The molar mass of $CaCl_2$ is

$$40.08 \text{ g mol}^{-1} \text{ Ca} + (2 \times 35.45 \text{ g mol}^{-1} \text{ Cl}) = 111 \text{ g mol}^{-1}$$

The mass of $CaCl_2$ you need is

$$(0.125 \text{ mol CaCl}_2)(111 \text{ g mol}^{-1} \text{ CaCl}_2) = 13.9 \text{ g CaCl}_2$$

**2.** Weigh out the 13.9 g of $CaCl_2$.

**3.** Add about 125 mL of water to a 250-mL volumetric flask, add the $CaCl_2$, and swirl.

**4.** Fill the flask with water to the 250-mL mark on the neck, and mix thoroughly.

**EXERCISE 13.10**

Outline the procedure for making up 500 mL of 0.200 *M* barium chloride.

This four-step procedure for preparing solutions is useful but not universal. Suppose, for instance, you are asked to prepare some 1 *M* hydrochloric acid. You will get stuck at the step in the procedure that tells you to weigh out the calculated amount of HCl, because pure HCl is a gas at room temperature. Chemists don't use HCl(*g*) for making aqueous solutions, but begin with a bottle of concentrated hydrochloric acid, 12 *M* HCl(*aq*), and prepare the 1 *M* solution using the *dilution method.* This method converts a concentrated solution into a more dilute solution by adding more solvent (see Figure 13.17). The solution with the higher concentration is sometimes called the *stock solution.* The number of moles of solute in a given amount of such a solution remains unchanged when more solvent is added to it. The newly prepared solution contains more solvent for the same amount of solute, so its concentration is always lower than that of the stock solution.

Chemists working in laboratories use the dilution method frequently. Many commercial reagents, particularly the common acids and ammonia, come from chemical companies as aqueous solutions that are too concentrated for most uses (see Table 13.3). Chemists typically dilute these solutions to the lower concentrations they need (see Figure 13.18).

**TABLE 13.3 Concentrations of Commercial Reagents**

| Reagent | Formula of Solute | Molarity of Commercial Solution |
|---|---|---|
| Hydrochloric acid | $HCl$ | 12 |
| Sulfuric acid | $H_2SO_4$ | 18 |
| Nitric acid | $HNO_3$ | 16 |
| Acetic acid | $CH_3CO_2H$ | 17 |
| Ammonia | $NH_3$ | 15 |

You can use a modified version of the four-step procedure to prepare solutions by the dilution method. As before, the moles of solute you need is equal to the molarity (moles solute per liter solution) times the amount of solution (in liters). Next, you calculate, not the mass (number of grams) of solute, but the volume (number of liters) of the starting solution.

Chemists almost always combine the two calculations, since the number of moles of solute is the same for the concentrated starting solution and the more dilute solution you're making. The combined equation is

$$V_{conc} M_{conc} = V_{dil} M_{dil}$$

where $V_{conc}$ and $M_{conc}$ are the volume and the concentration of the concentrated solution you begin with, and $V_{dil}$ and $M_{dil}$ are the volume and the concentration of the more dilute solution you are preparing. Rearranging the above equation gives you a way to calculate $V_{conc}$, the volume of concentrated solution you need:

$$V_{conc} = \frac{V_{dil} M_{dil}}{M_{conc}}$$

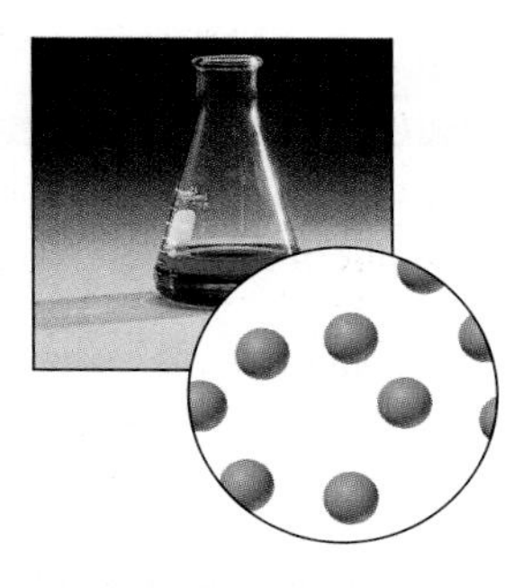

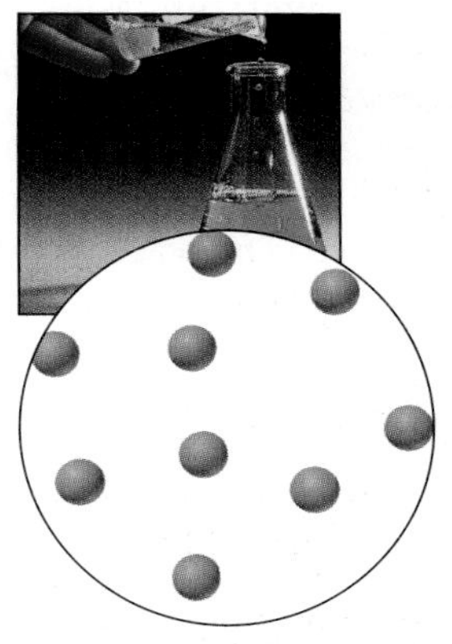

FIGURE 13.17
In dilution, the volume of the solution increases, while the number of solute particles remains the same.

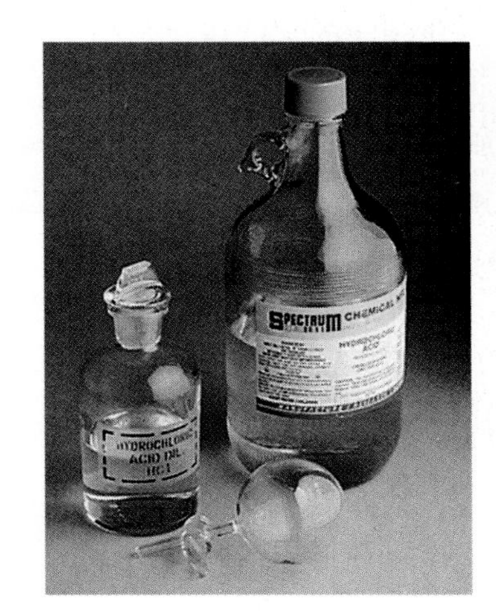


FIGURE 13.18
Manufacturers sell hydrochloric acid as 12 *M* solutions. This is too concentrated for many laboratory applications. Chemists usually prepare more dilute solutions (3 *M* is typical) and store them in reagent bottles until needed.

## CHAPTER BOX 13.3 PREPARING ACID SOLUTIONS

**Take extreme** care when preparing acid solutions by the dilution method. Concentrated laboratory reagents such as hydrochloric acid and especially sulfuric acid (see Table 13.3) are corrosive and very reactive. An extremely important laboratory safety rule is "Always Add Acid (AAA) to water." Never *(never!)* add water to concentrated acid.

When water and acid mix, an exothermic reaction occurs to form $H_3O^+(aq)$. The reaction can produce enough heat to cause the reaction mixture to boil out of the container carrying concentrated acid with it. These acid eruptions can travel a distance of several feet and can be quite dangerous. Adding acid *slowly to water,* with stirring, prevents the solution from generating heat rapidly enough to boil the water. There are two reasons why adding acid to water slowly prevents boilover. First, you have a large excess of water this way; it would take more heat to boil all of this water than to boil the small amount of water you would have added to the acid if you did it the other way around (see Chapter 12). Second, adding the acid slowly with stirring allows time for the heat to dissipate throughout the whole solution.

Don't underestimate the reactivity of concentrated acids. As an example of their strength, consider the effects on the organic molecules called carbohydrates that your body uses to generate the energy you need to live. Carbohydrates have the general formula $(CH_2O)_n$, and table sugar is one example. Adding concentrated sulfuric acid to table sugar removes water, leaving essentially pure carbon behind (as seen below). You must treat concentrated acids with respect.

If you mix concentrated sulfuric acid with table sugar, an exothermic reaction occurs that gets hot enough to boil water. The black substance left behind is essentially pure carbon. The reaction shown here is over in about 1 min.

Suppose you need to prepare 500 mL of 3.0 *M* $HCl(aq)$ starting with concentrated hydrochloric acid. Here is the revised four-step procedure:

1. Calculate the volume of concentrated acid you need to start with; $V_{dil}$ is 500 mL = 0.500 L, $M_{dil}$ is 3.0 *M*, and $M_{conc}$ is 12 *M*.

$$V_{conc} = \frac{V_{dil} M_{dil}}{M_{conc}}$$

$$= \frac{(0.500 \text{ L})(3.0\ M)}{12\ M} = 0.125 \text{ L} = 125 \text{ mL}$$

2. Measure this volume of concentrated acid. A graduated cylinder (see Figure 13.19) is often used for this purpose.
3. Add about half of the needed solvent (water) to a volumetric flask (a 500-mL one in this case), add the concentrated HCl, and swirl the flask to mix thoroughly.

4. Add more water until the liquid level comes exactly to the mark on the neck of the flask. Swirl the solution so that it will be uniform throughout.

FIGURE 13.19
Chemists frequently use graduated cylinders (shown here) to measure out amounts of solutions that do not match the volumes of volumetric flasks.

### EXAMPLE 13.11 Preparing Solutions by Dilution

Outline the procedure for preparing 1.0 L of 1.5 $M$ HCl($aq$) from concentrated hydrochloric acid.

**SOLUTION**

Use the revised four-step procedure:

1. Calculate the volume of concentrated acid you need to start with:

$$V_{\text{conc}} = \frac{V_{\text{dil}}M_{\text{dil}}}{M_{\text{conc}}} = \frac{(1.0\text{ L})(1.5\ M)}{12\ M} = 0.125\text{ L} = 125\text{ mL}$$

2. Measure out 125 mL of solute (concentrated HCl).
3. Put about half of the needed solvent (water) into a 1-L volumetric flask, add the 125 mL of concentrated HCl slowly and carefully, and swirl the flask to mix thoroughly.
4. Add more water until the liquid level comes exactly to the mark on the neck of the flask. Swirl the solution so that it will be uniform throughout.

**EXERCISE 13.11**

Outline the procedure for preparing 2.0 L of 6.0 $M$ $HNO_3$ from concentrated (16 $M$) nitric acid.

## 13.10 Concentration of Solutions: Percentages

**GOAL:** To determine the concentrations of solutions expressed as percentages.

You might think that the simplest concentration units to work with would be percentages, and in some ways they are. Health professionals often use aqueous solutions that contain a given percentage of some substance, such as a 0.9% saline (NaCl) solution. Of the different ways to calculate concentrations as percentages, at least three have widespread use: percent by mass, percent by volume, and mass-volume percent. Let's look at each of these.

**Percent by mass** is the mass of the solute divided by the total mass of the solution times 100%. In equation form, this is

$$\text{Percent by mass} = \frac{\text{Mass of solute}}{\text{Total mass of solution}} \times 100\%$$

**If you're rusty on how to calculate with percentages, see the Tool Box on this subject on page 34 in Chapter 2.**

The total mass of the solution, of course, is equal to the mass of the solute plus the mass of the solvent. The abbreviation %(m/m) is frequently used for percent by mass.

EXAMPLE 13.12 **Finding Percent by Mass**

**a.** You have made up a solution by dissolving 5.0 g of sodium chloride in 95 g of water. What is the %(m/m) of NaCl in this solution?
**b.** A solution contains 2.1 g of potassium hydroxide and 90 mL of ethanol. The density of ethanol is 0.79 g/mL. What is the %(m/m) of KOH in this solution?

SOLUTIONS

**a.** Here is the equation and the calculation:

$$\%(\text{m/m}) = \frac{\text{Mass of solute}}{\text{Mass of solute plus mass of solvent}} \times 100\%$$

$$= \frac{5.0 \text{ g NaCl}}{5.0 \text{ g NaCl} + 95 \text{ g H}_2\text{O}} \times 100\%$$

$$= 5.0\% \text{ (m/m)}$$

**b.** In this case you must first calculate the mass of the solvent from the density. Here's how that's done:

$$\text{Density} = \frac{\text{Mass}}{\text{Volume}}$$

$$\text{Mass} = \text{Volume} \times \text{Density}$$

$$\text{Mass} = (90 \text{ mL ethanol})(0.79 \text{ g mL}^{-1}) = 71 \text{ g ethanol}$$

Then you use the percent by mass equation:

$$\%(\text{m/m}) = \frac{\text{Mass of solute}}{\text{Mass of solute plus mass of solvent}} \times 100\%$$

$$= \frac{2.1 \text{ g KOH}}{2.1 \text{ g KOH} + 71 \text{ g ethanol}} \times 100\%$$

$$= 2.9\% \text{ (m/m)}$$

EXERCISE 13.12

Chemical companies often sell sodium hydroxide as a 50% (m/m) aqueous solution. Suppose you want to run a reaction that requires you to use 15 g of sodium hydroxide. How much of the 50% (m/m) solution should you use?

**Percent by volume** is the volume of the solute divided by the total volume of the solution times 100%. This way of expressing concentration is used most often when the solute and the solvent are either both liquids or both gases. The equation is

$$\text{Percent by volume} = \frac{\text{Volume of solute}}{\text{Total volume of solution}} \times 100\%$$

The total volume of the solution, of course, is usually equal to the volume of the solute plus the volume of the solvent. The abbreviation %(v/v) is used for percent by volume.

**There's no guarantee that the total volume of the final solution will be equal to the sum of the volumes of solute and solvent. Molecules of a liquid solute often fit in holes between the molecules of the solvent. In this case, the final volume of the solution will be slightly less than the volume of the two liquids added together. Volumes of gases, on the other hand, are always additive (unless a chemical reaction occurs).**

### EXAMPLE 13.13 Finding Percent by Volume

You have made up a solution by dissolving 12.5 mL of ethanol in 100 mL of water. What is the %(v/v) of ethanol in this solution, assuming that the volume of the final solution is equal to the sum of the individual volumes?

**SOLUTION**

The equation and the calculation look like this:

$$\%(\text{v/v}) = \frac{\text{Volume of solute}}{\text{Volume of solute plus volume of solvent}} \times 100\%$$

$$= \frac{12.5 \text{ mL ethanol}}{12.5 \text{ mL ethanol} + 100 \text{ mL } H_2O} \times 100\% = 11.1\%(\text{v/v})$$

**EXERCISE 13.13**

You need to make 12.5 L of a 10.0% (v/v) solution of oxygen in nitrogen. How much oxygen do you need?

*Normal saline* is the solution used to dissolve drugs for injection. Pharmacists and others who work with normal saline use a third kind of percentage to express the concentration of this solution. **Mass-volume percent,** abbreviated %(m/v), is equal to the mass of solute in grams divided by the volume of the solution in milliliters times 100%. The equation is

$$\%(\text{m/v}) = \frac{\text{Mass of solute (in g)}}{\text{Total volume of solution (in mL)}} \times 100\%$$

The result is not a true percentage. Dividing a mass by a volume doesn't give a unitless answer, as all percentage calculations should. This method is used only for aqueous solutions. Because the density of water (and of most dilute aqueous solutions) is 1.0 g $\text{mL}^{-1}$, the mass-volume percent is nearly the same as the percent by mass. For reasons we will discuss in Section 13.12, normal saline is 0.92% (m/v) salt water.

In dealing with mass percentages and volume percentages, you can use dimensional analysis to make sure you are keeping your units straight. With mass-volume percents, you must be a little more careful.

EXAMPLE 13.14 **Using Mass-Volume Percent**

Normal saline is 0.92% (m/v) NaCl in water. How many grams of NaCl do you need to make 150 mL of normal saline?

**SOLUTION**

Here are the equation and the calculation:

$$0.92\%\ (\text{m/v})\ \text{NaCl means}\ \frac{0.92\ \text{g NaCl}}{100\ \text{mL solution}}$$

This can be used as a unit factor

$$150\ \cancel{\text{mL solution}} \times \frac{0.92\ \text{g NaCl}}{100\ \cancel{\text{mL solution}}} = 1.4\ \text{g NaCl}$$

**EXERCISE 13.14**

How many grams of NaCl do you need to make 200 mL of normal saline?

The procedure you use for making up a solution having a concentration expressed as a percentage is very similar to the procedure you learned in Section 13.9. Here it is:

1. Calculate the mass or the volume of solute and solvent you need from the given percentage.
2. Measure the calculated mass or volume of solvent and put most of it in an appropriate flask. (In making %(m/m) and %(v/v) solutions you can put it all in.)
3. Measure and add the calculated mass or volume of solute.
4. For %(m/v) solutions, fill the flask to the mark. In all cases, mix completely.

The calculations for solutions of given percentage concentrations are easier than those for solutions of given molarity. This is why people who make up solutions on a routine basis often use percentages rather than molarity. The rest of the procedure is easier, too. You just have to mix the solute and solvent in an appropriate flask.

EXAMPLE 13.15 **Preparing Solutions with Given Percentage Concentrations**

Outline the procedure for making each of these solutions.

**a.** 200 g of 1% (m/m) KI in acetone (a common solvent)
**b.** 300 mL of 20% (v/v) ethanol in water (both are liquids)
**c.** 0.50 L of 5.0% (m/v) sodium bicarbonate in water

**SOLUTIONS**

**a.** The four-step procedure is straightforward:

**1.** 1% of 200 g is (200 g)(0.01) = 2 g. You need 2 g of KI. The rest of the solution is acetone. If the total is 200 g, you need 200 g − 2 g = 198 g of acetone.

2. Weigh 2 g of KI and put it in a flask.
3. Weigh 198 g of acetone and put it in a flask (see Figure 13.20).
4. Mix thoroughly.

**b.** The four-step procedure for a %(v/v) solution is similar:

1. 20% of 300 mL is (300 mL)(0.20) = 60 mL of ethanol. The rest of the solution is water. If the total is 300 mL, you need 300 mL − 60 mL = 240 mL of water.
2. Using a graduated cylinder, measure out 240 mL of water and put it in a flask.
3. Using a graduated cylinder, measure out 60 mL of ethanol and add it to the water.
4. Mix thoroughly.

**c.** The four-step procedure for a %(m/v) solution is more of the same:

1. In %(m/v) calculations you must use mL, and 0.50 L = 500 mL. 5.0% of 500 mL is (500 mL)(0.050) = 25 mL (which is equivalent to 25 g because the density of water is 1.0 g $mL^{-1}$). You need 25 g of $NaHCO_3$.
2. Put about 250 mL of water into a 500-mL volumetric flask. The solvent in %(m/v) solutions is always water.
3. Weigh out 25 g of $NaHCO_3$, put it in the volumetric flask with the water, and swirl.
4. Fill the flask to the 500-mL mark and mix thoroughly.

**EXERCISE 13.15**

Outline the procedure for making each of these solutions.

**a.** 150 g of 10%(m/m) KOH in ethanol
**b.** 50 mL of 15%(v/v) acetone in water (both are liquids)
**c.** 250 mL of 5.0%(m/v) sodium chloride in water

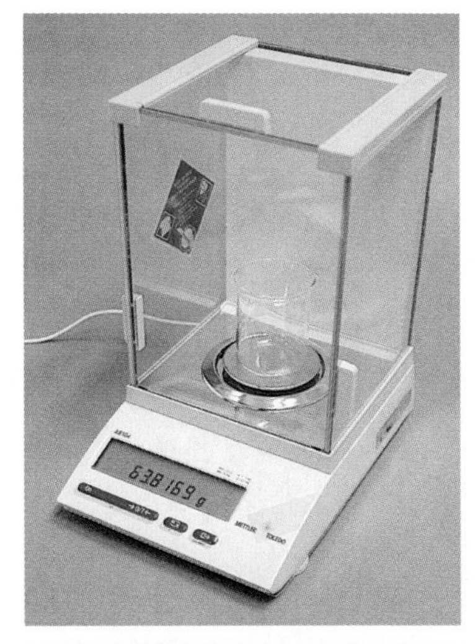

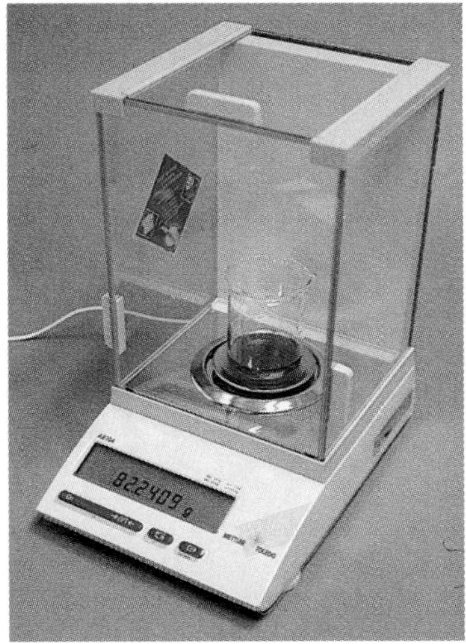

FIGURE 13.20
To weigh a liquid, you first weigh a container, which might be a flask or beaker. Then you add the liquid and weigh the combination. The mass of the liquid is equal to the mass of the flask-plus-liquid minus the mass of the flask. Some balances will subtract the mass of the flask automatically.

## 13.11 Concentrations: Molality and Mole Fraction

**GOAL:** To determine the concentrations of solutions expressed in terms of molalities and mole fractions.

There are so many different ways of measuring concentration because each way works best for some specific application or some particular kind of problem. This section covers two more concentration measures scientists sometimes use: molality and mole fraction.

The **molality** of a solution (abbreviated *m*) is the number of moles of solute per kilogram of solvent:

$$\text{Molality} = \frac{\text{Moles of solute}}{\text{kg of solvent}}$$

**The density of liquid water is 1.0 g $mL^{-1}$. Therefore, one may conclude that a 1.0 *m* aqueous solution is slightly less concentrated than a 1.0 *M* aqueous solution.**

This is different from molarity. The molarity tells you how much solute there is in a solution, but it doesn't tell you how much solvent is present. A 1 *M* aqueous solution of NaCl, for instance, contains 1 mol of NaCl per liter *of solution.* It contains less than a liter of water per liter of solution, and there is no simple way to tell how much less. A 1 *m* solution of aqueous NaCl, on the other hand, contains 1 mol of NaCl and 1 kg of water, so you know how much solvent there is. As with molarity, however, there is something you don't know: There is no simple way to tell what the total volume of solution is.

### EXAMPLE 13.16 Determining Molality

A saturated solution of potassium cyanide contains 50 g of KCN per 100 mL of water. What is the molality of this solution?

**SOLUTION**

To solve this problem you need moles of solute and kilograms of solvent. The solute is KCN. To get the number of moles equivalent to 50 g of KCN, you divide 50 g by the molar mass of KCN:

$$\text{Moles of KCN} = \frac{50\text{ g KCN}}{(39.1\text{ g mol}^{-1}\text{ K} + 12.0\text{ g mol}^{-1}\text{ C} + 14.0\text{ g mol}^{-1}\text{N})}$$

$$= \frac{50\text{ g KCN}}{65.1\text{ g mol}^{-1}\text{ KCN}} = 0.77\text{ mol KCN}$$

Next you need the mass of solvent in kilograms. Since the density of water is 1.0 g mL$^{-1}$, the mass of 100 mL of water is

$$100\ \cancel{\text{g H}_2\text{O}} \times \frac{1\text{ kg H}_2\text{O}}{1000\ \cancel{\text{g H}_2\text{O}}} = 0.10\text{ kg H}_2\text{O}$$

The molality is then

$$\frac{\text{Moles of KCN}}{\text{kg of H}_2\text{O}} = \frac{0.77\text{ moles}}{0.10\text{ kg}} = 7.7\ m$$

**EXERCISE 13.16**

What is the molality of a solution made by adding 35 g of $Ca(NO_3)_2$ to 150 mL of water?

The **mole fraction** of a solute in a solvent is the number of moles of the solute divided by the total number of moles in the solution. The equation is

$$\text{Mole fraction} = \frac{\text{Moles of solute}}{\text{Moles of solute} + \text{Moles of solvent}}$$

If the mole fraction (or any other fraction) were multiplied by 100%, it would be a percentage. The mole fraction is a unitless number.

### EXAMPLE 13.17 Determining Mole Fraction

**a.** What is the mole fraction of sodium phosphate in a solution that contains 0.0123 mol of $Na_3PO_4$ and 7.7 mol of water?
**b.** What is the mole fraction of calcium acetate in a solution made from 1.2 g of $(CH_3CO_2)_2Ca$ and 125 mL of water?

**SOLUTIONS**

**a.** The equation is

$$\text{Mole fraction} = \frac{\text{Moles of solute}}{\text{Moles of solute} + \text{Moles of solvent}}$$
$$= \frac{0.0123 \text{ mol } Na_3PO_4}{0.0123 \text{ mol } Na_3PO_4 + 7.7 \text{ mol } H_2O} = 0.0016$$

The mole fraction of $Na_3PO_4$ in this solution is 0.0016.

**b.** The numbers of moles aren't given here, so you have to calculate them first. The solution has

$$\frac{1.2 \text{ g}}{[4(12.0) + 6(1.0) + 4(16.0) + 40.1] \text{ g/mol}} = 0.0076 \text{ mol of calcium acetate}$$

Since the density of water is 1.0 g $mL^{-1}$, there are

$$\frac{125 \text{ g}}{(18.0 \text{ g mol}^{-1})} = 6.94 \text{ mol of } H_2O$$

The mole fraction of calcium acetate, calculated using the equation in part(a), is 0.0011.

**EXERCISE 13.17**

What is the mole fraction of each component in a solution prepared by mixing 10 g of acetone ($C_3H_6O$), 10 g of ethanol ($C_2H_6O$), and 10 g of water?

## 13.12 Colligative Properties

**GOAL:** To examine properties of solutions that depend only on the concentration of solute particles.

Doctors often advise patients with high blood pressure to reduce the sodium in their diets. People put the compound ethylene glycol ($HOCH_2CH_2OH$, the key ingredient in antifreeze) in their car radiators to prevent boiling in the summer and freezing in the winter. In areas with colder climates, road crews scatter salt on highways to melt ice in winter. These may seem to be a group of unconnected facts, but they arise from properties that are closely related. All of the properties depend only on the concentration of solute particles (whether molecules or ions), regardless of the chemical nature of the particles. Chemists' term for such properties is **colligative properties.**

We'll begin by examining one of the simpler colligative properties. When a solute dissolves in a solvent, the vapor pressure of the solvent decreases. Why does this happen? Figure 13.21 may help you visualize what happens to the molecules involved. For a molecule to escape the liquid phase, it must be at the surface (unless the liquid is boiling and thus bubbling). If you add a solute, the molecules or ions of the solute occupy some of the positions at the surface. Thus, fewer solvent molecules have a chance to escape to the vapor phase. This means, in turn, that the vapor pressure of the solvent—that is, the concentration of solvent molecules in the gas phase—decreases.

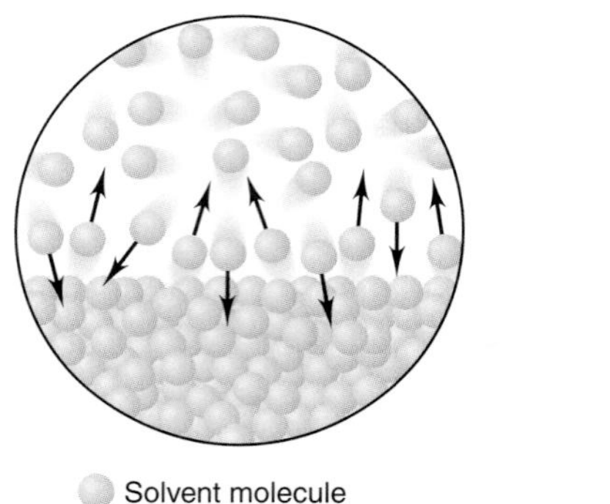


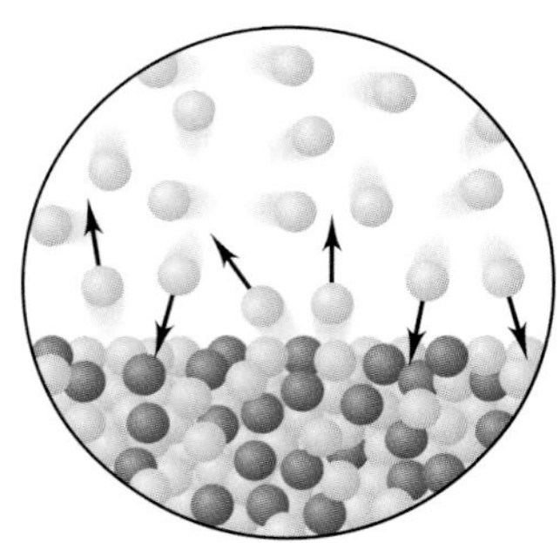

FIGURE 13.21
Only molecules at the surface can escape the liquid and move into the vapor phase. Solute molecules (as long as the solute has no appreciable vapor pressure itself) get in the way of solvent molecules. This means that there will be fewer solvent molecules in the gas phase above the solution, as shown here.

**The French physical chemist François Marie Raoult (1830–1901) discovered this relationship in 1886.**

Exactly how does the vapor pressure of a solution change as more and more solute is added? Once someone discovers the broad outline of a phenomenon, how to quantify it is the next question scientists will ask. The answer is almost always in the form of an equation. This time the equation is

$$P = \chi_{solvent} P^{\circ} \qquad \text{(Raoult's Law)}$$

where $P$ is the vapor pressure of the solvent above the solution, $\chi_{solvent}$ is the mole fraction of the solvent in the solution, and $P^{\circ}$ is the vapor pressure of the pure solvent. For Raoult's Law to hold true, the solute must have a vapor pressure of zero, or at least one much smaller than that of the solvent. As you might expect, Raoult's Law works very well when the solute is a solid, because solids usually have low vapor pressures compared to liquids.

Two other colligative properties are **boiling-point elevation** and **freezing-point depression.** These two are closely related. In freezing-point depression, the temperature at which a liquid solvent freezes to a solid decreases when a solute is added. In boiling-point elevation, the temperature at which a liquid solvent boils increases when a solute is added.

It is relatively easy to see how dissolved substances can affect the boiling and freezing points of solvents. For a solvent to boil, its vapor pressure must equal the external pressure (that is what defines a boiling point). Placing solute molecules in solution lowers the vapor pressure of the solvent. Therefore, heat must be added to make the vapor pressure of the solution equal to the external pressure (often atmospheric pressure).

At the freezing point of a solvent, the vapor pressures of the solute and the solvent are identical. One important implication of this is that their molecules enter and leave the solid phase at the same rate (see Figure 13.22). When a solution freezes, only the solvent molecules enter the solid phase and separate from the solution. This results in a solid block of pure solidified solvent, while the solute molecules remain in solution. At this point the vapor pressure of the solution is less than the vapor pressure of the solid (because the solution becomes more concentrated as solvent becomes solid). Therefore, the temperature must decrease to make the vapor pressures of the solid and solution identical. This means that the freezing point decreases.

The boiling point elevation and the freezing point depression are usually not very large. For example, a 1 *m* solution of a nonelectrolyte in water boils at 100.59°C and freezes at − 1.86°C (compared to 100°C and 0°C for pure water). Both boiling point elevation and freezing point depression are directly proportional to the concentration of solute particles in the solution. That is, the change in temperature equals a constant times the concentration of particles expressed as molality. The value of the constant depends on the identity of the solvent.

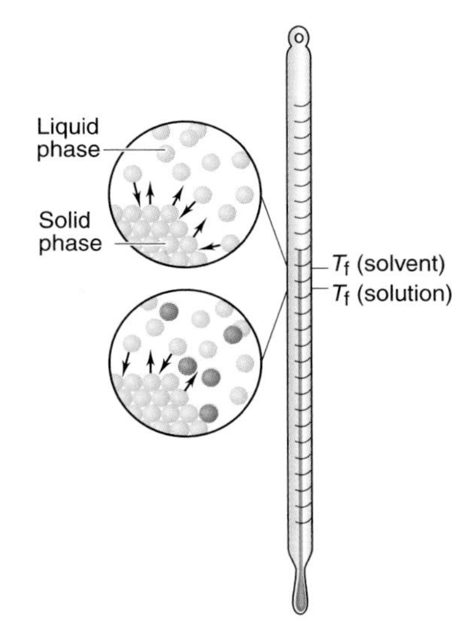


FIGURE 13.22
The vapor pressures of the solid and the liquid must be identical for a substance to freeze. When solutions freeze, only the solvent molecules enter the solid phase, and the vapor pressure of the solvent is always greater than that of the solution left behind. In order to lower the vapor pressure of the solid phase to that of the liquid phase, the temperature must decrease.

## EXAMPLE 13.18 Freezing-Point Depression and Boiling-Point Elevation

If these compounds are dissolved in water at equal molal concentrations, which one gives the largest decrease in the freezing point of the solution?
**A.** Table salt, NaCl **B.** Sugar, $C_{12}H_{22}O_{11}$ **C.** $NaNO_3$ **D.** $CaCl_2$

### SOLUTION

To answer this question you have to know what ions (if any) make up the compounds. NaCl (A), of course, divides into two particles. Sugar (B) is not an electrolyte and yields only one particle per molecule. $NaNO_3$ (C) also gives two particles per mole, a sodium ion ($Na^+$) and a nitrate ion ($NO_3^-$). $CaCl_2$ (D) gives three particles, a $Ca^{2+}$ ion and two $Cl^-$ ions. The answer is D.

### EXERCISE 13.18

If these compounds are dissolved in water at equal molal concentrations, which one gives the largest increase in the boiling point of the solution?
**A.** $Na_3PO_4$ **B.** $CaCl_2$ **C.** NaOH **D.** KI

**Some people knew that a pure solvent freezes out of solution, leaving solute in the unfrozen liquid, long before anyone did any theoretical work on colligative properties. Russians living near the Black Sea isolated salt from sea water by placing the water in large kettles and allowing some of it to freeze. They would periodically remove ice and add more sea water. They continued to do this until the water underneath the ice became saturated with salt. The salt then precipitated, and the Russians collected it.**

*Osmotic pressure* is another important colligative property, especially useful in biomedical fields and the health professions. The French physicist Jean Nollet discovered the phenomenon in 1784. He filled a pig bladder with a solution of alcohol and water. When he submerged the bladder in pure water, it expanded as water migrated into the bladder through the walls. This phenomenon occurs in a wide variety of situations. The pig bladder can be replaced by any other *semipermeable membrane* (a barrier that permits the migration of some molecules but not others), and the ethanol-water solution by

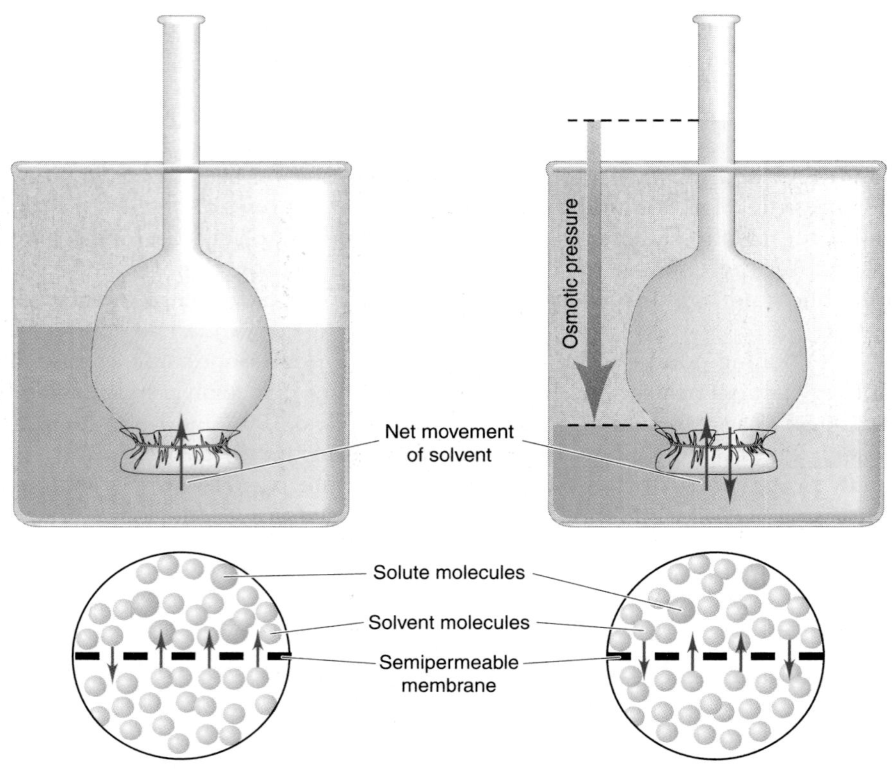


FIGURE 13.23
In this example of osmosis, a semipermeable membrane separates a pure solvent and a solution (left). More solvent molecules enter the solution than leave it. As a result (right), the volume of the solution increases as its concentration decreases. The osmotic pressure is defined as the pressure required to prevent this volume change.

almost any other solution. The pure solvent will migrate through the membrane into the solution. The process is termed **osmosis,** and the force that drives it is called **osmotic pressure.** More strictly speaking, the osmotic pressure is the pressure that must be applied to prevent the flow of a solvent through a semipermeable membrane (see Figure 13.23).

Osmotic pressure is one of the reasons doctors recommend low-salt diets for people with hypertension (high blood pressure). If your blood has a high concentration of salt in it, osmotic pressure forces water out of your cells into the blood. The more fluid there is in your circulatory system, the higher the total pressure will be. Eating less salt lowers the salt concentration and therefore the amount of fluid in your blood, reducing the pressure.

Medical teams often give fluids intravenously to burn patients. It is critical that the solution pumped into a patient have the same concentration of solute molecules as blood does, because blood cells can act like the bladder in Nollet's experiment. Adding pure water to the bloodstream causes blood cells to swell up and burst, which can have fatal consequences (see Figure 13.24). It's necessary to keep the osmotic pressure on the walls of blood cells approximately zero. Normal saline is 0.92% (m/v) salt, about the same concentration of salt as in blood.

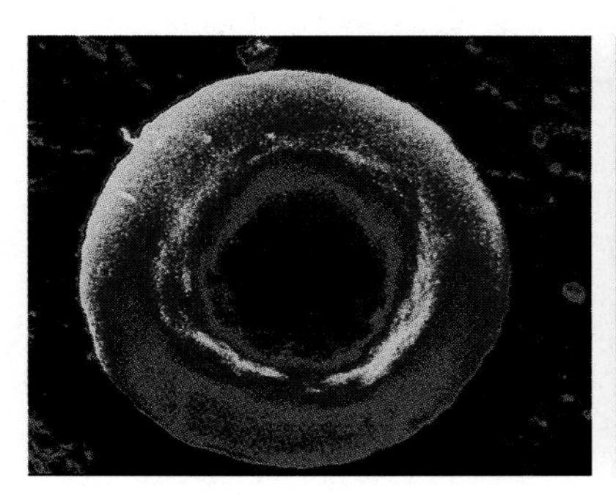

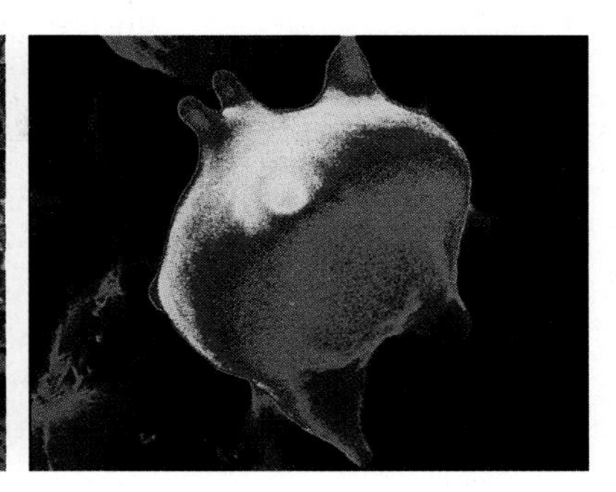

FIGURE 13.24
(Left) Blood cells placed in normal saline keep their usual shape, because there is no net flow of water across the cell membrane. (Center) The cells expand (as shown) and finally burst in solutions of low NaCl concentration, because water moves into the cells. (Right) In solutions of high NaCl concentration, cells shrink as water flows out of them.

## EXPERIMENTAL EXERCISE 13.1

When Svante Arrhenius (see Section 4.5, p. 128) was a student at the University of Uppsala in the early 1880s, almost all chemists held Dalton's view that atoms were structureless and indivisible. Arrhenius knew about experiments being done at the time with the freezing points of aqueous solutions, including Raoult's work showing that the amount the freezing point was lowered was proportional to the quantity of substance dissolved in water. Arrhenius knew that solutions of sodium chloride showed twice as large a freezing-point depression as solutions of table sugar of the same concentrations, and that solutions of sodium sulfate showed a freezing-point depression three times as large. He also knew that pure water and aqueous sugar solutions are poor conductors of electricity, and that solutions of sodium chloride and sodium sulfate are good conductors of electricity—that is, sodium chloride and sodium sulfate are electrolytes.

On the basis of this evidence, Arrhenius concluded that some molecules break down into separate particles when they dissolve in water. He further concluded that atoms could not be structureless indivisible species, because the sodium and chlorine atoms in table salt must enter into solution as charged particles; that is, as ions.

### QUESTIONS

1. What piece of experimental evidence led Arrhenius to the conclusion that atoms can form ions?
2. What piece of experimental evidence led Arrhenius to the conclusion that some substances divide into separate particles in aqueous solution?
3. Why didn't Arrhenius conclude that *all* substances divide into separate particles in aqueous solution?
4. Do facts that NaCl(*s*) is a nonconducting solid and NaCl(*aq*) is a good electrical conductor support Arrhenius's theory that ions exist in solution?
5. Solutions of acetic acid have freezing-point depressions very slightly greater than those of sugar solutions of the same concentrations, but much less than those of sodium chloride solutions of the same concentrations. Are these results consistent with the fact that acetic acid is a weak electrolyte?

Freezing-point depressions and boiling-point elevations are almost always small, but minor changes in the concentration of solute particles can result in very large changes in the osmotic pressure. The osmotic pressure is related to the number of moles of solute particles by an equation that resembles the gas law ($PV = nRT$):

$$\pi V = nRT \text{ or } \pi = \frac{nRT}{V} = MRT$$

Here $\pi$ is the osmotic pressure, $V$ is the volume of the solution, $n$ is the number of moles of particles in the solution, $M$ is the molarity of particles in solution, $T$ is the temperature in kelvins, and $R$ is the gas constant. Note that $M$ refers to the molarity of *particles* per liter of solution. A 0.1 $M$ solution of table salt has twice the osmotic pressure of a 0.1 $M$ solution of sugar, because it has twice as many particles in solution.

### EXAMPLE 13.19 Finding the Osmotic Pressure

What is the osmotic pressure of a 0.10 $M$ solution of table sugar at 300 K?

**SOLUTION**

You have everything needed to solve the problem: $M = 0.10 \text{ mol L}^{-1}$, $R = 0.082 \text{ L atm mol}^{-1} \text{ K}^{-1}$, and $T = 300$ K.

$$\pi = MRT = (0.10 \text{ mol L}^{-1})(0.082 \text{ L atm mol}^{-1} \text{ K}^{-1})(300 \text{ K}) = 2.5 \text{ atm}$$

**EXERCISE 13.19**

What is the osmotic pressure of a 0.1 $M$ solution of table salt at 300 K?

## CHAPTER SUMMARY

1. Solutions are homogeneous mixtures of solids, liquids, and/or gases.
2. Solutions can be prepared by dissolving solutes in solvents. When a solid or a gas dissolves in a liquid, the liquid is the solvent. When two liquids form a solution, the solvent is the liquid present in the larger amount.
3. Solutions can also be prepared by chemical reactions. For example, adding sodium metal to water produces an aqueous solution of NaOH and hydrogen gas.
4. For predicting the solubility of one substance in another, "like dissolves like" is a useful guideline.
5. Almost all ionic compounds are insoluble in water and other common solvents. Table 13.1 identifies the few ionic compounds that are water soluble. This is an important list, because these compounds are exceptions to a strong general trend.
6. A solution that is saturated contains the largest number of solute particles that that particular volume of solvent can accommodate at that temperature.

7. The solid that forms when two solutions of ionic compounds mix and react is called a precipitate. The reaction that produces it is a precipitation reaction.
8. Strong electrolytes are substances that dissociate completely in solution. They tend to be ionic compounds or strong acids, and their solutions are good electrical conductors.
9. Strong acids react completely with water to form the $H_3O^+$ ion and the anion of the acid. Weak acids react incompletely with water. Solutions of weak acids contain undissociated acid molecules, $H_3O^+$ ions, and the anion of the acid. The species present in the largest amount is the undissociated acid.
10. The concentration of a solution tells you how much solute is present relative to the amount of solution or solvent present.
11. Measures of concentration include molarity, various percentages such as %(m/m), %(v/v), and %(m/v), molality, and mole fraction. The best one to use in any given situation is the one most useful for the particular application or problem.
12. Colligative properties include vapor-pressure lowering, freezing-point depression, boiling-point elevation, and osmotic pressure. These properties depend only on the number of solute particles present in solution and not on the nature of those particles.

## IMPORTANT EQUATIONS

$$\text{Molarity } (M) = \frac{\text{Moles of solute}}{\text{Liters of solution}}$$

$$\text{Moles of solute} = V\,(\text{in liters}) \times M$$

$$V_{\text{conc}}M_{\text{conc}} = V_{\text{dil}}M_{\text{dil}}$$

Percent by mass, or %(m/m)

$$= \frac{\text{Mass of solute}}{\text{Total mass of solution}} \times 100\%$$

Percent by volume, or %(v/v)

$$= \frac{\text{Volume of solute}}{\text{Total volume of solution}} \times 100\%$$

Mass-volume percent, or %(m/v)

$$= \frac{\text{Mass of solute (in g)}}{\text{Total volume of solution (in mL)}} \times 100\%$$

$$\text{Molality } (m) = \frac{\text{Moles of solute}}{\text{kg of solvent}}$$

$$\text{Mole fraction (X)} = \frac{\text{Moles of solute}}{\text{Moles of solute} + \text{Moles of solvent}}$$

Raoult's Law: $P = X_{\text{solvent}}P°$

Osmotic pressure: $\pi V = nRT$ or $\pi = MRT$

## PROBLEMS

### SECTION 13.1 Solvents and Solutes

1. Identify the solvent and the solute in each of these solutions.
   a. A mixture of 100 mL of acetic acid and 120 mL of water
   b. A homogeneous mixture made from 120 g of solid ammonium acetate and 100 mL of water
   c. The solution obtained when sodium metal is added to water
   d. Household ammonia
   e. Alcoholic beverages

2. Identify the solvent and the solute in each of these solutions.
   a. A mixture of 100 mL of acetic acid and 20 mL of water
   b. A homogeneous mixture made from 20 g of solid ammonium chloride and 100 mL of water
   c. The solution obtained when calcium metal reacts with water
   d. Household bleach
   e. Carbonated beverages
3. For each solution, identify the solvent and the solute.
   a. A mixture of 10 mL of ethanol and 25 mL of water
   b. An alloy made from 95 g of aluminum and 2 g of manganese
   c. The solution obtained when 1 mol of $AgNO_3$ and 1 mol of NaCl are added to 1 L of water
   d. Vinegar
   e. Salt water
4. For each solution, identify the solvent and the solute.
   a. A mixture of 50 mL of acetone and 15 mL of water
   b. An alloy made from 2 g of carbon and 125 g of iron
   c. The solution obtained when 0.1 mol of $Ba(NO_3)_2$ and 0.1 mol of $Na_2SO_4$ are added to 100 mL of water
   d. Gasoline containing 10% ethanol
   e. A kamacite meteorite (6% nickel and 94% iron)
5. Name three aqueous solutions found around the house that are not mentioned in this chapter.
6. Give three examples of solid solutions that are not mentioned in this chapter.

## SECTION 13.2 Principles of Solubility

7. Which of these compounds is the most soluble in water? Explain your answer.
   A. $CH_3CH_2OH$ C. $CH_3CH_3$
   B. NaCl D. $Ca_3(PO_4)_2$
8. Which of these compounds is the least soluble in water? Explain your answer.
   A. $CH_3CH_2OH$ C. $CH_3CH_2CH_2CH_3$
   B. NaCl D. $(CH_3)_2C{=}O$
9. Which of these compounds is the most soluble in water? Explain your answer.
   A. $CH_3CH_3$ C. $CH_3CH_2OH$
   B. $CH_3CH_2CH_2CH_3$ D. $CH_3CH_2CH_2CH_2OH$
10. Which of these compounds is the least soluble in water? Explain your answer.
    A. $CH_3CH_2Cl$ C. $CH_3CH_2OH$
    B. $CH_3CH_2CH_2CH_2Cl$ D. $CH_3CH_2CH_2CH_2OH$
11. The atoms in hexane are arranged like this:

    $CH_3CH_2CH_2CH_2CH_2CH_3$

    What kind of intermolecular forces are present in compounds that dissolve readily in hexane?
12. The atoms in 1-octanol have this arrangement:

    $CH_3CH_2CH_2CH_2CH_2CH_2CH_2CH_2OH$

    What kind of intermolecular forces are present in compounds that dissolve readily in 1-octanol?
13. The atoms in ethylene glycol are arranged like this:

    $HOCH_2CH_2OH$

    What kind of intermolecular forces are present in compounds that dissolve readily in ethylene glycol?
14. The atoms in octane have this arrangement:

    $CH_3CH_2CH_2CH_2CH_2CH_2CH_2CH_3$

    What kind of intermolecular forces are present in compounds that dissolve readily in octane?
15. Give an example of a solvent in which water would be relatively insoluble.
16. Name two substances not mentioned in this chapter that are insoluble in water.
17. Explain why hydrogen gas forms a solution with any other gas.
18. Explain why formaldehyde ($H_2C{=}O$) dissolves well in water. (The biological preservative *formalin* is a solution of formaldehyde in water.)
19. Explain why HgS is insoluble in water.
20. Methylamine, $CH_3NH_2$, is a covalent molecular gas that is very soluble in water. Explain the reasons for this solubility.
21. Why is propane ($CH_3CH_2CH_3$) insoluble in water?
22. Explain why propane ($CH_3CH_2CH_3$) dissolves in butane ($CH_3CH_2CH_2CH_3$) in all proportions.
23. Draw a picture similar to the one on page 459 that shows solute-solvent interactions when formaldehyde ($H_2C{=}O$) dissolves in water.
24. Draw a picture similar to the one on page 459 that shows solute-solvent interactions when formaldehyde ($H_2C{=}O$) dissolves in trichloromethane ($HCCl_3$).
25. Draw a picture similar to the one on page 459 that shows solute-solvent interactions when methanol ($CH_3OH$) dissolves in ethanol ($CH_3CH_2OH$).

26. Draw a picture similar to the one on page 459 that shows solute-solvent interactions when pentane ($CH_3CH_2CH_2CH_2CH_3$) dissolves in hexane ($CH_3CH_2CH_2CH_2CH_2CH_3$).

## SECTION 13.3 Predicting the Solubility of Ionic Compounds in Water

27. Predict whether each compound is soluble in water.
a. $NH_4NO_3$
b. $Hg_2Cl_2$
c. $Fe(OH)_3$
d. NaOH

28. Predict whether each compound is soluble in water.
a. $NaHCO_3$
b. PbS
c. $K_2CO_3$
d. KOH

29. For each compound, tell whether it is water soluble.
a. AgBr
b. $CaSO_4$
c. $LiClO_4$
d. $NH_4Cl$

30. For each compound, tell whether it is water soluble.
a. $(CH_3CO_2)_2Ni$
b. $(NH_4)_2S$
c. LiOH
d. $Na_2CO_3$

31. Are these compounds soluble in water?
a. $CaCO_3$
b. $Na_2S$
c. $Na_2S_2O_3$
d. $PbCl_2$

32. Are these compounds soluble in water?
a. $CH_3CO_2Li$
b. $Ag_2O$
c. $(NH_4)_2CO_3$
d. $Mn(NO_3)_2$

33. Draw a picture showing how the $Ca^{2+}(aq)$ ion is stabilized by ion-dipole forces.

34. Draw a picture showing how the $OH^-(aq)$ ion is stabilized by hydrogen bonding.

35. Sodium iodide is somewhat soluble in the organic solvent acetone, $(CH_3)_2C{=}O$. Draw pictures showing how acetone solutions containing $Na^+$ and $I^-$ ions are stabilized by ion-dipole forces.

36. Potassium hydroxide is somewhat soluble in the organic solvent ethanol ($CH_3CH_2OH$). Draw pictures showing how ethanol solutions containing $K^+$ and $OH^-$ ions are stabilized by ion-dipole forces and hydrogen bonding, respectively.

37. Explain why measurable amounts of NaCl do not dissolve in hexane ($CH_3CH_2CH_2CH_2CH_2CH_3$).

38. Explain why most ionic compounds do not dissolve in water.

## SECTION 13.4 Saturated and Unsaturated Solutions

39. The solubility of ammonium chloride is 29.7 g per 100 mL of water at 0°C.
a. How much ammonium chloride will dissolve, and how much will remain undissolved, when 100 mL of water is added to 50 g of ammonium chloride at 0°C?
b. How much ammonium chloride will dissolve, and how much will remain undissolved, when 25 mL of water is added to 2.7 g of ammonium chloride at 0°C?

40. The solubility of ammonium chloride is 75.8 g per 100 mL of water at 100°C.
a. How much ammonium chloride will dissolve, and how much will remain undissolved, when 100 mL of water is added to 50 g of ammonium chloride at 100°C?
b. How much ammonium chloride will dissolve, and how much will remain undissolved, if the temperature of the solution in part (a) is reduced to 0°C? (See the previous problem for the solubility at 0°C.)

41. The solubility of copper(II) sulfate is 14.3 g per 100 mL of water at 0°C and 75.4 g per 100 mL at 100°C.
a. How much copper(II) sulfate will dissolve, and how much will remain undissolved, when 5.7 g of copper(II) sulfate is added to 21.1 mL of water at 0°C?
b. How much copper(II) sulfate will dissolve, and how much will remain undissolved, if the temperature of the solution in part (a) is raised to 100°C?

42. The solubility of copper(I) chloride is 0.0062 g per 100 mL of water at 0°C.
a. How much copper(I) chloride will dissolve, and how much will remain undissolved, when 1.00 g of copper(I) chloride is added to 1.0 L of water at 0°C?
b. How much copper(I) chloride will dissolve, and how much will remain undissolved, when 0.250 g of copper(I) chloride is added to 25 mL of water at 0°C?

43. The solubility of cadmium nitrate, $Cd(NO_3)_2$, is 109 g per 100 mL of water at 0°C.
a. How much cadmium nitrate will dissolve, and how much will remain undissolved, when 200 g of cadmium nitrate is added to 75 mL of water at 0°C?
b. How much cadmium nitrate will dissolve, and how much will remain undissolved, when 2.5 g of cadmium nitrate is added to 15 mL of water at 0°C?

44. The solubility of cobalt(II) chloride is 45 g per 100 mL of water at 7°C.
a. How much cobalt(II) chloride will dissolve, and how much will remain undissolved, when 1.00 g of cobalt(II) chloride is added to 1.0 mL of water at 7°C?
b. How much cobalt(II) chloride will dissolve, and how much will remain undissolved, when 25 g of cobalt(II) chloride is added to 45 mL of water at 7°C?

45. Sodium carbonate has a solubility of 7 g per 100 mL at 0°C and 46.0 g per 100 mL at 100°C. A chemist took 50 mL of a saturated solution of $Na_2CO_3$ at 100°C and cooled it to 0°C. How much $Na_2CO_3$ precipitated?

46. An inorganic salt has a water solubility of 2.55 g per 100 mL at 20°C. How much of the salt precipitates if 75 mL of a solution containing 12.44 g of the salt at 80°C is cooled to 20°C?

47. An inorganic salt has a water solubility of 0.19 g per 100 mL at 20°C. How much of the salt precipitates if 80 mL of a solution containing 1.99 g of the salt is cooled to 20°C?

48. An organic acid has a water solubility of 1.8 g per 100 mL at 0°C. How much of the acid precipitates if 75 mL of a solution containing 7.9 g of the acid is cooled to 0°C?

49. An organic acid has a water solubility of 0.80 g per 100 mL at 10°C. How much of the acid precipitates if 25 mL of a solution containing 1.99 g of the acid is cooled to 10°C?

50. An organic acid has a water solubility of 2.3 g per 100 mL at 5°C. How much of the acid precipitates if 150 mL of a solution containing 44.0 g of the acid is cooled to 5°C?

## SECTION 13.5 Precipitation Reactions

51. For the reaction in aqueous solution of each of these pairs of compounds, write both the balanced equation, including complete formulas of all compounds present, and a net ionic equation.
    a. $NH_4Cl$ and $AgNO_3$
    b. $(NH_4)_2S$ and $Pb(NO_3)_2$
    c. $(CH_3CO_2)_2Ca$ and $K_2SO_4$

52. For the reaction in aqueous solution of each of these pairs of compounds, write both the balanced equation, including complete formulas of all compounds present, and a net ionic equation.
    a. $Hg_2(NO_3)_2$ and KBr
    b. NaOH and $Al(NO_3)_3$
    c. $Li_2CO_3$ and $Fe(NO_3)_3$

53. For each reaction in aqueous solution, write both a balanced equation, including complete formulas of all compounds present, and a net ionic equation.
    a. $Ca(NO_3)_2$ and $Na_2CO_3$
    b. $Ba(NO_3)_2$ and $Na_2SO_4$
    c. $Hg_2(NO_3)_2$ and $CaCl_2$

54. For each reaction in aqueous solution, write both a balanced equation, including complete formulas of all compounds present, and a net ionic equation.
    a. $(CH_3CO_2)_2Pb$ and NaCl
    b. $FeCl_3$ and KOH
    c. $Sb(NO_3)_3$ and $(NH_4)_2S$

55. Each of these pairs of compounds reacts in aqueous solution. Write both a balanced equation, including complete formulas of all compounds present, and a net ionic equation.
    a. Ammonium chloride and lead(II) nitrate
    b. Sodium carbonate and iron(II) acetate
    c. Aluminum nitrate and potassium hydroxide

56. Each of these pairs of compounds reacts in aqueous solution. Write both a balanced equation, including complete formulas of all compounds present, and a net ionic equation.
    a. Copper(II) nitrate and sodium sulfide
    b. Sodium hydroxide and barium acetate
    c. Iron(II) chloride and lead nitrate

57. For the reaction in aqueous solution of each of these pairs of compounds, give both a balanced equation, including complete formulas of all compounds present, and a net ionic equation.
    a. Potassium hydroxide and nickel(II) nitrate
    b. Barium chloride and lithium carbonate
    c. Silver acetate and hydrogen iodide

58. For the reaction in aqueous solution of each of these pairs of compounds, give both a balanced equation, including complete formulas of all compounds present, and a net ionic equation.
    a. Mercurous(I) nitrate and ammonium bromide
    b. Bismuth(III) nitrate and ammonium sulfide
    c. Potassium chromate and lead nitrate

59. Draw a picture similar to Figure 13.9 (page 466) that illustrates what happens when aqueous solutions of NaOH and $Fe(NO_3)_2$ are mixed.

60. Draw a picture similar to Figure 13.9 (page 466) that illustrates what happens when aqueous solutions of $Pb(NO_3)_2$ and $FeCl_2$ are mixed.

## SECTION 13.6 Electrolytes: Dissociation in Solution

61. Which of these compounds is an electrolyte?
    A. $CH_3CH_2OH$
    B. $K_2CO_3$
    C. $CH_3CH_3$
    D. $Ca_3(PO_4)_2$

62. Which of these compounds is an electrolyte?
    A. $LiClO_4$
    B. AgI
    C. HgS
    D. $Al(OH)_3$

63. Draw a picture similar to Figure 13.13 (page 469) that shows what happens when an electric current passes through an aqueous solution of $CaCl_2$.

64. Draw a picture similar to Figure 13.13 (page 469) that shows what happens when an electric current passes through an aqueous solution of $FeCl_3$.

65. Explain why $NaCl(s)$ will not conduct electricity.

The battery is on, but no current flows through the NaCl(s).

66. Explain why $CH_3OH(aq)$ will not conduct electricity.

**SECTION 13.7 Strong and Weak Acids and Bases**

67. Classify each of these compounds as a strong electrolyte, a weak electrolyte, or a nonelectrolyte.
    a. $H_2SO_4$ b. $KClO_4$ c. $CH_3CH_2OH$

68. Classify each of these compounds as a strong electrolyte, a weak electrolyte, or a nonelectrolyte.
    a. $NH_3$ b. $K_2CrO_4$ c. $HClO_4$

69. Classify each compound as a strong electrolyte, a weak electrolyte, or a nonelectrolyte.
    a. $Al(NO_3)_3$ b. $CH_3CO_2H$ c. NaOCl

70. Classify each compound as a strong electrolyte, a weak electrolyte, or a nonelectrolyte.
    a. KOH b. $NaHCO_3$ c. KBr

71. Hydrogen fluoride is a gas at room temperature. It dissolves in water, forming a solution that gives a dim light with electrical conductivity testing apparatus. Which of these describes HF? Explain your answer.
    A. Reacts strongly with water and is a strong electrolyte
    B. Reacts weakly with water and is a weak electrolyte
    C. Reacts strongly with water and is a weak electrolyte
    D. Reacts weakly with water and is a strong electrolyte

72. A pigment called blue verdigris is a copper salt. It dissolves in water, forming a blue solution that gives a bright light with electrical conductivity testing apparatus. Which of these describes blue verdigris? Explain your answer.
    A. Dissociates completely in water and is a weak electrolyte
    B. Dissociates partially in water and is a strong electrolyte
    C. Dissociates completely in water and is a strong electrolyte
    D. Dissociates partially in water and is a weak electrolyte

**SECTION 13.8 Concentration of Solutions: Molarity**

73. Calculate the molarity of each of these solutions.
    a. 0.0775 mol of $H_2SO_4$ in enough water to give 150 mL of solution
    b. 0.500 mol of HCl in enough water to give 750 mL of solution
    c. $3.21 \times 10^{-3}$ mol of $Al(NO_3)_3$ in enough water to give 25 mL of solution

74. Calculate the molarity of each of these solutions.
    a. 1.5 mol of $CH_3OH$ in enough water to give 75 mL of solution
    b. 0.237 mol of $K_2CO_3$ in enough water to give 350 mL of solution
    c. $5.0 \times 10^{-3}$ mol of $AgNO_3$ in enough water to give 12.5 mL of solution

75. Calculate the molarity of each of these solutions.
    a. 0.015 mol of $H_3PO_4$ in enough water to give 100 mL of solution
    b. 1.75 mol of NaOH in enough water to give 2.0 mL of solution
    c. $1.00 \times 10^{-3}$ mol of $KAu(CN)_2$ in enough water to give 10 mL of solution

76. Calculate the molarity of each of these solutions.
    a. 0.044 mol of $Na_2Cr_2O_7$ in enough water to give 100 mL of solution
    b. 0.401 mol of NaCl in enough water to give 125 mL of solution
    c. $1.17 \times 10^{-3}$ mol of KI in enough water to give 22.5 mL of solution

77. Determine the molarity of each of these solutions.
    a. 0.52 mol of $Cd(NO_3)_2$ in enough water to give 250 mL of solution
    b. 7.1 mol of $HClO_4$ in enough water to give 12.5 L of solution
    c. $2.22 \times 10^{-2}$ mol of $Na_2CO_3$ in enough water to give 12.5 mL of solution

78. Determine the molarity of each of these solutions.
    a. 4.00 mol of NaBr in enough water to give 2.1 L of solution
    b. 0.065 mol of $(NH_4)_2S$ in enough water to give 750 mL of solution
    c. $2.0 \times 10^{-2}$ mol of $Hg_2(NO_3)_2$ in enough water to give 40 mL of solution

79. What is the molarity of each of these solutions?
   a. 1.95 g of $K_2CO_3$ in enough water to give 100 mL of solution
   b. 32 g of $Cu(NO_3)_2$ in enough water to give 500 mL of solution
   c. 77.5 g of NaBr in enough water to give 1.25 L of solution

80. What is the molarity of each of these solutions?
   a. 5.0 g of $FeCl_3$ in enough water to give 150 mL of solution
   b. 2.77 g of $CaBr_2$ in enough water to give 75 mL of solution
   c. 11.1 g of $Cd(NO_3)_2$ in enough water to give 125 mL of solution

81. What is the molarity of each of these solutions?
   a. 0.75 g of $Na_2CO_3$ in enough water to give 25 mL of solution
   b. 23.9 g of $KHSO_4$ in enough water to give 175 mL of solution
   c. 6.52 g of $CH_3CO_2H$ in enough water to give 400 mL of solution

82. What is the molarity of each of these solutions?
   a. 13.4 g of KOH in enough water to give 450 mL of solution
   b. 1.9 g of $Cr(NO_3)_3$ in enough water to give 200 mL of solution
   c. 125 g of NaCl in enough water to give 750 mL of solution

83. Calculate the molarity of each of these solutions.
   a. 1.04 g of silver nitrate in enough water to give 25 mL of solution
   b. 0.25 g of potassium permanganate in enough water to give 5.0 mL of solution
   c. 125 g of lithium chloride in enough water to give 1.50 L of solution

84. Calculate the molarity of each of these solutions.
   a. 0.78 g of barium bromide in enough water to give 20 mL of solution
   b. 2.25 g of aluminum nitrate in enough water to give 75 mL of solution
   c. 7.1 g of potassium hydroxide in enough water to give 200 mL of solution

85. Find the molarity of each of these solutions.
   a. 12.5 g of iron(II) nitrate in enough water to give 80 mL of solution
   b. 0.245 g of copper(II) acetate in enough water to give 50 mL of solution
   c. 8.42 g of acetic acid in enough water to give 150 mL of solution

86. Find the molarity of each of these solutions.
   a. 9.70 g of ammonium sulfide in enough water to give 200 mL of solution
   b. 5.43 g of calcium iodide in enough water to give 130 mL of solution
   c. 0.46 g of sodium bicarbonate (sodium hydrogen carbonate) in enough water to give 5.0 mL of solution

## SECTION 13.9 Preparing Solutions

87. How many moles of solute are present in each of these solutions?
   a. 125 mL of 2.2 $M$ NaCl
   b. 2.5 mL of 3.0 $M$ HCl
   c. 750 mL of 0.111 $M$ $AgNO_3$
   d. 35.45 mL of 0.2756 $M$ $BaCl_2$

88. How many moles of solute are present in each of these solutions?
   a. 10.0 L of $1.2 \times 10^{-2}$ $M$ $H_2SO_4$
   b. 23.5 mL of 12 $M$ HCl
   c. 2.55 L of 16 $M$ $HNO_3$
   d. 37.6 mL of 0.104 $M$ KCl

89. How many moles of solute are present in each of these solutions?
   a. 874 mL of $9.88 \times 10^{-2}$ $M$ $KMnO_4$
   b. 35.66 mL of $9.73 \times 10^{-2}$ $M$ NaOH
   c. 2.0 L of 2.5 $M$ $NH_3$
   d. 5.00 L of 4.55 $M$ $HNO_3$

90. How many moles of solute are present in each of these solutions?
   a. 34.22 mL of 0.5000 $M$ KOH
   b. 0.10 mL (one drop) of 18 $M$ $H_2SO_4$
   c. 11.25 mL of $2.001 \times 10^{-2}$ $M$ $CaCl_2$
   d. 5.00 L of 2.77 $M$ NaCl

91. How many moles of solute are present in each of these solutions?
   a. 5.00 mL of 0.355 $M$ potassium chloride
   b. 45.3 mL of 0.233 $M$ sodium hydroxide
   c. 32.7 mL of 0.897 $M$ ammonium bromide
   d. 766 mL of 0.0747 $M$ zinc nitrate

92. How many moles of solute are present in each of these solutions?
   a. 33.7 mL of 1.5 $M$ nitric acid
   b. 125 mL of 0.500 $M$ acetic acid
   c. 4.8 mL of 0.125 $M$ silver nitrate
   d. 7.55 mL of 0.0753 $M$ iron(III) nitrate

93. Give a procedure for making up each of these aqueous solutions, starting with water and the solid compound.
   a. 0.50 L of 1.0 $M$ KOH
   b. 250 mL of 0.25 $M$ $KMnO_4$
   c. 50 mL of 0.20 $M$ sodium hydrogen sulfate

94. Give a procedure for making up each of these aqueous solutions, starting with water and the solid compound.
    a. 75 mL of 0.125 *M* ammonium bromide
    b. 125 mL of 0.40 *M* $AgNO_3$
    c. 500 mL of 0.5000 *M* sodium hydroxide

95. Give a procedure for making up each of these aqueous solutions, starting with water and the solid compound.
    a. 1.25 L of 1.5 *M* NaCl
    b. 15 mL of 0.35 *M* $Ca(NO_3)_2$
    c. 25 mL of 0.40 *M* potassium carbonate

96. Give a procedure for making up each of these aqueous solutions, starting with water and the solid compound.
    a. 10 mL of 0.15 *M* aluminum acetate
    b. 0.5 L of 0.100 *M* sodium hydrogen carbonate (sodium bicarbonate)
    c. 50 mL of 0.125 *M* $Ca(NO_3)_2$

97. Describe the procedure for making up each of these aqueous solutions, starting with water and a stock solution of 2.00 *M* NaOH.
    a. 150 mL of 1.25 *M* NaOH
    b. 250 mL of 0.500 *M* NaOH
    c. 1.50 L of 0.250 *M* NaOH

98. Describe the procedure for making up each of these aqueous solutions, starting with water and a stock solution of 3.00 *M* NaOH.
    a. 500 mL of 0.400 *M* NaOH
    b. 750 mL of 1.50 *M* NaOH
    c. 25 mL of 0.250 *M* NaOH

99. Outline the procedure for making up each of these aqueous solutions from water and 16 *M* $HNO_3$.
    a. 250 mL of 2.5 *M* $HNO_3$
    b. 2.00 L of 3.00 *M* $HNO_3$
    c. 50.0 mL of 4.00 *M* $HNO_3$

100. Outline the procedure for making up each of these aqueous solutions from water and a stock solution of 5.00 *M* HI.
    a. 750 mL of 0.750 *M* HI
    b. 25 mL of 0.500 *M* HI
    c. 500 mL of 1.00 *M* HI

101. Give a procedure for preparing each of these aqueous solutions from water and 17 *M* acetic acid.
    a. 250 mL of 2.0 *M* acetic acid
    b. 50 mL of 0.25 *M* acetic acid
    c. 125 mL of 1.25 *M* acetic acid

102. Give a procedure for preparing each of these aqueous solutions from water and 18 *M* $H_2SO_4$. (Don't forget the word *carefully*.)
    a. 1 L of 3.0 *M* $H_2SO_4$
    b. 10 mL of 0.75 *M* $H_2SO_4$
    c. 750 mL of 0.20 *M* $H_2SO_4$

## SECTION 13.10 Concentration of Solutions: Percentages

103. Calculate %(m/m) for each solution. (The density of water is 1.00 g $mL^{-1}$.)
    a. 18 g of NaCl and 200 mL of water
    b. 4.5 g KOH and 120 mL of water
    c. 2.5 g of NaCl and 500 mL of water

104. Calculate %(m/m) for each solution. (The density of water is 1.00 g $mL^{-1}$.)
    a. 455 mg of KBr in 20 mL of water
    b. 128 mg of $KNO_3$ in 2.5 mL of water
    c. 8.7 g of aluminum nitrate in 75 mL of water

105. Calculate %(m/m) for each of these solutions. (The density of water is 1.00 g $mL^{-1}$.)
    a. 36 g of NaCl and 4.0 L of water
    b. 55.2 g of NaOH and 125 mL of water
    c. 5.5 g of NaCl in 250 mL of water

106. Calculate %(m/m) for each of these solutions. (The density of water is 1.00 g $mL^{-1}$.)
    a. 2.55 g of $K_2Cr_2O_7$ in 70 mL of water
    b. 22.4 g of $AgNO_3$ in 500 mL of water
    c. 550 mg of $(CH_3CO_2)_2Ca$ in 25 mL of water

107. Calculate %(m/m) for each of these solutions. (The density of water is 1.00 g $mL^{-1}$.)
    a. 60 g of $Ba(OH)_2$ in 1.50 L of water
    b. 10 g of $AlCl_3$ in 500 mL of water
    c. 2.7 g of $KMnO_4$ in 50 mL of water

108. Calculate %(m/m) for each of these solutions. (The density of water is 1.00 g $mL^{-1}$.)
    a. 22 g of $Al(NO_3)_3$ in 450 mL of water
    b. 35 g of $CuCl_2$ in 750 mL water
    c. 125 g of $CoSO_4$ in 1.25 L of water

109. You have a supply of 50%(m/m) NaOH solution in water, and a reaction calls for 22 g of sodium hydroxide. How much of the solution should you use?

110. You have a supply of 50%(m/m) NaOH solution in water, and you need 130 g of sodium hydroxide. How much of the solution should you use?

111. You have made up a solution by adding 250 mL of ethanol to 350 mL of water. What is the %(v/v) of ethanol in this solution?

112. You have made up a solution by adding 300 mL of ethanol to 1.00 L of water. What is the %(v/v) of ethanol in this solution?

113. For a scuba diving experiment you need to make 12,000 L of a 7.5%(v/v) solution of oxygen in helium. How much oxygen do you need? How much helium?

114. For a scuba diving experiment you need to make 20,000 L of a mixture that is 10%(v/v) oxygen, 45%(v/v) nitrogen, and 45%(v/v) helium. How much oxygen do you need? How much helium?

115. Normal saline is 0.92%(m/v) NaCl in water. How many grams of NaCl do you need to make 750 mL of normal saline?

116. Normal saline is 0.92%(m/v) NaCl in water. How many grams of NaCl do you need to make 5.5 L of normal saline?

117. Give a procedure for making up each of these solutions.
a. 25.0 g of 5.00%(m/m) KOH in ethanol
b. 750 mL of 45%(v/v) methanol in water (both are liquids)
c. 100 mL of 10%(m/v) sodium bicarbonate in water

118. Give a procedure for making up each of these solutions.
a. 150 g of 10%(m/m) KOH in ethanol
b. 75.00 mL of 1.00%(v/v) bromine in water (both are liquids)
c. 1.25 L of 0.92%(m/v) sodium chloride in water

119. Outline the procedure for preparing each of these solutions.
a. 100 g of 10%(m/m) NaOH in ethanol
b. 500 mL of 35%(v/v) water in isopropyl alcohol (both are liquids)
c. 5.0 L of 0.92%(m/v) sodium chloride in water

120. Outline the procedure for preparing each of these solutions.
a. 5.0 g of 2%(m/m) KI in acetone
b. 250 mL of 5%(v/v) 2-propanol in hexane (both are liquids)
c. 250 mL of 5%(m/v) sodium bicarbonate in water

## SECTION 13.11 Concentrations: Molality and Mole Fraction

121. A saturated solution of calcium iodide contains 209 g of $CaI_2$ per 100 mL of water at 20°C. What is the molality of this solution?

122. A saturated solution of calcium sulfate contains 0.209 g of $CaSO_4$ per 100 mL of water at 0°C. What is the molality of this solution?

123. Calculate the molality of each of these solutions.
a. 15 g of $CaCl_2$ in 120 mL of water
b. 1.50 g of $AgNO_3$ in 12.5 mL of water
c. 125 g of NaCl in 1.25 L of water

124. Calculate the molality of each of these solutions.
a. 3.5 g of sodium acetate in 25 mL of water
b. 159 g of $Na_2CO_3$ in 1.5 L of water
c. 208 g of $CH_3CH_2OH$ in 500 mL of water

125. What is the molality of each of these solutions?
a. 21 g of $NaH_2PO_4$ in 1.00 L of water
b. 2.7 g of $K_2CO_3$ in 37.5 mL of water
c. 2.12 kg of NaCl in 8.5 L of water

126. What is the molality of each of these solutions?
a. 2.22 g of magnesium chloride in 45 mL of water
b. 127 g of sodium hydrogen carbonate (sodium bicarbonate) in 550 mL of water
c. 37 g of KCl in 0.750 L of water

127. Calculate the mole fraction of the solute in each of these solutions.
a. 0.23 mol NaCl, 12.9 mol water
b. 12.5 g of $KClO_4$, 250 mL of water
c. 2.88 g of potassium hydroxide, 75 g of ethanol ($CH_3CH_2OH$)

128. Calculate the mole fraction of the solute in each of these solutions.
a. 20 g of benzene ($C_6H_6$), 75 g of cyclohexane ($C_6H_{12}$)
b. 1.2 mol of $H_2SO_4$, 1.7 mol of water
c. 15 g of $AgNO_3$, 250 g of ethanol ($CH_3CH_2OH$)

129. What is the mole fraction of the solute in each of these solutions?
a. 0.71 mol of $CuCl_2$, 21.0 mol of water
b. 7.67 g of $K_2CO_3$, 125 mL of water
c. 2.88 g of potassium iodide, 100 g of acetone, $(CH_3)_2C{=}O$

130. What is the mole fraction of the solute in each of these solutions?
a. 25.0 g of ethanol ($CH_3CH_2OH$), 75 mL of water
b. 9.2 g of NaCl, 1 L of water
c. 10.0 g of $Br_2(l)$, 150 g of $CCl_4$

131. Tincture of iodine was extensively used as an antiseptic during the first half of the 1900s. It is prepared by mixing 50 mL of water, 70 g of $I_2$, 50 g of KI, and 800 g of ethanol ($CH_3CH_2OH$). Calculate the mole fraction of each component.

132. A chemist performed experiments on a glass bulb that contained 2.0 mol of Ar, 2.5 mol of $N_2$, and 5.0 mol of CO. What is the mole fraction of each component of this mixture?

## SECTION 13.12 Colligative Properties

133. If these compounds are dissolved in water at equal molal concentrations, which one gives the largest decrease of the freezing point of the solution? Explain your answer.
A. $CH_3CH_2OH$
B. LiI
C. $Al(NO_3)_3$
D. $Ca(OH)_2$

134. If these compounds are dissolved in water at equal molal concentrations, which one gives the smallest decrease of the freezing point of the solution? Explain your answer.
A. $CH_3CH_2OH$
B. $Al(NO_3)_3$
C. LiI
D. $Ca(OH)_2$

135. When dissolved in water at equal molal concentrations, which of these substances gives the smallest increase of the boiling point of the solution? Explain your answer.
A. $NaHCO_3$ C. $CH_3OH$
B. $CuCl_2$ D. KOH

136. When dissolved in water at equal molal concentrations, which of these substances gives the largest increase of the boiling point of the solution? Explain your answer.
A. $NaHCO_3$ C. $CH_3OH$
B. $CuCl_2$ D. KOH

137. What is the osmotic pressure of each of these solutions?
a. 0.20 *M* $CH_3OH$ (covalent) at 300 K
b. 0.010 *M* NaCl at 27°C
c. 0.50 *M* $CaCl_2$ at 50°C

138. What is the osmotic pressure of each of these solutions?
a. 1.20 *M* $CH_3CH_2OH$ (covalent) at 290 K
b. 0.30 *M* table sugar ($C_{12}H_{22}O_{11}$) at 20°C
c. 0.050 *M* NaCl at 25°C

139. Determine the osmotic pressure of each of these solutions.
a. 0.15 *M* $Ca(NO_3)_2$ at 325 K
b. 1.00 *M* $CH_3CH_2OH$ at 10°C
c. 0.25 *M* $Al(NO_3)_3$ at 30°C

140. Determine the osmotic pressure of each of these solutions.
a. 1.35 *M* NaCl at 310 K
b. 0.015 *M* $KNO_3$ at 20°C
c. 0.43 *M* $(CH_3)_2CHOH$ (isopropyl alcohol, covalent) at 15°C

141. Find the osmotic pressure of each of these solutions.
a. 4.0 *M* $HNO_3$ at 295 K
b. 0.75 *M* NaCl at 22°C
c. 0.15 *M* $CH_3CH_2CH_2OH$ (covalent) at 350 K

142. Find the osmotic pressure of each of these solutions.
a. 1.0 *M* $(CH_3)_2C{=}O$ (acetone, covalent) at 310 K
b. 0.30 *M* $CaBr_2$ at 50°C
c. 0.375 *M* NaCl at 25°C

## SOLUTIONS TO EXERCISES

### Exercise 13.1

**a.** $NH_3(g)$ is the solute, $H_2O$ is the solvent.
**b.** $NaOCl(s)$ is the solute, $H_2O$ is the solvent. It is irrelevant that companies don't prepare bleach solutions by dissolving $NaOCl(s)$ in water, but rather by electrolysis (see Section 13.1). The final solution is $NaOCl(aq)$, which allows you to identify both the solvent and the solute.
**c.** $H_2O$ is the solute, and $CH_3CO_2H$ is the solvent.

### Exercise 13.2

**a.** "Like dissolves like." Water is a hydrogen bonding substance, and so are $CH_3OH$ (C) and $CH_3CH_2OH$ (D). These are very soluble in water, so neither is the answer. Both $CaF_2$ (A) and NaCl (B) are ionic. You know from your own experience that NaCl (table salt) is one of the few ionic compounds that dissolves in water, so the answer must be $CaF_2$.
**b.** The strongest intermolecular forces in $CH_3OH$ are hydrogen bonds, and $H_2O$ (A) is also a hydrogen-bonding substance. Of those given, it is the compound most soluble in $CH_3OH$.

### Exercise 13.3

**a.** $K_2S$ contains a Group 1A metal, so it should be soluble in water.
**b.** Neither of the ions in FeS is listed in Table 13.1, so this compound should be insoluble in water.
**c.** $PbCl_2$ is a listed exception to the general solubility of chloride salts. It is insoluble.
**d.** $FeCl_2$ is not one of the listed exceptions, so it should be soluble in water.
**e.** NaOH contains a Group 1A metal, and it is soluble in water.
**f.** Neither of the ions that make up $Fe(OH)_2$ is listed in Table 13.1, so the compound should be insoluble in water.
**g.** Table 13.1 mentions neither the mercury(II) ion nor the sulfide ion. Therefore, HgS follows the general rule that ionic compounds are insoluble in water.

### Exercise 13.4

**a.** The amount of calcium chloride being added, 50 g, is smaller than the amount that 100 mL of water can dissolve, which is 74 g. All 50 g of the $CaCl_2$ will dissolve, giving an unsaturated solution and no undissolved solid left over.
**b.** How much calcium chloride can dissolve in 15 mL of water? The answer can be found by using this proportional relationship:

$$15 \text{ mL} \times \frac{74 \text{ g } CaCl_2}{100 \text{ mL}} = 11 \text{ g } CaCl_2$$

The solution will become saturated when 11 g have dissolved, and the remaining 9 g will remain undissolved.

### Exercise 13.5

**a.** First, write the formulas of the starting materials. You learned how to write formulas from names in Chapter 7. Then you rearrange the ions to make the products, and decide which one is the precipitate (that is, the solid).

$$Hg(NO_3)_2(aq) + (NH_4)_2S(aq) \longrightarrow NH_4NO_3(aq) + HgS(s)$$

The equation isn't balanced, so you do that next. Ammonium ions and nitrate ions are the ones that need adjustment.

$$Hg(NO_3)_2(aq) + (NH_4)_2S(aq) \longrightarrow 2\ NH_4NO_3(aq) + HgS(s)$$

**b.** HgS is the insoluble product. The ions that make it up are $Hg^{2+}$ and $S^{2-}$. The net ionic equation is

$$Hg^{2+}(aq) + S^{2-}(aq) \longrightarrow HgS(s)$$

### Exercise 13.6

Compound A, HgS, is ionic, but is insoluble in water. It is not an electrolyte. $NH_4Cl$ (B) is a water-soluble ionic compound and therefore a good electrolyte. $C_2H_6$ (C) is a covalent compound and not an electrolyte. The answer is B.

### Exercise 13.7

$NH_3$ (A) is a weak base; so whatever it does, it does weakly. It reacts weakly with water to produce $NH_4^+$ and $OH^-$ ions. $H_2SO_4$ (B) is a covalent compound listed in Table 13.2 as a strong acid, so it must react strongly with water to produce $H_3O^+$ ions. $Na_2CO_3$ is an ionic compound. It produces separate ions in solution by dissociation. B is the answer.

### Exercise 13.8

**a.** $HNO_3$ reacts completely with water to form $H_3O^+$ and $NO_3^-$ ions; it is a strong electrolyte.
**b.** $K_2SO_4$ is a water-soluble salt; it is a strong electrolyte.
**c.** $CH_3OH$ dissolves in water, but it is a covalent nonelectrolyte.
**d.** $CH_3CO_2H$, acetic acid, is a weak acid and a weak electrolyte.
**e.** KBr is a water-soluble salt and a strong electrolyte.

### Exercise 13.9

**a.** Molarity is moles of solute per liter of solution:

$$\text{Molarity} = \frac{\text{Moles of solute}}{\text{Liters of solution}} = \frac{0.32\text{ mol }CH_3OH}{1.57\text{ L}} = 0.20\ M\ CH_3OH$$

**b.** This time you must calculate both the number of moles of solute and the number of liters of solution. The number of moles of $CaCl_2$ equals the number of grams divided by the molar mass. That is,

$$\frac{77.5\text{ g }CaCl_2}{(1 \times 40.08\text{ g mol}^{-1}\text{ Cl}) + (2 \times 35.45\text{ g mol}^{-1}\text{ Cl})} = 0.698\text{ mol }CaCl_2$$

The volume is given as 885 mL, which is 0.885 L. The equation is

$$\text{Molarity} = \frac{\text{Moles of solute}}{\text{Liters of solution}} = \frac{0.698\text{ mol }CaCl_2}{0.885\text{ L solution}} = 0.789\ M\ CaCl_2$$

### Exercise 13.10

Use the four-step procedure:

1. You need 0.500 L of 0.200 $M$ $BaCl_2$ (yes, you had to know the correct formula of barium chloride to work this problem). This means you need (0.500 L)(0.200 mol/L) = 0.100 mol of $BaCl_2$. The molar mass of $BaCl_2$ is 137 g $mol^{-1}$ Ba + (2 × 35.45 g $mol^{-1}$ Cl) = 208 g. The number of grams of $BaCl_2$ needed to prepare the solution is (0.100 mol $BaCl_2$)(208 g $mol^{-1}$ $BaCl_2$) = 20.8 g $BaCl_2$.
2. Weigh out the 20.8 g of $BaCl_2$.
3. Put about 250 mL of water into a 500-mL volumetric flask, add the $BaCl_2$, and swirl to dissolve.
4. Fill the flask to the 500-mL mark and mix thoroughly.

### Exercise 13.11

Once again, use the modified four-step procedure:

1. Calculate the volume of concentrated nitric acid you need to start with:

$$V_{conc} = \frac{V_{dil}M_{dil}}{M_{conc}}$$

$$= \frac{(2.0\ \text{L})(6.0\ M)}{16\ M} = 0.75\ \text{L} = 750\ \text{mL}$$

2. Measure out 750 mL of concentrated $HNO_3$.
3. Put about half of the needed water (about 1 L) into a 2.0-L volumetric flask, add the 750 mL of concentrated $HNO_3$ slowly and carefully, and swirl the flask to mix thoroughly.
4. Add more water until the liquid level comes exactly to the mark on the neck of the flask. Swirl the solution so that it will be uniform throughout.

### Exercise 13.12

The stock solution contains 50 g NaOH in every 100 g of solution. To get the amount of solution that contains 15 g of sodium hydroxide, simply use the unit conversion factor method:

$$15\ \cancel{\text{g NaOH}} \times \frac{100\ \text{g solution}}{50\ \cancel{\text{g NaOH}}} = 30\ \text{g solution}$$

### Exercise 13.13

The equation is the usual one. You'll have to do a little algebra after you set it up:

$$\%(\text{v/v}) = \frac{\text{Volume of oxygen } (V)}{\text{Volume of oxygen + volume of nitrogen}} \times 100\%$$

$$10.0\%(\text{v/v}) = \frac{V}{12.5\ \text{L}} \times 100\%$$

$$V = \frac{(10.0\%)(12.5\ \text{L})}{100\%} = 1.25\ \text{L of oxygen}$$

### Exercise 13.14

The equation and the calculation work like this:

$$\%(\text{m/v}) = \frac{\text{m}_{\text{NaCl}}}{\text{Volume of solution (in mL)}} \times 100\%$$

$$0.92\%(\text{m/v}) = \frac{\text{m}_{\text{NaCl}}}{200\ \text{mL}} \times 100\%$$

$$\text{m}_{\text{NaCl}} = \frac{[0.92\%(\text{m/v})](200\ \text{mL})}{100\%}$$

$$= 1.8\ \text{g of NaCl}$$

### Exercise 13.15

**a.** You use the usual procedure, modified for %(m/m):

1. 10% of 150 g is (150 g)(0.10) = 15 g. You need 15 g of KOH. The rest of the solution is ethanol. If the total is 150 g, you need 150 g − 15 g = 135 g of ethanol.
2. Weigh out the amount of KOH.
3. Weigh out the amount of ethanol.
4. Put them in a flask together and mix thoroughly.

**b.** The procedure for %(v/v) solutions is

1. 15% of 50 mL is (50 mL)(0.15) = 7.5 mL of acetone. The rest of the solution is water. If the total is 50 mL, you need 50 mL − 7.5 mL = 42.5 or 43 mL of water.

**2. and 3.** Using a graduated cylinder, measure out the amounts of the substances.

4. Put them in a flask together and mix thoroughly.

**c.** The procedure for %(m/v) solutions is

1. 5.0% of 250 mL is (250 mL)(0.05) = 12.5 mL; you need 12.5 g of NaCl.
2. Add about 125 mL of water to a 250-mL flask.
3. Weigh out the 12.5 g of NaCl, put it in the flask, and swirl.
4. Fill the flask to the 250-mL mark and mix thoroughly.

### Exercise 13.16

The mass of 150 mL of water is 150 g = 0.150 kg. The molar mass of $Ca(NO_3)_2$ is 40.1 g mol$^{-1}$ Ca + 2(14.0) g mol$^{-1}$ N + 6(16.0) g mol$^{-1}$ O = 164.1 g mol$^{-1}$ $Ca(NO_3)_2$. The number of moles of $Ca(NO_3)_2$ is 35 g/(164.1 g mol$^{-1}$) = 0.21 mol. Calculating the molality gives

$$\frac{0.21 \text{ mol Ca(NO}_3)_2}{0.150 \text{ kg H}_2\text{O}} = 1.4\ m$$

### Exercise 13.17

First, calculate the number of moles of each substance present. The molar masses are

$$C_3H_6O \text{ is } [(3 \times 12) + (6 \times 1) + 16] \text{ g mol}^{-1} = 58 \text{ g mol}^{-1}$$

$$C_2H_6O \text{ is } [(2 \times 12) + (6 \times 1) + 16)] \text{ g mol}^{-1} = 46 \text{ g mol}^{-1}$$

$$H_2O \text{ is } [(2 \times 1.0) + 16] \text{ g mol}^{-1} = 18 \text{ g mol}^{-1}$$

The number of moles of each component is

$$C_3H_6O\text{: } \frac{10 \text{ g}}{58 \text{ g mol}^{-1}} = 0.172 \text{ mol acetone}$$

$$C_2H_6O\text{: } \frac{10 \text{ g}}{46 \text{ g mol}^{-1}} = 0.217 \text{ mol ethanol}$$

$$H_2O\text{: } \frac{10 \text{ g}}{18 \text{ g mol}^{-1}} = 0.556 \text{ mol water}$$

The total number of moles present is 0.172 mol + 0.217 mol + 0.556 mol = 0.945 mol. The mole fraction of each component is therefore:

$$C_3H_6O\text{: } \frac{0.172 \text{ mol}}{0.945 \text{ mol}} = 0.18$$

$$C_2H_6O\text{: } \frac{0.217 \text{ mol}}{0.945 \text{ mol}} = 0.23$$

$$H_2O\text{: } \frac{0.556 \text{ mol}}{0.945 \text{ mol}} = 0.59$$

As a check, the sum of the three mole fractions should be 1.00, and it is.

### Exercise 13.18

All of these are strong electrolytes. Compound A gives four ions in solution, three $Na^+$ ions and one $PO_4^{3-}$ ion. The rest give fewer ions when they dissolve.

### Exercise 13.19

$M$ = 0.10 mol L$^{-1}$ × 2 particles = 0.20 mol L$^{-1}$ particles; $R$ = 0.082 L atm mol$^{-1}$ K$^{-1}$, and $T$ = 300 K.

$$\pi = MRT = (0.20 \text{ mol L}^{-1})(0.082 \text{ L atm mol}^{-1} \text{ K}^{-1})(300 \text{ K}) = 4.9 \text{ atm}$$

### Experimental Exercise 13.1

1. For a solution to conduct electricity, charges must somehow move in the solution. Ions such as $Na^+$ and $Cl^-$ provide a mechanism for charge transport.
2. If the amount of freezing-point depression depends only on the number of solute particles present, there must be twice as many particles in solutions of NaCl as there are in solutions of table sugar of identical concentrations. There must be three times as many particles present in solutions of sodium sulfate as there are in table sugar solutions. The freezing-point depression experiments convinced Arrhenius that sodium chloride and sodium sulfate dissociate in solution.
3. The fact that sugar solutions give the expected freezing-point depressions is evidence that sugar doesn't dissociate. It is easy to conclude from this that not all substances dissociate in water.
4. The fact that NaCl(*aq*) conducts electricity does support Arrhenius' contention that ions exist in solution, because ions provide the separated charges necessary for conduction. The fact that NaCl(*s*) does not conduct electricity neither supports nor contradicts the contention.
5. This kind of behavior demonstrates that acetic acid is only partially dissociated in solution—that is, it is a weak acid that reacts only incompletely with water. This is consistent with the behavior of weak electrolytes.

14

# THE DYNAMIC NATURE OF CHEMICAL REACTIONS: KINETICS AND EQUILIBRIUM



**KEY TERMS**

Since the beginning of this book, we have looked at and talked about chemical reactions. At the very beginning we considered the differences between chemical and physical changes. We went on to cover several types of reactions, including combustion, precipitation, and acid-base reactions. You can now predict the products of and write balanced equations for these types of reactions. Along the way we have explored the properties of substances on the molecular level and how they are influenced by the various types of bonding. You knew the physical differences between solids, liquids, and gases even before you took this course, but we have gone beyond what you can see (the macroscopic level) to explore what happens on microscopic and atomic levels.

In this chapter we combine knowledge about reactions with knowledge about chemical compounds at the molecular level. We will examine in detail what happens to atoms and molecules during chemical reactions. To use the language of chemists, this chapter introduces the topics of kinetics and equilibrium—how fast reactions go and what happens in reactions that do not go to completion.

FIGURE 14.1
The success or failure of any space mission depends on how fast reactions go in rocket motors and life support systems. As the nearly tragic Apollo 13 mission showed, knowledge of such reactions is neither complete nor perfect.

## 14.1 The Rates of Chemical Reactions

GOAL: To describe reaction rates in terms of what can be seen.

Suppose that you were to dissolve some bromine ($Br_2$) in an excess of the compound toluene ($C_7H_8$). At first the reaction mixture will be a reddish-brown color, the color of the bromine diluted by the toluene. Slowly, that bromine color will fade, and the solution becomes colorless (see Figure 14.2). The equation for the reaction is

$$C_7H_8 + Br_2 \longrightarrow C_7H_7Br + HBr$$

The products (benzyl bromide, $C_7H_7Br$, and HBr) are colorless, as is the reactant toluene ($C_7H_8$).

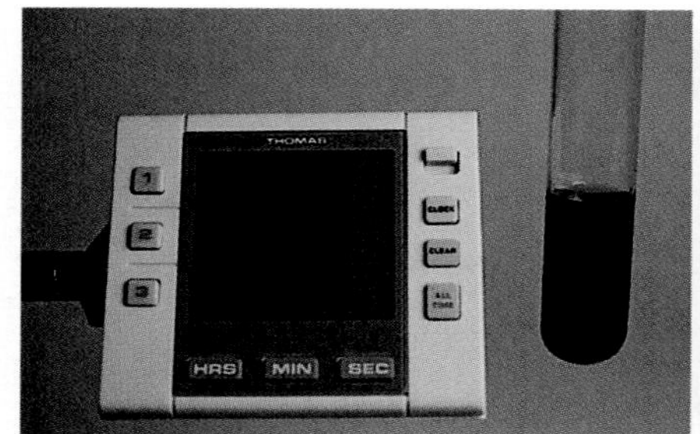


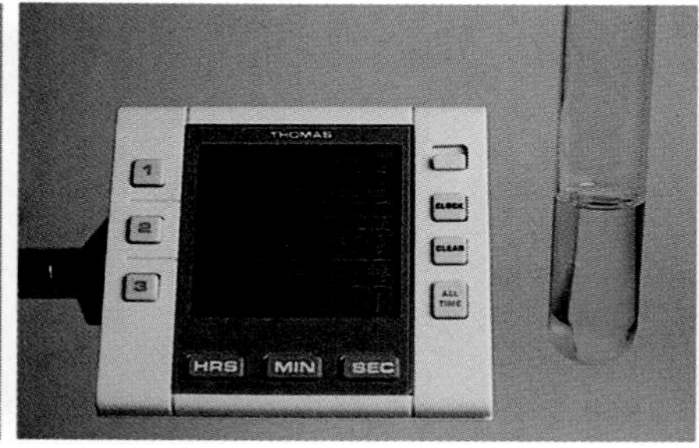


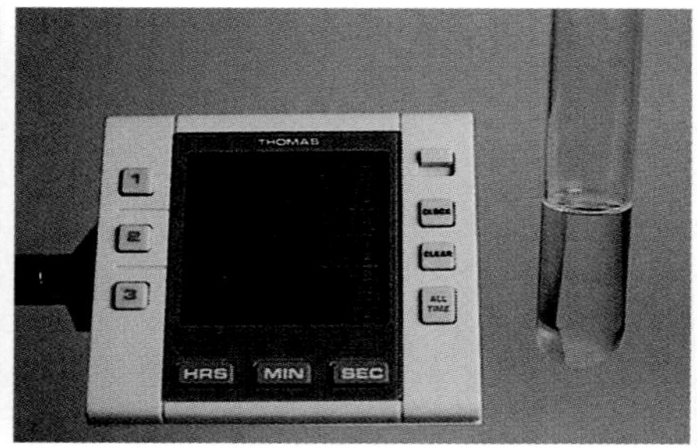


FIGURE 14.2
The reaction of reddish-brown bromine and colorless toluene takes time to go to completion. It is easy to see that the bromine is reacting because its characteristic color fades with time. After about 15 minutes, the reaction solution is colorless, which indicates that all the bromine has reacted. (This reaction must be run very carefully—bromine is corrosive and very dangerous. Don't try it without supervision.)

The reaction between bromine and toluene is not instantaneous. It needs time to finish—in chemists' language, to go to completion. All chemical reactions, in fact, require some finite amount of time to finish. You know, for example, that it can take years for a piece of iron to rust completely away. On the other hand, some reactions go very fast, reaching completion in about $10^{-9}$ s. Many reactions seem to finish immediately, but chemists can measure their exact rate of progress with modern equipment. Substances may react in less than a second, but in every case a reaction takes some finite, measureable amount of time.

**As we will discuss in Section 14.3, some reactions *never* go to completion.**

It isn't very informative to say that a reaction is fast or slow. As you might expect, chemists have worked out ways to express numerically the relative speeds at which reactions occur. For example, chemists use a *spectrometer* (see Figure 14.3) to follow the progress of the reaction of bromine and toluene by measuring the amount of red light the reaction mixture absorbs at fixed time intervals. The amount of light absorbed is related to the concentration of the bromine, so the absorption data provide an indirect measurement of how fast the bromine reacts—that is, it indicates the *rate* at which bromine reacts. Many ingenious methods have been devised for direct and indirect measurements of concentrations, sometimes under challenging conditions (such as during the controlled explosion inside a cylinder of an automobile engine).

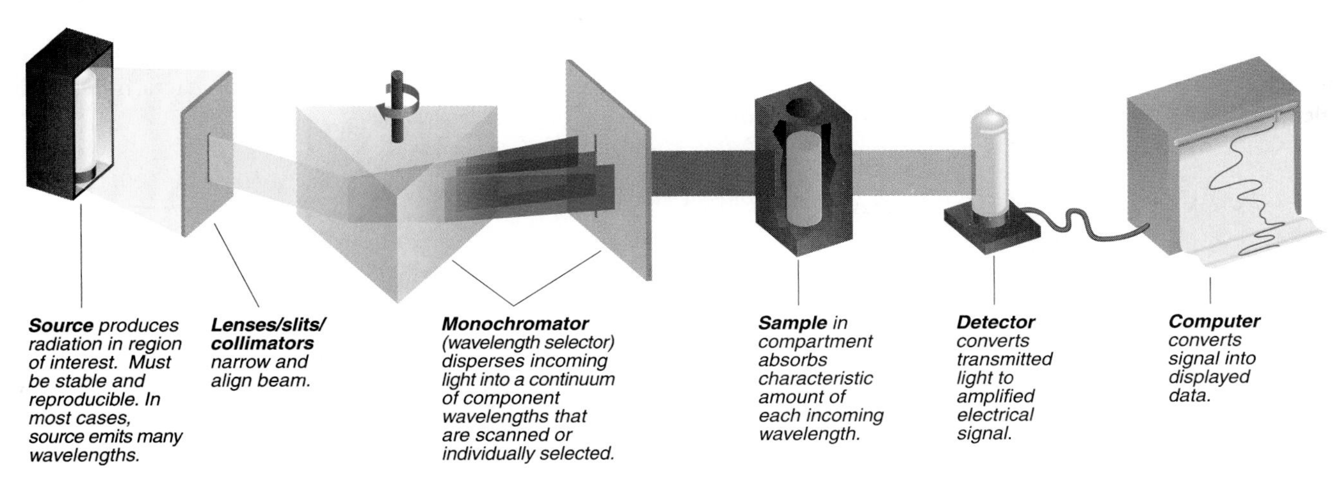
Source produces radiation in region of interest. Must be stable and reproducible. In most cases, source emits many wavelengths.
Lenses/slits/collimators narrow and align beam.
Monochromator (wavelength selector) disperses incoming light into a continuum of component wavelengths that are scanned or individually selected.
Sample in compartment absorbs characteristic amount of each incoming wavelength.
Detector converts transmitted light to amplified electrical signal.
Computer converts signal into displayed data.

FIGURE 14.3

Chemists use spectrometers to measure the amount of light (or other radiation) absorbed by a substance or mixture. Different types of spectrometers use radiation from different regions of the electromagnetic spectrum; visible, ultraviolet, and infrared are the most common.

Because reactant and product concentrations are intimately related to reaction rates, scientists have developed a shorthand notation for such concentrations. Square brackets around the chemical formula stand for the molarity (the number of moles per liter) of that substance in solution. Using this notation, $[CH_3Br]$ represents the molarity of methyl bromide.

Methyl bromide ($CH_3Br$) reacts with sodium hydroxide in solution to produce methyl alcohol (methanol) and sodium bromide. This is the equation:

**Methyl bromide ($CH_3Br$) is an important pesticide, used to fumigate soil before planting crops and to kill termites in homes. It is also implicated in the destruction of the ozone layer.**

$$CH_3Br + NaOH \longrightarrow CH_3OH + NaBr$$

FIGURE 14.4
You can find the average rate of speed of a car by dividing the distance traveled by the time the journey took. You can find the rate of speed at any given time by looking at the speedometer, but reaction rates are seldom that easy to determine.

Just as cars have rates of speed, reactions have rates. You can calculate the average speed of a car by dividing the distance it travels by the time it took to travel that distance (see Figure 14.4). Similarly, the rate of a chemical reaction is the change in the concentration of a reactant or product per unit time:

$$\textbf{Rate} = \frac{\textbf{Concentration change}}{\textbf{Time interval}}$$

Suppose you measure the concentration of $CH_3Br$ at two closely spaced times, as indicated by points 1 and 2 in Figure 14.5. Now you can calculate the average rate of the reaction during the time interval:

$$\text{Rate} = \frac{[CH_3Br]_2 - [CH_3Br]_1}{t_2 - t_1}$$

The closer together you can make the time measurements $t_1$ and $t_2$, the more closely you can estimate the rate at a given instant in time.

If you repeat such measurements several times during the course of a single experiment, you will find that the rate changes continuously as the reaction continues. The $CH_3Br$ disappears more quickly at the beginning of the reaction, when its concentration is high; the reaction rate slows down as its concentration decreases. Similarly, $CH_3OH$ forms more quickly at the beginning of the reaction than toward the end (see Figure 14.5). Such behavior is typical of most chemical reactions. In general, the rate of a reaction depends on the concentration of one or more reactants.

**The rate is the change in concentration per unit time.**

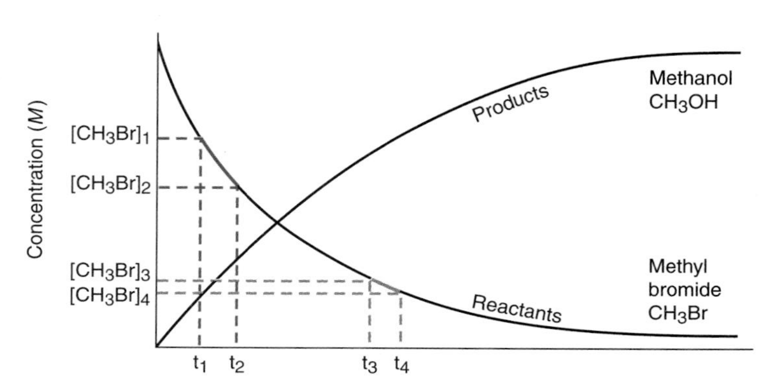


FIGURE 14.5
Concentration (molarity) versus time for methyl bromide and methanol in the reaction $CH_3Br + NaOH \rightarrow CH_3OH + NaBr$. The concentration of methyl bromide, a reactant, decreases, while the concentration of methanol, a product, increases. The concentrations change more rapidly at the beginning of the reaction (during the time interval from $t_1$ to $t_2$) than they do later (between $t_3$ and $t_4$). In other words, the rate is greater at the beginning than it is later. Most (but not all) reactions behave this way.

Chemists have done a lot of work in this area and can write equations to describe the relationship between rate and concentration for many reactions. We won't discuss this topic in detail, but we will present some observations on the methyl bromide–sodium hydroxide reaction illustrated in Figure 14.5 to give you an idea of how such relationships work. These observations are based on measurements from a series of experiments performed at a single temperature and in the same solvent.

1. If you double the concentration of $CH_3Br$, the rate of the reaction doubles. If you triple the concentration of $CH_3Br$, the reaction rate triples—and so on.
2. If you double the concentration of NaOH, the rate of the reaction doubles. If you triple the concentration of NaOH, the reaction rate triples—and so on.
3. If you add $NaNO_3$ to the reaction mixture, the rate does not change. Using KOH instead of NaOH also has no effect on the rate.
4. If you replace $CH_3Br$ with $CH_3Cl$, the products are $CH_3OH$ and NaCl, and the reaction is slower. If you use $CH_3I$ instead of $CH_3Br$, the reaction is faster. If you use $CH_3CH_2Br$ instead of $CH_3Br$, the reaction is slower.

**Original structure**

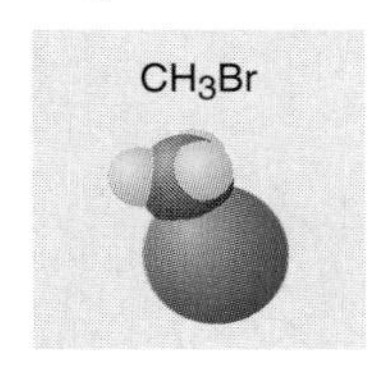


Replace with $CH_3Cl$

*Reaction rate decreases*

Replace with $CH_3I$

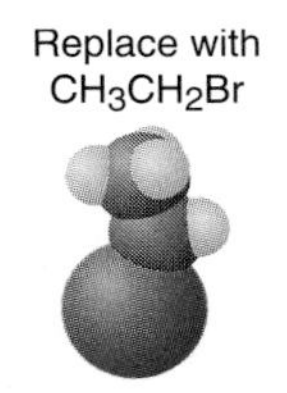

*Reaction rate increases*

Replace with $CH_3CH_2Br$

*Reaction rate decreases*

What do all these facts boil down to? Faced with a situation like this, chemists try to simplify things. We have begun this process by ignoring the effects of temperature and solvent.

Statements 1 and 2 above tell you that the rate of the reaction—the rate at which reactants become products—is directly proportional to the concentration of $CH_3Br$ and directly proportional to the concentration of NaOH. We can write an equation that says this:

$$\text{Rate} = k[CH_3Br][NaOH]$$

In words, the reaction rate at any time during the reaction is equal to a constant times the molar concentration of $CH_3Br$ times the molar concentration of NaOH. The proportionality constant $k$ in these equations is called the **rate constant.** The equation itself is called a **rate equation.**

Remember that NaOH is ionic, so it separates into $Na^+$ and $OH^-$ in solution. A chemist studying the reaction will ask whether it is the $Na^+$ ion, the $OH^-$ ion, or both together that influence the reaction rate. The experiments in statement 3 were designed to find out about the effect of the $Na^+$ ion. Adding $NaNO_3$ increases the concentration of $Na^+$ ions but does not change the rate of the reaction. This result suggests that the $Na^+$ ion has nothing to do with the reaction. Removing the sodium ions completely, by replacing NaOH with KOH, also has no influence on the reaction rate. Thus, it is the concentration of $OH^-$ rather than NaOH that has an effect on the rate. The rate equation can be modified to reflect this. The modified equation is

**The only way a chemist can find a rate equation is by experiment. It is impossible to look at the balanced equation for a reaction and predict with certainty what the rate expression will be—the best you can do is an educated guess.**

$$\text{Reaction rate} = k[CH_3Br][OH^-]$$

Since $CH_3Br$ consists of nonmetals only, you can predict that its five atoms are held together with covalent bonds. If you change the molecule, the rate of reaction will change. The experiments of statement 4 confirm this.

### EXAMPLE 14.1 Reactant Concentrations and Reaction Rates

**a.** Write the equation for the reaction between $CH_3Br$ and NaOH in net ionic form.
**b.** By what factor will the reaction rate, $k[CH_3Br][OH^-]$, increase if the concentration of NaOH is increased by a factor of 2.3?
**c.** By what factor will the reaction rate increase if the concentration of $CH_3Br$ is doubled and the concentration of NaOH tripled?

**SOLUTIONS**

**a.** The net ionic equation (see Section 8.8 for review) is

$$CH_3Br + OH^- \longrightarrow CH_3OH + Br^-$$

(The $Na^+$ ion is a spectator ion.)

**b.** If the concentration of NaOH increases by a factor of 2.3, so does the $OH^-$ concentration—and the rate.
**c.** If the concentration of $CH_3Br$ is increased by a factor of 2 and the concentration of NaOH by a factor of 3, the rate will increase by a factor of $2 \times 3 = 6$. The increase in the concentration of $CH_3Br$ doubles the rate and the increase in the concentration of $OH^-$ ions simultaneously triples the rate.

**EXERCISE 14.1**

A team of chemists studied this reaction:

$$(CH_3)_3CBr + NaOH \longrightarrow (CH_3)_3COH + NaBr$$

They discovered that doubling the concentration of $(CH_3)_3CBr$ doubled the reaction rate and that doubling the concentration of NaOH had no influence on the rate. What is the rate equation for this reaction?

## 14.2 Theory of Reaction Rates

**GOAL:** To understand what reaction rates mean on the molecular and atomic level.

In Section 14.1 we examined the reaction between $CH_3Br$ and NaOH in some depth. Chemists have performed many experiments on this and related reactions and have found that the reaction rate equals a constant times the concentration of $CH_3Br$ times the concentration of $OH^-$ (that is, the reaction rate = $k[CH_3Br][OH^-]$). This rate equation remains valid as long as neither the temperature nor the solvent changes. If you change the temperature or the solvent, the value of the "constant" $k$ changes, and so does the rate of the reaction. The rate constant $k$ is truly constant only for a restricted range of circumstances.

What are the effects of temperature and solvent on the rate of reaction? Here are two more statements to add to the list of four given in Section 14.1:

5. The four preceding statements hold true only if the temperature is constant. If the temperature increases, the rate increases; if the temperature decreases the rate decreases.
6. The reaction being considered takes place in solution, not in the gas phase. The nature of the solvent has an effect on the rate of this and all other reactions in solution. (Such effects are topics for more advanced courses in chemistry.)

Generations of chemists have contemplated what happens to atoms, molecules, and ions as reactions take place. They have concluded that there are three factors (collision frequency, energy factor, and probability factor) that influence the rates of reactions. Let's examine each of these factors in turn.

### Collision Frequency

The first factor that influences reaction rates is the **collision frequency.** This frequency might be expressed as the number of collisions between molecules that *might* react per second. It is only reasonable to think that two molecules must collide in order to react with one another. Further, it also seems reasonable to conclude that the more collisions there are per second, the more reactions there will be per second. That is, if the collision frequency increases, the reaction rate will increase. In the reaction of $CH_3Br$ with $OH^-$, the molecule and the ion must collide with each other (see Figure 14.6). The collision frequency for this reaction includes these collisions only, and not collisions between two $OH^-$ ions, between two $CH_3Br$ molecules, between reactant and solvent molecules, or with $Na^+$ ions.

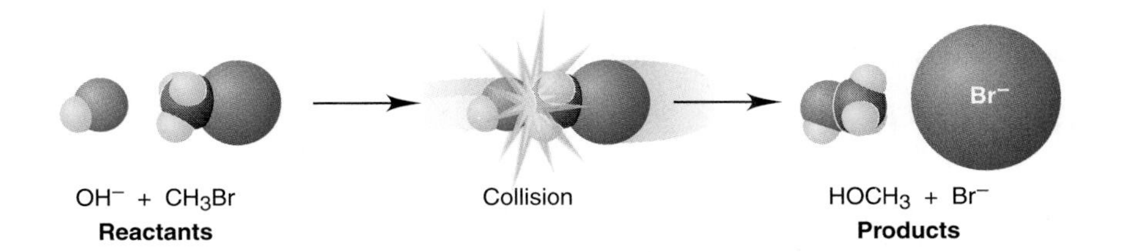


FIGURE 14.6
Reactant molecules must collide in order for product molecules to form.

There are several things a chemist can do to change the collision frequency for a given reaction. For example, you can change the concentrations of the particles involved in the reaction. If you double the number of $CH_3Br$ molecules in a given volume of solution (leaving everything else the same), the number of collisions doubles. This is why chemists use molar concentrations in equations such as this one:

**The number of moles per liter times Avogadro's number equals the number of molecules (or ions) per liter.**

$$\text{Rate} = k[CH_3Br][OH^-]$$

The temperature also influences the collision frequency. Molecules in a warmer sample have a higher kinetic energy than molecules in a colder one, which means that they move faster. If the molecules within a given volume move faster, they collide with one another more often. That is, the collision frequency increases as the temperature increases.

### Energy Factor

The second factor that influences reaction rates is called the **energy factor.** If a $CH_3Br$ molecule and an $OH^-$ ion glance off each other, nothing will happen. The reaction between $CH_3Br$ and $OH^-$ must disrupt the bonding within the $CH_3Br$ molecule, and some minimum amount of energy is required to do this. In the overall reaction the C—Br bond in $CH_3Br$ must break (and a C—OH bond in $CH_3OH$ forms). To be effective, a collision between the two species must have at least enough energy to get the bond-breaking process started. Collisions that don't have this much energy don't give the molecule and ion a chance to react. The energy factor is the fraction of collisions that have enough energy to cause a reaction to occur.

The actual amount of energy needed for a specific reaction to occur is something chemists can find out by experiment. They run the reaction at several different temperatures, find the rates of all these reactions, and perform some calculations. The minimum amount of energy that a collision must have for a reaction to happen is called the **activation energy.** One way to show what this means is to use an *energy diagram,* which is somewhat similar conceptually to the roller coaster profile in Section 4.1 (Figure 4.6). The vertical axis in both diagrams plots potential energy. The horizontal axis in the roller coaster diagram shows the progress of the coaster; the horizontal axis of an energy diagram shows the progress of the reaction (it is often called the *reaction coordinate*).

Figure 14.7 is the energy diagram for the reaction of $CH_3Br$ with $OH^-$. Note several things from this diagram. First, this reaction takes place in a single step. The C—O bond forms as the C—Br bond breaks; there is a continuity throughout the process. The activation energy is the amount of energy needed to carry the reactants to the high point (that is, the highest-energy point) on the curve in the diagram. The partially reacted species at this high point is called the **transition state.** Once the reaction is over the top, past the transition state, it goes quickly and readily to the products, just as a roller coaster slides to the bottom of the slope.

### Probability Factor

The third factor that affects reaction rates is called the **probability factor,** or **orientation factor.** This is the fraction of collisions in which the colliding species are aligned so that a reaction can take place.

Let's examine the reaction of $CH_3Br$ with $OH^-$ with respect to this factor. The product $CH_3OH$ has a bond between a carbon atom and an oxygen atom that was not present in the reactants. It seems likely that the oxygen atom (and not the hydrogen atom) of an $OH^-$ ion must come into direct contact with the carbon atom (not the bromine atom or a hydrogen atom) of a $CH_3Br$ molecule for the reaction to take place. In fact, other evidence indicates that the oxygen of the $OH^-$ ion must collide with the carbon of the $CH_3Br$ molecule from the side *opposite* the bromine atom for the reaction to take place (see Figure 14.8).

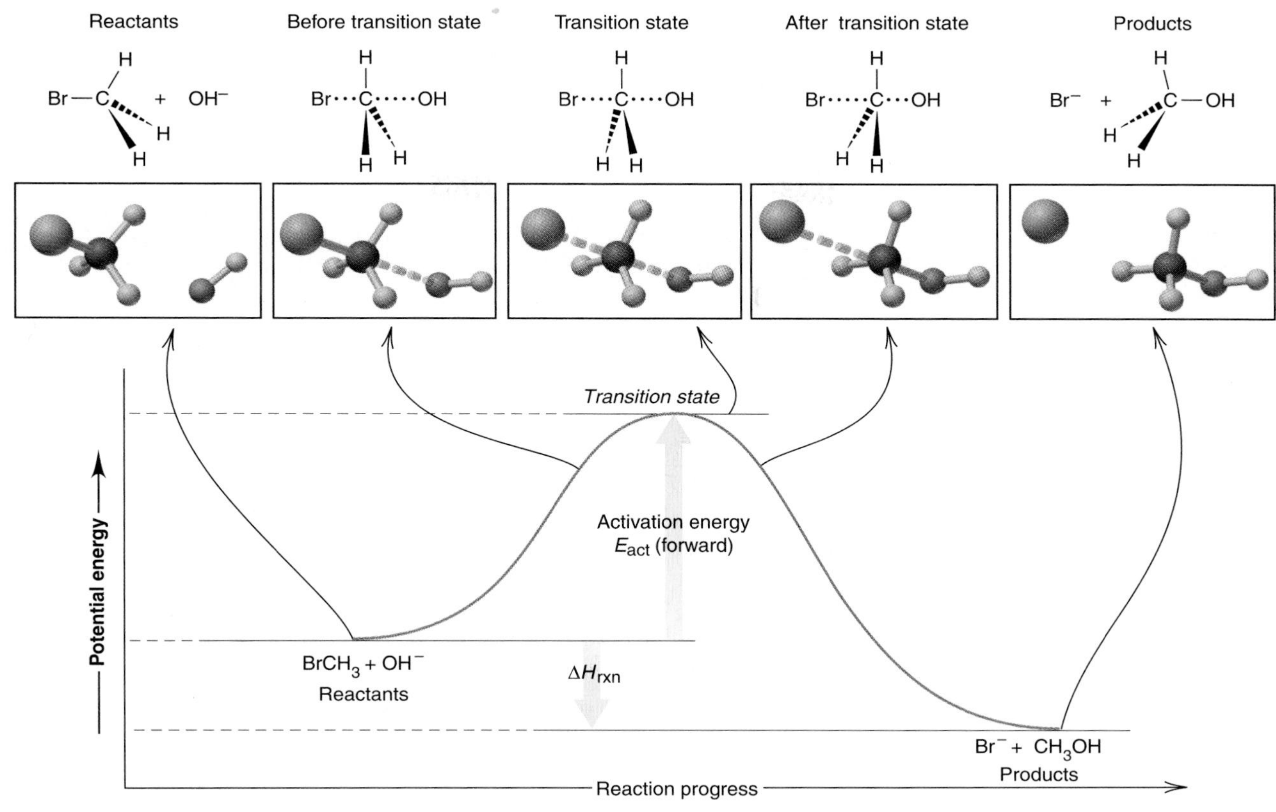


FIGURE 14.7
Energy diagram for the reaction between $CH_3Br$ and $OH^-$. The views at the molecular level show the actual progress of the reaction at five points on the diagram. The C—O bond forms gradually, and the C—Br bond breaks gradually.

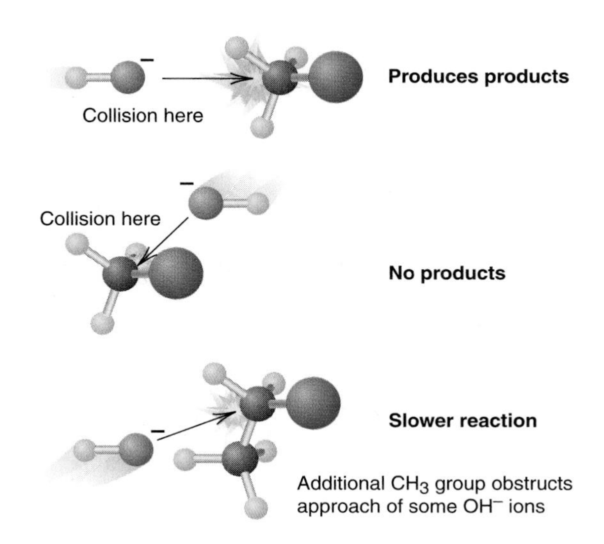


FIGURE 14.8
In the reaction $CH_3Br + NaOH \rightarrow CH_3OH + NaBr$, the oxygen atom of the hydroxide ion must collide with the side of the carbon atom opposite the bromine atom for the reaction to work (top). The fraction of collisions that are oriented this way is the *probability factor* for this reaction. Other orientations do not lead to product molecules (center). When $CH_3CH_2Br$ reacts instead of $CH_3Br$ (bottom), the extra carbon atom gets in the way of some $OH^-$ ions. The lower probability factor for $CH_3CH_2Br$ means a slower reaction.

## CHAPTER BOX 14.1 CATALYSTS

**A catalyst** is a substance that speeds up a reaction without being changed by the reaction (Section 8.7). The catalyst provides a different path that the reaction can take—one that has a lower activation energy ($E_{act}$) (see figure A below). Because a larger fraction of molecular collisions will have an energy greater than this new activation energy, the rate of the reaction will increase.

Most catalysts lower a reaction's activation energy by helping to break bonds. For example, a mixture of hydrogen and oxygen gases could stand at room temperature for centuries without reacting if left undisturbed. However, when a small amount of finely divided platinum is introduced into the mixture, the oxygen and hydrogen react explosively to form water. The surface of the metal adsorbs both hydrogen and oxygen molecules, breaking their bonds and forming bonds between the resulting atoms and the surface of the metal (see figure B below). The atoms migrate along the surface of the metal and react to form water. The water then leaves the surface of the platinum, making room for more hydrogen and oxygen reactant molecules.

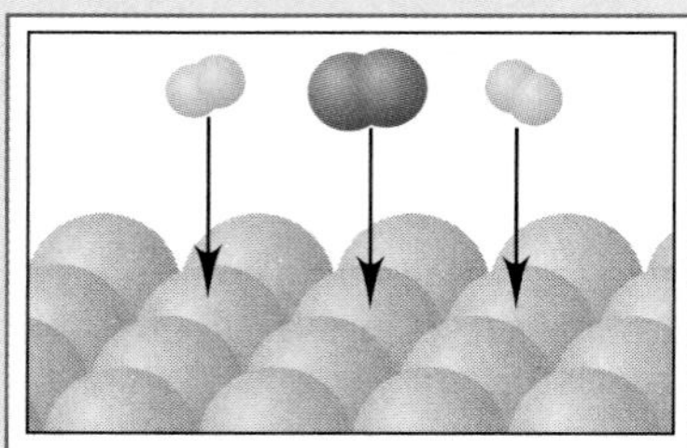

$H_2$ and $O_2$ are adsorbed on the surface of the platinum.

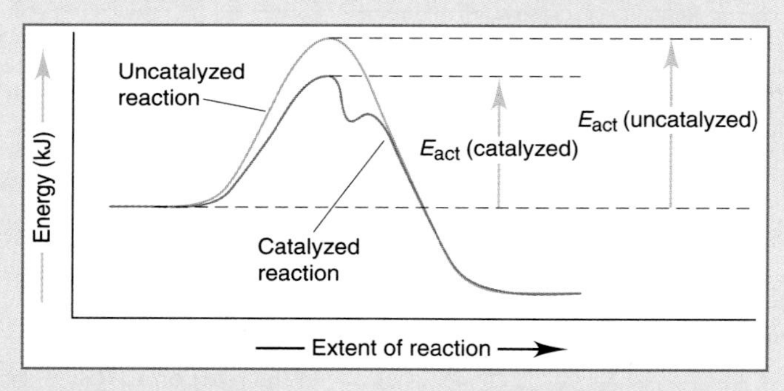


**FIGURE A**
The catalyst lowers the activation energy of a reaction by providing a different pathway for the reaction to take. The catalyzed reaction is faster because the energy barrier is lower.

The metal reacts with the $H_2$ and $O_2$, forming bonds between H and O atoms and atoms on the surface of the metal.

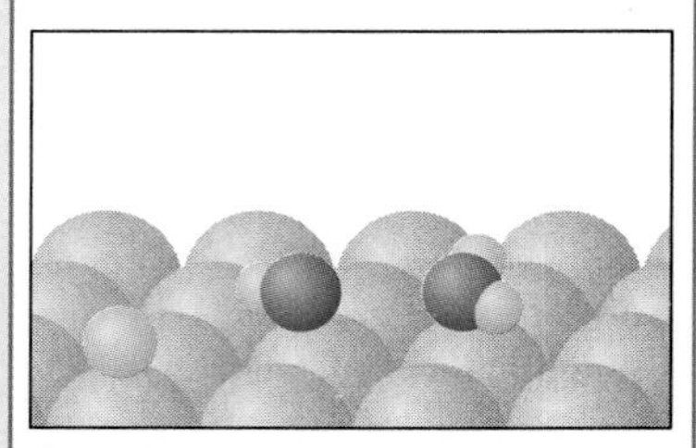

The atoms migrate toward each other, first forming OH and then $H_2O$.

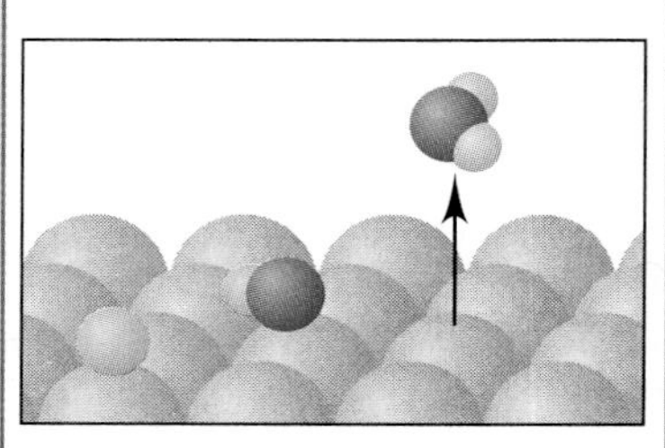

The water then desorbs from the metallic surface.

**FIGURE B**
Molecular view of platinum-catalyzed reaction of hydrogen and oxygen.

The fact that $CH_3CH_2Br$ reacts with $OH^-$ more slowly than does $CH_3Br$ under the same conditions (statement 4, Section 14.1) is mostly due to a probability factor. The extra carbon atom in $CH_3CH_2Br$, with its three hydrogen atoms, gets in the way of some of the approaching $OH^-$ ions and prevents the reaction from taking place. Figure 14.8 illustrates this.

In the rate equation for the reaction of $CH_3Br$ with NaOH,

$$\text{Rate} = k[CH_3Br][OH^-]$$

all three of the terms reflect the collision frequency. First, the more molecules per unit volume (that is, the higher the molar concentrations of $CH_3Br$ and $OH^-$), the more collisions there will be. Second, if the temperature increases while everything else remains constant, the number of collisions will increase. This increases the value of the rate constant $k$. Both the energy factor and the probability factor are unique for each reaction, and the rate constant $k$ reflects both of these factors. When the temperature increases, not only does the number of collisions increase, but the fraction of collisions with enough energy to yield product also increases.

There is an old rule of thumb that the rate of a reaction doubles with every 10°C rise in temperature. The collision frequency increases by only a small fraction when the temperature rises by 10°C, but the fraction of molecules that have enough energy to react increases dramatically. One practical application of the temperature dependency of reaction rates is the refrigeration of food. By cooling the food and thus slowing down the chemical reactions that cause it to spoil, you can keep food fresh for much longer periods of time.

**This rule of thumb is not exact, and it doesn't work for all reactions.**

## EXAMPLE 14.2 Theory of Reaction Rates

Which of the three factors (collision frequency, energy factor, and probability factor) is responsible for each of the following rate differences? Explain your answer in each case.

**a.** The rate constant for the reaction $CH_3Br + NaOH \rightarrow CH_3OH + NaBr$ is much larger than the rate constant for $(CH_3)_2CHBr + NaOH \rightarrow (CH_3)_2CHOH + NaBr$ when the two reactions are run under the same conditions of temperature and solvent.

**b.** The rate constant for the reaction $CH_3Cl + NaOH \rightarrow CH_3OH + NaCl$ is smaller than the rate constant for $CH_3Br + NaOH \rightarrow CH_3OH + NaBr$ when the two reactions are run under the same conditions of temperature and solvent.

### SOLUTIONS

**a.** The difference between the rate constants for the reaction of $CH_3Br$ and $(CH_3)_2CHBr$ is due to different probability or orientation factors. The replacement of two small hydrogen atoms by two much larger $CH_3$ groups on the carbon atom bonded to the bromine makes it less likely that the $OH^-$ ion will strike that carbon.

**b.** The collision frequency and the probability factor should be the same (or at least very similar) for the reaction of either $CH_3Cl$ or $CH_3Br$ with $OH^-$. The energy factors are different, because the bond strengths differ for the C—Cl and C—Br bonds.

### EXERCISE 14.2

Which of the three factors (collision frequency, energy factor, and probability factor) is responsible for each of the following rate differences? Explain your answer in each case.

**a.** The reaction $CH_3Br + NaOH \rightarrow CH_3OH + NaBr$ runs three times as fast when the concentration of $CH_3Br$ is 0.003 *M* than when it is 0.001 *M* and all other conditions are the same.

**b.** The rate of the reaction $CH_3CH_2Br + NaOH \rightarrow CH_3CH_2OH + NaBr$ is lower at 10°C than it is at 30°C.

## 14.3 Reactions at Equilibrium

**GOAL:** To understand what is happening when a chemical reaction reaches a state of dynamic equilibrium.

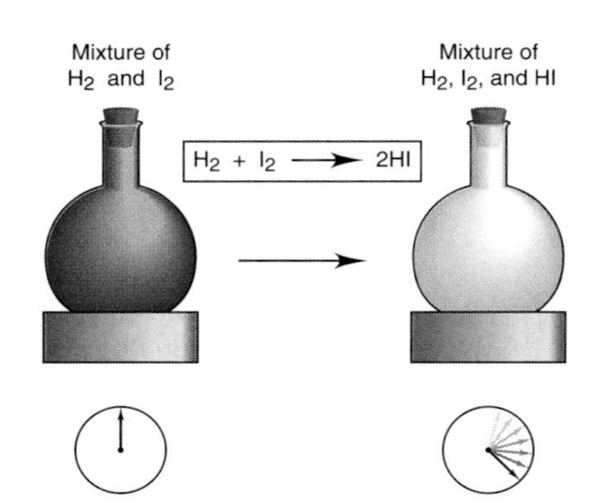


FIGURE 14.9
A mixture of $H_2$ and $I_2$ at 400°C reacts to form HI. The purple color of $I_2$ never completely disappears, indicating that the reaction does not go to completion.

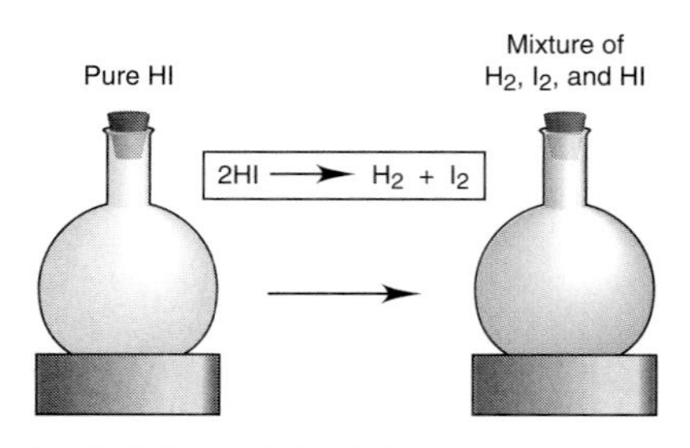


FIGURE 14.10
When pure HI is heated to 400°C, $H_2$ and $I_2$ form. If the total mass, temperature, and pressure are the same as they were for the experiment that started with $H_2$ and $I_2$ (Figure 14.9), the reaction flasks at the end of both experiments will have the *same* concentrations of HI, $I_2$, and $H_2$—the same *equilibrium concentrations.*

One type of problem introduced in the chapter on stoichiometry (Chapter 9) involved reactions that have converted some but not all of their reactants to products. Why should anyone think about reactions that aren't finished? The answer is that real reactions *don't* convert all of the reactant molecules to products—in other words, no reaction goes entirely to completion. When chemists say that a reaction goes to completion, they simply mean that the reaction works so well that unconverted reactants remain in quantities too small to be measured. Many reactions stop so far short of completion that chemists can measure the reactant concentrations easily. In such a case, the reactants give products for a while, but then the reaction seems to slow down and stop. At this point the reaction mixture contains both reactant and product molecules, and none of the concentrations change any more.

An example is the reaction of hydrogen with iodine in the gas phase. The product is hydrogen iodide, and the equation is

$$H_2(g) + I_2(g) \longrightarrow 2\ HI(g)$$

Suppose you put hydrogen and iodine gases in a flask. Because $I_2$ is a solid at room temperature, you must heat the flask to force it into the gas phase. A mixture of $H_2$ and $I_2$ gases at 400°C does interesting things. The mixture is purple at first because of the color of the iodine. The purple color fades, but it doesn't disappear. It decreases to a certain level and stays there. If you tested the mixture of gases, you would find that hydrogen, iodine, and hydrogen iodide are all present (see Figure 14.9). Did the hydrogen simply stop reacting with the iodine? It seems unlikely that this would happen. The hydrogen and iodine molecules are still colliding with each other, and they certainly reacted at the beginning of the experiment. There is no reason to think that all chemical reactivity has ceased just because the color stopped changing.

To figure out what is going on here, a chemist might perform some more experiments, such as heating pure, colorless HI gas in another flask to 400°C. The purple color of iodine will slowly appear and remain (see Figure 14.10). The reaction that happens here is exactly the reverse of the previous one:

$$2\ HI(g) \longrightarrow H_2(g) + I_2(g)$$

If you tested this second flask, you would find that hydrogen, iodine, and hydrogen iodide are all present. Did the hydrogen iodide molecules stop reacting with one another?

The answer is that *neither one* of the reactions ever stopped. If there are plenty of $H_2$ and $I_2$ molecules present and almost no HI molecules, the reaction of $H_2$ with $I_2$ produces HI faster than the reverse reaction consumes HI. If only HI molecules are present, the reaction of these molecules with each other is the only one that can happen at first. At some concentration of HI, $H_2$, and $I_2$ the two reactions occur at the same rate. Chemists show this by using a double arrow:

$$H_2(g) + I_2(g) \rightleftharpoons 2\ HI(g)$$

When the color of this reaction mixture stops changing, the reaction is said to have reached **equilibrium** (see Figure 14.11). More generally, equilibrium is the state at which there is no change in the concentrations of the reactants and products in a reaction. The fact that the concentrations aren't changing doesn't mean that no reactions are occurring. There are no concentration changes because the rate at which reactants are forming products is identical to the rate at which products are forming reactants. Chemists call this type of process a **dynamic equilibrium.**

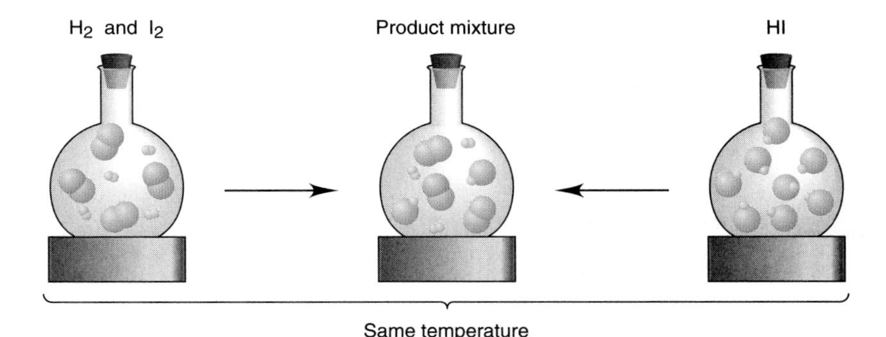


FIGURE 14.11
For the reaction $H_2(g) + I_2(g) \rightarrow 2\ HI(g)$, it doesn't matter whether you start with a mixture of $H_2$ and $I_2$ (left) or with pure HI (right). You will end up with the same mixture at equilibrium (center).

Thousands of observations show that once a reaction mixture reaches equilibrium, it does not spontaneously move to any nonequilibrium state. Some external factor (such as the actions of a chemist) may force the system to leave the equilibrium state; such a factor might be the addition of more of one of the reactants or products or a change in the temperature or the pressure. If something like that happens, the balance between the forward and reverse reactions will be disturbed, and the concentrations of reactants and products will change until a new equilibrium state is established.

It is important to recognize that it is the *rate* of conversion of reactants to products and the *rate* of conversion of products to reactants which are equal, not the concentrations. The concentrations of the reactants and products are almost always different, sometimes extremely so.

### EXAMPLE 14.3 Concentrations in a Reaction at Equilibrium

Two molecules of nitrogen dioxide react with each other to give one molecule of dinitrogen tetraoxide. The equation is

$$2\ NO_2(g) \rightleftharpoons N_2O_4(g)$$

Suppose some extra $NO_2(g)$ is added to an equilibrium mixture of $NO_2(g)$ and $N_2O_4(g)$. What will happen to the concentrations of nitrogen dioxide and dinitrogen tetraoxide after this addition?

**SOLUTION**

After the addition of more $NO_2(g)$, the rate of the forward reaction will exceed that of the reverse reaction. Some of the extra $NO_2(g)$ will react to give $N_2O_4(g)$, so the total $NO_2$ concentration will decrease and the $N_2O_4$ concentration will increase. When the reaction reaches the new equilibrium, both concentrations will be higher than they were before the extra $NO_2(g)$ was added.

**EXERCISE 14.3**

Sulfur trioxide ($SO_3$) is prepared by the reaction between sulfur dioxide ($SO_2$) and oxygen ($O_2$).

**a.** Write a balanced equation for this reaction.

**b.** If the reaction is at equilibrium, what happens to the concentrations of $SO_2$ and $SO_3$ if additional $O_2$ is added to the mixture?

## 14.4 Equilibrium Constant Expressions

**GOAL:** To write expressions that describe reactions at equilibrium.

The fact that forward and reverse reactions both have rate equations and constants means that chemists can write an equation describing each of these reactions at equilibrium. As an example, we'll use this familiar reaction:

$$H_2(g) + I_2(g) \rightleftharpoons 2\ HI(g)$$

The rate equation for the reaction between $H_2$ and $I_2$ is this: Rate = $k_f[H_2][I_2]$, where $k_f$ is the rate constant for the forward reaction (left to right). The reaction between two HI molecules has this rate equation: Rate = $k_r[HI][HI] = k_r[HI]^2$, where $k_r$ is the rate constant for the reverse reaction (right to left).

At equilibrium these two reaction rates are equal. This is another definition of the term *equilibrium,* in fact. At equilibrium, and only at equilibrium, the rates of the forward and reverse reactions are equal. Here's an equation that expresses this fact:

$$\text{Rate at equilibrium} = k_f[H_2]_{eq}[I_2]_{eq} = k_r[HI]^2_{eq}$$

The subscript eq indicates that the concentrations in this equation are the equilibrium concentrations. We can isolate the last two parts of this equation:

$$k_f[H_2]_{eq}[I_2]_{eq} = k_r[HI]^2_{eq}$$

And then we can rearrange to give:

$$\frac{k_f}{k_r} = \frac{[HI]^2_{eq}}{[H_2]_{eq}[I_2]_{eq}}$$

Because $k_f$ and $k_r$ are both constants (as long as the temperature doesn't change), their ratio is also a constant. We can use a new symbol for that constant:

$$K_{eq} = \frac{[HI]^2_{eq}}{[H_2]_{eq}[I_2]_{eq}}$$

This quantity, $K_{eq}$, is called the **equilibrium constant.** The equilibrium constant for any reaction is equal to the concentrations of the products divided by the concentrations of the reactants, each raised to the power of its coefficient in the balanced equation for the reaction. In general, for the reaction

$$aA + bB \rightleftharpoons cC + dD$$

the equilibrium constant expression is

$$\mathbf{K_{eq} = \frac{[C]^c[D]^d}{[A]^a[B]^b}} \quad \text{(Equilibrium constant expression)}$$

Equations involving $K_{eq}$ are valid only when the reaction under consideration is at equilibrium. Thus, chemists omit the subscript eq from the concentration terms; when writing these equations, they assume that those who will read them know this basic premise.

### EXAMPLE 14.4 Writing an Equilibrium Constant Expression

What is the equilibrium constant expression for this reaction?

$$2\,HI(g) \rightleftharpoons H_2(g) + I_2(g)$$

**SOLUTION**

The equilibrium constant expression should contain the concentrations of the products times each other divided by the concentration of the reactant squared. The answer is

$$K_{eq} = \frac{[H_2][I_2]}{[HI]^2}$$

Note that the reaction here is the reverse of the reaction discussed in the text above. This equilibrium constant expression is the *inverse* of the expression for the original reaction.

**EXERCISE 14.4**

What is the equilibrium constant expression for this reaction (see Figure 14.12)?

$$2\ NO_2(g) \rightleftharpoons N_2O_4(g)$$

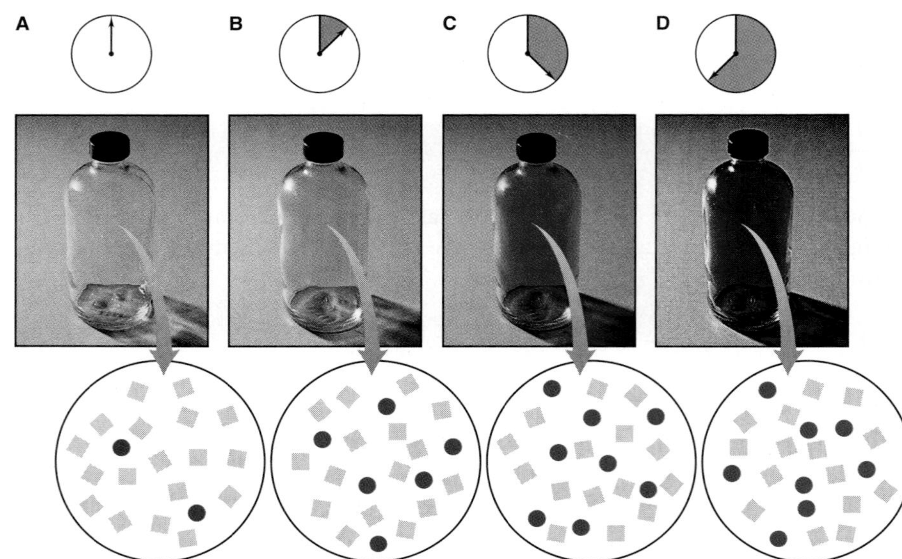


**FIGURE 14.12**
The bottle contains a mixture of $N_2O_4$ and $NO_2$. At left (A) the mixture consists mostly of colorless $N_2O_4$. When the mixture reaches equilibrium (C and D), you can see that there is much more brown $NO_2$.

The method for writing the equilibrium constant expression that we've been using works well if all of the reactants and products are in the same phase; that is, if the equilibrium is *homogeneous*. For *heterogeneous* equilibria (in which some of the compounds in the reaction are in different phases), the method is slightly different. Let's look at the following reaction as an example (see Figure 14.13):

$$CaCO_3(s) \rightleftharpoons CaO(s) + CO_2(g)$$

Equilibrium constant expressions are written in terms of the *concentrations* of reactants and products. In this reaction there are two solids, $CaCO_3$ and CaO. What are their concentrations? Well, it is certainly possible to calculate the molarity of a solid. Since molarity has units of mol $L^{-1}$, you could convert the density of a solid (in units of g $cm^{-3}$) to a molarity.

However, scientists never like to do unnecessary calculations, and here they are pointless. The concentrations of solids never change; they are always constant. Let's look at the implications of this fact for the equilibrium expression for the above reaction:

**FIGURE 14.13**
Heating limestone ($CaCO_3$) gives lime (CaO), which is then used in making both cement and glass.

$$K_{eq1} = \frac{[CaO(s)][CO_2(g)]}{[CaCO_3(s)]}$$

where $K_{eq1}$ is an equilibrium constant. The concentrations $[CaO(s)]$ and $[CaCO_3(s)]$ are both constants; their values don't change regardless of what happens during the reaction. If we lump all the constants together on the same side of the equation, we get

$$\frac{K_{eq1}[CaCO_3(s)]}{[CaO(s)]} = [CO_2(g)] = K_{eq}$$

where the new equilibrium constant $K_{eq}$ incorporates the concentrations of all the solids present.

Chemists use this same kind of analysis when solvents and other relatively pure liquids are involved in a reaction. For example, consider a reaction between a substance and the water it's dissolved in. The concentration of pure $H_2O(l)$ at 25°C is 55.5 *M*, and in dilute aqueous solutions this concentration never varies very much. Chemists simply include this concentration as part of $K_{eq}$, as we did for the concentrations of solids above.

Here are two examples of equilibrium constant expressions that incorporate this. The first reaction has a solid reactant and a liquid product, neither of which appears in the equilibrium constant expression:

$$P_4(s) + 6\,Cl_2(g) \rightleftharpoons 4\,PCl_3(l) \qquad K_{eq} = \frac{1}{[Cl_2(g)]^6}$$

This next equation is a reaction with water, which is the solvent for the reaction.

$$HF(aq) + H_2O(l) \rightleftharpoons H_3O^+(aq) + F^-(aq) \quad K_{eq} = \frac{[H_3O^+(aq)][F^-(aq)]}{[HF(aq)]}$$

### EXAMPLE 14.5 More Equilibrium Constant Expressions

Write the equilibrium constant expression for each of these reactions.

**a.** $N_2(g) + 3\,H_2(g) \rightleftharpoons 2\,NH_3(g)$
**b.** $NH_3(aq) + H_2O(l) \rightleftharpoons NH_4^+(aq) + OH^-(aq)$
**c.** $2\,H_2(g) + O_2(g) \rightleftharpoons 2\,H_2O(l)$

**SOLUTIONS**

**a.** All of the species in the reaction are gases, so you simply divide the concentrations of the products by the concentrations of the reactants. Don't forget exponents; two of the coefficients in the balanced equation are greater than 1. The answer is

$$K_{eq} = \frac{[NH_3(g)]^2}{[N_2(g)][H_2(g)]^3}$$

**b.** Water is the solvent, so you don't include it in the equilibrium constant expression:

$$K_{eq} = \frac{[NH_4^+(aq)][OH^-(aq)]}{[NH_3(aq)]}$$

**c.** The product, water, is a liquid and is not included in the equilibrium constant expression:

$$K_{eq} = \frac{1}{[H_2(g)]^2[O_2(g)]}$$

**EXERCISE 14.5**

Write the equilibrium constant expression for each of these reactions.

**a.** $P_4(s) + 5\ SO_2(g) \rightleftharpoons P_4O_{10}(s) + 5\ S(s)$

**b.** $2\ NaHCO_3(s) \rightleftharpoons Na_2CO_3(s) + CO_2(g) + H_2O(g)$

**c.** $OCl^-(aq) + H_2O(l) \rightleftharpoons HOCl(aq) + OH^-(aq)$

## 14.5 Simple Equilibrium Calculations

**GOAL:** To use data about reactions to calculate equilibrium constants and concentrations of reactants and products.

In Section 14.4 you learned to write equilibrium constant expressions. You might be wondering how to interpret such equations and what numerical values go with them.

Suppose you are working with this reaction:

$$CH_4(g) + 2\ O_2(g) \rightleftharpoons CO_2(g) + 2\ H_2O(g)$$

The equilibrium constant for this reaction is approximately $1 \times 10^{140}$ at 25°C. In an actual reaction mixture at this temperature, what would you expect to find? Which would be present in higher concentration, $CH_4$ or $H_2O$? To answer this kind of question, you might start from the equilibrium constant expression:

$$K_{eq} = \frac{[CO_2][H_2O]^2}{[CH_4][O_2]^2}$$

If the value of an equilibrium constant is greater than 1, the numerator (representing the product concentrations) is larger than the denominator (representing the reactant concentrations). In this case, the equilibrium constant is enormous: $1 \times 10^{140}$. Therefore, the equilibrium mixture consists almost entirely of $CO_2$ and $H_2O$, with almost none of the limiting reactant remaining.

As we noted in Example 14.4, when a chemical equation is written in reverse, the new equilibrium constant is the inverse of the original one. For the reaction written as $CO_2(g) + 2\ H_2O(g) \rightleftharpoons CH_4(g) + 2\ O_2(g)$, the equilibrium constant expression is

$$K_{eq\ reverse} = \frac{[CH_4][O_2]^2}{[CO_2][H_2O]^2}$$

and its value is

$$K_{eq\ reverse} = \frac{1}{K_{eq}} = 1 \times 10^{-140}$$

This value is much, much smaller than 1, and it indicates that (for the reverse of the above equation as written) the equilibrium mixture contains almost all reactants and almost no products. It's highly unlikely—at least at 25°C—that carbon dioxide and water will combine to form methane and oxygen.

## CHAPTER BOX 14.2 RATE VERSUS EQUILIBRIUM

**The equilibrium** constant for the combustion of methane at 25°C has the huge value of $1 \times 10^{140}$. You know from your own experience, however, that an oxygen-methane mixture does not *spontaneously* react to form water and carbon dioxide. For such a mixture to react, it is usually necessary to ignite it—the pilot light in a gas stove serves this purpose.

This and many similar observations have led scientists to conclude that, no matter how favorable the equilibrium constant is, it is not particularly related to how fast products

form. The *rate* of product formation depends on the magnitude of the activation energy (see Chapter Box 14.1 for the reaction between $H_2$ and $O_2$). Reactions with a large value of $E_{act}$ form products slowly, regardless of the value of $K_{eq}$. Reaction rates (kinetics) are related to collision frequency, collision energy, and orientation; equilibrium constants (thermodynamics) are primarily related to the energy difference between the reactants and the products. Reaction rates involve time and equilibrium constants do not.

In this section we will solve some equilibrium constant expressions. Chemists seldom give units with equilibrium constant values (contrary to our normal insistence that they be used). Recall that rate constants change when the temperature changes and that equilibrium constants are ratios of rate constants. Combining these two facts will tell you that equilibrium constants are temperature-dependent. When you give values of $K_{eq}$ you must specify the temperature.

You can calculate equilibrium constants from experimental data in the most straightforward way. The experimental data consist of measurements of the concentrations of the reactants and products at equilibrium. These values are simply substituted into the equilibrium constant expression. The method is illustrated in Example 14.6.

### EXAMPLE 14.6 Calculating Equilibrium Constants

**a.** For the reaction

$$2\ SO_2(g) + O_2(g) \rightleftharpoons 2\ SO_3(g)$$

at 1000°C, the concentrations of $SO_2(g)$, $O_2(g)$, and $SO_3(g)$ at equilibrium were found to be 3.2 *M*, 2.5 *M*, and 87.5 *M*, respectively. What is the equilibrium constant, $K_{eq}$, for this reaction?

**b.** A chemist found the concentrations of $H_2$, $I_2$, and HI to be $1.2 \times 10^{-3}$ *M*, $1.2 \times 10^{-3}$ *M*, and $8.4 \times 10^{-3}$ *M*, respectively, after the reaction involving them came to equilibrium at 425°C. What is the value of $K_{eq}$ for this reaction? The balanced equation is

$$H_2(g) + I_2(g) \rightleftharpoons 2\ HI(g)$$

**SOLUTIONS**

**a.** First, you write the equilibrium constant expression. Then you substitute the values for the equilibrium concentrations and calculate the result:

$$K_{eq} = \frac{[SO_3]^2}{[SO_2]^2[O_2]} = \frac{(87.5 \text{ mol L}^{-1})^2}{(3.2 \text{ mol L}^{-1})^2(2.5 \text{ mol L}^{-1})}$$
$$= 300$$

Since the product $SO_3$ was present in the highest concentration, it should be no surprise that the answer is greater than 1.

**b.** Write the equilibrium constant expression. Insert the given concentrations and calculate:

$$K_{eq} = \frac{[HI]^2}{[H_2][I_2]} = \frac{(8.4 \times 10^{-3} \text{ mol L}^{-1})^2}{(1.2 \times 10^{-3} \text{ mol L}^{-1})^2} = 49$$

Once again the concentration of the product is greater than that of the reactants, so the equilibrium constant is greater than 1.

**EXERCISE 14.6**

What is the equilibrium constant for the following reaction at 425°C? Use the concentration data given in Example 14.6.

$$2\ HI(g) \rightleftharpoons H_2(g) + I_2(g)$$

What do you do if you aren't given complete information about concentrations? You learned the methods to use in such cases in Chapter 9.

## EXAMPLE 14.7 Finding Equilibrium Constants without All Concentrations Given

Sulfur trioxide decomposes to give oxygen and sulfur dioxide in the gas phase. In one experiment, the initial concentration of $SO_3(g)$ was 0.20 *M*. After equilibrium was reached at 300°C, the concentration of $SO_2(g)$ was $2.18 \times 10^{-4}$ *M*. What is the value of the equilibrium constant?

**SOLUTION**

First, you need the balanced equation:

$$2\ SO_3(g) \rightleftharpoons 2\ SO_2(g) + O_2(g)$$

Next, you write the equilibrium constant expression.

$$K_{eq} = \frac{[SO_2]^2[O_2]}{[SO_3]^2}$$

Finally, you need the equilibrium concentrations—unfortunately you were given only one of those values. How can you find the other concentrations if you don't know the numbers of moles or the volume of the reaction container? The only numbers you have to work with are the initial concentration of $SO_3(g)$, which is 0.20 mol $L^{-1}$, and the equilibrium concentration of $SO_2(g)$, which is $2.18 \times 10^{-4}$ mol $L^{-1}$.

You will have to solve a problem related to the mole-mole problems of Chapter 9. To begin, construct a table like the ones in that chapter. With the given information filled in, it looks like this:

| Conc. (*M*) | $2[SO_3]$ $\rightleftharpoons$ | $2[SO_2]$ + | $[O_2]$ |
| --- | --- | --- | --- |
| At the beginning | 0.20 | 0 | 0 |
| Change | | | |
| At equilibrium | ? | $2.18 \times 10^{-4}$ | ? |

The change in the concentration of $SO_2$ is easy to calculate by subtracting the beginning concentration of 0 from the equilibrium concentration of $2.18 \times 10^{-4}$ *M*. You then calculate the changes in $[O_2]$ and $[SO_3]$ from the balanced equation using the unit factor method. The reaction forms 2 mol of $SO_2$ for every 1 mol of $O_2$. The change in $[SO_3]$ is the same as the change in $[SO_2]$, but $[SO_3]$ decreases as $[SO_2]$ increases. This gives you:

| Conc. (*M*) | $2[SO_3]$ | $\rightleftharpoons$ | $2[SO_2]$ | + | $[O_2]$ |
| --- | --- | --- | --- | --- | --- |
| At the beginning | 0.20 | | 0 | | 0 |
| Change | $-2.18 \times 10^{-4}$ | | $2.18 \times 10^{-4}$ | | $1.09 \times 10^{-4}$ |
| At equilibrium | ? | | $2.18 \times 10^{-4}$ | | ? |

You calculate the missing equilibrium by adding the changes to the initial concentrations:

| Conc. (*M*) | $2[SO_3]$ | $\rightleftharpoons$ | $2[SO_2]$ | + | $[O_2]$ |
| --- | --- | --- | --- | --- | --- |
| At the beginning | 0.20 | | 0 | | 0 |
| Change | $-2.18 \times 10^{-4}$ | | $2.18 \times 10^{-4}$ | | $1.09 \times 10^{-4}$ |
| At equilibrium | 0.20 | | $2.18 \times 10^{-4}$ | | $1.09 \times 10^{-4}$ |

Finally, you insert the values for all equilibrium concentrations into the equation for $K_{eq}$ and solve:

$$K_{eq} = \frac{[SO_2]^2[O_2]}{[SO_3]^2} = \frac{(2.18 \times 10^{-4}\ \text{mol L}^{-1})^2(1.09 \times 10^{-4}\ \text{mol L}^{-1})}{(0.20\ \text{mol L}^{-1})^2}$$
$$= 1.3 \times 10^{-10}$$

The equilibrium constant is $1.3 \times 10^{-10}$. Given the tiny product concentrations, the low value for $K_{eq}$ should be no surprise.

**EXERCISE 14.7**

A flask containing 2.00 *M* $H_2(g)$ and 1.00 *M* $I_2(g)$ was heated, producing 0.500 *M* HI$(g)$. What is the value of the equilibrium constant for the reaction $H_2(g) + I_2(g) \rightleftharpoons 2\ HI(g)$?

FIGURE 14.14
How large a number is $10^{90}$? Well, that many atoms would be enough to form a solid disk the size of an entire galaxy. The Andromeda galaxy is shown here.

As you can see, the possible values of equilibrium constants cover a very wide range. Values as small as $1 \times 10^{-70}$ and even smaller are possible, and so are values as large as $1 \times 10^{90}$ or more (this is an extremely large number—see Figure 14.14). Some reactions have equilibrium constants near 1, and in these the concentrations of both reactants and products at equilibrium are significant.

## 14.6 Solubility Equilibria

GOAL: To write solubility product expressions and use them to calculate equilibrium constants and solubilities.

Most of the reactions we have examined in this chapter take place in the gas phase. However, there are many reactions that yield a solid product in water (see Section13.5 for a discussion of these precipitation reactions). In this section we will take a closer look at the solubility of the "insoluble" salts. Throughout this discussion, assume that the solutions are saturated—that is, that no more salt can dissolve (see Section 13.4).

Consider silver chloride (AgCl). According to the simple rules for predicting the solubility of ionic compounds (see Section 13.3), AgCl is insoluble in water. However, AgCl is not completely, totally water-insoluble, and a few $Ag^+(aq)$ and $Cl^-(aq)$ ions are present in solution when AgCl(*s*) comes into contact with water (see Figure 14.15). Exactly how many ions go into solution depends on the solubility of AgCl(*s*).

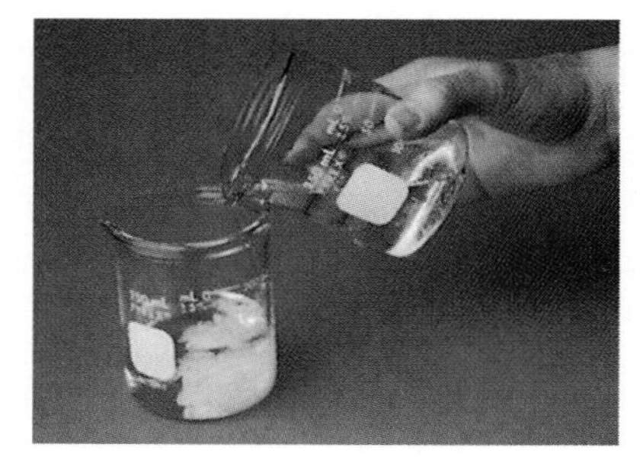

FIGURE 14.15
Although AgCl is insoluble in water, a few $Ag^+$ and $Cl^-$ ions will remain in the aqueous solution when $AgNO_3(aq)$ mixes with NaCl(*aq*).

To look at the solubility of an ionic compound in terms of an equilibrium, we need to start with a chemical equation. To get this, we write the formula for the solid on the left and the formulas for the ions that make it up on the right. Here is the equation for AgCl:

$$AgCl(s) \rightleftharpoons Ag^+(aq) + Cl^-(aq)$$

Now we can construct an equilibrium constant expression (remember that concentrations of solids do not appear in such an expression):

$$K_{eq} = [Ag^+(aq)][Cl^-(aq)]$$

Chemists call this type of equilibrium constant a **solubility product constant.** They use the symbol $K_{sp}$ for this kind of constant, rather than $K_{eq}$, and write the numerical value without units. Also, since every ion in this type of expression is in aqueous solution, the labels (*aq*) are usually omitted for convenience. This gives the same equation in a slightly different form:

$$K_{sp} = [Ag^+][Cl^-]$$

**STUDY ALERT**
**You will need to remember the formulas and charges of various ions to work the problems in this section. To review this, see Chapter 7.**

Once you know the formula of the precipitate that is involved, it's easy to write the expression for its $K_{sp}$. For instance, for AgBr you get $K_{sp} =$

$[Ag^+][Br^-]$. For $PbCl_2$ the balanced equation is $PbCl_2(s) \rightleftharpoons Pb^{2+}(aq) + 2\ Cl^-(aq)$, and the expression you get is

$$K_{sp} = [Pb^{2+}][Cl^-]^2$$

You won't get the right answer if you forget the exponents.

### EXAMPLE 14.8 Solubility Product Constant Expressions

Write the expression for $K_{sp}$ for each of these solids.

**a.** $CaCO_3$ **b.** $Ag_2SO_4$ **c.** $Al(OH)_3$ **d.** $Hg_2Cl_2$

**SOLUTIONS**

**a.** $CaCO_3$ is composed of the calcium ion ($Ca^{2+}$) and the carbonate ion ($CO_3^{2-}$). The expression is

$$K_{sp} = [Ca^{2+}][CO_3^{2-}]$$

**b.** $Ag_2SO_4$ is composed of two silver ions ($Ag^+$) and a sulfate ion ($SO_4^{2-}$). The expression is

$$K_{sp} = [Ag^+]^2[SO_4^{2-}]$$

**c.** $Al(OH)_3$ contains the aluminum ion ($Al^{3+}$) and three hydroxide ions ($OH^-$). The expression is

$$K_{sp} = [Al^{3+}][OH^-]^3$$

**d.** $Hg_2Cl_2$ consists of the mercurous ion ($Hg_2^{2+}$) and the chloride ion ($Cl^-$). The expression is

$$K_{sp} = [Hg_2^{2+}][Cl^-]^2$$

**The $Hg_2^{2+}$ ion is one of the few polyatomic cations.**

**EXERCISE 14.8**

Write the expression for $K_{sp}$ for each of these solids.

**a.** $Ag_2S$ **b.** $Fe_2S_3$ **c.** $BaSO_4$ **d.** $Ca_3(PO_4)_2$

**STUDY ALERT**

**It is necessary to realize that the solubility product expression is valid *only if the solution is saturated*—that is, only if there is too much solid for all of it to dissolve.**

Calculations using solubility product constants follow the usual pattern for equilibrium problems. You have to keep the balanced equation in mind.

### EXAMPLE 14.9 Calculations Involving Solubility Product Constants

**a.** The value of $K_{sp}$ for AgCl is $1.7 \times 10^{-10}$ at 25°C. What is the solubility of AgCl in water at that temperature?

**b.** The concentration of a saturated solution of $PbCl_2$ is $1.6 \times 10^{-2}$ *M* at 25°C. What is the value of $K_{sp}$ for lead(II) chloride at this temperature?

**SOLUTIONS**

**a.** The equation is

$$AgCl(s) \rightleftharpoons Ag^+(aq) + Cl^-(aq)$$

The solubility product expression for AgCl is therefore $K_{sp} = [Ag^+][Cl^-]$. If $x$ moles of AgCl dissolve per liter of solution, the concentrations of $Ag^+$ and $Cl^-$ will both be equal to $x$ at equilibrium. Thus you have:

$$K_{sp} = [x][x] = x^2 = 1.7 \times 10^{-10}$$
$$x = 1.3 \times 10^{-5}\ M$$

The solubility of AgCl in water is $1.3 \times 10^{-5}\ M$ at 25°C.

**b.** The equation is

$$PbCl_2(s) \rightleftharpoons Pb^{2+}(aq) + 2\ Cl^-(aq)$$

The solubility product expression for $PbCl_2$ is $K_{sp} = [Pb^{2+}][Cl^-]^2$. If $1.6 \times 10^{-2}$ mol of $PbCl_2$ dissolves per liter of solution, the concentration of $Pb^{2+}$ will be $1.6 \times 10^{-2}\ M$ and the concentration of $Cl^-$ will be twice that value, or $3.2 \times 10^{-2}\ M$. Substitution of these values gives:

$$K_{sp} = [1.6 \times 10^{-2}\ M][3.2 \times 10^{-2}\ M]^2 = 1.6 \times 10^{-5}$$

The $K_{sp}$ for $PbCl_2$ in water is $1.6 \times 10^{-5}$ at 25°C.

**EXERCISE 14.9**

**a.** What is the solubility product constant for $CaF_2$ if its solubility in water is $2.2 \times 10^{-4}\ M$?

**b.** What is the solubility of $CaCO_3$ in water if the $K_{sp}$ for $CaCO_3$ is $5 \times 10^{-9}$?

FIGURE 14.16
The low solubility of $CaCO_3$ in water is the driving force behind stalactite and stalagmite formation. (See Chapter Box 13.2.)

Note that the solubilities for many ionic compounds that earlier chapters said were insoluble in water have negative exponents and are therefore very small. But if these "insoluble" compounds actually dissolve in water, even to this small extent, what does that word really mean? Chemists have a practical rule of thumb for this. A compound that dissolves in water to the extent of $0.1\ M$ is called *soluble.* A compound with a solubility between $0.1\ M$ and $10^{-3}\ M$ is termed *slightly soluble.* If the solubility is less than $10^{-3}\ M$, the compound is said to be *insoluble.*

## 14.7 Le Chatêlier's Principle and Its Applications

**GOAL:** To predict what happens when a system of equilibrium is disturbed.

Suppose you are running a chemical reaction that has reached equilibrium. To all outward appearances nothing is happening, though you know that the action is continuous at the molecular level. Now you do something to upset the equilibrium. You might add one of the compounds involved in the reaction, increasing its concentration in the reaction mixture (see Example 14.3). You might allow another compound to evaporate out of the mixture. You might heat the mixture or decrease the pressure. What will happen to the concentrations of the reactants and products if you do any of these things?

This is a serious question for the practical chemist, because one of the most common problems chemists face is figuring out how to get the best possible yield from a reaction. A very broad and useful answer was proposed by the French chemist Henri Louis Le Chatêlier in 1888. His proposal is now known as **Le Chatêlier's Principle,** and it goes like this: *If you change one of the variables in a system at equilibrium, the other variables adjust in the direction that reduces the effect of the change.* The equilibrium shifts in the direction that minimizes the change made to the system.

Let's explore Le Chatêlier's Principle, beginning with the effects of concentration changes. Suppose you have a mixture of two gases in equilibrium with each other, such as the one in Example 14.3:

$$2\ NO_2(g) \rightleftharpoons N_2O_4(g)$$

Now you add some $NO_2(g)$ to the mixture, which increases the $NO_2$ concentration. Some of this $NO_2$ will react to give $N_2O_4$. This means that the $NO_2$ concentration will decrease below the high value it had after you made the addition. When the new equilibrium is established, however, the $NO_2$ concentration will be higher than it was before you added the extra $NO_2$. Another effect will be an increase in the $N_2O_4$ concentration. That isn't what you added, but its concentration will increase anyway, because the reaction will form some $N_2O_4$ from the added $NO_2$ (see Figure 14.17). This is the same conclusion as that reached in Example 14.3.

In this case the equilibrium has shifted to the right, toward the products. This is the direction that minimizes the change of adding more reactant. A different change would occur if you added or removed a product.

**STUDY ALERT**

**When chemists say an equilibrium shifts "to the right" or "to the left," they mean with respect to the double arrow in the balanced equation as written.**

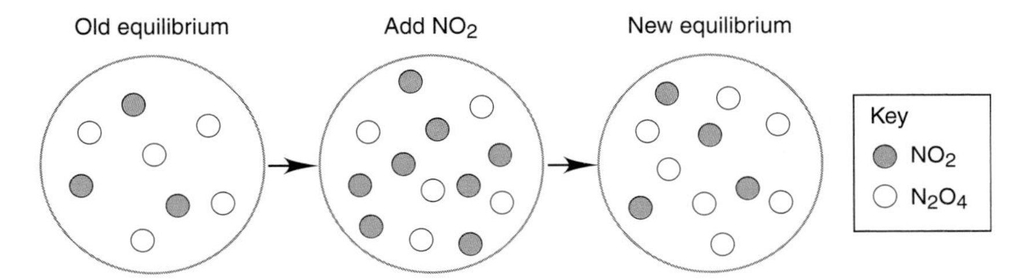


FIGURE 14.17

If you add $NO_2$ to an equilibrium mixture of $NO_2$ and $N_2O_4$, at first the concentration of $NO_2$ will increase (because that's what you added). Then the equilibrium shifts as the $NO_2$ reacts, decreasing the concentration of $NO_2$ and simultaneously increasing the concentration of $N_2O_4$, until the reaction reaches a new equilibrium state.

### EXAMPLE 14.10 Le Chatêlier's Principle and Reactant Concentrations

Suppose you have an equilibrium mixture of $H_2$, $I_2$, and HI. You remove some of the $I_2$. What will happen to the concentrations of $H_2$ and HI as the reaction establishes a new position of equilibrium?
The equation for the reaction is

$$2\,HI(g) \rightleftharpoons H_2(g) + I_2(g)$$

**SOLUTION**

If you remove some $I_2$, the equilibrium shifts to increase the concentration of $I_2$. As the concentration of $I_2$ increases, the concentration of $H_2$ will also increase, because $H_2$ is produced along with $I_2$. Since HI is the reactant, its concentration will decrease as the product concentrations increase.

**EXERCISE 14.10**

Suppose that you have an equilibrium mixture of $CH_3CO_2H$, $CH_3CH_2OH$, $H_2O$, and $CH_3CO_2CH_2CH_3$. This is the equation for the reaction:

$$CH_3CO_2H + CH_3CH_2OH \rightleftharpoons CH_3CO_2CH_2CH_3 + H_2O$$

You remove some of the $H_2O$. What will happen to the concentrations of the other three compounds as the reaction returns to equilibrium?

Le Chatêlier's Principle also applies to heat (or more precisely temperature changes), which can be thought of as a reactant or a product in many reactions (Section 8.9). Some reactions are exothermic—that is, they give off heat as they proceed. Adding heat to such a reaction mixture drives the equilibrium toward the reactant side. The effect of adding heat is the same as that of adding a product: The direction of reaction that minimizes the change is the one that goes back toward the reactants. The converse is true for reactions that are endothermic—reactions that absorb heat as they proceed. Adding heat to such a reaction mixture drives the equilibrium toward the product side.

Pressure changes during reactions that involve gases also give results that can be predicted from Le Chatêlier's Principle. Here's that familiar reaction between $NO_2$ and $N_2O_4$ again:

$$2\,NO_2(g) \rightleftharpoons N_2O_4(g)$$

Suppose you increase the pressure by compressing the reaction mixture with a piston. Remember that a mole of any gas exerts the same amount of pressure as a mole of any other gas. Thus, 2 mol of $NO_2(g)$ exert twice as much pressure as 1 mol of $N_2O_4(g)$, and the mixture will react to reduce the applied pressure by producing more $N_2O_4$ from $NO_2$, reducing the number of moles of gas. Once again the equilibrium shifts to the side that minimizes the change, this time a change in pressure.

**If the total number of moles of gases is the same on both sides of the balanced equation, pressure changes will have no effect on the equilibrium position. Pressure also has no measurable effect in reactions that involve only solids or liquids.**

## EXAMPLE 14.11 Le Chatêlier's Principle and Changes in Reaction Conditions

The reaction of $NO_2(g)$ to give $N_2O_4(g)$ is exothermic. If an equilibrium mixture of these gases is cooled, what will happen to the concentrations of $NO_2$ and $N_2O_4$?

**SOLUTION**

Since the reaction gives off heat, you can think of it this way:

$$2\ NO_2(g) \rightleftharpoons N_2O_4(g) + \text{heat}$$

If you cool the reaction mixture down, you remove heat. The reaction will move to counteract this heat loss by producing more heat. As heat is produced, so is $N_2O_4$; thus its concentration will increase. $NO_2$ will be consumed, and its concentration will decrease.

$NO_2$ is a brown gas and $N_2O_4$ is colorless. The equilibrium shifts to the right (toward $N_2O_4$) when the mixture cools in a low temperature bath.

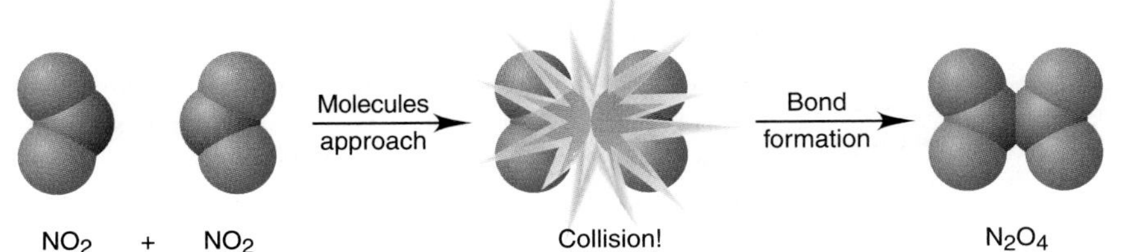


**EXERCISE 14.11**

Hydrogen reacts with nitrogen to produce ammonia. The supplier that provides your company with ammonia to make fertilizers has just quadrupled its prices, so you have decided to go into ammonia manufacturing. What pressure should you use—high or low—to get the greatest possible yield of ammonia from the reaction? Explain your answer.

## EXPERIMENTAL EXERCISE

### 14.1

A. L. Chemist headed up a team of scientists trying to prepare compound C using this reaction:

$$\text{Compound A} + \text{Compound B} \xrightarrow{\text{catalyst}} \underset{\text{(desired product)}}{\text{Compound C}} + H_2O$$

They looked up the properties of all these compounds and obtained the following information:

| Compound | Molar mass (g) | bp (°C) | Density (g mL$^{-1}$) | Water solubility (g 100 mL$^{-1}$) |
|---|---|---|---|---|
| Compound A | 116 | 205 | 0.97 | 1.0 |
| Compound B | 88 | 138 | 0.82 | 2.3 |
| Compound C | 186 | 226 | 0.86 | insoluble |
| $H_2O$ | 18 | 100 | 1.00 | — |

In their first experiment the scientific team mixed 1 mol of compound A with 1 mol of compound B and added a small amount of catalyst. They boiled the reaction mixture for 10 hours, using the apparatus shown in Figure A below. Only 18.6 g of the desired product were isolated from this experimental run. Other measurements showed that 18.6 g of compound C were present when the reaction was at equilibrium.

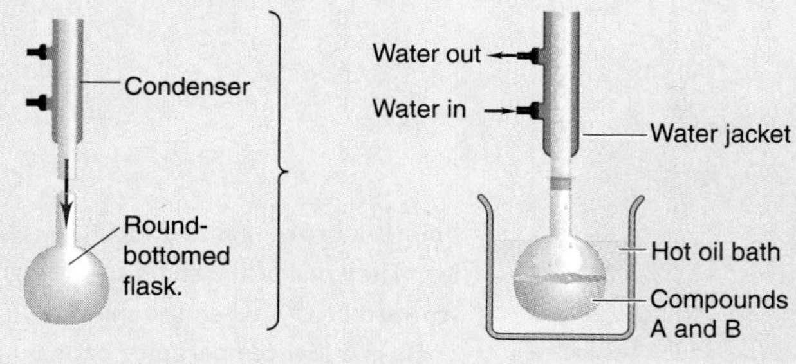


FIGURE A

The chemists put compounds A and B and the catalyst in a round-bottomed flask. They then fitted the flask with a condenser, placed it in a hot oil bath, and brought the mixture to a boil. The vapors produced by the boiling reaction mixture reconvert to the liquid state in the condenser and fall back down into the reaction mixture.

In a second experiment the team repeated the first procedure exactly, except that they boiled the mixture for 20 hours. Again they obtained only 18.6 g of the desired product. The team concluded that they needed to further modify their experimental procedure. They redesigned their apparatus to include a sidearm that could hold 20 mL of liquid (see Figure B above, right). The sidearm was kept at a constant temperature of 25°C.

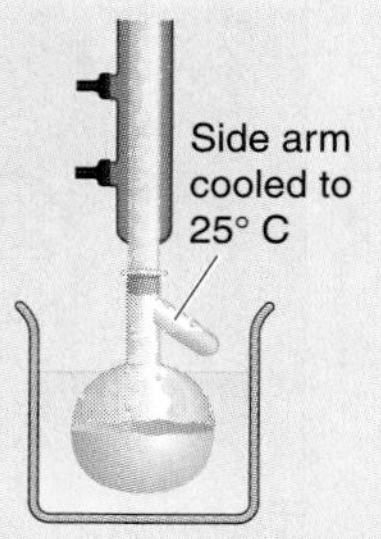


FIGURE B

The idea was to trap water in the sidearm as it dripped down the sides of the condenser. The other liquids (the reactants and products) that might be trapped in the sidearm are less dense than water. They floated on the water's surface and returned to the reaction mixture once the sidearm filled up, leaving the water behind. This effectively removed water from the reaction mixture and drove the equilibrium to the right, toward the product.

### QUESTIONS

1. What is the equilibrium constant for the reaction under the conditions used by A. L. Chemist and the scientific team?
2. Why would the setup shown in the figure above be ineffective for running the reaction in Exercise 14.10?
3. What is the theoretical yield, in grams, of compound C? Of water?
4. How do the water solubilities of compounds A and B influence the yield of this reaction?
5. Why was the second experiment the team carried out a waste of time?
6. Is a 20-mL sidearm large enough to maximize the yield of this reaction?

## CHAPTER SUMMARY

1. Chemical reactions can be fast, slow, or in between, but they all take a finite amount of time.
2. A rate equation gives the relationship between the reaction rate and reactant concentrations. The proportionality constant $k$ in the rate equation is called the rate constant. Each reaction at each temperature has its own rate constant.
3. The rate of a chemical reaction is influenced by three factors: (1) the frequency with which reactant molecules collide with each other, (2) the fraction of these collisions that have sufficient energy to give a reaction, and (3) the fraction of collisions that have a proper orientation to give a reaction.
4. The minimum amount of collision energy necessary to give products is known as the activation energy for a reaction. Each reaction has a unique activation energy.
5. Higher temperatures make reactions go faster. Increasing the temperature increases both the collision frequency and the fraction of collisions with an energy greater than the activation energy.
6. A reaction is at equilibrium when the rate of the forward reaction is identical to the rate of the reverse reaction. Chemical equilibria are dynamic—the forward and reverse reactions occur simultaneously and at the same rate.
7. The concentrations of solid and liquid reactants do not change and do not appear in equilibrium constant expressions in heterogeneous equilibria.
8. The solubility product constant is symbolized by $K_{sp}$. Values of these constants are typically given without units.
9. Le Chatêlier's Principle states that an equilibrium will shift in the direction that minimizes change in concentration, temperature, or pressure to the system.

## IMPORTANT EQUATIONS

Reaction rate equations often take this form (more complex equations are sometimes seen):

$$\text{Rate} = k\,[\text{Concentration Terms}]^x$$

The equilibrium constant for a general reaction such as

$$a\text{A} + b\text{B} \rightleftharpoons c\text{C} + d\text{D}$$

is $K_{eq} = \dfrac{[\text{C}]^c[\text{D}]^d}{[\text{A}]^a[\text{B}]^b}$

The solubility product constant $K_{sp}$ for the solubility of the general ionic compound $M_xN_y$ is:

$$K_{sp} = [\text{M}]^x[\text{N}]^y$$

## PROBLEMS

### SECTION 14.1 The Rates of Chemical Reactions

1. The rate equation for the reaction $(CH_3)_3CCl + H_2O \rightarrow (CH_3)_3COH + HCl$ is

   Rate $= k[(CH_3)_3CCl]$

   a. By what factor will the reaction rate change if the concentration of $(CH_3)_3CCl$ is doubled?
   b. For the same reaction, by what factor will the reaction rate change if the $H_2O$ concentration is doubled?

2. The rate equation for the reaction $CH_3CH_2Cl + CH_3CH_2ONa \rightarrow CH_3CH_2OCH_2CH_3 + NaCl$ is

   Rate $= k[CH_3CH_2Cl][CH_3CH_2O^-]$

   a. By what factor will the reaction rate change if the concentration of $CH_3CH_2Cl$ is tripled?
   b. For the same reaction, by what factor will the reaction rate change if $[CH_3CH_2Cl]$ is cut in half?

3. The rate equation for the reaction $CH_3CH_2CH_2Br + CH_3ONa \rightarrow CH_3CH_2CH_2OCH_3 + NaBr$ is

   Rate $= k[CH_3CH_2CH_2Br][CH_3O^-]$

   a. By what factor will the reaction rate change if the concentration of $CH_3O^-$ is cut in half?
   b. For the same reaction, by what factor will the reaction rate change if $CH_3OK$ is substituted for $CH_3ONa$ at the same concentration?

4. The rate equation for the reaction $(CH_3)_3CBr + H_2O \rightarrow (CH_3)_3COH + HBr$ is

   Rate $= k[(CH_3)_3CBr]$

   a. By what factor will the reaction rate change if the concentration of $(CH_3)_3CBr$ is cut to ⅓ of its original value?
   b. For the same reaction, by what factor will the reaction rate change if the $H_2O$ concentration is tripled?

5. The rate equation for the reaction $(CH_3)_3CI + OH^- \rightarrow (CH_3)_2C{=}CH_2 + I^- + H_2O$ is

   Rate $= k[(CH_3)_3CI][OH^-]$

   a. By what factor will the reaction rate change if the concentration of $OH^-$ is cut in half?
   b. For the same reaction, by what factor will the reaction rate change if KI is added to the reaction mixture?

6. The rate equation for the reaction $CH_3CH_2CH_2Cl + CH_3ONa \rightarrow CH_3CH_2CH_2OCH_3 + NaCl$ is

   Rate $= k[CH_3CH_2CH_2Cl][CH_3O^-]$

   a. By what factor will the reaction rate change if the concentration of $CH_3O^-$ is tripled?
   b. For the same reaction, by what factor will the reaction rate change if $[CH_3CH_2CH_2Cl]$ is cut in half and $CH_3OK$ is substituted for $CH_3ONa$ at the same concentration?

## EXPERIMENTAL PROBLEM

### 14.1

A. L. Chemist and his colleagues were studying a chemical reaction. Compound A decomposes to form compound B. A. L. Chemist dissolved 1.500 mol of compound A in 2.000 L of an inert solvent and measured the numbers of moles of reactant consumed and of product formed at 1.0-min intervals. The figure below is a graph of the data obtained.

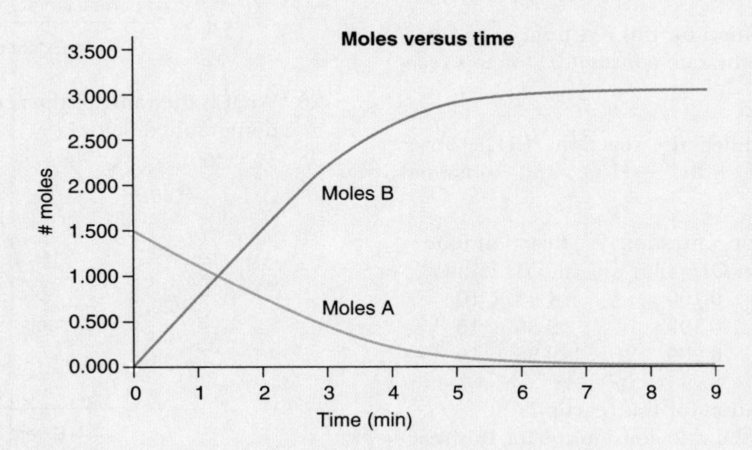


| Raw Data: | | | | | Moles | | | | | |
|---|---|---|---|---|---|---|---|---|---|---|
| A | 1.500 | 1.125 | 0.731 | 0.402 | 0.181 | 0.063 | 0.016 | 0.002 | 0.000 | 0.000 |
| B | 0.000 | 0.750 | 1.538 | 2.196 | 2.638 | 2.873 | 2.968 | 2.995 | 3.000 | 3.000 |
| time (min) | 0 | 1 | 2 | 3 | 4 | 5 | 6 | 7 | 8 | 9 |

**QUESTIONS**

**7.** What is the initial (starting) concentration of compound A?

8. Does this reaction essentially go to completion?

**9.** Write a balanced equation for the reaction A → B.

**10.** How many moles of compound B were present after 2.0 min?

**11.** How long does it take for the reaction to form the maximum number of moles of compound B?

12. Assume that the rate equation for this reaction is

Rate = $k$[A]

What are the units of the rate constant $k$?

13. A team of scientists studied the reaction $CH_3Cl + I^- \rightarrow CH_3I + Cl^-$ and obtained the data given to the right.

| Concentration of $CH_3Cl$ ($M$) | Concentration of $I^{-1}$ ($M$) | Reaction Rate (mol $L^{-1}$ $min^{-1}$) |
|---|---|---|
| 0.100 | 0.100 | $9.330 \times 10^{-2}$ |
| 0.200 | 0.100 | 0.1866 |
| 0.200 | 0.200 | 0.3732 |

a. What is the rate equation for this reaction?

b. What is the value of the rate constant, $k$, for this reaction?

14. A team of scientists studied the reaction $(CH_3)_3CBr + I^- \rightarrow (CH_3)_3CI + Br^-$ and obtained the data given below.

| Concentration of $(CH_3)_3CBr$ ($M$) | Concentration of $I^-$ ($M$) | Reaction Rate (mol $L^{-1}$ $min^{-1}$) |
|---|---|---|
| 0.100 | 0.100 | $4.112 \times 10^{-2}$ |
| 0.200 | 0.100 | $8.224 \times 10^{-2}$ |
| 0.200 | 0.200 | $8.224 \times 10^{-2}$ |

a. What is the rate equation for this reaction?
b. What is the value of the rate constant, *k*, for this reaction?

15. A team of scientists studied the reaction $(CH_3)_3CBr + OH^- \rightarrow (CH_3)_2C{=}CH_2 + Br^- + H_2O$ and obtained the data given below.

| Concentration of $(CH_3)_3CBr$ ($M$) | Concentration of $OH^-$ ($M$) | Reaction Rate (mol $L^{-1}$ $min^{-1}$) |
|---|---|---|
| 0.100 | 0.100 | $4.93 \times 10^{-3}$ |
| 0.200 | 0.100 | $9.86 \times 10^{-3}$ |
| 0.100 | 0.200 | $9.86 \times 10^{-3}$ |

a. What is the rate equation for this reaction?
b. What is the value of the rate constant, *k*, for this reaction?

16. A team of scientists studied the reaction $CH_3CH_2Br + H_2O \rightarrow CH_3CH_2OH + HBr$ and obtained the data given below.

| Concentration of $CH_3CH_2Br$ ($M$) | Concentration of $H_2O$ ($M$) | Reaction Rate (mol $L^{-1}$ $min^{-1}$) |
|---|---|---|
| 0.100 | 0.100 | $3.05 \times 10^{-2}$ |
| 0.100 | 0.200 | $6.10 \times 10^{-2}$ |
| 0.200 | 0.100 | $6.10 \times 10^{-2}$ |

a. What is the rate equation for this reaction?
b. What is the value of the rate constant, *k*, for this reaction?

## SECTION 14.2 Theory of Reaction Rates

17. For the reaction $C_6H_5ONa + CH_3I \rightarrow C_6H_5OCH_3 + NaI$, what two things could you do (in practical terms) to increase the collision frequency?

18. For the reaction in Problem 17, what could you do (in practical terms) to increase the energy factor?

19. What is the value of the activation energy for the reaction graphed below?

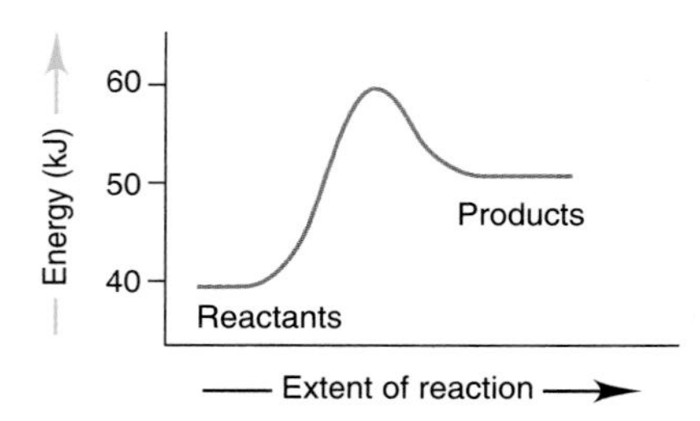


20. What is the value of the activation energy for the reaction graphed below?

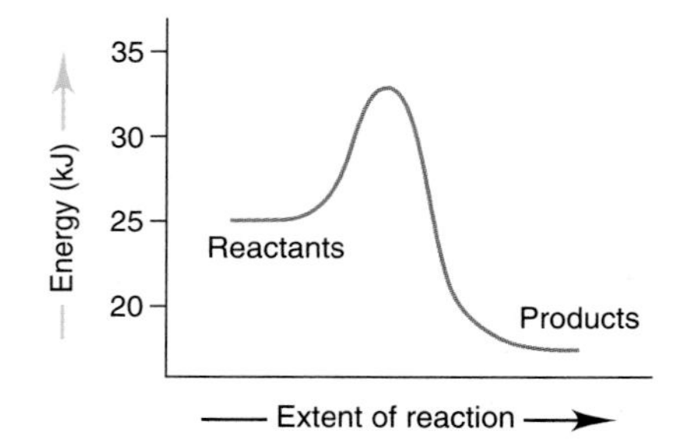


21. What is the value of the activation energy for the reaction graphed below?

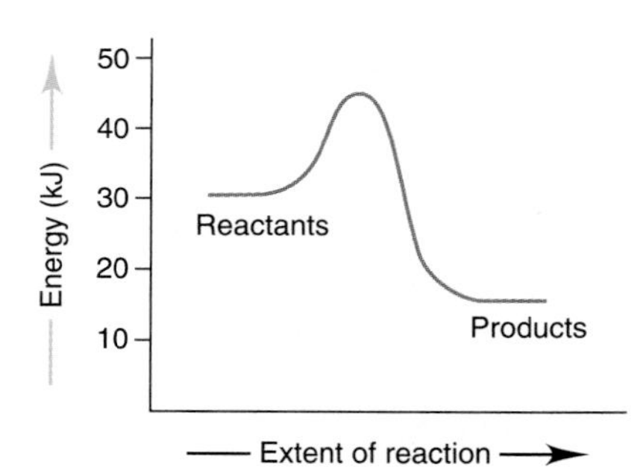


22. What is the value of the activation energy for the reaction graphed below?

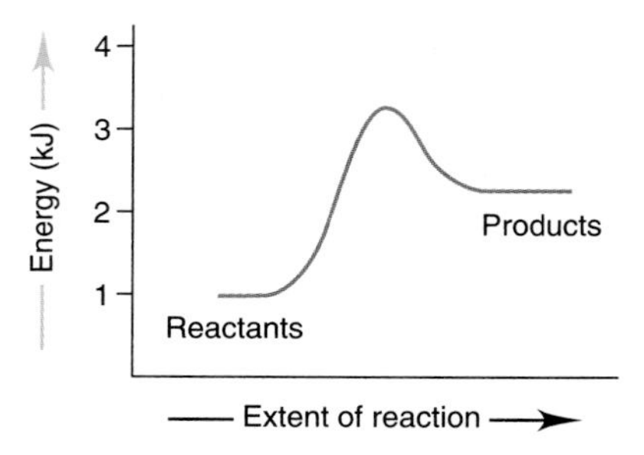

23. Which of the reactions (red, blue, purple, or green) shown on the graph below proceeds at the fastest rate, assuming that all reaction conditions are equal?

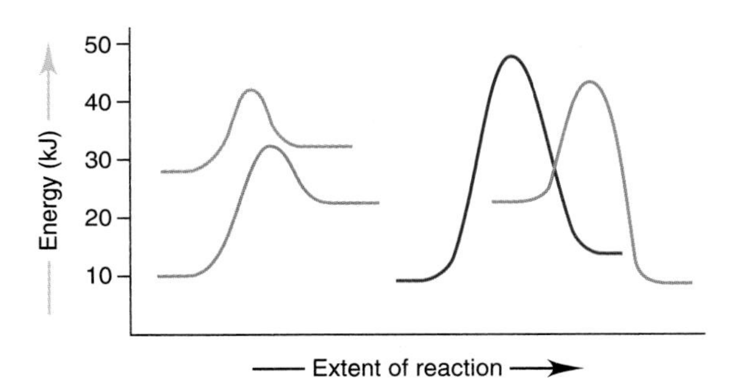


24. Which of the reactions (red, blue, purple, or green) shown on the graph in Problem 23 proceeds at the slowest rate, assuming that all reaction conditions are equal?

25. Under certain conditions reactant molecules collide $1 \times 10^5$ times per second. Of these collisions, 15% have an energy equal to or greater than the activation energy. Only 1 collision in 10 has the correct orientation for the reaction to occur. How many product molecules are produced per second?

26. Under certain conditions reactant molecules collide $5 \times 10^5$ times per second. Of these collisions, 12% have an energy equal to or greater than the activation energy. Only 1 collision in 45 has the correct orientation for the reaction to occur. How many product molecules are produced per second?

27. Under certain conditions molecules of a reactant make $2 \times 10^5$ collisions per second. Only 5% of these collisions have an energy equal to or greater than the activation energy. Only 1 out of every 120 collisions has the orientation necessary to produce the product. How many product molecules are produced per second?

28. Under certain conditions molecules of a reactant make $4 \times 10^5$ collisions per second. Only 2.4% of these collisions have an energy equal to or greater than the activation energy. Only 1 out of every 35 collisions has the orientation necessary to produce the product. How many product molecules are produced per second?

29. Reactant molecules collide $1 \times 10^6$ times per second under certain conditions. Of these collisions only 1.0% have energy equal to or greater than the activation energy. The molecules have the correct orientation in only 1 collision in 90. How many product molecules are produced per second?

30. Reactant molecules collide $2 \times 10^6$ times per second under certain conditions. Of these collisions only 2.9% have energy equal to or greater than the activation energy. The molecules have the correct orientation in only 1 collision in 55. How many product molecules are produced per second?

## SECTION 14.3 Reactions at Equilibrium

31. A team of chemists is making $SO_3$ using the gas phase reaction $2\ SO_2 + O_2 \rightleftharpoons 2\ SO_3$.
    a. Just after the chemists mix the reactants, which rate is faster, that for the forward reaction, $2\ SO_2 + O_2 \rightarrow 2\ SO_3$, or that for the reverse reaction, $2\ SO_3 \rightarrow 2\ SO_2 + O_2$?
    b. As the reaction approaches equilibrium, which slows down, the forward or the reverse reaction?

32. A team of chemists is making $SO_3$ using the gas phase reaction $2\ SO_2 + O_2 \rightleftharpoons 2\ SO_3$.
    a. When the reaction reaches equilibrium, what is the relationship between the rates of the forward and reverse reactions (that is, $2\ SO_2 + O_2 \rightarrow 2\ SO_3$ and $2\ SO_3 \rightarrow 2\ SO_2 + O_2$)?
    b. What is the relationship between the rate just after the reaction has begun and the rate at equilibrium, if the reaction runs at a constant temperature?

33. A team of chemists is examining the gas phase reaction $2\ NO_2 \rightleftharpoons N_2O_4$.
    a. Just after the chemists place pure $NO_2$ in the reaction vessel, is the forward reaction ($2\ NO_2 \rightarrow N_2O_4$) or the reverse reaction ($N_2O_4 \rightarrow 2\ NO_2$) faster?
    b. As the reaction approaches equilibrium, which slows down, the forward or the reverse reaction?

34. A team of chemists is examining the gas phase reaction $2\ NO_2 \rightarrow N_2O_4$.
    a. When the reaction reaches equilibrium, what is the relationship between the rates of the two reactions ($2\ NO_2 \rightarrow N_2O_4$ and $N_2O_4 \rightarrow 2\ NO_2$)?
    b. What is the relationship between the rate just after the reaction has begun and the rate at equilibrium, if the reaction runs at a constant temperature?

35. Ammonia can be prepared by running this gas phase reaction at high temperature and pressure in the presence of a catalyst:

    $$N_2(g) + 3\ H_2(g) \rightleftharpoons 2\ NH_3(g)$$

    a. At the beginning of the reaction, which reactant ($N_2$ or $H_2$) is consumed at the faster rate?
    b. As the reaction approaches equilibrium, which reaction (forward or reverse) runs at the faster rate?

36. a. As the reaction in Problem 35 approaches equilibrium, the chemists remove the catalyst. What happens to the rate of the reaction?
    b. What reaction(s) is (are) occurring at equilibrium, if any?

37. A team of chemists examined the reaction $N_2(g) + 3\ H_2(g) \rightleftharpoons 2\ NH_3(g)$ beginning with a sample of pure ammonia.
    a. Which of these reactions is faster immediately after the pure $NH_3$ is mixed with a catalyst?

    $$2\ NH_3(g) \longrightarrow N_2(g) + 3\ H_2(g)$$

    or

    $$N_2(g) + 3\ H_2(g) \longrightarrow 2\ NH_3(g)$$

    b. What relationship, if any, exists between the rate constants of the forward and reverse reactions at equilibrium?

38. A team of chemists examined the reaction $N_2(g) + 3\ H_2(g) \rightleftharpoons 2\ NH_3(g)$ beginning with a sample of pure ammonia.
    a. Which of these reactions, if either, is faster at equilibrium?

    $$2\ NH_3(g) \longrightarrow N_2(g) + 3\ H_2(g)$$

    or

    $$N_2(g) + 3\ H_2(g) \longrightarrow 2\ NH_3$$

    b. The chemists removed the catalyst after the reaction reached equilibrium. Now which of the two reactions, if either, is faster?

## SECTION 14.4 Equilibrium Constant Expressions

39. What is the equilibrium constant expression for each of these reactions?
    a. $2\ H_2(g) + O_2(g) \rightleftharpoons 2\ H_2O(g)$
    b. $N_2(g) + 2\ O_2(g) \rightleftharpoons 2\ NO_2(g)$
    c. $CO(g) + 2\ H_2(g) \rightleftharpoons CH_3OH(g)$

40. What is the equilibrium constant expression for each of these reactions?
    a. $COCl_2(g) \rightleftharpoons CO(g) + Cl_2(g)$
    b. $CO_2(g) + H_2O(g) \rightleftharpoons H_2CO_3(g)$
    c. $CO(g) + H_2O(g) \rightleftharpoons CO_2(g) + H_2(g)$

41. Write the equilibrium constant expression for each reaction.
    a. $2\ H_2CO(g) \rightleftharpoons CO_2(g) + CH_4(g)$
    b. $ClNO_2(g) + NO(g) \rightleftharpoons NO_2(g) + ClNO(g)$
    c. $PCl_3(g) + Cl_2(g) \rightleftharpoons PCl_5(g)$

42. Write the equilibrium constant expression for each reaction.
    a. $2\ NO(g) + O_2(g) \rightleftharpoons 2\ NO_2(g)$
    b. $Cl_2(g) + 3\ F_2(g) \rightleftharpoons 2\ ClF_3(g)$
    c. $N_2(g) + O_2(g) \rightleftharpoons 2\ NO(g)$

43. What is the equilibrium constant expression for each of these reactions?
    a. $P_4(s) + 6\ H_2(g) \rightleftharpoons 2\ P_2H_6(g)$
    b. $C(s) + O_2(g) \rightleftharpoons CO_2(g)$
    c. $3\ Fe(s) + 4\ H_2O(g) \rightleftharpoons Fe_3O_4(s) + 4\ H_2(g)$

44. What is the equilibrium constant expression for each of these reactions?
    a. $2\ Cu(s) + O_2(g) \rightleftharpoons 2\ CuO(s)$
    b. $CuO(s) + H_2(g) \rightleftharpoons Cu(s) + H_2O(g)$
    c. $C_5H_{10}(g) + Br_2(g) \rightleftharpoons C_5H_9Br(l) + HBr(g)$

45. Write the equilibrium constant expression for each reaction.
    a. $C_6H_{12}(g) + Cl_2(g) \rightleftharpoons C_6H_{11}Cl(l) + HCl(g)$
    b. $C_6H_{10}(l) + H_2(g) \rightleftharpoons C_6H_{12}(l)$
    c. $Ni(s) + 4\ CO(g) \rightleftharpoons Ni(CO)_4(g)$

46. Write the equilibrium constant expression for each reaction.
    a. $2\ Al(s) + 3\ Cl_2(g) \rightleftharpoons 2\ AlCl_3(s)$
    b. $4\ Al(s) + 3\ O_2(g) \rightleftharpoons 2\ Al_2O_3(s)$
    c. $CH_4(g) + Br_2(g) \rightleftharpoons CH_3Br(g) + HBr(g)$

47. Write the equilibrium constant expression for each of the reactions described.
    a. Copper metal reacts with HCl gas to form copper(II) chloride and hydrogen gas.
    b. Magnesium carbonate decomposes to magnesium oxide and carbon dioxide when heated to 400°C.
    c. Lead(II) hydroxide decomposes to lead(II) oxide and water at 400°C.

48. Write the equilibrium constant expression for each of the reactions described.
    a. Nitrogen reacts with hydrogen to form gaseous hydrazine ($N_2H_4$) at 350°C.
    b. Hydrogen peroxide, $H_2O_2(l)$, decomposes to form oxygen and water at room temperature.
    c. Calcium carbonate decomposes to give carbon dioxide and calcium oxide at 350°C.

## SECTION 14.5 Simple Equilibrium Calculations

49. The equilibrium constant for the reaction $2\ NO(g) + O_2(g) \rightleftharpoons 2\ NO_2(g)$ at 25°C is approximately $4 \times 10^{13}$. Which will be present in higher concentration at equilibrium, NO or $NO_2$?

50. The equilibrium constant for the reaction $N_2 + O_2 \rightleftharpoons 2\ NO$ at 25°C is approximately $5 \times 10^{-31}$. Which will be present in higher concentration at equilibrium, $N_2$ or NO?

51. A team of chemists ran this reaction at 1200 K:

$$CO(g) + 3\ H_2(g) \rightleftharpoons CH_4(g) + H_2O(g)$$

At equilibrium they measured these concentrations of reactants and products: CO, 0.30 *M*; $H_2$, 0.10 *M*; $CH_4$, 0.060 *M*; and $H_2O$, 0.020 *M*. What is the equilibrium constant for this reaction?

52. Under certain conditions, $NO(g)$ decomposes to form $N_2(g)$ and $O_2(g)$. A chemist placed 5.0 mol of $NO(g)$ in a 1-L flask and allowed it to decompose. At equilibrium, the flask contained 0.10 mol of $NO(g)$. What is the equilibrium constant for this reaction?

53. Under certain conditions in a 1.00-L container and at equilibrium, the concentrations of $NO_2$, NO, and $O_2$ are 0.200 *M*, $3.05 \times 10^{-2}$ *M*, and 0.600 *M*, respectively. Calculate the equilibrium constant for this reaction:

$$2\ NO_2(g) \rightleftharpoons 2\ NO(g) + O_2(g)$$

54. If $HF(g)$ is heated to 1300 K, a small amount of it decomposes to form $H_2(g)$ and $F_2(g)$. When 2.0 mol of pure $HF(g)$ is heated to this temperature in a 1.0-L reaction vessel and allowed to reach equilibrium, $6.3 \times 10^{-7}$ mol of $H_2(g)$ is formed. What is the value of the equilibrium constant for the decomposition of $HF(g)$ at 1300 K?

55. At high temperature and low pressure, carbon dioxide gas decomposes to form carbon monoxide and oxygen. When 2.00 mol of pure $CO_2(g)$ is placed in a 1.0-L reaction chamber and allowed to decompose under these conditions, 0.400 mol of $CO(g)$ is formed. What is the value of the equilibrium constant for this reaction?

56. If the flask representing the product mixture of the reaction between $H_2$ and $I_2$ to form HI in Figure 14.11 is an accurate reflection of the product distribution at equilibrium, calculate $K_{eq}$ for the reaction $H_2(g) + I_2(g) \rightleftharpoons 2\ HI(g)$.

57. $CuS(s)$ reacts with $O_2(g)$ to form $Cu(s)$ and $SO_2$ at high temperatures. A sample of $CuS(s)$ was placed in a closed 5.00-L container with 2.00 mol of $O_2(g)$ and heated, forming 1.2 mol of $SO_2(g)$ at equilibrium. What is $K_{eq}$ for this reaction?

58. Calcium carbonate decomposes to form $CaO(s)$ and $CO_2(g)$ at high temperatures. A sample of $CaCO_3(s)$ was placed in an evacuated 2.00-L reaction chamber and heated, forming 1.5 mol of $CO_2(g)$ at equilibrium. What is the value of $K_{eq}$ for this reaction?

## SECTION 14.6 Solubility Equilibria

59. Write the expression for $K_{sp}$ for each of these compounds.
a. $BaF_2$ b. $AlPO_4$ c. $Pb(OH)_2$

60. Write the expression for $K_{sp}$ for each of these compounds.
a. $Hg_2CO_3$ b. $Cd_3(PO_4)_2$ c. $Zn(NO_3)_2$

61. What is the expression for $K_{sp}$ for each of these compounds?
a. AgBr b. $Ag_2CO_3$ c. $Ni_3(PO_4)_2$

62. What is the expression for $K_{sp}$ for each of these compounds?
a. ZnS b. $Cd(OH)_2$ c. $Ag_3PO_4$

63. Calculate the water solubility in moles per liter of each of these compounds from the given value of the solubility product constant, $K_{sp}$, at 25°C.
a. Iron(II) fluoride, $FeF_2$, $K_{sp} = 2.36 \times 10^{-6}$
b. Lead sulfate, $PbSO_4$, $K_{sp} = 1.82 \times 10^{-8}$
c. Mercurous sulfate, $Hg_2SO_4$, $K_{sp} = 7.99 \times 10^{-7}$

64. Calculate the water solubility in moles per liter of each of these compounds from the given value of the solubility product constant, $K_{sp}$, at 25°C.
a. Mercury (II) iodide, $HgI_2$, $K_{sp} = 2.82 \times 10^{-29}$
b. Copper(II) phosphate, $Cu_3(PO_4)_2$, $K_{sp} = 1.39 \times 10^{-37}$
c. Calcium sulfate, $CaSO_4$, $K_{sp} = 7.10 \times 10^{-5}$

65. Calculate the solubility in moles per liter of water for each compound from the given value of the solubility product constant, $K_{sp}$, at 25°C.
a. Tin(II) hydroxide, $Sn(OH)_2$, $K_{sp} = 5.45 \times 10^{-27}$
b. Silver phosphate, $Ag_3PO_4$, $K_{sp} = 8.88 \times 10^{-17}$
c. Copper(I) sulfide, $Cu_2S$, $K_{sp} = 2.1 \times 10^{-47}$

66. Calculate the solubility in moles per liter of water for each compound from the given value of the solubility product constant, $K_{sp}$, at 25°C.
a. Iron(II) carbonate, $FeCO_3$, $K_{sp} = 3.1 \times 10^{-11}$
b. Zinc hydroxide, $Zn(OH)_2$, $K_{sp} = 1.2 \times 10^{-17}$
c. Silver carbonate, $Ag_2CO_3$, $K_{sp} = 8.45 \times 10^{-12}$

67. What is the solubility in moles per liter of each of these compounds in water at 25°C?
a. Nickel(II) hydroxide, $Ni(OH)_2$, $K_{sp} = 5.47 \times 10^{-16}$
b. Platinum(II) sulfide, PtS, $K_{sp} = 9.91 \times 10^{-74}$
c. Barium sulfate, $BaSO_4$, $K_{sp} = 1.07 \times 10^{-10}$

68. What is the solubility in moles per liter of each of these compounds in water at 25°C?
a. Iron(III) hydroxide, $Fe(OH)_3$, $K_{sp} = 2.64 \times 10^{-39}$
b. Manganese(II) carbonate, $MnCO_3$, $K_{sp} = 2.24 \times 10^{-11}$
c. Calcium carbonate, $CaCO_3$, $K_{sp} = 4.96 \times 10^{-9}$

69. Use the given value of $K_{sp}$ at 25°C to calculate the water solubility in moles per liter of each compound.
   a. Calcium fluoride, $CaF_2$, $K_{sp} = 1.46 \times 10^{-10}$
   b. Mercurous iodide, $Hg_2I_2$, $K_{sp} = 5.33 \times 10^{-29}$
   c. Calcium hydroxide, $Ca(OH)_2$, $K_{sp} = 4.68 \times 10^{-6}$

70. Use the given value of $K_{sp}$ at 25°C to calculate the water solubility in moles per liter of each compound.
   a. Lead (II) iodide, $PbI_2$, $K_{sp} = 8.49 \times 10^{-9}$
   b. Lead (II) sulfate, $PbSO_4$, $K_{sp} = 1.6 \times 10^{-8}$
   c. Cobalt(III) hydroxide, $Co(OH)_3$, $K_{sp} = 1 \times 10^{-43}$

71. From the given solubility at 25°C, calculate the value of $K_{sp}$ for each substance.
   a. Cadmium carbonate, $CdCO_3$, solubility = $2.4 \times 10^{-6}$ $M$
   b. Copper(I) chloride, CuCl, solubility = $4.1 \times 10^{-4}$ $M$
   c. Calcium hydroxide, $Ca(OH)_2$, solubility = $1.05 \times 10^{-2}$ $M$

72. From the given solubility at 25°C, calculate the value of $K_{sp}$ for each substance.
   a. Silver chromate, $Ag_2CrO_4$, solubility = $6.5 \times 10^{-5}$ $M$
   b. Bismuth iodide, $BiI_3$, solubility = $1.3 \times 10^{-5}$ $M$
   c. Barium carbonate, $BaCO_3$, solubility = $5.1 \times 10^{-5}$ $M$

73. The solubility at 25°C is given for each of these substances. Calculate $K_{sp}$.
   a. Lead (II) sulfide, PbS, solubility = $9.5 \times 10^{-15}$ $M$
   b. Zinc carbonate, $ZnCO_3$, solubility = $1.1 \times 10^{-5}$ $M$
   c. Magnesium hydroxide, $Mg(OH)_2$, solubility = $3.7 \times 10^{-5}$ $M$

74. The solubility at 25°C is given for each of these substances. Calculate $K_{sp}$.
   a. Copper(I) sulfide, $Cu_2S$, solubility = $8.3 \times 10^{-17}$ $M$
   b. Barium fluoride, $BaF_2$, solubility = $7.5 \times 10^{-3}$ $M$
   c. Barium sulfate, $BaSO_4$, solubility = $1.03 \times 10^{-5}$ $M$

75. For each of these substances, calculate $K_{sp}$ from the given solubility at 25°C.
   a. Cadmium hydroxide, $Cd(OH)_2$, solubility = $1.14 \times 10^{-5}$ $M$
   b. Cadmium sulfide, CdS, solubility = $8.8 \times 10^{-14}$ $M$
   c. Aluminum phosphate, $AlPO_4$, solubility = $9.9 \times 10^{-11}$ $M$

76. For each of these substances, calculate $K_{sp}$ from the given solubility at 25°C.
   a. Iron(II) hydroxide, $Fe(OH)_2$, solubility = $5.85 \times 10^{-10}$ $M$
   b. Iron(III) hydroxide, $Fe(OH)_3$, solubility = $2.0 \times 10^{-10}$ $M$
   c. Magnesium hydroxide, $Mg(OH)_2$, solubility = $1.44 \times 10^{-4}$ $M$

77. An aqueous solution of $AgNO_3$ is added slowly to an aqueous solution that contains both NaCl and $Na_2CrO_4$. The $K_{sp}$ for AgCl is $1.78 \times 10^{-10}$ and that for $Ag_2CrO_4$ is $2.45 \times 10^{-12}$. Which compound, AgCl or $Ag_2CrO_4$, precipitates from the solution first?

78. An aqueous solution of $Na_2S$ is added slowly to an aqueous solution that contains both $Cu(NO_3)_2$ and $AgNO_3$. The $K_{sp}$ for CuS is $9 \times 10^{-36}$ and that for $Ag_2S$ is $2 \times 10^{-49}$. Which compound, CuS or $Ag_2S$, precipitates from the solution first?

## SECTION 14.7 Le Châtelier's Principle and Its Applications

79. If each of these reactions is at equilibrium and the pressure is increased by depressing a piston does the concentration of reactants increase, decrease, or remain the same? Explain your answers.
   a. $CO_2(g) + CF_4(g) \rightleftharpoons 2\ COF_2(g)$
   b. $CO(g) + H_2O(g) \rightleftharpoons CO_2(g) + H_2(g)$
   c. $2\ NO(g) + Br_2(g) \rightleftharpoons 2\ NOBr(g)$

80. If each of these reactions is at equilibrium and the pressure is increased by depressing a piston does the concentration of reactants increase, decrease, or remain the same? Explain your answers.
   a. $N_2O_4(g) \rightleftharpoons 2\ NO_2(g)$
   b. $2\ H_2S(g) \rightleftharpoons 2\ H_2(g) + S_2(g)$
   c. $2\ NO(g) + O_2(g) \rightleftharpoons 2\ NO_2(g)$

81. If each of these reactions is at equilibrium and the pressure is decreased by depressing a piston does the concentration of reactants increase, decrease, or remain the same? Explain your answers.
   a. $MgCO_3(s) \rightleftharpoons MgO(s) + CO_2(g)$
   b. $2\ P(s) + 3\ Cl_2(g) \rightleftharpoons 2\ PCl_3(g)$
   c. $C(s) + O_2(g) \rightleftharpoons CO_2(g)$

82. If each of these reactions is at equilibrium and the pressure is decreased by depressing a piston does the concentration of reactants increase, decrease, or remain the same? Explain your answers.
   a. $2\ Fe(s) + O_2(g) \rightleftharpoons 2\ FeO(s)$
   b. $N_2(g) + 3\ H_2(g) \rightleftharpoons 2\ NH_3(g)$
   c. $2\ N_2O(g) \rightleftharpoons 2\ N_2(g) + O_2(g)$

83. Under certain conditions hydrogen reacts with oxygen to give hydrogen peroxide ($H_2O_2$) in the gas phase. The reaction is exothermic.
   a. Write a balanced equation for this reaction, showing heat as either a reactant or a product (whichever is appropriate).
   b. If you add hydrogen to the reaction mixture after it has reached equilibrium, what will be the effect on the concentration of oxygen? On the concentration of hydrogen peroxide? Explain your answer.

84. For the reaction of hydrogen with oxygen to give hydrogen peroxide in the gas phase (see Problem 83), answer these questions.
    a. If you remove oxygen from the reaction mixture after it has reached equilibrium, what will be the effect on the concentration of hydrogen? On the concentration of hydrogen peroxide? Explain your answer.
    b. If you add hydrogen peroxide to the reaction mixture after it has reached equilibrium, what will be the effect on the concentration of hydrogen? Explain your answer.

85. For the reaction of hydrogen with oxygen to give hydrogen peroxide in the gas phase (see Problems 83 and 84), answer these questions.
    a. If you increase the pressure by decreasing the volume of the reaction mixture, what will be the effect on the concentration of oxygen? On the concentration of hydrogen peroxide? Explain your answer.
    b. If you decrease the temperature of the reaction mixture, what will be the effect on the concentration of hydrogen? On the concentration of hydrogen peroxide? Explain your answer.

86. The usual reaction of hydrogen with oxygen in the gas phase yields water. The reaction is strongly exothermic.
    a. Write a balanced equation for this reaction, showing heat as either a reactant or a product (whichever is appropriate).
    b. If you add hydrogen to the reaction mixture for this reaction after it has reached equilibrium, what will be the effect on the concentration of oxygen? On the concentration of water? Explain your answer.

87. Consider the reaction of hydrogen with oxygen in the gas phase to yield water (see Problem 86).
    a. If you remove oxygen from the reaction mixture after it has reached equilibrium, what will be the effect on the concentration of hydrogen? On the concentration of water? Explain your answer.
    b. If you add water to the reaction mixture after it has reached equilibrium, what will be the effect on the concentration of hydrogen? Explain your answer.

88. If you increase the pressure by decreasing the volume of a reaction mixture of hydrogen, oxygen, and water (see Problems 86 and 87), what will be the effect on the concentration of hydrogen? On the concentration of water? Explain your answer.

89. If you decrease the temperature of a reaction mixture of hydrogen, oxygen, and water (see Problems 86–88), what will be the effect on the concentration of oxygen? On the concentration of water? Explain your answer.

90. The reaction between $H_2(g)$ and $N_2(g)$ to form $NH_3(g)$ is endothermic.
    a. Write a balanced equation for this reaction.
    b. Does heating the reaction mixture shift the equilibrium to the right or the left?
    c. Does increasing the pressure shift the equilibrium to the right or the left?

## ADDITIONAL PROBLEMS

91. For the reaction $2\ SO_2(g) + O_2(g) \rightleftharpoons 2\ SO_3(g)$, the value of $K_{eq}$ at 327°C is $2.8 \times 10^{12}$. If you place pure $SO_3$ in a flask at a concentration of 0.50 $M$ at this temperature, what will be the equilibrium concentration of $SO_2$?

92. For the reaction $2\ H_2(g) + S_2(g) \rightleftharpoons 2\ H_2S(g)$, the value of $K_{eq}$ at 1000 K is $5.0 \times 10^{-5}$. If you place pure $H_2S$ in a flask at a concentration of 1.5 $M$ at this temperature, what will be the equilibrium concentration of $H_2$?

## SOLUTIONS TO EXERCISES

### Exercise 14.1

Reaction rate $= k[(CH_3)_3CBr]$

### Exercise 14.2

**a.** The collision frequency is three times higher when the concentration of one of the reactant molecules is three times higher.

**b.** At lower temperatures the molecules move more slowly and collide with each other less often. Also, a smaller fraction of the collisions will have enough energy to give a reaction. Therefore, both the collision frequency and the energy factor are responsible for slower reactions at lower temperatures.

### Exercise 14.3

**a.** $2\ SO_2(g) + O_2(g) \rightleftharpoons 2\ SO_3(g)$

**b.** The oxygen reacts with $SO_2$ (decreasing its concentration) to form $SO_3$ (increasing its concentration) as a new position of equilibrium is attained.

### Exercise 14.4

The equilibrium constant is equal to the concentration of the product divided by the concentration of the reactant. Since $NO_2$ has a coefficient of 2 in the balanced equation, the concentration of $NO_2$ must be squared in the equilibrium constant expression.

$$K_{eq} = \frac{[N_2O_4]}{[NO_2]^2}$$

### Exercise 14.5

**a.** $K_{eq} = \dfrac{1}{[SO_2(g)]^5}$

**b.** $K_{eq} = [CO_2(g)][H_2O(g)]$ (Water is in the gas phase in this reaction and must be included in the equilibrium constant expression.)

**c.** $K_{eq} = \dfrac{[HOCl(aq)][OH^-(aq)]}{[OCl^-(aq)]}$

### Exercise 14.6

The reaction is the reverse of the one in Example 14.6, and the equilibrium constant expression is the inverse. You could go through the whole calculation, as shown below, or you could simply do this:

$$\frac{1}{\text{Previous answer}} = \text{Current answer}$$

$$K_{eq} = \frac{[H_2][I_2]}{[HI]^2} = \frac{(1.2 \times 10^{-3}\ \text{mol L}^{-1})^2}{(8.4 \times 10^{-3}\ \text{mol L}^{-1})^2} = 2.0 \times 10^{-2}$$

### Exercise 14.7

Here is the familiar equilibrium constant expression:

$$K_{eq} = \frac{[HI]^2}{[H_2][I_2]}$$

You of course have to calculate concentrations to put in the expression. You construct the usual table:

| Conc. ($M$) | $[H_2]$ + | $[I_2]$ $\rightleftharpoons$ | $2[HI]$ |
|---|---|---|---|
| At the beginning | 2.00 | 1.00 | 0 |
| Change | | | 0.500 |
| At equilibrium | ? | ? | 0.500 |

The balanced equation for the reaction tells you that the change in the concentration of $H_2$ (and $I_2$) is exactly half the change in the concentration of HI. The calculations are easy once you know that.

| | $[H_2]$ + | $[I_2]$ $\rightleftharpoons$ | $2[HI]$ |
|---|---|---|---|
| At the beginning | 2.00 | 1.00 | 0 |
| Change | − 0.250 | − 0.250 | 0.500 |
| At equilibrium | ? | ? | 0.500 |

| | $[H_2]$ + | $[I_2]$ $\rightleftharpoons$ | $2[HI]$ |
|---|---|---|---|
| At the beginning | 2.00 | 1.00 | 0 |
| Change | − 0.250 | − 0.250 | 0.500 |
| At equilibrium | 1.75 | 0.75 | 0.500 |

You can now calculate $K_{eq}$.

$$K_{eq} = \frac{[HI]^2}{[H_2][I_2]} = \frac{(0.500\ \text{mol L}^{-1})^2}{(1.75\ \text{mol L}^{-1})(0.75\ \text{mol L}^{-1})} = 0.19$$

### Exercise 14.8

**a.** $K_{sp} = [Ag^+]^2[S^{2-}]$

**b.** $K_{sp} = [Fe^{3+}]^2[S^{2-}]^3$

**c.** $K_{sp} = [Ba^{2+}][SO_4^{2-}]$

**d.** $K_{sp} = [Ca^{2+}]^3[PO_4^{3-}]^2$

### Exercise 14.9

**a.** $CaF_2(s) \rightleftharpoons Ca^{2+}(aq) + 2\ F^-(aq)$

$K_{sp} = [Ca^{2+}][F^-]^2 = (2.2 \times 10^{-4}\ M)(4.4 \times 10^{-4}\ M)^2$
$= 4.3 \times 10^{-11}$

**b.** $CaCO_3(s) \rightleftharpoons Ca^{2+}(aq) + CO_3^{2-}(aq)$

$K_{sp} = [Ca^{2+}][CO_3^{2-}] = 5 \times 10^{-9}$
$5 \times 10^{-9} = [x][x] = x^2$
$x = 7 \times 10^{-5}\ M$

### Exercise 14.10

If you remove $H_2O$, the reaction will shift to increase its concentration. As the concentration of $H_2O$ increases, the concentration of $CH_3CO_2CH_2CH_3$ will increase, too. The concentrations of $CH_3CO_2H$ and $CH_3CH_2OH$ will decrease as these compounds react to give more water.

**Exercise 14.11**

First, you need the balanced equation. Hydrogen and nitrogen are both diatomic molecules, and ammonia is $NH_3$. The balanced equation is

$$3\ H_2(g) + N_2(g) \rightleftharpoons 2\ NH_3(g)$$

There are more moles of gases on the left side of the equation (4) than on the right side (2). Higher pressure will push the reaction toward the side with fewer moles, that is, toward the desired product, $NH_3$.

**EXPERIMENTAL EXERCISE 14.1**

1. The equilibrium constant expression is:

$$K_{eq} = \frac{[C][H_2O]}{[A][B]}$$

At the end of the first experiment, 18.6 g of compound C was present at equilibrium. The reaction formed

$$\frac{18.6 \text{ g compound C}}{186 \text{ g mol}^{-1} \text{ compound C}} = 0.100 \text{ mol Compound C}$$

Since the balanced equation for the reaction says that 1 mol of compound A reacts with 1 mol of compound B to give 1 mol of compound C, there is 1.0 − 0.1 = 0.9 mol of each of compounds A and B present at equilibrium. The $K_{eq}$ is therefore

$$\frac{(0.100)(0.100)}{(0.900)(0.900)} = 0.0123$$

2. In Exercise 14.10, the reactants are acetic acid and ethanol. Both of these are completely soluble in water. The sidearm would contain a solution of $CH_3CO_2H$, $CH_3CH_2OH$, and water. Thus acetic acid and ethanol molecules would not return to the reaction mixture and leave the water behind as happens in the experiment described above.
3. Assuming the reaction was 100% complete, 1 mol of compound C (186 g) and 1 mol of water (18 g) would form.
4. Some of both reactants dissolve in the water in the sidearm and do not return to the reaction mixture. The yield is therefore lower than theoretical.
5. It had already been determined that 18.6 g of product was all that formed when the reaction was at equilibrium. Heating for a longer time cannot change this.
6. The maximum amount of water that can be obtained is 18 g, or 18 mL. A 20-mL sidearm is big enough.

15

# ACIDS AND BASES



**KEY TERMS**

FIGURE 15.1
Rain water picks up carbon dioxide and becomes weakly acidic. This acidic solution reacts with the base, $CaCO_3$, that makes up limestone (as it does in cave formation; see Chapter Boxes 13.2 on page 465 and 15.1 on page 550). Over millions of years this process has created the fantastic mountain towers of Guilin, which have inspired Chinese artists for millennia.

We introduced acids and bases in Section 8.4, though in a somewhat simplified way. As you know, these classes of compounds are among the most important in the study of chemistry. You encounter them in one form or another every day. From the citric acid in your breakfast orange juice to the sulfuric acid in acid rain, from the vinegar (acetic acid) in your salad dressing to the nucleic acids in your DNA, acids are everywhere. From the ammonia in window cleaners to the carbonates in antacid tablets, bases are universal too.

Acid-base reactions are a part of the molecular processes that make your body work and shape this planet's landscapes (see Figure 15.1). In terms of manufacturing output, sulfuric acid is first on the list of industrial chemicals, and ammonia is fourth. In other words, acid-base reactions are a vital and essential part of all our lives.

This chapter begins by discussing acids, bases, and their reactions, using a somewhat broader definition than the one Svante Arrhenius proposed. Next, equilibrium constant expressions are used to give a more precise idea of what happens when acids and bases dissolve in water. Finally, the chapter covers methods of measuring the concentrations of acids and bases in aqueous solutions.

## THE REACTIONS OF ACIDS AND BASES

### 15.1 Brønsted-Lowry Acids and Bases

**GOAL:** To define Brønsted-Lowry acids and bases, and to be able to identify them in reaction equations.

Before you can think about how acids and bases react, you must know how to identify members of these classes of compounds. In Section 8.4, we used Arrhenius's definitions of acids and bases. In these, an **Arrhenius acid** is a substance that produces $H^+$ ions in aqueous solutions, and an **Arrhenius base** is a substance that produces $OH^-$ ions in aqueous solutions. Chemists have known for some time that the $H^+$ ion doesn't exist in aqueous solutions as an independent species (see Chapter Box 8.4, page 282). The modernized Arrhenius definition of acids is that they are substances that produce $H_3O^+$ ions in aqueous solutions.

It did not take long for chemists to find other problems with the Arrhenius definitions. In practical terms, the $OH^-$ ion is far from being the only base. For example, we have already said that ammonia ($NH_3$), a common household chemical found in many cleaning products, is a base. It reacts as a base despite the fact that it contains neither an oxygen atom nor an OH group.

To see what is meant by the Brønsted-Lowry definitions of an acid and a base, consider the net ionic equation for any Arrhenius acid-base reaction in water:

$$H_3O^+ + OH^- \longrightarrow 2\,H_2O$$

When it reacts, the $H_3O^+$ ion gives up a proton, passing it to the $OH^-$ ion. This means that $H_3O^+$ is a **Brønsted-Lowry acid** in this reaction. The $OH^-$ in turn accepts the proton, making it a **Brønsted-Lowry base.**

The advantage of the Brønsted-Lowry definitions is that they cover acids and bases other than $H_3O^+$ and $OH^-$. Consider:

$$HSO_4^- + PO_4^{3-} \rightleftharpoons SO_4^{2-} + HPO_4^{2-}$$

During this reaction the $HSO_4^-$ ion donates a proton and is therefore an acid. You can tell this by looking at what happens to the sulfur-containing ions: The reaction turns the $HSO_4^-$ ion into the $SO_4^{2-}$ ion. The $PO_4^{3-}$ ion becomes the $HPO_4^{2-}$ ion during the same reaction. That is, the $PO_4^{3-}$ ion accepts a proton, which means that it is acting as a base. (See Figure 15.2.)

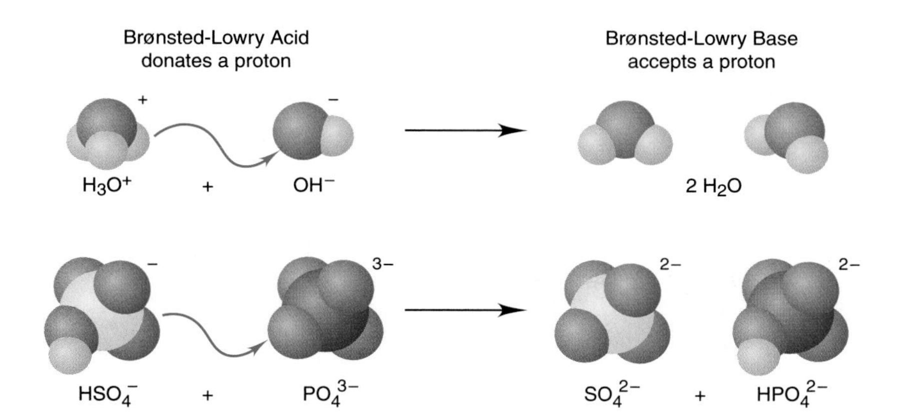


FIGURE 15.2
In Brønsted-Lowry acid-base reactions, a proton transfers from an acid to a base.

EXAMPLE 15.1

### Identifying Brønsted-Lowry Acids and Bases

For each of these acid-base reactions, first write the net ionic equation and then identify each reactant and each product as a Brønsted-Lowry acid or base.

**a.** $H_2SO_4 + NaH_2PO_4 \rightleftharpoons NaHSO_4 + H_3PO_4$
**b.** $NaOH + NaH_2PO_4 \rightleftharpoons H_2O + Na_2HPO_4$

**SOLUTIONS**

**a.** To write the net ionic equation, first separate the metal ion from each formula that contains one. The other ion contains the rest of the atoms and has a charge equal to and opposite that on the metal ion:

$$H_2SO_4 + Na^+ + H_2PO_4^- \rightleftharpoons Na^+ + HSO_4^- + H_3PO_4$$

Next cancel the spectator ions on both sides. This gives the net ionic equation:

$$H_2SO_4 + H_2PO_4^- \rightleftharpoons HSO_4^- + H_3PO_4$$

$H_2SO_4$ donates a proton in the reaction, so it must be an acid. (If you remembered its name, sulfuric acid, you might have guessed that.) $H_2PO_4^-$ accepts a proton, so it's a base. The $HSO_4^-$ ion must accept a proton to get back to $H_2SO_4$, and it is therefore a base. $H_3PO_4$ must give up a proton to get back to $H_2PO_4^-$, and it is therefore an acid. Note that there is always one acid and one base on each side of the equation.

**b.** To write the net ionic equation, the first step is

$$Na^+ + OH^- + Na^+ + H_2PO_4^- \rightleftharpoons 2\ Na^+ + HPO_4^{2-} + H_2O$$

Next cancel the spectator ions to get the net ionic equation:

$$OH^- + H_2PO_4^- \rightleftharpoons HPO_4^{2-} + H_2O$$

The $OH^-$ ion accepts a proton and is a base (if you remember the Arrhenius definition, you won't be surprised); $H_2PO_4^-$ donates a proton and is therefore an acid. On the right side, $HPO_4^{2-}$ must accept a proton to become $H_2PO_4^-$, and it is therefore a base. Similarly, $H_2O$ must donate a proton to become $OH^-$, so it is an acid.

**STUDY ALERT**

**You might have noticed that the $H_2PO_4^-$ ion was the base in part (a). Many acids can also serve as bases, and vice versa. (See Section 15.3 for more details.)**

**EXERCISE 15.1**

For each of the following acid-base reactions, identify each of the species present as a Brønsted-Lowry acid or base.

**a.** $H_2SO_4 + CH_3COOH \rightleftharpoons HSO_4^- + CH_3COOH_2^+$

**b.** $PO_4^{3-} + H_2PO_4^- \rightleftharpoons 2\ HPO_4^{2-}$

## 15.2 Brønsted-Lowry Acid-Base Reactions: Conjugate Acids and Bases

**GOAL:** To determine the formula of the conjugate acid or conjugate base of a given species.

**STUDY ALERT**

**When a proton adds to a base, the charge on the conjugate acid increases by 1 compared to the charge on the base. This must be true if charge is conserved. Adding a proton to $PO_3^{3-}$ gives $HPO_3^{2-}$; the charge increases from $-3$ to $-2$. Of course, the converse is also true. Removing a proton to form a conjugate base decreases the charge by 1. Removing a charge from $H_2SO_4$ to form $HSO_4^-$ decreases the charge from 0 to $-1$. It will be very useful to your academic career to pay serious attention to the charges on these conjugate acid-base pairs. Your test scores will improve, your instructor will be happy, you will feel good about yourself, and all will be right with the world. Trust us on this.**

Now you know the definitions of Brønsted-Lowry acids and bases and can use them to classify species in reactions. The next step is to learn how to predict the products of acid-base reactions.

Let's consider this reaction once again: $HSO_4^- + PO_4^{3-} \rightleftharpoons SO_4^{2-} + HPO_4^{2-}$. Relationships like the ones between $HSO_4^-$ and $SO_4^{2-}$ and between $PO_4^{3-}$ and $HPO_4^{2-}$ are so fundamental that they have their own names. If a proton (not a hydrogen atom or a "hydrogen" but a proton, $H^+$) is added to a molecule or ion, its **conjugate acid** forms. For example, $HPO_4^{2-}$ is the conjugate acid of the base $PO_4^{3-}$. You might keep in mind that the conjugate acid is always more acidic than the original species. If a proton ($H^+$) is removed from a species, its **conjugate base** forms. For example, $SO_4^{2-}$ is the conjugate base of the acid $HSO_4^-$. (See Figure 15.3.)

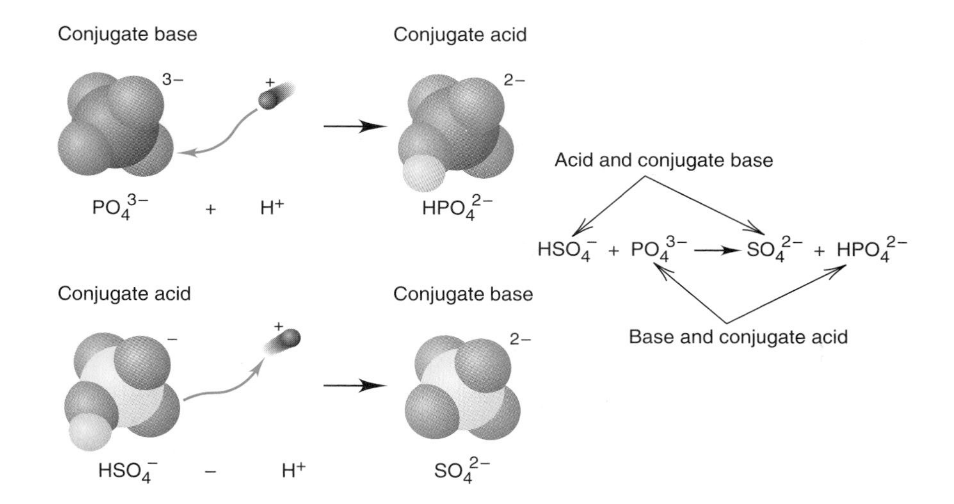


FIGURE 15.3
The relationships of conjugate acids and conjugate bases.

## EXAMPLE 15.2 Formulas of Conjugate Acids and Bases

What is the formula for each of these species?

**a.** The conjugate acid of $NH_3$
**b.** The conjugate base of $NH_3$
**c.** The conjugate acid of $OH^-$
**d.** The conjugate base of $OH^-$

### SOLUTIONS

**a.** To find the formula of the conjugate acid of $NH_3$, add $H^+$ to the formula. This gives $NH_4^+$. The charge increases by 1.

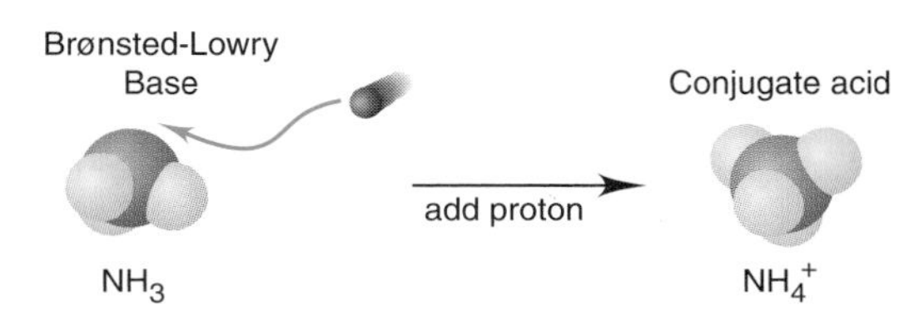


**b.** To find the conjugate base of $NH_3$, remove $H^+$. This gives $NH_2^-$. The charge decreases by 1.

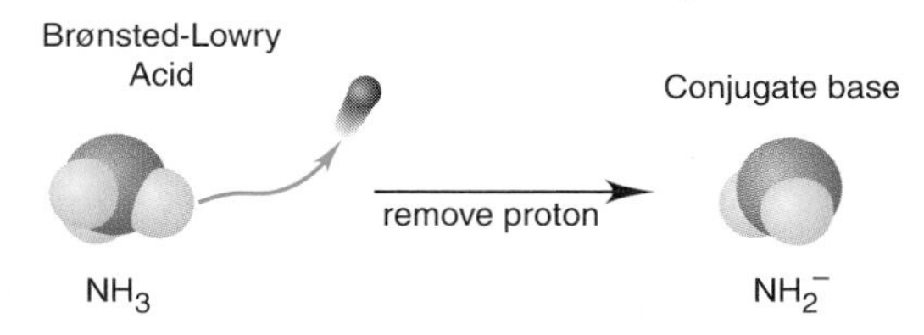


**c.** The conjugate acid of $OH^-$ is $H_2O$. When you add $H^+$ to $OH^-$, the $+1$ charge on the $H^+$ cancels out the $-1$ charge on the $OH^-$.

**d.** The conjugate base of $OH^-$ is $O^{2-}$.

### EXERCISE 15.2

What is the formula for each of these species?

**a.** The conjugate acid of $H_2S$
**b.** The conjugate base of $H_2S$
**c.** The conjugate acid of $CH_3^-$
**d.** The conjugate base of $CH_3^-$

## 15.3 Acid-Base Reactions: Reactants and Products

**GOAL:** To predict the products of Brønsted-Lowry acid-base reactions.

Now that you know how to find the formulas of conjugate acids and conjugate bases, you can begin to predict the products of acid-base reactions. For such a reaction, once you know what the reactants are, predicting what the products will be is relatively simple. First, you identify the reactant that acts as an acid. The conjugate base of this acid is one of the reaction products. The other reactant is the base, and its conjugate acid is the other product.

How do you know which of the reactants is the acid? Sometimes this is not a trivial question. In Example 15.1(a) $H_2PO_4^-$ acted as a base:

$$H_2SO_4 + H_2PO_4^- \rightleftharpoons HSO_4^- + H_3PO_4$$

In Example 15.1(b) it acted as an acid:

$$H_2PO_4^- + OH^- \rightleftharpoons HPO_4^{2-} + H_2O$$

A large number of molecules and ions exhibit this kind of behavior, acting as an acid in some reactions and as a base in others. Substances that can act as either Brønsted-Lowry acids or bases are said to be *amphiprotic*.

But we still have not answered the question "How do you know which of the reactants is the acid if you aren't given the equation?" The answer depends on the *strength* of the species as acids and bases. Strong acids lose their protons more easily than weak acids do, and strong bases attract protons more strongly than weak bases do. Given any pair of substances, the one that has the greater acid strength acts as the acid, and the other one acts as a base.

How can you tell which of two substances is the stronger acid? Here are some simple guidelines, which assume that you have no access to the lab facilities you would need to answer the question by experimenting. We will develop this topic further as we go along (Section 15.4).

- The more positive the charge of a species, the stronger its acidity. For example, $H_3O^+$ is a stronger acid than $H_2O$, which is a stronger acid than $OH^-$.
- The reverse works, too: The more negative the charge, the more basic the species. For example, $O^{2-}$ is a stronger base than $OH^-$, which is a stronger base than $H_2O$, and so on.
- Metal hydroxides, such as NaOH, $Fe(OH)_3$, etc., contain $OH^-$ and act as bases unless they are in the presence of a stronger base.
- An anion with no hydrogen will always act as a base.

### EXAMPLE 15.3 Predicting the Products of Acid-Base Reactions

Fill in the products and write the balanced net ionic equation for each of these reactions.

**a.** $H_2PO_4^- + OH^- \longrightarrow$
**b.** $H_3PO_4 + PO_4^{3-} \longrightarrow$
**c.** $HCl(aq) + NaOH(aq) \longrightarrow$

**SOLUTIONS**

**a.** Here the ions are given. Which reactant is the acid and which is the base? Well, you can predict that $OH^-$ is the base. One product is the conjugate acid of this base ($H_2O$), and the other product is the conjugate base of the acid (remove $H^+$ from $H_2PO_4^-$ to get $HPO_4^{2-}$). This gives you the reaction:

$$H_2PO_4^- + OH^- \longrightarrow H_2O + HPO_4^{2-}$$

**b.** Once again the ions are given. The acid is $H_3PO_4$, which you know either because you know its name (phosphoric acid) or because it's neutral while the other reactant has a $-3$ charge. The conjugate base of the acid $H_3PO_4$ is $H_2PO_4^-$, and the conjugate acid of the base $PO_4^{3-}$ is $HPO_4^{2-}$.

$$H_3PO_4 + PO_4^{3-} \longrightarrow H_2PO_4^- + HPO_4^{2-}$$

**c.** Here you must remember some chemistry you learned earlier. First, the species are given as being in aqueous solution, which means that they probably aren't what they seem (we introduced this topic in Chapter 8). $HCl(aq)$ is $H_3O^+(aq)$ and $Cl^-(aq)$, and $NaOH(aq)$ is $Na^+(aq)$ and $OH^-(aq)$. Thus the balanced net ionic equation is

$$H_3O^+(aq) + OH^-(aq) \longrightarrow 2\ H_2O(l)$$

**EXERCISE 15.3**

Fill in the products and write the balanced net ionic equation for each of these reactions.

**a.** $HBr(g) + H_2O(l) \longrightarrow$
**b.** $HClO_4(aq) + KCN(aq) \longrightarrow$
**c.** $CH_3OH_2^+ + OH^- \longrightarrow$

## 15.4 The Relative Strengths of Acids and Bases

**GOAL:** To predict the relative strengths of binary acids, oxyacids, metal hydroxides, and ammonia.

Section 15.3 stated you can decide which of two reactants acts as the acid by comparing their acid strengths. In this section we look at some additional ways to predict the relative strengths of acids.

First, let's examine the behavior of two hydrogen halides when they dissolve in water:

$$HBr(aq) + H_2O(l) \rightleftharpoons H_3O^+(aq) + Br^-(aq)$$
$$HF(aq) + H_2O(l) \rightleftharpoons H_3O^+(aq) + F^-(aq)$$

## CHAPTER BOX 15.1 ACIDS, BASES, AND CAVES

**Most cave** formation begins with rain. As water condenses into droplets and begins to fall, carbon dioxide dissolves in it. Because of the resulting carbonic acid ($H_2CO_3$), natural rain is slightly acidic. The rain water hits the ground and soaks into the soil. As vegetation decomposes in the soil, it produces more carbon dioxide, which also dissolves in the groundwater. In cave regions, this acidic groundwater percolates downward to the limestone below. There it finds cracks (see Figure A), where it reacts with the basic limestone (calcium carbonate).

The series of reactions that dissolves the limestone begins with carbon dioxide dissolving in water:

$$CO_2(g) \rightleftharpoons CO_2(aq)$$

The dissolved $CO_2$ reacts with water to give carbonic acid ($H_2CO_3$) which reacts slightly with water to give hydronium ion ($H_3O^+$) and bicarbonate (hydrogen carbonate) ion:

$$CO_2(aq) + H_2O(l) \rightleftharpoons H_2CO_3(aq)$$

$$H_2CO_3(aq) + H_2O(l) \rightleftharpoons H_3O^+(aq) + HCO_3^-(aq)$$

Finally, the $H_3O^+$ reacts with the calcium carbonate of the limestone and dissolves it. Calcium bicarbonate is more soluble in water than calcium carbonate is.

$$H_3O^+(aq) + CaCO_3(s) \rightleftharpoons Ca^{2+}(aq) + HCO_3^-(aq) + H_2O(l)$$

This is a slow process, but over tens and hundreds of thousands of years, huge caves can form.

Once the water that forms a cave drains away, other processes can begin. Air enters the cave, of course; most limestone caves are well-ventilated. This can trigger a series of reactions in the water that may drip from the roof of a cave. Sometimes this water becomes acidic in the soil far above, and then dissolves substantial quantities of calcium carbonate in narrow cracks in the limestone above the cave. Once it enters a cave passage with air low in carbon dioxide, the water begins to give off that gas.

Le Châtelier's Principle (see Chapter 14) does the rest. The reduced $CO_2$ concentration in the water causes $H_2CO_3$ to react, giving $CO_2$. The $H_2CO_3$ concentration decreases, and this causes $H_3O^+$ to react with $HCO_3^-$ to restore the equilibrium with $H_2CO_3$. As the $H_3O^+$ concentration decreases, $HCO_3^-$ begins to react with $Ca^{2+}$ and $H_2O$ to form $H_3O^+$. This last reaction also gives a precipitate of $CaCO_3$. This whole process is the reverse of the one that created the cave in the first place, the reverse of the equations above. The results of this prosaic-looking series of reactions can be spectacular (see Figure B).

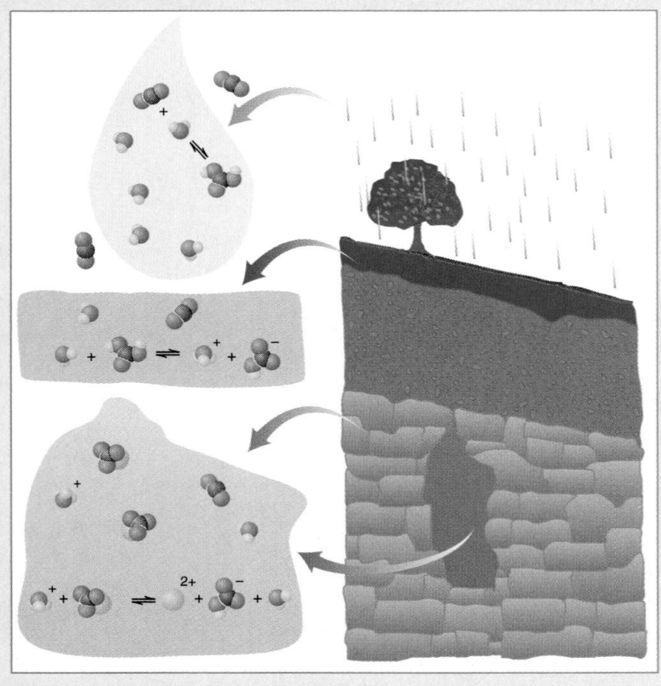

FIGURE A
Rain falls, percolates through the soil, and becomes acidic by dissolving carbon dioxide. The acidic water runs down through cracks in the limestone and reacts with the basic calcium carbonate of the limestone, making the cracks larger. Very slowly, caves form.

FIGURE B
These formations in Beth-shemesh Cave near Jerusalem were created by the process described in this Chapter Box.

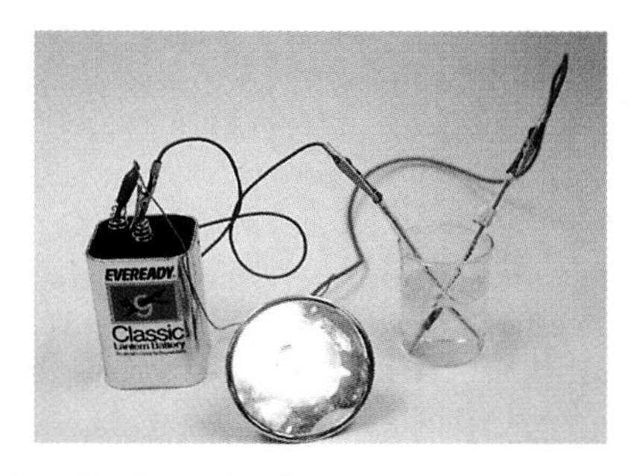


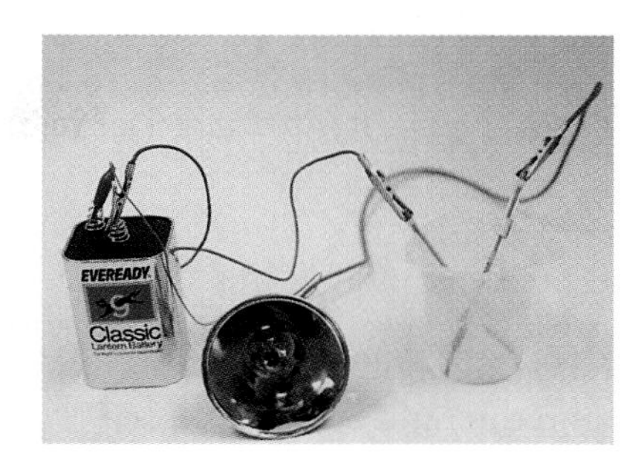


FIGURE 15.4
Conductivity of strong and weak acids. (A) An aqueous solution of HBr conducts electricity well, indicating that the solution contains many ions. (B) An aqueous solution of HF is a poor electrical conductor, indicating that there are few ions in solution. A plastic beaker is used to hold the HF solution because hydrofluoric acid is one of the few compounds that reacts with glass.

If you used the electrical conductivity apparatus described in Section 13.6, you'd discover that hydrogen bromide is a strong electrolyte; it reacts completely to form ions in aqueous solution. Hydrogen fluoride is a weak electrolyte; most of it remains in solution as undissociated HF(*aq*) molecules (see Figure 15.4). To indicate the position of the equilibrium for such reactions, chemists use arrows of unequal length:

**In a 1.0 *M* solution of HF(*aq*), about 97.4% of the HF molecules remain undissociated.**

$$HBr(aq) + H_2O(l) \rightleftharpoons H_3O^+(aq) + Br^-(aq)$$

$$HF(aq) + H_2O(l) \rightleftharpoons H_3O^+(aq) + F^-(aq)$$

Both of these reactions (and all Brønsted-Lowry acid-base reactions) can be represented like this:

$$\text{Stronger acid} + \text{Stronger base} \rightleftharpoons \text{Weaker acid} + \text{Weaker base}$$

The equilibrium position always lies toward the side of the weaker acid and the weaker base. From their electrical conductivities, you can conclude that HBr is a stronger acid than $H_3O^+$ and that HF is a weaker acid than $H_3O^+$. These facts let us begin a list of relative acid strengths:

$$HBr > H_3O^+ > HF$$

Imagine going into the laboratory and measuring the conductivity, or the equilibrium concentrations, for every one of the millions of possible acid-base reactions. It probably wouldn't be possible for anyone to accomplish this, even in a lifetime of work. Chemists needed ways to predict which of two possible acids in a reaction is the stronger.

Two classes of compounds are easy to work with—strong acids and nonelectrolytes. Section 8.8 presented the short list of strong acids, those compounds that dissociate completely in aqueous solution. You should memorize this list. The other class consists of covalently bonded compounds that don't dissociate at all in water—the nonelectrolytes. For example, the electrical conductivity of a solution of ethanol ($CH_3CH_2OH$) in water is slightly less than that of pure water. Ethanol is neither an acid nor a base in aqueous solution.

**You should know these strong acids in aqueous solution:**

**Perchloric Acid, $HClO_4$**
**Sulfuric Acid, $H_2SO_4$**
**Nitric Acid, $HNO_3$**
**Hydrochloric Acid, HCl**
**Hydrobromic Acid, HBr**
**Hydroiodic Acid, HI**

### EXAMPLE 15.4 Reactions of Acids with Water

Which of these compounds reacts most completely with water?
A. $CH_3CH_2OH$ B. $CH_3COOH$ C. $HNO_3$ D. HF

**SOLUTION**

The compound in this set that will react most completely with water is the strongest acid. Compounds B and D are weak acids, and A (ethanol) is a water-soluble nonelectrolyte. Compound C, nitric acid, is a strong acid and the right answer.

**EXERCISE 15.4**

Which of the following compounds reacts *least* with water?
A. $HClO_4$ B. $H_2SO_4$ C. HCl D. HF

### Acidity of Binary Hydrides

You can use the periodic table to predict the relative acid strengths of the **binary hydrides**—compounds made up of hydrogen and one other element. The general rule of thumb is that acid strength increases as you go across a row from left to right and increases as you go down in a group (see Figure 15.5). The strongest acids among the binary hydrides are those containing one of the elements to the right and toward the bottom of the periodic table.

Strength of binary acids

FIGURE 15.5

### EXAMPLE 15.5 Determining Relative Acidity of Binary Hydrides from the Periodic Table

Which of the following binary hydrides is the strongest acid?
A. $H_2O$ B. $H_2S$ C. $NH_3$ D. $PH_3$

**SOLUTION**

The strongest acid is the one whose element other than hydrogen is farthest down and to the right in the periodic table. The strongest acid in this set is $H_2S$. Sulfur is below oxygen, to the right of phosphorus, and both below and to the right of nitrogen in the periodic table. (Note: $H_2S$ is the strongest of these acids, but is only a weak acid in aqueous solution.)

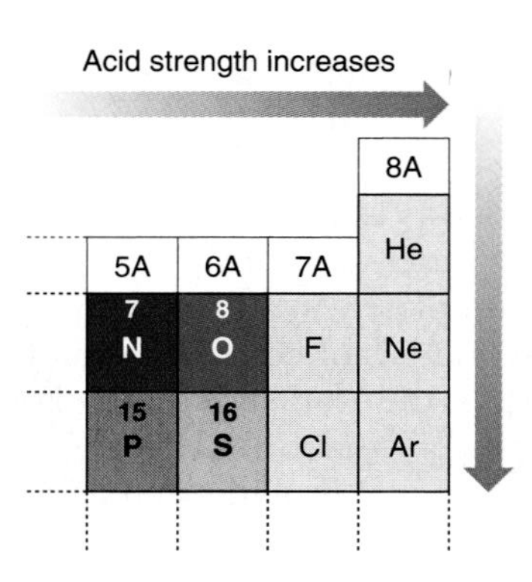

EXERCISE 15.5

Which of the following binary hydrides is the weakest acid?

A. HCl  B. HBr  C. $H_2S$  D. $H_2Se$

### Acidity of Oxyacids

Another method is useful for predicting the relative acidities of the oxyacids, acids containing one or more acidic OH groups bonded to a central atom. (Section 7.6 covers their nomenclature.) These acids have the general formula $O_xE(OH)_y$ where E is the central atom. Three of the strong acids you know, $HClO_4$, $H_2SO_4$, and $HNO_3$, are oxyacids. Note that their formulas can be rewritten as $O_3ClOH$, $O_2S(OH)_2$, and $O_2NOH$ to fit the general formula for an oxyacid. Here are Lewis diagrams of these acids:

```
      :O:                        :O:                     ..
      ||                  ..     ||    ..               :O:
 ..        ..                                            |    ..
 O=Cl—O—H            H—O—S—O—H                O=N—O—H
 ..        ..             ..     ||    ..          ..      ..
      ||                         :O:
      :O:
Perchloric acid             Sulfuric acid              Nitric acid
```

There are many other oxyacids; some examples are boric acid, $B(OH)_3$, phosphoric acid, $OP(OH)_3$, and carbonic acid, $OC(OH)_2$. (Usually you will see the formulas of these acids written as $H_3BO_3$, $H_3PO_4$, and $H_2CO_3$.) You can roughly predict the relative strengths of oxyacids by looking at the numbers of hydrogen and oxygen atoms in their molecules. In general, the strongest oxyacids have the greatest excess of oxygen over hydrogen atoms.

EXAMPLE 15.6 **Predicting Relative Acidity of Oxyacids**

Which of these oxyacids, $H_2SO_4$ (sulfuric acid) or $H_2SO_3$ (sulfurous acid), is the stronger acid? Explain your answer.

SOLUTION

$H_2SO_4$ is a stronger acid than $H_2SO_3$. You can predict this from the fact that $H_2SO_4$ has two more oxygens than hydrogens $(4 - 2 = 2)$, while $H_2SO_3$ has only one more oxygen $(3 - 2 = 1)$.

EXERCISE 15.6

Which of these oxyacids, HClO or $HClO_4$, is the weaker acid? Explain your answer.

### Strengths of Bases

Now you know some guidelines for predicting the relative strengths of acids. What are the guidelines for bases? For the purposes of this chapter they are relatively simple. Strong bases include the Group IA metal hydroxides (LiOH, NaOH, KOH, RbOH and CsOH):

$$NaOH(s) \xrightarrow{H_2O} Na^+(aq) + OH^-(aq)$$

Most of the Group 2A metal hydroxides, for instance $Ca(OH)_2$, are slightly soluble or insoluble in water. They produce aqueous solutions that are weakly basic because the $OH^-$ (*aq*) concentrations are low.

Many compounds react with water to form $OH^-(aq)$ in solution and thus are basic. The majority of these are weak bases because they react only slightly. The most common and best-known of these compounds is $NH_3$. You can identify it as a weak base because it forms $OH^-(aq)$ in solution, but not to a great extent:

$$NH_3(g) + H_2O(l) \rightleftharpoons NH_4^+(aq) + OH^-(aq)$$

A 1.0 *M* solution of $NH_3(aq)$ at room temperature has concentrations of 0.0042 *M* for both $NH_4^+(aq)$ and $OH^-(aq)$ and 0.996 *M* for $NH_3(aq)$. The equilibrium constant for the reaction at 25°C is $1.78 \times 10^{-5}$. This small number tells you that the equilibrium lies far to the left.

The guidelines for predicting base strength are closely related to those for acid strength. The most important guideline for base strength is this: The stronger an acid is, the weaker its conjugate base is. For example, HCl is a very strong acid, and therefore $Cl^-$ is a very weak base. Methane ($CH_4$) is so weak an acid it doesn't deserve to be called one, which means that $CH_3^-$ is an incredibly strong base. It's so strong, in fact, that its reaction with water to give $OH^-$ is explosive.

There is a second guideline for bases that often comes in handy: A base with any strength at all must have a pair of nonbonding electrons in its molecular structure. A base must accept a proton ($H^+$) from an acid. Since the $H^+$ ion has no electrons at all, both of the electrons for a covalent bond must be supplied by the base.

## EXAMPLE 15.7 Predicting Relative Basicity

Which of the following ions is the strongest base?

A. $SH^-$ B. $OH^-$ C. $Cl^-$ D. $F^-$

**SOLUTION**

Since all these ions have negative charges, they all must have at least one nonbonded pair of electrons. The strongest base is the one with the weakest conjugate acid. In the periodic table, the strongest acid has its central atom toward the right and bottom, and the weakest acid is toward the left and top. $H_2O$ is the weakest of the conjugate acids, so $OH^-$ is the strongest base in this group.

**EXERCISE 15.7**

Which of the following ions is the weaker base? Explain your answer.

A. $ClO^-$ B. $ClO_3^-$

## 15.5 Salts

**GOAL:** To use the formulas of salts to tell whether they are acidic, basic, or neutral.

This book has already covered some of the properties and characteristics of salts. For example, Chapter 7 deals with how to name them, and in Chapters 7 and 13 you learned how to predict their solubility in water. In Chapter 8 you found out how to predict what salts will form from Arrhenius acid-base reactions. Chapter 12 considered the physical properties of salts, including the fact that these ionic compounds are brittle solids at room temperature. In other words, salts are interesting, important, and useful substances in themselves. They are important constituents of your body, dissolved in body fluids, for example (see Table 15.1).

**TABLE 15.1 Relative Amounts of Various Ions in Body Fluids**

| Ion | % of Total Ions Present | | |
|---|---|---|---|
| | Blood plasma | Interstitial fluid | Intracellular fluid |
| **Cations** | | | |
| $Na^+$ | 92% | 95% | 8% |
| $K^+$ | 3% | 2.5% | 77% |
| $Ca^{2+}$ | 3% | 2% | 1% |
| $Mg^{2+}$ | 2% | 0.5% | 14% |
| **Anions** | | | |
| $Cl^-$ | 68% | 73% | 1% |
| $HCO_3^-$ | 16% | 19% | 5% |
| From proteins | 10% | — | 32% |
| From organic acids | 4% | 5% | — |
| $HPO_4^{2-}$ | 1% | 2% | 52% |
| $SO_4^{2-}$ | 1% | 1% | 10% |

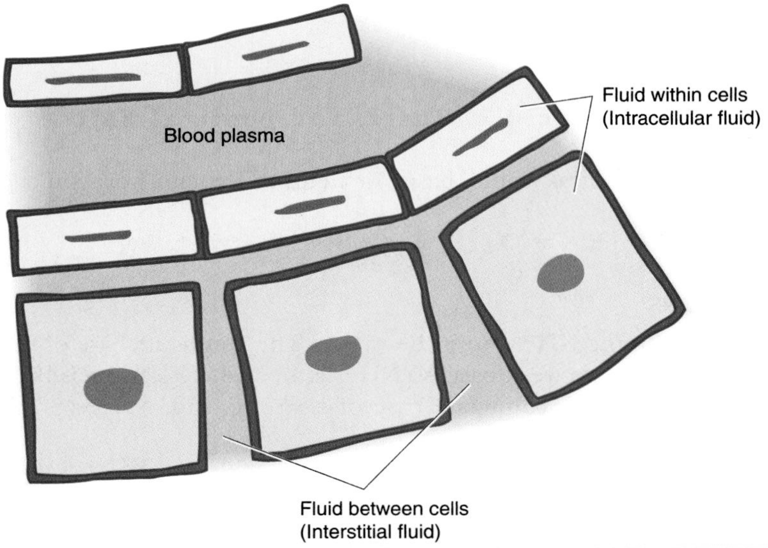

FIGURE 15.6
The word *salt* originally meant only NaCl, the substance used to flavor and preserve food (and it still does have that meaning, among others). The word *salt* comes from a Latin word, *sal,* that has the same meaning. The word *salary* comes from another Latin word, *salarium,* which meant money given to Roman soldiers to buy salt.

You have learned the Brønsted-Lowry definitions of acids and bases in this chapter. According to these definitions, not only the reactants but the products of many acid-base reactions can be salts that are acidic or basic. This section describes how to predict the acidity, basicity, or neutrality (that is, the lack of either acidity or basicity) of the salts you find among those products. (See Figure 15.6.)

The conclusion we will eventually come to is relatively simple. In the Arrhenius acid-base system, all reactions work the same way. Here is the general equation for such reactions and a familiar example:

$$\text{Acid} + \text{Base} \longrightarrow \text{Salt} + H_2O$$

$$HCl(aq) + NaOH(aq) \longrightarrow NaCl(aq) + H_2O(l)$$

When you eliminate the spectator ions to write the net ionic equation for this example (or for any other Arrhenius acid-base reaction), the salt disappears:

$$H^+ + OH^- \longrightarrow H_2O \qquad (\text{Better: } H_3O^+ + OH^- \longrightarrow 2\ H_2O)$$

Although the salt doesn't appear in the net ionic equation, it is what you would end up with if you evaporated the water after the reaction was finished. Before you can decide the acidity or basicity of such a salt, you must usually write the equation for the reaction with the complete formulas of the acid and the base.

**STUDY ALERT**
**Salts that can be prepared from a monoprotic (containing one acid proton) strong acid and a soluble metal hydroxide are always neutral. In this example NaCl is a neutral salt.**

Sodium acetate ($CH_3CO_2Na$) is an example. How can you make sodium acetate in aqueous solution by way of an acid-base reaction? Ordinarily, the cation of the salt (usually a metal ion, and sodium ion in this case) comes from its hydroxide. The anion (acetate ion in this case) comes from its conjugate acid (acetic acid in this case). This is the equation you get when you apply these guidelines to the example we're considering:

$$\underset{\text{Weak acid}}{CH_3CO_2H} + \underset{\text{Strong base}}{NaOH} \longrightarrow \underset{\text{Basic salt}}{CH_3CO_2Na} + \underset{\text{Water}}{H_2O}$$

Strong bases have relatively weak conjugate acids ($H_2O$), and weak acids have relatively strong conjugate bases ($CH_3CO_2^-$)—stronger than water, at least. You can therefore conclude that an aqueous solution of $CH_3CO_2Na$ is basic.

## EXAMPLE 15.8 Acidic, Basic, and Neutral Salts

Identify each salt as acidic, basic, or neutral. Explain how you arrived at the answers.

**a.** $NH_4Br$ **b.** $CaCO_3$ **c.** KI

**SOLUTIONS**

**a.** In $NH_4Br$, the $NH_4^+$ ion is the cation. The conjugate base of this ion is $NH_3$, which is a weak base; so $NH_4^+$ is an acid. The anion is $Br^-$, which is the very weak conjugate base of a strong acid. $NH_4Br$ is an acidic salt.

**b.** $CaCO_3$ contains the $Ca^{2+}$ ion. The hydroxide of $Ca^{2+}$ is $Ca(OH)_2$, a strong base; so $Ca^{2+}$ must be a very weak acid. $CaCO_3$ also contains $CO_3^{2-}$, the conjugate base of $HCO_3^-$, which in turn is the conjugate base of the weak acid $H_2CO_3$. $CaCO_3$ is a basic salt.

**c.** The anion $I^-$ is the conjugate base of the strong acid HI. The hydroxide KOH is a strong base. The conjugate bases of strong acids and the conjugate acids of strong bases are both very weak, so KI is a neutral salt.

**EXERCISE 15.8**

Identify each salt as acidic, basic, or neutral. Explain how you arrived at your answers.

**a.** $NaHCO_3$ **b.** $NH_4NO_3$ **c.** $(CH_3CO_2)_2Ba$

This simple method of predicting acidity or basicity of salts has wide application. Let's continue to look at sodium acetate ($CH_3CO_2Na$). The $Na^+$ ion contains no protons and is not acidic at all. The acetate ion ($CH_3CO_2^-$) is the conjugate base of acetic acid ($CH_3CO_2H$), which is a weak acid. Acetic acid reacts with water in this way:

$$CH_3CO_2H(aq) + H_2O(l) \rightleftharpoons CH_3CO_2^-(aq) + H_3O^+(aq)$$

The equilibrium constant is fairly small: $1.8 \times 10^{-5}$ at 25°C. This means that in a 1.0 *M* solution acetic acid is only about 0.4% dissociated.

Next let's consider a solution of sodium acetate in water, which contains $CH_3CO_2^-(aq)$ ions and $Na^+(aq)$ ions. A 1 *M* solution of sodium acetate is basic—the acetate ion reacts with water to form $OH^-(aq)$:

$$CH_3CO_2^-(aq) + H_2O(l) \rightleftharpoons CH_3CO_2H(aq) + OH^-(aq)$$

The equilibrium lies far to the left, as is usual with weak bases in water. The equilibrium constant is only $5.6 \times 10^{-10}$, in fact. In a 1 *M* solution of sodium acetate, the concentration of $CH_3CO_2^-$ is about 40,000 times larger than the concentrations of $OH^-$ and $CH_3CO_2H$.

The conjugate bases of all weak acids behave like this. They all yield basic solutions when their salts are dissolved in water. The conjugate bases of strong acids are so weakly basic that they will not react with water at all. For example, NaCl ($Cl^-$ is the conjugate base of HCl) gives neutral solutions when dissolved in water.

Conjugate acids of weak bases behave similarly. A solution of $NH_4Cl$ gives an acidic solution. The $NH_4^+$ ion is the conjugate acid of the weak base $NH_3$ and $Cl^-$ is the conjugate base of a strong acid. Dissolving $NH_4Cl$ in water gives this reaction:

$$NH_4^+(aq) + H_2O(l) \rightleftharpoons NH_3^+(aq) + H_3O^+(aq)$$

The resulting solution contains $H_3O^+$ ions and is therefore acidic.

## 15.6 Equilibria for the Reactions of Acids and Bases with Water

**GOAL:** To write chemical equations and equilibrium constant expressions that describe what happens when an acid or a base dissolves in water.

Let's review some of the important features of acids, bases, and salts we have covered up to now. Strong acids react essentially completely with water to produce a solution of $H_3O^+$ and the anion of the acid. It is impossible to detect unreacted acid molecules in solution. For example, in a 1.0 *M* solution of $HClO_4$ in water, the concentration is 1.0 *M* for both $H_3O^+(aq)$ and $ClO_4^-(aq)$. To use Brønsted-Lowry wording, $HClO_4$ is a stronger acid than $H_3O^+$ and water is a stronger base than $ClO_4^-$. The solution at equilibrium contains no detectable $HClO_4(aq)$; the only acid present is $H_3O^+(aq)$:

$$\begin{array}{c} HClO_4(l) + H_2O(l) \longrightarrow H_3O^+(aq) + ClO_4^-(aq) \\ \text{Stronger acid} + \text{Stronger base} \longrightarrow \text{Weaker acid} + \text{Weaker base} \end{array}$$

Reactions of weak acids with water are described by similar equations, except that the acids react much less strongly with water. Consider acetic acid. Once again its aqueous solution contains two acids, the unconverted $CH_3CO_2H(aq)$ and $H_3O^+(aq)$. The equilibrium lies to the left because $H_3O^+(aq)$ is a stronger acid than $CH_3CO_2H$. The equation can be written this way:

$$\begin{array}{c} CH_3CO_2H(aq) + H_2O(l) \rightleftharpoons H_3O^+(aq) + CH_3CO_2^-(aq) \\ \text{Weaker acid} + \text{Weaker base} \rightleftharpoons \text{Stronger acid} + \text{Stronger base} \end{array}$$

The fact that this equilibrium lies to the left is a defining characteristic of a weak acid.

You can construct equilibrium constant expressions for acid dissociations in water, just as you did for other reactions in Section 14.4. The equilibrium constant for the reaction of a weak acid with water is called an **acid dissociation constant** and is given the symbol $K_a$. For the reaction of a general weak acid HX with water, the equation is

$$HX(aq) + H_2O(l) \rightleftharpoons H_3O^+(aq) + X^-(aq)$$

**STUDY ALERT**

**For the general reaction $aA + bB \rightleftharpoons cC + dD$, the equilibrium constant expression is**

$$K_{eq} = \frac{[C]^c[D]^d}{[A]^a[B]^b} \quad \left(\text{see Sections 14.4 and 14.5}\right)$$

**When a reactant is the solvent, a pure liquid, or a solid, you omit its concentration from the equilibrium constant expression.**

The equilibrium constant expression for this reaction, with the label (*aq*) omitted as always, is

$$K_a = \frac{[H_3O^+][X^-]}{[HX]}$$

This is the specific expression for the acid dissociation constant for acetic acid in water:

$$K_a = \frac{[H_3O^+][CH_3CO_2^-]}{[CH_3CO_2H]}$$

For weak acids, $K_a < 1$ and the equilibrium lies to the left. The $K_a$ for acetic acid is only $1.8 \times 10^{-5}$, and in a 1.0 *M* solution acetic acid is only about 0.4% dissociated. For strong acids, $K_a > 1$ and the equilibrium lies to the right with essentially no molecular HX present. That's the only real difference between the two types of acids.

Note that for either strong or weak acids, the strongest acid that can persist in aqueous solution is the hydronium ion, $H_3O^+$. This is always true. Any acids stronger than $H_3O^+$ will react with $H_2O$ to form $H_3O^+$ and the anion of the acid. Look at the reaction of perchloric acid ($HClO_4$) with water, at the top of this page. Essentially no $HClO_4$ exists in solution at equilibrium; the only acid present is $H_3O^+$.

## EXAMPLE 15.9 Writing $K_a$ Expressions

Write the $K_a$ expression for an aqueous solution of each of these acids.
**a.** HF **b.** $NH_4Cl$ **c.** $H_2S$ (one reaction for each proton)

**SOLUTIONS**

**a.** First, you write the equation for the reaction of the acid with water. Then you can write the equilibrium constant expression in the usual way.

$$HF + H_2O \rightleftharpoons H_3O^+ + F^-$$

$$K_a = \frac{[H_3O^+][F^-]}{[HF]}$$

**b.** The process is the same as the one in part (a). You ignore the spectator ion $Cl^-$.

$$NH_4^+ + H_2O \rightleftharpoons H_3O^+ + NH_3$$

$$K_a = \frac{[H_3O^+][NH_3]}{[NH_4^+]}$$

**c.** Here are two reactions. Each must be considered separately.

$$H_2S + H_2O \rightleftharpoons H_3O^+ + HS^-$$
$$HS^- + H_2O \rightleftharpoons H_3O^+ + S^{2-}$$

Each of the reactions has its own $K_a$. These can be designated as $K_{a1}$ and $K_{a2}$, respectively. The two expressions are

$$K_{a1} = \frac{[H_3O^+][HS^-]}{[H_2S]}$$

$$K_{a2} = \frac{[H_3O^+][S^{2-}]}{[HS^-]}$$

**EXERCISE 15.9**

Write the $K_a$ expression for an aqueous solution of each of these acids.
**a.** HCN **b.** NaHS **c.** $H_2CO_3$ (one reaction for each proton)

The strong bases we need to consider consist of the Group 1A metal hydroxides and the Group 2A metal hydroxides that dissolve in water. The only base in these aqueous solutions is $OH^-(aq)$. For example, in aqueous solution NaOH is 100% dissociated; the equilibrium is completely to the right:

$$NaOH(s) \xrightarrow{H_2O} Na^+(aq) + OH^-(aq)$$

Weak bases react incompletely with water to form hydroxide ion, and the equilibrium lies toward the left. As we have seen, this is the case in aqueous solutions of $NH_3$ and acetate ion ($CH_3CO_2^-$), for example:

$$NH_3(g) + H_2O(l) \rightleftharpoons NH_4^+(aq) + OH^-(aq)$$
$$CH_3CO_2^-(aq) + H_2O(l) \rightleftharpoons CH_3CO_2H(aq) + OH^-(aq)$$

These equations (and the aqueous solutions they represent) contain two bases at equilibrium, just as equations for acids contain two acids. Aqueous solutions of ammonia contain unconverted $NH_3(aq)$ and $OH^-$. Aqueous solutions of acetate ion contain $CH_3CO_2^-$ and $OH^-$. Since the equilibrium lies to the left in both of these reactions, the stronger base in either solution is the $OH^-$ ion.

In fact, the strongest base that can exist in water is the hydroxide ion, $OH^-$. Bases stronger than the $OH^-$ ion (such as the hydride ion, $H^-$, or the oxide ion, $O^{2-}$) react with water to form $OH^-$. The ionic compound $Na_2O$ contains the strong base $O^{2-}$, but when $Na_2O$ is placed in water, the $O^{2-}$ ion reacts to form $OH^-$.

**Even though weak bases do not *dissociate* in aqueous solution (they react with water), $K_b$ is called the base dissociation constant for historical reasons.**

The equilibrium constant for the reaction of a weak base with water is called the **base dissociation constant** and has the symbol $K_b$. For acetate ion the equation for the reaction is:

$$CH_3CO_2^- + H_2O \rightleftharpoons CH_3CO_2H + OH^-$$

and the $K_b$ expression is:

$$K_b = \frac{[CH_3CO_2H][OH^-]}{[CH_3CO_2^-]}$$

For the reactions of all weak bases with water, $K_b < 1$ and the equilibrium lies toward the left.

### EXAMPLE 15.10 Writing $K_b$ Expressions

Write the $K_b$ expression for an aqueous solution of each of these bases:

**a.** NaF **b.** $CH_3NH_2$ **c.** $Na_2CO_3$ (both reactions)

**SOLUTIONS**

**a.** As you did for an acid, you write the equation for the reaction of the base with water, and derive the equilibrium constant expression from that

$$F^- + H_2O \rightleftharpoons HF + OH^-$$

$$K_b = \frac{[HF][OH^-]}{[F^-]}$$

**b.** $CH_3NH_2 + H_2O \rightleftharpoons CH_3NH_3^+ + OH^-$

$$K_b = \frac{[CH_3NH_3^+][OH^-]}{[CH_3NH_2]}$$

**c.** There are two reactions, and you treat them separately, ignoring the spectator ions.

$$CO_3^{2-} + H_2O \rightleftharpoons HCO_3^- + OH^-$$
$$HCO_3^- + H_2O \rightleftharpoons H_2CO_3 + OH^-$$

Each of the reactions has its own $K_b$ expression. The constants are $K_{b1}$ for the first reaction and $K_{b2}$ for the second.

$$K_{b1} = \frac{[HCO_3^-][OH^-]}{[CO_3^{2-}]} \quad \text{and} \quad K_{b2} = \frac{[H_2CO_3][OH^-]}{[HCO_3^-]}$$

**EXERCISE 15.10**

Write the $K_b$ expression for an aqueous solution of each of these bases.
**a.** NaCN **b.** $NaHCO_3$ **c.** $Na_2S$ (both reactions)

## MEASURING ACIDS AND BASES

In the first part of this chapter we looked at acid-base reactions from the point of view of what happens when acids and bases are mixed. Simply dissolving an acid in water involves a chemical reaction between the acid and the water (which acts as a base). Knowing generally what happens in such aqueous solutions is only part of the story, however. In a wide range of situations, scientists need to know the concentrations of ions in acidic or basic solutions, especially the concentration of $H_3O^+(aq)$. For example, a medical researcher might need to know how much the $H_3O^+(aq)$ concentrations in patients' blood samples change when they are given a particular medication. Another scientist might need to know what concentration of $H_3O^+(aq)$ in acid rain will cause irreversible harm to forests, rivers, or lakes.

In order to answer these and a host of other questions, chemists have developed methods for measuring the concentrations of ions in aqueous solution. For example, they use electronic instruments called *pH meters* to measure concentrations of $H_3O^+$ directly (see Figure 15.7). An older experimental technique called *titration* is used to measure the total amount of acid or base present in a solution. If less exact results are sufficient, chemists can estimate concentrations by doing calculations using the $K_a$ and $K_b$ expressions.

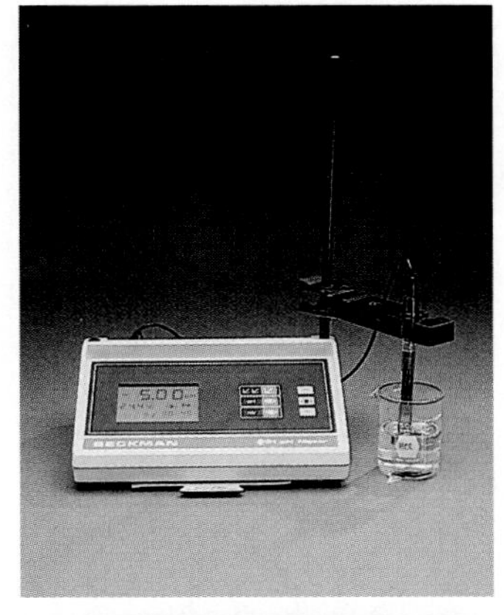


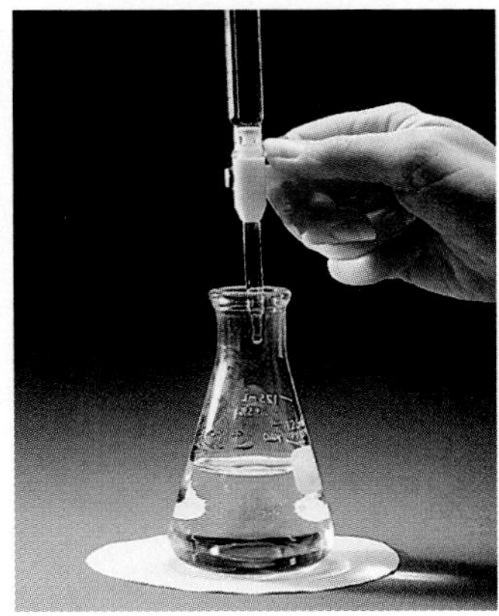

FIGURE 15.7
Chemists use pH meters (top) to measure the concentration of $H_3O^+$ ions in aqueous solutions. Titration techniques (bottom) are used to measure the total amount of acid or base present in a sample.

## 15.7 Hydronium and Hydroxide Ion Concentrations

**GOAL:** To calculate the concentration of $H_3O^+$ ion given the concentration of $OH^-$ ion, and vice versa.

We have often looked at the acid-base reaction for the formation of water:

$$H_3O^+(aq) + OH^-(aq) \rightleftharpoons 2\,H_2O(l)$$

We first saw it as the net ionic equation defining any Arrhenius acid-base reaction (Section 8.8). Now we are interested in the reverse of this reaction, which is:

$$2\,H_2O(l) \rightleftharpoons H_3O^+(aq) + OH^-(aq)$$

As usual, the unequal lengths of the double arrows show that the acids and bases on one side of the equation are stronger than those on the other. The reason we need to look at this equation is that this reaction must occur in pure water. In other words, pure water must contain some $H_3O^+$ and some $OH^-$.

After the reaction reaches equilibrium, there is much more $H_2O$ than $H_3O^+$ and $OH^-$. This should not surprise you, for two reasons. First, from an experimental point of view, pure water does not conduct electricity well at all, so it cannot contain a high concentration of ions. Second, according to the guidelines for predicting acid strength, $H_3O^+$ has a larger positive charge than $H_2O$ and must therefore be the stronger acid. For this reason, the equilibrium must lie to the left. Both of these reasons lead to the correct answer. The concentrations of $H_3O^+$ and $OH^-$ in pure water are definitely small, but exactly what are they?

The answer is that they are *very* small. At equilibrium at 25°C, the concentrations of $H_3O^+$ and $OH^-$ are equal to 0.000000100 *M*, or $1.00 \times 10^{-7}$ *M*. This allows us to calculate a constant for the reaction:

$$2\,H_2O(l) \rightleftharpoons H_3O^+(aq) + OH^-(aq)$$

$$K_w = [H_3O^+][OH^-] = (1.00 \times 10^{-7}\text{ mol L}^{-1})(1.00 \times 10^{-7}\text{ mol L}^{-1})$$
$$K_w = 1.00 \times 10^{-14}$$

This equilibrium constant has a special name and symbol. It is called the **ionization constant for water,** is symbolized by $K_w$, and is conventionally written without units:

$$K_w = 1.00 \times 10^{-14} = [H_3O^+][OH^-]$$

This is one constant whose value you need to memorize. With that value you can calculate the concentration of either $H_3O^+$ or $OH^-$ in any dilute aqueous solution if you know the concentration of the other. Since $K_w$ is an equilibrium constant, the product of $[H_3O^+]$ and $[OH^-]$ at equilibrium must always be equal to $1.00 \times 10^{-14}$.

Let's look at a couple of important points about the $K_w$ expression. First, a **neutral solution** is one in which $[H_3O^+]$ equals $[OH^-]$. In this case the $K_w$ expression is:

$$K_w = 1.00 \times 10^{-14} = [H_3O^+][OH^-]$$
$$= [H_3O^+][H_3O^+]$$
$$[H_3O^+]^2 = 1.00 \times 10^{-14}$$
$$\mathbf{[H_3O^+] = 1.00 \times 10^{-7}\ \mathit{M} = [OH^-]} \quad \textbf{in all neutral solutions}$$

From this result we can conclude that an **acidic solution** is one in which the $H_3O^+$ concentration is greater than the $OH^-$ concentration (and also greater than $1.00 \times 10^{-7}$ *M*). Similarly, a **basic solution** is one in which the $OH^-$ concentration is greater than the $H_3O^+$ concentration (and also greater than $1.00 \times 10^{-7}$ *M*).

### EXAMPLE 15.11 Concentrations of $H_3O^+$ and $OH^-$ Ions

**a.** The concentration of $H_3O^+$ in a certain aqueous solution is $5.0 \times 10^{-3}$ *M*. What is the concentration of $OH^-$? Is this solution acidic or basic?

**b.** The concentration of $OH^-$ in an aqueous solution is $2.0 \times 10^{-9}$ *M*. What is the concentration of $H_3O^+$? Is this solution acidic or basic?

**SOLUTIONS**

**a.** The equation to use is $[H_3O^+][OH^-] = 1.00 \times 10^{-14}$. With the given $H_3O^+$ concentration put in, this becomes

$$(5.0 \times 10^{-3})[OH^-] = 1.00 \times 10^{-14}$$

$$[OH^-] = \frac{1.00 \times 10^{-14}}{5.0 \times 10^{-3}} = 0.20 \times 10^{-11}\ M$$

The $OH^-$ concentration is therefore $0.20 \times 10^{-11}$ *M*, or, in correct scientific notation, $2.0 \times 10^{-12}$ *M*. The given value for $[H_3O^+]$ is greater than $1 \times 10^{-7}$ *M*, so the solution is acidic.

**b.** The equation to use is, once again, $[H_3O^+][OH^-] = 1.00 \times 10^{-14}$. Substituting in the given $OH^-$ concentration gives

$$[H_3O^+](2.0 \times 10^{-9}) = 1.00 \times 10^{-14}$$

$$[H_3O^+] = \frac{1.00 \times 10^{-14}}{2.0 \times 10^{-9}} = 0.50 \times 10^{-6}\ M$$

The $H_3O^+$ concentration is therefore $0.50 \times 10^{-5}$ *M*, or, in correct scientific notation, $5.0 \times 10^{-6}$ *M*. Since $[H_3O^+]$ is higher than $1 \times 10^{-7}$ *M*, this solution is acidic.

**EXERCISE 15.11**

**a.** The concentration of $H_3O^+$ in a certain aqueous solution is $5.0 \times 10^{-7}$ *M*. What is the concentration of $OH^-$? Which is this solution, acidic or basic?

**b.** The concentration of $OH^-$ in an aqueous solution is $2.0 \times 10^{-1}$ *M*. What is the concentration of $H_3O^+$? Which is this solution, acidic or basic?

## TOOL BOX: LOGARITHMS

If you can handle scientific notation and memorize some definitions, you can use logarithms. Since you will be using logarithms in the next section, we are reviewing this topic now.

Logarithms are exponents. The logarithms (called *logs* for short) that you will use in this course are exponents of the base 10. That is, the log of $10^{23}$ is 23, and the log of $10^{-7}$ is $-7$. In general, $\log 10^x = x$. Your next question might be "What do I do about a number like $6.02 \times 10^{23}$ or $2 \times 10^{-7}$?" The answer in theoretical terms is, you convert the whole thing into 10 to an exponent.

TOOLBOX

FIGURE 15.8
To find the logarithm of a number with most scientific calculators, enter the number and hit the "log" key.

The main point of developing decimal exponents in the first place was to use them as logarithms, because this made pen-and-paper calculations easier to do. In practical terms, you will use a scientific calculator to work with logarithms. (You might want to get out your calculator and its instruction book now—go ahead; we'll wait.) We will round off logs to four decimal places, although a calculator will show more. (See Figure 15.8.)

Let's find the log of $6.02 \times 10^{23}$. First, the log of 6.02 is 0.7796. You can get this answer with most calculators by punching in 6.02 and hitting the log button. (Some calculators work a little differently.) Remember, this result means that $6.02 = 10^{0.800}$, which is reasonable, since the log of 1 is zero ($10^0 = 1$) and the log of 10 is 1 ($10^1 = 10$). The log of 6.02 must be between the values 0 and 1. The log of $10^{23}$ is, of course, 23.

To multiply two numbers you add their exponents; for example, $10^3 \times 10^8 = 10^{11}$. To get the log of $6.02 \times 10^{23}$, you add the log of 6.02 and the log of $10^{23}$. In other words, $10^{0.800} \times 10^{23} = 10^{23.800}$. This means that 23.800 is the log of $6.02 \times 10^{23}$. If you can enter numbers (such as $6.02 \times 10^{23}$) in exponential notation into your calculator, you can get this answer directly.

What about the log of $2 \times 10^{-7}$? The log of 2 is 0.3 ($2 = 10^{0.3}$), and the log of $10^{-7}$ is $-7$. The log of $2 \times 10^{-7}$ is therefore $-6.6990$ (because $10^{0.3} \times 10^{-7} = 10^{-6.7}$). Don't be surprised to see that the log of a number that has a negative exponent doesn't begin with that exponent.

Sometimes you will need to convert logarithms back to the original numbers (these are called *antilogs*). Suppose that the logarithm of a certain number is $-2.3$. What is that number? To answer this question, you need to find the value of $10^{-2.3}$. To do this with most calculators, you enter 2.3 and change the sign so that $-2.3$ shows on the calculator display. Then you push the $10^x$ button. Your calculator will then display the value 0.0050, which you can easily translate to scientific notation as $5.0 \times 10^{-3}$.

Finally, what do you do about significant figures when you use logarithms? The number of significant figures in the answer should be the same as the number of significant figures after the decimal point of the number. You will note that we have been following this rule above.

### TOOL BOX EXAMPLE 1 **Working with Logarithms**

**a.** What is the logarithm of each of these numbers?
(1) 3.0 (2) $3.0 \times 10^{21}$ (3) $3.0 \times 10^{-9}$

**b.** What is the antilog of each of these logarithms?
(1) 0.60 (2) 7.60 (3) $-3.60$

#### SOLUTIONS

**a.** (1) 0.48 (2) 21.48 (3) $-8.53$

**b.** (1) 4.0 (2) $4.0 \times 10^7$ (3) $2.5 \times 10^{-4}$

### TOOL BOX EXERCISE 1

**a.** What is the logarithm of each of these numbers?
(1) 2.718 (2) $5 \times 10^{-21}$ (3) $7.2 \times 10^9$

**b.** What is the antilog of each of these logarithms?
(1) 2.57 (2) 7.57 (3) $-3.43$

## 15.8 A Measure of Acidity or Basicity: pH

**GOAL:** To estimate relative acidity or basicity from a given pH, and to calculate the pH of a solution from either $[H_3O^+]$ or $[OH^-]$ and vice versa.

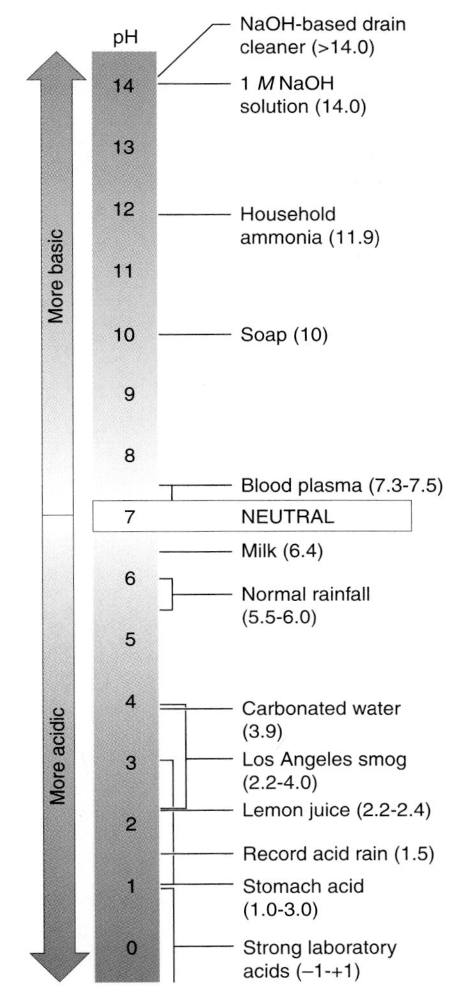


FIGURE 15.9
The pH values for some common solutions. Remember: If the pH is less than 7, the solution is acidic. If the pH equals 7, it's neutral. If the pH is over 7, it's basic.

The $H_3O^+$ concentrations of aqueous solutions are useful to many people. Scientists studying such diverse topics as industrial processes, pollution, acid rain, and blood and body chemistry need this kind of information. It's so useful and important, in fact, that many people who are not scientists also know and use it from time to time.

Since you have almost certainly seen acidities expressed on the pH scale, we will start by trying to convey a feel for what such numbers mean. A neutral solution, one that is neither acidic nor basic, has a pH of 7. Lower pH values are acidic, and higher pH values are basic.

Figure 15.9 shows some common pH values. You should become familiar with these. Later they will help you decide whether an answer you have calculated is reasonable. If you calculate a pH of 12 for a strongly acidic solution, for instance, you'll know that you have to go back and try again.

### EXAMPLE 15.12 Determining Basicity or Acidity from pH

Which of these is the pH of a weakly basic solution?
A. 1.5 B. 4.5 C. 9.5 D. 13.5

**SOLUTION**

Basic solutions have pH values above 7. The closer the pH is to 7, the weaker the base is. The pH of 9.5 (C) is that of a weakly basic solution. (The pH of 13.5 is for a strongly basic solution.)

**EXERCISE 15.12**

Which of these is the pH of a *strongly* acidic solution?
A. 1.0 B. 6.5 C. 7.5 D. 13.0

Now that you've learned what the numbers on the pH scale mean, the time has come to try some calculations. Scientists of a century ago often had difficulty in presenting data on changes in $[H_3O^+]$, even to others in the scientific community. It is not unusual for the concentration of $H_3O^+$ to increase or decrease by a factor of 1 million or more during the course of some process. Suppose, for instance, that you neutralized a 1 *M* HCl(*aq*) solution by adding NaOH(*aq*). The $H_3O^+$ concentration is 1.0 *M* to start and ends up at $1.0 \times 10^{-7}$ *M*. In other words, $[H_3O^+]$ decreases by a factor of 10 million in this simple experiment. Scientists once found concentration changes this large to be clumsy and inconvenient to work with.

The difficulty clearly lies in the fact that concentration changes tend not to occur on a convenient linear scale, but exponentially—that is, concentrations change by many powers of 10. However, after reading the preceding Tool Box, you know that scientists can express very small and very large numbers by using logarithms. Scientists developed the pH scale to report $H_3O^+$ concentrations logarithmically.

To work with pH, let's begin by looking at its precise scientific definition:

$$\mathbf{pH = -\log[H_3O^+]}$$

As usual, $[H_3O^+]$ represents the concentration of $H_3O^+$ in moles per liter. For example, suppose that $[H_3O^+]$ is $1.0 \times 10^{-7}$ *M*. First, we take the logarithm of the concentration and then change the sign of the logarithm:

$$\begin{aligned} pH &= -\log[H_3O^+] = -\log(1.0 \times 10^{-7}) = -(\log 1.0 + \log 10^{-7}) \\ &= -(0 - 7) = 7.00 \end{aligned}$$

The pH of a neutral solution, a solution in which $[H_3O^+] = [OH^-] = 1.0 \times 10^{-7}$, is 7.00. Each change of 1 on the pH scale represents a change of a factor of 10 in $[H_3O^+]$.

EXAMPLE 15.13 **Calculating the pH from $[H_3O^+]$**

A chemist found that a sample of orange juice had an $H_3O^+$ concentration of $5.1 \times 10^{-3}$ *M*. What is the pH of this juice?

SOLUTION

The equation is $pH = -\log[H_3O^+]$. Therefore,

$$pH = -\log(5.1 \times 10^{-3}) = 2.29$$

EXERCISE 15.13

What is the pH of a solution for which $[H_3O^+] = 4.0 \times 10^{-11}$ *M*?

Now you can see where the pH values in Figure 15.9 come from. It's important to keep in mind that, although the pH scale is a direct measure of the $H_3O^+$ concentration, the *lower* the pH value, the *higher* the concentration of $H_3O^+$. If $[H_3O^+]$ is $2 \times 10^{-2}$ *M*, the pH is 1.7. Since $[H_3O^+]$ is much higher than $[OH^-]$ in this case, the solution is acidic. If $[H_3O^+]$ drops to $2 \times 10^{-12}$, then the pH is 11.7. In this case $[H_3O^+] < 1 \times 10^{-7}$, so the solution is basic.

Converting a reported pH value back to the $H_3O^+$ concentration is also relatively simple. If the pH is 3.2, for example, the value of $\log[H_3O^+]$ is $-3.2$, and the concentration of $H_3O^+$ is the antilog of $-3.2$, or $6 \times 10^{-4}$ *M*:

$$pH = 3.2 = -\log[H_3O^+]$$

Therefore

$$\log[H_3O^+] = -3.2$$

and

$$[H_3O^+] = 10^{-3.2} = 6 \times 10^{-4}\ M \qquad \text{(Don't forget the units.)}$$

**Some calculators have a $\log^{-1}$ key, and others have a $10^x$ key. To solve problems like this, you'll need to know how to use the appropriate function key on your calculator. If all else fails, you might try reading the directions.**

### EXAMPLE 15.14 Calculating $[H_3O^+]$ from the pH

A detergent labeled "mild" has a pH of 7.80. What is the concentration of $H_3O^+$ in this detergent? Do you think that the detergent deserves the label "mild"?

**SOLUTION**

If the pH is 7.80, the $H_3O^+$ concentration is

$$[H_3O^+] = \text{antilog}(-\text{pH}) = 1.6 \times 10^{-8}\ M$$

This detergent has a pH closer to neutral than most soaps do, so it does deserve to be called "mild."

**EXERCISE 15.14**

A drain cleaner has a pH of $-0.2$. What is the concentration of $H_3O^+$ in this cleaner? Most drain cleaners use NaOH as their active ingredient. Is this one of them?

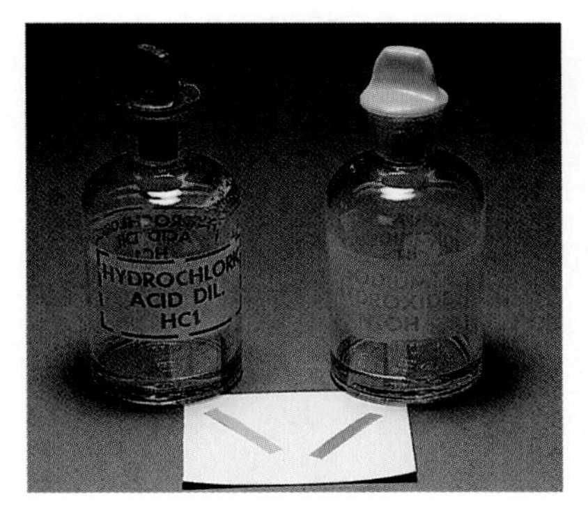


FIGURE 15.10
To simply find out whether a solution is acidic or basic, chemists still use litmus paper. Acids turn litmus paper red (as at left above), and bases turn litmus paper blue (as at right above), leading to the mnemonic "Acids are red and bases are blue." Pure water and solutions that are at or close to neutrality don't change the color of litmus paper at all.

## 15.9 Experimental Determination of the pH of a Solution

**GOAL:** To show how chemists determine the pH of a solution experimentally by using indicators and pH meters.

There are a variety of ways to find the pH of a sample experimentally, and you are probably familiar with at least some of them. You may have used a simple kit to test the water in a fish tank or the soil in a garden for acidity or basicity.

The oldest method of measuring the pH is through the use of **indicators,** which are substances whose color changes depending on the pH. Many naturally occurring substances can be used as indicators. If you extract the color from a red rose with alcohol, the solution you get is an indicator that changes color in strongly basic solution. You can dry strips of purple cabbage and use them as indicators. The most famous indicator is **litmus.** Isolated from certain lichens, this compound turns pink in the presence of acid and blue in the presence of base (see Figure 15.10).

**Chemists use litmus paper to answer a simple question: Is this solution acidic or basic? A "litmus test" in politics also answers a simple yes-or-no question.**

Compounds that act as acid-base indicators have two defining characteristics. First, acid-base indicators are almost always acids themselves (at least at low pH values). Obviously, an indicator needs to be sensitive to pH changes, and what species are more sensitive to changes in $H_3O^+$ concentration than acids and their conjugate bases? Second, indicators and their conjugate bases must have distinctly different colors. Indicators have one color at low pH and a different color at high pH.

Let's see exactly how an indicator works. For a general indicator we will use the abbreviation HIn (the H in HIn stands for the proton that the indicator donates). For litmus, the acid form of the indicator, HIn, is red and its conjugate base, $In^-$, is blue. If you put HCl(*aq*) solution on paper containing the acid form of litmus, the indicator stays protonated and the paper stays red. If you add $OH^-$ to the HCl solution until all of the acid (that is, all of the HCl and HIn combined) reacts, the following transformation occurs:

$$\underset{\text{Red}}{\text{HIn}} + \text{OH}^- \longrightarrow \text{H}_2\text{O} + \underset{\text{Blue}}{\text{In}^-}$$

In other words, the $OH^-$ removes a proton from the indicator to make its conjugate base. When the indicator is litmus, this reaction changes the color of the paper from red to blue.

Over the years, chemists have discovered indicators that change colors over many different pH ranges. Table 15.2 lists several of the most common of these indicators.

Chemists and many others use the color changes of indicators to get a rough idea of the acidity or basicity of a sample. To get a better idea of the pH of a sample, you can use a series of indicators. Example 15.15 demonstrates how it is done.

**TABLE 15.2 Indicators**

| Indicator | pH Range | Color at Lower pH | Color at Higher pH |
|---|---|---|---|
| Methyl orange | 3.1–4.4 | Red | Yellow |
| Bromocresol green | 4.0–5.6 | Yellow | Blue |
| Methyl red | 4.4–6.2 | Red | Yellow |
| Bromocresol purple | 5.2–6.8 | Yellow | Purple |
| Phenol red | 6.4–8.0 | Yellow | Red |
| Phenolphthalein | 8.0–10.0 | Colorless | Red |
| Thymolphthalein | 9.4–10.6 | Colorless | Blue |

### EXAMPLE 15.15 Using Indicators to Determine pH

What is the approximate pH of a solution that gives a yellow color when tested with methyl orange and a red color when tested with methyl red?

**SOLUTION**

You need to look at Table 15.2. Note that methyl orange changes from red to yellow in the pH range from 3.1 to 4.4, and methyl red changes from red to yellow in the pH range from 4.4 to 6.2. The methyl orange results tell you the pH of the solution is 4.4 or greater, and the methyl red results tell you the pH is 4.4 or less. Together the results suggest that the pH of the solution is around 4.4.

**EXERCISE 15.15**

What is the approximate pH of a solution that gives a purple color when tested with bromocresol purple and is colorless when tested with phenolphthalein?

Testing a solution with a variety of indicators is a clumsy and time-consuming process. Chemists use *pH paper* (see Figure 15.11) to measure pH faster (though the result is still approximate). This type of paper is made by combining a variety of indicators on a paper strip.

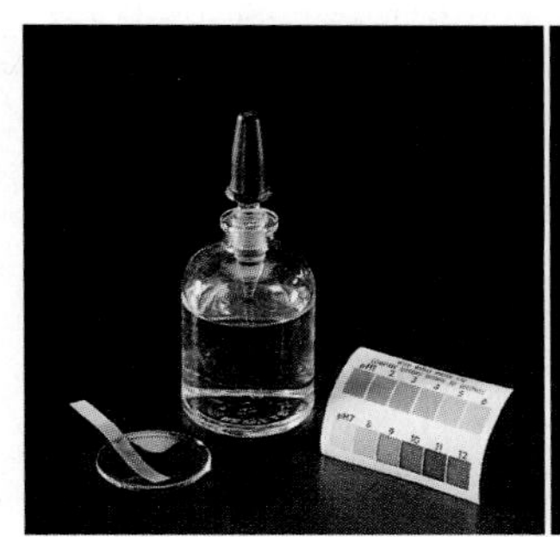
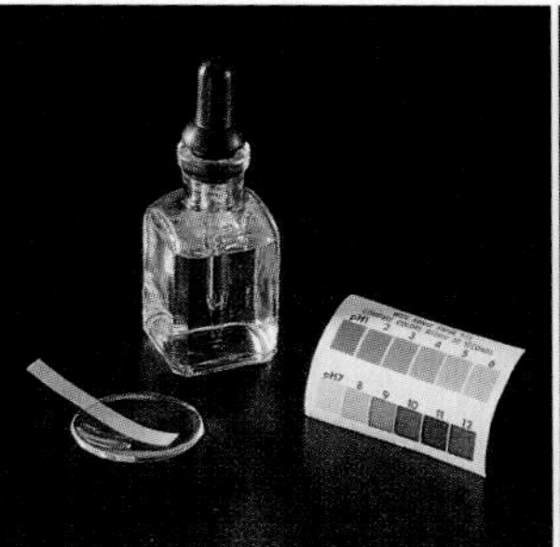
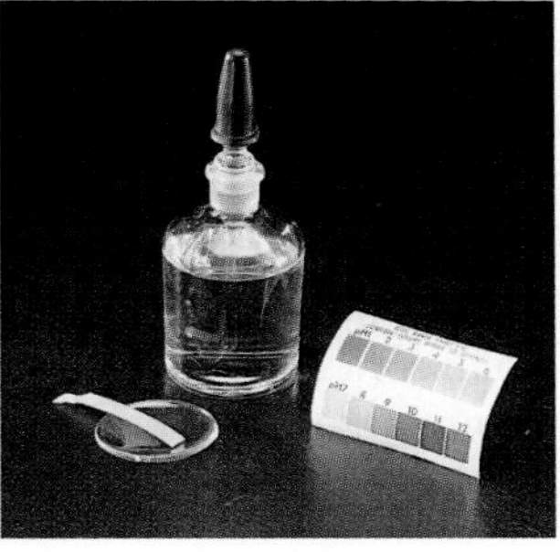

FIGURE 15.11
For somewhat more precise measurements of pH, chemists use pH paper. This paper has a range of colors covering a relatively broad range of pH values. The first solution tested (left) contains $CH_3CO_2Na$ and is basic. $NaCl(aq)$ (center) is neutral. The solution on the right is $NH_4Cl(aq)$ and is acidic.

When chemists need a more accurate pH value than they can get with indicator paper, they usually use an electronic device called a *pH meter.* Not only does this kind of device give more precise measurements (see Figure 15.12), but it is often more convenient to use; however, pH meters are expensive.

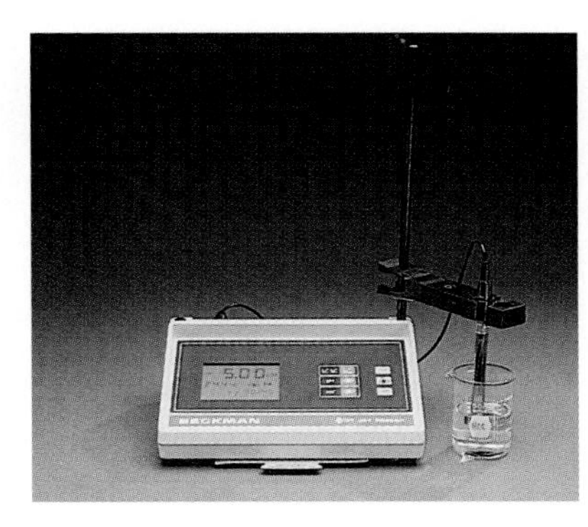

FIGURE 15.12
Chemists in the lab usually measure pH using electronic pH meters like this one. These instruments give more precise measurements than pH paper does. When properly calibrated, pH meters are also more accurate than pH paper.

## 15.10 Concentrations from $K_a$ Expressions

**GOAL:** To use $K_a$ expressions to calculate the concentrations of compounds and ions in solution.

Reactions of acids with water reach equilibrium almost instantly. We can use the equilibrium calculation techniques of Chapter 14 to predict the pH of a weak acid solution, and we can also calculate the concentrations of the other species present. The theory behind equilibrium constant expressions predicts experimental results very well.

Let's consider a 1.0 *M* solution of acetic acid. The equation for the reaction at equilibrium is

$$CH_3CO_2H + H_2O \rightleftharpoons CH_3CO_2^- + H_3O^+$$

The equilibrium constant for this reaction is the $K_a$ for acetic acid, which is $1.8 \times 10^{-5}$. The $K_a$ expression is

$$K_a = 1.8 \times 10^{-5} \text{ mol L}^{-1} = \frac{[CH_3CO_2^-][H_3O^+]}{[CH_3CO_2H]}$$

The equilibrium constant is small, and the equilibrium lies far to the left. In other words, most of the $CH_3CO_2H$ in the solution is undissociated.

To solve for $[H_3O^+]$ directly, you will need to do some mathematical manipulations. From the balanced equation, $[CH_3CO_2^-] = [H_3O^+] = x$, as usual. For each $H_3O^+$ ion or $CH_3CO_2^-$ ion formed, one $CH_3CO_2H$ molecule is destroyed; so in a 1.0 *M* solution $[CH_3CO_2H]$ is $1.0 - x$. Entering these terms into the $K_a$ expression gives:

$$1.8 \times 10^{-5} = \frac{(x)(x)}{1.0 - x}$$

$$x^2 + (1.8 \times 10^{-5})x - (1.8 \times 10^{-5}) = 0$$

While it's possible to solve this equation, the work will be much easier if we can make a simplifying assumption. This assumption is that *the equilibrium concentration of acetic acid is the same as its initial concentration.* You can make the same assumption for any weak acid in aqueous solution in which the initial concentration of the acid is much greater than the value of $K_a$. (By "much greater," we mean "larger by a factor of at least $10^3$.") Example 15.16 shows how to use this method.

## EXAMPLE 15.16 Calculations Using $K_a$ Expressions

Calculate the concentrations of $CH_3CO_2^-$, $H_3O^+$, and $OH^-$ in a 1.0 *M* aqueous solution of acetic acid. What is the pH of this solution? ($K_a$ for acetic acid is $1.8 \times 10^{-5}$ mol $L^{-1}$.)

**SOLUTION**

The equation for the reaction at equilibrium is

$$CH_3CO_2H + H_2O \rightleftharpoons CH_3CO_2^- + H_3O^+$$

The $K_a$ expression is

$$K_a = 1.8 \times 10^{-5} \text{ mol L}^{-1} = \frac{[CH_3CO_2^-][H_3O^+]}{[CH_3CO_2H]}$$

First, is the assumption valid? That is, is the initial concentration of acetic acid much greater than the $K_a$? Yes, $1.0 \gg 1.8 \times 10^{-5}$. This means that you can use 1.0 *M* for $[CH_3CO_2H]$ in the $K_a$ expression. Let $x = [CH_3CO_2^-] = [H_3O^+]$. Next, enter the known and unknown concentrations into the $K_a$ expression and solve for $x$:

$$1.8 \times 10^{-5}\ \text{mol L}^{-1} = \frac{[CH_3CO_2^-][H_3O^+]}{[CH_3CO_2H]} = \frac{(x)(x)}{1.0\ M} = \frac{x^2}{1.0\ M}$$

$$(1.8 \times 10^{-5}\ M)(1.0\ M) = 1.8 \times 10^{-5}\ M^2 = x^2$$

$$x = 4.2 \times 10^{-3}\ M$$

That is, $[CH_3CO_2^-] = [H_3O^+] = 4.2 \times 10^{-3}\ M$. To calculate the remaining concentration needed, $[OH^-]$, use the expression for $K_w$.

$$K_w = 1.0 \times 10^{-14} = [H_3O^+][OH^-]$$

$$1.0 \times 10^{-14} = (4.2 \times 10^{-3})[OH^-]$$

$$[OH^-] = \frac{1.0 \times 10^{-14}}{4.2 \times 10^{-3}} = 2.4 \times 10^{-12}\ M$$

Finally, here's the calculation of the pH:

$$pH = -\log[H_3O^+] = -\log(4.2 \times 10^{-3}) = 2.38$$

Now that you know the equilibrium value of $[H_3O^+]$, you can test whether the simplifying assumption is valid. That is, is $1.0 - x$ approximately the same as 1.0? Since $x = 4.2 \times 10^{-3}$, $1.0 - x = 1.0$ (to the correct number of significant figures).

**EXERCISE 15.16**

Calculate the concentrations of $HS^-$, $H_3O^+$, and $OH^-$ in a 0.10 $M$ aqueous solution of $H_2S$ ($K_a = 9.1 \times 10^{-8}$ mol L$^{-1}$ for the dissociation of the first proton). What is the pH of this solution?

As you think further about the equilibria present in a 1.0 M solution of acetic acid, you might wonder whether the calculations in Example 15.16 are valid. Two equilibria were used to solve the problem in that example:

$$CH_3CO_2H + H_2O \rightleftharpoons CH_3CO_2^- + H_3O^+$$

$$2\ H_2O \rightleftharpoons H_3O^+ + OH^-$$

Both reactions occur at the same time in the aqueous solution. The $H_3O^+$ ion is common to both reactions, and it appears in both the $K_a$ and the $K_w$ expressions. The equilibrium constant for the first reaction is the $K_a$ of acetic acid, and that of the second reaction is $K_w$ for water.

$$K_a = 1.8 \times 10^{-5}\ M = \frac{[CH_3CO_2^-][H_3O^+]}{[CH_3CO_2H]}$$

$$K_w = 1.0 \times 10^{-14} = [H_3O^+][OH^-]$$

At equilibrium, of course, the solution (because it is homogeneous) can have *only one value* of $[H_3O^+]$. How can you solve for the concentration of $H_3O^+$ if the ion is being formed in two different reactions? In fact, you can solve the $K_a$ and $K_w$ expressions simultaneously using long involved algebraic techniques. However, chemists as a rule are quite lazy, and never want to do any extra work. A close look at the chemical processes involved here shows the path to an easier solution.

First, look at the two equilibrium constants, and note that $K_a \gg K_w$. Since $10^{-5}$ is a billion times bigger than $10^{-14}$, you can conclude that the acid dissociation equilibrium lies much further to the right than the water dissociation equilibrium. As a rough approximation, about $10^5$ more $H_3O^+$ ions are produced by the reaction between acetic acid and water than by the dissociation of water itself. Also, the acetic acid dissociation will make the solution acidic by producing $H_3O^+$. This will drive the water dissociation reaction to the left, according to Le Châtelier's Principle.

These arguments lead us to the conclusion that the $K_w$ equilibrium can be ignored for calculations of $[H_3O^+]$ and $[CH_3CO_2^-]$, because $[H_3O^+]$ from the dissociation of water will be very small compared to $[H_3O^+]$ from the dissociation of acetic acid. This approximation can almost always be made.

Almost always isn't all the time. When does this approximation not work? For the approximation to lead to a valid result, two criteria must be met. The first you have already seen: $K_a \gg K_w$. As long as $K_a$ is at least a thousand times larger than $K_w$, you are on safe ground in ignoring $K_w$. Second, remember that most of the $H_3O^+$ ions present in solution came from acetic acid dissociation. This can only be true if you start out with a relatively large concentration of acetic acid.

Consider an extreme case. Suppose you made a $1 \times 10^{-10}$ *M* solution of acetic acid. If 100% of the acetic acid dissociated (it will not), you would get $[H_3O^+] = 10^{-10}$ *M* from the acid dissociation. The dissociation of water, however, produces $[H_3O^+] = 10^{-7}$ *M*, or about 1000 times as much $H_3O^+$ in this particular example. You can see that it is necessary to have a certain minimum amount of acetic acid present for the basic assumption that $K_w$ can be ignored to work. How much is this minimum? As long as the initial concentration of weak acid is at least $1 \times 10^{-4}$ *M* and $K_a \gg K_w$, the assumption is valid.

## 15.11 Buffer Solutions

**GOAL:** To learn how solutions that resist changes in pH are made and how they work.

Both acids and bases can be very corrosive and dangerous (see Chapter Box 13.3). This section considers a phenomenon used by both natural systems and chemists to protect themselves and the things they work with from damage from too much acid or base.

The obvious thing to do when you spill a strong acid is to neutralize it with a base. A spill of hydrochloric acid, for example, might be neutralized with sodium bicarbonate ($NaHCO_3$), which is a relatively weak base. Thus, if you add too much of it, you don't give rise to a new danger. You could use sodium hydroxide (NaOH) to neutralize the acid, but if you add an excess of that, you get a different, equally serious problem. Sodium hydroxide is as dangerous and corrosive as hydrochloric acid is, and in emergencies you don't have the time or opportunity to measure things carefully.

**Suppose you spill some HCl or NaOH on *yourself.* Immediately wash yourself off with large amounts of water. Don't just splash water on—drench yourself with water. Never mind if it makes a mess. Your health is more important.**

Suppose you want to protect against *both* acids and bases. To neutralize strong acids you need a weak base, and to neutralize strong bases you need a weak acid. A mixture of a weak base and a weak acid can protect against both strong acids and strong bases. Such a mixture is called a **buffer.** In ordinary English, the word *buffer* is used for anything that neutralizes or cushions the shock of opposing forces. The use of the word in chemistry is not much different. A *buffer* is a solution that resists pH changes when either acid or base is added. Formally, a buffer solution is an aqueous solution that contains both a weak acid and its conjugate base (or, alternatively, both a weak base and its conjugate acid).

There are two ways to make an effective buffer solution. One is to dissolve both a weak acid and its sodium salt in water. For example, you can make an effective buffer solution by dissolving 0.5 mol of acetic acid and 0.5 mol of sodium acetate in 1 L of water. An alternative method that gives the same result is to neutralize some (but not all) of a weak acid solution with a strong base. For example, if you add 0.5 mol of NaOH to 1 L of a 1.0 *M* solution of acetic acid, exactly half of the acetic acid reacts to give sodium acetate:

**You can use a potassium salt instead of a sodium salt. The point is that the salt is soluble in water.**

$$CH_3CO_2H + NaOH \longrightarrow CH_3CO_2Na + H_2O$$

These two methods result in exactly the same solution. Either way you end up with a mixture of the weak acid $CH_3CO_2H$ and its conjugate base $CH_3CO_2^-$ in solution.

### EXAMPLE 15.17 Identifying Buffer Solutions

Which of these mixtures is a buffer solution?

A. $HNO_3$ + excess NaOH
B. Excess $HNO_3$ + NaOH
C. Excess HCl + $NH_3$
D. HCl + excess $NH_3$

**SOLUTION**

In solutions A and B a strong acid is mixed with a strong base. That won't give a buffer. Solution C (after mixing) contains $H_3O^+$, $NH_4^+$, and $Cl^-$ ions. No weak base is present, so this solution will not act as a buffer. Solution D contains $NH_4^+$, $NH_3$, and $Cl^-$. That is, the solution contains both a weak acid and its conjugate base, and is therefore a buffer solution. D is the answer.

**EXERCISE 15.17**

Which of these mixtures is a buffer solution?

A. Excess HCl + $CH_3CO_2H$
B. Excess NaOH + $CH_3CO_2H$
C. Excess $CH_3CO_2Na$ + $HNO_3$
D. $CH_3CO_2Na$ + excess $HNO_3$

Let's look at how the acetic acid–sodium acetate buffer works. You have a solution that contains a large amount of both $CH_3CO_2H$ and $CH_3CO_2^-$. Suppose you add a strong acid, for instance, hydrochloric acid, to the buffer. The hydrochloric acid (or any other strong aqueous acid) dissociates completely to give $H_3O^+(aq)$. Its reaction with the buffer is

$$H_3O^+(aq) + CH_3CO_2^-(aq) \longrightarrow H_2O(l) + CH_3CO_2H(aq)$$

In other words, the buffer neutralizes most of the $H_3O^+$, and the pH change of the solution is relatively small (see Figure 15.13). The other product of the reaction is acetic acid, which is already present in the buffer.

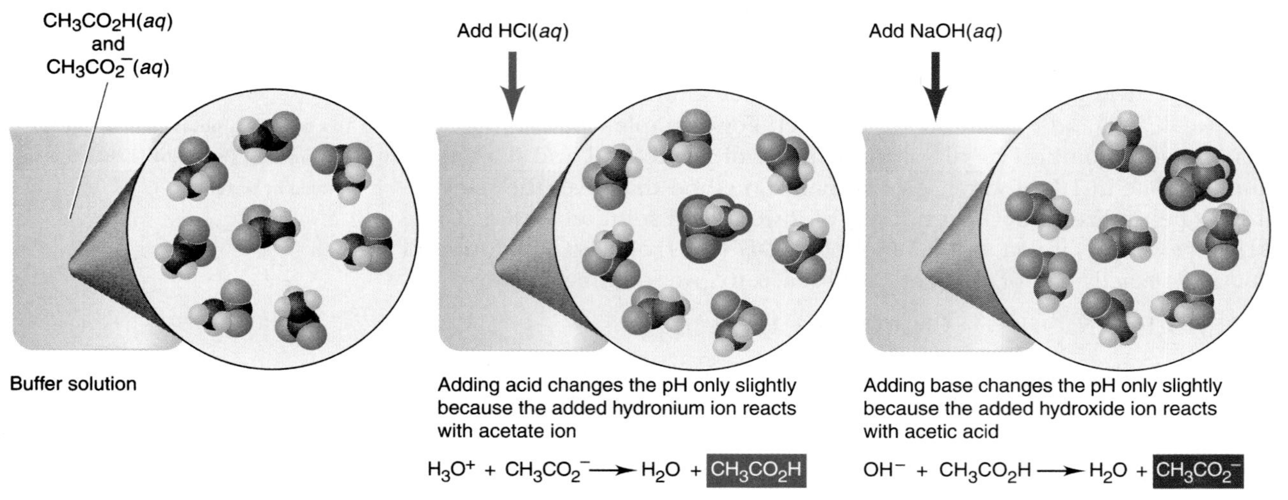
CH3CO2H(aq)
and
CH3CO2−(aq)
Buffer solution
Add HCl(aq)
Adding acid changes the pH only slightly because the added hydronium ion reacts with acetate ion
H3O+ + CH3CO2− → H2O + CH3CO2H
Add NaOH(aq)
Adding base changes the pH only slightly because the added hydroxide ion reacts with acetic acid
OH− + CH3CO2H → H2O + CH3CO2−

FIGURE 15.13
**How an acetic acid–sodium acetate buffer works.**

You can also add a strong base, sodium hydroxide, for example, to this buffer. The reaction that occurs is

$$OH^-(aq) + CH_3CO_2H(aq) \longrightarrow CH_3CO_2^-(aq) + H_2O(l)$$

The buffer neutralizes $OH^-$ as it did $H_3O^+$. Once again the pH change is much smaller than it would be if the solution was not a buffer.

The pH of a buffer solution is easy to calculate if you know the concentrations of acid and conjugate base used to make the buffer. Buffer solutions contain both a weak acid, which we'll designate HA, and its conjugate base, $A^-$. The $K_a$ expression for HA must be valid for the buffer solution:

$$K_a = \frac{[H_3O^+][A^-]}{[HA]} = [H_3O^+]\frac{[A^-]}{[HA]}$$

If you take the negative log of both sides of this equation, you get:

$$-\log(K_a) = -\log[H_3O^+] - \log\left(\frac{[A^-]}{[HA]}\right)$$

The term $-\log[H_3O^+]$ is the pH, so you can substitute to get:

$$-\log(K_a) = \text{pH} - \log\left(\frac{[A^-]}{[HA]}\right)$$

Then rearranging gives:

$$\text{pH} = -\log(K_a) + \log\left(\frac{[A^-]}{[HA]}\right)$$

The term $-\log(K_a)$ is called the $pK_a$, just as $-\log(H_3O^+)$ is called the pH. For any buffer solution, then:

$$\mathbf{pH = pK_a + \log\left(\frac{[A^-]}{[HA]}\right)} \qquad \textbf{(Henderson-Hasselbalch Equation)}$$

This equation is known as the **Henderson-Hasselbalch Equation** after the biochemists who developed it. Note that when $[A^-] = [HA]$, the last term becomes $\log(1) = 0$, and the pH equals the $pK_a$.

### EXAMPLE 15.18 Using the Henderson-Hasselbalch Equation

A buffer solution was prepared by dissolving 0.500 mol of $NH_3$ and 0.500 mol of $NH_4Cl$ ($K_a = 5.6 \times 10^{-10}$) in enough water to give 1.00 L of solution.

**a.** What is the pH of this buffer solution?
**b.** What is the pH of the solution after $1.0 \times 10^{-2}$ mol of HCl is added?
**c.** What is the pH of the solution after $1.0 \times 10^{-3}$ mol of $OH^-$ is added?

**SOLUTIONS**

**a.** The equation for the reaction is:

$$NH_4^+(aq) + H_2O(l) \rightleftharpoons NH_3(aq) + H_3O^+(aq)$$

The acid HA in the Henderson-Hasselbalch Equation is $NH_4^+$, and the base $A^-$ is $NH_3$. Since $[NH_4^+] = [NH_3]$ in the original solution, pH = $pK_a$.

$$pH = pK_a = -\log(K_a) = -\log(5.6 \times 10^{-10}) = 9.25$$

**b.** The extra acid added to the solution gives the following reaction:

$$NH_3 + H_3O^+ \longrightarrow NH_4^+ + H_2O$$

This reaction increases the concentration of $NH_4^+$ by an amount equal to the added $H_3O^+$ and decreases the concentration of $NH_3$ by the same amount. You can determine the exact changes from the usual table:

| | $[NH_3]$ | $[NH_4^+]$ |
|---|---|---|
| Initial | 0.500 | 0.500 |
| Change | $-1.0 \times 10^{-2}$ | $+1.0 \times 10^{-2}$ |
| Final | 0.490 | 0.510 |

You can now determine the pH of the final solution from the Henderson-Hasselbalch Equation:

$$pH = pK_a + \log\left(\frac{[NH_3]}{[NH_4^+]}\right) = 9.25 + \log\left(\frac{0.490}{0.510}\right)$$
$$= 9.25 + \log(0.961) = 9.25 + (-0.017) = 9.23$$

Note that there has been a very small change in pH from that in part (a) and that the pH has decreased, indicating that the solution is slightly more acidic. If $1 \times 10^{-2}$ mol of HCl were added to 1 L of pure water, the pH would change from 7 to 2, and the $[H_3O^+]$ increases by a factor of $10^5$! Buffer solutions resist changes in pH very effectively.

**c.** The extra base added to the solution gives the following reaction:

$$NH_4^+ + OH^- \longrightarrow NH_3 + H_2O$$

This reaction increases the concentration of $NH_3$ by an amount equal to the added $OH^-$ and decreases the concentration of $NH_4^+$ by an equal amount. Making a table gives:

| | $[NH_4^+]$ | $[NH_3]$ |
|---|---|---|
| Initial | 0.500 | 0.500 |
| Change | $-1.0 \times 10^{-3}$ | $+1.0 \times 10^{-3}$ |
| Final | 0.499 | 0.501 |

$$pH = pK_a + \log\left(\frac{[NH_3]}{[NH_4^+]}\right) = 9.25 + \log\left(\frac{0.501}{0.499}\right)$$

$$pH = 9.25 + \log(1.004) = 9.25 + 0.0017 = 9.25$$

There is virtually no change in the pH of this buffer. For comparison, adding $1 \times 10^{-3}$ mol of $OH^-$ to 1 L of pure water would change the pH from 7 to 11. Once again you can see that buffer solutions are very good at keeping pH constant.

FIGURE 15.14
Buffers resist pH changes. At the top right is a beaker of pure water containing phenol red indicator. The yellow color of the indicator shows that the solution has a pH between 6.4 and 8. At the bottom left is a beaker of a buffer made with $H_2PO_4^-$ and $HPO_4^{2-}$ and containing the same indicator. The buffer solution and the water have similar pH values (see Exercise 15.18). If you add 1 mL of a 1 *M* NaOH solution to the pure water, it becomes basic, as shown by the red color of the indicator in the beaker on the top left. If you add the same amount of NaOH to the buffer solution, the color of the indicator doesn't change significantly, as the bottom right beaker shows.

**EXERCISE 15.18**

A buffer solution was prepared by dissolving 0.50 mol of $NaH_2PO_4$ ($K_a = 6.2 \times 10^{-8}$) and 0.50 mol of $Na_2HPO_4$ in enough water to give 1.0 L of solution.

**a.** What is the pH of this buffer solution?

**b.** What is the pH of the solution just after addition of $1 \times 10^{-2}$ mol of $H_3O^+$?

**c.** What is the pH of the solution just after addition of $1 \times 10^{-2}$ mol of $OH^-$?

The acetic acid–sodium acetate buffer is only one of several buffers that chemists commonly use. Buffers made from the ions $H_2PO_4^-$ and $HPO_4^{2-}$ are common. The $H_2PO_4^-$ ion is weakly acidic and the $HPO_4^{2-}$ ion is weakly basic in solutions that are near neutrality. (See Figure 15.14.)

Your body is largely made up of liquids, including blood and other fluids inside and outside cells. Small changes in the pH of these fluids can be life-threatening. For example, blood has a pH of 7.4. If the pH of blood drops below 7 or rises much above 7.8, many of the enzymes in body tissues don't work, and other vital functions go haywire. Extreme cases are fatal. The natural buffer that does most of the job of keeping the blood pH near 7.4 is made up of carbonic acid ($H_2CO_3$) and bicarbonate ion ($HCO_3^-$). (Phosphate buffers and certain proteins also serve as buffers in body fluids.)

The fact that carbonic acid in water can give off carbon dioxide has some health effects. Suppose you breathe heavily (hyperventilate), greatly reducing the partial pressure of $CO_2(g)$ in your lungs. Because of Le Châtelier's

## EXPERIMENTAL EXERCISE 15.1

A. L. Chemist and members of a scientific team were given a solution (Solution 1) containing a substance to analyze. They were informed that the substance was either an acid or a base, and they were asked to determine its $K_a$ or $K_b$.

A small sample of Solution 1 gave a blue color when tested with bromocresol green, and another sample gave a yellow color when tested with bromocresol purple. The team members concluded that the substance was a weak acid, which they referred to as HA.

Since the concentration of Solution 1 was unknown, the team members decided to use chemistry to determine the $K_a$ of the acid. They took 10.0 mL of Solution 1 and added just enough NaOH(*aq*) to completely neutralize the acid present. They then added the neutralized solution to a second 10.0-mL portion of Solution 1 to prepare a new solution (Solution 2). They assumed that Solution 2 was a buffer solution in which $[HA] = [A^-]$, and therefore its pH was equal to the $pK_a$. The measured pH of Solution 2 was 5.20, meaning that the $K_a$ was $6.3 \times 10^{-6}$.

**QUESTIONS**

1. Other than water, what species is present in the largest amount in Solution 1?
2. What color would Solution 1 give if it were tested with the indicator phenolphthalein?
3. What reaction occurred when Solution 1 was neutralized?
4. Other than water, what species is present in the largest amount in the neutralized solution?
5. Justify the assumption that Solution 2 is a buffer.
6. Is the team's conclusion that Solution 1 contains a weak acid justified on the basis of the indicator results?
7. Is the $K_a$ value reported by A. L. Chemist and the team consistent with those typical of weak acids?
8. What base other than NaOH could have been used to neutralize Solution 1?
9. Had analysis shown that Solution 1 contained a weak base rather than a weak acid, what modification would you make in the experimental procedure used?

Principle, the following reactions occur (note that the product of each is a reactant in the previous one):

$$CO_2(aq) \rightleftharpoons CO_2(g)$$
$$H_2CO_3(aq) \rightleftharpoons H_2O(l) + CO_2(aq)$$
$$HCO_3^-(aq) + H_3O^+(aq) \rightleftharpoons H_2CO_3(aq) + H_2O(l)$$

As you breathe heavily, you exhale $CO_2$, and this decreases the $CO_2$ concentration in your blood. Then, $H_2CO_3$ dissociates to give $CO_2$, and $HCO_3^-$ reacts with $H_3O^+$ to restore the blood concentration of $H_2CO_3$. This loss of $H_3O^+$ makes your blood more basic (alkaline), a condition called *alkalosis.* The opposite condition, called *acidosis,* can be caused by breathing too little.

FIGURE 15.15
We have mentioned acid rain earlier in this book (see Chapter 8, for example). This lake in the Adirondack Mountains is acidic because of acid rain. The rock in the area is mostly granite, which is almost completely insoluble in water and therefore has essentially no buffering action. Lakes in limestone regions are protected against acid rain by the $CaCO_3$ in that rock. This isn't really buffering; limestone wouldn't protect a lake against "basic rain" if that were a problem.

## 15.12 Measuring Amounts of Acids and Bases: Titration

GOAL: To understand how titration works, and to calculate the amounts of acids and bases in samples from titration data.

FIGURE 15.16
Here a chemist is using titration to find out the amount of acid present in an unknown sample.

Preceding sections have talked about how to measure the acid or base strength, the *intensity* of the acidic or basic quality, of a given substance. This section covers how to measure the *amount* of an acid or base. The classical method chemists use to do this is called **titration** (see Figure 15.16) To measure the amount of acid present in a sample, add a base of known concentration until all of the acid is neutralized. Stop when the reaction is exactly complete. The amount of acid that was originally in the sample is the same as the amount of the base that reacted with it. To find out when the reaction is complete, chemists use either an indicator or a pH meter. Figure 15.17 shows the process of titration.

The key to the concentration calculation lies in the number of molecules or ions that react. One $H_3O^+$ ion reacts with one $OH^-$ ion to give water. The same is true of moles, of course; 1 mol of $H_3O^+$ ions reacts with 1 mol of $OH^-$ ions.

The process of calculating from titration data works like this. If you are titrating an unknown acid with a known base, you know what the original concentration of base is (that is, you know its molarity exactly). You don't know the concentration of the acid, but you measure the exact volume of acid solution before you begin. You use a buret to measure the volume of base that reacts exactly with the measured volume of acid. Next, you can calculate the number of moles of base that reacted (volume of base times molarity of base equals moles of base; the units are mol $L^{-1} \times L = mol$). The number of moles of base used is equal to the number of moles of acid. If you need to find the concentration of the acid and not just the number of moles, then you divide the number of moles of acid by the volume of acid solution to get the concentration.

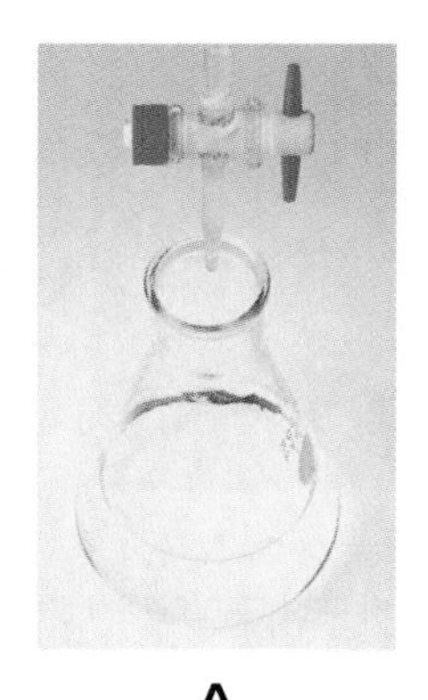

A

B

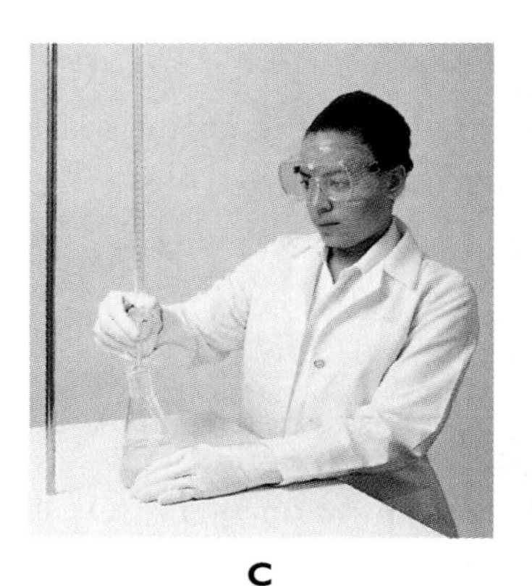

C

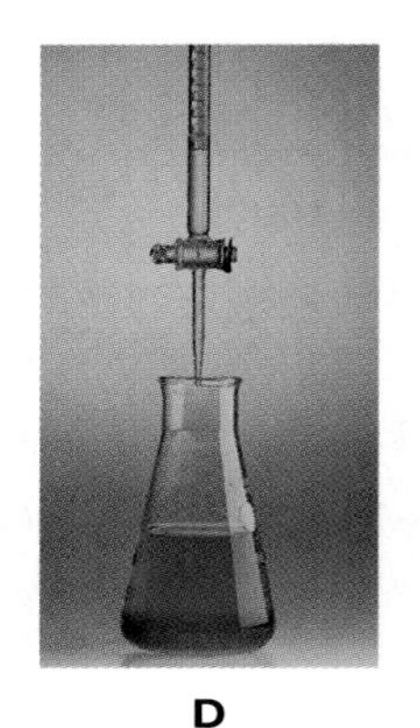

D

E

FIGURE 15.17
Titration of an acid with a base using phenolphthalein as the indicator.
(A) The acid solution is in the flask (with a small amount of indicator) and the base solution is in the buret above. (B) At first the chemist can run the base into the acid fairly quickly. The pink cloud shows that the solution is basic where the base hits the acid.
(C) When the chemist swirls the flask, the pink color disappears. As the titration approaches the *end point,* where the reaction is complete and no acid is unreacted and no excess base has been added, the chemist adds base more carefully and washes the sides of the flask with water. (D) When the end point is reached, the solution will be a very pale pink. (E) If the chemist adds too much base, the solution will turn a darker pink.

### EXAMPLE 15.19 Titrating an Acid with a Known Base

A 25.00-mL sample of HCl(*aq*) required 27.25 mL of 0.112 *M* NaOH(*aq*) to be completely neutralized. What is the molarity of the HCl(*aq*)?

**SOLUTION**

In order to find the molarity of the HCl(*aq*), you first need to know the number of moles of HCl in the 25.00-mL sample. The number of moles of HCl in the sample is exactly equal to the number of moles of base it takes to neutralize the HCl. The number of moles of $OH^-$ is the molarity of the NaOH(*aq*) multiplied by the volume used (27.25 mL = 0.02725 L):

$$\text{Moles of HCl} = \text{moles of NaOH} = 0.02725\ \text{L} \times 0.112\ \text{mol L}^{-1}$$
$$= 3.05 \times 10^{-3}\ \text{mol}$$

Now you can divide the number of moles of HCl by the volume (25.00 mL = 0.025 L) to get the molarity:

$$\text{Molarity of HCl} = \frac{\text{moles of HCl}}{\text{liter of solution}} = \frac{3.05 \times 10^{-3}\ \text{mol HCl}}{0.02500\ \text{L sol'n}}$$
$$= 0.122\ \text{mol L}^{-1}\ \text{HCl}$$
$$= 0.122\ M$$

**EXERCISE 15.19**

A 20.00-mL sample of HCl(*aq*) required 33.26 mL of 0.102 *M* NaOH(*aq*) to be neutralized completely. What is the molarity of the HCl(*aq*)?

The calculations are done in exactly the same way when a base is titrated with a known acid. You calculate the number of moles of acid used, which is equal to the number of moles of base.

### EXAMPLE 15.20 Titrating a Base with a Known Acid

A 25.00-mL sample of $Ca(OH)_2$(*aq*) was titrated with 12.76 mL of 0.0199 *M* HCl. What is the concentration of the $Ca(OH)_2$(*aq*)?

**SOLUTION**

Here the calculations require you to keep track of the stoichiometry of the reaction. The number of moles of $H_3O^+$ still equals the number of moles of $OH^-$, but there are two moles of $OH^-$ for every mole of $Ca(OH)_2$. You must keep track of this in your calculation.

$$Ca(OH)_2(aq) + 2\ HCl(aq) \longrightarrow CaCl_2(aq) + 2\ H_2O(l)$$

Moles of $OH^-$ = moles of $H_3O^+$ = (0.01276 L)(0.0199 mol $L^-$) = $2.54 \times 10^{-4}$ mol

$$(2.54 \times 10^{-4} \text{ mol OH}^-)\left(\frac{1 \text{ mol Ca(OH)}_2}{2 \text{ mol OH}^-}\right) = 1.27 \times 10^{-4} \text{ mol Ca(OH)}_2$$

$$\text{Molarity of Ca(OH)}_2 = \frac{1.27 \times 10^{-4} \text{ mol}}{0.0250 \text{ L}} = 5.08 \times 10^{-3}\ M$$

**EXERCISE 15.20**

A chemist titrated 20.0 mL of NaOH(*aq*) with 19.87 mL of 1.23 *M* HCl. What is the molarity of the NaOH(*aq*)?

How do chemists find out the concentration of the "known" base or acid solution they use in titrations? They usually determine this concentration by titrating a solution of a substance called a **primary standard** with the base or acid. A primary standard is a solid substance of known molar mass and acid-base properties that can be purchased in pure form. One such substance is potassium hydrogen phthalate (abbreviated KHP), which has the formula $C_8H_5O_4K$ and a molar mass of 204.23 g. It's a monoprotic acid (that is, it releases only one proton in acid-base reactions), and it's relatively easy to purify at room temperature. Chemists use KHP to standardize (that is, find the concentration of) solutions of bases.

The standardization procedure is relatively simple. You dry a sample of the KHP, weigh it, and dissolve it in water. It doesn't matter how much water you add; you calculate the number of moles of KHP from the mass. You then titrate the KHP solution with the aqueous solution of base until neutralization is achieved. The calculations are similar to those we've already done; the number of moles of acid (KHP) originally present equals the number of moles of base added to reach the neutralization point.

### EXAMPLE 15.21 Calculations Involving a Primary Standard

A 0.8975-g sample of KHP was weighed out and dissolved in about 50 mL of water. The KHP solution was then titrated with an aqueous solution of NaOH. After 37.25 mL of the base was added, the solution was neutral. What is the molarity of the NaOH solution?

**SOLUTION**

$$\text{Moles of NaOH} = \text{moles of KHP} = \frac{0.8975 \text{ g}}{204.23 \text{ g mol}^{-1}} = 4.395 \times 10^{-3} \text{ mol}$$

$$\text{Molarity} = \frac{4.395 \times 10^{-3} \text{ mol}}{0.03725 \text{ L}} = 0.1180\ M$$

**EXERCISE 15.21**

A 1.000-g sample of KHP was weighed out and dissolved in water. The KHP solution was then neutralized by the addition of 46.93 mL of an aqueous solution of NaOH of unknown concentration. What was the molarity of the NaOH solution?

## CHAPTER SUMMARY

1. A Brønsted-Lowry acid is a proton-donating species, and a Brønsted-Lowry base is a proton-accepting species.
2. A base forms its conjugate acid by accepting a proton from a Brønsted-Lowry acid. An acid forms its conjugate base by donating a proton to a Brønsted-Lowry base.
3. For binary hydrides, acid strength is a function of the position on the periodic table of the element other than hydrogen. When comparing binary hydrides whose central elements are in the same row on the periodic table, the element farther to the right makes the most acidic hydride. When comparing binary hydrides whose central elements are in the same periodic table group, the element closer to the bottom makes the most acidic hydride.
4. In all acid-base reactions, the stronger acid reacts with the stronger base to form a weaker base and a weaker acid. In acid-base reactions, the equilibrium always lies toward the side of the weaker acid and the weaker base.
5. The stronger an acid is, the weaker its conjugate base will be, and vice versa. A salt formed from a strong acid and a strong base yields a neutral aqueous solution. A salt formed from a weak acid and a strong base is always weakly basic, and a salt formed from a weak base and a strong acid is weakly acidic.
6. Strong acids react completely with water. The only acid present in an aqueous solution of a strong acid is $H_3O^+$. The only base present in an aqueous solution of a strong base is $OH^-$.
7. Acid dissociation constants are symbolized by $K_a$. Base dissociation constants are symbolized by $K_b$. Both are equilibrium constants.
8. The constant for water is given the symbol $K_w$ and has the value $1 \times 10^{-14}$.
9. The acidity of an aqueous solution is frequently expressed as the pH, defined as $-\log[H_3O^+]$. A low pH (less than 7) indicates acidity, a pH of 7 indicates neutrality, and a high pH (greater than 7) indicates that the tested solution is basic.
10. Indicators are acids and bases that have distinctly different colors at different pH values. They can be used to determine the approximate pH of aqueous solutions and the end points of titrations.
11. A buffer solution contains significant amounts of an acid and its conjugate base. Buffer solutions resist changes in pH.

## IMPORTANT EQUATIONS

Acid dissociation constant for the weak acid HA:

$$K_a = \frac{[H_3O^+][A^-]}{[HA]}$$

Base dissociation constant for the weak base B:

$$K_b = \frac{[HB^+][OH^-]}{[B]}$$

Water dissociation constant: $K_w = 1.0 \times 10^{-14} = [H_3O^+][OH^-]$

$$pH = -\log[H_3O^+]$$
$$pK_a = -\log(K_a)$$

Henderson-Hasselbalch Equation: $pH = pK_a + \log\left(\frac{[A^-]}{[HA]}\right)$

## PROBLEMS

### SECTION 15.1 **Brønsted-Lowry Acids and Bases**

1. Identify each substance in these reactions as either an acid or a base.
   a. $H_2SO_4 + HNO_3 \rightleftharpoons HSO_4^- + H_2NO_3^+$
   b. $AsH_3 + H_3O^+ \rightleftharpoons AsH_4^+ + H_2O$
   c. $CH_3O^- + H_2O \rightleftharpoons CH_3OH + OH^-$
2. Identify each substance in these reactions as either an acid or a base.
   a. $HClO_4 + CH_3CO_2H \rightleftharpoons ClO_4^- + CH_3CO_2H_2^+$
   b. $H_2O + CN^- \rightleftharpoons OH^- + HCN$
   c. $CH_3CH_2OH + H_3PO_4 \rightleftharpoons CH_3CH_2OH_2^+ + H_2PO_4^-$
3. For each of these Brønsted-Lowry acid-base reactions in aqueous solution, write the balanced net ionic equation and identify each reactant and each product as either an acid or a base.
   a. $HNO_3 + Na_2HPO_4 \rightleftharpoons NaNO_3 + NaH_2PO_4$
   b. $KOH + H_2SO_4 \rightleftharpoons KHSO_4 + H_2O$
   c. $NH_3 + H_3PO_4 \rightleftharpoons NH_4^+ + H_2PO_4^-$
4. For each of these Brønsted-Lowry acid-base reactions, write the balanced net ionic equation and identify each of the substances present as either an acid or a base.
   a. $NaHSO_4 + NaOH \rightleftharpoons Na_2SO_4 + H_2O$
   b. $LiOH + NH_4Cl \rightleftharpoons LiCl + H_2O + NH_3$
   c. $Na_2CO_3 + NaHSO_4 \rightleftharpoons Na_2SO_4 + NaHCO_3$
5. For each of these Brønsted-Lowry acid-base reactions in aqueous solution, write the balanced net ionic equation and identify each of the substances present as either an acid or a base.
   a. $LiOH + LiH_2PO_4 \rightleftharpoons Li_2HPO_4 + H_2O$
   b. $HCl + Na_3PO_4 \rightleftharpoons H_3PO_4 + NaCl$
   c. $H_2SO_4 + Na_3PO_4 \rightleftharpoons H_3PO_4 + Na_2SO_4$
6. For each of these Brønsted-Lowry acid-base reactions in aqueous solution, write the balanced net ionic equation and identify each of the substances present as either an acid or a base.
   a. $K_2SO_4 + H_2SO_4 \rightleftharpoons KHSO_4$
   b. $HClO_4 + KH_2PO_4 \rightleftharpoons H_3PO_4 + KClO_4$
   c. $NH_3 + HBr \rightleftharpoons NH_4Br$

### SECTION 15.2 **Brønsted-Lowry Acid-Base Reactions: Conjugate Acids and Bases**

7. Give the formulas of both the conjugate acid and the conjugate base of each of these compounds.
   a. $CH_3NH_2$ b. $H_2O$ c. $HNO_3$
8. Give the formulas of both the conjugate acid and the conjugate base of each of these compounds.
   a. $H_2S$ b. $CH_3OH$ c. $CH_3NH_2$
9. Write the formulas of both the conjugate acid and the conjugate base for each of these compounds.
   a. $N_2H_4$ b. $H_2SO_4$ c. $PH_3$
10. Write the formulas of both the conjugate acid and the conjugate base for each of these compounds.
    a. HOCl b. $H_3BO_3$ c. $N_2H_2$
11. What are the formulas of both the conjugate acid and the conjugate base of each of these ions?
    a. $HSO_4^-$ b. $H_3O^+$ c. $HBO_3^{2-}$
12. What are the formulas of both the conjugate acid and the conjugate base of each of these ions?
    a. $HS^-$ b. $H_2PO_4^-$ c. $HCO_3^-$
13. Write the formulas for both the conjugate acid and the conjugate base of each of these ions.
    a. $HPO_4^{2-}$ b. $NH_2^-$ c. $N_2H_3^-$

14. Write the formulas for both the conjugate acid and the conjugate base of each of these ions.
a. $CH_3OH_2^+$ b. $H_2BO_3^-$ c. $OH^-$

## SECTION 15.3 Acid-Base Reactions: Reactants and Products

15. Write the balanced net ionic equation, including the formulas of the products, for each of these reactions.
a. $H_3PO_4 + OH^- \rightarrow$
b. $HCl(aq) + NaOH(aq) \rightarrow$
c. $H_2SO_4$ + large excess $Na_2CO_3 \rightarrow$

16. Write the balanced net ionic equation, including the formulas of the products, for each of these reactions.
a. Large excess $H_3BO_3 + LiOH \rightarrow$
b. $NH_3(g) + HCl(g) \rightarrow$
c. $NH_3(aq) + HCl(aq) \rightarrow$

17. For each reaction, give the balanced net ionic equation, including the formulas of the products.
a. $H_3BO_3$ + large excess $LiOH \rightarrow$
b. $H_3PO_4(aq) + HPO_3^{2-}(aq) \rightarrow$
c. $HBr(g) + H_2O(l) \rightarrow$

18. For each reaction, give the balanced net ionic equation, including the formulas of the products.
a. $H_3O^+(aq) + CaCO_3(s) \rightarrow$
b. $NaHCO_3(aq) + NaOH(aq) \rightarrow$
c. $OH^- + HOCl \rightarrow$

19. What is the complete balanced net ionic equation for each of these reactions?
a. $KOH(aq) + HClO_4(aq) \rightarrow$
b. $K_2CO_3 + H_3PO_4 \rightarrow$ (assume equimolar amounts)
c. $NaHCO_3 + Ca(OH)_2 \rightarrow$

20. What is the complete balanced net ionic equation for each of these reactions?
a. $HBr(g) + CH_3NH_2(g) \rightarrow$
b. $HBr(aq) + CH_3NH_2(aq) \rightarrow$
c. $H_2SO_4(aq) + SO_4^{2-}(aq) \rightarrow$
(assume equimolar amounts)

21. Write the complete balanced net ionic equation for each of these reactions.
a. $HI(g) + H_2O(l) \rightarrow$
b. $H_3O^+(aq) + MgCO_3(s) \rightarrow$
(assume equimolar amounts)
c. $NaHSO_4(aq) + KOH(aq) \rightarrow$

22. Write the complete balanced net ionic equation for each of these reactions.
a. $KHSO_4 + HSO_3F \rightarrow$
b. $H_2SO_4 + NaH_2BO_3 \rightarrow$
c. $Na_2O(s) + H_2O(l) \rightarrow$

## SECTION 15.4 The Relative Strengths of Acids and Bases

23. Which of these compounds reacts most completely with water? Explain your answer.
A. $H_2SO_4$ C. $CH_3CH_2OH$
B. HF D. $CH_3CO_2H$

24. Which of these compounds reacts the least with water? Explain your answer.
A. $NH_3$ C. $HNO_3$
B. $CH_3COOH$ D. $CH_3CH_2OH$

25. Which of these compounds is the strongest acid? Explain your answer.
A. HCl B. HF C. $H_2S$ D. $H_2O$

26. Which of these compounds is the weakest acid? Explain your answer.
A. HCl B. HF C. $H_2S$ D. $H_2O$

27. Which of these compounds is the strongest base? Explain your answer.
A. HCl B. HF C. $H_2S$ D. $H_2O$

28. Which of these compounds is the weakest base? Explain your answer.
A. HCl B. HF C. $H_2S$ D. $H_2O$

29. Which of the species in each pair is the stronger acid? Explain your answers.
a. $HCO_2H$ or $CH_3OH$
b. $H_2S$ or $H_3S^+$
c. $HS^-$ or $S^{2-}$

30. Which of the species in each pair is the stronger acid? Explain your answers.
a. $H_2SO_4$ or $H_2SO_3$
b. $NH_3$ or $NH_4^+$
c. $NH_3$ or $NH_2^-$

31. Which of the species in each pair is the stronger acid? Explain your answers.
a. $H_3PO_2$ or $H_3PO_4$
b. $AsH_3$ or $AsH_4^+$
c. $CH_4$ or $CH_3^-$

32. Which of the species in each pair is the stronger acid? Explain your answers.
a. HF or NaH
b. $HNO_3$ or $HNO_2$
c. $OH^-$ or $O^{2-}$

33. Which of the species in each pair is the stronger base? Explain your answers.
a. $HCO_3^-$ or $CO_3^{2-}$
b. $OH^-$ or $I^-$
c. $HSO_4^-$ or $HSO_3^-$

34. Which of the species in each pair is the stronger base? Explain your answers.
   a. $O_2^-$ or $O_2^{2-}$
   b. $Cl^-$ or $OCl^-$
   c. $Cl^-$ or $F^-$

35. Which of the species in each pair is the stronger base? Explain your answers.
   a. $H_2SO_4$ or $CH_3CO_2H$
   b. $H_2PO_4^-$ or $HPO_4^{2-}$
   c. $OH^-$ or $SH^-$

36. Which of the species in each pair is the stronger base? Explain your answers.
   a. $HSO_4^-$ or $HCO_3^-$
   b. $HS^-$ or $S^{2-}$
   c. $NH_2^-$ or $F^-$

## SECTION 15.5 Salts

37. Identify each salt as acidic, basic, or neutral. Explain your answers.
   a. $NaHCO_3$ b. LiBr c. $NH_4I$

38. Identify each salt as acidic, basic, or neutral. Explain your answers.
   a $Ca_3(PO_4)_2$ b. $KNO_3$ c. $NH_4Br$

39. Is each of these salts acidic, basic, or neutral? Explain your answers.
   a. $Ca(NO_3)_2$ b. $Na_2SO_4$ c. LiCl

40. Is each of these salts acidic, basic, or neutral? Explain your answers.
   a. $K_2S$ b. $NH_4NO_3$ c. $CaCO_3$

41. Write the equation for the reaction that occurs when 100 mL of an aqueous solution of each of these acids is slowly added to 100 mL of NaOH(*aq*). Assume that the concentration of both aqueous solutions is exactly 1.0 *M*. Predict whether the final solution is acidic, basic, or neutral.
   a. HCl b. $H_2SO_4$ c. $NH_4Cl$

42. Write the equation for the reaction that occurs when 100 mL of an aqueous solution of each of these acids is slowly added to 100 mL of NaOH(*aq*). Assume that the concentration of both aqueous solutions is exactly 1.0 *M*. Predict whether the final solution is acidic, basic, or neutral.
   a. $CH_3CO_2H$ b. $NH_4Cl$ c. HOCl

43. Write the equation for the reaction that occurs when 100 mL of an aqueous solution of each of these bases is slowly added to 100 mL of HBr(*aq*). Assume that the concentration of both aqueous solutions is exactly 1.0 *M*. Predict whether the final solution is acidic, basic, or neutral.
   a. $NH_3$ b. NaOCl c. KOH

44. Write the equation for the reaction that occurs when 100 mL of an aqueous solution of each of these is slowly added to 100 mL of HCl(*aq*). Assume that the concentration of both aqueous solutions is exactly 1.0 *M*. Predict whether the final solution is acidic, basic, or neutral.
   a. $NaNO_3$ b. $CH_3CO_2Na$ c. $NaHCO_3$

45. Explain exactly how you would prepare 1.0 L of a 0.50 *M* aqueous solution of each substance, given the starting materials.
   a. $Na_2SO_4$ from 1.0 *M* $H_2SO_4$(*aq*) and 50% (*m*/*m*) NaOH
   b. $NH_4Cl$ from 2.0 *M* $NH_3$(*aq*) and 3.0 *M* HCl(*aq*)

46. Explain exactly how you would prepare 0.50 L of a 0.25 *M* aqueous solution of each substance, given the starting materials.
   a. $NaClO_4$ from 1.0 *M* $HClO_4$(*aq*) and 2.0 *M* NaOH(*aq*)
   b. $CH_3CO_2Na$ from 2.0 *M* $CH_3CO_2H$(*aq*) and 3.0 *M* NaOH(*aq*)

## SECTION 15.6 Equilibria for the Reactions of Acids and Bases with Water

47. Write the $K_a$ expression for an aqueous solution of each of these acids.
   a. $Na_2HPO_4$
   b. $NH_4NO_3$
   c. $H_2CO_3$ (one for each proton)

48. Write the $K_a$ expression for an aqueous solution of each of these acids.
   a. HF
   b. HCN
   c. $NaH_2PO_4$ (one for each proton)

49. What is the $K_a$ expression for an aqueous solution of each of these acids?
   a. $CH_3CO_2H$
   b. $H_3PO_4$ (one for each proton)
   c. $NaHSO_4$

50. What is the $K_a$ expression for an aqueous solution of each of these acids?
   a. $NaHCO_3$
   b. $H_3BO_3$ (first proton only)
   c. $HCO_2H$ (one acidic proton)

51. For each of these acids in aqueous solution, write the $K_a$ expression.
   a. $HO_2CCO_2H$ (one for each proton)
   b. $NH_4Br$
   c. $H_3PO_2$ (first proton only)

52. For each of these acids in aqueous solution, write the $K_a$ expression.
   a. HOCl b. $KHSO_4$ c. $CH_3CH_2CO_2H$

53. Write the $K_b$ expression for an aqueous solution of each of these bases.
 a. $CH_3CO_2K$
 b. $Na_3PO_4$ (all three reactions)
 c. $NH_3$

54. Write the $K_b$ expression for an aqueous solution of each of these bases.
 a. KOCl  b. KCN  c. $HCO_2Li$

55. What is the $K_b$ expression for an aqueous solution of each of these bases?
 a. $Li_2HPO_4$ (both reactions)
 b. $Na_2SO_4$ (one reaction only)
 c. KF

56. What is the $K_b$ expression for an aqueous solution of each of these bases?
 a. $CH_3NH_2$  b. $KBrO_2$  c. $NaH_2BO_3$

57. Write a balanced equation for the reaction that occurs when $Mg_3N_2(s)$ is added to water.

58. Write a balanced equation for the reaction that occurs when $KH(s)$ is added to water.

59. One method for preparing a saturated aqueous solution of $Ca(OH)_2$ is to add 5.0 g of CaO to 1.0 L of water, stir vigorously, and then filter the solution. Write a balanced equation for the reaction between $CaO(s)$ and water.

60. In old miners' lamps, calcium carbide ($CaC_2$) reacts with water to give acetylene ($C_2H_2$) and calcium hydroxide. Write a balanced equation for this reaction.

## SECTION 15.7 Hydronium and Hydroxide Ion Concentrations

61. The $H_3O^+$ concentrations of three solutions are given. For each of these solutions, calculate the $OH^-$ concentration and classify the solutions as either acidic or basic.
 a. $9.1 \times 10^{-7}$ *M*
 b. 2.10 *M*
 c. $7.27 \times 10^{-11}$ *M*

62. The $H_3O^+$ concentrations of three solutions are given. For each of these solutions, calculate the $OH^-$ concentration and classify the solutions as either acidic or basic.
 a. $3.9 \times 10^{-3}$ *M*
 b. $1.51 \times 10^{-11}$ *M*
 c. $5.2 \times 10^{-5}$ *M*

63. The $H_3O^+$ concentrations of three solutions are given. For each of these solutions, calculate the $OH^-$ concentration and classify the solutions as either acidic or basic.
 a. $2.22 \times 10^{-9}$ *M*
 b. $4.2 \times 10^{-4}$ *M*
 c. $6.2 \times 10^{-13}$ *M*

64. The $H_3O^+$ concentrations of three solutions are given. For each of these solutions, calculate the $OH^-$ concentration and classify the solutions as either acidic or basic.
 a. $5.9 \times 10^{-11}$ *M*
 b. $9.44 \times 10^{-3}$ *M*
 c. $2 \times 10^{-6}$ *M*

65. The $H_3O^+$ concentrations of three solutions are given. For each of these solutions, calculate the $OH^-$ concentration and classify the solutions as either acidic or basic.
 a. 0.122 *M*  b. $4.55 \times 10^{-13}$ *M*  c. 0.450 *M*

66. The $H_3O^+$ concentrations of three solutions are given. For each of these solutions, calculate the $OH^-$ concentration and classify the solutions as either acidic or basic.
 a. $1.58 \times 10^{-3}$ *M*
 b. $1.05 \times 10^{-8}$ *M*
 c. 1.05 *M*

67. The $OH^-$ concentrations of three solutions are given. For each of these solutions, calculate the $H_3O^+$ concentration and classify the solutions as either acidic or basic.
 a. $4.07 \times 10^{-2}$ *M*
 b. $7.98 \times 10^{-12}$ *M*
 c. $1.94 \times 10^{-3}$ *M*

68. The $OH^-$ concentrations of three solutions are given. For each of these solutions, calculate the $H_3O^+$ concentration and classify the solutions as either acidic or basic.
 a. $9.00 \times 10^{-8}$ *M*
 b. $6.33 \times 10^{-5}$ *M*
 c. 1.2 *M*

69. The $OH^-$ concentrations of three solutions are given. For each of these solutions, calculate the $H_3O^+$ concentration and classify the solutions as either acidic or basic.
 a. $1.234 \times 10^{-5}$ *M*
 b. $7.73 \times 10^{-9}$ *M*
 c. $8.19 \times 10^{-11}$ *M*

70. The $OH^-$ concentrations of three solutions are given. For each of these solutions, calculate the $H_3O^+$ concentration and classify the solutions as either acidic or basic.
 a. $5.45 \times 10^{-7}$ *M*
 b. $7.29 \times 10^{-11}$ *M*
 c. $2.4 \times 10^{-3}$ *M*

71. The $OH^-$ concentrations of three solutions are given. For each of these solutions, calculate the $H_3O^+$ concentration and classify the solutions as either acidic or basic.
a. 0.0554 $M$
b. $1.33 \times 10^{-13}$ $M$
c. $6.05 \times 10^{-1}$ $M$

72. The $OH^-$ concentrations of three solutions are given. For each of these solutions, calculate the $H_3O^+$ concentration and classify the solutions as either acidic or basic.
a. 2.11 $M$
b. $4.66 \times 10^{-3}$ $M$
c. $4.2 \times 10^{-9}$ $M$

### TOOL BOX **Logarithms**

73. Write the logarithm for each of these numbers.
a. 9.052 b. $9.2 \times 10^{31}$ c. $5.05 \times 10^{-8}$

74. Write the logarithm for each of these numbers.
a. $2 \times 10^{7}$ b. $3.1 \times 10^{-12}$ c. 115

75. What is the logarithm of each of these numbers?
a. $7.23 \times 10^{-14}$ b. $8 \times 10^{7}$ c. 59.3

76. What is the logarithm of each of these numbers?
a. $8.5 \times 10^{-22}$ b. $4.9 \times 10^{7}$ c. 9.05

77. Give the logarithm of each of these numbers.
a. $4 \times 10^{-9}$ b. $1.32 \times 10^{17}$ c. $7 \times 10^{37}$

78. Give the logarithm of each of these numbers.
a. $2.48 \times 10^{-14}$ c. $4.14 \times 10^{-21}$
b. $8.1 \times 10^{-4}$

79. Write the antilog for each of these logarithms.
a. 7.91 b. $-0.5507$ c. $-23.77$

80. Write the antilog for each of these logarithms.
a. $-2.07$ b. 12.9 c. 2.3101

81. What is the antilog of each of these logarithms?
a. $-12.19$ b. 5.42 c. 3.447

82. What is the antilog of each of these logarithms?
a. 3.405 b. $-12.77$ c. $-15.690$

83. Give the antilog of each of these logarithms.
a. $-13.459$ b. 6.1237 c. $-5.556$

84. Give the antilog of each of these logarithms.
a. 23.301 b. $-5.55$ c. $-27.4353$

### SECTION 15.8 **A Measure of Acidity or Basicity: pH**

85. Identify each of these solutions as acidic, basic, or neutral from the pH given.
a. 4.7 c. 2.3 e. 7.0
b. 12.9 d. 8.7

86. Identify each of these solutions as acidic, basic, or neutral from the pH given.
a. 7.9 c. 11.1 e. 2.2
b. 5.1 d. 9.3

87. Tell whether each of these solutions is acidic, basic, or neutral from the pH given.
a. 13.9 c. 7.75 e. 3.8
b. 4.77 d. 9.32

88. Tell whether each of these solutions is acidic, basic, or neutral from the pH given.
a. 1.4 c. 6.89 e. 10.4
b. 11.1 d. 8.4

89. Find the pH of each substance from the given $H_3O^+$ concentration.
a. A stomach acid, $[H_3O^+] = 8.8 \times 10^{-3}$ $M$
b. Another stomach acid, $[H_3O^+] = 2.2 \times 10^{-2}$ $M$
c. Sea water, $[H_3O^+] = 2.5 \times 10^{-8}$ $M$
d. Milk of magnesia, $[H_3O^+] = 3.1 \times 10^{-11}$ $M$

90. Find the pH of each substance from the given $H_3O^+$ concentration.
a. Soft drink, $[H_3O^+] = 1.5 \times 10^{-4}$ $M$
b. An acid rain sample, $[H_3O^+] = 1.7 \times 10^{-4}$ $M$
c. Another rain sample, $[H_3O^+] = 3.1 \times 10^{-6}$ $M$
d. Perspiration, $[H_3O^+] = 2.5 \times 10^{-6}$ $M$

91. Find the pH of each of these samples from the given $[H_3O^+]$.
a. Lemon juice, $[H_3O^+] = 5.1 \times 10^{-3}$ $M$
b. Household ammonia, $[H_3O^+] = 1.25 \times 10^{-12}$ $M$
c. A brand of beer, $[H_3O^+] = 1.0 \times 10^{-4}$ $M$
d. A different beer, $[H_3O^+] = 2.2 \times 10^{-4}$ $M$

92. Find the pH of each of these samples from the given $[H_3O^+]$.
a. Milk, $[H_3O^+] = 1.1 \times 10^{-7}$ $M$
b. Saliva, $[H_3O^+] = 3.2 \times 10^{-7}$ $M$
c. Drain cleaner, $[H_3O^+] = 1.2 \times 10^{-15}$ $M$
d. A detergent, $[H_3O^+] = 5.77 \times 10^{-9}$ $M$

93. For a solution having each of these $H_3O^+$ concentrations given, calculate the pH.
a. $9.1 \times 10^{-7}$ $M$ c. $7.27 \times 10^{-11}$ $M$
b. 2.10 $M$ d. $3.9 \times 10^{-3}$ $M$

94. For a solution having each of these $H_3O^+$ concentrations given, calculate the pH.
a. $5.66 \times 10^{-10}$ $M$ c. 0.0123 $M$
b. $8.1 \times 10^{-7}$ $M$ d. $7.78 \times 10^{-4}$ $M$

95. For a solution with each of these values for $[H_3O^+]$, calculate the pH.
a. $2.22 \times 10^{-9}$ $M$ c. $6.2 \times 10^{-13}$ $M$
b. $4.2 \times 10^{-4}$ $M$ d. $5.9 \times 10^{-11}$ $M$

96. For a solution with each of these values for $[H_3O^+]$, calculate the pH.
a. $9.22 \times 10^{-6}$ $M$ c. $2.1 \times 10^{-3}$ $M$
b. $9.79 \times 10^{-11}$ $M$ d. $3.6 \times 10^{-8}$ $M$

97. Find both $[H_3O^+]$ and $[OH^-]$ for each of these samples from the given pH.
 a. Stomach acid (heartburn-inducing), pH = 1.5
 b. Vinegar, pH = 3.1
 c. Sea water, pH = 8.3
 d. HCl(*aq*), pH = 2.23

98. Find both $[H_3O^+]$ and $[OH^-]$ for each of these samples from the given pH.
 a. Acid rain sample, pH = 3.7
 b. A detergent, pH = 8.9
 c. A drain cleaner in water, pH = 14.3
 d. KOH(*aq*), pH = 11.22

99. Find both the $H_3O^+$ and the $OH^-$ concentration of each of these solutions from the given pH.
 a. Orange juice, pH = 2.4
 b. Soda water, pH = 4.0
 c. Baking soda in water, pH = 8.4
 d. NaOH(*aq*), pH = 12.75

100. Find both the $H_3O^+$ and the $OH^-$ concentration of each of these solutions from the given pH.
 a. An industrial detergent, pH = 9.6
 b. A borax solution, pH = 9.8
 c. $Ca(OH)_2(aq)$, pH = 11.8
 d. $H_2SO_4(aq)$, pH = 0.55

101. Find both $[H_3O^+]$ and $[OH^-]$ for each of these samples from the given pH.
 a. 3.8 b. 6.6 c. 8.98 d. 11.7

102. Find both $[H_3O^+]$ and $[OH^-]$ for each of these samples from the given pH.
 a. 10.1 b. 2.25 c. 5.38 d. 7.45

### SECTION 15.9 Experimental Determination of the pH of a Solution

103. Using Table 15.2 in Section 15.9 (p. 568), give the approximate pH of a solution that gives a yellow color with methyl orange and a yellow color with bromocresol purple.

104. Using Table 15.2 in Section 15.9, give the approximate pH of a solution that gives a blue color with bromocresol green and a yellow color with phenol red.

105. Using Table 15.2 in Section 15.9, give the approximate pH of a solution that gives a red color with phenol red and is colorless with thymolphthalein.

106. Using Table 15.2 in Section 15.9, give the approximate pH of a solution that gives a purple color with bromocresol purple and a faint red color with phenol red.

107. Using Table 15.2 in Section 15.9, give the approximate pH of a solution that gives a colorless solution with phenolphthalein and a red color with phenol red.

108. Using Table 15.2 in Section 15.9, give the approximate pH of a solution that gives a yellow color with phenol red and a yellow color with methyl red.

109. Some fellow chemists need a more precise and accurate value for the pH of a solution than they can get using indicators. What technique would you recommend?

110. You don't care what the pH of a certain sample is, you only want to know whether it is acidic or basic. What technique will give you this information reliably and quickly?

### SECTION 15.10 Concentrations from $K_a$ Expressions

111. Calculate the concentrations of $HS^-$, $H_3O^+$, and $OH^-$ in a 0.050 *M* solution of $H_2S$ in water ($K_{a1}$ = $9.1 \times 10^{-8}$ *M* for the first proton; you can ignore $K_{a2}$). What is the pH of this solution? (Note: $H_2S$ is toxic and smells terrible; do not handle any but the smallest amounts of it in the real world.)

112. Calculate the concentrations of $HPO_4^{2-}$, $H_3O^+$, and $OH^-$ in a 0.20 *M* solution of $NaH_2PO_4$ in water ($K_{a2}$ = $6.23 \times 10^{-8}$ *M* for the first proton in $NaH_2PO_4$; you can ignore other equilibria). What is the pH of this solution?

113. Calculate the concentrations of $H_2BO_3^-$, $H_3O^+$, and $OH^-$ in a 1.0 *M* solution of $H_3BO_3(aq)$ ($K_{a1}$ = 7.3 $\times 10^{-10}$ *M* for the first proton; you can ignore $K_{a2}$). What is the pH of this solution?

114. What are the concentrations of $F^-$, $H_3O^+$, and $OH^-$ in a 0.50 *M* solution of HF(*aq*) ($K_a = 3.5 \times 10^{-4}$ *M*). What is the pH of this solution?

115. Calculate the pH of 0.50 *M* $NH_4Cl(aq)$, $K_a$ = $5.6 \times 10^{-10}$ *M*.

116. Calculate the pH of $1.5 \times 10^{-3}$ *M* $HIO_3(aq)$, $K_a$ = 0.20 *M*.

117. Calculate the pH of 0.25 *M* $HNO_2(aq)$, $K_a$ = $5.1 \times 10^{-4}$ *M*.

118. Calculate the pH of 0.0100 *M* benzoic acid(*aq*), $K_a$ = $6.3 \times 10^{-5}$ *M*.

119. A chemist dissolved 0.25 mol of a weak acid in 500 mL of water. The pH of the solution was 2.3. What is the $K_a$ of this acid?

120. A chemist dissolved 0.100 mol of a weak acid in 200 mL of water. The pH of the solution was 3.1. What is the $K_a$ of this acid?

121. A solution prepared by dissolving 0.051 mol of a weak acid in 250 mL of water has a pH of 2.1. What is the $K_a$ of this acid?

122. A solution prepared by dissolving 0.12 mol of a weak acid in 400 mL of water has a pH of 2.9. What is the $K_a$ of this acid?

123. A solution prepared by dissolving 0.25 mol of a weak acid in 800 mL of water has a pH of 4.5. What is the $K_a$ of this acid?

124. A solution prepared by dissolving 0.18 mol of a weak acid in 500 mL of water has a pH of 3.7. What is the $K_a$ of this acid?

## SECTION 15.11 Buffer Solutions

125. Decide whether each of these mixtures yields a buffer in aqueous solution. Explain your answers.
   a. $NH_4Br + NH_3$
   b. Excess $HClO_4$ + LiOH
   c. $HClO_4$ + excess LiOH

126. Decide whether each of these mixtures yields a buffer in aqueous solution. Explain your answers.
   a. $H_2SO_4$ + excess $HNO_3$
   b. Excess KOH + $CH_3CO_2H$
   c. NaOH + excess $CH_3CO_2H$

127. Does each of these mixtures produce a buffer in aqueous solution? Explain your answers.
   a. HF + excess $HNO_3$
   b. Excess HF + LiOH
   c. HF + excess LiOH

128. Does each of these mixtures produce a buffer in aqueous solution? Explain your answers.
   a. $NH_3$ + excess NaOH
   b. Excess $CH_3CO_2H$ + HBr
   c. KOH + excess $CH_3CO_2H$

129. A buffer solution was prepared by dissolving 0.20 mol of $CH_3CO_2H$ ($K_a = 1.8 \times 10^{-5}$ *M*) and 0.20 mol of $CH_3CO_2Na$ in enough water to give 1 L of solution.
   a. What was the pH of this buffer solution?
   b. What was the pH of the buffer solution after $2 \times 10^{-2}$ mol of HCl was added?
   c. What was the pH of the buffer solution after $5 \times 10^{-3}$ mol of $OH^-$ was added?

130. A buffer solution was prepared by dissolving 0.10 mol of HF ($K_a = 3.5 \times 10^{-4}$ *M*) and 0.10 mol of NaF in enough water to give 1 L of solution.
   a. What was the pH of this buffer solution?
   b. What was the pH of the buffer solution after $2 \times 10^{-2}$ mol of $HNO_3$ was added?
   c. What was the pH of the buffer solution after $8 \times 10^{-3}$ mol of $OH^-$ was added?

131. A 5.00-mL portion of 0.100 *M* NaOH was added to 45.00 mL of 0.100 *M* $CH_3CO_2H$ ($K_a = 1.75 \times 10^{-5}$ *M*). What was the pH of the resulting solution?

132. A 10.00-mL portion of 0.200 *M* NaOH was added to 90 mL of 0.125 *M* $CH_3CO_2H$ ($K_a = 1.75 \times 10^{-5}$ *M*). What was the pH of the resulting solution?

133. If 5.00 mL of 0.250 *M* NaOH was added to 45.00 mL of 0.250 *M* $KH_2PO_4$ ($K_{a2} = 6.23 \times 10^{-8}$ *M*), what was the pH of the resulting solution?

134. If 5.00 mL of 0.50 *M* NaOH was added to 45.00 mL of 0.250 *M* $KH_2PO_4$ ($K_{a2} = 6.23 \times 10^{-8}$ *M*), what was the pH of the resulting solution?

135. What was the pH of the resulting solution when 10.00 mL of 0.100 *M* HCl was added to 90.00 mL of 0.100 *M* $NaNO_2$ ($K_a = 5.1 \times 10^{-4}$ *M*)?

136. What was the pH of the resulting solution when 5.00 mL of 0.250 *M* HCl was added to 45.00 mL of 0.200 *M* sodium benzoate ($K_a = 6.3 \times 10^{-5}$ *M*)?

## SECTION 15.12 Measuring Amounts of Acids and Bases: Titration

137. A 15.00-mL sample of $HNO_3(aq)$ required 21.78 mL of 0.205 *M* NaOH(*aq*) to be completely neutralized. What is the molarity of the $HNO_3(aq)$?

138. A 20.00-mL sample of HBr(*aq*) required 17.20 mL of 0.177 *M* NaOH(*aq*) to be completely neutralized. What is the molarity of the HBr(*aq*)?

139. A 10.00-mL sample of $H_2SO_4(aq)$ required 31.23 mL of 0.232 *M* NaOH(*aq*) to be completely neutralized. What is the molarity of the $H_2SO_4(aq)$?

140. A 22.00-mL sample of $H_2SO_4(aq)$ required 24.29 mL of 0.105 *M* NaOH(*aq*) to be completely neutralized. What is the molarity of the $H_2SO_4(aq)$?

141. A 25.00-mL sample of NaOH(*aq*) required 17.81 mL of 0.112 *M* HCl(*aq*) to be completely neutralized. What is the molarity of the NaOH(*aq*)?

142. A 30.00-mL sample of KOH(*aq*) required 41.34 mL of 0.345 *M* HCl(*aq*) to be completely neutralized. What is the molarity of the KOH(*aq*)?

143. A 1.2397-g sample of KHP was very carefully weighed out and dissolved in about 50 mL of water. The KHP solution was then titrated with an aqueous solution of NaOH. When 24.94 mL of the NaOH solution had been added, the resulting solution became neutral. What is the molarity of the NaOH(*aq*)?

144. A 2.8734-g sample of KHP was very carefully weighed out and dissolved in about 50 mL of water. The KHP solution was then titrated with an aqueous solution of NaOH. When 31.06 mL of the NaOH solution had been added, the resulting solution became neutral. What is the molarity of the NaOH(*aq*)?

145. A 2.0034-g sample of KHP was very carefully weighed out and dissolved in about 50 mL of water. The KHP solution was then titrated with an aqueous solution of NaOH. When 28.46 mL of the NaOH solution had been added, the KHP was neutralized. What is the molarity of the NaOH(*aq*)?

146. A 1.9282-g sample of KHP was very carefully weighed out and dissolved in about 50 mL of water. The KHP solution was then titrated with an aqueous solution of NaOH. When 41.12 mL of the NaOH solution had been added, the KHP was neutralized. What is the molarity of the NaOH(*aq*)?

### ADDITIONAL PROBLEMS

147. Suppose you have a 0.100 *M* aqueous solution of the weak base hydroxylamine ($NH_2OH$, whose $K_b = 9.1 \times 10^{-9}$). Calculate the equilibrium concentrations of $NH_3OH^+$, $H_3O^+$, and $OH^-$.

$$NH_2OH(aq) + H_2O(l) \rightleftharpoons NH_3OH^+(aq) + OH^-(aq)$$

148. What is the pH of a 0.250 *M* solution of $NH_3(aq)$ ($K_b = 1.8 \times 10^{-5}$ *M*)? Calculate the equilibrium concentrations of $NH_3$, $NH_4^+$, $H_3O^+$, and $OH^-$.

$$NH_3(aq) + H_2O(l) \rightleftharpoons NH_4^+(aq) + OH^-(aq)$$

## SOLUTIONS TO EXERCISES

### Exercise 15.1

**a.** The net ionic equation was given:

$$H_2SO_4 + CH_3COOH \rightleftharpoons HSO_4^- + CH_3COOH_2^+$$

$H_2SO_4$ donates a proton in the reaction, so it must be a Brønsted-Lowry acid. $HSO_4^-$ is a Brønsted-Lowry base. $CH_3COOH$ accepts a proton, so it acts as a base in this reaction. (This is true despite its name, acetic acid.) $CH_3COOH_2^+$ is an acid.

**b.** The net ionic equation was given:

$$PO_4^{3-} + H_2PO_4^- \rightleftharpoons 2\ HPO_4^{2-}$$

$PO_4^{3-}$ accepts a proton and is a base. $H_2PO_4^-$ donates a proton in this reaction and is therefore an acid here. On the right side, one $HPO_4^{2-}$ ion is an acid and the other is a base.

### Exercise 15.2

**a.** Add $H^+$ to $H_2S$ to get the formula of the conjugate acid, $H_3S^+$.

**b.** To find the formula of the conjugate base, subtract $H^+$. This gives $HS^-$.

**c.** The conjugate acid of $CH_3^-$ is $CH_4$.

**d.** The conjugate base of $CH_3^-$ is $CH_2^{2-}$.

### Exercise 15.3

**a.** HBr is made up of nonmetals and is a covalent gas. The net ionic equation is

$$HBr(g) + H_2O(l) \longrightarrow Br^-(aq) + H_3O^+(aq)$$

**b.** $HClO_4$ is a strong acid (see Section 8.8) and reacts completely with water to form $H_3O^+$. HCN is not a strong acid. The net ionic equation is therefore:

$$H_3O^+ + CN^- \longrightarrow H_2O + HCN$$

**c.** $CH_3OH_2^+ + OH^- \longrightarrow CH_3OH + H_2O$

### Exercise 15.4

First, classify all the alternatives. Compounds A, B, and C are strong acids. The compound that is least reactive with water is the weakest acid, HF (D).

### Exercise 15.5

The strongest acid is the compound with the nonhydrogen element farthest toward the right and bottom of the periodic table, so the weakest acid has the nonhydrogen element toward the left and top. The weakest acid in this group is therefore $H_2S$.

### Exercise 15.6

HClO is the weaker acid. Both oxyacids contain one hydrogen atom, but $HClO_4$ has four oxygen atoms to one for HClO.

### Exercise 15.7

The conjugate acids of these two ions are HClO and $HClO_3$, respectively. $HClO_3$ has more oxygens per hydrogen atom, so it's the stronger acid. Its conjugate base, $ClO_3^-$, is therefore the weaker base.

### Exercise 15.8

**a.** $NaHCO_3$ contains the $HCO_3^-$ ion, the conjugate base of $H_2CO_3$. $NaHCO_3$ is a basic salt. It is used in over-the-counter medicines to neutralize stomach acid.

**b.** $NH_4NO_3$ contains the $NH_4^+$ ion, which is the conjugate acid of $NH_3$. $NH_4NO_3$, an important fertilizer and potentially dangerous explosive, is acidic.

**c.** $Ba(O_2CCH_3)_2$ contains the acetate ion, which is the conjugate base of the weak acid acetic acid. Barium acetate is a basic salt.

### Exercise 15.9

**a.** $HCN + H_2O \rightleftharpoons H_3O^+ + CN^-$

$$K_a = \frac{[H_3O^+][CN^-]}{[HCN]}$$

**b.** $HS^- + H_2O \rightleftharpoons H_3O^+ + S^{2-}$

$$K_a = \frac{[H_3O^+][S^{2-}]}{[HS^-]}$$

This is the same as the $K_{a2}$ expression for $H_2S$ (you ignore the spectator ion $Na^+$).

**c.** $H_2CO_3 + H_2O \rightleftharpoons H_3O^+ + HCO_3^-$
$HCO_3^- + H_2O \rightleftharpoons H_3O^+ + CO_3^{2-}$

$$K_{a1} = \frac{[H_3O^+][HCO_3^-]}{[H_2CO_3]}$$

$$K_{a2} = \frac{[H_3O^+][CO_3^{2-}]}{[HCO_3^-]}$$

### Exercise 15.10

**a.** $CN^- + H_2O \rightleftharpoons HCN + OH^-$

$$K_b = \frac{[HCN][OH^-]}{[CN^-]}$$

**b.** $HCO_3^- + H_2O \rightleftharpoons H_2CO_3 + OH^-$

$$K_b = \frac{[H_2CO_3][OH^-]}{[HCO_3^-]}$$

**c.** There are two reactions, and as usual you treat them separately.

$S^{2-} + H_2O \rightleftharpoons HS^- + OH^-$
$HS^- + H_2O \rightleftharpoons H_2S + OH^-$

Each reaction has its own $K_b$ expression. As usual, you use $K_{b1}$ for the first reaction and $K_{b2}$ for the second.

$$K_{b1} = \frac{[HS^-][OH^-]}{[S^{2-}]} \quad \text{and} \quad K_{b2} = \frac{[H_2S][OH^-]}{[HS^-]}$$

### Exercise 15.11

**a.** Use the equation $[H_3O^+][OH^-] = 1.00 \times 10^{-14}$. With the given concentration substituted in, this becomes

$$(5.0 \times 10^{-7})[OH^-] = 1.00 \times 10^{-14}$$

$$[OH^-] = \frac{1.00 \times 10^{-14}}{5.0 \times 10^{-7}} = 2.0 \times 10^{-8}\ M$$

The given value of $[H_3O^+]$ is greater than $1 \times 10^{-7}\ M$, so the solution is acidic.

**b.**

$$[H_3O^+][OH^-] = 1.00 \times 10^{-14}$$

$$[H_3O^+](2.0 \times 10^{-1}) = 1.00 \times 10^{-14}$$

$$[H_3O^+] = \frac{1.00 \times 10^{-14}}{2.0 \times 10^{-1}} = 5.0 \times 10^{-14} M$$

The value of $[H_3O^+]$ is much smaller than $1 \times 10^{-7}$, so $[OH^-]$ must be much larger; the solution is basic.

### Exercise 15.12

Acidic pH values are below 7, and the more acidic the solution, the farther below 7 is its pH. The lowest pH value is A, which is the pH of the most acidic solution.

### Exercise 15.13

$$pH = -\log[H_3O^+] = -\log(4.0 \times 10^{-11}) = 10.40$$

### Exercise 15.14

If the pH is $-0.2$, the $H_3O^+$ concentration is

$$[H_3O^+] = \text{antilog}(-pH) = 10^{-(-0.2)} = 1.6\ M$$

From the pH this drain cleaner is strongly acidic rather than basic. It can't contain NaOH, which is a strong base. (This cleaner probably contains $H_2SO_4$. Handle *any* drain cleaner with *extreme* care! Just because you can buy these chemicals in a grocery store, that doesn't mean they're safe.)

### Exercise 15.15

Bromocresol purple is purple at pH values of 6.8 or higher and phenolphthalein is colorless at pH values of 8.0 or lower. The pH of the solution is between 6.8 and 8.0.

### Exercise 15.16

The equilibrium is

$$H_2S + H_2O \rightleftharpoons HS^- + H_3O^+$$

The equilibrium constant expression is:

$$K_a = 9.1 \times 10^{-8}\ M = \frac{[HS^-][H_3O^+]}{[H_2S]}$$

The initial concentration of $H_2S$ is much greater than the $K_a$, so you can use $0.10\ M$ for $[H_2S]$ in the equilibrium constant expression. Let $x = [HS^-] = [H_3O^+]$. The equilibrium constant expression is then:

$$9.1 \times 10^{-8}\ \text{mol L}^{-1} = \frac{[HS^-][H_3O^+]}{[H_2S]} = \frac{(x)(x)}{0.10\ M} = \frac{x^2}{0.10\ M}$$

$$(9.1 \times 10^{-8}\ M)(0.10\ M) = 9.1 \times 10^{-9}\ M^2 = x^2$$

$$x = 9.5 \times 10^{-5}\ M$$

$[HS^-] = [H_3O^+] = 9.5 \times 10^{-5}$ $M$. To calculate $[OH^-]$, use the expression for $K_w$.

$$K_w = 1.0 \times 10^{-14} = [H_3O^+][OH^-]$$
$$1.0 \times 10^{-14} = (9.5 \times 10^{-5})[OH^-]$$
$$[OH^-] = \frac{1.0 \times 10^{-14}}{9.5 \times 10^{-5}} = 1.0 \times 10^{-10}\ M$$

Finally, the pH is

$$pH = -\log[H_3O^+] = -\log(9.5 \times 10^{-5}) = 4.02$$

### Exercise 15.17

Mixture A contains two acids, which won't buffer at all. Mixture B yields a solution containing two bases, NaOH and $CH_3CO_2Na$, and that won't work either. Mixture D has the same problem as A; you end up with a solution containing two acids. Mixture C is the answer because the resulting solution contains $CH_3CO_2H$ and $CH_3CO_2Na$.

### Exercise 15.18

**a.** You recognize that $H_2PO_4^-$ is the acid and $HPO_4^{2-}$ is the base. Since $[H_2PO_4^-] = [HPO_4^{2-}]$ in the original solution, $pH = pK_a$.

$$pH = pK_a = -\log(K_a) = -\log(6.2 \times 10^{-8}) = 7.21$$

**b.** Adding acid to the solution gives the reaction

$$HPO_4^{2-} + H_3O^+ \longrightarrow H_2PO_4^- + H_2O$$

This increases the concentration of $H_2PO_4^-$ and decreases the concentration of $HPO_4^{2-}$. You can determine the exact changes using the usual table:

| | $[H_2PO_4^-]$ | $[HPO_4^{2-}]$ |
|---|---|---|
| Initial | 0.50 | 0.50 |
| Change | $+1 \times 10^{-2}$ | $-1 \times 10^{-2}$ |
| Final | 0.51 | 0.49 |

The Henderson-Hasselbalch Equation gives the final answer:

$$pH = pK_a + \log\left(\frac{[HPO_4^{2-}]}{[H_2PO_4^-]}\right) = 7.21 + \log\left(\frac{0.49}{0.51}\right)$$
$$= 7.21 + \log(0.96) = 7.21 + (-0.017) = 7.19$$

**c.** Adding base to the solution means that the following reaction occurs:

$$H_2PO_4^- + OH^- \longrightarrow HPO_4^{2-} + H_2O$$

This reaction increases the concentration of $HPO_4^{2-}$ and decreases the concentration of $H_2PO_4^-$. You make the table:

| | $[H_2PO_4^-]$ | $[HPO_4^{2-}]$ |
|---|---|---|
| Initial | 0.50 | 0.50 |
| Change | $-1 \times 10^{-2}$ | $+1 \times 10^{-2}$ |
| Final | 0.49 | 0.51 |

$$pH = 7.21 + \log\left(\frac{0.51}{0.49}\right) = 7.21 + \log(1.048)$$
$$= 7.21 + 0.017 = 7.23$$

### Exercise 15.19

$$\text{Moles of HCl} = \text{Moles of NaOH}$$
$$= (0.03326\ \text{L})(0.102\ \text{mol L}^{-1})$$
$$= 3.39 \times 10^{-3}\ \text{mol}$$
$$\text{Molarity of HCl} = \frac{3.39 \times 10^{-3}\ \text{mol}}{0.02000\ \text{L}} = 0.170\ M$$

### Exercise 15.20

$$\text{Moles of NaOH} = \text{Moles of } H_3O^+$$
$$= (0.01987\ \text{L})(1.23\ \text{mol L}^{-1})$$
$$= 2.44 \times 10^{-2}\ \text{mol}$$
$$\text{Molarity of NaOH} = \frac{2.44 \times 10^{-2}\ \text{mol}}{0.0200\ \text{L}} = 1.22\ M$$

### Exercise 15.21

$$\text{Moles of NaOH} = \text{Moles of KHP}$$
$$= \frac{1.000\ \text{g}}{204.23\ \text{g mol}^{-1}} = 4.896 \times 10^{-3}\ \text{mol}$$
$$\text{Molarity} = \frac{4.896 \times 10^{-3}\ \text{mol}}{0.04693\ \text{L}} = 0.1043\ M$$

### Tool Box Exercise 1

**a.** (1) 0.4342 (2) −20.3 (3) 9.86
**b.** (1) $3.7 \times 10^2$ (2) $3.7 \times 10^7$ (3) $3.7 \times 10^{-4}$

### Experimental Exercise 15.1

1. Solution 1 is most likely a solution of the weak acid, which the team called HA. The dominant species in the solution (other than water) is HA itself, which reacts only slightly with water to form $H_3O^+$ and $A^-$.
2. Phenolphthalein is colorless below pH 8.0, so Solution 1 would remain colorless.
3. The reaction is $HA(aq) + OH^-(aq) \rightleftharpoons A^-(aq) + H_2O(l)$.
4. The major species in the neutralized solution is $A^-$. A large amount of the spectator ion $Na^+$ is also present.
5. A 10-mL portion of Solution 1 contains some specific amount of HA. When neutralized, all the HA is converted to $A^-$. Mixing the neutralized solution with a 10-mL portion of Solution 1 gives a solution that contains equal amounts of HA and $A^-$. This is a classic buffer solution.
6. No, it isn't. A dilute solution of a strong acid would give the same results. Indicators only measure $H_3O^+$ concentrations, not acid strength.
7. Yes, it is. Weak acids have $K_a$ values less than 1 and greater than $10^{-14}$, generally speaking.
8. Any water-soluble hydroxide could be used. It is the $OH^-$ ion that neutralized the acid, so any source of that ion will work.
9. If Solution 1 contained a weak base instead of a weak acid, it would have to be neutralized with a strong acid (such as HCl) instead of a base. The rest of the procedure would work fine.

16

# OXIDATION-REDUCTION REACTIONS



KEY TERMS

This chapter introduces another class of reactions vitally important both to industry and to life on earth—oxidation-reduction (often called redox) reactions. Utility companies use redox reactions to convert the potential energy of fossil fuels to electrical energy. Your body's cells use other oxidation-reduction reactions to convert food energy to a form of energy you can use to move and think. Redox chemistry shares a quality common to many of the topics in this book—it is important to you personally, for your very existence. (See Figure 16.1.)

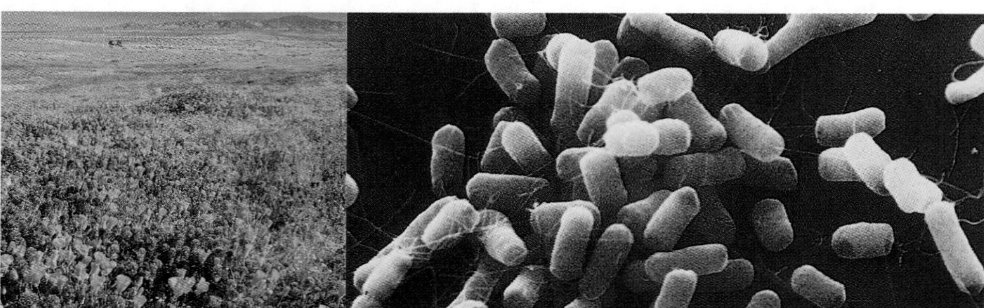

FIGURE 16.1
A power plant and the people working in it, flowers, and bacteria all use oxidation-reduction reactions to convert fuel to usable energy.

Early chemists thought that the processes now called redox reactions were transfers of phlogiston. Alchemists theorized that when metal samples were heated in air, they "lost phlogiston." Lavoisier pointed out that the samples gained mass, and that a theory proposing that a gain in mass is caused by a loss of something else was not likely to be a good one. He proposed instead that the metals reacted with a substance present in the air, a substance he called *oxygen*. The metals gained mass in their reaction with air, he said, because oxygen added to the metal. Lavoisier's proposal matched well with observed reality. Chemists' present understanding of oxidation-reduction reactions begins from that point. Also, Lavoisier's name for the substance in the air, oxygen, stuck. Joseph Priestley was the actual discoverer of oxygen, but his name, "dephlogisticated air," fell by the wayside.

**See Section 8.5 for our previous treatment of oxidation and reduction reactions.**

## 16.1 Oxidation and Reduction: Definitions

GOAL: To define and give examples of oxidation, reduction, oxidizing agents, and reducing agents.

The term *oxidation* originally referred to reactions with oxygen that gave oxygen-containing products. The reaction of aluminum metal provides one example. Aluminum is relatively corrosion-proof because its surface oxidizes to give a thin, almost impervious film of aluminum oxide. The reaction is

$$4\,Al(s) + 3\,O_2(g) \longrightarrow 2\,Al_2O_3(s)$$

The oxide covers the surface of the metal and prevents metal below the surface from oxidizing. This is the main reason why aluminum can be used in cooking utensils. The oxide coating is relatively inert and prevents the reactive metal itself from coming into contact with the acids (vinegar, for example) and bases (baking soda) used in cooking. Otherwise, the aluminum itself would react with your food. (See Figure 16.2.)

FIGURE 16.2
From foil through cans and cookware to airplane fuselages, aluminum metal is ubiquitous in modern society.

Iron objects become rusty when the metallic iron oxidizes to iron(III) oxide. The equation for the oxidation of iron is similar to that for aluminum, though rusting requires water and is much different in detail:

$$4\,\mathrm{Fe}(s) + 3\,\mathrm{O_2}(g) \longrightarrow 2\,\mathrm{Fe_2O_3}(s)$$

The equations may look alike, but the products are quite different. The rust that forms on the surface of iron and some of its steel alloys isn't impervious. It flakes off, exposing more metal below the surface, which continues to oxidize. This process can continue until the metal is all gone (see Figure 16.3). This kind of practical consequence can make all the difference, both in the laboratory and in the everyday world.

Just as chemists originally defined *oxidation* as the addition of oxygen, they defined *reduction* as the loss of oxygen. Refiners convert most iron ore (and recycled rust) to the metal by a reduction reaction with carbon monoxide:

$$\mathrm{Fe_2O_3}(s) + 3\,\mathrm{CO}(g) \longrightarrow 2\,\mathrm{Fe}(s) + 3\,\mathrm{CO_2}(g)$$

Copper metal is produced from the oxide similarly by reduction with either hydrogen or carbon:

$$\mathrm{CuO}(s) + \mathrm{H_2}(g) \longrightarrow \mathrm{Cu}(s) + \mathrm{H_2O}(g)$$
$$2\,\mathrm{CuO}(s) + \mathrm{C}(s) \longrightarrow 2\,\mathrm{Cu}(s) + \mathrm{CO_2}(g)$$

FIGURE 16.3
Given enough time, this metal wheel will rust completely away.

As chemists continued to investigate useful reactions like these, it became clear that the gain and loss of oxygen as a definition of oxidation and reduction was too limiting. As with acid-base reactions, thinking about the chemistry led to a new definition of oxidation and reduction. According to the

FIGURE 16.4
You can remember which name goes with which process by using this (somewhat cutesy) mnemonic about Leo the Lion: "LEO says GER." Loss of Electrons = Oxidation, Gain of Electrons = Reduction.

modern definitions, **oxidation** is the loss of electrons by an atom, molecule, or ion, and **reduction** is the gain of electrons (see Figure 16.4).

Electrons do not float free in ordinary matter; they're always associated with an atom or group of atoms. Chemists therefore concluded that every time an atom, molecule, or ion loses an electron (is oxidized), some other species must gain an electron (be reduced). Conversely, whenever something gains an electron, something else must have lost it. In other words, *oxidation and reduction always occur together.* Therefore, this type of reaction is often called an **oxidation-reduction reaction** (or **redox reaction,** for short).

To appreciate the usefulness of the modern definition, consider an oxidation-reduction reaction that doesn't involve oxygen, such as the one illustrated in parts (A) and (B) of Figure 16.5. If you put a piece of copper wire into a colorless aqueous solution of silver nitrate, the solution soon shows the blue color characteristic of copper(II) ions. At the same time, the copper wire becomes coated with whiskers of metallic silver. The net ionic equation for this reaction is

$$2\,Ag^{+}(aq) + Cu(s) \longrightarrow 2\,Ag(s) + Cu^{2+}(aq)$$

Comparing the charges of the reactants with those of the products, we see that each silver ion gains an electron ($Ag^{+} + e^{-} \rightarrow Ag$). The silver ions are *reduced* to silver atoms. At the same time, each copper atom loses two electrons ($Cu \rightarrow Cu^{2+} + 2\,e^{-}$) and is *oxidized.*

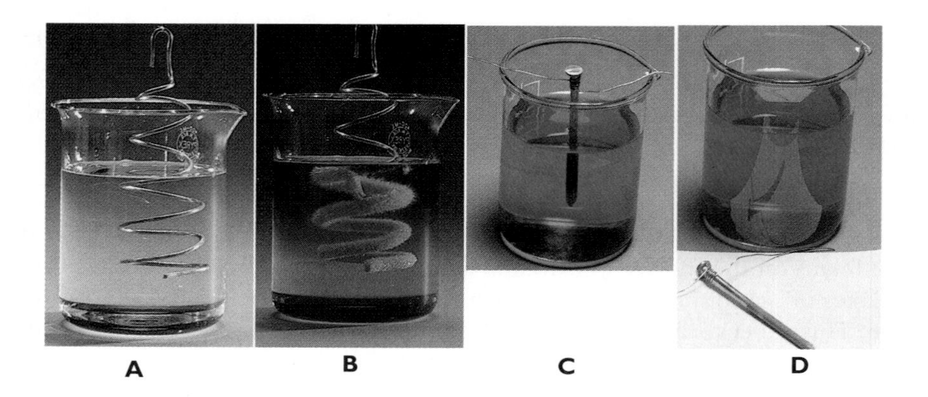


FIGURE 16.5
Redox reactions that do not involve oxygen. (A) A solution of $AgNO_3(aq)$ in which a copper wire has just been immersed. (B) After a few minutes you can see a blue color in the solution, indicating the presence of $Cu^{2+}(aq)$ ions. The copper wire has become coated with silver metal. (C) A solution of $CuSO_4(aq)$ just after an iron nail was dropped into it. (D) The color intensity of the solution has decreased, and the nail is coated with copper metal.

The definitions of oxidation and reduction focus on the atom that is acted upon. We can also think in terms of one species causing the other to be oxidized or reduced. From this point of view, each copper atom in the reaction illustrated in Figure 16.5 causes two silver ions to be reduced. Consequently, we can say that copper is the **reducing agent** in this reaction. Similarly, the silver ions cause the oxidation of the copper atom, so silver ion is the **oxidizing agent.** Note that an oxidizing agent is reduced, and a reducing agent is oxidized.

**STUDY ALERT**
**It is very easy to get oxidized, reduced, oxidizing agent, and reducing agent confused. Take a few moments *right now,* and get it down pat.**

Figure 16.5 shows another redox reaction in parts (C) and (D). Here an iron nail has been suspended in an aqueous solution of copper(II) sulfate. After a few minutes, the nail becomes completely coated with a reddish-brown layer of metallic copper, and the blue color of the dissolved copper ions is less intense. The net ionic equation for this reaction is

$$Cu^{2+}(aq) + Fe(s) \longrightarrow Cu(s) + Fe^{2+}(aq)$$

This time copper ion is being reduced and is the oxidizing agent, causing the oxidation of iron. What makes copper switch roles in the two reactions? This aspect of redox reactions is another point of similarity to acid-base reactions. Oxidizing and reducing agents have varying strengths relative to one another, just as acids and bases do. This idea will be pursued further in Section 16.2.

### EXAMPLE 16.1 Identifying Oxidizing and Reducing Agents

In each of these reactions, identify the oxidizing agent, the reducing agent, the substance that is oxidized, and the substance that is reduced.

**a.** $4\,Al(s) + 3\,O_2(g) \rightarrow 2\,Al_2O_3(s)$
**b.** $2\,Na(s) + Cl_2(g) \rightarrow 2\,NaCl(s)$

**SOLUTIONS**

**a.** According to the old definition of oxidation, aluminum gains oxygen and so should be oxidized. Let's check this conclusion against the modern definition. Aluminum goes from neutral Al atoms to $Al^{3+}$ ions, so it loses electrons and is oxidized. Aluminum, $Al(s)$, is therefore the reducing agent. Oxygen, $O_2(g)$, is reduced and is the oxidizing agent.
**b.** Sodium is transformed from neutral Na atoms to $Na^+$ ions. It loses electrons and is oxidized; $Na(s)$ is the reducing agent. Chlorine goes from $Cl_2$ to $2\,Cl^-$, gaining electrons; $Cl_2(g)$ is reduced and is the oxidizing agent.

**EXERCISE 16.1**

In each of these reactions, identify the oxidizing agent, the reducing agent, the substance that is oxidized, and the substance that is reduced.

**a.** $CuO(s) + H_2(g) \rightarrow Cu(s) + H_2O(g)$
**b.** $2\,Na(s) + 2\,H_2O(l) \rightarrow 2\,NaOH(aq) + H_2(g)$

It is sometimes difficult to tell whether a covalent substance is oxidized or reduced. Let's consider two common reactants, hydrogen and oxygen gases.

Hydrogen gas is a covalent molecule, $H_2$. Hydrogen is more electronegative than metals but has a lower electronegativity than most other nonmetals. Whenever $H_2$ is a reactant, forming a covalent compound as a product, chemists almost always consider it to be oxidized. To aid in explaining why, here is the classic reaction between $H_2$ and $O_2$ to form water:

$$2\,H_2(g) + O_2(g) \longrightarrow 2\,H_2O(l)$$

You know that oxygen atoms are more electronegative than hydrogen atoms, so it is reasonable to expect that the hydrogens give up their electrons to the oxygens in the formation of water. Of course, this isn't strictly true. Water is a covalent molecule and contains no ions at all; the hydrogen atoms do not truly give up their electrons. However, a close look at the structure of water is revealing (see Figure 16.6). Both bonds in the water molecule are polar; the

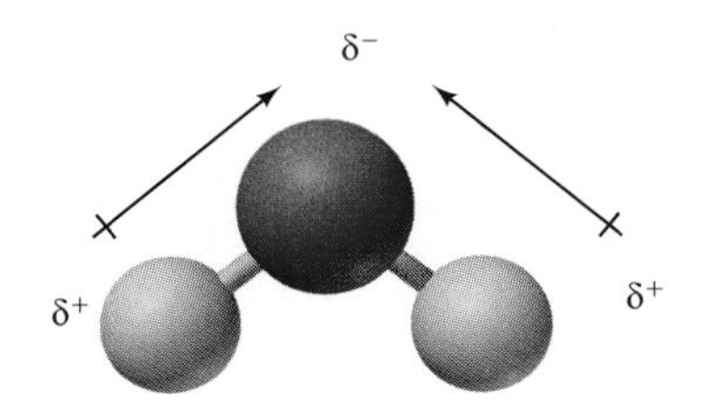


FIGURE 16.6

hydrogen atoms have considerable positive character. The hydrogen atoms that had no charge at all in $H_2$ now have some positive charge in $H_2O$. Thus, for the purposes of redox chemistry, hydrogen is considered to be oxidized in this and all similar reactions. The opposite is also true. When $H_2(g)$ is formed as a product from a covalent compound, that compound is reduced.

The arguments for oxygen are similar. Oxygen is one of the most electronegative elements, and in almost all of its covalent compounds (including water) it has a partial negative charge. When $O_2(g)$ is a reactant and forms a covalent product, the oxygen is reduced and, of course, is the oxidizing agent. As with hydrogen, the opposite is also true. When $O_2(g)$ is formed as a product from a covalent compound, that compound is oxidized. Since oxygen is very electronegative, such reactions do not occur easily, but chemists can force them to occur. Here's an example:

$$6\,H_2O(l) + 2\,Cu^{2+}(aq) \xrightarrow{\text{electricity}} O_2(g) + 4\,H_3O^+ + 2\,Cu(s)$$

In this reaction the water is oxidized and the copper ions are reduced.

## EXAMPLE 16.2 Identifying Covalent Oxidizing and Reducing Agents

In each of these reactions, identify the oxidizing agent, the reducing agent, the substance that is oxidized, and the substance that is reduced.

**a.** $CH_3CHO + H_2 \xrightarrow{\text{Ni}} CH_3CH_2OH$

**b.** $CH_4 + 2\,O_2 \rightarrow CO_2 + 2\,H_2O$

### SOLUTIONS

**a.** Hydrogen reacts to form a covalent compound. Therefore, $H_2$ is oxidized and is the reducing agent. The hydrogen atoms have partial positive charges in the product. Something else must be the oxidizing agent (the compound reduced), and $CH_3CHO$ is the only other substance around.

**b.** Oxygen reacts to form covalent products, and under those circumstances it is almost always the oxidizing agent and the substance reduced. Since methane ($CH_4$) is the only other compound present, it must be the reducing agent and the substance oxidized.

### EXERCISE 16.2

In each of these reactions, identify the oxidizing agent, the reducing agent, the substance that is oxidized, and the substance that is reduced.

**a.** $CH_3CH_2SH + H_2 \xrightarrow{\text{Ni}} CH_3CH_3 + H_2S$

**b.** $2\,CH_2{=}CH_2 + O_2 \xrightarrow{\text{Ag, 250°C}} 2\,CH_2{-}CH_2$ (with O bridging the two carbons, forming a three-membered ring)

## EXPERIMENTAL EXERCISE

## 16.1

Ernst G. Stahl was a famous German physician in the early part of the 1700s. His medical lectures at the University of Halle were well attended, and by 1716 he was the personal physician of the Prussian king, Frederick William I.

Around 1700 Stahl turned his attention to the study of combustion, the first-known of all redox reactions (of course, they weren't called that then). Combustion was a particularly important reaction in the 1700s (and still is), although it was not well understood. Metals could not be isolated from their ores unless they were burned in charcoal. A brand-new invention, the steam engine, required combustion to make the steam that ran the engine. Artificial lighting, particularly in the cities, was a major need and could only be supplied by burning something.

Stahl proposed that combustible objects were rich in a substance called *phlogiston* (from the Greek, "to set on fire"). Burning an object caused it to lose phlogiston, either to the air or to another object. For example, wood and charcoal were rich in phlogiston, and the ashes remaining after burning them were not. When a metallic ore was roasted in charcoal, the phlogiston transferred from the charcoal to the metal. Stahl also proposed that the rusting of iron was simply a type of combustion. This brilliant observation is essentially correct and was far from obvious.

The problem with Stahl's theory was that when pure metals are heated in air, they gain mass (Boyle had showed this 50 years earlier), even though they supposedly lose phlogiston. This is exactly the opposite of what happens when charcoal burns. In that case the remaining ashes have less mass than the original charcoal. This apparent flaw in the phlogiston theory was pretty much ignored until Antoine Lavoisier got around to looking at combustion.

Lavoisier placed a diamond and some air in a large sealed glass vessel and weighed the entire setup. He burned the diamond until it completely disappeared and reweighed the apparatus. Although no diamond was present in the flask, the total mass was the same before and after the experiment. When Lavoisier examined the contents of the flask, he found that the remaining air contained carbon dioxide. Lavoisier made three major conclusions from these and other observations: (1) Diamonds consist of pure carbon, (2) air is necessary to support combustion, and (3) the phlogiston theory is worthless—part of the air is transferred from one substance to another.

Lavoisier placed a diamond in a sealed flask and burned it. The total mass of the flask and contents was the same before and after combustion. The gas inside the flask contained carbon dioxide at the end of the experiment. These and other facts led Lavoisier to conclude that air is necessary to support combustion.

About this time Joseph Priestley came to Paris and, like everyone who was anyone in the field of chemistry, paid a visit to Lavoisier's laboratory. Priestley told Lavoisier about his experiments in isolating "dephlogisticated air." Lavoisier immediately recognized that air contains two components (an idea vaguely advanced by others), one of which supports combustion and one of which does not. He named the combustion-supporting substance oxygen (the other substance was eventually named nitrogen).

### QUESTIONS

1. Give an argument that Stahl might make to support his idea that metals are rich in phlogiston and metallic ores are low in phlogiston.
2. Was Lavoisier's experiment with the diamond sufficient to conclude that air was necessary to support combustion?
3. Give a modern argument supporting Stahl's proposal that the rusting of iron is a type of combustion.
4. Lavoisier also performed the experiments described below. Which of them helped him conclude that air contains two substances?
   - **A.** Lavoisier showed that $CO_2(g)$ was exhaled by animals when placed in a chamber filled with air or in one filled with oxygen.
   - **B.** He heated a large amount of metal in a small amount of air and observed that not all the air was consumed by the metal.
   - **C.** He heated mercury oxide in an evacuated vessel and produced oxygen gas.

## 16.2 Oxidizing and Reducing Agents

GOAL: To use half-reactions to investigate properties of some common oxidizing and reducing agents.

Oxidation-reduction reactions have many things in common with acid-base reactions. For example, we can say that oxidizing agents are substances that react with reducing agents (and vice versa), just as acids are substances that react with bases (and vice versa). The conjugate bases of strong acids are weak, and the reducing agents that form from strong oxidizing agents are weak. This analogy can be taken further. Brønsted-Lowry acid-base reactions involve the transfer of protons; the reaction of acetic acid with ammonia is an example:

$$CH_3CO_2H + NH_3 \longrightarrow CH_3CO_2^- + NH_4^+$$

Although the proton, $H^+$, does not actually exist as an independent species in chemical reactions, you can think of this reaction as taking place in two steps, which can be called half-reactions:

First: $CH_3CO_2H \longrightarrow CH_3CO_2^- + H^+$

Then: $NH_3 + H^+ \longrightarrow NH_4^+$

Adding the two steps gives the overall reaction (since $H^+$ occurs on both sides, it cancels out). You can express any acid-base reaction in the same way, using separate half-reactions for the acid and the base.

The equations for oxidation-reduction reactions are often more complex than acid-base reactions, although those we have looked at so far in this chapter haven't been. This complexity is why chemists often use a half-reaction approach to balance redox equations. (We'll look at the full procedure later in the chapter.) As a simple example to illustrate the process, let's consider the reaction between calcium metal and oxygen. The balanced equation is

$$2\,Ca(s) + O_2(g) \longrightarrow 2\,CaO(s)$$

Calcium metal loses electrons and is the reducing agent. Its half-reaction is

$$2\,Ca \longrightarrow 2\,Ca^{2+} + 4\,e^-$$

Oxygen is the oxidizing agent; $O_2$ gains electrons. Its half-reaction is:

$$O_2 + 4\,e^- \longrightarrow 2\,O^{2-}$$

When we add the two half-reactions and then combine the $Ca^{2+}$ and $O^{2-}$ ions to give CaO, we get the overall reaction (see also Figure 16.7):

$$\begin{array}{lllll} & 2\,Ca & & \longrightarrow & 2\,Ca^{2+} + \cancel{4\,e^-} \\ + & & O_2 + \cancel{4\,e^-} & \longrightarrow & 2\,O^{2-} \\ \hline & 2\,Ca & + O_2 & \longrightarrow & 2\,Ca^{2+} + 2\,O^{2-} \\ \text{or} & 2\,Ca(s) & + O_2(g) & \longrightarrow & 2\,CaO(s) \end{array}$$

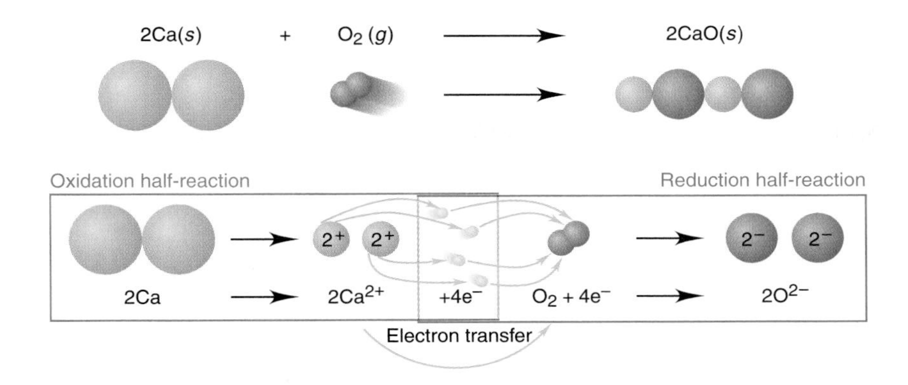


FIGURE 16.7

The more readily an oxidizing agent gains electrons, the more powerful it is. Electronegative elements and cations of weakly electropositive metals will accept electrons readily.

These half-reactions make several things about redox reactions clear. You can easily see that oxidizing agents gain electrons and reducing agents lose them, for instance. Many of the more common oxidizing agents (but by no means all of them) are electronegative elements and oxygen-containing molecules and ions (see Figure 16.8). Many cations are useful oxidizing agents. Cations of weakly electropositive metals having high positive charges (for example, $Sn^{4+}$, $Fe^{3+}$, and $Co^{3+}$) can be quite strong oxidizing agents.

Electropositive elements (that is, metals) are often useful as reducing agents, just as electronegative elements (oxygen, for example) are useful as oxidizing agents. Elements such as sodium readily give up valence electrons to form ions that have noble gas electron configurations. For this reason, the metals are strong reducing agents. The metals of Groups 1A and 2A, aluminum, and iron are examples.

FIGURE 16.8
Potassium permanganate, $KMnO_4$, is a powerful oxygen-containing oxidizing agent used in water purification. Solutions of $KMnO_4$ have this distinctive purple color.

### EXAMPLE 16.3 Identifying Oxidizing and Reducing Agents

In each of these pairs, predict which of the two species is the more powerful oxidizing agent. Explain your answers.

**a.** $Br_2$ or $Cl_2$ **b.** $Fe^{3+}$ or $Fe^{2+}$ **c.** $NO_2^-$ or $NO_3^-$

**SOLUTIONS**

**a.** $Cl_2$ is the more powerful oxidizing agent. The more electronegative element is more powerful as an oxidizing agent.
**b.** $Fe^{3+}$ is more powerful. With a larger positive charge it attracts electrons more strongly.
**c.** $NO_3^-$ is more powerful. It contains more oxygen atoms.

**EXERCISE 16.3**

In each of these pairs, predict which of the two species will be the more powerful oxidizing agent. Explain your answers.

**a.** $F_2$ or $I_2$ **b.** $Sn^{2+}$ or $Sn^{4+}$ **c.** $MnO_4^-$ or $MnO_3^-$

You can find half-reactions for a number of common oxidizing and reducing agents in Table 16.1. By convention, all half-reactions are written as reductions in these tables. The strongest reducing agents appear as the *products* of the reactions at the top of our table, and the strongest oxidizing agents are the *reactants* in the reactions at the bottom. (Not all tables use this order; some put the strongest oxidizing agents at the top.)

One of the most useful things about Table 16.1 is that it allows you to predict the products and the direction of redox reactions. All redox reactions involve two half-reactions. The *half-reaction closer to the top of the table is always the oxidation half-reaction* for a spontaneous reaction (and you must reverse it when copying it from the table). The *half-reaction nearer the bottom is the reduction half-reaction.*

**TABLE 16.1 Half-Reactions for Some Common Oxidizing and Reducing Agents**

Half-reactions that take place in acidic solution. In all cases the oxidizing agent is a reactant and the reducing agent is a product. The closer the half-reaction is to the top of the table, the weaker the oxidizing agent and the stronger the reducing agent.

Strength of reactant as an oxidizing agent (increases downward) / Strength of product as a reducing agent (increases upward)

| Half-reaction |
|---|
| $Na^+ + e^- \rightarrow Na(s)$ |
| $Al^{3+} + 3\ e^- \rightarrow Al(s)$ |
| $2\ H_2O + 2\ e^- \rightarrow H_2(g) + 2\ OH^-$ |
| $Zn^{2+} + 2\ e^- \rightarrow Zn(s)$ |
| $Fe^{2+} + 2\ e^- \rightarrow Fe(s)$ |
| $Sn^{2+} + 2\ e^- \rightarrow Sn$ |
| $2\ H_3O^+ + 2e^- \rightarrow H_2(g) + 2\ H_2O(l)$ |
| $Cu^{2+} + 2\ e^- \rightarrow Cu(s)$ |
| $I_2(s) + 2\ e^- \rightarrow 2\ I^-$ |
| $Fe^{3+} + e^- \rightarrow Fe^{2+}$ |
| $Ag^+ + e^- \rightarrow Ag(s)$ |
| $2\ NO_3^- + 4\ H_3O^+ + 2\ e^- \rightarrow N_2O_4(g) + 6\ H_2O$ |
| $Br_2 + 2\ e^- \rightarrow 2\ Br^-$ |
| $O_2(g) + 4\ H_3O^+ + 4\ e^- \rightarrow 6\ H_2O$ |
| $Cr_2O_7^{2-} + 14\ H_3O^+ + 6\ e^- \rightarrow 2\ Cr^{3+} + 21\ H_2O$ |
| $Cl_2 + 2\ e^- \rightarrow 2\ Cl^-$ |
| $MnO_4^- + 8\ H_3O^+ + 5\ e^- \rightarrow Mn^{2+} + 12\ H_2O$ |
| $H_2O_2 + 2\ H_3O^+ + 2\ e^- \rightarrow 4\ H_2O$ |
| $Co^{3+} + e^- \rightarrow Co^{2+}$ |
| $F_2 + 2\ e^- \rightarrow 2\ F^-$ |

Half-reactions that take place in basic solutions. Again, the oxidizing agent is a reactant and the reducing agent is a product. The stronger oxidizing agents and the weaker reducing agents are at the bottom of the table.

| Half-reaction |
|---|
| $2\ H_2O + 2\ e^- \rightarrow H_2(g) + 2\ OH^-$ |
| $O_2(g) + 2\ H_2O + 4\ e^- \rightarrow 4\ OH^-$ |

## CHAPTER BOX 16.1 HISTORICAL OXIDIZING AND REDUCING AGENTS

**Oxidizing agents** have been key ingredients for many centuries in the manufacture of explosives. It is not known exactly when the Chinese discovered the formula for gunpowder, but it was common in China by the time Marco Polo arrived in 1275. The Chinese used gunpowder primarily to make fireworks.

The classic recipe for gunpowder consists of a mixture of potassium nitrate, sulfur, and charcoal (carbon). When detonated, gunpowder does not actually explode; it simply burns very quickly to give large volumes of gases. The products of the burning include $N_2$, CO, and $CO_2$, along with thick smoke that contains small particles of $K_2CO_3$, $K_2SO_4$, and $K_2S$. Blasting powder used in the construction and mining industries contained sodium nitrate in place of potassium nitrate.

Blasting powder and gunpowder aren't used very widely today. They have been largely replaced by stronger explosives such as ammonium nitrate, whose reaction is:

$$2\,NH_4NO_3(s) \longrightarrow 2\,N_2(g) + O_2(g) + 4\,H_2O(g) + \text{heat}$$

A mixture of ammonium nitrate and fuel oil is frequently used as an explosive in mining. Ammonium nitrate is also used in fertilizers.

Gunpowder, blasting powder, and ammonium nitrate all contain the nitrate ion, $NO_3^-$. The nitrate ion acts as the oxidizing agent in its reactions. For explosives to work well, they must burn very rapidly. The oxygen needed for the rapid combustion cannot come from the air—the rapidly expanding gases that are products of the reaction push the atmospheric oxygen away. A strong oxidizing agent, in this case $NO_3^-$, is a necessary component of the mixture.

The reactions of all explosions are highly exothermic (that is, they give off plenty of heat). These reactions are also redox reactions, and most of them form large volumes of gaseous products. The destructive power of an explosion comes from the shock wave caused by the rapid increase in volume of the hot gases produced by the reaction. High explosives such as TNT (trinitrotoluene, $C_7H_5O_6N_3$) or nitroglycerin ($C_3H_5O_9N_3$, used in dynamite) produce shock waves that travel at over 16,000 km $h^{-1}$. Both TNT and nitroglycerin contain three nitro groups. The nitro group is related to the nitrate ion and acts as the oxidizing agent in the reaction of these high explosives.

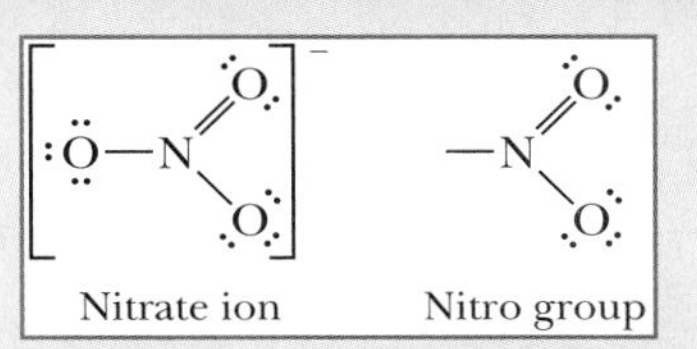


For a short time in U.S. history, the reducing properties of aluminum metal had great industrial importance. Until almost 1900 aluminum metal was extremely expensive, more costly than gold. It was very difficult to extract the metal from its oxide, $Al_2O_3$, even though this oxide is found very commonly as the mineral bauxite. In 1886 (while he was an undergraduate student at Oberlin College), Charles Martin Hall developed an inexpensive way to manufacture aluminum in his garage. Hall made a pile of money as the founder of Alcoa, and aluminum became relatively inexpensive (see Chapter Box 13.1). The French chemist Paul Heroult discovered the same process in the same year. It is now called the Hall-Heroult process.

At about the time of Hall's discovery, the steel industry started to boom, and engineers began to build bridges and buildings with steel beams and girders. How to connect the steel beams was a practical problem—an obvious modern choice, welding, had to wait for the development of the acetylene torch. At this point the reducing properties of aluminum came into play. If a rivet is placed between two steel beams, the surrounding hole is filled with a mixture of iron(III) oxide and powdered aluminum, and the mixture is then ignited, this reaction occurs:

$$Fe_2O_3(s) + 2\,Al(s) \longrightarrow Al_2O_3(s) + 2\,Fe(l)$$

The molten iron formed in this reaction quickly freezes and holds the rivet and beam in place. This is called a *thermite reaction.*

Ammonium nitrate fertilizer was the main ingredient in the bomb that destroyed the Alfred Murrah Federal Building in Oklahoma City in 1995.

Figure 16.5 shows that silver whiskers form on a copper wire dipped in aqueous silver nitrate solution, while copper(II) ions go into solution at the same time. Let's see how we can use half-reactions to predict this. Two half-reactions are involved. As they appear in Table 16.1, these half-reactions are

$$Cu^{2+} + 2\ e^- \longrightarrow Cu(s) \quad \text{(closer to the top of Table 16.1)}$$
$$Ag^+ + e^- \longrightarrow Ag(s) \quad \text{(closer to the bottom of Table 16.1)}$$

Having identified the two half-reactions involved, we rewrite the one closer to the top of the table as an oxidation by reversing it, but leave the second one written as a reduction. This gives:

$$Cu(s) \longrightarrow Cu^{2+} + 2\ e^-$$
$$Ag^+ + e^- \longrightarrow Ag(s)$$

In other words, we predict that the copper metal will be oxidized and the silver ion will be reduced to silver metal. As Figure 16.5 shows, that's exactly what happens. We will discuss how to use half-reactions to balance redox equations in the next section.

### EXAMPLE 16.4 Predicting the Direction of Redox Reactions

**a.** What, if any, is the oxidation half-reaction when $F_2$ is bubbled through a solution of $Cl^-$ in water?
**b.** What, if any, is the oxidation half-reaction when $Fe^{2+}(aq)$ is added to a solution of $I^-(aq)$?

**SOLUTIONS**

**a.** According to Table 16.1, $F_2$ is an extremely powerful oxidizing agent, so you can be fairly certain that something will be oxidized. Chloride ion, the other species present, has this oxidation half-reaction: $2\ Cl^- \rightarrow Cl_2(aq) + 2\ e^-$. The reduction half-reaction is $F_2 + 2\ e^- \rightarrow 2\ F^-(aq)$. (It's always worthwhile to write both half-reactions, just to make sure that you have one oxidation and one reduction rather than two of one kind.)
**b.** There are two half-reactions in Table 16.1 that involve the $Fe^{2+}$ ion—one in which $Fe^{3+}$ is reduced to $Fe^{2+}$, and one in which $Fe^{2+}$ is reduced to metallic Fe. The only half-reaction in the table that involves iodide ion is the reduction of $I_2$ to $I^-$. Putting these three in the order in which they appear in the table gives:

$$Fe^{2+} + 2\ e^- \longrightarrow Fe$$
$$I_2 + 2\ e^- \longrightarrow 2\ I^-$$
$$Fe^{3+} + e^- \longrightarrow Fe^{2+}$$

Remember that the half-reaction closer to the top of the table must be the oxidation and must be reversed, while the lower one is the reduction. One possible combination of two of the above half-reactions is

$$Fe^{2+} + 2\ e^- \longrightarrow Fe$$
$$2\ I^- \longrightarrow I_2 + 2\ e^-$$

The other possible combination is

$$2\,I^- \longrightarrow I_2 + 2\,e^-$$
$$Fe^{3+} + e^- \longrightarrow Fe^{2+}$$

In the first of these combinations, the oxidation is the lower reaction in the table. In the second combination, $Fe^{2+}$ is a product and not a reactant. Neither reaction will occur.

**EXERCISE 16.4**

**a.** What, if any, is the reduction half-reaction when an iron nail is placed in a solution of $SnCl_2(aq)$ ?
**b.** What, if any, is the reduction half-reaction when $Fe^{3+}(aq)$ is added to a solution of $I^-(aq)$?

## 16.3 Balancing Redox Equations: The Half-Reaction Method

**GOAL:** To balance equations for redox reactions using the half-reaction method.

You first learned how to balance simple chemical equations in Chapter 8. That method, which chemists call balancing by inspection, works well for relatively simple equations. The equations for oxidation-reduction reactions are often far from simple, however. They can include six or more chemical species instead of three or four, for instance. To balance equations like this, you must simplify them first.

In Section 16.2 you learned to use half-reactions to predict the direction and products of reactions. Here you will use them to balance redox equations. In this section the importance of the electrons shown in half-reactions will become obvious.

In the half-reaction method of balancing equations, you separate the reaction into two parts, the oxidation and the reduction half-reactions. Then you balance the half-reactions by inspection, using electrons to balance the charges. Finally you combine the two halves in such a way that the electrons cancel out. The final result is a complete, balanced chemical equation. Here is the step-by-step procedure for the half-reaction method:

**Step 1** Write the unbalanced equation for the reaction. For example, let's consider the reaction of permanganate ion with chloride ion in acidic solution. (You sometimes need to know whether the reaction takes place in acidic or basic solution.)

$$MnO_4^- + Cl^- \longrightarrow Mn^{2+} + Cl_2 \qquad \text{(in acidic solution)}$$

**Step 2** Separate the unbalanced equation into two unbalanced half-reactions, using common sense and reason to divide up the chemical symbols. In this example, manganese-containing reactants give manganese-containing products and chlorine-containing reactants give chlorine-containing products:

$$MnO_4^- \longrightarrow Mn^{2+}$$
$$Cl^- \longrightarrow Cl_2$$

**Step 3** Balance each half-reaction. To get the correct answer every time, you should follow this sequence in strict order:

**a.** In one of your half-reactions, use inspection to balance everything except oxygen, hydrogen, and charge. Let's start with the manganese half-reaction:

$$MnO_4^- \longrightarrow Mn^{2+}$$

Here the Mn is already balanced.

**b.** Balance oxygen by adding $H_2O$ to the side of the half-reaction that needs oxygen. In this half-reaction there are four oxygen atoms on the left side and none on the right. To get four oxygen atoms on the right side, you add four water molecules:

$$MnO_4^- \longrightarrow Mn^{2+} + \mathbf{4\,H_2O}$$

**c.** Balance hydrogen atoms by adding $H^+$ ions. The half-reaction has no hydrogen atoms on the left and eight on the right. You need to add eight $H^+$ ions to the left side:

$$MnO_4^- + \mathbf{8\,H^+} \longrightarrow Mn^{2+} + \mathbf{4\,H_2O}$$

**d.** Finally, balance the charge by adding electrons. Electrons have a negative charge, so you must add them to the side of the half-reaction that has more positive charges. The left side of the half-reaction has a total charge of $+7$ ($-1$ for $MnO_4^-$ and $+1$ for each of the eight $H^+$'s). The right side has a total charge of $+2$ (from $Mn^{2+}$). In order to get the same charge on both sides, you must add 5 electrons (with a combined charge of $-5$) to the left side:

$$\mathbf{5\,e^-} + MnO_4^- + 8\,H^+ \longrightarrow Mn^{2+} + 4\,H_2O$$

This half-reaction is now balanced for both atoms and charges.

Now repeat the sequence for the other half-reaction:
Step 3a balances chlorine:

$$Cl^- \longrightarrow Cl_2$$
$$\mathbf{2}\,Cl^- \longrightarrow Cl_2 \qquad \text{(balancing chlorine)}$$

Steps 3b and 3c are unnecessary because there is no oxygen or hydrogen in this half-reaction.
Step 3d balances the charge:

$$2\,Cl^- \longrightarrow Cl_2$$
$$2\,Cl^- \longrightarrow Cl_2 + \mathbf{2\,e^-} \qquad \text{(balancing charge)}$$

**Step 4** If both half-reactions have the same number of electrons, go to Step 5. If the half-reactions contain different numbers of electrons, then multiply each balanced half-reaction by the coefficient of the electrons in the *other* half-reaction. The result will be that the number of electrons on the left side of one half-reaction equals the number of electrons on the right side of the other half-reaction. There are five electrons on the left side of the reduction half-reaction:

$$5\,e^- + MnO_4^- + 8\,H^+ \longrightarrow Mn^{2+} + 4\,H_2O$$

There are two electrons on the right side of the oxidation half-reaction:

$$2\,Cl^- \longrightarrow Cl_2 + 2\,e^-$$

To get the same number of electrons in both half-reactions, you multiply the reduction half-reaction by 2 and the oxidation half-reaction by 5:

$$2(5\,e^- + MnO_4^- + 8\,H^+ \longrightarrow Mn^{2+} + 4\,H_2O) = \mathbf{10\,e^-} + 2\,MnO_4^- + 16\,H^+ \longrightarrow 2\,Mn^{2+} + 8\,H_2O$$

$$5(2\,Cl^- \longrightarrow Cl_2 + 2\,e^-) = 10\,Cl^- \longrightarrow 5\,Cl_2 + \mathbf{10\,e^-}$$

**Step 5** Add the half-reactions together. Cancel out electrons (and anything else you can).

$$\begin{array}{r} 10\,e^- + 2\,MnO_4^- + 16\,H^+ \longrightarrow 2\,Mn^{2+} + 8\,H_2O \\ 10\,Cl^- \longrightarrow 5\,Cl_2 + 10\,e^- \\ \hline \cancel{10\,e^-} + 2\,MnO_4^- + 16\,H^+ + 10\,Cl^- \longrightarrow 2\,Mn^{2+} + 8\,H_2O \\ + 5\,Cl_2 + \cancel{10\,e^-} \end{array}$$

$$2\,MnO_4^- + 16\,H^+ + 10\,Cl^- \longrightarrow 2\,Mn^{2+} + 8\,H_2O + 5\,Cl_2$$

**Step 6** If the reaction takes place in acidic solution, add $H_2O$ to *both sides* so that the final equation has $H_3O^+$ instead of $H^+$ (remember than $H^+$ doesn't actually exist). If the reaction takes place in basic solution, add $OH^-$ to *both sides* so that the final equation has $H_2O$ instead of $H^+$. Cancel anything you can. In the example, you add 16 $H_2O$ to each side and then combine 16 $H^+$ and 16 $H_2O$ on the left to get 16 $H_3O^+$. This gives the complete balanced equation:

$$2\,MnO_4^- + 16\,H_3O^+ + 10\,Cl^- \longrightarrow 2\,Mn^{2+} + 24\,H_2O + 5\,Cl_2$$

**Step 7** Check to make sure that the equation is balanced for both atoms and charges.

### EXAMPLE 16.5 Balancing Redox Equations using Half-Reactions

Balance each of these equations.

**a.** $F_2 + Cl^- \longrightarrow Cl_2 + F^-$

**b.** $MnO_4^- + CN^- \longrightarrow MnO_2 + CNO^-$ (in basic solution)

**STUDY ALERT**

**The half-reaction method is a powerful way to balance equations. Chemists use it when nothing else works easily. You can balance every redox equation correctly if you take the steady, meticulous approach this technique requires.**

**SOLUTIONS**

**a.** First, separate the given equation into appropriate half-reactions:

$F_2 \longrightarrow F^-$ and $Cl^- \longrightarrow Cl_2$

Balance each of the half-reactions. Here's the first:

$F_2 \longrightarrow F^-$

$F_2 \longrightarrow \mathbf{2}\,F^-$ (balancing fluorine; no oxygen or hydrogen to worry about)

$F_2 + \mathbf{2\,e^-} \longrightarrow 2\,F^-$ (balancing charge)

Here's the other half-reaction:

$Cl^- \longrightarrow Cl_2$

$\mathbf{2}\,Cl^- \longrightarrow Cl_2$ (balancing chlorine; no oxygen or hydrogen)

$2\,Cl^- \longrightarrow Cl_2 + \mathbf{2\,e^-}$ (balancing charge)

Combine the balanced half-reactions. Since there are two electrons in each of them, the electrons are already balanced. Add the balanced half-reactions and cancel out the electrons:

$$F_2 + 2\,e^- \longrightarrow 2\,F^-$$
$$2\,Cl^- \longrightarrow Cl_2 + 2\,e^-$$
$$F_2 + 2\,Cl^- + \cancel{2\,e^-} \longrightarrow 2\,F^- + Cl_2 + \cancel{2\,e^-}$$

The balanced equation is

$F_2 + 2\,Cl^- \longrightarrow 2\,F^- + Cl_2$

**b.** The first step is to separate the equation into two half-reactions:

$MnO_4^- \longrightarrow MnO_2$ and $CN^- \longrightarrow CNO^-$

Next, balance everything in each half-reaction, following the sequence given in the text. For one of the half-reactions:

$MnO_4^- \longrightarrow MnO_2$ (manganese is balanced)

$MnO_4^- \longrightarrow MnO_2 + \mathbf{2\,H_2O}$ (balancing oxygen)

$MnO_4^- + \mathbf{4\,H^+} \longrightarrow MnO_2 + 2\,H_2O$ (balancing hydrogen)

$\mathbf{3\,e^-} + MnO_4^- + 4\,H^+ \longrightarrow MnO_2 + 2\,H_2O$ (balancing charge)

For the other half-reaction:

$CN^- \longrightarrow CNO^-$

$CN^- + \mathbf{H_2O} \longrightarrow CNO^-$ (balancing oxygen)

$CN^- + H_2O \longrightarrow CNO^- + \mathbf{2\,H^+}$ (balancing hydrogen)

$CN^- + H_2O \longrightarrow CNO^- + 2\,H^+ + \mathbf{2\,e^-}$ (balancing charge)

Next, multiply each balanced half-reaction by the coefficient of the electrons in the other half-reaction:

$$\mathbf{2}(3\,e^- + MnO_4^- + 4\,H^+ \longrightarrow MnO_2 + 2\,H_2O) = 6\,e^- + 2\,MnO_4^- + 8\,H^+ \longrightarrow 2\,MnO_2 + 4\,H_2O$$

$$\mathbf{3}(CN^- + H_2O \longrightarrow CNO^- + 2\,H^+ + 2\,e^-) = 3\,CN^- + 3\,H_2O \longrightarrow 3\,CNO^- + 6\,H^+ + 6\,e^-$$

Next, add the balanced half-reactions together, canceling the electrons and anything else you can.

$$6\,e^- + 2\,MnO_4^- + 8\,H^+ \longrightarrow 2\,MnO_2 + 4\,H_2O$$
$$3\,CN^- + 3\,H_2O \longrightarrow 3\,CNO^- + 6\,H^+ + 6\,e^-$$

$$6\,e^- + 2\,MnO_4^- + 2\,H^+ + 3\,CN^- + 3\,H_2O \longrightarrow 2\,MnO_2 + 4\,H_2O + 3\,CNO^- + 6\,H^+ + 6\,e^-$$

which reduces to:

$$2\,MnO_4^- + 2\,H^+ + 3\,CN^- \longrightarrow 2\,MnO_2 + H_2O + 3\,CNO^-$$

This reaction takes place in basic solution, so you must add $OH^-$ to both sides to change $H^+$ to $H_2O$. Then cancel anything you can.

$$2\,MnO_4^- + 2\,H^+ + \mathbf{2\,OH^-} + 3\,CN^- \longrightarrow 2\,MnO_2 + H_2O + \mathbf{2\,OH^-} + 3\,CNO^-$$

$$2\,MnO_4^- + \cancel{2}\,H_2O + 3\,CN^- \longrightarrow 2\,MnO_2 + \cancel{H_2O} + 2\,OH^- + 3\,CNO^-$$

This is the answer:

$$2\,MnO_4^- + H_2O + 3\,CN^- \longrightarrow 2\,MnO_2 + 2\,OH^- + 3\,CNO^-$$

**EXERCISE 16.5**

Balance each of these equations.

**a.** $Fe^{2+} + MnO_4^- \rightarrow Fe^{3+} + Mn^{2+}$ (in acid solution)

**b.** $Fe + Sn^{2+} \rightarrow Fe^{2+} + Sn$

## 16.4 Oxidation Numbers

**GOAL:** To determine the oxidation numbers of elements in simple molecules and ions.

The methods of identifying oxidizing and reducing agents introduced in Sections 16.1 and 16.2 may not always be practical. What if the formulas of the compounds involved in a reaction are more complex and unfamiliar? How do you decide where the electrons come from and where they go? You can do it by writing and balancing the half-reactions, of course. Another, shorter way is to use somewhat artificial constructs called oxidation numbers.

An **oxidation number** is equal to the charge that an atom or ion *would have* if all of its polar bonds were ionic. Of course, the bonding in covalent molecules is different from that in ionic compounds because of electron sharing, but the oxidation number system ignores this. Chemists sometimes use oxidation numbers to keep track of the loss and gain of electrons in redox reactions, but these numbers do not necessarily correspond to any real charges of atoms or ions.

Chemists use this set of guidelines to assign oxidation numbers.

1) The oxidation number of any atom of a *pure element* is zero. For example, the oxidation numbers of hydrogen, oxygen, and sulfur in $H_2$, $O_2$, and $S_8$ are all zero, and so are the oxidation numbers of atoms in metal samples (Zn, Al, Na, etc.).
2) The oxidation number of an atom in any *monatomic ion* is equal to the charge of the ion. For example, the oxidation number of sodium in $Na^+$ is $+1$; that of sulfur in $S^{2-}$ is $-2$.
3) Fluorine has an oxidation number of $-1$ in all its compounds.
4) Oxygen almost always has an oxidation number of $-2$. There are a couple of exceptions to this rule. The most important of these are $O_2$, in which oxygen's oxidation number is zero, and $H_2O_2$ (hydrogen peroxide), in which the oxidation number of oxygen is $-1$. Other exceptions occur when oxygen is bonded to fluorine (for example, in $OF_2$) and in certain polyatomic ions (such as $O_2^-$). The latter two exceptions are somewhat rare, and you can determine the oxidation number of oxygen in compounds of this type by using the rest of these rules.
5) Hydrogen has an oxidation number of $+1$ when bonded to nonmetals and $-1$ when bonded to metals.
6) The sum of the oxidation numbers of all the atoms in a neutral molecule is zero. For a polyatomic ion the sum of the oxidation numbers equals the charge on the ion.
7) The preceding rules make it possible to assign oxidation numbers to other elements in compounds. Electronegative elements almost always have negative oxidation numbers, and electropositive elements similarly tend to have positive oxidation numbers.

### EXAMPLE 16.6 Assigning Oxidation Numbers

What is the oxidation number of each element in these species?

**a.** $I_2$ **b.** KBr **c.** $KClO_4$ **d.** $H_4SiO_4$ **e.** $Cr_2O_7^{2-}$

**SOLUTIONS**

**a.** Iodine has an oxidation number of zero in elemental iodine (from the first guideline).

**b.** KBr is composed of the ions $K^+$ and $Br^-$. (See Chapter 7.) The oxidation number of potassium is therefore $+1$ and that of bromine is $-1$ (both from the second guideline).

**c.** $KClO_4$ is composed of the $K^+$ and $ClO_4^-$ ions. The oxidation number of potassium in $K^+$ is $+1$. The sum of the oxidation numbers of the elements in the $ClO_4^-$ ion is $-1$. Remember that the sum of the oxidation numbers must equal the charge of the ion.

4(oxidation number of O) + (oxidation number of Cl) = $-1$

Each oxygen has an oxidation number of $-2$. Therefore:

4($-2$) + (oxidation number of Cl) = $-1$ =

($-8$) + (oxidation number of Cl) = $-1$

Oxidation number of Cl $= -1 + 8 = +7$

A moment's reflection will tell you that there is no such thing in the real world as a chlorine cation with a charge of +7. For one thing, it takes a lot of energy to remove even one electron from a neutral chlorine atom—chlorine is highly electronegative. Removing seven electrons by chemical means would be impossible. Furthermore, in this case the electrons would have to be taken by oxygen atoms, which are only slightly more electronegative than chlorine. Instead, you must recognize that the internal bonding in $ClO_4^-$ is covalent (as it is in all polyatomic ions), and that the oxidation numbers of chlorine and oxygen do not represent real charges.

**d.** Each hydrogen atom has an oxidation number of +1 and each oxygen atom has an oxidation number of −2. Also, you know that the sum of all the oxidation numbers is zero. Therefore:

$$4(\text{oxidation number of H}) + 4(\text{oxidation number of O}) + (\text{oxidation number of Si}) = 0$$

$$4(+1) + 4(-2) + (\text{oxidation number of Si}) = 0$$

$$\text{Oxidation number of Si} = +8 - 4 = +4$$

**e.** The sum of all the oxidation numbers must equal the charge on the ion, in this case −2. You know that oxygen has an oxidation number of −2. Therefore:

$$2(\text{oxidation number of Cr}) + 7(-2) = -2$$

$$2(\text{oxidation number of Cr}) = -2 + 14 = +12$$

$$\text{Oxidation number of Cr} = \frac{+12}{2} = +6$$

Again, the idea of a chromium cation with a +6 charge is absurd. The $Cr_2O_7^{2-}$ ion is covalently bonded, and the oxidation numbers are not charges.

**EXERCISE 16.6**

What is the oxidation number of each element in these species?

**a.** CaS **c.** $Na_2SO_4$ **e.** $Fe_2O_3$
**b.** $HNO_3$ **d.** $S_8$ **f.** $OF_2$

One of the reasons to look at oxidation numbers is that it can sometimes be tricky to tell exactly what is being oxidized and reduced. Chapter Box 16.1 presented this equation for an explosion:

$$2\,NH_4NO_3(s) \longrightarrow 2\,N_2(g) + O_2(g) + 4\,H_2O(g)$$

How can we identify the oxidizing and reducing agents? A quick look at the equation shows that this is not simple, but an analysis of the oxidation numbers gives us a clue.

Ammonium nitrate, $NH_4NO_3$, consists of two ions, $NH_4^+$ and $NO_3^-$. Using the techniques in Example 16.6, we can determine that the oxidation number of nitrogen in $NH_4^+$ is −3 and the oxidation number of nitrogen in $NO_3^-$ is +5. On the product side of the equation, nitrogen in $N_2$ has an oxidation number of zero. The nitrogen in $NH_4^+$ is oxidized to $N_2$ because the oxidation number increases from $-3$ to 0—suggesting that the nitrogen in $NH_4^+$ loses electrons and increases its oxidation number. Similarly, the nitrogen in $NO_3^-$ changes its oxidation number from $+5$ to 0; it is reduced.

That is not all we can determine, however. One of the products is $O_2$, in which the oxidation number of oxygen is zero. The only source of oxygen on the reactant side of the equation is the nitrate ion, $NO_3^-$. At least some of the oxygens in the nitrate ion change oxidation number from $-2$ to 0, and in the overall reaction oxygen is oxidized. In this reaction $NH_4^+$ is a reducing agent and $NO_3^-$ is both an oxidizing (the nitrogen) and a reducing (the oxygen) agent.

## EXAMPLE 16.7 Oxidation-Reduction Reactions and Oxidation Numbers

For this reaction, determine all the oxidation numbers and identify the oxidizing agent, the substance oxidized, the reducing agent, and the substance reduced.

$$2\,K(s) + 2\,NH_3(l) \longrightarrow 2\,KNH_2(s) + H_2(g)$$

**SOLUTION**

Here is a small table that organizes the information on **oxidation numbers:**

Oxidation Numbers

| Reactants | | Products | |
|---|---|---|---|
| K | $NH_3$ | $KNH_2$ | $H_2$ |
| 0 | N, $-3$<br>H, $+1$ | K, $+1$<br>N, $-3$<br>H, $+1$ | 0 |

Potassium loses electrons (its oxidation number changes from 0 to $+1$); it is the reducing agent and is oxidized. The hydrogen in $NH_3$ gains electrons so $NH_3$ is reduced and is the oxidizing agent.

**EXERCISE 16.7**

For this reaction, determine all the oxidation numbers, and identify the oxidizing agent, the substance oxidized, the reducing agent, and the substance reduced.

$$Fe_2O_3(s) + 6\,Na(s) \longrightarrow 3\,Na_2O(s) + 2\,Fe(s)$$

## 16.5 Electricity from Redox Reactions

**GOAL:** To explain and predict the behavior of species involved in the redox reactions that take place in electrochemical cells and batteries.

One of the two half-reactions of a redox reaction produces electrons. It is possible to run the two half-reactions of an overall reaction in separate containers and cause a current to flow in a wire from one container to the other. This may seem odd to you—how can half-reactions run in separate containers? But in fact it happens all around you all the time. Batteries work on this principle, so every time you start your car, use your portable CD player, or turn on a flashlight, you are running half-reactions this way. (See Figure 16.9.)

A

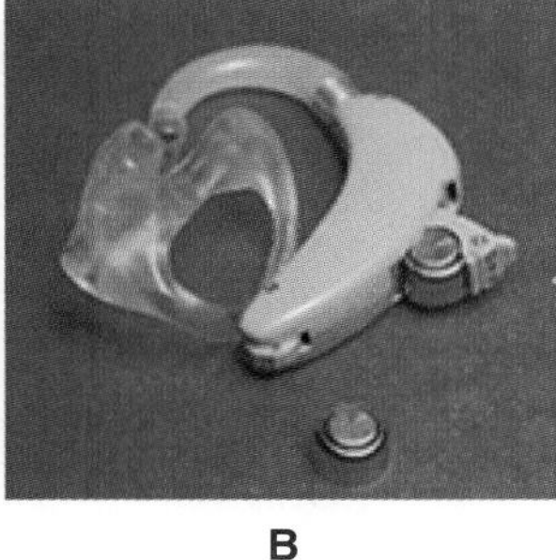
B

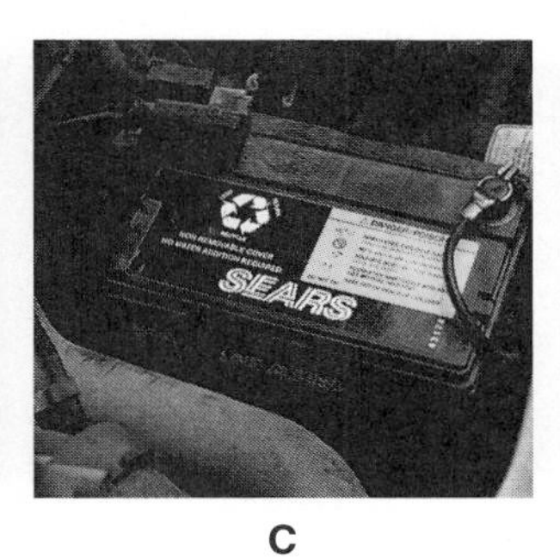

C

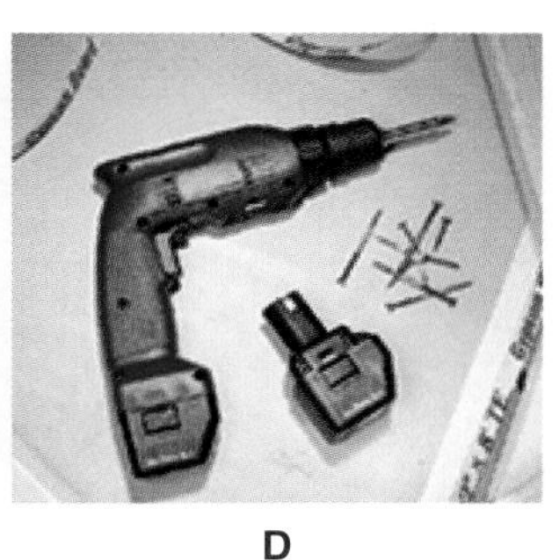
D

FIGURE 16.9
You use batteries all the time; they are a familiar part of daily life. (A) Dry cell batteries, (B) silver battery, (C) lead storage battery, and (D) a nicad (nickel-cadmium) rechargeable battery.

None of this was obvious (or even thought of) before the late 1700s. The Italian anatomist Luigi Galvani (1732–1798) noticed that the muscles in frogs' legs twitched when an electric spark touched them (see Figure 16.10). He experimented further with the effects of electricity on muscle tissue. He finally found that he could get the twitching effect just by placing muscle tissue in contact with two different metals. He decided that the effect was caused by the tissue.

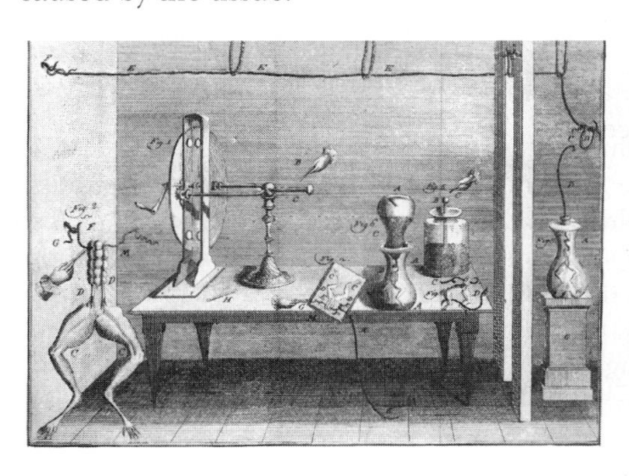

FIGURE 16.10
Galvani made the connection between electricity and biology in his experiments with the muscles in frogs' legs. Ultimately, this led to the development of the pacemaker, which helps some cardiac patients keep a regular heartbeat.

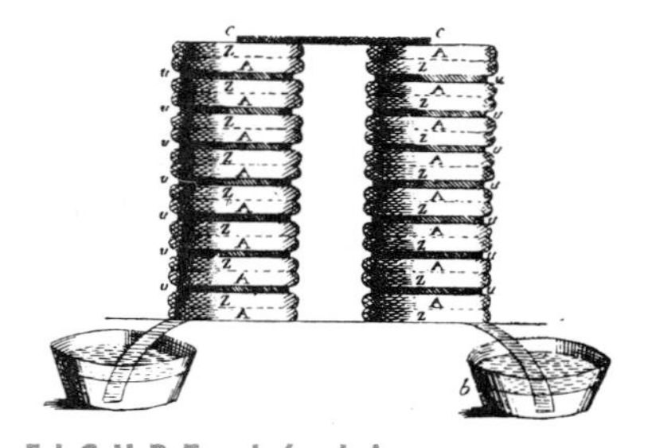

FIGURE 16.11
Volta's electrochemical cell was similar to the one shown here.

Alessandro Volta (1745–1827), another Italian, decided that the muscle tissue was just a wet and salty medium, and it was such a medium that was important in generating an electric current. He separated some plates of copper and zinc with pieces of cardboard soaked in salt water. He immediately found that an electrical current flowed even though there was no tissue in the system. His first electrochemical cell, made in 1800, was similar to the one shown in Figure 16.11, though Volta didn't have a light bulb to use as a current detector. His cell exploited this simple oxidation-reduction reaction

$$Cu^{2+}(aq) + Zn(s) \longrightarrow Cu(s) + Zn^{2+}(aq)$$

The half-reactions are

$$Zn(s) \longrightarrow 2\ e^- + Zn^{2+}(aq) \qquad \text{and} \qquad Cu^{2+}(aq) + 2\ e^- \longrightarrow Cu(s)$$

To get the electrons to pass through the wire in an electrochemical cell is a little bit tricky. For example, in Figure 16.12, no current can pass through the wire because the copper(II) ions and the zinc metal are in direct contact. This means that the electrons pass directly from the zinc metal to the copper(II) ions without passing through the wire.

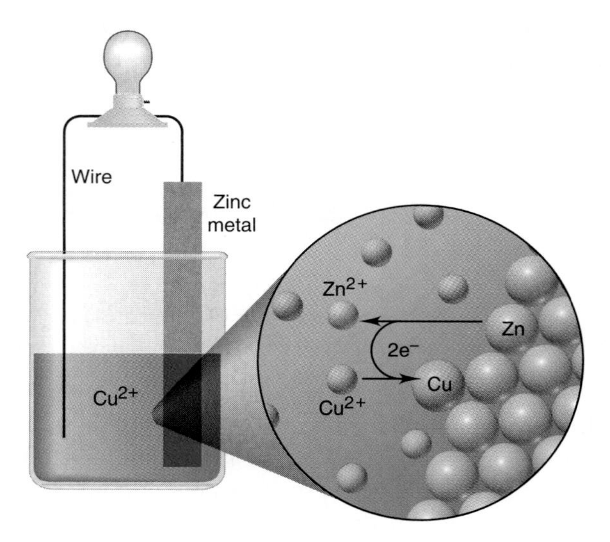


FIGURE 16.12
The reaction $Cu^{2+}(aq) + Zn(s) \rightarrow Cu(s) + Zn^{2+}(aq)$ occurs in this apparatus, but the electrons transfer from zinc to copper at the surface of the zinc strip and not through the wire. The light bulb stays dark. Unless the copper(II) ions are separated from the zinc metal the electrons won't pass through the wire.

Simply separating the two solutions as shown in Figure 16.13 doesn't work either, at least not for long. The problem is that this setup makes it impossible to keep the individual solutions electrically neutral. In the left-hand cell there is a strip of zinc metal in a solution of $ZnCl_2(aq)$; on the right is copper metal in $CuCl_2(aq)$. As soon as the wire connects the metal strips, a few Zn atoms in the cell on the left give up their electrons to form $Zn^{2+}$ ions, which dissolve. The electrons zip through the wire to the copper metal side.

However, no negative charges form in the left-hand cell to counterbalance the buildup of positive charge (from the newly formed $Zn^{2+}$ ions—remember that the electrons left through the wire). This buildup of positive charge attracts the electrons and prevents them from leaving the left-hand cell. The electron flow and the reaction stop.

A similar effect happens in the cell on the right. At first a few electrons arrive through the wire and a few copper(II) ions are reduced to copper metal. However, there is no positively charged species present to replace the $Cu^{2+}$ ions that react. This leaves extra $Cl^-$ in solution with no counterbalancing positive charge, which means that an overall negative charge begins to develop in the right-hand cell. This prevents electrons from moving from left to right through the wire and helps to stop the reaction.

Before electrons can flow freely through the wire, something must prevent the buildup of charges in both cells. One solution is a *salt bridge,* which is a gel or paste that contains ions, often enclosed in a glass tube. The salt bridge connects the two cells just as the wire does and, when the current flows through the wire, releases ions of the appropriate charge into each solution. Figure 16.14 illustrates the effect of a salt bridge. When the current isn't flowing, the ions stay immobile in the gel. Thus the wires can be disconnected, and the whole apparatus can be put in storage. Practical batteries use salt bridges or more modern equivalents to achieve significant shelf life.

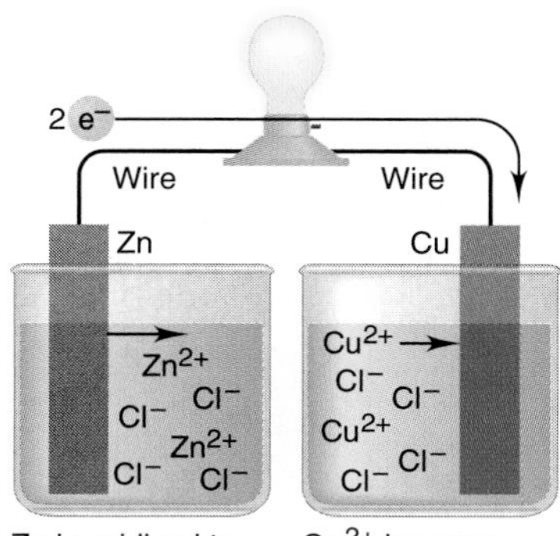


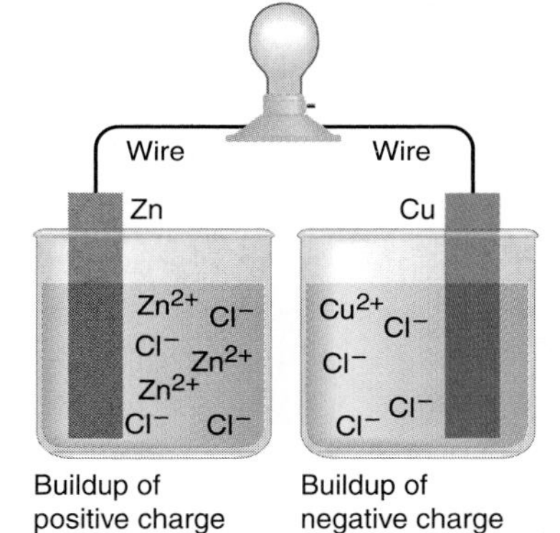


**FIGURE 16.13**
(Top) Here the only connection between the copper(II) ions and the zinc is the wire through the light bulb. In theory the $Zn^{2+}$ ions could go into solution from the zinc strip: $Zn(s) \rightarrow Zn^{2+}(aq) + 2\ e^-$. This half-reaction would release electrons to flow through the wire. They would light the bulb and then react with $Cu^{2+}$ ions at the copper strip: $Cu^{2+}(aq) + 2\ e^- \rightarrow Cu(s)$. (Bottom) However, those reactions quickly produce a positive charge in the zinc beaker and a negative charge in the copper beaker, stopping the electron flow very quickly. The bulb actually remains unlit.

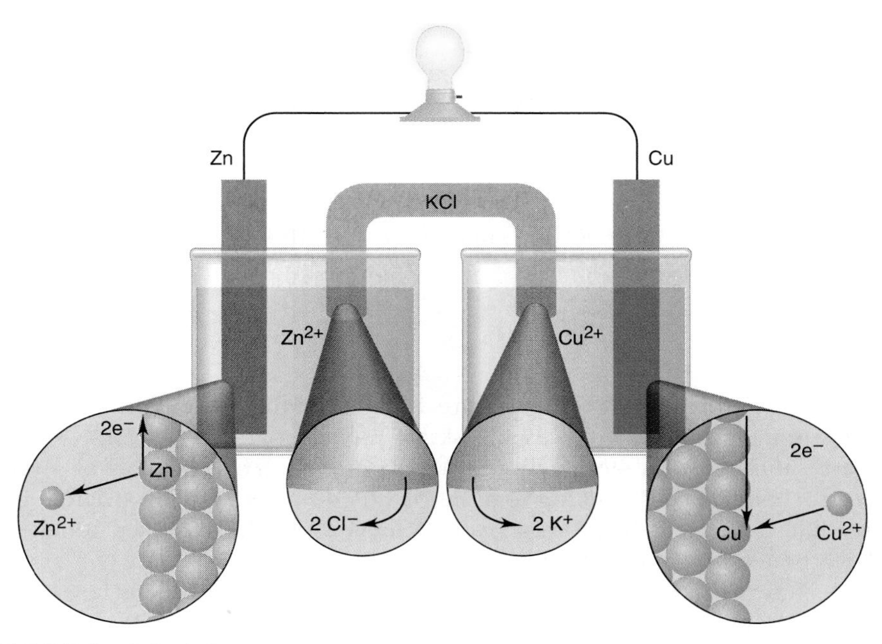


**FIGURE 16.14**
In a working electrochemical cell, as the half-cell on the left produces $Zn^{2+}$ ions, electrons move through the wire and light the bulb, and $Cl^-$ ions move in the opposite direction out of the salt bridge into the solution. This keeps the charges balanced in the cell on the left. Similarly, potassium ions move out of the salt bridge and into the cell on the right to balance the charges there.

EXAMPLE 16.8 **Direction of Electron Flow in an Electrochemical Cell**

In what direction do the electrons flow in the wires in this electrochemical cell? In the salt bridge, in what direction do the $Cl^-$ ions move? Explain your answers.

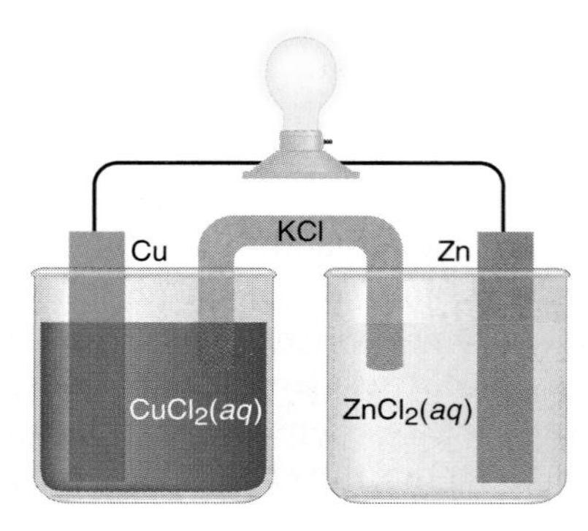


**SOLUTION**

Zinc metal reacts in the right-hand cell, producing $Zn^{2+}$ ions in solution and electrons flowing into the zinc metal strip. These electrons flow from the zinc metal through the wire to the copper metal strip in the left-hand cell. Like electrons, $Cl^-$ ions have a negative charge. As the electrons leave the zinc by way of the wire, $Cl^-$ ions flow through the salt bridge into the zinc-containing cell (on the right) to replace the lost negative charge.

**EXERCISE 16.8**

In what direction do $K^+$ ions migrate in the salt bridge in Example 16.8? Explain your answer.

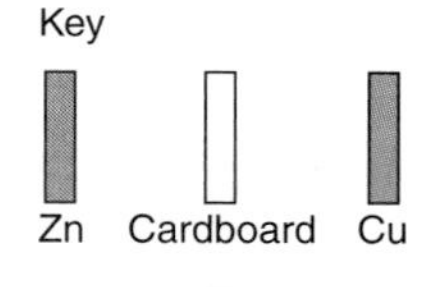


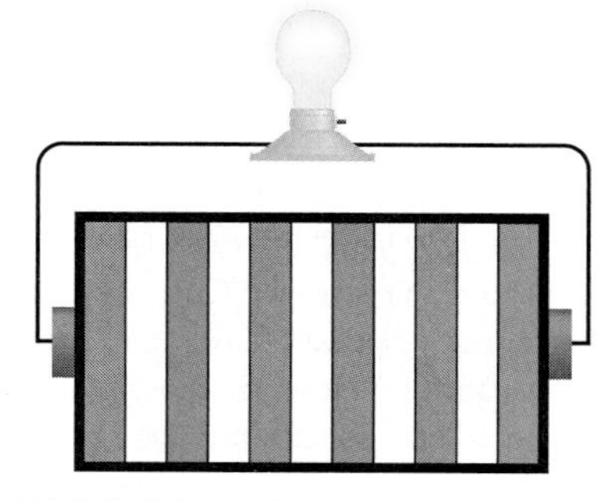

FIGURE 16.15
This series of alternating disks of zinc, cardboard saturated with salt water, and copper is a *battery.* Note that the saturated cardboard serves as a salt bridge.

The electrochemical cell in Figure 16.14 is cumbersome. To make something more portable, Volta used disks of metal separated by disks of cardboard saturated with salt, as shown in Figure 16.15. This assembly of several electrochemical cells working together is called a **battery.** This original battery had a short life and wasn't very reliable. The designs used today are much better.

A battery is an electrochemical cell or simply a series of cells having potential chemical energy. Nothing happens until the circuit is completed, which is usually done by flipping a switch. This allows current to flow, which means that the redox reaction that converts the stored energy to active electrical current can begin to run. This electrical energy can then be used to light a bulb, play a cassette tape, or whatever.

The battery most often used in small electrical and electronic appliances is the **alkaline dry cell**—this is what you usually use in flashlights, boom boxes, remote control devices, and so forth. These batteries contain a number of parts, as depicted in Figure 16.16. Zinc metal and manganese dioxide are the substances that react to produce electrical energy.

Commercial batteries are marked with a ⊕ at one end and a ⊖ at the other end. These ends are called the **cathode** and **anode,** respectively. The oxidation half-reaction occurs at the anode, and the reduction half-reaction at the cathode. These definitions of anode and cathode are valid for all electrochemical cells, whether or not they are used in a battery.

In a typical alkaline dry cell battery, the reactions that occur are

Anode: $Zn(s) + 2\,OH^-(aq) \longrightarrow Zn(OH)_2(s) + 2\,e^-$ (oxidation)

Cathode: $2\,e^- + 2\,MnO_2(s) + H_2O(l) \longrightarrow Mn_2O_3(s) + 2\,OH^-(aq)$ (reduction)

Overall: $2\,MnO_2(s) + H_2O(l) + Zn(s) \longrightarrow Mn_2O_3(s) + Zn(OH)_2(s)$

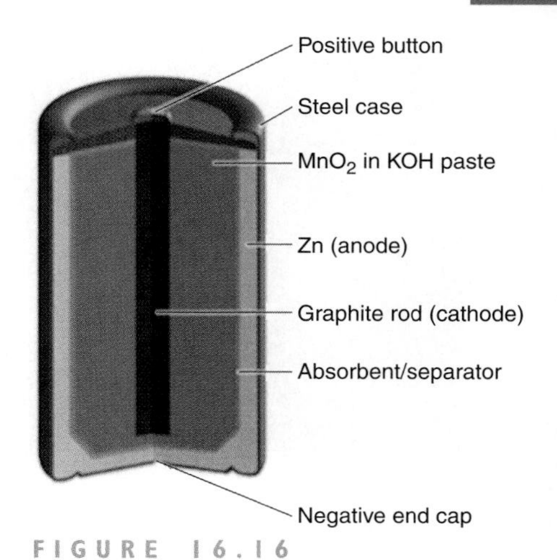


FIGURE 16.16
In an alkaline dry cell battery, the graphite cathode is chemically inert and acts as a wire to carry the electrical current. The half-reactions are actually complex multi-step processes approximated by the overall reaction $2\,MnO_2(s) + H_2O(l) + Zn(s) \rightarrow Mn_2O_3(s) + Zn(OH)_2(s)$.

### EXAMPLE 16.9 Redox Reactions in Alkaline Batteries

**a.** In the reaction that takes place in alkaline batteries, given below, what is the reducing agent?

**b.** Does this reaction take place in an acidic medium, or is a basic medium required?

$$2\,MnO_2(s) + H_2O(l) + Zn(s) \longrightarrow Mn_2O_3(s) + Zn(OH)_2(s)$$

**SOLUTIONS**

**a.** Metals are usually reducing agents, which suggests that $Zn(s)$ is the reducing agent here. You can check by using oxidation numbers. In $Zn(s)$, zinc has an oxidation number of 0, while in $Zn(OH)_2$ the oxidation number is +2. The substance being oxidized (the reducing agent) gives up electrons when it reacts, as zinc does here.

**b.** $Zn(OH)_2$ is zinc hydroxide, which can't exist in acidic solution. The reaction medium here must be basic. That is why KOH is present in the paste, and why the battery is called "alkaline."

**EXERCISE 16.9**

In the reaction that takes place in alkaline batteries (see Example 16.9), what is the oxidizing agent?

FIGURE 16.17
Galvanized garbage cans (left) are being replaced by lighter plastic ones (right). The latter are often made from polyethylene recycled from grocery bags and milk jugs.

**_Electrochemical cell_ is a general term. The kind of cell that produces electricity from chemical reactions is called a _voltaic_ (or _galvanic_) cell. The kind that uses electricity to make nonspontaneous redox reactions occur is an _electrolytic cell._ Both are electrochemical cells.**

## 16.6 Electroplating and Electrolysis

GOAL: To understand the process of electroplating, and to calculate the amount of metal deposited by a given amount of current in this process.

The copper-zinc electrochemical cell described in Section 16.5 produces electricity, but it also does something else: It plates additional copper on the copper metal electrode. Cells made with metals and their salts can produce electricity, and you can use electricity to produce metals from their salts. This process, called **electroplating,** is fairly widespread in industry. It's used to purify copper, galvanize metal garbage cans, and plate metals of all kinds (see Figure 16.17).

Copper for use in electrical wires and circuits should be very pure. Copper when it is first refined from copper ore contains other metals as impurities. Electricity is the key to the purification process. By means of an external electrical source, plates of impure copper can be given a positive charge while plates of pure copper are simultaneously given a negative charge. In a practical setup, plates of pure and impure copper alternate in solution. (See Figure 16.18.) When the electricity flows, the impure copper plates put $Cu^{2+}$ ions in solution in order to dump their excess positive charge. These $Cu^{2+}$ ions are then reduced back to copper metal on the negatively charged plate, as electrons react with $Cu^{2+}$ ions in the solution to produce pure copper. The solution never builds up an excess of either charge, because the $Cu^{2+}$ ions are consumed just as fast as they are formed.

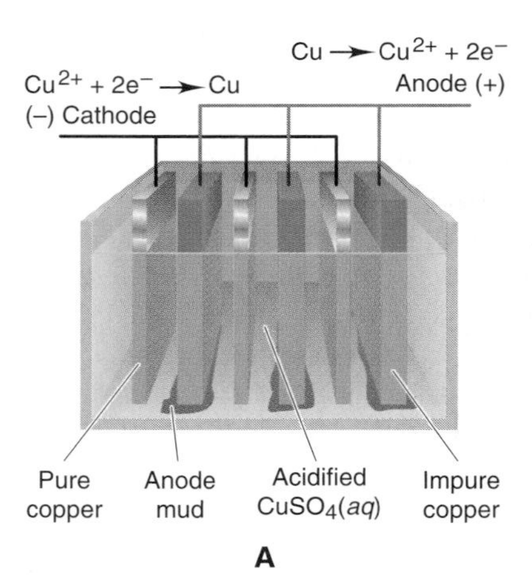

FIGURE 16.18
(A) Purifying copper by the electrolytic method. (B) At the start thin plates of pure copper alternate with thick plates of impure copper. As the purification proceeds, the copper in the impure plates is forced into solution by the electrical current. The $Cu^{2+}$ ions in solution are then reduced back to copper metal on the pure plates, which grow in size as the process continues. Some of the impurities present in the impure copper fall to the bottom. These impurities include silver and sometimes gold.

Impurities less active than copper (those below copper in Table 16.1), including silver and gold, fall to the bottom of the cell as the copper dissolves around them. Plant engineers adjust the voltage so that only the copper and not the more active metals plate out on the purified copper slab.

Electroplating is not the only technique that uses an external electrical source to run a chemical reaction. In the early 1800s Humphry Davy was intrigued by Volta's battery. Chemists of the day didn't fully understand the connection between chemical and electrical processes. In a truly magnificent experiment Davy prepared sodium and potassium metals by passing an electric current through their molten carbonate salts. He generated the current by using a huge battery consisting of 250 metal plates (look at Figure 16.15 and think bigger). Industries today use this process, which is called **electrolysis,** to prepare many metals, including potassium, sodium, and magnesium.

Electrolysis procedures are quite simple in concept. They use an electrolytic cell, which consists of a container that holds the molten metallic salt and two electrodes made of a solid conducting material (often graphite) connected to an electrical source. The metal ions migrate to the cathode, where they are reduced, and the anions migrate to the anode, where they are oxidized. Sodium metal is made this way from molten sodium chloride, as illustrated in Figure 16.19.

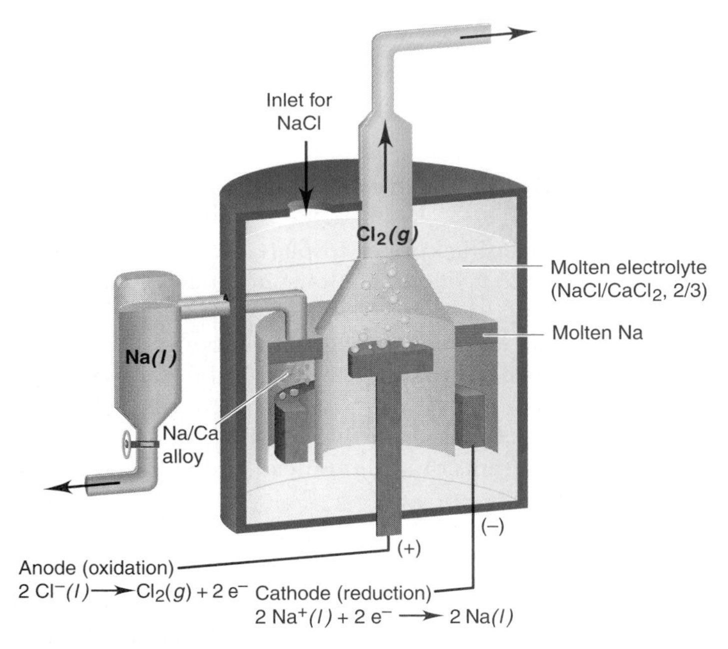


FIGURE 16.19
In a Downs cell for producing Na(*l*), when the engineer turns on the electricity, $Cl^-$ ions move to the anode where they are oxidized: $2\ Cl^- \rightarrow Cl_2(g) + 2\ e^-$. The $Na^+$ ions are reduced at the cathode: $Na^+ + e^- \rightarrow Na(l)$. Care must be taken to prevent the chlorine gas from contacting the molten sodium—the resulting reaction is violent.

You can see that it would be useful to know exactly how much current is needed to plate out a specific amount of metal or prepare a specific amount of an element. For example, if you were electroplating a metal with gold to improve its electrical conductivity for use in a computer, you certainly would want to use the minimum amount of gold necessary to accomplish your task.

Let's look more closely at the preparation of sodium by the electrolysis of sodium chloride. Passing 1 mol of electrons through the apparatus in Figure 16.19 will produce 1 mol of sodium, since the half-reaction is simply $Na^+ + e^- \rightarrow Na$. However, 2 mol of electrons must pass through the system to make 1 mol of chlorine gas; that half-reaction is $2\ Cl^- \rightarrow Cl_2 + 2\ e^-$.

The charge on an electron is $1.60219 \times 10^{-19}$ C. Therefore, the charge on 1 mol of electrons is

$$(6.02205 \times 10^{23} \text{ electrons mol}^{-1})(1.60219 \times 10^{-19} \text{ C electron}^{-1}) = 96{,}485 \text{ C mol}^{-1}$$

Michael Faraday did a great deal of work in developing the quantitative laws of electrolysis in the early part of the 1800s, and this number is called the **Faraday constant,** $F$, to honor him. The value of the constant is usually given with three significant figures:

$$F = 96{,}500 \text{ C mol}^{-1} = 9.65 \times 10^4 \text{ C mol}^{-1} \qquad \text{(Faraday constant)}$$

This allows us to compute the charge of $n$ moles of electrons easily enough:

$$\text{Charge } (Q) = nF$$

This is only part of the picture, however. Chemists and engineers usually determine charges by measuring the current flow for a given time. The unit for current is the *ampere* (A). A charge of 1 coulomb passes a point when a current of 1 ampere flows there for 1 second.

$$1 \text{ C} = (1 \text{ A})(1 \text{ s}) = 1 \text{ A s}$$

With this information, a variety of calculations are possible, as illustrated in Example 16.10.

## EXAMPLE 16.10 Stoichiometry of Copper Purification

**a.** A current flows, causing copper to plate out from a solution containing $Cu^{2+}$ ions. If 125 C of charge flows, how many grams of copper will be deposited?

**b.** A current of 15.0 A passes through a solution of $Cr(NO_3)_3$ for 1 h. How many moles of chromium metal will be deposited on an inert metal plate?

**c.** A current of 50.0 A is passed through NaCl($l$) in an electrolysis experiment. How many grams of sodium metal and chlorine gas will be produced in 30.0 min?

**SOLUTIONS**

**a.** The half-reaction is $Cu^{2+} + 2\ e^- \rightarrow Cu(s)$. This means that two moles of electrons are needed to deposit each mole of copper. The equation is

$$125 \text{ C}\left(\frac{1 \text{ mol e}^-}{9.65 \times 10^4 \text{ C}}\right)\left(\frac{1 \text{ mol Cu}(s)}{2 \text{ mol e}^-}\right)\left(\frac{63.55 \text{ g Cu}(s)}{1 \text{ mol Cu}(s)}\right) = 0.0412 \text{ g Cu}(s)$$

**b.** The half-reaction is $Cr^{3+} + 3\ e^- \rightarrow Cr$. To work this problem you first need to convert amperes (the product of current and time) to coulombs. Then you continue as in part (a).

$$15.0\ \text{A}\left(\frac{1\ \text{C}}{1\ \text{A s}}\right)\left(\frac{60\ \text{min}}{1\ \text{hr}}\right)\left(\frac{60\ \text{s}}{1\ \text{min}}\right) = 5.40 \times 10^4\ \text{C}$$

$$n = \frac{Q}{F} = \frac{5.40 \times 10^4\ \text{C}}{9.65 \times 10^4\ \text{C mol}^{-1}} = 0.560\ \text{mol electrons}$$

$$(0.560\ \text{mol electrons})\left(\frac{1\ \text{mol Cr}}{3\ \text{mol electrons}}\right) = 0.187\ \text{mol Cr}$$

**c.** The half-reactions are $Na^+ + e^- \rightarrow Na$ and $2\ Cl^- \rightarrow Cl_2 + 2e^-$.

$$(50.0\ \text{A})\left(\frac{1\ \text{C}}{1\ \text{A s}}\right)(30.0\ \text{min})\left(\frac{60\ \text{s}}{1\ \text{min}}\right) = 9.00 \times 10^4\ \text{C}$$

$$n = \frac{Q}{F} = \frac{9.00 \times 10^4\ \text{C}}{9.65 \times 10^4\ \text{C mol}^{-1}} = 0.933\ \text{mol electrons}$$

$$(0.933\ \text{mol electrons})\left(\frac{1\ \text{mol Na}}{1\ \text{mol electrons}}\right)\left(\frac{23.0\ \text{g Na}}{1\ \text{mol Na}}\right) = 21.5\ \text{g Na}$$

$$(0.933\ \text{mol electrons})\left(\frac{1\ \text{mol Cl}_2}{2\ \text{mol electrons}}\right)\left(\frac{70.9\ \text{g Cl}_2}{1\ \text{mol Cl}_2}\right) = 33.1\ \text{g Cl}_2$$

**EXERCISE 16.10**

**a.** A 15-A current flows for 1 h during copper purification. How many grams of copper will be purified by this current?

**b.** How long must a 25-A current pass through an electrolytic cell to produce 2 mol of sodium metal in an electrolysis experiment?

## CHAPTER SUMMARY

1. Chemical species are oxidized when they lose electrons and are reduced when they gain electrons. (In earlier theory, substances were oxidized when they added oxygen and reduced when they lost oxygen.)
2. Oxidation and reduction always occur simultaneously. Oxidation cannot take place without an accompanying reduction. Reactions in which these linked processes occur are called redox reactions.
3. Oxidizing agents oxidize some other species; they are reduced in redox reactions. Reducing agents reduce other species; they are oxidized in redox reactions.
4. Half-reactions represent the oxidation and reduction portions of a redox reaction separately. In oxidation half-reactions the electrons appear on the right side of the equation; in reduction half-reactions the electrons are on the left side.
5. In Table 16.1 the strongest oxidizing agents appear as reactants nearer the bottom of the table, and the strongest reducing agents appear as products nearer the top of the table.
6. In balancing redox reactions, it is vital that the number of electrons lost in the oxidation half-reaction equals the number of electrons gained in the reduction half-reaction.

7. Oxidation numbers are artificial in the sense that they seldom have any relation to actual charges of elements in molecules and polyatomic ions. However, oxidation numbers are a useful tool for keeping track of the total electron count.
8. Redox reactions can be used to generate electricity, a useful form of energy. Batteries are a classic example of this use of redox reactions.
9. Electroplating is a technique in which an electric current passes through a solution of a metallic salt, in such a way that a metal plates out of solution onto solid electrodes.
10. Electrolysis is a technique in which an electric current passes through a solution or a molten metallic salt. Metal ions are reduced to the metal at the anode and anions are oxidized at the cathode.

## PROBLEMS

### SECTION 16.1 Oxidation and Reduction: Definitions

1. Define *oxidizing agent.*
2. Define *reducing agent.*
3. Explain how to identify the substance that is reduced in a chemical equation.
4. Explain how to identify the substance that is oxidized in a chemical equation.
5. In each of these reactions, identify the oxidizing agent, the reducing agent, the substance that is oxidized, and the substance that is reduced.
   a. $2\ C_2H_2(g) + 5\ O_2(g) \rightarrow 4\ CO_2(g) + 2\ H_2O(g)$
   b. $Ca(s) + Br_2(l) \rightarrow CaBr_2(s)$
   c. $C_2H_4(g) + H_2(g) \xrightarrow{Pt} C_2H_6(g)$
6. In each of these reactions, identify the oxidizing agent, the reducing agent, the substance that is oxidized, and the substance that is reduced.
   a. $CuS(s) + O_2(g) \rightarrow Cu(s) + SO_2(g)$
   b. $2\ C_6H_5CHO + OH^- \rightarrow C_6H_5CO_2^- + C_6H_5CH_2OH$
   c. $Zn(s) + Sn^{2+}(aq) \rightarrow Zn^{2+}(aq) + Sn(s)$
7. In each of these reactions, identify the oxidizing agent, the reducing agent, the substance that is oxidized, and the substance that is reduced.
   a. $Mg(s) + I_2(g) \rightarrow MgI_2(s)$
   b. $S_8(s) + 8\ O_2(g) \rightarrow 8\ SO_2(g)$
   c. $2\ CH_3CHO(l) + O_2(g) \rightarrow 2\ CH_3CO_2H(l)$
8. In each of these reactions, identify the oxidizing agent, the reducing agent, the substance that is oxidized, and the substance that is reduced.
   a. $N_2(g) + O_2(g) \rightarrow 2\ NO(g)$
   b. $2\ Fe^{3+}(aq) + 2\ I^-(aq) \rightarrow 2\ Fe^{2+}(aq) + I_2(aq)$
   c. $CH_4 + Cl_2 \rightarrow CH_3Cl + HCl$
9. In each of these reactions, identify the oxidizing agent, the reducing agent, the substance that is oxidized, and the substance that is reduced.
   a. $PCl_3(l) + Cl_2(g) \rightarrow PCl_5(s)$
   b. $C(s) + H_2O(g) \rightarrow CO(g) + H_2(g)$
   c. $2\ C_2H_2(g) + O_2(g) \rightarrow 4\ C(s) + 2\ H_2O(g)$
10. In each of these reactions, identify the oxidizing agent, the reducing agent, the substance that is oxidized, and the substance that is reduced.
    a. $C(s) + 2\ S(s) \rightarrow CuS_2(s)$
    b. $Cu^{2+}(aq) + Cu(s) \rightarrow 2\ Cu^+(aq)$
    c. $Hg_2^{2+}(aq) + Br_2(l) \rightarrow 2\ Hg^{2+}(aq) + 2\ Br^-(aq)$
11. In each of these reactions, identify the oxidizing agent, the reducing agent, the substance that is oxidized, and the substance that is reduced.
    a. $CO(g) + H_2O(g) \rightarrow CO_2(g) + H_2(g)$
    b. $FeO(s) + CO(g) \rightarrow Fe(s) + CO_2(g)$
    c. $2\ Na(l) + CuCl_2(s) \rightarrow 2\ NaCl(s) + Cu(l)$
12. In each of these reactions, identify the oxidizing agent, the reducing agent, the substance that is oxidized, and the substance that is reduced.
    a. $Zn(s) + Cu^{2+}(aq) \rightarrow Zn^{2+}(aq) + Cu(s)$
    b. $Fe_2O_3(s) + 2\ Al(s) \rightarrow 2\ Fe(l) + Al_2O_3(s)$
    c. $SnCl_4(aq) + Sn(s) \rightarrow 2\ SnCl_2(aq)$

### SECTION 16.2 Oxidizing and Reducing Agents

13. In each of these pairs, which is the more powerful oxidizing agent? Explain your answers.
    a. $MnO_2$ or $MnO_4^-$
    b. $Cu^+$ or $Cu^{2+}$
    c. $O_2$ or $S_8$

14. In each pair, which is the more powerful oxidizing agent? Explain your answers.
a. $Sn^{2+}$ or $Sn^{4+}$ b. $I_2$ or $I^-$ c. $N_2$ or $O_2$

15. In each of these pairs, which is the more powerful oxidizing agent? Explain your answers.
a. Mn(III) or Mn(II) c. $Cl_2$ or $Cl^-$
b. $VO^{2+}$ or $VO_2^+$

16. In each of these pairs, which is the more powerful oxidizing agent? Explain your answers.
a. $MnO_4^-$ or $MnO_3^-$ c. $Cl_2$ or $I_2$
b. Na or $Cl_2$

17. Which species in each pair is the more powerful reducing agent? Explain your answers.
a. Li or K c. Mn or $Mn^{2+}$
b. $Cl_2$ or $Br_2$

18. Which species in each pair is the more powerful reducing agent? Explain your answers.
a. Na or Mg c. $Fe^{3+}$ or $Fe^{2+}$
b. CO or $CO_2$

19. Give the half-reaction for oxidation (if any) for the reaction that occurs (if one does) in each of these experiments.
a. Sodium metal is heated with solid zinc chloride ($ZnCl_2$).
b. Liquid bromine is added to an aqueous solution of sodium chloride.
c. Chlorine gas is bubbled through a solution containing iron (II) ions.

20. Give the half-reaction for oxidation (if any) for the reaction that occurs (if one does) in each of these experiments.
a. Fluorine gas is bubbled through an aqueous solution of $KMnO_4$.
b. Solid $I_2$ is heated with tin metal.
c. An aqueous solution containing $Fe^{2+}$ ions is added to silver metal.

21. Give the half-reaction for reduction (if any) for the reaction that occurs (if one does) in each of the experiments described in Problem 19.

22. Give the half-reaction for reduction (if any) for the reaction that occurs (if one does) in each of the experiments described in Problem 20.

23. For each of the experiments described, write the half-reaction for oxidation (if any) for the reaction that occurs (if one does).
a. Chlorine gas is bubbled through a solution containing $Fe^{3+}$ ions.
b. Solid aluminum is heated strongly with $Fe_2O_3$.
c. Oxygen gas is bubbled through an aqueous solution containing $I^-$.

24. For each of the experiments described, write the half-reaction for oxidation (if any) for the reaction that occurs (if one does) in each of these experiments.
a. Aqueous solutions of hydrogen peroxide ($H_2O_2$) and $Ag^+$ ion are mixed.
b. Iron filings, Fe(*s*), are added to an aqueous solution containing $Br^-$.
c. Solutions containing $MnO_4^-(aq)$ and $Fe^{2+}(aq)$ are mixed.

25. Write the half-reaction for reduction (if any) for the reaction that occurs (if one does) in each of the experiments described in Problem 23.

26. Write the half-reaction for reduction (if any) for the reaction that occurs (if one does) in each of the experiments described in Problem 24.

## SECTION 16.3 Balancing Redox Equations: The Half-Reaction Method

27. Balance each equation using the half-reaction method.
a. $Cr + MnO_2 \rightarrow Mn^{2+} + Cr^{3+}$ (acidic conditions)
b. $H_2O_2 + MnO_4^- \rightarrow O_2 + Mn^{2+}$ (acidic conditions)
c. $H_2SO_4 + Cu \rightarrow SO_2 + Cu^{2+}$

28. Balance each equation using the half-reaction method.
a. $As + NO_3^- \rightarrow H_3AsO_3 + NO$ (acidic conditions)
b. $H_2 + AgCl \rightarrow HCl + Ag$
c. $MnO_4^- + H_2CO \rightarrow MnO_2 + CO_2$ (basic conditions)

29. Balance each equation using the half-reaction method.
a. $HNO_3 + Cu \rightarrow Cu^{2+} + NO_2$
b. $Zn + BrO_4^- \rightarrow Zn^{2+} + Br^-$ (basic conditions)
c. $MnO_4^- + S^{2-} \rightarrow MnO_2 + S$ (basic conditions)

30. Balance each equation using the half-reaction method.
a. $CrO_3 + CH_3OH \rightarrow CO_2 + Cr^{2+}$ (acidic conditions)
b. $MnO_4^- + NO_2^- \rightarrow MnO_2 + NO_3^-$ (basic conditions)
c. $OsO_4 + Ag \rightarrow Os + Ag^+$ (acidic conditions)

31. Use the half-reaction method to balance each of these equations.
a. $H_2O_2 + NO \rightarrow H_2O + HNO_2$ (acidic conditions)
b. $H_2O_2 + Hg \rightarrow Hg^{2+} + OH^-$
c. $CrO_4^{2-} + Fe \rightarrow Fe^{3+} + Cr^{3+}$ (acidic conditions)

32. Use the half-reaction method to balance each of these equations.
a. $H_2S + MnO_2 \rightarrow Mn^{2+} + S$ (acidic conditions)
b. $CuI + Ag \rightarrow AgI + Cu$
c. $AuCl_2 \rightarrow Au + AuCl_3$

### SECTION 16.4 Oxidation Numbers

33. What is the oxidation number of each of the elements in these species?
a. $CuF$
b. $CrO_4^{2-}$
c. $KClO_3$
d. $IF_5$

34. What is the oxidation number of each of the elements in these species?
a. $[VO_3(OH)]^{2-}$
b. $K_2CO_3$
c. $BrF_3$
d. $MnO_4^{2-}$

35. Determine the oxidation number of each of the elements in these species.
a. $Al_2O_3$
b. $K_3VO_4$
c. $TiF_6^{2-}$
d. $HOCl$

36. Determine the oxidation number of each of the elements in these species.
a. $BiCl_4^-$
b. $CaCO_3$
c. $CaO$
d. $BH_4^-$

37. In each of these species what is the oxidation number of each of the elements?
a. $PdCl_6^{2-}$
b. $NpO_2^{1+}$
c. $ZnS$
d. $CaCr_2O_7$

38. In each of these species what is the oxidation number of each of the elements?
a. $UO_2Cl_2$
b. $Na_2CrO_4$
c. $P_4$
d. $NaOH$

39. Explain how to use oxidation numbers to identify an oxidizing agent in a chemical equation.

40. Explain how to use oxidation numbers to identify a reducing agent in a chemical equation.

41. For each of these reactions, give the oxidation number of all elements, and identify the oxidizing agent, the substance oxidized, the reducing agent, and the substance reduced.
a. $2\,CuS(s) + 3\,O_2(g) \rightarrow 2\,CuO(s) + 2\,SO_2(g)$
b. $2\,Al(s) + 6\,HCl(aq) \rightarrow 2\,AlCl_3(aq) + 3\,H_2(g)$
c. $CO_2(g) + 2\,H_2(g) \rightarrow C(s) + 2\,H_2O(g)$

42. For each of these reactions, give the oxidation number of all elements, and identify the oxidizing agent, the substance oxidized, the reducing agent, and the substance reduced.
a. $2\,K(s) + 2\,H_2O(l) \rightarrow 2\,KOH(aq) + H_2(g)$
b. $Cl_2(aq) + 2\,KBr(aq) \rightarrow 2\,KCl(aq) + Br_2(aq)$
c. $2\,C_2H_6(g) + 7\,O_2(g) \rightarrow 6\,H_2O(l) + 4\,CO_2(g)$

43. For each of the elements in these reactions, give the oxidation number. Also, identify the oxidizing agent, the substance oxidized, the reducing agent, and the substance reduced in each case.
a. $2\,Fe_2O_3(s) + 3\,C(s) \rightarrow 3\,CO_2(g) + 4\,Fe(s)$
b. $Mg(s) + 2\,HNO_3(aq) \rightarrow Mg(NO_3)_2(aq) + H_2(g)$
c. $C_4H_6 + 2\,H_2 \xrightarrow{Pt} C_4H_{10}$

44. For each of the elements in these reactions, give the oxidation number. Also, identify the oxidizing agent, the substance oxidized, the reducing agent, and the substance reduced in each case.
a. $2\,Fe(s) + 2\,H_2SO_4(aq) \rightarrow 2\,FeHSO_4(aq) + H_2(g)$
b. $C(s) + 2\,H_2(g) \rightarrow CH_4(g)$
c. $Zn(s) + 2\,HBr(aq) \rightarrow ZnBr_2(aq) + H_2(g)$

45. Determine all oxidation numbers and identify the oxidizing agent, the substance oxidized, the reducing agent, and the substance reduced in each reaction.
a. $2\,Cs(l) + H_2O(l) \rightarrow 2\,CsOH(aq) + H_2(g)$
b. $2\,H_2S(g) + SO_2(g) \rightarrow 3\,S(s) + 2\,H_2O(g)$
c. $N_2(g) + O_2(g) \rightarrow 2\,NO(g)$

46. Determine oxidation numbers and identify the oxidizing agent, the substance oxidized, the reducing agent, and the substance reduced in each reaction.
a. $H_2CO(aq) + 5\,H_2O(l) + 4\,Ag^+(aq) \rightarrow 4\,Ag(s) + CO_2(g) + 4\,H_3O^+(aq)$
b. $2\,H_2S(g) + 3\,O_2(g) \rightarrow 2\,SO_2(g) + 2\,H_2O(g)$
c. $3\,Mg(s) + N_2(g) \rightarrow Mg_3N_2(s)$

### SECTION 16.5 Electricity from Redox Reactions

47. What half-reaction occurs at the cathode in the cell shown? In what direction do electrons flow through the wires and light bulb? In the salt bridge, in what direction do the $K^+$ ions move?

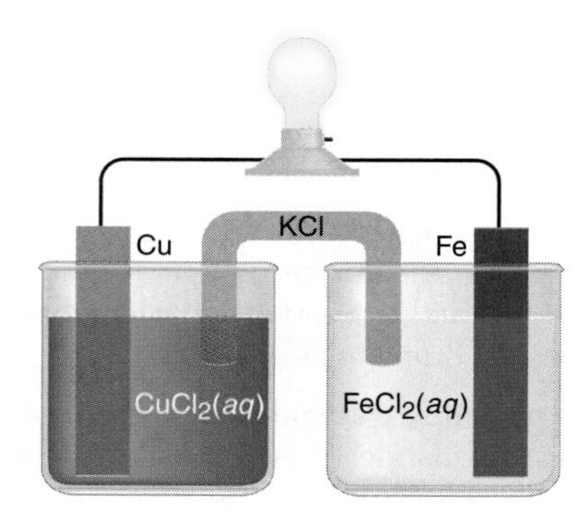


48. What half-reactions occur at the anode in the cell in Problem 47? In the salt bridge, in what direction do the $Cl^-$ ions move?

49. In the cell shown below, the platinum wire electrode does not react. Instead, $H_2$ gas bubbles evolve. What half-reaction occurs at the anode in this cell? In the salt bridge, in what direction do $Cl^-$ ions move?

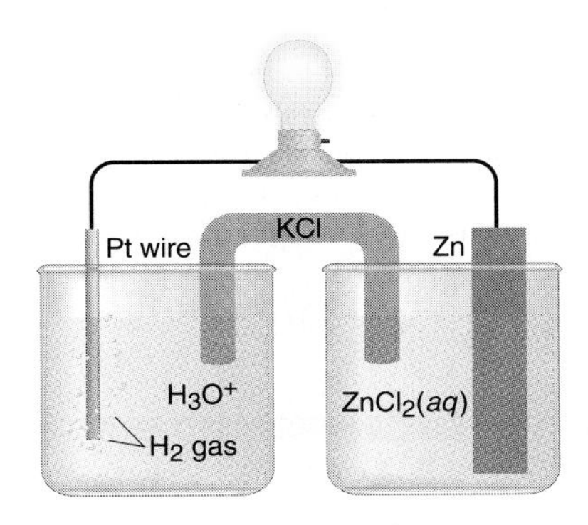


50. What half-reaction occurs at the cathode in the cell in Problem 49? In what direction do electrons flow through the wires and light bulb? In the salt bridge, in what direction do the $K^+$ ions move?

51. Lead storage batteries (Figure 16.9 shows one) are the kind used in almost all cars.
    a. In the reaction that takes place in these batteries (given below), what is the reducing agent?

$$Pb(s) + PbO_2(s) + 2\,H_2SO_4(aq) \longrightarrow 2\,PbSO_4(s) + 2\,H_2O(l)$$

    b. What is the balanced half-reaction for the reduction?

52. The reaction that takes place in lead storage batteries is given in Problem 51.
    a. What is the oxidizing agent?
    b. What is the balanced half-reaction for the oxidation?

53. Mercury batteries are used to power some electric watches and other small appliances. The overall cell reaction is

$$HgO(s) + Zn(s) \longrightarrow ZnO(s) + Hg(l)$$

    a. What is the oxidizing agent?
    b. What species is reduced?
    c. Write the balanced oxidation half-reaction that occurs in basic conditions.

54. The overall reaction that takes place in mercury batteries is given in Problem 53.
    a. What is the reducing agent?
    b. What species is oxidized?
    c. Write the balanced reduction half-reaction that occurs in basic conditions.

55. Silver batteries (Figure 16.9 shows one) are used in pacemakers and hearing aids. The unbalanced cell reaction is

$$Ag_2O(s) + Zn(s) \longrightarrow Ag(s) + ZnO(s)$$

    a. Write a balanced equation for the oxidation half-reaction in basic conditions.
    b. What reaction occurs at the cathode?

56. The unbalanced equation for the reaction that takes place in silver batteries is given in Problem 55.
    a. Write a balanced equation for the reduction half-reaction in basic conditions.
    b. Write a balanced equation for the overall cell reaction.
    c. What reaction occurs at the anode?

## SECTION 16.6 Electroplating and Electrolysis

57. In electroplating the goal is to put a layer of metal on an object. Is the object the cathode or the anode? Explain your answer.

58. In the purification of copper the metal dissolves from one electrode and redeposits on the other. Which is the anode, and which is the cathode? Explain your answer.

59. A current flows, causing zinc to plate out from a solution of $Zn^{2+}$ ions. If 478 C of charge flows, how many grams of zinc are deposited?

60. Some jewelers are electroplating silver onto a platter. If $2.53 \times 10^3$ C of charge flows through the solution of $Ag^+$ ions, how many grams of silver are deposited?

61. A group of auto customizers passed a current of 45.0 A through a solution of $Cr(NO_3)_3$ for 2 h and 15 min. How many moles of chromium metal will be deposited on their van decoration?

62. Some auto customizers passed a current of 30.0 A through a solution of $Cu(NO_3)_2$ for 1 h and 35 min. How many moles of copper metal will be deposited on the van decoration?

63. A group of chemists passed a current of 50.0 A through CsCl in an electrolysis experiment. How many grams of cesium metal will they produce in 1 h and 30 min?

64. Certain jewelers passed a current of 11.5 A through a solution of gold(I) ions to add gold plating to a statuette. How many grams of gold will be plated in 48 min?

65. How long must a current of 15 A pass through a solution of $Cu^{2+}$ ions to deposit 1.00 g of copper metal on a plate?

66. How long must a current of 30 A pass through a solution of $Cr^{3+}$ ions to deposit 1.50 g of chromium?

67. How long must a current of 25 A be passed through a solution of $Ag^{+}$ ions to deposit 0.50 g of silver?

68. How long must a current of 30 A be passed through a solution of $Zn^{2+}$ ions to deposit 2.50 g of zinc?

**ADDITIONAL PROBLEMS**

69. A group of chemists prepared a sample of $F_2$ by passing a current of 8.4 A through a mixture of KF and HF for 27 min. (a) How many moles of fluorine gas did they produce? (b) What volume did this sample of $F_2$ occupy at STP?

70. A group of chemists prepared a sample of $Cl_2$ by passing a current of 35 A through molten NaCl for 44 min. (a) How many moles of chlorine gas did they produce? (b) What volume did this sample of $Cl_2$ occupy at STP?

## EXPERIMENTAL PROBLEM 16.1

A. L. Chemist and a team of scientists were studying redox reactions. In the first experiment they slowly added $FeCl_3(aq)$ to NaI(*aq*), resulting in the formation of a colored solution (Solution 1). In a second experiment they bubbled a small amount of $Cl_2(g)$ through a saturated solution of KI(*aq*). This also gave a colored solution (Solution 2). The team attempted to isolate the compound responsible for the colors of Solutions 1 and 2, and obtained a black solid that gave off purple fumes when heated.

**QUESTIONS**

**71.** What reaction occurs when $FeCl_3(aq)$ is added to NaI(*aq*)?
**72.** What reaction occurs in the second experiment?
**73.** What substance was responsible for the purple color of the fumes from the sample obtained from Solutions 1 and 2?
**74.** What evidence supports the conclusion that a redox reaction occurs in both experiments?
**75.** What is the oxidation half-reaction in the first experiment?
**76.** What is the reduction half-reaction in the second experiment?
**77.** What ionic compound is present in Solution 2 besides KI?
**78.** What is the oxidizing agent in the first experiment?
**79.** What is the reducing agent in the second experiment?
**80.** According to Table 16.1, which of the following is the best reducing agent? Which is the best oxidizing agent?

$Fe^{3+}$ $Na^{+}$ $K^{+}$ $Cl^{-}$ $I^{-}$ $Cl_2$ $I_2$

## SOLUTIONS TO EXERCISES

**Exercise 16.1**

**a.** According to the old definition, copper loses oxygen, so it should be reduced. Let's check. Copper goes from $Cu^{2+}$ ions to neutral Cu atoms. It gains electrons, is reduced, and is the oxidizing agent. Hydrogen, $H_2$, is oxidized and is the reducing agent.
**b.** Sodium goes from neutral atoms to $Na^{+}$ ions. It loses electrons, is oxidized, and is the reducing agent. Water is reduced and is the oxidizing agent.

**Exercise 16.2**

**a.** Hydrogen, $H_2$, reacts to form a covalent compound. Therefore it is oxidized and is the reducing agent. The compound $CH_3CH_2SH$ is the oxidizing agent, the species reduced.
**b.** Oxygen is the oxidizing agent, the species reduced. The other compound present, $CH_2{=}CH_2$, is the reducing agent and the species oxidized.

**Exercise 16.3**

**a.** Fluorine is more electronegative than iodine and is therefore reduced more easily. Thus $F_2$ is the more powerful oxidizing agent.
**b.** $Sn^{4+}$ has the larger positive charge and should be able to add electrons more easily than $Sn^{2+}$ can. $Sn^{4+}$ is probably the stronger oxidizing agent.
**c.** $MnO_4^-$ has more oxygen atoms and is therefore the better oxidizing agent.

**Exercise 16.4**

**a.** The $Fe^{2+} \rightarrow Fe$ half-reaction is higher in Table 16.1 than the $Sn^{2+} \rightarrow Sn$ half-reaction. This means that $Sn^{2+}$ will be reduced to Sn by iron metal. The reduction half-reaction is $Sn^{2+} + 2\,e^- \rightarrow Sn$. The oxidation half-reaction is $Fe(s) \rightarrow Fe^{2+} + 2\,e^-$.
**b.** Here $I^-$ (higher in Table 16.1) is oxidized to $I_2$ and $Fe^{3+}$ (lower in the table) is reduced to $Fe^{2+}$. The reduction half-reaction is $Fe^{3+} + e^- \rightarrow Fe^{2+}$. The oxidation half-reaction is $2\,I^- \rightarrow I_2 + 2\,e^-$.

**Exercise 16.5**

**a.** Separate the given equation into half-reactions and balance each one separately. For one of the half-reactions:

$Fe^{2+} \longrightarrow Fe^{3+}$ (iron is already balanced; there is no oxygen or hydrogen)

$Fe^{2+} \longrightarrow Fe^{3+} + \mathbf{e^-}$ (balancing charge)

For the other half-reaction:

$MnO_4^- \longrightarrow Mn^{2+}$ (manganese is already balanced)

$MnO_4^- \longrightarrow Mn^{2+} + \mathbf{4\,H_2O}$ (balancing oxygen)

$MnO_4^- + \mathbf{8\,H^+} \longrightarrow Mn^{2+} + 4\,H_2O$ (balancing hydrogen)

$MnO_4^- + 8\,H^+ + \mathbf{5\,e^-} \longrightarrow Mn^{2+} + 4\,H_2O$ (balancing charge)

Multiply the first half-reaction by 5 and add the two half-reactions:

$$5(Fe^{2+} \longrightarrow Fe^{3+} + \mathbf{e^-}) = 5\,Fe^{2+} \longrightarrow 5\,Fe^{3+} + 5\,\mathbf{e^-}$$
$$MnO_4^- + 8\,H^+ + \mathbf{5\,e^-} \longrightarrow Mn^{2+} + 4\,H_2O$$

$$5\,Fe^{2+} + MnO_4^- + 8\,H^+ + \cancel{5\,e^-} \longrightarrow 5\,Fe^{3+} + Mn^{2+} + 4\,H_2O + \cancel{5\,e^-}$$

$$5\,Fe^{2+} + MnO_4^- + 8\,H^+ \longrightarrow 5\,Fe^{3+} + Mn^{2+} + 4\,H_2O$$

Because the reaction is in acid solution, convert $H^+$ to $H_3O^+$ by adding 8 $H_2O$ to both sides of the equation:

$$5\,Fe^{2+} + MnO_4^- + 8\,H^+ + \mathbf{8\,H_2O} \longrightarrow 5\,Fe^{3+} + Mn^{2+} + 4\,H_2O + \mathbf{8\,H_2O}$$

This gives the answer:

$$5\,Fe^{2+} + MnO_4^- + 8\,H_3O^+ \longrightarrow 5\,Fe^{3+} + Mn^{2+} + 12\,H_2O$$

**b.** Separate the overall equation into half-reactions. One half-reaction is

$Fe \longrightarrow Fe^{2+}$ (iron is already balanced; oxygen and hydrogen are not involved)

$Fe \longrightarrow Fe^{2+} + 2\,e^-$ (balancing charge)

The second half-reaction is

$Sn^{2+} \longrightarrow Sn$ (tin is balanced; no need to worry about oxygen and hydrogen)

$Sn^{2+} + 2\,e^- \longrightarrow Sn$ (balancing charge)

It's easy to see that the final balanced equation is

$$Fe + Sn^{2+} \longrightarrow Fe^{2+} + Sn$$

**Exercise 16.6**

**a.** Ca, +2; S, −2 (for the charges on monatomic ions, see Chapter 7)
**b.** H, +1; O, −2; N, +5 (hydrogen and oxygen from the guidelines, nitrogen by calculation)
**c.** Na, +1; O, −2; S, +6 (sodium and oxygen from the guidelines, sulfur by calculation)
**d.** S, 0 (first guideline)
**e.** Fe, +3; O, −2 (oxygen from the fourth guideline, iron by calculation)
**f.** F is always −1 in its compounds; this means that oxygen is +2 here.

**Exercise 16.7**

Here is the table of formulas and oxidation numbers:

| Reactants | | Products | |
|---|---|---|---|
| $Fe_2O_3$ | Na | $Na_2O$ | Fe |
| Fe, +3<br>O, −2 | 0 | Na, +1<br>O, −2 | 0 |

The Fe in $Fe_2O_3$ gains electrons; $Fe_2O_3$ is the oxidizing agent and the substance reduced. Sodium loses electrons, and is therefore the reducing agent and the substance oxidized.

**Exercise 16.8**

As the $Cu^{2+}$ ions are reduced to Cu metal in the cell on the left, the $K^+$ ions migrate toward the left to replace the positive charge.

**Exercise 16.9**

Zinc is the reducing agent, and nothing seems to be happening to the water. This leaves $MnO_2(s)$. Check it using oxidation numbers. In $MnO_2$ the oxidation number of Mn is +4, and in $Mn_2O_3$ it's +3. $MnO_2$ is reduced and is the oxidizing agent.

**Exercise 16.10**

**a.** The half-reaction is $Cu^{2+} + 2\ e^- \rightarrow Cu(s)$. This means that 2 moles of electrons are needed to deposit each mole of copper. The equation is:

$$\left(\frac{15\ C}{1\ s}\right)(3600\ s)\left(\frac{1\ mol\ e^-}{9.65 \times 10^4\ C}\right)\left(\frac{1\ mol\ Cu(s)}{2\ mol\ e^-}\right)\left(\frac{63.55\ g\ Cu(s)}{1\ mol\ Cu(s)}\right) = 17.8\ g\ Cu(s)$$

**b.** To get 2 mol of Na metal requires 2 mol of electrons because the half-reaction is

$$Na^+ + e^- \longrightarrow Na$$

Therefore the charge that passes through the cell needs to be

$$(2\ mol\ electrons)(9.65 \times 10^4\ C\ mol\ electrons^{-1}) = 1.93 \times 10^5\ C$$

$$1.93 \times 10^5\ C = 25\ A \times s$$

or $$\frac{1.93 \times 10^5\ C}{25\ A} = 7.72 \times 10^3\ s$$

$$(7.72 \times 10^3\ s)\left(\frac{1\ min}{60\ s}\right)\left(\frac{1\ h}{60\ min}\right) = 2.14\ h$$

**Experimental Exercise 16.1**

1. According to Stahl, charcoal loses phlogiston to the air when it burns. Since metallic ores yield metals when heated with charcoal, it is reasonable to conclude that the metals gain the phlogiston lost by the charcoal. Since the metallic ores accept phlogiston from the charcoal, they must be lacking in phlogiston. The reaction product, the metal, must be rich in phlogiston.
2. Not at all—this experiment only shows that combustion can occur if air is present. Lavoisier recognized this and ran a second experiment. He heated a diamond in an evacuated bulb (containing no air) and no reaction was observed. It was only after Lavoisier completed this second experiment that he concluded that air was necessary for combustion to occur.
3. When things such as wood and charcoal burn, they add oxygen to form oxides, specifically $CO_2$. When iron rusts, it adds oxygen to form the oxide $Fe_2O_3$. The reactions are sufficiently similar to support the contention.
4. Only experiment B, in which only part of the air can combine with the metal.

17

# NUCLEAR CHEMISTRY



## KEY TERMS

In all the chemical reactions we have looked at so far in this book, the changes that occur have been due to electrons moving within or between atoms. No atomic nuclei have changed in any of the reactions you have seen up to now. In this chapter we will look at processes that change the nuclei of atoms. Chemists who study these processes call their work **nuclear chemistry.** Physicists study the same phenomena, of course, and they call the field *nuclear physics*.

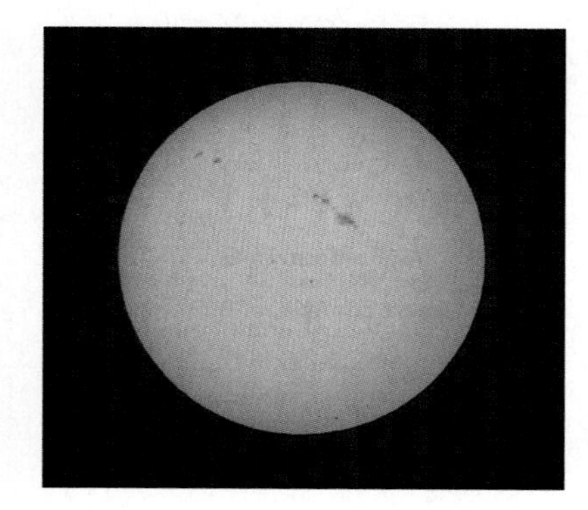

FIGURE 17.1
The sun's energy comes from nuclear changes.

Dalton's early atomic theory proposed that atoms were unchangeable. Arrhenius (Section 4.5) and chemists who followed him showed how the electronic structures of atoms and molecules can change. Eventually chemists and physicists demonstrated that even the nuclei of atoms can alter. In fact, without changes in nuclei, the universe couldn't exist in its present form.

This chapter considers transformations that occur in the nuclei of atoms. The sun and the stars are enormous nuclear furnaces. Nuclear reactions in the sun convert millions of tons of matter into energy each and every second, transforming even more millions of tons of hydrogen atoms into helium atoms in the process (see Figure 17.1). These are typical results of nuclear reactions. Atoms of one element usually change into atoms of another element, and the reactions typically release enormous amounts of energy.

Heat from radioactive processes in the earth's core has played a major role in this planet's geological development. Here at the earth's surface, people use electrical energy from nuclear power plants in their homes and at work. Many people worry about nuclear warfare, though perhaps not as much as a few years ago, and many benefit from nuclear medicine. Changes in the nuclei of atoms have a profound influence on human life.

FIGURE 17.2
Using the natural radioactivity of an isotope of carbon from cooking fires, archaeologists were able to establish the date of construction of these Native American cliff dwellings.

## 17.1 Radioactivity

GOAL: To learn about the historical origins of knowledge of radioactivity.

For more information about cathode rays, see Chapter Box 4.3, p.129.

The main output of a cathode ray tube is a beam of electrons, and these rays played a major role in the discovery that atoms are made up of smaller particles. This discovery about atomic structure is not the only one that came from the study of cathode rays. As the 1800s drew to a close, the physicist

Wilhelm Roentgen (see Figure 17.3) turned his attention toward cathode rays. He knew from the work of others that these rays cause some substances to give off faint light, and he performed experiments to try to understand what was happening.

Roentgen performed most of these experiments in a darkened room. He blocked his cathode ray tube with cardboard so that the light the tube gave off wouldn't affect his night vision. His targets, samples of substances that might glow if exposed to cathode rays, were located across the room. Roentgen noticed that one of his targets, a piece of paper coated with a complex platinum salt, was glowing (see Figure 17.4). It was doing this despite the cardboard that blocked the line of sight between the cathode ray tube and the target. Roentgen knew that cathode rays don't penetrate cardboard. More amazingly, his piece of coated paper glowed even in the next room. When he turned the tube off, the glow disappeared; when he turned it back on, the glow came back.

Roentgen performed more experiments. He found that his new rays didn't bend in the presence of either electric or magnetic fields. (Cathode rays are moving electrons, and they are bent by both). He had discovered a new kind of rays. He didn't know what they were, so he called them X rays (X is commonly used as a symbol for an unknown quantity in mathematics). He announced his discovery of X rays in 1895, but refused to patent the rays, allowing them to be used freely for the good of humanity.

FIGURE 17.3
The German physicist Wilhelm Konrad Roentgen (1845–1923). In 1901 he became the first winner of the Nobel Prize in physics.

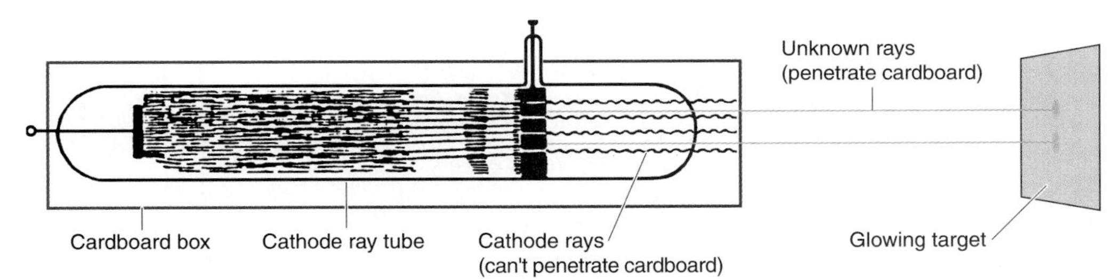


FIGURE 17.4
Roentgen's experiment. He called his previously undiscovered rays "X rays" after the commonly used symbol for an unknown in mathematics.

In his first public demonstration Roentgen took a picture of the bone structure of a volunteer's hand. Medical X rays came into use almost immediately. Roentgen's discovery triggered an intense wave of excitement among scientists. Dozens of physicists set out to study the new rays. They eventually learned that **X rays** are a form of high-energy electromagnetic radiation (see Chapter 6).

One of the scientists excited by Roentgen's discovery was the French physicist Antoine-Henri Becquerel (1852–1908). Some substances give off low-frequency light when high-frequency light hits them, and he was studying this phenomenon. Becquerel wondered whether substances would give off X rays if they sat in sunlight. His experiment was a simple one. He took a carefully wrapped photographic plate, put a uranium salt sample on top of it, and let it sit in the sun. If the substance gave off X rays, the rays would penetrate the wrapping and produce a blurred image on the photographic plate. If no X rays were emitted, nothing would happen to the plate, Becquerel thought.

**Sunlight cannot cause anything at all to give off X rays. The photons that reach the earth from the sun do not have enough energy to trigger the process that gives X rays.**

**In those days the chemicals that produced photographic images were coated on glass plates. Today they are imbedded in plastic film.**

He developed his plate and found an image. Then he set out to repeat his experiment. He prepared another wrapped photographic plate with the uranium salt on top and waited for a sunny day. After several cloudy days he decided to develop the plate anyway, just to make sure that sunlight was necessary for image formation. This is what scientists call a *control experiment.* To Becquerel's surprise the plate showed an image.

Becquerel studied this new phenomenon intensely, and so did others. Marie Curie developed a device for measuring the amount of radiation given off by a sample. She found that the amount of radiation from most of her samples was proportional to the amount of uranium in the sample, and that another element (thorium) gives off the same general kind of rays. She coined the name **radioactivity** for this phenomenon.

The Curies and others observed a connection between radiation and energy very early on. In 1901 Pierre Curie observed that a gram of radium, the most strongly radioactive element known at the time, gave off a tremendous amount of heat along with the radiation. In 1905 the American geologist Clarence Edward Dutton suggested that local concentrations of radioactivity provide the heat needed to drive volcanos and other geothermal processes in the interior of the planet. Today it seems more likely that heating of the earth's core by radioactivity results in continental drift, which has volcanism as a side effect (see Figure 17.5). Dutton's suggestion was a first step in the right direction.

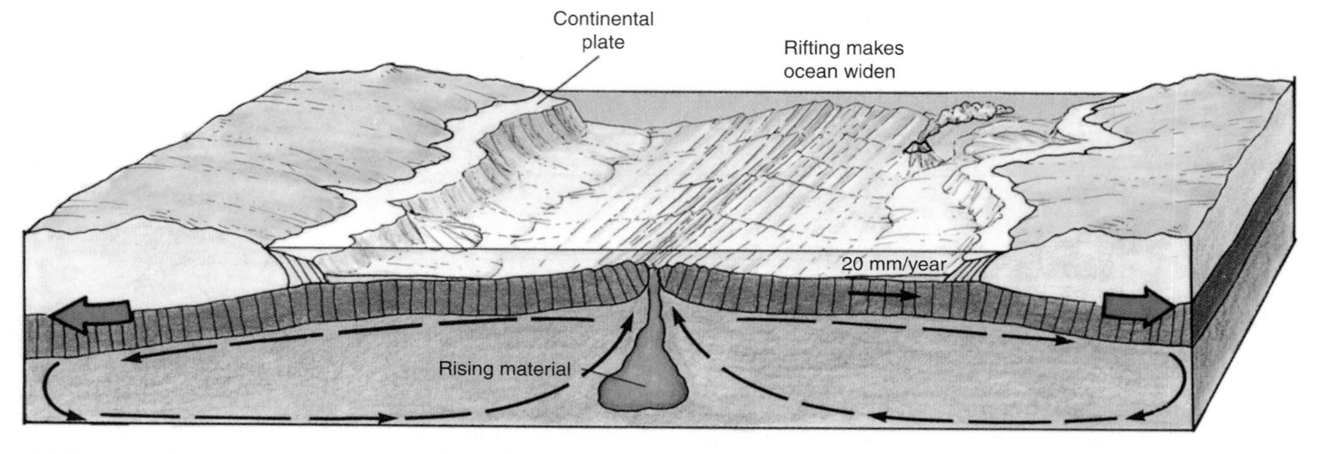


FIGURE 17.5
Rock in the lower mantle of the earth's crust, heated by radioactivity in the earth's core, expands and rises through cooler, more dense rock. As this rock cools near the surface, it becomes more dense and sinks. This is the internal geological process that causes continental drift.

The American chemist William Draper Harkins noted in 1915 that the mass of four hydrogen atoms is slightly greater than the mass of a helium atom. By that time Albert Einstein had published his famous equation $E = mc^2$, and Harkins used it to calculate the amount of energy equivalent to the mass difference. He pointed out that the amount of energy that would be given off by the *fusion* (that is, the combination) of four hydrogen atoms to produce one helium atom was enormous. He proposed that the sun and the stars generate energy this way—that is, by fusion. He was approximately correct, although the process requires several intermediate steps and is more complicated than his simple idea suggested. This topic is covered in more detail later in this chapter.

By 1925 the English chemist and physicist Francis William Aston had carefully determined the atomic masses of most of the stable isotopes of the known elements. This meant that he could use Einstein's equation to calculate the amount of energy obtainable from *nuclear fission* (the splitting of nuclei, described in Section 17.6). He even predicted some of the possible dangers associated with fission. The atomic age was on its way, whether humanity was ready for it or not.

CHAPTER BOX **17.1 MARIE SKLODOWSKA CURIE**

**Marie Sklodowska Curie** was born in Poland during a period of Russian occupation. She worked hard to support a brother and sister who had emigrated to France and to save enough money to make the journey herself. In the meantime she learned as much as she could. Eventually she was able to move to Paris, where she enrolled in a university.

The work she did proved to be a vitally important part of the scientific revolution that swept physics and chemistry beginning in 1895. Her husband Pierre, a well-established scientist in his own right, gave up his own lines of research to work with her. This was good judgment on his part; her work was far more important.

Pure samples of uranium compounds have a reliable, rather low level of radioactivity. The Curies found some samples of uranium ores that gave off more radiation than could be accounted for by the amount of uranium they contained. The two scientists used radioactivity as an indicator to isolate two new elements, polonium (named for Marie's native country) and radium (for its intense radioactivity). The Curies and Becquerel won the Nobel Prize in physics in 1903 for their pioneering work with radioactivity. In 1911 Marie Curie won a second Nobel Prize, in chemistry, for her discovery of the two elements. She was the first person to win two Nobel Prizes and is considered one of the greatest scientists of this century. Her daughter Irène was also a Nobel Prize winner. The Curie family is the only one that has ever won four Nobel Prizes.

Following the lead of the Curies, other scientists discovered several other radioactive elements. Radium quickly found a use in treating certain forms of cancer. Radiation is a two-edged sword, however. Although it has been invaluable in medicine, Marie Curie died of a leukemia that was almost certainly caused by her exposure to radiation in the course of her research.

Marie Sklodowska Curie (1867–1934)

CHAPTER BOX **17.2 N RAYS: A CAUTIONARY TALE**

**The year** was 1903, and excitement about the new radiation discovered by Becquerel was sweeping the scientific community. A reputable French physicist, P. Blondlot of the University of Nancy, announced the discovery of another kind of radiation. He called it N rays after his home city and university. As they had for the other new kinds of radiation, scientists performed research and published scores of papers on the new phenomenon.

Some of the resulting evidence seemed strange, and some researchers were unable to replicate Blondlot's experimental results at all. The American physicist Robert Wood visited Blondlot's lab and witnessed some of his experiments. Blondlot and his assistants called out the locations of spectral lines during one experimental run. Taking advantage of the darkness of the lab, Wood removed a vital prism from the experimental apparatus. None of the experimenters noted the difference. They continued to call out "results" as before.

If he had anything fraudulent to hide, Blondlot would not have invited Wood and others to his laboratory. He was not a conscious faker. Led on by the temptations of fame, Blondlot deluded himself. The lesson is this: If you believe something strongly enough, you are likely to see evidence to support your belief, whether the evidence is there or not. This principle has many applications, in both science and daily life. Some techniques scientists use to avoid this kind of self-delusion are illustrated in Example 17.1.

### EXAMPLE 17.1 Control Experiments

A person with a cold takes a vitamin pill, and the cold goes away. The person then declares that the vitamin is a cure for colds. Propose a suitable control experiment.

**SOLUTION**

The next time the person catches a cold, he or she might take another kind of pill, perhaps an aspirin or a salt tablet. If the cold goes away (and they often do), the cure is due to some other means (most probably the person's immune system) and not the vitamin.

Knowing the power of human self-deception, the person might ask a friend to pick a pill (either vitamin or not) without telling the person what kind it was. This procedure is called a *blind experiment.* To be even more careful, the person might ask two friends to help. One friend would select the pill to be taken and give it to the other friend without telling which kind it was. Neither the person with the cold nor the friend giving the person the pill would know what kind of pill was being tested. This procedure is called a *double blind experiment.* In tests of new drugs, when both patient and researcher may have strong personal hopes for the success of a treatment, this method is considered necessary for reliable results.

Finally, there is the question of *statistical significance,* which is outside the scope of this book. How many patients does a medical researcher need to test a drug on before the researcher can be sure that the drug is responsible for the cure (or lack of it)? Whole courses are devoted to questions such as this.

**EXERCISE 17.1**

In the tale of N rays in Chapter Box 17.2, what was the control experiment? Was it blind, double blind, or neither?

## 17.2 Natural Radioactivity

**GOAL:** To review the structure of the atom in order to understand how radioactivity arises.

Some elements are naturally radioactive. As physicists found out within a few years of Becquerel's discovery, these elements decay in several different ways. Their atoms emit rays with at least three different sets of properties, and physicists called these three kinds of radiation alpha particles, beta particles, and gamma rays. In this section we consider how these radioactive processes work.

Let's begin by reviewing the structure of the atom (see Figure 17.6). Each atom has a nucleus at its center, and the nuclear volume is much smaller than the volume occupied by the atom as a whole. If an atom were the size of a large football stadium, for example, its nucleus would be the size of a small green pea. Despite its tiny size the nucleus contains virtually all the mass of an atom.

All nuclei contain protons and neutrons—except the nucleus of an ordinary hydrogen atom, which consists of a single proton. The **mass number** of a given atom is equal to the sum of the number of protons and the number of neutrons in the nucleus. All nuclei of atoms of a given element have the same number of protons, which is identical to the element's atomic number. Isotopes of an element all have the same number of protons but differing numbers of neutrons, and consequently different mass numbers. A cloud of electrons surrounds the nucleus. Ordinary chemical changes, the topic of most of this book, involve changes in the valence electrons. This chapter examines changes in the nucleus. All of the reactions introduced in this chapter involve nuclei, and they are all collectively called **nuclear reactions.**

**You can find a more detailed treatment of the structure of the atom in Chapter 4.**

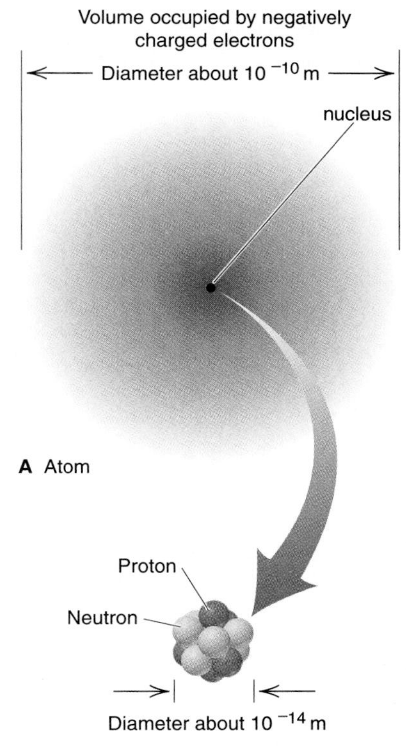


FIGURE 17.6
A typical atom

## EXAMPLE 17.2 The Composition of Atoms and Ions

How many protons, neutrons, and electrons are present in each of these atoms or ions? (You may need to consult a periodic table.)

**a.** A tritium atom (a hydrogen atom with a mass number of 3)
**b.** A sodium ion, $Na^+$, with a mass number of 23
**c.** An iodide ion, $I^-$, with a mass number of 127

### SOLUTIONS

**a.** Hydrogen has the atomic number 1, so the nucleus contains 1 proton. The number of neutrons is equal to the mass number minus the number of protons, or $3 - 1 = 2$. The tritium atom is a neutral species, so it has the same number of electrons as protons—that is, 1 electron.
**b.** Sodium has the atomic number 11, so the nucleus contains 11 protons. The number of neutrons is $23 - 11 = 12$. A positively charged ion has one fewer electron than a neutral atom, so the number of electrons is 10.
**c.** Iodine has the atomic number 53, so the nucleus contains 53 protons. The number of neutrons is $127 - 53 = 74$. With a negative charge, this ion has one more electron than a neutral iodine atom, for a total of 54 electrons.

### EXERCISE 17.2

How many protons, neutrons, and electrons are present in each of these atoms or ions? (You may consult a periodic table if you wish.)

**a.** A uranium atom with a mass number of 238
**b.** A fluoride ion, $F^-$, with a mass number of 19
**c.** A barium ion, $Ba^{2+}$, with a mass number of 138

Now that we have reviewed the basics of atomic structure, let's move on to explore the types of radiation emitted by unstable nuclei. Scientists observe three basic types of emissions from radioactive nuclei: alpha particles, beta particles, and electromagnetic radiation (usually in the form of X rays or gamma rays).

**Even though they now know that alpha particles are helium nuclei and beta particles are electrons, scientists still use these terms to signify the nuclear origins of these species.**

An **alpha particle** consists of two protons and two neutrons. It is the nucleus of a helium atom, in fact. Ernest Rutherford fired alpha particles into a closed space and found helium gas there afterwards. This was part of the evidence that he used to pin down the nature of the alpha particle.

**Beta particles** are high-speed electrons that are emitted from the nuclei of radioactive atoms. Becquerel made a close study of their properties and was the first to conclude this. Beta particles differ from other electrons only in that they originate in atomic nuclei, not because they are fundamentally different in any way.

**Gamma rays** are high-energy photons. They behave as other photons do and are not deflected by electric or magnetic fields. The wavelengths of gamma rays are shorter than those of X rays, and their energies are thus even higher.

Some radioactive species emit particles called **positrons.** A positron has the same mass as an electron but possesses a positive rather than a negative charge. The positron is the *antimatter* version of the electron, an antielectron. This particle has a +1 charge but otherwise is like an electron. When a positron and an electron collide, they annihilate each other, converting their masses into a burst of high-energy photons (almost always gamma rays). Physicists hardly ever observe positron emission directly. The positron annihilates the first electron it encounters after leaving the nucleus, producing a high-energy photon. The experimenters detect this photon. (See Figure 17.7.)

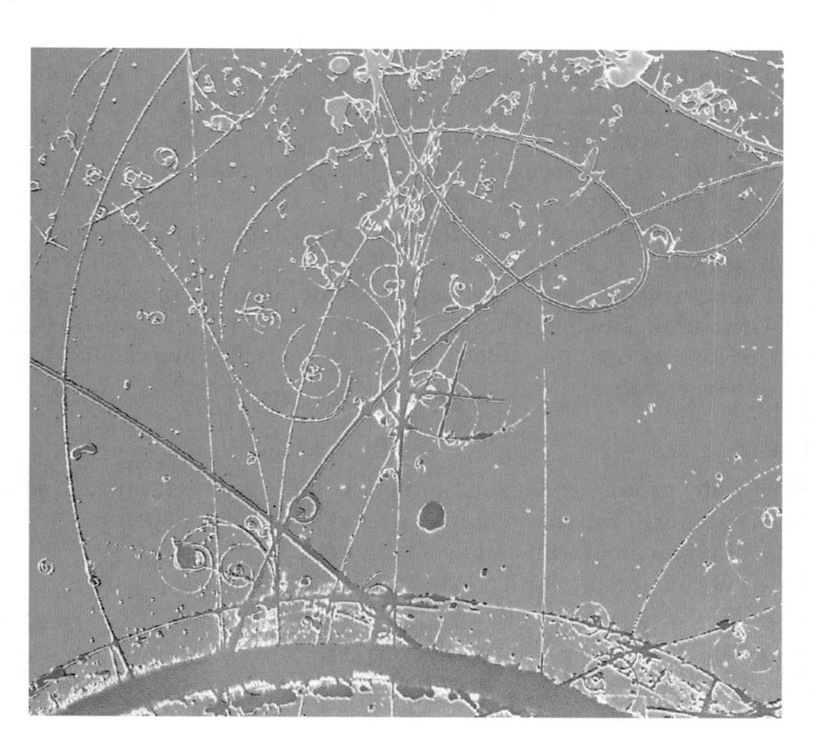

FIGURE 17.7
This image shows the tracks of particles produced by highly energetic nuclear reactions, as well as electron-positron annihilations.

Alpha and beta particles and gamma rays also differ in their ability to penetrate solid objects (see Figure 17.8). Gamma rays are most penetrating; they pass through matter even better than X rays do. Alpha particles are the least penetrating of the three. A positron can't pass through any ordinary matter—it will annihilate an electron first. However, the radiation given off by this annihilation is strongly penetrating.

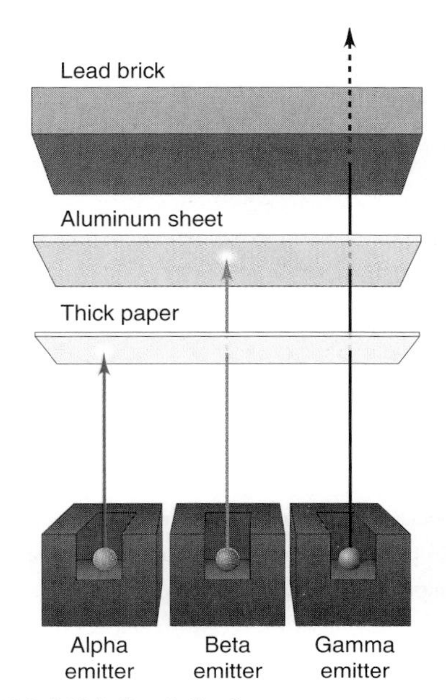


FIGURE 17.8
The penetrating power of the three kinds of radiation varies: A thick sheet of paper will block alpha particles, but both beta particles and gamma rays will pass through. An aluminum sheet stops beta particles but not gamma rays. A thick wall of lead bricks will absorb almost all gamma rays.

## EXAMPLE 17.3 Alpha, Beta, and Gamma Radiation

The illustration below shows three possible paths for radiation. Which of the three paths, A, B, or C, will an alpha particle take? Explain your answer. (Note: Ernest Rutherford used this kind of apparatus in his 1897 investigation of alpha and beta particles.)

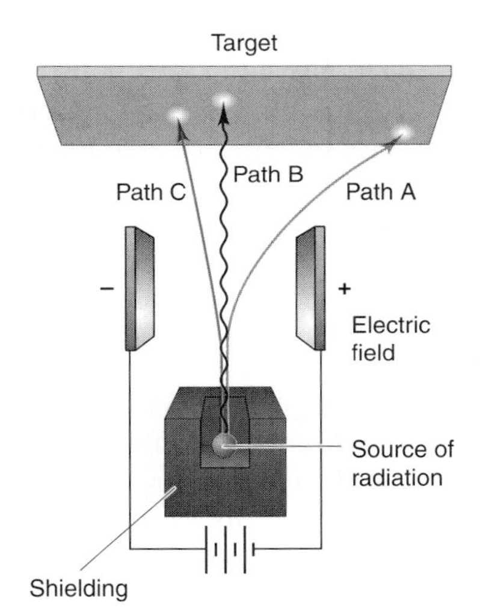


FIGURE 17.9

**SOLUTION**

Like charges repel and unlike charges attract. Alpha particles have positive charges and will take path C, because they will be drawn to the negative charge in the electric field. Also, alpha particles are more massive than beta particles, and thus are harder to deflect. Path C is less strongly deflected than path A.

**EXERCISE 17.3**

Which of the three common types of radiation will take path B in Figure 17.9?

Alpha particles, beta particles, gamma rays, and X rays have enough energy to convert any atoms or molecules they hit into ions. All four of these forms of radiation are called *ionizing radiation* because of this effect. All types of ionizing radiation are both dangerous and useful. Such radiation can kill the cancerous tissue in a tumor when concentrated on it. The ions formed by such radiation can also change the makeup of the body's cells, causing cancer and mutations.

## 17.3 Symbols and Equations for Radioactive Decay

**GOAL:** To write equations for radioactive decay processes of given isotopes.

Exactly the same nuclear reactions occur in pure uranium, uranium oxide, and uranium fluoride, for example.

Before you can write equations for nuclear reactions, you have to know what symbols to use and how the laws of conservation of matter and energy apply to nuclear changes. Remember that these reactions involve nuclear changes only, so it doesn't matter whether the radioactive species is an element or a compound.

We will discuss factors that influence the stability of nuclides later in this chapter.

Any particular nucleus, defined by its numbers of protons and neutrons, is called a **nuclide.** Some nuclides are stable. That is, they do not undergo spontaneous radioactive decay. Many more are unstable. An unstable nuclide will decompose to give a different nuclide plus radiation. (Still more nuclides are in what are called *excited states;* such a nuclide will decompose to give a high-energy photon and the same nuclide in a less excited state.) This section looks at the decomposition of unstable nuclides.

To write the symbol for a specific nuclide, you begin with the symbol for the element. Ordinarily you write the atomic number of the atom to the lower left and the mass number to the upper left. Figures 17.10 and 17.11 show how it's done, with both the general form and specific examples. Note the difference in meaning between the words *isotope* and *nuclide.* All five of the examples in Figure 17.11 are nuclides, but only two forms of the same element can be isotopes of each other.

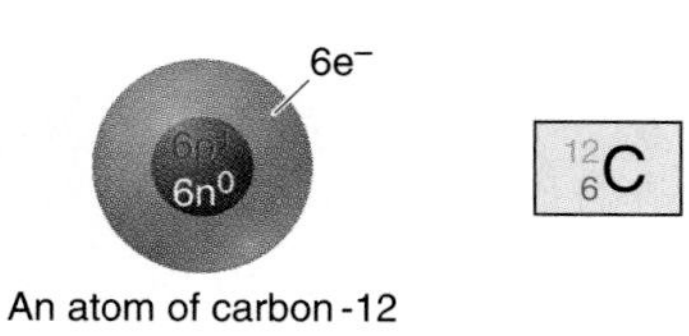


FIGURE 17.10

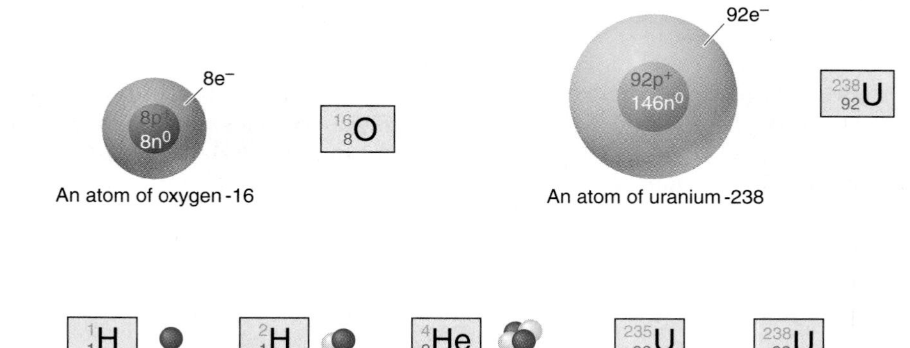


FIGURE 17.11

Examples of symbols for nuclides—from left to right are the symbols for the nuclides of ordinary hydrogen, "heavy hydrogen" or *deuterium,* ordinary helium, and two isotopes of uranium. The nuclei are not drawn to scale.

You can calculate the number of neutrons from the information presented by the symbol for a nuclide, just as you did in Example 17.2 from information given in sentence form. From the symbols in Figure 17.10, an ordinary hydrogen nuclide has no neutrons (1 − 1), deuterium has 1 (2 − 1), helium has 2 (4 − 2), and $^{238}U$ has 146 (238 − 92).

You also need to know the symbols for some subatomic particles. The alpha particle and the most common isotope of helium have the same symbol, though sometimes scientists use the Greek letter alpha ($\alpha$) instead.

Alpha particle = $^{4}_{2}He$ or $\alpha$ or $^{4}_{2}\alpha$

The proton and the simplest isotope of hydrogen also have the same symbol, although the letter p is sometimes used to symbolize a proton.

Proton = $^{1}_{1}H$ or p or $^{1}_{1}p$

Finally, here are the symbols for the electron, the positron, and the neutron:

Electron = $^{0}_{-1}e$ or $^{0}_{-1}\beta$ Positron = $^{0}_{1}e$ or $^{0}_{1}\beta$ Neutron = $^{1}_{0}n$

Chemists and physicists sometimes use the symbol $\beta$ (the Greek letter beta) instead of e. (See Table 17.1.)

**STUDY ALERT**

**There are at least two other ways in which the mass number is given. If you have the name or symbol of an element, you know the atomic number (or at least you can look it up on a periodic table). In these situations you can simply give the name of the element and follow it with the mass number (for example, tin-112 or uranium-238). You can also give the mass number as a superscript to the left of the symbol of the element ($^{112}Sn$ or $^{238}U$), omitting the atomic number (which is implied by the element's symbol).**

**TABLE 17.1 Symbols for Elemental Particles**

Symbols for Elemental Particles

| Alpha particle | Proton | Electron | Positron | Neutron |
|---|---|---|---|---|
| $^{4}_{2}He$ | $^{1}_{1}H$ $^{1}_{1}p$ | $^{0}_{-1}e$ $^{0}_{-1}\beta$ | $^{0}_{1}e$ $^{0}_{1}\beta$ | $^{1}_{0}n$ |

It may seem strange that an electron is ejected from an atomic nucleus that contains only protons and neutrons. The −1 for the electron arises from the fact that charge is conserved in nuclear reactions as it is in all other chemical processes. The overall reaction that produces the electron is probably this:

$$^{1}_{0}n \longrightarrow {}^{0}_{-1}e + {}^{1}_{1}p$$

The charge or subscript on the left side of this reaction must equal the sum of the charges or subscripts on the right (see Figure 17.12). The proton remains in the nucleus, while the electron (a beta particle) ejects at high speed.

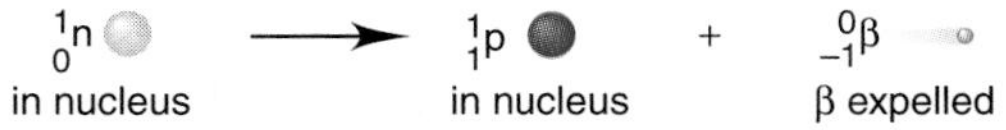

$$^{1}_{0}n \longrightarrow {}^{1}_{1}p + {}^{0}_{-1}\beta$$

in nucleus — in nucleus — β expelled

FIGURE 17.12

You may have concluded from the discussion in Section 17.1 that mass and energy are not conserved separately, and in fact they are not. In nuclear reactions it is the total mass-energy, mass and energy linked by $E = mc^2$, that doesn't change. Nuclear reactions can change small amounts of mass into gigantic amounts of energy. Not only is the overall quantity of mass-energy conserved, the amount of mass converted to energy isn't enough to change a mass number.

Let's consider **beta decay** first to illustrate these points. Nuclides with excess neutrons often undergo beta decay, because this process converts a neutron into a proton. One of the best-known radioactive isotopes is carbon-14, or $^{14}C$. Archaeologists use it to find the age of plant materials, artifacts, and charcoal from ancient fires. What is the nuclide formed by the radioactive decay of carbon-14?

**For more details about radioactive dating, see Section 17.4**

To write an equation for any radioactive decay process, you begin by writing down what you already know. In this case, we know that an electron (the beta particle) is a product. Carbon-14, the only nuclide on the left side of the equation, has the atomic number 6. Chemists and physicists symbolize the unknown nuclide, the other product, by the notorious letter X. This gives us a start:

$$^{14}_{6}C \longrightarrow {}^{A}_{Z}X + {}^{0}_{-1}e$$

Next, we'll calculate the mass number, $A$, of the unknown nuclide X. Mass number is conserved in these reactions—that is, the sum of the mass numbers or subscripts on the left side of the equation must equal the sum of the mass numbers or subscripts on the right side. The unknown mass number can thus be calculated using this simple equation:

$$14 = A + 0$$

We can solve this just by looking at it; the mass number of X is 14.

Atomic number, $Z$, is also conserved in these reactions. Here's the equation that will give us $Z$ for the unknown nuclide resulting from the decay of carbon-14:

$$6 = Z + (-1)$$

Solving this equation gives us an atomic number of 7 for X. This means that the unknown nuclide is

$$^{14}_{7}X$$

A glance at the periodic table reveals that the element with atomic number 7 is nitrogen. We replace X with N and get the final equation:

$$^{14}_{6}C \longrightarrow {}^{14}_{7}N + {}^{0}_{-1}e$$

Equations for **positron emission** are very similar to those for beta decay. Nuclides that have relatively few neutrons often decay by this route, because it converts a proton into a neutron. Neon-19 is one such nuclide. What is the equation for the decay of neon-19 by positron emission? What nuclide is the product?

**Positrons are annihilated as soon as they encounter an electron from any source. Often a positron emitted from a nucleus annihilates an electron in an electron shell of the same atom, producing gamma rays.**

We know that the equation has a positron as a product, and the periodic table tells us that the atomic number of the reactant neon nuclide is 10. That gives us this preliminary equation:

$$^{19}_{10}Ne \longrightarrow {}^{A}_{Z}X + {}^{0}_{1}e$$

The unknown product nuclide has a mass number of $19 - 0 = 19$ and an atomic number of $10 - 1 = 9$. According to the periodic table, the element with an atomic number of 9 is fluorine. The final equation is

$$^{19}_{10}Ne \longrightarrow {}^{19}_{9}F + {}^{0}_{1}e$$

The isotope $^{19}F$, the only stable isotope of fluorine, is the decay product of positron emission by $^{19}Ne$.

Relatively few nuclei decay by **electron capture.** In this process the nucleus absorbs a 1*s* electron. The lightest isotope to decay this way is beryllium-7. What is the balanced equation for the electron capture decay of beryllium-7?

The atomic number of beryllium is 4. The beryllium nuclide is a reactant, and since the nucleus absorbs a 1*s* electron, that electron is a reactant, too. The starting equation is

$$^{7}_{4}\text{Be} + {}^{0}_{-1}\text{e} \longrightarrow {}^{A}_{Z}\text{X}$$

By now you're probably getting the hang of this. This isotope of beryllium decays through electron capture to the most common isotope of lithium, and the final equation is

$$^{7}_{4}\text{Be} + {}^{0}_{-1}\text{e} \longrightarrow {}^{7}_{3}\text{Li}$$

Finally, let's look at **alpha decay.** Uranium was the first radioactive element to be identified, and its most common isotope is uranium-238. This isotope is an alpha emitter, as many radioactive isotopes of the heaviest elements are. To write the equation for this alpha decay process, first we put the uranium-238 nuclide on the left and the alpha particle on the right. There is also another nuclide as product, once again indicated by the letter X.

$$^{238}_{92}\text{U} \longrightarrow {}^{A}_{Z}\text{X} + {}^{4}_{2}\text{He}$$

To find out what the product nuclide is, we balance this equation, remembering that mass number and charge are conserved in a nuclear transformation. First we compute the mass number *A* of the unknown:

$$238 = A + 4$$
$$A = 238 - 4 = 234$$

The total number of protons in the unknown nuclide is

$$92 = Z + 2$$
$$Z = 92 - 2 = 90$$

This means the unknown nuclide has 90 protons and a total mass number of 234. The equation becomes:

$$^{238}_{92}\text{U} \longrightarrow {}^{234}_{90}\text{X} + {}^{4}_{2}\text{He}$$

Element 90 on the periodic table is thorium. The equation for the alpha decay of uranium-238 is therefore:

$$^{238}_{92}\text{U} \longrightarrow {}^{234}_{90}\text{Th} + {}^{4}_{2}\text{He}$$

You may be wondering what happened to gamma rays. Many of the equations we have balanced in this section produce a gamma ray. Just as we ignored the energy produced in chemical reactions in earlier chapters, we ignored the gamma rays here. You already know that gamma rays are emitted when positrons annihilate electrons. Many nuclides give off gamma rays directly. The nuclei involved are in an unstable state—they have extra energy and are said to be *excited* or *metastable*. These nuclides lose their excess energy by giving off a photon of energy in the gamma-ray region of the electromagnetic spectrum. (See Figure 17.13.)

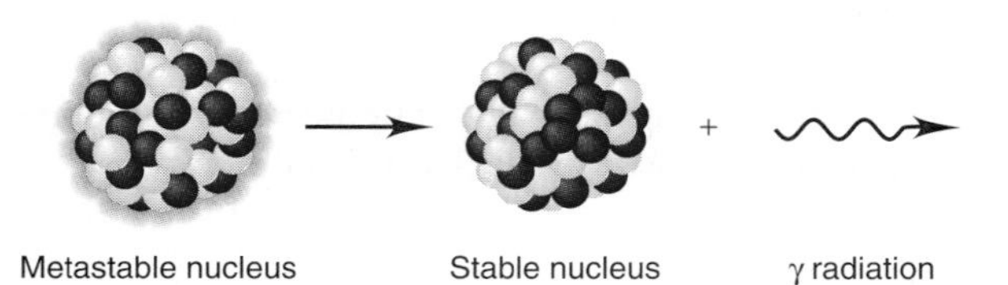


FIGURE 17.13

Metastable nucleus Stable nucleus $\gamma$ radiation

Photons have energy but lack both mass and charge. When a nucleus in an excited state gives off a gamma ray, neither the atomic number nor the mass number changes. In situations where it is important to keep track of such things, excited (metastable) nuclei are identified by including the letter m after the value of the mass number. The decay of radium-226 is a classic example of gamma emission. The decay process first gives a metastable radon-222 nuclide that loses its excess energy by emission of a gamma ray. The equations are (the Greek letter gamma, $\gamma$, symbolizes a gamma ray):

$$^{226}_{88}\text{Ra} \longrightarrow {}^{222\text{m}}_{86}\text{Rn} + {}^{4}_{2}\text{He}$$
$$^{222\text{m}}_{86}\text{Rn} \longrightarrow {}^{222}_{86}\text{Rn} + \gamma$$

**EXAMPLE 17.4** **Writing Equations for Radioactive Decay**

Write an equation for each of these decay processes.

**a.** The beta decay of potassium-40, a weakly radioactive, naturally occurring isotope of an essential mineral nutrient

**b.** The alpha decay of radon-222, the most stable isotope of a noble gas implicated in indoor air pollution

**c.** The electron capture decay of argon-37

**SOLUTIONS**

In each case you begin by writing the symbol for the nuclide given as a starting material. For beta decay you put an electron on the product side of the equation; for alpha decay you have an alpha particle (helium nucleus) as one product; for electron capture you include an electron as a second starting material. In all three equations you need to identify the product nuclide. The sums of the mass numbers on the left and the right sides of the equation must be the same, as well as the sums of the charges. The answers are

**a.** $^{40}_{19}\text{K} \longrightarrow {}^{40}_{20}\text{Ca} + {}^{0}_{-1}\text{e}$

**b.** $^{222}_{86}\text{Rn} \longrightarrow {}^{218}_{84}\text{Po} + {}^{4}_{2}\text{He}$

**c.** $^{37}_{18}\text{Ar} + {}^{0}_{-1}\text{e} \longrightarrow {}^{37}_{17}\text{Cl}$

**EXERCISE 17.4**

Write an equation for each of these decay processes.

**a.** The beta decay of tritium ($^3_1H$)

**b.** The alpha decay of radium-222, an isotope of an element discovered by Marie and Pierre Curie

**c.** Positron emission by sodium-20

## 17.4 The Rate of Radioactive Decay: Half-Lives

**GOAL:** To understand what half-life means, and to perform some simple and useful calculations based on half-lives.

The amount of time it takes for radioactive decay to occur depends on processes in the nucleus alone. You can burn compounds containing radioactive isotopes, freeze them, bury them in the ground, or shoot them into orbit. They will decay in their own good time, and that time depends only on what happens inside their nuclei.

As far as an individual radioactive nuclide is concerned, decay is unpredictable. Scientists can calculate probabilities, but they can't be sure whether a particular nuclide will decay in the next second or the next million years. Large samples of radioactive nuclides are another story. In large samples scientists can predict what fraction of the nuclides is likely to decay in any given time period.

The fraction of a sample of radioactive atoms that decays in any given time period depends only on the identity of the nuclide. Most notably, it does not depend on the number of atoms you start with. For any given type of nuclide, half of the individual nuclides present in a sample will decay in a length of time that is the same for any sample of that type of nuclide, regardless of the size of the sample or the state of chemical combination. This length of time required for half of the nuclides to decay is called the **half-life.** From their observations, physicists and chemists have learned the half-lives of a large number of types of nuclides.

Figuring out how long it will take a third, or a seventh, or an eleventh of a given sample to decay requires mathematical methods that are beyond the scope of this book. The calculation of how long it takes for $\frac{1}{4}$ or $\frac{1}{8}$ or any other related fraction of a sample to decay is a rather simple calculation once you know the half-life. Example 17.5 illustrates such a calculation.

### EXAMPLE 17.5 Simple Half-Life Calculations

The radioactive isotope radon-222 has a half-life of 3.80 days. How long will it take the number of radioactive atoms in a sample of radon-222 to decrease to $\frac{1}{8}$ of its original value?

FIGURE 17.14
All plants, including trees such as this ancient bristlecone pine, absorb carbon dioxide from the atmosphere and incorporate it into their tissues. The exact age of a living tree can be determined by counting the growth rings. The oldest living bristlecone pines were seedlings about 4,000 years ago, but *dendrochronologists* (scientists who study tree rings) have used the wood of dead bristlecone pines to take the scale back about 10,000 years. Using this data scientists have calibrated the $^{14}C$ scale and can adjust calculated ages of plant tissues for the small changes in cosmic ray intensity that have occurred over the past 10,000 years.

**SOLUTION**

It will take 3.80 days (1 half-life) to reduce the number of atoms to ½ of its original value. After 3.80 more days the number will be down to ½ of ½, or ¼, of the original value. Another 3.80 days reduces the number down to ⅛, or $(\frac{1}{2})^3$, of its original value. The total time is thus 3(3.80 days) = 11.4 days.

**EXERCISE 17.5**

The half-life of carbon-14 is 5730 years. A group of archaeologists studying the remains of an Ice Age settlement in France find that a piece of wood from a house in the settlement contains $\frac{1}{16}$ of its original number of $^{14}C$ atoms. How old is this sample?

Exercise 17.5 suggests one use that scientists have found for their knowledge of half-lives. The American chemist Bertram Boltwood suggested the general idea first, in 1907, and Willard Libby developed it more fully in 1946. Knowing how much of a radioactive isotope there is in a sample both at its origins and today, or measuring the amounts of both the radioactive isotope and its decay product in a sample, allows scientists to calculate the age of the sample.

Radiocarbon dating is based on this idea. Cosmic rays bombard nitrogen atoms in the upper atmosphere, giving radioactive carbon-14 atoms. (Cosmic rays consist mostly of high-energy particles, including fast-moving protons and neutrons, that originate elsewhere in the universe.) The reaction of a nitrogen-14 atom with a high-energy neutron produces carbon-14:

$$^{14}_{7}N + ^{1}_{0}n \longrightarrow ^{14}_{6}C + ^{1}_{1}H$$

Carbon-14 undergoes beta decay and has a half-life of 5730 years.

Cosmic ray bombardment produces carbon-14 in the upper atmosphere at a rate that is nearly constant, and the rate of decay of the carbon-14 atoms is constant. Thus there is a small, nearly constant amount of carbon-14 in the atmosphere all the time, almost entirely in the form of $^{14}CO_2$. That is, the fraction of atmospheric carbon dioxide that was $^{14}CO_2$ a thousand years ago, five thousand years ago, or ten thousand years ago is nearly the same as it is today (see Figure 17.14).

As plants grow, they incorporate this carbon-14 into their tissues. Once the plants die, they absorb no more carbon-14, and the atoms of this isotope in their tissues slowly decay and disappear. Scientists can use radiocarbon dating to find the ages of pieces of wood, other plant remains, and charcoal from wood fires; the limits to the ages that can be determined are between a few hundred and over 50,000 years old (see Exercise 17.5 and Figure 17.15).

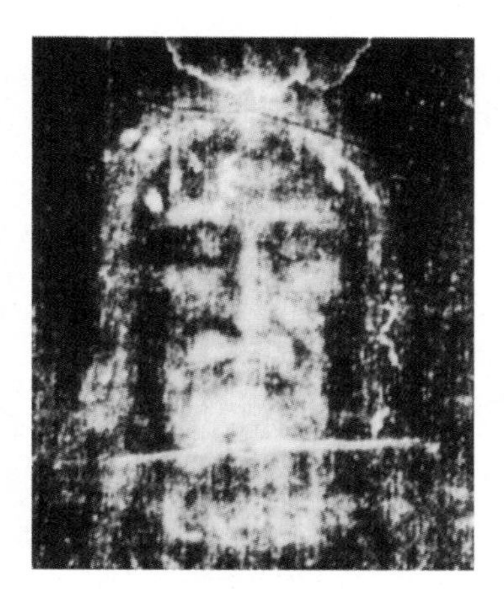

FIGURE 17.15
Some people believe that the Shroud of Turin is the burial shroud of Jesus Christ. Scientists have used carbon-14 dating to examine the Shroud. The plant fibers from which it was made grew over a thousand years after the crucifixion of Christ.

Uranium-series dating is useful for finding the ages of rocks, almost all of which are much older than the samples that can be dated by carbon-14 analysis. The most common uranium isotope, uranium-238, has a half-life of $4.1 \times 10^9$ years, just slightly less than the age of the earth. The final product of the decay of uranium-238 is the stable isotope lead-206. The overall decay process has several steps, but the radioactive nuclides involved have small half-lives compared to that of uranium-238. Scientists can sometimes determine the age of a rock sample by measuring the ratio of $^{238}U$ to $^{206}Pb$ in it.

A third useful method based on half-lives is potassium-argon dating. This makes use of the naturally occurring radioactive isotope potassium-40, which decays into the isotope argon-40. If a lava flow contains potassium and the rock structure is tight enough to keep the argon gas from escaping, radioactive dating is possible. Because potassium-40 has a long half-life ($1.3 \times 10^9$ years), geologists can date only very old lava flows this way. Enough time must pass for a detectable amount of argon-40 to accumulate.

FIGURE 17.16
Dating of rocks collected by Apollo astronauts has revealed that the moon is about 4.2 billion years old.

## 17.5 Nuclear Reactions and Artificial Elements

**GOAL:** To describe how nuclear reactions are used to produce artificial elements, and to write equations for such reactions.

Once the Curies and others established the existence of radioactivity, the paths of transformation for certain elements into others were worked out fairly quickly. For instance, uranium-238 decays by a series of steps to give a stable lead isotope as the final product. Physicists and chemists worked out this string of nuclear reactions (which includes the radium and radon decays described in Section 17.4) by the early 1900s.

Scientists then took a more active role in causing nuclear reactions. Ernest Rutherford was the first person to do this, though he was trying to do something else at the time. He was bombarding targets with alpha particles; in fact, he was doing an experiment similar to the experiments with gold foil that led to his discovery about the structure of the atom (see Section 4.6 for details). Nitrogen gas was Rutherford's target this time, and his detector produced a response typical of high-energy protons. Rutherford balanced the equation. If nitrogen and alpha particles are the reactants, and a proton is one of the products, what else is a product? You know how to figure this out by now. First, you write the starting equation this way:

$$^{14}_{7}N + ^{4}_{2}He \longrightarrow ^{1}_{1}H + ^{A}_{Z}X$$

Next, you balance the equation.

$$^{14}_{7}N + ^{4}_{2}He \longrightarrow ^{1}_{1}H + ^{17}_{8}X$$

The unknown nuclide has a mass number of 17 and contains 8 protons. A glance at the periodic table fills in the blank.

$$^{14}_{7}N + ^{4}_{2}He \longrightarrow ^{1}_{1}H + ^{17}_{8}O$$

Rutherford was the first person to observe a nuclear reaction resulting from the collision of two nuclides (as opposed to a radioactive decay reaction), and he was also the first to transmute one element into another.

**Transmutation—changing one kind of matter into another—was one of the goals of the medieval alchemists. Their hope was to convert "base" metals such as lead into "noble" ones such as gold. Nuclear chemists could carry out this particular transmutation, but the cost would be far greater than the value of the resulting gold.**

Rutherford's experiments were only a modest beginning, and a number of other reactions were soon being studied. The team of Frédéric and Irène Joliot-Curie bombarded aluminum-27 atoms with alpha particles. (Irène was the daughter of Marie and Pierre Curie.) They found that sometimes an aluminum atom absorbed an alpha particle and gave off a neutron. The nuclide formed in this process then emits a positron, giving a stable nuclide as a final product. This was the first observed example of positron decay. The intermediate product of this two-reaction sequence was the first artificially produced radioactive nuclide. This work earned the Joliot-Curies the Nobel Prize. The equations for their reaction sequence are the focus of the following example and exercise.

### EXAMPLE 17.6 Artificial Radioactivity

As the Joliot-Curies discovered, an aluminum-27 atom (the most abundant isotope) absorbs an alpha particle and emits a neutron. What is the equation for this reaction?

**SOLUTION**

Aluminum has an atomic number of 13, and the isotope used by the Joliot-Curies has a mass number of 27. The equation is therefore

$$^{27}_{13}\text{Al} + ^{4}_{2}\text{He} \longrightarrow ^{1}_{0}\text{n} + ^{A}_{Z}\text{X}$$
$$^{27}_{13}\text{Al} + ^{4}_{2}\text{He} \longrightarrow ^{1}_{0}\text{n} + ^{30}_{15}\text{P}$$

The Joliot-Curies won their Nobel Prize for producing phosphorus-30.

**EXERCISE 17.6**

The phosphorus-30 nuclide produced in the reaction in Example 17.6 emits a positron. What is the product of this radioactive decay?

These experiments and others clearly showed that it was possible to prepare isotopes of known elements. Some of these reaction products were stable, but a great many more were not. Artificial radioactivity had arrived. This was an interesting development for its own sake. Most unstable nuclides have half-lives much shorter than the age of the earth. Even if they were present when the earth formed from interstellar dust and gas, they would have decayed into stable nuclides long ago. Today's scientists can't study the properties of these isotopes unless they make them themselves (or they are formed from the radioactive decay of other radionuclides–see end of chapter problem 49).

**The elements technetium (atomic number 43), promethium (61), and astatine (85) had not yet been isolated because they do not exist naturally on the earth in more than the most minute quantities. The half-lives of their most stable isotopes are much shorter than the age of this planet.**

The ability to prepare isotopes of known elements also led directly to an interesting question. Could scientists use these techniques to prepare samples of elements previously unknown? When this question was first considered seriously, no sample of any element beyond uranium, which has atomic number 92, had ever been isolated. Furthermore, three elements whose atomic numbers were less than 92—the elements with the atomic numbers 43, 61, and 85—had yet to be observed on earth.

Attempts to make elements beyond uranium produced another discovery as a side effect. The Italian-American physicist Enrico Fermi bombarded uranium with neutrons, but the results were confusing. He and other physicists studied the reaction further. Some of these researchers finally concluded that a process called **nuclear fission** had occurred (see Figure 17.20). In this process a nucleus splits into two fragments, each with an atomic number lower (usually much lower) than that of the starting nucleus. We will cover this topic much more fully in Section 17.6.

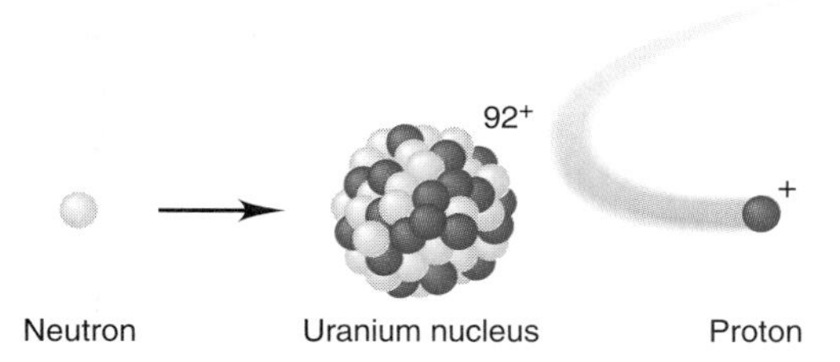


FIGURE 17.17
Why did the physicists bombard the uranium nucleus with neutrons when proton bombardment would be a more direct way to increase the atomic number, and thus make a new element? Protons and nuclei both have positive charges, and they repel each other strongly. To force a proton into a nucleus, the collision must have enough energy to overcome the very strong forces of repulsion between the two positively charged species. Neutrons are not repelled by a nucleus because they are uncharged (neutral). They react with nuclei at much lower energies.

In the search for new elements, physicists looked more closely at the debris of the nuclear reaction between uranium and neutrons. Edwin McMillan and Philip Abelson found traces of a suspicious beta-emitting isotope, tracked it down, and isolated neptunium-239, with atomic number 93. Glenn T. Seaborg (see Figure 17.18) was instrumental in the next step. If neptunium-239 is a beta emitter, Seaborg reasoned, there must be yet another new element beyond neptunium. For more information about this, consider the following example and exercise.

FIGURE 17.18
Glenn T. Seaborg (born 1912) is a pioneer in the field of nuclear chemistry. He has prepared several new elements and many new isotopes, including iodine-131, which is used to treat certain thyroid conditions. In 1951, he and Edwin McMillan were awarded the Nobel Prize for their work.

## EXAMPLE 17.7 Producing Artificial Elements

Neptunium-239 requires two steps to be made. First, a uranium-238 nucleus absorbs a neutron to produce another isotope of uranium. This product then emits a beta particle to produce neptunium-239. Give the equations for these two processes.

**SOLUTION**

$$^{238}_{92}\text{U} + {}^{1}_{0}\text{n} \longrightarrow {}^{239}_{92}\text{U}$$

$$^{239}_{92}\text{U} \longrightarrow {}^{239}_{93}\text{Np} + {}^{0}_{-1}\text{e}$$

**EXERCISE 17.7**

Write the equation for the beta decay of neptunium-239. (Seaborg and his colleagues could write this equation, which is how they knew there was another new element to be found.)

Since the 1940 discovery of artificial plutonium (atomic number 94), minute traces of plutonium-244 have been found in rocks in California; plutonium actually does occur naturally on earth. The half-life of plutonium-244 is $8.2 \times 10^7$ years, almost a hundred million years. This is not a large fraction of the age of the earth, which is approximately $4.5 \times 10^9$ years, but it is apparently long enough to allow a tiny amount of plutonium-244 to survive.

By 1945 researchers had filled all the holes in the periodic table, prepared dozens of isotopes of existing elements, and made a few new elements with atomic numbers larger than that of uranium. By then plutonium was being produced in substantial quantities for use in nuclear weapons.

There are only so many new elements that can be made by the neutron bombardment of uranium. For example, curium (atomic number 96) can be prepared in only very small quantities (micrograms) this way. Neutron bombardment of uranium is not effective at all in making elements with atomic numbers larger than 100. To prepare these elements, scientists must bombard nuclides that have atomic numbers between 95 and 98 with high-energy beams of light nuclei. Scientists at the University of California at Berkeley prepared the elements nobelium (No, atomic number 102) and lawrencium (Lr, atomic number 103) using these reactions:

$$^{246}_{96}\mathrm{Cm} + ^{12}_{6}\mathrm{C} \longrightarrow ^{(258-x)}_{102}\mathrm{No} + x\,\mathrm{n}$$
$$^{252}_{98}\mathrm{Cf} + ^{11}_{5}\mathrm{B} \longrightarrow ^{(263-x)}_{103}\mathrm{Lr} + x\,\mathrm{n}$$

In these reactions the number of neutrons given off varies from 5 to 10. It appears that only very small numbers of atoms of elements having atomic numbers larger than 100 can be prepared. The nuclei of these elements are so unstable that they decay within seconds or minutes. For example, on February 6, 1996 element 112 with 165 neutrons was produced by fusing zinc and lead at the Heavy-Ion Research Center in Germany. It decayed after 280 microseconds ($2.80 \times 10^{-4}$ sec!).

## 17.6 Nuclear Fission

**GOAL:** To examine the processes involved in the splitting of atoms.

FIGURE 17.19
Lise Meitner (1878–1968). Inspired by the Curies' discovery of radium, she earned her doctorate in 1906. Forced to flee from Hitler's Germany to Sweden in 1938, she published the first report of uranium fission from Stockholm in 1939.

The discussion in Section 17.5 moved past the topic of nuclear fission almost as if it were a mere side issue in the quest for new elements. It wasn't, of course. Few scientific discoveries have had so much impact on world affairs so quickly.

As you recall, Enrico Fermi bombarded uranium with neutrons in the early 1930s, and the results were confusing. The German physical chemist Otto Hahn and his colleagues studied the problem. They finally discovered, much to their surprise, that a radioactive isotope of barium was one of the products. Hahn's colleague Lise Meitner, who was by then a refugee from Adolf Hitler's Germany (see Figure 17.19), finally proposed in clear language what had happened. Under bombardment by neutrons some of the uranium atoms had split into pieces (see Figure 17.20).

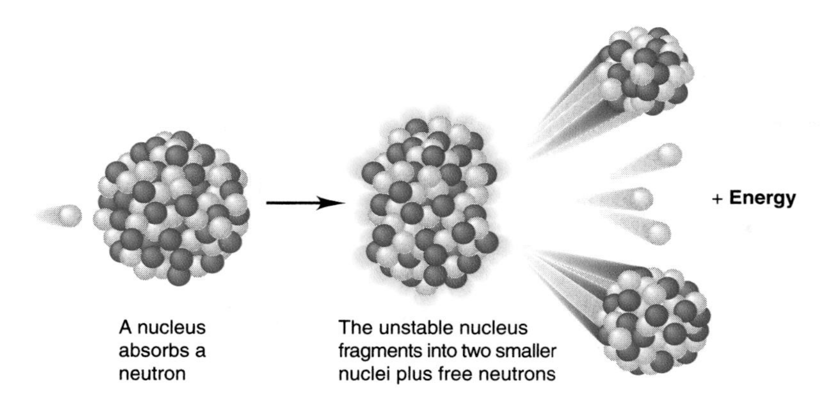
+ Energy
A nucleus absorbs a neutron
The unstable nucleus fragments into two smaller nuclei plus free neutrons

FIGURE 17.20
The first known instance of nuclear fission occurred when a uranium-235 nucleus absorbed a neutron to form an unstable nuclide. This nuclide split into two smaller fragments. The atomic numbers of the fragments add up to 92. This type of fission always releases neutrons.

When neutrons bombard the mixture of isotopes that comprise naturally occurring uranium, the nuclear chemistry is quite complex. An uncommon isotope of uranium, uranium-235, undergoes fission. The much more common uranium-238 absorbs neutrons to give beta-emitting nuclei, as described in Section 17.5. Plutonium also undergoes fission.

Here are the equations for several of the fission routes that begin with uranium-235. As you can see, the fission process can yield several products.

$$^{235}_{92}\text{U} + {}^{1}_{0}\text{n} \longrightarrow {}^{236m}_{92}\text{U} \longrightarrow \begin{cases} {}^{103}_{42}\text{Mo} + {}^{131}_{50}\text{Sn} + 2\,{}^{1}_{0}\text{n} \\ {}^{142}_{56}\text{Ba} + {}^{91}_{36}\text{Kr} + 3\,{}^{1}_{0}\text{n} \\ {}^{81}_{32}\text{Ge} + {}^{152}_{60}\text{Nd} + 3\,{}^{1}_{0}\text{n} \end{cases}$$

These equations have several implications. First, they ignore energy, as many chemical equations do. However, the energy is the most significant product of a fission reaction from a human point of view. Recall that Aston found precise atomic masses in the 1920s. He and others were then able to calculate possible energy yields from nuclear reactions using $E = mc^2$. How were these calculations performed, and what were their implications?

Nuclei are made up of protons and neutrons, which collectively are called **nucleons.** You might expect that the mass of each nucleus would be equal to the sum of the masses of the nucleons. Aston noticed that the masses for all the elements except hydrogen were less than that. The "missing" mass is equivalent to the amount of energy given off when each nucleus was assembled. The difference between the mass of the nucleons and the mass of the nucleus is the mass value used in the equation $E = mc^2$. The energy value from the equation divided by the number of nucleons for a nucleus is called the **binding energy** for that nucleus. Scientists who do these calculations typically divide the binding energy of the nucleus by the total number of nucleons and report the results as *binding energy per nucleon.*

Figure 17.21 shows a plot of binding energy per nucleon versus mass number for the most stable isotopes of the elements. The dip in the center of the plot shows that energy will be released if very heavy nuclei split to form nuclei of medium masses, or if nuclei with low mass numbers fuse to form nuclei of medium mass. In other words, heavy elements tend to undergo fission, while light elements tend to undergo fusion. Iron-56 is the most stable isotope of all.

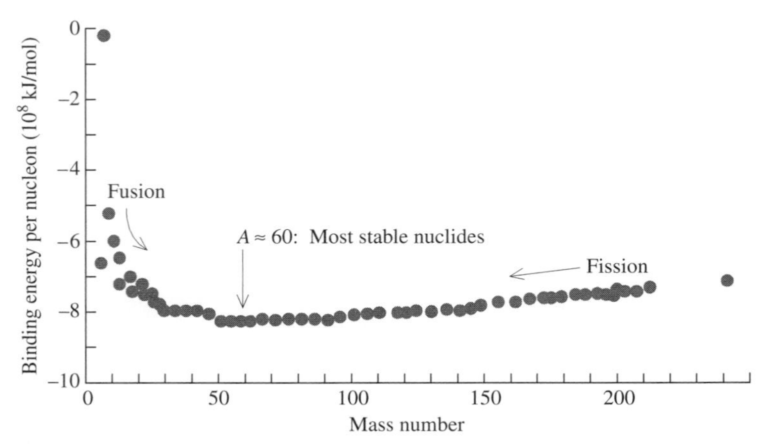


FIGURE 17.21
A plot of binding energy per nucleon versus mass number shows that elements nearest the top of the graph, especially hydrogen, have high potential energy. Hydrogen and other elements on the left side of the graph will give off energy when they undergo fusion. Elements on the right side give off energy when they undergo fission. The isotope iron-56 and those near it on the graph are the most stable.

**Proposals about a "nuclear winter" arose from studies of the planet Mars, where clouds of dust from planet-wide dust storms can lower the temperature many degrees. The nuclear winter proposal suggests that if a number of nuclear bombs (less than 100) were detonated on the surface of the earth within a period of a few hours or days, a large cloud of smoke and dust would enter the atmosphere. This dust would block the sunlight, lowering the temperature of the planet to below the freezing point of water. This could very easily lead to the destruction of all or at least most life on the planet. There is a considerable amount of empirical evidence that supports this proposal.**

The second important implication of the equations for the fission of uranium-235 involves the neutrons. One neutron starts the process, and two or three neutrons are given off. Two or three other uranium-235 atoms can then absorb these neutrons, and break up to give six to nine more neutrons. These neutrons can cause more fission reactions, giving more energy and producing 18 to 27 neutrons—and so on.

When a product of a reaction triggers another identical step, as it does here, the result is a *chain reaction.* If a chain reaction expands as this one does, and if energy is produced from each step of the reaction, the potential for an explosion exists. Physicists and chemists knew decades before World War II that energy might be available from nuclear reactions. They just didn't know how to tap this energy.

There were several important engineering problems to be solved before a fission bomb could be built. If there aren't enough uranium atoms in a small enough volume, most of the neutrons produced will escape from the surface of the sample rather than cause fission. (The smallest amount of fissionable material needed to produce an explosion is called the **critical mass.**) If the reaction builds up slowly enough, the explosion may blow the uranium apart or perhaps merely melt it, stopping the reaction before it produces a high yield of energy. The Manhattan Project during World War II solved these practical problems (see Figure 17.22). Since that time scientists and engi-

FIGURE 17.22
The left picture shows the nuclear bomb blast that leveled the Japanese city of Hiroshima. The picture on the right shows one result of the blast. If the nuclear winter proposals are correct, a more intense nuclear war might end human civilization and even cause the extinction of all species.

neers in several nations, knowing only that there is a way to make the process work, have produced fission weapons.

After the horrors of Hiroshima and Nagasaki many people turned with hope to the prospect of peaceful power from the atom. Nuclear engineers can run the same fission reactions that produce explosions in a controlled way, producing heat that can be converted to electricity. In one design, masses of uranium too small and impure to form a critical mass are formed into rods. Operators can move other rods containing such neutron-absorbing elements as cadmium or boron in and out between the fuel-containing rods. This movement controls the reaction rate.

Nuclear power is a controversial public policy issue. For a simplified look at some of the questions involved, see Chapter Box 17.3.

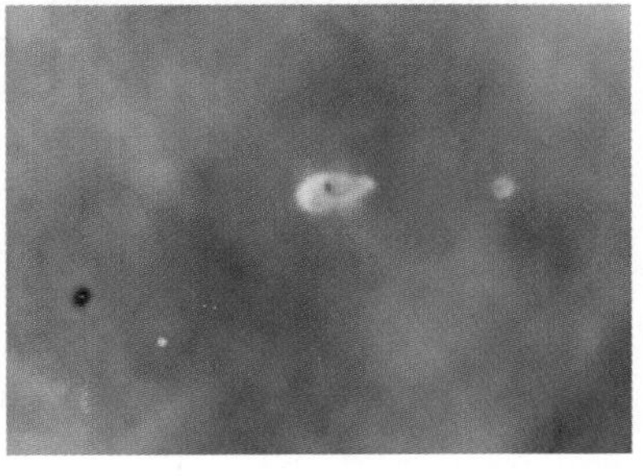

FIGURE 17.23
The Orion Nebula (upper photo) is a huge cloud of interstellar dust and gas in which stars are forming. Its stars are 500,000 years old or less; one of them is thought to be about 2,000 years old. The lower photo, taken by the Hubble Space Telescope in 1994, shows disks of dust and gas around some of these young stars.

## 17.7 Nuclear Fusion

**GOAL:** To examine the nuclear process that powers the sun and the stars (as well as the hydrogen bomb).

To people who have heard about it only in connection with the hydrogen bomb, **nuclear fusion** can seem terribly threatening. Without both present and past nuclear fusion, however, humans wouldn't and couldn't exist. The sun is a nuclear fusion reactor, and the earth is made up of the ashes of the fusion fires of stars that burned out long ago. This section looks at the processes that make the sun and the other stars what they are, summarizing the work of astronomers and physicists over the centuries.

Stars begin as huge clouds of hydrogen gas with other gases and dust mixed in. Such clouds can begin to collapse, drawn by their own gravity or pushed in by interstellar explosions nearby. As the gas atoms and molecules and the dust particles fall toward the center, they speed up, and this faster motion translates into heat. When enough mass has accumulated in the center of the cloud, the temperature can rise into the millions of degrees. At these temperatures protons in the cloud can move rapidly enough to collide with one another, overcoming the electrostatic repulsion between their positive charges. Fusion reactions begin to occur. Figure 17.23 shows a nebula in which this is happening, the Orion Nebula. If you use binoculars, you can make out this nebula as a fuzzy spot in the southern sky on a clear winter evening (see Figure 17.24).

For fusion reactions to ignite a star, the mass of the cloud apparently must be quite large, at least 20% of the mass of the sun. This is the size of the smallest currently burning stars, at any rate. Several reaction sequences are possible, and this one seems likely:

$$
\begin{aligned}
{}^{1}_{1}\mathrm{H} + {}^{1}_{1}\mathrm{H} &\longrightarrow {}^{2}_{1}\mathrm{H} + {}^{0}_{1}\mathrm{e} \\
{}^{1}_{1}\mathrm{H} + {}^{2}_{1}\mathrm{H} &\longrightarrow {}^{3}_{2}\mathrm{He} \\
{}^{3}_{2}\mathrm{He} + {}^{3}_{2}\mathrm{He} &\longrightarrow {}^{4}_{2}\mathrm{He} + 2\,{}^{1}_{1}\mathrm{H}
\end{aligned}
$$

These reactions convert four protons into one helium, a positron, and a vast amount of energy. This energy, in the form of heat, stops the gravitational

FIGURE 17.24
The constellation Orion is prominent in the winter skies in the Northern Hemisphere. Bright stars form the constellation's outline, with the Orion Nebula forming part of the sword. You can see the nebula, a small area of fuzzy brightness, with binoculars or a small telescope.

## CHAPTER BOX 17.3 NUCLEAR POWER: WHAT NOW?

**Conventional nuclear** power plants produce electricity from a fluid (either water or a mixture of molten potassium and sodium metals) heated by nuclear fission. Hundreds of such plants around the world function quietly and effectively (as pictured in Figure A). The Chernobyl incident (pictured in Figure B) is a graphic example of how badly things can go wrong. Citizens need to decide what to do about nuclear power. Do you think it should be used more than it is now? About the same? Much less? Here are some things to think about.

1. Reactors in First World or Western industrialized nations, are all much safer than the Chernobyl reactor was. They are not perfectly safe. Nothing is perfectly safe.

2. If you live near a nuclear power plant, your safety is not in your own hands. You are far more likely to die in a car accident than from anything that might happen in a nuclear power plant, but you use cars by your own choice, and this gives you somewhat more control over your own fate.

3. Unlike fossil fuel (gas, oil, coal) power plants, nuclear plants produce no carbon dioxide along with their energy, and therefore don't contribute to global warming.

4. Much of the slowdown in building new nuclear plants in the United States has been caused by the fact that they are very expensive. Conservation strategies can reduce the demand for power, and other power-producing technologies can make power available much more cheaply.

5. Where and how to store the radioactive waste that nuclear plants produce is an unsolved problem. Furthermore, at the end of their useful lives nuclear power plants become large masses of radioactive waste themselves.

What's important to you? Chemists and physicists can provide useful information to help you move toward an informed decision. Beyond that point, science can't answer questions about values. Scientists can and should participate in the debate, because they are citizens too. Scientists can often lay out what *can* be done, but they have no special ability to decide what *should* be done.

FIGURE B
The Chernobyl nuclear power plant, near Kiev in Ukraine, after the 1986 disaster. The graphite core of the reactor burned for more than three days, and all of the people who lived near the plant had to move away to avoid the dangerous levels of radioactivity in the area.

FIGURE A
Nuclear power plant in Grand Gulf, Mississippi.

collapse of the original cloud. There is a balancing of forces here. Gravity, which seems reasonable on earth, is a huge force in such giant bodies as stars. It is opposed by the equally enormous kinetic energy of atoms heated by nuclear reactions. To counterbalance the force of gravity, the sun (see Figure 17.25) must run $1 \times 10^{38}$ nuclear reactions per second, converting 5 million tons of mass into energy to counter the gravitational force.

Here on earth, hydrogen is a rather uncommon element. It forms part of water, but on the whole there isn't very much hydrogen in the planet's crust. The opposite is true for the universe as a whole. It has been estimated that 90% of the mass of the universe is hydrogen, 9% is helium, and all the rest of the elements put together total about 1%. The sun contains a greater proportion of helium and less hydrogen, but the other elements still make up a very small percentage of the total.

After several billion years of life, most stars begin to run out of hydrogen. The time required for this to happen varies with the brightness of the star. A giant blue-white star like Rigel in Orion will burn out in a few hundred million years, while a small and dim star might last many billions of years. (The sun apparently has several billion years of life to go, so you shouldn't get worried.) When a star begins to run out of hydrogen, it changes in several ways (see Figure 17.26). The nuclear reactions slow down, the star cools, and the force of gravity begins to win out. The inner core collapses, the outer layers expand, and the star becomes a red giant. The collapsing core of the star becomes hotter. At these higher temperatures three helium nuclei can fuse to

**Scientists know what elements are present in the sun and other stars because excited atoms of each element give off characteristic line spectra (see Chapter 6). The light from even very dim stars can be sensed by a telescope, analyzed in a spectrometer, and compared to laboratory spectra of the elements.**

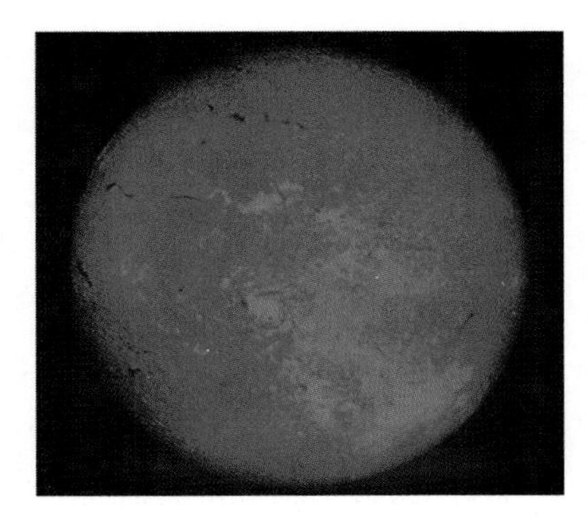

FIGURE 17.25
The sun

Coalescing cosmic dust
Cosmic dust
① Primitive coalescing star burns hydrogen at $10^7$K.
② Star begins burning He, causes expansion into red giant. Core of red giant burns He to form $^{12}C$, $^{16}O$, $^{20}Ne$, $^{24}Mg$ at $2 \times 10^8$ K.
③ Red supergiant core burns carbon and oxygen to form nuclei through $^{40}Ca$ at $7 \times 10^8$ K. Further heating to $3 \times 10^9$ K forms nuclei through $^{56}Fe$ and $^{58}Ni$.
④ Core implodes to form a neutron star, while outer layers explode into a supernova. Neutron-capture processes form heavier nuclei.
⑤ Debris from supernova eventually ends up in second-generation stars.
Neutron star

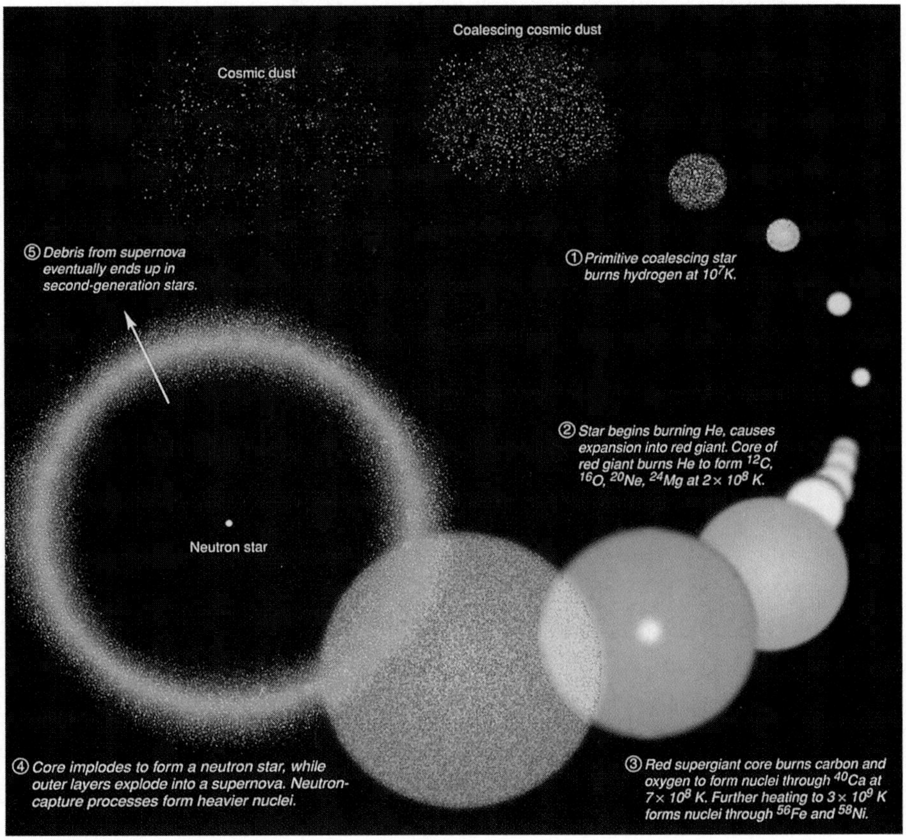

FIGURE 17.26
The life cycle of a star

give carbon. Carbon can react with helium to give oxygen, and oxygen can react with helium to give neon. At even higher temperatures these elements can fuse to give even heavier elements, all the way up to iron and its neighbors, which have the most stable nuclei.

### EXAMPLE 17.8 Equations for Fusion Reactions

Write the equation for the fusion of three helium nuclei to give carbon. (A gamma ray is another product of the reaction.)

**SOLUTION**

$$3\,{}^{4}_{2}\text{He} \longrightarrow {}^{12}_{6}\text{C} + \gamma$$

**EXERCISE 17.8**

Write the equation for the fusion of carbon-12 with helium-4 to give oxygen. (A gamma ray is another product of this reaction as well.)

Most of the energy a star can give off comes from the conversion of hydrogen to helium. As you can see from the graph of binding energy per nucleon in Figure 17.21, the energy difference between hydrogen and helium is much greater than that between helium and iron.

Where do elements heavier than iron come from? When most of a star's fuel is gone, it begins a final collapse. If the star is large enough, this collapse can cause an enormous explosion, resulting in what is called a **supernova** (see Figure 17.27). The energies available, enormous even compared to the normal energies of stars, can drive the formation of the heavier elements, including very unstable ones. All of the heavy elements in the universe, including silver, gold, uranium, and the iodine you require for the proper function-

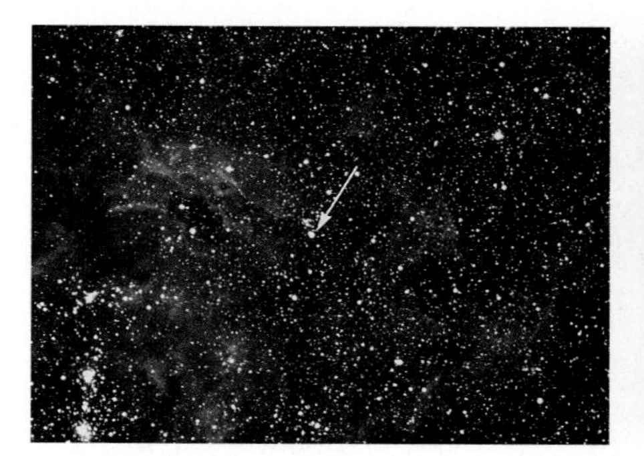

FIGURE 17.27
In 1987 a supernova appeared in a small galaxy near the Milky Way, about 160,000 light-years from earth. The arrow in the photo on the left shows the star that exploded. You can see in the photo on the right that the supernova is enormously brighter than the other stars in the cloud.

ing of your thyroid gland, were made in the furnaces of supernova explosions. (See Figure 17.28.)

In the end all stars collapse to form what might be called cinders. Smaller stars end up as dwarfs that have enormous density. The electronic structures of the atoms that make them up are crushed into a very small space, forming what is called *degenerate matter.* Larger stars can collapse even further, until the nuclei of the atoms are in contact. The result is a *neutron star.* The largest stars apparently can collapse beyond that, to form *black holes.*

FIGURE 17.28
The Crab Nebula is the remnant of a supernova that was observed by Chinese astronomers in the year 1054 AD.

### EXAMPLE 17.9 Practical Fusion Reactions

**a.** Several reactions occur in a hydrogen bomb explosion. One is between lithium-6 and a neutron to give helium-4 and another product. Write the equation for this reaction.

**b.** In the early 1990s some chemists claimed that fusion took place in a certain design of electrochemical cell at room temperature. This process was called *cold fusion.* Most physicists thought that these claims were very unlikely. Why did they think this?

**SOLUTIONS**

**a.** As usual, you write down what is given and then simply balance the equation. The second product is a radioactive isotope of hydrogen called a triton (a tritium nucleus).

$${}^{6}_{3}\text{Li} + {}^{1}_{0}\text{n} \longrightarrow {}^{4}_{2}\text{He} + {}^{3}_{1}\text{H}$$

**b.** As nuclei get closer and closer to one another, their positive charges will repel each other more and more strongly. No known process could make nuclei come in contact with one another at room temperature.

**EXERCISE 17.9**

**a.** Another reaction that occurs in a hydrogen bomb is between a tritium nucleus (triton) (${}^{3}$H) and a deuterium nucleus (deuteron) (${}^{2}$H) to give helium-4 and another product. Write the equation for this reaction.

**b.** Nuclear fusion is a promising source of energy, but making its use practical has been difficult. What is the main scientific obstacle to the design of practical fusion power plants?

## CHAPTER SUMMARY

1. In ordinary chemical reactions valence electrons are transferred, but the nuclei of atoms do not change at all. Nuclear chemists study reactions that change the nuclei of atoms. These reactions typically involve energy changes far greater than those involved in chemical reactions.

2. A radioactive nucleus can decay by emitting an alpha particle (helium nucleus), a beta particle (high-energy electron), a gamma ray (high-energy photon), or a positron (antielectron) or by capturing an inner-shell electron.
3. Equations for nuclear reactions are balanced by balancing the superscripts (mass numbers, the numbers of protons plus neutrons) and the subscripts (atomic numbers or charges) of the reactants and products.
4. No one can predict when any individual radioactive atom will decay. The probability of an individual nucleus of a given radioactive isotope decaying depends only on what happens in that nucleus, not on anything outside it. The rate of decay of a sample containing many radioactive atoms is predictable. The typical measure of the rate of decay of an isotope is the half-life, the time it takes half of the atoms in a given sample to decay. Half-lives are useful in determining the dates of past events.
5. Nuclear chemists and physicists have made samples of artificial isotopes of known elements and of previously unknown elements by bombarding samples of known elements with high-energy subatomic particles or small nuclei.
6. Tremendous amounts of energy are available from nuclear fission, in which the nuclei of heavy elements (mainly uranium and plutonium) split to give smaller nuclei. Nuclear weapons give off this energy in uncontrolled form; nuclear reactors produce it in a much more controlled manner.
7. Even greater amounts of energy are available from nuclear fusion, which is the process that fuels the sun. In this process small nuclei (typically hydrogen and other very light nuclei) combine to give larger nuclei. Elements near iron in the periodic table do not give energy from either fusion or fission.

## PROBLEMS

### SECTION 17.1 Radioactivity

1. X rays represent a type of radiation you have already studied. What type is that?
2. Give two major differences between ordinary chemical reactions and nuclear reactions.
3. Briefly summarize the major discovery of Wilhelm Roentgen.
4. Briefly summarize the work of Antoine-Henri Becquerel.
5. Briefly summarize the discoveries of Marie and Pierre Curie.
6. Briefly summarize the ideas of William Draper Harkins.
7. What is a source of the earth's internal heat?
8. Toward the end of the 1800s there was a major scientific controversy about the age of the earth. Geologists maintained that millions of years of time were needed to shape the earth, while physicists calculated that the earth and the sun would cool completely within a few tens of thousands of years. The discovery of radioactivity effectively ended this debate. Why?
9. Which of these statements about chemical and nuclear reactions is *false?*
   A. The energy changes in nuclear reactions are larger than those in chemical reactions.
   B. One element cannot be changed to another by way of a chemical reaction.
   C. Different isotopes of an element behave the same way in a nuclear reaction.
   D. The chemical state of an atom has no effect on the course of its nuclear reactions.

10. Which of these statements about chemical and nuclear reactions is *true*?
    A. The energy changes in nuclear reactions are smaller than those in chemical reactions.
    B. One element cannot be changed to another by way of a nuclear reaction.
    C. Different isotopes of an element behave differently in chemical reactions.
    D. The chemical state of an atom has no effect on the course of its nuclear reactions.

11. In Becquerel's experiments with a uranium salt and photographic film, suppose he had wrapped his photographic plates with clear plastic wrap. Would his experiments have worked? Why or why not?

12. In Becquerel's experiments with a uranium salt and photographic film, suppose he had wrapped his photographic plates with heavy lead foil. Would his experiments have worked? Why or why not?

## SECTION 17.2 Natural Radioactivity

13. Where is the positive charge of an atom located?

14. Where is most of the mass of an atom located?

15. Where are the neutrons of an atom located?

16. Where is the negative charge in an atom?

17. Which of these types of radiation penetrates other objects *most strongly*?
    A. Alpha particles C. Gamma rays
    B. Beta particles D. Positrons

18. Which of these types of radiation penetrates other objects *least effectively*?
    A. Alpha particles C. Gamma rays
    B. Beta particles D. Positrons

19. How many protons, neutrons, and electrons are present in each of these atoms or ions?
    a. A deuterium atom (a hydrogen atom with a mass number of 2)
    b. An iron(III) ion, $Fe^{3+}$, with a mass number of 56
    c. An oxide ion, $O^{2-}$, with a mass number of 17

20. How many protons, neutrons, and electrons are present in each of these atoms or ions?
    a. A radon atom with a mass number of 222
    b. A phosphide ion, $P^{3-}$, with a mass number of 31
    c. A calcium ion, $Ca^{2+}$, with a mass number of 40

21. How many protons, neutrons, and electrons are present in each of these atoms or ions?
    a. An oxygen atom with a mass number of 18
    b. A copper(I) ion, $Cu^{+}$, with a mass number of 63
    c. A chloride ion, $Cl^{-}$, with a mass number of 37

22. How many protons, neutrons, and electrons are present in each of these atoms or ions?
    a. An argon atom with a mass number of 40
    b. A nitride ion, $N^{3-}$, with a mass number of 14
    c. A radium ion, $Ra^{2+}$, with a mass number of 226

23. How many protons, neutrons, and electrons are present in each of these atoms or ions?
    a. An iodine atom with a mass number of 127
    b. A mercury(II) ion, $Hg^{2+}$, with a mass number of 198
    c. A sulfide ion, $S^{2-}$, with a mass number of 32

24. How many protons, neutrons, and electrons are present in each of these atoms or ions?
    a. A xenon atom with a mass number of 132
    b. A bromide ion, $Br^{-}$, with a mass number of 81
    c. A scandium ion, $Sc^{3+}$, with a mass number of 45

25. How many protons, neutrons, and electrons are present in each of these atoms or ions?
    a. A xenon atom with a mass number of 126
    b. A manganese(II) ion, $Mn^{2+}$, with a mass number of 55
    c. An oxide ion, $O^{2-}$, with a mass number of 18

26. How many protons, neutrons, and electrons are present in each of these atoms or ions?
    a. A radon atom with a mass number of 224
    b. An iodide ion, $I^{-}$, with a mass number of 126
    c. A barium ion, $Ba^{2+}$, with a mass number of 136

27. What are the mass number and the charge of an alpha particle?

28. What are the mass number and the charge of a beta particle? Of a positron?

## SECTION 17.3 Symbols and Equations for Radioactive Decay

29. Two isotopes used in blood analyses for medical diagnoses are phosphorus-32 and sodium-24, both of which decay by emitting a beta particle. Write equations for both of these nuclear reactions.

30. Isotopes of very heavy elements tend to emit alpha particles after very short half-lives. Radioactive isotopes that do this include $^{250}Fm$, $^{254}No$, $^{256}Lr$, and $^{257}Rf$ (Rf has been suggested as the symbol for element 106). Write equations for these alpha decays.

31. Most naturally occurring vanadium is the stable isotope $^{51}V$, but a small amount is the weakly radioactive $^{50}V$, which can decay by either electron capture or beta emission. Write equations for both of these nuclear reactions.

32. Both of the isotopes used in nuclear weapons and nuclear power plants, $^{235}U$ and $^{239}Pu$, normally decay by alpha emission. Write equations for both of these decay processes.

33. Two isotopes used in diagnostic medicine, $^{131}Ba$ (used to detect bone tumors) and $^{51}Cr$ (used to study blood processes), decay by electron capture. Write equations for these decay processes.

34. Four radioactive isotopes used in cancer therapy are $^{60}Co$, $^{90}Y$, $^{131}I$, and $^{198}Au$. All of these undergo beta decay. Write equations for all of these decay processes.

35. Write equations for these decay processes of mercury isotopes: alpha decay of $^{184}Hg$, positron emission by $^{186}Hg$, electron capture by $^{197}Hg$, and beta decay of $^{204}Hg$.

36. Write equations for these decay processes of gold isotopes: positron emission by $^{176}Au$, alpha decay of $^{177}Au$, electron capture by $^{193}Au$, and beta decay of $^{200}Au$.

37. Write equations for these decay processes of isotopes of osmium: positron emission by $^{180}Os$, electron capture by $^{185}Os$, alpha decay of $^{186}Os$, and beta decay of $^{196}Os$.

38. Write equations for these decay processes of isotopes of lead: alpha decay of $^{184}Pb$, positron emission by $^{195}Pb$, electron capture by $^{197}Pb$, and beta decay of $^{214}Pb$.

39. Write equations for the decay processes of isotopes of radon: electron capture by $^{200}Rn$, positron emission by $^{209}Rn$, alpha decay of $^{218}Rn$, and beta decay of $^{224}Rn$.

40. Write equations for the decay processes of isotopes of astatine: alpha decay of $^{198}At$, positron emission by $^{202}At$, and electron capture by $^{206}At$.

41. Write equations for these decay processes of isotopes of platinum: positron emission by $^{173}Pt$, electron capture by $^{176}Pt$, alpha decay of $^{178}Pt$, and beta decay of $^{201}Pt$.

42. Write equations for these decay processes of isotopes of radium: alpha decay of $^{211}Ra$, electron capture by $^{213}Ra$, and beta decay of $^{228}Ra$.

### SECTION 17.4 The Rate of Radioactive Decay: Half-Lives

43. The buried ruins of a small village in Tanzania yield a piece of wood that contains $1/2$ as much carbon-14 as a modern sample of the same wood. How old is this sample? The half-life of carbon-14 is 5730 years.

44. A sample of charcoal from an ancient firepit in northern Mexico had $1/4$ as much carbon-14 as a modern sample of charcoal. How old is this sample?

45. A sample of charcoal used as a black pigment in ancient paintings of wild animals found in a French cave had $1/32$ as much carbon-14 as a modern sample of charcoal. How old is this sample?

46. A sample of bone from a Neanderthal site in Turkey had $1/256$ as much carbon-14 as a modern sample of human bone. How old is this sample?

47. A geologist sent a sample from a very ancient lava bed to a dating laboratory and was told that the amounts of $^{40}K$ and $^{40}Ar$ were equal. What is the age of this sample? The half-life of $^{40}K$ is $1.3 \times 10^9$ years.

48. A geologist sent a sample of an igneous rock (formed in a volcano) to a dating laboratory and was told that the ratio of $^{40}K$ to $^{40}Ar$ was 1:3. What is the age of this sample?

49. Radium-226 has a half-life of approximately 1600 years. Despite the fact that the half-life of this isotope is far shorter than the age of the earth, Marie Curie isolated it from mineral samples. This radium did not survive since the earth's formation. Where, then, did it come from?

50. Radioisotopes used in nuclear medicine are employed in low concentrations and have short half-lives. Why?

51. The half-life of $^{99m}Tc$, used in diagnosing a variety of health problems, is approximately 6 hours. How many half-lives does it take for the number of radioactive nuclides in a sample of $^{99m}Tc$ to decrease to $1/128$ of its original value? How long is this in hours?

52. The half-life of $^{131}I$, used in the diagnosis and treatment of thyroid problems, is approximately 8 days. How many days does it take for the number of radioactive nuclides in an $^{131}I$ sample to decrease to $1/128$ of its original value?

53. The half-life of $^{51}Cr$, used in the determination of blood volume and red blood cell lifetimes, is 27.8 days. How many days does it take for the radioactivity of a $^{51}Cr$ sample to decrease to $1/128$ of its original value?

54. The half-life of $^{42}K$, used in nuclear medicine, is 12.4 hours. How long does it take for the radioactivity of a $^{42}K$ sample to decrease to $1/128$ of its original value?

### SECTION 17.5 Nuclear Reactions and Artificial Elements

55. Elements with higher atomic numbers than those currently known will be more and more difficult to isolate, for two main reasons. What are those reasons?

56. The most abundant artificial element on earth was made for use in nuclear weapons and for the production of nuclear energy. What is this element?

57. Technetium (atomic number 43) has no stable isotopes and does not exist in nature. It was first made by bombarding $^{96}Mo$ with $^{2}H$; this reaction also gives a neutron as a product. Write the equation for this reaction.

58. Physicists at the University of California bombarded $^{209}Bi$ with alpha particles and obtained three neutrons and another element that does not exist in nature. What is this element? Write an equation for this reaction.

59. In 1949 University of California physicists bombarded a sample of $^{241}Am$ with alpha particles. They obtained two neutrons and an element that had never been made before. What element was it? Write the equation for this reaction.

60. University of California physicists also bombarded a sample of californium-249 with carbon-12 nuclei. They obtained four neutrons and another element. Write an equation for this reaction.

61. University of California physicists also bombarded a sample of californium-249 with oxygen-18 nuclei. They obtained four neutrons and another element. Write an equation for this reaction.

62. Bombarding $^{142}Nd$ with neutrons produces some promethium (Pm) and a beta particle. (This is how Pm, which has no stable isotopes, was first made.) Write the equation for this reaction.

63. Mlle. Marguerite Perey of the Curie Institute in Paris discovered a new element. This element can be made by bombarding $^{230}Th$ with protons; the products are two alpha particles and the element. What element is it? Write the equation for this reaction.

64. Bombarding $^{249}Bk$ with neutrons produces a beta particle and another highly radioactive element. What element is it? Write the equation for this reaction.

## SECTION 17.6 Nuclear Fission

65. An accident in a nuclear power plant can have more serious consequences than an accident in a coal-fired power plant. What features of radioactivity are responsible for this difference?

66. The radioactive nuclides strontium-90 (a fission product found in the fallout from nuclear testing) and radium-226 both accumulate in the bones of people who are exposed to them, thus making these nuclides more dangerous than they would be otherwise. Why do these substances accumulate in bone? (Hint: Look at the periodic table.)

67. Several of the heaviest known radioactive nuclides decay by spontaneous fission. For instance, $^{252}Cf$ can decay to give $^{140}Xe$, four neutrons, and another atom. What is the other atom formed by this reaction?

68. The isotope $^{235}U$ can undergo fission by more than one route. For example, it can undergo spontaneous fission to give $^{139}Ba$, two neutrons, and another atom. What is the other atom formed by this reaction?

## SECTION 17.7 Nuclear Fusion

69. Which of the following processes gives the *most* energy?
    A. The reaction of hydrogen with oxygen to give water
    B. The fusion of four hydrogen nuclei to give helium
    C. The decay of $^{238}U$ to give an alpha particle
    D. The fission of $^{235}U$ to give two smaller nuclei plus neutrons

70. Which of the following processes gives the *least* energy?
    A. The reaction of hydrogen with oxygen to give water
    B. The fusion of four hydrogen nuclei to give helium
    C. The decay of $^{238}U$ to give an alpha particle
    D. The fission of $^{235}U$ to give two smaller nuclei plus neutrons

71. As the cores of red giant stars collapse, oxygen nuclei can react with alpha particles to give gamma rays and a new type of nuclei. Give the equation for this process.

72. At very high stellar temperatures, carbon nuclei can react with oxygen nuclei to give gamma rays and another kind of nuclei. Give the equation for this process.

73. Second-generation stars such as the sun (that is, stars formed long after the beginning of the universe and young enough to include material from older stars) can make helium from four hydrogen nuclei by a six-step process catalyzed by carbon. Give the equation for the first step of this process, the reaction of a $^{12}C$ nucleus with a proton to give a gamma ray and another nucleus.

74. Give the equation for the second step of the process in Problem 73, in which the new nucleus, the product of the first step, decays by emitting a positron.

75. In the third step of the process in Problem 73, the nucleus produced by the second step reacts with a proton to give another gamma ray and a $^{14}N$ nucleus. Give the equation for the third step.

76. In the fourth step of the process in Problem 73, the $^{14}N$ produced in the third step reacts with a proton to give a new nucleus and a gamma ray. Give the equation for the fourth step.

77. In the fifth step, the nucleus formed by the reaction in the fourth step decays by emitting a positron. Give the equation for the fifth step.

78. In the sixth and final step, the nucleus produced by the reaction of the fifth step reacts with a proton to give a $^{12}C$ nucleus and an alpha particle. Give the equation for this step.

## SOLUTIONS TO EXERCISES

**Exercise 17.1**

Wood's removal of the prism was the control experiment. It was a blind experiment. Wood knew that he had removed the prism, but Blondlot and his colleagues did not. (Mistakes by chemists and physicists are typically revealed when experimenters in other labs fail to repeat results; this sort of control experiment is very unusual in the physical sciences.)

**Exercise 17.2**

**a.** The nucleus of the uranium atom contains 92 protons (the atomic number) and 146 neutrons (the mass number minus the number of protons). This is a neutral atom, so it has the same number of electrons as protons, or 92.
**b.** The nucleus of the fluoride ion has 9 protons and 10 neutrons. This is a negative ion, and thus has 10 electrons.
**c.** All barium nuclei contain 56 protons. The number of neutrons in this isotope is 138 − 56 = 82. With a 2+ charge, this ion has 2 fewer electrons than a neutral barium atom does, or 54 electrons.

**Exercise 17.3**

Path B has no deflection and can be followed only by radiation that has no charge. Beta particles have a negative charge, and their path would bend toward the positive side of the electric field. Gamma rays have no charge and therefore take path B.

**Exercise 17.4**

**a.** $^{3}_{1}\text{H} \longrightarrow {}^{3}_{2}\text{He} + {}^{0}_{-1}\text{e}$
**b.** $^{222}_{88}\text{Ra} \longrightarrow {}^{218}_{86}\text{Rn} + {}^{4}_{2}\text{He}$
**c.** $^{20}_{11}\text{Na} \longrightarrow {}^{20}_{10}\text{Ne} + {}^{0}_{1}\text{e}$

**Exercise 17.5**

The fraction $\frac{1}{16}$ is $\frac{1}{2} \times \frac{1}{2} \times \frac{1}{2} \times \frac{1}{2} = (\frac{1}{2})^4$. The sample of wood has therefore been around for 4 half-lives of $^{14}$C, or is about 23,000 years old.

**Exercise 17.6**

$$^{30}_{15}\text{P} \longrightarrow {}^{0}_{1}\text{e} + {}^{A}_{Z}\text{X}$$
$$^{30}_{15}\text{P} \longrightarrow {}^{0}_{1}\text{e} + {}^{30}_{14}\text{Si}$$

**Exercise 17.7**

$$^{239}_{93}\text{Np} \longrightarrow {}^{239}_{94}\text{Pu} + {}^{0}_{-1}\text{e}$$

Plutonium was the new element Seaborg expected to be discovered.

**Exercise 17.8**

$$^{12}_{6}\text{C} + {}^{4}_{2}\text{He} \longrightarrow {}^{16}_{8}\text{O} + \gamma$$

**Exercise 17.9**

**a.** Once again you balance the equation.

$$^{3}_{1}\text{H} + {}^{2}_{1}\text{H} \longrightarrow {}^{4}_{2}\text{He} + {}^{1}_{0}\text{n}$$

**b.** All nuclei have positive charges, and they therefore repel one another. Before fusion can take place, two nuclei must come into contact with each other. In stars (or bombs) this requires a temperature of millions of degrees. The problem is how to keep a reaction at this temperature confined and under control.

18

# ORGANIC CHEMISTRY



**KEY TERMS**

One of the oldest divisions in the field of chemistry is between **organic chemistry** and **inorganic chemistry.** (See Chapter Box 7.1 for an introduction to these terms.) In the early years of the science, organic chemicals were defined as those that come from living organisms. Inorganic substances were considered to come primarily from rocks and minerals. The two kinds of compounds seemed very different then, and in many ways they still do. Most of the organic compounds known then were liquids (such as ethyl alcohol) or solids with relatively low melting points (such as candle wax). Most inorganic compounds were solids with high melting points such as salts and minerals. (Now chemists know that most organic compounds contain only covalent bonds, while many inorganic ones contain ionic bonds; see Figure 18.1.)

FIGURE 18.1
It's easy to understand why early chemists thought that organic and inorganic compounds are fundamentally different in some way. Could it be possible that the compounds that make up a kitten and those that make up a rock follow the same scientific principles?

Many of the differences between organic and inorganic compounds arise from the singular properties of the element carbon. Carbon atoms readily form covalent bonds among themselves. As you already know, several of the elements are diatomic, and some other elements (including sulfur and phosphorus) occur in molecules having more than two atoms (see Figure 18.2). These elements form a few structures that have extended chains of atoms surrounded by atoms of other elements. However, only carbon forms such an enormous variety of extended covalent structures. For instance, a simple plastic grocery bag contains chains of carbon atoms many thousands of atoms long (with more thousands of hydrogen atoms attached). As for paper bags, their long carbon chains contain an occasional oxygen atom, with many hydrogen and oxygen atoms to the sides.

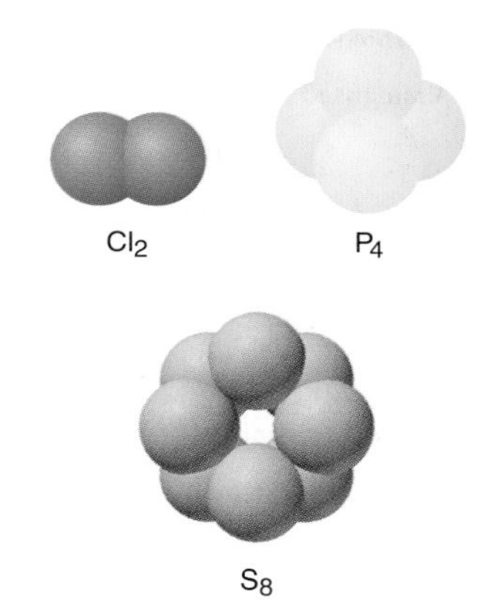


FIGURE 18.2
The structures of elemental chlorine, phosphorus, and sulfur are all covalent.

Faced with the fact that large molecules are common among organic compounds, organic chemists have developed efficient methods of depicting structures, and this chapter begins by introducing these. Faced with literally millions of different carbon-based compounds, organic chemists also had to come up with a way to organize their study. Modern chemists classify organic compounds according to certain structural features, and the scheme for naming organic compounds begins with those classifications. Much of this chapter describes the structures, names, and properties of the most common members of these classes of organic compounds.

**STUDY ALERT**
**In many ways learning organic chemistry is like learning a foreign language. The written language of organic chemistry is not complex. As with any language, the best way to learn to use it is to practice, practice, practice until you can do it right.**

## 18.1 Lewis Diagrams in Organic Chemistry

**GOAL:** To draw and interpret two kinds of Lewis diagrams widely used to depict organic compounds.

Lewis diagrams for organic compounds can be tedious to draw. In a structure like this one, all of the carbon and hydrogen atoms and the bonds joining them take time to write down:

```
    H  H  H  H  H  H  H  H  H  H  H  H  H  H  H
    |  |  |  |  |  |  |  |  |  |  |  |  |  |  |
 H—C—C—C—C—C—C—C—C—C—C—C—C—C—C—C—H
    |  |  |  |  |  |  |  |  |  |  |  |  |  |  |
    H  |  H  H  H  H  H  H  H  H  H  H  H  H  H
    H—C—H
       |
       H
```

This section introduces two of the conventions that organic chemists use to save time and space, called *condensed structures* and *bond-line formulas.* Both condensed structures and bond-line formulas take advantage of the fact that carbon forms four covalent bonds in stable molecules with rare exceptions. In structures without charges (that is, in almost all organic molecules), nitrogen almost always has three bonds and one pair of nonbonding electrons. Oxygen has two bonds and two nonbonding pairs, while the halogens form one bond and have three nonbonding pairs of electrons. Hydrogen atoms form one bond and have no nonbonding electron pairs. (See Table 18.1.)

**TABLE 18.1 Common Bonding Arrangements in Organic Molecules**

| Carbon | Nitrogen | Oxygen | Halogens | Hydrogen |
|---|---|---|---|---|
| —C— (with single bonds above and below) | —N̈— (with single bond below) | :O: (with single bonds above and below) | :C̈l: (with single bond below) | H (with single bond below) |
| \C= (two single bonds and one double bond) | —N̈= | :O: (with double bond below) | :F̈: (with single bond below) | |
| =C= | :N≡ | | :B̈r: (with single bond below) | |
| —C≡ | | | | |

**STUDY ALERT**
**Condensed structures and bond-line formulas (discussed later in this section) do not *look* like complete Lewis diagrams, but they *are* the same. That's the key to understanding what the conventions mean.**

Let's see how to turn a Lewis diagram for a large molecule into a condensed structure. We'll use the molecule shown at the beginning of this section (an alkane; see Section 18.9).

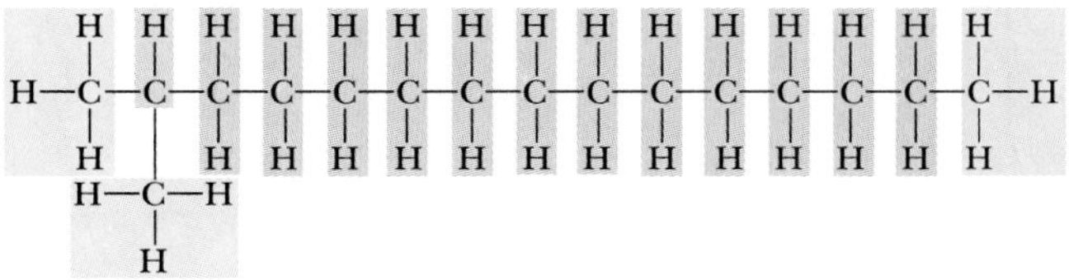

```
    H  H  H  H  H  H  H  H  H  H  H  H  H  H  H
    |  |  |  |  |  |  |  |  |  |  |  |  |  |  |
 H—C—C—C—C—C—C—C—C—C—C—C—C—C—C—C—H
    |  |  |  |  |  |  |  |  |  |  |  |  |  |  |
    H  |  H  H  H  H  H  H  H  H  H  H  H  H  H
    H—C—H
       |
       H
```

The ends are $CH_3$ groups, and there are many $CH_2$ units in the middle. Here's what we get if we rewrite the structure in a first stage of condensation:

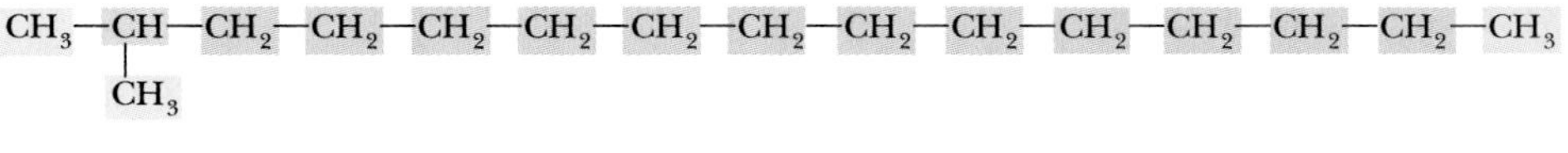

$$CH_3—CH—CH_2—CH_2—CH_2—CH_2—CH_2—CH_2—CH_2—CH_2—CH_2—CH_2—CH_2—CH_2—CH_3$$
(with a $CH_3$ branch bonded below the CH)

This structure shows 34 fewer bonds than the original. The letter H appears 18 fewer times, though we do have to write 15 subscript numbers. We can omit almost all of the carbon-carbon bonds, too, to get this:

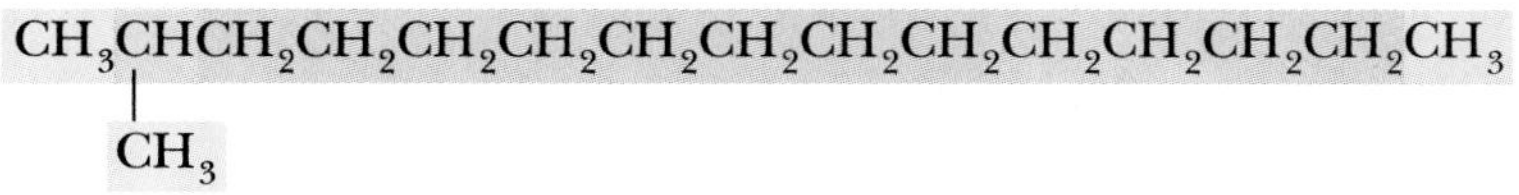

$$CH_3CHCH_2CH_2CH_2CH_2CH_2CH_2CH_2CH_2CH_2CH_2CH_2CH_2CH_3$$
(with a $CH_3$ branch bonded below the CH)

**By convention, these structures show only the bonds between carbon atoms. They assume that you know that each hydrogen atom makes only one bond, to a carbon atom.**

Generally, you need to show bonds explicitly only where connections are not obvious.

This is still the same molecule. The same information is there, but in condensed form. Can we go further? Well, there are two $CH_3$ groups on the left end of the molecule, and twelve consecutive $CH_2$ groups in the middle. We can save even more time and space by writing the structure of the molecule this way:

$$(CH_3)_2CH(CH_2)_{12}CH_3$$

It's still the same structure if you turn it around: $CH_3(CH_2)_{12}CH(CH_3)_2$

This is the **condensed structure** of the original molecule. If you keep the bonding arrangements for carbon and hydrogen shown in Table 18.1 in mind, you can reconstruct the original complete Lewis diagram from this formula. (Be careful not to condense the structure so far that you can't do this. For example, the molecular formula $C_{16}H_{34}$ doesn't tell you how the atoms are connected.) Organic chemists use condensed and bond-line structures extensively.

## EXAMPLE 18.1 Interpreting Condensed Structures

Draw the complete Lewis diagram from each of these condensed structures.

**a.** $(CH_3)_2CHCH_2OH$ **b.** $(CH_3)_3C(CH_2)_2CH_3$

**SOLUTIONS**

**a.** Each carbon atom has four bonds to it. The carbon in each of the first two $CH_3$ groups already has three bonds to hydrogen atoms, so it can be bonded to only one other carbon, the carbon in the CH group that is next in the condensed structure:

```
   H  H
   |  |
H—C—C
   |  |
   H  |
 H—C—H
      |
      H
```

After you draw that much, you attach the next carbon with its two hydrogen atoms.

```
   H  H  H
   |  |  |
H—C—C—C
   |  |  |
   H  |  H
 H—C—H
      |
      H
```

Finally, you add the OH group. The oxygen atom has two pairs of nonbonding electrons:

```
   H  H  H
   |  |  |   ..
H—C—C—C—O—H
   |  |  |   ..
   H  |  H
 H—C—H
      |
      H
```

**b.** When there are multiple $CH_3$ groups, they fan out at the end of the molecule. Multiple $CH_2$ groups string together in the middle of the molecule. That's all there is to it, really.

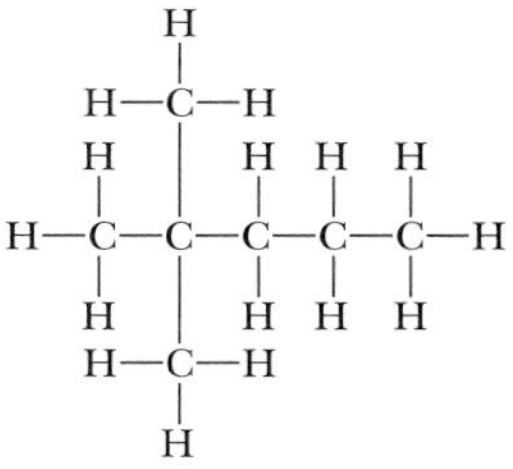

```
        H
        |
     H—C—H
   H    |    H  H  H
   |    |    |  |  |
H—C—C—C—C—C—H
   |    |    |  |  |
   H    |    H  H  H
     H—C—H
        |
        H
```

**EXERCISE 18.1**

Draw the complete Lewis diagram from each of these condensed structures.

**a.** $CH_3OCH(CH_3)_2$ **b.** $HOCH_2C(CH_3)_3$

The other common type of abbreviated formula used in organic chemistry is called a **bond-line formula.** Compounds with atoms connected in a circle—a *ring*—occur often in organic chemistry, and the atoms and bonds in them are especially clumsy to draw and interpret using ordinary Lewis diagrams. Bond-line formulas are a quick and easy way to draw such structures, but you have to know what they mean. These special rules for bond-line formulas apply to the carbon and hydrogen atoms in the structures:

- Bends in lines, ends of lines, and intersections of lines all represent carbon atoms.

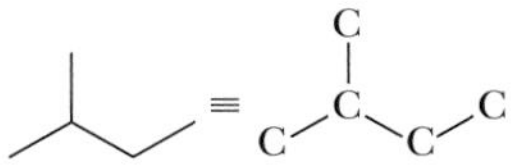

- Hydrogen atoms bonded to carbon atoms aren't shown at all in bond-line formulas. You decide how many hydrogen atoms are attached to any given carbon atom by counting the number of bonds shown to that carbon and subtracting that number from 4.

Two bonds: therefore this carbon atom is bonded to 4 − 2 = 2 hydrogen atoms.

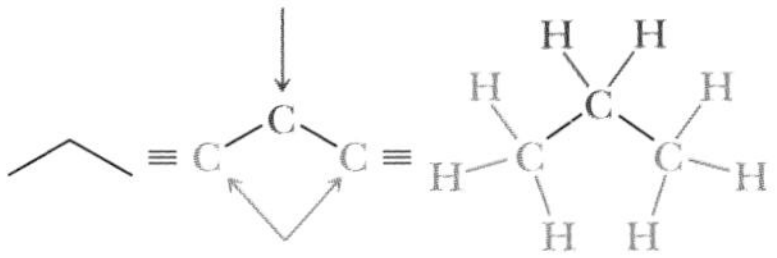

One bond: therefore these carbon atoms are bonded to 4 − 1 = 3 hydrogen atoms.

- Chemists always indicate atoms other than carbon and hydrogen by the usual letter symbols. Hydrogen atoms bonded to these non-carbon atoms are also shown.

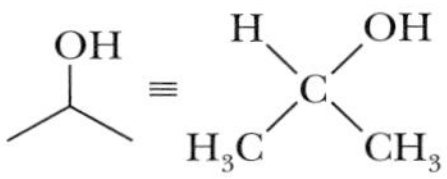

- Nonbonding electrons are rarely shown in these structures.

Chemists frequently use bond-line formulas for molecules that contain long chains or rings of atoms. For example, here are the condensed structure and bond-line formula of the compound called 2-methylcyclohexanol:

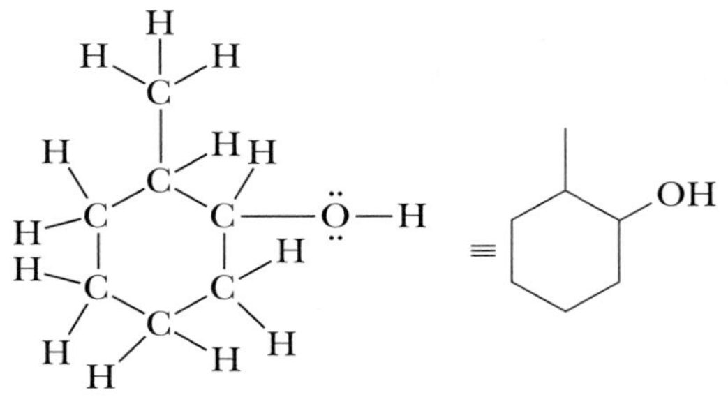

EXAMPLE 18.2 **Using Bond-Line Formulas**

**a.** Draw the bond-line formula for compound A.
**b.** What is the molecular formula of compound B?

A. 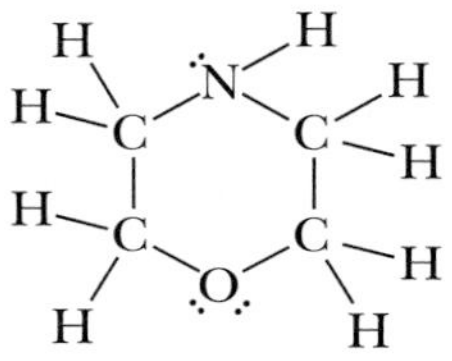 B.

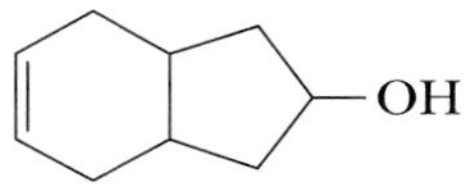

**SOLUTIONS**

**a.** Each carbon atom is one corner of a six-membered ring, which you draw as a hexagon. Two corners of the ring are occupied by the nitrogen and oxygen. You show the hydrogen bonded to the nitrogen as you usually do in a Lewis diagram. This is the bond-line formula for compound A:

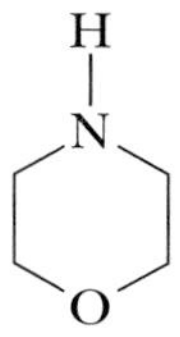

**b.** To find the molecular formula of compound B, first count the corners and intersections in the bond-line formula. There are 9, which means there are 9 carbon atoms.

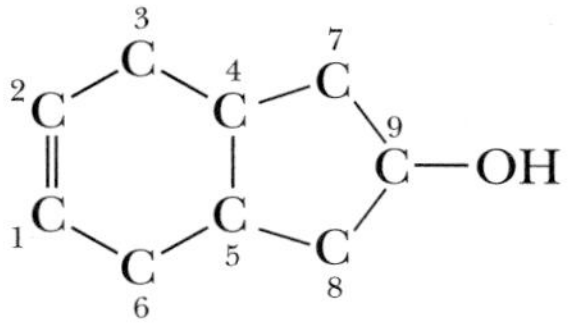

Each of the carbons has four bonds to it, whether they are shown or not; the missing bonds are to hydrogen atoms. The two carbons at the left side of the molecule each have three bonds shown, two in the double bond and one in a single bond. One more bond, to a hydrogen, gives four apiece for those carbons. Count your way through the whole structure this way and you'll see that there are 13 hydrogen atoms that aren't shown in the bond-line formula:

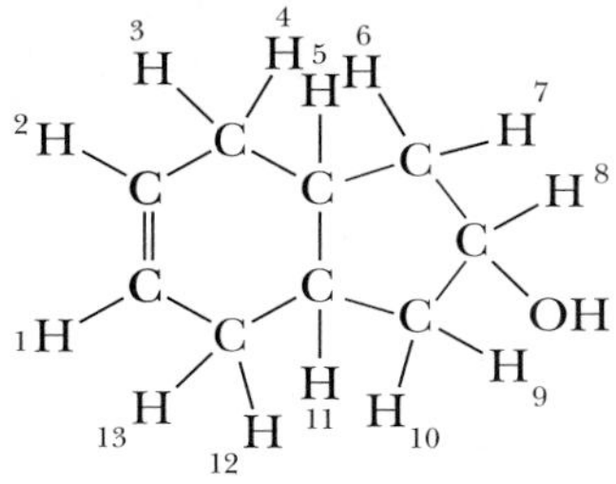

Finally, don't forget to add the hydrogen and oxygen that are shown in the bond-line formula as the OH group. The molecular formula of the compound is $C_9H_{14}O$.

**EXERCISE 18.2**

**a.** Draw the complete Lewis diagram for compound C.
**b.** What is the molecular formula of compound D?

**C.** 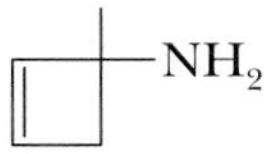 $NH_2$ **D.** 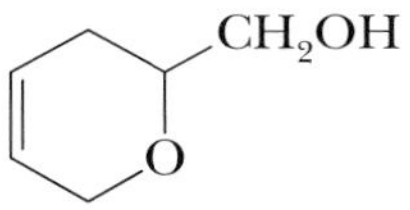 $CH_2OH$ O

## 18.2 Organic and Inorganic Compounds

**GOAL:** To define *organic* and *inorganic* compounds, and to show how the classification originated.

Organic chemicals are all around you, and your body functions only because you are largely constructed from them. This does not make organic chemistry easy. Plants and animals contain amazingly complex mixtures of compounds, and the compounds themselves seldom have simple formulas or structures. Figure 18.3 on page 672 shows Lewis diagrams of a few of the first organic compounds to be isolated in pure form. Simple inorganic compounds have formulas such as NaCl, $CaSO_4$, and $KNO_3$ and were much easier for early chemists to understand and deal with.

**Isolating a single compound from a tissue sample can be a difficult and time-consuming project.**

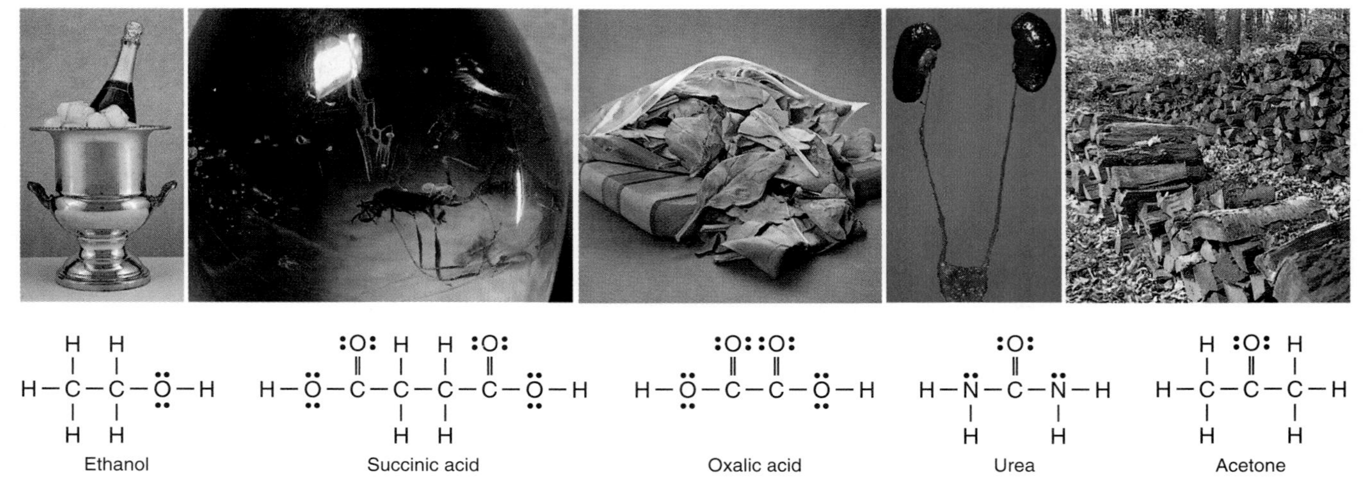


FIGURE 18.3
These are Lewis diagrams of some of the first organic compounds to be isolated in pure form. Ethyl alcohol was isolated from fermentation mixtures, succinic acid from amber heated to a high temperature, oxalic acid from several plants including spinach, urea from urine, and acetone from wood that was strongly heated without being burned.

In the late 1700s Antoine Lavoisier studied both inorganic and organic compounds. He made enough progress with inorganic compounds to be able to propose a nomenclature system for them (see Chapter Box 7.3, page 233), and chemists now use a system based on his. All Lavoisier could accomplish in the organic area was to suggest that organic compounds (by which he meant compounds from plants or animals) were ones that include carbon, hydrogen, and oxygen. Compounds from animal sources might also contain nitrogen and phosphorus, he decided.

Lavoisier's proposal was objective and testable, and therefore scientific. Most scientists of his time believed the *vital force* concept, which held (among other things) that organic compounds cannot be synthesized from inorganic starting materials. This philosophy had elements of the mystical about it. Vitalism (as it was called) had a strong hold on the scientists of the time.

**For more information about naming inorganic compounds, see Chapter 7, especially Figures 7.10 and 7.11.**

The Swedish chemist Jöns Jakob Berzelius (1779–1848) analyzed more than 2,000 compounds in the early 1800s. His results established the Law of Definite Proportions (and indirectly Dalton's atomic theory) beyond reasonable doubt. His great mass of findings drove him to develop a shorthand system for expressing formulas of compounds. Chemists still use his letter symbols for the elements and his formulas for compounds.

**The differences in properties between silver fulminate (an explosive) and silver cyanate (a stable compound) are due, of course, to differences in structure. The fulminate ion is $CNO^-$, and the cyanate ion is $NCO^-$. Berzelius didn't know this and had no way to find it out. He was so attached to the idea of a vital force that he didn't look for any other explanations for the differences between the two compounds. (There is a lesson here.) The basic concept that structure determines properties took another 75 years to take hold.**

Berzelius analyzed organic compounds as well as inorganic ones. He concluded that it was probably impossible to make organic compounds from inorganic ones, but that both types of compounds follow the same rules of chemical combination. He recognized that organic molecules and their formulas were often much larger and more complicated than those of the inorganic compounds known at the time.

The vital force concept provides easy explanations for inconvenient facts, and Berzelius used it extensively. For example, in the early 1820s Justus von Liebig (1803–1873) analyzed silver fulminate while Friedrich Wöhler (1800–1882) studied silver cyanate. These two compounds have very different properties but the same formula. Berzelius explained away the difference by saying that silver fulminate and silver cyanate had different amounts of vital force.

Then in 1828 Wöhler ran a key chemical reaction. He began with ammonium cyanate, a fairly well-known inorganic compound whose molecular formula is $CH_4N_2O$. Wöhler simply heated this compound, and the product he got was urea (see Figure 18.4). Not only is urea definitely organic (it is an end product of protein metabolism), but urea and ammonium cyanate have exactly the same molecular formula. Critics said that Wöhler's ammonium cyanate must have contained vital force.

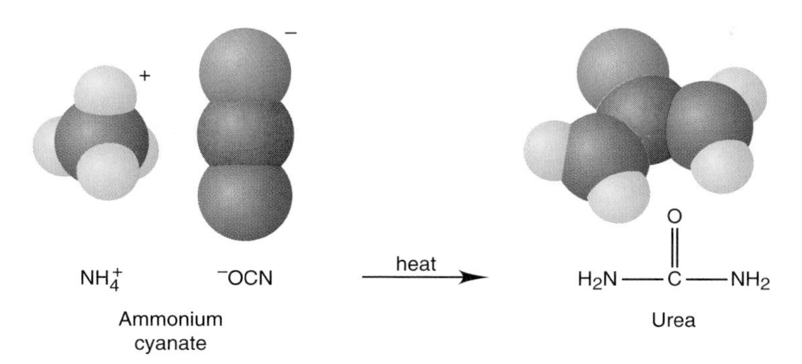


FIGURE 18.4

By the mid-1840s chemists had synthesized several organic compounds, not even from inorganic compounds, but from the elements that make them up. Despite this development it took a long time for vitalism to wither away. Vitalism seems to give a special status to the compounds that make up human beings, and ideas that flatter the human ego usually take a long time to die out, even when someone clearly shows that they are unreasonable.

In the two centuries since Lavoisier, chemists have crossed the line between organic compounds (from living sources) and inorganic ones (from nonliving sources) so often that the original distinction between the two has become meaningless. Not only have chemists made organic compounds from inorganic ones, but geologists have found that many rocks and minerals originate from living organisms. As you probably know, limestone is a rock widely used in construction, and it consists principally of calcium carbonate ($CaCO_3$). Most limestone is "organic," formed from the shells of ancient organisms (see Figure 18.5).

Today chemists use a variation of Lavoisier's original idea: Organic compounds are those composed principally of carbon and hydrogen, but which often also contain the elements oxygen, nitrogen, sulfur, phosphorus, and/or halogens. Advocates of organic foods and fertilizers adhere to the old definition of the word *organic*, which can lead to confusion. Table 18.2 contrasts the two definitions by applying them to several compounds.

FIGURE 18.5
Limestone and mollusk shells are both primarily calcium carbonate. (This is a photo of Coeymans limestone mudcrack fossils.)

**TABLE 18.2 Organic and Inorganic Compounds**

| Material | Formula | Organic (from a living source)? | Organic (primarily C and H)? |
|---|---|---|---|
| Bone meal | $Ca_5(PO_4)_3OH$ | Yes | No |
| DDT (a pesticide) | $C_{14}H_9Cl_5$ | No | Yes |
| Table sugar | $C_{12}H_{22}O_{11}$ | Yes | Yes |
| Nickel-iron meteorite | Ni and Fe (a mixture) | No | No |

## CHAPTER BOX 18.1 SPONTANEOUS GENERATION

**Where does** life come from? The vital force debate among chemists in the early 1800s clearly arose from this question. Can life come spontaneously from nonliving material? Early biologists, chemists, and even physicists confronted this related issue of *spontaneous generation.*

Some early thinkers believed that many forms of life were created by spontaneous generation. It seemed that tadpoles grew from mud, flies from rotting meat, and rats from garbage, for instance. By the 1700s educated people realized that even insects had to have parents, that they hatch from eggs and do not arise directly from their food sources. The origin of one-celled organisms, which were discovered in 1677, remained an open question.

The Italian biologist Lazzaro Spallanzani (1729–1799) performed experiments in 1768 that appeared to settle the matter. Microorganisms seem to appear and grow anywhere there is food for them. Spallanzani sealed meat broth in flasks and boiled the flasks, not just for a few seconds, but for as much as an hour. Nothing grew in the flasks after this. The prolonged boiling, he thought, killed all the microorganisms in the solution, in the air above it, and on the walls of the flask. Sealing kept new organisms out. Only spontaneous generation could produce new microorganisms, and since none appeared, there was no spontaneous generation.

Vitalism (which included the idea of spontaneous generation) was a persistent system of belief. Was it possible that the boiling destroyed not only existing microorganisms but also some vital force? In about 1860, the famous French chemist Louis Pasteur (1822–1895) took the next step.

Pasteur was a religious man. He believed that the creation of life was done only by God, and he set out to disprove the existence of spontaneous generation. Instead of sealing his flasks completely, he left long, thin, looping necks on them (as shown below, right). Air could enter his flasks, but dust could not. No microbes grew in Pasteur's flasks after he boiled them. Then Pasteur let some of his broth slosh into the necks of his flasks and back down again. Microbes grew after he did that. Pasteur's experiments were repeatable by religious believers and nonbelievers alike, so they were accepted.

The Irish physicist John Tyndall (1820–1893) closed a few final loopholes by demonstrating that dust contains microorganisms, and spontaneous generation was no longer a viable theory in its original form. For the surprising twentieth-century sequel, see Chapter Box 19.3.

Italian biologist Lazzaro Spallanzani (1729–1799)

(Top, right) French chemist Louis Pasteur (1822–1895). (Bottom) After Pasteur had boiled the growth medium in a flask like these, bacteria didn't grow in it, despite the connection with the outside air. This key experiment helped Pasteur show that bacteria, like all other living organisms, arise from other bacteria rather than by spontaneous generation.

## EXAMPLE 18.3 Differentiating Organic and Inorganic Compounds

Chemical companies synthesize many tons of vitamin C (ascorbic acid, $C_6H_8O_6$) in factories every year. Is this vitamin C organic? Explain your answer.

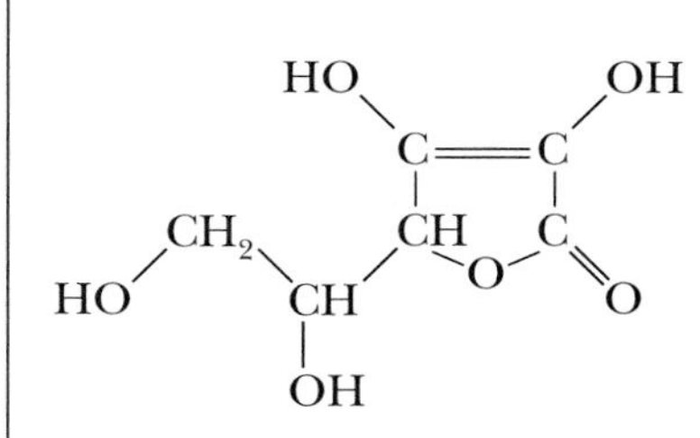

FIGURE 18.6
Vitamin C (ascorbic acid) is an essential vitamin that is found in fruit, particularly citrus fruits.

**SOLUTION**

The answer depends on which definition of *organic* you use. Vitamin C contains principally carbon and hydrogen, and any sample of it is certainly considered by chemists to be organic. This particular sample is not organic to a user of the definition that organic compounds must come from living sources.

Vitamin C molecules manufactured in factories and vitamin C molecules isolated from plant and animal sources are identical. There are no chemical (or nutritional) differences between synthetic (manufactured) vitamin C and the organic (isolated from citrus fruits or rose hips) vitamin C that is sold in health food stores. You will usually pay more for the organic vitamin C, and you get no extra benefits for the extra money.

**EXERCISE 18.3**

Decaying vegetation in marshes gives off methane ($CH_4$) and hydrogen sulfide ($H_2S$) gases. Are these gases organic? Explain your answer.

## 18.3 Functional Groups

**GOAL:** To trace the development of early organic chemistry to the unifying idea of fundamental classes of compounds.

What are the structures of organic compounds like? This was an important research question for at least two or three generations after the structures of the simplest inorganic salts were known. Before they could draw structures, chemists had to know the molecular formulas, and even these were in question for decades.

In the midst of considerable confusion, chemists performed experiments, proposed theories, and made progress. To the chemists of the time, organic chemistry was confusing and almost impossible to understand. In 1835 Wöhler said, "Organic chemistry today almost drives me mad. To me it appears like a primeval tropical forest full of the most remarkable things, a dreadful endless jungle into which one does not dare enter for there seems to be no way out."

With the clarity of vision that comes with hindsight, some results and ideas of the early chemists seem better and more advanced than others. We'll look at a few of these here. This may give you the idea that organic chemistry made constant and coherent advances in the first half of the 1800s, instead of being the muddle it usually seemed to the chemists of the time.

You have already had a hint of the basic problem in the early chapters of this book. Chemists were unsure of the formula of even the fundamental compound water for years after Dalton proposed his atomic theory. Eight grams of oxygen and one gram of hydrogen formed nine grams of water; this much they knew. If the atomic mass of hydrogen were 1 and that of oxygen were 8 (or if masses were 2 and 16, respectively), the formula of water would be HO. If the atomic masses were 1 and 16 (or 2 and 32), the formula would be $H_2O$. Similarly, methane, the simplest organic compound, has the formula $CH_4$; but one German chemist thought it had the formula $C_4H_8$. Berzelius and other chemists found out the ratios of the masses of the elements in many organic compounds with good accuracy in the early 1800s. There was confusion about the actual atomic masses for a half-century after that.

**The word *oxygen* comes from Greek words meaning "to give rise to acids." Lavoisier renamed Priestly's "dephlogisticated air" oxygen because he believed it to be a necessary component of all acids. Even the best of scientists can be wrong occasionally.**

A key error by Berzelius led to a great deal of confusion. He thought that the formula for silver oxide was AgO, but, in fact, it is $Ag_2O$. He and others had worked out many of the formulas of organic acids from the formulas of their silver salts, so the acid formulas were inevitably wrong.

Other mistakes in organic formulas came from confusion over the formulas of inorganic acids. Many (if not most) chemists of the early 1800s believed that oxygen rather than hydrogen was the essential element in acids, and that units of $H_2O$ in formulas represented relatively unimportant "water of composition." Berzelius thought that the formula of acetic acid ($CH_3CO_2H$) was $C_4H_6O_3$—double the true molecular formula of $C_2H_4O_2$ minus one $H_2O$.

**Michel Eugène Chevreul (1786–1889) took the acids that he isolated from soap, made them into candles, and (with Gay-Lussac) patented the process. This was an important invention in a society that relied on candles for lighting. The new candles burned brighter and longer than the old tallow candles, and they also smelled much better. Chevreul remained an active scientist through a very long life. He published his last scientific paper at the age of 102.**

Schemes of organization are often key to advances in a science. The idea that organic compounds come in groups or families is a fundamental one in organic chemistry. Fats, oils, soaps, and waxes were compounds that chemists recognized early on as being members of groups. The remarkable French chemist Chevreul worked on these substances from 1810 to 1823 (see Figure 18.7). He found that soaps are the sodium salts of organic acids and that both fats and oils are constructed from organic acids and glycerol.

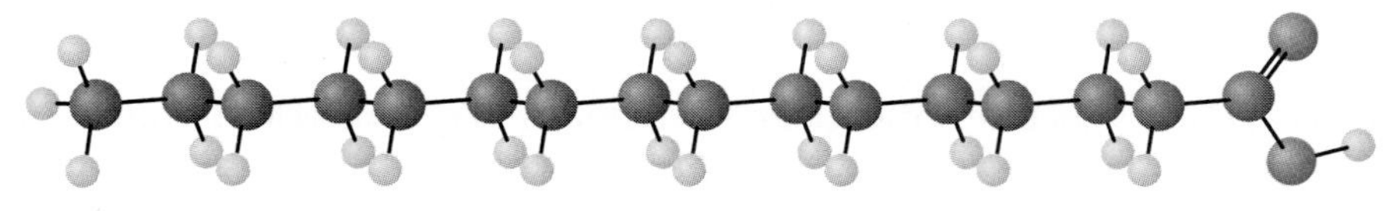

```
     H   H   H   H   H   H   H   H   H   H   H   H   H   H   H  :O:
     |   |   |   |   |   |   |   |   |   |   |   |   |   |   |   ||   ..
H—C—C—C—C—C—C—C—C—C—C—C—C—C—C—C—C—O—H
     |   |   |   |   |   |   |   |   |   |   |   |   |   |   |        ..
     H   H   H   H   H   H   H   H   H   H   H   H   H   H   H
```

FIGURE 18.7
These are representations of palmitic acid, one of Chevreul's discoveries. He isolated this fatty acid from fats and used it to make candles.

As you already know, bonding is a vital part of the study of chemistry. Berzelius based an early theory about bonding on inorganic compounds. He proposed that atoms are held together in molecules by static electricity and that compounds are made by combining "positive" and "negative" elements. (Molecules are held together by the electrostatic forces, yes, but not in the way Berzelius meant.) Then, in the early 1830s, the French chemist Jean Baptiste André Dumas ran some reactions that replaced the "positive" element hydrogen with the "negative" element chlorine in organic molecules. This experiment created problems for Berzelius's theory.

Berzelius was a great authority by that time and had become oversensitive to attacks on his theories during his later years. He rejected Dumas's ideas, especially since Dumas could propose no force of attraction that would hold molecules together other than static electricity. Chemical bonds must exist in some form or another, chemists knew, but the nature of those bonds remained a puzzle until the early 1900s. (Keep in mind that, at this point in this book, you know much more about bonding than Berzelius ever did.)

Data and theories accumulated. Finally, the German chemist August Kekulé proposed that carbon atoms were tetravalent (that is, that each carbon atom forms four bonds to other atoms). He helped to assemble the first international conference in chemistry, at which the Italian chemist Cannizzaro ended the confusion between atomic and molecular weights. Chemists worked out the main principles of organic structures fairly quickly after that.

Inorganic compounds turn up as recognizable units in many chemical reactions, especially precipitation reactions. In contrast, if a chemist only knows the identity of the elements present in an organic compound, the chemist can predict very little of its chemistry. It is much more important to know exactly how the atoms in an organic molecule are connected because that is what determines how they react.

Organic chemists' experimentation and theorizing during the 1800s finally resulted in a useful unifying idea. Its earliest name, put forth by the French chemist Charles Gerhardt in 1853, was the "second-type theory." Organic chemists now call it the *functional group concept*. A **functional group** is an association of atoms that largely determines the physical and chemical properties of organic molecules that contain it. Alcohols are one functional group, and we'll look at them next.

## EXPERIMENTAL EXERCISE

### 18.1

A. L. Chemist was drinking coffee with a colleague, and the talk wandered around to Berzelius's theories. Suppose, they wondered, a chemist believes (correctly) that 1 mol of silver oxide has a mass of 231.7 g. Suppose the chemist also believes that the atomic mass of oxygen is 16 and that the formula of silver oxide is AgO.

**QUESTION**

1. What atomic mass of silver would this chemist calculate using these assumptions?

## 18.4 Alcohols and Ethers

**GOAL:** To recognize and give examples of alcohols and ethers.

FIGURE 18.8
People have made alcohol by fermentation for many thousands of years. The process of distillation to separate the alcohol from the water, as is being done in this distillery, is a much more recent invention.

To most people the word *alcohol* means the compound called ethyl alcohol, grain alcohol, or ethanol, which is the active ingredient in alcoholic beverages (see Figure 18.8). Even chemists sometimes use the word *alcohol* by itself to mean this single compound. Chemists in the 1800s found that there are large numbers of organic compounds that contain one atom of oxygen in their molecular formula and that react with other compounds in the same way ethanol does. The structures of these compounds all contain a grouping of carbon and hydrogen atoms (called an **alkyl group**) attached to an OH group by a covalent bond. The functional group that identifies **alcohols** is this covalently bonded—OH group.

Figure 18.9 shows the Lewis diagrams, names, and structures of several alcohols. Note that all of the names of these alcohols end in the two letters *ol.* Chemists use suffixes such as this one in the names to indicate the functional groups of most compounds, as you will see in subsequent sections.

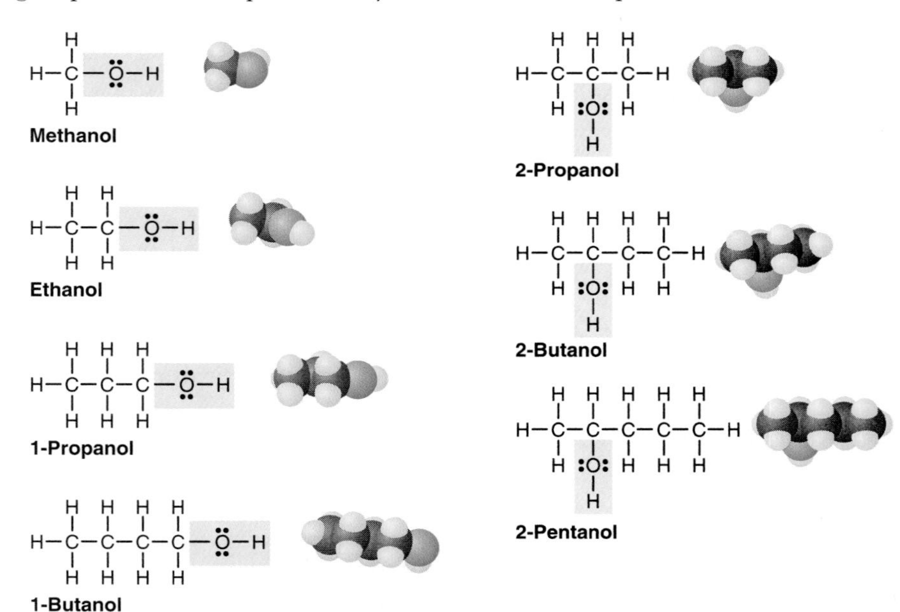


FIGURE 18.9
All seven of these compounds contain an alkyl group and an OH group, and they are therefore all alcohols. The four structures on the left are the first four members of the series of unbranched alcohols, and the three on the right begin another series.

Some care must be taken when working with Lewis diagrams and condensed structures of organic molecules. Keep in mind that these diagrams and formulas are not really representing *structures;* they merely show how the atoms are connected to each other. All of the representations in Figure 18.10 are ethanol. Similarly, for propanol, you get the same structure by placing the OH group on either one of the end carbon atoms of the three-carbon chain (that is, $HOCH_2CH_2CH_3$ is identical to $CH_3CH_2CH_2OH$). However,

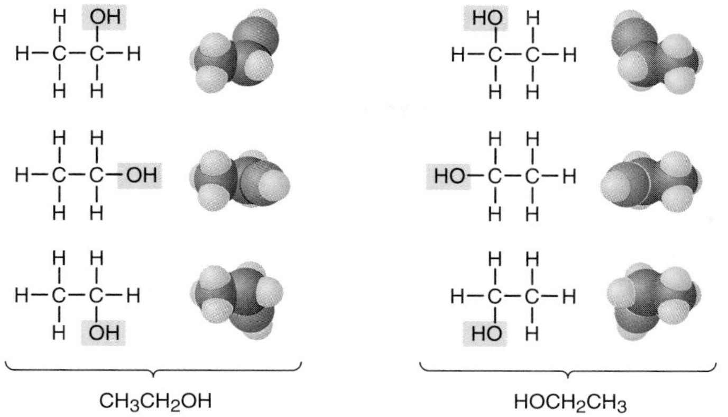

$CH_3CH_2OH$ $HOCH_2CH_3$

FIGURE 18.10

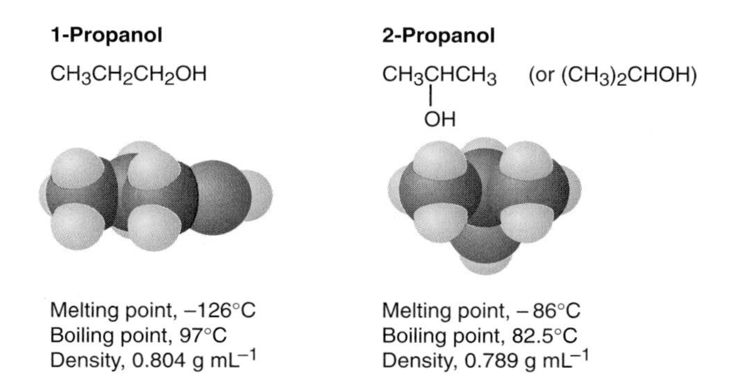


FIGURE 18.11

placing the OH group on the central carbon gives a different compound, one with different chemical and physical properties (see Figure 18.11).

To reflect this kind of difference in structure, the names of alcohols are given numerical prefixes according to the following rules:

- Number the carbon atoms in the longest continuous chain of carbon atoms. Start with the end that assigns the lowest possible number to the carbon atom bearing the OH group.
- Use the number of the carbon atom bearing the OH group as the prefix.

Chemists around the world have adopted an official method for assigning *systematic names* to organic compounds. This method uses standard names for alkyl groups, standard suffixes for functional groups, and rules such as the preceding ones for numbering and for naming compounds containing more than one functional group. The names given in Figure 18.9 are systematic names.

Many compounds have *common names* that were in wide use before the systematic nomenclature was invented. We have already mentioned several names for ethanol. Another name for 2-propanol is *isopropyl alcohol,* and you will find it by that name on drugstore shelves.

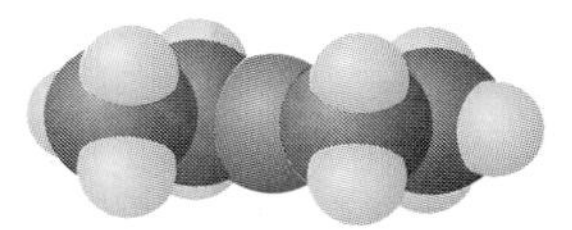


Diethyl Ether

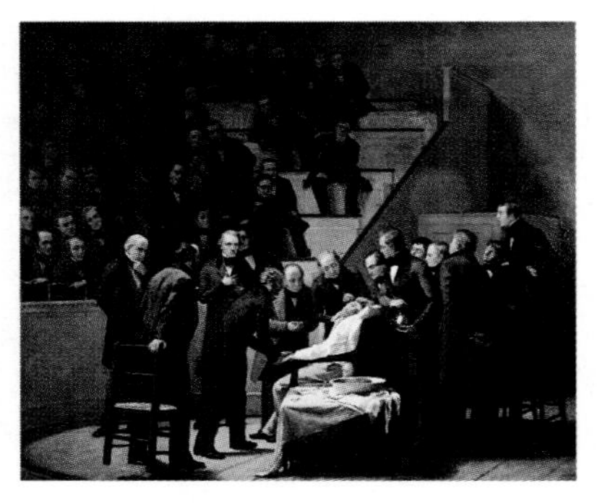

FIGURE 18.12
The American dentist W. T. G. Morton's demonstration of the anesthetic properties of ether changed surgery forever.

### EXAMPLE 18.4 Identifying Alcohols

Which of these compounds is most likely to be an alcohol?
A. Cocaine B. Cholesterol C. Octane D. Aspirin

**SOLUTION**

Without looking up the structures in a reference book, you can't know for sure. However, cholesterol (compound B) has a name that ends in the letters *ol*, and therefore is likely to be an alcohol (and it is).

**EXERCISE 18.4**

Which of these compounds is most likely to be an alcohol?
A. Insulin B. Polypropylene C. Acetone D. Menthol

There is another functional group whose members contain one singly bonded oxygen atom per molecule. These are the **ethers.** In these compounds the oxygen atom has two alkyl groups attached to it, instead of one as in the alcohols. The most common member of this group of compounds is diethyl ether, commonly called simply *ether,* just as ethanol is referred to as *alcohol.* The structure of diethyl ether is shown in Figure 18.12.

## 18.5 Aldehydes and Ketones

GOAL: To recognize and give examples of aldehydes and ketones.

**Aldehydes and ketones are not the only kinds of compounds that contain carbonyl groups. This pair of atoms forms part of groups that also contain another oxygen atom or a nitrogen atom.**

There are four different families of organic compounds whose functional groups contain one oxygen atom in addition to carbon and hydrogen. Two are the alcohols and the ethers discussed in Section 18.4, which contain carbon-oxygen single bonds. The other two are the aldehydes and the ketones, which contain a carbon-oxygen double bond. Chemists call this C=O group a **carbonyl group.**

An **aldehyde** has only one carbon atom attached to the carbonyl group. The only exception is the smallest of the family, formaldehyde ($H_2C{=}O$), which has two hydrogen atoms and no carbon atoms bonded to the carbonyl group. Figure 18.13 shows the structures of formaldehyde and another simple aldehyde. Formaldehyde is an important (and somewhat dangerous) in-

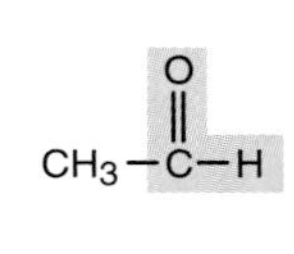


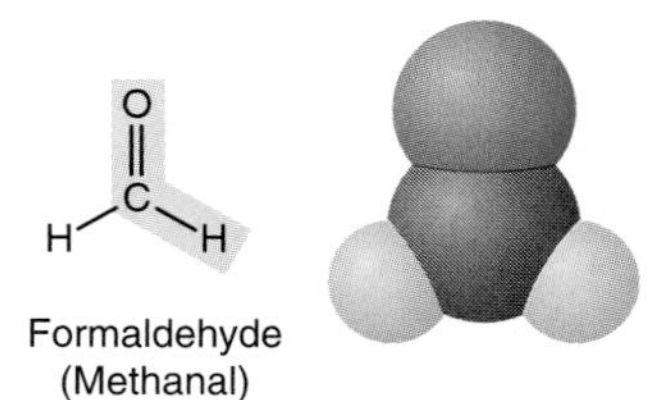

Formaldehyde
(Methanal)

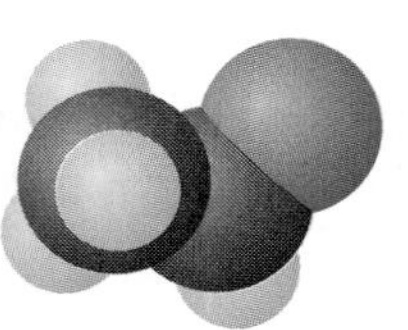



Acetaldehyde
(Ethanal)

FIGURE 18.13

## CHAPTER BOX 18.2 THE ANESTHETIC PROPERTIES OF DIETHYL ETHER

**By the** 1840s, scientists' knowledge of gross human anatomy was extensive, but this knowledge didn't help physicians and surgeons very much. There were no anesthetics, and operating was almost always more dangerous than doing anything else. The risk of the patient dying of shock during an operation was extremely high. The only common surgical operation at the time was the amputation of limbs. The greatness of surgeons was gauged by the speed with which they could remove a damaged arm or leg.

In the meantime, chemists investigating the properties of their newly discovered compounds found some that could ease pain and produce a "high" as early as the late 1700s. This knowledge spread. A group of men and women got together to sniff ether (and other compounds) in search of a high. One fellow breathed in too much ether, passed out, fell down, and got a severe laceration on his head. To everyone's surprise he wasn't in any pain at all, and only knew he was injured because he could see blood running down when he came to. Events such as these gave rise to the idea that diethyl ether and other compounds might be useful as anesthetics.

It took time for this sort of knowledge to penetrate from the happy world of parties (and chemists) to the grimmer world of the surgeon. In 1842 Crawford W. Long of Georgia used ether as an anesthetic during an operation to remove a tumor. Word of Morton's 1846 demonstration (see Figure 18.12) at Massachusetts General Hospital spread rapidly among surgeons, transforming their practices radically. As a general anesthetic, diethyl ether has a great many advantages. It is inexpensive, easy to administer, and a very good muscle relaxant. Furthermore, it has only a slight influence on blood pressure, pulse rate, and the rate of respiration. The major disadvantages are an irritating effect on the respiratory passages, and nausea as an aftereffect. (For this reason diethyl ether is seldom used today as a general anesthetic; anesthesiologists use similar ethers that have fluorine atoms replacing some of the hydrogen in the structure.)

Diethyl ether has another disadvantage, one that was especially important in the second half of the 1800s. This compound has a relatively low boiling point, and mixtures of ether vapor and atmospheric oxygen are explosive. Surgery at night was a very dangerous enterprise when the only source of light was a kerosene lamp. Surgeons during the Civil War were frequently caught between the proverbial rock and hard place. Battles were almost always fought during the day, since much less ammunition was wasted when soldiers could see their target. The wounded were brought to army doctors all day long, but there were so many casualties that they could not possibly all be treated during daylight hours. Army surgeons on both sides were forced to do tricky operations at night, using candles or kerosene lamps to see by. They operated in the open as much as they could, but wind shifts caused more than one explosion when ether-oxygen mixtures blew into lamps.

dustrial chemical. Dissolved in water and methanol, it forms the preservative solution called *formalin,* used by biologists.

The systematic names of aldehydes end with the suffix *al* attached to the name of the alkyl root as *ol* is attached in naming alcohols. As always, there are compounds whose common names are more familiar than their systematic ones. Vanillin, the compound responsible for the odor and flavor of vanilla, is an aldehyde. Cinnamaldehyde is the compound primarily responsible for the distinctive smell of cinnamon. Here are these and several other aldehydes:

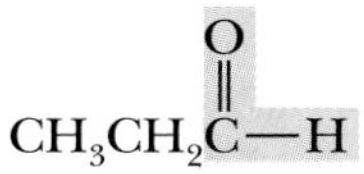

$CH_3CH_2C(=O)H$ — Propan**al**

$CH_3CH(CH_3)CH_2C(=O)H$ — 3-Methylbutan**al**

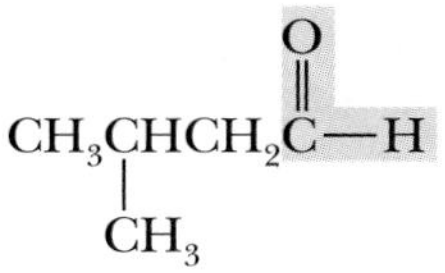

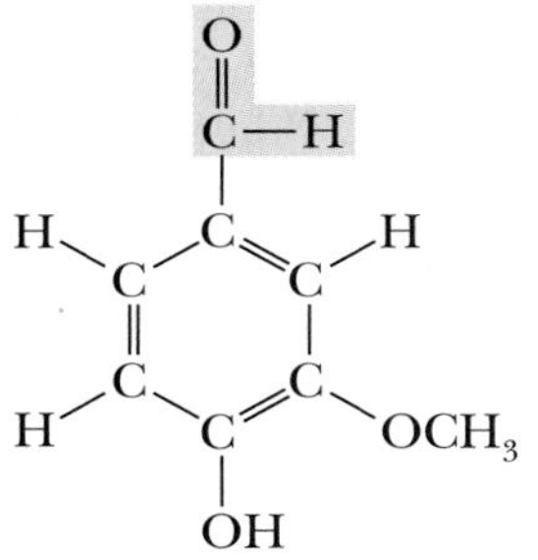

Vanillin: benzene ring bearing C(=O)H, OH, and $OCH_3$ — Vanillin

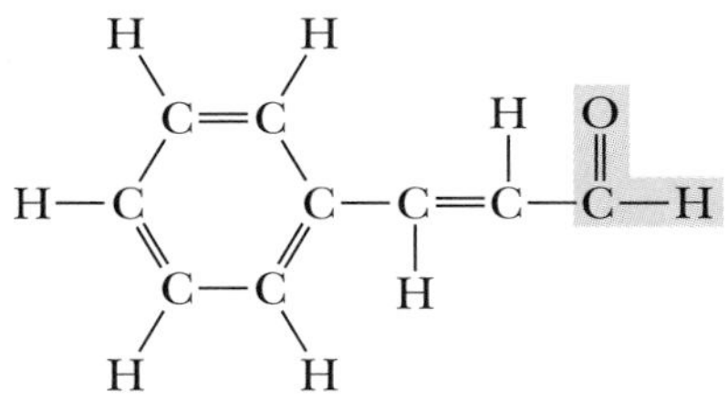

$C_6H_5CH=CHC(=O)H$ — Cinnamaldehyde

A **ketone** has two carbon atoms attached to the carbonyl group. You can see that ketones are closely related to aldehydes. Chemists put them in a different category because they react differently with oxidizing agents. Aldehydes are relatively easy to oxidize; a mild oxidizing agent such as $Ag^+$ will do the job (see Table 16.1). Ketones, on the other hand, will oxidize only under extreme conditions. The systematic names of ketones end in the suffix *one* (pronounced like the word *own*). Here are three of the smaller and simpler members of the ketone family:

**Acetone is the solvent in fingernail polish remover. Like many ketones, it has a distinctive odor.**

O
‖
C
$H_3C$ $CH_3$

Acet**one**

O

Cyclohexan**one**

O
‖
C
$H_3C$ $CH_2CH_3$

Butan**one**

EXAMPLE 18.5 **Identification of Oxygen-Containing Functional Groups**

**a.** Which of these compounds is a ketone?

**A.**
```
   H  O  H
   |  ‖  |
H—C—C—C—H
   |     |
   H     H
```

**C.**
```
   O     H
   ‖     |
H—C—O—C—H
         |
         H
```

**B.**
```
   H  O
   |  ‖
H—C—C—O—H
   |
   H
```

**D.**
```
   H  O
   |  ‖
H—C—C—H
   |
   H
```

**b.** Which of these compounds is an aldehyde?

A. Acetophenone
B. Piperonal
C. Tetrahydrofuran
D. Pinene

**SOLUTIONS**

**a.** Compound A has two carbon atoms attached to its carbonyl group and is therefore a ketone. Compounds B and C have two oxygen atoms attached to the same carbon atom, functional groups we have not yet studied. Compound D has a hydrogen attached to its carbonyl group and is therefore an aldehyde.

**b.** The name piperonal (B) ends in *al* and therefore is the aldehyde of this group.

**EXERCISE 18.5**

**a.** Which of these compounds is an alcohol?

```
     H     H                      O  H
     |     |                      ‖  |
A. H—C—O—C—H         C.  H—O—C—C—H
     |     |                         |
     H     H                         H

     H                           H  O—H
     |                           |  |
B. H—C—C=O           D.  H—C—C—H
     |  |                        |  |
     H  H                        H  H
```

**b.** Which of these compounds is a ketone?

A. Acetophenone  C. Tetrahydrofuran
B. Decanal  D. Pinene

## 18.6 Carboxylic Acids

**GOAL:** To recognize and give examples of carboxylic acids.

Acetic acid ($CH_3CO_2H$), the compound that gives vinegar its distinctive taste and odor, is a **carboxylic acid.** So is palmitic acid, Chevreul's discovery (see Figure 18.7). Carboxylic acids are the only organic compounds (with few exceptions) that form acidic solutions when dissolved in water. They contain an alkyl group bonded to a group usually abbreviated as $CO_2H$. The $CO_2H$ group is called a **carboxyl group.** For examples of simple carboxylic acids, see Figure 18.14.

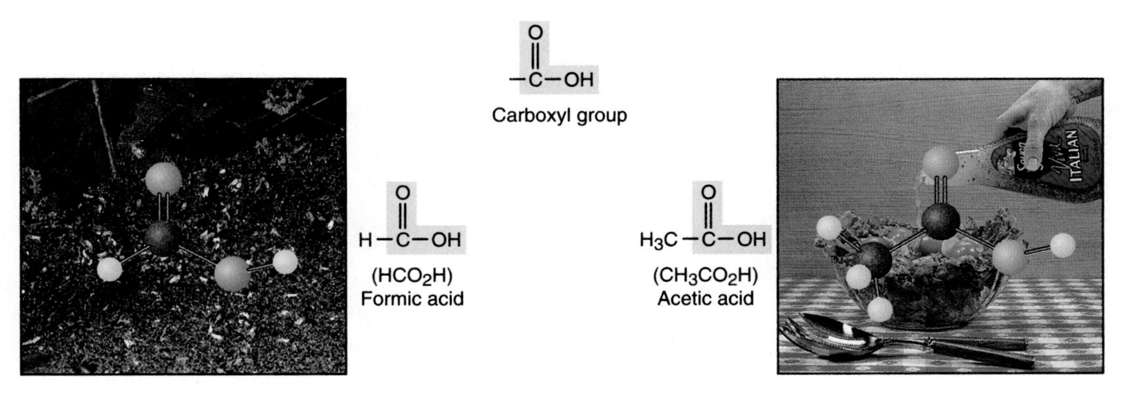


FIGURE 18.14
Formic acid puts the sting into the bites of some ants; the word *formic* comes from *formica,* the Latin word for "ant." Acetic acid gives vinegar (and oil and vinegar salad dressing) its distinctive sour taste.

## EXAMPLE 18.6 Identifying Carboxylic Acids

Which of these compounds is a carboxylic acid?

A. $(CH_3)_2CHOH$ C. $CH_3CO_2CH_3$
B. $(CH_3)_2CHCO_2H$ D. $CH_3OCH_2CH_3$

**SOLUTION**

Compound B contains a $CO_2H$ group and is therefore a carboxylic acid.

**EXERCISE 18.6**

Which of these compounds is a carboxylic acid?

**A.** $H{-}CH_2{-}O{-}CH_2{-}H$ (H–C(H)(H)–O–C(H)(H)–H)

**B.** H–C(H)(H)–C(H)=O

**C.** H–O–C(=O)–C(H)(H)–H

**D.** H–C(H)(H)–C(H)(O–H)–H

## 18.7 Esters and Amides

**GOAL:** To recognize and give examples of esters and amides.

**Chemists have discovered that the singly bonded oxygen atom in an ester almost always comes from the alcohol, not the acid. To discover this, they used oxygen isotopes as *tracers*.**

Chemists often use carboxylic acids to make compounds that contain other functional groups. **Esters,** in which the hydrogen of the carboxyl group is replaced by an alkyl group, are one type of compounds made from carboxylic acids. The acid reacts with an alcohol, as in this example reaction:

$$H_3C{-}\overset{O}{\overset{\|}{C}}{-}O{-}H + CH_3CH_2{-}O{-}H \xrightarrow{\text{catalyst}} H_3C{-}\overset{O}{\overset{\|}{C}}{-}O{-}CH_2CH_3 + H_2O$$

Acid    Alcohol    Ester

This OH and this H form $H_2O$, and the remaining parts of the acid and the alcohol form the ester.

Many fruits and flowers owe their pleasant smells to esters (see Figure 18.15). Other members of the ester family are all around you and inside you. Fats and oils are made up of fatty acids (relatively large carboxylic acids) and glycerol. Glycerol is an alcohol, and fats and oils are members of the ester family. For more details about the role of such compounds in living things, see Chapter 19.

```
   H  H      :O:  H  H  H  H
   |  |      ||   |  |  |  |
H—C—C—Ö—C—C—C—C—C—H
   |  |      ¨    |  |  |  |
   H  H           H  H  H  H
```

Ethyl butanoate

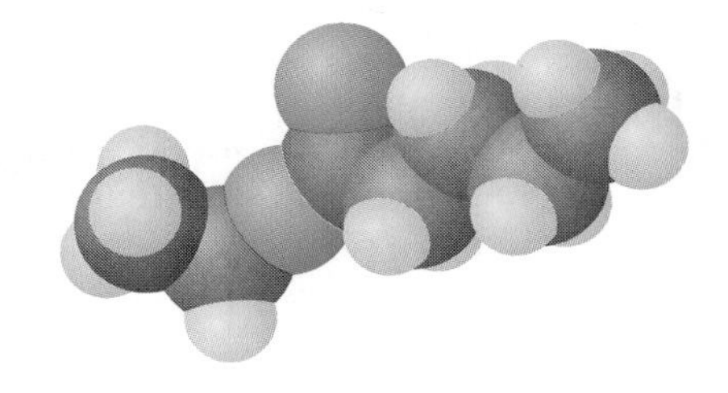

FIGURE 18.15
The ester ethyl butanoate is partly responsible for the pleasant and distinctive odor of pineapples. Similar esters lend their odors to other fruits.

An **amide** contains an alkyl group bonded to a carbonyl group to which a nitrogen atom is also singly bonded. There are probably large amounts of amides nearby as you read this book. Many foods and all animals, including yourself, contain proteins, and the most common functional group in protein molecules is the amide group. Also, nylon is made by linking together a series of smaller units using amide groups. For the structure of a simple amide, see Figure 18.16; for more about both proteins and nylon, see Chapter 19.

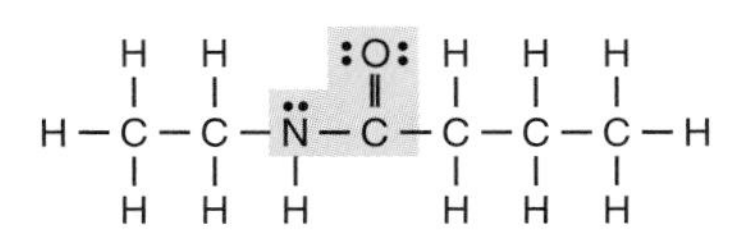

An amide, N-ethylbutanamide

FIGURE 18.16
Nylon fibers consist of long series of segments connected by amide functional groups. The photo shows nylon-66 being made in a chemistry laboratory.

EXAMPLE 18.7 **Carboxylic Acid, Ester, and Amide Functional Groups**

**a.** Which of these compounds is an ester?

```
      H  O  H                  O     H
      |  ‖  |                  ‖     |
A.  H—C—C—C—H         C.  H—C—O—C—H
      |     |                        |
      H     H                        H

      H  O                     H  O
      |  ‖                     |  ‖
B.  H—C—C—O—H         D.  H—C—C—H
      |                        |
      H                        H
```

**b.** Draw the structure of an ester that has the molecular formula $C_3H_6O_2$.

**SOLUTIONS**

**a.** Compound C is the ester because it contains the $CO_2R$ group. Compound A is a ketone, B is a carboxylic acid, and D is an aldehyde.

**b.** There are two esters that have the formula $C_3H_6O_2$, and either one of these is a correct answer.

```
   H  O       H                     O       H  H
   |  ‖       |                     ‖       |  |
H—C—C—O—C—H       and       H—C—O—C—C—H
   |          |                             |  |
   H          H                             H  H
```

**EXERCISE 18.7**

**a.** Which of these compounds is an amide?

```
       H                           H
       |                           |
A.  H—C—C=N—H            C.  H—C—N—H
       |  |                        |  |
       H  H                        H  H

       H     H                       O  H
       |     |                       ‖  |
B.  H—C—N—C—H            D.  H—N—C—C—H
       |  |  |                     |     |
       H  H  H                     H     H
```

**b.** Draw the structure of a carboxylic acid that has the molecular formula $C_3H_6O_2$.

## 18.8 Amines

**GOAL:** To recognize and give examples of compounds containing the amine functional group.

**Another key feature of amines is that the nitrogen has a nonbonding pair of electrons.**

Organic compounds containing three single bonds to nitrogen, none to an adjacent carbonyl group, are **amines.** Relatively small amine molecules are partly responsible for the unpleasant odor of dead fish. The Polish-American chemist Casimir Funk thought that all members of a certain class of nutrients were amines, so he called them *vitamines* (from "vital amines"). The name was shortened to *vitamin* when others found that not all of these compounds contain an amine group.

**Historians of science sometimes call synthetic dye manufacturing the first industry created by scientific investigators working in chemistry.**

*Aniline* is perhaps the most important amine in the history of technology (see Figure 18.17). The British chemist Henry Perkin used it to make the first artificial dye in 1856, when he was still a teenager. He produced a color called *mauve.* Aniline and its relatives are still the starting materials for about half of all synthetic dyes produced today.

```
     H       H
      \     /
       C — C
      //     \\        ..
H — C          C — N — H
      \      /         |
       C = C           H
      /     \
     H       H
```

Aniline

FIGURE 18.17
Aniline is still important in the dye industry.

## EXAMPLE 18.8 Structures of Amines

Draw the structure of an amine that contains one carbon atom and one nitrogen atom in addition to hydrogen atoms.

**SOLUTION**

First, connect one carbon atom to one nitrogen atom, and then surround them with hydrogen atoms until the carbon has four bonds connected to it and the nitrogen has three. Place a nonbonding pair of electrons on the nitrogen. This compound, methylamine (or aminomethane), is the simplest amine.

```
    H
    |   ..
H — C — N — H
    |   |
    H   H
```

**EXERCISE 18.8**

There are two different amines that contain two carbon atoms and one nitrogen atom. Draw the structures of both.

## 18.9 Hydrocarbons

**GOAL:** To recognize and give examples of various subclasses of hydrocarbons.

The **hydrocarbons** form what is perhaps the most fundamental of the organic functional group families. These compounds contain carbon and hydrogen, and nothing else. Fuels such as natural gas, gasoline, kerosene, and heating oil consist almost entirely of hydrocarbons (see Figure 18.18).

FIGURE 18.18
The major use of alkanes is as fuels. The stove (top) burns methane. The cigarette lighters on the right burn butane, which is a gas at room temperature and normal atmospheric pressure, but liquefies at higher pressures. The cooking unit inside the recreational vehicle (left) uses propane, which is stored in the tank shown.

In Chapter 8 you learned how to balance combustion reactions, and you read about some of the problems that arise from the burning of hydrocarbons.

Hydrocarbons are divided into four subclasses. The **alkanes** contain only single bonds, and their names end with the suffix *ane.* The four smallest alkanes are gases named methane, ethane, propane, and butane (see Figure 18.19). Methane is the major constituent of natural gas, and propane and butane are also well-known fuels. Larger alkane molecules are liquids at room temperature. The octane scale that indicates the extent to which gasoline will burn in car engines without knocking is named for a liquid alkane. Even larger alkane molecules are soft solids at room temperature. Paraffin (a wax) is a mixture of relatively large alkane molecules.

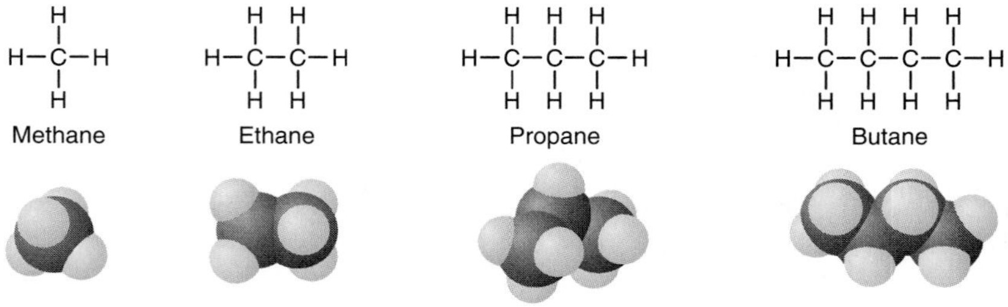

Methane Ethane Propane Butane

FIGURE 18.19
The structures of four small alkanes.

Virtually all organic molecules contain alkyl groups, portions that could be called alkanes if nothing else were attached to them. The alkyl group forms the "backbone" of the molecule, to which other functional groups are attached.

Despite the "ene" in their names, neither ethylene glycol nor polyethylene contain double bonds. These are common names, not derived from the system now used by chemists.

The **alkenes,** sometimes called *olefins* in the chemical industry, are compounds that contain carbon-carbon double bonds. The names of alkenes end in the suffix *ene.* The smallest alkene is ethene, often called ethylene, $CH_2{=}CH_2$. This compound is the starting point for synthesizing hundreds of other compounds, including the plastic polyethylene.

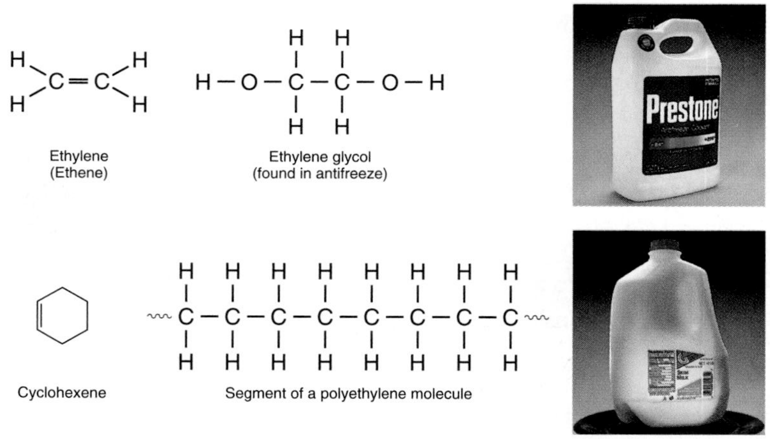

Ethylene (Ethene)

Ethylene glycol (found in antifreeze)

Cyclohexene

Segment of a polyethylene molecule

FIGURE 18.20
Alkenes and substances made from them. Chemical companies make ethylene glycol by oxidizing ethylene. The best-quality polyethylene has hundreds of thousands or even millions of connected $CH_2$ groups in each molecule.

Hydrocarbons containing carbon-carbon triple bonds are called **alkynes.** The systematic names of alkynes have the suffix *yne.* The simplest of these is ethyne, which has the more common but somewhat misleading name of acetylene. The structure of ethyne is $H{-}C{\equiv}C{-}H$. The three-carbon member of the family is propyne, which has the condensed structure $CH_3C{\equiv}CH$. Here are structures of some other alkynes:

$H_3C{-}CH_2{-}C{\equiv}C{-}H$ (1-Butyne)   $H_3C{-}C{\equiv}C{-}CH_3$ (2-Butyne)   Ethynylcyclopentane (cyclopentyl$-C{\equiv}C{-}H$)

**Aromatic hydrocarbons** contain a closed circle of six carbon atoms called a benzene ring. In many ways they look like alkenes, since the ring contains what appear to be carbon-carbon double bonds. Aromatic hydrocarbons don't react very much like alkenes, however, and thus were given their own class name. The name arose when chemists in the mid-1800s discovered compounds that were both fragrant and chemically similar to benzene. Since those early discoveries chemists have found vast numbers of "aromatic" compounds that have all sorts of odors. This is another word, like "organic," that has a different meaning for chemists than it has for the general public.

**The low reactivity of carbon-carbon double bonds in aromatic compounds is due to resonance, which we discussed in Section 10.3. Resonance lowers the potential energy of the ring, greatly increasing its stability.**

The fundamental member of the aromatic hydrocarbons is *benzene.* Figure 18.21 shows the structures of benzene and several other aromatic compounds.

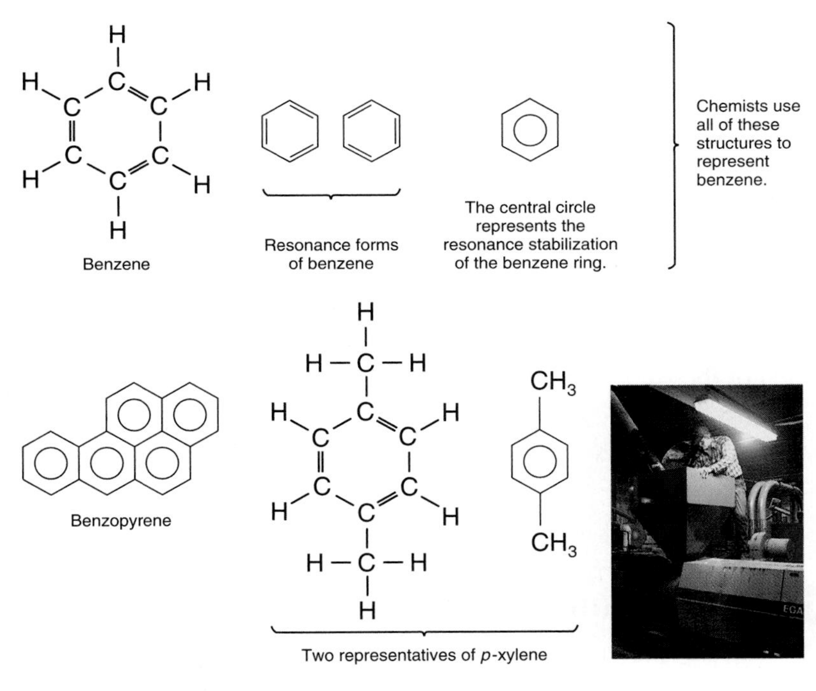
Benzene
Resonance forms of benzene
The central circle represents the resonance stabilization of the benzene ring.
Chemists use all of these structures to represent benzene.
Benzopyrene
Two representatives of p-xylene

FIGURE 18.21
Aromatic hydrocarbons. Chemists represent benzene rings in several ways, depending on the situation; the bond-line formula is the most common. Benzopyrene (bottom left) has the dubious distinction of being the first compound identified as a carcinogen (cancer-causing compound) for humans. It's part of the soot that is produced by almost all combustion processes. The aromatic hydrocarbon *p*-xylene (bottom center) is a starting material (along with ethylene glycol) for making the plastic polyethylene terephthalate (PETA). This plastic, used in soft drink bottles and carpets, is highly recyclable. The photo shows PETA being shredded for recycling.

## EXAMPLE 18.9 Hydrocarbon Structures

**a.** Which of these hydrocarbons is an alkane?

A. $CH_3CH{=}CH_2$   C. $CH_3CH_2CH_3$

B. $H_2C{=}C{=}CH_2$   D. $HC{\equiv}CCH_3$

**b.** What is the molecular formula of this compound?

**SOLUTIONS**

**a.** Alkanes contain no double or triple bonds. The only one to fit this definition is compound C.

**b.** First, count the carbons:

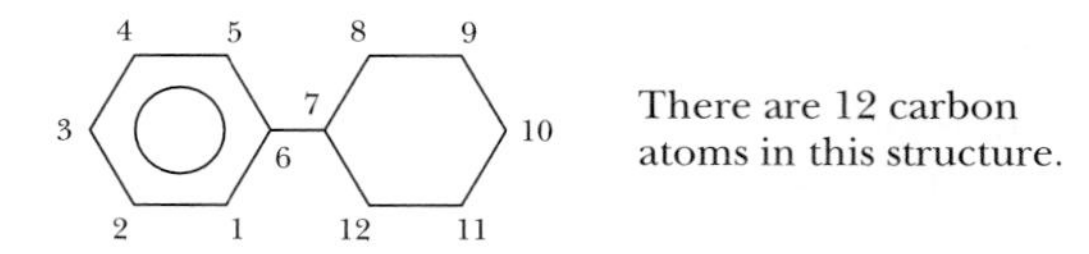

There are 12 carbon atoms in this structure.


To find the number of hydrogens, first draw the three double bonds of the benzene ring.

Then count the hydrogens, remembering that carbon atoms in organic compounds have four bonds to them. You may want to write in the carbon-hydrogen bonds not shown in the bond-line formula. The locations of all hydrogens are shown here:

The formula is $C_{12}H_{16}$.

**EXERCISE 18.9**

Which of these compounds is an aromatic hydrocarbon?

A. O B. O C. $-CH_3$ D. $-CH_3$

## 18.10 Nucleophilic Substitution

**GOAL:** To gain some understanding of organic reactions in general by focusing on one important type.

Organic compounds take part in many kinds of reactions. Alkanes are best known for their combustion (oxidation) reactions, and carboxylic acids for their reactions with bases. This section focuses on a particular type of reaction called **nucleophilic substitution.** Such a reaction is a substitution because one functional group is replaced by another. (We'll get to the "nucleophilic" part in a moment.)

You saw an example of this kind of reaction in Chapter 14. Here it is:

$$CH_3Br + NaOH \longrightarrow CH_3OH + NaBr$$

In net ionic form, the equation is

$$CH_3Br + OH^- \longrightarrow CH_3OH + Br^-$$

The rate equation for this reaction at any constant temperature is

$$\text{Reaction rate} = k[CH_3Br][OH^-]$$

The bromine atom leaves $CH_3Br$ as $Br^-$, and the OH group replaces it.

Many similar compounds react the same way. Chemists can change the alkyl group from $CH_3$ to something else or use other halogens (and even some nonhalogens). They can use many other attacking groups besides $OH^-$ and still get the same kind of reaction. The simple form of the rate equation suggests that the molecular mechanism of the reaction (the pathway it follows) is probably simple as well. Chemists who studied the reaction came up with the following mental picture, which explains the body of observed facts.

The hydroxide ion has a negative charge, which is mostly concentrated on the electronegative oxygen atom. The $CH_3Br$ molecule is polar. The bromine atom is the most electronegative one in the molecule, so there is a partial negative charge on the Br end and a partial positive charge on the $CH_3$ end. As the two species collide in solution, some of them line up as shown in Figure 18.22; this is the only orientation that leads to reaction.

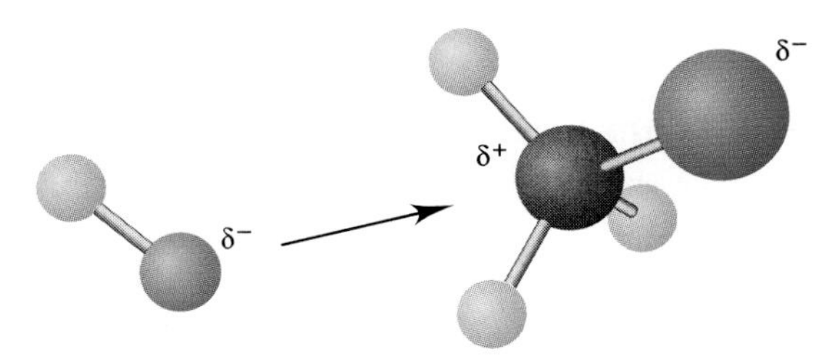


FIGURE 18.22
In the reaction $CH_3Br + NaOH \rightarrow CH_3OH + NaBr$, the electrostatic force of attraction between positive and negative charges helps bring the reactants together.

Because the negative hydroxide ion is electrostatically attracted to the positive nucleus, this ion is called a **nucleophile,** from Greek words that mean "nucleus-loving." From the opposite point of view, the carbon atom in $CH_3Br$ could be called an *electrophile,* for "electron-loving," but organic chemists more often use the term **substrate** for the species being attacked by the nucleophile. The product, the $Br^-$ ion, is the **leaving group,** because it leaves the substrate during the reaction. Based on the evidence they have gathered, chemists have become convinced that the structure shown in Figure 18.23 represents the transition state for this reaction. In the transition state, the oxygen atom is in the process of forming a bond to the carbon atom, while the carbon-bromine bond is simultaneously breaking. The three carbon-hydrogen bonds move from left to right, like the ribs of an umbrella turning inside out.

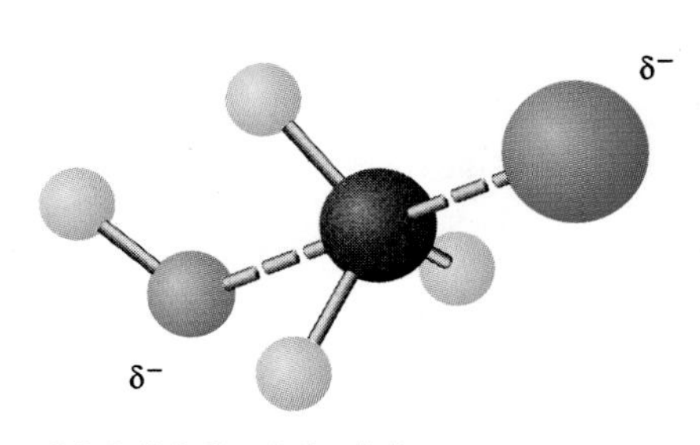


FIGURE 18.23

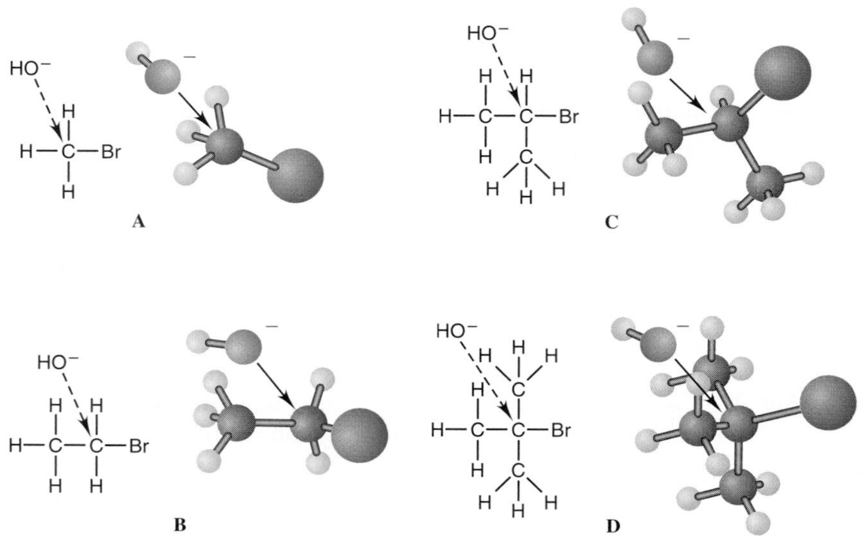

**FIGURE 18.24**
Compound A undergoes nucleophilic substitution reactions the fastest, and compound D is the slowest. It becomes more and more difficult for the nucleophile to reach the carbon atom to which the bromine is attached as more hydrogen atoms are replaced with $CH_3$ groups.

## EXAMPLE 18.10 Terminology of Nucleophilic Substitution

Identify the nucleophile and the leaving group in this reaction.

$$CH_3S^- + CH_3CH_2Cl \longrightarrow CH_3CH_2SCH_3 + Cl^-$$

**SOLUTION**

The nucleophile is the reactant with the negative charge, which is $CH_3S^-$. The leaving group is the halide ion, $Cl^-$, that breaks away from the alkyl group.

**EXERCISE 18.10**

Identify the substrate and the nucleophile in this reaction.

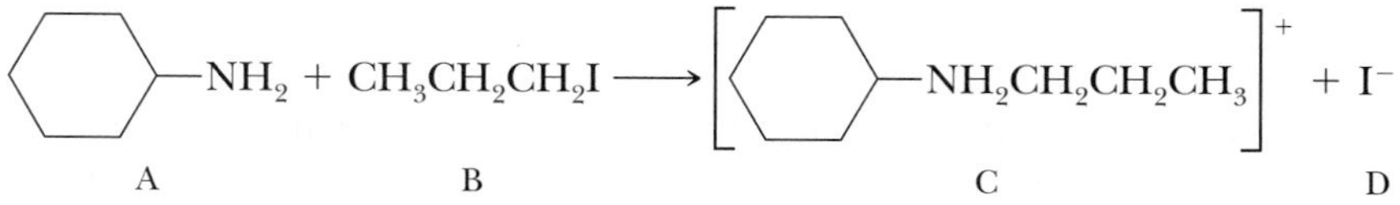

$$\underset{A}{C_6H_{11}\text{—}NH_2} + \underset{B}{CH_3CH_2CH_2I} \longrightarrow \underset{C}{[C_6H_{11}\text{—}NH_2CH_2CH_2CH_3]^+} + \underset{D}{I^-}$$

Other kinds of reactions occur, but those are beyond the scope of this book.

What evidence did organic chemists use to decide that the nucleophile attacks the carbon atom of the substrate in this nucleophilic substitution reaction? Most obviously, the nucleophile is bonded to that carbon atom after the reaction is over. Furthermore, the rate law suggests that the nucleophile and the substrate collide at some time during the reaction. And there is yet other evidence: The rate of the reaction changes when a different alkyl group is part of the substrate. The rate slows down drastically if a hydrogen atom in $CH_3Br$ is replaced with a methyl group ($CH_3$), as in compound B in Figure 18.24. The rate is even slower if two of the hydrogens are replaced with methyl groups (compound C). If methyl groups are swapped for all three of the hydrogens, giving compound D in Figure 18.24, the nucleophilic substitution reaction we are discussing here doesn't occur at all. These facts show that the nucleophile's access to the central carbon atom grows more and more difficult as methyl groups are added.

There is even more direct evidence. Suppose a chemist attempts a substitution reaction on a ring compound like this one, using $OH^-$ to replace the Br:

**Note that the bromine and the methyl group are on the same side of the ring.**

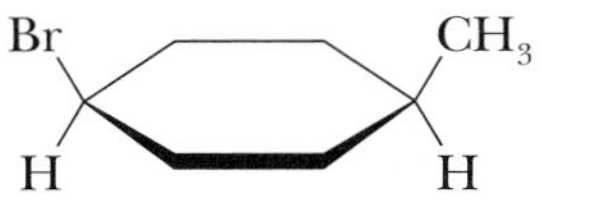

Br CH$_3$
H H

There are two possible substitution products. In one of them the $OH^-$ has exactly the same position as the Br atom did, and in the other the $OH^-$ is bonded to the same carbon atom but on the opposite side of the molecule. Here are the structures of these two possible products:

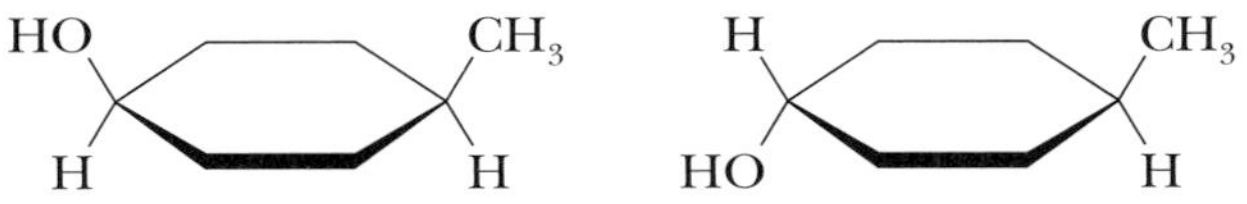

HO CH$_3$ H CH$_3$
H H HO H

Of these two possible products, only the compound on the right forms. This is compelling evidence that nucleophiles really do attack the side of the carbon atom opposite the leaving group.

Even though no organic chemist has ever actually watched the interactions of molecules as a reaction takes place, chemists have figured out in remarkable detail what happens during many reactions.

## CHAPTER SUMMARY

1. Originally the term organic was used by chemists to refer to compounds that came from living matter; it now refers to compounds composed primarily of carbon and hydrogen. The older definition of the word has persisted among nonchemists, which is a possible source of confusion.
2. Organic chemists use two important types of shorthand to represent molecules: condensed structures and bond-line formulas. Interpreting them requires the knowledge that carbon forms four covalent bonds in essentially all stable organic compounds and hydrogen forms one.
3. Organic chemists use the functional group concept to organize their subject matter. A functional group is a collection of atoms that confers a distinctive set of chemical and (sometimes) physical properties on organic compounds that contain it.
4. There are four functional groups that contain a single oxygen atom in addition to carbon and hydrogen. Molecules of all alcohols contain a carbon atom covalently bonded to an OH group. In ethers, a carbon atom bonds to an oxygen atom, which in turn bonds to another carbon atom. In aldehydes the carbon atom of a C=O (carbonyl) group bonds to at least one hydrogen atom. In ketones the carbon of the C=O group bonds to two other carbon atoms.

5. Carboxylic acids contain a carbonyl group whose carbon is bonded to an OH group; in condensed structures the carboxyl group is written as $CO_2H$. These compounds are among the few organic compounds that are acidic in aqueous solution. In esters, a carbon atom replaces the hydrogen in the $CO_2H$ group of a carboxylic acid. In amides, a nitrogen atom is singly bonded to a carbonyl group.
6. Amines contain nitrogen atoms singly bonded to carbons that have no oxygens bonded to them.
7. Hydrocarbons are organic compounds that contain only carbon and hydrogen. Subclasses of hydrocarbons include alkanes (which contain only single bonds), alkenes (which have a carbon-carbon double bond), alkynes (a carbon-carbon triple bond), and aromatic compounds (a benzene ring). The main use of hydrocarbons is as fuels.
8. In nucleophilic substitution reactions, a base (nucleophile) can attack a compound that includes a carbon-containing group (substrate) to replace a leaving group (often a halide ion).

## PROBLEMS

### SECTION 18.1 Lewis Diagrams in Organic Chemistry

1. Draw a Lewis diagram from each of these condensed structures.
   a. $(CH_3)_2CH(CH_2)_3CH_3$
   b. $CH_3(CH_2)_4CHCl_2$
   c. $(CH_3)_4C$
2. Draw a Lewis diagram from each of these condensed structures.
   a. $(CH_3)_2CHOH$
   b. $H_2N(CH_2)_6C(CH_3)_3$
   c. $CF_3(CH_2)_2CH(CH_3)_2$
3. For each of these condensed structures, draw the Lewis diagram.
   a. $Cl_2CH(CH_2)_2CH_2Br$
   b. $(CH_3)_2C(OH)CH_3$
   c. $(CH_3)_2CHOCH_3$
4. For each of these condensed structures, draw the Lewis diagram.
   a. $HOCH_2CH_2OH$
   b. $CH_3CH(NH_2)CH_2CH_3$
   c. $ClCH_2CHClCH_2Cl$
5. Write a condensed structure from each of these Lewis diagrams.

```
      H     H
      |     |
a. H—C—Ö—C—H
      |  ..  |
      H     H
```

```
      H  OH
      |  |
b. H—C—C—H
      |  |
      H  H
```

```
      H  H  H  H  H  H
      |  |  |  |  |  |
c. H—C—C—C—C—C—C—H
      |  |  |  |  |  |
      H  H  H  H  H  H
```

6. Write a condensed structure from each Lewis diagram.

```
       H  Cl  H
       |  |   |
a. Cl—C—C—C—Cl
       |  |   |
       H  H   H
```

```
          H
          |
        H—C—H
      H   |  H  H  H
      |   |  |  |  |
b. H—C———C—C—C—C—H
      |   |  |  |  |
      H   |  H  H  H
        H—C—H
          |
          H
```

```
      H     H  H
      |     |  |
c. H—C—O—C—C—H
      |     |  |
      H     |  H
          H—C—H
            |
            H
```

7. For each of these Lewis diagrams, write the condensed structure.

a. 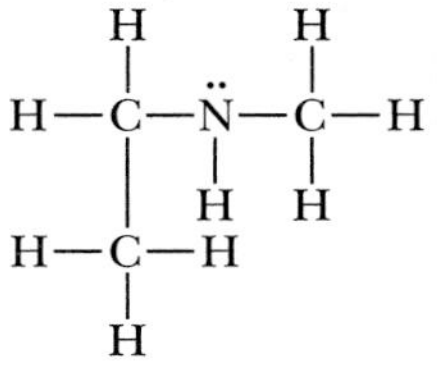 H—C—N̈—C—H (with H above and H below each C; H—C—H with H below, under the first C)

b. 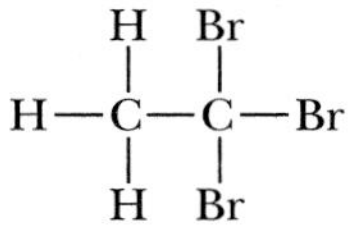 H—C—C—Br (H and H on the first C, Br and Br on the second C)

c. HO—C—C—C—C—C—C—OH (H above and H below each C)

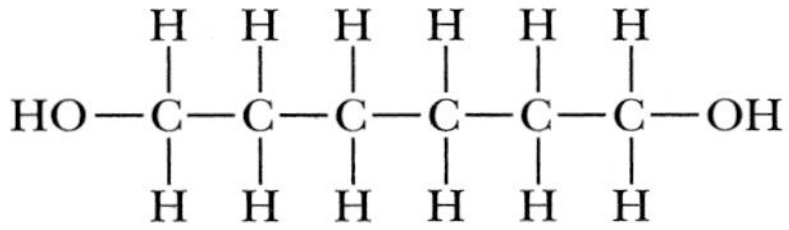

8. For each of these Lewis diagrams, write the condensed structure.

a. 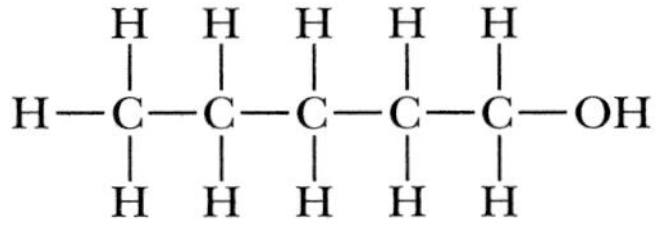 H—C—C—C—C—C—OH (H above and H below each C)

b. 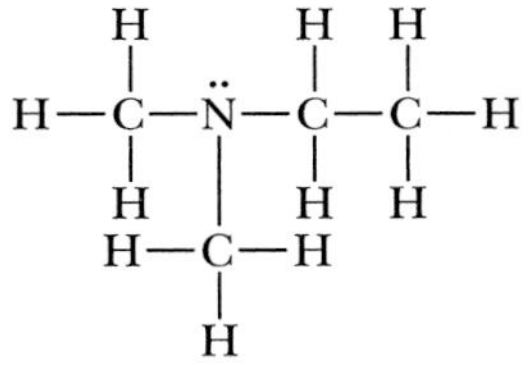 H—C—N̈—C—C—H (H on each C; H—C—H with H below, under N)

c. H—C—C—Ö—C—C—C—H (H on each C; H—C—H with H below, under the third C)

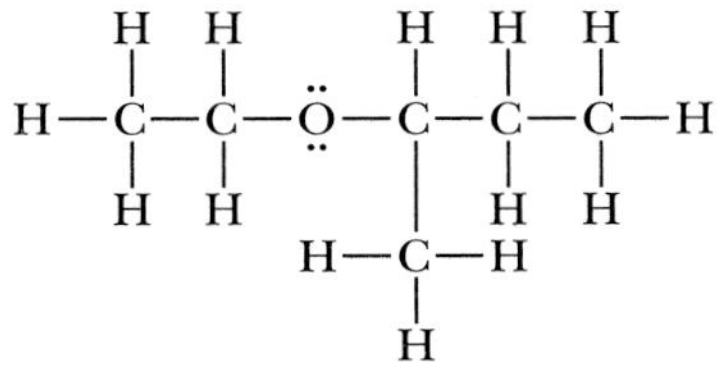

9. Draw a Lewis diagram, showing all nonbonding electrons, for each of these compounds.

a.  (ring containing O)   b. 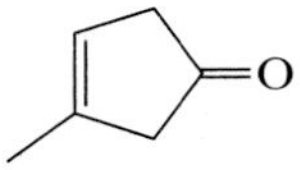 =O   c. 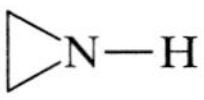 N—H

10. Draw a Lewis diagram, showing all nonbonding electrons, for each of these compounds.

a. 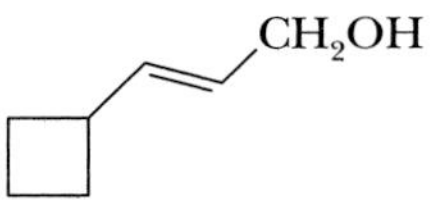 $CH_2OH$   b. 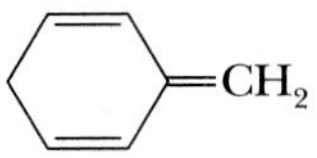 $=CH_2$

c. 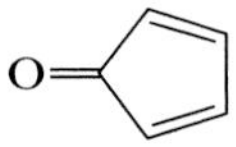 O=

11. For each of these compounds, draw a Lewis diagram, with all nonbonding electrons shown.

a. 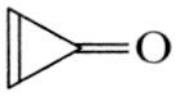 =O   b. 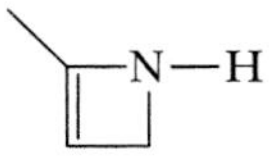 N—H   c. 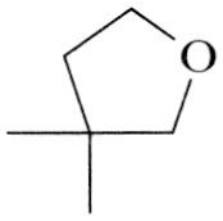 O

12. For each of these compounds, draw a Lewis diagram, with all nonbonding electrons shown.

a. 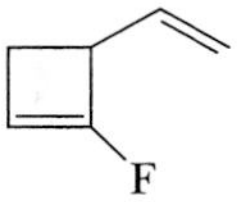 F   b. 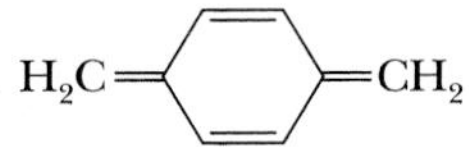 $H_2C=$ $=CH_2$

c. 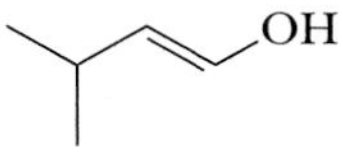 OH

13. What is the molecular formula of each of these compounds?

a. 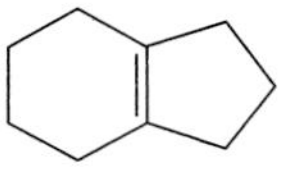   b. 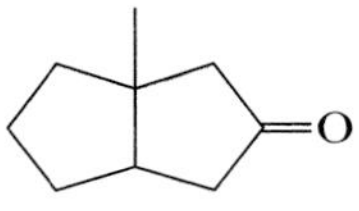 =O

c. 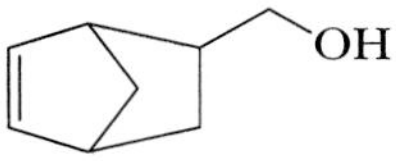 OH

14. What is the molecular formula of each of these compounds?

a. 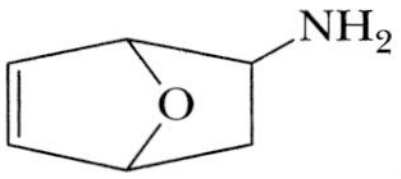 $NH_2$, O   b. 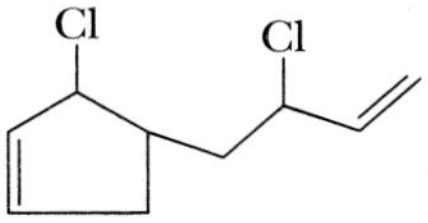 Cl, Cl

c.

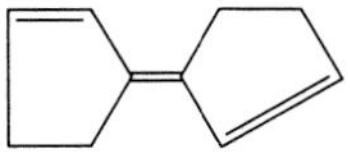

15. Give the molecular formula for each of these compounds.

a. 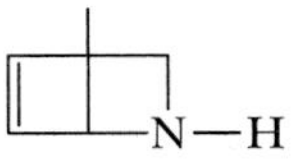 N—H   b. 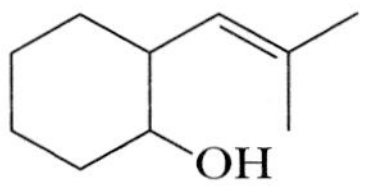 OH

c. 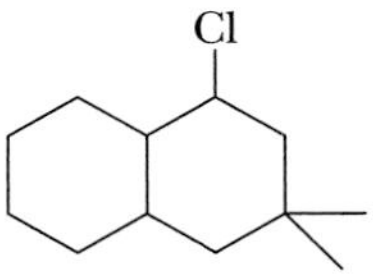 Cl

16. Give the molecular formula for each of these compounds.

a. 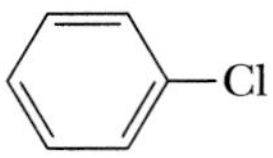 Cl   b. 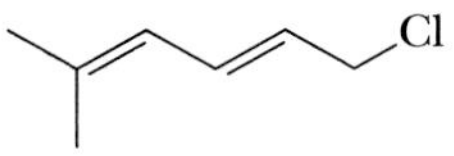 Cl

c.

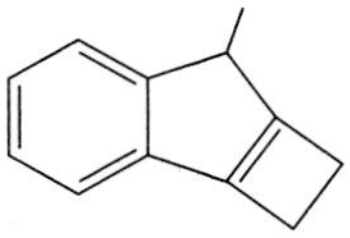

## SECTION 18.2 Organic and Inorganic Compounds

17. Which of these substances are organic compounds? Explain your answer in each case.
   a. Diatomaceous earth ($SiO_2$), from skeletons of microscopic animals called *diatoms*
   b. Quartz sand ($SiO_2$) eroded from volcanic deposits
   c. Ethylene ($H_2C{=}CH_2$) from natural gas

18. Which of these substances are organic compounds? Explain your answer in each case.
   a. Silica ($SiO_2$) isolated from a meteorite
   b. Atropine ($C_{17}H_{23}NO_3$), a poison isolated from the deadly nightshade plant
   c. Ethylene ($H_2C{=}CH_2$) from a tomato

19. State whether each of these substances is organic. Explain your answers.
   a. The straw in a basket of the Anasazi people (circa 1200 AD)
   b. Cocaine ($C_{12}H_{21}NO_4$)
   c. Spring water

20. State whether each of these substances is organic. Explain your answers.
   a. A diamond
   b. Polystyrene plastic (empirical formula $C_8H_8$)
   c. Amber (fossilized tree sap)

21. Prepare a table similar to Table 18.2 that contains these entries.
   a. Formaldehyde, prepared from the reaction between carbon monoxide and hydrogen ($H_2$)
   b. Pure carbon, prepared from the incomplete combustion of acetylene
   c. Pure carbon, prepared from the incomplete combustion of wood

22. Prepare a table similar to Table 18.2 that contains these entries.
   a. Sodium chloride isolated from bovine blood
   b. Vitamin C isolated from lemons
   c. Calcium oxide prepared by heating oyster shells

## SECTION 18.4 Alcohols and Ethers

23. Which of these compounds is an ether?
   A. $(CH_3CH_2)_2O$  B. $(CH_3CO)_2O$
   C. $CH_3CH_2OH$  D. $CH_3CO_2H$

24. Which of the compounds in Problem 23 is an alcohol?

25. Which of these compounds is an alcohol?

```
      H  O                  H  O
      |  ||                 |  ||
   H—C—C                 H—C—C
A.    |  |            B.    |  |
   H—C—C—H               H—C—O
      |  |                  |
      H  H                  H

       H  H                    H
       |  |                    |
    H—C—C—H             H3C—C—O
C.     |  |            D.      |  |
  H3C—C—C—H               H—C—C—H
       |  |                    |  |
       H  OH                   H  H
```

26. Which of the compounds in Problem 25 is an ether?

27. Draw the Lewis diagrams of these alcohols.
   a. An alcohol that contains two carbon atoms
   b. Two alcohols having the molecular formula $C_3H_8O$

28. Draw the Lewis diagrams of these alcohols.
   a. An alcohol that contains four carbon atoms and no double bonds
   b. A different alcohol that contains four carbon atoms and no double bonds.

29. Draw the Lewis diagrams of these ethers.
   a. An ether that contains four carbon atoms and no double bonds
   b. A different ether that contains four carbon atoms and no double bonds

30. Draw the Lewis diagrams of these compounds.
   a. An ether that contains a ring
   b. An alcohol that contains a ring

31. Which of these compounds is not identical to the other three?

```
      H  OH H  H  H
      |  |  |  |  |
A. H—C—C—C—C—C—H
      |  |  |  |  |
      H  H  H  H  H

      H  H  H  OH H
      |  |  |  |  |
B. H—C—C—C—C—C—H
      |  |  |  |  |
      H  H  H  H  H

      H  H  H  H  H
      |  |  |  |  |
C. H—C—C—C—C—C—H
      |  |  |  |  |
      H  OH H  H  H

      H  H  OH H  H
      |  |  |  |  |
D. H—C—C—C—C—C—H
      |  |  |  |  |
      H  H  H  H  H
```

32. Which of these compounds is not identical to the other three?

```
      H   H   H   H
      |   |   |   |
A. H—C—C—C—C—Cl
      |   |   |   |
      H   H   H   H

      H   H   H   H
      |   |   |   |
B. H—C—C—C—C—H
      |   |   |   |
      Cl  H   H   H

      H   H   H   H
      |   |   |   |
C. Cl—C—C—C—C—H
      |   |   |   |
      H   H   H   H

      H   H   H   H
      |   |   |   |
D. H—C—C—C—C—H
      |   |   |   |
      H   H   Cl  H
```

33. A chemist believed that the molecular formula of methane was $C_4H_8$. Draw a Lewis diagram of a compound having this formula in which each carbon atom forms four single covalent bonds and each hydrogen atom forms one covalent bond.

34. Berzelius thought that the molecular formula of acetic acid was $C_4H_6O_3$. Draw a Lewis diagram of a compound having this formula in which all bonds are covalent and each carbon atom forms four bonds, each oxygen atom forms two bonds, and each hydrogen atom forms one bond.

## EXPERIMENTAL PROBLEM 18.1

Once again A. L. Chemist and colleagues were talking about Berzelius and his theories. Suppose, one of them asked, a chemist had the mistaken impression that the molar mass of silver is 216 g mol$^{-1}$ (see Experimental Exercise 18.1). Suppose the chemist isolated an unknown carboxylic acid and prepared its silver salt, whose formula is $C_xH_yO_2Ag$ (where $x$ and $y$ are unknown) and whose molar mass is known to be 363 g mol$^{-1}$.

**QUESTIONS**

**35.** What are the values of $x$ and $y$ in $C_xH_yO_2Ag$, if 216 g mol$^{-1}$ is the molar mass of silver?

**36.** Suppose the chemist burned 363 mg of the silver salt and isolated 385.4 mL of $CO_2(g)$ at STP. Does this experimental result support your answer to Question 1?

**SECTION 18.5 Aldehydes and Ketones**

37. Which of these compounds is an aldehyde?

```
      H      H                  H
      |      |                  |
A. H—C—O—C—H         B. H—C—C=O
      |      |                  |   |
      H      H                  H   H

      O   H                     H   O—H
      ‖   |                     |   |
C. H—O—C—C—H         D. H—C—C—H
          |                     |   |
          H                     H   H
```

38. Which of these compounds is a ketone?

```
         O
         ‖
         C                      H
        / \                     |
A. H—C—C—H           B. H—C—C=O
      |   |                     |   |
      H   H                     H   O—H

        O—H
         |
         C—H                    H       H
        / \                     |       |
C. H—C—C—H           D. H—C—O—C—H
      |   |                     |       |
      H   H                     H       H
```

39. Draw the Lewis diagram of each of these aldehydes.
    a. An aldehyde that contains three carbon atoms and no carbon-carbon double bond.
    b. An aldehyde that contains three carbon atoms and one carbon-carbon double bond.

40. Draw the Lewis diagram of each of these ketones.
    a. A ketone that contains four carbon atoms and no carbon-carbon double bonds.
    b. A ketone containing four carbon atoms that is different from the answer to (a).

41. From the condensed structure, identify each of these compounds as an alcohol, ether, aldehyde, or ketone.
    a. $CH_3CHO$
    b. $CH_3OCH_2CH_2CH_3$
    c. $CH_3(CH_2)_2CH_2OH$

42. From the condensed structure, identify each of these compounds as an alcohol, ether, aldehyde, or ketone.
    a. $(CH_3CH_2)_2O$
    b. $CH_3COCH_2CH_3$
    c. $(CH_3)_2CHOCH(CH_3)_2$

43. State whether each of these compounds is an alcohol, ether, aldehyde, or ketone.
    a. $CH_3CH_2CHOHCH_2CH_3$
    b. $CH_3CH_2CHO$
    c. $(CH_3)_2CHCOCH(CH_3)_2$

44. State whether each of these compounds is an alcohol, ether, aldehyde, or ketone.
    a. $(CH_3)_2CHCOCH_2CH_3$
    b. $CH_3CHOHCH_2CH_3$
    c. $CH_3CH_2CH_2CHO$

## SECTION 18.6 Carboxylic Acids

45. Identify each of the functional groups present in this molecule.

```
     H      H  O  H  H  O
     |      |  ‖  |  |  ‖
  H—C—O—C—C—C—C—C—O—H
     |      |     |  |
     H      H     H  H
```

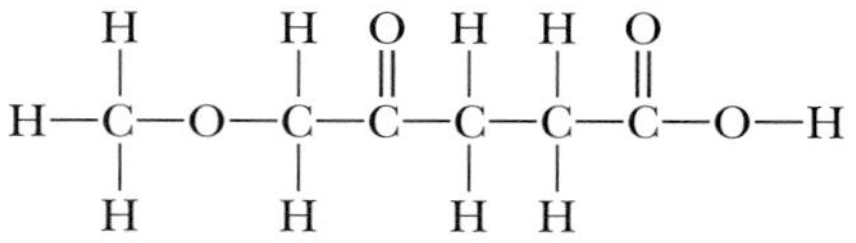

46. Identify each of the functional groups present in this molecule.

```
        O  H  H  H  H  H  H  O
        ‖  |  |  |  |  |  |  ‖
  H—O—C—C—C—C—C—C—C—C—H
           |  |  |  |  |  |
           H  H  H  H  H  O—H
```

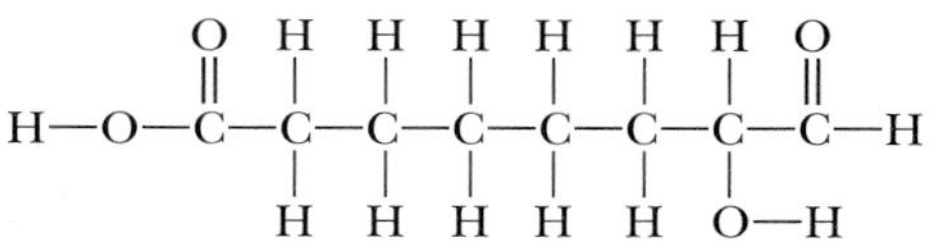

47. Draw a Lewis diagram for each of these compounds.
    a. $CH_3CH_2CO_2H$
    b. $CH_3CH_2CO_2^-\ K^+$

48. Draw a Lewis diagram for each of these compounds.
    a. $CH_3CO_2H$
    b. $CH_3CO_2^-\ Na^+$

49. What is the structural difference between alcohols and carboxylic acids?

50. What is the structural difference between aldehydes and ketones?

## SECTION 18.7 Esters and Amides

51. Which of these represents an amide? (Hint: The one you can draw a Lewis diagram for is the answer.)

    A. $CH_3\overset{O}{\overset{\|}{C}}NCH_3$ C. $CH_3\overset{O}{\overset{\|}{C}}NH_2CH_3$

    B. $CH_3\overset{O}{\overset{\|}{C}}NHCH_3$ D. $CH_3\overset{O}{\overset{\|}{C}}NH_3CH_3$

52. Which of these represents an ester? (Hint: The one you can draw a Lewis diagram for is the answer.)

    A. $CH_3\overset{O}{\overset{\|}{C}}OCH_3$ C. $CH_3\overset{O}{\overset{\|}{C}}OH_2CH_3$

    B. $CH_3\overset{O}{\overset{\|}{C}}OHCH_3$ D. $CH_3\overset{O}{\overset{\|}{C}}OH_3CH_3$

53. Which of these compounds is an ester?

    A. 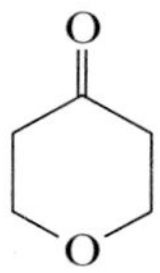 (six-membered ring with C=O and ring O) B. (cyclohexanone with O—$CH_3$)

    C. (six-membered ring with C=O and ring O) D. 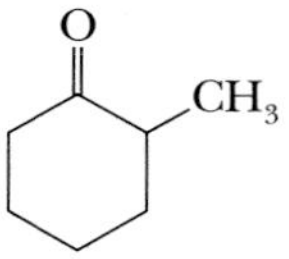 (cyclohexanone with $CH_3$)

54. Which of these compounds is an amide?

    A. 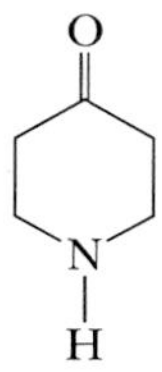 (six-membered ring with C=O and ring N—H) B. (cyclohexanone with $NH_2$)

    C. (cyclohexane with =NH) D. 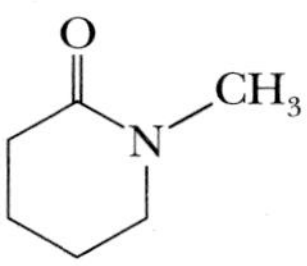 (six-membered ring with C=O and ring N—$CH_3$)

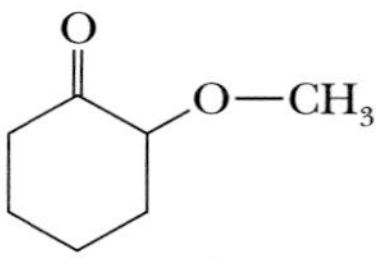

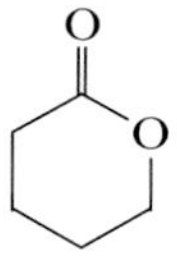

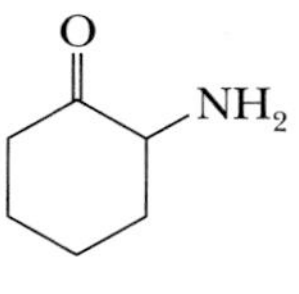

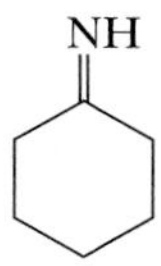

55. Draw a Lewis diagram for each of these compounds.
   a. $CH_3CH_2CO_2CH_2CH_3$
   b. $CH_3CONH_2$
   c. $CH_3O_2CCH_3$

56. Draw a Lewis diagram for each of these compounds.
   a. $(CH_3)_2CHCOOCH_3$
   b. $(CH_3)_2CHCON(CH_3)_2$
   c. $CH_3O_2CCH_2CH_3$

57. What is the structural difference between esters and carboxylic acids?

58. What is the structural difference between esters and ethers?

## SECTION 18.8 Amines

59. Two of these molecules contain amine groups. Identify one of them.

A. 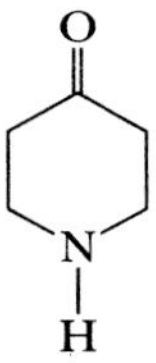 (six-membered ring with C=O and N–H)
B. (cyclohexanone with $NH_2$)
C. (cyclohexane with =NH)
D. 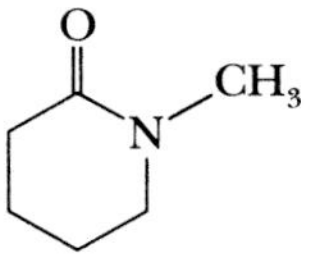 (six-membered ring with C=O and N–$CH_3$)

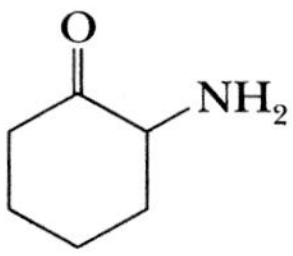

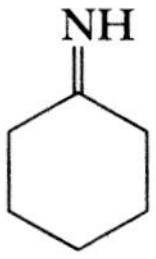

60. Identify the other molecule containing an amine group in Problem 59.

61. There are two different amines that contain two carbon atoms. Draw the Lewis diagrams of both of them.

62. What is the major difference between amines and amides?

## SECTION 18.9 Hydrocarbons

63. Which of these compounds is an alkene?

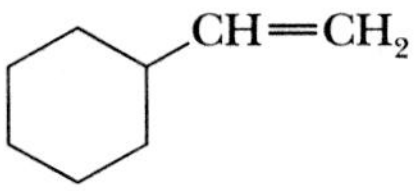

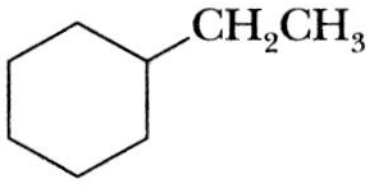

A. 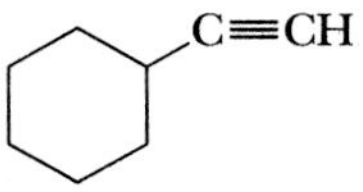 (cyclohexane–C≡CH)
B. (cyclohexane–$CH{=}CH_2$)
C. (cyclohexane–$CH_2CH_3$)
D. 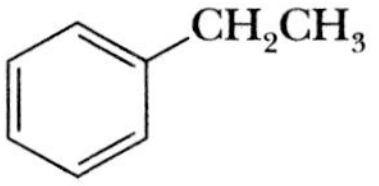 (benzene–$CH_2CH_3$)

64. Of the compounds in Problem 63, which is an alkane?

65. Of the compounds in Problem 63, which is an alkyne?

66. Of the compounds in Problem 63, which is an aromatic hydrocarbon?

67. Draw the Lewis diagram of compound B in Problem 63.

68. Draw the Lewis diagram of compound D in Problem 63.

69. What is the basic structural difference between an alkane and an alkene?

70. What is the basic structural difference between an alkane and an alkyne?

71. What is the basic structural difference between an alkene and an alkyne?

72. What is the basic structural difference between an alkene and an aromatic hydrocarbon?

## SECTION 18.10 Nucleophilic Substitution

73. Identify the nucleophile, the substrate, and the leaving group in this reaction.

$$C_6H_5{-}CH_2CH_2Cl + NH_3 \longrightarrow Cl^- + C_6H_5{-}CH_2CH_2\overset{+}{N}H_3$$

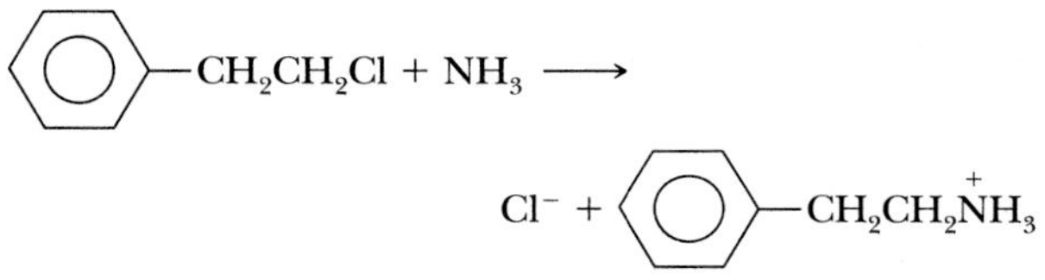

74. Identify the nucleophile, the substrate, and the leaving group in this reaction.

$$CH_3CH_2I + CH_3NH_2 \longrightarrow I^- + [CH_3CH_2NH_2CH_3]^+$$

75. For this reaction, identify the nucleophile, the substrate, and the leaving group.

$$[CH_3OH_2]^+ + Br^- \longrightarrow CH_3Br + H_2O$$

76. For this reaction, identify the nucleophile, the substrate, and the leaving group.

$$C_6H_5{-}CH_2CH_2\overset{+}{O}H_2 + Cl^- \longrightarrow H_2O + C_6H_5{-}CH_2CH_2Cl$$

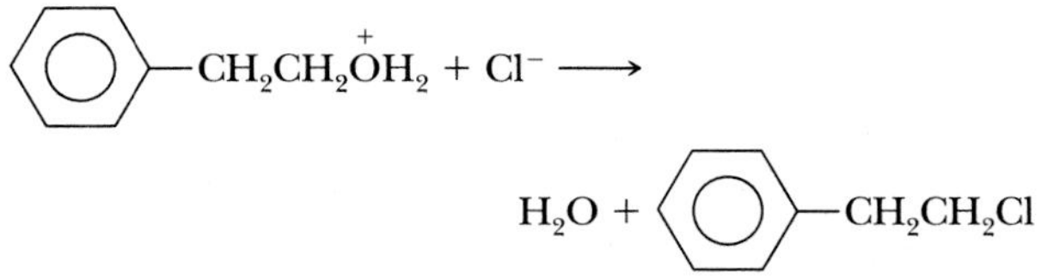

## ADDITIONAL PROBLEMS

77. Which of the functional groups discussed in this chapter is most likely to be found in the molecules that make up each of these substances?
   a. Bacon (the lean part)
   b. The compounds that give a banana its smell
   c. Natural gas

78. Which of the functional groups discussed in this chapter is most likely to be found in the molecules of each of these substances?
   a. Whiskey (the active ingredient)
   b. Cholestanone (a steroid)
   c. 2,3-dimethyldecanal

79. Which of the functional groups covered in this chapter is most likely to be part of the molecules of each of these substances?
   a. Bacon (the fat part)
   b. Cyclodecene
   c. Home heating oil

80. Which of the functional groups covered in this chapter is most likely to be part of the molecules of each of these substances?
   a. The compounds that give a fish its smell
   b. An organic compound that forms an acidic solution in water
   c. 2,3-dimethyl-4-octanone

## EXPERIMENTAL PROBLEM

### 18.2

A. L. Chemist and a team of colleagues studied this reaction between acetic acid and ethanol to form an ester:

$$CH_3\overset{\overset{\displaystyle O}{\|}}{C}{-}OH + CH_3CH_2OH \longrightarrow CH_3\overset{\overset{\displaystyle O}{\|}}{C}{-}OCH_2CH_3 + H_2O$$

They considered two possible series of steps (pathways) by which the reaction might occur. In the first pathway the oxygen atom in the OH group in acetic acid makes a bond to the $CH_3CH_2$ group in the final product:

$$CH_3\overset{\overset{\displaystyle O}{\|}}{C}{-}\boxed{O}{-}H + CH_3CH_2OH \longrightarrow CH_3\overset{\overset{\displaystyle O}{\|}}{C}{-}\boxed{O}{-}CH_2CH_3 + H_2O$$

In the second pathway the oxygen in the OH group in ethanol makes the bond in the final product.

$$CH_3\overset{\overset{\displaystyle O}{\|}}{C}OH + CH_3CH_2{-}\boxed{O}{-}H \longrightarrow CH_3\overset{\overset{\displaystyle O}{\|}}{C}{-}\boxed{O}{-}CH_2CH_3 + H_2O$$

In order to determine which pathway is correct, the team ran the reaction using ethanol that contained the isotope oxygen-18 instead of ordinary oxygen (that is, they used $CH_3CH_2{-}^{18}O{-}H$).

**QUESTIONS**

**81.** The reaction being studied has a relatively small equilibrium constant. As a consultant to the team, what suggestion would you offer to drive the equilibrium to the right?

82. Which of the products contains the $^{18}O$ isotope if the first pathway is correct?

**83.** Which of the products contains the $^{18}O$ isotope if the second pathway is correct?

## SOLUTIONS TO EXERCISES

### Exercise 18.1

**a.** The carbon of the first $CH_3$ group already has three bonds to hydrogens, so it can be bonded to only one other atom, the oxygen that follows it in the condensed structure. After you draw that, you come to the next carbon with one bond to a hydrogen and two bonds to $CH_3$ groups, which finishes the structure. The oxygen atom has two pairs of nonbonding electrons.

```
    H       H   H
    |   ..  |   |
H — C — O — C — C — H
    |   ..  |   |
    H       |   H
        H — C — H
            |
            H
```

**b.** First comes a hydrogen, then an oxygen, then a $CH_2$ group, then a carbon. This last carbon has three $CH_3$ groups attached to it. The structure looks like this:

```
            H
            |
        H — C — H
        H   |   H
    ..  |   |   |
H — O — C — C — C — H
    ..  |   |   |
        H   |   H
        H — C — H
            |
            H
```

### Exercise 18.2

**a.** The rules are the usual ones, and the Lewis diagram for compound C is

```
        H
        |
    H — C — H
        |   ..
H — C — C — N — H
    ||  |   |
H — C — C   H
        |\
        H H
```

**b.** The molecular formula of compound D is $C_6H_{10}O_2$.

### Exercise 18.3

Methane contains only carbon and hydrogen, and this sample of methane also originated from living (or once-living) tissues by a biological process. It is organic by any definition. The hydrogen sulfide gas also originated from living tissues and is organic by that definition. However, it contains no carbon, so it is an inorganic compound to a chemist.

### Exercise 18.4

Menthol (compound D) has a name that ends in the letters *ol*, and therefore you can predict once again that it is the alcohol.

### Exercise 18.5

**a.** Compound D has an alkyl group attached to an OH group and is therefore an alcohol.

**b.** Acetophenone (compound A) is a ketone.

### Exercise 18.6

Compound C has an alkyl group attached to a $CO_2H$ group (written backward). This compound is therefore a carboxylic acid (it's acetic acid, in fact).

### Exercise 18.7

**a.** Compound D has a carbonyl group whose carbon is attached to a nitrogen atom, and it is therefore an amide.

**b.** There is only one carboxylic acid that has the molecular formula $C_3H_6O_2$. Here it is:

```
    H   H   O
    |   |   ||
H — C — C — C — O — H
    |   |
    H   H
```

### Exercise 18.8

The structures of these two amines are

```
    H   H                      H       H
    |   |                      |       |
H — C — C — N — H    and   H — C — N — C — H
    |   |   |                  |   |   |
    H   H   H                  H   H   H
    Ethylamine                 Dimethylamine
```

### Exercise 18.9

A hydrocarbon is defined as a compound consisting of only carbon and hydrogen. This means that either C or D is the answer, because compounds A and B both contain oxygen. Only compound D contains a benzene ring (see Figure 18.21), so it is the aromatic hydrocarbon.

### Exercise 18.10

It looks like compound B is the substrate. Compound A is the nucleophile, because it replaces the departing $I^-$ ion.

### Experimental Exercise 18.1

1. If 1 mol of silver oxide has a mass of 231.7 g and contains 1 mol of O, which has a mass of 16 g, then 1 mol of silver oxide contains 231.7 − 16.0 = 215.7 g of silver. If the formula of silver oxide is AgO, this 215.7 g of silver is 1 mol of silver, which means that the atomic mass of silver is 215.7.

19

# BIOCHEMISTRY



## KEY TERMS

In the early 1800s chemists called the substances they obtained from living organisms *organic compounds*. Since then the field of organic chemistry has expanded to include vast areas of industrial and theoretical chemistry and has almost completely lost the early focus on living systems. A new field, **biochemistry**, has grown up to deal with the chemistry of life (see Figure 19.1).

In this chapter we first consider the relationships among the physical, biological, and social sciences. Then we will look at the major classes of biomolecules, and briefly consider the structures and some of the functions of these compounds in living organisms.

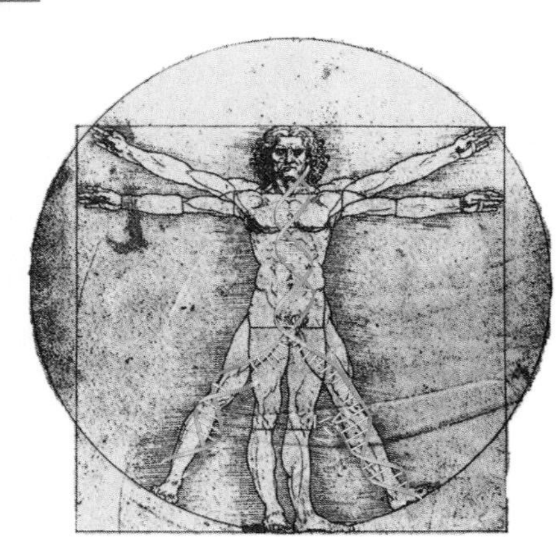

FIGURE 19.1
Biochemistry is the study of the chemistry of life.

## 19.1 From the Organism to the Molecule

GOAL: To outline the relationships among the scientific disciplines that study living organisms, especially human beings.

FIGURE 19.2
People are complex subjects for scientific study.

Humans have many aspects, and scientists and scholars are actively studying all of those aspects in one way or another (see Figure 19.2). Those who study the interactions of people in societies, for example, are *sociologists*. *Psychologists* focus more on the individual human being, on mind and motivation, but clearly the border between these two social sciences is somewhat fuzzy. We will not deal with the behavioral sciences in this chapter—though it is well known that some chemical substances can influence human behavior in a profound way.

Human beings have a physical structure, and its study falls in the realm of *biology*. Muscles and blood, skin and bones, brain and heart—people are made up of organs and tissues. Such subfields of biology as *anatomy* and *physiology* study what these systems are and how they work. (See Figure 19.3.)

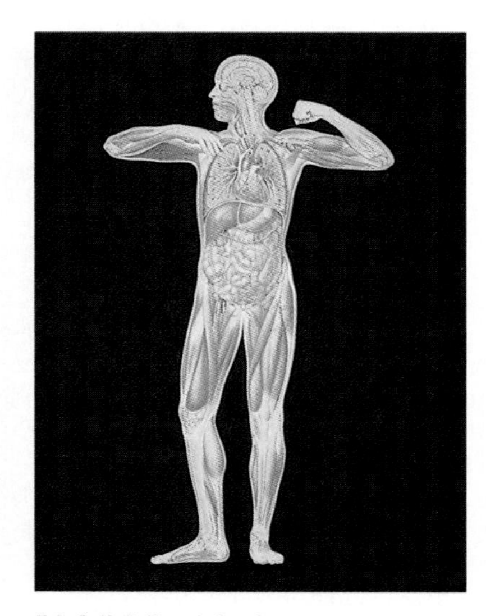

FIGURE 19.3
Biologists use models like this, showing muscles and organs, to teach their students about human anatomy.

By now you have probably realized that this section is examining the human organism on a progressively smaller scale. Organs are clearly visible with the naked eye. The next step down in scale from organs is tissues. You can see some details of muscle tissue, stomach lining, and bone structure with the unaided eye, but the view through a microscope is more informative about how tissues do their jobs. All organs and tissues are made up of cells, which always have fluids and sometimes have solid materials between them. Each cell, in turn, has a cell wall that contains structures within it. Study at the level of cells is the area called *cell biology*. Finally, scientists examine the molecules that make up cells and their substructures. This level is the focus of *biochemistry* or *molecular biology*.

Some human cells contain stores of fat. Biochemists know that an important function of fat is energy storage. Fats react with oxygen in the body to provide the motive power you need to live and work. This kind of reaction is part of the process called *metabolism*. Fats are, of course, chemical compounds, and they are members of a larger class called **lipids** (see Figure 19.4). We will consider lipids in more detail in Section 19.2. Not all lipids are energy-storing molecules. Cell walls consist largely of lipids, for example, and vegetable oils are also lipids.

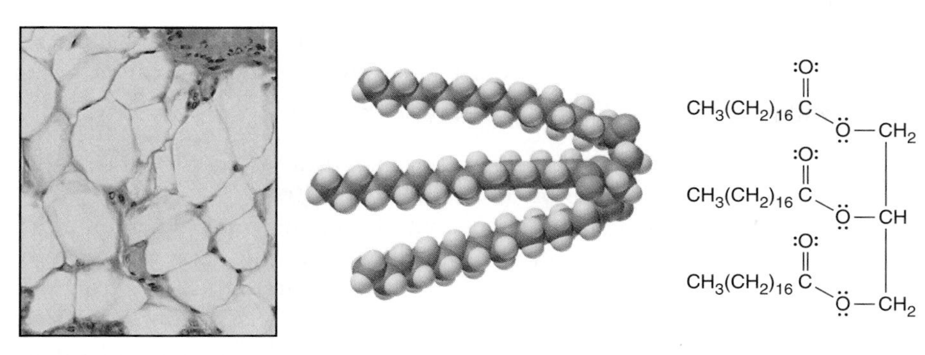


FIGURE 19.4
Lipids. (Left) Cells contain stored fat, a source of energy for the body. (Center) Tristearin, a typical fat molecule. (Right) Lewis diagram of tristearin.

Some human and animal tissues also contain a substance called *glycogen.* Molecules of glycogen are made up of linked sugar molecules. Plant tissues contain starch, which is also made of linked sugar molecules. These compounds are energy suppliers, like fats. Sugars and their polymers are members of the class of compounds called **carbohydrates,** which we will consider in Section 19.3. Again, not all carbohydrates are energy-storing molecules. Both cartilage in people and animals and cellulose in plants provide structural stiffness. Both of these substances consist of carbohydrates.

**Polymers are long chain molecules consisting of a basic structural unit repeated many times (hundreds to many thousands of times depending on the polymer and how it was prepared). For example, polyvinyl chloride, known as PVC, has a basic structural unit of —$CH_2CHCl$— and a formula of —$(CH_2CHCl)_n$—.**

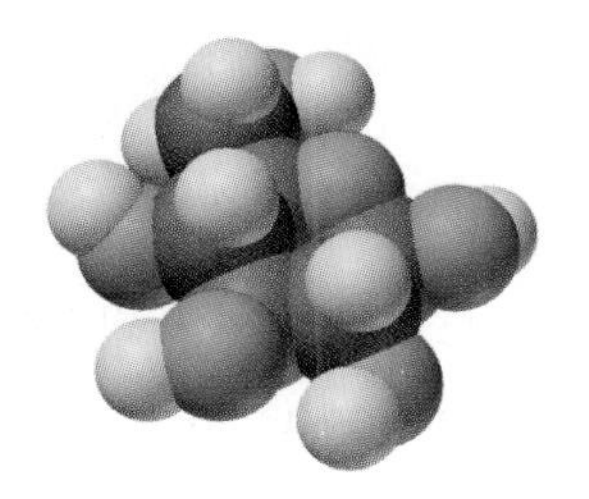

FIGURE 19.5
Carbohydrates. (Left) These plant fibers contain cellulose, a compound that provides rigidity to wood and other plant tissue. (Center) Many of these single glucose molecules link together to form cellulose. (Right) Lewis diagram of glucose.

All animal cells also contain compounds called **proteins.** These substances perform a remarkable number of functions within human cells (see Figure 19.6). In muscle cells protein molecules contract and relax, and as a result you move (or don't). Other proteins, called *enzymes,* catalyze the reactions that comprise the body's metabolism. Like fats and carbohydrates, proteins can also serve as an important source of metabolic energy. Like glyco-

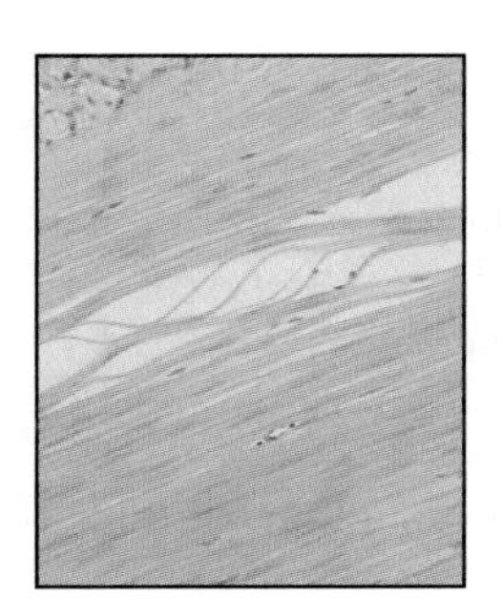

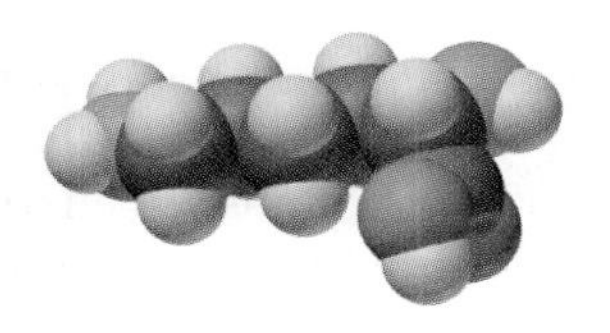

FIGURE 19.6
Proteins. (Left) This is muscle tissue, high in protein. (Center) A lysine molecule. This amino acid is present in most protein molecules. (Right) Lewis diagram of lysine.

gen and starch, proteins are *polymers,* long molecules made from simple repeating units. The repeating units of proteins are compounds called *amino acids.* Human enzyme systems are powerful, but human bodies cannot make all of the amino acids they need. People must get some amino acids in the foods they eat, and these are called the *essential amino acids.* We will examine amino acids and proteins in more detail in Section 19.4.

Large polymer molecules can be made by linking many smaller units. Chemists make nylon, for example, by reacting carboxylic acids with amines. Your body makes proteins by a similar reaction. (Despite this, the structures of proteins and nylon are different in important ways; see Section 19.4.) Your body makes glycogen, and potato plants make starch, by linking sugar molecules with one another.

The nucleus is a key structure within each cell. The nucleus governs the inheritance of traits when the cell divides, among other things. Human cells contain 48 chromosomes, and each chromosome contains DNA, the molecule that governs inheritance. DNA molecules, like proteins and starch molecules, are polymers. We will consider the nucleic acids DNA and RNA in Section 19.5 (see Figure 19.7).

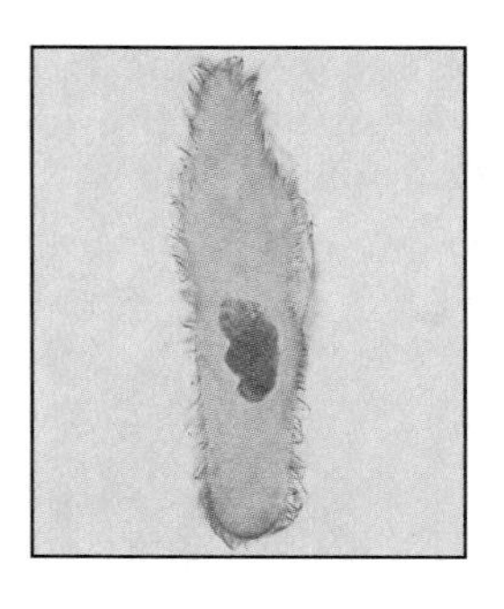

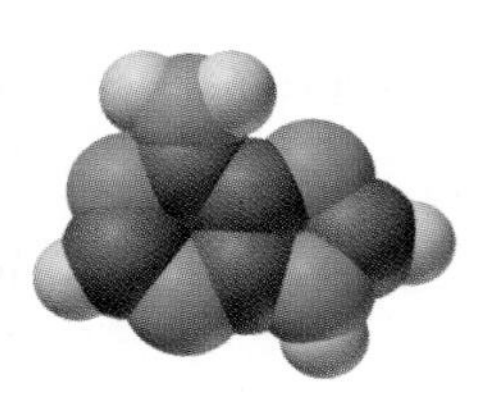

FIGURE 19.7
Nucleic acids. (Left) The nucleus of this paramecium consists largely of nucleic acids. (Middle) Adenine is one of several building blocks that make up DNA (for more detail, see Section 19.5). (Right) Lewis diagram of adenine.

Finally, other substances regulate the speeds of biological processes and transfer information from one part of the body to another, or even from one organism to another. These substances include hormones, vitamins, and minerals, which may or may not be members of the four major classes mentioned earlier in this section. We will cover a few of these substances in Section 19.6.

## EXAMPLE 19.1 Important Classes of Biochemical Substances

Four major classes of biochemical compounds are lipids, carbohydrates, proteins, and nucleic acids.

**a.** In which of these classes of compounds does table sugar (sucrose) belong?
**b.** Of which of these classes of compounds are enzymes members?
**c.** Which of these classes is *not* an important source of energy in metabolism?
**d.** For which of these classes of compounds is pasta (spaghetti, etc.) important as a dietary source?

**SOLUTIONS**

**a.** Carbohydrates
**b.** Proteins
**c.** Nucleic acids
**d.** Carbohydrates

**EXERCISE 19.1**

Four major classes of biochemical compounds are lipids, carbohydrates, proteins, and nucleic acids.

**a.** For which of these classes of compounds is meat an important dietary source?
**b.** Of which of these classes of compounds is DNA a member?
**c.** Of which of these classes of compounds is starch a member?
**d.** Of which of these classes of compounds is RNA a member?

## 19.2 Lipids: Fats and Oils

GOAL: To identify the structural features that characterize lipids, including triglycerides and cholesterol.

**Fundamentally, lipids are biomolecules that can be extracted from cells by organic solvents. There are many kinds of lipids and all of them have biological importance. Fats and steroids are among the most abundant lipids in the body.**

People tend to think about fat in terms of its influence on their own appearance and that of others. In recent years, dietary fats of various kinds have been implicated as possible factors in heart disease. This section covers several of the structural features thought to be important in this respect.

The **triglycerides** are the fats found most often in the human diet. Here is the structure of a typical triglyceride:

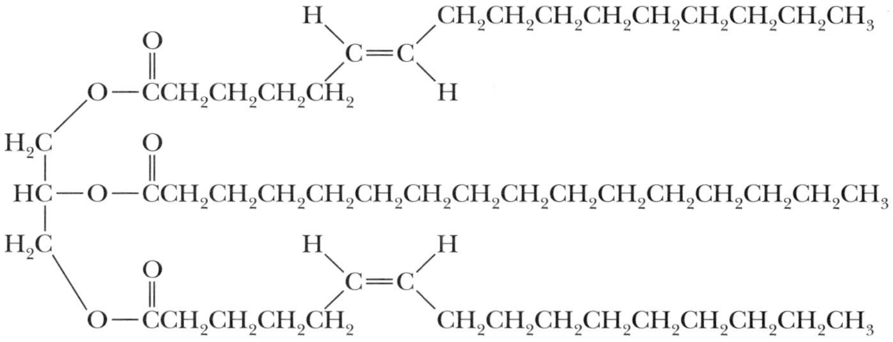

$$H_2C-O-\overset{O}{\overset{\|}{C}}CH_2CH_2CH_2CH_2-CH{=}CH-CH_2CH_2CH_2CH_2CH_2CH_2CH_2CH_2CH_3 \quad (\text{trans: H, H on opposite sides})$$

$$HC-O-\overset{O}{\overset{\|}{C}}CH_2CH_2CH_2CH_2CH_2CH_2CH_2CH_2CH_2CH_2CH_2CH_2CH_2CH_2CH_3$$

$$H_2C-O-\overset{O}{\overset{\|}{C}}CH_2CH_2CH_2CH_2-CH{=}CH-CH_2CH_2CH_2CH_2CH_2CH_2CH_2CH_2CH_3 \quad (\text{cis: H, H on same side})$$

Note that the molecule contains two different types of functional groups. The three ester groups are responsible for the prefix *tri* (which means "three") in the name *triglyceride.* There are also two carbon-carbon double bonds in this compound. In other words, alkene functional groups are also present. Triglycerides differ from one another in the lengths of the carbon chains, the positions of the double bonds, and the numbers of double bonds.

**The *glyceride* portion of the name *triglyceride* comes from *glycerol:***

$$\begin{array}{l} H_2COH \\ \;| \\ HCOH \quad \text{Glycerol} \\ \;| \\ H_2COH \end{array}$$

If a triglyceride molecule contains one or more carbon-carbon double bonds, the compound is *unsaturated.* If there are two or more such bonds, the compound can be called *polyunsaturated.* If there are no carbon-carbon double bonds at all, the compound is *saturated.* Unsaturated fats are thought to be more healthful as part of your diet than saturated ones.

Normally, fats come from animals and oils from plants, but both are triglycerides. Fats are simply more solid than oils at room temperature; that is, they have higher melting points. This difference is related to the degree of unsaturation. The more saturated a fat is, the higher its melting point is likely to be. Oils, because they are unsaturated, are likely to be healthier than fats.

Recently another factor has become a topic in the health debate. Fats and oils described as *trans* are apparently less healthful than ones described as *cis*. These labels are based on the arrangement of atoms around a carbon-carbon double bond. When two identical atoms or groups are on the same side of a line running along a double bond and the other points of attachment to the doubly bonded carbon atoms are different, the molecule is *cis*. When identical atoms or groups are on opposite sides, the molecule is *trans*. There are examples of *cis* and *trans* configurations in the triglyceride structure on the previous page and in Figure 19.8.

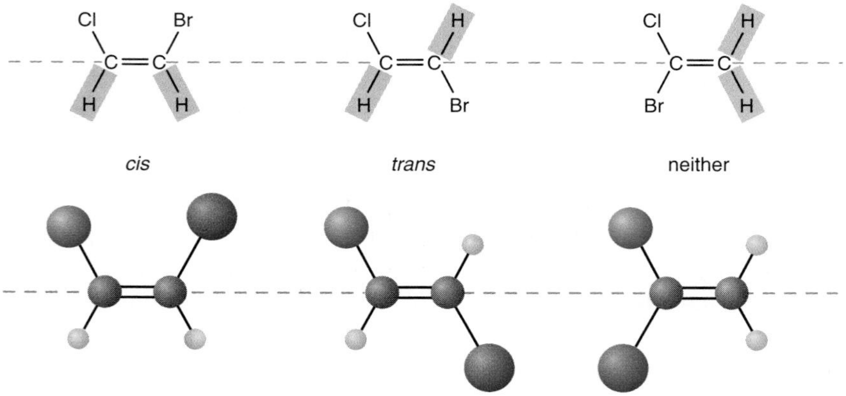

Cl Br / C=C / H H — *cis*

Cl H / C=C / H Br — *trans*

Cl H / C=C / Br H — neither

**FIGURE 19.8**
In the *cis* configuration at left, the two hydrogens are identical to each other and on the same side of a line connecting the doubly bonded carbons. In the center, the *trans* alkene has two hydrogens on opposite sides of the double bond. On the right, the alkene has two hydrogens on the same carbon and is neither *cis* nor *trans*.

**EXAMPLE 19.2** ***cis* and *trans* Alkenes**

Which of the following is a *trans* alkene? Explain your answer.

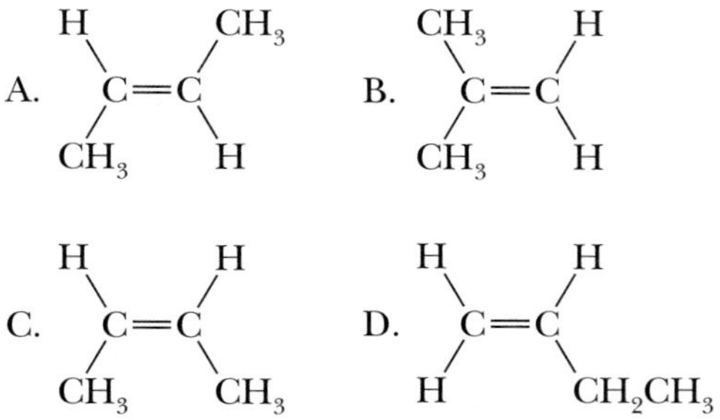

A. H, $CH_3$ (top); C=C; $CH_3$, H (bottom)

B. $CH_3$, H (top); C=C; $CH_3$, H (bottom)

C. H, H (top); C=C; $CH_3$, $CH_3$ (bottom)

D. H, H (top); C=C; H, $CH_2CH_3$ (bottom)

**SOLUTION**

Alkene A has two hydrogens (identical) on opposite sides of a line connecting the doubly bonded carbons. Therefore, it is *trans*. (It also has two identical $CH_3$ groups on opposite sides of the line connecting the doubly bonded carbons.) Alkenes B and D have two hydrogen atoms on the same carbon, so they are neither *cis* nor *trans*. Alkene C is *cis*.

**EXERCISE 19.2**

Which of the following alkenes is neither *cis* nor *trans*? Explain your answer.

A. H, $CH_3$ on left carbon; $CH_3$, H on right carbon: (H)($CH_3$)C=C($CH_3$)(H)

B. (CH$_3$)(CH$_3$)C=C(H)(CH$_3$)

C. (H)(CH$_3$)C=C(H)(CH$_3$)

D. (H)(CH$_3$)C=C(H)(CH$_2$CH$_3$)

Cholesterol is a famous chemical compound; as a source of serious health problems, it is the focus of continuing debate. As Figure 19.9 shows, it is a complex molecule. The key to dealing with such large structures is to focus on the parts of the molecule that are important functionally and ignore the rest. For example, the cholesterol molecule contains two functional groups. It is both an alcohol and an alkene. The rest of the molecule is much less reactive.

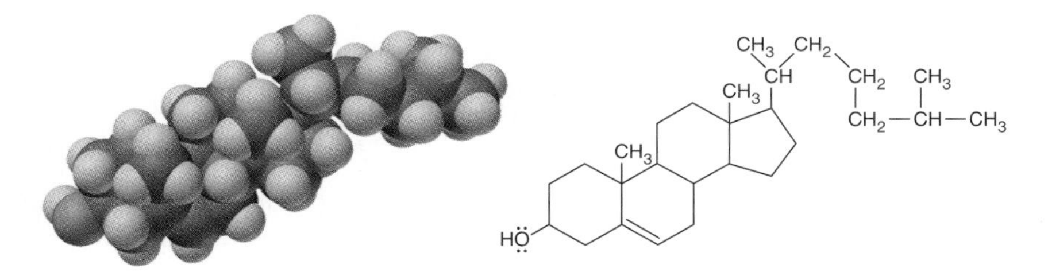


FIGURE 19.9
Cholesterol

Another important structural feature of cholesterol is its system of rings. The molecule contains four rings, arranged like this:

Any compound with this ring system is classed as a *steroid*. Cholesterol is an important starting molecule in biochemistry. The body uses it to make many other steroid molecules that are vital to body functions. Among these are cortisone, testosterone, and estradiol.

## EXAMPLE 19.3 Cholesterol and Steroids

**a.** The molecular structure of the male sex hormone testosterone is shown. How many carbon atoms does a molecule of testosterone contain?

**b.** What functional groups does this molecule contain?

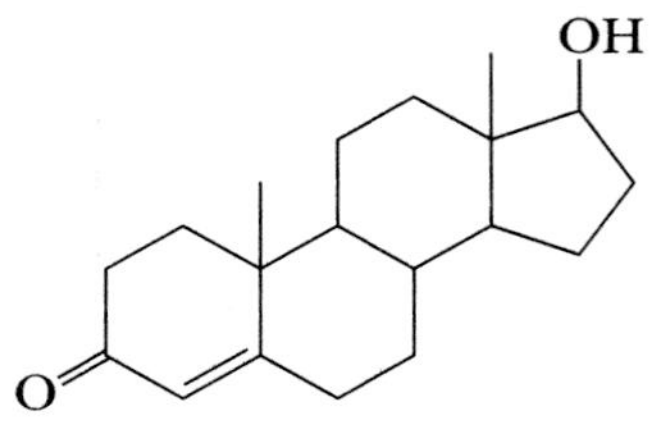

### SOLUTIONS

**a.** In structures of this kind, bends in lines, intersections of three or four lines, and the ends of single lines represent carbon atoms (see Section 18.1 for these rules). A testosterone molecule contains 19 carbon atoms, as shown in this Lewis diagram:

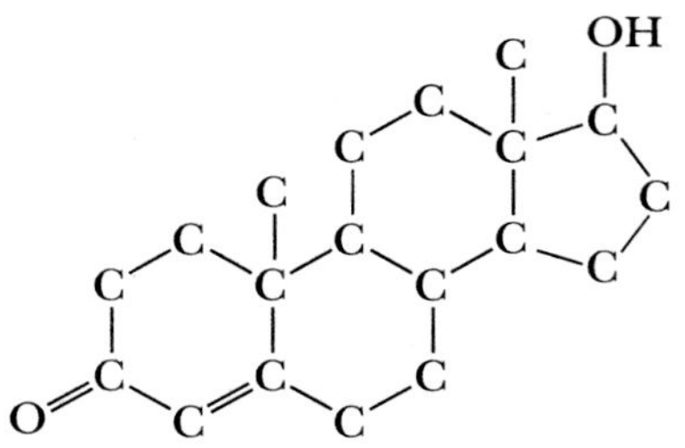

**b.** It has a carbonyl (C═O) group with two carbons attached to the carbon of that group, an OH group, and a carbon-carbon double bond. Thus it is a ketone, an alcohol, and an alkene.

### EXERCISE 19.3

**a.** The molecular structure of cortisone, an adrenal hormone, is shown below. What functional groups are present in this molecule?

**b.** How many carbon atoms does the molecule contain? How many hydrogen atoms? What is the molecular formula of cortisone?

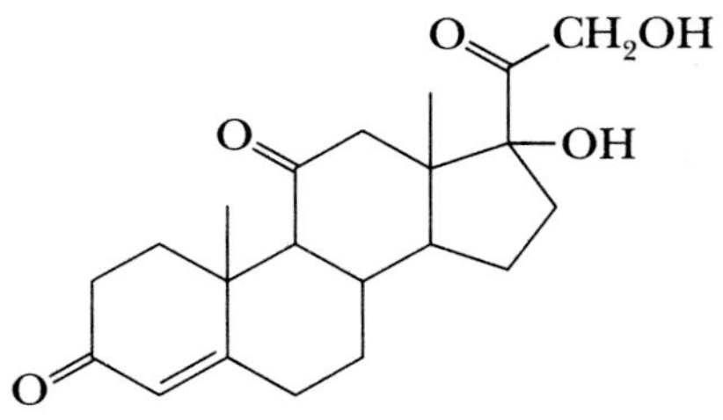

## CHAPTER BOX 19.1 SOAP

**Heating fats** and oils with wood ashes and water produces soap, along with a compound called glycerol. Human beings have been doing this (in some societies at least) for thousands of years. You may remember the long-lived French chemist Michel Eugène Chevreul from Chapter 18; he studied this reaction. The reaction between a triglyceride and a base gives the salt of a carboxylic acid (that is, a soap) and glycerol. Isolating this soap and adding a strong acid to it yields long-chain carboxylic acids called *fatty acids.* Chevreul made fine candles from such acids. The reactions look like this:

$$\begin{array}{l} H_2C{-}O{-}\overset{O}{\overset{\|}{C}}CH_2(CH_2)_{15}CH_3 \\ HC{-}O{-}\overset{O}{\overset{\|}{C}}CH_2(CH_2)_{15}CH_3 \\ H_2C{-}O{-}\overset{O}{\overset{\|}{C}}CH_2(CH_2)_{15}CH_3 \end{array} + 3\,NaOH \longrightarrow \underset{\text{Glycerol}}{\begin{array}{l} H_2C{-}OH \\ HC{-}OH \\ H_2C{-}OH \end{array}} + 3\,\underset{\text{Soap}}{NaO{-}\overset{O}{\overset{\|}{C}}CH_2(CH_2)_{15}CH_3}$$

$$\underset{\text{Soap}}{CH_3(CH_2)_{15}CH_2COONa} + H_3O^+ \longrightarrow \underset{\text{Fatty acid}}{CH_3(CH_2)_{15}CH_2COOH} + Na^+ + H_2O$$

Here's how soap works to clean things. Think of a soap molecule as containing a water-soluble salt bonded to a long, oil-soluble alkyl group. When you dissolve the soap in water, the ionic salt dissolves well, but the alkyl group does not. Large numbers of the soap molecules group together, with their water-soluble ends oriented outward and their oil-soluble tails making something like a droplet of oil. Oily and greasy dirt doesn't normally dissolve in water. Its molecules adhere to one another through van der Waals forces (see Section 13.2) and do not form hydrogen bonds with the water molecules. Oil and grease molecules can dissolve, however, in the tangle of alkyl groups in soapy water (see Figure below). "Like dissolves like," as you know, and the alkyl groups of soap molecules attract the grease molecules through van der Waals forces. Thus you can rinse them down the drain.

Suppose you are an old-time soap maker. What do you do with the leftover glycerol after you've made your soap and candle wax? This was a practical problem for some time. Then chemists found that glycerol reacts with nitric acid to make nitroglycerin. The Swedish chemist Alfred Nobel invented a practical explosive, dynamite, by mixing the unstable nitroglycerin with an inert filler. He used the fortune he made from his invention to fund the Nobel Prizes.

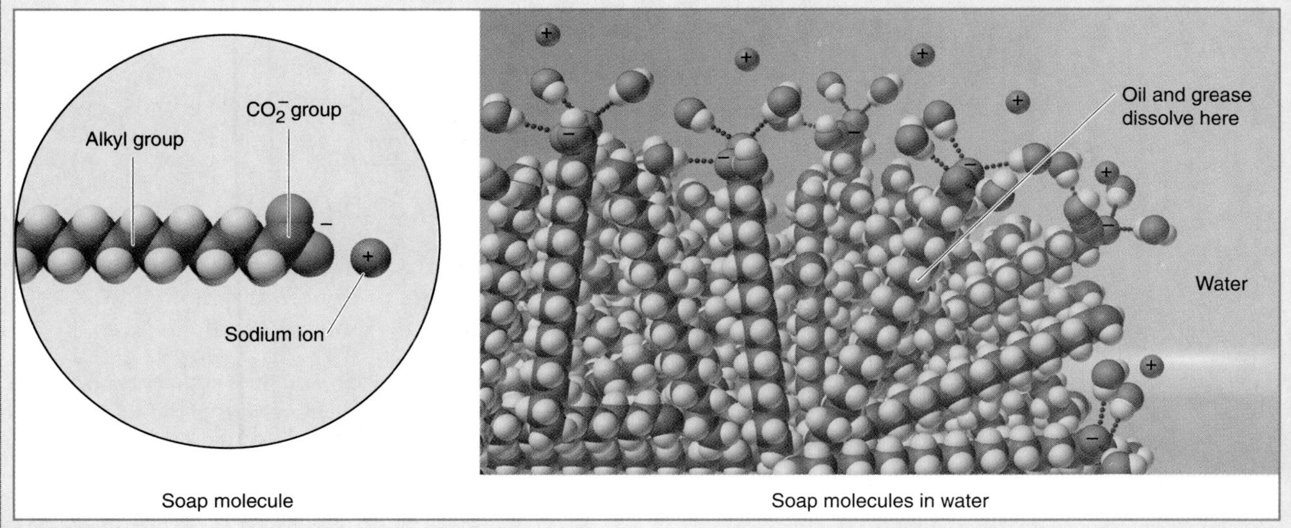


Soap in Action

FIGURE 19.10
Sugar cane

## 19.3 Carbohydrates: Sugars and Their Relatives

**GOAL:** To identify the main structural features of carbohydrate molecules.

There are many different kinds of sugar molecules, and chemists have been fascinated by them as long as there have been chemists. Gay-Lussac (whom you met in Section 5.7), along with his colleague Thénard, was the first to work out the composition of sugar. Another French chemist, Ansleme Payen, discovered that he could use activated charcoal to remove colored impurities when refining sugar cane to produce table sugar (see Figure 19.10). Activated charcoal is used today in aquarium filters, gas masks, and a multitude of other applications.

Sugar molecules contain carbon, hydrogen, and oxygen. Many sugar molecules contain one oxygen atom and two hydrogen atoms for each carbon atom. The empirical formula $CH_2O$ inspired the family name *carbohydrate*—from *carbo,* "carbon," and *hydrate,* "containing water." (Your instructor might or might not appreciate bad puns about C water.)

**For a review of empirical formulas, see Section 9.8.**

The simplest member of the carbohydrate family is called D-glyceraldehyde. The structure of this compound is shown four different ways in Figure 19.11. You can see from the figure that the molecular formula of D-glyceraldehyde is $C_3H_6O_3$, which gives it the empirical formula $CH_2O$. The molecule contains three functional groups: one carbonyl group with a hydrogen bonded to its carbon and two OH groups. Thus it is an aldehyde and an alcohol. The third structure in Figure 19.11 is a *Fischer projection.* It represents exactly the same molecule as the structure to its right.

**All simple sugars contain either an aldehyde or a ketone functional group and multiple OH groups.**

```
   O
   ‖
   C—H              O⸌ C ⸍H          O⸌ C ⸍H
   |                   |                |
H—C—O—H           H▻C◅OH          H—+—OH
   |                   |                |
H—C—O—H             CH₂OH            CH₂OH
   |
   H
```

FIGURE 19.11
At left is a conventional Lewis diagram of D-glyceraldehyde; it does not convey any three-dimensional information. The two structures in the center are ways of showing the tetrahedral arrangement of the atoms around the central carbon atom. The H and OH to the left and right of the central carbon atom are also closer to the viewer. (The triangles mimic artistic perspective; the point of the triangle seems to recede into the distance.) The top and bottom carbon atoms are farther from the viewer than the central carbon. At right is a three-dimensional (ball-and-stick, space-filling) model of the molecule.

## CHAPTER BOX 19.2 STEREOISOMERS

**Most of** the compounds you have studied so far in this book have had unique formulas: NaCl is sodium chloride; $CH_4$ is methane, $C_2H_6$ is ethane. A few have been isomers, compounds having the same formula but different arrangements of atoms. Figure A below shows two structures that are examples of isomers. These compounds are called 2-methylpentane and 3-methylpentane, and they differ only in the location of the methyl group on the five-carbon chain. Compounds like these are *structural,* or *constitutional, isomers,* which differ in the order of attachment of the atoms in the molecule.

The discussion of *cis* and *trans* fat molecules earlier in this chapter introduced isomers of another kind. In these isomers, called *stereoisomers,* the atoms are attached in exactly the same order. The only difference is the spatial arrangement of the atoms. Figure B shows *cis* and *trans* isomers named *cis*-2-butene and *trans*-2-butene.

Pairs of stereoisomers come in two varieties: those that are mirror images of each other and those that are not. You can see that the *cis* and *trans* isomers in Figure B are not mirror images. What do mirror-image isomers look like? For an example of mirror-image objects that are not the same, consider a person's two hands. (Ignore the things that make any hand unique, such as small scars.) In a mirror your right hand looks like your left hand, and vice versa. In contrast, most noses in a mirror look pretty much like the real noses, with no right-left differences. Figure C below illustrates molecules that are mirror-image isomers.

FIGURE B

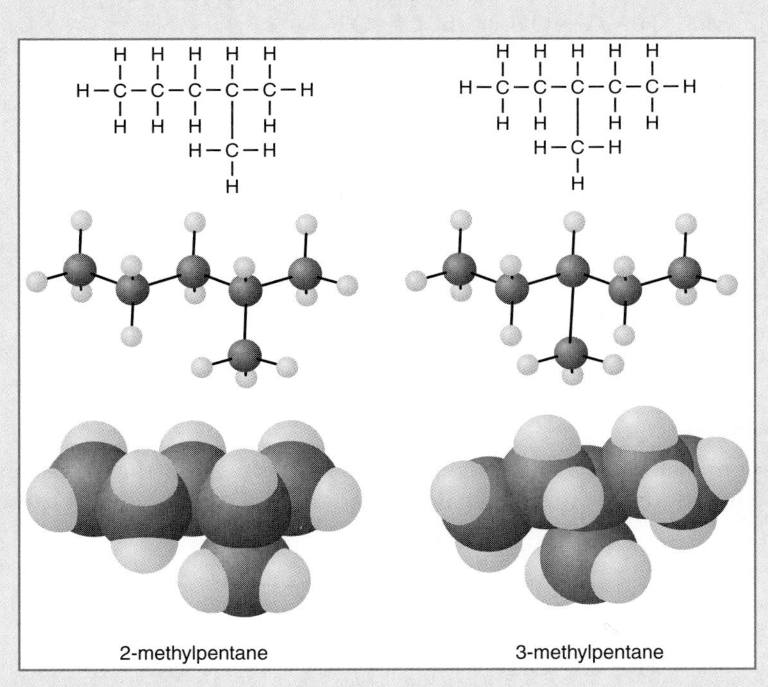


FIGURE A

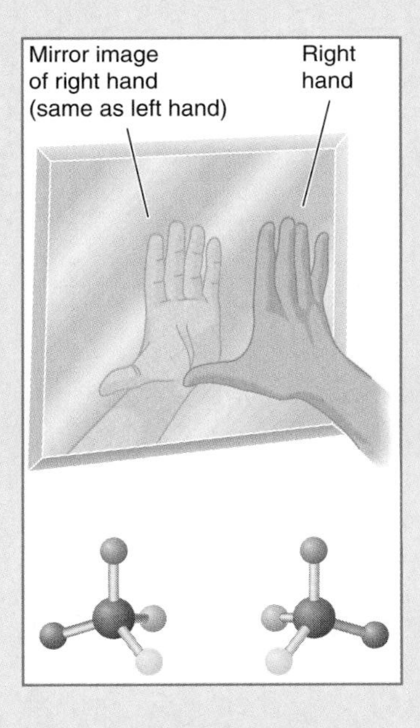


FIGURE C
Mirror-image isomers. (Top) The mirror image of a right hand is a left hand, but the two hands are not the same. (Bottom) If a compound has four different groups attached to a central carbon atom, two molecules that are mirror images of each other will not be the same. Many compounds found in biochemical systems have this property.

The prefixes here are the same ones used in naming binary covalent compounds. See Table 7.4, Section 7.5.

The names of sugars and other carbohydrates usually end in the letters *ose* (for example, glucose, sucrose, and cellulose). Since D-glyceraldehyde is an aldehyde and a carbohydrate, it can be classified as an **aldose.** If the compound is a ketone and a carbohydrate, it is termed a **ketose.** Number prefixes based on the Greek numbers are widely used in chemistry, as you already know. Since D-glyceraldehyde has three carbon atoms, it is also referred to as a *triose.* A four-carbon sugar is a *tetrose,* one with five carbons is a *pentose,* and one with six carbons is a *hexose.* All three of these conventions are often combined: D-glyceraldehyde is thus an *aldotriose.*

## EXAMPLE 19.4 Classification of Simple Sugars

Assign to each of the following sugars a classification that indicates both the functional group (aldehyde or ketone) and the number of carbons.

**a.**
$$\begin{array}{rcl} & CH_2OH & \\ & | & \\ & C{=}O & \\ & | & \\ H- & C & -OH \\ & | & \\ HO- & C & -H \\ & | & \\ & CH_2OH & \end{array}$$

**b.**
$$\begin{array}{c} CH_2OH \\ | \\ C{=}O \\ | \\ CH_2OH \end{array}$$

**c.**
$$\begin{array}{rcl} H- & C & {=}O \\ & | & \\ H- & C & -OH \\ & | & \\ HO- & C & -H \\ & | & \\ H- & C & -OH \\ & | & \\ H- & C & -OH \\ & | & \\ & CH_2OH & \end{array}$$

**SOLUTIONS**

**a.** This compound is a ketone, contains five carbons, and is a carbohydrate (as all sugars are). It is therefore a ketopentose.
**b.** Ketone, three carbons, carbohydrate: ketotriose.
**c.** Aldehyde, six carbons, carbohydrate: aldohexose (This one is *glucose,* the commonest simple sugar in the human body.)

**EXERCISE 19.4**

Assign each of the following sugars to a classification that indicates both the functional group (aldehyde or ketone) and the number of carbons.

**a.**
$$\begin{array}{rcl} H- & C & {=}O \\ & | & \\ H- & C & -OH \\ & | & \\ HO- & C & -H \\ & | & \\ & CH_2OH & \end{array}$$

**b.**
$$\begin{array}{rcl} H- & C & {=}O \\ & | & \\ H- & C & -OH \\ & | & \\ H- & C & -OH \\ & | & \\ H- & C & -OH \\ & | & \\ & CH_2OH & \end{array}$$

**c.**
$$\begin{array}{rcl} & CH_2OH & \\ & | & \\ & C{=}O & \\ & | & \\ HO- & C & -H \\ & | & \\ H- & C & -OH \\ & | & \\ H- & C & -OH \\ & | & \\ & CH_2OH & \end{array}$$

## 19.4 Proteins

**GOAL:** To identify the key structural features of all proteins, including enzymes.

As you know, proteins are a vital part of everyone's diet, and **amino acids** are the chemical building blocks of proteins. Approximately 20 amino acids are found in every human body. Ten of these are the *essential amino acids*. The body's biochemical systems are incapable of making these compounds, and you must eat foods containing them if you are to survive and be healthy.

Chemists have made many amino acids, defined as compounds that contain both an amino group ($NH_2$) and a carboxyl group ($CO_2H$). The amino acids present in the body have several features in common. First, they are $\alpha$-amino acids, meaning that the $NH_2$ group is attached to the carbon atom directly adjacent to the C=O group of the carboxylic acid, as in this structure of the amino acid alanine:

$$CH_3CH(NH_2)C(=O)—OH$$ Alanine, an α-amino acid

Another important feature of alanine is the fact that the $\alpha$-carbon, the one with the $NH_2$ group attached to it, has four different groups (H, $CH_3$, $CO_2H$, and $NH_2$) attached to it. Yes, alanine and almost all of the other amino acids in the human body can exist as mirror-image isomers. As with carbohydrates, only one of the two forms commonly exists in your body (and the bodies of other organisms). For amino acids, it is the L form that commonly exists in nature. Figure 19.12 shows the two forms of alanine.

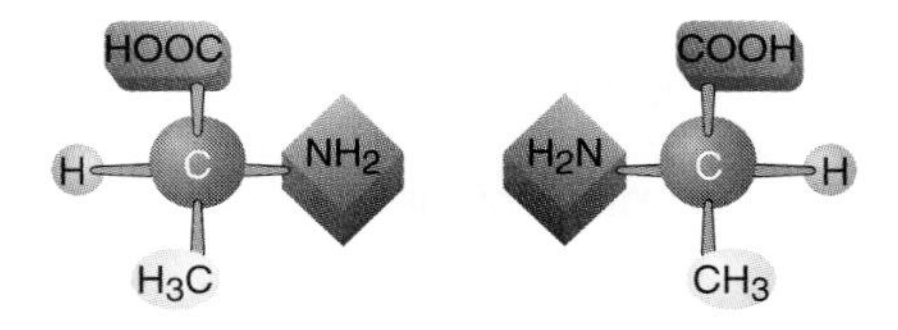


**FIGURE 19.12**
The D and L forms of alanine. The L form (on the right) is very common in nature.

In all the other amino acids that are found in your body, there are groups other than the $CH_3$ group of L-alanine. The hydrogen and the amino and carboxyl groups, as well as the central carbon atom, are the same in all amino acids. Figure 19.13 shows the structures of ten essential amino acids.

**The only exception to the $\alpha$-carbon having four different groups attached to it is glycine, $H_2NCH_2CO_2H$.**

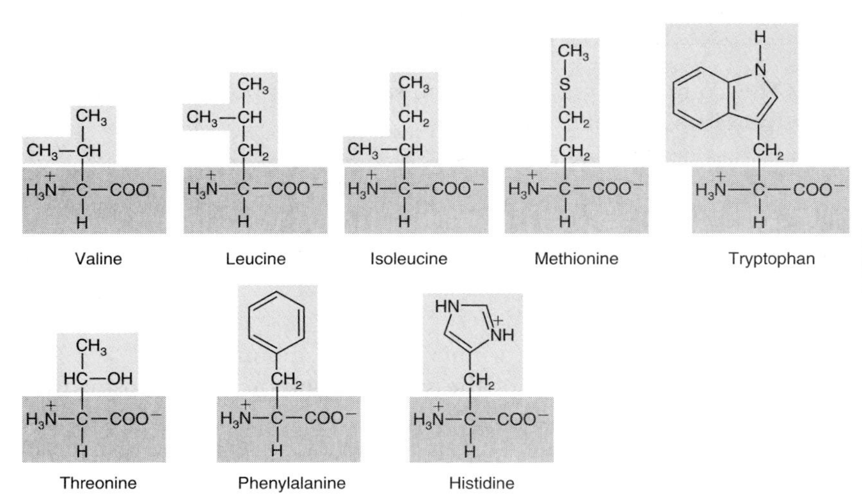


**FIGURE 19.13**
These "essential" amino acids are essential for a healthy diet.

FIGURE 19.14
This hot-air balloon is made of nylon.

Amino acids react with one another in complex processes in the body to yield proteins. In these reactions, carboxyl groups react with amino groups to give amide groups plus water. For a simple carboxylic acid and a simple amine group, the reaction goes like this:

$$CH_3\overset{\overset{\displaystyle O}{\|}}{C}{-}OH + CH_3NH_2 \xrightarrow{\text{catalyst}} CH_3\overset{\overset{\displaystyle O}{\|}}{C}{-}NHCH_3 + H_2O$$

Industrial chemists use a related process to make nylon (see Figure 19.14). In both processes—the production of proteins in the body and the production of nylon in a factory—the starting functional groups are the same. The products, of course, are very different. The similarities and differences are shown in Figure 19.15.

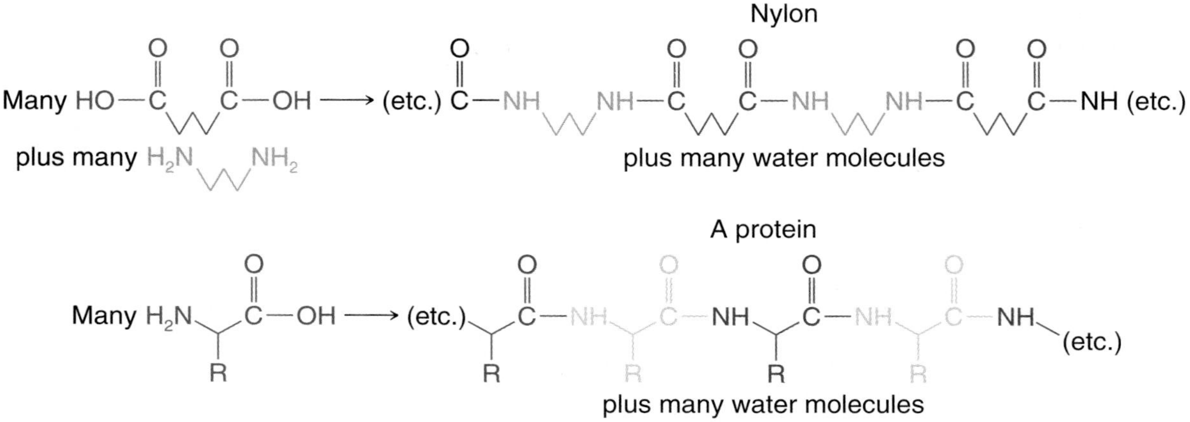

Nylon

Many $HO{-}\overset{\overset{\displaystyle O}{\|}}{C}\!\sim\!\sim\!\overset{\overset{\displaystyle O}{\|}}{C}{-}OH$ plus many $H_2N\!\sim\!\sim\!NH_2$ $\longrightarrow$ (etc.) $\overset{\overset{\displaystyle O}{\|}}{C}{-}NH\!\sim\!\sim\!NH{-}\overset{\overset{\displaystyle O}{\|}}{C}\!\sim\!\sim\!\overset{\overset{\displaystyle O}{\|}}{C}{-}NH\!\sim\!\sim\!NH{-}\overset{\overset{\displaystyle O}{\|}}{C}\!\sim\!\sim\!\overset{\overset{\displaystyle O}{\|}}{C}{-}NH$ (etc.)

plus many water molecules

A protein

Many $H_2N{-}\underset{\underset{\displaystyle R}{|}}{CH}{-}\overset{\overset{\displaystyle O}{\|}}{C}{-}OH$ $\longrightarrow$ (etc.)$-\underset{\underset{\displaystyle R}{|}}{CH}{-}\overset{\overset{\displaystyle O}{\|}}{C}{-}NH{-}\underset{\underset{\displaystyle R}{|}}{CH}{-}\overset{\overset{\displaystyle O}{\|}}{C}{-}NH{-}\underset{\underset{\displaystyle R}{|}}{CH}{-}\overset{\overset{\displaystyle O}{\|}}{C}{-}NH{-}\underset{\underset{\displaystyle R}{|}}{CH}{-}\overset{\overset{\displaystyle O}{\|}}{C}{-}NH-$(etc.)

plus many water molecules

FIGURE 19.15
Both nylon and protein molecules consist of smaller units linked together by amide functional groups. These amide groups, in turn, form by way of reactions that link amino groups with carboxyl groups, as shown here (in simplified form). In these amino acid and protein structures, R represents the fourth group attached to the $\alpha$-carbon.

## EXAMPLE 19.5 Nylon and Protein Structures

This compound was made from smaller compounds containing $NH_2$ and $CO_2H$ functional groups:

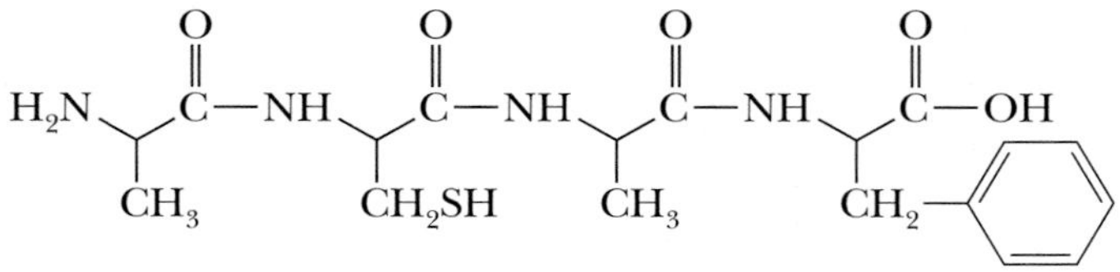

$$H_2N{-}\underset{\underset{\displaystyle CH_3}{|}}{CH}{-}\overset{\overset{\displaystyle O}{\|}}{C}{-}NH{-}\underset{\underset{\displaystyle CH_2SH}{|}}{CH}{-}\overset{\overset{\displaystyle O}{\|}}{C}{-}NH{-}\underset{\underset{\displaystyle CH_3}{|}}{CH}{-}\overset{\overset{\displaystyle O}{\|}}{C}{-}NH{-}\underset{\underset{\displaystyle CH_2{-}C_6H_5}{|}}{CH}{-}\overset{\overset{\displaystyle O}{\|}}{C}{-}OH$$

**a.** Give the molecular formula of the compound.

**b.** Draw the structure of the smaller compound that forms the leftmost unit of the compound.

**c.** Is this a small protein-like molecule (a *polypeptide*) or a nylon-like molecule? Explain your answer.

**SOLUTION**

**a.** First, count the atoms symbolized by letters in the structure. Then, each intersection of three lines represents a CH unit, and the ring to the lower right has the formula $C_6H_5$. The total formula is $C_{18}H_{26}N_4O_5S$.

**b.** Convert the first amide group from the left end of the molecule to the $CO_2H$ and $NH_2$ groups from which it was formed. The structure of the smaller compound is

O
||
$H_2N$ C—OH
$CH_3$

**c.** Since this compound was made from amino acids, it is a polypeptide rather than a nylon molecule.

**EXERCISE 19.5**

This compound was made from smaller compounds containing $NH_2$ and $CO_2H$ functional groups.

O O O O
|| || || ||
$H_2N$ NH—C C—NH NH—C C—OH
$CH_3$ $CH_2SH$ $CH_3$ $CH_2$—

**a.** How many molecules of water were given off when this molecule was formed from smaller units?

**b.** Draw the structure of the smaller compound that forms the rightmost unit of the compound.

**c.** Is this molecule a polypeptide, or is it related to nylon? Explain your answer.

One of the best-known types of proteins is the group serving as *enzymes*. These remarkable substances catalyze almost every reaction that occurs in any living organism. (Almost all enzymes are proteins, though a few RNA molecules can also serve as biological catalysts.) Enzymes, like all catalysts, make reactions go faster. Biochemical reactions catalyzed by enzymes usually run $10^6$ to $10^{12}$ times faster than the same reactions without any enzyme. Enzymes are also very *specific*. That means that an enzyme will catalyze one and only one reaction and will not influence the rates of other reactions.

**See Section 19.5 for some information about RNA.**

A seemingly extreme example of the specificity of enzymes was discovered a century ago. Starch and cellulose are very similar carbohydrates. They not only have the same empirical formula, but they are both polymers of glucose, and the individual glucose molecules are linked via the same functional group. Starch and cellulose are isomers of each other, in fact; the direction of the linkage is the only difference between them. This seemingly minor difference has an important consequence, however. Humans and other animals eat and digest starch for energy, but cannot digest cellulose at all. The German chemist Emil Fischer worked on figuring out why this was the case in the 1890s. He found that normal digestive enzymes can cut the linkage between one glucose unit and another in starch but do not react with the almost identical linkage in cellulose.

Biochemists continue to work out the details of the structures of enzyme molecules and the reactions involving them. Before biochemists can work out all the reaction details, they must first discover the structures of the enzymes and other proteins they study. A fundamental approach is to find out what amino acids make up a protein molecule. To learn this chemists use a method called *hydrolysis.* Proteins react with water in the presence of a catalyst, breaking apart to form amino acids. This reaction is the reverse of the second reaction shown in Figure 19.15. Biochemists usually use chromatography to identify the separated amino acids.

**See Section 2.3 for information about chromatography.**

Once the identities of the amino acids are known, the next task is to learn their sequence. In the protein the amino acids are linked together to form a chain, but in what order? To begin with, biochemists can use several chemical methods to identify the N-terminal residue (that is, the amino acid that has a free $NH_2$ group, at one of the two ends of a protein chain). Other methods identify the amino acid at the other end, the C-terminal residue (the one at the end with a free $CO_2H$ group). You can learn the details in more advanced books on chemistry or biochemistry.

In practical terms, repeated analyses of the ends can identify the order of no more than about 20 of the amino acids in a sequence. Most proteins in real organisms are chains of amino acids considerably longer than this. Thus methods involving *partial hydrolysis* are useful; these break a long protein chain into pieces. Then the biochemists find the order of the amino acids in each piece.

**TABLE 19.1 Common Amino Acids and their Abbreviations**

| Amino Acid | Abbreviation |
|---|---|
| Alanine | Ala |
| Arginine | Arg |
| Asparagine | Asn |
| Aspartic acid | Asp |
| Cysteine | Cys |
| Glutamic acid | Glu |
| Glutamine | Gln |
| Glycine | Gly |
| Histidine | His |
| Isoleucine | Ile |
| Leucine | Leu |
| Lysine | Lys |
| Methionine | Met |
| Phenylalanine | Phe |
| Proline | Pro |
| Serine | Ser |
| Threonine | Thr |
| Tryptophan | Trp |
| Tyrosine | Tyr |
| Valine | Val |

Biochemists have a shorthand method of writing the sequence of amino acids in a protein. Lewis diagrams or even bond-line formulas are difficult to write and interpret for any protein made up of more than a few amino acids. In the shorthand system every amino acid has a three-letter abbreviation (see Table 19.1). Alanine is Ala, valine is Val, tyrosine is Tyr, and histidine is His, for example. Also, the N-terminal residue is always written at the left end of a chain, and the C-terminal residue is always at the right end. Consider this tetrapeptide (polypeptide consisting of four amino acids):

$$H_2N\text{–}CH(CH_3)\text{–}C(=O)\text{–}NH\text{–}CH(CH_2SH)\text{–}C(=O)\text{–}NH\text{–}CH(CH_3)\text{–}C(=O)\text{–}NH\text{–}CH(CH_2C_6H_5)\text{–}C(=O)\text{–}OH$$

The four amino acids, from left to right, are alanine, cysteine, alanine again, and phenylalanine. The shorthand system represents the sequence of amino acids in this polypeptide as Ala-Cys-Ala-Phe. Clearly the system saves biochemists considerable time and space when writing the structures of polypeptides and proteins.

EXAMPLE 19.6

## Sequences of Amino Acids in Small Polypeptides

Total hydrolysis of an unknown tetrapeptide gave leucine, lysine, tyrosine, and tryptophan. Partial hydrolysis gave the dipeptides Tyr-Lys, Trp-Leu, and Leu-Tyr. What is the sequence of this tetrapeptide?

### SOLUTION

Lysine and tyrosine occur in the order Tyr-Lys, and no other dipeptide includes lysine. Tyrosine also occurs in Leu-Tyr, so the order so far is Leu-Tyr-Lys. Finally, tryptophan precedes leucine in Trp-Leu. The full sequence is Trp-Leu-Tyr-Lys.

### EXERCISE 19.6

Total hydrolysis of an unknown pentapeptide gave alanine, lysine, valine, and histidine. Partial hydrolysis gave the dipeptides His-His, His-Lys, Val-His, and Ala-Val. What is the sequence of this pentapeptide?

The sequence of amino acids in a protein is called the *primary structure.* How the protein chain coils is the *secondary structure* and how it folds is the tertiary structure (see Figure 19.16). Some protein molecules associate with other protein molecules to make stable aggregations. These are the *quaternary structures.*

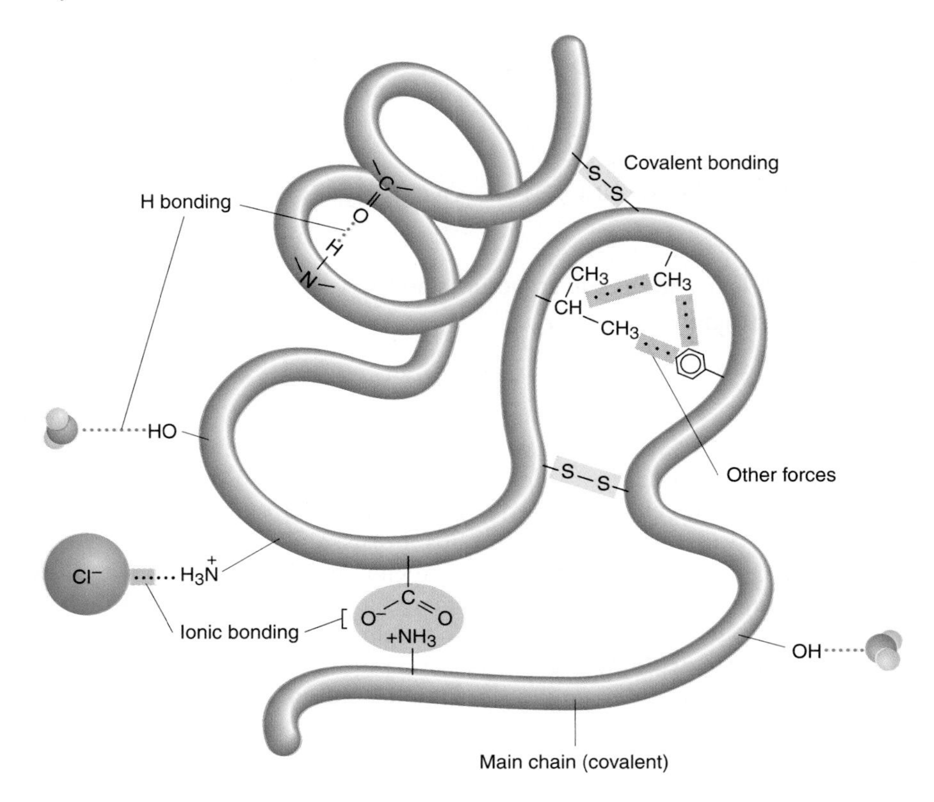


FIGURE 19.16
A combination of covalent bonding, ionic bonding, and other forces is responsible for proteins' shapes (and therefore for their functions).

## CHAPTER BOX 19.3 THE MILLER-UREY EXPERIMENT

**Chapter Box 18.1** told how Louis Pasteur showed that once living organisms, even microorganisms, had been killed in a flask of nutrient medium, life did not arise again. This doesn't show that life might not arise given a different chemical starting point, a longer time, and an outside energy source.

Harold C. Urey won the Nobel Prize early in his career for the discovery of deuterium ($^{2}H$). Many years later, after that discovery had played a crucial part in the development of nuclear weapons, Urey decided to work in the area of geochemistry. He thought (correctly so far) that his discoveries in that field wouldn't have military applications. Urey thought that in its early years the earth had an atmosphere much different from the one it has now, one much more like that of the planet Jupiter.

Urey's student Stanley Miller ran an important experiment. He mixed water, ammonia, methane, and hydrogen together in a flask. As energy sources for chemical reactions, the early earth would have had ultraviolet light from the sun and lightning, so Miller added an electric discharge to his apparatus (see the figure at right). He let his experiment run for a week. Then he isolated what compounds he could from the reaction mixture, which was something of a mess by then.

Miller found a number of different reaction products. Among them were some α-amino acids, including several of those important to life on earth today. Chemists and others have followed up this line of research extensively, and it seems clear that most of the compounds that make up living systems today can be made by experiments similar to this one.

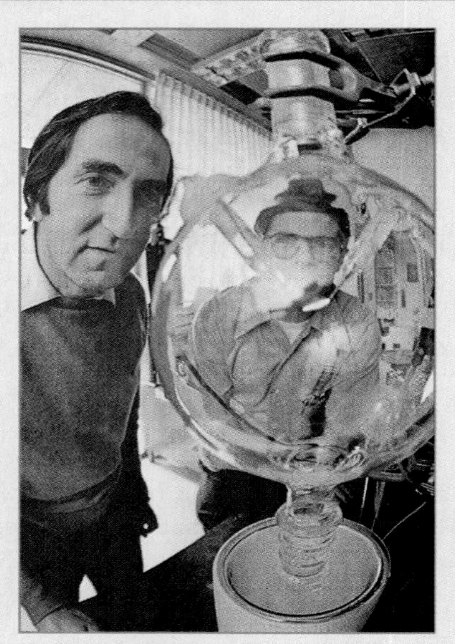

Miller used this experimental setup to investigate the chemistry that might have occurred in the atmosphere and oceans early in the history of this planet.

## 19.5 Nucleic Acids: DNA and RNA

**GOAL:** To learn about the structures and functions of the molecules that govern heredity.

The first discovery of what is now called **DNA** was made in 1869. The year before that Johann Miescher, a young Swiss chemist, went to Germany to work in biochemistry. He got some pus-soaked bandages from a hospital, extracted white blood cells from them, and isolated a substance from the cells. This substance was acidic, contained phosphorus, and consisted of very large molecules. White blood cells have large nuclei, and Miescher knew that the substance came from them. He later isolated a purer form of the substance from salmon sperm, available from fish in the rivers near his Swiss home. The name **nucleic acid** was later coined by a pupil of Miescher's.

The first step in understanding a complex substance is almost always to cut it into pieces and analyze the pieces. By the 1920s biochemists knew that there were two types of nucleic acids, and that each of their molecules had three parts. In deoxyribonucleic acid (DNA) the parts are a phosphate group, the five-carbon sugar deoxyribose, and one of four compounds called *bases* that contain mostly carbon and nitrogen. The four bases are named adenine, cytosine, thymine, and guanine. In ribonucleic acid (RNA) ribose replaces deoxyribose. Three of the four RNA bases are the same as the ones in DNA. Figure 19.17 shows how the elements of DNA and RNA chains bond to one another.

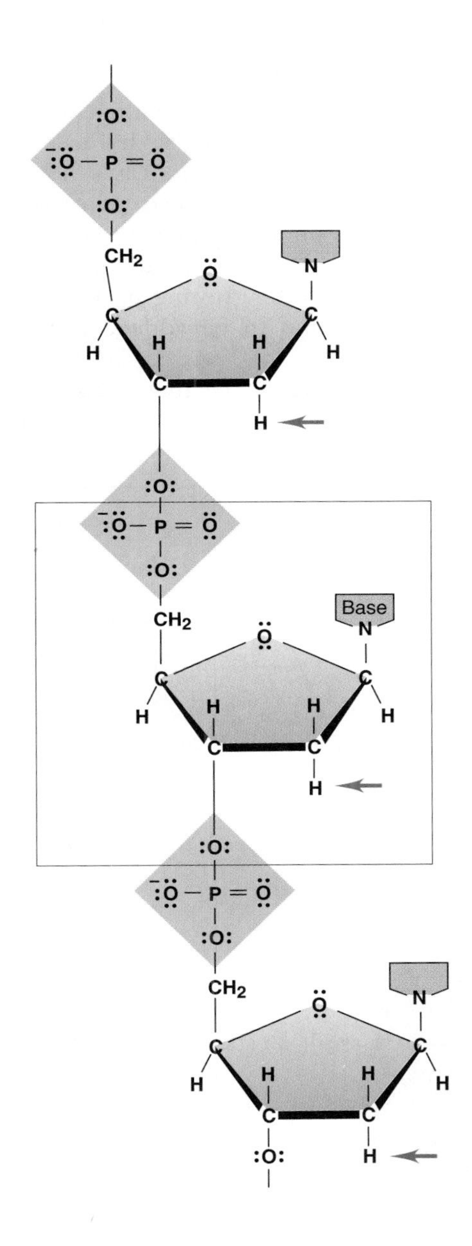


FIGURE 19.17
DNA backbone. A DNA molecule is made from deoxyribose (rings highlighted in red), phosphate groups (purple diamonds), and bases (above and to the right of each deoxyribose ring). RNA contains ribose (in which an OH group replaces an H in the five-membered ring). Also, one of the four RNA bases is different from one of the DNA bases: otherwise single chains of DNA and RNA are the same.

The connections between the pieces took years to work out. In each long DNA chain, deoxyribose and phosphate pieces alternate from one end of the molecule to another. One base hangs off each deoxyribose. The unit of one phosphate, one ribose or deoxyribose, and one base is called a **nucleotide.** You can see how they link up in the DNA molecule in Figure 19.17.

James Watson and Francis Crick worked out the double helix structure of DNA in 1953 (see Figure 19.18). In this structure, each long chain of DNA is paired with another. The chains are joined by hydrogen bonds between bases. Of the six possible pairings of two bases, only adenine-thymine and guanine-cytosine pairs actually form. When a cell divides, the pairs of DNA chains separate, and the dividing cell makes a new chain to match each old one.

The sequence of the bases in DNA codes the genetic information that the molecule carries. Certain enzymes use the DNA code to build RNA molecules in the cell nucleus. This RNA then moves to other parts of the cell, where other enzymes use its coded information in building protein molecules. Biochemists and molecular biologists have worked out the code that DNA uses: A set of three specific bases in a DNA molecule codes for a set of three bases in the RNA molecule, which in turn triggers the addition of a specific amino acid to a growing protein chain. For example, if an RNA molecule contains the sequence cytosine-uracil-adenine, the cellular mechanisms add the amino acid leucine to the growing protein. The sequence adenine-cytosine-uracil causes threonine to be added, and uracil-adenine-adenine means "stop."

The portion of the DNA molecule that encodes a complete protein is called a **gene.** All of the DNA in a particular cell is called the **genome,** and this codes for every protein the cell can make—and thus how the cell be-

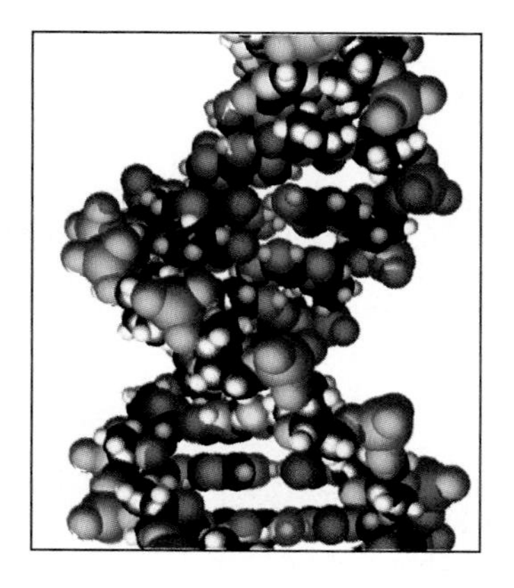

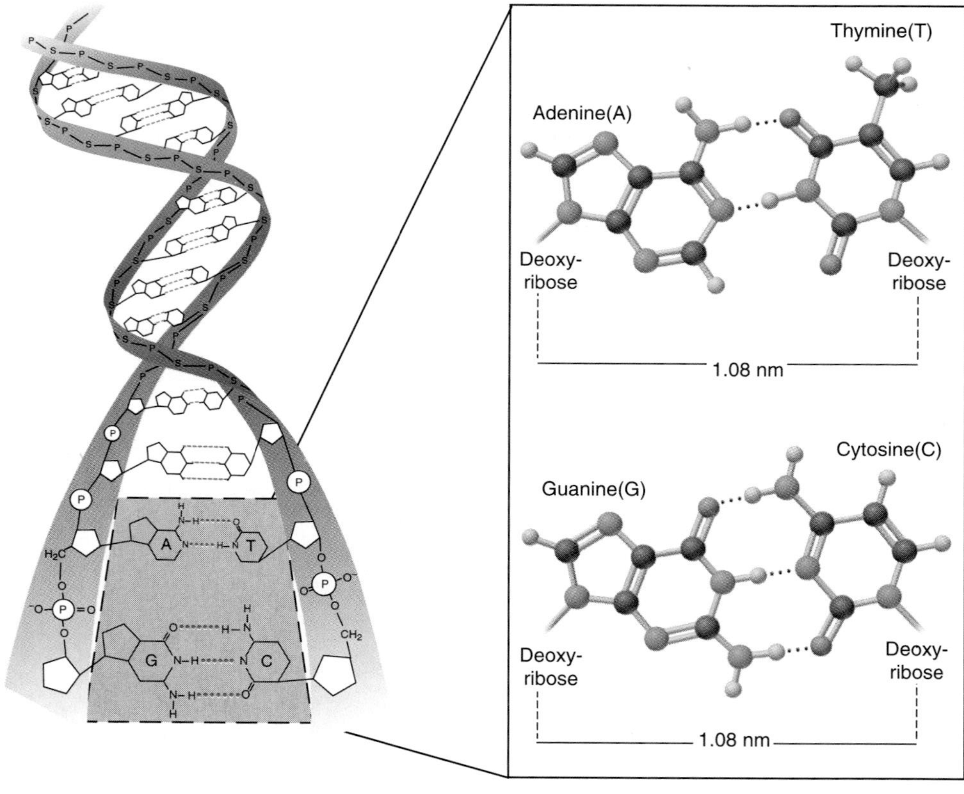


FIGURE 19.18

The double helix of DNA. At left, a space-filling model. Center, an expanded schematic diagram showing how the chains are put together. The structures at right show how the bases pair with one another.

haves. Biochemists and molecular biologists are now working on the Human Genome Project, an effort to map the complete structure of human DNA.

### EXAMPLE 19.7 Nucleic Acids

Many simple sugars have the empirical formula $CH_2O$, which is the same as the molecular formula of formaldehyde.

**a.** What are the molecular and empirical formulas of the base adenine found in DNA and RNA? The structure of adenine is given in Figure 19.18.

**b.** What compound has a molecular formula that is the same as the empirical formula of adenine?

**SOLUTIONS**

**a.** Adenine has the molecular formula $C_5H_5N_5$ and the empirical formula CHN.

**b.** The empirical formula of adenine is the same as the molecular formula of hydrogen cyanide ($H—C{\equiv}N$:). (Yes, it is possible to make adenine from hydrogen cyanide if the conditions are right.)

**EXERCISE 19.7**

**a.** What are the molecular and empirical formulas of cytosine?

**b.** If you were trying to make cytosine using an experimental procedure like that of Stanley Miller, what starting materials might you use?

## 19.6 Other Molecules and Ions in Biological Systems

**GOAL:** To examine some other biomolecules: ATP, hormones, vitamins, minerals, and alkaloids.

### The Energy Molecule

Nucleotides are vital for life as a part of DNA and RNA, but they serve other important functions. One of them, adenosine triphosphate, or **ATP,** is the molecule that carries energy around in the body and passes it on to other molecules, allowing for movement and for the production of other high-energy molecules. Here is the structure of ATP:

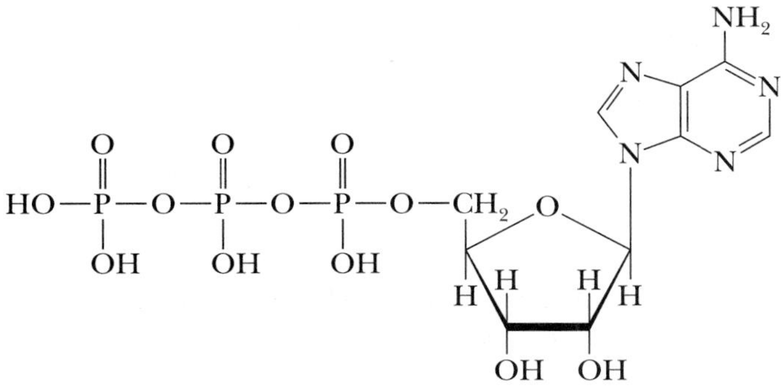

ATP

ATP reacts with water in the presence of an enzyme to release an inorganic phosphate group and ADP (adenosine diphosphate). The reaction also releases about 35 kJ of energy per mole in a form that can drive other reactions.

### Hormones

**Hormones** are molecules that act as messengers, controlling chemical processes in the body, including metabolism. Members of this diverse group are defined by their functions and not by their structures. Insulin, the hormone that regulates glucose and is used to treat diabetes mellitus, is a polypeptide. Epinephrine (adrenaline) is a phenol derivative:

OH

Phenol is an aromatic alcohol.

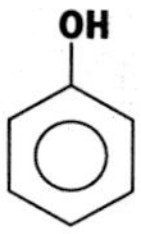

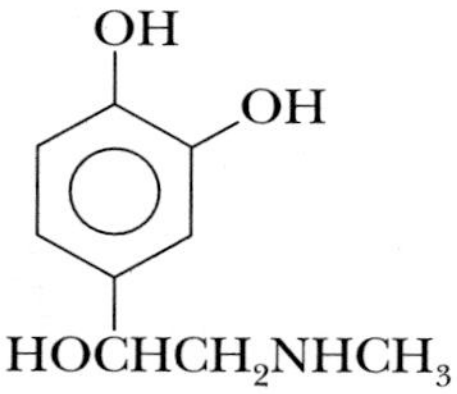

OH
OH
HOCHCH$_2$NHCH$_3$

Epinephrine
(Adrenaline)

Other hormones are amino acid derivatives or steroids. You can find the structures of two hormones, testosterone and cortisone, in Example 19.3 and Exercise 19.3. These compounds are steroids, as are many of the hormones associated with the processes of growth and reproduction.

### Vitamins

Like hormones, **vitamins** are not classified by structure. They are substances that must be part of the diet if people are to remain in good health. Vitamin A provides the pigment that captures light in the receptors of the eye. Niacin and riboflavin are important in almost all oxidation-reduction reactions that take place in cells. The specific functions of vitamin C (ascorbic acid) and vitamin E are somewhat obscure to biochemists. Many health claims, some of them exaggerated, have grown up around these two substances.

It seems likely that Vitamin E's job is to react with oxidizing agents, preventing other important compounds from being oxidized prematurely.

Here is the structure of Vitamin A:

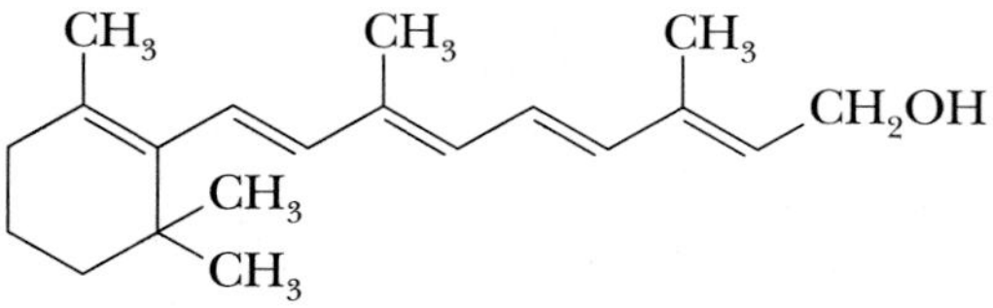

CH$_3$ CH$_3$ CH$_3$
CH$_2$OH
CH$_3$
CH$_3$

Vitamin A

You can see that molecules of this compound will be attracted to each other by van der Waals forces. As a result, Vitamin A is much more soluble in fat than it is in water. Vitamin C, on the other hand, contains several OH groups and is more soluble in water than in fat.

### Minerals

To geologists, minerals are relatively pure crystalline substances. Rocks, on the other hand, are complex mixtures and are often not crystalline. In nutrition, **minerals** are cations and anions, and people usually ingest them as part of mixtures more complex than any that geologists ever study (see Figure 19.19).

Like vitamins, minerals have many biochemical functions. Calcium (as calcium ion, $Ca^{2+}$) and phosphorus (as phosphate ion, $PO_4^{3-}$) form the solid part of bones. Iron (as $Fe^{2+}$ ion), is part of the proteins hemoglobin and myoglobin, working to transport oxygen in the blood. Biochemists have found that zinc ($Zn^{2+}$) is vital to the functioning of over a hundred enzymes and also to the senses of taste and smell.

FIGURE 19.19

Left is a specimen of the mineral calcite (calcium carbonate). Milk (right) is a *very* complex mixture and a major source of dietary calcium for many people.

Iodine (which you ingest as iodide ion, $I^-$, not as $I_2$) is part of the thyroid hormone thyroxin, the structure of which is shown in Figure 19.20. Lack of iodine in the diet leads to a condition called *goiter*, in which the thyroid glands in the neck (which make thyroxin) become enlarged.

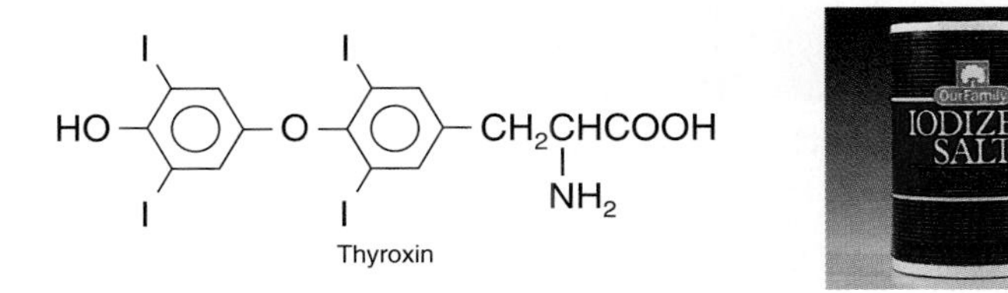


FIGURE 19.20

If you use iodized salt, a goiter will not be a problem. Goiter is still common among less medically aware peoples who live far from the sea.

#### Alkaloids: Chemical Self-Defense

Human beings and other animals have many ways to defend themselves from predators. They can fight, hide, or flee, as circumstances dictate. Plants can do none of these things, and you might wonder how they survive in such large numbers. All of them resort to chemical self-defense in one form or another.

One mechanism is the use of indigestible material in plant tissues. Animals can't digest cellulose, the main structural material of plants. Even termites don't digest the cellulose in the wood they eat. Bacteria in their stomachs do this.

Some plants use a much more active and dangerous form of self-defense. Their tissues produce compounds that poison any animal that eats them. Almost all of the members of an entire family of compounds, the alkaloids, are toxic in one way or another. Extracting the tissues of certain plants with acid separates these compounds from the others. The fact that these compounds react with acids means that they are bases, and this alkaline nature is reflected in the name **alkaloid.** All of the alkaloids are amines, and many contain other functional groups.

Coniine can be isolated from hemlock and was part of the poison used in the execution of the Greek philosopher Socrates. Nicotine is an addictive drug in small quantities; in larger quantities it is a deadly poison, and farmers sometimes use it as an insecticide. Atropine is another dangerous poison, isolated from a plant called the *deadly nightshade.* Cocaine (from the South American coca plant) is a chemical relative of atropine (see Figure 19.21).

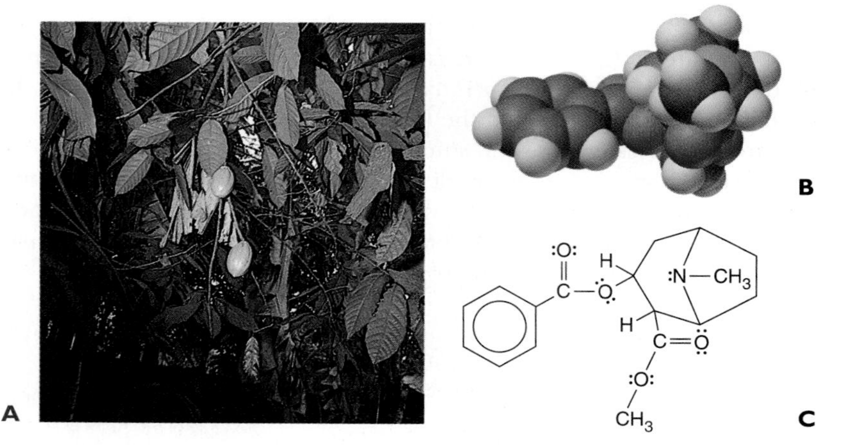


FIGURE 19.21
Cocaine. (A) The coca plant makes cocaine, probably for self-defense. (B) Space-filling model of a cocaine molecule. (C) Lewis diagram of the same substance.

EXAMPLE 19.8 **Functional Groups and Classification of Biomolecules**

**a.** What functional groups are present in vitamin A? (You will find the structure of this compound earlier in this section.)
**b.** Of which of the following families is vitamin A a member?
A. Lipids C. Carbohydrates
B. Proteins D. Nucleic acids

SOLUTIONS

**a.** Vitamin A contains an OH group (at the far right of the structure as shown) and several carbon-carbon double bonds; it is an alcohol and an alkene.
**b.** Vitamin A does not contain amine, carboxyl, or amide groups, as proteins do. It is not a polyhydroxy aldehyde or ketone, as carbohydrates are. It does not contain a pentose, phosphate, or a base, as nucleic acids do. Vitamin A is a lipid (answer A), with a polar OH group at the end of a long nonpolar chain.

EXERCISE 19.8

**a.** What functional groups are present in thyroxin? The structure of thyroxin is shown in Figure 19.20.
**b.** Of which of the following families is thyroxin a member?
A. Nucleic acids C. Carbohydrates
B. Amino acids D. Lipids

## CHAPTER SUMMARY

1. Biochemistry is the study of the chemistry of living systems.
2. Lipids are most prominent in the body as energy-storing compounds, though specialized members of the family serve other purposes. You can find lipids in the form of fats and oils in your local supermarket. The most common fats in the human diet are triglycerides, which may be saturated or unsaturated.
3. Carbohydrates are so named because they have the empirical formula $CH_2O$ (or something close to that). Simple sugars are aldoses or ketoses, polyhydroxy aldehydes or ketones. More complex carbohydrates, including starch and cellulose, consist of many molecules of simple sugars linked together. Starches and glycogen store metabolic energy, while cellulose and cartilage serve as structural stiffeners.
4. Amino acids have this structure: $H_2N—CHR—CO_2H$, where R indicates a group that contains carbons, hydrogens, and sometimes other atoms. These compounds link together in long chains to form proteins, which are the main component of muscle tissue. Most enzymes (which are biochemical catalysts) are proteins.
5. Nucleic acids, including DNA and RNA, are long chains of nucleotides. These nucleotides are in turn made up of inorganic phosphate, the sugar deoxyribose (in DNA) or ribose (in RNA), and various bases. Nucleic acids are the molecules of genetic inheritance.
6. The bodies of human beings and other organisms contain a number of biochemically active compounds. These include hormones, vitamins, minerals, and alkaloids (the last of which are found in plants).

## PROBLEMS

### SECTION 19.1 From the Organism to the Molecule

1. Four major classes of biochemical compounds are lipids, carbohydrates, proteins, and nucleic acids. Of which of these classes of compounds are fats members?
2. Of which of the four classes of compounds in Problem 1 is glycogen a member?
3. Of which of the four classes of compounds in Problem 1 are plant cell walls made?
4. Of which of the classes of compounds in Problem 1 does muscle tissue consist?
5. Which of the classes in Problem 1 are formed by amino acids linking together?
6. Which of the classes in Problem 1 are formed by simple sugars linking together?
7. Which of the classes of compounds in Problem 1 is most important in genetic inheritance?
8. Which of the classes of compounds in Problem 1 serve as a store of energy?
9. Of which of the classes of compounds in Problem 1 is vegetable oil a dietary source?
10. Of which of the classes of compounds in Problem 1 are potatoes a dietary source?

### SECTION 19.2 Lipids: Fats and Oils

11. Which have the higher melting points, fats or oils? Explain your answer.
12. Explain the meaning of the word *triglyceride*.
13. What structural feature is responsible for the word *saturated* in saturated fats?
14. What structural feature do polyunsaturated fats have in their molecules?

15. Which of the following compounds is a *cis* alkene? Explain your answer.

A. $BrHC{=}CHBr$ (Br / H on the left carbon; Br / H on the right carbon)

B. $Br_2C{=}CHBr$ (Br / Br on the left carbon; Br / H on the right carbon)

C. $BrHC{=}CHBr$ (Br / H on the left carbon; H / Br on the right carbon)

D. $Br_2C{=}CH_2$ (Br / Br on the left carbon; H / H on the right carbon)

16. Which of the compounds in Problem 15 is a *trans* alkene? Explain your answer.

17. Which two compounds in Problem 15 are neither *cis* nor *trans* alkenes? Explain your answer.

18. Classify this compound as a *cis* alkene, a *trans* alkene, or neither. Explain your answer.

19. What is the molecular formula of cholesterol? (The structure of cholesterol is given in Section 19.2.)

20. What is the molecular formula of the female sex hormone estradiol? This is the structure of estradiol:

OH
HO
Estradiol

21. Another female sex hormone is estrone. What is its molecular formula? This is its structure:

O
HO
Estrone

22. Ergosterol is a precursor of vitamin D. What is its molecular formula? Its structure is

$CH_3$ $CH_3$
$CHCH{=}CHCHCH(CH_3)_2$
HO
Ergosterol

## SECTION 19.3 Carbohydrates: Sugars and Their Relatives

23. Explain the origin of the word *carbohydrate.*

24. Explain how carbohydrates and hydrocarbons differ.

25. Classify each of these sugars, indicating both the carbonyl functional group present and the number of carbons.

a.
$CH_2OH$
$C{=}O$
$HO{-}C{-}H$
$CH_2OH$

b.
$CH_2OH$
$C{=}O$
$HO{-}C{-}H$
$H{-}C{-}OH$
$H{-}C{-}OH$
$CH_2OH$

c.
$H{-}C{=}O$
$H{-}C{-}OH$
$H{-}C{-}OH$
$H{-}C{-}OH$
$CH_2OH$

26. Classify each of these sugars, indicating both the carbonyl functional group present and the number of carbons.

a.
$H{-}C{=}O$
$H{-}C{-}OH$
$HO{-}C{-}H$
$CH_2OH$

b.
$CH_2OH$
$C{=}O$
$H{-}C{-}OH$
$H{-}C{-}OH$
$HO{-}C{-}H$
$H{-}C{-}OH$
$CH_2OH$

c.
$H{-}C{=}O$
$HO{-}C{-}H$
$CH_2OH$

27. Classify each of these sugars, indicating both the carbonyl functional group present and the number of carbons.

a.
$CH_2OH$
$H{-}C{-}OH$
$H{-}C{-}OH$
$H{-}C{-}OH$
$C{=}O$
$CH_2OH$

b.
$CH_2OH$
$HO{-}C{-}H$
$C{=}O$
$CH_2OH$

c.
$CH_2OH$
$H{-}C{-}OH$
$HO{-}C{-}H$
$H{-}C{-}OH$
$H{-}C{=}O$

28. Classify each of these sugars, indicating both the carbonyl functional group present and the number of carbons.

a.
```
    H—C=O
      |
    H—C—OH
      |
   HO—C—H
      |
   HO—C—H
      |
   HO—C—H
      |
    H—C—OH
      |
      CH2OH
```

b.
```
      CH2OH
      |
      C=O
      |
   HO—C—H
      |
      CH2OH
```

c.
```
      CH2OH
      |
    H—C—OH
      |
   HO—C—H
      |
    H—C=O
```

## SECTION 19.4 Proteins

29. What functional group do both nylon and proteins have in common?

30. Protein molecules must have two different functional groups ($NH_2$ and $CO_2H$) at their ends, but nylon molecules can have identical terminal groups. Explain the reason for this fact.

31. This molecule was made from smaller molecules containing $NH_2$ and $CO_2H$ functional groups.

```
                  O          O               O
                  ||         ||              ||
H2N\/\NH—C\/\C—NH\/\C—OH
```

a. Draw the structures of the smaller molecules that were used to make this molecule.
b. Is this a polypeptide or a nylon-like molecule? Explain your answer.
c. How many water molecules were produced when this molecule was made?

32. This molecule was made from smaller molecules containing $NH_2$ and $CO_2H$ groups.

```
       O          O          O          O          O
       ||         ||         ||         ||         ||
H2N\/C—NH\/C—NH\/C—NH\/C—NH\/C—OH
```

a. Draw the structures of the smaller molecules that were used to make this molecule.
b. Is this a polypeptide or a nylon-like molecule? Explain your answer.
c. How many water molecules were produced when this molecule was made?

33. An unknown pentapeptide was completely hydrolyzed and found to contain aspartic acid (Asp), lysine (Lys), proline (Pro), and methionine (Met). After partial hydrolysis the dipeptides Lys-Asp, Met-Pro, Lys-Met, and Pro-Lys were identified. What is the sequence of amino acids in this pentapeptide?

34. An unknown tetrapeptide was completely hydrolyzed and found to contain aspartic acid (Asp), lysine (Lys), proline (Pro), and methionine (Met). After partial hydrolysis the dipeptides Lys-Asp, Met-Pro, and Pro-Lys, were identified. What is the sequence of amino acids in this tetrapeptide?

35. An unknown pentapeptide was completely hydrolyzed and found to contain glutamic acid (Glu), histidine (His), proline (Pro), and glycine (Gly). After partial hydrolysis the dipeptides His-Gly, His-Pro, Glu-His, and Gly-His were identified. What is the sequence of amino acids in this pentapeptide?

36. An unknown pentapeptide was completely hydrolyzed and found to contain valine (Val), tyrosine (Tyr), histidine (His), and alanine (Ala). After partial hydrolysis the dipeptides Tyr-Ala, Tyr-His, Val-Tyr, and Ala-Tyr were identified. What is the sequence of amino acids in this pentapeptide?

37. An unknown pentapeptide was completely hydrolyzed and found to consist of glutamine (Gln), tyrosine (Tyr), arginine (Arg), and lysine (Lys). After partial hydrolysis the dipeptides Arg-Gln, Lys-Lys, Tyr-Arg, and Lys-Tyr were identified. What is the sequence of amino acids in this pentapeptide?

38. An unknown pentapeptide was completely hydrolyzed and found to consist of glutamine (Gln), tyrosine (Tyr), arginine (Arg), and lysine (Lys). After partial hydrolysis the dipeptides Gln-Arg, Tyr-Lys, Lys-Gln, and Lys-Lys were identified. What is the sequence of amino acids in this pentapeptide?

39. An unknown pentapeptide was completely hydrolyzed and found to consist of glutamine (Gln), tyrosine (Tyr), arginine (Arg), and lysine (Lys). After partial hydrolysis the dipeptides Lys-Arg, Lys-Tyr, Gln-Lys, and Tyr-Lys were identified. What is the sequence of amino acids in this pentapeptide?

40. An unknown pentapeptide was completely hydrolyzed and found to consist of glutamine (Gln), tyrosine (Tyr), arginine (Arg), and lysine (Lys). After partial hydrolysis the dipeptides Lys-Lys, Arg-Gln, Tyr-Lys, and Gln-Tyr were identified in the product. What is the sequence of amino acids in this pentapeptide?

## SECTION 19.5 Nucleic Acids: DNA and RNA

41. What are the molecular and empirical formulas of thymine and guanine (See Figure 19.18 for the necessary structures.)

42. What are the molecular and empirical formulas of ribose and deoxyribose? (See Figure 19.17 for the necessary structures.)

43. The DNA bases that contain two rings are called purine bases. Which are these? (See Figure 19.18 for the necessary structures.)

44. The DNA bases that contain one ring are called pyrimidine bases. Which are these? (See Figure 19.18 for the necessary structures.)

45. Analyses of nucleic acids invariably find equal amounts of purine and pyrimidine bases. Explain why. (See problems 43 and 44 for the definition of purine and pyrimidine.)

46. Analyses of nucleic acids that have less thymine than cytosine always find less adenine than guanine. Explain why.

### SECTION 19.6 Other Molecules and Ions in Biological Systems

47. What is ATP?

48. What are hormones?

49. What are alkaloids?

50. What are vitamins?

51. What does the word *minerals* mean to geologists?

52. What does the word *minerals* mean to biochemists?

53. What functional groups are present in epinephrine (adrenaline)? (For the structure of this compound, see Section 19.6.)

54. Of the compounds shown in Section 19.6, which one does not contain an alcohol functional group?

### ADDITIONAL PROBLEMS

55. What is the most abundant functional group in proteins?

56. What is the most abundant functional group in carbohydrates?

57. Which of the following functional groups is most common in lipids discussed in this chapter?
A. Aldehyde C. Alcohol
B. Ketone D. Ester

58. Which of the following classes of compounds contains the highest proportion of ester groups?
A. Proteins C. Carbohydrates
B. Amino acids D. Lipids

59. Which of the following classes of compounds contains the highest proportion of amide groups?
A. Amino acids C. Carbohydrates
B. Proteins D. Lipids

60. Which of the following classes of compounds contains the highest proportion of alcohol groups?
A. Proteins C. Carbohydrates
B. Amino acids D. Lipids

## SOLUTIONS TO EXERCISES

### Exercise 19.1

**a.** Proteins **c.** Carbohydrates
**b.** Nucleic acids **d.** Nucleic acids

### Exercise 19.2

Alkene A has two hydrogens on the opposite sides of a line connecting the doubly bonded carbons; therefore it is *trans.* Alkenes C and D have two hydrogens on the same side; they are *cis.* Alkene B has two identical $CH_3$ groups on the same carbon, so it is neither *cis* nor *trans.*

### Exercise 19.3

**a.** A cortisone molecule contains three ketone groups, two alcohol groups, and one alkene group.

**b.** Cortisone contains 21 carbon atoms, as shown in the structure at right. To count hydrogen atoms, remember that each of the carbon atoms must have four bonds connected to it. A carbon with three bonds already shown has one hydrogen bonded to it, one with two bonds shown has two hydrogens, and the carbons at the ends of lines have three hydrogens. The total number of hydrogen atoms in a cortisone molecule is 28. (Don't forget to count the OH hydrogens and the hydrogens in the $CH_2$ group at the upper right of the structure as shown.) The molecular formula of cortisone is $C_{21}H_{28}O_5$. The ungainly structure sketched below is only for your count of the bonds to the carbon atoms, of course; it isn't suitable for any formal purpose.

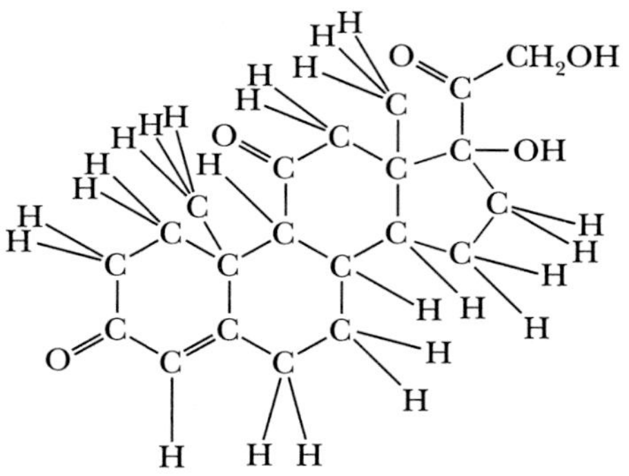

**Exercise 19.4**

**a.** Four carbons, aldehyde: aldotetrose
**b.** Five carbons, aldehyde: aldopentose (this is *ribose,* the R of RNA)
**c.** Six carbons, ketone: ketohexose (this is *fructose,* found in honey and fruit juices)

**Exercise 19.5**

**a.** The structure contains three amide functional groups. One molecule of water was given off to form each amide group, giving a total of three molecules of water.
**b.** This time convert the rightmost amide group to the $CO_2H$ and $NH_2$ groups that formed it. The structure is

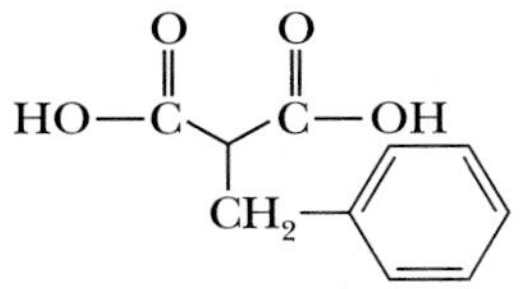

O O
‖ ‖
HO—C—C—OH
CH$_2$—(benzene ring)

A dicarboxylic acid

**c.** Since this molecule was made from molecules with two $NH_2$ and two $CO_2H$ groups apiece, it is similar to nylon.

**Exercise 19.6**

The pentapeptide contains two histidine units, as shown by the first dipeptide listed. The sequence is Ala-Val-His-His-Lys.

**Exercise 19.7**

**a.** The molecular and empirical formulas of cytosine are both $C_4H_5N_3O$.
**b.** Cytosine might be made by combining three molecules of HCN and one of $CH_2O$ (formaldehyde).

**Exercise 19.8**

**a.** Thyroxin is a phenol (the OH group on the left), an ether (the oxygen between the two rings), an amine (from the $NH_2$ group), and a carboxylic acid (from the $CO_2H$ group).
**b.** From the list of functional groups in part (a), thyroxin is an amino acid (answer B).

# *Appendix A*
# Algebra and Solving Numerical Problems

From your experience with the courses you have taken up to now, undoubtedly you know that scientists frequently work with numbers. Sometimes the numbers come in chart or table form, but in chemistry and physics you will often need to work with equations. This section will refresh your memory on how to rearrange equations to get a numerical answer from them. (Presumably all readers of this book have taken an algebra course at one point or other in their careers.)

As you work through the numerical problems in this book, you will see that problem solvers typically *rearrange* mathematical equations (algebraic expressions) before solving them. This means that they place the unknown quantity in the expression on one side of an equation and all known quantities on the other side. Then the problem solver does the arithmetic needed to get the numerical answer. While this is not the only technique you can use, it is almost universal, and it has the merit of being straightforward, uncomplicated, and relatively easy.

By way of illustration, suppose that in the expression below $a = 10$ and $c = 5$ and you want to calculate the value of b.

$$ab = c$$

The basic idea is to rearrange this equation so that the unknown quantity, b, is isolated on the left and that the known quantities, a and c, are placed on the right. We will see why this particular left/right order was chosen later in the appendix.

There are two simple mathematical tools that problem solvers use to rearrange equations. The first is that *any quantity divided by itself is equal to 1.* For example, $a/a = 1$, (won)/(won) = 1, and $(345\pi^2)/(345\pi^2) = 1$. The second is that if you do *any mathematical operation* (addition, subtraction, multiplication, division, taking the square root, cubing, etc.) *to both sides of an equation, this leaves the validity of the equation unchanged.* For example, if your equation is $2 = 2$, adding 5 to both sides gives $(2 + 5) = (2 + 5)$, or $7 = 7$, which is still a valid expression.

To isolate b on the left hand side of our equation $ab = c$, divide both sides of the equation by a.

$$\frac{ab}{a} = \frac{c}{a}$$

Since $ab/a = b$ because $a/a = 1$, the expression above reduces to:

$$b = \frac{c}{a}$$

You can now solve the equation by substituting the known values of a and c. Since a = 10 and c = 5, this gives:

$$b = \frac{c}{a} = \frac{5}{10} = 0.5$$

In the next example, we will rearrange and solve an equation that has a different form than the previous one. Again we will solve for b, and again we will use values of a = 10 and c = 5. Since the equation is different, the value of b may well be different—we'll just have to solve the equation to find out. Here's the new equation:

$$\frac{a}{b} = c$$

Rearranging this equation requires two steps. First, multiply both sides of the equation by b.

$$\frac{ab}{b} = bc$$

Since b/b = 1, this gives you:

$$a = bc$$

In the second step, divide both sides of the expression by c to isolate b on the right.

$$\frac{a}{c} = \frac{bc}{c}$$

Since c/c = 1, this gives you:

$$\frac{a}{c} = b$$

Now you can plug the numbers into the equation and solve. Using the values a = 10 and c = 5, solving for b gives us:

$$\frac{10}{5} = b = 2$$

You will notice that in this last example we isolated b on the right side of the equation, and in the first example b was isolated on the left side. It really doesn't matter where the unknown appears, and the side that is more convenient is usually easier to solve.

What do you do if the equation is more complicated? The answer here is that you solve it one step at a time (which is how to solve complicated problems of almost any kind). Here's an example. Once again we will use the values a = 10 and c = 5 and solve for b.

$$\frac{a}{b^2c} - 10 = 1$$

Where do you start? You see that you have several things to do to get an equation in which b is equal to some combination of everything else. You solve this problem as you should solve all problems, one step at a time. Generally it's best to add or subtract before multiplying or dividing, so we'll add ten to both sides of the equation first.

$$\frac{a}{b^2c} - 10 + 10 = 1 + 10, \text{ which gives } \frac{a}{b^2c} = 11$$

Next let's get $b^2$ away from a and c. To do this, multiply both sides of the equation by $b^2$.

$$\frac{b^2a}{b^2c} = 11\ b^2, \text{ which gives } \frac{a}{c} = 11\ b^2$$

How do you get the 11 away from the $b^2$? Divide both sides by 11.

$$\frac{a}{11\ c} = \frac{11\ b^2}{11}, \text{ which gives } \frac{a}{11\ c} = b^2$$

Finally we need to take the square root of both sides of the equation. This gives:

$$\sqrt{\frac{a}{11\ c}} = b$$

Finally (whew!), we plug in the values a = 10 and c = 5 and do the arithmetic.

$$\sqrt{\frac{10}{11 \times 5}} = b, \text{ which gives } \sqrt{\frac{10}{55}} = b, \text{ or } b = 0.4$$

When you're solving any type of problem, remember to stop at the end and check to see if your solution is reasonable, that it makes sense. Chapter Box 3.2 contains a short discussion on the reasonableness of answers. For the problems in this appendix, you couldn't do that, since the equations here have no units and no reference to anything in the real world.

# *Appendix B*
# Answers to Selected Problems

## CHAPTER 1

**2.** Three specific activities are:
**a.** Dyeing your hair **b.** Cooking **c.** Eating

**4.** Three things that consist exclusively of chemicals are:
**a.** Wood **b.** Plastics **c.** Food

**6. a.** Chemistry and astronomy
**b.** Physics and chemistry
**c.** Chemistry and geology

**8. b.** Extraction of minerals
**d.** Faults in the earth's crust

**10. a.** Burning candles
**b.** Iron manufacture
**c.** Gunpowder manufacture

**14. b.** The older thread of mixing human characteristics with chemistry

**16.** Put some ketchup on a plate and see if it gets runny.

**20.** Blow air into a balloon. Without letting go of the mouth of the balloon, squeeze the bottom and see what happens.

**(The term *falsifiable* as used here means that the proposed theory can be tested by experiment or observation.)**

**22.** This idea cannot be regarded as being scientific because it is not falsifiable.

**24.** This idea can be regarded as being scientific because the elements can be counted.

**30.** This idea can be regarded as being scientific because it is falsifiable.

**32.** Scientific, because it is falsifiable.

**36.** This idea can be regarded as being scientific because it is falsifiable.

**40.** Scientific, because it is falsifiable.

**42. b.** Data; this is the raw information that was obtained from the experiment.
**d.** Conclusion; this is a deduction made from the results of the experiment.

**44. b.** Conclusion; this is a deduction made from the results of the experiment.
**d.** Data; this is the raw information that was obtained from the experiment.

**46.** Vinegar is acidic.

**50.** Weigh the sample of metal being used. Heat the metal strongly in an open container. Compare the weight of the metal before and after the experiment.

**52.** Relativity and evolution are scientific facts. The Theories of Evolution and Relativity are attempts to interpret these observed facts.

**58.** Aluminum is more difficult to extract than iron.

**64.** Li–lithium, Na–sodium, K–potassium, Rb–rubidium, Cs–cesium, Fr–francium

**68.** His hypothesis was that air was essential for combustion.

**70.** The total mass of the reaction mixture would have remained the same. The air would not have rushed in at the end.

## CHAPTER 2

**2. b.** Pancake syrup **d.** Apple juice

**4. a.** Homogeneous **d.** Homogeneous

**6. b.** Heterogeneous **d.** Homogeneous

**8.** 17% of the sample is tin.

**12.** 91.7% of the alloy consists of gold.

**16. b.** 27.4% of the shots attempted were successful.
**d.** 31.0% of the shots attempted were successful.

**18. b.** 48.0% of the shots attempted were successful.
**d.** 36.2% of the shots attempted were successful.

**20. b.** Percentage of elephant seals = 22.2%
Percentage of fur seals = 77.7%
Percentage of crabeater seals = 0.183%
**d.** Percentage of elephant seals = 12.8%
Percentage of fur seals = 87.0%
Percentage of crabeater seals = 0.198%

**22. b.** Percentage of copper = 57.1%
Percentage of tin = 28.6%
Percentage of zinc = 14.3%
**d.** Percentage of copper = 44.4%
Percentage of tin = 44.4%
Percentage of zinc = 11.1%

**24.** Percentage of iron = 85.7%
Percentage of nickel = 7.62%
Percentage of chromium = 3.81%
Percentage of carbon = 2.86%

**28. b.** 25 grams of aluminum hydroxide
25 grams of magnesium carbonate
50 grams of calcium carbonate
**d.** 5 grams of aluminum hydroxide
45 grams of magnesium carbonate
50 grams of calcium carbonate

**30. b.** 225 grams of aluminum hydroxide, 150 grams of magnesium carbonate, and 375 grams of calcium carbonate.
**d.** 300 grams of calcium carbonate, 150 grams of magnesium carbonate, and 300 grams of aluminum hydroxide.

**34. a.** Honey **b.** Apple juice **c.** Eggnog

**38. b.** Gas **d.** Solid

**44. a.** Extraction. The salt will dissolve in water and the mixture can be filtered, leaving the insoluble pepper behind.
**b.** Filtration. The sand is insoluble in the seawater. The sand will be left behind.
**c.** Winnowing. The popcorn kernels are much lighter than the seeds and they will be blown away.

**50.** The more volatile

**51.** C—this part of the tube is cooler—condensation occurs there

**52.** The boiling point of the mixture increases.

**56. b.** Physical
**d.** Chemical

**58. b.** The color of the liquid
**d.** The phase of the sample

**60. b.** The color of the substance
**d.** Its solubility in water

**64.** Yes; it is possible to have a pure mixture. As long as all of the components of the mixture are known then the mixture is pure.

**66.** "Pure mountain spring water" is also an incorrect statement since all water that is obtained from the ground contains some amount of dissolved salts and this makes it a mixture.

**74.** Four elements that exist in nature as single atoms:
Krypton, Kr
Neon, Ne
Argon, Ar
Helium, He

**76. b.** Mn **d.** Pb **f.** Ne **h.** N

**78. b.** Potassium **d.** Silver **f.** Strontium **h.** Tin **j.** Uranium

**80. b.** $H_2SO_4$ **d.** $NaNO_3$ **f.** $N_2O_4$

**82. b.** $Br_2$ **d.** $As_4$

**84. b.** N:3, H:12, P:1, O:4
**d.** N:2, H:4, O:3
**f.** K:1, Mn:1, O:4
**h.** Co:1, C:1, O:3

**90.** Molecules are arranged closer together in a solid than in a liquid.

**92.** Wood

**94.** $CH_4$; 4 H atoms are present in a molecule of methane

**98. b.** The scent of perfume travels through a room.
**d.** Ammonia is dissolved in liquid household cleaners.

**100. b.** It is flammable. **d.** Nitrogen ($N_2$)

## CHAPTER 3

**4.** The magnitude denotes the size of a measurement. The unit compares the measurement to a standard. The only way that a measurement makes sense is if both the magnitude and the unit are included.

**6.** In a space shuttle the objects are "weightless"—that is, the acceleration due to gravity present is used to keep them in orbit. It does not contribute to their weight—but the object still has mass, since its mass is independent of *g*.

**10. a.** Lowers accuracy only
**b.** Lowers accuracy only
**c.** Lowers neither accuracy nor precision
**d.** Lowers accuracy and precision

**12. b.** Average = 9.8 kg
**d.** Average = 47 s
**f.** Average = 1081.1 kg
**h.** Average = 328.0 s
**j.** Average = 0.065 g

**16. a.** Average time = 66 s
**b.** Both the accuracy and the precision of these results are quite low.

**18. b.** 4 significant figures **d.** 1 significant figure
**f.** 5 significant figures **h.** Exact **j.** 5 significant figures

**30.** Grams of nickel = 0.146 g; to 3 significant figures

**34.** Volume remaining = 30 mL; to zero decimal places

**38. b.** $1 \times 10^{-1}$ mg
**d.** $4 \times 10^{12}$y
**f.** $3.88 \times 10^{-4}$ km
**h.** $5.111 \times 10^{1}$ mL
**j.** $2.387 \times 10^{6}$ pm

**40. b.** 8001 $m^3$
**d.** 0.077371 kg
**f.** 0.01414 s
**h.** 978000 $ft^3$
**j.** 1110000000000 s

**42. b.** $1.0 \times 10^{2}$ $m^4$; to 2 significant figures
**d.** $1.69 \times 10^{2}$ $m^3$; to 3 significant figures
**f.** $7.5 \times 10^{31}$; to 2 significant figures

**44. b.** $3.1 \times 10^{33}$ m; to 2 significant figures
**d.** $1.0 \times 10^{-1}$ s; to 2 significant figures
**f.** $1.9 \times 10^{1}$

**46. a.** $10^{-3}$ **b.** $10^{+3}$ **c.** $10^{-6}$

**54. a.** A box of cereal **b.** A turkey **c.** A small tin of tomato paste **d.** A bag of flour

**58. b.** −328°F **f.** °F = 1832 **j.** −40°F
**d.** 100°C **h.** 373 K
**60. a.** 2000 $cm^3$ **c.** 950 $cm^3$ **e.** 4000 $cm^3$
**b.** 0.25 $m^3$ **d.** 50 $cm^3$
**64.** 0.893 g/mL; to 3 significant figures
**68.** The fact that a piece of lead will sink in water.
**72.** 8940 g; to 3 significant figures
**74. b.** 53.4 g; to 3 significant figures
**d.** 174 g; to 3 significant figures
**78.** $3.00 \times 10^{10}$ cm/s
**82.** 1 wk = 604,800 s
**86. b.** $\mu m \rightarrow m: \dfrac{1\ m}{1 \times 10^{+6}\ \mu m}$
**d.** $m \rightarrow cm: \dfrac{100\ cm}{1\ m}$
**88. b.** $7.21 \times 10^{6}$ m **h.** $7.50 \times 10^{3}$ cm
**d.** $4.54 \times 10^{-4}$ g **j.** 6.5 mg
**f.** 783 m = $7.83 \times 10^{2}$ m
**90. b.** $6.78 \times 10^{1}$ mL **h.** $5.6 \times 10^{4}$ kg
**d.** $9.81 \times 10^{-4}$ kg **j.** $9.88 \times 10^{3}$ pm
**f.** $7.4 \times 10^{-2}$ ms
**94. b.** 9,800 $kg/m^2$ **h.** 10.06 m
**d.** 805 km/h **j.** 2.77 kg
**f.** $1.2 \times 10^{3}$ mi
**100.** 55,000 m/h
**104.** $1.96 \times 10^{12}$ mi/y
**108. b.** 3 **d.** 6
**110. b.** mass of copper 4.1 kg mass of zinc = 0.6 kg
mass of tin = 0.3 kg
**d.** mass of zinc = 1.7 kg mass of copper = 3.3 kg

## CHAPTER 4

**4. b.** = $1.27 \times 10^{5}$ J **d.** $1.55 \times 10^{-19}$ J
**6. b.** $3.09 \times 10^{4}$ kg **d.** $1.76 \times 10^{-3}$ kg
**8. b.** $7.93 \times 10^{5}$ J = $7.93 \times 10^{2}$ kJ
**d.** $3.3 \times 10^{5}$ J = $3.3 \times 10^{2}$ kJ
**12.** $2.7 \times 10^{4}$ J
**14. b.** Position 3 **d.** Position 4
**16. b.** 725°C **d.** Yes
**20.** It was converted into kinetic energy as the book fell and it was also converted into sound energy and heat as the book hit the ground.
**24. b.** Energy stored by an object
**d.** Energy of an object's position
**28.** The kinetic energy of liquid water molecules at 10°C is less than that of liquid water molecules at 90°C.
**32.** A substance is hot relative to another substance because its particles have more kinetic energy than the particles of the colder substance. When the two substances are brought into contact, the particles of the different substances collide and the particles of the hotter substance, which are moving faster, transfer their energy to the colder substance.

As these colder particles acquire this energy, their kinetic energy (and thus their temperature) increases. In this way heat "flows" from a hot substance to a cold substance.
**34. b.** 580 J **d.** $5.4 \times 10^{3}$ J
**42.** Du Fay proposed the existence of positive vitreous fluid and a negative resinous fluid. The gold leaves contained an excess of either vitreous or resinous fluid, which made the leaves repel each other.
**46.** *Insulating parts:* cork and glass
*Conducting parts:* brass rod and the tin coating
**50.** Thomson's model of the atom makes more sense because the positively charged particles are very close to the negatively charged particles thus making for a very stable structure. It is expected that the positively charged particles would "draw" the negatively charged particles to themselves. It is hard, at first, to understand Rutherford's model. After all, one question that the novice chemist must ask is "why don't the electrons fall into the mass of a positive charge?"
**54. b.** 47 **d.** 3
**58. b.** K **d.** Br
**60. b.** Sr **d.** Sn
**62.**

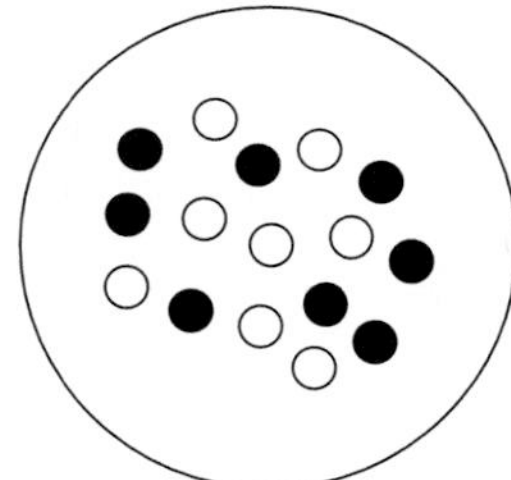

8 electrons surround the nucleus

**64. b.** 18 **d.** 18 **f.** 54 **h.** 18 **j.** 28

**66.**

| | Symbol | # of Protons | # of Electrons | Charge | Cation/ Anion/Neutral |
|---|---|---|---|---|---|
| b. | Ca | 20 | 18 | +2 | cation |
| d. | Ne | 10 | 10 | 0 | neutral |
| f. | H | 1 | 2 | −1 | anion |
| h. | Al | 13 | 10 | +3 | cation |
| j. | N | 7 | 10 | −3 | anion |

**68. b.** 3 **d.** 4

**76. a.** C **b.** A

**80.** (+ −) (+ −) (+ −)

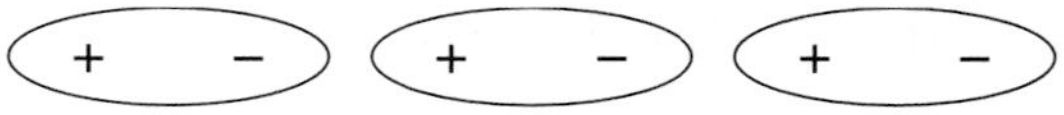

## CHAPTER 5

**2. b.** Semimetal **f.** Metal **j.** Nonmetal
**d.** Metal **h.** Metal

**4. b.** The element is a nonmetal because it is a poor thermal and electrical conductor.
**d.** The element is a nonmetal because it is a gas.

**6. b.** C **d.** At **f.** P **h.** Ag **j.** B

**8. b.** Be **d.** Cl **f.** Ca **h.** At **j.** Ba

**10. b.** $SrCl_2$ **d.** $MgI_2$

**14. b.** Main group element **h.** Main group element
**d.** Main group element **j.** Actinide
**f.** Main group element

**16. b.** 4 **d.** 6 **f.** 2 **h.** 6 **j.** 4

**18. b.** 2 **d.** 4 **f.** 8 **h.** 2 **j.** 4

**20. b.** F **d.** Cl **f.** Sr **h.** As **j.** O

**24. b.** $[Ne]3s^23p^4$ **f.** $[Ne]3s^23p^3$ **j.** $[Ne]3s^1$
**d.** $1s^2$ **h.** $[He]2s^22p^3$

**26. b.** $X$ **d.** 3 **f.** $XCl_3$

**28. b.** $[He]2s^22p^6$ **f.** $[He]2s^22p^6$
**d.** $[Kr]5s^24s^{10}5p^6$ **h.** $[Ar]4s^23d^{10}4p^6$

**30. b.** $Ca^{2+}$ **d.** $S^{2-}$ **f.** $P^{3-}$ **h.** $K^+$

**32. b.** 1 **d.** 1 **f.** 2 **h.** 1 **j.** 3

**34. b.** $GaBr_3$ **f.** $Na_2S$ **j.** $H_2S$
**d.** $Al_2O_3$ **h.** $Cs_3P$

**36. b.** $4.82 \times 10^{24}$ C atoms $mol^{-1}$
**d.** $1.6 \times 10^{25}$ atoms $mol^{-1}$

**38. b.** $1.3 \times 10^{25}$ H atoms **d.** $8.4 \times 10^{24}$ H atoms

**40. b.** $8.5 \times 10^{24}$ O atoms **d.** $8.5 \times 10^{24}$ O atoms

**42. b.** 80.9 g $mol^{-1}$ HBr **h.** 209 g $mol^{-1}$ $PCl_5$
**d.** 100 g $mol^{-1}$ $KHCO_3$ **j.** 65.1 g $mol^{-1}$ KCN
**f.** 134 g $mol^{-1}$ $AlCl_3$

**44. b.** $2.99 \times 10^{-22}$ g $molecule^{-1}$
**d.** $2.18 \times 10^{-22}$ g $molecule^{-1}$

**46. b.** $1.59 \times 10^{-11}$ mol of H **d.** 25 mol S

**48.** 24.3 amu

**54. a.** 42.0 g **b.** 126 g **c.** 84.0 g

**56. a.** 2 **b.** 4 **c.** 2 **d.** 0

**58. b.** 2 **d.** 3 **f.** 1

**60.** 35.46 g, the smallest obtainable

**64.** $2.41 \times 10^{24}$ atoms of $Y$ in 1 mol of compound 5

**66.** $Y$ is Cl because it has a molar mass of 35.46 g/mol

## CHAPTER 6

**2.** B

**4. b.** $2.24 \times 10^4\ s^{-1}$ **d.** $5.6 \times 10^2\ s^{-1}$

**6. b.** $3.1 \times 10^{-12}$ m

**10.** AM—lower frequency ($8.80 \times 10^5$ Hz), therefore longer wavelength.
FM—higher frequency ($9.55 \times 10^7$ Hz).

**12.** $4 \times 10^{14}\ s^{-1}$

**14. b.** $3.279 \times 10^{-6}$ m **d.** $2.985 \times 10^{-6}$ m

**16. b.** $6.65 \times 10^{14}$ Hz
**d.** $7.35 \times 10^{14}$ Hz (1 Hz = $s^{-1}$)

**20. b.** $6 \times 10^3\ s^{-1}$ **d.** $2.1 \times 10^{15}\ s^{-1}$

**22. b.** $4 \times 10^{-11}$ m **d.** 3.00 m

**28. a.** $5.83 \times 10^{-28}$ $photon^{-1}$
**b.** $3.51 \times 10^{-7}$ kJ $mol^{-1}$

**31. b.** $3.46 \times 10^{-18}$ J $photon^{-1}$
**d.** $1.29 \times 10^{-28}$ J $photon^{-1}$

**34. b.** $4.86 \times 10^{19}\ s^{-1}$ $photon^{-1}$
**d.** $9.10 \times 10^{18}\ s^{-1}$ $photon^{-1}$

**36. b.** $3.52 \times 10^{10}$ J $mol^{-1}$
**d.** $1.57 \times 10^{13}$ J $mol^{-1}$

**38. b.** $7.19 \times 10^9\ s^{-1}$ $photon^{-1}$
**d.** $1.57 \times 10^{16}\ s^{-1}$ $photon^{-1}$

**42. b.** $2.95 \times 10^2$ nm **d.** $2.58 \times 10^2$ nm

**44.** K, Na, Ba, Ca, Mg

**45. b.** $3s$ orbital **d.** $4d$ orbital

**48. b.** $n = 2, \ell = 0$ **d.** $n = 2, \ell = 3$ **f.** $n = 4, \ell = 0$

**50.** $5s$ $5p$ $5d$ and $5f$

**54.** Total number of orbitals is 16

**56. a.** $n = 3, \ell = 1, m = 0$ **b.** $n = 1, \ell = 0, m = 0$

**62. b.** 2 **d.** 6

**64. b.** [↑↓] $4s$ [ ][ ][ ] $4p$ Ca
**d.** [↑↓] $3s$ [↑↓][↑][↑] $3p$ S

**66. b.** [↑][↑][ ][ ][ ] $3d$ $V^{3+}$
**d.** [↑↓][↑↓][↑][↑][↑] $3d$ $Co^{2+}$

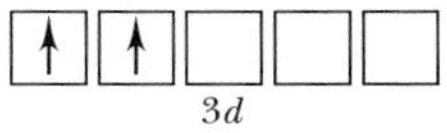

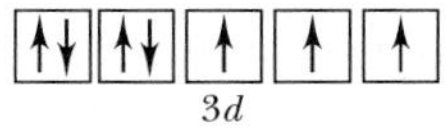

**68. b.** 1 **d.** 1

**70. a.** Forbidden state **c.** Excited state
**b.** Ground state **d.** Forbidden state

**72. b.** Excited **d.** Excited

**76. a.** The photoelectric effect
**b.** The obtaining of an interference pattern

## CHAPTER 7

2. **b.** lithium ion **d.** radium ion

All of the ions in problem 4 form only one cation when they react, and are found in Table 7.1.

4. **b.** $Al^{3+}$ **d.** $Sr^{2+}$

Most of the elements in problems 6 and 8 form different cations when they react. Roman numerals must be included when naming them.

6. **b.** tin(II) **d.** gold(III)
8. **b.** manganese(VII) **d.** chromium(VI)

Problems 10 and 12 can be answered by writing the symbol of the element and the charge of the ion which is the positive number denoted by the Roman numeral in parenthesis (except for the mercurous ion).

10. **b.** $Cr^{2+}$ **d.** $Ti^{4+}$
12. **b.** $V^{5+}$ **d.** $Ni^{2+}$
16. **b.** $Se^{2-}$ **d.** $F^-$
18. **b.** $ClO_3^-$ **d.** $CO_3^{2-}$
22. **b.** dichromate ion **d.** nitrite ion
26. **b.** $CaH_2$ **d.** $Sr(HSO_4)_2$
28. **b.** $HgSO_4$ **d.** $NaClO_4$
32. **b.** $KClO$ **d.** $NH_4HCO_3$
34. **b.** $Co(NO_3)_3$ **d.** $CrSO_3$
36. **b.** Correct formula: $(NH_4)_2SO_4$
    **d.** Correct formula: $Al_2O_3$
    **f.** Correct formula: $AlCl_3$
38. **b.** mercurous chloride
    **d.** sodium dichromate
42. **b.** mercury(II) chloride
    **d.** lead(II) bromide
46. **b.** tungsten(VI) chloride
    **d.** magnesium permanganate
48. **b.** copper(II) nitrite **d.** tin(II) perchlorate
50. **b.** boron trichloride **f.** disulfur decafluoride
    **d.** dinitrogen monoxide
54. **b.** disulfur dioxide **f.** xenon trioxide
    **d.** diselenium dibromide
56. **b.** $AsCl_5$ **d.** $P_2I_4$
58. **b.** $NI_3$ **d.** $P_4O_6$
60. **b.** hypobromous acid **d.** perchloric acid
64. **b.** $HNO_2$ **d.** $HCl(aq)$
68. **b.** covalent: binary compound **d.** ionic
74. **b.** sodium hydride **d.** copper(I) hydroxide
78. **b.** strontium hydride **d.** copper(I) iodide
80. **b.** $AgBr$ **d.** $CH_3CO_2Na$
82. **b.** $Ti_3(PO_4)_4$ **d.** $NH_4HCO_3$
86. **b.** $Na_3P$ **d.** $NH_4HSO_4$
88. **b.** covalent **d.** covalent
92. **b.** covalent **d.** covalent
94. **b.** covalent **d.** ionic
96. **b.** soluble **d.** insoluble
98. **b.** soluble **d.** insoluble

## CHAPTER 8

2. **b.** $NaOH + H_3PO_4 \rightarrow Na_3PO_4 + 3H_2O$
   **d.** $2CuO + C \rightarrow 2Cu + CO_2$
6. **b.** $6H_2S + 2SO_3 \rightarrow 6H_2O + S_8$
   Reactants: $H_2S + SO_3$ Products: $H_2O + S_8$
   **d.** $2S + Cl_2 \rightarrow S_2Cl_2$
   Reactants: $S + Cl_2$ Products: $S_2Cl_2$
8. **b.** One mole of copper(I) sulfide reacts with one mole of oxygen to form two moles of copper metal and one mole of sulfur dioxide.
   **d.** One mole of copper sulfide reacts with one mole of oxygen to form one mole of copper metal and one mole of sulfur dioxide.
12. **b.** balanced **d.** balanced
14. **b.** not balanced **d.** not balanced
18. **b.** $Mg(OH)_2 + 2H_2SO_4 \rightarrow Mg(HSO_4)_2 + 2H_2O$
    **d.** $Na_2CO_3 + 2C \rightarrow 2Na + 3CO$
20. **b.** $2NO + 2H_2 \rightarrow N_2 + 2H_2O$
    **d.** $P_4O_{10} + 6H_2O \rightarrow 4H_3PO_4$
22. **b.** $2NaHCO_3 \rightarrow Na_2CO_3 + H_2O + CO_2$
    **d.** $Cu_2O + CO \rightarrow 2Cu + CO_2$
24. **b.** $Al_2O_3 + 3Cl_2 + 3C \rightarrow 2AlCl_3 + 3CO$
    **d.** $4PH_3 + 8O_2 \rightarrow P_4O_{10} + 6H_2O$
28. **a.** balanced
    **b.** $MgO + 2HCl \rightarrow MgCl_2 + H_2O$
    **c.** balanced
    **d.** $Pb(OH)_2 + 2HCl \rightarrow PbCl_2 + 2H_2O$
    **e.** missing reactant
30. **b.** $2KHSO_4 + Ba(OH)_2 \rightarrow BaSO_4 + K_2SO_4 + 2H_2O$
    **d.** $H_2SO_4 + 2CuOH \rightarrow Cu_2SO_4 + 2H_2O$
32. **b.** $2HNO_3 + Ca(OH)_2 \rightarrow Ca(NO_3)_2 + 2H_2O$
    **d.** $H_2SO_4 + 2RbOH \rightarrow Rb_2SO_4 + 2H_2O$
34. **b.** $2HClO_4 + Fe(OH)_2 \rightarrow Fe(ClO_4)_2 + 2H_2O$
    **d.** $2HBr + Zn(OH)_2 \rightarrow ZnBr_2 + 2H_2O$
38. **b.** The products will be aluminum sulfate and water.

$$3H_2SO_4 + 2Al(OH)_3 \longrightarrow Al_2(SO_4)_3 + 6H_2O$$

    **d.** The products will be calcium acetate and water.

$$Ca(OH)_2 + 2CH_3COOH \longrightarrow (CH_3COO)_2Ca + 2H_2O$$

40. **b.** The products will be potassium perchlorate, carbon dioxide, and water.

$$2HClO_4 + K_2CO_3 \longrightarrow 2KClO_4 + CO_2 + H_2O$$

**d.** The products will be iron(III) sulfate and water.

$$3H_2SO_4 + 2Fe(OH)_3 \longrightarrow Fe_2(SO_4)_3 + 6H_2O$$

**42. b.** $MnO_2$ is reduced and Al is oxidized.
**d.** CO was oxidized and $O_2$ was reduced.

**46. b.** $2Ba + O_2 \rightarrow 2BaO$
**d.** $C_5H_{12} + 8O_2 \rightarrow 5CO_2 + 6H_2O$

**48. b.** $2Cl_2 + O_2 \rightarrow 2Cl_2O$
**d.** $2C_4H_{10}O + 12O_2 \rightarrow 8CO_2 + 10H_2O$

**50. b.** $2K + 2H_2O \rightarrow 2KOH + H_2$

**52. b.** One mole of liquid sulfuric acid reacts with one mole of solid sodium chloride to yield one mole of hydrogen chloride gas and one mole of solid sodium hydrogen sulfate.
**d.** Passing an electric current through a heated mole of liquid magnesium chloride yields one mole of solid magnesium and one mole of chlorine gas.

**54. b.** Four moles of solid silver metal react with one mole of gaseous oxygen upon heating to produce two moles of solid silver oxide.
**d.** One mole of methane gas reacts with one mole of gaseous ammonia at 1200°C in the presence of platinum to produce one mole of hydrogen cyanide gas and three moles of hydrogen gas.

**58. b.** $2HCl(aq) + Fe(s) \rightarrow H_2(g) + FeCl_2(aq)$
**d.** $S(l) + O_2(g) \xrightarrow{\text{heat}} SO_2(g)$

**60. b.** $2NaCl(l) \xrightarrow{\text{electric current}} 2Na(l) + Cl_2(g)$
**d.** $CO_2(g) + Ca(OH)_2(s) \rightarrow CaCO_3(s) + H_2O(l)$

**64. b.** Balanced equation:

$$3H_3PO_4(l) + 6NaOH(aq) \longrightarrow 3Na_2HPO_4(aq) + 6H_2O(l)$$

Net ionic equation:

$$3H_3PO_4(l) + 6OH^-(aq) \longrightarrow 3HPO_4^{2-}(aq) + 6H_2O(l)$$

**d.** Balanced equation:

$$NaHSO_4(aq) + NaHCO_3(aq) \longrightarrow Na_2SO_4(aq) + CO_2(g) + H_2O(l)$$

Net ionic equation:

$$HSO_4^-(aq) + HCO_3^-(aq) \longrightarrow CO_2(g) + SO_4^{2-}(aq) + H_2O(l)$$

**66. b.** Balanced equation:

$$CsHSO_4(aq) + CsOH(aq) \longrightarrow Cs_2SO_4(aq) + H_2O(l)$$

Net ionic equation:

$$HSO_4^-(aq) + OH^-(aq) \longrightarrow SO_4^{2-}(aq) + H_2O(l)$$

**d.** Balanced equation:

$$NaHCO_3(aq) + HCl(aq) \longrightarrow NaCl(aq) + CO_2(g) + H_2O(l)$$

Net ionic equation:

$$HCO_3^-(aq) + H^+(aq) \longrightarrow H_2O(l) + CO_2(g)$$

**70. b.** The products will be solid aluminum chloride and aqueous sodium chloride.
Balanced equation:

$$AlCl_3(aq) + 3NaOH(aq) \longrightarrow 3NaCl(aq) + Al(OH)_3(s)$$

Net ionic equation:

$$Al^{3+}(aq) + 3OH^-(aq) \longrightarrow Al(OH)_3(s)$$

**d.** The products will be aqueous sodium nitrate, water and gaseous carbon dioxide.
Balanced equation:

$$HNO_3(aq) + NaHCO_3(aq) \longrightarrow NaNO_3(aq) + H_2O(l) + CO_2(g)$$

Net ionic equation:

$$H^+(aq) + HCO_3^-(aq) \longrightarrow H_2O(l) + CO_2(g)$$

**72. b.** One mole of nitrogen gas reacts with two moles of oxygen gas to yield two moles of gaseous nitrogen dioxide and 66 kJ of heat.
**d.** One mole of solid sulfur and one mole of oxygen gas react to yield one mole of sulfur dioxide gas and 297 kJ of heat.

**78. b.** $2NH_4^+(aq) + Mg(OH)_2(s) \rightarrow 2NH_3(aq) + 2H_2O(l) + Mg^{2+}(aq)$
**d.** $C_4H_8(l) + 6O_2(g) \xrightarrow{\text{heat}} 4CO_2(g) + 4H_2O(g)$

**80. b.** One mole of aqueous barium ions reacts with one mole of aqueous sulfate ions to yield one mole of solid barium sulfate.
**d.** One mole of liquid ethanol reacts with one mole of oxygen gas when heated in the presence of silver metal to yield one mole of liquid ethanoic acid and one mole of water vapor.

**86.** The data shown in Experimental Problem 8.1 clearly indicate that a small sample size does not necessarily provide enough evidence to support generalized conclusions.

## CHAPTER 9

**1. b.** For every two moles of Fe, one mole of $Al_2O_3$ is produced from a complete reaction.
**d.** For every mole of $Fe_2O_3$, 2 moles of Al are needed for a complete reaction.

**3. a.** $NO + NO_2 \rightarrow N_2O_3$

$$\frac{1 \text{ mol NO}}{1 \text{ mol NO}_2}, \frac{1 \text{ mol NO}_2}{1 \text{ mol NO}}, \frac{1 \text{ mol N}_2\text{O}_3}{1 \text{ mol NO}}, \frac{1 \text{ mol NO}}{1 \text{ mol N}_2\text{O}_3}$$

$$\frac{1 \text{ mol NO}_2}{1 \text{ mol N}_2\text{O}_3}, \frac{1 \text{ mol N}_2\text{O}_3}{1 \text{ mol NO}_2}$$

**b.** $2Al + 6HCl \rightarrow 2AlCl_3 + 3H_2$

$$\frac{2\text{ mol Al}}{6\text{ mol HCl}}, \frac{6\text{ mol HCl}}{2\text{ mol Al}}, \frac{2\text{ mol Al}}{2\text{ mol AlCl}_3}, \frac{2\text{ mol AlCl}_3}{2\text{ mol Al}},$$
$$\frac{2\text{ mol Al}}{3\text{ mol H}_2}, \frac{3\text{ mol H}_2}{2\text{ mol Al}}, \frac{6\text{ mol HCl}}{2\text{ mol AlCl}_3}, \frac{2\text{ mol AlCl}_3}{6\text{ mol HCl}},$$
$$\frac{6\text{ mol HCl}}{3\text{ mol H}_2}, \frac{3\text{ mol H}_2}{6\text{ mol HCl}}, \frac{2\text{ mol AlCl}_3}{3\text{ mol H}_2}, \frac{3\text{ mol H}_2}{2\text{ mol AlCl}_3}.$$

**5. b.** $2C_3H_6 + 9O_2 \xrightarrow{\text{heat}} 6CO_2 + 6H_2O$

$$\frac{2\text{ mol C}_3\text{H}_6}{9\text{ mol O}_2}, \frac{9\text{ mol O}_2}{2\text{ mol C}_3\text{H}_6}, \frac{2\text{ mol C}_3\text{H}_6}{6\text{ mol CO}_2},$$
$$\frac{6\text{ mol CO}_2}{2\text{ mol C}_3\text{H}_6}, \frac{2\text{ mol C}_3\text{H}_6}{6\text{ mol H}_2\text{O}}$$
$$\frac{6\text{ mol H}_2\text{O}}{2\text{ mol C}_3\text{H}_6}, \frac{9\text{ mol O}_2}{6\text{ mol CO}_2}, \frac{6\text{ mol CO}_2}{9\text{ mol O}_2},$$
$$\frac{6\text{ mol H}_2\text{O}}{9\text{ mol O}_2}, \frac{9\text{ mol O}_2}{6\text{ mol H}_2\text{O}}, \frac{6\text{ mol CO}_2}{6\text{ mol H}_2\text{O}}, \frac{6\text{ mol H}_2\text{O}}{6\text{ mol CO}_2}.$$

**c.** $2NaOH + H_2SO_4 \rightarrow Na_2SO_4 + 2H_2O$

$$\frac{2\text{ mol NaOH}}{1\text{ mol H}_2\text{SO}_4}, \frac{1\text{ mol H}_2\text{SO}_4}{2\text{ mol NaOH}}, \frac{2\text{ mol NaOH}}{1\text{ mol Na}_2\text{SO}_4},$$
$$\frac{1\text{ mol Na}_2\text{SO}_4}{2\text{ mol NaOH}}, \frac{2\text{ mol NaOH}}{2\text{ mol H}_2\text{O}}, \frac{2\text{ mol H}_2\text{O}}{2\text{ mol NaOH}},$$
$$\frac{1\text{ mol H}_2\text{SO}_4}{1\text{ mol Na}_2\text{SO}_4}, \frac{1\text{ mol Na}_2\text{SO}_4}{1\text{ mol H}_2\text{SO}_4}, \frac{1\text{ mol H}_2\text{SO}_4}{2\text{ mol H}_2\text{O}},$$
$$\frac{2\text{ mol H}_2\text{O}}{1\text{ mol H}_2\text{SO}_4}, \frac{1\text{ mol Na}_2\text{SO}_4}{2\text{ mol H}_2\text{O}}, \frac{2\text{ mol H}_2\text{O}}{1\text{ mol Na}_2\text{SO}_4}.$$

**7. b.** $2C_6H_6 + 15O_2 \xrightarrow[\text{Pd}]{\text{heat}} 12CO_2 + 6H_2O$

$$\frac{2\text{ mol C}_6\text{H}_6}{15\text{ mol O}_2}, \frac{15\text{ mol O}_2}{2\text{ mol C}_6\text{H}_6}, \frac{2\text{ mol C}_6\text{H}_6}{12\text{ mol CO}_2},$$
$$\frac{12\text{ mol CO}_2}{2\text{ mol C}_6\text{H}_6}, \frac{2\text{ mol C}_6\text{H}_6}{6\text{ mol H}_2\text{O}}, \frac{6\text{ mol H}_2\text{O}}{2\text{ mol C}_6\text{H}_6},$$
$$\frac{6\text{ mol H}_2\text{O}}{15\text{ mol O}_2}, \frac{15\text{ mol O}_2}{6\text{ mol H}_2\text{O}}, \frac{15\text{ mol O}_2}{12\text{ mol CO}_2},$$
$$\frac{12\text{ mol CO}_2}{15\text{ mol O}_2}, \frac{12\text{ mol CO}_2}{6\text{ mol H}_2\text{O}}, \frac{6\text{ mol H}_2\text{O}}{12\text{ mol CO}_2}.$$

**11. b.** 0.75 mol of $H_2$ is required to produce 0.5 mol of $NH_3$.

**13. a.** 10 mol of P are required to produce 2.5 mol $P_4O_6$.
**b.** 3.75 mol of $Br_2$ are required to produce 2.5 mol of $FeBr_3$.

**15. b.** 0.495 mol of $H_2O$ are required to produce 0.33 mol $H_3PO_4$.

**17. b.** 0.0875 mol of $C_{12}H_{24}$ are needed to produce 1.05 mol of $CO_2$.

**21. a.** $Na_2CO_3 + CaCl_2 \rightarrow 2NaCl + CaCO_3$
0.23 mol of $CaCO_3$ is formed when 0.23 mol of $Na_2CO_3$ is reacted with excess $CaCl_2$.
**b.** $Al(s) + 6HCl(aq) \rightarrow 2AlCl_3(s) + 3H_2(g)$
0.99 mol of hydrogen gas is produced when 0.66 mol of Al is reacted with excess hydrochloric acid.
**c.** The equation is balanced.
0.50 mol of $Br_2$ is required to react completely with 0.50 mol of Ca.

**23. b.** $9.03 \times 10^4$ mol of $CO_2$ are produced when $3.01 \times 10^4$ mol of $C_3H_8$ is completely reacted with $O_2$.

**24. b.** $4.02 \times 10^{-2}$ mmol of $CO_2$ is produced when $2.01 \times 10^{-2}$ mmol of $C_2H_6$ reacts with $O_2$ in excess.

**27. a.** 0 mmol of $H_2$ and 125 mmol of $O_2$ remain unreacted.
**b.** 190 mol of $H_2$ and 4730 mol of $N_2$ remain unreacted.

**31.** At the end of the reaction 0.3 mol of $ZnBr_2$ is produced and 0.9 mol of Zn remain.

**33.** 0.60 mol of magnesium metal remained at the end and 0.50 mol of $MgCl_2$ was produced.

**37.** 1.7 mol of methane and 26.7 mol of ammonia remain unreacted. 23.3 mol of hydrogen cyanide are formed.

**39. b.** Mass of 0.030 mol of MgSO = 3.61 g
**d.** Mass of 0.50 mol of $K_2CO_3$ = 69.1 g

**41. b.** 1.6 g $CaF_2$ **d.** 0.536 g AgCN

**45. b.** 8034 g **d.** 504 g

**47. b.** $3.42 \times 10^{-4}$ mol NaCl **d.** 7.5 mol $Fe_2(CO_3)_3$

**49. b.** $2.7 \times 10^6$ mol $CO_2$ **d.** $4 \times 10^{-8}$ mol PbS

**51. b.** $8.8 \times 10^{-2}$ mol $C_{12}H_{22}O_{11}$
**d.** $1.27 \times 10^4$ mol $CaCO_3$

**53. b.** 60.19 mol $Na_2CrO_4$

**55. b.** 180 mol $CaCl_2$

**59. a.** 5.07 mol $C_6H_6$ **b.** $4.1 \times 10^{-2}$ mol $C_{10}H_{10}O_4$

**65. b.** 131 g $H_2$

**67.** 206 g $NaHSO_4$

**75.** 0.82 g KOCl

**79.** 3030 g $CO_2$

**83.** $SiO_2 + 4HF \rightarrow SiF_4 + 2H_2O$
2 mol of $SiF_4$ are formed.

**87.** $2Fe + 3Br_2 \rightarrow 2FeBr_3$
4.83 mol of $FeBr_3$ are formed.

**93. b.** CuCl is the limiting reactant.
16.9 g $CuCl_2$ are formed.
**d.** $N_2$ is the limiting reactant.
15.3 g of $NH_3$ are formed.

**95.** $2Al + 3Br_2 \rightarrow 2AlBr_3$
$Br_2$ is the limiting reactant.
2.22 g of $AlBr_3$ are formed.

**101.** $Na_2CO_3(aq) + CaCl_2(aq) \rightarrow CaCO_3(s) + 2NaCl(aq)$
The limiting reactant is $Na_2CO_3$.
47.2 g $CaCO_3$ is formed.

**105.** The limiting reactant is $AgNO_3$.
4.21 g AgCl is formed.

**109.** The limiting reactant is $Pb(NO_3)_2$
1.05 g $PbCl_2$ are formed.

**111. b.** 25.5% Cl **d.** 81.6% Cl

**115. b.** 4.20% H **d.** 63.6% N

**117. b.** 52.0% Cr 48.0% O **d.** 23.6% K 76.4% I
**119. b.** 84.1% C 15.9% H
**d.** 24.7% K 34.8% Mn 40.5% O
**121. b.** The sample contains 14 kg of nickel.
**d.** The sample contains 8.63 mg of barium.
**125. b.** 46.7 mg N **d.** 142.8 mg Cl
**127. b.** 4.6 g Al **d.** 11 kg P
**131. b.** Empirical formula = CuCl
**135. b.** Empirical formula = $N_2O_3$
**139. b.** Empirical formula = KClO
**141. b.** Empirical formula = $C_3H_5O_2$
**d.** Empirical formula = HO
**145. b.** Molecular formula = $C_6H_9$
**d.** Molecular formula = $C_6H_3N_3O_6$
**147. b.** Molecular formula = CO
**d.** Molecular formula = $C_8H_8$
**151. b.** 230 g $CO_2$ **c.** 10 kg $H_2O$
**153. b.** 5 mol $CH_4$ contains more carbon.
**d.** 1 mol of $Ca(HCO_3)_2$ contains 2 mol of C. One mol of $CO_2$ contains 1 mol of C. One mol of $Ca(HCO_3)_2$ contains a larger mass of carbon.
**155.** $4AsH_3(g) + 3O_2(g) \rightarrow 6H_2O(g) + 4As(s)$

## CHAPTER 10

**5.** Covalent bonding is present in $PbCl_4$, which is why its melting point is so low and it is easily decomposed.

**7. b.** $Li^+ :\ddot{O}:^{2-} Li^+$

**9. b.** $CaCl_2$

$:\ddot{Cl}:^- Ca^{2+} :\ddot{Cl}:^-$

**11. b.** $:\ddot{Cl}—\ddot{Cl}:$

**d.** H—$\ddot{S}$—H

**13. b.** $NF_3$

:F—N—F:
   |
  :F:

**d.** $H_2N—NH_2$

H  H
|  |
H—N—N—H

**15. b.** C≡O

H
|
**d.** H—C=N—H

**19.** [H—O—C(=O)—O:]$^-$ ⟷ [H—O—C(—O:)=O]$^-$

| | | | | |
|---|---|---|---|---|
| **27. b.** | :O: ‖ C with :F: and :F: | 120 | > ideal angle | The repulsive forces of the C=O are greater than those of the C—F bonds so the F—C—F angle is less than 120°. If the F—C—F angle decreases then the F—C—O angle must increase. |

**31. a.**

| | Lewis Diagram | Shape | Approx. Bond Angle/° | Explanation |
|---|---|---|---|---|
| **a.** | :Cl—O—Cl: | bent | 109.5 | This is a tetrahedron of electron pairs. Two of these are unbonded and thus "invisible" when determining molecular geometry. |
| **b.** | $SiH_4$ (H—Si with three H) | tetrahedral | 109.5 | The silicon atom has four groups of electrons around it and there four atoms attached to it. |
| **c.** | $:\ddot{N}$ with three :Cl: | trigonal pyramidal | 109.5 | There is a tetrahedron of electron pairs, one pair is unbonded, which means that it is "invisible," similar to those in part (a). |

**35. b.** $[:\ddot{O}—C≡N:]^-$ linear 180

**37. b.** $[H—\ddot{C}(H)—H]^-$ trigonal pyramidal 109.5 See question 31c.

**41. a.** :Br—F: 180°

**b.** H—C(=O)—H 120°

c. 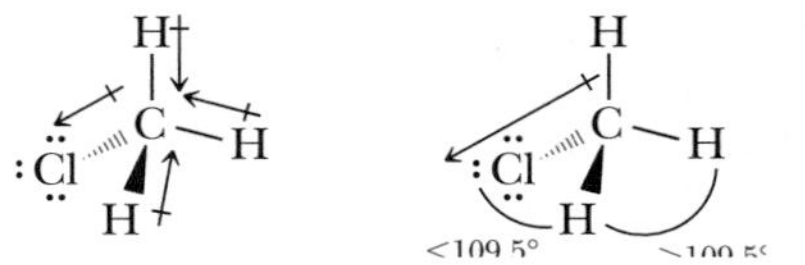 (3-D structures of $CH_2Cl$ ... bond angles $<109.5°$, $>109.5°$)

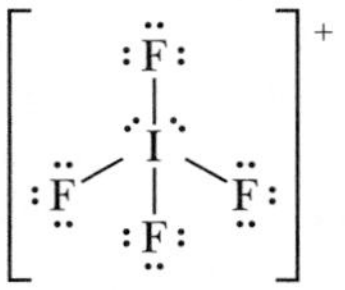

49. a. 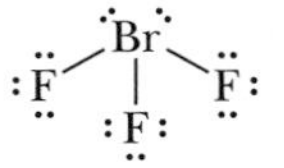 $BrF_3$ Lewis diagram
b. 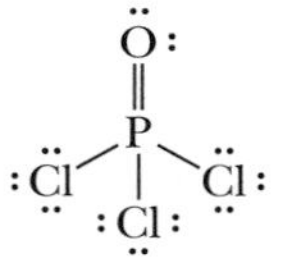 $POCl_3$ Lewis diagram
c. $[IF_4]^+$ Lewis diagram
d. 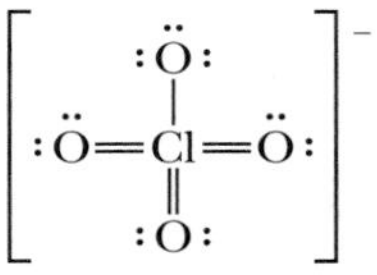 $[ClO_4]^-$ Lewis diagram

53. b.

| Lewis Diagram | Shape of the Molecule |
|---|---|
| 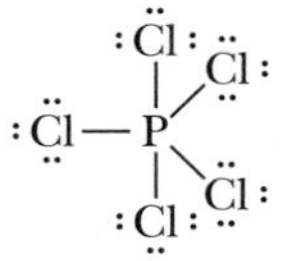 $PCl_5$ | trigonal bipyramidal |

59. Resonance structures that obey the octet rule.
The double bond can be formed with any of the three oxygen atoms.

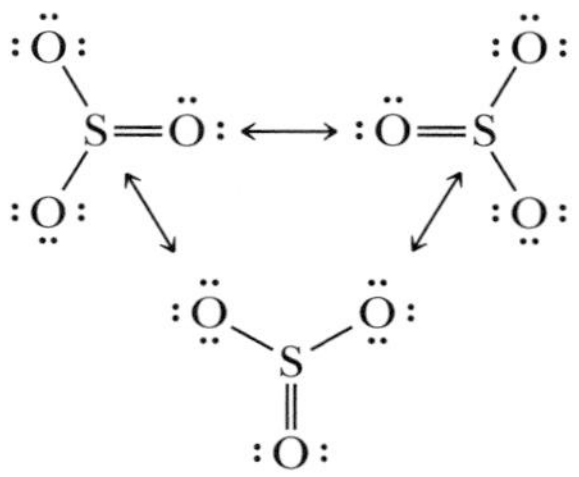

Diagram of $SO_3$ that does not obey the octet rule.
Three double bonds are made to the sulfur atom.

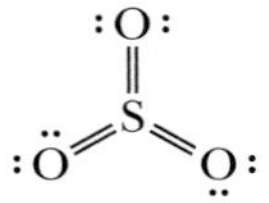

61. Lewis diagram of $ClO_4^-$ that obeys the octet rule and violates the rule that states that oxygen should have two bonds.

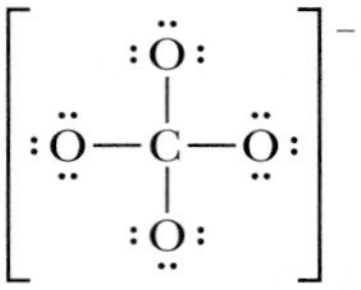

Lewis diagram of $ClO_4^-$ that violates the octet rule but obeys the rule that says that each oxygen should have two bonds to it.

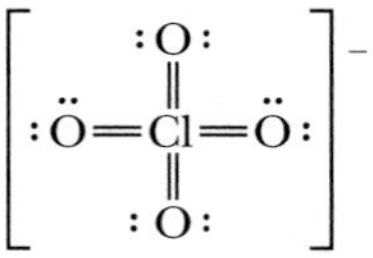

67. b. The molecule is ammonia.

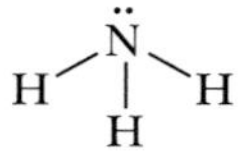

d. The compound is $AlI_3$, which is ionic.

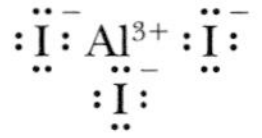

## CHAPTER 11

2. $1.43 \times 10^{20}$ mol $N_2$
6. Earth has $O_2$ present in its atmosphere whereas Venus and Mars have none because on earth there are plants that produce $O_2$.
10. D. Ozone absorbs high-energy radiation as it approaches the planet and prevents it from reaching earth.
12. C. Will decrease the pressure inside of the container.
16. $2.59 \times 10^4$ mm Hg
34.0 atm

19.

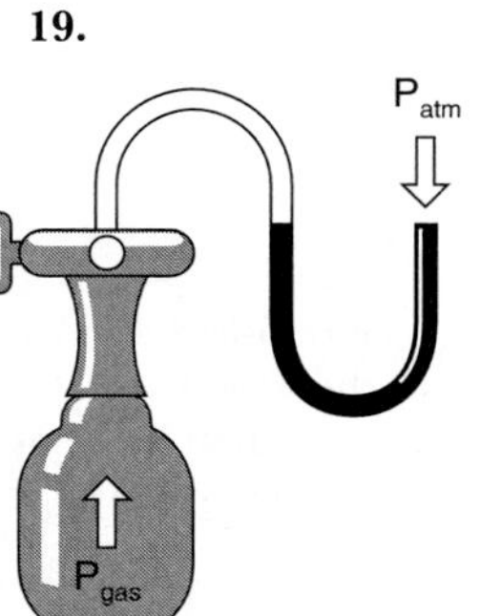


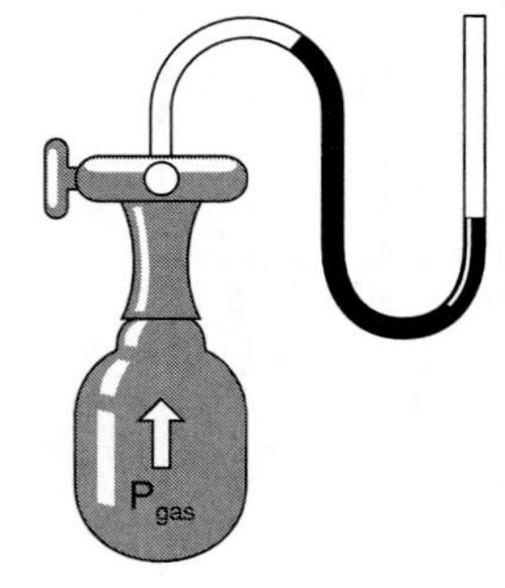


Original
Level of Mercury

Since the pressure of the gas is less than atmospheric pressure, it will support a shorter column of mercury than the atmosphere does.

**20.** 815 mm Hg
**22.** $a = m\ s^{-2}$

**24. b.** $0.998\ \text{atm}\ \dfrac{(14.7\ \text{psi})}{(1\ \text{atm})} = 14.7\ \text{psi}$

**d.** $740\ \text{torr}\ \dfrac{(1.013 \times 10^5\ \text{Pa})}{(760\ \text{torr})}\ \dfrac{(1\ \text{kPa})}{(1000\ \text{Pa})} = 98.6\ \text{kPa}$

**26. b.** $45\ \text{atm}\ \dfrac{(1\ \text{atm})}{(14.7\ \text{psi})} = 3.1\ \text{atm}$

**d.** $1.21 \times 10^3\ \text{torr}\ \dfrac{(1.013 \times 10^5\ \text{Pa})}{(760\ \text{torr})}\ \dfrac{(1\ \text{kPa})}{(1000\ \text{Pa})} = 1.61 \times 10^2\ \text{kPa}$

**30.** $1.6\ \text{lb ft}^{-2}$ 0.57 mm Hg
**34.** $8.3 \times 10^{-5}$ psi $5.7 \times 10^{-6}$ atm
**36. b.** 184 mm Hg
**38. b.** 1.59 atm **c.** 48 atm
**42. b.** 15.2 mL
**46.** $8.5 \times 10^5\ m^3$
**50.** When the volume increases the molecules now have more room in which to move. They collide with the walls less frequently and thus the resultant force that they exert on the walls decreases. Pressure = force/area so if the force decreases then the pressure will also decrease.
**54. b.** 11.2 mL
**56. b.** 22.7 mL **c.** $4.5 \times 10^5$ L
**60.** 17.3 mL
**64.** 446 mL
**68. b.** $4.61 \times 10^{-4}$ L **c.** $8.22 \times 10^{-2}$ mL
**70. a.** 1.6 mL
**72. b.** 0.41 atm
**74. b.** 0.85 atm
**76.** 4900 L
**82. a.** Volume = 22.4 L **c.** Volume = $2.80 \times 10^{-2}$ L
**b.** Volume = 560 L
**86. b.** $3.52 \times 10^{-3}$ L
**90.** Volume of $Cl_2$ needed = 126 L
**96. a.** $6.1 \times 10^{-5}$ mol $SO_2$ **c.** $4.5 \times 10^{-4}$ mol $H_2O$
**100. b.** 3.85 L
**104. b.** $1.01 \times 10^{-3}$ atm
**108.** 0.606 L
**114. a.** $2.97\ g\ L^{-1}$ **b.** $2.86\ g\ L^{-1}$ **c.** $0.541\ g\ L^{-1}$
**118. a.** $63.9\ g\ mol^{-1}$
**122.** Molecular formula = $C_2H_2$
**126.** % of water in the air = 0.317%
**130. b.** 1.6 atm **c.** Flask I
**134.** 0.936
**138.** 1.87
**144.** At low pressures and high temperatures the molecules of a gas are far apart and thus the volume occupied by the gas molecules is inconsequential when compared to the actual volume of the gas.
**148.** The production of water vapor expels all of the air inside the can. When the can is stoppered and cooled, all of the water vapor condenses and there is very little air in the can. The pressure inside of the can is much less than the pressure outside of the can. This pressure difference causes the can to collapse.
**152.** Molar mass = $70.00\ g\ mol^{-1}$
Empirical formula = $CH_2$
Molecular formula = $C_5H_{10}$
**154. b.** 12.3 atm
**156.** The flask in step 3 will contain more sample if $P_{atm}$ = 765 mm Hg because the increased atmospheric pressure prevents some of the gas from escaping the flask when compared to $P_{atm}$ = 752 mm Hg.
**162.** The result would have been identical because if more liquid was used then more gas would have been expelled from the container so that the pressure inside of the container would be equal to that outside of the container. This would have resulted in their obtaining the same mass of liquid remaining in the flask upon its cooling.

## CHAPTER 12

**4.** A
**8.** Helium
**12.** Chlorine, hydrogen, bromine, water, and hydrogen chloride gas.
**18.** D
**22.**

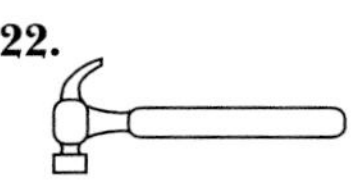

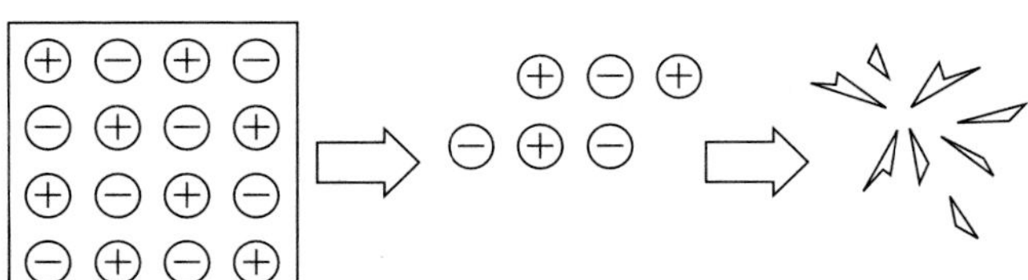

Block of Sodium Chloride

The Crystal Shatters, Producing Small Pieces of Sodium Chloride

**26. b.** $H_2N—NH_2$ can form hydrogen bonds because the hydrogen atoms are attached to nitrogen atoms and the nitrogen atom has an unbonded pair of electrons.

**d.** $CH_3CH_2CH_2CH_3$ cannot form hydrogen bonds because the hydrogen atoms are not bonded to electronegative elements.

**28. b.** $(CH_3CH_2)_2O$ can engage in hydrogen bonding with water molecules.

$CH_3CH_2$ and $CH_3CH_2$ bonded to O, which is hydrogen-bonded to $\overset{\delta+}{H}$ of two water molecules ($\overset{\delta-}{O}$ with $\overset{\delta+}{H}$ $\overset{\delta+}{H}$).

**d.** $CH_3CH_2OH$ can participate in hydrogen bonding with water molecules.

$CH_3CH_2$—O—$\overset{\delta+}{H}$, with O hydrogen-bonded to $\overset{\delta+}{H}$ of one water molecule and $\overset{\delta+}{H}$ hydrogen-bonded to another ($\overset{\delta+}{H}$—$\underset{\delta-}{O}$—$\overset{\delta+}{H}$).

**32.** All have induced dipole-induced dipole forces. Hydrogen bonding can occur in (a); (b) and (c) are polar and have dipole-dipole forces.

**36.** *(The electron cloud of the argon atom is deformed away from the* $I^-$ *ion.)*

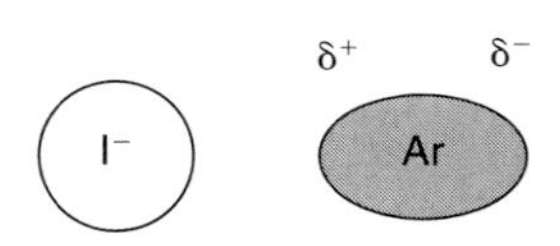


**40.** $CH_3CH_2CH_2CH_2CH_2CH_2CH_2CH_2CH_2CH_2CH_2—OH$

**44.**

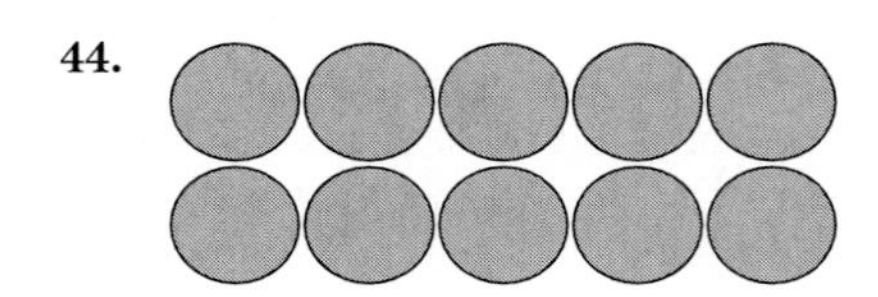

**46. b.** Heat absorbed = 57 kJ

**d.** Heat absorbed = 4.03 J

**48. b.** Heat absorbed = 400 kJ

**d.** Heat absorbed = 8.4 kJ

**52. b.** Heat required = $1.7 \times 10^4$ kJ

**d.** Heat required = 280 kJ

**54. b.** Heat required = 340 kJ

**d.** Heat required = 14.1 J

**58. b.** Heat required = $7.8 \times 10^7$ kJ

**d.** Heat required = $1.7 \times 10^4$ kJ

**62.** The perspiration is evaporated by the heat from the body. This loss of heat lowers the body's temperature and keeps you cool.

**64.** As the strength of the intermolecular forces present in a liquid increases, the molar heat of vaporization of the liquid increases.

**68.** Yes it does, because the molar mass obtained from Experiment 1 is approximately one and a half times the expected molar mass of acetic acid (60 g $mol^{-1}$).

**70.** These experimental results do support his hypothesis because they show that at different temperatures and conditions the concentration of dimers present in a sample of acetic acid differs. The temperature at which Experiment 2 was conducted is higher than that of Experiment 1. This would cause some of the hydrogen bonding between the molecules present in the sample in Experiment 1 to break so that the concentration of dimers present decreases. This is reflected by the decrease in the molar mass obtained in Experiment 2. The key thing to note is that the molar mass obtained in Experiment 2 is still higher than the expected molar mass of 60 g $mol^{-1}$, which shows that some dimers are present in the sample even under these conditions.

## CHAPTER 13

**2. b.** Water is the solvent and ammonium chloride is the solute.

**d.** Sodium hypochlorite is the solute and water is the solvent.

**4. b.** Iron is the solvent and carbon is the solute.

**d.** Gasoline is the solvent and ethanol is the solute.

**8.** C; $CH_3CH_2CH_2CH_3$ is the least soluble in water.

**12.** The strongest intermolecular forces present in 1-octanol are van der Waals forces of attraction. Compounds that dissolve readily in 1-octanol also have van der Waals forces as their strongest intermolecular forces.

**16.** Carbon–graphite or diamond.
Vegetable oil

**22.** The strongest intermolecular forces of attraction between propane molecules and butane molecules are induced dipole-induced dipole forces so they dissolve in one another in all proportions.

**26.** $CH_3CH_2CH_2CH_2CH_2CH_3$
〰〰〰 van der Waals forces
$CH_3CH_2CH_2CH_2CH_3$

**28. b.** insoluble **d.** soluble

**32. b.** insoluble **d.** soluble

**38.** Most ionic compounds do not dissolve in water because the ionic bonds between the ions in the crystal are much stronger than any ion-dipole forces that can be formed between water molecules and the ions.

**40. b.** At 0°C solubility = 29.7 g per 100 mL

Amount of ammonium chloride that does not dissolve = 50.0 g − 29.7 g = 20.3 g

**42. a.** Dissolved Cu = 0.062 g

Amount that remains undissolved = 0.94 g

**48.** Amount that precipitates = 6.5 g

**52. b.** $3NaOH(aq) + Al(NO_3)_3(aq) \rightarrow 3NaNO_3(aq) + Al(OH)_3(s)$

Net ionic equation:

$$Al^{3+}(aq) + 3OH^-(aq) \longrightarrow Al(OH)_3(s)$$

**54. b.** $FeCl_3(aq) + 3KOH(aq) \rightarrow Fe(OH)_3(s) + 3KCl(aq)$
Net ionic equation:

$$Fe^{3+}(aq) + 3OH^-(aq) \longrightarrow Fe(OH)_3(s)$$

**58. b.** $2Bi(NO_3)_3(aq) + 3(NH_4)_2S(aq) \rightarrow 6NH_4NO_3(aq) + Bi_2S_3(s)$

Net ionic equation:

$$2Bi^{3+}(aq) + 3S^{2-}(aq) \longrightarrow Bi_2S_3(s)$$

**64.**

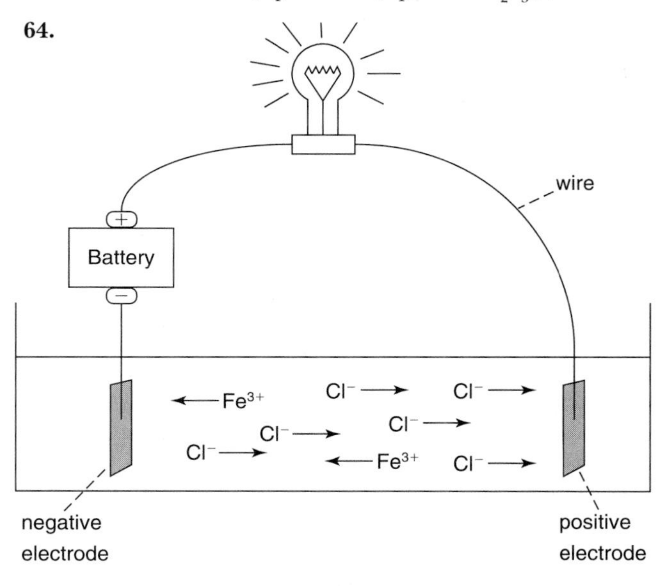


**70. a.** strong electrolyte **c.** strong electrolyte
**b.** strong electrolyte

**72.** The answer is C. Blue verdigris is a strong electrolyte because it dissolves in water to form a solution that is a good conductor of electricity. In order for it to do this it must fully dissolve.

**76. a.** 0.44 *M* **b.** 3.21 *M* **c.** $5.20 \times 10^{-2}$ *M*

**78. b.** $8.67 \times 10^{-2}$ *M*

**80. b.** 0.185 *M*

**82. b.** $3.99 \times 10^{-2}$ *M*

**84. b.** 0.141 *M*

**86. b.** 0.142 *M*

**88. b.** 0.28 mol **d.** $3.91 \times 10^{-3}$ mol

**92. b.** $6.25 \times 10^{-2}$ mol
**d.** $5.69 \times 10^{-4}$ mol

**96. b.** $m = (0.5\ L)(0.100\ mol\ L^{-1})\dfrac{(84.10\ g\ NaHCO_3)}{(1\ mol)}$

$= 4\ g\ NaHCO_3$

Add the 4 g of $NaHCO_3$ to 0.25 L of water. Swirl the flask until all of the solute has dissolved. Add water until the volume of the solution is 0.5 L. Swirl the flask until the solution is uniform.

**98. b.** $V_{conc} = \dfrac{V_{dil}M_{dil}}{M_{conc}}$

$= \dfrac{(0.750\ L)(1.50\ M)}{(3.00\ M)}$

$= 0.375\ L = 375\ mL$

Measure out 375 mL of 3.00 *M* NaOH. Add it slowly to 375 mL of water. Swirl the flask to mix thoroughly. Add water until the volume of the solution is exactly 750 mL. Swirl the flask to mix thoroughly.

**102. b.** Volume of 18 *M* $H_2SO_4$ needed:

$$V_{conc} = \frac{V_{dil}\,M_{dil}}{M_{conc}}$$

$$= \frac{(0.010\ L)(0.75\ M)}{(18\ M)}$$

$$= 4.2 \times 10^{-4}\ L = 0.42\ mL$$

Measure out 0.42 mL of 18 M $H_2SO_4$. Add it slowly and carefully to 5.0 mL of water. Swirl the flask to mix thoroughly. Add water until the volume of the solution is exactly 10 mL. Swirl the flask to mix thoroughly.

**104. b.** 4.9% $KNO_3$

**108. b.** 4.5% $CuCl_3$ **c.** 9.09% $CoSO_4$

**110.** 260 g of the solution are needed.

**116.** 51 g

**120. b.** 1. Volume of 2-propanol = (0.05)(250 mL) = 12.5 mL

Volume of hexane = 250 mL – 12.5 mL = 238 mL

2. Measure out 12.5 mL of 2-propanol into a flask.
3. Measure out 238 mL of hexane.
4. Mix thoroughly.

**124. b.** 1.0 m

**126. b.** 2.75 m

**128. b.** 0.414

**130. b.** $2.9 \times 10^{-3}$

**134.** $CH_3CH_2OH$ gives the smallest decrease, because when it is dissolved in solution it gives the smallest number of particles per molecule. (See answer to Problem 133.)

**138. b.** 7.2 atm

**140. b.** 0.72 atm

**142. b.** 23.9 atm

## CHAPTER 14

**2. b.** The rate of the reaction will be halved.
**6. b.** The reaction rate decreases by a factor of one half.
**8.** Yes
**12.** $min^{-1}$
**13. b.** $k = 9.33\ mol^{-1}\ L^2\ min^{-1}$
**16. b.** $k = 3.05\ mol^{-1}\ L^2\ min^{-1}$
**18.** The energy factor can be increased by an increase in temperature.
**22.** Activation energy = 2.2 kJ
**26.** Number of product molecules produced in 1 second = $1.33 \times 10^3$
**30.** Number of product molecules formed in 1 second = $1.05 \times 10^3$
**32. b.** As long as the temperature remains constant the rate constant remains the same.
**34. b.** The rate constant remains the same at the beginning of the reaction and at the end.
**36. a.** When the catalyst is removed both the forward and the reverse rates of reaction decrease.

**40. a.** $K_{eq} = \frac{[Cl_2(g)][CO(g)]}{[COCl_2(g)]}$

**b.** $K_{eq} = \frac{[H_2CO_3(g)]}{[CO_2(g)][H_2O(g)]}$

**c.** $K_{eq} = \frac{[CO_2(g)][H_2(g)]}{[CO(g)][H_2O(g)]}$

**46. a.** $K_{eq} = \frac{1}{[Cl_2(g)]^2}$

**b.** $K_{eq} = \frac{1}{[O_2(g)]^2}$

**c.** $K_{eq} = \frac{[HBr(g)][CH_3Br(g)]}{[CH_4(g)][Br_2(g)]}$

**50.** $N_2$ is present at a higher concentration.
**54.** $2HF(g) \rightleftharpoons H_2(g) + F_2(g)$

$K_{eq} = 2.0 \times 10^{-13}$

**58.** $CaCO_3(s) \rightleftharpoons CaO(s) + CO_2(g)$

$K_{eq} = 0.75$

**60. b.** $K_{sp} = [Cd^{2+}]^3[PO_4^{3-}]^2$
**64. b.** $1.67 \times 10^{-8}\ mol\ L^{-1}$
**66. b.** $1.4 \times 10^{-6}\ mol\ L^{-1}$
**68. b.** $4.73 \times 10^{-6}\ mol\ L^{-1}$
**70. b.** $1.3 \times 10^{-4}\ mol\ L^{-1}$
**72. b.** $K_{sp} = 7.7 \times 10^{-19}$
**76. b.** $K_{sp} = 4.3 \times 10^{-38}$
**78.** CuS precipitates out of the solution first.
**80. b.** There are three moles of product gases and two moles of reactant gases. If the pressure is increased then the equilibrium shifts towards the reactants. The concentration of $H_2S$ increases and the concentrations of $H_2$ and $S_2$ decrease.
**82. b.** There are four moles of reactant gases and two moles of product gases. If the pressure is decreased the equilibrium shifts towards the reactants. The concentrations of $N_2$ and $H_2$ increase while the concentration of $NH_3$ decreases.
**84. a.** If the concentration of oxygen is decreased the equilibrium shifts towards the side of the reactants. This means that the concentration of hydrogen peroxide decreases while the concentrations of hydrogen and oxygen increase.
**b.** If the concentration of hydrogen peroxide increases the equilibrium shifts towards the reactants. The concentration of hydrogen peroxide decreases while the concentrations of hydrogen and oxygen increase.
**90. b.** Heating the mixture drives the equilibrium to the right. This is done to remove the added heat from the system.
**92.** Since the equilibrium constant of the forward reaction is so much less than 1, it can be assumed that all of the $H_2S$ is used up to form $H_2$ and $S_2$. This means that the concentration of $H_2$ at equilibrium is 1.5 *M*.

## CHAPTER 15

**2. b.** $H_2O$—acid
$CN^-$—base
$OH^-$—base
HCN—acid
**4. b.** Net ionic equation:

$$OH^- + NH_4^+ \rightleftharpoons NH_3 + H_2O$$

$OH^-$—base
$NH_4^+$—acid
$NH_3$—base
$H_2O$—acid

**6. b.** Net ionic equation:

$$HClO_4 + H_2PO_4^- \rightleftharpoons H_3PO_4 + ClO_4^-$$

$HClO_4$—acid
$H_2PO_4^-$—base
$H_3PO_4$—acid
$ClO_4^-$—base
**8. b.** conjugate acid—$CH_3OH_2^+$
conjugate base—$CH_3O^-$
**10. b.** conjugate acid—$H_4BO_3^+$
conjugate base—$H_2BO_3^-$

**12. b.** conjugate acid—$H_3PO_4$
conjugate base—$HPO_4^{2-}$
**14. b.** conjugate base—$HBO_3^{2-}$
conjugate acid—$H_3BO_3$
**16. b.** $NH_3(g) + HCl(g) \rightarrow NH_4Cl(s)$
**20. b.** $H_3O^+(aq) + CH_3NH_2(aq) \rightarrow H_2O(l) + CH_3NH_3^+(aq)$
**22. b.** $H_2SO_4 + H_2BO_3^- \rightarrow H_3BO_3 + HSO_4^-$
**26.** D—$H_2O$, because oxygen is to the left of fluorine and chlorine, and it is above sulfur in the periodic table.
**30. b.** $NH_4^+$ is a stronger acid than $NH_3$ because it has a positive charge whereas $NH_3$ is uncharged.
**32. b.** $HNO_3$ is the stronger acid because it has three oxygen atoms whereas $HNO_2$ only has two.
**34. b.** $OCl^-$ is the stronger base. Its conjugate acid, HOCl, is a weaker acid than the conjugate acid of $Cl^-$, HCl.
**36. b.** $S^{2-}$ is the stronger base because it has a higher negative charge than $HS^-$.
**40. b.** $NH_4NO_3$ is acidic. $NH_4^+$ is acidic and $NO_3^-$ is neutral.
**42. b.** $NH_4Cl + NaOH(aq) \rightarrow$
$NH_3(g) + H_2O(l) + NaCl(aq)$
The final solution is basic because $NH_3(g)$ is very soluble in water. It reacts according to the following equation:

$$NH_3(g) + H_2O(l) \longrightarrow NH_4^+(aq) + OH^-(aq)$$

**46. a.** For 0.50 L to have a concentration of 0.25 *M* it must contain 0.125 mol of solute. One mole of $HClO_4$ reacts with one mole of NaOH to produce one mole of $NaClO_4$. 0.125 L of 1.0 *M* $HClO_4$ contains 0.125 mol of $HClO_4$. 62.5 mL of 2.0 *M* NaOH solution contains 0.125 mol of NaOH. Measure out 0.125 L of 1.0 *M* $ClO_4$ solution and 62.5 mL of 2.0 *M* NaOH in separate flasks. Slowly, and with constant stirring, add the acid to the base and swirl until the mixture is uniform. Then add water until the total volume of the solution is 0.50 L. Mix thoroughly.
**b.** For a 0.50 L solution to have a concentration of 0.25 *M* it must contain 0.125 mol of solute. One mole of $CH_3CO_2H$ reacts with one mole of NaOH to produce one mole of $CH_3CO_2Na$.
Measure out 0.0625 L of 2.0 *M* $CH_3CO_2H$ and 0.042 L of 3.0 *M* NaOH into separate flasks. Carefully add the acid to the base and swirl until the mixture is uniform. Add water until the volume of the mixture is 0.50 L. Mix thoroughly.

**48. b.** $HCN(aq) + H_2O(l) \rightleftharpoons H_3O^+(aq) + CN^-(aq)$

$$K_a = \frac{[CN^-][H_3O^+]}{[HCN]}$$

**52. b.** $HSO_4^-(aq) + H_2O(l) \rightleftharpoons H_3O^+(aq) + SO_4^{2-}(aq)$

$$K_a = \frac{[SO_4^{2-}][H_3O^+]}{[HSO_4^-]}$$

**56. b.** $K_b = \dfrac{[HBrO_2][OH^-]}{[BrO_2^-]}$

**58.** $KH(s) + H_2O(l) \rightarrow KOH(aq) + H_2(g)$
**62. b.** $[H_3O^+][OH^-] = 1.00 \times 10^{-14}$

$$[OH^-] = \frac{1.00 \times 10^{-14}}{1.51 \times 10^{-11}} = 6.62 \times 10^{-4}\ M$$

The solution is basic.
**64. b.** $[H_3O^+][OH^-] = 1.00 \times 10^{-14}$

$$[OH^-] = \frac{1.00 \times 10^{-14}}{9.44 \times 10^{-3}} = 1.06 \times 10^{-12}\ M$$

The solution is acidic.
**66. b.** $[H_3O^+][OH^-] = 1.00 \times 10^{-14}$

$$[OH^-] = \frac{1.00 \times 10^{-14}}{1.05 \times 10^{-8}} = 9.52 \times 10^{-7}\ M$$

The solution is basic.
**70. b.** $[H_3O^+][OH^-] = 1.00 \times 10^{-14}$

$$[H_3O^+] = \frac{1.00 \times 10^{-14}}{7.29 \times 10^{-11}} = 1.37 \times 10^{-4}\ M$$

The solution is acidic.
**72. b.** $[H_3O^+][OH^-] = 1.00 \times 10^{-14}$

$$[H_3O^+] = \frac{1.00 \times 10^{-14}}{4.66 \times 10^{-3}} = 2.15 \times 10^{-12}\ M$$

The solution is basic.
**74. b.** $\log(3.1 \times 10^{-12}) = -11.51$
**78. b.** $\log(8.1 \times 10^{-4}) = -3.09$
**80. b.** $10^{12.9} = 7.94 \times 10^{12}$
**84. c.** $10^{-27.4353} = 3.67029 \times 10^{-28}$
**86. b.** acidic **d.** basic
**88. b.** basic **d.** basic
**92. b.** $pH = -\log(3.2 \times 10^{-7}) = 6.5$
**d.** $pH = -\log(5.77 \times 10^{-9}) = 8.24$
**94. b.** $pH = -\log(8.1 \times 10^{-7}) = 6.1$
**d.** $pH = -\log(7.78 \times 10^{-4}) = 3.10$
**98. b.** $[H_3O^+] = 10^{-8.9} = 1.3 \times 10^{-9}\ M$

$$[OH^-] = \frac{1.00 \times 10^{-14}}{1.3 \times 10^{-9}} = 7.9 \times 10^{-6}\ M$$

**d.** $[H_3O^+] = 10^{-11.22}$
$= 6.026 \times 10^{-12}\ M$

$$[OH^-] = \frac{1.00 \times 10^{-14}}{6.026 \times 10^{-12}}$$
$= 1.660 \times 10^{-3}\ M$

**102. b.** $[H_3O^+] = 10^{-2.25}$
$= 5.62 \times 10^{-3}\ M$

$$[OH^-] = \frac{1.00 \times 10^{-14}}{5.62 \times 10^{-3}}$$
$= 1.78 \times 10^{-12}\ M$

**d.** $[H_3O^+] = 10^{-7.45}$
$= 3.55 \times 10^{-8}\ M$

$$[OH^-] = \frac{1.00 \times 10^{-14}}{3.55 \times 10^{-8}}$$
$= 2.82 \times 10^{-7}\ M$

**106.** 7.0–8.0

**114.** $x = [F^-] = [H_3O^+]$
$x = 1.3 \times 10^{-2}\ M$

$$[OH^-] = \frac{1.00 \times 10^{-14}}{1.3 \times 10^{-2}}$$
$= 7.7 \times 10^{-13}$
$pH = -\log(1.3 \times 10^{-2})$
$= 1.9$

**118.** pH = 3.1

**124.** $K_a = 1.1 \times 10^{-7}\ M$

**126. b.** This mixture does not yield a buffer solution. The solution obtained contains only $CH_3CO_2^-$ ions, $K^+$ ions and $OH^-$ ions. It does not contain the conjugate acid of $CH_3CO_2^{2}$, $CH_3CO_2H$.

**128. b.** This mixture does not produce a buffer. Both $CH_3CO_2H$ and HBr are acids.

**132.** pH = 4.1

**138.** 0.152 *M*

**142.** 0.475 *M*

**146.** 0.230 *M*

**148.** pH = 11.3
$[OH^-] = [NH_4^+] = 2.12 \times 10^{-3}\ M$, $[NH_3] = 0.250\ M$

## CHAPTER 16

**5. b.** Ca is the **reducing agent.** The bromine has been converted to $Br^-$ ions. For this to occur it must have gained electrons. The bromine has been **reduced.** The bromine is the **oxidizing agent.**

**6. b.** $C_6H_5CHO$—reducing agent; it has been oxidized.
$C_6H_5CHO$—oxidizing agent; it has been reduced.

**8. b.** $Fe^{3+}$—oxidizing agent; it has been reduced.
$I^-$—reducing agent; it has been oxidized.

**10. b.** $Cu^{2+}$—oxidizing agent; it has been reduced.
Cu(*s*)—reducing agent; it has been oxidized.

**12. b.** $Fe_2O_3$—oxidizing agent; it has been reduced.
Al—reducing agent; it has been oxidized.

**14. b.** $I_2$ is the more powerful oxidizing agent. It is neutral and can accept electrons more readily than $I^-$.

**16. b.** $Cl_2$ is the more powerful oxidizing agent. It is more electronegative than Na.

**18. b.** CO is the more powerful reducing agent because it has less oxygen than $CO_2$.

**20. b.** $Sn(s) \rightarrow Sn^{2+} + 2e^-$

**26. a.** No reaction occurs.
**b.** No reaction occurs.
**c.** $MnO_4^- + 8H_3O^+ + 5e^- \rightarrow Mn^{2+} + 12H_2O$

**28. b.** Unbalanced equation:

$$H_2 + AgCl \longrightarrow HCl + Ag$$

Half reactions:

$$H_2 \longrightarrow HCl(1)$$
$$AgCl \longrightarrow Ag(2)$$

Half reaction (1)

$H_2 + \mathbf{2Cl^-} \longrightarrow \mathbf{2HCl}$
(balancing Cl atoms and H atoms)
$H_2 + 2Cl^- \longrightarrow 2HCl + \mathbf{2e^-}$ (balancing charge)

Half reaction (2)

$AgCl + \mathbf{e^-} \longrightarrow Ag + Cl^-$
(balancing charge and Cl atoms)

Multiply half reaction (2) by 2

$$\mathbf{2}AgCl + 2e^- \longrightarrow 2Ag + 2Cl^-$$

Add half reactions:

$$2AgCl + 2e^- \longrightarrow 2Ag + 2Cl^-$$
$$H_2 + 2Cl^- \longrightarrow 2HCl + 2e^-$$
$$2AgCl + H_2 \longrightarrow 2Ag + 2HCl$$

Complete balanced equation:

$$2AgCl + H_2 \longrightarrow 2Ag + 2HCl$$

**30. b.** Unbalanced equation:

$$MnO_4^- + NO_2^- \longrightarrow MnO_2 + NO_3^-$$

Half reactions:

$$MnO_4^- \longrightarrow MnO_2(1)$$
$$NO_2^- \longrightarrow NO_3^-(2)$$

Half reaction (1):

$$MnO_4^- + \mathbf{4H^+} + \mathbf{3e^-} \longrightarrow MnO_2 + \mathbf{2H_2O}$$

(balancing charge, H, and O atoms)

Since the reaction takes place in basic conditions:

$$MnO_4^- + 4H_2O + 3e^- \longrightarrow MnO_2 + 2H_2O + 4OH^-$$

Canceling $2H_2O$ leaves:

$$MnO_4^- + 2H_2O + 3e^- \longrightarrow MnO_2 + 4OH^-$$

Half reaction (2):

$$NO_2^- + \mathbf{H_2O} \longrightarrow NO_3^- + \mathbf{2H^+} + \mathbf{2e^-}$$

(balancing charge, H, and O atoms)

Again, because of basic conditions:

$$NO_2^- + H_2O + 2OH^- \longrightarrow NO_3^- + 2H_2O + 2e^-$$

Canceling $H_2O$:

$$NO_2^- + 2OH^- \longrightarrow NO_3^- + H_2O + 2e^-$$

Balancing the electrons in both half reactions and adding them gives the complete balanced equation.

$$2MnO_4^- + \cancel{4H_2O} + 6e^- \longrightarrow 2MnO_2 + \cancel{8OH^-}$$
$$3NO_2^- + \cancel{6OH^-} \longrightarrow 3NO_3^- + \cancel{3H_2O} + \cancel{6e^-}$$
$$3NO_2^- + 2MnO_4^- + H_2O \longrightarrow 3NO_3^- + 2MnO_2 + 2OH^-$$

Final equation:

$$3NO_2^- + 2MnO_4^- + H_2O \longrightarrow 3NO_3^- + 2MnO_2 + 2OH^-$$

**32. b.** Unbalanced equation:

$$CuI + Ag \longrightarrow AgI + Cu$$

Half reactions:

$$CuI \longrightarrow Cu \ (1)$$
$$Ag \longrightarrow AgI \ (2)$$

Half reaction (1):

$$CuI + \mathbf{e^-} \longrightarrow Cu + \mathbf{I^-}$$

(balancing I atoms and charge)

Half reaction (2):

$$Ag + \mathbf{I^-} \longrightarrow AgI + \mathbf{e^-}$$

(balancing I atoms and charge)

Adding both half reactions gives the complete balanced equation:

$$CuI + \cancel{e^-} \longrightarrow Cu + \cancel{I^-}$$
$$Ag + \cancel{I^-} \longrightarrow AgI + \cancel{e^-}$$
$$CuI + Ag \longrightarrow AgI + Cu$$

**36. b.** oxidation number of Ca = +2
oxidation number of C = +4
oxidation number of oxygen = −2

**d.** oxidation number of boron = +3
oxidation number of hydrogen = −1

**38. b.** oxidation number of sodium = +1
oxidation number of oxygen = −2
oxidation number of chromium = +6

**d.** oxidation number of sodium = +1
oxidation number of oxygen = −2
oxidation number of hydrogen = +1

**42. b.**

| Compound | Element | Oxidation number |
|---|---|---|
| $Cl_2$ | Cl | 0 |
| KBr | K | +1 |
| KBr | Br | −1 |
| KCl | K | +1 |
| KCl | Cl | −1 |
| $Br_2$ | Br | 0 |

oxidizing agent = $Cl_2$
substance oxidized = $Br^-$
reducing agent = $Br^-$
substance reduced = $Cl_2$

**44. b.**

| Compound | Element | Oxidation number |
|---|---|---|
| C | C | 0 |
| $H_2$ | H | 0 |
| $CH_4$ | C | −4 |
| $CH_4$ | H | +1 |

oxidizing agent = C
substance oxidized = $H_2$
reducing agent = $H_2$
substance reduced = C

**46. b.**

| Compound | Element | Oxidation number |
|---|---|---|
| $H_2S$ | H | +1 |
| $H_2S$ | S | −2 |
| $O_2$ | O | 0 |
| $SO_2$ | S | +4 |
| $SO_2$ | O | −2 |
| $H_2O$ | H | +1 |
| $H_2O$ | O | −2 |

oxidizing agent = $O_2$
substance oxidized = S in $H_2S$
reducing agent = S in $H_2S$
substance reduced = $O_2$

**50.** The reaction that occurs at the cathode is: $2H_3O^+ + 2e^- \rightarrow H_2(g) + 2H_2O$. Electrons flow from the zinc electrode to the platinum electrode. The $K^+$ ions enter the $H_3O^+(aq)$ solution.

**54. b.** Zn is oxidized

**56. b.** $Ag_2O(s) + Zn(s) \rightarrow ZnO(s) + 2Ag(s)$

**60.** Mass of Ag deposited = 2.86 g Ag

**64.** Mass of copper deposited = 11.0 g

**70.** Moles of $Cl_2(g)$ produced = 0.479 mol
Volume of $Cl_2(g)$ formed = 10.7 L

**72.** $2KI(aq) + Cl_2(g) \rightarrow 2KCl(aq) + I_2(l)$

**76.** $Cl_2(g) + 2e^- \rightarrow 2Cl^-(aq)$

**80.** The best reducing agent is $I^-$. The best oxidizing agent is $Cl_2$.

## CHAPTER 17

**4.** Antoine-Henri Becquerel conducted experiments that showed that uranium emitted X rays both in the presence and the absence of sunlight.

**8.** The discovery of radioactivity enabled the age of the earth to be determined by measuring the amount of radioactivity present in rocks.

**12.** Becquerel's experiment would not have worked because lead foil absorbs xrays. The emission of xrays would not have been detected in either case.

**16.** The negative charge in an atom is found in the shells of electrons surrounding the nucleus.

**22. b.** number of protons = 7
number of neutrons = 7
number of electrons = 10

**24. b.** number of protons = 35
number of neutrons = 46
number of electrons = 36

**26. b.** number of protons = 53
number of neutrons = 73
number of electrons = 54

**30.** ${}^{250}_{100}Fm \rightarrow {}^{246}_{98}Cf + {}^{4}_{2}He$
${}^{254}_{102}No \rightarrow {}^{250}_{100}Fm + {}^{4}_{2}He$
${}^{256}_{103}Lr \rightarrow {}^{252}_{101}Md + {}^{4}_{2}He$
${}^{257}_{106}Rf \rightarrow {}^{253}_{104}Db + {}^{4}_{2}He$

**36.** ${}^{176}_{79}Au \rightarrow {}^{176}_{78}Pt + {}^{0}_{1}e$
${}^{177}_{79}Au \rightarrow {}^{173}_{77}Ir + {}^{4}_{2}e$
${}^{193}_{79}Au + {}^{0}_{-1}e \rightarrow {}^{193}_{78}Pt$
${}^{200}_{79}Au \rightarrow {}^{200}_{80}Hg + {}^{0}_{-1}e$

**42.** ${}^{211}_{88}Ra \rightarrow {}^{207}_{86}Rn + {}^{4}_{2}He$
${}^{213}_{88}Ra + {}^{0}_{-1}e \rightarrow {}^{213}_{87}Fr$
${}^{228}_{88}Ra \rightarrow {}^{228}_{89}Ac + {}^{0}_{-1}e$

**44.** Age of charcoal = 11460 years

**48.** $2.6 \times 10^{9}$ years

**52.** 56 days

**58.** ${}^{209}_{83}Br + {}^{4}_{2}He \rightarrow {}^{210}_{85}At + 3\,{}^{1}_{0}n$
The element is ${}^{210}_{85}At$.

**60.** ${}^{249}_{98}Cf + {}^{12}_{6}C \rightarrow {}^{257}_{104}Db + 4\,{}^{1}_{0}n$

**64.** ${}^{249}_{97}Bk + {}^{1}_{0}n \rightarrow {}^{250}_{98}Cf + {}^{0}_{-1}e$
The element is ${}^{250}_{98}Cf$.

**68.** ${}^{235}_{92}U \rightarrow {}^{139}_{56}Ba + {}^{94}_{36}Kr + 2\,{}^{1}_{0}n$
The other atom is ${}^{94}_{36}Kr$.

**72.** ${}^{12}_{6}C + {}^{16}_{8}O \rightarrow {}^{2}_{14}Si + \gamma$

**76.** ${}^{14}_{7}N + {}^{1}_{1}H \rightarrow {}^{15}_{8}O + \gamma$

## CHAPTER 18

**2. b.**

```
                                  H
                                  |
                               H—C—H
       H   H   H   H   H   H      |    H
   ..  |   |   |   |   |   |      |    |
H—N—C—C—C—C—C—C—C—C—H
   |   |   |   |   |   |   |      |    |
   H   H   H   H   H   H   H      |    H
                               H—C—H
                                  |
                                  H
```

**c.** $CH_3CH_2OCHCH_2CH_3$ (with $CH_3$ on the CH carbon)

```
CH3CH2OCHCH2CH3
         |
        CH3
```

**4. b.**

```
   H   H   H   H
   |   |   |   |
H—C—C—C—C—H
   |   |   |   |
   H   |   H   H
     :N—H
       |
       H
```

**8. a.** $CH_3(CH_2)_3CH_2OH$

**b.** $CH_3NCH_2CH_3$ (with $CH_3$ on N)

```
CH3NCH2CH3
    |
   CH3
```

**14. a.** $C_6H_9NO$
**b.** $C_9H_{12}Cl_2$
**c.** $C_{10}H_{12}$

**18. b.** B and C

**20. b.** B and C

**22. b.** Vitamin C    yes    yes

**24.** C

**28. b.**

```
   H   H   H   H
   |   |   |   |
H—C—C—C—C—H
   |   |   |   |
   H   |   H   H
     :Ö—H
```

**32.** D

**34.**

```
   H  :Ö:  H      Ö:
   |   ||  |     //
H—C———C———C———C
   |       |     \
   H       H      Ö—H
```

**36.** If formula of silver salt is $C_8H_{15}O_2Ag$, then burning $1 \times 10^{-3}$ mol (363 mg) should produce $8 \times 10^{-3}$ mol of $CO_2$. And $8 \times 10^{-3}$ mol $CO_2$ at STP should equal $8 \times 10^{-3}$ mol

$$\frac{(22.4\ \text{L})}{(1\ \text{mol})}\ \frac{(1000\ \text{mL})}{(1\ \text{L})} = 179.2\ \text{mL}.$$

So the experimental result given does not support the answer to question 35.

**40. a.**

```
     H  :O:  H   H
     |   ||  |   |
  H—C—C—C—C—H
     |       |   |
     H       H   H
```

**b.** (cyclobutanone ring)=Ö:

**42. b.** Ketone

**44. b.** Alcohol

**46.** Carboxyl, alcohol and aldehyde carbonyl groups

**50.** Both aldehydes and ketones contain the (C=O) carbonyl group. The difference is that in aldehydes the carbonyl group is attached to a carbon and a hydrogen whereas in a ketone the carbonyl group is attached to two alkyl groups.

**54.** D

**56. b.**

```
      H   H    :O:        H
      |   |    ||    ..   |
  H—C—C——C——N—C—H
      |   |          |    |
      H   |          |    H
    H—C—H   H—C—H
          |          |
          H          H
```

**60.** A or B

**64.** C

**70.** An alkane contains only C—C single bonds, whereas an alkyne contains C≡C triple bonds

**74.** $CH_3NH_2$ is the nucleophile
$CH_3CH_2I$ is the substrate
$I^-$ is the leaving group

**78. b.** R—C=O ketone (with R attached below the C)

```
R—C=O
  |
  R
```

**80. b.** $CO_2H$ carboxylic acid

**82.** Water

## CHAPTER 19

**2.** Glycogen is a carbohydrate

**6.** Simple sugars link together to form more complex members of the class carbohydrates

**10.** Potatoes are a dietary source of carbohydrate

**12.** The prefix tri- means that there are three ester groups in the molecule. The suffix -glyceride reflects the fact that the alcohol part of the ester comes from the alcohol glycerol.

**16.** C is a *trans* alkene because the bromine atoms and the hydrogen atoms are on opposite sides of the double bond.

**20.** $C_{18}H_{24}O_2$

**22.** $C_{28}H_{44}O$

**30.** Proteins are formed by peptide bonds between the amino group of one $\alpha$-amino acid and the carboxylic group of another. Thus one terminus of the molecule has an amino group and the other a carboxylic acid group. Nylon is made up of monomers with two identical functional groups, diamines and dicarboxylic acids. Both functional groups can bond, creating a molecule with identical terminal groups.

**32. b.** This is a polypeptide-like molecule because it is made of alpha amino acids.

**34.** Met-Pro-Lys-Asp

**36.** Val-Tyr-Ala-Tyr-His

**40.** Arg-Gln-Tyr-Lys-Lys

**44.** Thymine and cytosine

**48.** Hormones are molecules that act as chemical messengers to control certain chemical processes in the body.

**54.** B

**60.** C

# Glossary

## a

**Absolute zero** The temperature at which a sample of an ideal gas would reach a volume of zero. This is 0 K or −273°C. (*p. 390*)
**Accuracy** Closeness of a measurement to the correct value. (*p. 7*)
**Acid** See Arrhenius acid, Brønsted-Lowry acid. (*p. 265*)
**Acid dissociation constant** For the reaction of a proton donor with water, the product of factors expressing the concentrations of the products divided by the product of factors expressing the concentrations of the reactants. (*p. 558*)
**Acidic solution** Aqueous solution in which $[H_3O^+]$ is greater than $1.00 \times 10^{-7}$ M. (*p. 562*)
**Acidosis** A condition in which the blood is more acidic than normal. (*p. 577*)
**Acids** Substances that produce $H^+$ (Arrhenius definition). (*p. 265*)
**Actinides** Elements with atomic numbers from 90 to 103 inclusive. (*p. 157*)
**Activation energy** The amount of energy needed before a given reaction can take place. (*p. 510*)
**Alcohol** (1) Any compound containing an alkyl group bonded to an OH group. (2) The compound ethyl alcohol or ethanol, $C_2H_5OH$. (*p. 678*)
**Aldehyde** Compound containing a C=O (carbonyl) group, in which the carbon atom of the carbonyl is attached to one or two hydrogen atoms and one carbon atom or none at all. (*p. 680*)
**Aldose** Carbohydrate containing an aldehyde group. (*p. 714*)
**Alkali metals** Members of Group 1A in the periodic table. (*p. 153*)
**Alkaline dry cell** Device that stores energy in a chemically alkaline medium and releases this energy in the form of electricity. (*p. 617*)
**Alkaline earth metals** Members of Group 2A in the periodic table. (*p. 154*)
**Alkaloid** Nitrogen containing bases that occur in plants; they are often poisonous when ingested. (*p. 725*)
**Alkalosis** A condition in which the blood is more basic (alkaline) than normal. (*p. 577*)
**Alkane** Compound containing carbon atoms and hydrogen atoms, connected to one another by single bonds only. (*p. 688*)
**Alkene** Compound containing only C and H, along with a carbon-carbon double bond. (*p. 688*)
**Alkyl group** Assemblage of singly bonded carbon and hydrogen atoms with an empty bond for something else to attach to. (*p. 678*)
**Alkyne** Compound containing only C and H, along with a carbon-carbon triple bond. (*p. 688*)
**Allotropes** Forms of solid compounds having different crystal structures. (*p. 429*)
**Alloy** A mixture of two or more elements that has metallic properties. (*p. 32*)
**Alpha decay** Form of radioactive decay in which the nucleus of an atom gives off an alpha particle (a helium nucleus that consists of two protons and two neutrons). (*p. 643*)
**Alpha particle** Particle consisting of two protons and two neutrons given off by the nuclei of some radioactive elements. (*p. 638*)
**Amide** Compound containing an alkyl group or H atom attached to a C=O group that is in turn attached to a nitrogen atom. (*p. 685*)
**Amine** Compound containing a nitrogen atom bonded with single bonds to one or more carbon atoms. (*p. 686*)
**Amino acid** Compound containing an amine group and a carboxylic acid group on the same carbon; molecule used to make proteins. (*p. 715*)
**Amplitude** Height of a wave. (*p. 191*)
**Amorphous solid** One that has no long-range order in its molecular structure. (*p. 56*)
**Amorphous solids** Having no definite form; used for solid compounds having no long-range. (*p. 426*)
**Amphiprotic** A substance that can act as either an acid or a base. (*p. 548*)
**Angular quantum number (l)** Quantum number describing the shape of an orbital. (*p. 204*)
**Angular structure** Structure with three points (or nuclei) that are not in a straight line. (*p. 360*)

**Anion** A chemical species with a negative charge. (*p. 134*)
**Anode** Electrode where oxidation occurs; it is the negative electrode in an electrochemical cell. (*p. 617*)
**Aqueous** A solution in water. (*p. 276*)
**Aqueous solution** Homogeneous mixture in which water is the solvent. (*p. 456*)
**Aromatic hydrocarbons** Compounds that consist of carbon and hydrogen only, and which contain rings of carbon atoms with unusually stable electronic structures. (*p. 689*)
**Arrhenius acid** A substance that produces $H^+$ (or, in the modern form, $H_3O^+$) in solution. (*p. 544*)
**Arrhenius base** A substance that produces $OH^-$ in solution. (*p. 544*)
**Atmosphere** (atm) Unit of pressure equal to 760 mm Hg. (*p. 385*)
**Atom** The smallest particle an element can be divided into and still retain its characteristics as an element. (*p. 49*)
**Atomic mass unit** One-twelfth of the mass of a carbon atom of the isotope $^{12}C$. (*p. 130*)
**Atomic number** Equal to the number of protons in the nucleus of an atom of a given element. (*p. 131*)
**Atomic radius** Distance from the center of an atom to its outer surface. (*p. 165*)
**ATP** Adenosine triphosphate is the molecule that carries energy around in the body and passes it on to other molecules, providing energy for movement and making other high-energy molecules. (*p. 723*)
**Average** The sum of a series of measurements divided by the number of measurements made. (*p. 71*)

$$\text{Average} = \frac{\text{Measurement 1} + \text{Measurement 2} + \cdots}{\text{Number of measurements made}}$$

**Average atomic mass** A weighted average of the mass of the isotopes present under natural conditions. (*p. 179*)
**Avogadro's Hypothesis** Equal volumes of any gases under the same conditions of temperature and pressure contain the same number of particles. (*p. 396*)
**Avogadro's number** The number of particles in a mole of any substance, $6.02 \times 10^{23}$. (*p. 177*)

## b

**Balanced equation** All the correct coefficients are given. (*p. 256*)
**Balancing equations by inspection** A directed trial-and-error approach to balancing simple chemical equations. Details of the method are given in Chapter 8. (*p. 259*)
**Base** See Arrhenius base, Brønsted-Lowry base. (*p. 265*)
**Base dissociation constant** For the reaction of a proton acceptor with water, the product of factors expressing the concentrations of the products divided by the product of factors expressing the concentrations of the reactants. (*p. 560*)
**Bases** Substances that produce $OH^-$ in water (Arrhenius definition). (*p. 265*)
**Basic solution** Aqueous solution in which $[OH^-]$ is greater than $1.00 \times 10^{-7}$ M. (*p. 562*)
**Battery** Device for converting oxidation-reduction reactions to usable electricity. (*p. 616*)
**Bent structure** See angular. (*p. 360*)
**Beta decay** Form of radioactive decay in which the nucleus of an atom gives off an energetic electron. (*p. 642*)
**Beta particle** Energetic electron given off by the nucleus of certain radioactive elements. (*p. 638*)
**Beta Rays** Electrons with high energy, emitted by certain radioactive elements. (*p. 194*)
**Binary compound** Compound consisting of two elements. (*p. 234*)
**Binary hydrides** Compound consisting of two elements, one of which is hydrogen. (*p. 552*)
**Binding energy** Calculated by subtracting the actual mass of the nucleus from the mass of the nucleons that make it up and calculating the energy equivalent to that mass using the equation $E = mc^2$. (*p. 651*)
**Biochemistry** Area of chemistry in which the molecular makeup of living systems is studied. (*p. 704*)
**Body-centered cubic structure** Arrangement of atoms in a crystal, in which eight atoms form the corners of a cube, and another atom occupies the center of the cube. (*p. 443*)
**Boiling point elevation** The increase in the boiling temperature of a solvent when a solute is added. (*p. 484*)
**Bond** Attraction of two or more atoms for one another. (*p. 340*)
**Bond angle** The angle between the lines connecting two nuclei to a single central atom. (*p. 356*)
**Bond length** The average distance between two atoms bonded to each other in a molecule or ion. (*p. 344*)
**Bond-line formula** Variant of a Lewis diagram that symbolizes bonds and carbon atoms by lines, and which does not show hydrogen atoms attached to carbon atoms at all. These show the connections between atoms in organic molecules in a more concise form than conventional Lewis diagrams can achieve. (*p. 669*)
**Boyle's Law** The pressure of a sample of gas is inversely proportional to its volume at constant temperature (that is, $P_1V_1 = P_2V_2$). (*p. 387*)
**Brittle** Having the property of fracturing and breaking apart under stress. (*p. 151*)
**Brønsted-Lowry acid** A substance that donates protons during chemical reactions. (*p. 545*)

**Brønsted-Lowry base** A substance that accepts protons during chemical reactions. (*p. 545*)
**BTU** ("British Thermal Unit") A unit of energy equal to 1055.06 joules. (*p. 115*)
**Buffer** A solution that resists pH changes when either acid or base is added. (*p. 573*)

## C

**Calorie** A unit of energy equal to 4.148 joules. Originally it was defined as the amount of heat needed to raise the temperature of 1 g of water 1°C. The "dieter's calorie" is the kilocalorie. (*p. 115*)
**Carbohydrate** Compound containing an aldehyde or ketone functional group and several OH groups, resulting in an empirical formula $CH_2O$ or something close to it. (*p. 705*)
**Carbonyl group** A carbon atom doubly bonded to an oxygen atom, a C=O group. (*p. 680*)
**Carboxyl group** C=O group, an oxygen atom bonded to a carbon atom which in turn is bonded to two other atoms. (*p. 683*)
**Carboxylic acid** Compound containing a hydrogen atom or alkyl group bonded to a $CO_2H$ group. (*p. 683*)
**Catalysts** Substances that increase the rates of reactions without being changed by them. (*p. 276*)
**Cathode** Electrode where reduction occurs; it is the positive electrode in an electrochemical cell. (*p. 617*)
**Cation** A chemical species with a positive charge. (*p. 134*)
**Celsius scale** Temperature scale in which the freezing point of water is defined as 0°C and the boiling point of water is defined as 100°C. (*p. 88*)
**Centigrade scale** Another name for the Celsius scale. (*p. 88*)
**Charge** The number of excess protons or neutrons a species has. (*p. 280*)
**Charles' Law** The volume of a sample of gas is directly proportional to its temperature at constant pressure (that is, $V_1/T_1 = V_2/T_2$ or $V_1/V_2 = T_1/T_2$). (*p. 389*)
**Chemical energy** The potential energy possessed by a chemical (not a nuclear) fuel. (*p. 122*)
**Chemical formula** A list of the elements present in a substance and their numerical ratios. (*p. 53*)
**Chemical property** Pertaining to reactions, in which one substance is transformed into another. (*p. 46*)
**Chemistry** The study of matter and the changes it undergoes. (*p. 2*)
**Chromatography** A separation method in which a moving phase transports substances along a stationary phase that absorbs the substances to different extents. (*p. 44*)
**Colligative properties** Characteristics of a solution that depend only on the number of solute particles in a solution, not on their chemical nature. (*p. 484*)
**Collision frequency** The number of collisions between reacting molecules per unit time. (*p. 509*)
**Combustion** The reaction that occurs when a substance is burned in oxygen (or air). (*p. 270*)
**Compound** A pure substance that contains atoms of more than one element. (*p. 52*)
**Conclusions** The deductions and inferences drawn from results. (*p. 11*)
**Condensation** The transformation of a gas into a liquid or solid. (*p. 42*)
**Condensed structure** Abbreviated variant of a Lewis diagram that expresses connections between atoms in a molecule in a concise yet complete form by grouping atoms. (*p. 668*)
**Conductor** Substance that transfers electricity well. (Also a substance that transfers heat well.) (*p. 126*)
**Conservation of Electrical Charge** Electrical charge is neither created not destroyed. (*p. 280*)
**Conservation of Energy** States that energy is neither created nor destroyed. It is not exactly true, since energy can be converted to mass. (*p. 281*)
**Conjugate acid** Species formed by adding a proton ($H^+$) to a compound or ion. (*p. 546*)
**Conjugate base** Species formed by removing a proton from a compound or ion. (*p. 546*)
**Conversion factor** An arithmetic value that converts one unit into another unit. (*p. 94*)
**Cosmic Rays** High-energy radiation that rains down on Earth from space. Over 99% of this radiation consists of particles. (*p. 195*)
**Coulomb** The amount of electrical charge carried by a current of one ampere in one second. (*p. 131*)
**Covalent bond** A bond formed when two atoms share one or more pairs of electrons. (*p. 344*)
**Covalent compound** Compound in which electrons are shared (in ionic compounds electrons are not shared). Covalent compounds usually contain nonmetals only, while ionic compounds usually contain both metals and nonmetals. (*p. 234*)
**Critical mass** Quantity of fissionable material (uranium or plutonium) sufficient to sustain a nuclear chain reaction and give an explosion. (*p. 652*)
**Crystalline solids** Solid materials with long-range order in their molecular or ionic structures. (*p. 426*)
**Crystallography** The study of the structure of atoms and molecules in crystals. (*p. 443*)
**Cubic closed-packed structure** Arrangement of atoms in a crystal, in which each atom touches six

atoms in its own plane and three in each adjacent plane. The location of atoms in one plane matches the location of gaps between the atoms in the third plane. (*p. 442*)

## d

**Dalton's Law of Partial Pressures** Scientific law which states that the sum of the partial pressures of all the gases in a mixture equals the total pressure. (*p. 402*)
**Data** The raw information obtained from experiments. (*p. 11*)
**Density** Mass per unit volume. (*p. 91*)
**Diatomic** Containing two atoms. (*p. 51*)
**Diffusion** The rate at which gases mix. (*p. 404*)
**Dipole-dipole forces** The electrostatic attraction between the negatively-charged end of a polar molecule and the positively-charged end of another. (*p. 439*)
**Dipole-induced dipole forces** The electrostatic attraction between a charged end of a polar molecule and an ordinarily nonpolar molecule or atom that has had its centers of positive and negative charge separated under the influence of the nearby charge. (*p. 439*)
**Dipole moment** It is defined as the amount of charge separated times the distance between the positive and negative centers of a molecule. It is a vector quantity having both a magnitude and a direction. (*p. 362*)
**Dissociates** To separate, as ions do when they go into solution. (*p. 341*)
**Dissociation** The separation of ions that occurs when ionic compounds dissolve. (*p. 469*)
**Dissolves** Forms a homogeneous mixture with another substance. (*p. 456*)
**Distillation** A separation method in which a substance is evaporated and then condensed elsewhere. (*p. 42*)
**DNA** Deoxyribonucleic acid, large polymeric molecules made up of connected bases, deoxyribose, and phosphate units that carry the codes for the molecules that express the inherited features of organisms. (*p. 720*)
**Double bond** Covalent bond in which four electrons are shared by two atoms. (*p. 351*)
**Ductile** Having the ability to be drawn into wires. (*p. 151*)
**Dynamic equilibrium** Reaction in which the concentration of reactants and products does not change, but in which reactant molecules produce product molecules and vice versa at equal rates. (*p. 515*)

## e

**Effective nuclear charge** The atomic number of an element minus the number of inner shell electrons present. (*p. 165*)
**Effusion** The rate at which a gas escapes from a tiny hole. (*p. 404*)
**Electrolysis** A technique in which an electric current passes through a solution or a molten metallic salt. Metal ions are reduced to the metal at the anode and anions are oxidized at the cathode. (*p. 619*)
**Electrolyte** Compound that dissolves in water to give a solution that conducts electricity. (*p. 469*)
**Electron** Particle with a negative charge and a mass 1/1837 times that of a hydrogen atom. Electrons surround the nuclei in atoms. (*p. 128*)
**Electron capture** Form of radioactive decay in which the nucleus of an atom takes in an electron, converting a proton into a neutron in the process. (*p. 643*)
**Electron configuration** A list of the arrangement of electrons in the shells and subshells of an atom or ion. (*p. 160*)
**Electromagnetic radiation** Light, both visible and invisible (for example, infrared, ultraviolet, X rays, radio waves, etc.) (*p. 193*)
**Electronegative** Able to attract electrons to itself in a compound, a characteristic of atoms of nonmetals. (*p. 170*)
**Electronegativity** Ability of an element to attract electrons to itself in a compound. (*p. 169*)
**Electrophile** Chemical species that attacks a source of electrons. (*p. 691*)
**Electroplating** A technique in which an electric current passes through a solution of a metallic salt, in such a way that metals are plated out of solution onto solid electrodes. (*p. 618*)
**Electropositive** Not able to attract electrons to itself in a compound, a characteristic of metal atoms. (*p. 170*)
**Element** A substance that cannot be broken down into simpler substances by chemical means. (*p. 15*)
**Empirical formula** The simplest formula for a compound, one that gives the number of atoms present in terms of whole-number ratios. (*p. 318*)
**Endothermic reactions** Reactions which absorb heat. (*p. 281*)
**Energy** The ability to move an object through a distance. This is the same as the ability to do work. (*p. 113*)
**Energy factor** Fraction of collisions in which the molecules colliding have sufficient energy for a reaction to take place. (*p. 510*)
**Equilibrium** (1) The state of a reaction at which there is no change in the concentration of the reactants and products. (2) The state of a reaction in which the rates of the forward and reverse processes are equal. (*p. 515*)
**Equilibrium constant** The concentrations of the products of a reac-

tion at equilibrium divided by the concentrations of the reactants. See the Important Equations at the end of Chapter 14 for the equation form. (*p. 517*)
**Ester** Compound containing a hydrogen atom or alkyl group bonded to a C=O group that is bonded to an oxygen atom that is bonded to an alkyl group. (*p. 684*)
**Ether** (1) Compound containing two alkyl groups bonded to an oxygen atom. (2) The compound diethyl ether, $C_4H_{10}O$. (*p. 680*)
**Evaporation** The transformation of a liquid into a gas. (*p. 42*)
**Excess** A quantity of a given reactant that is more than enough to react with the others. (*p. 304*)
**Excited state** Any electron configuration that has a higher potential energy that the ground state electron configuration. (*p. 209*)
**Exothermic reactions** Reactions which evolve heat. (*p. 281*)
**Experiment** A test or a trial. (*p. 8*)
**Exponent** The power to which a base unit (usually 10 or e) is raised. (*p. 79*)
**Extensive properties** Properties that depend on the amount of a substance in the sample. (*p. 46*)
**Extraction** A method for separating substances based on differences in solubility. A liquid is passed over a sample, and parts of the sample dissolve in the liquid. (*p. 41*)

## f

**Fahrenheit scale** Temperature scale in which the freezing point of water is defined as 0°F and the boiling point of water is defined as 212°F. (*p. 88*)
**Falsifiable** An idea or concept that can be shown to be false by experiment. (*p. 9*)
**Faraday constant** The amount of charge on one mole of electrons. It has a value of $9.65 \times 10^4$ C mol$^{-1}$. (*p. 620*)
**Filtration** A method of separation in which a liquid or gas is passed through a barrier with holes to remove solid or liquid particles from it. (*p. 41*)
**First Law of Thermodynamics** Energy can be neither created nor destroyed. Another statement is that energy is conserved. (*p. 119*)
**Fission** The splitting apart of large atomic nuclei (for example uranium or plutonium) to produce less massive nuclei, neutrons, an energy. (*p. 634*)
**Force** A physical agency that causes acceleration. (*p. 137*)
**Freezing-point depression** The decrease in the temperature at which a solvent freezes when a solute is added. (*p. 484*)
**Frequency** The number of waves that pass a certain location per unit time. (*p. 190*)
**Functional group** A grouping of atoms that largely determines the physical and chemical properties of an organic molecule. (*p. 677*)
**Fusion** The combining of small atomic nuclei (usually hydrogen) to produce a larger nucleus plus energy. (*p. 634*)

## g

**Gamma ray** High-energy photon given off by the nuclei of some radioactive elements. (*p. 638*)
**Gamma rays** Very high-energy electromagnetic radiation emitted by certain radioactive elements. (*p. 194*)
**Gene** The portion of the DNA molecule that encodes a complete protein. (*p. 722*)
**Genome** All of the DNA in a cell. (*p. 722*)
**Ground state** The electron configuration that has the lowest possible potential energy. (*p. 209*)
**Group** Column of elements in the periodic table. (*p. 152*)

## h

**Half-life** For a given radioisotope, the length of time during which half of the atoms in any given sample will decay. (*p. 645*)
**Halides** The elements of Group 7A on the periodic table. These are fluorine, chlorine, bromine, iodine, and astatine. (*p. 154*)
**Halogens** Members of Group 7A in the periodic table. (*p. 154*)
**Heat** A form of kinetic energy in which molecules move or vibrate. (*p. 116*)
**Heisenberg Uncertainty Principle** The conclusion that neither the position nor the momentum of an electron in an atom can be pinpointed at a particular instant in time. (*p. 203*)
**Heterogeneous mixture** One with boundaries between two or more phases. (*p. 32*)
**Hexagonal closed-packed structure** Arrangement of atoms in a crystal, in which each atom touches six atoms in its own plane and three in each adjacent plane. The location of atoms in one plane matches the location of the atoms in the third plane. (*p. 442*)
**Homogeneous mixture** One with uniform properties throughout, at any level much larger than the molecular. (*p. 32*)
**Hormone** Chemical messengers that control chemical processes in the body, including metabolism. (*p. 724*)
**Hund's Rule** Electrons occupy a subshell in a manner that maximizes the total spin. (*p. 209*)
**Hydrocarbons** Compounds consisting of carbon and hydrogen only. (*p. 687*)
**Hydrogen bonding** Force of attraction between hydrogen atoms

attached to electronegative atoms and nonbonding pairs of electrons on other electronegative atoms. (*p. 436*)
**Hydrolysis** Adding the elements of water to a compound. (*p. 718*)
**Hypothesis** An idea put forth for testing, one which requires more careful examination. (*p. 12*)

## i

**Ideal Gas Equation** PV = nRT where P is the pressure, V is the volume, n is the number of moles, T is the temperature of a gas and R is the ideal gas constant. (*p. 397*)
**Indicator** A substance that changes color at a given pH. (*p. 567*)
**Induced dipole-induced-dipole forces** The electrostatic attraction between nonpolar molecules or atoms, caused by the fact that the centers of positive and negative charge in each species separate under the influence of the other. (*p. 440*)
**Inert** Unreactive. (*p. 155*)
**Infrared radiation** Electromagnetic radiation having slightly less energy than visible light. (*p. 193*)
**Inorganic chemistry** Archaic, the study of compounds from nonliving sources. Modern, the study of the elements other than carbon and hydrogen. (*p. 666*)
**Insulator** Substance that does not conduct electricity well. (Also, a substance that does not conduct heat well.) (*p. 126*)
**Intensive property** Independent of the amount of substance present. (*p. 46*)
**Ion** An electrically charged object that consists of an atom or group of atoms. (*p. 128*)
**Ion-dipole forces** The forces of attraction between a charged species and the oppositely-charged end of a polar molecule. (*p. 462*)
**Ionic bond** A bond that forms when oppositely charged species (ions) attract one another. (*p. 341*)
**Ionic compound** Electrically neutral species combining cations and anions. The charges of the ions cancel out. (*p. 229*)
**Ionization constant for water** $K_w = [H_3O^+][OH^-] = 1 \times 10^{-14}$. (*p. 562*)
**Ionizing radiation** High-energy photons or particles capable of removing electrons from atoms and molecules, thereby producing ions. (*p. 639*)
**Isotopes** Two atoms having the same number of protons but a different number of neutrons. (*p. 136*)

## j

**Joule** The SI derived unit of energy which has units of kg $m^2$ $s^{-2}$. (*p. 114*)

## k

**Kelvin** The base unit of temperature used by scientists. The zero temperature on the Kelvin scale is “absolute zero,” the temperature at which all molecular motion ceases, and 1 K equals 1°C. (*p. 90*)
**Kelvin scale** Temperature scale in which 0.0 K is defined as the lowest possible temperature; the temperature at which molecular motion ceases. (*p. 90*)
**Ketone** Compound containing a C=O (carbonyl) group, in which the carbon atom of the carbonyl is attached to two carbon atoms. (*p. 682*)
**Ketose** Carbohydrate containing a ketone group. (*p. 714*)
**Kilowatt-hour** Unit of energy, equal to $3.6 \times 10^6$ joules. (*p. 115*)
**Kinetic energy** The energy an object has by virtue of its motion. (*p. 114*)

## l

**Lanthanides** Elements with atomic numbers from 58 to 71 inclusive. (*p. 157*)
**Law of Conservation of Mass** States that matter is neither created nor destroyed. It is not exactly true, since mass can be converted to energy. (*p. 256*)
**Laws** Observations or experimental results that have been confirmed repeatedly. They are usually simple verbal statements or equations. (*p. 12*)
**Leaving group** Portion of a molecule that breaks off from the rest during a reaction, to be replaced by a *nucleophile*. (*p. 691*)
**Le Chatêlier’s Principle** If you change one of the variables in a system at equilibrium, the other variables will shift to counteract the effects of this change. (*p. 527*)
**Lewis diagram** A diagram showing the number of bonding and nonbonding electrons associated with each atom in a molecule or ion. (*p. 347*)
**Limiting reactant** Reactant not in excess, the one that will be used up first as a reaction proceeds. (*p. 312*)
**Linear structure** Structure with three points (or nuclei) in a straight line. (*p. 356*)
**Lipid** Fat or oil molecule. (*p. 704*)
**Liter** Metric system unit of volume equal to $10^3$ cubic centimeters, a little more than an English system quart. (*p. 91*)
**Litmus** An indicator that is red when acidic and blue when basic. (*p. 567*)

## m

**Magnetic quantum number (m)** Quantum number that describes the orientation in space of an orbital. (*p. 205*)
**Magnitude** The size of a measurement, a number. (*p. 70*)

**Main group elements** The elements in Groups 1A-8A on the periodic table. They have either *s*-electrons or both *s*- and *p*-electrons in their valence shells; *d* and *f* subshells are either completely empty or completely full. (*p. 157*)
**Malleable** Having the ability to be beaten into flat sheets. (*p. 151*)
**Mass number** The sum of the numbers of protons and neutrons in a given nucleus. (*p. 637*)
**Mass spectrometer** An instrument that generates positive ions and employs electromagnetic fields to deflect them in such a way that their mass-to-charge ratios can be found. (*p. 135*)
**Mass-volume percent** The mass of solute in grams divided by the volume of the solution in milliliters. The solution must be aqueous. (*p. 479*)
**Measurement** The act of determining the size or amount of something. (*p. 70*)
**Melting point** Temperature at which a substance changes from a solid to a liquid. (*p. 430*)
**Metabolism** The process of converting oxygen plus fats, carbohydrates, and proteins into usable energy by the body. (*p. 704*)
**Metal** A substance that has the typical properties of metals, such as metallic luster, good conductivity of heat and electrical current, malleability and ductility, etc. (*p. 151*)
**Metallic bond** A "sea" of shared electrons that holds together samples of metals. (*p. 345*)
**Metalloid** Semimetal. (*p. 152*)
**Meter** Metric system unit of length, a little more than an English system yard. (*p. 86*)
**Metric System** A system of measurement that uses the meter, gram, second, kelvin, mole, ampere, and candela as units. Other units are made by combining these units, and all units are sized using prefixes that represent factors of 10. (*p. 84*)
**Microwaves** A type of electromagnetic radiation, with less energy than infrared but more energy than radio waves. (*p. 194*)
**Mineral** In nutrition, minerals are cations and anions. (*p. 725*)
**Mixture** A sample that is composed of more than one pure substance. (*p. 31*)
**Molality** The number of moles of solute per 1 kg of solvent. (*p. 481*)
**Molar heat of fusion** The amount of heat absorbed when one mole of a substance freezes, or the amount of heat released when one mole of a substance melts. (*p. 445*)
**Molar heat of vaporization** The amount of energy needed to convert one mole of a given substance from a liquid state to a gas state without a change in temperature. (*p. 447*)
**Molar mass** The mass (usually in grams) of one mole of a substance. (*p. 176*)
**Molarity** The number of moles of substance per liter of solution. (*p. 472*)
**Mole** Avogadro's number ($6.02 \times 10^{23}$) of anything. (*p. 175*)
**Mole fraction** The number of moles of a solute divided by the total number of moles of the substances in the solution. (*p. 482*)
**Molecular biology** Area of biology in which the molecular makeup of living systems is studied. (*p. 704*)
**Molecular formula** List of the elements present in a compound, including the numbers of each kind of atom present. (*p. 320*)
**Molecule** A particle containing more than one atom bonded together. (*p. 50*)
**Monatomic** Consisting of one atom only. (*p. 225*)

## n

**Net ionic equations** Equations which show all solids, liquids, gases, and ions that participate in the reactions. Spectator ions are omitted. (*p. 280*)
**Network solid** Material made up of one or more very large molecules. (*p. 426*)
**Neutral** When electric charge is involved, having a total charge of zero; containing an equal number of protons and neutrons. (*p. 132*)
**Neutral atom** An atom in which the number of protons equals the number of electrons; that is, an atom with a total charge of zero. (*p. 132*)
**Neutral solution** Aqueous solution in which $[H_3O^+]$ equals $[OH^-]$ at $1.00 \times 10^{-7}$ M. (*p. 562*)
**Neutralization** A complete reaction between an acid and a base. (*p. 266*)
**Neutron** An electrically neutral particle with a mass approximately equal to that of a hydrogen atom. Neutrons are found in the nuclei of atoms. (*p. 131*)
**Noble gases** Members of Group 8A in the periodic table. (*p. 154*)
**Nomenclature** Area of chemistry (or any other discipline) dealing with names. (*p. 222*)
**Nonelectrolyte** Substance that does not conduct electricity when dissolved in water. (*p. 469*)
**Nonmetal** A substance that has the typical properties of nonmetals (and lacks the typical properties of metals). Typical properties of nonmetals include poor reflectivity, low conductivity of heat and electrical current, brittleness in the solid state, etc. (*p. 151*)
**Nuclear chemistry** Field of study in which reactions that change the nuclei of atoms are studied. Also called *nuclear physics.* (*p. 632*)
**Nuclear fission** See fission. (*p. 649*)

**Nuclear fusion** See fusion. (*p. 653*)
**Nuclear reaction** The interaction of a nucleus with another particle or nucleus at high energy to produce one or more different nuclei. (*p. 637*)
**Nucleic acids** Large polymeric molecules, including DNA and RNA, that perform functions of coding for the structures of molecules inherited from generation to generation in organisms. (*p. 720*)
**Nucleons** Subatomic particles found in an atomic nucleus, including protons and neutrons. (*p. 651*)
**Nucleophile** Substance that attacks an atom that has a partial positive charge. (*p. 691*)
**Nucleophilic substitution** Type of reaction in which an electron-rich species attacks a center of positive charge in another species (the *substrate*), displacing another electron-rich group. (*p. 690*)
**Nucleotide** A unit containing one phosphate, one ribose or deoxyribose, and one base, as found in RNA, DNA, and the energy molecule ATP. (*p. 721*)
**Nuclide** Atom or ion that contains a given number of protons and neutrons. (*p. 640*)

## O

**Octahedral structure** Structure with eight sides (all equilateral triangles) and six corners. (*p. 367*)
**Octet rule** Guideline which states that atoms tend to assume noble gas configurations with eight electrons in their outer shells as they form molecules and ions. (*p. 347*)
**Orbital** Region of space in an atom in which electrons can be found. (*p. 201*)
**Orbital (box) diagram** A method of writing electron configurations of atoms or ions that places electrons in boxes that represent individual atomic orbitals. (*p. 208*)
**Organic chemistry** Archaic, the study of compounds originating from living systems. Modern, the chemistry of compounds consisting largely of carbon and hydrogen. (*p. 666*)
**Orientation factor** Fraction of collisions in which the molecules that collide that are aligned in such a way as to permit a reaction to take place. (*p. 510*)
**Osmosis** The migration of a solvent through a semipermeable membrane to an area of higher solute concentration. (*p. 486*)
**Osmotic pressure** The pressure that must be applied to prevent the flow of solvent though a semipermeable membrane. (*p. 486*)
**Oxidation** In older terms, the gain of oxygen in a reaction. In more modern terms, the gain of electrons in an oxidation-reduction reaction. (*p. 596*)
**Oxidation number** The charge that an atom or ion would have if all of its bonds were ionic. (*p. 609*)
**Oxidation-reduction reaction** Chemical change in which one reactant transfers oxygen atoms to another reactant (old definition) or removes electrons from another reactant (new definition). (*p. 596*)
**Oxidized** The process of gaining, or the state of having gained, oxygen in a chemical reaction. Also, increase in oxidation number. (*p. 270*)
**Oxidizing agent** Species that gains electrons in an oxidation-reduction reaction. (*p. 596*)
**Oxyacid** A member of a class of compounds whose molecules contain an oxyanion and enough hydrogens to balance the negative charge. (*p. 237*)
**Oxyanion** Charged species consisting of a central atom with one or more oxygen atoms attached to it. (*p. 227*)

## P

**Paired spin** Two electrons occupy an orbital or a subshell with opposite spins. (*p. 208*)
**Parallel spin** Two or more electrons occupy a subshell with the same spin. (*p. 209*)
**Partial pressure** The pressure exherted by an individual gas in a mixture of gases. (*p. 402*)
**Pascal (Pa)** SI unit of pressure; 101,325 Pa = 1 atm. (*p. 385*)
**Pauli Exclusion Principle** No two electrons in an atom can have four quantum numbers ($n$, $\ell$, $m$, $m_s$) that are the same. (*p. 207*)
**Percent by mass** The mass of the solute divided by the total mass of the solution times 100%. (*p. 477*)
**Percent by volume** The volume of the solute divided by the total volume of the solution times 100%. (*p. 478*)
**Percentages** Percentages are defined as Part/Whole × 100%. They are used to indicate the relative amount of individual components, the "Parts," in a collection, the "Whole." (*p. 34*)
**Period** Row of elements in the periodic table. The properties of the elements in a period vary in a systematic way. (*p. 152*)
**Phase** A state that matter can assume.
**Photon** Packet of light energy. (*p. 198*)
**Physical properties** Ones that can be measured without changing the formula of the substances present. (*p. 46*)
**Polar covalent bond** Covalent bond in which the electrons are not shared equally; in which the center of positive charge and the center of negative charge are separated. (*p. 344*)
**Polarity** Separation of charges in a molecule. (*p. 361*)
**Polarizable** Not polar, but capable of developing a dipole when a

charged object approaches. (*p. 440*)
**Polyatomic** Species consisting of two or more atoms. (*p. 227*)
**Polycrystalline** Consisting of a large number of small crystals. (*p. 429*)
**Positron** Elemental particle having a positive charge and the mass and other properties of an electron; the antimatter version of the electron. (*p. 638*)
**Positron emission** Form of radioactive decay in which a particle with a positive charge, but otherwise with properties identical to those of an electron, is given off. (*p. 642*)
**Potential energy** The energy an object has by virtue of its position in relation to an object that exerts a force on it. (*p. 114*)
**Power** Energy per unit time. (*p. 115*)
**Precipitate** An insoluble solid product resulting from the mixture of soluble compounds. (*p. 466*)
**Precipitation reaction** Chemical change that gives an insoluble solid product from the mixing of two soluble compounds. (*p. 466*)
**Precision** The closeness of several measurements of the same thing to each other. (*p. 71*)
**Pressure** Force per unit area. (*p. 384*)
**Primary standard** A substance of known molecular weight and acid-base properties used to determine the concentration of solutions used in titration. (*p. 580*)
**Principal Quantum Number (n)** Quantum number used in calculating the energy of an electron in an orbital. (*p. 203*)
**Probability factor** See Orientation factor. (*p. 510*)
**Products** Chemical species formed in a chemical reaction. (*p. 255*)
**Properties** The typical characteristics of substances. (*p. 45*)
**Protein** Large molecule made up of many amino acids which have linked together by forming amide functional groups. (*p. 705*)
**Proton** A positively charged particle with a mass approximately equal to that of a hydrogen atom. Protons form part of atomic nuclei. (*p. 130*)
**Pure substance** One that contains only one kind of molecule or atom. (*p. 47*)

## q

**Quantum number** Countable number used in calculating properties of electrons in atoms. (*p. 203*)

## r

**Radiation** High-energy photons or particles emitted from a source. (*p. 638*)
**Radioactivity** The property of elements whose nuclei give off alpha particles, beta particles, or gamma rays. (*p. 634*)
**Radio waves** Electromagnetic radiation having low frequency, low energy, and long wavelengths. (*p. 194*)
**Rate constant** The rate of a reaction is equal to the rate constant multiplied by terms that express the concentrations of certain reactants in that reaction. (*p. 507*)
**Rate equation** Mathematical relationship in which the rate of a reaction is equal to the rate constant multiplied by terms that express the concentrations of certain reactants in that reaction. (*p. 507*)
**Reactants** Chemical species that undergo chemical transformations. (*p. 255*)
**Redox reaction** Short term for oxidation-reduction. (*p. 596*)
**Reduced** The process of losing, or the state of having lost, oxygen or oxidation number in a chemical reaction. (*p. 270*)
**Reducing agent** Species that loses electrons in an oxidation-reduction reaction. (*p. 596*)
**Reduction** In older terms, the loss of oxygen in a reaction. In more modern terms, the loss of electrons in an oxidation-reduction reaction. (*p. 596*)
**Relative humidity** The amount of water actually in an air sample divided by the amount of water that the same amount of air at that temperature can hold times 100%. (*p. 402*)
**Resonance hybrid** The weighted average of all the resonance structures of a given molecule or ion. (*p. 355*)
**Resonance structure** A Lewis diagram that shows one of two or more possible ways of arranging the electrons for a given skeleton of atoms. (*p. 354*)
**Results** Data presented in organized form, the significant outcomes of experiments. (*p. 11*)

## s

**Saturated solution** Homogeneous mixture in which no more of a given solute will dissolve when it is added. (*p. 464*)
**Scientific notation** A system of writing numbers consisting of two parts. The first part is a number with a value between 1 and 10. This is then multiplied by $10^x$, giving the magnitude of the number. (*p. 78*)
**Second Law of Thermodynamics** Energy may be changed from one useful form to another, or from more useful forms to less useful forms, but never from less useful to more useful forms. (*p. 119*)
**Semimetal** Substance having properties intermediate between those of metals and nonmetals. (*p. 152*)
**Semipermeable membrane** Barrier that allows molecules of a solvent to migrate from regions of lower to higher solute concentration, but prevents the migration of solute molecules. (*p. 485*)

**Shell** Layer of electrons around the nucleus of an atom. (*p. 159*)
**SI units** A standard set of units based on the metric system that is used extensively but not exclusively by practicing scientists. The system uses the meter, kilogram, second, kelvin, mole, ampere, and candela as base units. (*p. 88*)
**Significant figures** All of those that are known exactly plus the one that is not. (*p. 73*)
**Simple cubic structure** Arrangement of atoms in a crystal, in which eight atoms occupy the corners of a cube. (*p. 443*)
**Single bond** Covalent bond in which two electrons are shared by two atoms. (*p. 351*)
**Solubility** The amount of a solute that will dissolve in a standard amount of a given solvent. (*p. 464*)
**Solubility product constant** The product of terms that express the concentrations of the ions that make up a given ionic compound in solution. See the Important Equations at the end of Chapter 14 for the equation form. (*p. 524*)
**Solute** A substance that has dissolved in another. (*p. 456*)
**Solution** A homogeneous liquid or solid mixture. (*p. 456*)
**Solvent** A substance capable of dissolving another, or that has already dissolved another. (*p. 456*)
**Spectator ion** Ion that is present in a solution but doesn't participate in the reaction that occurs. (*p. 280*)
**Spin quantum number** ($m_s$) This quantum number indicates the direction of the spin of an electron. (*p. 206*)
**Stable** Unlikely to react, low in potential energy. (*p. 160*)
**Standard** A known value which is used for comparison purposes against experimentally determined values. (*p. 74*)
**Standard molar volume** The volume occupied by one mole of an ideal gas at STP, 22.4 L. (*p. 396*)
**Standard temperature and pressure (STP)** 0°C and 1 atm. Also called Standard. (*p. 396*)
**State** A form or phase that matter can take. (*p. 37*)
**Stoichiometry** The branch of chemistry that considers how much of one substance reacts with another, or how much is produced when another substance reacts. (*p. 300*)
**Strong acid** Species which, in aqueous solution, donates virtually all of its protons to form $H_3O^+$. (*p. 278*)
**Strong acids** Acids that react completely in aqueous solution to give $H_3O^+$ and a spectator ion. (*p. 278*)
**Strong base** Species which contains the $OH^-$ ion or which accepts a proton from $H_2O$ in aqueous solution to give $OH^-$ in high yield. (*p. 279*)
**Strong bases** Substances which dissolve in water to give a solution of metal ions and $OH^-(aq)$ ions. (*p. 279*)
**Strong electrolyte** Substance that conducts electricity strongly when dissolved in water. (*p. 470*)
**Sublimation** The transformation of a gas into a solid. (*p. 42*)
**Subshell** Subdivision of a layer of electrons around the nucleus of an atom. (*p. 161*)
**Substitution reaction** Chemical transformation in which one group replaces another on an alkyl group. (*p. 690*)
**Substrate** In a nucleophilic substitution reaction, the molecule or ion containing a center of positive charge that is attacked by an electron-rich species. (*p. 691*)
**Supernova** Vast explosion that takes place late in the evolution of very large and bright stars. (*p. 656*)

## t

**Temperature** A numerical measure of warmth or coldness. (*p. 88*)
**Tetrahedral structure** An arrangement of four atoms around a central atom in which the outer atoms are separated from one another as far as possible, at the corners of a tetrahedron. (*p. 358*)
**Tetrahedron** Four-sided pyramid-like geometrical figure. The sides are equilateral triangles. (*p. 358*)
**Theories** More general constructions than laws, and are put forth to explain laws or law-like behavior. Generally laws tell what happens, and theories explain why. (*p. 12*)
**Titration** The process of reacting an acid with a base for the purpose of measuring the amount of one or the other. (*p. 578*)
**Torr** A unit of pressure equal to 1.00 mm Hg. (*p. 385*)
**Transition metal** (or transition element) Member of one of the B groups in the periodic table. (*p. 157*)
**Transition state** Points of maximum energy in a reaction diagram. (*p. 510*)
**Triglyceride** Lipid (fat or oil) molecule consisting of glycerol and three fatty acid molecules bonded together by way of ester functional groups. (*p. 707*)
**Trigonal** Three-sided. (*p. 357*)
**Trigonal bipyramid** In this structure three outer atoms are arranged in a planar triangle around a central atom. Another outer atom is directly above the central atom, with a fifth atom directly below the central atom. (*p. 366*)
**Trigonal planar structure** Flat three-sided structure. (*p. 357*)
**Trigonal pyramid** Structure with three sides and a center point elevated above the plane of the sides. (*p. 359*)
**Triple bond** Covalent bond in which six electrons are shared by two atoms. (*p. 351*)

## U

**Ultraviolet radiation** Electromagnetic radiation having slightly more energy than visible light. (*p. 193*)

**Unit** Standard to which the magnitude of a measurement is compared. (*p. 70*)

**Unit factor** Mathematical tool used to convert measurements from one unit to another. As suggested by the name, all unit factors are equal to one. (*p. 94*)

**Unsaturated solution** Homogeneous mixture that has the capacity to dissolve more of a given solute than is already present. (*p. 464*)

## V

**Valence** The number of chemical bonds an atom of a given element usually forms. (*p. 171*)

**Valence electrons** Electrons in the outermost occupied shell of an atom. (*p. 160*)

**Velocity** Distance covered per unit time in a specified direction. (In mathematical terms, velocity is a vector quantity and speed is a scalar measure?) (*p. 190*)

**Viscosity** The resistance of a fluid to flow. (*p. 429*)

**Vitamin** Substances that must be part of our diets if we are to remain in good health. (*p. 724*)

**Volatility** The ease with which a liquid is converted into a gas. (*p. 42*)

**VSEPR theory** Valence Shell Electron Pair Repulsion theory. This predicts that the electrons around a central atom will repel one another, producing an arrangement in which they are as far apart as possible. (*p. 356*)

## W

**Watts** An SI unit of power defined as one $J\ s^{-1}$. (*p. 115*)

**Wavelength** The distance from the peak of one wave to the peak of the next. (*p. 190*)

**Weak acid** Species which, in aqueous solution, donates relatively few of its protons to form $H_3O^+$. (*p. 278*)

**Weak acids** Acids that react incompletely in aqueous solution. (*p. 278*)

**Weak base** Species which accepts protons reluctantly in aqueous solution. (*p. 554*)

**Weak electrolyte** Substance that conducts electricity weakly when dissolved in water. (*p. 474*)

## X

**X ray** Photon of electromagnetic radiation having less energy than gamma rays but more than ultraviolet radiation. (*p. 633*)

**X rays** A type of high-energy electromagnetic radiation. (*p. 194*)

# Credits

**CHAPTER 1**

Opening Illustration, Tony Stone Images, Inc.
Figure 1.1, *Left,* Photo Researchers, Inc. *Middle,* Chris Sorensen *Right,* The Rodin Museum: Philadelphia: Gift of Jules Mastbaum
Figure 1.2, *Left,* Joe Dimaggio/Peter Arnold *Right,* Horst Schaffer/Peter Arnold
Figure 1.3, North Wind Picture Archives
Figure 1.6, The Science Museum/ Science and Society Picture Library
Figure 1.8, Patrick Watson
Figure 1.9, Francois Gohier/Photo Researchers, Inc.
Figure 1.13, The Bettmann Archive
Figure 1.14, Patrick Watson
Figure 1.15, CERN/Science Photo Library
Figure 1.18, Stephen Frisch
Figure 1.19, The Bettmann Archive
Figure 1.20, Chris Sorensen
Figure 1.21, Chris Sorensen
Figure 1.22, Chris Sorensen
Figure 1.23, Chris Sorensen

**CHAPTER 2**

Opening Illustration, Photo Disk, Inc.
Figure 2.1, *A,* Stephen Frisch; *Inset,* Carla Montgomery *B,* Patrick Watson.
Figure 2.2, *A,* Patrick Watson *B,* Phototake
Figure 2.3, Werner Forman Archive/Art Resource
Figure 2.4, The Bettmann Archive
Figure 2.5, *Top,* Chris Sorensen; *Inset,* Ken Eward/Science Source
Figure 2.7, Bob Pool/Tom Stack and Assoc.
Figure 2.8, *Left,* Stephen Frisch *Right,* Chris Sorensen
Figure 2.9, Robert Auerbach
Figure 2.10, Tony Eveans/Marcel Sire
Figure 2.11, Chris Sorensen
Figure 2.12, Stephen Frisch
Figure 2.13, Richard Megna/ Fundamental Photographs
Figure 2.14, Peter Arnold, Inc.
Figure 2.15, *Left,* Patrick Watson *Right,* Chris Sorensen
Figure 2.20, Chris Sorensen
Figure 2.21, Patrick Watson
Figure 2.22, *Left,* Art Resource/Erich Lessing *Right,* JLM Visuals/Breck P. Kent
Figure 2.24, Tom Way/IBM Electronics/TOPO METRIX
Figure 2.29, Stephen Frisch
Figure 2.31, *Left,* Stephen Frisch *Center,* Stephen Frisch *Right,* Ward's Science Catalog
Figure 2.33, Patrick Watson
Figure 2.34, Stephen Frisch

**CHAPTER 3**

Opening Illustration, Tony Stone Images, Inc.
Figure 3.1, NASA
Figure 3.4, *Left,* Donald Brownlee/University of Washington *Right,* Anglo-Australian Telescope, Photography by David Malin
Figure 3.7, Chris Sorensen
Figure 3.8, The Granger Collection
Figure 3.9, Patrick Watson
Figure 3.11, Erich Lessing/Art Resource
Figure 3.12, Stephen Frisch
Figure 3.13, Anglo-Australian Telescope, Photography by David Malin
Figure 3.16, Chris Sorensen
Figure 3.18, Patrick Watson
Figure 3.19, Patrick Watson
Figure 3.20, Bill O'Connor/Peter Arnold, Inc.

**CHAPTER 4**

Opening Illustration, Tony Stone Images, Inc.
Figure 4.1, *Left, Middle, & Right,* Chris Sorensen
Figure 4.2, Paul Reuben
Figure 4.3, Chris Sorensen
Figure 4.4, Patrick Watson Inset, Patrick Watson
Figure 4.5, Chris Sorensen
Figure 4.7, *Left,* Ann Duncan/Tom Stack & Associates *Right,* David Hoffman/TSW
Figure 4.8, Stephen Frisch
UNEX4.2, Chris Sorensen
Figure 4.10, U.S. Army Corps of Engineering
Figure 4.13, Richard Kaylin/TSW
Figure 4.14, *Left,* Richard Day, Animals/Animals *Right,* Raymond A. Mendez, Animals/Animals, Earth Scenes
Figure 4.15, Patrick Watson
Figure 4.16, Art Resource/Erich Lessing
Figure 4.17, Art Resource/Erich Lessing

**CHAPTER 5**

Opening Illustration, Tony Stone Images, Inc.
Figure 5.2, Patrick Watson
Figure 5.3, Patrick Watson
Figure 5.4, Patrick Watson
Figure 5.5, Stephen Frisch
Figure 5.6, Stephen Frisch
Figure 5.7, *Left,* Tom Way/IBM Electronics *Right,* Peter Arnold, Inc.
Figure 5.9, Yaov Levy/Phototake
Figure 5.10, Patrick Watson
Figure 5.11, *Left,* Patrick Watson *Right,* Patrick Watson
Figure 5.12, Patrick Watson
Figure 5.30, Chris Sorensen
Figure 5.31, Patrick Watson
Figure 5.32, Patrick Watson

### CHAPTER 6

Opening Illustration, Tony Stone Images, Inc. & PhotoDisk, Inc.
Figure 6.1, Martin Dohrn/Science Photo Library
Figure 6.5, Thomas T. Arny
Figure 6.6, *Top,* Michael Marten/Science Photo Library *Bottom,* Patrick Watson
Figure 6.7, Chris Sorensen *Inset,* Chris Sorensen
Figure 6.8, F. S. Judd Research Laboratories, Kodak Ltd. *Negative,* Patrick Watson
Figure 6.10, Kodak
Figure 6.11, The Bettmann Archive, Kodak
Figure 6.12, *Left,* Stephen Frisch *Center,* Stephen Frisch *Right,* Washington, D.C. Tourism Conventions Bureau
Figure 6.13, Kodak
Figure 6.15, The Bettmann Archive
Figure 6.31, B, Stephen Frisch
Figure 6.32, PSSC Physics, ed2, Newton, MA 1965, D.C. Heath & Company and Educational Development Center, Inc.

### CHAPTER 7

Opening Illustration, Tony Stone Images, Inc.
Figure 7.2, Patrick Watson
Figure UN1(*A,B*), *Left & Right,* Chris Sorensen
Chapter Box 7.3, Culver Pictures
Figure 7.12, William St. George
Figure 7.16, Patrick Watson
Figure 7.17, Stephen Frisch

### CHAPTER 8

Opening Illustration, Tony Stone Images, Inc.
Figure 8.4, Richard Megna/Fundamental Photographs
Figure 8.6, Bassam Z. Shakhashiri
Figure 8.7, Stephen Frisch
Figure 8.8, Chris Sorensen
Figure 8.9, Stephen Frisch
Figure 8.10, AIP Emilio Segre Visual Archives
Figure 8.12, Patrick Watson
Figure 8.13, Patrick Watson
Figure 8.15, *Left,* NASA *Right,* NASA/JPL
Figure 8.18, *Top; Left,* Chris Sorensen *Top; Right,* Patrick Watson *Bottom; Left,* Photo Researchers *Bottom; Right,* Chris Sorensen
Figure 8.19, John Shaw/Tom Stack & Associates
Figure 8.20, Patrick Watson
Figure 8.24, Jon Brennis
Figure 8.29, Stephen Frisch
Figure 8.30, Patrick Watson

### CHAPTER 9

Opening Illustration, Photo Disk, Inc.
Figure 9.1, M. W. Kellogg Company
Figure 9.3, *Top,* Paul Brierly and Michael Freeman, by Courtesy of Charles Moore, Science Museum London *Bottom,* Patrick Watson
Figure 9.4, Patrick Watson
Figure 9.5, S. Allen/Peter Arnold
Figure 9.6, Stephen Frisch
Figure 9.11, *Left,* Bassam Shakhashiri *Right,* NASA
Figure 9.12, Patrick Watson
Figure 9.13, Patrick Watson
Figure 9.19, Jeffrey Scovil
Figure 9.21, Spencer Grant/Photo Researchers
Figure 9.23, Patrick Watson
Figure 9.24, Chris Sorensen
Figure 9.26, Chris Sorensen
Figure 9.27, Stephen Frisch
Figure 9.28, Stephen Frisch
Figure 9.29, Photo Researchers
Figure 9.30, Ward's Earth Science Catalog

### CHAPTER 10

Opening Illustration, Tony Stone Images, Inc.
Figure 10.1, Patrick Watson
Figure 10.2, The Granger Collection
Figure 10.11, Stephen Frisch
Figure 10.12, Francis Simon/American Institute of Physics Emilio Segre Visual Archives
Figure 10.14, Stephen Frisch
Figure 10.16, Stephen Frisch
Figure 10.19, Stephen Frisch
Figure 10.27, Stephen Frisch
Figure 10.30, Stephen Frisch

### CHAPTER 11

Opening Illustration, Tony Stone Images, Inc.
Figure 11.1, Harvey Lloyd/The Stock Market
Figure 11.2, Museum of Comparative Zoology, Harvard University
Figure 11.3, Byron Augustin/Tom Stack & Associates
Figure 11.4, Frank Rossotto/Stock Trek
Figure 11.5, Grant Heilman Photography
Figure 11.6, *Left & Right,* Patrick Watson
Figure 11.8, *Left,* Frank Rossotto/Stock Trek *Right,* NASA
Figure 11.13, Patrick Watson
UNExa11.3, Chris Sorensen
Chapter Box 11.2, The Bettmann Archive
Figure 11.16, *Left,* Northwind *Right,* Grant Heilman
Chapter Box 11.13, Figure A, The Bettmann Archive
Figure 11.21, Frank Rossotto/Stock Trek
Figure 11.22, Chris Sorensen
Figure 11.23, Patrick Watson
Figure 11.24. Patrick Watson
Figure 11.25, The Granger Collection
UNP11.147, Stephen Frisch
UNP11.191, The Granger Collection

### CHAPTER 12

Opening Illustration, Tony Stone Images, Inc.
Figure 12.2, Stephen Frisch
Figure 12.3, *Left,* Jeffrey Scovil *Center,* Patrick Watson *Right,* Patrick Watson
Figure 12.4, *Left & Right,* Dr. E. R. Degginger
Figure 12.6, *Left & Right,* Stephen Frisch
Figure 12.8, *Left,* James L. Amos/Peter Arnold, Inc. *Right,* James L. Amos/Peter Arnold, Inc.
Figure 12.9, Carla Montgomery
Figure 12.10, The Bettmann Archive
Figure 12.11, *Left image,* Chris Sorensen *Right image,* Chris Sorensen
Figure 12.12, Professor. J. F. Allen/University of St. Andrews, School of Physics and Astronomy
CB12.1*A, B,* Culver Pictures *C,* Gemological Institute
CB12.1*B, B,* Patrick Watson *C,* Patrick Watson
Figure 12.15, Darek Karp/Animals/Animals, Earth Scenes
Figure 12.20, Chris Sorensen
Figure 12.21, Chris Sorensen
Figure 12.27, David Scharf/Peter Arnold, Inc.
Figure 12.29, Stephen Frisch

## CHAPTER 13

Opening Illustration, Tony Stone Images, Inc.
Figure 13.1, Patrick Watson
Figure 13.2, Patrick Watson
Figure 13.3, Photo Courtesy Mass Bay Brewing Co.
Figure 13.7, Stephen Frisch
Figure 13.8, Zig Leszczynski, Animals/Animals. Earth Scenes
Chapter Box 13.2, *Top,* Rich Buzzelli/Tom Stack & Associates *Bottom,* Richard Megna/Fundamental Photographs
Figure 13.10, Patrick Watson
Figure 13.11, Stephen Frisch
Figure 13.12, Stephen Frisch
Figure 13.14, Stephen Frisch
Figure 13.15, Stephen Frisch
Figure 13.16, Patrick Watson
Figure 13.17, Stephen Frisch
Figure 13.18, Patrick Watson
Chapter Box 13.3, Stephen Frisch
Figure 13.19, Patrick Watson
Figure 13.20, Patrick Watson
Figure 13.24, Photo Researchers
UNPB13.65, Stephen Frisch

## CHAPTER 14

Opening Illustration, PhotoDisk, Inc.
Figure 14.1, Frank Rosotto/Stock Trek
Figure 14.2, Patrick Watson
Figure 14.4, Chris Sorensen
Figure 14.12, Stephen Frisch
Figure 14.13, PhotoDisk, Inc.
Chapter Box 14.2, Chris Sorensen
Figure 14.14, JPL
Figure 14.15, Patrick Watson
Figure 14.16, Richard Megna/Fundamental Photographs
UNEX14.11Sol, Stephen Frisch

## CHAPTER 15

Opening Illustration, PhotoDisk, Inc.
Figure 15.1, Top & Bottom Dennis Cox/China Stock
Chapter Box 15.1B, JLM Visuals
Figure 15.4, Stephen Frisch
Figure 15.6, Patrick Watson
Figure 15.7, Stephen Frisch
Figure 15.8, Chris Sorensen
Figure 15.10, Patrick Watson
Figure 15.11, Stephen Frisch
Figure 15.12, Stephen Frisch
Figure 15.14, Patrick Watson
Figure 15.15, Robert Winslow/Tom Stack & Associates
Figure 15.16, Stephen Frisch
Figure 15.17, Fundamental Photographs

## CHAPTER 16

Opening Illustration, Tony Stone Images, Inc.
Figure 16.1, Tom Stack & Associates, Inga Spence/Tom Stack & Associates, David Phillips/Science Source/Photo Researchers
Figure 16.2, AP/World Wide Photos
Figure 16.3, Photo Researchers
Figure 16.5, Stephen Frisch
Figure 16.8, Patrick Watson
Chapter Box 16.1, AP/World Wide Photos/David Glass
Figure 16.9, *A,* Chris Sorensen *B,* Patrick Watson *C&D,* Chris Sorensen
Figure 16.10, *Left,* Burndy Library *Right,* Patrick Watson
Figure 16.11, The Granger Collection
Figure 16.16, Patrick Watson
Figure 16.17, Chris Sorensen
Figure 16.18*B,* Tom Hollyman/Photo Researchers

## CHAPTER 17

Opening Illustration, Tony Stone Images, Inc.
Figure 17.1, Ward's Natural Science Establishment
Figure 17.2, Richard J. Green/Photo Researchers
Figure 17.3, Culver Pictures
Chapter Box 17.1, AIP Emilio Segre Visual Archives
Figure 17.7, CERN
Figure 17.14, Ray Richardson Animals/Animals, Earth Scenes
Figure 17.15, Catholic News Service
Figure 17.16, NASA
Figure 17.18, Culver Pictures
Figure 17.19, Culver Pictures
Figure 17.22, AP/Wide Word Photos
Figure 17.23, Hubble Space Telescope
Figure 17.24, StarLight
Chapter Box 17.3, *Left,* Gary D. McMichael/Photo Researchers *Top; Right and Bottom,* Bettmann News Photos
Figure 17.25, Ward's Natural Science Establishment
Figure 17.27, Royal Observatory, Edinburgh
Figure 17.28, Palomar Observatory/California Institute of Technology

## CHAPTER 18

Opening Illustration, PhotoDisk, Inc.
Figure 18.1, Chris Sorensen
Figure 18.3, Chris Sorensen; Raymond A. Mendez Animals/Animals Earth, Scenes; Chris Sorensen; Peter Arnold, Inc.; Chris Sorensen
Figure 18.5, Catherine Ursillo/Photo Researchers
Chapter Box 18.1, *Lower; Left,* Bettmann *Top; Right,* Snark International *Lower; Right,* Institut Pasteur
Figure 18.6, Chris Sorensen
Figure 18.8, Jed DeKalb/Courtesy of Jack Daniel Distillery, Lynchburg
Figure 18.12, The Granger Collection/Robert Hinckley
Figure 18.14, *Left,* C. C. Lockwood Animals/Animals, Earth Scenes *Right,* Chris Sorensen
Figure 18.15, Chris Sorensen
Figure 18.16, Monsanto Chemical Co.
Figure 18.17, Chris Sorensen
Figure 18.18, Chris Sorensen
Figure 18.20, Chris Sorensen
Figure 18.21, Steve Elmore/Tom Stack & Associates

## CHAPTER 19

Opening Illustration, Tony Stone Images, Inc. & PhotoDisk, Inc.
Figure 19.1, Literary Work of Leonardo Da Vinci. Compiled and edited from the original manuscripts by Jean-Paul Richter, 3rd ed. Published in London by Phaidon, 1970, vol. I/Reproduced Courtesy The Trustees of The Boston Public Library
Figure 19.2, Chris Sorensen
Figure 19.3, John Bavosi/Science Photo Library
Figure 19.4, Bruce Iverson
Figure 19.5, Bruce Iverson
Figure 19.6, Bruce Iverson
Figure 19.7, Bruce Iverson
Figure 19.10, Bruno J. Zehender/Peter Arnold, Inc.
Figure 19.14, B. Taylor/H. Armstrong Roberts
CB19.3, Jon Brenneis
Figure 19.18, BioGrafx/Kenneth Eward
Figure 19.19, *Left,* Jeffrey Scovil *Right,* Chris Sorensen
Figure 19.20, Chris Sorensen
Figure 19.21, Walter Schmidt/Peter Arnold, Inc.

# Index